A COMPETITIVE APPROACH TO

Linear
Algebra

Useful for

- Undergraduate and Postgraduate Courses for All Indian Universities
- CSIR-JRF, NET, SET, GATE and All Entrance Examinations for Admission in M.Sc., M.Phil. and Ph.D. Programmes

Dr. SUDHIR KUMAR PUNDIR

M.Sc., M.Phil., NET, Ph.D.
Associate Professor
Department of Mathematics
S.D. (P.G.) College, Muzaffarnagar (U.P.)

CBSPD

CBS Publishers & Distributors Pvt Ltd

New Delhi • Bengaluru • Chennai • Kochi • Kolkata • Lucknow • Mumbai
Hyderabad • Jharkhand • Nagpur • Patna • Pune • Uttarakhand

A Competitive
Approach to
Linear Algebra

ISBN: 978-81-239-2791-6

First Edition: 2015
Reprint: 2019, 2023

Published by **Satish Kumar Jain** and produced by **Varun Jain** for

CBS Publishers & Distributors Pvt Ltd

4819/XI Prahlad Street, 24 Ansari Road, Daryaganj, New Delhi 110 002, India.
Ph: 011-23289259, 23266861, 23266867
Fax: 011-23243014

Website: www.cbspd.com
e-mail: delhi@cbspd.com

Corporate Office: 204 FIE, Industrial Area, Patparganj, Delhi 110 092
Ph: 011-4934 4934   Fax: 011-4934 4935

e-mail: publishing@cbspd.com; publicity@cbspd.com

Branches

- **Bengaluru:** Seema House 2975, 17th Cross, KR Road, Banasankari 2nd Stage, Bengaluru 560 070, Karnataka, India
  Ph: +91-80-26771678/79   Fax: +91-80-26771680   e-mail: bangalore@cbspd.com
- **Chennai:** 7, Subbaraya Street, Shenoy Nagar, Chennai 600 030, Tamil Nadu, India
  Ph: +91-44-26680620, 26681266   Fax: +91-44-42032115   e-mail: chennai@cbspd.com
- **Kochi:** 42/1325, 1326, Power House Road, Opp KSEB, Power House, Ernakulum Kochi 682 018, Kerala, India
  Ph: +91-484-4059061-65,67   Fax: +91-484-4059065   e-mail: kochi@cbspd.com
- **Kolkata:** 147, Hind Ceramics Compound, 1st Floor, Nilgunj Road, Belghoria, Kolkata-700056, West Bengal, India
  Ph: +033-25633055, 033-25633056   e-mail: kolkata@cbspd.com
- **Lucknow:** Basement, Khushnuma Complex, 7 Meerabai Marg (Behind Jawahar Bhawan),Lucknow-226001, UP, India
  Ph: +0522-4000032   e-mail: tiwari.lucknow@cbspd.com
- **Mumbai:** PWD Shed, Gala no 25/26, Ramchandra Bhatt Marg, Next to JJ Hospital Gate no. 2, Opp. Union Bank of India,
  Noorbaug, Mumbai-400009, Maharashtra, India
  Ph: 022-66661880/89
  e-mail: mumbai@cbspd.com

Representatives

• Hyderabad	0-9885175004	• Jharkhand	0-9811541605	• Nagpur	0-9421945513
• Patna	0-9334159340	• Pune	0-9923910676	• Uttarakhand	0-9716462459

Printed at Glorious Printers, Delhi, India

Preface

The book entitled 'A Competitive Approach to LINEAR ALGEBRA' is meant for UG and PG students of all Indian Universities. Besides, it will also be very useful for students preparing for various competitive examinations like CSIR-JRF/NET, SET, GATE and various entrance examinations for admission in M.Sc., M.Phil. and Ph.D. programmes.

Special and conscious efforts have been made to keep the writing style simple. Students who are tired of complex concepts and abstract presentation styles, will find this book simple and straightforward. It is a collection and compilation work from various sources and has been endeavoured to include as much as information as could be possible. The book's objective is to provide a conceptual understanding of the fundamentals of linear algebra. Different concepts have been explained with the help of examples. A large number of problems with solutions have been provided to assist one get a firm grip on the ideas developed. There is plenty of scope in the form of exercise for the reader to try and solve the problem on his own. To make the book self-contained and competition-oriented a chapter review of basic terms, results and questions have been given at the end of each chapter. Also, at the end of each section, graded examples, illustrations and facts have been given with the name of Competition Corner. These help the students to grasp the thing better. These include problems on the entire section and are carefully selected to represent variety.

I express my gratitude to the authors and publishers of various books I consulted.

I wish to sincerely thank Sh S.K. Jain, Managing Director, CBS Publishers & Distributors Pvt. Ltd., New Delhi for his encouragement and help in bringing out this publication in a present nice form.

My special thanks to Sh. Y.N. Arjuna, Senior Director, Publishing, Editorial and Publicity, CBS Publishers & Distributors, New Delhi whose encouragement and unstinted support enabled me to complete my book. Sh. B.M. Singh and Sh. Sunil Dutt, CBS Publisher & Distributors deserve special mention for their kind help and support. Mr. Peeyush Goel of M/s Dreamshapers also deserves special mention for nice typesetting.

I must also record my appreciation due to my wife Dr. Rimple, daughter Rijuta and son Shreesh for their understanding and love during the long period that I have taken to complete this book.

Above all, I am thankful to The Almighty God, without whose grace nothing is possible for anyone.

Readers are welcome to point out errors, if any, and send their valuable suggestions for improving the quality of the book.

Dr. Sudhir Kumar Pundir
email: skpundir05@yahoo.co.in

Contents

Chapter 1   Introduction — 1–62

1.1	Introduction	1
1.2	Number System	1
1.3	Interval	2
1.4	Concept of Sets	3
1.5	Type of Sets	6
1.6	Subset	7
1.7	Universal Set	9
1.8	Venn Diagrams	13
1.9	Operations on Sets	13
1.10	Some Results on Venn Diagram	21
1.11	Ordered Pair	23
1.12	Relation	28
1.13	Classification of Relations	30
1.14	Functions	39
1.15	Type of Functions	40
1.16	Binary Operation	47
1.17	Number of Binary Operations	47
1.18	Properties of Binary Operation	47
1.19	Algebraic Structure	48
1.20	Identity Element	52
1.21	Inverse of an Element	53
1.22	Composition Table for Binary Operation on Finite Sets	55

Chapter 2   Matrix Algebra — 63–136

2.1	Introduction	63
2.2	Type of Matrices	63
2.3	Operation on Matrices	65
2.4	Properties of Matrix Addition	67
2.5	Properties of Multiplication of Matrix by a Scalar	68
2.6	Multiplication of Matrices	72
2.7	Properties of Matrix Multiplication	73
2.8	Determinant of a Square Matrix	75
2.9	Properties of Determinants	76
2.10	Evolution of a Determinant by Sarrus Diagram	76
2.11	Minor and Cofactors	76
2.12	Singular and Non-Singular Matrix	77
2.13	Transpose of a Matrix	77
2.14	Properties of Transpose of a Matrix	78
2.15	Symmetric Matrix	78
2.16	Skew-Symmetric Matrix	79
2.17	Properties of Symmetric and Skew-Symmetric Matrix	79
2.18	Complex Matrix	80
2.19	Submatrix of a Matrix	81
2.20	Minors of a Matrix	86
2.21	Rank of a Matrix	87
2.22	Echleon From of a Matrix	87
		88

2.23 Elementary Transformations (or E-Transformations) of a Matrix   92
2.24 Elementary Matrices   93
2.25 Invariance of Rank under E-Transformations   95
2.26 Normal Form   96
2.27 Equivalence of Matrices   98
2.28 Row and Column Equivalence of Matrices   99
2.29 Inverse of a Matrix   116
2.30 Inverse of a Matrix by Elementary Transformations   121

Chapter 3   System of Linear Equations   137–176

3.1 Introduction   137
3.2 Vectors and their Dependence and Independence   137
3.3 Homogeneous Linear Equations   140
3.4 Nature of the Solution of the Equation $AX = 0$ 142
3.5 Non-Homogeneous Equations   148
3.6 Condition for Consistency   148
3.7 Gauss Elimination Method   155
3.8 LU Decomposition Method   163

Chapter 4   Eigen Values and Eigen Vectors of a Matrix   177–234

4.1 Introduction   177
4.2 The Characteristic Equation of a Matrix   177
4.3 Characteristic Vectors or Eigenvectors of a Matrix   178
4.4 Relation Between Eigenvalues and Eigenvectors   178
4.5 Eigenvalues of Special Type of Matrices   180
4.6 The Cayley-Hamilton Theorem   202
4.7 Diagonalisation of a Matrix   214
4.8 Algebraic and Geometric Multiplicity of an Eigenvalue   222

Chapter 5   Vector Spaces   235–348

5.1 Introduction   235
5.2 Prologue to Vector Space   235
5.3 Vector Spaces   237
5.4 Elementary Properties of Vector Spaces   251
5.5 Vector Subspaces: Vector Space within Vector Space   253
5.6 Elementary Properties of Vector Subspaces   253
5.7 Algebra of Subspaces   254
5.8 Linear Sum of Two Subspaces   264
5.9 Direct Sum of Vector Subspaces   265
5.10 Linear Combination of Vectors   273
5.11 Linear Dependence and Independence of Vectors   275
5.12 Basis of a Vector Space   290
5.13 Finite Dimensional Vector Space   290
5.14 Dimension of Subspace of a Vector Space   294
5.15 Cosets   317
5.16 Quotient Spaces   318
5.17 Isomorphism   321

Chapter 6   Linear Transformations   349–418

6.1 Introduction   349
6.2 Some Definitions   350
6.3 Properties of Linear Transformations   350
6.4 Algebra of Linear Transformations   352

6.5	Linear Operator	356
6.6	Algebra of Linear Operators	356
6.7	Range and Null Space of a Linear Transformation	357
6.8	Product of Linear Transformations	358
6.9	Polynomials in a Linear Operator	360
6.10	Invertible Linear Transformation	361
6.11	Non-Singular Linear Transformations	362
6.12	Coordinate Vector	370
6.13	Matrix Representation of a Linear Transformation	370
6.14	Change of Basis	375
6.15	Similarity of Matrices	377
6.16	Similarity of Linear Transformations	378
6.17	Determinant of a Linear Transformation on a Finite Dimensional Vector Space	381
6.18	Scalar Transformation	381
6.19	Trace of a Matrix	381
6.20	Trace of a Linear Transformation on a Finite Dimensional Vector Space	382

Chapter 7 Linear Functionals — 419–470

7.1	Introduction	419
7.2	Dual Spaces	420
7.3	Dual Basis	423
7.4	Second Dual Space: Bidual Space	426
7.5	Natural Mapping	429
7.6	Annihilator	429
7.7	Annihilator of an Annihilator	431
7.8	Eigenvalues and Eigenvectors of a Linear Transformation	445
7.9	Minimal Polynomial	448
7.10	Invariance of Linear Operator	449
7.11	Diagonalization	453

Chapter 8 Inner Product Spaces — 471–554

8.1	Introduction	471
8.2	Orthogonality and Orthonormality	474
8.3	Orthogonal Expansion	480
8.4	The Adjoint of a Linear Transformation	489
8.5	Properties of the Adjoint	490
8.6	Self-Adjoint Transformation	491
8.7	Congruent Operators	498
8.8	Inner Product Vector Space Isomorphism	499
8.9	Unitary Operators	502
8.10	Normal Operators	507
8.11	Positive Operators	509
8.12	Perpendicular Projection	514
8.13	Properties of a Perpendicular Projection	516
8.14	Invariance and Reducibility in Inner Product Space	518
8.15	Orthogonal Projections	521
8.16	Characterization of Spectra	526
8.17	The Spectral Theorem for Normal Operators	530
8.18	Spectral Theorem for Self-Adjoint Operators	535

Chapter 9 Bilinear, Quadratic and Hermitian Forms — 555–644

9.1	Introduction	555
9.2	Bilinear Forms	555
9.3	Bilinear Forms and Matrices	557

9.4	Quadratic Forms	575
9.5	Real Symmetric Bilinear and Quadratic Forms: Law of Inertia	576
9.6	Orthogonal Diagonalization of the Quadratic Form	578
9.7	Quadratic Forms and Matrices	585
9.8	Matrix of Quadratic Form	586
9.9	Conversion of a Symmetric Matrix into Quadratic Form	587
9.10	Congruence Operation on a Square Matrix	593
9.11	Congruence of Quadratic Forms	593
9.12	Equivalence of Real Quadratic Forms	593
9.13	The Linear Transformation of a Quadratic Form	594
9.14	Congruent Reduction of a Symmetric Matrix	594
9.15	Rank of a Quadratic Form	596
9.16	Reduction of a Real Quadratic Form over Real Field	599
9.17	Normal (or Canonical) Form of a Real Quadratic Matrix	600
9.18	Signature and Index of a Real Quadratic Form	602
9.19	Reduction of a Real Quadratic Form over the Field of Complex Numbers	603
9.20	Orthogonal Reduction of a Real Quadratic Form	604
9.21	Classification of Real Symmetric Matrices	616
9.22	Positive-Definiteness of a Quadratic Form X′AX in Terms of Leading Principal Minors of A	622
9.23	Hermitian Forms	634
9.24	Matrix Representation of a Hermitian Form	635

Chapter 10   Canonical Forms — 645–668

10.1	Introduction	645
10.2	Similarity of Linear Transformations	645
10.3	Invariant Subspace	645
10.4	Invariant Direct-Sum Decompositions	646
10.5	Normal Form	647
10.6	Triangular Form	648
10.7	Nilpotent Transformation	651
10.8	Jordan Canonical Form	651
10.9	Rational Canonical Form	653
10.10	Raw and Column Space of a Matrix	653

Chapter 11   Modules — 669–705

11.1	Introduction	669
11.2	Modules	669
11.3	Coset R-Module	670
11.4	General Properties of Modules	670
11.5	Submodules	670
11.6	Linear Sum of Two Modules	672
11.7	Homomorphism of Modules (Linear Transformations)	673
11.8	Quotient Modules	675
11.9	Cyclic Module	675
11.10	Simple and Semi-Simple Modules	678
11.11	Free Modules	681
11.12	Noetherian and Artinian Modules	685
11.13	Filtered and Graded Modules	690
11.14	Smith Normal Form over a PID and Rank	693
11.15	Finitely Generated Modules over a PID	697

Bibliography — 707

Index — 709–712

1 INTRODUCTION

1.1 INTRODUCTION

The concept of set is fundamental in all branches of mathematics. It was developed by German mathematician George Cantor. This chapter introduces the notations and terminology of set theory. Classical which is fundamental to the study of pure mathematics and esential for all study of pure mathematics.

1.2 NUMBER SYSTEM

The number system plays a key role in mathematics. The real number system R is one of the most important and beautiful mathematical system. There are different ways of introducing the real number system, but the most common way is to start with Peano's Axioms for the natural numbers. The axioms for natural numbers, discovered by the Italian Mathematician Peano are:

   (i)   1 is a natural number
   (ii)   Each natural number n has a successor $(n+1)$.
   (iii)   Two natural numbers are equal if their successors are equal.
   (iv)   Except 1, each natural number is a successor of a natural number.
   (v)   Any set of natural numbers contains 1 and the successor $(k+1)$ of every natural number k whenever it contains k in the set N of natural numbers.

REMARKS

- Axiom (v) is commonly known as the axiom of induction or principle of finite induction.
- The above axioms completely define the set of natural numbers.

Definition: *The numbers 1, 2, 3, … are called natural numbers.* We represent the set of natural numbers by N. *i.e.,*   N = {1,2,3, …}

The Peano's axioms can be used to extend the set N of natural numbers to another large system, known as the set of integers.

Definition: *The numbers … , –3, –2, –1, 0, 1, 2, 3… are called integers. We represent the set of integers by Z.* *i.e.,*   Z = { … –3, –2, –1, 0,1,2,3,.. }

Integers are used to define the rational numbers.

Definition: *Any number of the form p/q, where* $p,q \in$ *Z, q ≠ 0 and p,q have no common factor (except ± 1) is called a rational number.*

The set of rational numbers is denoted by Q.

$$Q = \left\{ \frac{p}{q} ; p,q \in Z, q \neq 0 \right\}$$

REMARK

- The set of rational numbers consists of integers and fractions.

Definition (1). *Any number which is not rational, is called an irrational number.*

For example, $\sqrt{2}, \sqrt{3}$ etc. It should be noted that every rational number can be expressed as a terminating or recurring decimal whereas every irrational number can be expressed as a non-terminating infinite decimal.

Definition (2). *A number which is either rational or irrational is called a real number.* The set of real numbers is denoted by R.

Definition (3). *Let* $a \in R$ *and n be any positive integer then we can define* $a^n = a.a.a...n$ *times. In particular* $\quad a = a$

$$a^2 = a.a$$
$$a^3 = a.a.a = a^2.a \text{ and so on.}$$

Also, if n is any negative integer, then we have $x^{-n} = (x^n)^{-1} = (x^{-1})^n$

Definition (4). *A real number a is called positive, if* $a > 0$ *and the set of all positive real numbers, denoted by* R^+, *is given by* $R^+ = \{x : x \in R, x > 0\}$.

Definition (5). *A real number a is called negative if* $a < 0$ *and the set of all negative real numbers, denoted by* R^-, *is given by* $R^- = \{x : x \in R, x < 0\}$

1.3 INTERVAL

A subset S of R is called an interval if $a, b \in S, x \in R$ such that $a < x < b$ implies $x \in S$. There are following four type of intervals.

(i) $a \circ\!\!-\!\!-\!\!-\!\!-\!\!-\!\!-\!\!\circ b \quad \Rightarrow \quad]a, b[= \{x : a < x < b\}$

(ii) $a \bullet\!\!-\!\!-\!\!-\!\!-\!\!-\!\!-\!\!\bullet b \quad \Rightarrow \quad [a, b] = \{x : a \leq x \leq b\}$

(iii) $a \circ\!\!-\!\!-\!\!-\!\!-\!\!-\!\!-\!\!\bullet b \quad \Rightarrow \quad]a, b] = \{x : a < x \leq b\}$

(iv) $a \bullet\!\!-\!\!-\!\!-\!\!-\!\!-\!\!-\!\!\circ b \quad \Rightarrow \quad [a, b[= \{x : a \leq x < b\}$

Observations

➠ The set $]a, b[$ in which the end points are not included, is called an open interval.

➠ The set $[a, b]$ also contains both its end points, is called a closed interval.

➠ The sets $[a, b[$ and $]a, b]$ are called half open (or half closed) intervals or semi-open (or semi-closed) as they contain only one end point.

REMARKS

- If S is any interval and if c and d are two elements of S, then all numbers lying between c and d are also elements of S.

- The proper use of a bracket, for example, parenthesis' for open and square brackets for closed and end points, itself specifies the interval. As such, to emphasize the nature of an interval, we shall drop the used 'description' and shall simply express the interval by using the appropriate brackets.

1.3.1 LENGTH OF AN INTERVAL

The number $b - a$ is called length of the intervals $]a, b[, [a, b[,]a, b]$ and $[a, b]$. If the length of the interval is finite, the interval is said to be finite and if the length is infinite, then it is known as infinite interval.

1.3.2 ABSOLUTE VALUE OF A REAL NUMBER

The absolute value of a real number a denoted by $|a|$ is the real number a, – a or 0 according as a is positive, negative or zero, i.e.,

$$|a| = \begin{cases} a & if \quad a \geq 0 \\ -a & if \quad a < 0 \end{cases}$$

From the above definition, it is clear that

(i) $|a| = \max\{a, -a\}$　　(ii) $-|a| = \min\{a, -a\}$　　(iii) $|a| \geq a \geq -|a|$

1.3.3 SOME USEFUL RESULTS

(i) $|xy| = |x| \cdot |y|$　　　　　　　　　　(ii) $|x+y| \leq |x| + |y|$

(iii) $|x - y| \geq ||x| - |y||$　　　　　　(iv) $|x - y| \leq |x| + |y|$

(v) $\left|\dfrac{x}{y}\right| = \dfrac{|x|}{|y|}$

(vi)　If $\in > 0$, then $|x - y| < \varepsilon \Leftrightarrow y - \in < x < y + \in$

1.4 CONCEPT OF SETS

The theory of sets is one of the most important tools of pure mathematics. Pure mathematics is the study of sets equipped with assigned structures, known as mathematical systems. In this section, we shall study some fundamental concept of set theory.

Definition: 'A set is a well defined collection of objects'.

The objects of a set are called the elements or members of that set and their membership is defined by certain conditions.

The basic concept used can be defined. Suppose, for example, one defines the set by "A set is well defined collection of objects" then what is meant by a collection. Perhaps, then one defines "A collection is an aggregate of things". What then is an aggregate? How our language is finite, so after sometime we will run out of new words to use and have to repeat some words already questioned. The definition is then circular and obviously worthless.

Mathematicians realize that there must be some undefined or primitive concept at the moment they have agreed that set shall be such a primitive concept.

The sets are usually denoted by the capital letters of English alphabets, say *A,B,C,... , X, Y, Z.*

For example :

(1)　The collection of the letters *a, b, c, d,...*

(2)　The collection of all natural numbers denoted by N.

(3)　The students of M.Sc., Mathematics in C.C.S. University, Meerut.

(4)　The collection of vowels in English alphabet. This set containing only five elements, namely *a, e, i, o, u.*

(5)　The collection of all states in Indian union.

If *S* is a set, an object *a* in the collection *S* is called an element of *S*. This fact is expressed in symbol as $a \in S$ (read as *a* is in *S* or *a* belongs to *S*). If *a* is not in *S*, we write $a \notin S$. For example, $4 \in R$, the set of real numbers, but $\sqrt{-2} \notin R$.

Here, Greek letter $\in$ denotes 'belongs to'. It is the abbreviation of the Greek word meaning 'is'.

REMARKS

- By the term 'well defined' we mean that we are given a collection of objects, with certain definite property, so that we are able to determine whether a given object belongs to our collection or not. Thus, every collection of objects is not a set.
- Set and aggregate both have the same meaning.
- The elements of a set must be distinguished from one another. For example, the collection of sand particles does not form a set.
- The collection of rich persons of a city is not a set. However the collection of those persons of city whose wealth exceeds, a fixed amount, say ruppes fifty thousands, is a set.
- The order is not preserved in case of a set, whereas order is necessarily preserved in case of sequence. That is to say, each of the sets {1,2,3}, {3,2,1}, {1,3,2} denotes the same sets.
- The repetition of an element does not change the nature of a set, i.e., each of the sets {1,2,3}, {1,2,2,3}, {1,3,3,2} denotes the same sets.

1.4.1 REPRESENTATION OF A SET

There are two ways of representing a set:

   (i)   Roster or tabulation method

   (ii) Set-builder or rule method

Roster Method: In this method, the elements of the set are listed within braces, and separated by comma.

For example:

(1)   A= {1,2,3,4,5,6}

(2)   The set of vowels of English alphabet may be represent as {a, e, i, o, u},

(3)   The set of a natural numbers from 1 to 100 may be written as N = {1,2,3, ..., 100}. We use three dots in the middle to include the missing elements.

(4)   The set of positive integers, which is a non-ending set may be written as Z^+ = {1,2,3,4,5, ...}. The three dots in the end means that the elements continue in the same manner.

(5)   The set of prime number is written as P = {2,3,5,7,11,13,17,19....}

Set-Builder Method: In this method, we first try to find a property which characterizes the elements of a set, that is, a property P, which all the elements of the set possess and which no other objects possess. Then, we describe the set as {$x : x$ has property P}.

This is to be read as "the set of all x such that x has property P".

For example:

(1)   The set of all integers can be written as Z= {$x : x$ is an integer}

(2)   The set A = {1, 2, 3,4, 5} can be written as A = {$x \in N : x \leq 5$}.

(3)   The set of complex numbers can be written as C = {$a + ib : a, b \in R$}

(4)   The set A = {1,8,27, ...} can be written as A = {$x^3 : x \in Z^+$}.

Solved Examples

Example 1. *Use the Roster method to identify each set:*

(a) *The set of possible integers greater than 8 and less than 14.*

(b) *The set of numbers whose elements are the first five positive odd integers.*

(c) *The set of even positive integers.*

(d) *The set of even positive integers that are divisible by 10.*

(e) *The set of all vowels in English alphabets which precedes r.*

Solution. (a) {9, 10, 11, 12, 13} (b) {1,3,5,7,9}

(c) {2, 4, 6, 8, 10 ...} (d) {10,20,30,40,50 ...}

(e) {a, e, i, o}

Example 2. *Use the set-builder method, identify the following sets :*

(a) $A = \{1,3,5,7,9,...\}$ (b) $B = \left\{1, \dfrac{1}{4}, \dfrac{1}{9}, \dfrac{1}{16}, \dfrac{1}{25}, ...\right\}$

(c) $C = \{0,1, 2, 3, ...\}$ (d) $D = \left\{\dfrac{1}{2}, \dfrac{2}{3}, \dfrac{3}{4}, \dfrac{4}{5}, ...\right\}$

Solution. (a) The set of odd positive integers.

(b) Here, elements of the set B are the reciprocals of the squares of the natural numbers.

So, the set $B = \left\{\dfrac{1}{n^2} : n \in N\right\}$

(c) The set of whole numbers.

(d) Here, each element in the given set has the denominator one more than the numerator. Hence,

$$D = \left\{x : x = \dfrac{n}{n+1} : n \in N\right\}$$

Example 3. *Write the set* $\left\{\dfrac{1}{2}, \dfrac{2}{5}, \dfrac{3}{10}, \dfrac{5}{26}, ...\right\}$ *in the set-builder form.*

Solution. We observe that each element in the given set has the denominator one more than the square of the numerator. Also, the numerator begins with 1. Hence, in the set builder form, the given set can be written as

$$\left\{x : x = \dfrac{n}{n^2 + 1} : n \in N\right\}$$

❧ EXERCISE 1.1

1. Which of the following collections are sets?

(i) All mathematics students in your college.

(ii) All poor hockey players in the college.

(iii) All odd numbers less than 20.

(iv) The collection of good teachers in your college.

(v) All successful and rich people in your city.

(vi) The people in your immediate family (father, mother, sister, brother).

2. Write the members of each of following sets by the Roster method.
   (i) {$x : x$ is an odd whole number less than 14}
   (ii) {$x : x^2 < 36$ and $x \in N$}
   (iii) {$x :$ squares of all whole numbers less than 8}
   (iv) {$x : x$ is a prime number, $10 < x < 20$}
   (v) {$x: x$ is a composite number less than 20}

3. Rewrite the following sets using set-builder method.
   (i) $A = \{2, 4, 6, 8, \ldots\}$
   (ii) $B = \left\{1, \dfrac{1}{2}, \dfrac{1}{3}, \dfrac{1}{4}, \ldots\right\}$
   (iii) $C = \{0, 3, 6, 9, 12, \ldots\}$
   (iv) $D = \{0, 4, 6, 8, 10, \ldots\}$

4. List the elements of the following sets.
   (i) $A = \{x : x^2 \leq 16 : x \in Z\}$
   (ii) $B = \{x : 1 \leq x \leq 5$ and $x \in N\}$
   (iii) $C = \{x : x \in N$ and x is a proper factor of 5}
   (iv) $D = \{x : x$ is a month of year having 31 days}

5. Use the appropriate symbols $\in$ or $\notin$ to fill in the blanks below:
   (i) 12 … the set of all numbers dividing 84.
   (ii) K … the set of all vowels of the English alphabets.
   (iii) $\dfrac{1}{2}$ … the set of natural number.
   (iv) India … the set of members of UNO.
   (v) $\sqrt{2}$ …. The set of rational number
   (vi) 15 … the set of multiples of 3.

Answers

1. (i), (iii), (vi)   **2.** (i) {1, 3, 5, 7, 9, 11, 13} (ii) {1, 2, 3, 4, 5} (iii) {0, 1, 4, 9, 16, 25, 36, 49}
   (iv) {11, 13, 17, 19} (v) {1, 4, 6, 8, 9, 10, 12, 14, 15, 16, 18} **3.** (i) $A = \{x : x = 2n : n \in N\}$
   (ii) {$1/n : n \in N$} (iii) {$x : x = 3n, n$ is the whole number (iv){$x : x = 2n, n$ is the whole number}
4. (i) {–4, –3, –2, –1, 0, 1, 2, 3, 4} (ii) {1, 2, 3, 4, 5} (iii) {3, 5}
   (iv) {Jan, March, May, July, August, October, December}
5. (i) $\in$ (ii) $\notin$ (iii) $\notin$ (iv) $\in$ (v) $\notin$ (vi) $\in$

1.5 TYPE OF SETS

(i) **Empty Set:** A set containing no elements is called empty set and is denoted by the symbol ϕ.

For example:
(1) $\phi = \{x : x$ is a negative integer whose square is –1}
(2) $\phi = \{x : x$ is a natural number lying between 2 and 3}
(3) $\phi = \{$the set of such persons, who never die}
(4) $\phi = \{x : x$ is a real number, $x^2 < 0\}$
(5) $\phi = \{x : x$ is an even prime number greater than five}
(6) $\phi = \{$the set of real numbers which are the solution of equation $x^2 + 1 = 0\}$
(7) $\phi = \{x : x$ is a straight ling passing through three distinct points on a circle}

REMARKS

- The empty set is also known as null set or void set.
- The Roster method, the empty set is denoted by {}.
- To describe the null set, we can use any property, which is not true for any element.
- It is wrong to use the expression 'an empty' or 'a null set' as there is one and only one empty set through, it may have many-many descriptions. We shall always call 'The empty or the null set.'
- A set consisting of at least one element is called a non-empty or non-void set.
- {ϕ} is not a null set.

(ii) **Singleton Set:** Set containing only one element is a singleton set. The set $\{a\}$ is a singleton set.

REMARKS

- $\{0\}$ is not a null set, since it contains 0 as its member. It is a singleton set.
- A room containing only one man is not same thing as a man. In a similar way, the singleton set $\{a\}$ is not the same thing as the element a.

(iii) **Finite Set:** A set is said to be finite if it consists of only finite number of elements. Here, the process of counting the different elements comes to an end.

    For example :

    (1) Set of natural numbers less than 50.

    (2) Set of all persons in a city.

    (3) Set of English alphabets.

    (4) Set of all persons on the earth.

(iv) **Infinite Set:** A set which is not finite, *i.e.,* it contains infinite number of elements. Here, process of counting the different elements never comes to an end.

    For example :

    (1) Set of natural numbers N $= \{1,2,3, \ldots\}$

    (2) Set of all points of plane.

    (3) Set of all even integers.

    (4) Set of rational numbers lying between two integers.

(v) **Equal Sets:** Two sets are said to be equal if they contain exactly the same elements.

    For example:

                  $A = \{x : x$ is a letter in the word 'Area'$\}$, *i.e.,* $A = \{a, r, e\}$

    And $B = \{y : y$ is a letter in the word 'ear'$\}$, *i.e.,* $B = \{a, r, e\}$

    Here A and B are equal sets.

1.5.1 CARDINAL NUMBER OF A SET

The number of distinct elements contained in a finite set A is called cardinal number of A and is denoted by $n(A)$.

1.5.2 EQUIVALENT SETS

Two finite sets are said to be equivalent if they have the same cardinal number.

REMARKS

- Equivalent sets are not always equal but equal sets are always equivalent.
- The number of distinct elements in a finite set is also called the order of the set. If the order of a set is zero, the set is empty.
- If the order of a set is one, the set is singleton.
- The order of an infinite set is never defined.

1.6 SUBSET

Let A and B be two sets. Then set A is said to be a subset of the set B if every element of A is also an element of B. Symbolically, we write $A \subseteq B$.

When A is subset of B, it means that 'A is contained in B' or 'B contains A'. Here B is called superset of A and is written as $B \supset A$.

REMARKS

- Every set is a subset of itself.
- Empty set is a subset of every set.
- If A is not a subset of B, we write $A \nsubseteq B$.
- An element cannot be a subset of a set, only a set can be subset of a set.

1.6.1 PROPER SUBSET

We know that for A to be a subset of B all that is needed is that every element of A is in B. It is possible that every element of B may or may not be in A. If it so happens that every element of B is also in A, then we will have $B \subset A$. Obviously, then A and B are the same set, so that we have $A \subset B$ and $B \subset A \Leftrightarrow A = B$.

If every element of A is in B, but every element of B is not in A , i.e., if $A \subset B$ and $B \not\subset A$, then A is said to be a proper subset of B.

For example :

(1) $\{a, b\}$ is a proper subset of $\{a, b, c\}$.

(2) Set of natural numbers N is a proper subset of set Z of integers.

REMARKS

- Here, it follows that every element of A is an element of B and B contains at least one element which does not belong to A.
- If the subset is not proper, it is called *improper subset*. For example, $A \subseteq A$ and $\phi \subseteq A$ are improper subsets.

1.6.2 NUMBER OF SUBSETS OF A SET

Let A contains n distinct elements such that $0 < r \leq n$. If we consider those subsets of A that have r elements each, then the number of ways in which r elements can be choosen out of n elements is ^nC_r. Therefore, the number of subsets of A having r elements each is ^nC_r.

Hence, the total number of subsets of A is equal to

$$^nC_0 + {}^nC_1 + {}^nC_2 + ... + {}^nC_n = (1+1)^n = 2^n$$

For example:

(1) If a set A has one element, then it has $2^1 = 2$ subsets.

(2) If a set A has two elements, then it has $2^2 = 4$ subsets.

REMARKS

- The number of proper subsets of a set with n elements is $2^n - 2$.
- The collection of all possible subsets of a given set A is called power set. It is denoted by $P(A)$. For example : If $A = \{1,2,3\}$ then the power set $P(A) = \{\phi, \{1\}, \{2\}, \{3\}, \{1,2\}, \{1,3\}, \{2,3\}, \{1,2,3\}\}$.
- $P(\phi) = \{\phi\}$
- The power set of any given set is always non-empty.

1.7 UNIVERSAL SET

In any discussion , we are given particular set and we consider different subsets of the given set. This given set is called Universal Set. It is denoted by U.

For Example :

(1) The universal set is of real numbers R, while considering the set of natural numbers, whole numbers, integers and rational numbers.

(2) The set of alphabets is the universal set from which the letters of any word may be chosen to form a set.

(3) In geometry, we discuss set of lines, triangles and circles, then the universal set is the plane, in which the lines, triangles and circles lie.

REMARKS

- Universal set is a super set of each of the given sets.
- The universal set is not unique.

1.7.1 COMPLEMENT OF A SET

Let U be the universal set and the set $A \subseteq U$. Complement of set A with respect to the universal set U is the set of all those elements of U which are not the elements of A and is denoted by A' or A^c, i.e.,
$$A' = \{x : x \in U \text{ and } x \notin A\}$$

For example :

(i) If $U = \{1, 2, 3, 4, 5, 6, 7, 8, 9, 11\}$ and $A = \{1, 2, 3\}$ then $A' = \{4, 5, 6, 7, 8, 9, 11\}$.

REMARKS

- Complement of the universal set is the null set and *vice-versa*.
- $(A')' = A$
- If $A \subseteq B$, then $B' \subseteq A'$.
- $x \in A' \Leftrightarrow x \notin A$

Solved Examples

Based on the following Results

➠ A set containing no element is called empty set.

➠ Set containing finite number of elements is called finite otherwise infinite.

➠ The number of distinct elements contained in a finite set is called its cardinality.

➠ Two finite sets are said to be equivalent if they have same cardinality.

➠ A set A is said to be subset of a set B if every elements of A belongs to B.

➠ Total number of subsets of a set A of n elements is 2^n.

➠ $A' = \{x : x \in A' \text{ and } x \notin A\}$

Example 1. *Let $A = \{1,2,3\}$, then find $P(A)$.*

Solution. Since $A = \{1, 2, 3\}$ then,
$$P(A) = \{\phi, \{1\}, \{2\}, \{3\}, \{1,2\}, \{1,3\}, \{2,3\}, \{1,2,3\}\}$$

Example 2. Let $A = \{a,b,c,d\}$, $B = \{a,b,c\}$ and $C = \{b,d\}$, find all sets X such that

(i) $X \subset B$ and $X \subset C$ (ii) $X \subset A$ and $X \not\subset B$

Solution. (i) Here, we have

$$P(B) = \{\phi, \{a\}, \{b\}, \{c\}, \{a, b\}, \{a,c\}, \{b, c\}, \{a, b, c\}\}.$$

And $P(C) = \{\phi, \{b\}, \{d\}, \{b, d\}\}$, then $X \subset B$ and $X \subset C$ implies

$$X \in P(B) \text{ and } X \in P(C)$$

$$X = \{\phi, \{b\}\}$$

(ii) Here, we have, $X \subset A$ and $X \not\subset B$, which implies that

$$X \in P(A) \text{ and } X \notin P(B)$$

Therefore

$$X = \{\{d\}, \{a,b,d\}, \{b,c,d\}, \{a,c,d\}, \{a,d\}, \{b,d\}, \{c,d\}, \{a,b,c,d\}\}$$

Example 3. Write down all the subsets of the following sets.

(i) $\{a\}$ (ii) $\{a,b\}$ (iii) $\{a,b,c\}$ (iv) ϕ

Solution. (i) Let $A = \{a\}$. Since A contains only one element, therefore, the total number of subsets is $2^1 = 2$, which are given by ϕ and $\{a\}$.

(ii) Here, total number of subsets, $= 2^2 = 4$, which are given by ϕ, $\{a\}, \{b\}$, $\{a, b\}$

(iii) Here, total number of subsets $= 2^3 = 8$, and given by

$$\phi, \{a\}, \{b\}, \{c\}, \{a,b\}, \{a,c\}, \{b,c\}, \{a,b,c\}$$

(iv) since ϕ contains no element therefore the number of subsets $= 2^0 = 1$. The only subset is ϕ.

Example 4. Which of the following sets are empty. Also, give the reason.

(i) $A = \{x : x \neq x, \text{ is a real number}\}$.

(ii) $B = \{x : x + 4 = 4\}$

(iii) $C = \{x : x^3 - 3 = 0 \text{ and } x \text{ is rational number}\}$

Solution. (i) Here, $A = \{x : x \neq x, x \text{ is a real number}\}$. Since $x \neq x$ is not true

$$\Rightarrow \quad A = \phi$$

(ii) $B = \{x : x + 4 = 4\} = \{x : x = 0\} = \{0\}$

$$\Rightarrow \quad B \text{ has one element } 0, \text{ therefore } B \neq \phi.$$

(iii) Since there is no rational number whose square is 3, so $x^3 - 3 = 0$ is not satisfied for any rational numbers. Therefore, C is an empty set.

Example 5. Which of the following sets are finite and which are infinite.

(i) The set of natural numbers divisible by 2.

(ii) The set of natural numbers less then 8.

(iii) The set of integers whose square is even.

(iv) The set of integers greater than −18.

(v) The set of lines passing through a point.

(vi) *The set of points of a plane at a fixed distance from a given point in the plane.*

(vii) *The set of points common to two given parallel lines.*

(viii) *The set of the roots of a polynomial of nth degree.*

Solution.

(i) The given set is $\{2, 4, 6, 8, ...\}$. It has an infinite number of elements, therefore it is an infinite set.

(ii) The given set is $\{1,2,3,4,5,6,7\}$. It has seven elements, *i.e.*, finite number of elements. Hence, it is a finite set.

(iii) The given set is $\{..., -8, -6, -4, -2, 0, 2, 4, 6, 8,...\}$. It has infinite number of elements, therefore it is an infinite set.

(iv) Here, the given set is $\{-17, -16, ..., 0, 1, 2 ...\}$. It has infinite number of elements therefore, it is an infinite set.

(v) Since infinite number of lines can pass throught a fixed point, therefore the given set is an infinite set.

(vi) Since the points in a plane at a fixed distance from a given point in the plane lie on a circle with the given point as center and the number of points on a circle is infinite. Therefore, the given set is an infinite set.

(vii) Since two parallel lines cannot meet anywhere, therefore, the set of points common to two given parallel lines is empty, therefore the given set cannot be infinite. Hence, it is a finite set.

(viii) Since, a polynomial of nth degree always have atmost n roots.

Therefore, the given set is always a finite set.

Example 6. *Which of the following sets are equivalent ϕ, $\{0\}$ and $\{\phi\}$.*

Solution. Since ϕ has no element. Also, $\{0\}$ and $\{\phi\}$, each contains one element namely 0 and ϕ respectively. Hence, $\{0\}$ and $\{\phi\}$ are equivalent.

Example 7. *Which of the following sets are equal ?*
$$A = \{1,2,3\}, B = \{2,3,4\}, C = \{3,2,1\}, D = \{2,3,5\}$$

Solution. Since $1 \in A$ but $1 \notin B$, therefore $A \neq B$. A and C have exactly the same element, therefore $A = C$.

Also,
$$1 \in C \text{ but } 1 \notin D \Rightarrow C \neq D$$
$$4 \in B \text{ but } 4 \notin C \Rightarrow B \neq C$$
$$4 \in B \text{ but } 4 \notin D \Rightarrow B \neq C$$
$$1 \in A \text{ but } 1 \notin D \Rightarrow A \neq D$$

Hence, only A and C are equal sets.

❧ EXERCISE 1.2

1. Fill in the blanks:

(i) A set which contains no element is called ... set.

(ii) If $A = \{1,2,3\}$ and $B = \{3,2,1\}$ then they are said to be ...

(iii) If $A = \{a, b, c\}$ and $B = \{c, d, e\}$ then they are said to be ...

(iv) If every element of a set B is also an element of A, then B is said to be ... of A.

(v) The empty set is a ... of every set.

(vi) Every set is a .... of itself.

(vii) The set Z of integers is a ... of set of natural numbers N.

2. Which of the followings sets are equal?

   (i) $A = \{1,2,3\}$

   (ii) $B = \{1,2,2,3\}$

   (iii) $C = \{x \in R : x^3 - 6x^2 + 11x - 6 = 0)$

3. Which of the following sets are equivalent to the set $\{4,7,11,17,20\}$?

   (i) $\{5,1,2,3,4\}$

   (ii) {all odd numbers less then 10}

   (iii) {the months of 30 days of a year}

   (iv) {all the prime numbers which lie between 10 and 25}.

4. Which of the following sets are finite and which are infinite ?

   (i) $\{x \in N : x > 10\}$

   (ii) $\{x \in N : x < 100\}$

   (iii) $\{x \in R : 1 \le x \le 2\}$

   (iv) Set of vowels in English alphabets.

   (v) The set of prime numbers less than 100.

   (vi) The set of multiple of 8.

5. Which of the following statements are true? Give the reason.

   (i) For any two sets A and B either $A \subseteq B$ or $B \subseteq A$

   (ii) Every subset of a finite set is finite.

   (iii) A subset of an infinite set may be finite.

   (iv) Every set has a proper subset.

   (v) A set containing n elements have 2^n subsets.

   (vi) If $A = \{1,2,3,4,5,6\}$ and $B = \{$whole numbers less than 6$\}$, then $A = B$.

(vii) The empty set has no proper subset.

6. Examine which of the following sets are empty?

   (i) The set of tigers in your class.

   (ii) The set of triangles having three equal sides.

   (iii) The set of all numbers which, when added to zero, yield sum greater than the original.

   (iv) The set of odd numbers which are divisible by 2.

   (v) The set of men, who never die.

7. Which of the following statements are true?

   (i) If $x \in A$ and $A \subset B$, then $x \in B$

   (ii) If $A \subset B$ and $B \subset C$, then $A \subset C$

   (iii) If $A \not\subset B$ and $B \not\subset C$, then $A \not\subset C$

   (iv) If $x \in A$ and $A \not\subset B$, then $x \in B$

   (v) If $A \subset B$ and $x \notin B$, then $x \notin A$

8. Are the following sets, *i.e.*, (A and B) are equal.

   (i) $A = \{x : x$ is a letter of the word 'LITTLE'$\}$
     $B = \{x : x$ is a letter in the word 'TITLE'$\}$

   (ii) $A = \{x : x$ is a letter in the word 'FOLLOW'$\}$
     $B = \{x : x$ is a letter in the word 'WOLF'$\}$

   (iii) $A = \{x : x$ is a letter in the word 'LOYAL'$\}$
     $B = \{x : x$ is a letter In the word 'ALLOY'$\}$

9. Write down all possible subsets of each of the following sets.

   (i) $\{a\}$ (ii) $\{0,1\}$ (iii) $\{a, b, c\}$

   (iv) $\{1, \{1\}\}$ (v) ϕ

10. Which of the following statements are true?

   (i) $\{a, \phi\} \in \{a, \{a, \phi\}\}$

   (ii) If $A \subseteq B$ and $B \subseteq C$, then $A \subseteq C$

   (iii) If $a \in B$ and $B \subseteq C$, then $a \in C$

Answers

1. (i) Empty (ii) equal (iii) equivalent (iv) subset (v) subset (vi) subset (vii) super set.

2. $A = B = C$ 3. (i), (ii), (iv)   4. (ii), (iv), (v) are finite sets and (i), (iii), (vi) are infinite.

5. (i) F   (ii) T   (iii) T   (iv) F   (v) T   (vi) F   (vii) T   6. (i), (iii), (iv), (v)

7. (i), (ii), (v)   8. (i) Equal, (ii) Equal, (iii) Equal

9. (i) ϕ, $\{a\}$; (ii) ϕ,$\{0\}$,$\{1\}$,$\{0,1\}$; (iii) ϕ,$\{a\}$,$\{b\}$,$\{c\}$, $\{a, b\}$,$\{b,c\}$,$\{a,c\}$,$\{a,b,c\}$

   (iv) $\{1\}$; $\{1\}$, $\{\{1\}\}$, $\{1,\{1\}\}$;

10. (i) , (ii), (iii)

1.8 VENN DIAGRAMS

A set can be represented by closed figures like circles, triangles, rectangles, etc. The point in the interior of the figure represents the elements of the set. Such a representations is called a Venn diagram. In Venn diagram, the universal set is usually represented by a rectangular region and its subset by closed bounded regions inside the rectangular region. For example, if A is a subset of B, i.e., $A \subset B$. This is shown in figure 1.

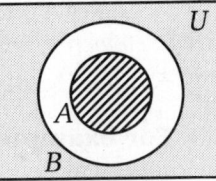

Fig. (1)

REMARKS

- The diagrams drawn to represent sets are called Venn diagram or Venn-Euler diagrams, after the name of British mathematician **Venn.**
- If A and B are two sets, which are not equal, but have common elements, then to represent A and B, We draw two intersecting circles.
- Two disjoint sets are represented by two non-intersecting circles.
- Venn diagrams are to be used for clarity and are no substitute for precise proof.

1.9 OPERATIONS ON SETS

1.9.1 UNION AND INTERSECTION OPERATIONS

(i) Union of Two sets

Let A and B be two sets. Then Union of A and B, denoted by $A \cup B$ is the set of all those elements, which either belongs to A or B or to both A and B.

It should be noted that the common elements are to be taken only once.

Symbolically: $A \cup B = \{x : x \in A \text{ or } x \in B\}$ It is shown in the adjoining figure 2.

For example :

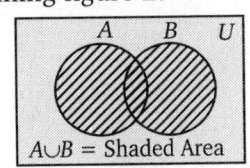

$A \cup B$ = Shaded Area

Fig. (2)

(1) Let $A = \{3,4,5,6,7\}$ and $B = \{5,6,7,8,9\}$

Then $A \cup B = \{3, 4, 5, 6, 7, 8, 9\}$

(2) Let $A = \{x : x = 2n, n = 1, 2, 3, ...\} = \{2, 4, 6, 8, ...\}$

$B = \{x : x = 3n, n = 1, 2, 3, ...\} = \{3, 6, 9, 12, ...\}$

Then $A \cup B = \{x : x \text{ is multiple of } 2 \text{ or a multiple of } 3\}$
$= \{2, 3, 4, 6, 8, 9, 10, 12, ...\}$

(3) Let A = set of even natural numbers = $\{2, 4, 6, 8,...\}$
and B = set of natural numbers = $\{1, 2, 3, 4, 5,...\}$

Then $A \cup B = \{1, 2, 3, 4,...\}$

REMARKS

- $x \in (A \cup B) \Leftrightarrow x \in A \text{ or } x \in B$.
- $x \notin (A \cup B) \Leftrightarrow x \notin A \text{ and } x \notin B$
- $A \cup B = B \cup A$, i.e., union of sets is commutative.
- $A \cup A' = U \text{ and } A \cup U = U$
- $A \cup \phi = A$
- If $A, B, C, D, ..., Z$ is a finite family of sets, then their union is denoted by $A \cup B \cup C \cup D ... \cup Z$.
- $(A \cup B) \cup C = A \cup (B \cup C)$, i.e., a union of sets is associative.

(ii) Intersection of Two sets

Let A and B be two sets. Then intersection of A and B, denoted by $A \cap B$ is the set of all those elements, which belongs to both A and B.

Symbolically: $A \cap B = \{x : x \in A \text{ and } x \in B\}$ It is shown in the adjoining figure 3.

For example:

(1) Let $\quad A = \{2, 4, 6, 8, 10\}$ and $B = \{1, 2, 3, 4, 5\}$

Then $A \cap B = \{2, 4\}$

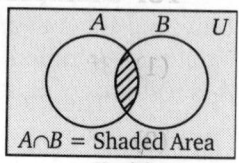

$A \cap B$ = Shaded Area

Fig. (3)

(2) If $\quad A = \{x : x = 3n, n \in Z\}$

$\qquad B = \{x : x = 4n, n \in Z\}$

Then $A \cap B = \{x : x \text{ is multiple of } 3 \text{ and } x \text{ is a multiple of } 4\}$

$\qquad = \{x : x \text{ is multiple of } 3 \text{ and } 4 \text{ both})$

$\qquad = \{x : x = 12n, n \in Z\}$

REMARKS

- $x \in (A \cap B) \Leftrightarrow x \in A \text{ and } x \in B$.
- $x \notin (A \cap B) \Leftrightarrow x \notin A \text{ or } x \notin B$
- $A = A \cap A$, i.e., intersection of sets is idempotent.
- $A \cap \phi = \phi$
- $A \cap U = A$, where U is a universal set.
- $A \cap B = B \cap A$, i.e., intersection of sets is commutative.
- $(A \cap B) \cap C = A \cap (B \cap C)$ intersection of sets is associative.
- If $A, B, C, D, \ldots, Z$ is a finite family of sets, then their intersection is denoted by $A \cap B \cap C \ldots \cap Z$.

(iii) Distributive Property of Union and Intersection

(i) $A \cup (B \cap C) = (A \cup B) \cap (A \cup C)$　(ii) $A \cap (B \cup C) = (A \cap B) \cup (A \cap C)$

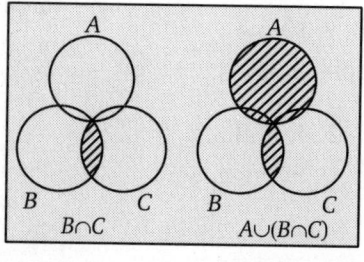

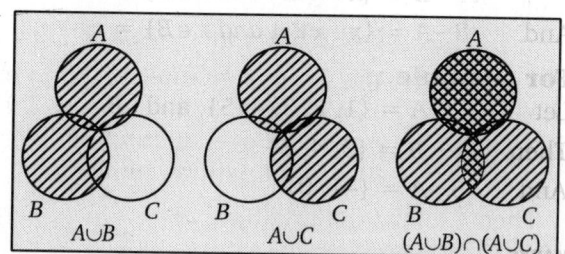

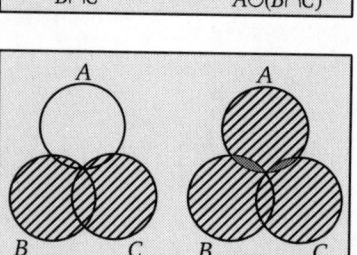

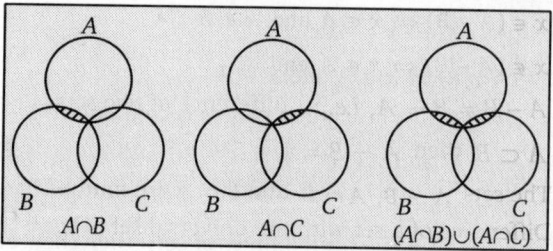

Fig. (4)

1.9.2 DISJOINT SETS

When two sets have no common elements, they are called disjoint sets. Thus, if $A \cap B = \phi$, then A and B are disjoint. It is shown in the adjoining figure 5.

For example :

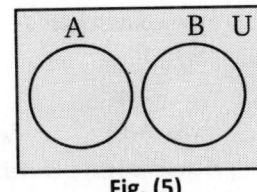
Fig. (5)

(1) If $A = \{2, 4, 6, 8\}$ and $B = \{1, 3, 5, 7, 9\}$
Then, $A \cap B = \phi$

(2) If A = Boys in school
B = Girls in school
Then, $A \cap B = \phi$

REMARKS

- If $A \cap B \neq \phi$, , then A and B are said to be intersecting or overlapping sets.
- A family of sets is said to be pairwise disjoint family of sets if and only if any two sets of this family are disjoint. For example, classes of A_2, A_3, A_5 and A_7 defined as

  $A_2 = \{2, 2^2, 2^3, ...\}$; $A_3 = \{3, 3^2, 3^3, ...\}$; $A_5 = \{5, 5^2, 5^3, ...\}$ and $A_7 = \{7, 7^2, 7^3, ...\}$ are pairwise disjoint.

- $\phi \cap A = \phi, i.e.,$ null set is disjoint from every subset.

1.9.3 DIFFERENCE OF TWO SETS

If A and B are two sets, then the set of all elements which belong to A but do not belong to B is called the difference of sets A and B and is denoted by $A \sim B$. The set of all elements which belong to B but do not belong to A is called the difference of sets B and A and is denoted by $B \sim A$.

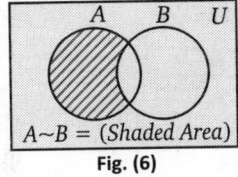
$A \sim B$ = (Shaded Area)
Fig. (6)

Therefore,

$A \sim B = \{x : x \in A \ and \ x \notin B\} = A \cap B'$

And $B \sim A = \{x : x \notin A \ and \ x \in B\} = B \cap A'$

For example :
Let $A = \{1, 2, 3, 4, 5\}$ and $B = \{-1, 0, 1, 2\}$
Then, $A \sim B = \{3, 4, 5\}$
And $B \sim A = \{-1, 0\}$

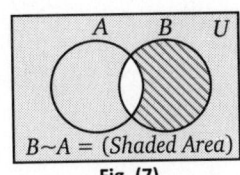
$B \sim A$ = (Shaded Area)
Fig. (7)

REMARKS

- $x \in (A - B) \Leftrightarrow x \in A$ and $x \notin B$.
- $x \notin (A - B) \Leftrightarrow x \notin A$ and $x \in B$
- $A - B \neq B \sim A$, i.e., , difference of two sets is not commutative.
- $A \subset B$ then $A \sim B = \phi$
- The sets $A \sim B$, $A \cap B$ and $B \sim A$ are mutually disjoint.
- Difference of a set with the universal set is known as complementation.
- $A \sim B$ is a subset of A and $B \sim A$ is a subset of B.

1.9.4 SYMMETRIC DIFFERENCE OF TWO SETS

If A and B are two sets, then the symmetric difference of two sets A and B is denoted by AΔB is given by $A \Delta B = (A \sim B) \cup (B \sim A)$

Symbolically: $A \Delta B = \{x : (x \in A \text{ and } x \notin B) \text{ or } (x \in B \text{ and } x \notin A)\}$

For example :

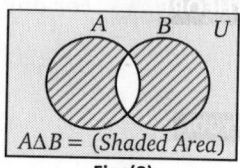

(i) If $\qquad A = \{1,2,3,4,5,6,7,8\}$ and $B = \{1,3,5,6,7,8,9\}$

 Then $\qquad A \sim B = \{2, 4\}$ and $B \sim A = \{9\}$

 and $\qquad A \Delta B = \{2, 4, 9\}$

Equivalent Sets: Two finite sets A and B are said to be equivalent if their cardinal numbers are same , *i.e.,* $n(A) = n(B)$.

$A\Delta B = $ (Shaded Area)

Fig. (8)

1.9.5 LAW OF EXCLUDED MIDDLE AND LAW OF CONTRADICTION

Two special properties of set operations are known as the excluded middle axioms and law of contradiction. The excluded middle axioms are very important because they are the only set operations described here that are not valid for both classical sets and fuzzy sets. Let A be any subset of universal set X. Then , we define.

(i)   Axiom of the excluded middle: $A \cup A' = U$

(ii) Axiom of the contradiction: $A \cup A' = \phi$

THEOREM 1. (*i*) $A \cup \phi = A$ (*ii*) $A \cap \phi = \phi$ (*iii*) $A \cup A = A$
(*iv*) $A \cap A = A$ (*v*) $A \cup B = B \cup A$ (*vi*) $A \cap B = B \cap A$

Proof.   (i) Let x be an arbitrary element of $A \cup \phi$ i.e., $x \in A \cup \phi$

Then, by definition $x \in A \cup B \Leftrightarrow x \in A$ or $x \in B$ i.e., $x \in A \cup \phi$

$\Rightarrow \qquad x \in A$ or $x \in \phi$

$\Leftrightarrow \qquad x \in A$ $\qquad\qquad\qquad\qquad\qquad$ ($\because \phi$ is a null set $\Rightarrow x \notin \phi$)

Therefore, $A \cup \phi = A$

(ii) Let x be an arbitrary element of $A \cap \phi$.

$x \in A \cap \phi \Leftrightarrow x \in A$ and $x \in \phi$ $\qquad\qquad\qquad$ ($\because \phi$ is a null set)

Therefore, $A \cap \phi = \phi$

(iii) Let x be an arbitrary element of $A \cup A$,

$x \in A \cup A \Leftrightarrow x \in A$ and $x \in B$ $\qquad\qquad$ (Repeated statement)

$\Leftrightarrow x \in A$

Therefore, $A \cup A = A$

(iv) Let x be an arbitrary element of $A \cap A$,

$x \in A \cap A \qquad \Leftrightarrow x \in A$ and $x \in A$ $\qquad\qquad$ (Repeated statement)

$\Leftrightarrow x \in A$

Therefore, $A \cap A = A$

(v) Let x be an arbitrary element of $A \cup B$,

$x \in A \cup B \qquad\qquad \Leftrightarrow x \in A$ and $x \in B$ $\qquad\quad$ (Writing in reverse order)

$\Leftrightarrow x \in B$ or $x \in A \Leftrightarrow x \in B \cup A$

Therefore, $A \cup B = B \cup A$

(vi) Let x be an arbitrary element of $A \cap B$

$x \in A \cap B \qquad \Leftrightarrow x \in A$ and $x \in B$ \qquad (Writing in reverse order)

$\Leftrightarrow x \in B$ and $x \in A \Leftrightarrow x \in B \cap A$

Therefore, $A \cap B = B \cap A$

THEOREM 2. *For any three sets A, B and C*

(i) $A \cup (B \cup C) = (A \cup B) \cup C$ \qquad (ii) $A \cap (B \cap C) = (A \cap B) \cap C$

(iii) $A \cup (B \cap C) = (A \cup B) \cap (A \cup C)$ \qquad (iv) $A \cap (B \cup C) = (A \cap B) \cup (A \cap C)$

Proof. (i) Let x be an arbitrary element of $A \cup (B \cup C)$, then

$$x \in A \cup (B \cup C)$$

$\Leftrightarrow \qquad x \in A$ or $x \in (B \cup C) \Leftrightarrow x \in A$ or $(x \in B$ or $x \in C)$

$\Leftrightarrow \qquad (x \in A$ or $x \in B)$ or $x \in C$ \qquad (By associativity)

$\Leftrightarrow \qquad x \in (A \cup B)$ or $x \in C \Leftrightarrow x \in A \cup (B \cup C)$

Therefore, $A \cup (B \cup C) = (A \cup B) \cup C$

(ii) Let x be an arbitrary element of $A \cap (B \cap C)$, then $x \in A \cap (B \cap C)$

$\Leftrightarrow \qquad x \in A \qquad$ and $x \in (B \cap C) \Leftrightarrow x \in A$ and $(x \in B$ and $x \in C)$

$\Leftrightarrow \qquad (x \in A \qquad$ and $x \in B)$ and $x \in C$ \qquad (By associativity)

$\Leftrightarrow \qquad x \in (A \cap B)$ and $x \in C \Leftrightarrow x \in (A \cap B) \cap C$

Therefore, $A \cap (B \cap C) = (A \cap B) \cap C$

(iii) Let x be an arbitrary element of $A \cup (B \cap C)$, then

$$x \in A \cup (B \cap C)$$

$\Leftrightarrow \qquad x \in A$ or $x \in (B \cap C) \Leftrightarrow x \in A$ or $(x \in B$ and $x \in C)$

$\Leftrightarrow \qquad (x \in A$ or $x \in B)$ and $(x \in A$ or $x \in C) \Leftrightarrow x \in (A \cup B)$ and $x \in (A \cup C)$

$\Leftrightarrow \qquad x \in (A \cup B) \cap (A \cup C)$

Therefore, $A \cup (B \cap C) = (A \cup B) \cap (A \cup C)$

(iv) Let x be an arbitrary element of $A \cap (B \cup C)$, then

$$x \in A \cap (B \cup C)$$

$\Leftrightarrow \qquad x \in A$ and $x \in (B \cup C) \Leftrightarrow x \in A$ and $(x \in B$ or $x \in C)$

$\Leftrightarrow \qquad (x \in A$ and $x \in B)$ or $(x \in A$ and $x \in C) \Leftrightarrow x \in (A \cap B)$ or $x \in (A \cap C)$

Therefore, $A \cap (B \cup C) = (A \cap B) \cup (A \cap C)$

THEOREM 3. (i) $(A')' = A$ \qquad (ii) $A \cup A' = U$, *where U is the universal set.*

(iii) $A \cap A' = \phi$ \qquad (iv) $(A \cup B)' = A' \cap B'$ *(De' Morgan's Law)*

(v) $(A \cap B)' = A' \cup B'$ *(De' Morgan's Law)*

Proof. (i) Let x be an arbitrary element of $(A')'$, then

$$x \in (A')' \Leftrightarrow x \notin A' \Leftrightarrow x \in A$$

Therefore, \qquad $(A')' = A$

(ii) Let x be an arbitrary element of $(A \cup A')$,

$x \in (A \cup A') \qquad \Leftrightarrow x \in A$ or $x \in A' \Leftrightarrow x \in A$ or $x \in U - A$

$\Leftrightarrow \qquad x \in A$ or $(x \in U, x \notin A) \Leftrightarrow x \in U$

Therefore, \qquad $A \cup A' = U$

(iii) Let x be an arbitrary element of $(A \cap A')$,

$$x \in (A \cap A') \qquad \Leftrightarrow x \in A \text{ and } x \in A' \text{ but if } x \in A \text{ then } x \notin A'$$

Therefore, $\qquad A \cap A' = \phi$

(iv) Let x be an arbitrary element of $(A \cup B)'$,

$$x \in (A \cup B)' \qquad \Leftrightarrow x \notin (A \cup B) \qquad \Leftrightarrow x \notin A \text{ and } x \notin B$$

$$\Leftrightarrow \qquad x \in A' \text{ and } x \in B' \quad \Leftrightarrow x \in A' \cap B'$$

Therefore, $\qquad (A \cup B)' = A' \cap B'$

(v) Let x be an arbitrary element of $(A \cap B)'$,

$$x \in (A \cap B)' \qquad \Leftrightarrow x \notin (A \cap B) \qquad \Leftrightarrow x \notin A \text{ or } x \notin B$$

$$\Leftrightarrow \qquad x \in A' \text{ or } x \in B' \quad \Leftrightarrow x \in A' \cup B'$$

Therefore, $\qquad (A \cap B)' = A' \cup B'$

Solved Examples

Based on the following Results

➡ $A \cup B = \{x : x \in A \text{ or } x \in B\}$	➡ $A \cap B = \{x : x \in A \text{ and } x \in B\}$
➡ $x \notin A \cap B \Leftrightarrow x \notin A \text{ or } x \notin B$	➡ $x \notin A \cup B \Leftrightarrow x \notin A \text{ and } x \notin B$
➡ $x \in A - B \Leftrightarrow x \in A \text{ and } x \notin B$	➡ $A \Delta B = (A \sim B) \cup (B \sim A)$
➡ $A \cup \phi = A$	➡ $A \cap \phi = \phi$
➡ $(A')' = A$	➡ $A \cup A' = U$
➡ $A \cap A' = \phi$	➡ $(A \cup B)' = (A' \cap B')$
➡ $(A \cap B)' = A' \cup B'$	

Example 1. *Show that (i)* $A \subset (A \cup B)$, *(ii)* $(A \cap B) \subset A$.

Solution. (i) Let $x \in A$ be arbitrary then $x \in A$ certainly but may or may not belong to B.

$$\Rightarrow \qquad\qquad x \in A \cup B \quad \Rightarrow \quad A \subset A \cup B$$

Therefore, $\qquad x \in A \qquad \Rightarrow \quad x \in A \cup B$ gives $A \subset A \cup B$

(ii) Let $\qquad\qquad x \in A \cap B \qquad$ where x is arbitrary

$$x \in A \cap B \quad \Rightarrow \quad x \in A \text{ and } x \in B$$

In particular, $\quad x \in A \cap B \quad \Rightarrow \quad x \in A$

Therefore, $\qquad (A \cap B) \subset A$

REMARK

• Similarly we can show that (i) $B \subset (A \cup B)$ and (ii) $A \cap B \subset B$.

Example 2. *Let A and B be two sets, if $A \cap X = B \cap X = \phi$ and $A \cup X = B \cup X$ for some set X, prove that $A = B$.*

Solution. Given that $A \cup X = B \cup X$

$$\Rightarrow \quad A \cap (A \cup X) = A \cap (B \cup X) \qquad \text{(taking intersection by } A \text{ on both sides)}$$

$$\Rightarrow \quad A = A \cap (B \cup X) \qquad\qquad (\because A \cap (A \cup X) = A)$$

$$\Rightarrow \quad A = (A \cap B) \cup (A \cap X) \qquad\qquad \text{(By distributive law)}$$

$\Rightarrow \quad A = (A \cap B) \cup \phi \quad \Rightarrow \quad A = A \cap B$

$\Rightarrow \quad A \subset (A \cap B) \quad \Rightarrow \quad A \subseteq B \qquad \qquad \dots(1)$

Again consider, $A \cup X = B \cup X$

$\Rightarrow \quad B \cap (A \cup X) = B \cap (B \cup X) \qquad$ (taking intersection with B)

$\Rightarrow \quad B \cap (A \cup X) = B$

$\Rightarrow \quad (B \cap A) \cup (B \cap X) = B \qquad$ (By distributive law)

$\Rightarrow \quad (B \cap A) \cup \phi = B \qquad$ (Given $B \cap X = \phi$)

$\Rightarrow \quad (B \cap A) = B \qquad$ $(\because A \cap B = B \cap A)$

$\Rightarrow \quad A \cap B = B \qquad \Rightarrow \quad B \subset A \cap B \Rightarrow \quad B \subseteq A \qquad \dots(2)$

Hence, (1) and (2) gives $A \subseteq B$ and $B \subseteq A$.

$\Rightarrow \quad A = B$

Example 3. *For any two sets A and B, show that*

(i) $P(A \cap B) = P(A) \cap P(B)$, *(ii)* $P(A) \cup P(B) \subset P(A \cup B)$

Solution. (i) Let $\quad X \in P(A \cap B) \quad \Rightarrow X \subseteq A \cap B$

$\Rightarrow \quad X \subseteq A$ and $X \subset B \Rightarrow X \in P(A)$ and $X \in P(B)$

$\Rightarrow \quad X \in P(A) \cap P(B)$

Therefore, $\quad P(A \cap B) \subset P(A) \cap P(B) \qquad \dots(1)$

Now, let $X \in P(A) \cap P(B) \Rightarrow X \in P(A)$ and $X \in P(B)$

$\Rightarrow \quad X \subseteq A$ and $X \subseteq B \Rightarrow X \subseteq A \cap B$

$\Rightarrow \quad X \in P(A \cap B)$

Therefore, $\quad P(A) \cap P(B) \subset P(A \cap B) \qquad \dots(2)$

From (1) and (2), we conclude that

$\qquad P(A \cap B) = P(A) \cap P(B)$

(ii) Let $\quad X \in P(A) \cup P(B) \quad \Rightarrow X \in P(A)$ and $X \in P(B)$

$\Rightarrow \quad X \subseteq A$ or $X \subseteq B \quad \Rightarrow X \subseteq A \cup B$

$\Rightarrow \quad X \in P(A \cup B)$

Therefore, $\qquad P(A) \cup P(B) \subset P(A \cup B)$

REMARK

- Converse of the result (ii) is not necessarily true. For example, let $A = \{1,2\}$ and $B = \{3,5,6\}$, then we find that $X = \{1,2,3,5\}$ which is a subset of $A \cup B$. Therefore, $X \in P(A \cup B)$. But $X \notin P(A), X \notin P(B)$. So,
$$X \notin P(A) \cup P(B) \quad \Rightarrow P(A \cup B) \not\subset P(A) \cup P(B)$$

1.9.6 SOME MORE RESULTS

1. If A and B are any two sets, then

    (i) $A - B = A \cap B'$ (ii) $A - B = A \Leftrightarrow A \cap B = \phi$

    (iii) $(A - B) \cup B = A \cup B$ (iv) $A \subset B \Leftrightarrow B' \subset A'$

    (v) $(A - B) \cup (B - A) = (A \cup B) - (A \cap B)$

2. If A and B are any two sets, then

(i) $A - (B \cap C) = (A - B) \cup (A - C)$ (ii) $A - (B \cup C) = (A - B) \cap (A - C)$

(iii) $A \cap (B - C) = (A \cap B) - (A \cap C)$

❧ EXERCISE 1.3

1. Let $A = \{a, b\}$, $B = \{a, b, c\}$. Is $A \subset B$? Find $A \cup B$ and $A \cap B$.

2. If $A = \{1,2,3,4\}$, $B = \{2,4,6,8\}$, $C = \{3,4,5,6\}$ and universal set $U = \{1,2,3,4,...9\}$. Verify that $A \cap (B \cup C) = (A \cap B) \cup (A \cap C)$.

3. If A, B, C are subsets of a set X, then show that $A \subseteq B$ and $B \subseteq C \Rightarrow A \subseteq C$.

4. Find the union of the following sets:

(i) $A = \{x : x$ is an even integer$\}$,

    $B = \{x : x$ is an odd integer$\}$.

(ii) $A = \{x : x$ is a multiple of 2$\}$,

    $B = \{x : x$ is a multiple of 3$\}$.

(iii) $A = \{x : x$ is a rational number $\}$,

    $B = \{x : x$ is an irrational number$\}$.

(iv) $A = \{x : x$ is a negative integer$\}$,

    $B = \{x : x$ is a non- negative integer$\}$

5. Find the intersection of the following sets.

(i) $A = \{x : x$ is an even integer$\}$,

    $B = \{x : x$ is an odd integer$\}$

(ii) $A = \{x : x$ is a rational number $\}$,

    $B = \{x : x$ is an irrational number$\}$.

(iii) $A = \{x : x$ is a multiple of 5$\}$,

    $B = \{x : x$ is a multiple of 2$\}$

(iv) $A = \{x : x$ is a rational number $\}$,

    $B = \{x : x$ is a real number$\}$

6. If $A = \{1,2,3,4\}$, $B = \{2,4,6,8\}$ and $C = \{3,4,5,6\}$, find

(i) $(A \cup B) \cap C$

(ii) $A \cup (B \cap C)$

7. Write T for true and F for false statement.

(i) $A \subset (A \cup B)$ (T/F)

(ii) $(A \cup B) \subset B$ (T/F)

(iii) $(A \cap B) \subset A$ (T/F)

(iv) $A \cup A = A$ and $A \cap A = A$ (T/F)

(v) If $A \cap B = \phi$, then $A \cap \phi = B$ (T/F)

(vii) If A and B are disjoint sets, then intersection of their union and intersection is the null set.   (T/F)

(viii) If A is the proper subset of U, then the union of A and A' is U.   (T/F)

(ix) $U' = \phi$ and $\phi' = U$ (T/F)

(x) $(A \cup B)' = A' \cap B'$ (T/F)

(xi) $A \cap A'$ is always empty   (T/F)

(xii) $(A \cap B)' = A' \cup B'$ (T/F)

8. If $A = \{1, 2, 3, 4, 5, 6, 7, 8\}$ and $B = \{1, 3, 5, 6, 7, 8, 9\}$, then show that

$$A \Delta B = \{2, 4, 9\}$$

9. Let $A = \{x : x \in N\}$,

    $B = \{x : x = 2n : n \in N\}$,

    $C = \{x : x = 2n-1 : n \in N\}$

and $D = \{x : x$ is a prime natural number$\}$. Find

(i) $A \cap B$ (ii) $A \cap C$

(iii) $A \cap D$ (iv) $B \cap C$

(v) $B \cap D$ (vi) $C \cap D$

10. For any two sets A and B, prove that $P(A) = P(B)$ implies that $A = B$

11. For any two sets A and B, show that

(i) $A \cup (A \cap B) = A$

(ii) $A \cap (A \cup B) = A$

(iii) $(A \cup B) \cap (A \cap B') = A$

(iv) $A' \cup B = U \Rightarrow A \subset B$

(v) $A \subset B \Leftrightarrow B' \subset A'$

12. Let $A = \{1, 2, 3, 4\}$, $B = \{2, 3, 4, 5\}$ and $C = \{4, 5, 6, 7\}$. Verify that

(i) $A \cup (B \cap C) = (A \cup B) \cap (A \cup C)$

(ii) $A \cap (B \cup C) = (A \cap B) \cup (A \cap C)$

(iii) $A \cap (B - C) = (A \cap B) - (A \cap C)$

(iv) $A - (B \cup C) = (A - B) \cap (A - C)$

(v) $A - (B \cap C) = (A - B) \cup (A - C)$

13. Show that

(i) If a sets has only even elements, then it has 2 subsets.

(ii) If $B \subset A$ and B has one element less than that of A, show that A has twice as many subset as B has.

(iii) A set with 2 element has 2^2 subsets, a set with 3 elements has 2^3 subsets and so on.

14. If $X = \{4^n - 3n - 1 : n \in N\}$ and $Y = \{9(n - 1) : n \in N\}$, show that $X \subset Y$.

15. Show that $A - B$, $A \cap B$ and $B - A$ are pairwise disjoint.

16. Show that $A \cup B \subseteq A \cap B$ implies that $A = B$.

Answers

1. (i) Yes. $\{a, b, c\}$, $\{a, b\}$;

4. (i) $A \cup B = \{x : x \text{ is non-zero integer}\}$ (ii) $A \cup B = \{x : x \text{ is a multiple of 2 or 3}\}$

(iii) $A \cup B = \{x : x \text{ is a real number}\}$ (iv) $A \cup B = \{x : x \text{ is an integer}\}$

5. (i) ϕ (ii) ϕ (iii) ϕ (iv) $\{x : x \text{ is a rational number}\}$

6. (i) $\{3, 4, 6\}$,   (ii) $\{1, 2, 3, 4, 6\}$

7. (i) T   (ii) F   (iii) T   (iv) T   (v) F   (vi) T   (vii) T

(viii) T   (ix) T   (x) T   (ix) T   (x) T   (xi) T   (xii) T

9. (i) B   (ii) C   (iii) D   (iv) ϕ (v) 2   (vi) $D - \{2\}$

1.10 SOME RESULTS ON VENN DIAGRAM

If A is a finite set, then $n(A) = $ No. of elements in the set A.

The following results may be remembered for direct application:

(1) $n(A \cup B) = n(A) + n(B) - n(A \cap B)$

(2) $n(A \cup B) = n(A) + n(B)$, provided A and B are disjoint, i.e., if $n(A \cap B) = 0$

Fig. (9)

(3) $n(A \cap B') = n(A) - n(A \cap B)$

(4) $n(B \cap A') = n(B) - n(A \cap B)$

(5) $n(A \cup B) = n(A \cap B') + n(B \cap A') + n(A \cap B)$

(6) $n(A \Delta B) = n(A) + n(B) - 2n(A \cap B)$

(7) $n(A' \cup B') = n[(A \cap B)'] = n(U) - n(A \cap B)$

(8) $n(A' \cap B') = n[(A \cup B)'] = n(U) - n(A \cup B)$

(9) $n(A - B) = n(A) - n(A \cap B) \Rightarrow n(A - B) + n(A \cap B) = n(A)$

(10) $n(A \cup B \cup C) = n(A) + n(B) + n(C) - n(A \cap B) - n(B \cap C)$
$$- n(A \cap C) + n(A \cap B \cap C)$$

Solved Examples

Example 1. *In a group of athletic teams in a school, 21 are in the basket ball, 26 in the hockey team and 29 in the football team. If 14 play hockey and basket ball, 12 play football and basket ball, 15 play hockey and football and 8 play all the three games. Find (i) how many players are there in all (ii) how many play football only.*

Solution. Let A, B and C denote the set of players, who play basket ball, hockey and football respectively. Then, according to question, we have

$$n(A) = 21, n(B) = 26, n(C) = 29$$
$$n(A \cap B) = 14, n(A \cap C) = 12, n(B \cap C) = 15 \text{ and } n(A \cap B \cap C) = 8$$

Therefore, $n(A \cap B \cap C) = [n(A) + n(B) + n(C) + n(A \cap B \cap C)]$
$$- [n(A \cap B) + n(A \cap C) + n(B \cap C)]$$
$$= [21 + 26 + 29 + 8] - [14 + 12 + 15] = 43$$

Hence, the total number of players is 43. Now, the number of players playing football only is $[29 - (7 + 8 + 4)] = 10$.

Example 2. *In a canteen, out of 123 students, 42 students buy ice-cream, 36 buy burst and 10 buy cakes, 15 students buy ice-cream and 11 buy ice-cream and buns but no cakes. Draw Venn diagram to illustrate the above information and find (i) how many students buy nothing at all (ii) how many students buy at least two items. (iii) how many students buy all three items.*

Solution. Define the sets A, B and C such that

A = Set of students who buy cakes

B = Set of students who buy ice-cream

C = Set of students who buy buns

According to question, we have,

$n(A) = 10$; $n(B) = 42$; $n(C) = 36$; $n(B \cap C) = 15$;

$n(A \cap B) = 10$; $n[(A \cap C) - B] = 4$;

$n[(B \cap C) - A] = 11$ and $n[A - B \cup C] = 10$

Now we have $n(B \cup C) = n(B) + n(C) - n(B \cap C)$

$= 42 + 36 - 15 = 63$

$n(B \cup C) - n(B) = 63 - 42 = 21$

and $n(B \cup C) - n(C) = 63 - 36 = 27$

The above distribution of the students can be illustrated by Venn diagram (Figure 10). Now, total number of students buying something.

$= 10 + 6 + 21 + 4 + 4 + 11 + 17 = 73$

Fig. (10)

(i) Number of students who did not buy anything $= 123 - 73 = 50$

(ii) Number of students buying at least two items $= 6 + 4 + 4 + 11 = 25$

and (iii) Number of students buying all three items $= 4$

☙ EXERCISE 1.4

1. Out of 80 students who secured first class marks in Mathematics or in Physics, 50 obtained first class marks in Mathematics, 10 in both Physics and Mathematics. How many students secured first class marks in Physics only?

2. The Mathematics club in a school held an open house on three afternoons 115, 110 and 135 students attended both the first, second and third afternoons respectively. 25 attended just the first, 30 attended both the first and second days, 80 attended both the first and third days, and 60 attended both the second and third days. How many attended (i) all three days (ii) just the second day (iii) just the third day?

3. In a school of 250 pupils, 100 are girls, and 200 pupils stay at school for lunch. If 40 girls go home for lunch. Find the number of boys who go home for lunch.

4. In a class of 150 students, the following results were obtained in a certain examination. 45 students failed in Maths; 50 students failed in Physics, 48 students failed in Chemistry, 35 failed in both Maths and Chemistry, 25 failed in the three subject. Find the number of students who have failed in at least one subject.

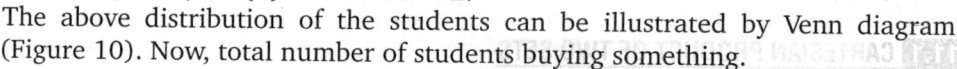

Answers			
1. 30	**2.** 20, 30, 15	**3.** 10	**4.** 71

1.11 ORDERED PAIR

Sometimes, there are situations in which order is very important. Some results may be affected by order and other are not.

Definition: *An ordered pair is a pair of entries whose components occur in a specific order. It is written by listing the two components in the specific order, separating them by a comma and enclosing the pair in parentheses.*

Symbolically: If A and B are two sets, then by ordered pair of elements, we must mean a pair (a,b): $a \in A$, $b \in B$ in that order.

REMARKS

- It may be noted that (a, b) is not the same as $\{a, b\}$. The former denotes an ordered pair whereas the latter denotes a set.
- $(a, b) \neq (b, a)$ unless $a = b$.
- Ordered pair may have the same first and second components, *i.e.*, two elements of an ordered pair need not be distinct.
- Two ordered pairs are said to be equal when both the first components are equal and their second components are also equal.

1.11.1 CARTESIAN PRODUCT OF TWO SETS

The set of all ordered pairs of elements (a,b), $a \in A$, $b \in B$ is called the cartesian product of two sets A and B. It is denoted by $A \times B$.

Symbolically : $A \times B = \{(a, b) : a \in A, b \in B\}$

For example :

If $A = \{2, 3\}$ and $B = \{4, 5, 6\}$, then

$$A \times B = \{(2, 4), (2, 5), (2, 6), (3, 4), (3, 5), (3, 6)\}$$

REMARKS

- $A \times B = \phi \Leftrightarrow A = \phi$ or $B = \phi$
- If A and B are finite sets, then $n(A \times B) = n(A) . n(B)$
- If either A or B is infinite, then $A \times B$ is an infinite set.

1.11.2 ORDERED TRIPLET

If A, B, C are three sets, then by ordered triple product of elements, we mean a triplet (a, b, c) : $a \in A$, $b \in B$, $c \in C$ in that order.

This is also called ordered 3-tuple.

The set of all ordered triplets (a, b, c): $a \in A$, $b \in B$, $c \in C$ is also called the cartesian triple product of three sets A, B and C and is denoted by $(A \times B \times C)$

Symbolically: $A \times B \times C = \{(a, b, c): a \in A, b \in B, c \in C\}$

REMARK

- In general, the cartesian product on n sets $A_1, A_2, ..., A_n$ is a ordered n tuples $(a_1, a_2, ..., a_n)$, where $a_1 \in A_1, a_2 \in A_2, ..., a_n \in A_n$. It is denoted by $A_1 \times A_2 ... \times A_n$ or briefly by $\prod_{i=1}^{n} A_i$ where $\prod$ stands for the product.

Solved Examples

Example 1. *If A= {1, 2} and B= {a, b, c}, find the value of A×B, B×A, A×A, B×B.*

Solution. We have $A = \{1, 2\}$ and $B = \{a, b, c\}$.

Therefore,

$A \times B = \{(1, a), (1, b), (1, c), (2, a), (2, b), (2, c)\}$

$B \times A = \{(a, 1), (a, 2), (b, 1), (b, 2), (c, 1), (c, 2),\}$

$A \times A = \{(1, 1), (1, 2), (2, 1), (2, 2)\}$

$B \times B = \{(a, a), (a, b), (a, c), (b, a) (b, b), (b, c), (c, a), (c, b), (c, c)\}$

Example 2. *If $A = \{1, 2, 3\}$, $B = \{a, b, c, d\}$ and $C = \{-1, -2\}$, find $A \times B$, $B \times A$ and $C \times (B \cup C)$.*

Solution. Given that $A = \{1, 2, 3\}, B = \{a, b, c, d\}$ and $C = \{-1, -2\}$.

Therefore,

$A \times B = \{(1, a), (1, b), (1, c), (1, d), (2, a), (2, b), (2, c), (2, d),$
$\qquad\qquad (3, a), (3, b), (3, c), (3, d)\}$

$B \times A = \{(a, 1), (b, 1), (c, 1), (d, 1), (a, 2), (b, 2), (c, 2), (d, 2), (a, 3),$
$\qquad\qquad (b, 3), (c, 3), (d, 3)\}$

Also, $B \cup C = \{a, b, c, d, -1, -2\}$

Therefore,

$C \times (B \cup C) = \{(-1, a), (-1, b), (-1, c), (-1, d), (-1, -1), (-1, -2), (-2, a),$
$\qquad\qquad (-2, b), (-2, c), (-2, d), (-2, -1), (-2, -2)\}$

Example 3. *Find the values of a and b if (4a–2, b+4) = (2a, 4).*

Solution. Since we know that two ordered pairs (a_1, b_1) and (a_2, b_2) are said to be equal if $a_1 = a_1$ and $b_1 = b_2$. Therefore, for the equality of two given ordered pairs, we have

$4a - 2 = 2a$ and $b + 4 = 4$

Therefore, $4a - 2a = 2 \Rightarrow a = 1$ and $b + 4 = 4 \Rightarrow b = 0$

Example 4. *If $A = \{1, 2, 3, 4\}$ and $B = \{4, 5\}$, represent $A \times B$, $B \times A$ and $B \times B$ pictorially and find their values.*

Solution. Given $A= \{1, 2, 3, 4\}$ and $B = \{4, 5\}$

$A \times B = \{(1, 4), (1, 5), (2, 4), (2, 5), (3, 4), (3, 5), (4, 4), (4, 5)\}$

$B \times A = \{(4, 1), (5, 1), (4, 2), (5, 2), (4, 3), (5, 3), (4, 4), (5, 4)\}$

And $B \times B = \{(4, 4), (4, 5), (5, 4), (5, 5)\}$

Pictorially, $A \times B$, $B \times B$ and $B \times A$ can be represented as shown in figure 11.

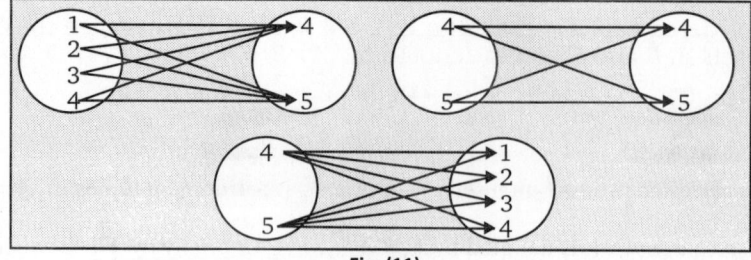

Fig. (11)

Example 5. *Let A = {1, 2, 3, 4} and B = {5, 7, 9}. Determine (i) A×B, (ii) B×A. Also represent A×B and B×A graphically.*

Solution.   (i) Given   A = {1, 2, 3, 4} and B = {5, 7, 9}. Then,

$$A×B = \{(1, 5), (1, 7), (1, 9), (2, 5), (2, 7), (2, 9)\ (3, 5), (3, 7), (3, 9),$$
$$(4, 5), (4, 7), (4, 9)\}$$

Graphically, it can be represented as shown in Figure 12.

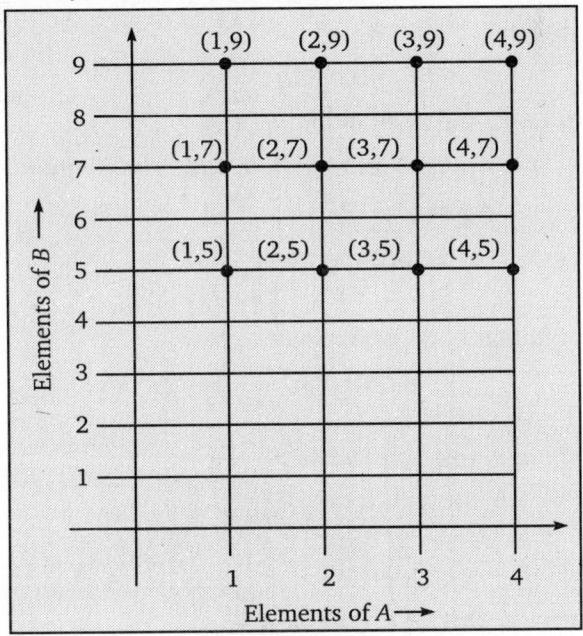

Fig. (12) : A × B

Now, B×A = {(5, 1), (5, 2), (5, 3) (5, 4) (7, 1), (7, 2), (7, 3) (7, 4) (9, 1), (9, 2), (9 3) (9, 4)}

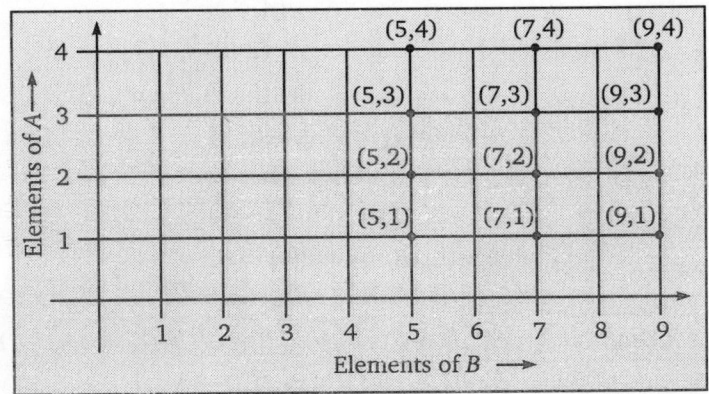

Fig. (13) : B × A

THEOREM 1. *For any three subsets A, B and C , we have.*

(i) A × (B ∩ C) = (A×B) ∩ (A×C)   (ii) A × (B ∪ C) = (A×B) ∪ (A × C)

Proof. (i) Let $(x, y) \in A \times (B \cap C)$

$\Rightarrow$ Then, $x \in A$ and $y \in (B \cap C)$

$\Rightarrow$ $x \in A$ and $y \in B$ and $y \in C \Rightarrow x \in A, y \in B$ and $x \in A, y \in C$

$\Rightarrow$ $(x, y) \in A \times B$ and $(x, y) \in (A \times C) \Rightarrow (x, y) \in (A \times B) \cap (A \times C)$

But (x, y) is arbitrary, therefore

$$A \times (B \cap C) \subset (A \times B)(A \times C) \qquad \dots (1)$$

Conversely,

If $(x, y) \in (A \times B) \cap (A \times C)$

Then, $(x, y) \in A \times B$ and $(x, y) \in A \times C$

$\Rightarrow$ $x \in A, y \in B$ and $x \in A, y \in C \Rightarrow x \in A, y \in B$ and $y \in C$

$\Rightarrow$ $x \in A$ and $y \in (B \cap C)$ $\Rightarrow$ $(x, y) \in A \times (B \cap C)$

But (x, y) is arbitrary, therefore

$$(A \times B) \cap (A \times C) \subseteq A \times (B \cap C) \qquad \dots (2)$$

From (1) and (2), we conclude that

$$A \times (B \cap C) = (A \times B) \cap (A \times C)$$

(ii) $(x, y) \in A \times (B \cup C)$

Then, $x \in A$ and $y \in (B \cup C)$

$\Rightarrow$ $x \in A$ and $y \in B$ or $y \in C$

$\Rightarrow$ $(x \in A$ and $y \in B)$ or $(x \in A$ and $y \in C)$

$\Rightarrow$ $\{(x, y) \in (A \times B)\}$ or $\{(x, y) \in (A \times C)\}$

$\Rightarrow$ $(x, y) \in (A \times B) \cup (A \times C)$

Since (x, y) is arbitrary, therefore

$$A \times (B \cup C) \subseteq (A \times B) \cup (A \times C) \qquad \dots (1)$$

Conversely,

If $(x, y) \in (A \times B) \cup (A \times C)$

Then, $(x, y) \in (A \times B)$ or $(x, y) \in (A \times C)$

$\Rightarrow$ $(x \in A$ and $y \in B)$ or $(x \in A$ and $y \in C)$

$\Rightarrow$ $x \in A$ and $(y \in B$ or $y \in C) \Rightarrow (x, y) \in A \times (B \cup C)$

But (x, y) is arbitrary, therefore

$$(A \times B) \cup (A \times C) \subseteq A \times (B \cup C) \qquad \dots (2)$$

From (1) and (2), we conclude that

$$A \times (B \cup C) = (A \times B) \cup (A \times C).$$

THEOREM 2. *For any sets A, B, C, D we have* $(A \times B) \cap (C \times D) = (A \cap C) \times (B \cap D)$

Proof. If $(a, b) \in (A \times B) \cap (C \times D)$, then

$\Rightarrow$ $(a, b) \in (A \times B)$ and $(a, b) \in (C \times D)$

$\Rightarrow$ $(a \in A$ and $b \in B)$ and $(a \in C$ and $(b \in D)$

$\Rightarrow$ $(a \in A$ and $a \in C)$ and $(b \in B$ and $b \in D)$

$\Rightarrow$ $a \in (A \cap C)$ and $b \in (B \cap D) \Rightarrow (a, b) \in (A \cap C) \times (B \cap D)$

Since (a, b) is arbitrary, therefore

$$(A \times B) \cap (C \times D) \subseteq (A \cap C) \times (B \cap D) \qquad \dots (1)$$

Now, let $(a, b) \in (A \cap C) \times (B \cap D)$

$\Rightarrow \quad a \in (A \cap C)$ and $b \in (B \cap D) \Rightarrow (a \in A$ and $a \in C)$ and $(b \in B$ and $b \in D)$

$\Rightarrow \quad (a \in A$ and $b \in B)$ and $(a \in C$ and $b \in D)$

$\Rightarrow \quad (a, b) \in (A \times B) \cap (C \times D)$

Since, (a, b) is arbitrary, therefore

$$(A \cap C) \times (B \cap D) \subseteq (A \times B) \cap (C \times D) \qquad \dots (2)$$

From (1) and (2), we conclude that

$$(A \times B) \cap (C \times D) = (A \cap C) \times (B \cap D)$$

REMARKS

- $(A \times B) \cap (B \times A) = (A \cap B) \times (B \cap A)$
- $A \times (B' \cup C')' = A \times (B \cap C) = (A \times B) \cap (A \times C)$
- $A \times (B' \cap C')' = A \times (B \cup C) = (A \times B) \cup (A \times C)$

THEOREM 3. *If A and B are two non-empty sets having n elements in common, then $A \times B$ and $B \times A$ have n^2 elements in common.*

Proof.   We know that $(A \times B) \cap (C \times D) = (A \cap C) \times (B \cap D)$

$$(A \times B) \cap (B \times A) = (A \cap B) \times (B \cap A)$$
$$(A \times B) \cap (B \times A) = (A \cap B) \times (A \cap B)$$

Since $(A \times B)$ has n elements, therefore $(A \cap B) \times (B \cap A)$ has n^2 elements.

$$(A \times B) \cap (B \times A) = (A \cap B) \times (B \cap A) \text{ has } n^2 \text{ elements.}$$

Hence, $(A \times B)$ and $(B \times A)$ have n^2 elements in common.

REMARKS

- For any three sets, A, B, C, we have $A \times (B - C) = (A \times B) - (A \times C)$
- If A and B are any two non-empty sets, then $A \times B = B \times A$ iff $A = B$.
- If $A \subseteq B$, then $A \times A \subseteq (A \times B) \cap (B \times A)$
- If $A \subseteq B$, then $A \times C \subseteq B \times C$ for any set C.
- If $A \subseteq B$ and $C \subseteq D$, then $A \times C \subseteq B \times D$.
- $A \times B = A \times C \Rightarrow B = C$

⚡ EXERCISE 1.5

1. If $A = \{a, b, c\}$, $B = \{d\}$, $C = \{2\}$, then verify

  (i) $A \times (B \cup C) = (A \times B) \cup (A \times C)$

  (ii) $A \times (B \cap C) = (A \times B) \cap (A \times C)$

  (iii) $A \times (B - C) = (A \times B) - (A \times C)$

  (iv) $(A \cap B) \times C = (A \times C) \cap (B \times C)$

2. If $A = \{2, 3\}$, $B = \{1, 2, 3\}$, $C = \{2, 3, 4\}$, show that $A \times A = (B \times B) \cap (C \times C)$.

3. If $A = \{1, 2, 3\}$, $B = \{4, 5\}$ and $C = \{1, 2, 3, 4, 5\}$, then show that $(C \times B) - (A \times B) = B \times B$.

4. The ordered pairs $(2,7)$, $(4, 8)$ and $(5, 9)$ and among nine elements of the set $A \times B$. Determine the other six elements of $A \times B$.

5. Let $A = \{2, 3, 5, 7\}$, $B = \{1, 12, 13, 15\}$. How many elements are there in $A \times B$? In $B \times A$? Is $A \times B = B \times A$? Is $n(A \times B) = n(B \times A)$?

6. Let A and B be two sets. Show that sets $A \times B$ and $B \times A$ have an element in common if and only if the sets A and B have an element in common.

7. Some elements of $A \times B$ are (a, x), (a, y), (d, z). If $A : \{a, b, c\ d\}$, find the remaining elements of $A \times B$ such that $n(A \times B)$ is least.

8. If A and B are two sets having 3 elements in common. If $n(A) = 5$, $n(B) = 4$, find $n(A \times B)$ and $n\{(A \times B) \cap (B \times A)\}$.

9. The ordered pairs $(1, 1)$, $(2, 2)$ and $(3, 3)$ are among the 3 elements in the set $A \times B$. If A and B have 3 elements each, how many elements in all does the set $A \times B$ have? Also find the remaining elements.

10. If A and B are two sets such that $n(A) = 3$ and $n(B) = 2$. If $(x, 1)$, $(y, 2)$, $(z, 1)$ are in $A \times B$, find A and B, where x, y, z are distinct.

11. Write 'T' for true and 'F' for false statement:
    (a) If $A = (a, b)$ and $B = (b, a)$, then $A \times B = \{(a, b)\ (b, a)\}$ **(T/F)**
    (b) $\{(a, x), (a, y), (b, x), (b, y)\}$ is product set. **(T/F)**
    (c) If $n(A) = x$ and $n(B) = y$ and $A \cap B = \phi$, then $n (A \times B) = xy$ **(T/F)**
    (d) If A and B are non-empty sets, then $A \times B$ is a non-empty set of ordered pairs (x, y) such that $x \in A$ and $y \in A$. **(T/F)**

12. (a) If $A = \{1, 2, 3\}$, $B = \{4, 5\}$ and $C = \{1, 2, 3, 4, 5\}$. Find
    (i) $A \times B$,   (ii) $C \times B$,   (iii) $B \times B$
    (b) If $A = \{1, 2, 3, 4\}$ and $B = \{5, 7, 9\}$, find $(A \times B) \cap (A \cap B)$.

Answers

4. $(2, 8)$, $(2, 9)$, $(4, 7)$, $(4, 9)$, $(5, 7)$, $(5, 8)$ **5.** 16, 16, No, yes

7. (a, y), $(a, 2)$, (b, x), (b, y), (b, z), (c, x), (c, z), (d, x), (d, y) **8.** 20, 9

9. 9, $(1, 2)$, $(1, 3)$, $(2, 1)$, $(2, 3)$, $(3, 1)$, $(3, 2)$

10. (i) $A = \{x, y, z\}$, $B = \{1, 2\}$, (ii) (a) F   (b) T,   (c) T   (d) F

12. (a) (i) $A \times B = (1, 4)$, $(1, 5)$, $(2, 4)$, $(2, 5)$, $(3, 4)$, $(3, 5)$
    (ii) $C \times B = \{(1, 4), (1, 5), (2, 4), (2, 5), (3, 4), (3, 5), (4, 4), (4, 5), (5, 4), (5, 5), \}$
    (iii) $B \times B = \{(4, 4), (4, 5), (5, 5)\}$ (b) ϕ

1.12 RELATION

Let us take two sets of natural numbers N_1 and N_2. We define R as a relation between them such that N_1 is a square of N_2. Then we can write $1R1$, $2R4$, $3R9$, ...

In terms of ordered pair, we can write
$$R = \{(1, 1), (2, 4), (3, 9), (4, 16), ...\} = \{(x, y : x, y \in N \text{ and } y = x^2\}$$
The relation from set N to N is a subset of $N \times N$ such that $y = x^2$.

Definition: *Let A and B be two sets. Then a relation R from A to B is a subset of $A \times B$.*

Symbolically: R is a relation from A to $B \Leftrightarrow R \subseteq A \times B$.

REMARKS

- If R is a relation from A to B, then A is called the domain and B the range of R.
- If R is a relation from a non-empty set A to a non-empty set B and if $(a, b) \in R$, then we write aRb, read as "a is related to b by the relation R." On the other hand, if $(a, b) \notin R$, we write $a\bar{R}b$ and say that 'a is not related to b by the relation R'.
- In particular, any subset $A \times A$ defined a relation in A, known as Binary relation.

☞ ILLUSTRATIONS

(1) If $a, b \in N$ and R is defined as "a is divisor of b" then R is relation on N. The subset $N \times N$, which corresponds to the relation R is $S = \{(n, r): n \in N, r \in N\}$ Here, it is clear that $(1, 3)$, $(2, 4)$, $(3, 9)$ $(4, 8)$, $(4, 4)$, are in S, whereas $(2, 3)$, $(4, 5)$, $(5, 6)$ are not in S.

(2)   If R is a relation from set $A = \{1,2,3\}$ to the set $B = \{-1, -2\}$ defined by $x + y = 0$, then $R = \{(1, -1), (2, -2)\}$

Here, domain of R is $\{1, 2\}$ and Range $= \{-1, -2\}$.

(3)   If $A = \{a, b, c, d, e\}$ and $B = \{f, g, h, i\}$ and let $R = \{(a, g), (a, i), (d, h), (e, f)\}$ by a relation from A to B then

Domain of $R = \{a, d, e\}$ and Range of $R = \{g, i, h, f\}$

(4)   If $a, b \in R$, the set of real numbers and R is "$|a - b|$ is a rational number" then R is a relation on R. The subset S of $R \times R$ which corresponds to the relation is

$$S = \{(a, b + a): a \in R, b \in Q\}$$

It is observed that $\left(1, 2\dfrac{1}{2}\right), \left(\pi, \pi - \dfrac{1}{2}\right)$ belongs to S, while $(\sqrt{2}, \pi + \sqrt{2}) \notin S$.

(5)   If $A = \{2, 3, 4\}$ and $B = \{a, b, c\}$, then $R = \{(2, b), (3, c), (2, a), (4, a)\}$ being a subset of $A \times B$, is a relation from $A \times B$. Here $(2, b), (3, c), (2, a), (4, a) \in R$, so we may write $2Rb, 3Rc, 2Ra, 4Ra$. But $(3, b) \notin R$ therefore, $3 \not{R} b$.

(6)   If $a, b \in N$ and R is defined by "$a - b$ is divisible by a number $n \in N$", then R is a relation on N . The subset S of $N \times N$ corresponding to the relation by

$$S = \{n, n + rm : n \in N, r \in N\}$$

Here, $m = 3, (2, 8), (5, 11) \in S$ [$\because 2 - 8 = 6$, which is divisible by 3]

While $(3, 8) \in S$ [$\because 3 - 8 = 5$, which is not divisible by 3]

1.12.1 TOTAL NUMBER OF RELATIONS

Let A and B be two non-empty finite sets consisting p and q elements respectively, then $A \times B$ consists of $p q$ ordered pairs. Therefore, total number of subset of $A \times B$ is 2^{pq}.

REMARKS

- For a non-empty set A, $\phi \in A \times A$, therefore it is a relation on A, called void or empty relation on A.
- The void relation ϕ and the universal relation $A \times B$ are called trivial relations from A to B.
- The void and universal relation on set A respectively the smallest and the largest relation on A.

1.12.2 IDENTITY RELATION

Let A be a set. The identity relation on A is the relation $I_A = \{(x, x) : x \in A\}$ on A.

For example : If $A = \{a, b, c\}$ then the relation $I_A = \{(a, a), (b, b), (c, c)\}$ is the identity relation. $R = \{(a, a), (b, b)\}$ is not an identity relation as $(c, c) \notin R$.

1.12.3 INVERSE OF A RELATION

Let A, B be two non-empty sets and R be a relation from a set A to B and let (x, y), number of the subset D of $A \times B$ corresponding to the relation R from A to B.

To the relation R from the set A to the set B, there corresponds a relation from the set B to the set A called the inverse of the relation, denoted by R^{-1} such that the subset $B \times A$ corresponding to the relation R^{-1} is $= \{(y, x): (x, y) \in D\}$.

i.e., $yR^{-1}x \Leftrightarrow xRy$

☞ **ILLUSTRATIONS**

(1) Let $A = \{a, b, c\}$ and $B = \{1, 2, 3\}$ be two sets and let $R = \{(a, 1), (a, 2), (b, 1),$ $(b, 2)\}$ be a relation from A to B then $R^{-1} = \{(1, a), (2, a), (1, b), (2, b)\}$

(2) If $A = \{1, 2, 3\}$, $B = \{5, 6, 7\}$ and let $R = \{(1, 5), (2, 5), (2, 7)\}$ be a relation from A to B.

Then $\qquad R^{-1} = \{(5, 1), (5, 2), (7, 2)\}$ which is a relation from B to A.

Also, Domain $(R) = \{1, 2\} = $ Range (R^{-1})

And, Range $(R) = \{5, 7\} = $ Domain (R^{-1})

(3) The inverse of the relation *"is less than"* In R *"is greater than"*.

REMARK

• Sometimes, the inverse of a relation coincides with the relation itself.
For example, the inverse of the relation "perpendicular to" in the set of straight lines coincides with itself.

1.13 CLASSIFICATION OF RELATIONS

(a) **Reflexive Relation:** Let R be a relation on a set A.

"A relation R is said to be reflexive if $(x, x) \in R \; \forall \; x \in A$" i.e., $x \, R \, x \; \forall \; x \in A$

☞ **ILLUSTRATIONS**

(1) In a set of integers, a relation R defined by $x \, R \, y$ iff $x - y$ is divisible by 4, then R is a reflexive relation because $x - x = 0$ which is a divisible by 4.

(2) The universal relation on a non-empty set A is reflexive.

(3) The relation "is less than," i.e., '$<$' in the set of rational number is not reflexive, because no member have the relation is less than to itself.

(4) The relation "is a factor of" in the set of rational number is reflexive, since every rational number is a factor of itself.

(5) The relation "is less than or equal to." i.e., $\leq$ in the set of natural number is reflexive.
$$n \leq n \; \forall \; n \in N$$

(b) **Symmetric Relation.** *A relation R on a set A is said to be symmetric if*
$$(y, x) \in R \text{ whenever } (x, y) \in R \; \forall \; x, y \in R$$
i.e., $\qquad x \, R \, y \Leftrightarrow y \, R \, x \; \forall \; x, y \in R$

☞ **ILLUSTRATIONS**

(1) Let l_1, l_2 be two lines such that l_1 is perpendicular to l_2, i.e., $l_1 \perp l_2$. Then $l_1 \perp l_2 \Rightarrow l_2 \perp l_1$. Therefore the relation $\perp$ is symmetric.

(2) The identity and the universal relation on a non-empty set are symmetric relations.

(3) Consider the set N of natural numbers and the relation 'is less than'. This relation is not symmetric. Since if $2 < 3$ then $3 \nless 2$.

Let $A = \{1, 2, 3\}$ and relations R_1 and R_2 defined by
$$R_1 = \{(1, 2), (1, 3), (3, 1), (2, 1)\} \text{ and } R_2 = \{(1, 2), (2, 3), (3, 1)\}$$
Then R_1 is a symmetric relation, but R_2 is not symmetric.

(c) **Transitive Relation:** *A relation R on a set A is said to be transitive iff $(x, y) \in R$ and $(y, z) \in R \Rightarrow (x, z) \in R \; \forall \; x, y, z \in A$,* i.e., $x \, R \, y, \; y \, R \, z \Rightarrow x R z$.

☛ **ILLUSTRATIONS**

(1) Let a, b, c be three numbers such that a is a factor of b and b is a factor of c, then obviously a is a factor of c. Therefore, 'is a factor of' is a transitive relation.

(2) If l_1, l_2, l_3 are three lines such that $l_1 \perp l_2$ and $l_2 \perp l_3$ then it is obvious that l_1 is parallel to l_3. Therefore the relation " $\perp$ " is not transitive.

(3) The identity and universal relation on a non-empty set are transitive.

(4) Let l_1, l_2, l_3 be three straight lines, such that l_1 is parallel to l_2 and l_2 is parallel to l_3 then it is clear that l_1 is parallel to l_3. Therefore, 'is parallel to' is a transitive relation.

(d) **Anti-symmetric Relation.** *A relation R on a non-empty set A is said to be an anti-symmetric relation iff* $(x, y) \in R$ *and* $(y, x) \in R \Rightarrow x = y \ \forall \ x, y \in R$

REMARKS

• The identity relation R on a set A is an anti – symmetric relation.

• If $(x, y) \in R$ and $(y, x) \notin R$, then it may be noted that $x = y$.

• The universal relation on a set A containing at least two elements is not anti – symmetric.

1.13.1 EQUIVALENCE RELATIONS

A relation R on a set E is said to be equivalence if it is

(i) Reflexive,   (ii) Symmetric and   (iii) Tansitive

☛ **ILLUSTRATIONS**

(1) In a set of integers, a relation R is defined by $x R y$ if and only if $x - y$ is divisible by 4. Then R is an equivalence relation. Since

(a) For $x R x$, $x - x = 0$ is divisible by 4. Therefore, it is reflexive.

(b) For $x R y$. Let $x - y = 4m$ so $y - x = 4m$, which is also divisible by 4. Therefore, it   is symmetric.

(c) For $x R y$, let $x - y = 4m$; for $y R z$, let $y - z = 4n$. By adding these two equations,   we get $x - z = -4(m + n)$,

which is divisible by 4. Therefore it is transitive.

(2) Let R be a relation on the set of all lines in a plane L defined by $(l_1, l_2) \in R$ if and only if line l_1 is parallel to l_2, then R is an equivalence relation because

(a) For each line $l \in L$, we have l is parallel to l.

$\Rightarrow lRl \Rightarrow R$ is reflexive.

(b) Let $l_1, l_2 \in L$ such that $(l_1, l_2) \in R$, then

$\Rightarrow (l_1, l_2) \in R \Rightarrow l_1$ is parallel to $l_2 \Rightarrow l$ is symmetric.

(c) Let $l_1, l_2, l_3 \in L$ such that (l_1, l_2) and $(l_2, l_3) \in R$, then obviously $(l_1, l_3) \in R$ because

if l_1 is parallel to l_2 and l_2 is parallel to l_3, then l_3 should be parallel to l_1.

1.13.2 CONGRUENCE MODULO 'm'

Let m be an arbitrary but fixed integer. If $x - y$ is divisible by m, then two integers x and y are said to be congruence modulo m of one another.

Symbolically : $x \equiv y \pmod{m}$ if $x - y$ divisible by m.

For example : $32 \equiv 2 \pmod{3}$, as $32 - 2 = 30$ which is divisible by 3.

1.13.3 COMPOSITION OF RELATIONS

Let R_1 and R_2 be two relations from sets A to B and B to C respectively, then we can define a relation $R_1 \, o \, R_2$ from A to C, such that $(x, z) \in R_1 \, o \, R_2$ if and only if there exist $y \in Y$ such that $(x, y) \in R_1$ and $(y, z) \in R_2$.

This relation is called composition of R_1 and R_2.

REMARKS

- $R_1 o R_2 \neq R_2 o R_1$
- $(R_2 o R_1)^{-1} = R_1^{-1} o R_2^{-1}$

For example : Let A, B, C be three sets such that

$A = \{-1, -2\}$, $B = \{p, q, r\}$ and $C = \{\alpha, \beta, \gamma\}$

Also, $\qquad R_1 = \{(-1, p), (-1, r), (-2, q)\}$ is a relation from A and B and

$\qquad\qquad R_2 = \{(p, \alpha), (q, \beta), (r, \gamma)\}$ and is a relation from set to B to C.

Then $R_2 o R_1$ is a relation from A to C given by

$\qquad\qquad R_2 o R_1 = \{(-1, \alpha), (-1, \gamma), (-z, \beta)\}$

THEOREM 4. *The intersection of two equivalence relations on a set is an equivalence relation.*

Proof. Let R_1, R_2 be two equivalence relation on a set A. To show $(R_1 \cap R_2)$ also an equivalence relation.

(i) Let $a \in A$ be arbitrary.

Since R_1 and R_2 both are reflexive on A.

$\therefore \qquad\qquad (a, a) \in R_1$ and $(a, a) \in R_2 \Rightarrow (a, a) \in R_1 \cap R_2$

Therefore, $(R_1 \cap R_2)$ is reflexive.

(ii) Let $a, b \in A$ such that $(a, b) \in R_1 \cap R_2$

$\qquad\qquad (a, b) \in R_1 \cap R_2 \Rightarrow (a, b) \in R_1$ and $(a, b) \in R_2$

Also, R_1 and R_2 both are symmetric on A.

Therefore, $(b, a) \in R_1$ and $(b, a) \in R_2 \Rightarrow (b, a) \in R_1 \cap R_2 \Rightarrow (R_1 \cap R_2)$ is symmetric on A.

(iii) Let $a, b, c \in A$ such that $(a, b) \in R_1 \cap R_2$, $(b, c) \in R_1 \cap R_2$

Then, $(a, b) \in R_1 \cap R_2$ and $(b, c) \in R_1 \cap R_2$

$\Rightarrow \quad \{(a, b) \in R_1$ and $(a, b) \in R_2$ and $\{(b, c) \in R_1$ and $(b, c) \in R_2\}$

$\Rightarrow \quad \{(a, b) \in R_1, (b, c) \in R_1\}$ and $\{(a, b) \in R_2, (b, c) \in R_2\}$

$\Rightarrow \quad (a, c) \in R_1$ and $(a, c) \in R_2 \qquad\qquad [\because R_1$ and R_2 both are transitive.]

$\Rightarrow \quad (a, c) \in R_1 \cap R_2$

Therefore, $(R_1 \cap R_2)$ is transitive on A.

From (i), (ii) and (iii), we have $R_1 \cap R_2$ is reflexive, symmetric and transitive, and hence $R_1 \cap R_2$ is an equivalence relation.

REMARK

- The union of two equivalence relations on a set is not necessarily an equivalence relation.

THEOREM 5. *If R is an equivalence relation, then R^{-1} is also an equivalence relation.*

Proof. Let R be an equivalence relation on a set A. Then by definition of relation on a set, we have

$$R \subseteq A \times A \Rightarrow R^{-1} \subseteq A \times A$$

Therefore, R^{-1} is a relation on A.

Now, to show R^{-1} is an equivalence relation.

(i) Let $a \in A$, then $(a, a) \in R$ ($\because R$ is an equivalence relation and hence reflexive)

$\Rightarrow \quad (a, a) \in R^{-1}$

Thus, $(a, a) \in R^{-1} \forall a \in R \Rightarrow R^{-1}$ is reflexive on A.

(ii) Let $(a, b) \in R^{-1}$, then $(a, b) \in R^{-1} \Rightarrow (b, a) \in R$

$\Rightarrow \quad (a, b) \in R$ ($\because R$ is symmetric)

$\Rightarrow \quad (b, a) \in R^{-1}$

Therefore R^{-1} is symmetric .

(iii) Let $(a, b) \in R^{-1}$ and $(b, c) \in R^{-1}$ then $(a, b) \in R^{-1} \Rightarrow (b, a) \in R$

and $(b, c) \in R^{-1} \Rightarrow (c, b) \in R$

Now, $(c, b) \in R$ and $(b, a) \in R$

$(c, a) \in R$ ($\because R$ is transitive)

$(a, c) \in R^{-1}$

Therefore R^{-1} is transitive .

From (i), (ii) and (iii), we conclude that R^{-1} is an equivalence relation.

Solved Examples

Based on the following Results

➠ If $n(A) = p$, $n(B) = q$ then total number of subsets of $A \times B = 2^{pq}$.

➠ **Reflexive Relation:** $xRx, \forall x \in A$

➠ **Symmetric relation:** $xRy \Leftrightarrow yRx \forall x, y \in R$

➠ **Transitive relation:** $xRy, yRz \Rightarrow xRz$

➠ **Anti-symmetric relation:** $xRy \Rightarrow yRx \Leftrightarrow x = y$

➠ **Equivalence relation:** Reflexive, symmetric and transitive **(RST).**

➠ **Partial ordered relation:** Reflexive, anti-symmetric and transitive **(RAT)**

➠ R is equivalence $\Rightarrow R^{-1}$ is equivalence.

➠ Intersection of two equivalence relations on a set is again equivalence.

Example 1. *Let Z be the set of integers. Define a relation R on Z such that x R y holds if and only if x − y is divisible by 5, $x \in Z$, $y \in Z$. Show that it is an equivalence relation.*

Solution. (i) For each $x \in Z$, $x - x$ i.e., 0 is divisible by 5.

Therefore, for all $x \in Z$, $x R x \Rightarrow x$ is reflexive.

(ii) Let $x R y \Rightarrow x - y$ is divisible by 5.

$\Rightarrow y - x$ is divisible by 5.

Thus $xRy = yRx$

Therefore R is symmetric.

(iii) Let us suppose xRy and yRz, then $(x - y)$ and $(y - z)$ are both divisible by 5. Hence, 5 is also a divisor of $(x - y) + (y - z)$.

5 is a divisor of $(x - z)$.

Therefore, $xRy, yRz \Rightarrow xRz \Rightarrow R$ is transitive.

From (i), (ii) and (iii), we conclude that R is an equivalence relation.

Example 2. *Let* $N \times N$ *be the set of ordered pairs of natural numbers. Also, let R be the relation in* $N \times N$, *defined by (a, b) R (c, d) if and only if* $a + d = b + c$. *Show that R is an equivalence relation.*

Solution. (i) For all $(a, b) \in N \times N$, we have $a + b = b + a$, i.e., $(a, b) R (b, a)$.

Therefore, R is reflexive.

(ii) Let $(a, b) R (c, d)$, then, by definition of R

$$(a + d) = (b + c) \text{ or } (c + b) = (d + a)$$

$(c, d) R (a, b) \Rightarrow R$ is symmetric.

(iii) Let us suppose $(a, b) R (c, d)$ and $(c, d) R (e, f)$, then

$$a + d = b + c \text{ and } c + f = d + e$$

$\Rightarrow \quad (a + d) + (c + f) = (b + c) + (d + e) \Rightarrow a + f = b + e$

$\Rightarrow \quad (a, b) R (e, f)$

Therefore, R is transitive.

Hence, from (i), (ii) and (iii), we conclude that R is an equivalence relation.

Example 3. *If R is the relation for natural number defined by* $x + 4y = 20$. *Find the domain and range of the relation R.*

Solution. Let $x + 4y = 20 \quad \Rightarrow \qquad y = \dfrac{20 - x}{4}$

For $x = 4, y = 4$ and for $x = 8, y = 3$.

For $x = 16, y = 1$ and for $x = 12, y = 2$

Therefore, Domain = {4, 8, 12, 16} and range = {4, 3, 2, 1}

Example 4. *A relation R defined on the set of integers Z, as follows*

$$(x, y) \in R \Rightarrow x^2 + y^2 = 25$$

Express R and R^{-1} *as the sets of ordered pairs and hence find their respective domains.*

Solution. Since $(x, y) \in R \Leftrightarrow x^2 + y^2 = 25 \quad \Rightarrow \quad y = \pm\sqrt{25 - x^2}$

If $x = 0 \quad \Rightarrow \quad y = 5$.

Therefore, $(0, 5) \in R$ and $(0, -5) \in R$

Now, $x = 3 \quad \Rightarrow \quad y = \sqrt{25 - 9} = \pm 4$

$(3, 4) \in R, (-3, 4) \in R, (3, -4) \in R$ and $(-3, -4) \in R$

$x = \pm 4 \quad \Rightarrow \quad y = \pm 3$

Therefore, $(4, 3) \in R$, $(-4, 3) \in R$, $(4, -3) \in R$ and $(-4, -3) \in R$

$x = \pm 5 \Rightarrow y = \sqrt{25 - 25} = 0 \quad \therefore \quad (5, 0) \in R$ and $(-5, 0) \in R$

Here, it is clear that for any other integral value of x, y is not an integer. Therefore,

$R = \{(0, 5), (0, -5), (3, 4), (-3, 4), (3, -4), (-3, -4), (4, 3), (-4, 3), (4, -3),$
$(-4, -3), (5, 0), (-5, 0)\}$

and $R^{-1} = \{(5, 0), (-5, 0), (4, 3), (4, -3), (-4, 3), (-4, -3), (3, 4), (3, -4),$
$(-3, 4), (-3, -4), (0, 5), (0, -5)\}$

Also, Domain $(R) = \{0, 3, -3, 4, -4, 5, -5\} =$ domain of (R^{-1}).

Example 5. *Consider the set $A = \{a, b, c\}$. Give an example of a relation R on A which is*

(i) *reflexive and symmetric but not transitive.*

(ii) *symmetric and transitive, but not reflexive.*

(iii) *reflexive and transitive, but not symmetric.*

Solution. (i) Given $A = \{a, b, c\}$

Let $R = \{(a, a), (a, b), (b, a), (b, c), (c, b), (b, b), (c, c)\}$ on A.

Clearly, R is reflexive and symmetric but not transitive.

(ii) Let $R = \{(a, a), (a, b), (b, a), (b, b)\}$ on A.

Here, R is symmetric and transitive but not reflexive.

(iii) Let $R = \{(a, a), (b, b), (c, c), (a, b)\}$ on A.

Here, R is reflexive, transitive but not symmetric.

Example 6. *If R is a relation on $N \times N$, show that the relation R defined by $(a, b)\ R\ (c, d)$ if and only if $ad = bc$ is an equivalence relation.*

Solution. (i) Since $ab = ba\ \forall\ a, b \in N$.

Therefore, $(a, b)\ R\ (a, b) \forall\ a, b \in N \Rightarrow R$ is reflexive.

(ii) We have $(a, b)\ R\ (c, d)$ iff $ad = bc\ \forall\ a, b, c, d \in N$

Now, $(c, d)\ R\ (a, b)$ iff $cb = da\ \forall\ a, b, c, d \in N \Rightarrow R$ is symmetric.

(iii) We have $(a, b)\ R\ (c, d)$ iff $ad = bc\ \forall\ a, b, c, d \in N$

Therefore, $(a, b)\ R\ (c, d), (c, d)\ R\ (e, f) \Rightarrow (a, b)\ R\ (e, f)\ \forall\ a, b, c, d \in N$

Using $(a, d), (c, f) = (b, c)(d, e)$

$\Rightarrow \qquad (a, f) = (b, e) \Rightarrow R$ is transitive

Hence, from (i), (ii) and (iii), we conclude that R is an equivalence relation.

Example 7. *Let R_1 and R_2 be two relations on a set A, where $A = \{1, 2, 3, 5\}$ such that*

$R_1 = \{(1, 1), (1, 2), (1, 5), (2, 1), (2, 5)\}$

and $\qquad R_2 = \{(3, 3), (3, 2), (2, 3), (1, 2), (2, 1)\}$

Then, which of the following statement is false :

(i) $R_1 \cup R_2$ *is symmetric*   (ii) $R_1 \cap R_2$ *is transitive*

(iii) $R_1 \cap R_2$ *is symmetric*   (iv) $R_1 \cup R_2$ *is transitive.*

Solution. (i) As $(1, 2) \in R_1$, also $(2, 1) \in R_1$, therefore, it is symmetric and as $(1, 2) \in R_2$, also $(2, 1) \in R_2 \Rightarrow R_2$ is symmetric.

Now, $R_1 \cup R_2 = \{(1, 1), (1, 2), (1, 5), (2, 1), (2, 5), (3, 3), (3, 2), (2, 3)\}$

In $R_1 \cup R_2$, as $(1, 2) \in R_1 \cup R_2$, also $(2, 1) \in R_1 \cup R_2 \Rightarrow R_1 \cup R_2$ is symmetric

Therefore, (i) is true.

(ii) We have $R_1 \cap R_2 = \{(1, 2), (2, 1)\}$

$\Rightarrow$ (1, 1) should also belong to $R_1 \cap R_2$.

But in this case $(1, 1) \notin R_1 \cap R_2$. Hence, $R_1 \cap R_2$ is not transitive.

Therefore, (ii) is false.

(iii) We have, $R_1 \cap R_2 = \{(1, 2), (2,1)\}$

$\qquad (1, 2) \in R_1 \cap R_2$ and also $(2, 1) \in R_1 \cap R_2$.

Therefore, (iii) is true.

(iv) In $R_1 \cup R_2$, $(1, 2) \in R_1 \cup R_2$

and $(2, 5) \in R_1 \cup R_2$, also $(1, 5) \in R_1 \cup R_2$

$\Rightarrow$ $R_1 \cup R_2$ is transitive

Therefore, (iv) is true.

Example 8. *If A be the set of all triangles in a plane and R = $\{(a, b) : \Delta a = \Delta b\}$, i.e.,*

aRb $\Leftrightarrow$ Area of triangle a = Area of triangle b, then show that R is an equivalence relation.

Solution. (i) Since, for all $a \in A$ we have $\Delta a = \Delta a$

Therefore, $aRa \Rightarrow R$ is reflexive.

(ii) For any $a, b \in A$, we have $(a, b) \in R \Rightarrow \Delta a = \Delta b$

$\Rightarrow \qquad \Delta b = \Delta a \Rightarrow (b, a) \in R$

Therefore, $(b, a) \in R$, i.e., $bRa \Rightarrow R$ is symmetric.

(iii) For all $a, b, c \in A$, we have $(a, b) \in R$, $(b, c) \in R$

$\qquad \Delta a = \Delta b$ and $\Delta b = \Delta c \Rightarrow \Delta a = \Delta c \Rightarrow (a, c) \in R$

Therefore, R is transitive.

Hence, from (i), (ii) and (iii), we conclude that R is an equivalence relation.

Example 9. *Let Z be a set of non-zero integers and a relation R defined by $xRy \Rightarrow x^y = y^x \ \forall \ x, y \in Z$, then show that R is not an equivalence relation on Z.*

Solution. (i) Let $x \in Z$, then $x^x = x^x, \ \forall \ x \in Z$

$\Rightarrow \qquad xRx, \ \forall \ x \in Z$

Therefore, R is reflexive.

(ii) Let $x, y \in Z$, such that xRy, i.e., $x^y = y^x$

$\Rightarrow \qquad x^y = y^x \Rightarrow y^x = x^y$

Therefore, $xRy \Rightarrow yRx, \ \forall \ x, y \in Z$

$\Rightarrow \qquad R$ is symmetric.

(iii) Let $x, y, z \in Z$ such that xRy and yRz

i.e., $\quad x^y = x^y$ and $y^z = z^y$ which does not give $x^z = z^x$

$\Rightarrow \qquad R$ is not transitive.

Hence, we conclude that R is not an equivalence relation.

Example 10. *Let $A = R \times R$ (R is the set of real numbers) and define the following relation on A : (a, b) R (c, d) iff $a^2 + b^2 = c^2 + d^2$*

    (i) *verify that R is an equivalence relation on A.*

    (ii) *describe geometrically what the equivalence classes are for this reason.*

Solution. (i) we have $(a, b)R(c, d)$ $\Rightarrow a^2 + b^2 = c^2 + d^2$

    $\Rightarrow$ $c^2 + d^2 = a^2 + b^2 \Rightarrow (c, d)R(a, b)$ ...(1)

    $\Rightarrow$ R is symmetric.

Now, $(a, b)R(c, d)$ and $(c, d)R(x, y) \Rightarrow a^2 + b^2 = c^2 + d^2$

and $c^2 + d^2 = x^2 + y^2$

    $\Rightarrow$ $a^2 + b^2 = x^2 + y^2 \Rightarrow (a, b)R(x, y)$ ...(2)

    $\Rightarrow$ R is transitive.

Again $(a, b)R(a, b) \Leftrightarrow a^2 + b^2 = a^2 + b^2$ ...(3)

    $\Rightarrow$ R is reflexive.

Hence, from (1), (2) and (3), we conclude that R is an equivalence relation.

    (ii) For any point (a, b), the sum $a^2 + b^2$ is the square of the distance from the origin. The equivalence classes are, therefore, the set of points in the plane which have the same distance from the origin. Hence, the equivalence classes are concentric circles centered at the origin.

Example 11. *Let R be the binary relation defined as $R = \{(a, b) \in R^2 : a - b \le 3\}$. Determine whether R is reflexive, symmetric, anti symmetric and transitive.*

Solution. We have $(a, b) \in R^2 : a - b \le 3$.

    $\Rightarrow$ $(a, a) \in R^2 : a - a \le 3$ i.e., $0 \le 3$, which is true. So, R is reflexive.

In a similar way, we can easily show that R is neither symmetric, anti symmetric nor transitive.

1.13.4 RELATIONS OTHER THAT EQUIVALENCE

Let R be a given relation on the set X. Then R is

    (1)   non-reflexive if $\exists x$, such that $(x, x) \notin R$.

    (2)   anti-reflexive or reflexive if $i_x \cap R = \phi$ (where i_x is the identity relation on X or $\forall x \in X : (x, x) \notin R$

    (3)   non-symmetrical if for some $(x, y) \in R$, we have $(y, x) \notin R$

    (4)   anti-symmetric if $R \cap R^{-1} = i$, i.e., $(x, y) \in R$ and $(y, x) \in R \Rightarrow x = y$

    (5)   asymmetric if $R \cap R^{-1} = \phi$, i.e., $(x, y) \in R \Rightarrow (y, x) \notin R$

    (6)   non-transitive if $R \circ R \not\subset R$

    (7)   anti-transitive if $(R \circ R) \cap R = \phi$

    (8)   A reflexive and symmetric, but not transitive relation is called a tolerance relation.

    (9)   A non-symmetric transitive relation is called an ordered relation.

    (10)   A reflexive, anti-symmetric and transitive relation is called partial-ordered relation.

☞ EXERCISE 1.6

1. If R is the relation 'is less than' from $A = \{1, 2, 3, 4, 5\}$ to $B = \{1, 4, 5\}$, find the set of ordered pairs corresponding to R. Also find R^{-1}.

2. A relation R defined from a set $A = \{2, 3, 4, 5\}$ to a set $B = \{3, 6, 7, 10\}$ as follows : $(x, y) \in R \Rightarrow x$ divides y. Write R as a set of ordered pairs and determine the domain and range of R. Also find R^{-1}.

3. Find the domain and range of $A = \{1, 2, 3, 4, 5, 6\}$ when the relation are defined as
   (i) $x R_1 y$ if and only if $x - y > 0$
   (ii) $x R_2 y$ if and only if $x + y < 0$

4. Two sets A and B are given by $A = \{1, 2, 8, 9\}$ and $B = \{2, 3, 4, 6, 7\}$ and if R is the relation form A to B given by $\{(1,2), (1,3), (2,4), (2,6)\}$, then which of the following statement is true?
   (i) Domain (R) = Range (R^{-1}) and Range (R) = Domain (R^{-1})
   (ii) Domain (R) = Domain (R^{-1}) and Range (R) = Range (R^{-1})
   (iii) Domain (R) = Range (R^{-1}) and Range (R) = Domain (R^{-1})
   (iv) Domain (R) = Range (R)

5. If R is a relation on a set A, then which of the following statement is not true?
   (i) If R is reflexive then R^{-1} is reflexive.
   (ii) If R is symmetric then R^{-1} is symmetric.
   (iii) If R is transitive, then R^{-1} is transitive.
   (iv) None of these

6. Find the domain and range of the following relations:
   (i) $R = \{(x + 1, x + 5)\} : x \in \{0, 1, 2, 3, 4, 5\}$
   (ii) $R = \{(x, x^3) : x$ is a prime number, less than 10$\}$
   (iii) $R = \{(a, b) : a \in N, a < 5, b = 4\}$
   (iv) $R = \{(a, b) : b = |a - l|, a \in Z, $ and $|a| \le 3\}$

7. Let R_1 be the relation defined on the set of reals R such as $(a, b) \in R_1$ if and only if $1 + ab > 0$ for all $a, b \in R$. Show that R_1 is reflexive, symmetric but not transitive.

8. Let R be relation on $N \times N$, defined by $(a, b) R (c, d)$ if and only if $ad \, (b + c) = bc \, (a + d)$. Show that R is an equivalence relation.

9. Show that the relation 'congruence modulo m' on the set of integers is an equivalence relation.

10. Let R_1 be a relation on the set of reals defined by $R_1 = \{(a, b) \in R \times R : a^2 + b^2 = 1\}$ Show that R_1 is not an equivalence relation on R.

11. In a set L of all straight lines in a plane, discuss which of the following two relations are equivalence relations L.
    (i) $R_1 = \{(x, y) : x, y \in L$ and x is parallel to $y\}$
    (ii) $R_2 = \{(x, y) : x, y \in L$ and x is perpendicular to $y\}$.

12. Show that the relation $R = \{(a, b) : a - b = $ even integer $\forall \, a, b \in Z\}$, i.e., $a R b \Leftrightarrow a - b = $ even integer, is an equivalence relation.

13. Show that the relation R in N, the set of natural numbers, defined by $x R y$ if $x^2 - 4xy + 3y^2 = 0$, $(x, y \in N)$ is reflexive, not symmetric and not transitive.

14. For the given relation R on a set S, determine which are equivalence relations:
    (i) S is the set of all rational numbers, $a R b$ if and only if $a = b$
    (ii) S is the set of all real numbers iff
        (a) $|a| = |b|$ (b) $a \ge b$
    (iii) S is the set of all triangles in a plane, $a R b$ iff a is congruent to b.
    (iv) S is the set of all triangles in a plane, $a R b$ iff a and b have equal perimeters.

15. An integer m is said to be related to another integer n if m is a multiple of n. Show that this relation is reflexive and transitive but not symmetric.

16. Let R be a relation defined on the set of natural number N as $R = \{(x, y) : x, y \in N, 2x + y = 41\}$. Find the domain and range of R.

17. Let O be the origin. Define a relation between two points P and Q in a plane if $PO = OQ$. Show that the relation is an equivalence relation.

18. Given the relation $R = \{(1, 2), (2, 3)\}$ on the set of natural number N, add a minimum of ordered pairs so that the enlarged relation is symmetric, transitive and reflexive.

19. Let N denote the set of all natural numbers and R be the relation on N×N defined by $(a,b)R(c,d) \Leftrightarrow ad(b+c) = bc(a+d)$. Show that R is an equivalence relation.

20. Show that the relation, which is symmetric and transitive, is not necessarily reflexive.

Answers

1. $aRb = \{(1,4),(1,5),(2,4),(3,4),(2,5),(3,5),(4,5)\}$,
$R^{-1} = \{(4,1),(5,1),(4,2),(5,2),(4,3),(5,3),(5,4)\}$

2. Domain $(R) = \{2, 3, 5\}$, Range $(R) = \{3, 6, 10\}$, $R^{-1} = \{(6,2),(10,2),(3,3),(6,3),(10,5)\}$

3. (i) $\{2, 3, 4, 5, 6\}$, $\{1, 2, 3, 4, 5\}$, (ii) ϕ, ϕ **4.** (iii) **5.** (iv) **6.** (i) Domain $(R) = \{1, 2, 3, 4, 5, 6\}$, Range $(R) = \{5, 6, 7, 8, 9, 10\}$ (ii) Domain $(R) = \{2, 3, 5, 7\}$, Range $(R) = \{8, 27, 125, 243\}$ (iii) Domain $(R) = \{1,2,3,4\}$, Range $(R) = \{4\}$ (iv) Domain $(R) = \{0,-1,-2,-3,1,2,3\}$, Range $(R) = \{1, 2, 3, 4, 0, 1, 2\}$ **11.** $R_1 = $ Equivalence relation, $R_2 = $ Not equivalence **14.** (i), (ii) **16.** Domain $(R) = \{1, 2, ..., 19, 20\}$, Range $(R) = \{39, 37, 35, ..., 5, 3, 1\}$ **18.** $\{(1, 2), (2, 1), (2, 3), (3, 2), (1, 3), (3, 1), (1, 1), (2, 2), (3, 3), (4, 4), ...\}$

1.14 FUNCTIONS

Definition: *Let A and B be two sets, then a rule or correspondence, which associates each element of A to a unique element to B, is called a function from set A to set B.*

If a general element of set A is denoted by x, and of set B is denoted by y, then we say that y is a function of x if, for every $x \in A$, one and only one value of $y \in B$ can be determined.

Symbolically: If f is a function from a set A to a set B, then we write $f : A \to B$, read as f is a function from A to B or f maps A to B.

1.14.1 RANGE AND DOMAIN OF A FUNCTION

Let an element $y \in B$ be corresponded by an element $x \in A$, then y is called the image of x and is denoted by $f(x)$. Here, x is defined as the pre-image of y.

The set A is called the domain and the set B is called the co-domain of the function f.

The set of all f-images of the elements of A, is called image set or the range of f and is denoted by

$$f(A) \quad \text{or} \quad \{f(x) : x \in A\}$$

Evidently, $f(A) \subseteq B$.

Thus, a mapping $f : A \to B$ is the set of ordered pairs $\{(a, b) : a \in A, b \in B\}$, so that no two ordered pairs have the same finite element.

$$f = \{(a, b): a \in A, b \in B, b = f(x) \, \forall \, a \in A\}$$

For example : Let $A = \{-2, -1, 0, 1, 2\}$ and B is the set of natural numbers for every $x \in A$, $f(x) \in B$ and $f(x) = x^2$.

Here, A is the domain and B is the co-domain.

$f(a)$ is the value of the function $f(x)$, when x takes the value a, *i.e.,* when x is replaced by a.

The elements of the co-domain which is equal to $f(x)$ form the range.

When $x = -2$, $f(-2) = (-2)^2 = 4$

When $x = -1$, $f(-1) = 1$

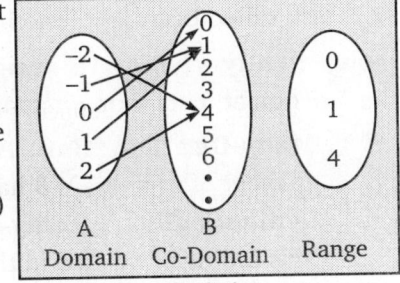

Fig. (14)

When $x = 0, f(0) = 0$

When $x = 1, f(1) = 1$

When $x = 2, f(2) = 4$.

Which can be illustrated in the figure (14).

REMARKS

- If $f : A \rightarrow B$ then a single element in A cannot have more than one image in B. However, two or more elements in A may have the same image in B.
- Every element in A must have its image in B, but every element in B may not have it pre-image in A.
- To each element x in A, there exists a unique element y in B such that $y = f(x)$.
- The unique element y of B is called the value of f at x (the image of f under x), and written as $y = f(x)$.
- The range of f consist of those elements in B which appear as the image of at least one element in A.
- Range of a function is the image of its domain.
- Range is a subset of co-domain.

1.15 TYPE OF FUNCTIONS

(a) **One-One function:** A function f from A to B, *i.e.,* $f : A \rightarrow B$ is said to be one-one (or injective) iff distinct elements of A have distinct images.

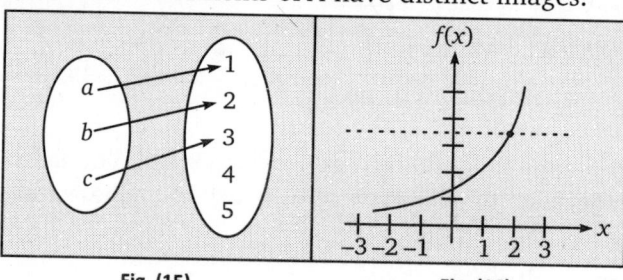

Fig. (15)   Fig. (16)

Symbolically: f is one-one if for $x_1, x_2 \in A$, we have

$$x_1 \neq x_2 \quad \Rightarrow \quad f(x_1) \neq f(x_2) \; \forall \; x_1, x_2 \in A$$

or

$$f(x_1) = f(x_2) \Rightarrow \quad x_1 = x_2 \; \forall \; x_1, x_2 \in A$$

It is also called Univalent function.

Graphically, a function is one-one if and only if no line parallel to x-axis meets the graph of the function at more than one point.

(b) **Many-One Function:** A function $f : A \rightarrow B$ is called many-one, if at least one element of co-domain B has two or more than two pre-images in domain A.

Symbolically: f is many-one if for $x_1, x_2 \in A$, we have $x_1 \neq x_2 \Rightarrow f(x_1) = f(x_2)$

This can be illustrated in the following figures.

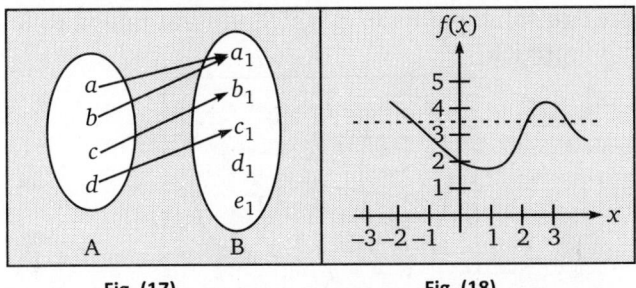

| Fig. (17) | Fig. (18) |

Graphically, a function is many-one if and only if a line parallel to x-axis meets the graph of the function at more than one point.

REMARK

- One-many function does not exist.

(c) Onto function: A function $f : A \rightarrow B$ is called an onto function, if there is no element of B which is not an image of some element of A, *i.e.*, every element of B appears as the image of at least one element of A. This is illustrated in Figure 20.

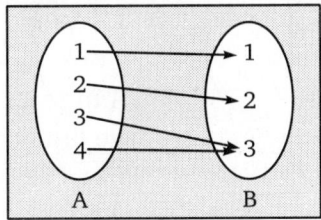

Fig. (19): Onto Function

REMARKS

- In an onto function, Range = Co-domain
- Onto function is also called surjective.

(d) Into function: A function $f : A \rightarrow B$ is called an into function, *i.e.*, if there is at least one element of set B which has no pre-image in the set A. This is illustrated in Figure 20.

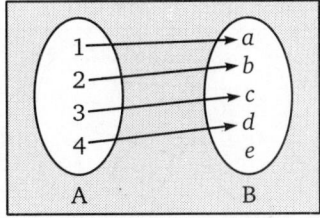

Fig. (20): Into Function

REMARK

- In an into function, Range $\subset$ Co-domain.

(e) One-One Into Function: A function $f : A \rightarrow B$ is called a one-one into function, if it is both one-one and into, *i.e.,* the different points in A are joined to different points in B and there are some points in B which are not joined to any point in A. This is illustrated in Figure 21.

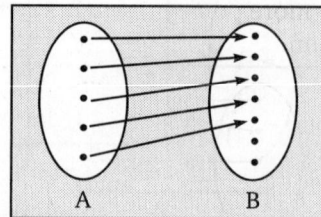

Fig. (21): One-One Into Function

Symbolically : One-one into function is defined as

    (i) Range $\subset$ Co-domain.

    (ii) $f(x_1) \neq f(x_2) \Rightarrow x_1 \neq x_2$.

(f) **One-One Onto Function:** A function $f : A \rightarrow B$ is both one-one and onto, *i.e.,* the different points in A are joined to different points in B and no point in B is left vacant. This is illustrated in Figure 22.

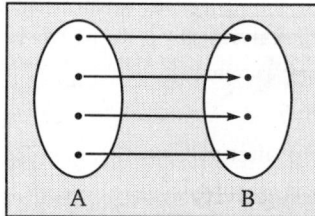

Fig. (22): One-one Onto Function

REMARKS

- One-one onto mapping is also known as bijective or one-to-one.
- For a one-one onto function

    Range = Co-domain, and $x_1 \neq x_2 \Rightarrow f(x_1) \neq f(x_2)$ or $f(x_1) = f(x_2) \Rightarrow x_1 = x_2$

(g) **Many-One Into Function:** A function $f : A \rightarrow B$ which is both many-one and into function is called a many-one into function, *i.e.,* two or more points in A are joined to some points in B and there are some point in B which are not joined to any point in A. Therefore, for many-one into function.

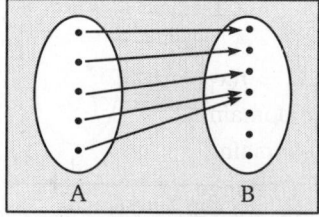

Fig. (23): Many-One Into Function

    (i) Range $\subset$ Co-domain.

    (ii) $x_1 \neq x_2$

         $\Rightarrow f(x_1) = f(x_2)$

(h) Many-One Onto Function: If function $f : A \to B$ is both many-one and onto function is called a many one onto function, *i.e.,* in B one point is joined to at least one point in A and two or more points in A are joined to some points in B. Therefore, for many-one onto function.

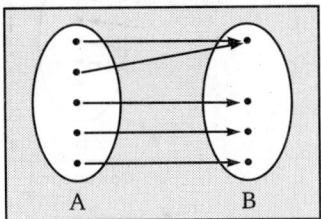

Fig. (24): Many-One Onto Function

(i) Range = Co-domain.

(ii) $x_1 \ne x_2 \quad \Rightarrow f(x_1) = f(x_2)$

WORKING RULE

1.	**For checking the Injectivity (One-One) of the function**	
	Let x and y be two arbitrary elements in the domain of f.	
STEP 1.	Take $f(x) = f(y)$	
STEP 2.	If we get $x = y$, after solving $f(x) = f(y)$. Then, $f : A \to B$ is one-one.	
2.	**For checking the surjectivity (Onto) of a function**	
STEP 1.	Take an arbitrary element y in the co-domain.	
STEP 2.	Put $f(x) = f(y)$	
STEP 3.	Solve $f(x) = y$ for x and obtain x in terms of y.	
STEP 4.	Get the equation of the form $x = g(y)$	
STEP 5.	If $x = g(y)$ belongs to domain f, for all values of y, then f is onto.	

Solved Examples

Based on the following Results

➡ For a function $f : A \to B, A =$ domain, $B =$ co-domain.

➡ **For one-one function:** $x_1 \ne x_2 \Rightarrow f(x_1) \ne f(x_2) \ \forall \ x_1, x_2 \in A$

➡ or $f(x) = f(x_2) \Rightarrow x_1 = x_2 \ \forall \ x_1, x_2 \in A$

➡ **For many-one function:** $x_1 \ne x_2 \Rightarrow f(x_1) = f(x_2), \ x_1, x_2 \in A$

➡ **For onto function:** Range = co-domain

➡ **For into function:** Range $\subseteq$ co-domain

➡ **For one-one into function:** (i) Range $\subseteq$ co-domain
(ii) $f(x_1) \ne f(x_2) \Rightarrow x_1 \ne x_2$

➡ **For one-one onto function:** (i) Range = codomain
(ii) $x_1 \ne x_2 \Rightarrow f(x_1) \ne f(x_2)$ or $f(x_1) = f(x_2) \Rightarrow x_1 = x_2$

➡ **For many-one into function:** (i) Range = co-domain   (ii) $x_1 = x_2 \Rightarrow f(x_1) = f(x_2)$

➡ **For many-one onto function:** (i) Range = co-domain (ii) $x_1 \ne x_2 \Rightarrow f(x_1) = f(x_2)$

Example 1. *Let $f : \mathrm{R} \to \mathrm{R}$ be a function defined by*

$$f(x) = \begin{cases} 3x - 1 \text{ when } & x > 3 \\ x^2 - 1 \text{ when } & -2 \le x \le 3 \\ x + 3 \text{ when } & x < -2 \end{cases}$$

Find (i) $f(2)$, (ii) $f(4)$, (iii) $f(-1)$, (iv) $f(-3)$

Solution. (i) $f(2) = (2)^2 - 2 = 4 - 2 = 2$

(ii) $f(4) = 3(4) - 1 = 12 - 1 = 11$

(iii) $f(-1) = (-1)^2 - 2 = 1 - 2 = -1$

(iv) $f(-3) = 2(-3) + 3 = -6 + 3 = -3$

Example 2. *For $y = +\sqrt{x}$, say whether it is a function or not. If it is a function, find its domain and range.*

Solution. Here we have $y = +\sqrt{x}$ ...(1)

Since y is real if $x \ge 0$ and is unique and finite for each $x \ge 0$.

Therefore, (1) is a function with domain $[0, \infty[$.

Again from (1), $y \ge 0 \ \forall \ x \ge 0$

Hence, range $= [0, \infty[$

Example 3. *Find the domain of $f(x) = \dfrac{x^3 - x^2 + 4x + 2}{3x + 11}$*

Solution. Since f is defined for all real values of x except when $3x + 11 = 0$

i.e., when, $x = -\dfrac{11}{3}$

Hence, domain of $f = \mathrm{R} - \left\{-\dfrac{11}{3}\right\}$

Example 4. *Let $f : \mathrm{N} - \{1\} \to \mathrm{N}$ be defined by $f(n) =$ the highest prime factor of n. Show that f is neither one-one nor onto. Also, find the range f.*

Solution. Since we have

$f(6) =$ the highest prime factor of $6 = 3$

$f(9) =$ the highest prime factor of $9 = 3$

$f(12) =$ the highest prime factor of $12 = 3$

Therefore, f is a many-one function.

Clearly, image of any $n \in \mathrm{N} - \{1\}$ is the largest prime number that divides n. So the range of f consists of prime number only. Consequently, range of $f \ne \mathrm{N}$ (Co-domain)

$\Rightarrow$ f is not onto function.

Hence, f is neither one-one nor onto. The range of f is the set of all prime numbers.

Example 5. *Let $A = \{1, 2\}$. Find all one-to-one function from A to A.*

Solution. Let $f : A \to A$ be a one-one function.

Then, for $f(1)$, there are two choices, *i.e.,* 1 or 2.

Let us first suppose $f(1) = 1$.
As $f : A \to A$ is one-one, $f(2) = 2$
Therefore, we have $f(1) = 1, f(2) = 2$
Now, let $f(1) = 2$
Since, $f : A \to A$ is one-one, therefore $f(2) = 1$.
Therefore, we have $f(1) = 2$ and $f(2) = 1$.
Hence, we have two one-one function say f and g form A and A given by
$$f(1) = 1, f(2) = 2 \text{ and } f(2) = 1 \text{ and } f(1) = 2.$$

Example 6. *Let $\{x \in R : -1 \le x \le 1\} = B$. Show that $f : A \to B$ given by $f(x) = x |x|$ is one-one and onto.*

Solution. Let x, y be any two elements in A, then
$$x \ne y \Rightarrow x|x| \ne y|y| \Rightarrow f(x) \ne f(y).$$
Therefore, f is one-one.
Since, range of $f = f(A) = B$ so $f : A \to B$ is onto mapping. Hence f is one-one and onto.

Example 7. *Find the domain and range of the function.*
$$f(x) = -\sqrt{-5 - 6x - x^2}$$

Solution. Given that, $f(x) = -\sqrt{-5 - 6x - x^2}$
For f to be real, $-5 - 6x - x^2 \ge 0$ $\Rightarrow$ $x^2 + 6x + 5 \le 0$
$\Rightarrow$ $x^2 + 6x \le -5$ $\Rightarrow$ $x^2 + 6x + 9 \le -5 + 9$
$\Rightarrow$ $(x + 3)^2 \le 4$ $\Rightarrow$ $|x + 3|^2 \le 4$
$\Rightarrow$ $|x + 3| \le 2$ $\Rightarrow$ $-2 \le x + 3 \le 2$
$\Rightarrow$ $-2 -3 \le x \le 2 -3$ $\Rightarrow$ $-5 \le x \le -1$
Therefore, domain of $f(x) = [-5, -1]$
To find the range of $f(x)$,   put $y = f(x)$
Therefore, $f(x) = -\sqrt{-5 - 6x - x^2}, y \le 0$
$\Rightarrow$ $y^2 = -5 -6x -x^2$ $\Rightarrow$ $x^2 + 6x + (y^2 + 5) = 0$
For real x, discriminant ≥ 0,   *i.e.*, $(6)^2 - 4 \times 1 \times (y^2 + 5) \ge 0$
$\Rightarrow$ $36 - 4y^2 - 20 \ge 0$ $\Rightarrow$ $-4y^2 \ge -16$
$\Rightarrow$ $y^2 \le 4$ $\Rightarrow$ $|y|^2 \le 4$
$\Rightarrow$ $|y| \le 2$ *i.e.*, $-2 \le y \le 2$
But $y \le 0$ therefore, $-2 \le y \le 0$.
Hence,   Range of $f = [-2, 0]$

Example 8. *For a finite set A, if $f : A \to A$ is a one-one function, show that f is onto.*

Solution. Let $A = \{a_1, a_2, ..., a_n\}$ be a finite set.
Since $f : A \to A$ is one-one function, therefore $f(a_1), f(a_2), ..., f(a_n)$ are distinct elements of the set A, but A has only n elements. Therefore,
$$A = \{f(a_1), f(a_2), ..., f(a_n)\}$$
$\Rightarrow$ Co-domain = Range
Hence, every element in A (co-domain) has its pre-image in the domain A.
$\Rightarrow$ $f : A \to A$ is onto.

REMARK

- For a finite set A, if $f : A \rightarrow A$ is onto function, then f is one-one.

Example 9. If $f : R \rightarrow R$ be a function defined by $f(x) = 4x^3 - 7$, show that the function f is bijective.

Solution. Given that $f(x) = 4x^3 - 7;\ x \in R$

(i) f is one-one : Let $x_1, x_2 \in R$

Now, $f(x_1) = f(x_2)$

$\Rightarrow \quad 4x_1^3 - 7 = 4x_2^3 - 7 \qquad\qquad \Rightarrow 4x_1^3 = 4x_2^3$

$\Rightarrow \quad x_1^3 = x_2^3 \qquad\qquad\qquad \Rightarrow x_1^3 - x_2^3 = 0$

$\Rightarrow \quad (x_1 - x_2)(x_1^2 + x_1 x_2 + x_2^2) = 0$

$\Rightarrow \quad (x_1 - x_2)\left[\left(x_1 + \dfrac{x_2}{2}\right)^2 + \dfrac{3x_2^2}{4}\right] \qquad \left\{\because \left[\left(x_1 + \dfrac{x_2}{2}\right)^2 + \dfrac{3x_2^2}{4} \neq 0\right]\right\}$

$\Rightarrow \quad (x_1 - x_2) = 0 \qquad\qquad\qquad \Rightarrow x_1 = x_2$

Therefore, f is one-one.

(ii) f is onto : Let $c \in R$

$f(x) = c \quad \Rightarrow 4x^3 - 7 = c \qquad \Rightarrow x = \left(\dfrac{c+7}{4}\right)^{1/3}$

Now, $\left(\dfrac{c+7}{4}\right)^{1/3} \in R$ and $f\left\{\left(\dfrac{c+7}{4}\right)^{1/3}\right\} = 4\left[\left(\dfrac{c+7}{4}\right)^{1/3}\right]^3 - 7 = c + 7 - 7 = c$

Which implies that c is the image of $\left(\dfrac{c+7}{4}\right)^{1/3}$

Therefore, f is onto. Hence, f is bijective function.

Example 10. Let A and B be two sets. Prove that $f : A \times B \rightarrow B \times A$ difined by $f(a, b) = (b, a)$ is one-one and onto.

Solution. **(i) f is one-one :** Let (a_1, b_1) and $(a_2, b_2) \in A \times B$ such that

$f(a_1, b_1) = f(a_2, b_2)$

$\Rightarrow \qquad\qquad (b_1, a_1) = (b_2, a_2)$

$\Rightarrow \qquad\qquad b_1 = b_2$ and $a_1 = a_2$

Therefore, $(a_1, b_1) = (a_2, b_2)$

Thus, $f(a_1, b_1) = f(a_2, b_2)$

$\Rightarrow \qquad\qquad (a_1, b_1) = (a_2, b_2)\ \forall\ (a_1, b_1), (a_2, b_2) \in A \times B$

$\Rightarrow \quad f$ is one-one.

(ii) f is onto : Let $(b, a) \in B \times A$ such that $b \in B$ and $a \in A$.

$\Rightarrow \qquad\qquad (a, b) \in A \times B$

Therefore, for all $(b, a) \in B \times A$, there exist $(a, b) \in A \times B$ such that $f(a, b) = (b, a)$

$\Rightarrow \quad f$ is onto. Hence f is one-one and onto.

1.16 BINARY OPERATION

Let S be a non-empty set. Then any function from $S \times S$ to S is called binary operation. It is usually denoted by any of the following symbols : *, o, O, $\oplus$, +, . etc.

That is, any function $* : S \times S \to S$ is said to be binary operation, where $S \times S = \{(a, b): a, b \in S\}$.

Instead of writting $*(a, b)$, we write $a * b$ for $a, b \in S$. Thus a binary operation $*$ on S is a rule which assign to a pair $a, b \in S$ another element $a*b \in S$.

☞ **ILLUSTRATIONS**

(1) The addition $+: Z \times Z \to Z$ is a binary operation, because for all $a, b \in Z$, we have $a+b \in Z$.

(2) The multiplication $'.'$: $Z \times Z \to Z$ is binary operation, as $a.b \in Z$ for all $a, b \in Z$.

(3) The subtraction $'-'$: $N \times N \to N$ is not a binary operation, as $3 \in N$, $5 \in N$ but $3-5 = -2 \notin N$.

(4) The subtraction $'-'$: $Z \times Z \to Z$ is binary operation, as $a-b \in Z$ for all $a, b \in Z$.

(5) Let S be a non-empty set and $P(S)$ be its power set. Then the union opeartion on $P(S)$ is a binary opaeration as $A \cup B \in P(S)$ for all $A, B \in P(S)$.

(6) Addition of the set S of all irrationals is not a binary operation as $2+\sqrt{3} \in S$ and $2-\sqrt{3} \in S$ but $2 + \sqrt{3} + 2 - \sqrt{3} \notin S$.

(7) Multiplication on the set S of all irrationals is not a binary operation as
$$\sqrt{2} \in S, -\sqrt{2} \in S \text{ but } \left(\sqrt{2}\right)\left(-\sqrt{2}\right) = -2 \notin S.$$

REMARKS

- A binary operation combines any two elements a and b of a set to give another element of the same set, this fact is also known as closure property or we can say that the set is closed with respect to binary opeartion.
- Binary operation is never one-one.

1.17 NUMBER OF BINARY OPERATIONS

Let $f : A \to B$ be a function, where $n(A) = p$ and $n(B) = q$. Then, the total number of functions from A to B is given by $[n(B)]^{n(A)}$.

Let S be a finite set containing n elements, then $S \times S$ will have n^2 elements. Since the binary operation is a function form $S \times S$ to S, so that total number of binary operations on S is $(n)^{n^2}$.

☞ **ILLUSTRATION**

(1) If $S = \{1, 2\}$, then the total number of binary operations on $S = (2)^{2^2} = 2^4 = 16$.

1.18 PROPERTIES OF BINARY OPERATION

Let $'*'$: $S \times S \to S$ and $'o'$: $S \times S \to S$ be any two binary operations, where S is any non-empty set. Then we define the following properties.

(i) Associative property : A binary operation $*$ on S is said to be associative, if
$$(a*b)*c = a*(b*c) \text{ for all } a, b, c \in S.$$

If '*' denotes addition (+) and multiplication (×), then '*' is always associative on S but if '*' denotes subtraction (–), then '*' is not associative on S.

For example : Addition and multiplication on Z, the set of all integers are always associative, *i.e.,*

$$(a + b) + c = a + (b + c)$$

and
$$(a \times b) \times c = a \times (b \times c) \text{ for all } a, b, c \in Z.$$

But subtraction is not associative on Z, *i.e.,*

$$3 - (5 - 4) \neq (3 - 5) - 4$$

(ii) Commutative property : A binary opeartion '*' on a set S is said to be *commutative* if

$$a * b = b * a \text{ for all } a, b \in S.$$

If '*' denotes additon (+) and multiplication (×), then '*' is always *commutative* on S but if * denotes subtraction (–), then '*' is not commutative on S.

For example : If $S = Z$, the set of all integers, then

$$a + b = b + a$$

and
$$a \times b = b \times a \text{ for all } a, b \in Z.$$

But subtraction is not commutative as $3 - 5 \neq 5 - 3$ for $3, 5 \in Z$.

(iii) Distributive property : A binary opeartion '*' is said to be *distributive* over another binary operation ' o' on S if

$$a * (b \text{ o } c) = (a * b) \text{ o } (a * c) \text{ for all } a, b, c \in S$$

and
$$(b \text{ o } c) * a = (b * a) \text{ o } (c * a) \text{ for all } a, b, c \in S$$

Here the first distribution is known as *left distribution* whereas the second is known as *right distribution*. That is,

Left distributive property :

$$a * (b \text{ o } c) = (a * b) \text{ o } (a * c) \text{ for all } a, b, c \in S$$

Right distributive property :

$$(b \text{ o } c) * a = (b * a) \text{ o } (c * a) \text{ for all } a, b, c \in S$$

For example : If $S = Z$, the set of all integers and '*' denotes multiplication, 'o' denotes addition, then '*' is distributive over 'o' on Z, *i.e.,*

$$a \times (b + c) = a \times b + a \times c$$

and
$$(b + c) \times a = b \times a + c \times a$$

But if '*' denotes addition and 'o' denotes multiplication, then '*' is not distributive over 'o' on Z, *i.e.,*

$$5 + (3 \times 2) \neq (5 + 3) \times (5 + 2)$$

1.19 ALGEBRAIC STRUCTURE

Definition. *Let S be a non-empty set and '*' be a binary operation on S, then $(S, *)$ is known as algebraic structure.*

Solved Examples

Example 1. *Determine whether the following operations are binary on the given set or not:*

(i) '*' *on* N *defined by* $a * b = a^b$ *for all* $a, b \in N$

(ii) '*' *on* N *defined by* $a * b = a + b - 2$ *for all* $a, b \in N$

(iii) '*' on N defined by $a*b = a^b + b^a$ for all a, b ∈ N

(iv) '*' on Z defined by $a*b = \sqrt{|ab|}$ for all a, b ∈ Z

(v) '*' on Z defined by $a*b = \sqrt{a^2 + b^2}$ for all a, b ∈ Z

(vi) '*' on Z defined by $a*b = \dfrac{1}{a-b}$ for all a, b ∈ Z

Solution.   (i)   For all a, b ∈ N, a^b ∈ N.

∴   a * b ∈ N

Hence, * is binary operation on N.

(ii) For all a, b ∈ N, a + b ∈ N

But a+b–2 ∉ N for all a, b ∈ N,

For example if a = b = 1 then a+b–2=0 ∉ N.

Hence, * is not binary on N, where a * b = a+b–2.

(iii) For all a, b ∈ N, a^b ∈ N and b^a ∈ N, then

$$a^b + b^a \in N \Rightarrow a * b \in N$$

Thus a * b ∈ N for all a, b ∈ N. Hence, * is binary operation on N.

(iv) If a = 1 and b = 2, then $|ab| = |1 \times 2| = 2$.

Now $\sqrt{|ab|} = \sqrt{2}$

∴ $\sqrt{|ab|}$ ∉ Z for all a, b ∈ Z

⇒   a * b ∉ Z for all a, b ∈ Z

Hence * is not binary operation on Z.

(v) Since $a * b = \sqrt{a^2 + b^2}$, for all a, b ∈ Z

If a = b = 1, then $\sqrt{a^2 + b^2} = \sqrt{1^2 + 2^2} = \sqrt{2}$ ∉ Z

∴ $\sqrt{a^2 + b^2}$ ∉ Z for all a, b ∈ Z

⇒   a * b ∉ Z for all a, b ∈ Z

Hence * is not binary operation on Z .

(vi) Since $a * b = \dfrac{1}{a-b}$ for all a, b ∈ Z.

If a = 3 and b = 1, then $\dfrac{1}{a-b} = \dfrac{1}{3-1} = \dfrac{1}{2}$ ∉ Z

∴ $\dfrac{1}{a-b}$ ∉ Z for all a, b ∈ Z

⇒ $a * b = \dfrac{1}{a-b}$ ∉ Z for all a, b ∈ Z

Hence * is not binary operation on Z.

Example 2.   *Let S be a set having more than one element. Let '*' S × S→S be binary operation defined by a * b = a for all a, b ∈ S. Is (S, *) associative or commutative?*

Solution.   Since S has more than one elenment, so if a, b ∈ S, then a ≠ b.

Now,　　　　　　$a * b = a$ and $b * a = b$

$\therefore$　　　　　　$a + b \neq b * a$　　　　for all a, b

Thus $*$ is not commutative on S.

Again,　　　　$a * (b * c) = a * b = a$

and　　　　$(a * b) * c = a * c = a$

$\therefore$　　　　$a * (b * c) = (a * b) * c$　　for all $a, b, c \in S$

Thus $*$ is associative on S.

Example 3. *Let '*' be a binary opeuration on* Q *the set of all rational numbers. Find which of the binary opeurtions are commutative.*

(i) $a * b = a - b$　　*for all* $a, b \in$ Q

(ii) $a * b = a^2 + b^2$　*for all* $a, b \in$ Q.

Solution.　(i)　$a * b = a - b = -(b - a) = -b * a$

$\therefore$　　　　　　　$a * b \neq b * a$ for all $a, b \in$ Q.

Thus $*$ is not commutative on Q.

(ii)　$a^2 \in$ Q, $b^2 \in$ Q for all $a, b \in$ Q. Then

$a * b = a^2 + b^2$

$= b^2 + a^2$ [$\because$ Addition is always commutative on Q]

$= b * a$

$\therefore$　　　　$a * b = b * a$ for all $a, b \in$ Q

Thus $*$ is commutative on Q.

Example 4. *Discuss the commutativity and associativity of the binary opeurtion '*' on* R *defined by* $a * b = a + b + ab$ *for all* $a, b \in$ R *where on R.H.S. we have usual addition and multiplication of real numbers.*

Solution.　Since $a * b = a + b + ab \in$ R for all $a, b \in$ R.

(i) Commutativity : Let $a, b \in$ R. Then

$a * b = a + b + ab$

$= b + a + ba$

[$\because$ Usual addition and multipplication are always commutative on R]

$= b * a$

$\therefore$　　　　$a * b = b * a$ for all $a, b \in$ R

So '$*$' is commutative on R.

(ii) Associativity : Let $a, b, c \in$ R. Then

$(a * b) * c = (a + b + ab) * c$

$= a + b + ab + c + (a + b + ab)c$

$= a + b + ab + c + ac + bc + abc$

[$\because$ Multiplication is distributive over addition on R]

$= a + b + c + bc + ab + ac + abc$

[$\because$ Addition is commutative on R]

$= a + b + c + bc + a(b + c + bc)$

$$= a * (b + c + bc)$$
$$= a * (b * c) \qquad \text{[By definition of '*']}$$
$$\therefore \qquad (a * b) * c = a * (b * c) \text{ for all } a, b, c \in R.$$

Hence, '*' is associative on R.

Example 5. *Discuss the commutativity and associativity of the binary opeartion '*' defined on Q by the rule $a * b = a - b + ab$ for all $a, b \in Q$.*

Solution. Since $a * b = a - b + ab \in Q$ for all $a, b \in Q$.

(i) **Commutativity :** Let $a, b \in Q$. Then
$$a * b = a - b + ab$$
$$\neq b - a + ba$$
$$\neq b * a \qquad [\because a - b \neq b - a]$$
$$\therefore \qquad a * b \neq b * a \text{ for all } a, b \in Q.$$

(ii) **Associativity :** Let $a, b, c \in Q$. Then
$$(a * b) * c = (a - b + ab) * c$$
$$= a - b + ab - c + (a - b + ab)c$$
$$= a - b + ab - c + ac - bc + abc \qquad \ldots(1)$$

[$\because$ Multiplication is distributive over addition and subtraction on Q]

Also, $\qquad a * (b * c) = a * (b - c + bc)$
$$= a - (b - c + bc) + a(b - c + bc)$$
$$= a - b + c - bc + ab - ac + abc$$
$$= a - b + ab + c - ac - bc + abc \qquad \ldots(2)$$

From eqn. (1) and (2), we have
$$(a * b) * c \neq a * (b * c) \text{ for all } a, b, c \in Q.$$

So, '*' is not associative on Q.

Example 6. *Discuss the commutativity and associativity of the binary opeartion '*' on R defined by*

$$a * b = \frac{ab}{4} \text{ for all } a, b \in R.$$

Solution. Since $a * b = \frac{ab}{4} \in R$ for all $a, b \in R$.

(i) **Commutatively:** Let $a, b \in R$. Then
$$a * b = \frac{ab}{4}$$
$$= \frac{ba}{4} \quad [\because \text{Multiplication is always commutative on R}]$$
$$= b * a$$
$$\therefore \qquad a * b = b * a \text{ for all } a, b \in R.$$

So, '*' is commutative on R.

(ii) Associatively: Let $a, b, c \in R$. Then

$$(a * b) * c = \frac{ab}{4} * c$$

$$= \frac{(ab/4)c}{4} = \frac{abc}{16} \qquad \dots (1)$$

Now, $\quad a * (b * c) = a * \dfrac{bc}{4}$

$$= \frac{a(bc/4)}{4} = \frac{abc}{16} \qquad \dots(2)$$

From eqn. (1) and (2), we get

$\quad (a * b) * c = a * (b * c)$ for all $a, b, c \in R$.

So. '*' is associative on R.

Example 7. *Let $A = N \times N$ and '*' be a binary operation on A defined by*

$(a, b) * (c, d) = (ac, bd)$ for all $a, b, c, d \in N$.

Show that '' is commutative and associative on A*

Solution. Since $\quad (a, b) * (c, d) = (ac, bd) \in A$ for all $(a, b), (c, d) \in A$

(i) Commutativity: Let $(a, b), (c, d) \in A$. Then

$$(a, b) * (c, d) = (ac, bd)$$

$$= (ca, db)$$

$$[\because \text{Multiplication is always commutative on } N]$$

$$= (c, d) * (a, b)$$

$\therefore (a, b) * (c, d) = (c, d) * (a, b)$ for all $(a, b), (c, d) \in A$.

So, '*' is commutative on A.

(ii) Associativity: Let $(a, b), (c, d), (e, f) \in A$. Then

$$\{(a,b) * (c,d) * (e,f) = (ac,bd) * (e,f)\}$$

$$= (ace, bdf) \qquad \dots (1)$$

$$(a,b) * \{(c,d) * (e,f)\} = (a,b) * (ce,df)$$

$$= (ace, bdf)$$

$\therefore \{(a,b) * (c,d)\} * (e,f) = (a,b) * \{(c,d) * (e,f)\}$ for all $(a,b),(c,d),(e,f) \in A$

So, '*' is associative on A.

1.20 IDENTITY ELEMENT

Let $(S, *)$ be an algebraic structure. If there exists an element $e \in S$ such that $a * e = a = e * a$ for all $a \in S$ then e is called on identity element of S with respect to binary operation '*'.

☞ ILLUSTRATIONS

(1) Let $S = Z$, the set of all integers and addition (+) is a binary operation on Z. Then we know that $0 \in Z$ such that

$$0 + a = a = a + 0 \text{ for all } a \in Z.$$

Thus, 0 is the identity element for addition on Z.

(2) If $S = N$, the set of all natural numbers and multiplication '*' is binary operation on N. Then we know that $1 \in N$ such that

$$1.a = a = a.1 \text{ for all } a \in N.$$

Thus, 1 is the identity element for multiplication on N.

(3) Let $P(S)$ denote the power set of a non-empty set S. Then we know that

$$A \cup \phi = A = \phi \cup A \quad \text{for all } A \in P(S)$$

Thus, ϕ is the identity element for the union of sets on $P(S)$.

(4) Let $P(S)$ denotes the power set of a non-empty set S. Then we know that

$$A \cap S = A = S \cap A \quad \text{for all } A \in P(S)$$

REMARK

- '0' is not an identity element for addition on N.

1.21 INVERSE OF AN ELEMENT

Let $(S, *)$ be an algebraic structure, and e be the identity element of S. Then an element $a \in S$ is said to be invertible element, if there exists an element $b \in S$ such that

$$a * b = e = b * a$$

The element b is called an inverse of a.

☞ ILLUSTRATIONS

(1) Let $(Z, +)$ be an algebraic structure. Then $0 \in Z$ is the identity element for addition. Now corresponding to each element $a \in Z$, there exists $-a \in Z$ such that

$$a + (-a) = 0 = (-a) + a$$

Thus, $-a$ is the additive inverse of a.

(2) Let $(R, *)$ be an algebraic structure. Then $1 \in R$ is the multiplicative identity. If $a \neq 0 \in R$, then correspondig to each non-zero element $a \in R$, then exists an element $\dfrac{1}{a} \in R$, such that

$$a.\frac{1}{a} = 1 = \frac{1}{a}.a$$

Thus $\dfrac{1}{a}$ is the multiplicative inverse of a.

REMARK

- The inverse of an element a (if it exists) with respect to the addition (or multiplication) binary operations is generally is denoted by $a^{-1}\left(\text{or } \dfrac{1}{a} \right)$

THEOREM 1. *Let (S,*) be an algebraic structure. If S has the identity element for '*', then it is unique.*

Proof. Let e_1 and e_2 be two identities for '*' on S. Then by definition of identity, we have,
$$e_1 * e_2 = e_2, \text{ as } e_1 \text{ is the identity} \qquad \ldots (1)$$
and
$$e_1 * e_2 = e_1, \text{ as } e_2 \text{ is the identity} \qquad \ldots (2)$$
From equations (1) and (2), we get
$$e_1 = e_2$$
Hence, the identity, if exists, is unique.

THEOREM 2. *Let (S, *) be algebraic structure with identity element e. Then every element $a \in S$ has unique inverse in S, if '*' is associative.*

Proof. Let b and c be two inverses of an element $a \in S$ in S. Then we have
$$a*b = e = b*a$$
and
$$a*c = e = c*a$$
Now,
$$(b*a)*c = e*c \qquad\qquad (\because b*a = e)$$
$$= c \qquad\qquad \text{[By the definition of identity]}$$
and
$$b*(a*c) = b*e \qquad\qquad (\because a*c = e)$$
$$= b \qquad\qquad \text{[By the definition of identity]}$$
Since '*' is associative on S, so that $(b*a)*c = b*(a*c)$ for all $a, b, c \in S$
$$\Rightarrow \qquad c = b$$
Hence, $a \in S$ has unique inverse in S.

THEOREM 3. *Let (S,*) be an algebraic structure and '*' be an associative binary operation of S. If a be an invertible element of S, then*
$$(a^{-1})^{-1} = a$$

Proof. Let e be the identity element of S for '*' and a be an invertible element of S, then
$$a * a^{-1} = e = a^{-1} * a$$
$$\Rightarrow \qquad a^{-1} * a = e = a * a^{-1}$$
$$\Rightarrow \qquad a \text{ is the inverse of } a^{-1}$$
$$\Rightarrow \qquad (a^{-1})^{-1} = a$$

REMARK

- For identity element e, we have $e = e^{-1}$

Theorem 4. *Let (S, *) be an algebraic structure and * be an associative binary operation on S. If each element of S has inverse in S. Then $(a * b)^{-1} = b^{-1} * a^{-1}$.*

Proof. Let e be the identity element of S for '*'. Now for $a, b \in S$, we have
$$(a * b)*(b^{-1} * a^{-1}) = a* (b * b^{-1})*a^{-1} \qquad\qquad \text{(By associativity)}$$
$$= a* e * a^{-1} \qquad\qquad (\because b*b^{-1} = e)$$
$$= a* a^{-1} \qquad\qquad (\because a*e = a)$$
$$= e \qquad\qquad (\because a * a^{-1} = e)$$
Similarly,
$$(b^{-1} * a^{-1}) * (a * b) = e$$
$$\therefore \qquad (a *a)*(b^{-1} *a^{-1}) = e = (b^{-1} *a^{-1})*(a*b)$$
$$\Rightarrow \quad b^{-1} * a^{-1} \text{ is the inverse of } a*b$$
$$\Rightarrow \qquad (a * b)^{-1} = b^{-1} * a^{-1}$$

1.22 COMPOSITION TABLE FOR BINARY OPERATION ON FINITE SETS

Let $S = \{a_1, a_2, \dots a_n\}$ be a finite set and '*' be a binary operation on S. Then we construct a composition table by means of following instructions:

First we write down the elements $a_1, a_2, \dots, a_n$ of S in a top horizontal row and in a left vertical column as shown below. Now we write down the element $a_i * a_j$ at the intersection of row headed by a_i $(1 \le i \le n)$ and the column headed by a_i $(1 \le j \le n)$ to get the following table:

*	a_1	a_2	$\cdots$	a_j	$\cdots$	a_n
a_1	$a_1 * a_1$	$a_1 * a_2$	$\cdots$	$a_1 * a_j$	$\cdots$	$a_1 * a_n$
a_2	$a_2 * a_1$	$a_2 * a_2$	$\cdots$	$a_2 * a_j$	$\cdots$	$a_2 * a_n$
$\vdots$	$\vdots$	$\vdots$	$\vdots$	$\vdots$	$\vdots$	$\vdots$
a_i	$a_i * a_1$	$a_i * a_2$	$\cdots$	$a_i * a_j$	$\cdots$	$a_i * a_n$
$\vdots$	$\vdots$	$\vdots$	$\vdots$	$\vdots$	$\vdots$	$\vdots$
a_n	$a_n * a_1$	$a_n * a_2$	$\cdots$	$a_n * a_j$	$\cdots$	$a_n * a_n$

From above table, we conclude the following properties:

  (i) Closure property: If all the entries in the table are the elements of S, then S is closed under binary operation '*'.

  (ii) Commutative property: If the entries in every row coincide with the corresponding entries in the coerresponding column, we say that the composition is commutative, otherwise it is said to be non-commutative.

  (iii) Existence of identity: If the row headed by an element a_i of S just coincides with the top row of the table and the column headed by a_i coincides with column on extreme left of the table, then a_j is the identity element of S for the composition '*'.

  (iv) Existence of inverse: Look at the position of the identity element e anywhere in the table except in the top row and in the extreme column. If e is placed at the intersection of the row headed by a_i and the column headed by a_j, then a_i and a_j are the inverse of each other.

Solved Example

Example 1. *Let $S = \{1,2,3,4\}$ and '*' be a binary operation in S defined by $a*b = r$, where r is the least non-negative remainder when ab is divided by 5. Construct the composition table for '*' on S.*

Solution. We have

             $1*1$ = least non-negative remainder, when 1 is divided by 5

                 = 1

             $2*3$ = least non-negative remainder, when 6 is divided by 5

                 = 1

Similarly,

             $1*2 = 2,\ 1*3 = 3,\ 1*4 = 4,$

2*1 = 2, 2*2 =4, 2*3 =1, 2*4 =3,
3*1 = 3, 3*2 =1, 3*3 =4, 3*4 =2,
4 *1 =4, 4*2 =3, 4*3=2, 4*4 =1

Now, we construct the composition tabel for '*' on S as follows:

*	1	2	3	4
1	1	2	3	4
2	2	4	1	3
3	3	1	4	2
4	4	3	2	1

From above composition table, we observe that

(i) All the entries in the table are the elements of S, so S is closed under '*'.

(ii) The binary operation '*' is commutative on S, as the composition table is symmetrical about the diagonal starting at the upper left corner and ending at the lower right corner.

(iii) 1 is the identity element for '*' because the row headed by 1 coincides with the top row and the column headed by 1 coincides with extreme left column of the table.

(iv) Every element of *S* is invertible with respect to '*' because the identity element 1 appears in each row headed by 2 and column headed by 3, so 2 and 3 are the inverse of each other. Similarly 4 is the inverse of itself.

Example 2. *Prepare the composition table for multiplication on the set S = {1, ω, ω²} of cube root of unity. Show that multiplication on S satisfies closure property, the associative law and commutative law. What is the identity element? Write down the multiplicative inverse of each element of S.*

Solution. Since ω is a cube root of unity, we we have $\omega^3 =1$, $\omega^4 =\omega$,

Now we construct the composition table for '.' as follows:

•	1	ω	ω²
1	1	ω	ω²
ω	ω	ω²	1
ω²	ω²	1	ω

From above table we observe that

(i) All the entries and the elements of S, so S is closed under multiplication.

(ii) Since, 1, ω, ω² are complex numbers and multiplication is always associative on complex numbers. Hence, multiplication is associative on *S*.

(iii) The composition table is symmetrical about the diagonal starting at the upper left corner and ending at the lower right corner, so multiplication is commutative on *S*.

(iv) Since the row headed by 1 coincides with the top row, so 1 is the identity element in *S* for multiplication.

(v) 1 is the identity element so 1 is the inverse of itself. Also, 1 is the intersection of row headed by ω and column headed by ω², so ω and ω² are the inverse of each other.

Example 3. *Construct the composition table for the composition of functions, defined on the set $S = \{f_1, f_2, f_3, f_4\}$ of four functions from C, the set of all complex numbers to itself, given by*

$$f_1(z) = z, f_2(z) = -z, f_3(z) = \frac{1}{z} \text{ and } f_4(z) = -\frac{1}{z} \text{ for all } z \in C.$$

Solution. Since each function is defined from C to C, so $f_1 \circ f_1, f_1 \circ f_2, f_1 \circ f_3, f_1 \circ f_4$ etc, exist.

So, $(f_1 \circ f_1)(z) = f_1(f_1(z)) = f_1(z)$

$$(f_1 \circ f_2)(z) = f_1(f_2(z)) = f_1(-z) = -z = f_2(z)$$

$$(f_1 \circ f_3)(z) = f_1(f_3(z)) = f_1\left(\frac{1}{z}\right) = \frac{1}{z} = f_3(z)$$

$$(f_1 \circ f_4)(z) = f_1(f_4(z)) = f_1\left(-\frac{1}{z}\right) = -\frac{1}{z} = f_4(z)$$

$$(f_2 \circ f_1)(z) = f_2(f_1(z)) = f_2(z) = -z = f_2(z)$$

$$(f_2 \circ f_2)(z) = f_2(f_1(z)) = f_2(z) = -z = f_2(z)$$

$$(f_2 \circ f_2)(z) = f_2(f_2(z)) = f_2(-z) = -(-z) = z = f_1(z)$$

Similarly, other compositions can be obtained.

Now we construct the composition table for the composition of functions 'o' as follows:

o	f_1	f_2	f_3	f_4
f_1	$f_1 \circ f_1 = f_1$	$f_1 \circ f_2 = f_2$	$f_1 \circ f_3 = f_3$	$f_1 \circ f_4 = f_4$
f_2	$f_2 \circ f_1 = f_2$	$f_2 \circ f_2 = f_1$	$f_2 \circ f_3 = f_4$	$f_2 \circ f_4 = f_3$
f_3	$f_3 \circ f_1 = f_3$	$f_3 \circ f_2 = f_4$	$f_3 \circ f_3 = f_1$	$f_3 \circ f_4 = f_2$
f_4	$f_4 \circ f_1 = f_4$	$f_4 \circ f_2 = f_3$	$f_4 \circ f_3 = f_2$	$f_4 \circ f_4 = f_1$

❧ EXERCISE 1.7

1. Let $A = \{-2, -1, 0, 1, 2\}$ and $f : A \to Z$ given by $f(x) = x^2 - 2x - 3$. Find :

   (i) the range of f,

   (ii) pre-image of 6, –3 and 5.

2. Find the domain and range of the following function

$$f(x) = \sqrt{(x-1)(3-x)}$$

3. Find the range of the following function

$$f(x) = \frac{1}{(2x-3)(x+1)}$$

4. Find the domain and range of the following functions :

   (i) $f(x) = \dfrac{x^2 - 1}{x - 1}$ (ii) $y = -|x|$

   (iii) $f(x) = \dfrac{|x-1|}{x-1}$ (iv) $y = \sqrt{x-3}$

5. If $A = \{-1, 0, 2, 5, 6, 11\}$,

          $B = \{-2, -1, 0, 18, 25, 108\}$

   and $f(x) = x^2 - x - 2$, find $f(A)$.

6. Let A be the set of two positive integers. Let $f : A \to Z^+$, set of positive integers be defined by $f(n) = p$, where p is the highest prime factor of n. If range of $f = \{3\}$, find A.

7. Find the domain for which the function $f(x) = 2x^2 - 1$ and $g(x) = 1 - 3x$ are equal.

8. Let $f_1 : R \to R$ and $f_2 : C \to C$ be two functions defined as $f_1(x) = x^3$ and $f_2(x) = x^3$. Show that they are not equal.

9. Let $A = \{p, q, r, s\}$ and $B = \{1, 2, 3\}$. Which of the following relations from A to B not a function?

(i) $R_1 = \{(p, 1), (q, 2), (r, 1), (s, 2)\}$

(ii) $R_2 = \{(p, 1), (q, 1), (r, 1), (s, 1)\}$

(iii) $R_3 = \{(p, 1), (q, 2), (r, 2), (s, 3)\}$

(iv) $R_4 = \{(p, 2), (q, 3), (r, 2), (s, 2)\}$

10. Write the following relations as sets of ordered pairs and find which of them are functions :

(i) $\{(x, y) : y = 3x, x \in (1, 2, 3),$
$$y \in (3, 6, 9, 12)\}$$

(ii) $\{(x, y) : y > x + 1, x = 1, 2$ and $y = 2, 4, 6\}$

(iii) $\{(x, y) : x + y = 3$
$$x, y \in (0, 1, 2, 3)\}$$

11. Express the following functions as sets of ordered pairs, and find their range :

(i) $f_1 : A \to R : f_1(x) = x^2 + 1$
where $A = \{-1, 0, 2, 4\}$

(ii) $f_2 : A \to N : f_2(x) = 2x$
where $A = \{x : x \in N, x \le 10\}$

12. Let $f : R \to R$ be a function such that $f(x) = 2^x$. Determine :

(i) range of f

(ii) $\{x : f(x) = 1\}$

(iii) whether $f(x + y) = f(x) \cdot f(y)$ holds

13. Let $f : R^+ \to R$, be a function such that $f(x) = \log x$. Determine :

(i) the image set of domain of f

(ii) $\{x : f(x) = -2\}$

(iii) whether $f(xy) = f(x) + f(y)$ holds

14. Give an example of a map which is :

(i) one-to-one but not onto

(ii) not one to one, but onto

(iii) neither one-to-one nor onto

Answers

1. i) $f(A) = \{-4, -3, 0, 5\}$, (ii) $\phi, \{1, 2\}, -2$ **2.** Domain $= [1, 3]$, Range $= [-1, 1]$

3. $\left] -\infty, \dfrac{-8}{25} \right] \cup [0, \infty[$

4. (i) $R - \{1\}$, $R - \{2\}$, (ii) $R : R - R^+$, (iii) $R - \{1\}$, $\{-1, 1\}$, (iv) $[3, \infty[$, $[0, \infty]$

5. $f(A) = \{1, -2, 18, 28, 108\}$ **6.** $A = \{3, 6\}$ or $(3, 9)$ or $[3, 12]$ etc. **7.** $(-2, 1/2)$ **9.** (iii)

10. (i) $\{(1, 3), (2, 6), (3, 9)\}$, function, (ii) $\{(1, 4), (1, 6), (3, 4), (3, 6)\}$, not function

(iii) $\{(0, 3), (1, 2), (2, 1), (3, 0)\}$, function

11. (i) $f_1 = \{x, f(x) : x \in A\} = \{(-1, 2), (0, 1), (2, 5), (4, 17)\}$

(ii) $f_2 = \{(x, g(x)) : x \in A\} = \{(1,2),(2,4), (3, 6), ..., (10, 20)\}$

12. (i) Range of $f = R^+$, the set of positive real numbers, (ii) $(x : f(x) = 1) = \{0\}$,

(iii) $f(x+y) = f(x) \cdot f(y)$ holds for all $x, y \in R$

14. (i) $n \mapsto n^2 : N \to N$ (ii) $n \to |n| : Z \to N \cup \{0\}$ (iii) $n \to |n|^2 : Z \to N \cup \{0\}$

Selected Terms and Results

TERMS

- **Interval :** A subset S of $\mathbf{R}$ is called an interval if $a, b \in S$; $x \in \mathbf{R}$ such that $a < x < b$ implies $x \in S$.
- **Set:** A set is a well defined collection of objects.
- **Empty set :** A set containing no element is called empty set.
- **Finite set :** A set is said to be finite if it consists of only finite number of elements. Otherwise it is said to be infinite.
- **Equivalent set :** Two finite sets are said to be equivalent if they have the same cardinal number.
- **Union of two sets :** The union of two sets A and B denoted by $A \cup B$ is the set of those elements which either belong to A or B or to both.
- **Intersection of two sets :** The intersection of two sets A and B, denoted by $A \cap B$ is the set of all those elements which belongs to both A and B.
- **Disjoint sets :** When two sets have no common element, they are called disjoint sets.
- **Ordered pair :** An ordered pair is a pair of entries whose components occur in a specific order.
- **Cartesian product of two sets :** The set of all ordered pairs of elements (a, b) : $a \in A$, $b \in B$ is called the cartesian product of two sets A and B.
- **Relation :** Let A and B be two sets. Then a relation R from A to B is a subset of $A \times B$.
- **Reflexive Relation :** A relation R is said to be reflexive if $(x, x) \in R \; \forall \, x \in A$.
- **Symmetric Relation :** A relation R on a set A is said to be symmetric if $(y, x) \in R$ whenever $(x, y) \in R$.
- **Transitive Relation :** A relation R on a set A is said to be transitive if $(x, y) \in R$, $(y, z) \in R$ $\Rightarrow (x, z) \in R$.

- **Anti-symmetric Relation :** A relation R on a non-empty set A is said to be an anti-symmetric iff $(x, y) \in R$ and $(y, x) \in R$ $\Rightarrow x = y \; \forall \, x, y \in R$.
- **Equivalence Relation :** A relation R on a set E is said to be equivalence if it is
  (i) reflexive
  (ii) anti-symmetric and
  (iii) transitive
- **Partial Ordered Relation :** A relation R on a set E is said to be partial ordered relation if it is
  (i) reflexive
  (ii) anti-symmetric and
  (iii) transitive
- **Function :** Let A and B be two sets, then the rule or correspondence which associates each element of A to a unique element of B is called a function or mapping.
- **Range and domain of a function:** Let an element $y \in B$ be corresponded by an element $x \in A$, then y is called the image of x and is denoted by $f(x)$. The set A is called the domain and the set B is called the co-domain of the function f.
- **One-One function :** A function $f : A \rightarrow B$ is said to be one-one iff distinct elements of A have distinct images.
- **Onto function :** A function $f : A \rightarrow B$ is called an onto function if there is no element of B which is not the image of some element A, *i.e.*, every element of B appears as the image of at least one element of A.
- **Into function :** A function $f : A \rightarrow B$ is called an into function, if there is at least one element of set B which has no pre-image in the set A.
- **Even function :** A function $f : A \rightarrow B$ is said to be an even function if $f(-x) = f(x) \forall x \in A$.
- **Odd function :** A function $f : A \rightarrow B$ is said to be an odd function if $f(-x) = -f(x) \forall x \in A$.

RESULTS

- Total number of subsets of a set A is equal to 2^n where n is the number of elements of A.
- The number of proper subsets of a set with n elements is $2^n - 2$.
- Union of sets is commutative, associative and idempotent.
- Difference of two sets is not commutative.
- Difference of a set with the universal set is called complementation.
- The identity and universal relations on a non-empty sets are transitive.
- The intersection of two equivalence relations on a set is equivalence relation.
- The union of two equivalence relations on a set is not necessarily an equivalence relation.
- If gof and fog both exist, they may not be equal.
- The composition of function is associative but not commutative.
- The composition of any function with the identity function is the function itself.
- The inverse of bijective function is unique.
- The inverse of bijective function is again bijective.

Review Questions and Project Work

1. Define union, intersection, difference and symmetric difference of two sets.
2. Define the power set of a set
3. How many element does the power set of a set S with n elements have?
4. Define what it mean for a function from the set of positive integers to the set of positive integer to be one to one.
5. Define the inverse of a function.
6. Let $f(n)$ be the function from the set of integers to the set of integers such that $f(n) = n^2 + 1$. What are the domain, co-domain and range of this function.
7. Give an example of a function from the set of positive integers to the set of positive integers that is :

   (a) both one-one and onto.
   (b) one-one but not onto.
   (c) neither one-one nor onto.
   (d) not one-one but is onto.
8. When the empty set the power set of a set?
9. (a) Define what is means for two sets to be equal.

   (b) Describe the ways to show that two sets are equal.
10. Let A and B be sets in a finite universal set U. List the following in order of increasing size :

    (a) $|A|$, $|A \cup B|$, $|A \cap B|$, $|U|$, $|\phi|$
    (b) $|A - B|$, $|A \oplus B|$, $|A| + |B|$, $|A \cup B|$, $|\phi|$
11. Research where the concept of a function first arose and describe how this concept was first used.

Objective type Questions

FILL IN THE BLANKS

1. A relation R on a set A is symmetric iff $R =$ _____ .
2. Let R be an anti-symmetric relation on a set A such that $(a, b) \in R$ and $(b, a) \in R$. Then _____ .
3. Let R be a relation on a set A such that $R = R^{-1}$. Then R is _____ .
4. Let $A = \{1, 2, 3\}$, then the smallest equivalence relation on A is _____ .
5. Let A be a finite set. Then the smallest equivalence relation on A is the _____
6. The void relation on a set is _____ and _____ but not _____ .
7. Let R be a relation defined by $R = \{(4, 5), (1, 4), (4, 6), (7, 6), (3, 7)\}$ on N. Then $R \circ R^{-1}$ = _____ .
8. Let $R = \{(a, a), (b, c), (a, b)\}$ be a relation on a set $A = \{a, b, c\}$. Then the minimum number of ordered pairs which when added to R make it transitive is _____ .

TRUE/FALSE

Write T for True and F for false statement.

1. A binary relation is a set. **(T/F)**
2. A void set defines a relation. **(T/F)**
3. The total number of relations from a set containing m elements to a finite set containing n elements is 2^{mn}. **(T/F)**
4. Every relation is a function. **(T/F)**
5. Every function is a relation. **(T/F)**
6. The total number of bijections from a set containing n elements to a set containing n elements is n^n. **(T/F)**
7. Every equivalence relation is symmetric. **(T/F)**
8. Every symmetric relation is equivalence. **(T/F)**
9. Every anti-symmetric relation is symmetric. **(T/F)**
10. The composition of functions is commutative. **(T/F)**
11. Reflexivity is redundant in the definition of an equivalence relation on a set A, because by symmetry $(a, b) \in R \Rightarrow (b, a) \in R$ and by transitivity $(a, b) \in R$ and $(b, a) \in R \Rightarrow (a, a) \in R$ **(T/F)**
12. The relation $R = \{(1, 2), (1, 3)\}$ is a transitive relation on a set $A = \{1, 2, 3\}$. **(T/F)**
13. The identity relation on a finite set A is the smallest equivalence relation on A. **(T/F)**

MULTIPLE CHOICE QUESTIONS

Choose the most appropriate one.

1. Let R_1 and R_2 be two equivalence relation on a set. Consider the following assertion :
   (i) $R_1 \cup R_2$ is an equivalence relation.
   (ii) $R_1 \cap R_2$ is an equivalence relation.
   Which of the following is correct?
   (a) Both assertions are true.
   (b) Assertion (i) is true but assertion (ii) is not true.
   (c) Assertion (ii) is true but assertion (i) is not true.
   (d) Neither (i) nor (ii) is true.

2. The 'subset' relation on a set of set is :
   (a) a partial ordering
   (b) an equivalence relation
   (c) transitive and symmetric only
   (d) transitive and anti-symmetric only

3. Let R be a symmetric and transitive relation on a set A, then :
   (a) R is reflexive and hence an equivalence relation.
   (b) R is reflexive and hence a partial order.
   (c) R is not reflexive and hence is not an equivalence relation .
   (d) None of the above

4. The number of equivalence relations of the set $\{1, 2, 3, 4\}$ is :
   (a) 4   (b) 15
   (c) 16   (d) 24

5. Suppose A is a finite set with n elements. The number of elements in the large equivalence relation of A is :
   (a) 1   (b) n
   (c) $n + 1$ (d) n^2

6. The binary relation $S = \phi$ on the set $A = \{1, 2, 3\}$ is :
   (a) neither reflexive nor symmetric
   (b) symmetric and reflexive
   (c) transitive and reflexive
   (d) transitive and symmetric

7. Let $f(x) = x^2 + x$ and $g(x) = x + 1$ then fog is :
   (a) $x^2 + 3x + 2$ (b) $x^2 + x + 1$
   (c) $(x+1)^2 + (x+1)$ (d) None of these

8. Let A and B be sets with cardinalities m and n respectively. The number of one-to-one mapping from A to B where $m < n$ is :
   (a) m^n (b) ^nP_m
   (c) ^mC_n (d) ^nC_m

9. The number of functions from m element set to n element set is :
   (a) $m + n$ (b) m^n
   (c) n^m (d) $m * n$

10. _____ is an unordered collection of elements where an element can occur as a member more than once :
    (a) Multiset   (b) Ordered set
    (c) Set   (d) None of these

11. The number of substrings of all lengths that can be formed from a character string of length $n =$ _____

   (a) n (b) n^2

   (c) $\dfrac{n(n-1)}{1}$ (d) $\dfrac{n(n+1)}{2}$

12. In a room containing 28 females, there are 18 females who speak English, 15 females speak French and 22 speak German. 9 females speak both English and French, 11 Females speak both French and German whereas 13 speak both German and English. How many females speak all the three languages?

   (a) 9   (b) 8

   (c) 7   (d) 6

13. Consider the following statements :

   S_1 : There exist infinite set A, B and C such that $A \cap (B \cap C)$ is finite.

   S_2 : There exist two irrational numbers x and y such that $(x + y)$ is rational.

   Which of the following is True about S_1 and S_2?

   (a) Only S_1 is correct.

   (b) only S_2 is correct.

   (c) Both S_1 and S_2 are correct.

   (d) None of the S_1 and S_2 is correct.

14. The power set 2^S of the set $S = \{3, (1, 4), 5\}$ is:

   (a) $\{5, 3, 1, 4, (1, 3, 5), (1, 4, 5), (3, 4), \phi\}$

   (b) $\{5, 3 (1, 4), 5\}$

   (c) $\{5, (3), [3, (1,4)], (3, 5), \phi\}$

   (d) None of the above.

15. Let A be a finite set of size n, the number of elements in the power set of $A \times A$ is :

   (a) 2^n (b) 2^{n^2}

   (c) $(2^n)^2$ (d) None of these

16. Let S be an infinite set and $S_1, S_2, S_3, \ldots S_n$ be the sets such that $S_1 \cup S_2 \cup S_3 \cup \ldots \cup S_n = S$. Then :

   (a) at least one of the set S_i is a finite set.

   (b) not more than one of the set S_i can be finite.

   (c) at least one of the sets S_i is an infinite set.

   (d) None of the above

17. The number of elements in the power set $P(S)$ of the set $S = \{\{\phi\}, 1, \{2, 3\}\}$ is :

   (a) 2   (b) 4

   (c) 8   (d) None of these

18. Let A and B be sets and A' and B' denote the complements of the sets A and B. The set $(A - B) \cup (B - A) \cup (A \cap B)$ is equal to :

   (a) $A \cup B$ (b) $A' \cup B'$

   (c) $A \cap B$ (d) $A' \cap B'$

19. Let $P(S)$ denote the power set of set S which of the following is always TRUE?

   (a) $P(P(S)) = P(S)$

   (b) $P(S) \cap S = P(S)$

   (c) $P(S) \cap P(P(S)) = (\phi)$

   (d) $S \in P(S)$

Answers

FILL IN THE BLANKS

1. $R = R$ **2.** $a = b$ **3.** Many-One into **4.** Symmetric **5.** $\{(1, 1), (2, 2), (3, 3)\}$ **6.** The identity relation on A. **7.** Symmetric, transitive but not reflexive **8.** $\{(5, 5), (4, 4), (6, 5), (6, 6), (5, 6), (7, 7)\}$

TRUE/ FALSE

1. T	2. T	3. F	4. F	5. T	6. F	7. T	8. F	9. F
10. F	11. F	12. T	13. T					

MULTIPLE CHOICE QUESTIONS

1. (c)	2. (a)	3. (d)	4. (b)	5. (d)	6. (d)	7. (a)	8. (b)	9. (c)
10. (a)	11. (d)	12. (d)	13. (c)	14. (d)	15. (b)	16. (c)	17. (c)	18. (a)
19. (c)								

MATRIX ALGEBRA

2.1 INTRODUCTION

A set of mn numbers either real or complex arranged in the form of a reactangular array in which there are m rows and n columns, rectangular arrrangement is called a matrix of order $m \times n$ which is denoted by $[a_{ij}]_{m \times n}$ where $i = 1, 2, 3, ..., m$ represents the number of rows and $j = 1, 2, 3, ..., n$ represents the number of columns and thus a matrix of order $m \times n$ is usually written as

$$[a_{ij}]_{m \times n} = \begin{bmatrix} a_{11} & a_{12} & \cdots & a_{1n} \\ a_{21} & a_{22} & \cdots & a_{2n} \\ \vdots & \vdots & \vdots & \vdots \\ a_{m1} & a_{m2} & \cdots & a_{mn} \end{bmatrix}_{m \times n}$$

REMARK

- Sometimes, a matrix is a rectangular array of numbers enclosed in double straight lines shown as '$\| \ \|$' or enclosed in parenthesis '( )'.

2.2 TYPE OF MATRICES

(i) **Null matrix (or zero matrix):** A matrix of order $m \times n$ is called a *null matrix* if it contains all mn elements zero. It is denoted by O and is usually written as

$$O = \begin{bmatrix} 0 & 0 & \cdots & 0 \\ 0 & 0 & \cdots & 0 \\ \vdots & \vdots & \vdots & \vdots \\ 0 & 0 & \cdots & 0 \end{bmatrix}_{m \times n}$$

(ii) **Row matrix:** A matrix having only one row and n columns is called a *row matrix* of order $1 \times n$.

For example : $A = \begin{bmatrix} a_{11} & a_{12} & a_{13} & \cdots & a_{1n} \end{bmatrix}_{1 \times n}$

(iii) **Column matrix :** A matrix having m rows and only one column is called a *column matrix* of order $m \times 1$.

For example : $A = \begin{bmatrix} a_{11} \\ a_{21} \\ a_{31} \\ \vdots \\ a_{m1} \end{bmatrix}_{m \times 1}$

(iv) **Horizontal matrix :** A matrix having more columns than the the number of its rows, is called *Horizontal matrix.*

For example : $A = \begin{bmatrix} a_{11} & a_{12} & a_{13} \\ a_{21} & a_{22} & a_{23} \end{bmatrix}_{2 \times 3}$

(v) **Vertical matrix :** *A matrix having more number of rows than its columns, is called vertical matrix.* For exmaple :

$$A = \begin{bmatrix} a_{11} & a_{12} \\ a_{21} & a_{22} \\ a_{31} & a_{32} \end{bmatrix}_{3 \times 2}$$

REMARK

- Row matrix is also a horizontal matrix and column matrix is also a vertical matrix.

(vi) **Square matrix :** A matrix having a number of rows equal to number of columns, is called *square matrix.*

For example : $A = \begin{bmatrix} a_{11} & a_{12} & a_{13} \\ a_{21} & a_{22} & a_{23} \\ a_{31} & a_{32} & a_{33} \end{bmatrix}_{3 \times 3}$

Here, the matrix A has 3 rows and 3 columns, so it is a square matrix. Also the elements a_{11}, a_{22}, a_{33} are placed in the diagonal, so these elements are known as *diagonal elements.*

(vii) **Diagonal matrix :** A matrix of order $n \times n$ is called a *diagonal matrix* if it contains all its off diagonal elements equal to zero.

Suppose $A = [a_{ij}]_{n \times n}$ and if $a_{ij} = 0$ for all $i \neq j$, then A is a diagonal matrix. Diagonal matrix of order $n \times n$ is usually written as

$$\text{Diag} [a_{11} \quad a_{22} \quad a_{33} \quad ... \quad a_{nn}]$$

For example :

$$A = \begin{bmatrix} 1 & 0 & 0 \\ 0 & 2 & 0 \\ 0 & 0 & 3 \end{bmatrix}_{3 \times 3} = \text{Diag} [1 \ 2 \ 3]$$

(viii) **Scalar matrix :** A diagonal matrix whose diagonal elements are all equal but not equal to 1 is called a *scalar matrix.*

For example : $A = \begin{bmatrix} k & 0 & 0 \\ 0 & k & 0 \\ 0 & 0 & k \end{bmatrix}, k \neq 1$

(ix) **Unit matrix :** A square matrix of order $n \times n$ having all off diagonal elements equal to zero and each of the diagonal elements equal to 1, is called a *unit matix.* It is usually denoted by I_n and is written as

$$I_n = \begin{bmatrix} 1 & 0 & ... & 0 \\ 0 & 1 & ... & 0 \\ 0 & 0 & ... & 0 \\ \vdots & \vdots & \vdots & \vdots \\ 0 & 0 & ... & 1 \end{bmatrix}_{n \times n}$$

REMARK

- Unit matrix can also be denoted by I.

  (x) **Triangular matrix :** A matrix in which the elements lying above or below principal diagonal are all zero, is called a *triangular matrix*.

There are two kinds of triangular matrix.

  (a) **Upper triangular matrix :** A matrix of order $n \times n$ is called an *upper triangular matrix* if it contains all its elements below the diagonal elements equal to zero.

Suppose $A = [a_{ij}]_{n \times n}$ and if $a_{ij} = 0$ for all $i > j$, then A is an upper triangular matrix.

For example :
$$A = \begin{bmatrix} 2 & 3 & 4 \\ 0 & 1 & 5 \\ 0 & 0 & 3 \end{bmatrix}_{3 \times 3}$$

is an upper triangular matrix of order 3×3.

  (b) **Lower triangular matrix :** A matrix of order $n \times n$ is called a *lower triangular matrix* if it contains all its elements above the diagonal elements equal to zero.

Suppose $A = [a_{ij}]_{n \times n}$ and if $a_{ij} = 0$ for all $i < j$, then A is called lower triangular matrix.

For example :
$$A = \begin{bmatrix} 1 & 0 & 0 \\ 3 & 4 & 0 \\ 5 & 6 & 7 \end{bmatrix}_{3 \times 3}$$

is a lower triangular matrix of order 3×3.

2.3 OPERATION ON MATRICES

2.3.1 ADDITION OF MATRICES

Suppose A and B are two matrices of same order, then the addition of these two matrices is obtained by adding corresponding elements of A and B. It is denoted by $A + B$. If the order of A and B is $m \times n$, then the order of $A+B$ will be $m \times n$.

Suppose $A = [a_{ij}]_{m \times n}$ and $B = [b_{ij}]_{m \times n}$

then $A + B = [a_{ij} + b_{ij}]_{m \times n}$

For example: If $A = \begin{bmatrix} 1 & 2 & 3 \\ 5 & 1 & 4 \\ 7 & 8 & 9 \end{bmatrix}$ and $B = \begin{bmatrix} 1 & 3 & 5 \\ 5 & 0 & 1 \\ 3 & 2 & 12 \end{bmatrix}$

then
$$A + B = \begin{bmatrix} 1 & 2 & 3 \\ 5 & 1 & 4 \\ 7 & 8 & 9 \end{bmatrix} + \begin{bmatrix} 1 & 3 & 5 \\ 5 & 0 & 1 \\ 3 & 2 & 12 \end{bmatrix}$$

$$= \begin{bmatrix} 1+1 & 2+3 & 3+5 \\ 5+5 & 1+0 & 4+1 \\ 7+3 & 8+2 & 9+12 \end{bmatrix}$$

$$= \begin{bmatrix} 2 & 5 & 8 \\ 10 & 1 & 5 \\ 10 & 10 & 21 \end{bmatrix}$$

REMARK

REMARK

- If the orders of the matrices are different, then they are not conformable for addition.

2.3.2 SUBSTRACTION OF MATRICES

Suppose A and B are two matrices of same order, then the substraction of A and B, i.e., A–B is obtained by substracting each element of B from the corresponding element of A. If A and B are of order $m \times n$, then the order of $A - B$ will be of order $m \times n$.

Let
$$A = [a_{ij}]_{m \times n} \text{ and } B = [b_{ij}]_{m \times n}$$

then
$$A - B = [a_{ij} - b_{ij}]_{m \times n}$$

For example: If
$$A = \begin{bmatrix} 1 & 2 & 3 \\ 3 & 4 & 5 \\ 5 & 6 & 7 \end{bmatrix} \text{ and } B = \begin{bmatrix} 0 & 5 & 2 \\ 3 & -2 & 2 \\ 5 & 7 & 8 \end{bmatrix}$$

then
$$A - B = \begin{bmatrix} 1 & 2 & 3 \\ 3 & 4 & 5 \\ 5 & 6 & 7 \end{bmatrix} - \begin{bmatrix} 0 & 5 & 2 \\ 3 & -2 & 2 \\ 5 & 7 & 8 \end{bmatrix}$$

$$= \begin{bmatrix} 1-0 & 2-5 & 3-2 \\ 3-3 & 4-(-2) & 5-2 \\ 5-5 & 6-7 & 7-8 \end{bmatrix}$$

$$= \begin{bmatrix} 1 & -3 & 1 \\ 0 & 6 & 3 \\ 0 & -1 & -1 \end{bmatrix}$$

REMARK

- If the order of matrices are different, then they are not conformable for subtraction.

2.3.3 MULTIPLICATION OF A MATRIX BY A SCALAR

Suppose A is a matrix of order $m \times n$ and k is a scalar, then the multiplication of A by k, i.e. kA is obtained by multiplying each element of A by k.

Let
$$A = [a_{ij}]_{m \times n} \ \forall \ 1 \le i \le m \text{ and } i \le j \le m$$

For example : If
$$A = \begin{bmatrix} 1 & 2 & 3 \\ 4 & 5 & 6 \\ 7 & 8 & 9 \end{bmatrix} \text{ and } k = 3,$$

then
$$3A = 3 \begin{bmatrix} 1 & 2 & 3 \\ 4 & 5 & 6 \\ 7 & 8 & 9 \end{bmatrix}$$

$$= \begin{bmatrix} 3 \times 1 & 3 \times 2 & 3 \times 3 \\ 3 \times 4 & 3 \times 5 & 3 \times 6 \\ 3 \times 7 & 3 \times 8 & 3 \times 9 \end{bmatrix} = \begin{bmatrix} 3 & 6 & 9 \\ 12 & 15 & 18 \\ 21 & 24 & 27 \end{bmatrix}$$

2.3.4 EQUALITY OF MATRICES

Two matrices are said to be equal if both have same order and having same corresponding elements.

For example : The matrices $A = \begin{bmatrix} 1 & 2 \\ -3 & 4 \end{bmatrix}$ and $B = \begin{bmatrix} x & y \\ z & 4 \end{bmatrix}$ are said to be equal if $x = 1$, $y = 2$ and $z = -3$.

2.4 PROPERTIES OF MATRIX ADDITION

2.4.1 COMMUTATIVE LAW

If A and B are two matrices of same order $m \times n$, then
$$A + B = B + A$$

Proof. Let $A = [a_{ij}]_{m \times n}$ and $B = [b_{ij}]_{m \times n}$ where $1 \le i \le m$ and $1 \le j \le n$. Then

$$\begin{aligned}
A + B &= [a_{ij}]_{m \times n} + [b_{ij}]_{m \times n} \\
&= [a_{ij} + b_{ij}]_{m \times n} \quad \text{(By definition of addition)} \\
&= [b_{ij} + a_{ij}]_{m \times n} \quad \text{(Real numbers are always commutative)} \\
&= [b_{ij}]_{m \times n} + [a_{ij}]_{m \times n} \\
&= B + A
\end{aligned}$$

Hence, $A + B = B + A$

2.4.2 ASSOCIATIVE LAW

If A ,B and C are three matrices of same order $m \times n$, then
$$(A + B) + C = A + (B + C)$$

Proof. Let $A = [a_{ij}]_{m \times n}$ and $B = [b_{ij}]_{m \times n}$ where $1 \le i \le m$ and $1 \le j \le n$. Then

$$\begin{aligned}
(A+B)+C &= ([a_{ij}]_{m \times n} + [b_{ij}]_{m \times n}) + [c_{ij}]_{m \times n} \\
&= [a_{ij} + b_{ij}]_{m \times n} + [c_{ij}]_{m \times n} \\
&= [(a_{ij} + b_{ij}) + (c_{ij})]_{m \times n} \quad \text{(Numbers are always associative)} \\
&= [a_{ij}]_{m \times n} + ([b_{ij} + c_{ij}]_{m \times n}) \\
&= [a_{ij}]_{m \times n} + ([b_{ij}]_{m \times n} + [c_{ij}]_{m \times n}) \\
&= A + (B + C)
\end{aligned}$$

Hence, $(A+B) +C = A + (B+C)$

2.4.3 ADDITIVE IDENTITY

If A is a matrix of order $m \times n$ and O is a null matrix of the same order $m \times n$, then
$$A + O = A = O + A$$

Proof. Let $A = [a_{ij}]_{m \times n}$ and $O = [0]_{m \times n}$, then

$$\begin{aligned}
A + O &= [a_{ij}]_{m \times n} + [0]_{m \times n} \\
&= [a_{ij} + 0]_{m \times n} \\
&= [a_{ij}]_{m \times n} \\
&= A
\end{aligned}$$

Also $$\begin{aligned}
O + A &= [0]_{m \times n} + [a_{ij}]_{m \times n} \\
&= [0 + a_{ij}]_{m \times n} \\
&= [a_{ij}]_{m \times n} \\
&= A
\end{aligned}$$

Hence $A + O = A = O + A$

Therefore, the null matrix O is treated as an additive identity.

2.4.4 ADDITIVE INVERSE

If A is a matrix of order $m \times n$ and $-A$ is the negative of A, so its order is also $m \times n$, then

$$-A + A = O \quad \text{(null matrix)}$$

Hence, $-A$ *is the* additive inverse *of* A.

2.4.5 CANCELLATION LAW

If A, B and C are three matrices of order $m \times n$ then

(i) $A + B = A + C \Rightarrow B = C$ (Left cancellation law)

(ii) $B + A = C + A \Rightarrow B = C$ (Right cancellation law)

Proof.

(i) It is given that

$$A + B = A + C \qquad \qquad \qquad \text{...(1)}$$

Adding $-A$ to the left of both sides, we get

$$-A + (A + B) = -A + (A + C)$$

$\Rightarrow \qquad (-A + A) + B = (-A + A) + C \qquad$ (From associative law)

$\Rightarrow \qquad \qquad O + B = O + C \qquad \qquad$ (By additive inverse)

$\Rightarrow \qquad \qquad \qquad B = C \qquad \qquad \qquad$ (By additive inverse)

Similarly, we can prove that if $B + A = C + A$, then $B = C$.

2.5 PROPERTIES OF MULTIPLICATION OF MATRIX BY A SCALAR

(i) **Distribution law of scalar multiplication over matrix addition :** If A and B are two matrices of order $m \times n$ and k is any scalar, then

$$k(A + B) = kA + kB$$

Proof. Let $A = [a_{ij}]_{m \times n}$ and $B = [b_{ij}]_{m \times n}$, then

$$
\begin{aligned}
k(A + B) &= k([a_{ij}]_{m \times n} + [b_{ij}]_{m \times n}) \\
&= k([a_{ij} + b_{ij}]_{m \times n}) \\
&= [k(a_{ij} + b_{ij})]_{m \times n} \\
&= [ka_{ij} + kb_{ij}]_{m \times n} \\
&= [ka_{ij}]_{m \times n} + [kb_{ij}]_{m \times n} \\
&= k[a_{ij}]_{m \times n} + k[b_{ij}]_{m \times n} \\
&= kA + kB
\end{aligned}
$$

Hence $\qquad \qquad k(A + B) = kA + kB$.

(ii) If A is a matrix of order $m \times n$ and a, b are two scalars, then

$$(a + b)A = aA + bA$$

Proof. Let $\qquad \qquad A = [a_{ij}]_{m \times n}$, then

$$
\begin{aligned}
(a + b)A &= (a + b)[a_{ij}]_{m \times n} \\
&= [(a + b)a_{ij}]_{m \times n} \qquad \text{(By scalar multiplication)} \\
&= [aa_{ij} + ba_{ij}]_{m \times n} \qquad \text{(Numbers are distributive)} \\
&= [aa_{ij}]_{m \times n} + [ba_{ij}]_{m \times n} \\
&= a[a_{ij}]_{m \times n} + b[a_{ij}]_{m \times n}
\end{aligned}
$$

$$= aA + bA$$

Hence $\qquad (a + b)A = aA + bA$

(iii)   If A is a matrix of order $m \times n$ and a, b are two scalars, then

$$a(bA) = (ab)A$$

Proof. Let $A = [a_{ij}]_{m \times n}$, then

$$
\begin{aligned}
a(bA) &= a(b[a_{ij}]_{m \times n}) \\
&= a(b[a_{ij}]_{m \times n}) && \text{(By scalar multiplication)} \\
&= [a(ba_{ij})]_{m \times n} \\
&= [(ab)a_{ij}]_{m \times n} && \text{(Numbers are associative)} \\
&= (ab)[a_{ij}]_{m \times n} \\
&= (ab)\, A
\end{aligned}
$$

Hence $\qquad a(bA) = (ab)A.$

(iv)   If A is a matrix of order $m \times n$ and k is any scalar, then

$$(-k)A = -(kA) = k(-A)$$

Proof. Let $A = [a_{ij}]_{m \times n}$, then

$$
\begin{aligned}
(-k)A &= (-k)[a_{ij}]_{m \times n} \\
&= [(-k)a_{ij}]_{m \times n} && \text{(By scalar multiplication)} \\
&= [-ka_{ij}]_{m \times n} \\
&= -[ka_{ij}]_{m \times n} \\
&= kA
\end{aligned}
$$

Now

$$
\begin{aligned}
(-k)A &= (-k)[a_{ij}]_{m \times n} \\
&= [(-k)a_{ij}]_{m \times n} \\
&= [k(-a_{ij})]_{m \times n} \\
&= k[-a_{ij}]_{m \times n} \\
&= k(-A)
\end{aligned}
$$

Hence $\qquad (-k)A = -(kA) = k(-A)\,.$

Solved Examples

Example 1. *Find the number of rows and columns in the following matrices :*

(i) $[1 \ \ 2 \ \ 3 \ \ 4]$
(ii) $\begin{bmatrix} -1 \\ 3 \\ 0 \\ 4 \end{bmatrix}$
(iii) $\begin{bmatrix} -1 & 0 & 3 & 4 \\ 5 & 6 & 7 & 8 \end{bmatrix}$

Solution.   (i)   In the matrix $[1 \ \ 2 \ \ 3 \ \ 4]$, there is one row and four columns.

(ii)   In the matrix $\begin{bmatrix} -1 \\ 3 \\ 0 \\ 4 \end{bmatrix}$ there are four rows and one column.

(iii) In the given matrix $\begin{bmatrix} -1 & 0 & 3 & 4 \\ 5 & 6 & 7 & 8 \end{bmatrix}$.

There are 2 rows and 4 columns.

Example 2. *Find the order of the following matrix :*

$$\begin{bmatrix} -1 & 0 & 3 & 4 \\ 5 & 6 & 7 & 8 \\ 0 & 3 & -2 & 4 \end{bmatrix}$$

Solution. Here the given matrix has 3 rows and 4 columns, so its order is 3×4.

Example 3. *For what values of x, y and z, the matrices A and B are equal :*

$$A = \begin{bmatrix} x & 1 & 2 \\ 3 & 4 & z \\ -1 & 0 & 2 \end{bmatrix} \text{ and } B = \begin{bmatrix} 3 & 1 & 2 \\ 3 & 4 & 5 \\ -1 & y & 2 \end{bmatrix}$$

Solution. Since the matrices A and B are given equal, then comparing the corresponding elements, we get

$$x = 3, y = 0 \text{ and } z = 5.$$

Example 4. If $A = \begin{bmatrix} 1 & 0 \\ 2 & -1 \end{bmatrix}, B = \begin{bmatrix} 3 & 7 \\ 4 & 8 \end{bmatrix}, C = \begin{bmatrix} -1 & 1 \\ 0 & 0 \end{bmatrix}$ then find :

(i) 7 A;   (ii) –3B   (iii) 2C   (iv) A–5B   (v) 4A+3C.

Solution. (i)

$$7 A = 7 \begin{bmatrix} 1 & 0 \\ 2 & -1 \end{bmatrix} = \begin{bmatrix} 7 \times 1 & 7 \times 0 \\ 7 \times 2 & 7 \times -1 \end{bmatrix}$$

$$= \begin{bmatrix} 7 & 0 \\ 14 & -7 \end{bmatrix}$$

(ii)

$$-3B = -3 \begin{bmatrix} 3 & 7 \\ 4 & 8 \end{bmatrix} = \begin{bmatrix} -3 \times 3 & -3 \times 7 \\ -3 \times 4 & -3 \times 8 \end{bmatrix}$$

$$= \begin{bmatrix} -9 & -21 \\ -12 & -24 \end{bmatrix}$$

(iii)

$$2C = 2 \begin{bmatrix} -1 & 1 \\ 0 & 0 \end{bmatrix} = \begin{bmatrix} 2 \times -1 & 2 \times 1 \\ 2 \times 0 & 2 \times 0 \end{bmatrix}$$

$$= \begin{bmatrix} -2 & 2 \\ 0 & 0 \end{bmatrix}$$

(iv)

$$A - 5B = \begin{bmatrix} 1 & 0 \\ 2 & -1 \end{bmatrix} - 5 \begin{bmatrix} 3 & 7 \\ 4 & 8 \end{bmatrix}$$

$$= \begin{bmatrix} 1 & 0 \\ 2 & -1 \end{bmatrix} - \begin{bmatrix} 15 & 35 \\ 20 & 40 \end{bmatrix}$$

$$= \begin{bmatrix} 1-15 & 0-35 \\ 2-20 & -1-40 \end{bmatrix} = \begin{bmatrix} -14 & -35 \\ -18 & -41 \end{bmatrix}$$

(iv) $\qquad 4A + 3C = 4\begin{bmatrix} 1 & 0 \\ 2 & -1 \end{bmatrix} + 3\begin{bmatrix} -1 & 1 \\ 0 & 0 \end{bmatrix}$

$$= \begin{bmatrix} 4 & 0 \\ 8 & -4 \end{bmatrix} + \begin{bmatrix} -3 & 3 \\ 0 & 0 \end{bmatrix}$$

$$= \begin{bmatrix} 4-3 & 0+3 \\ 8+0 & -4+0 \end{bmatrix}$$

$$= \begin{bmatrix} 1 & 3 \\ 8 & -4 \end{bmatrix}$$

Example 5. *Find the additive inverse of the matrix :*

$$A = \begin{bmatrix} 2 & -3 & -1 & -1 \\ -3 & 1 & -2 & 2 \\ 1 & -2 & -8 & 7 \end{bmatrix}$$

Solution. The additive inverse of $A = -A$

$\therefore \qquad\qquad -A = -\begin{bmatrix} 2 & -3 & -1 & -1 \\ -3 & 1 & -2 & 2 \\ 1 & -2 & -8 & 7 \end{bmatrix}$

$$= \begin{bmatrix} -2 & 3 & 1 & 1 \\ 3 & -1 & 2 & -2 \\ 1 & 2 & 8 & -7 \end{bmatrix}$$

Example 6. *If $A = \begin{bmatrix} 1 & 2 \\ 3 & 4 \end{bmatrix}, \quad B = \begin{bmatrix} -1 & 5 \\ 5 & 7 \end{bmatrix}$*

then prove that $5(A+B) = 5A + 5B.$

Solution. $\qquad\qquad A + B = \begin{bmatrix} 1 & 2 \\ 3 & 4 \end{bmatrix} + \begin{bmatrix} -1 & 5 \\ 5 & 7 \end{bmatrix}$

$$= \begin{bmatrix} 1-1 & 2+5 \\ 3+5 & 4+7 \end{bmatrix}$$

$$= \begin{bmatrix} 0 & 7 \\ 8 & 11 \end{bmatrix}$$

Now $\qquad 5(A+B) = 5\begin{bmatrix} 0 & 7 \\ 8 & 11 \end{bmatrix} = \begin{bmatrix} 5\times0 & 5\times7 \\ 5\times8 & 5\times11 \end{bmatrix} = \begin{bmatrix} 0 & 35 \\ 40 & 55 \end{bmatrix}$

and $\qquad\qquad 5A = 5\begin{bmatrix} 1 & 2 \\ 3 & 4 \end{bmatrix} = \begin{bmatrix} 5\times1 & 5\times2 \\ 5\times3 & 5\times4 \end{bmatrix} = \begin{bmatrix} 5 & 10 \\ 15 & 20 \end{bmatrix}$

$\qquad\qquad 5B = 5\begin{bmatrix} -1 & 5 \\ 5 & 7 \end{bmatrix} = \begin{bmatrix} 5\times-1 & 5\times5 \\ 5\times5 & 5\times7 \end{bmatrix} = \begin{bmatrix} -5 & 25 \\ 25 & 35 \end{bmatrix}$

Now $\qquad\qquad 5A + 5B = \begin{bmatrix} 5 & 10 \\ 15 & 20 \end{bmatrix} + \begin{bmatrix} -5 & 25 \\ 25 & 35 \end{bmatrix}$

$$= \begin{bmatrix} 5-5 & 10+25 \\ 15+25 & 20+35 \end{bmatrix}$$

$$= \begin{bmatrix} 0 & 35 \\ 40 & 55 \end{bmatrix}$$

Hence $\qquad 5(A+B) = 5A + 5B$

2.6 MULTIPLICATION OF MATRICES

Let A and B be two matrices of order $m \times n$ and $n \times p$ respectively. Then a matrix C of order $m \times p$ is obtained by multiplying each row of A to each column of B.

Suppose $A = [a_{ij}]_{m \times n}$, $B = [b_{jk}]_{n \times p}$, then $C = [c_{ik}]_{m \times p}$ is known as the multiplication of A and B if

$$c_{ik} = \sum_{j=1}^{n} a_{ij} b_{jk}$$

and hence we can write $\qquad C = AB$

WORKING RULE

First we check whether the matrices are conformable for multiplication or not. For this we check that if the number of columns of first matrix is equal to the number of rows of the second matrix, then the matrices can be multiplied. Multiplication is operated by the rule (row × Column). In this rule, we first put the first row of the first matrix next to the first column of the second matrix and the corresponding elements are now multiplied and then summed up which gives the first element of the first row of the product matrix. This process runs till the first row of the first matrix is operated to all columns of the second matrix. After that the first process is applied to the second, third etc. rows of the first matrix.

For example : If $\qquad A = \begin{bmatrix} 2 & 1 & 5 \\ 6 & 2 & 3 \end{bmatrix}_{2 \times 3}$ and $B = \begin{bmatrix} 3 & 4 \\ 5 & 6 \\ 7 & 8 \end{bmatrix}_{3 \times 2}$, then

$$AB = \begin{bmatrix} 2 & 1 & 5 \\ 6 & 2 & 3 \end{bmatrix} \begin{bmatrix} 3 & 4 \\ 5 & 6 \\ 7 & 8 \end{bmatrix}$$

$$= \begin{bmatrix} 2 \times 3 + 1 \times 5 + 5 \times 7 & 2 \times 4 + 1 \times 6 + 5 \times 8 \\ 6 \times 3 + 2 \times 5 + 3 \times 7 & 6 \times 4 + 2 \times 6 + 3 \times 8 \end{bmatrix}$$

$$= \begin{bmatrix} 6 + 5 + 35 & 8 + 6 + 40 \\ 18 + 10 + 21 & 24 + 12 + 24 \end{bmatrix}$$

$$= \begin{bmatrix} 49 & 54 \\ 49 & 60 \end{bmatrix}$$

REMARKS

- If the number of columns of the matrix A is equal to the number of rows of matrix B, then A and B are conformable for the multiplication AB but not for BA.
- Square matrices are always conformable for multiplication both ways.

2.7 PROPERTIES OF MATRIX MULTIPLICATION

(i) Associative law : If A, B and C are the matrices of order $m \times n$, $n \times p$ and $p \times q$, then

$$AB)C = A(BC)$$

Proof. Let $A = [a_{ij}]_{m \times n}$, $B = [b_{jk}]_{n \times p}$ and $C = [c_{kl}]_{p \times q}$, then

$$AB = [a_{ij}]_{m \times n} \cdot B = [b_{jk}]_{n \times p} = [x_{ik}]_{m \times p}$$

where

$$x_{ik} = \sum_{j=1}^{n} a_{ij} b_{jk} \qquad \qquad ...(1)$$

$\therefore$

$$(AB)C = [x_{ik}]_{m \times p} \cdot [c_{kl}]_{p \times q} = [u_{il}]_{m \times q}$$

where

$$u_{il} = \sum_{k=1}^{p} x_{ik} c_{kl} \qquad \qquad ...(2)$$

Now from (1) and (2), we get

$$u_{il} = \sum_{k=1}^{p} \left(\sum_{j=1}^{n} a_{ij} b_{jk} \right) c_{kl}$$

$$= \sum_{j=1}^{n} a_{ij} \left(\sum_{k=1}^{p} b_{jk} c_{kl} \right) \qquad \qquad ...(3)$$

Now

$$BC = [b_{jk}]_{m \times p} \cdot [c_{kl}]_{p \times q} = [v_{jl}]_{m \times q}$$

where

$$v_{jl} = \sum_{k=1}^{p} b_{jk} c_{kl} \qquad \qquad ...(4)$$

From equations (3) and (4), we get

$$u_{il} = \sum_{j=1}^{n} a_{ij} v_{jl} \qquad \qquad ...(5)$$

Equation (2) implies that u_{il} is the (i, l)th element in $(AB)C$ and the equation (5) implies that u_{il} is (i, l)th element in $A(BC)$. Therefore by the equality of two matrices, we obtain

$$(AB)C = A(BC).$$

(ii) Matrix multiplication satisfies the distributive law over the matrix addition.

If A, B and C are the matrices of order $m \times n$, $n \times p$ and $n \times p$ respectively, then

$$A(B+C) = AB + AC$$

Proof. Let $A = [a_{ij}]_{m \times n}$, $B = [b_{jk}]_{n \times p}$ and $C = [c_{jk}]_{n \times p}$, then

$$B+C = [b_{jk}]_{n \times p} + [c_{jk}]_{n \times p}$$

$$= [b_{jk} + c_{jk}]_{n \times p}$$

$\therefore$

$$A(B+C) = [a_{ij}]_{m \times n} + [b_{jk} + c_{jk}]_{n \times p}$$

$$= [x_{ik}]_{m \times p} \qquad \qquad ...(1)$$

where
$$x_{ik} = \sum_{j=1}^{n} a_{ij}(b_{jk} + c_{jk})$$

$$= \sum_{j=1}^{n} (a_{ij}b_{jk} + a_{ij}c_{jk})$$

$$= \sum_{j=1}^{n} a_{ij}b_{jk} + \sum_{j=1}^{n} a_{ij}c_{jk}$$

$= (i, k)$ the element of AB + (i, k) the element of AC

$= (i, k)$ the element of $(AB + BC)$

But equation (1) implies that x_{ik} is the (i, k) the element of $A(B+C)$, therefore, by the definition of equality of two matrices, we must have

$A(B + C) = AB + AC.$

REMARK

- Matrix multiplication is not commutative in general.

Solved Examples

Example 1. If $A = \begin{bmatrix} 2 & 3 & 4 \\ 3 & 2 & 3 \\ -1 & 1 & 2 \end{bmatrix}, B = \begin{bmatrix} 1 & 3 & 0 \\ -1 & 2 & 1 \\ 1 & 0 & 2 \end{bmatrix}$

then find AB and BA and show that $AB \neq BA$.

Solution. Since A and B are square matrices of order 3×3 so that multiplication of AB and BA is possible.

Now,

$$AB = \begin{bmatrix} 2 & 3 & 4 \\ 3 & 2 & 3 \\ -1 & 1 & 2 \end{bmatrix}\begin{bmatrix} 1 & 3 & 0 \\ -1 & 2 & 1 \\ 1 & 0 & 2 \end{bmatrix}$$

$$= \begin{bmatrix} 2\times1+3\times(-1)+4\times0 & 2\times3+3\times2+4\times0 & 2\times0+3\times1+4\times2 \\ 1\times1+2\times(-1)+3\times0 & 1\times3+2\times2+3\times0 & 1\times0+2\times1+3\times2 \\ -1\times1+1\times(-1)+2\times0 & -1\times3+1\times2+2\times0 & -1\times0+1\times1+2\times2 \end{bmatrix}$$

$$= \begin{bmatrix} 2-3+0 & 6+6+0 & 0+3+8 \\ 1-2+0 & 3+4+0 & 0+2+6 \\ -1+1+0 & -3+2+0 & 0+1+4 \end{bmatrix}$$

$$= \begin{bmatrix} -1 & 12 & 11 \\ -1 & 7 & 8 \\ -2 & -1 & 5 \end{bmatrix}_{3\times3}$$

and $BA = \begin{bmatrix} 1 & 3 & 0 \\ -1 & 2 & 1 \\ 0 & 0 & 2 \end{bmatrix}\begin{bmatrix} 2 & 3 & 4 \\ 1 & 2 & 3 \\ -1 & 1 & 2 \end{bmatrix}$

$$= \begin{bmatrix} 1\times2+3\times1+0\times(-1) & 1\times3+3\times2+0\times1 & 1\times4+3\times3+0\times2 \\ -1\times2+2\times1+1\times(-1) & -1\times3+2\times2+1\times1 & -1\times4+2\times3+1\times2 \\ 0\times2+0\times1+2\times(-1) & 0\times3+0\times2+2\times1 & 0\times4+0\times3+2\times2 \end{bmatrix}$$

$$= \begin{bmatrix} 2+3-0 & 3+6+0 & 4+9+0 \\ -2+2-1 & -3+4+1 & -4+6+2 \\ 0+0-2 & 0+0+2 & 0+0+4 \end{bmatrix}$$

$$= \begin{bmatrix} 5 & 9 & 13 \\ -1 & 2 & 4 \\ -2 & 2 & 4 \end{bmatrix}_{3\times3}$$

Here AB and BA have same order but different corresponding elements. Hence $AB \neq BA$.

Example 2.   If $A = \begin{bmatrix} 7 & 0 & 0 \\ 0 & 7 & 0 \\ 0 & 0 & 7 \end{bmatrix}$ and $B = \begin{bmatrix} a & b & c \\ d & e & f \\ g & h & i \end{bmatrix}$ then prove that $AB = 7B$.

Solution. $AB = \begin{bmatrix} 7 & 0 & 0 \\ 0 & 7 & 0 \\ 0 & 0 & 7 \end{bmatrix}\begin{bmatrix} a & b & c \\ d & e & f \\ g & h & i \end{bmatrix}$

$$= \begin{bmatrix} 7\times a+0\times d+0\times g & 7\times b+0\times e+0\times h & 7\times c+0\times f+0\times i \\ 0\times a+7\times d+0\times g & 0\times b+7\times e+0\times h & 0\times c+7\times f+0\times i \\ 0\times a+0\times d+7\times g & 0\times b+0\times e+7\times h & 0\times c+0\times f+7\times i \end{bmatrix}$$

$$= \begin{bmatrix} 7a+0+0 & 7b+0+0 & 7c+0+0 \\ 0+7d+0 & 0+7e+0 & 0+7f+0 \\ 0+0+7g & 0+0+7h & 0+0+7i \end{bmatrix}$$

$$= \begin{bmatrix} 7a & 7b & 7c \\ 7d & 7e & 7f \\ 7g & 7h & 7i \end{bmatrix} = 7\begin{bmatrix} a & b & c \\ d & e & f \\ g & h & i \end{bmatrix} = 7B$$

2.8 DETERMINANT OF A SQUARE MATRIX

Let A be a square matrix. Then the determinant which is formed by the elements of matrix A is usually denoted by $|A|$.

For example : If $A = \begin{bmatrix} a_{11} & a_{12} & a_{13} \\ a_{21} & a_{22} & a_{23} \\ a_{31} & a_{32} & a_{33} \end{bmatrix}$, then its determinant is

$$A = \begin{vmatrix} a_{11} & a_{12} & a_{13} \\ a_{21} & a_{22} & a_{23} \\ a_{31} & a_{32} & a_{33} \end{vmatrix}$$

REMARK

- The determinant of a matrix is reduced to a number.

2.9 PROPERTIES OF DETERMINANTS

(1) The value of a determinant is zero if all the elements of a row or column are zero.

(2) The value of a determinant remain unchanged when rows are changed into corresponding columns.

(3) If any two rows or columns of a determinant are interchanged the sign of the determinant is changed.

(4) If any two rows or columns of a determinant are identical, then the value of the determinant is zero.

(5) If every element of same columns or row is the sum of two terms then determinant is equal to the sum of two determinant are containing only the first term and other the second term only in place of each sum.

(6) If each element of a row (or column) is multiplied by a constant k, then then value of the new determinant will be k times the value of original determinant.

(7) If each element of a row (or column) of a determinant by a constant k and then added to the corresponding elements of some other row (or column) then the value of the determinant remain same.

(8) If the elements of the determinant are the polynomial in a variable x and if by putting $x = a$, the determinant vanishes then $(x - a)$ will be a factor of determinant.

2.10 EVOLUTION OF A DETERMINANT BY SARRUS DIAGRAM

$$\begin{vmatrix} a_{11} & a_{12} & a_{13} \\ a_{21} & a_{22} & a_{23} \\ a_{31} & a_{32} & a_{33} \end{vmatrix} = a_{11}(a_{22}a_{33} - a_{32}a_{23}) - a_{12}(a_{21}a_{33} - a_{31}a_{23}) + a_{13}(a_{21}a_{32} - a_{31}a_{22})$$
$$= a_{11}a_{22}a_{33} + a_{12}a_{31}a_{23} + a_{13}a_{21}a_{32} - (a_{11}a_{32}a_{23} + a_{11}a_{21}a_{33} + a_{13}a_{31}a_{22})$$

**WORKING RULE**

Write the columns of the determinant and again write the first and second columns on the right side and draw the lines as shown in the following figure :

$$
\begin{array}{ccccc}
a_{11} & a_{12} & a_{13} & a_{11} & a_{12} \\
a_{21} & a_{22} & a_{23} & a_{21} & a_{22} \\
a_{31} & a_{32} & a_{33} & a_{31} & a_{32}
\end{array} \quad + -
$$

For example :

Let

$$A = \begin{vmatrix} 1 & 2 & 3 \\ 2 & 3 & 4 \\ 2 & 0 & 5 \end{vmatrix}$$

Then we have

$$
\begin{array}{ccccc}
1 & 2 & 3 & 1 & 2 \\
2 & 3 & 4 & 2 & 3 \\
2 & 0 & 5 & 2 & 0
\end{array}
$$

$$|A| = 1 \cdot 3 \cdot 5 + 2 \cdot 4 \cdot 2 + 3 \cdot 2 \cdot 0 - (2 \cdot 3 \cdot 3 + 0 \cdot 4 \cdot 1 + 5 \cdot 2 \cdot 2)$$
$$= 15 + 16 + 0 - (18 + 0 + 20)$$
$$= 31 - 38 = -7$$

2.11 MINOR AND COFACTORS

In determinant, $\quad \Delta = \begin{vmatrix} a_{11} & a_{12} & a_{13} \\ a_{21} & a_{22} & a_{23} \\ a_{31} & a_{32} & a_{33} \end{vmatrix} \qquad \qquad \dots(1)$

If we leave the row and column passing through the element a_{ij} then we obtained the second order determinant, which is called the minor of the element a_{ij}. It is denoted by M_{ij}. Therefore, in a determinant of order 3, we may get 9 minors corresponding to the 9 elements of the determinant.

For example, in determinant (1)

$$\text{Minor of } a_{21} = \begin{vmatrix} a_{12} & a_{13} \\ a_{32} & a_{33} \end{vmatrix} = M_{21}$$

and

$$\text{Minor of } a_{32} = \begin{vmatrix} a_{11} & a_{13} \\ a_{21} & a_{23} \end{vmatrix} = M_{32}$$

If we expand the determinant along the first row, then

$$\Delta = (-1)^{1+1} a_{11} M_{11} + (-1)^{1+2} a_{12} M_{12} + (-1)^{1+3} a_{13} M_{13}$$
$$= a_{11} M_{11} - a_{12} M_{12} + a_{13} M_{13}$$

Similarly, along second column, we can write

$$\Delta = -a_{12} M_{12} + a_{22} M_{22} - a_{32} M_{32}$$

Cofactor: If we multiply the minor M_{ij} by $(-1)^{i+j}$. Then resulting value is called cofactor of the element a_{ij}. If A_{ij} is the cofactor of then we write

$$\text{Cofactor of } a_{ij} = A_{ij} = (-1)^{i+j} M_{ij}$$

$$\text{Cofactor of } a_{21} = A_{21} = (-1)^{2+1} M_{21} = -\begin{vmatrix} a_{12} & a_{13} \\ a_{32} & a_{33} \end{vmatrix}$$

$$\text{Cofactor of } a_{32} = A_{32} = (-1)^{3+2} M_{32} = -\begin{vmatrix} a_{11} & a_{13} \\ a_{21} & a_{23} \end{vmatrix}$$

Hence, cofactor of $a_{ij} = (-1)^{i+j}$ determinant obtained by leaving row and column passing through that element. Therefore, we can write

$$\Delta = a_{11} A_{11} + a_{12} A_{12} + a_{13} A_{13}$$
$$\Delta = a_{21} A_{21} + a_{22} A_{22} + a_{23} A_{23}$$
$$\Delta = a_{31} A_{31} + a_{32} A_{32} + a_{33} A_{33}$$

and

$$a_{11} A_{21} + a_{12} A_{22} + a_{13} A_{23} = 0$$
$$a_{11} A_{31} + a_{12} A_{32} + a_{13} A_{33} = 0$$

2.12 SINGULAR AND NON-SINGULAR MATRIX

Definition. *A matrix whose determinant value is zero, is said to be singular matrix.*

If the matrix is not singular then it is said to be non-singular.

For example : If $A = \begin{bmatrix} 2 & 3 \\ 6 & 9 \end{bmatrix}$, then its determinant value.

$$|A| = \begin{bmatrix} 2 & 3 \\ 6 & 9 \end{bmatrix} = 2 \times 9 - 3 \times 6 = 18 - 18 = 0$$

Thus the matrix A is singular.

2.13 TRANSPOSE OF A MATRIX

Definition. Consider a matrix $A = [a_{ij}]_{m \times n}$. Then a matrix which is obtained by interchanging the rows and columns of A is called the transpose of A. It is denoted by A' or A^T.

That is , if $A = [a_{ij}]_{m \times n}$, then $A' = [a_{ji}]_{n \times m}$.

For example : If $\qquad A = \begin{bmatrix} 2 & 3 & 5 \\ 1 & 6 & 7 \end{bmatrix}_{2 \times 3}$, then its transpose is

$$A' = \begin{bmatrix} 2 & 3 & 5 \\ 1 & 6 & 7 \end{bmatrix}'$$

$$= \begin{bmatrix} 2 & 1 \\ 3 & 6 \\ 5 & 7 \end{bmatrix}_{3 \times 2}$$

REMARKS

- Transpose of row matrix is a column matrix and transpose of a column matrix is a row matrix.
- If a matrix is square then its transpose will be a square matrix of same order.

2.14 PROPERTIES OF TRANSPOSE OF A MATRIX

THEOREM 1. If A' and B' are the transpose of the matrix A and B respectively, then :

(i) $(A')' = A$

(ii) $(A+B)' = A' + B'$, here A and B must be of same order.

(iii) $(kA)' = kA'$, here k is any scalar.

(iv) $(AB)' = B'A'$, here AB and $B'A'$ are conformable for multiplication.

Proof.

(i) Let $\quad A = [a_{ij}]_{m \times n}$, then $\quad A' = [a_{ji}]_{n \times m}$ since

$\qquad (i, j)$th element in $(A')' = (j, i)$th element in A'

$\qquad\qquad\qquad\qquad\qquad = (i, j)$th element in A

Thus by the definition of equality of matrices, we must have $(A')' = A$.

(ii) Let $A = [a_{ij}]_{m \times n}$, $B = [b_{ij}]_{m \times n}$. So, $A' = [a_{ji}]_{n \times m}$ and $B' = [b_{ji}]_{n \times m}$, then

$\qquad (i, j)$th element in $(A+B)' = (j, i)$th element in $(A+B)$

$\qquad\qquad\qquad\qquad\qquad = (j, i)$th element in $A + (j, i)$th element in B

$\qquad\qquad\qquad\qquad\qquad = (i, j)$th element in $A' + (i, j)$th element in B'

$\qquad\qquad\qquad\qquad\qquad = (i, j)$th element in $(A'+B')$

Thus by the definition of equality of matrices, we get

$$(A'+B') = A' + B'$$

(iii) Let $A = [a_{ij}]_{m \times n}$ so that $A' = [a_{ji}]_{n \times m}$ and k be a scalar, then

(i, j)th element in $(kA)' = (j, i)$th element in (kA)

$= (i, j)$th element in kA'

Thus by the defintion of equality of matrices, we get

$$(kA)' = kA'$$

(iv) Let $A = [a_{ij}]_{m \times n}$ and $B = [b_{ij}]_{n \times p}$ then AB is conformable for multiplication and having the order $m \times p$. Therefore, the order of $(AB)'$ is $p \times m$. Since the orders of A' and B' are respectively $n \times m$ and $p \times n$ so $B'A'$ is conformable for multiplication and having the order $p \times m$.

Now $k, i)$th element in $(AB)' = (i, k)$th element in AB

$$= \sum_{j=1}^{n} a_{ij} b_{jk}$$

[By the definition of multiplication of matrices]

But (k, i)th element in $B'A' = \sum_{j=1}^{n} b_{kj} b_{ji}$

$$= \sum_{j=1}^{n} a_{ji} b_{kj}$$

$= (i, k)$th elemetn in AB

$\therefore$ (k, i)th element in $A'B' = (k, i)$th element in $B'A'$

Thus by the definition of equality of matrices, we must have

$$(AB) = B'A'$$

2.15 SYMMETRIC MATRIX

A matrix 'A' is said to be a symmetric matrix if $A' = A$, that is, the transpose of a matrix is equal to the matrix itself.

For exmaple : If $A = \begin{bmatrix} 1 & 2 & 3 \\ 2 & 4 & 5 \\ 3 & 5 & 6 \end{bmatrix}$, then

$A' = \begin{bmatrix} 1 & 2 & 3 \\ 2 & 4 & 5 \\ 3 & 5 & 6 \end{bmatrix}$ so that $A' = A$

Hence, A is symmetric.

2.16 SKEW-SYMMETRIC MATRIX

A matrix 'A' is said to be a skew-symmetric matrix if $A' = -A$.

For exmaple : If $A = \begin{bmatrix} 0 & 2 & 3 \\ -2 & 0 & 4 \\ -3 & -4 & 0 \end{bmatrix}$, then

$$A' = \begin{bmatrix} 0 & -2 & -3 \\ 2 & 0 & -4 \\ 3 & 4 & 0 \end{bmatrix}$$

$$= \begin{bmatrix} 0 & 2 & 3 \\ -2 & 0 & 4 \\ 3 & -4 & 0 \end{bmatrix} = -A$$

Hence A is skew-symmetric matrix.

REMARK

- The diagonal elements of a skew-symmetric matrix are all zero.

2.17 PROPERTIES OF SYMMETRIC AND SKEW-SYMMETRIC MATRIX

(i) *If A is a symmetric (skew-symmetric) matrix, then kA is symmetric (skew-symmetric) matrix, where k is any scalar.*

Proof. Let A be a symmetric matrix, then $A' = A$, since we have

$$(kA)' = kA'$$
$$= kA \qquad (\because A' = A)$$
$$\Rightarrow \qquad (kA)' = (kA)$$

$\Rightarrow$ kA is symmetric matrix

Also, if A is a skew-symmetric matrix, then $A' = -A$.

Since we have $\qquad (kA)' = kA'$

$$= k(-A) \qquad (\because A' = -A)$$
$$= -(kA)$$
$$\therefore \qquad (kA') = -(kA)$$

$\Rightarrow$ kA is skew-symmetric matrix.

(ii) *If A and B are symmetric (skew-symmetric) matrices then A+ B is symmetric (skew-symmetric) matrix.*

Proof. Let A and B be symmetric matrices, then $A' = A$ and $B' = B$.

Since we have

$$(A+B)' = A' + B'$$
$$\Rightarrow \qquad (A+B)' = (A+B) \qquad (\because A' = A, B' = A)$$

$\Rightarrow$ $A+B$ is symmetric.

Similarly, if A and B are skew-symmetric matrices, then $A' = -A$ and $B' = -B$.

But $\qquad (A+B)' = A' + B'$

$$\Rightarrow \qquad (A+B)' = -A - B \qquad (\because A' = -A, B' = -B)$$
$$\Rightarrow \qquad (A+B)' = -(A+B)$$

$\Rightarrow$ $A+B$ is skew-symmetric matrix.

(iii) *If A is any matrix, then AA' and A'A both are symmetric matrices.*

Proof. If $\qquad (AA')' = (A')'A' \qquad [\because (AB)' = B'A']$

$$= AA'$$

$\Rightarrow$ AA' is symmetric.

Also $(A'A)' = A'(A')'$ [$\because$ $(AB)' = B'A'$]

$$= A'A$$ [$\because$ $(A')' = A$]

$\Rightarrow$ AA' is symmetric.

(iv) *If A is any square matrix, then A + A' is symmetric and A − A' is skew-symmetric.*
    Proof. We have $(A + A')' = A' + (A')'$ [$\because$ $(A+B)' = A' + B'$]

$$= A' + A$$ [$\because$ $(A')' = A$]

$$= A + A'$$ [Matrix addition is commutative.]

$\Rightarrow$ $A + A'$ is symmetric.

Similarly, $(A − A')' = A' − (A')'$

$$= A' − A$$

$$= − (A − A')$$

$\Rightarrow$ $A − A'$ is skew symmetric.

(v) *All positive integral powers of a symmetric matrix are symmetric.*
    Proof. Let A be a symmetric matrix and m be any positive integer, then

$$A_m = A . A . A . \ldots . A \ (m \text{ times})$$

$\Rightarrow$ $(A^m)' = [A . A . A . \ldots A \ (m \text{ times})]'$

$$= A' . A' . A' \ldots A' \ (m \text{ times}) \quad (\because A' = A)$$

$$= A . A . A . \ldots A \ (m \text{ times})$$

$$= A^m$$

$\Rightarrow$ A^m is symmetric matrix.

2.18 COMPLEX MATRIX

A matrix 'A' *is said to be* complex matrix if it contains some of its elements equal to a complex number.

2.18.1 CONJUGATE OF A COMPLEX MATRIX

Let A be a complex matrix, then a matrix which is obtained by replacing all the complex elements of A by their complex conjugate number, is called conjugate of a matrix. It is denoted by $\overline{A}$.

For example: If $A = \begin{bmatrix} 1+2i & 3i & 6 \\ 7 & 2+4i & 1+i \end{bmatrix}$, then

$$\overline{A} = \begin{bmatrix} 1-2i & -3i & 6 \\ 7 & 2-4i & 1-i \end{bmatrix}$$

2.18.2 TRANSPOSE CONJUGATE OF A MATRIX

The transpose of the conjugate matrix is called the transposed conjugate of that matrix. It is denoted by A^θ.

Therefore, we must have $(A^\theta) = (\overline{A})'$.

2.18.3 PROPERTIES OF TRANSPOSE CONJUGATE OF MATRIX

(1) $(A^\theta) = A$

(2) $(A + B)^\theta = A^\theta + B^\theta$, A, B being of same order

(3) $(kA)^\theta = \bar{k} A^\theta$, k being a complex number

(4) $(AB)^\theta = B^\theta A^\theta$, A, B being conformable for multiplication.

2.18.4 HERMITIAN AND SKEW-HERMITIAN MATRICES

(1) **Hermitian matrix :** A matrix A is said to be Hermitian if $A^\theta = A$.

(2) **Skew-Hermitian matrix :** A matrix A is said to be skew-Hermitian if $A^\theta = -A$.

2.18.5 ORTHOGONAL AND UNITARY MATRICES

(i) **Orthogonal matrix :** A matrix A is said to be orthogonal if $A'A = I$, where I is a unit matrix of order same as of order A.

(ii) **Unitary Matrix :** A square matrix A is said to be unitary if $A^\theta A = I$.

For example : If $A = \begin{bmatrix} \cos\theta & \sin\theta \\ -\sin\theta & \cos\theta \end{bmatrix}$, then

$$A' = \begin{bmatrix} \cos\theta & -\sin\theta \\ \sin\theta & \cos\theta \end{bmatrix}$$

$$A'A = \begin{bmatrix} \cos\theta & -\sin\theta \\ \sin\theta & \cos\theta \end{bmatrix}\begin{bmatrix} \cos\theta & \sin\theta \\ -\sin\theta & \cos\theta \end{bmatrix}$$

$$= \begin{bmatrix} \cos^2\theta + \sin^2\theta & 0 \\ 0 & \sin^2\theta + \cos^2\theta \end{bmatrix}$$

$$= \begin{bmatrix} 1 & 0 \\ 0 & 1 \end{bmatrix} = I$$

$\Rightarrow$ A is orthogonal matrix.

For example : If $A = \begin{bmatrix} 0 & -i \\ i & 0 \end{bmatrix}$, then $\bar{A} = \begin{bmatrix} 0 & i \\ -i & 0 \end{bmatrix}$

$$A^\theta = (\bar{A})' = \begin{bmatrix} 0 & i \\ i & 0 \end{bmatrix}$$

$$A^\theta A = \begin{bmatrix} 0 & -i \\ i & 0 \end{bmatrix}\begin{bmatrix} 0 & -i \\ i & 0 \end{bmatrix} = \begin{bmatrix} 1 & 0 \\ 0 & 1 \end{bmatrix} \qquad (\because i^2 = -1)$$

$$= I$$

$\Rightarrow$ A is unitary.

2.18.6 PROPERTIES OF HERMITIAN AND SKEW-HERMITIAN MATRICES

(1) If A is Hermitian (skew-Hermitian) matrix, then $i\,A$ is Hermitian (skew-Hermitian) matrix.

(2) If A, B are Hermitian or skew-Hermitian, then $A + B$ is Hermitian or skew-Hermitian.

(3) If A is Hermitian or skew-Hermitian, then $\bar{A}$ is Hermitian or skew-Hermitian.

Solved Examples

Based on the following Results

➡ $A^\theta = (\overline{A})'$
➡ For Hermitian matrix $A^\theta = A$
➡ For skew-Hermitian $A^\theta = -A$
➡ For orthogonal matrix $AA' = I$
➡ For unitary matrix $A^\theta A = I$

Example 1. *If A be any square matrix, prove that $A + A^\theta$, AA^θ, $A^\theta A$ are all Hermitian and $A - A^\theta$ is skew-Hermitian.*

Solution. (i) $(A+A^\theta)^\theta = A^\theta + (A^\theta)^\theta$ $[\because (A+B)^\theta = A^\theta + B^\theta]$

$= A^\theta + A$ $[\because (A^\theta)^\theta = A]$

$= A + A^\theta$

$\Rightarrow$ $A + A^\theta$ is Hermitian.

(ii) $(AA^\theta)^\theta = (A^\theta)^\theta A^\theta$ $[\because (AB)^\theta = B^\theta A^\theta]$

$= AA^\theta$ $[\because (A^\theta)^\theta = A]$

$\Rightarrow$ $A A^\theta$ is Hermitian.

(iii) $(A^\theta A)^\theta = A^\theta (A^\theta)^\theta = A^\theta A$

$\Rightarrow$ $A^\theta A$ is Hermitian.

(iv) $(A - A^\theta) = A^\theta - (A^\theta)^\theta$

$= A^\theta - A = -(A - A^\theta)$

$\Rightarrow$ $A - A^\theta$ is skew-Hermitian.

Example 2. *Prove that the matrix*

$$\begin{bmatrix} \dfrac{1+i}{2} & \dfrac{-1+i}{2} \\ \dfrac{1+i}{2} & \dfrac{1-i}{2} \end{bmatrix}$$

is unitary.

Solution.   Let us suppose $A = \begin{bmatrix} \dfrac{1+i}{2} & \dfrac{-1+i}{2} \\ \dfrac{1+i}{2} & \dfrac{1-i}{2} \end{bmatrix}$

$\therefore$ $A' = \begin{bmatrix} \dfrac{1+i}{2} & \dfrac{1+i}{2} \\ \dfrac{-1+i}{2} & \dfrac{1-i}{2} \end{bmatrix}$

$A^\theta = \overline{A'} = \begin{bmatrix} \dfrac{1-i}{2} & \dfrac{1-i}{2} \\ \dfrac{-1-i}{2} & \dfrac{1+i}{2} \end{bmatrix}$

Now
$$A^\theta A = \begin{bmatrix} \dfrac{1-i}{2} & \dfrac{1-i}{2} \\ \dfrac{-1-i}{2} & \dfrac{1-i}{2} \end{bmatrix} \begin{bmatrix} \dfrac{1+i}{2} & \dfrac{-1+i}{2} \\ \dfrac{1+i}{2} & \dfrac{1-i}{2} \end{bmatrix}$$

$$= \begin{bmatrix} \dfrac{1}{4}(1-i^2)+\dfrac{1}{4}(1-i)^2 & -\dfrac{1}{4}(1-i^2)+\dfrac{1}{4}(1-i)^2 \\ -\dfrac{1}{4}(1+i^2)+\dfrac{1}{4}(1+i)^2 & \dfrac{1}{4}(1-i^2)+\dfrac{1}{4}(1-i)^2 \end{bmatrix}$$

$$= \begin{bmatrix} \dfrac{1}{4}(1+1)+\dfrac{1}{4}(1+1) & 0 \\ 0 & \dfrac{1}{4}(1+1)+\dfrac{1}{4}(1+1) \end{bmatrix}$$

$$= \begin{bmatrix} 1 & 0 \\ 0 & 1 \end{bmatrix} = I$$

Hence A is unitary.

📑 EXERCISE 2.1

---

1. If $A = \begin{bmatrix} 1 & 0 \\ 2 & -1 \end{bmatrix}$, $B = \begin{bmatrix} 3 & 7 \\ 4 & 8 \end{bmatrix}$, $C = \begin{bmatrix} -1 & 1 \\ 0 & 0 \end{bmatrix}$,

then prove that
$$A + (B+C) = (A+B) + C.$$

2. If $A = \begin{bmatrix} 2 & 3 \\ 4 & -5 \end{bmatrix}$, $B = \begin{bmatrix} 8 & 9 \\ 6 & 7 \end{bmatrix}$, then find

(i) $2A + 3B$

(ii) $5A - 3B$

3. Find $A + B$ and show that $A+B = B+A$, when

(i) $A = \begin{bmatrix} 7 & 8 \\ 9 & 2 \\ 3 & 4 \end{bmatrix}$, $B = \begin{bmatrix} 2 & 3 \\ 4 & 5 \\ 6 & 7 \end{bmatrix}$

(ii) $A = \begin{bmatrix} 1 & 2 & -3 \\ 4 & 1 & 5 \\ -3 & -2 & 2 \end{bmatrix}$, $B = \begin{bmatrix} 3 & -1 & 2 \\ 4 & 2 & 5 \\ 2 & 0 & 3 \end{bmatrix}$

4. If $A = \begin{bmatrix} \cos\alpha & -\sin\alpha \\ \sin\alpha & \cos\alpha \end{bmatrix}$,

$B = \begin{bmatrix} \cos\beta & -\sin\beta \\ \sin\beta & \cos\beta \end{bmatrix}$,

then show that $AB = BA$.

5. If $A = \begin{bmatrix} 1 & 1 & -1 \\ 2 & 0 & 3 \\ 3 & -1 & 2 \end{bmatrix}$, $B = \begin{bmatrix} 1 & 3 \\ 0 & 2 \\ -1 & 4 \end{bmatrix}$ and

$C = \begin{bmatrix} 1 & 2 & 3 & -4 \\ 2 & 0 & -2 & 1 \end{bmatrix}$, then find $A(BC)$,

$(AB)C$ and show that $A(BC) = (AB)C$.

6. (i) If $A = \begin{bmatrix} 1 & 2 \\ 3 & 4 \end{bmatrix}$, $B = \begin{bmatrix} 2 & 1 \\ 4 & 2 \end{bmatrix}$,

$C = \begin{bmatrix} 5 & 1 \\ 7 & 4 \end{bmatrix}$, then verify that :

$$A(B+C) = AB + BC.$$

(ii) For the matrices :

$A = \begin{bmatrix} 3 & 2 \\ 1 & 5 \end{bmatrix}$, $B = \begin{bmatrix} 2 & -4 \\ -3 & 1 \end{bmatrix}$, $C = \begin{bmatrix} 1 & 2 \\ 3 & -1 \end{bmatrix}$

Verify $(A+B) C = AC + BC$.

7. If $\begin{bmatrix} 4 & 1 & 2 \\ 0 & 5 & 3 \end{bmatrix} \begin{bmatrix} 3 & 4 & 5 \\ -1 & 0 & -2 \\ 3 & 4 & 7 \end{bmatrix}$

$= \begin{bmatrix} 8x + 3y & 6z & 32 \\ 4 & 12 & 26x - 5y \end{bmatrix}$

find the values of x, y, z.

8. For what values of a, b, c, d, e the following matrices are same ?

$A = \begin{bmatrix} a & 1 & 2 \\ c & b & 3 \\ 1 & -1 & 0 \end{bmatrix}$, $B = \begin{bmatrix} 0 & e & 2 \\ 7 & 9 & d \\ e & -1 & a \end{bmatrix}$

9. If $A = \begin{bmatrix} 2 & 3 \\ 1 & -4 \end{bmatrix}$, $B = \begin{bmatrix} 1 & 0 \\ 0 & 1 \end{bmatrix}$, then find

(i) $10B$ (ii) $4A$

(iii) $A+B$ (iv) $A-B$

(v) $2A-3B$ (vi) $5A+3B$.

10. Find the product of

(i) $A = \begin{bmatrix} 1 & 2 \\ 3 & 4 \end{bmatrix}$, $B = \begin{bmatrix} 1 & 7 \\ 2 & 3 \end{bmatrix}$

(ii) $A = \begin{bmatrix} 1 & 2 & 1 \\ 4 & 0 & 2 \end{bmatrix}$, $B = \begin{bmatrix} 3 & -4 \\ 1 & 5 \\ -2 & 2 \end{bmatrix}$

(iii) $A = \begin{bmatrix} x \\ y \\ z \end{bmatrix}$, $B = \begin{bmatrix} x & y & z \end{bmatrix}$

11. If $A = \begin{bmatrix} 1 & 0 & 0 \\ 0 & 1 & 0 \\ 0 & 0 & 1 \end{bmatrix}$, then prove that $A^3 = A$.

12. Find the number of rows and columns in the following matrices :

(i) $\begin{bmatrix} 2 & 6 & 7 & 8 \\ 2 & 5 & 11 & 6 \end{bmatrix}$ (ii) $\begin{bmatrix} 1 & 1 & 1 & 1 \end{bmatrix}$

(iii) $\begin{bmatrix} 2 & -1 & 3 \\ 0 & 3 & 4 \\ 2 & 3 & 7 \\ 2 & 5 & 11 \end{bmatrix}$ (iv) $\begin{bmatrix} x \\ y \\ z \end{bmatrix}$

13. Find the order of the following matrices :

(i) $\begin{bmatrix} 1 & 1 & 1 & 1 \end{bmatrix}$ (ii) $\begin{bmatrix} 1 & a & b & 0 \\ 0 & c & d & 1 \\ 1 & a & b & 0 \end{bmatrix}$

(iii) $\begin{bmatrix} 2 \\ 3 \\ 4 \\ 5 \end{bmatrix}$ (iv) $\begin{bmatrix} 2 & 3 \\ 3 & 4 \\ 5 & 6 \\ 7 & 8 \end{bmatrix}$

14. Are the following matrices conformable for addition ?

(i) $\begin{bmatrix} 1 & 2 & 3 \\ 4 & 5 & 6 \\ 7 & 8 & 9 \end{bmatrix}$, $\begin{bmatrix} 6 & 4 \\ 7 & 4 \\ 7 & 3 \end{bmatrix}$

(ii) $\begin{bmatrix} 5 & 6 \\ 7 & 8 \end{bmatrix}$, $\begin{bmatrix} 9 & 10 \\ 11 & 12 \end{bmatrix}$

(iii) $\begin{bmatrix} 3 & 2 & -1 \\ 0 & 4 & 0 \end{bmatrix}$, $\begin{bmatrix} 2 & 1 \\ 5 & 2 \\ 7 & 8 \end{bmatrix}$

(iv) $\begin{bmatrix} x \\ y \\ z \end{bmatrix}$, $\begin{bmatrix} a \\ b \\ c \end{bmatrix}$

15. Find the additive inverse of matrix :

$$A = \begin{bmatrix} 1 & 2 & -7 & 5 \\ 0 & 5 & 0 & 8 \\ 0 & 0 & 0 & -8 \end{bmatrix}$$

16. Are the following matrices conformable for the product AB ?

(i) $A = \begin{bmatrix} 5 & 7 \\ 8 & 9 \end{bmatrix}$, $B = \begin{bmatrix} 3 & 5 & 7 \\ 1 & 0 & 1 \end{bmatrix}$

(ii) $A = \begin{bmatrix} -1 & 2 \\ 3 & 4 \end{bmatrix}$, $B = \begin{bmatrix} 3 \\ 4 \end{bmatrix}$

(iii) $A = \begin{bmatrix} 2 & 1 & 0 \\ 3 & 2 & 1 \\ 1 & 0 & 1 \end{bmatrix}$, $B = \begin{bmatrix} 1 & 2 & 3 & 4 \\ 2 & 0 & 1 & 2 \\ 3 & 1 & 0 & 5 \end{bmatrix}$

(iv) $A = \begin{bmatrix} 1 & 1 & 1 & 1 \end{bmatrix}$, $B = \begin{bmatrix} 1 \\ 1 \\ 1 \\ 1 \end{bmatrix}$

17. Find the values of x and y if :

$$\begin{bmatrix} x & 3 \\ 5 & x-y \end{bmatrix} = \begin{bmatrix} 2 & 3 \\ 5 & 1 \end{bmatrix}$$

18. Transpose the matrix :

$$A = \begin{bmatrix} 1 & 3 & 5 \\ -2 & 5 & 6 \\ 7 & 0 & 3 \end{bmatrix}.$$

19. If $A = \begin{bmatrix} 2 & 1 & 4 \\ 5 & -3 & 7 \end{bmatrix}$, $B = \begin{bmatrix} 1 & 2 & 3 \\ 2 & -4 & 6 \end{bmatrix}$

Find $2A - 3B$ and $B - \dfrac{A}{2}$. Also write down the unit matrix of order 3.

20. Prove that the matrix $\dfrac{1}{3} \begin{bmatrix} 1 & 2 & 2 \\ 2 & 1 & -2 \\ -2 & 2 & 1 \end{bmatrix}$ is orthogonal.

Answers

2. (i) $\begin{bmatrix} 28 & 33 \\ 26 & 11 \end{bmatrix}$ (ii) $\begin{bmatrix} -14 & -12 \\ 2 & -46 \end{bmatrix}$ **3.** (i) $\begin{bmatrix} 9 & 11 \\ 13 & 7 \\ 9 & 11 \end{bmatrix}$ (ii) $\begin{bmatrix} 4 & 1 & -1 \\ 8 & 3 & 10 \\ -1 & -2 & 5 \end{bmatrix}$

7. $x = 1, y = 3, z = 4$

8. $a = 0, b = 9, c = 7, d = 3, e = 1$

9. (i) $\begin{bmatrix} 10 & 0 \\ 0 & 10 \end{bmatrix}$ (ii) $\begin{bmatrix} 8 & 12 \\ 4 & -16 \end{bmatrix}$ (iii) $\begin{bmatrix} 3 & 3 \\ 1 & -3 \end{bmatrix}$

(iv) $\begin{bmatrix} 1 & 3 \\ 1 & -5 \end{bmatrix}$ (v) $\begin{bmatrix} 1 & 6 \\ 2 & -11 \end{bmatrix}$ (vi) $\begin{bmatrix} 13 & 15 \\ 5 & -17 \end{bmatrix}$

10. (i) $\begin{bmatrix} 5 & 13 \\ 11 & 33 \end{bmatrix}$ (ii) $\begin{bmatrix} 3 & 8 \\ 8 & -12 \end{bmatrix}$ (iii) $\begin{bmatrix} x^2 & xy & xz \\ yx & y^2 & yz \\ zx & zy & z^2 \end{bmatrix}$

12. (i) Rows = 2, Columns = 4   (ii) Row = 1, Columns = 4

(iii) Rows = 4, Columns = 3   (ii) Rows = 3, Column = 1.

13. (i) 1×4 (ii) 3×4 (iii) 4×1 (iv) 4×2

14. (i) No   (ii) Yes   (iii) No   (iv) Yes

15. $\begin{bmatrix} -1 & -2 & 7 & -5 \\ 0 & -5 & 0 & -8 \\ 0 & 0 & 0 & 8 \end{bmatrix}$ **16.** (i) Yes   (ii) Yes   (iii) Yes   (iv) Yes

17. $x = 2, y = 1$ **18.** $\begin{bmatrix} 1 & -2 & 7 \\ 3 & 5 & 0 \\ 5 & 6 & 3 \end{bmatrix}$ **19.** $\begin{bmatrix} 1 & -4 & -1 \\ 4 & 6 & -4 \end{bmatrix}, \begin{bmatrix} 0 & \frac{3}{2} & 1 \\ -\frac{1}{2} & -\frac{5}{3} & -\frac{5}{2} \end{bmatrix}, \begin{bmatrix} 1 & 0 & 0 \\ 0 & 1 & 0 \\ 0 & 0 & 1 \end{bmatrix}$

2.19 SUBMATRIX OF A MATRIX

Let A be a matrix of order $m \times n$, then a matrix obtained from A by removing some rows and columns, is called a submatrix of the matrix A.

For Example : (i) Consider a matrix $A = \begin{bmatrix} 1 & 2 & 3 & 6 \\ 5 & 7 & 9 & 9 \\ 4 & 5 & 6 & 12 \end{bmatrix}$ of order 3×4, then a matrix $\begin{bmatrix} 1 & 2 & 6 \\ 5 & 7 & 9 \end{bmatrix}$ is a submatrix of A, which is obtained from A by removing third column and fourth row.

(ii) Consider a matrix $A = \begin{bmatrix} 1 & 5 & 2 \\ 0 & 1 & 3 \\ 0 & 0 & 1 \end{bmatrix}$ of order 3×3, then a matrix

$$\begin{bmatrix} 1 & 5 \\ 0 & 1 \\ 0 & 0 \end{bmatrix}$$

is submatrix of A, which is obtained from A by removing third column.

REMARK

- If the given matrix A is a square matrix, then a square submatrix of A is known as principal submatrix.

2.20 MINORS OF A MATRIX

Let A be a matrix of order $m \times n$, then the determinant of every square submatrix of A is called a minor of A. If the order of the determinant of square submatrix of A is $r \times r$, then it is denoted as r-minor of A.

For example : (i) Consider a matrix

$$A = \begin{bmatrix} 1 & 2 & 3 \\ 2 & 4 & 6 \end{bmatrix}$$

Then all the 2-minors of A are

$$\begin{vmatrix} 1 & 2 \\ 2 & 4 \end{vmatrix}, \begin{vmatrix} 2 & 3 \\ 4 & 6 \end{vmatrix}, \begin{vmatrix} 1 & 3 \\ 2 & 6 \end{vmatrix}$$

(ii) Consider a matrix

$$A = \begin{bmatrix} 1 & 2 & 3 \\ 5 & 7 & 9 \\ 4 & 5 & 6 \\ 6 & 9 & 12 \end{bmatrix}$$

Then all the 3-minors of A are

$$\begin{vmatrix} 1 & 2 & 3 \\ 5 & 7 & 9 \\ 4 & 5 & 6 \end{vmatrix}, \begin{vmatrix} 1 & 2 & 3 \\ 5 & 7 & 9 \\ 6 & 9 & 12 \end{vmatrix}, \begin{vmatrix} 1 & 2 & 3 \\ 4 & 5 & 6 \\ 6 & 9 & 12 \end{vmatrix}, \begin{vmatrix} 5 & 7 & 9 \\ 4 & 5 & 6 \\ 6 & 9 & 12 \end{vmatrix}$$

2.21 RANK OF A MATRIX

Let A be a matrix of order $m \times n$, then a non-negative integer r is said to the rank of matrix A if it possesses the following two properties :

(i) There exists at least one r-minor of A which is not equal to zero.

(ii) Every s-minor of A for all $s > r$ is zero.

We denote the rank of A by $\rho(A)$.

In other words, the rank of a matrix is the order of any highest order of a non-zero minor of the matrix.

REMARKS

- If the order of a matrix A is $m \times n$, then $\rho(A) \leq$ min. $\{m, n\}$
- A is a null matrix iff $\rho(A) = 0$.
- If A is any non-zero matrix, then $\rho(A) \geq 1$.
- $\rho(A) \geq r$, if there exists a non-zero r-minor of A.
- For any square matrix A of order n, $\rho(A) = n$ iff A is non-singular.
- For any square matrix A of order n, $\rho(A) < n$ iff A is singular.
- $r(A) \leq r$ if every s-minor of A is zero, where $s > r$.
- Every $(r+1)$- rowed minor of A can be expressed as a linear combination of its r-rowed minors, therefore if every r-minor of A is zero, then its every $(r+1)$-minor is also zero.

2.22 ECHLEON FROM OF A MATRIX

A matrix A is said to be in Echelon form if :

  (i) every row of A has all its entries 0 which occurs below every row having a non-zero entry. and

  (ii) the number of zeros before the first non-zero entry in a row is less than the number of such zeros in the next row.

REMARK

- The rank of a matrix is equal to the number of non-zero rows in Echelon form of that matrix.
  For example: Consider a matrix

$$A = \begin{bmatrix} 0 & 2 & 3 & 5 \\ 0 & 0 & 3 & 2 \\ 0 & 0 & 0 & 0 \end{bmatrix}$$

Clearly, A is in Echelon form which has 2 non-zero rows, hence the rank of A is 2.

THEOREM 1. *The rank of the transpose of a matrix is equal to the rank of that matrix.*

Proof. Let A be a marix, then A' is its transpose and let $\rho(A) = r$, then there exists an r-rowed minor of A which is not equal to zero and all s-rowed minors of A are zero, where $s > r$. Let $|B|$ be a r-rowed minor of A such that $|B| \neq 0$. Since A' is the transpose of A, then $|B'|$ is the r-rowed minor of A' but $|B'| = |B| \neq 0$, therefore $\rho(A') \geq r$. Suppose there is an s-minor $|C|$ of A' such that $|C| \neq 0$, where $s > r$, then $|C'|$ will be an s-minor of A such that $|C'| = |C| \neq 0$, therefore $\rho(A) > r$ which is a contradiction, hence $\rho(A') = r$.

Solved Examples

Based on the following Results

➠ The rank of a matrix is the order of any highest order of a non-zero minor of the matrix.

➠ A matrix A is said to be in Echelon form if
  (i) every row of A has all its entries 0 which occurs below every row having a non-zero entry. and (ii) the number of zeros before the first non-zero entry in a row is less than the number of such zeros in the next row.

➠ The rank of a matrix is equal to the number of non-zero rows in Echelon form.

Example 1. *Find the rank of the following matrices :*

  (i) $\begin{bmatrix} 3 & 0 & 0 \end{bmatrix}$

  (ii) $\begin{bmatrix} 1 & 2 & 3 \\ 2 & 4 & 5 \end{bmatrix}$

  (iii) $\begin{bmatrix} 1 & 2 & 3 \\ 3 & 4 & 5 \\ 4 & 5 & 6 \end{bmatrix}$

  (iv) $\begin{bmatrix} 1 & 5 & 2 & 4 \\ 0 & 1 & 3 & 1 \\ 0 & 0 & 1 & 3 \end{bmatrix}$

Solution. (i) Let $A = \begin{bmatrix} 3 & 0 & 0 \end{bmatrix}$, then A is the non-zero rowed matrix, thereofore $\rho(A) \geq 1$.
  Also A is a matrix of order of 1×3, then $\rho(A) \leq 1$, hence $\rho(A) = 1$.

  (ii) Let $A = \begin{bmatrix} 1 & 2 & 3 \\ 2 & 4 & 5 \end{bmatrix}$

  The order of A is 2×3, then $\rho(A) \leq 2$.

Also there is a 2-minor $\begin{vmatrix} 2 & 3 \\ 4 & 5 \end{vmatrix}$ of A which is not equal to zero, then $\rho(A) \geq 2$, hence $\rho(A) = 2$.

(iii) Let $A = \begin{bmatrix} 1 & 2 & 3 \\ 3 & 4 & 5 \\ 4 & 5 & 6 \end{bmatrix}$. The order of A is 3×3, then $\rho(A) \leq 3$.

Now $|A| = \begin{vmatrix} 1 & 2 & 3 \\ 3 & 4 & 5 \\ 4 & 5 & 6 \end{vmatrix} = 1(24 - 25) - 2(18 - 20) + 3(15 - 16) = 0$ '

$\therefore$ The only 3-minor $|A|$ of A is zero, thus $\rho(A) < 3$. Further, there is a 2-minor $\begin{vmatrix} 1 & 2 \\ 3 & 4 \end{vmatrix}$ of A which is not equal to zero, hence $\rho(A) = 2$.

(iv) Let $A = \begin{bmatrix} 1 & 5 & 2 & 4 \\ 0 & 1 & 3 & 1 \\ 0 & 0 & 1 & 3 \end{bmatrix}$

The order of A is 3×4, then $\rho(A) \leq 3$.

Now there is a 3-minor $\begin{vmatrix} 1 & 5 & 2 \\ 0 & 1 & 3 \\ 0 & 0 & 1 \end{vmatrix}$ of A which is not equal to zero, then $\rho(A) \geq 3$.

Hence $\rho(A) = 3$.

Example 2. *Let A and B be two square matrices of order n. If $\rho(A) = \rho(B) = n$, then prove that $\rho(AB) = n$ and conversely.*

Solution. Suppose $\rho(A) = \rho(B) = n$, then both A and B are non-singular.

$\therefore$ $\qquad\qquad\qquad\qquad |A| \neq 0$ and $|B| \neq 0$

$\Rightarrow$ $\qquad\qquad\qquad\qquad |AB| = |A||B| \neq 0$

Since the order of AB is n and $|AB| \neq 0$, therefore $\rho(AB) = n$.

Conversely, suppose that $\qquad \rho(AB) = n$

$\therefore$ $\qquad\qquad\qquad\qquad |AB| \neq 0$

$\Rightarrow$ $\qquad\qquad\qquad\qquad |A||B| \neq 0$

$\Rightarrow$ $\qquad\qquad\qquad\qquad |A| \neq 0$ and $|B| \neq 0$

$\Rightarrow$ $\qquad\qquad\qquad\qquad \rho(A) = n, \rho(B) = n.$

Example 3. *Prove that every skew-symmetric matrix of odd order has rank less than its order.*

Solution. Let A be a skew-symmetric matrix of order n, where n is an odd natural number, then $A' = -A$.

Now $\qquad\qquad\qquad\qquad A' = -A$

$\Rightarrow$ $\qquad\qquad\qquad |A'| = |-A|$

$\Rightarrow$ $\qquad\qquad\qquad |A| = |(-1)A|$ $\qquad\qquad [\because |A'| = |A|]$

$\Rightarrow$ $\qquad\qquad\qquad |A| = (-1)^n |A|$

$\Rightarrow$ $\qquad\qquad\qquad |A| = -|A|$ $\qquad\qquad\qquad [\because n$ is odd$]$

$$\Rightarrow \qquad 2|A| = 0$$

$$\Rightarrow \qquad |A| = 0$$

$$\Rightarrow \qquad \rho(A) \neq n$$

But $\rho(A) \leq n$, hence $\qquad \rho(A) < n.$

Example 4. *If A be a non-zero column and B is a non-zero row matrix, then show that $\rho(AB)=1$.*

Solution. Let
$$A = \begin{bmatrix} a_{11} \\ a_{21} \\ \vdots \\ a_{m1} \end{bmatrix} \text{ and } B = \begin{bmatrix} b_{11} & b_{12} & \cdots & b_{1n} \end{bmatrix}$$

be two non-zero column and row matrices respectively.

Then we have
$$AB = \begin{bmatrix} a_{11} \\ a_{21} \\ \vdots \\ a_{m1} \end{bmatrix} \begin{bmatrix} b_{11} & b_{12} & \cdots & b_{1n} \end{bmatrix}$$

$$= \begin{bmatrix} a_{11}b_{11} & a_{11}b_{12} & \cdots & a_{11}b_{1n} \\ a_{21}b_{11} & a_{21}b_{12} & \cdots & a_{21}b_{1n} \\ \vdots & \vdots & \vdots & \vdots \\ a_{m1}b_{11} & a_{m1}b_{12} & \cdots & a_{m1}b_{1n} \end{bmatrix}$$

Clearly, AB is a matrix of order $m \times n$ and AB is a non-zero matrix since A and B are non-zero matrices,

then $\qquad \rho(AB) \geq 1 \qquad \qquad ...(1)$

also every 2-minor of AB vanishes, then

$$\rho(AB) \leq 1 \qquad \qquad ...(2)$$

From (1) and (2) we have

$$\rho(AB) = 1.$$

Example 5. *If A is n-rowed square matrix of rank $n-1$, then show that adj A is non-zero matrix.*

Solution. Since the rank of A is $n-1$, i.e., $\rho(A) = n-1$, then there exists a non-zero $(n-1)$ minor of A, therefore there exists at least one element of $adj A$ which is non-zero, hence $adj A$ is a non-zero matrix.

Example 6. *Let A be a square matrix of order n. Show that $\rho(adj. A)$ is n or 0 in accordance with $\rho(A)$ is n or less than $n-1$.*

Solution. Suppose $\rho(A) = n$, then $|A| \neq 0$

But we know that

$$|adj. A| = |A|^{n-1}$$

$$\Rightarrow \qquad |adj. A| \neq 0 \qquad \qquad [\because |A| \neq 0]$$

$$\therefore \qquad \rho(adj. A) = n, \text{ since the order of } adj A = n.$$

Next, when $\rho(A) < n - 1$, then $|A| = 0$ and every r-minor of A is zero, where $r \geq n-1$, therefore every element of $adj\ A$ is zero so that $adj.\ A$ is a null matrix, hence

$$\rho(adj.\ A) = 0.$$

Example 7. *Find the value of x so that $\rho(A) \leq 2$, where A is the matrix given by*

$$\begin{bmatrix} 3x-8 & 3 & 3 \\ 3 & 3x-8 & 3 \\ 3 & 3 & 3x-8 \end{bmatrix}$$

Solution. Since $\rho(A) \leq 2$, then $|A| = 0$ because A is a square matrix of order 3×3.

Now $|A| = 0$

$$\Rightarrow \quad \begin{bmatrix} 3x-8 & 3 & 3 \\ 3 & 3x-8 & 3 \\ 3 & 3 & 3x-8 \end{bmatrix}$$

$\Rightarrow \quad (3x-8)\{(3x-8)(3x-8)-9\} - 3\{3(3x-8)-9\} + 3\{9 - 3(3x-8)\} = 0$

$\Rightarrow \quad (3x-8)^3 - 9(3x-8) - 9(3x-8) + 27 + 27 - 9(3x-8) = 0$

$\Rightarrow \quad (3x-8)^3 - 27(3x-8) + 54 = 0$

$\Rightarrow \quad (3x-5)^2(3x-2) = 0$

$\Rightarrow \quad 3x-2 = 0 \text{ or } 3x-5 = 0$

$\Rightarrow \quad x = \dfrac{2}{3} \text{ or } x = \dfrac{5}{3}$

When $x = \dfrac{2}{3}$, then $A = \begin{bmatrix} -6 & 3 & 3 \\ 3 & -6 & 3 \\ 3 & 3 & -6 \end{bmatrix}$, clearly there is a 2-minor $\begin{vmatrix} -6 & 3 \\ 3 & -6 \end{vmatrix}$ of A,

which is non-zero, hence $\rho(A) = 2$.

Again, when $x = \dfrac{5}{3}$, then $A = \begin{bmatrix} -3 & 3 & 3 \\ 3 & -3 & 3 \\ 3 & 3 & -3 \end{bmatrix}$

Clearly, there is a 2-minor $\begin{vmatrix} -3 & 3 \\ 3 & -3 \end{vmatrix}$ of A, which is non-zero, hence $\rho(A) = 2$.

❧ EXERCISE 2.2

1. Find the rank of the following matrices :

(i) $\begin{bmatrix} 0 & 0 \\ 0 & 0 \end{bmatrix}$

(ii) $\begin{bmatrix} 5 & 10 \\ 3 & 6 \end{bmatrix}$

(iii) $\begin{bmatrix} 1 & -3 & 4 & 7 \\ 9 & 1 & 2 & 0 \end{bmatrix}$

(iv) $\begin{bmatrix} 1 & 2 & 3 \\ 2 & 1 & 0 \\ 0 & 1 & 2 \end{bmatrix}$

(v) $\begin{bmatrix} 1 & 2 & -7 & 5 \\ 0 & 5 & 0 & 8 \\ 0 & 0 & 0 & -3 \end{bmatrix}$

(vi) $\begin{bmatrix} 0 & 1 & 2 & 1 \\ 1 & 2 & 3 & 2 \\ 3 & 1 & 1 & 3 \end{bmatrix}$

(vii) $\begin{bmatrix} 1 & 2 & 3 & 4 \\ 2 & 4 & 6 & 8 \\ 3 & 6 & 9 & 12 \end{bmatrix}$

(viii) $\begin{bmatrix} 1 & 5 & 4 & 6 \\ 2 & 7 & 5 & 9 \\ 3 & 9 & 6 & 12 \end{bmatrix}$

(ix) $\begin{bmatrix} 1 & x & x^2 \\ 1 & y & y^2 \\ 1 & z & z^2 \end{bmatrix}$

(x) $\begin{bmatrix} 1 & 1 & 1 & 1 \\ 1 & 1 & 1 & 1 \\ 1 & 1 & 1 & 1 \\ 1 & 1 & 1 & 1 \end{bmatrix}$

(xi) $\begin{bmatrix} 1 & 0 & 0 & 0 \\ 0 & 1 & 0 & 0 \\ 0 & 0 & 1 & 0 \\ 0 & 0 & 0 & 1 \end{bmatrix}$

2. If $A = \begin{bmatrix} 0 & 1 & 0 & 0 \\ 0 & 0 & 1 & 0 \\ 0 & 0 & 0 & 1 \\ 0 & 0 & 0 & 0 \end{bmatrix}$,

find $\rho(A)$, $\rho(A^2)$, $\rho(A^3)$ and $\rho(A^4)$.

3. Show that the rank of a matrix does not alter on affixing any number of additional rows or columns of zeros.

4. Show that the rank of a matrix is greater than or equal to the rank of its every sub-matrix.

5. Find the rank of A, B, $A+B$ and AB, where

$$A = \begin{bmatrix} 1 & 1 & -1 \\ 2 & -3 & 4 \\ 3 & -2 & 3 \end{bmatrix} \text{ and } B = \begin{bmatrix} -1 & -2 & -1 \\ 6 & 12 & 6 \\ 5 & 10 & 5 \end{bmatrix}$$

Answers

1. (i) 0 (ii) 1 (iii) 2 (iv) 2 (v) 3 (vi) 3 (vii) 1 (viii) 2 (ix) Rank $= 3$ if $x \neq y \neq z$; Rank $= 2$ if only two of x, y, z are different; Rank $= 1$ if $x = y = z$. (x) 1 (xi) 4

2. $\rho(A) = 3$, $\rho(A^2) = 2$, $\rho(A^3) = 1$, $\rho(A^4) = 0$

5. $r(A) = 2$, $\rho(B) = 1$, $\rho(A+B) = 2$, $\rho(AB) = 0$

2.23 ELEMENTARY TRANSFORMATIONS (OR E-TRANSFORMATIONS) OF A MATRIX

Consider the matrices

$$A = \begin{bmatrix} 1 & 2 & -3 \\ 3 & 0 & 1 \end{bmatrix}, B = \begin{bmatrix} 3 & 0 & 1 \\ 1 & 2 & -3 \end{bmatrix}, C = \begin{bmatrix} -3 & 2 & 1 \\ 1 & 0 & 3 \end{bmatrix}, D = \begin{bmatrix} 4 & 8 & -12 \\ 3 & 0 & 1 \end{bmatrix}$$

$$E = \begin{bmatrix} -3 & 2 & 7 \\ 1 & 0 & 21 \end{bmatrix}, F = \begin{bmatrix} 1 & 2 & -3 \\ 6 & 6 & -8 \end{bmatrix} \text{ and } G = \begin{bmatrix} 7 & 0 & 1 \\ -11 & 0 & -3 \end{bmatrix}$$

From above matrices, we observe that :

(1) B can be obtained from A by interchanging the first and second row.

(2) C can be obtained from A by interchanging the first and third column.

(3) D can be obtained from A by multiplying each element of the first row by 4.

(4) E can be obtained from C by multiplying each element of third column by 7.

(5) F can be obtained from A by adding to the elements of second row, 3 times the corresponding elements of the first row.

(6) G can be obtained from B by adding to the elements of first column, 4 times the corresponding elements of the third column.

Such transformations as performed above are known as elementary transformations (or E-operations or E-transformations).

Elementary transformations on rows are known as elementary row transformations whereas the transformations on columns are known as elementary column transformations.

Thus we may define E-transformations as follows:

Definition: *An elementary transformation (or E-transformation) is an operation of any of the following types:*

(i) *The interchange of any two rows (or columns).*

(ii) *The multiplication of any row (or column) by any non-zero number.*

(iii) *The addition of non-zero scalar multiple of any row (or column) to another row (or column).*

2.23.1 NOTATIONS FOR E-TRANSFORMATIONS

E-transformations can be denoted by the following notations:

   (i)   The transformation of interchanging i^{th} and j^{th} row of a matrix is denoted by $R_i \leftrightarrow R_j$.

   (ii) The transformation of interchanging i^{th} and j^{th} column is denoted by $C_i \leftrightarrow C_j$.

   (iii) The transformation of multiplication of ith row of a matrix by non-zero scalar k is denoted by $R_i \rightarrow kR_i$.

   (iv) The transformation of multiplication of jth column by a non-zero scalar k is denoted by $C_{ij} \rightarrow kC_j$.

   (v)   The transformation of addition of a non-zero scalar k multiple of jth row to another i^{th} row of a matrix is denoted by $R_i \rightarrow R_i + kR_j \ (i \neq j)$.

   (vi) The transformation of addition of a non-zero scalar k multiple of jth column to another i^{th} column of a matrix is denoted by $C_i \rightarrow C_i + kC_j \ (i \neq j)$.

2.24 ELEMENTARY MATRICES

A matrix which is obtained from a unit (identity) matrix by a single E-transformation is known as an elementary matrix.

For example : Consider the matrices

$$\begin{bmatrix} 0 & 0 & 1 \\ 0 & 1 & 0 \\ 1 & 0 & 0 \end{bmatrix}, \begin{bmatrix} 1 & 0 & 0 \\ 0 & 1 & 0 \\ 3 & 0 & 1 \end{bmatrix}, \begin{bmatrix} 1 & 0 & 0 \\ 0 & 3 & 0 \\ 0 & 0 & 1 \end{bmatrix}$$

Clearly, these matrices are elementary matrices because these are obtained from a unit matrix I_3 (the identity matrix of order 3×3) by performing the E-transformations $C_1 \rightarrow C_3$, $R_3 \rightarrow R_3 + 3R_1$ and $R_2 \rightarrow 3R_2$.

The elementary matrices of different types can be denoted by the following notations:

   (i)   The elementary matrix obtained by interchanging i^{th} and j^{th} rows (or columns) of a unit matrix is denoted by E_{ij}.

   (ii) The elementary matrix obtained by multiplying i^{th} row (or column) of a unit matrix by a non-zero scalar k is denoted by $E_i(k)$.

   (iii) The elementary matrix obtained by adding a non-zero scalar k multiple of jth row (or column) to i^{th} row (or column) of a unit matrix is denoted by $E_{ij}(k)$.

Obviously, $|E_{ij}| = -1, \ | E_i(k) | = k \neq 0$ and $| E_{ij}(k) | = 1$

Hence we can say that all the elementary matrices are non-singular and hence they possess their inverse.

THEOREM 1. *Every E-row (column) transformation of a matrix can be obtained by pre-multiplication (post-multiplication) with the corresponding elementary matrix.*

Proof.   Let A be a matrix of order $m \times n$, then we can write $A = I_m A I_n$

So that any elementary row transformation can be obtained by subjecting the pre-factor I_m and the same elementary column transformation can be obtained by subjecting the post-factor I_n.

In order to prove this result we shall first prove that any E-row transformation a product AB can be obtained by subjecting the pre-factor A to the same E-row transformation and any E-column transformation of a product AB can be obtained by subjecting the post-factor B to the same E-column transformation, where B is a matrix of order $n \times p$.

Let $A = [a_{ij}]_{m \times n}$ and $B = [b_{ij}]_{n \times p}$, then the product AB is conformable.

We can write A and B as follows :

$$A = \begin{bmatrix} R_1 \\ R_2 \\ \vdots \\ R_m \end{bmatrix} \text{ and } B = \begin{bmatrix} C_1 & C_2 & \cdots & C_p \end{bmatrix}$$

where $R_1, R_2, ..., R_m$ are row vectors of A and $C_1, C_2, ..., C_p$ are column vectors of B.

Now
$$AB = \begin{bmatrix} R_1C_1 & R_1C_2 & \cdots & R_1C_p \\ R_2C_1 & R_2C_2 & \cdots & R_2C_p \\ \vdots & \vdots & & \vdots \\ R_mC_1 & R_mC_2 & \cdots & R_mC_p \end{bmatrix}_{m \times p}$$

If σ be any E-row transformation, then
$$(\sigma A)B = \sigma(AB)$$

[**Note:** If σ denotes the operation $R_1 \leftrightarrow R_2$, then

$$\sigma A = \begin{bmatrix} R_2 \\ R_1 \\ \vdots \\ R_m \end{bmatrix}$$

$\therefore \qquad (\sigma A)B = \begin{bmatrix} R_2 \\ R_1 \\ \vdots \\ R_m \end{bmatrix} \begin{bmatrix} C & C_2 & \cdots & C_p \end{bmatrix} = \begin{bmatrix} R_2C_1 & R_2C_2 & \cdots & R_2C_p \\ R_1C_1 & R_1C_2 & \cdots & R_1C_p \\ \vdots & & & \\ R_mC_1 & R_mC_2 & \cdots & R_mC_p \end{bmatrix}$

Clearly $\qquad (\sigma A)B = \sigma(AB)$

Similarly, if the columns $C_1, C_2, ..., C_p$ of B be subjected to any E-column transformation, the columns of AB are also subjected to the same E-column transformation.

i.e., $\qquad \sigma(\sigma B) = \sigma(AB)$ where σ denotes $C_1 \leftrightarrow C_2$.

Now we move to main theorem, if A is a matrix of order $m \times n$, then we can write
$$A = I_m A \text{ where } I_m \text{ is a unit matrix of order } n \times m.$$

If σ be any E-row transformation, then $\quad \sigma(I_m A) = (\sigma I_m)A = EA$

where E is the elementary matrix corresponding to the same row transformation σ.

Similarly, we can also write $A = AI_m$

and if σ dentoes the E-column transformation, then
$$\sigma(AI_m) = A(\sigma I_m) = AE_1$$

where E_1 is the elementary matrix corresponding to the same column transformation.

Hence the theorem.

2.25 INVARIANCE OF RANK UNDER E-TRANSFORMATIONS

THEOREM 1. *Elementary transformation (E-transformation) do not change the rank of matrix.*

Proof. Since we know that *E*- transformations are of three types. Therefore, we shall prove this theorem for three cases.

Case I. *Interchanging the rows (or columns) does not change the rank.*

Let A be a matrix of order $m \times n$ of rank r and let B be a matrix obtained from A by interchanging the rows R_i and R_j, *i.e.*, by *E*-transformation $R_i \leftrightarrow R_j$. Let the rank of B be s. Then we shall prove that

$$r = s.$$

Since rank of $A = r$. This implies that A contains at least one. r-rowed square submatrix with non-zero determinant. Let it be R *i.e.*, det $R \neq 0$. Let us suppose S to be the r-rowed square submatrix of B having the same rows as are in R though these rows may be in different positions. Then either

$$\det S = \det R \text{ or } \det S = - \det R$$

But $$\det R \neq 0 \Rightarrow \det S \neq 0$$

$\therefore$ Rank of $B \geq r \Rightarrow s \geq r.$ ...(1)

Further, since the matrix A can also be obtained from B by *E*-transformation $R_i \leftrightarrow R_j$. Then we have

$$r \geq s \qquad \qquad \qquad ...(2)$$

Hence from (1) and (2) we conclude that $r = s$.

Case II. *Multiplication of the elements of a row by a non-zero number does not change the rank.*

Let A be a matrix of order $m \times n$ of rank r and let B be a matrix obtained from A by *E*- transformation $R_i \rightarrow kR_i$ where $k \neq 0$ and let rank of B be s. Therefore we shall prove that $s = r$. Suppose B_0 is an $(r + 1)$-rowed square submatrix of B, then there exists A_0 of $(r+1)$-rowed square submatrix of A such that either

$$\det B_0 = \det A_0 \text{ or } \det B_0 = k \det A_0$$

But rank of $A = r$ this means that every $(r+1)$-rowed square submatrix of A has zero determinant.

$\therefore$ $|A_0| = 0 \Rightarrow |B_0| = 0$

$\Rightarrow$ Every $(r+1)$-rowed square submatrix will have zero determinant.

$\Rightarrow$ Rank of B cannot exceed the rank of A

$\Rightarrow$ $s \leq r.$ ...(1)

Further, since the matrix A can also be obtained from B by *E*-transformation

$$R_i \rightarrow \left(\frac{1}{k}\right) R_i .$$

Thus we have

$$r \leq s \qquad \qquad \qquad ...(2)$$

Hence from (1) and (2) we conclude that

$$r = s.$$

Case III. *Addition of any row to the product of any number k and other row does not change the rank.*

Let the rank of a matrix A of order $m \times n$ be r and let B is obtained by the E-transformation $R_i \to R_i + kR_j$ and let rank of B be s. Then we shall prove that

$$s = r.$$

Now, if B_0 is an $(r + 1)$-rowed square submatrix of B, there exists uniquely A_0 an $(r+1)$-rowed square submatrix of A.

Since we know that any E-transformation does not change the determinant value. Therefore if no row of A_0 is a part of ith row of A, or if two rows of A_0 are the parts of the ith and j^{th} rows of A, then det B_0 = det A_0.

But   the rank of $A = r \Rightarrow$ det $A_0 = 0 \Rightarrow$ det $B_0 = 0$.

Now suppose if a row of A_0 is a part of ith row of A and not row is a part of jth row, then

$$\det B_0 = \det A_0 + k \det C_0$$

where C_0 is an $(r + 1)$-rowed square submatrix which is obtained from A_0 by E-transformation $R_i \to R_i + kR_j$.

Clearly, all the $(r +1)$ rows of C_0 are exactly same as the rows of some $(r + 1)$-rowed square submatrix of A, though in some different position. Therefore det C_0 is ± 1 times det of some $(r+1)$-rowed square submatrix A. But the rank of A is r. This implies every $(r+ 1)$-rowed square submatrix will have zero determinant.

$\therefore$ det $A_0 = 0$, det $C_0 = 0 \Rightarrow$ det $B_0 = 0$

hence rank of B cannot exceed the rank of A

$\therefore$ $s \le r$ ... (1)

Further, since A can also be obtained from B by E-transformation $R_i \to R_i - KR_j$, therefore we have

$$r \le s$$ ... (2)

From (1) and (2) we conclude that

$$r = s.$$

REMARKS

- The rank of a matrix does not change by a series of E-transformation.
- The rank of a matrix does not change by a column-transformation.

2.26 NORMAL FORM

Definition: *If a matrix is reduced to the form* $\begin{pmatrix} I_r & O \\ O & O \end{pmatrix}$ *. Then this form is called normal form of the given matrix.*

THEOREM 1. *Every matrix of order $m \times n$ of rank r can be reduced to the form* $\begin{pmatrix} I_r & O \\ O & O \end{pmatrix}$ *by a finite number of E-transformations, where I_r is the unit matrix of order $r \times r$.*

Proof. Let $A = [a_{ij}]_{m \times n}$ be a matrix of order $m \times n$ and of rank r. If A is a zero matrix, then its rank is zero and thus A can be written as $\begin{pmatrix} I_r & O \\ O & O \end{pmatrix}$.

Let us suppose A is a non-zero matrix. It means that it has at least one of its elements non-zero. Let this non-zero element be $a_{ij} = k \neq 0$.

Let B be a matrix which is obtained from A by E-transformations $R_1 \leftrightarrow R_i$ and and $C_1 \leftrightarrow C_j$ and whose leading element is k. Again using the E-transformation $R_1 \rightarrow \dfrac{1}{K} R_1$ on B we get a matrix C whose leading element becomes 1. Let this matrix C be

$$C = \begin{bmatrix} 1 & c_{12} & c_{13} & \cdots & c_{1n} \\ c_{21} & c_{22} & c_{23} & \cdots & c_{2n} \\ c_{31} & c_{32} & c_{33} & \cdots & c_{3n} \\ \cdots & \cdots & \cdots & \cdots & \cdots \\ c_{m1} & c_{m2} & c_{m3} & \cdots & c_{mn} \end{bmatrix}_{m \times n}$$

Now subtracting first column after multiplying by suitable number from remaining columns of C and subtracting first row after multiplying by suitable number from remaining rows of C, we obtain a matrix D whose elements of the first row and first column are zero except the leading element. Let D be given as

$$D = \begin{bmatrix} 1 & 0 & 0 & \cdots & 0 \\ 0 & & & & \\ 0 & & A_1 & & \\ \vdots & & & & \\ 0 & & & & \end{bmatrix}_{m \times n}$$

where A_1 is a matrix of order $(m-1) \times (n-1)$.

If this matrix A_1 is non-zero matrix, then we shall apply above process on A_1. Since we know that E-transformation will not effect the first row and first column of D, so that we shall apply E-transformations on D and there is no need to take A_1 separately. Continuing this process finitely we obtain a matrix M such that

$$M = \begin{pmatrix} I_k & O \\ O & O \end{pmatrix}$$

This implies that matrix M has a rank k. But M is obtained from A by a finite number of E-transformations and we know that E-tansformations do not change the rank, therefore k must be equal to r.

Hence the matrix A of order $m \times n$ of rank r can be reduced to the form $\begin{pmatrix} I_r & O \\ O & O \end{pmatrix}$ by a finite number of E-transformations.

REMARK

- The form $\begin{pmatrix} I_r & O \\ O & O \end{pmatrix}$ of A is also called first canonical form.

Corollary 1. *The rank of a matrix of order $m \times n$ is r if and only if it can be reduced to the form $\begin{bmatrix} I_r & O \\ O & O \end{bmatrix}$ by a finite chain of E-transformations.*

Proof. The condition is necessary : The proof is the same as that of above theorem.

The condition is sufficient: Suppose the matrix of order $m \times n$ is reduced to the form $\begin{bmatrix} I_r & O \\ O & O \end{bmatrix}$ by a finite chain of E-transformations, since we know that E-transformations do not change the rank of the matrix and the rank of the matrix $\begin{bmatrix} I_r & O \\ O & O \end{bmatrix}$ is r, hence the rank of the given matrix is r.

Corollary 2. *If A is a matrix of order $m \times n$ of rank r, then there exist two non-singular matrices P and Q such that*

$$PAQ = \begin{bmatrix} I_r & O \\ O & O \end{bmatrix}$$

Proof. The matrix A can be reduced to the form $\begin{bmatrix} I_r & O \\ O & O \end{bmatrix}$ by a finite chain of E-transformations and E-row (column) transformations are equivalent to pre (post) multiplication by the corresponding elementary matrices $P_1, P_2,..., P_s, Q_1, Q_2,..., Q_t$, such that

$$P_s P_{s-1} ... P_2 P_1 A \, Q_1 \, Q_2 ... Q_{t-1} \, Q_t = \begin{bmatrix} I_r & O \\ O & O \end{bmatrix}$$

Let $P = P_s P_{s-1} ... P_2 P_1$ and $Q_1, Q_2,..., Q_t$, then

$$PAQ = \begin{bmatrix} I_r & O \\ O & O \end{bmatrix}$$

2.27 EQUIVALENCE OF MATRICES

Let A be a matrix of order $m \times n$. If a matrix B of order $m \times n$ is obtained from A by a finite chain of elementary transformation on A, then A is said to be equivalent to B. We write symbolically as $A \sim B$ which is read as 'A is equivalent to B'.

THEOREM 1. *The relation '$\sim$' in the set of all $m \times n$ matrices is an equivalent relation.*

Proof. We shall prove that the relation '$\sim$' is

    (i) reflexive   (ii) symmetric   (iii) transitive.

  (i) Reflexivity: Let A be an $m \times n$ matrix, then A can be obtained from A itself by the elementary transformation $R_i \rightarrow R_i$, for all $i = 1, 2, ... , m$

$$\therefore \qquad\qquad A \sim A$$

  (ii) Symmetry: Let A and B be any two $m \times n$ matrices such that $A \sim B$.

    Now $A \sim B \Rightarrow B$ can be obtained from A by a finite chain of elementary transformations on A

    $\Rightarrow A$ can also be obtained from B by a finite chain of elementary transformations on B.

$$\Rightarrow \qquad\qquad B \sim A$$

    i.e., If $A \sim B$, then $\qquad B \sim A$.

  (iii) Transitivity: Let A, B and C be any three $m \times n$ matrices such that $A \sim B$ and $B \sim C$.

    $A \sim B \Rightarrow B$ can be obtained from A by a finite chain of elementary transformations on A.

    $B \sim C \Rightarrow C$ can be obtained from B by a finite chain of elementary transformations on B.

On combining these two statements we can say that C can also be obtained from A by a finite chain of elementary transformations on A.

$$\therefore \qquad\qquad A \sim C$$

i.e., $\qquad\qquad A \sim B, B \sim C \Rightarrow A \sim C$

Hence the relation '$\sim$' is an equivalence relation.

2.28 ROW AND COLUMN EQUIVALENCE OF MATRICES

(i) **Row equivalence of matrix :** A matrix A is said to be row equivalent to a matrix B, if B can be obtained from A by a finite chain of elementary row transformations on A and we write $A \overset{R}{\sim} B$.

(ii) **Column equivalence of matrix :** A matrix A is said to be column equivalent to a matrix B, if B can be obtained from A by a finite chain of elementary column transformations on A, we write $A \overset{C}{\sim} B$.

Theorem 1. **(Employment of only row transformations) :**

Let A be a matrix of order $m \times n$ of rank r, then there exists a non-singular matrix P such that

$$PA = \begin{bmatrix} G \\ O \end{bmatrix}$$

where G is a matrix of order $r \times n$ of rank r and O is a null matrix of order $(m-r) \times n$.

Proof. Since A is a matrix of order $m \times n$ of rank r, then there exist two non-singular matrices P and Q such that

$$PAQ = \begin{bmatrix} I_r & O \\ O & O \end{bmatrix} \qquad\qquad ...(1)$$

Further, since every non-singular matrix can be expressed as the product of elementary matrices.

So let $\qquad\qquad Q = Q_1 Q_2 \ldots Q_t$

where $Q_1, Q_2, ..., Q_t$ are elementary matrices. Now (1) can be written as

$$PAQ_1 Q_2 \ldots Q_t = \begin{bmatrix} I_r & O \\ O & O \end{bmatrix} \qquad\qquad ...(2)$$

Again, every elementary column transformation of a matrix is equivalent to post-multiplication with the corresponding elementary matrix. Since no column transformation can affect the last $(m - r)$ rows of RHS of (2), therefore post-multiplying the LHS of (2) by elementary matrices $Q_t^{-1}, Q_{t-1}^{-1}, ..., Q_2^{-1}, Q_1^{-1}$ successively and effecting the corresponding column transformations in RHS of (2), we get

$$PA = \begin{bmatrix} G \\ O \end{bmatrix}$$

Since the elementary transformations do not change the rank, therefore the rank of PA is the same as the rank of A which is r, thus the rank of $\begin{bmatrix} G \\ O \end{bmatrix}$ is r. Hence the rank of G is r as G has r rows and the last $(m - r)$ rows of the matrix $\begin{bmatrix} G \\ O \end{bmatrix}$ consist only zero entries, i.e., O is a null matrix of order $(m - r) \times n$.

THEOREM 2. (Employment of only column transformations)

Let A be a matrix of order m × n of rank r, then there exists a non-singular matrix Q such that

$$AQ = [H \quad O]$$

where H is a matrix of order m × n and O is a null matrix of order m × (n − r).

Proof. Since A is a matrix of order $m \times n$, then there exist two non-singular matrices P and Q such that

$$PAQ = \begin{bmatrix} I_r & O \\ O & O \end{bmatrix} \qquad \text{...(1)}$$

Further, since every non-singular matrix can be expressed as the product of elementary matrices.

So let $\qquad P = P_1 P_2 \dots P_s$

where $P_1 P_2 \dots P_s$ are elementary matrices.

Now (1) can be written as

$$P_1 P_2 \dots P_s\, AQ = \begin{bmatrix} I_r & O \\ O & O \end{bmatrix} \qquad \text{...(2)}$$

Again, every elementary row transformation of a matrix is equivalent to pre-multiplication with the corresponding elementary matrix. Since no row transformation can affect the last $(n - r)$ columns of RHS of (2), therefore premultiplying the LHS of (2) by elementary matrices $P_1^{-1}, P_2^{-1}, \dots, P_s^{-1}$ successively and effecting the corresponding row transformations in RHS of (2), we get

$$AQ = [H \quad O]$$

Since the elementary transformations do not change the rank, therefore the rank of AQ is the same as the rank of A, which is r, thus the rank of $[H \quad O]$ is r, hence the rank of H is r as H has r columns and the last $(n - r)$ column of the matrix $[H \quad O]$ consists only zero entries *i.e.*, O is a null matrix of order $m \times n - r$.

THEOREM 3. (Rank of product of matrices)

The rank of a product of two matrices cannot exceed the rank of either matrix, i.e., if A and B be two matrices conformable for the product AB, then ρ(AB) ≤ ρ(A), ρ(AB) ≤ ρ(B), i.e., ρ(AB) ≤ min {ρ(A), ρ(B).

Proof. Let A be a matrix of order $m \times n$ of rank r_1 and B be a matrix of order $n \times p$ of order r_2, then AB is conformable and let r be the rank of AB.

We shall prove that

$$r \le r_1, r \le r_2$$

Since A is a matrix of order $m \times n$ of rank r_1, then there exists a non-singular matrix P such that

$$PA = \begin{bmatrix} G \\ O \end{bmatrix} \qquad \text{...(1)}$$

where G is a matrix of order $r_1 \times n$ of rank r_1 and O is a null matrix of order $(m - r_1) \times n$.

Now post-multiplying both sides of (1) by B, we get

$$PAB = \begin{bmatrix} G \\ O \end{bmatrix} B \qquad \qquad ...(2)$$

Since the rank of matrix does not change by pre-multiplying it by a non-singular, therefore

$$\rho(PAB) = \rho(AB) = r$$

$\therefore$ Rank of the matrix $\begin{bmatrix} G \\ O \end{bmatrix} B = r$

Since the rank of G is r_1 so it contains r_1 non-zero rows, therefore the matrix $\begin{bmatrix} G \\ O \end{bmatrix}$ B cannot have more than r_1 rows, hence

$$\text{Rank of } \begin{bmatrix} G \\ O \end{bmatrix} B \leq r_1$$

$\therefore$ $\qquad\qquad r \leq r_1$ $\qquad\qquad\qquad\qquad\qquad ...(3)$

i.e., $\qquad$ Rank of $AB \leq$ rank of the prefactor A

Again, since $\quad \rho(AB) = \rho((AB)')$

$\Rightarrow \qquad\qquad \rho(AB) = \rho(B'A')\rho(B') = \rho(B)$

$\therefore \qquad\qquad r \leq r_2$ $\qquad\qquad\qquad\qquad\qquad ...(4)$

i.e., $\qquad$ rank of $AB \leq$ rank of the post factor B

From (3) and (4) we conclude that

$$r \leq \min \{r_1, r_2\}$$

i.e., $\qquad\qquad \rho(AB) \leq \min \{\rho(A), \rho(B)\}$.

THEOREM 4. *Every non-singular matrix is row equivalent to a unit matrix.*

Proof. Let $A = [a_{ij}]$ be a non-singular matrix of order $n \times n$. We shall prove the theorem by mathematical induction on n.

Suppose $n = 1$, then $A = [a_{11}]$, therefore in this case the theorem is trivially proved. Thus we assume that the theorem holds for all non-singular matrices of order $n-1$. For a non-singular matrix $A = [a_{ij}]$ of order $n \times n$ there must be at least one non-zero element in the first column of A otherwise $|A| = 0$. Suppose that $a_{i1} = s \neq 0$.

Now (if necessary), interchanging the ith and first row, we get a matrix B whose leading element is s and which is not equal to zero.

$$B = \begin{bmatrix} s & a_{i2} & a_{i3} & \cdots & a_{in} \\ a_{21} & a_{22} & a_{23} & \cdots & a_{2n} \\ \vdots & \vdots & \vdots & \vdots & \vdots \\ a_{(i-1)1} & a_{(i-1)2} & a_{(i-3)3} & \cdots & a_{(i-1)n} \\ a_{11} & a_{12} & a_{13} & \cdots & a_{1n} \\ a_{(i+1)1} & a_{(i+1)2} & a_{(i+1)3} & \cdots & a_{(i+1)n} \\ \vdots & \vdots & \vdots & \vdots & \vdots \\ a_{n1} & a_{n2} & a_{n3} & \cdots & a_{nn} \end{bmatrix}$$

Multiplying each element of the first row of B by $1/s$, we get a matrix C whose leading element is equal to unity.

Let

$$C = \begin{bmatrix} 1 & c_{12} & c_{13} & \cdots & c_{1n} \\ c_{21} & c_{22} & c_{23} & \cdots & c_{2n} \\ c_{31} & c_{32} & c_{33} & \cdots & c_{3n} \\ \vdots & \vdots & \vdots & \vdots & \vdots \\ c_{n1} & c_{n2} & c_{n3} & \cdots & c_{nn} \end{bmatrix}$$

Now applying $R_2 \to R_2 - c_{21}R_1$, $R_2 \to R_2 - c_{31}R_1$ etc, we get a matrix D in which all elements of the first column except the leading element are equal to zero.

Let

$$D = \begin{bmatrix} 1 & d_{12} & d_{13} & \cdots & d_{1n} \\ 0 & & & & \\ 0 & & A_1 & & \\ \cdots & & & & \\ 0 & & & & \end{bmatrix}$$

where A_1 is a non-singular matrix, otherwise $|A_1| = 0 \Rightarrow |D| = 0 \Rightarrow |A| = 0$.
By hypothesis A_1, a matrix of order $(n-1) \times (n-1)$, can be transformed to a unit matrix I_{n-1} of order $(n-1) \times (n-1)$ by elementary row transformations. If these two row transformations are applied to the matrix D, they will not effect the first row and the first column of D, and we, therefore, get a matrix M such that

$$M = \begin{bmatrix} 1 & d_{12} & d_{13} & \cdots & d_{1n} \\ 0 & 1 & 0 & \cdots & 0 \\ 0 & 0 & 1 & \cdots & 0 \\ \vdots & \vdots & \vdots & \vdots & \vdots \\ 0 & 0 & 0 & 0 & 0 \end{bmatrix}$$

Again applying $R_2 \to R_2 - d_{12}R_1$, $R_3 \to R_3 - d_{13}R$, etc. we get a matrix I_n, the unit matrix of order $n \times n$.

Hence A is reduced to I_n by elementary row transformations only and hence the theorem.

THEOREM 5. *If A is a non-singular matrix of order $n \times n$, there exist elementary matrices E_1, E_2, ..., E_s such that*

$$E_s E_{s-1} \ldots E_2 E_1 A = I_n.$$

Proof. Since A is a non-singular matrix of order $n \times n$ then by above theorem, it can be reduced to I_n by elementary row transformations only. Further since every elmentary row transformation is equivalent to pre-multiplication by the corresponding elementary matrix, therefore there exists elementary matrices E_1, E_2, ..., E_s such that

$$E_s E_{s-1} \ldots E_2 E_1 A = I_n$$

THEOREM 2. *Every non-singular matrix A is expressible as the product of elementary matrices.*

Proof. Let A be a non-singular matrix of order $n \times n$, then by corollory 1, we have

$$E_s E_{s-1} \ldots E_2 E_1 A = I_n \qquad \ldots(1)$$

Pre-multiplying both sides of (1) by $(E_s E_{s-1} \ldots E_2 E_1)^{-1}$, we get

$$(E_s E_{s-1} \ldots E_2 E_1)^{-1} (E_s E_{s-1} \ldots E_2 E_1)A = (E_s E_{s-1} \ldots E_2 E_1)^{-1}I_n$$

$$\Rightarrow \qquad I_n A = E_1^{-1}E_2^{-1} \ldots E_{s-1}^1 E_s^{-1}$$

$$\therefore \qquad A = E_1^{-1}E_2^{-1} \ldots E_{s-1}^1 E_s^{-1}$$

Since each $E_1^{-1}(i = 1, 2, ..., s)$ is elementary matrix.

THEOREM 7. *The rank of a matrix does not change by pre-multiplication or postmultiplication with a non-singular matrix.*

Proof. We know that elementary transformations do not change the rank of a matrix and element row (column) transformations are equivalent to pre(post)-multiplication with the corresponding elementary matrices.

Since every non-singular matrix can be expressed as the product of elementary matrices.

Solved Examples

Example 1. *Show that the matrices $\begin{bmatrix} 1 & 2 & 3 \\ 2 & 4 & 6 \end{bmatrix}$ and $\begin{bmatrix} 0 & 3 & 2 \\ 0 & 6 & 4 \end{bmatrix}$ are equivalent.*

Solution. Let
$$A = \begin{bmatrix} 1 & 2 & 3 \\ 2 & 4 & 6 \end{bmatrix} \text{ and } B = \begin{bmatrix} 0 & 3 & 2 \\ 0 & 6 & 4 \end{bmatrix}$$

Applying $R_2 \to R_2 - 2R_1$ on A
$$A \sim \begin{bmatrix} 1 & 2 & 3 \\ 0 & 0 & 0 \end{bmatrix}$$

Again applying $C_1 \to C_1 - \dfrac{1}{2} C_2$
$$A \sim \begin{bmatrix} 0 & 2 & 3 \\ 0 & 0 & 0 \end{bmatrix}$$

Again applying $R_2 \to R_2 + 2R_1$
$$A \sim \begin{bmatrix} 0 & 2 & 3 \\ 0 & 4 & 6 \end{bmatrix}$$

Again applying $C_2 \leftrightarrow C_3$
$$A \sim \begin{bmatrix} 0 & 3 & 2 \\ 0 & 6 & 4 \end{bmatrix} = B$$

$$A \sim B$$

Thus B can be obtained from A by a finite number of elementary transformations on A. Hence A and B are equivalent.

Example 2. *If A and B be two equivalent matrices, then show that $\rho(A) = \rho(B)$.*

Solution. Since A and B are equivalent, therefore B can be obtained from A by a finite chain of elementary transformations on A and elementary transformations do not change the rank of the matrices, hence $\rho(A) = \rho(B)$.

Example 3. *Show that if two matrices A and B have the same size and the same rank they are equivalent.*

Solution. Let A and B be two matrices of order $m \times n$ and $\rho(A) = \rho(B) = r$. Then we have
$$A \sim \begin{bmatrix} I_r & O \\ O & O \end{bmatrix} \text{ and } B \sim \begin{bmatrix} I_r & O \\ O & O \end{bmatrix}$$

Since '$\sim$' is symmetric, then
$$B \sim \begin{bmatrix} I_r & O \\ O & O \end{bmatrix} \Leftrightarrow \begin{bmatrix} I_r & O \\ O & O \end{bmatrix} \sim B$$

Again '~' is transitive, then

$$A \sim \begin{bmatrix} I_r & O \\ O & O \end{bmatrix} \text{ and } \begin{bmatrix} I_r & O \\ O & O \end{bmatrix} \sim B$$

$$\Rightarrow \qquad A \sim B$$

Hence A and B are equivalent.

Example 4. *Use E-transformations to reduce the following matrices to triangular form and hence find their rank.*

$$(i) \begin{bmatrix} 5 & 3 & 14 & 4 \\ 0 & 1 & 2 & 1 \\ 1 & -1 & 2 & 0 \end{bmatrix} \qquad (ii) \begin{bmatrix} 8 & 1 & 3 & 6 \\ 0 & 3 & 2 & 2 \\ -8 & -1 & -3 & 4 \end{bmatrix}$$

Solution. (i) Let

$$A = \begin{bmatrix} 5 & 3 & 14 & 4 \\ 0 & 1 & 2 & 1 \\ 1 & -1 & 2 & 0 \end{bmatrix}$$

Applying $R_1 \leftrightarrow R_3$

$$A \sim \begin{bmatrix} 1 & -1 & 2 & 0 \\ 0 & 1 & 2 & 1 \\ 5 & 3 & 14 & 4 \end{bmatrix}$$

Again applying $R_3 \to R_3 - 5R_1$

$$A \sim \begin{bmatrix} 1 & -1 & 2 & 0 \\ 0 & 1 & 2 & 1 \\ 0 & 8 & 4 & 4 \end{bmatrix}$$

Again applying $R_3 \to R_3 - 8R_2$

$$A \sim \begin{bmatrix} 1 & -1 & 2 & 0 \\ 0 & 1 & 2 & 1 \\ 0 & 0 & -12 & -4 \end{bmatrix}$$

The last equivalent matrix is in Echelon form (or triangular form) which has three non-zero rows. Hence $\rho(A) = 3$.

(ii) Let $$A = \begin{bmatrix} 8 & 1 & 3 & 6 \\ 0 & 3 & 2 & 2 \\ -8 & -1 & -3 & 4 \end{bmatrix}$$

Applying $C_1 \to \dfrac{1}{8} C_1$

$$A \sim \begin{bmatrix} 1 & 1 & 3 & 6 \\ 0 & 3 & 2 & 2 \\ -1 & -1 & -3 & 4 \end{bmatrix}$$

Again applying $R_3 \to R_3 + R_1$

$$A \sim \begin{bmatrix} 1 & 1 & 3 & 6 \\ 0 & 3 & 2 & 2 \\ 0 & 0 & 0 & 10 \end{bmatrix}$$

The last equivalent matrix is in Echelon form (or triangular form) which has three non-zero rows, hence $\rho(A) = 3$.

Example 5. *Is the matrix* $\begin{bmatrix} 1 & 2 & 1 \\ -1 & 0 & 2 \\ 2 & 1 & -3 \end{bmatrix}$ *equivalent to* I_3 *?*

Solution. Let $$A = \begin{bmatrix} 1 & 2 & 1 \\ -1 & 0 & 2 \\ 2 & 1 & -3 \end{bmatrix}$$

Then $\qquad\qquad |A| = 1(0-2) - 2(3-4) + 1(-1-0) = -2 + 2 + 1 = 1 \neq 0$

Therefore, A is a non-singular matrix of order 3×3, so it is row equivalent to a unit matrix. Hence $A \sim I_3$.

Example 6. *If A and B are two matrices of the same type, then* $\rho(A+B) \leq \rho(A) + \rho(B)$

Solution. Let A and B be two matrices of order $m \times n$ and let $\rho(A) = r_1$ and $\rho(B) = r_2$.

Now $\qquad\qquad \rho(A) = r_1 \Rightarrow A$ contains r_1 linearly independent rows

and $\qquad\qquad \rho(B) = r_2 \Rightarrow B$ contains r_2 linearly independent rows.

Therefore $A + B$ will contain at most $r_1 + r_2$ linearly independent rows

$\therefore \qquad\qquad\qquad \rho(A+B) \leq r_1 + r_2$

Hence $\qquad\qquad \rho(A+B) \leq \rho(A) + \rho(B)$.

Example 7. *If A and B are two n-rowed square matrices, then* $\rho(AB) \geq \rho(A) + \rho(B) - n$

Solution. Let $\rho(A) = r$, then there exists two non-singular matrices P and Q such that

$$PAQ = \begin{bmatrix} I_r & O \\ O & O \end{bmatrix} \qquad\qquad ...(1)$$

Pre-multiplying by P^{-1} and post-multiplying by Q^{-1} we get

$$A = P^{-1} \begin{bmatrix} I_r & O \\ O & O \end{bmatrix} Q^{-1} \qquad\qquad ...(2)$$

Consider a matrix

$$C = P^{-1} \begin{bmatrix} O_r & O \\ O & I_{n-r} \end{bmatrix} Q^{-1}$$

Now $$A + C = P^{-1} \left\{ \begin{bmatrix} I_r & O \\ O & O \end{bmatrix} + \begin{bmatrix} O_r & O \\ O & I_{n-r} \end{bmatrix} \right\} Q^{-1}$$

$\Rightarrow \qquad\qquad A + C = P^{-1} \begin{bmatrix} I_r & O \\ O & I_{n-r} \end{bmatrix} Q^{-1}$

$\Rightarrow \qquad\qquad A + C = P^{-1} I_n Q^{-1}$

$\Rightarrow \qquad\qquad A + C = P^{-1} Q^{-1}$

$\therefore$ $A+C$ is non-singular matrix of order $m \times n$, since $P^{-1}Q^{-1}$ is non-singular of order $n \times n$.

Therefore, $\qquad \rho(A+C) = n$

But by the definition of C, we have

$$\rho(C) = n - r$$

$\Rightarrow \qquad\qquad \rho(C) = n - \rho(A) \qquad\qquad\qquad\qquad \because \rho(A) = r$

Since the rank of a matrix does not change on pre-multiplying it with a non-singular matrix, then

$$\rho((A+C))\,B = \rho(B) \qquad\qquad [\because A+C \text{ is non-singular matrix.}]$$

$$\Rightarrow \qquad \rho(B) = \rho(AB + CB)$$

$$\Rightarrow \qquad \rho(B) \leq \rho(AB) + \rho(CB) \qquad\qquad [\because \rho(A+B) \leq \rho(A)+\rho(B)]$$

$$\Rightarrow \qquad \rho(B) \leq \rho(AB) + \rho(C) \qquad\qquad [\because \ \rho(CB) \leq \min\{\rho(C), \rho(B)\}]$$

$$\Rightarrow \qquad \rho(B) \geq \rho(AB) + n - \rho(A) \qquad\qquad [\because \rho(C) = n - \rho(A)]$$

$$\therefore \qquad \rho(AB) \geq \rho(A) + \rho(B) - n.$$

Example 8. *If A be any non-singular matrix and B a matrix such that AB exists, then show that*
$$\rho(AB) = \rho(B).$$

Solution. Since A is a non-singular matrix, then $B = A^{-1}(AB)$

We know that

$$\rho(AB) \leq \rho(B) \qquad\qquad\qquad …(1)$$

Now $$\rho(B) = \rho(A^{-1}(AB)) \leq \rho(AB)$$

or $$\rho(B) \leq \rho(AB) \qquad\qquad\qquad …(2)$$

From (1) and (2), we have

$$\rho(AB) = \rho(B).$$

Example 9. *If A is a square matrix of order $n \times n$ such that $A^2 = A$, then show that*
$$\rho(A) + \rho(I_n - A) = n.$$

Solution. We have $$A^2 = A$$

$$\Rightarrow \qquad A - A^2 = 0$$

$$\Rightarrow \qquad AI_n - A^2 = 0 \qquad\qquad\qquad [\because AI_n = A]$$

$$\Rightarrow \qquad A(I_n - A) = 0$$

$$\therefore \qquad P(A(I_n - A)) = 0$$

Also we know that

$$\rho(A(I_n - A)) \geq \rho(A) + \rho(I_n - A) - n$$

$$\Rightarrow \qquad \rho(A) + \rho(I_n - A) - n \leq 0$$

$$\therefore \qquad \rho(A) + \rho(I_n - A) \leq n \qquad\qquad\qquad …(1)$$

Again we know that

$$\rho(A + I_n - A) = \rho(A) + \rho(I_n - A)$$

$$\Rightarrow \qquad \rho(I_n) \leq \rho(A) + \rho(I_n - A)$$

$$\therefore \qquad \rho(A) + \rho(I_n - A) \geq n \qquad\qquad [\because \rho(I_n) = n] \qquad …(2)$$

From (1) and (2), we get

$$\rho(A) + \rho(I_n - A) = n$$

Example 10. *If A is a square matrix of order $n \times n$ and $\rho(A) = n - 1$, show that $\rho(adj.\ A) = 1$.*

Solution. Since A is an $n \times n$ matrix and $\rho(A) = n - 1$,

then we have $\qquad\qquad |A| = 0$

But we know that $A(adj.\ A) = |A|I_n \Rightarrow A(adj.\ A) = O$

$\therefore \qquad\qquad \rho(A\ adj.\ A) = 0$

Also $\qquad\qquad \rho(A\ adj.\ A) \geq \rho(A) + \rho(adj.\ A) - n$

$\Rightarrow \qquad \rho(A) + \rho(adj.\ A) - n \leq O$

$\Rightarrow \qquad \rho(A) + \rho(adj.\ A) \leq n$

$\Rightarrow \qquad \rho(adj.\ A) \leq n - \rho(A)$

$\Rightarrow \qquad \rho(adj.\ A) \leq n - (n - 1)$

$\therefore \qquad \rho(adj.\ A) \leq 1 \qquad\qquad\qquad\qquad ...(1)$

Since $\rho(A) = n - 1$, then there exists at least one minor of order $n - 1$ of A not equal to zero, therefore there exists at least one element of *adj. A* which is non-zero, it follows that

$$\rho(adj.\ A) > 0 \qquad\qquad\qquad ...(2)$$

From (1) and (2), we get

$$\rho(adj.\ A) = 1.$$

Example 11. *Find the rank of the matrix*

$$A = \begin{bmatrix} 1 & 3 & 4 & 3 \\ 3 & 9 & 12 & 9 \\ -1 & -3 & -4 & -3 \end{bmatrix}$$

Solution. We have $\qquad A = \begin{bmatrix} 1 & 3 & 4 & 3 \\ 3 & 9 & 12 & 9 \\ -1 & -3 & -4 & -3 \end{bmatrix}$

Applying $R_2 \to R_2 - 3R_1$ and $R_3 \to R_3 + R_1$

$$A \sim \begin{bmatrix} 1 & 3 & 4 & 3 \\ 0 & 0 & 0 & 0 \\ 0 & 0 & 0 & 0 \end{bmatrix}$$

The last equivalent matrix is in Echelon form which has one non-zero row.

Hence $\qquad \rho(A) = 1.$

Example 12. *Determine the rank of the following matrices :*

(i) $\begin{bmatrix} 2 & -1 & 3 & 4 \\ 0 & 3 & 4 & 1 \\ 2 & 3 & 7 & 5 \\ 2 & 5 & 11 & 6 \end{bmatrix}$ (ii) $\begin{bmatrix} -2 & -1 & -3 & -1 \\ 1 & 2 & 3 & -1 \\ 1 & 0 & 1 & 1 \\ 0 & 1 & 1 & -1 \end{bmatrix}$

Solution. (i) Let $\qquad\qquad A = \begin{bmatrix} 2 & -1 & 3 & 4 \\ 0 & 3 & 4 & 1 \\ 2 & 3 & 7 & 5 \\ 2 & 5 & 11 & 6 \end{bmatrix}$

Applying $R_3 \to R_3 - R_1$, $R_4 \to R_4 - R_1$

$$A \sim \begin{bmatrix} 2 & -1 & 3 & 4 \\ 0 & 3 & 4 & 1 \\ 0 & 4 & 4 & 1 \\ 0 & 6 & 8 & 2 \end{bmatrix}$$

Again applying $R_3 \to R_3 - \dfrac{4}{3} R_2$, $R_4 \to R_4 - 2R_2$

$$A \sim \begin{bmatrix} 2 & -1 & 3 & 4 \\ 0 & 3 & 4 & 1 \\ 0 & 0 & -4/3 & -1/3 \\ 0 & 0 & 0 & 0 \end{bmatrix}$$

Again applying $R_3 \to 3R_3$

$$A \sim \begin{bmatrix} 2 & -1 & 3 & 4 \\ 0 & 3 & 4 & 1 \\ 0 & 0 & -4 & -1 \\ 0 & 0 & 0 & 0 \end{bmatrix}$$

The last equaivalent matrix is in Echelon form which has 3 non-zero rows, hence $\rho(A) = 3$.

(ii) Let $A = \begin{bmatrix} -2 & -1 & -3 & -1 \\ 1 & 2 & 3 & -1 \\ 1 & 0 & 1 & 1 \\ 0 & 1 & 1 & -1 \end{bmatrix}$

Applying $R_1 \to R_2$

$$A \sim \begin{bmatrix} 1 & 2 & 3 & -1 \\ -2 & -1 & -3 & -1 \\ 1 & 0 & 1 & 1 \\ 0 & 1 & 1 & -1 \end{bmatrix}$$

Again applying $R_2 \to R_2 + 2R_1$, $R_3 \to R_3 - R_1$

$$A \sim \begin{bmatrix} 1 & 2 & 3 & -1 \\ 0 & 3 & 3 & -3 \\ 0 & -2 & -2 & 2 \\ 0 & 1 & 1 & -1 \end{bmatrix}$$

Again applying $R_2 \to R_2 + R_3$

$$A \sim \begin{bmatrix} 1 & 2 & 3 & -1 \\ 0 & 1 & 1 & -1 \\ 0 & -2 & -2 & 2 \\ 0 & 1 & 1 & -1 \end{bmatrix}$$

Again applying $R_3 \to R_3 + 2R_2$, $R_4 \to R_4 - R_2$

$$A \sim \begin{bmatrix} 1 & 2 & 3 & -1 \\ 0 & 1 & 1 & -1 \\ 0 & 0 & 0 & 0 \\ 0 & 0 & 0 & 0 \end{bmatrix}.$$

The last equivalent matrix is in Echelon form which has 2 non-zero rows, hence $\rho(A) = 2$.

Example 13. *Find the rank of the matrix*

$$A = \begin{bmatrix} 6 & 1 & 3 & 8 \\ 4 & 2 & 6 & -1 \\ 10 & 3 & 9 & 7 \\ 16 & 4 & 12 & 15 \end{bmatrix}$$

Solution. We have

$$A = \begin{bmatrix} 6 & 1 & 3 & 8 \\ 4 & 2 & 6 & -1 \\ 10 & 3 & 9 & 7 \\ 16 & 4 & 12 & 15 \end{bmatrix}$$

Applying $R_1 \rightarrow R_1 - R_2$

$$A \sim \begin{bmatrix} 2 & -1 & -3 & 9 \\ 4 & 2 & 6 & -1 \\ 10 & 3 & 9 & 7 \\ 16 & 4 & 12 & 15 \end{bmatrix}$$

Again applying $R_2 \rightarrow R_2 - 2R_1$, $R_3 \rightarrow R_3 - 5R_1$, $R_4 \rightarrow R_4 - 8R_1$

$$A \sim \begin{bmatrix} 2 & -1 & -3 & 9 \\ 0 & 4 & 12 & -19 \\ 0 & 8 & 24 & -38 \\ 0 & 12 & 0 & -57 \end{bmatrix}$$

Again applying $R_2 \rightarrow R_2 - 2R_1$, $R_3 \rightarrow R_3 - 5R_1$, $R_4 \rightarrow R_4 - 8R_2$

$$A \sim \begin{bmatrix} 2 & -1 & -3 & 9 \\ 0 & 4 & 12 & -19 \\ 0 & 8 & 0 & 0 \\ 0 & 12 & 0 & 0 \end{bmatrix}$$

The last equivalent matrix is Echelon form which has 2 non-zero rows, hence
$$\rho(A) = 2.$$

Example 14. *Find the rank of the matrix*

$$A = \begin{bmatrix} 1 & a & b & 0 \\ 0 & c & d & 1 \\ 1 & a & b & 0 \\ 0 & c & d & 1 \end{bmatrix}$$

Solution. We have

$$A = \begin{bmatrix} 1 & a & b & 0 \\ 0 & c & d & 1 \\ 1 & a & b & 0 \\ 0 & c & d & 1 \end{bmatrix}$$

Applying $R_3 \rightarrow R_3 - R_1$, $R_4 \rightarrow R_4 - R_2$

$$A \sim \begin{bmatrix} 1 & a & b & 0 \\ 0 & c & d & 1 \\ 0 & 0 & 0 & 0 \\ 0 & 0 & 0 & 0 \end{bmatrix}$$

The last equivalent matrix is in Echelon form which has 2 non-zero rows, hence
$$\rho(A) = 2.$$

Example 15. *Reduce the matrix* $A = \begin{bmatrix} 1 & -1 & 2 & -3 \\ 4 & 1 & 0 & 2 \\ 0 & 3 & 0 & 4 \\ 0 & 1 & 0 & 2 \end{bmatrix}$ *to the normal form* $\begin{bmatrix} I_r & O \\ O & O \end{bmatrix}$ *and hence determine its rank.*

Solution. We have $A = \begin{bmatrix} 1 & -1 & 2 & -3 \\ 4 & 1 & 0 & 2 \\ 0 & 3 & 0 & 4 \\ 0 & 1 & 0 & 2 \end{bmatrix}$

Applying $R_2 \rightarrow R_2 - 4R_1$

$$A \sim \begin{bmatrix} 1 & -1 & 2 & -3 \\ 0 & 5 & -8 & 14 \\ 0 & 3 & 0 & 4 \\ 0 & 1 & 0 & 2 \end{bmatrix}$$

Applying $C_2 \rightarrow C_2 + C_1$, $C_3 \rightarrow C_3 - 2C_1$, $C_4 \rightarrow C_4 + 3C_1$

$$A \sim \begin{bmatrix} 1 & 0 & 0 & 0 \\ 0 & 5 & -8 & 14 \\ 0 & 3 & 0 & 4 \\ 0 & 1 & 0 & 2 \end{bmatrix}$$

Applying $R_2 \leftrightarrow R_4$

$$A \sim \begin{bmatrix} 1 & 0 & 0 & 0 \\ 0 & 1 & 0 & 2 \\ 0 & 3 & 0 & 4 \\ 0 & 5 & -8 & 14 \end{bmatrix}$$

Applying $R_3 \rightarrow R_3 - 3R_2$, $R_4 \rightarrow R_4 - 5R_2$

$$A \sim \begin{bmatrix} 1 & 0 & 0 & 0 \\ 0 & 1 & 0 & 2 \\ 0 & 0 & 0 & -2 \\ 0 & 0 & -8 & 4 \end{bmatrix}$$

Applying $C_4 \rightarrow C_4 - 2C_2$

$$A \sim \begin{bmatrix} 1 & 0 & 0 & 0 \\ 0 & 1 & 0 & 0 \\ 0 & 0 & 0 & -2 \\ 0 & 0 & -8 & 4 \end{bmatrix}$$

Applying $C_3 \leftrightarrow C_4$

$$A \sim \begin{bmatrix} 1 & 0 & 0 & 0 \\ 0 & 1 & 0 & 0 \\ 0 & 0 & -2 & 0 \\ 0 & 0 & 4 & -8 \end{bmatrix}$$

Applying $R_4 \rightarrow R_4 + 2R_3$

$$A \sim \begin{bmatrix} 1 & 0 & 0 & 0 \\ 0 & 1 & 0 & 0 \\ 0 & 0 & -2 & 0 \\ 0 & 0 & 0 & -8 \end{bmatrix}$$

Applying $R_3 \to \dfrac{1}{2} R_3, R_4 \to -\dfrac{1}{8} R_4$

$$A \sim \begin{bmatrix} 1 & 0 & 0 & 0 \\ 0 & 1 & 0 & 0 \\ 0 & 0 & 1 & 0 \\ 0 & 0 & 0 & 1 \end{bmatrix}$$

$\therefore$ $\qquad\qquad\qquad A \sim I_4$

Hence $\qquad\qquad \rho(A) = 4$

Example 16. *Find the rank of the matrix*

$$A = \begin{bmatrix} 2 & -2 & 0 & 6 \\ 4 & 2 & 0 & 2 \\ 1 & -1 & 0 & 3 \\ 1 & -2 & 1 & 2 \end{bmatrix}$$

by reducing it to normal form.

Solution. We have $\qquad\qquad A = \begin{bmatrix} 2 & -2 & 0 & 6 \\ 4 & 2 & 0 & 2 \\ 1 & -1 & 0 & 3 \\ 1 & -2 & 1 & 2 \end{bmatrix}$

Applying $R_1 \to \dfrac{1}{2} R_1$

$$A \sim \begin{bmatrix} 1 & -1 & 0 & 3 \\ 4 & 2 & 0 & 2 \\ 1 & -1 & 0 & 3 \\ 1 & -2 & 1 & 2 \end{bmatrix}$$

Applying $R_2 \to R_2 - 4R_1$, $R_3 \to R_3 - R_1, R_4 \to R_4 - R_1$

$$A \sim \begin{bmatrix} 1 & -1 & 0 & 3 \\ 0 & 6 & 0 & -10 \\ 0 & 0 & 0 & 0 \\ 0 & -1 & 1 & -1 \end{bmatrix}$$

Applying $C_2 \to C_2 + C_1$, $C_4 \to C_4 - 3C_1$

$$A \sim \begin{bmatrix} 1 & -1 & 0 & 0 \\ 0 & 6 & 0 & -10 \\ 0 & 0 & 0 & 0 \\ 0 & -1 & 1 & 1 \end{bmatrix}$$

Applying $R_2 \leftrightarrow R_4$

$$A \sim \begin{bmatrix} 1 & 0 & 0 & 0 \\ 0 & -1 & 1 & 1 \\ 0 & 0 & 0 & 0 \\ 0 & 6 & 0 & -10 \end{bmatrix}$$

Applying $R_4 \to R_4 + 6R_2$

$$A \sim \begin{bmatrix} 1 & 0 & 0 & 0 \\ 0 & -1 & 1 & 1 \\ 0 & 0 & 0 & 0 \\ 0 & 0 & 6 & -4 \end{bmatrix}$$

Applying $C_3 \to C_3 + C_2$, $C_4 \to C_4 + C_2$

$$A \sim \begin{bmatrix} 1 & 0 & 0 & 0 \\ 0 & -1 & 0 & 0 \\ 0 & 0 & 0 & 0 \\ 0 & 0 & 6 & -4 \end{bmatrix}$$

Applying $C_3 \to \dfrac{1}{6}C_3$, $C_4 \to \dfrac{-1}{4}C_4$, $C_2 \to (-1)C_2$

$$A \sim \begin{bmatrix} 1 & 0 & 0 & 0 \\ 0 & 1 & 0 & 0 \\ 0 & 0 & 0 & 0 \\ 0 & 0 & 1 & 1 \end{bmatrix}$$

Applying $R_3 \leftrightarrow R_4$

$$A \sim \begin{bmatrix} 1 & 0 & 0 & 0 \\ 0 & 1 & 0 & 0 \\ 0 & 0 & 1 & 1 \\ 0 & 0 & 0 & 0 \end{bmatrix}$$

Applying $C_4 \to C_4 - C_3$

$$A \sim \begin{bmatrix} 1 & 0 & 0 & 0 \\ 0 & 1 & 0 & 0 \\ 0 & 0 & 1 & 0 \\ 0 & 0 & 0 & 0 \end{bmatrix}$$

$$A \sim \begin{bmatrix} I_3 & O \\ O & O \end{bmatrix}$$

Hence $\rho(A) = 3$.

Example 17. *Find two non-singular matrices P and Q such that PAQ is in the normal form where*

$$A = \begin{bmatrix} 1 & 1 & 1 \\ 1 & -1 & -1 \\ 3 & 1 & 1 \end{bmatrix}$$

Also find the rank of the matrix A.

Solution. We write $A = I_3 A I_3$

or $$\begin{bmatrix} 1 & 1 & 1 \\ 1 & -1 & -1 \\ 3 & 1 & 1 \end{bmatrix} = \begin{bmatrix} 1 & 0 & 0 \\ 0 & 1 & 0 \\ 0 & 0 & 1 \end{bmatrix} A \begin{bmatrix} 1 & 0 & 0 \\ 0 & 1 & 0 \\ 0 & 0 & 1 \end{bmatrix}$$ ...(1)

In order to find P and Q such that $PAQ = \begin{bmatrix} I_r & O \\ O & O \end{bmatrix}$ we shall reduce the matrix on LHS of (1) by using elementary transformations, while in doing so we shall apply elementary row transformation to pre-factor of A and elementary-column transformation to post-factor of A on RHS of (1).

Now applying $R_2 \to R_2 - R_1$, $R_3 \to R_3 - 3R_1$

$$\begin{bmatrix} 1 & 1 & 1 \\ 1 & -2 & -2 \\ 0 & -2 & -2 \end{bmatrix} = \begin{bmatrix} 1 & 0 & 0 \\ -1 & 1 & 0 \\ -3 & 0 & 1 \end{bmatrix} A \begin{bmatrix} 1 & 0 & 0 \\ 0 & 1 & 0 \\ 0 & 0 & 1 \end{bmatrix}$$

Applying $C_2 \to C_2 - C_1$, $C_3 \to C_3 - C_1$

$$\begin{bmatrix} 1 & 0 & 0 \\ 0 & -2 & -2 \\ 0 & -2 & -2 \end{bmatrix} = \begin{bmatrix} 1 & 0 & 0 \\ -1 & 1 & 0 \\ -3 & 0 & 1 \end{bmatrix} A \begin{bmatrix} 1 & -1 & -1 \\ 0 & 1 & 0 \\ 0 & 0 & 1 \end{bmatrix}$$

Applying $R_2 \to \left(-\dfrac{1}{2} \right) R_2$

$$\begin{bmatrix} 1 & 0 & 0 \\ 0 & 1 & 1 \\ 0 & -2 & -2 \end{bmatrix} = \begin{bmatrix} 1 & 0 & 0 \\ 1/2 & -1/2 & 0 \\ -3 & 0 & 1 \end{bmatrix} A \begin{bmatrix} 1 & -1 & -1 \\ 0 & 1 & 0 \\ 0 & 0 & 1 \end{bmatrix}$$

Applying $R_3 \to R_3 + 2R_2$

$$\begin{bmatrix} 1 & 0 & 0 \\ 0 & 1 & 1 \\ 0 & 0 & 0 \end{bmatrix} = \begin{bmatrix} 1 & 0 & 0 \\ 1/2 & -1/2 & 0 \\ -2 & -1 & 1 \end{bmatrix} A \begin{bmatrix} 1 & -1 & -1 \\ 0 & 1 & 0 \\ 0 & 0 & 1 \end{bmatrix}$$

Applying $C_2 \to C_3 - C_2$

$$\begin{bmatrix} 1 & 0 & 0 \\ 0 & 1 & 0 \\ 0 & 0 & 0 \end{bmatrix} = \begin{bmatrix} 1 & 0 & 0 \\ 1/2 & -1/2 & 0 \\ -2 & -1 & 1 \end{bmatrix} A \begin{bmatrix} 1 & -1 & 0 \\ 0 & 1 & -1 \\ 0 & 0 & 1 \end{bmatrix}$$

or $$\begin{bmatrix} I_2 & O \\ O & O \end{bmatrix} = PAQ$$

where $$P = \begin{bmatrix} 1 & 0 & 0 \\ 1/2 & -1/2 & 0 \\ -2 & -1 & 1 \end{bmatrix} \text{ and } Q = \begin{bmatrix} 1 & -1 & 0 \\ 0 & 1 & -1 \\ 0 & 0 & 1 \end{bmatrix}$$

$$A \sim \begin{bmatrix} I_2 & O \\ O & O \end{bmatrix}$$

Hence $$\rho(A) = 2.$$

Example 18. *Determine non-singular matrices P and Q such PAQ is in the normal form* $\begin{bmatrix} I_r & O \\ O & O \end{bmatrix}$, *where*

$$A = \begin{bmatrix} 3 & 2 & -1 & 5 \\ 5 & 1 & 4 & -2 \\ 1 & -4 & 11 & -19 \end{bmatrix}$$

Solution. Since A is a matrix of order 3×4, therefore we write

$$A = I_3 A I_4$$

or $$\begin{bmatrix} 3 & 2 & -1 & 5 \\ 5 & 1 & 4 & -2 \\ 1 & -4 & 11 & -19 \end{bmatrix} = \begin{bmatrix} 1 & 0 & 0 \\ 0 & 1 & 0 \\ 0 & 0 & 1 \end{bmatrix} A \begin{bmatrix} 1 & 0 & 0 & 0 \\ 0 & 1 & 0 & 0 \\ 0 & 0 & 1 & 0 \\ 0 & 0 & 0 & 1 \end{bmatrix}$$

Applying $R_1 \leftrightarrow R_4$

$$\begin{bmatrix} 1 & -4 & 11 & -19 \\ 5 & 1 & 4 & -2 \\ 3 & 2 & -1 & 5 \end{bmatrix} = \begin{bmatrix} 0 & 0 & 1 \\ 0 & 1 & 0 \\ 1 & 0 & 0 \end{bmatrix} A \begin{bmatrix} 1 & 0 & 0 & 0 \\ 0 & 1 & 0 & 0 \\ 0 & 0 & 1 & 0 \\ 0 & 0 & 0 & 1 \end{bmatrix}$$

Applying $R_2 \to R_2 - 5R_1$, $R_3 \to R_3 - 3R_1$

$$\begin{bmatrix} 1 & -4 & 11 & -19 \\ 0 & 21 & -51 & -93 \\ 0 & 14 & -34 & 62 \end{bmatrix} = \begin{bmatrix} 0 & 0 & 1 \\ 0 & 1 & -5 \\ 1 & 0 & -3 \end{bmatrix} A \begin{bmatrix} 1 & 0 & 0 & 0 \\ 0 & 1 & 0 & 0 \\ 0 & 0 & 1 & 0 \\ 0 & 0 & 0 & 1 \end{bmatrix}$$

Applying $C_2 \to C_2 + 4C_1$, $C_3 \to C_3 - 11C_1$, $C_4 \to C_4 + 19C_1$

$$\begin{bmatrix} 1 & 0 & 0 & 0 \\ 0 & 21 & -51 & 93 \\ 0 & 14 & -34 & 62 \end{bmatrix} = \begin{bmatrix} 0 & 0 & 1 \\ 0 & 1 & -5 \\ 1 & 0 & -3 \end{bmatrix} A \begin{bmatrix} 1 & 4 & -11 & 19 \\ 0 & 1 & 0 & 0 \\ 0 & 0 & 1 & 0 \\ 0 & 0 & 0 & 1 \end{bmatrix}$$

Applying $C_2 \to \dfrac{1}{7}C_2$, $C_3 \to -\dfrac{1}{17}C_3$, $C_4 \to \dfrac{1}{31}C_4$

$$\begin{bmatrix} 1 & 0 & 0 & 0 \\ 0 & 3 & 3 & 3 \\ 0 & 2 & 2 & 2 \end{bmatrix} = \begin{bmatrix} 0 & 0 & 1 \\ 0 & 1 & -5 \\ 1 & 0 & -3 \end{bmatrix} A \begin{bmatrix} 1 & 4/7 & 11/7 & 19/31 \\ 0 & 1/7 & 0 & 0 \\ 0 & 0 & -1/7 & 0 \\ 0 & 0 & 0 & 1/31 \end{bmatrix}$$

Applying $R_2 \to \dfrac{1}{3}R_2$, $R_3 \to \dfrac{1}{2}R_3$

$$\begin{bmatrix} 1 & 0 & 0 & 0 \\ 0 & 1 & 1 & 1 \\ 0 & 1 & 1 & 1 \end{bmatrix} = \begin{bmatrix} 0 & 0 & 1 \\ 0 & 1/3 & -5/3 \\ 1/2 & 0 & -3/2 \end{bmatrix} A \begin{bmatrix} 1 & 4/7 & 11/7 & 19/31 \\ 0 & 1/7 & 0 & 0 \\ 0 & 0 & -1/7 & 0 \\ 0 & 0 & 0 & 1/31 \end{bmatrix}$$

Applying $R_3 \to R_3 - R_2$

$$\begin{bmatrix} 1 & 0 & 0 & 0 \\ 0 & 1 & 1 & 1 \\ 0 & 0 & 0 & 0 \end{bmatrix} = \begin{bmatrix} 0 & 0 & 1 \\ 0 & 1/3 & -5/3 \\ 1/2 & -1/3 & 1/6 \end{bmatrix} A \begin{bmatrix} 1 & 4/7 & 11/7 & 19/31 \\ 0 & 1/7 & 0 & 0 \\ 0 & 0 & -1/7 & 0 \\ 0 & 0 & 0 & 1/31 \end{bmatrix}$$

Applying $C_3 \to C_3 - C_2$, $C_4 \to C_4 - C_2$

$$\begin{bmatrix} 1 & 0 & 0 & 0 \\ 0 & 1 & 0 & 0 \\ 0 & 0 & 0 & 0 \end{bmatrix} = \begin{bmatrix} 0 & 0 & 1 \\ 0 & 1/3 & -5/3 \\ 1/2 & -1/3 & 1/6 \end{bmatrix} A \begin{bmatrix} 1 & 4/7 & 9/119 & 19/217 \\ 0 & 1/7 & -1/7 & -1/7 \\ 0 & 0 & -1/17 & 0 \\ 0 & 0 & 0 & 1/31 \end{bmatrix}$$

or

$$\begin{bmatrix} I_2 & O \\ O & O \end{bmatrix} = PAQ$$

where

$$P = \begin{bmatrix} 0 & 0 & 1 \\ 0 & 1/3 & -5/3 \\ 1/2 & -1/3 & 1/6 \end{bmatrix}$$

and
$$Q = \begin{bmatrix} 1 & 4/7 & 9/119 & 19/217 \\ 0 & 1/7 & -1/7 & -1/7 \\ 0 & 0 & -1/17 & 0 \\ 0 & 0 & 0 & 1/31 \end{bmatrix}$$

$$\therefore \qquad A \sim \begin{bmatrix} I_2 & O \\ O & O \end{bmatrix}$$

Hence $\qquad \rho(A) = 2.$

❧ EXERCISE 2.3

1. Are the followings pairs of matrices equivalent ?

(i) $\begin{bmatrix} 4 & 0 & 2 \\ 3 & 1 & 0 \\ 5 & 2 & 0 \end{bmatrix}, \begin{bmatrix} 3 & 9 & 0 & 2 \\ 7 & -2 & 0 & 1 \\ 8 & 1 & 1 & 5 \end{bmatrix}$

(ii) $\begin{bmatrix} 2 & -1 & 3 & 4 \\ 0 & 3 & 4 & 1 \\ 2 & 3 & 7 & 5 \\ 2 & 5 & 11 & 5 \end{bmatrix}, \begin{bmatrix} 1 & 0 & -5 & 6 \\ 3 & -2 & 1 & 2 \\ 5 & -2 & -9 & 14 \\ 4 & -2 & -4 & 8 \end{bmatrix}$

Determine the rank of the following matrices:

2. $\begin{bmatrix} 1 & 1 & 1 \\ 2 & 2 & 2 \\ 3 & 3 & 3 \end{bmatrix}$

3. $\begin{bmatrix} 2 & 1 & 3 \\ 4 & 7 & 13 \\ 4 & -3 & -1 \end{bmatrix}$

4. $\begin{bmatrix} 4 & 5 & 6 \\ 5 & 6 & 7 \\ 7 & 8 & 9 \end{bmatrix}$

5. $\begin{bmatrix} 1 & 2 & 3 \\ 2 & 3 & 4 \\ 3 & 5 & 7 \end{bmatrix}$

6. $\begin{bmatrix} 2 & 3 & 7 \\ 3 & -2 & 4 \\ 1 & -3 & -1 \end{bmatrix}$

7. $\begin{bmatrix} 3 & -1 & 2 \\ -6 & 2 & -4 \\ -3 & 1 & -2 \end{bmatrix}$

8. $\begin{bmatrix} 1 & 2 & 3 & 1 \\ 2 & 4 & 6 & 2 \\ 1 & 2 & 3 & 2 \end{bmatrix}$

9. $\begin{bmatrix} 1 & 3 & 4 & 3 \\ 3 & 9 & 12 & 9 \\ 1 & 3 & 4 & 1 \end{bmatrix}$

10. $\begin{bmatrix} 1 & 2 & -1 & 4 \\ 2 & 4 & 3 & 5 \\ -1 & -2 & 6 & -7 \end{bmatrix}$

11. $\begin{bmatrix} 1 & 2 & -4 & 5 \\ 2 & -1 & 3 & 6 \\ 8 & 1 & 9 & 7 \end{bmatrix}$

12. $\begin{bmatrix} 1 & -1 & 3 & 6 \\ 1 & 3 & -3 & -4 \\ 5 & 3 & 3 & 11 \end{bmatrix}$

13. $\begin{bmatrix} 1 & 2 & 3 & 0 \\ 2 & 4 & 3 & 2 \\ 3 & 2 & 1 & 3 \\ 6 & 8 & 7 & 5 \end{bmatrix}$

14. $\begin{bmatrix} 2 & 3 & -1 & -1 \\ 1 & -1 & -2 & -4 \\ 3 & 1 & 3 & -2 \\ 6 & 3 & 0 & -7 \end{bmatrix}$

15. $\begin{bmatrix} 1 & 2 & 1 & 2 \\ 1 & 3 & 2 & 2 \\ 2 & 4 & 3 & 4 \\ 3 & 7 & 4 & 6 \end{bmatrix}$

16. $\begin{bmatrix} 3 & -2 & 0 & -1 \\ 0 & 2 & 2 & 1 \\ 1 & -2 & -3 & 2 \\ 0 & 1 & 2 & 1 \end{bmatrix}$

17. $\begin{bmatrix} 0 & 1 & -3 & -1 \\ 1 & 0 & 1 & 1 \\ 3 & 1 & 0 & 2 \\ 1 & 1 & -2 & 0 \end{bmatrix}$

18. $\begin{bmatrix} 1 & 2 & -1 & 3 \\ 4 & 1 & 2 & 1 \\ 3 & -1 & 1 & 2 \\ 1 & 2 & 0 & 1 \end{bmatrix}$

19. $\begin{bmatrix} 1 & 0 & 2 & 1 \\ 0 & 1 & -2 & 1 \\ 1 & -1 & 4 & 0 \\ -2 & 2 & 8 & 0 \end{bmatrix}$

20. $\begin{bmatrix} 8 & 0 & 0 & 1 \\ 1 & 0 & 8 & 1 \\ 0 & 0 & 1 & 8 \\ 0 & 1 & 1 & 8 \end{bmatrix}$

21. $\begin{bmatrix} 6 & 1 & 3 & 8 \\ 4 & 2 & 6 & -1 \\ 10 & 3 & 9 & 7 \\ 16 & 4 & 12 & 15 \end{bmatrix}$

22. Reduce the matrix

$$\begin{bmatrix} 0 & 1 & -3 & -1 \\ 1 & 0 & 1 & 1 \\ 3 & 1 & 0 & 2 \\ 1 & 1 & -2 & 0 \end{bmatrix}$$

to normal form and find its rank.

23. Find the rank of the matrix :

$$A = \begin{bmatrix} 1 & 2 & 3 \\ 2 & 3 & 4 \\ 3 & 5 & 7 \end{bmatrix}$$

after reducing it to normal form.

24. Reduce the matrix

$$A = \begin{bmatrix} 9 & 7 & 3 & 6 \\ 5 & -1 & 4 & 1 \\ 6 & 8 & 2 & 4 \end{bmatrix}$$

to normal form and find its rank.

25. Use elementary row or column transformations to find the rank of the matrix

$$\begin{bmatrix} 1 & 1 & 2 & 3 \\ 1 & 3 & 0 & 3 \\ 1 & -2 & -3 & -3 \\ 1 & 1 & 2 & 3 \end{bmatrix}$$

26. Find the rank of A, B, A+B, AB and BA where

$$A = \begin{bmatrix} 1 & 1 & -1 \\ 2 & -3 & 4 \\ 3 & -2 & 3 \end{bmatrix}, B = \begin{bmatrix} -1 & -2 & -1 \\ 6 & 12 & 6 \\ 5 & 10 & 5 \end{bmatrix}$$

27. Find two non-singular matrices P and Q such that PAQ is in the normal form where

$$A = \begin{bmatrix} 1 & -1 & 2 & -1 \\ 4 & 2 & -1 & 2 \\ 2 & 2 & -2 & 0 \end{bmatrix}$$

Also find the rank of the matrix A.

28. Show that if A and B are equivalent matrices, then there exists non-singular matrices P and Q such that $B = PAQ$.

29. Show that the rank of a matrix is not altered if a column of a matrix is multiplied by a non-zero scalar.

30. Find matrices P and Q such that

$$P \begin{bmatrix} 2 & 2 & -6 \\ -1 & 2 & 2 \end{bmatrix} Q \text{ is in the normal form.}$$

Answers

1. (i) Not equivalent   (ii) Not equivalent

2. 1	**3.** 2	**4.** 2	**5.** 2	**6.** 1	**7.** 2	**8.** 2	**9.** 2	**10.** 3
11. 3	**12.** 3	**13.** 3	**14.** 3	**15.** 4	**16.** 2	**17.** 3	**18.** 3	**19.** 4
20. 2	**21.** 3	**22.** 2	**23.** 3	**24.** 3				

25. $\rho(A) = 2, \rho(B) = 1, \rho(A+B) = 2, \rho(AB) = 0, \rho(BA) = 1$

26. $\rho(A) = 2.29, R = \begin{bmatrix} 1 & 1 \\ 1/2 & 0 \end{bmatrix}, S = \begin{bmatrix} 1 & 4 & 8 \\ 0 & 0 & 1 \\ 0 & 1 & 3 \end{bmatrix}$

2.29 INVERSE OF A MATRIX

Let A be a non-singular matrix of order $n \times n$. Then it is said to be invertible if there exists a non-singular square matrix of order $n \times n$ such that

$$AB = I_n = BA$$

where I_n is the unit matrix of order $n \times n$.

The matrix B is the inverse of A, we write $B = A^{-1}$.

THEOREM 1. *The inverse of a matrix, if it exists, is unique.*

Proof. Let A be a non-singular matrix of order $n \times n$ and if possible, let B and C be its inverses, then we have

$$AB = I_n = BA \qquad \qquad \qquad ...(1)$$

and $$AC = I_n = CA \qquad \qquad \qquad ...(2)$$

From (1) and (2) we get

$$AB = AC$$

$$\Rightarrow \qquad B(AB) = B(AC)$$

$$\Rightarrow \qquad (BA)B = (BA)C \qquad \qquad \text{(By associative Law)}$$

$$\Rightarrow \qquad I_n B = I_n C \qquad \qquad \text{[Using (1)]}$$

$$\Rightarrow \qquad B = C$$

THEOREM 2. *A square matrix is invertible if and only if it is non-singular.*

Proof. Let A be a square matrix of order $n \times n$ and suppose that A is invertible, then there exists a matrix B of order n such that

$$AB = I_n = BA$$

$$\Rightarrow \qquad |AB| = |I_n|$$

$$\Rightarrow \qquad |A||B| = 1$$

$$\Rightarrow \qquad |A| \neq 0$$

Thus, A is non-singular.

Conversely, suppose that A is non-singular matrix, then we have

$$A(adj.\,A) = |A|\,I_n = (adj.\,A)A$$

$$\Rightarrow \qquad A\left(\frac{adj.\,A}{|A|}\right) = I_n = \left(\frac{adj.\,A}{|A|}\right)A \qquad \left[\because |A| \neq 0 \Rightarrow \frac{1}{|A|} \text{exists}\right]$$

$$\Rightarrow \qquad AB = I_n = BA, \text{ if } B = \frac{adj.\,A}{|A|}$$

Thus, A is invertible.

THEOREM 3. *If A is an invertible matrix, then $(A^{-1})^{-1} = A$.*

Proof. Since A is invertible, then we have

$$AA^{-1} = I = AA^{-1}$$

$$\Rightarrow \qquad A \text{ is the inverse of } A^{-1}$$

$$\therefore \qquad (A^{-1})^{-1} = A$$

THEOREM 4. **(Reversal law):** *If A and B are invertible matrices of the same order, then AB is invertible and $(AB)^{-1} = B^{-1}A^{-1}$.*

Proof. Since A and B are invertible, therefore we have

$$|A| \neq 0, \ |B| \neq 0$$

$\Rightarrow \qquad |AB| = |A| \ |B|$

$\Rightarrow \qquad |AB| \neq 0$

$\therefore \quad AB$ is invertible.

Now $\qquad (AB)(B^{-1}A^{-1}) = A(BB^{-1})A^{-1}$ [By associative law]

$\Rightarrow \qquad (AB)(B^{-1}A^{-1}) = A(I)A^{-1}$ $[\because BB^{-1} = I]$

$\Rightarrow \qquad (AB)(B^{-1}A^{-1}) = (AI)A^{-1}$ [By associative law]

$\Rightarrow \qquad (AB)(B^{-1}A^{-1}) = AA^{-1}$ $[\because AI = A]$

$\Rightarrow \qquad (AB)(B^{-1}A^{-1}) = I$ $[\because AA^{-1} = I]$

Also, $\qquad (B^{-1}A^{-1})(AB) = B^{-1}(A^{-1}A)B$ [By associative law]

$\Rightarrow \qquad (B^{-1}A^{-1})(AB) = B^{-1}(I)B$ $[\because A^{-1}A = I]$

$\Rightarrow \qquad (B^{-1}A^{-1}) \ (AB) = B^{-1}(IB)$ [By associative law]

$\Rightarrow \qquad (B^{-1}A^{-1})(AB) = B^{-1}B$ $[\because IB = B]$

$\Rightarrow \qquad (B^{-1}A^{-1})(AB) = I$ $[\because B^{-1}B = I]$

$\therefore \qquad (AB)(B^{-1}A^{-1}) = I = (B^{-1}A^{-1})(AB)$

$\Rightarrow \qquad (AB)^{-1} = B^{-1}A^{-1}$

REMARK

- If A, B and C are three invertible matrices of the same order, then $(ABC)^{-1} = C^{-1}B^{-1}A^{-1}$.

THEOREM 5. *If A is an invertible square matrix, then A' is also invertible and $(A')^{-1} = (A^{-1})'$, where A' is the transpose of A.*

Proof. Since A is an invertible matrix, then we have

$$|A| \neq 0$$

Now $\qquad |A| = |A'|$

$\Rightarrow \qquad |A'| \neq 0$

$\Rightarrow A'$ is invertible.

Also $\qquad AA^{-1} = I = A^{-1}A$

$\Rightarrow \qquad (AA^{-1})' = (I)' = (A^{-1}A)'$

$\Rightarrow \qquad (A^{-1})' A' = I = A' (A^{-1})'$ (By reversal rule of transpose)

$\Rightarrow (A^{-1})'$ is the inverse of A'

$\Rightarrow \qquad (A')^{-1} = (A^{-1})'$

THEOREM 6. *The inverse of an invertible matrix is a symmetric matrix.*

Proof. Let A be an invertible symmetric matrix, then

$$|A| \neq 0 \text{ and } A' = A$$

Now by above theorem.

$$(A')^{-1} = (A^{-1})'$$

$$\Rightarrow \qquad (A^{-1})' = A^{-1} \qquad\qquad\qquad [\because A' = A]$$

$$\Rightarrow \quad A^{-1} \text{ is a symmetric matrix.}$$

THEOREM 7. *If A is an invertible matrix, then* $(adj. A)' = adj (A')$

Proof. Since A is an invertible matrix, then

$$|A| \neq 0$$

Now $\qquad\qquad |A'| = |A|$

$$\Rightarrow \qquad\qquad |A'| \neq 0$$

$\Rightarrow A'$ is invertible

$\Rightarrow (A')^{-1}$ exists.

We have $\qquad A(adj. A) = |A|I$

$$\Rightarrow \qquad (A \; adj. \; A) = (|A|I)' = |A|I' = |A|I$$

$$\Rightarrow \qquad (adj. A') A' = |A|I \qquad\qquad\qquad\qquad ...(1)$$

Also $\qquad (adj. A)' A' = |A'| I$

$$\Rightarrow \qquad (adj. A)'A' = |A|I \qquad [\because |A'| = |A|] \qquad ... (2)$$

From (1) and (2), we get

$$(adj. A)' A' = (adj. A')A'$$

$$\Rightarrow \quad (adj. A)' A' (A')^{-1} = (adj. A') A' (A')^{-1}$$

$$\Rightarrow \qquad (adj. A)' I = (adj. A')I$$

$$\Rightarrow \qquad (adj. A)' = adj. (A)$$

THEOREM 8. *The adjoint of a symmetric matrix is also a symmetric matrix,*

Proof. Let A be a symmetric matrix of order $n \times n$, then

$$A' = A$$

Now by above theorem

$$(adj. A)' = adj(A')$$

$$\Rightarrow \qquad (adj. A)' = adj A \qquad\qquad\qquad [\because A' = A]$$

$$\Rightarrow \quad adj A \text{ is a symmetric matrix.}$$

THEOREM 9. *If A is a non-singular matrix, then* $|A^{-1}| = |A|^{-1}$.

Proof. Since A is a non-singular matrix, then

$$|A| \neq 0$$

$\Rightarrow$ A^{-1} exists.

Also $$AA^{-1} = |I| = A^{-1}A$$

$\Rightarrow$ $$|AA^{-1}| = |I| = 1$$

$\Rightarrow$ $$|A|\,|A^{-1}| = 1$$

$\Rightarrow$ $$|A^{-1}| = \frac{1}{|A|} = |A|^{-1}$$

THEOREM 10. *If A and B are non-singular matrices of the same order, then*

$$adj.\ (AB) = (adj.\ B)\ (adj.\ A)$$

Proof. Since A and B are non-singular matrices of the same order, then AB exists.

Also $$|A| \neq 0, |B| \neq 0$$
$\Rightarrow$ $$|AB| = |A||B| \neq 0$$
$\Rightarrow$ $(AB)^{-1}$ exists.

Now we have

$$A(adj.\ A) = |A|\ I \qquad \qquad ...(1)$$
and $$B(adj.\ B) = |B|\ I \qquad \qquad ...(2)$$
Also $$AB\ (adj.\ B) = |AB|\ I \qquad \qquad ...(3)$$
We have

$$(AB)\ (adj.\ B\ adj.\ A) = A(\ adj.\ B)\ adj.\ A \qquad \text{(By associative law)}$$

$\Rightarrow$ $$(AB)\ (adj.\ B\ adj.\ A) = A(\ |B|\ I\ )\ adj.\ A \qquad \text{[Using (1)]}$$

$\Rightarrow$ $$(AB)\ (adj.\ B\ adj.\ A) = |B|(AI\ )\ adj.\ A$$

$\Rightarrow$ $$(AB)\ (adj.\ B\ adj.\ A) = |B|(A\ adj.\ A)$$

$\Rightarrow$ $$(AB)\ (adj.\ B\ adj.\ A) = |B||A\ |\ I$$

$\therefore$ $$(AB)\ (adj.\ B\ adj.\ A) = |AB|\ I \qquad \qquad ...(4)$$

From (3) and (4), we get

$$(AB)\ (adj.\ AB) = (AB)\ (adj.\ B\ adj.\ A)$$

$\Rightarrow$ $$(AB)^{-1}\ (AB)(adj.\ AB) = (AB)^{-1}\ (AB)\ (adj.\ B\ adj.\ A)$$

$$[\because (AB)^{-1}\ \text{exists.}]$$

$\Rightarrow$ $$I(adj.\ AB) = I(adj.\ B\ adj.\ A)$$

$\Rightarrow$ $$adj.\ (AB) = (adj.\ B)\ (adj.\ A)$$

THEOREM 11. **(Cancellation laws) :** *Let A, B and C be three square matrices of the same order. If A is a non-singular matrix, then*

(i) $AB = AC \Rightarrow B = C$ [Left Cancellation law]

(ii) $BA = CA \Rightarrow B = C$ [Right Cancellation law]

Proof. Since A is a non-singular matrix, then

$$|A| \neq 0 \Rightarrow A^{-1}\ \text{exists.}$$

(i)　We have

$$AB = AC$$

$\Rightarrow$　　　$A^{-1}(AB) = A^{-1}(AC)$　　　　　　　[$\because A^{-1}$ exists.]

$\Rightarrow$　　　$(A^{-1}A)B = (A^{-1}A)C$　　　　　　　[By associative law]

$\Rightarrow$　　　　　$IB = IC$　　　　　　　　[$\because A^{-1}A = I$]

$\Rightarrow$　　　　　$B = C$　　　　　　　　[$\because IB = B, IC = C$]

(ii)　We have

$$BA = CA$$

$\Rightarrow$　　　$(BA)\, A^{-1} = (CA)\, A^{-1}$　　　　　　[$\because A^{-1}$ exists.]

$\Rightarrow$　　　$B\,(AA^{-1}) = C(AA^{-1})$　　　　　　[By associative law]

$\Rightarrow$　　　　　$BI = CI$　　　　　　　　[$\because AA^{-1} = I$]

$\Rightarrow$　　　　　$B = C$　　　　　　　　[$\because BI = B, CI = C$]

THEOREM 12. *If the product of two non-null square matrices is a null matrix, then both of them must be singular.*

Proof.　　Let A and B be two non-null matrices of the same order $n \times n$ such that

$$AB = O \qquad\qquad\qquad ...(1)$$

where O is a null matrix of order $n \times n$.

Let, if possible B be a non-singular matrix, then B^{-1} exists.

From (1) we have

$$AB = O$$

$\Rightarrow$　　　　$(AB)B^{-1} = OB^{-1}$

$\Rightarrow$　　　　$A(BB^{-1}) = O$　　　　　[By associative law and $OB^{-1} = O$]

$\Rightarrow$　　　　$AI_n = O$　　　　　　　[$\because BB^{-1} = I_n$]

$\Rightarrow$　　　　$A = O$　　　　　　　　[$\because AI_n = A$]

which is a contradiction because A is a non-null matrix.

Therefore, B is a singular matrix.

Similarly, we can prove that A is a singular matrix.

2.30　INVERSE OF A MATRIX BY ELEMENTARY TRANSFORMATIONS

Let A be a non-singular matrix of order $n \times n$, then there exists a finite number of elementary matrices $E_1, E_2, ..., E_3, ..., E_s$ such that

$$E_s E_{s-1} ... E_2 E_1 A = I_n$$

$\Rightarrow$　　　$E_s E_{s-1} = E_2 E_1 A A^{-1} = I_n A^{-1}$　　　　[$\because |A| \neq 0 \Rightarrow A^{-1}$ exists.]

$\Rightarrow$　　$(E_s E_{s-1} ... E_2 E_1)\,(AA^{-1}) = I_n A^{-1}$　　　　[By associative law]

$\Rightarrow$　　$(E_s E_{s-1} ... E_2 E_1)\, I_n = A^{-1}$　　　　[$\because AA^{-1} = I_n, A^{-1} = A^{-1}$]

Hence,　　　　$A^{-1} = (E_s E_{s-1} ... E_2 E_1)\, I_n$　　　　　　...(1)

We know that every non-singular matrix of order $n \times n$ can be reduced to the unit matrix I_n by a finite chain of elementary row-transformations only and each elementary row-transformation of a matrix is equivalent to pre-multiplication by the corresponding elementary matrix.

From (1) it follows that if a non-singular matrix A of order $n \times n$ is reduced to the unit matrix I_n by a finite chain of elementary row-transformations only, then the same chain of elementary row-transformations applied to the unit matrix I_n gives the inverse of A.

WORKING RULE

Let A be a non-singular matrix of order $n \times n$, then we follow the following steps :	
STEP 1.	Write $A = I_n A$ ...(1)
STEP 2.	Apply elementary row-transformations on A on L.H.S. of (1) and reduce it to I_n and apply corresponding elementary row-transformations on the pre-factor I_n on R.H.S. of (1) till we obtain $I_n = BA$.
STEP 3.	Finally, we write $A^{-1} = B$.

Solved Examples

Based on the following Results

➡ A matrix B is said to be the inverse of A if $AB = BA = I$, where I is the unit matrix if same order.
➡ A square matrix is invertiable if and only if A is non-singular.
➡ $(A^{-1})^{-1} = A$

Example 1. *By using elementary row-transformations find the inverse of the following matrices:*

(i) $\begin{bmatrix} 1 & 2 \\ 3 & 7 \end{bmatrix}$ (ii) $\begin{bmatrix} 1 & 2 \\ 2 & -1 \end{bmatrix}$

Solution. (i) We write

$$A = I_2 A$$

or

$$\begin{bmatrix} 1 & 2 \\ 3 & 7 \end{bmatrix} = \begin{bmatrix} 1 & 0 \\ 0 & 1 \end{bmatrix} A$$

Applying $R_2 \to R_2 - 3R_1$, we get

$$\begin{bmatrix} 1 & 2 \\ 0 & 1 \end{bmatrix} = \begin{bmatrix} 1 & 0 \\ -3 & 1 \end{bmatrix} A$$

Again applying $R_1 \to R_1 - 2R_2$, we get

$$\begin{bmatrix} 1 & 0 \\ 0 & 1 \end{bmatrix} = \begin{bmatrix} 7 & -2 \\ -3 & 1 \end{bmatrix} A$$

$$\Rightarrow \qquad I_2 = BA$$

$$\Rightarrow \qquad A^{-1} = B = \begin{bmatrix} 7 & -2 \\ -3 & 1 \end{bmatrix}.$$

(ii) We write

$$A = I_2 A$$

or

$$\begin{bmatrix} 1 & 2 \\ 2 & -1 \end{bmatrix} = \begin{bmatrix} 1 & 0 \\ 0 & 1 \end{bmatrix} A$$

Applying $R_2 \to R_2 - 2R_1$, we get

$$\begin{bmatrix} 1 & 2 \\ 0 & -5 \end{bmatrix} = \begin{bmatrix} 1 & 0 \\ -2 & 1 \end{bmatrix} A$$

Applying $R_2 \to -\dfrac{1}{5} R_2$, we get

$$\begin{bmatrix} 1 & 2 \\ 0 & 1 \end{bmatrix} = \begin{bmatrix} 1 & 0 \\ 2/5 & -1/5 \end{bmatrix} A$$

Applying $R_1 \to R_1 - 2R_2$, we get

$$\begin{bmatrix} 1 & 0 \\ 0 & 1 \end{bmatrix} = \begin{bmatrix} 1/5 & 2/5 \\ 2/5 & -1/5 \end{bmatrix} A$$

$$\Rightarrow \qquad I_2 = BA$$

$$\Rightarrow \qquad A^{-1} = B = \begin{bmatrix} 1/5 & 2/5 \\ 2/5 & -1/5 \end{bmatrix}$$

Example 2. *Find the inverse of the matrix*

$$A = \begin{bmatrix} 1 & 2 & 1 \\ 3 & 2 & 3 \\ 1 & 1 & 2 \end{bmatrix}$$

by using elementary row-transformation.

Solution. We write $\qquad A = I_3 A$

or

$$\begin{bmatrix} 1 & 2 & 1 \\ 3 & 2 & 3 \\ 1 & 1 & 2 \end{bmatrix} = \begin{bmatrix} 1 & 0 & 0 \\ 0 & 1 & 0 \\ 0 & 0 & 1 \end{bmatrix} A$$

Applying $R_2 \to R_2 - 3R_1$, $R_3 \to R_3 - R_1$, we get

$$\begin{bmatrix} 1 & 2 & 1 \\ 0 & -4 & 0 \\ 0 & -1 & 1 \end{bmatrix} = \begin{bmatrix} 1 & 0 & 0 \\ -3 & 1 & 0 \\ -1 & 0 & 1 \end{bmatrix} A$$

Applying $R_2 \to \dfrac{-1}{4} R_2$, we get

$$\begin{bmatrix} 1 & 2 & 1 \\ 0 & 1 & 0 \\ 0 & -1 & 1 \end{bmatrix} = \begin{bmatrix} 1 & 0 & 0 \\ 3/4 & -1/4 & 0 \\ -1 & 0 & 1 \end{bmatrix} A$$

Applying $R_3 \to R_3 + R_2$, we get

$$\begin{bmatrix} 1 & 2 & 1 \\ 0 & 1 & 0 \\ 0 & 0 & 1 \end{bmatrix} = \begin{bmatrix} 1 & 0 & 0 \\ 3/4 & -1/4 & 0 \\ -1/4 & -1/4 & 1 \end{bmatrix} A$$

Applying $R_1 \rightarrow R_1 - 2R_2$, we get

$$\begin{bmatrix} 1 & 0 & 1 \\ 0 & 1 & 0 \\ 0 & 0 & 1 \end{bmatrix} = \begin{bmatrix} -1/2 & 1/2 & 0 \\ 3/4 & -1/4 & 0 \\ -1/4 & -1/4 & 1 \end{bmatrix} A$$

Applying $R_1 \rightarrow R_1 - R_3$, we get

$$\begin{bmatrix} 1 & 0 & 0 \\ 0 & 1 & 0 \\ 0 & 0 & 1 \end{bmatrix} = \begin{bmatrix} -1/4 & 3/4 & -1 \\ 3/4 & -1/4 & 0 \\ -1/4 & -1/4 & 1 \end{bmatrix} A$$

$$\Rightarrow \qquad I_3 = BA$$

$$\Rightarrow \qquad A^{-1} = B = \begin{bmatrix} -1/4 & 3/4 & -1 \\ 3/4 & -1/4 & 0 \\ -1/4 & -1/4 & 1 \end{bmatrix}$$

Example 3. *Using elementary transformations, find the inverse of the following matrix :*

$$A = \begin{bmatrix} 1 & 2 & 3 \\ 2 & 5 & 7 \\ -2 & -4 & -5 \end{bmatrix}$$

Solution. We write

$$A_3 = I_3 A$$

or

$$\begin{bmatrix} 1 & 2 & 3 \\ 2 & 5 & 7 \\ -2 & -4 & -5 \end{bmatrix} = \begin{bmatrix} 1 & 0 & 0 \\ 0 & 1 & 0 \\ 0 & 0 & 1 \end{bmatrix} A$$

Applying $R_2 \rightarrow R_2 - 2R_1$, $R_3 \rightarrow R_3 + 2R_1$, we get

$$\begin{bmatrix} 1 & 2 & 3 \\ 0 & 1 & 1 \\ 0 & 0 & 1 \end{bmatrix} = \begin{bmatrix} 1 & 0 & 0 \\ -2 & 1 & 0 \\ 2 & 0 & 1 \end{bmatrix} A$$

Applying $R_1 \rightarrow R_1 - 2R_2$, we get

$$\begin{bmatrix} 1 & 0 & 1 \\ 0 & 1 & 1 \\ 0 & 0 & 1 \end{bmatrix} = \begin{bmatrix} 5 & -2 & 0 \\ -2 & 1 & 0 \\ 2 & 0 & 1 \end{bmatrix} A$$

Applying $R_1 \rightarrow R_1 - R_3$, $R_2 \rightarrow R_2 - R_3$, we get

$$\begin{bmatrix} 1 & 0 & 0 \\ 0 & 1 & 0 \\ 0 & 0 & 1 \end{bmatrix} = \begin{bmatrix} 3 & -2 & -1 \\ -4 & 1 & -1 \\ 2 & 0 & 1 \end{bmatrix} A$$

$$\Rightarrow \qquad I_3 = BA$$

$$\Rightarrow \qquad A^{-1} = B = \begin{bmatrix} 3 & -2 & -1 \\ -4 & 1 & -1 \\ 2 & 0 & 1 \end{bmatrix}$$

Example 4. *Find the inverse of the matrix*

$$A = \begin{bmatrix} 0 & 1 & 2 & 2 \\ 1 & 1 & 2 & 3 \\ 2 & 2 & 2 & 3 \\ 2 & 3 & 3 & 3 \end{bmatrix}$$

by using elementary transformations.

Solution. We write

$$A = I_4 A$$

or

$$\begin{bmatrix} 0 & 1 & 2 & 2 \\ 1 & 1 & 2 & 3 \\ 2 & 2 & 2 & 3 \\ 2 & 3 & 3 & 3 \end{bmatrix} = \begin{bmatrix} 1 & 0 & 0 & 0 \\ 0 & 1 & 0 & 0 \\ 0 & 0 & 1 & 0 \\ 0 & 0 & 0 & 1 \end{bmatrix} A$$

Applying $R_2 \leftrightarrow R_1$, we get

$$\begin{bmatrix} 1 & 1 & 2 & 3 \\ 0 & 1 & 2 & 2 \\ 2 & 2 & 2 & 3 \\ 2 & 3 & 3 & 3 \end{bmatrix} = \begin{bmatrix} 0 & 1 & 0 & 0 \\ 1 & 0 & 0 & 0 \\ 0 & 0 & 1 & 0 \\ 0 & 0 & 0 & 1 \end{bmatrix} A$$

Applying $R_3 \to R_3 - 2R_1$, $R_4 \to R_4 - 2R_1$, we get

$$\begin{bmatrix} 1 & 1 & 2 & 3 \\ 0 & 1 & 2 & 2 \\ 0 & 0 & -2 & -3 \\ 0 & 1 & -1 & -3 \end{bmatrix} = \begin{bmatrix} 0 & 1 & 0 & 0 \\ 1 & 0 & 0 & 0 \\ 0 & -2 & 1 & 0 \\ 0 & -2 & 0 & 1 \end{bmatrix} A$$

Applying $R_1 \to R_1 - R_2$, $R_4 \to R_4 - R_2$, we get

$$\begin{bmatrix} 1 & 0 & 0 & 1 \\ 0 & 1 & 2 & 2 \\ 0 & 0 & -2 & -3 \\ 0 & 0 & -3 & -5 \end{bmatrix} = \begin{bmatrix} -1 & 1 & 0 & 0 \\ 1 & 0 & 0 & 0 \\ 0 & -2 & 1 & 0 \\ -1 & -2 & 0 & 1 \end{bmatrix} A$$

Applying $R_3 \to -\dfrac{1}{2} R_3$, we get

$$\begin{bmatrix} 1 & 0 & 0 & 1 \\ 0 & 1 & 2 & 2 \\ 0 & 0 & 1 & 3/2 \\ 0 & 0 & -3 & -5 \end{bmatrix} = \begin{bmatrix} -1 & 1 & 0 & 0 \\ 1 & 0 & 0 & 0 \\ 0 & 1 & -1/2 & 0 \\ -1 & -2 & 0 & 1 \end{bmatrix} A$$

Applying $R_2 \to R_2 - 2R_3$, $R_4 \to R_4 + 3R_3$, we get

$$\begin{bmatrix} 1 & 0 & 0 & 1 \\ 0 & 1 & 0 & -1 \\ 0 & 0 & 1 & 3/2 \\ 0 & 0 & 0 & -1/2 \end{bmatrix} = \begin{bmatrix} -1 & 1 & 0 & 0 \\ 1 & -2 & 1 & 0 \\ 0 & 1 & -1/2 & 0 \\ -1 & 1 & -3/2 & 1 \end{bmatrix} A$$

Applying $R_4 \rightarrow -2R_4$, we get

$$\begin{bmatrix} 1 & 0 & 0 & 1 \\ 0 & 1 & 0 & -1 \\ 0 & 0 & 1 & 3/2 \\ 0 & 0 & 0 & 1 \end{bmatrix} = \begin{bmatrix} -1 & 1 & 0 & 0 \\ 1 & -2 & 1 & 0 \\ 0 & 1 & -1/2 & 0 \\ 2 & -2 & 3 & -2 \end{bmatrix} A$$

Applying $R_1 \rightarrow R_1 - R_4, R_2 \rightarrow R_2 + R_4, R_3 \rightarrow R_3 - \dfrac{3}{2}R_4$, we get

$$\begin{bmatrix} 1 & 0 & 0 & 0 \\ 0 & 1 & 0 & 0 \\ 0 & 0 & 1 & 0 \\ 0 & 0 & 0 & 1 \end{bmatrix} = \begin{bmatrix} -3 & 3 & -3 & 2 \\ 3 & -4 & 4 & -2 \\ -3 & 4 & -5 & 3 \\ 2 & -2 & 3 & -2 \end{bmatrix} A$$

$\Rightarrow \qquad\qquad I_4 = BA$

$\Rightarrow \qquad\qquad A^{-1} = B = \begin{bmatrix} -3 & 3 & -3 & 2 \\ 3 & -4 & 4 & -2 \\ -3 & 4 & -5 & 3 \\ 2 & -2 & 3 & -2 \end{bmatrix}$

EXERCISE 2.4

Using elementary row-transformations, find the inverse of each of the following matrices, if it exists :

1. $\begin{bmatrix} 5 & 2 \\ 2 & 1 \end{bmatrix}$

2. $\begin{bmatrix} 2 & 3 \\ 0 & 1 \end{bmatrix}$

3. $\begin{bmatrix} 1 & 6 \\ -3 & 5 \end{bmatrix}$

4. $\begin{bmatrix} 1 & 2 & 3 \\ 2 & 4 & 5 \\ 3 & 5 & 6 \end{bmatrix}$

5. $\begin{bmatrix} 1 & 2 & -1 \\ -1 & 1 & 2 \\ 2 & -1 & 1 \end{bmatrix}$

6. $\begin{bmatrix} 1 & -1 & 0 \\ 1 & -3 & 9 \\ 8 & 9 & 2 \end{bmatrix}$

7. $\begin{bmatrix} 0 & 1 & 2 \\ 1 & 2 & 3 \\ 3 & 1 & 1 \end{bmatrix}$

8. $\begin{bmatrix} 2 & 3 & 1 \\ 2 & 4 & 1 \\ 3 & 7 & 2 \end{bmatrix}$

9. $\begin{bmatrix} 1 & -3 & 2 \\ 2 & 0 & 0 \\ 1 & 4 & 1 \end{bmatrix}$

10. $\begin{bmatrix} 2 & -1 & 3 \\ 1 & 2 & 4 \\ 3 & 1 & 1 \end{bmatrix}$

11. $\begin{bmatrix} 1 & 1 & 1 \\ 2 & 2 & 3 \\ 2 & 4 & 9 \end{bmatrix}$

12. $\begin{bmatrix} 2 & 0 & -1 \\ 5 & 1 & 0 \\ 0 & 1 & 3 \end{bmatrix}$

13. $\begin{bmatrix} 1 & 2 & 0 \\ 2 & 3 & -1 \\ 1 & -1 & 3 \end{bmatrix}$

14. $\begin{bmatrix} 1 & 1 & 2 \\ 3 & 1 & 1 \\ 2 & 3 & 1 \end{bmatrix}$

15. $\begin{bmatrix} 1 & 2 & 1 \\ 3 & 2 & 3 \\ 1 & 1 & 2 \end{bmatrix}$

16. $\begin{bmatrix} 3 & 0 & -1 \\ 2 & 3 & 0 \\ 0 & 4 & 1 \end{bmatrix}$

17. $\begin{bmatrix} -1 & -3 & 3 & -1 \\ 1 & 1 & -1 & 0 \\ 2 & -5 & 2 & -3 \\ -1 & 1 & 0 & 1 \end{bmatrix}$

18. $\begin{bmatrix} 1 & 1 & 2 & 0 \\ 0 & 1 & 1 & -1 \\ 2 & 1 & 2 & 1 \\ 3 & -2 & 1 & 6 \end{bmatrix}$

19. $\begin{bmatrix} 1 & 2 & -1 \\ 3 & 8 & 2 \\ 4 & 9 & -1 \end{bmatrix}$

Answers

1. $\begin{bmatrix} 1 & -2 \\ -2 & 5 \end{bmatrix}$

2. $\begin{bmatrix} 1 & -3 \\ 0 & 2 \end{bmatrix}$

3. $\dfrac{1}{23}\begin{bmatrix} 5 & -6 \\ 3 & 1 \end{bmatrix}$

4. $\begin{bmatrix} 1 & -3 & 2 \\ -3 & 3 & -1 \\ 2 & -1 & 0 \end{bmatrix}$

5. $\dfrac{1}{14}\begin{bmatrix} 3 & -1 & 5 \\ 5 & 3 & -1 \\ -1 & 5 & 3 \end{bmatrix}$

6. $\dfrac{1}{157}\begin{bmatrix} 87 & -2 & 9 \\ -70 & -2 & 9 \\ -33 & 17 & 2 \end{bmatrix}$

7. $\dfrac{1}{2}\begin{bmatrix} 1 & -1 & 1 \\ -8 & 6 & -2 \\ 5 & -3 & 1 \end{bmatrix}$

8. $\begin{bmatrix} 1 & 1 & -1 \\ -1 & 1 & 0 \\ 2 & -5 & 2 \end{bmatrix}$

9. $\dfrac{1}{22}\begin{bmatrix} 0 & 11 & 0 \\ -2 & -1 & 4 \\ 8 & -7 & 6 \end{bmatrix}$

10. $\dfrac{1}{30}\begin{bmatrix} 2 & -4 & 10 \\ -11 & 7 & 5 \\ 5 & 5 & -5 \end{bmatrix}$

11. $\dfrac{1}{3}\begin{bmatrix} -6 & 5 & -1 \\ 15 & -8 & 1 \\ -6 & 3 & 0 \end{bmatrix}$

12. $\begin{bmatrix} 3 & -1 & 1 \\ -15 & 6 & -5 \\ 5 & -2 & 2 \end{bmatrix}$

13. $\dfrac{1}{6}\begin{bmatrix} -8 & 6 & 2 \\ 7 & -3 & -1 \\ 5 & -3 & 1 \end{bmatrix}$

14. $\dfrac{1}{11}\begin{bmatrix} -2 & 5 & -1 \\ -1 & -3 & 5 \\ 7 & -1 & -2 \end{bmatrix}$

15. $\dfrac{1}{4}\begin{bmatrix} -1 & 3 & -4 \\ 3 & -1 & 0 \\ -1 & -1 & 4 \end{bmatrix}$

16. $\begin{bmatrix} 3 & -4 & -3 \\ -2 & 3 & 2 \\ 8 & -12 & 9 \end{bmatrix}$

17. $\begin{bmatrix} 0 & 2 & 1 & 3 \\ 1 & 1 & -1 & -2 \\ 1 & 2 & 0 & 1 \\ -1 & 1 & 2 & 6 \end{bmatrix}$

18. $\begin{bmatrix} 2 & -1 & 1 & -1 \\ -5 & -3 & 1 & 1 \\ 2 & 3 & -1 & 0 \\ -3 & -1 & 0 & 1 \end{bmatrix}$

Selected Terms and Results

TERMS

- **Matrix :** A set of *mn* numbers either real or complex arranged in the form of a rectangular array in which there are *m* rows and *n* columns is called matrix of order $m \times n$.

- **Null matrix:** If all elements are zero, then matrix is called null matrix.

- **Row matrix :** Matrix having only one row and *n* column.

- **Column marix :** Matrix having *m* rows and one column.

- **Horizontal matrix :** Matrix having more columns then the number of its rows.

- **Vertical matrix :** Matrix having more number of rows then the columns.

- **Square matrix :** Matrix having equal number of rows and columns.

- **Diagonal matrix :** Contains all its off-diagonal elements equal to zero.

- **Scalar matrix :** A diagonal matrix whose diagonal elements are equal (but not equal to 1) is called scalar matrix.

- **Unit matrix :** A matrix having all off diagonal elements equal to zero and each of the diagonal element equal to 1 is called unit or identity matrix.

- **Triangular matrix :** A matrix in which the element lying above or below the principal diagonal are all zero, is called a triangular matrix.

- **Upper triangular matrix :** A square matrix containing all its elements below the diagonal elements equal to zero.

- **Lower triangular matrix :** A square matrix containing all its elements above the diagonal elements equal to zero.

- **Determinant of a square matrix :** Let *A* be a square matrix. Then the determinant is formed by the elements of matrix *A* and is denoted by $|A|$.

- **Singular and Non-singular matrix :** A matrix whose determinant value is zero is called singular otherwise non-singular.

- **Transpose of a matrix :** Transpose of a matrix *A* is obtained by interchanging the rows and columns of the given matrix. It is denoted by *A'*.

- **Symmetric matrix :** If $A = A'$

- **Skew-symmetric matrix :** If $A = -A'$

- **Complex matrix :** A matrix which contains some of its elements equal to complex number.

- **Conjugate of a complex matrix:** The conjugate of a complex matrix *A* is obtained by replacing all the complex elements of *A* by their complex conjugate number.

- **Transpose conjugate of a matrix :** The transpose of the conjugate matrix is called the transposed conjugate of that matrix. It is denoted by A^θ.

- **Hermitian matrix :** If $A^\theta = A$

- **Skew Hermitian matrix :** If $A^\theta = -A$

- **Orthogonal matrix :** If $AA' = I$

- **Unitary matrix :** If $AA^\theta = I$

- **Submatrix of a matrix :** A matrix obtained from the given matrix by removing some rows and columns is called submatrix.

- **Minor of a matrix :** Let *A* be a matrix of order $m \times n$, then the determinant of every square submatrix of *A* is called a minor of *A*.

- **Rank of a matrix :** The rank of a matrix is the order of any highest order of a non-zero minor of the matrix.

- **Elementry matrix :** A matrix which is obtained from a unit matrix by a single E-transformation is called elementary matrix.

- **Inverse of a matrix :** Let *A* be a non-singular matrix of order $n \times n$ then *A* is said to be invertible if there exists a non-singular square matrix of order $n \times n$ such that $AB = BA = I$.

RESULTS

- Matrices addition is associative and commutative.
- Scalar multiplication is distributive over matrix addition.
- Matrix multiplication is associative but not commutative in general.
- Matrix multiplication satisfies the distributive law over matrix addition.
- The determinant of a matrix is reduced to a number.
- Transpose of a column matrix is a row matrix and *vice-versa*.
- Transpose of a square matrix is square matrix.
- The rank of the transpose of a matrix is equal to the rank of that matrix.
- The rank of a matrix is equal to the number of non-zero rows in Echelon form of that matrix.
- E-transformation do not change the rank of a matrix.
- Multiplication of the elements of a row by a non-zero number does not change the rank.

- Addition of any row to the product of any number k and other row does not change the rank.
- The rank of a product of two matrices cannot exceed the rank of either matrix.
- Every non-singular matrix A is expressible as the product of elementary matrices.
- The rank of a matrix does not change by pre-mutliplication or post-multiplication with a non-singular matrix.
- Two matrices of same size and same rank are equivalent.
- The inverse of a matrix if exists, is unique.
- A square matrix is invertible iff it is non-singular.
- $(A')^{-1} = (A^{-1})'$
- $(AB)^{-1} = B^{-1}A^{-1}$
- If A is non-singular matrix then $|A^{-1}| = |A|^{-1}$
- $Adj. (AB) = (Adj. B)(Adj. A)$

Review Questions and Project Work

1. Find the values of a, b, c, d if

$$3\begin{bmatrix} a & b \\ c & d \end{bmatrix} = \begin{bmatrix} a & 6 \\ -1 & 2d \end{bmatrix} + \begin{bmatrix} 4 & a+b \\ c+d & 3 \end{bmatrix}$$

   (**Ans.** $a=2, b=4, c=1, d=3$)

2. Find the matrices A and B such that $AB = 0$

   (**Ans.** $A = \begin{bmatrix} 1 & 2 \\ 2 & 4 \end{bmatrix}, B = \begin{bmatrix} 6 & 2 \\ -3 & -1 \end{bmatrix}$)

3. Find a 2×2 matrix A such that A^2 is diagonal but not A.   (**Ans.** $A = \begin{bmatrix} 1 & 2 \\ 3 & -1 \end{bmatrix}$)

4. Find an upper triangular matrix A such that $A^3 = \begin{bmatrix} 8 & -57 \\ 0 & 27 \end{bmatrix}$.   (**Ans.** $A = \begin{bmatrix} 2 & -3 \\ 0 & 3 \end{bmatrix}$)

5. (i) If A has a zero row then show that AB has a zero row.

   (ii) If B has a zero column then show that AB has a zero column.

6. If $A = [a_{ij}]$ and $B = [b_{ij}]$ be upper triangular matrices. Prove that AB is upper triangular with diagonal $a_{11}b_{11}, a_{22}b_{22}, \dots a_{nn}b_{mm}$.

7. Show that for $x=3, y=0, z=3$, the matrix

$$A = \begin{bmatrix} 3 & x+2i & yi \\ 3-2i & 0 & 1+2i \\ yi & 1-xi & -1 \end{bmatrix}$$ is Hermitian.

8. If A is a complex matrix, show that AA^θ and $A^\theta A$ are Hermitian.

9. If A is a square matrix, show that

   (i) $A + A^T$ is symmetric.

   (ii) $A - A^T$ is skew-symmetric.

10. Let $f(x), g(x)$ be polynomials and A be any square matrix then show that

   (i) $(f+g)(A) = f(A) + g(A)$

   (ii) $(f \cdot g)(A) = f(A) \cdot g(A)$

   (iii) $f(A) g(A) = g(A) f(A)$

Objective type Questions

FILL IN THE BLANKS

1. If the rank of a square matrix is not equal to its order, then the matrix is _____ .

2. The rank of every non-zero matrix is always greater than or equal to _____ .

3. The rank of I_n is _____ where I_n is a unit matrix of order n .

4. The rank of a matrix is $\leq r$. If all $(r+1)$-rowed minors of the matrix _____ .

5. The rank of A and A^T are _____ .

6. The rank of $A = \begin{bmatrix} 1 & 2 & 3 \\ 2 & 4 & 5 \end{bmatrix}$ is _____ .

7. The rank of $A = \begin{bmatrix} 1 & 1 & 1 & 1 \\ 1 & 1 & 1 & 1 \\ 1 & 1 & 1 & 1 \\ 1 & 1 & 1 & 1 \end{bmatrix}$ is _____ .

8. If A is a non-zero column matrix and B is a non-zero row matrix, then rank of $(AB) =$ _____ .

9. All the elementary matrices are _____ .

10. Elementary transformation does not change the _____ of the matrix.

11. If $A \sim \begin{pmatrix} I_{10} & 0 \\ 0 & 0 \end{pmatrix}$, then the rank of $A =$ _____ .

12. The ranks of two equivalent matrices are _____ .

13. The rank of a matrix $\begin{bmatrix} 1 & 2 & 3 \\ 0 & 1 & 4 \\ 0 & 0 & 0 \end{bmatrix}$ is _____ .

14. The matrix $\begin{bmatrix} 1 & 2 & 3 \\ 0 & 1 & 4 \\ 0 & 0 & 0 \end{bmatrix}$ is an _____ .

15. The rank of the product of two matrices cannot exceed the _____ of either matrix.

16. The rank of the matrix diag. $[\lambda_1 \ \lambda_2 \ \lambda_3]$ is _____ .

17. The rank of the matrix $A = \begin{bmatrix} 1 & 2 & 0 & 0 \\ 0 & 0 & 1 & -1 \\ 0 & 0 & 0 & 0 \end{bmatrix}$ is _____ .

18. Non-square matrix has _____ inverse.

19. A matrix A is said to be singular if $|A| =$ _____ .

20. A matrix is said to be _____ if it is square and non-singular.

21. If $|A| \neq 0$, then matrix is said to be _____ .

22. The inverse of a matrix, if exist is _____ .

23. If A and B be two non-singular matrices of the same order, then AB is _____ .

24. $(AB)^{-1} =$ _____ .

25. The transpose of the matrix of cofactors is known as _____ .

26. The necessary and sufficient condition that a square matrix may possess an inverse is that it be _____ .

27. The inverse of the inverse of a matrix A is equal to _____ .

TRUE/FALSE

Write T for true and F for false statement.

1. $\begin{bmatrix} 3 & 5 \\ 7 & 9 \end{bmatrix} = \begin{bmatrix} 3 & 7 \\ 5 & 9 \end{bmatrix}$ **(T/F)**

2. If $A = \begin{bmatrix} 1 & 0 \\ 0 & 1 \end{bmatrix}$, $B = \begin{bmatrix} 3 & 4 \\ 6 & 7 \end{bmatrix}$, then

   $A + B = \begin{bmatrix} 4 & 4 \\ 6 & 8 \end{bmatrix}$ **(T/F)**

3. If $A = \begin{bmatrix} -1 & 2 & 3 \\ 5 & 6 & 7 \end{bmatrix}$, then

   $5A = \begin{bmatrix} -5 & -10 & -15 \\ -25 & 30 & 35 \end{bmatrix}$. **(T/F)**

4. Square matrices are always conformable for multiplication. **(T/F)**

5. A matrix A is symmetric if $A' = -A$. **(T/F)**

6. The diagonal elements of a skew-symmetric matrix are all zero. **(T/F)**

7. The rank of zero matrix is 1. **(T/F)**

8. The rank of I_4 is 4. **(T/F)**

9. The rank of a matrix, if it is reduced to an Echelon form, is equal to the number if non-zero rows in Echelon form. **(T/F)**

10. If $A = [a_{ij}]_{m \times n}$ and $|A| = 0$, then the rank of $A \geq n$. **(T/F)**

11. If rank $A = 3$, then the rank of its transpose is 3. **(T/F)**

12. The rank of the matrix $\begin{bmatrix} 0 & 1 & 2 & 3 \\ 0 & 0 & 1 & -1 \\ 0 & 0 & 0 & 0 \end{bmatrix}$ is 3. **(T/F)**

13. The rank of the matrix $\begin{bmatrix} 1 & 1 & 1 & 1 \\ 1 & 1 & 1 & 1 \\ 1 & 1 & 1 & 1 \end{bmatrix}$ is 4. **(T/F)**

14. If $A = \begin{bmatrix} a_{11} \\ a_{21} \\ \vdots \\ a_{m1} \end{bmatrix}$ and $B = [b_{11} \quad b_{12} \quad \dots \quad b_{nn}]$ then the rank of $AB = 1$. **(T/F)**

15. If the rank of a square matrix of order n is $n - 1$, then $Adj\ A \neq 0$. **(T/F)**

16. The rank of a matrix is always greater than or equal to the rank of its every submatrix. **(T/F)**

17. The rank of $(AB) \geq$ rank of A. **(T/F)**

18. The elementary transofrmation changes the rank of a matrix. **(T/F)**

19. If $A \sim \begin{pmatrix} I_r & O \\ O & O \end{pmatrix}$, then rank of $A = r - 1$. **(T/F)**

20. Then rank of a diag. $[1, 2, 3, ..., n]$ is $\dfrac{n(n+1)}{2}$. **(T/F)**

21. If $A = [a_{ij}]_{m \times n}$ and rank of $A = r$, there exist two singular matrices such that $PAQ = \begin{pmatrix} I_r & O \\ O & O \end{pmatrix}$. **(T/F)**

22. Every square matrix possesses inverse. **(T/F)**

23. Every non-singular matrix possesses inverse. **(T/F)**

24. If A, B are any two $n \times n$ matrices such that $BA = 0$, where 0 is the null matrix. Then at least one of them is non-singular. **(T/F)**

25. The inverse of an orthogonal matrix is not necessarily orthogonal. **(T/F)**

26. $Adj.\ (AB) = Adj.\ (A)\ Adj.\ (B)$ **(T/F)**

27. The inverse of matrix A exists if A is singular. **(T/F)**

MULTIPLE CHOICE QUESTIONS

Choose the most appropriate one :

1. The unit matrix of order 2 is :

(a) $\begin{bmatrix} 1 & 0 \\ 0 & -1 \end{bmatrix}$ (b) $\begin{bmatrix} 1 & 0 \\ 0 & 1 \end{bmatrix}$

(c) $\begin{bmatrix} -1 & 0 \\ 0 & -1 \end{bmatrix}$ (d) $\begin{bmatrix} 0 & 1 \\ 1 & 0 \end{bmatrix}$

2. A matrix A is skew-symmetric matrix of :

(a) $A' = -A$ (b) $A' = A$

(c) $A' = A^2$ (d) $A' = -A^2$

3. A matrix A is Hermitian if :

(a) $A^\theta = -A$ (b) $A^\theta = 0$

(c) $A^\theta = A$ (d) $A^\theta = A'$

4. $(AB)' = ?$

(a) $A'B'$ (b) $B'A'$

(c) $-A'B'$ (d) $-B'A'$

5. $(AB)^\theta = ?$

(a) $A^\theta B^\theta$ (b) A^θ

(c) B^θ (d) $B^\theta A^\theta$

6. If $A = \begin{bmatrix} 1 & -1 \\ -1 & 1 \end{bmatrix}$, $B = \begin{bmatrix} -1 & 1 \\ 1 & -1 \end{bmatrix}$, then $A + B = ?$

(a) $\begin{bmatrix} 0 & 0 \\ 0 & 0 \end{bmatrix}$ (b) $\begin{bmatrix} 0 & 0 \\ 0 & 1 \end{bmatrix}$

(c) $\begin{bmatrix} 1 & 0 \\ 0 & 0 \end{bmatrix}$ (d) $\begin{bmatrix} 1 & 1 \\ 1 & 1 \end{bmatrix}$

7. The order of the matrix $\begin{bmatrix} 2 & 3 & 4 & 5 \\ 6 & 7 & 8 & 9 \end{bmatrix}$ is :

(a) 2×3 (b) 2×4

(c) 3×4 (d) 4×2

8. If the rank $A = \begin{bmatrix} a_{11} & a_{12} \\ a_{21} & a_{22} \end{bmatrix}$ is 2, then rank of

$\begin{bmatrix} a_{11} & a_{21} \\ a_{12} & a_{22} \end{bmatrix}$ is :

(a) 3   (b) 2

(c) 1   (d) none of these

9. If A is a null matrix, then its rank is :

(a) 0   (b) 1

(c) 2   (d) none of these

10. The rank of I_6 is :

(a) 2   (b) 3

(c) 5   (d) 6

11. If A and B are equivalent, and rank $A = r$, then rank of B is :

(a) $r - 1$ (b) $r + 1$

(c) r (d) 0

12. The rank of the matrix $\begin{bmatrix} 2 & 3 & 4 & 5 \\ 0 & 3 & 6 & 7 \\ 0 & 0 & -1 & 0 \\ 0 & 0 & 0 & 1 \end{bmatrix}$ is :

(a) 3   (b) 4

(c) 2   (d) 1

13. If $A = \begin{bmatrix} 1 & 1 & 1 \\ 1 & 1 & 1 \\ 1 & 1 & 1 \end{bmatrix}$, then the rank of A^2 is :

(a) 3   (b) 2

(c) 1   (d) 0

14. If $A = \begin{bmatrix} 1 \\ 2 \\ 3 \end{bmatrix}$, $B = [2 \quad 3 \quad 4]$, then the rank of AB is :

(a) 2   (b) 3

(c) 1   (d) 0

15. If the rank of a matrix $\geq r$, then there is at least one r-rowed minor of the matrix, whose rank is :

(a) 1   (b) 0

(c) 2   (d) r

16. If the rank of $A = m$ and rank of $B = n$, then :

(a) rank $(AB) \geq$ rank (A)

(b) rank of $(AB) = mn$

(c) rank $(AB) \geq$ rank (B)

(d) rank $(AB) \leq \min\{$rank (A), rank $(B)\}$

17. If A is a matrix such that there exists a square submatrix of order r which is non-singular

and every square submatrix of order $r + 1$ is singular, then the rank of A is :

(a) $r + 1$ (b) r

(c) $r - 1$ (d) $r + 2$

18. The transpose of the matrix of cofactors is known as :

(a) inverse   (b) adjoint

(c) transpose   (d) none of these

19. For the inverse of a matrix A it is necessary that A must be :

(a) singular   (b) non-singular

(c) diagonal   (d) none of these

20. The $\dfrac{(Adj.\,A)}{|A|}$ is known as :

(a) A^{-1} (b) A^2

(c) A (d) none of these

21. For a matrix A, which one of the following is a number?

(a) A^{-1} (b) $adj.\,A$

(c) Rank of A (d) None of these

22. If a non-singular matrix A is symmetric, then A^{-1} is :

(a) Skew-symmetric   (b) Hermitian

(c) diagonal   (d) symmetric

23. If A is non-singular matrix, then $(A^{-1})^{-1}$ is :

(a) I (b) A^{-1}

(c) A (d) AA^{-1}

24. The necessary and sufficient condition that a square matrix amy possess an inverse is that it be :

(a) non-singular   (b) singular

(c) triangular   (d) none of these

25. The diagonal elements of Hermitian matrix are:

(a) Complex number   (b) Real number

(c) Natural numbers   (d) none of these

26. The diagonal elements of skew-Hermitian matrix are :

(a) Pure real numbers or zero

(b) Pure imaginary or zero

(c) Compelx numbers

(d) none of these

27. Let $A = [a_{ij}]_{m \times n}$, and $\tilde{A} = [a_{ij}, b_i]_{m \times (n+1)}$ for $AX = \bar{b}$, then :

   (a) A is coefficient matrix and $\tilde{A}$ the augmented matrix

   (b) $\tilde{A}$ is coefficient matrix and A is augmented matrix.

   (c) A, $\tilde{A}$ are coefficient matrix

   (d) none of these

28. Let $A = 0$, and B and C are matrices such that $AB = AC$, then—

   (a) $B = C$ (b) $B \neq C$

   (c) $B \neq A$ (d) $C \neq A$

Answers

FILL IN THE BLANKS

1. singular	**2.** 1	**3.** n	**4.** vanish	**5.** same	**6.** 2	**7.** 1	**8.** 1
9. Non singular	**10.** rank	**11.** 10	**12.** same	**13.** 2	**14.** Echelon form	**15.** rank
16. 3	**17.** 2	**18.** no	**19.** 0	**20.** invertiable	**21.** non-singular
22. unique	**23.** Non-singular	**24.** $B^{-1}A^{-1}$	**25.** adjoint
26. non-singular	**27.** A, itself

TRUE/ FALSE

1. T	**2.** T	**3.** F	**4.** T	**5.** F	**6.** T	**7.** F	**8.** T	**9.** T
10. F	**11.** T	**12.** F	**13.** F	**14.** T	**15.** T	**16.** T	**17.** F	**18.** F
19. F	**20.** F	**21.** F	**22.** F	**23.** T	**24.** T	**25.** F	**26.** T	**27.** F

MULTIPLE CHOICE QUESTIONS

1. (b)	**2.** (a)	**3.** (c)	**4.** (b)	**5.** (d)	**6.** (a)	**7.** (b)	**8.** (b)	**9.** (a)
10. (d)	**11.** (c)	**12.** (b)	**13.** (c)	**14.** (c)	**15.** (b)	**16.** (d)	**17.** (b)	**18.** (b)
19. (b)	**20.** (a)	**21.** (c)	**22.** (d)	**23.** (c)	**24.** (a)	**25.** (b)	**26.** (b)	**27.** (a)
28. (a)

COMPETITION CORNER
for JRF, NET/SET, GATE Aspirants

SOME FASCINATING FACTS

- Let F be a field and m,n be positive integers. A mapping $A : m \times n \to F$ associating each ordered pair $(i,j) \in m \times n$ to be scalars $a_{ij} \in F$ is called an $m \times n$ matrix (matrix as a mapping).

- The null matrix and identity matrix are examples of matrices in row canonical form.

- Let $f(x) = a_0 x^n + a_1 x^{n-1} + a_2 x^{n-2} + ... + a_{n-1} x + a_n$ be a polynomial over a field F and A be a square matrix over F. Then
  $$f(A) = a_0 A^n + a_1 A^{n-1} + a_2 A^{n-2} + ... + a_{n-1} A + a_n I$$
  is called a matrix polynomial.

- Let there be a system of simultaneous linear equations in n unknowns in Echelon form, then
  (i) the system has a unique solution if $m = n$, i.e. the system is in triangular form.
  (ii) the system has an infinite number of solutions if $m < n$ i.e., there are more variable than the number of equations.

- If the Echelon form of simultaneous linear equation contains more variables than equation, then each of the remaining $n - m$ free variables may take any value, so the system has infintely many solutions. The general solution of such a system may be obtained in either of the following two equivalent ways:
  (i) Arbitrarily assign values to $n-m$ free variables and solve uniquely for m to obtain a solution of the system.
  (ii) Find the values of m pivot variables in terms of $(n-m)$ frce variables to find the general solution of the system.

- r is the rank of an $m \times n$ matrix if
  (i) every square submatrix of order $(r+1)$ or more is singular.
  (ii) there exist at least one square submatrix of order r which is non-singular.

- The system of linear equations $AX = B$ is consistent iff the rank of argumented matrix $[A : B]$ is equal to the rank of the coefficient matrix A.

- The identity matrix I can be defined in the form of Kroncker delta as follows :
  $$I = [\delta_{ij}] \text{ where } \delta_{ij} = \begin{cases} 0 \text{ if } i \neq j \\ 1 \text{ if } i = j \end{cases}$$

- Let A be a real matrix, then following are equivalent :
  (i) A is orthogonal.
  (ii) The rows of A form an orthonormal set.
  (iii) The columns of A form an orthonormal set.

- Every diagonal entry of a Hermitian matrix is real.

- Every diagonal element of a skew-Hermitian matrix is either zero or purely imaginary.

- The maximum number of linearly independent row vectors of a matrix A is called row-rank and the maximum number of linearly independent columns vectors of a matrix A is called its columns rank.

- The rank of a skew-symmetric matrix cannot be one.

- The rank of any matrix is greater than or equal to the rank of its every submatrix.

SOME IMPORTANT ILLUSTRATIONS

1. A^2 is symmetric if either A is symmetric or A is skew-symmetric.
2. If A is skew symmetric then $AA' = A'A$ and A^2 are symmetric.
3. All positive integral powers of a symmetric matrix are symmetric.
4. If A and B are symmetric then $AB + BA$ is symmetric and $AB - BA$ is skew symmetric.
5. If A is skew-symmetric matrix and C is any column matrix then $C'AC = 0$
6. If A is symmetric then $B'AB$ is symmetric.
7. If A is skew-symmetric then $B'AB$ is skew-symmetric.
8. If A is Hermitian matrix then iA is skew Hermitian and *vice-versa*.
9. If A and B are orthogonal matrices then AB and BA are orthogonal.

10. If A and B are unitary matrices then AB and BA are unitary.

11. If A is any matrix then AA^θ and $A^\theta A$ are both Hermitian.

12. If A is a non-zero column matrix and B is a non-zero row matrix then $\rho(AB) = 1$.

13. If A is a non-zero matrix of order $n \times 1$ and B is a non-zero matrix of order $1 \times n$ then rank$(AB) = 1$.

14. rank (AA^θ) = rank (A)

15. All matrices cannot be reduced to their normal forms by using only row operations.

16. If A be $n \times n$ matrix, then
    $$adj.(adj.\ A) = |A|^{n-2}.\ A.$$

17. A is a zero of the polynomial
    (i) $f(x) = x^2 + 2x - 11$
    (ii) $g(x) = x^2 - x - 8$

18. If B has a zero column, then B is not invertiable.

19. If A and B are invertiable, then $A + B$ need not be invertiable.

20. A invertible if and only if A' is invertible.

21. A diagonal matrix $D = $ diag $(a_1, a_2, ..., a_n)$ is invertible if and only if no $a_i = 0$.

22. A is row equivalent to B if and only if there exist elementary matrices $E_1, E_2, ..., E_r$ such that
    $$E_r.\ E_{r-1}, ..., E_2 E_1 A = B.$$

23. B is row equivalent to A if and only if there exists an invertible matrix P such that $B = PA$.

24. B is equivalent to A if and only if there exists an invertible matrix P and Q such that $B = PAQ$.

25. B is column equivalent to A if an only if there exists an invertible matrix Q such that $B = AQ$.

26. If A is an square upper triangular matrix with a zero on its diagonal then A is not invertible.

27. All square Echelon matrices are upper triangular matrices.

28. All upper triangular matrices are in Echolon form.

29. A is lower triangular if and only if A' is upper triangular.

30. The diagonal elements of a skew-symmetric matrix must be zero.

31. AB need not be symmetric even though A and B are symmetric.

32. Let A and B be symmetric matrices then AB is symmetric if and only if A and B commutes.

33. A is symmetric then $A^2, A^3, ..., A^n$ are symmetric.

34. A is symmetric then $f(A)$ is symmetric for any polynomial $f(x)$.

35. If A is Hermitian and skew-Hermitian then $A = 0$.

36. The matrix $A = \begin{pmatrix} a_1 & a_2 & a_3 \\ b_1 & b_2 & b_3 \\ c_1 & c_2 & c_3 \end{pmatrix}$ is orthogonal if and only if its rows $u_1 = (a_1, a_2, a_3)$, $u_2 = (b_1, b_2, b_3)$, $u_3 = (c_1, c_2, c_3)$ form an orthonormal set.

37. $u_1, u_2, ..., u_r$ is an orthonormal set of vectors in $\mathbf{C}^n$ if and only if $\bar{u}_1, \bar{u}_2, ..., \bar{u}_r$ is an orthonormal set.

38. The determinant of matrix A and its transpose A' are equal.

39. If A is singular then AB is also singular.

40. If A is orthogonal then $|A| = \pm 1$.

41. If A is a square matrix and $n > 0$ then
    $$A^{-n} = (A^{-1})^n.$$

42. The matrix $A = \begin{bmatrix} a & b \\ c & d \end{bmatrix}$ will be invertible if $ad - bc \neq 0$ and singular if $ad - bc = 0$ also
    $$A^{-1} = \frac{1}{ad - bc} \begin{bmatrix} d & -b \\ -c & a \end{bmatrix}.$$

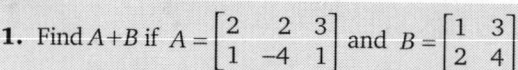

Self Assessment Test

1. Find $A+B$ if $A = \begin{bmatrix} 2 & 2 & 3 \\ 1 & -4 & 1 \end{bmatrix}$ and $B = \begin{bmatrix} 1 & 3 \\ 2 & 4 \end{bmatrix}$

2. Show that for any matrix $A -(-A) = 2A$.

3. Find the size of the product of AB if $A = (a_{ij})_{2\times3}$ and $B = (b_{ij})_{3\times4}$.

4. If A be an $m\times n$ matrix such that $m > 1$, $n > 1$ find the condition under which $A.u$ and $v.A$ are defined (u and v are vectors.)

5. Show that the matrix AA' and $A'A$ are defined for any matrix A.

6. Evaluate $f(A)$ for the polynomial $f(x) = 2x^2 - 4x + 5$.

7. Show that if $AB=A$ and $BA=B$ then A and B are idempotent.

8. Show that the elementary matrices are invertible and their inverses are also elementary matrices.

9. Show that the matrix
$$A = \begin{pmatrix} 1/9 & 8/9 & -4/9 \\ 4/9 & -4/9 & -7/9 \\ 8/9 & 1/9 & 4/9 \end{pmatrix}$$ is orthogonal.

10. If $A = \begin{pmatrix} 1/\sqrt{5} & 2/\sqrt{5} \\ x & y \end{pmatrix}$ is orthogonal. Find the value of x and y.

11. Show that $A = \begin{pmatrix} \frac{1}{3} - \frac{2}{3}i & \frac{2}{3}i \\ -\frac{2}{3}i & -\frac{1}{3} - \frac{2}{3}i \end{pmatrix}$ is unitary.

12. Find the inverse of $A = \begin{bmatrix} 3 & 1 & 0 \\ -1 & 2 & 2 \\ 5 & 0 & -1 \end{bmatrix}$.

13. Prove that $A = \begin{bmatrix} 3 & 7-4i & -2+5i \\ 7+4i & -2 & 3+i \\ -2-5i & 3-i & 4 \end{bmatrix}$ is a Hermitian matrix.

14. Reduce the following matrices to row reduced Echelon form and find their rank

(i) $\begin{bmatrix} 1 & 2 & -1 & 0 \\ 2 & 1 & -1 & 1 \\ 3 & 4 & 2 & 2 \end{bmatrix}$ (ii) $\begin{bmatrix} 1 & 1 & 1 & 1 \\ 1 & 3 & -2 & 1 \\ 2 & 0 & -3 & 2 \\ 3 & 3 & -3 & 3 \end{bmatrix}$

15. Show that the vectors
$$\begin{pmatrix} 1 & 0 \\ 0 & 0 \end{pmatrix}, \begin{pmatrix} 1 & 1 \\ 0 & 0 \end{pmatrix}, \begin{pmatrix} 1 & 1 \\ 1 & 0 \end{pmatrix}, \begin{pmatrix} 0 & 0 \\ 0 & 1 \end{pmatrix}$$
are linearly independent.

16. If $A = \begin{bmatrix} 1 & 2 & 3 \\ 3 & 2 & 1 \\ 1 & 3 & 2 \\ 2 & 1 & 3 \end{bmatrix}$, then find non-singular matrices P and Q such that PAQ is in the normal form.

17. Reduce the following matrices to their normal forms and find their rank.

(i) $A_1 = \begin{bmatrix} 1 & 1 & 1 & -1 \\ 1 & 2 & 3 & 4 \\ 3 & 4 & 5 & 2 \end{bmatrix}$

(ii) $A_2 = \begin{bmatrix} 1 & 4 & 3 & 2 \\ 1 & 2 & 3 & 4 \\ 2 & 6 & 7 & 5 \end{bmatrix}$

18. Find the inverse of the following matrices :

(i) $A_1 = \begin{bmatrix} 1 & 2 & 1 \\ 3 & 2 & 3 \\ 1 & 1 & 2 \end{bmatrix}$ (ii) $A_2 = \begin{bmatrix} 1 & 1 & 1 \\ 1 & 2 & 3 \\ 1 & 3 & 6 \end{bmatrix}$

19. Using elementary operations, find the inverse of the matrix
$$A = \begin{bmatrix} 1 & -1 & 0 & 2 \\ 0 & 1 & 1 & -1 \\ 2 & 1 & 2 & 1 \\ 3 & -2 & 1 & 6 \end{bmatrix}$$

20. If $A = \begin{bmatrix} \cos\theta & \sin\theta & 0 \\ \sin\theta & \cos\theta & 0 \\ 0 & 0 & 0 \end{bmatrix}$ then show that $AA^{-1} = A^{-1}A = I$.

SYSTEM OF LINEAR EQUATIONS

3.1 INTRODUCTION

In this chapter we shall study the nature of solutions of a system of linear equations with the help of the theory of matrices discussed in previous chapters. Before going into details of solutions of linear equations we shall try to understand the concepts of linearly dependent and independent set of vectors.

3.2 VECTORS AND THEIR DEPENDENCE AND INDEPENDENCE

An ordered set of n numbers $(x_1, x_2, x_3, ,..., x_n)$ is known as a vector of order n.

The n numbers $x_1, x_2, x_3, ,..., x_n$ are called the components of the vector. We denote this vector by a single letter X. Conveniently, we may write the components of a vector X in the form of a row or in the form of a column.

Therefore, we may write
$$X = [x_1, x_2, x_3, ,..., x_n]$$
which is known as a n-dimensional row vector or it may be written as

$$X = \begin{bmatrix} x_1 \\ x_2 \\ x_3 \\ \vdots \\ x_n \end{bmatrix}$$

which is known as an n-dimensional column vector.

If we consider an $m \times n$ matrix, then it contains m-row vectors and n-column vectors, each row vector consists of the components of an n-vector and each column vector consists of the components of an m-vector.

For Example: Consider a matrix of order 3×4

$$A = \begin{bmatrix} 1 & -1 & 4 & 5 \\ 2 & 3 & 0 & -7 \\ 3 & 2 & 2 & 6 \end{bmatrix}$$

Then,
$$R_1 = [1 \quad -1 \quad 4 \quad 5]$$
$$R_2 = [2 \quad 3 \quad 0 \quad -7]$$
$$R_3 = [3 \quad 2 \quad 2 \quad 6]$$

$$C_1 = \begin{bmatrix} 1 \\ 2 \\ 3 \end{bmatrix}, C_2 = \begin{bmatrix} -1 \\ 3 \\ 2 \end{bmatrix}$$

$$C_3 = \begin{bmatrix} 4 \\ 0 \\ 2 \end{bmatrix}, C_4 = \begin{bmatrix} 5 \\ -7 \\ 6 \end{bmatrix}$$

Thus A can be written as

$$A = \begin{bmatrix} R_1 \\ R_2 \\ R_3 \end{bmatrix}$$

or
$$A = \begin{bmatrix} C_1 & C_2 & C_3 & C_4 \end{bmatrix}$$

Definiton : *If all the components of a vector are zero, then it is known as a null vector or a zero vector. It is usually denoted by capital letter O.*

For Example : The vectors

$$[0 \quad 0 \quad 0 \quad 0] \text{ and } \begin{bmatrix} 0 \\ 0 \\ 0 \\ 0 \end{bmatrix} \text{ are both null vectors.}$$

3.2.1 SUM OF TWO VECTORS

Let $X = (x_1, x_2, x_3, ..., x_n)$ and $Y = (y_1, y_2, y_3, ..., y_n)$ be two vectors, then $X + Y$ is obtained by adding their corresponding components. Thus

$$X + Y = (x_1 + y_1, x_2 + y_2, ..., x_n + y_n)$$

REMARK

- If two vectors are of different dimensions then they cannot be added up.

3.2.2 MULTIPLICATION OF A VECTOR BY A SCALAR

Let $X = (x_1, x_2, ..., x_n)$ be an n-vector and λ be a scalar, then λX can be obtained on multiplication of each component of X by λ Thus

$$\lambda X = (\lambda x_1, \lambda x_2, ..., \lambda x_n)$$

3.2.3 LINEAR DEPENDENCE AND INDEPENDENCE OF VECTORS

Let $x_1, x_2, x_3, ..., x_m$ be m vectors. Then they are said to be linearly independent if

$$\lambda_1 x_1 + \lambda_2 x_2 + ... + \lambda_m x_m = O \qquad \text{(Zero vector)}$$
$$\Rightarrow \qquad \lambda_1 = \lambda_2 = \lambda_3 = ... = \lambda_m = 0$$

If none of $\lambda_1, \lambda_2, ..., \lambda_m$ is zero, then the vectors $x_1, x_2, ..., x_m$ are called linearly dependent.

3.2.4 LINEAR COMBINATION OF VECTORS

A vector X is said to be a linear combination of the vectors $x_1, x_2, ..., x_m$ if there exists scalars $\lambda_1, \lambda_2, ..., \lambda_m$ such that

$$X = \lambda_1 x_1 + \lambda_2 x_2 + ... + \lambda_m x_m$$

Suppose that the vectors $x_1, x_2, ..., x_m$ are linearly dependent, then in the equation.

$$\lambda_1 x_1 + \lambda_2 x_2 + ... + \lambda_m x_m = O \qquad \qquad ... (1)$$

there is at least one of $\lambda_1, \lambda_2, \ldots \lambda_m$ is non-zero, let it be λ_r, then equation (1) can be written as

$$\lambda_r X_r = -\lambda_1 x_1 - \lambda_2 x_2 - \ldots \lambda_{r-1} x_{r-1} - \lambda_{r+1} x_{r+1} - \ldots - \lambda_m x_m$$

$$\Rightarrow \qquad X_r = \left(-\frac{\lambda_1}{\lambda_r}\right) x_1 + \left(-\frac{\lambda_2}{\lambda_r}\right) x_2 + \ldots + \left(-\frac{\lambda_{r-1}}{\lambda_r}\right) x_{r-1}$$

$$+ \left(-\frac{\lambda_{r+1}}{\lambda_r}\right) x_{r+1} + \ldots + \left(-\frac{\lambda_m}{\lambda_r}\right) x_m$$

$$\Rightarrow \qquad X_r = k_1 x_1 + k_2 x_2 + \ldots + k_{r-1} x_{r-1} + k_{r+1} x_{r+1} + \ldots + k_m x_m$$

It follows that X_r is a linear combinations of vectors $x_1, x_2, \ldots, x_{r-1}, x_{r+1}, \ldots, x_m$.

Hence if a set of vectors is linearly dependent, then at least one member of the set can be expressed as a linear combination of the remaining vectors.

3.2.5 LINEAR DEPENDENCE OF THE ROWS AND COLUMNS OF A SQUARE MATRIX

Consider a square matrix of order 3×3, namely

$$A = \begin{bmatrix} a_{11} & a_{12} & a_{13} \\ a_{21} & a_{22} & a_{23} \\ a_{31} & a_{32} & a_{33} \end{bmatrix}$$

or

$$A = \begin{bmatrix} C_1 & C_2 & C_3 \end{bmatrix}$$

where

$$C_1 = \begin{bmatrix} a_{11} \\ a_{21} \\ a_{31} \end{bmatrix}, C_2 = \begin{bmatrix} a_{12} \\ a_{22} \\ a_{32} \end{bmatrix} \text{ and } C_3 = \begin{bmatrix} a_{13} \\ a_{23} \\ a_{33} \end{bmatrix}$$

The columns C_1, C_2, C_3 are linearly dependent if there exist scalars k_1, k_2, k_3 not all zero such that

$$k_1 C_1 + k_2 C_2 + k_3 C_3 = O$$

$$\Rightarrow \qquad k_1 \begin{bmatrix} a_{11} \\ a_{21} \\ a_{31} \end{bmatrix} + k_2 \begin{bmatrix} a_{12} \\ a_{22} \\ a_{32} \end{bmatrix} + k_3 \begin{bmatrix} a_{13} \\ a_{23} \\ a_{33} \end{bmatrix} = \begin{bmatrix} 0 \\ 0 \\ 0 \end{bmatrix}$$

$$\therefore \qquad \begin{aligned} k_1 a_{11} + k_2 a_{12} + k_3 a_{13} = 0 \\ k_1 a_{21} + k_2 a_{22} + k_3 a_{23} = 0 \\ k_1 a_{31} + k_2 a_{32} + k_3 a_{33} = 0 \end{aligned}$$

i.e., if $\qquad |A| = 0$

Hence, the columns of A are linearly dependent if $|A| = 0$. Since $|A'| = |A|$, if $|A| = 0$, then $|A'| = 0$. Now if $|A'| = 0$, then the columns of A' are linearly dependent but the columns of A' are the rows of A. Hence if $|A| = 0$, then both the rows and columns of A are linearly dependent. It follows that if $|A| \neq 0$, then its rows and columns are linearly independent and vice-versa.

3.2.6 LINEAR DEPENDENCE AND INDEPENDENCE OF ANY MATRIX

Consider a matrix of order $m \times n$, given by

$$A = \begin{bmatrix} a_{11} & a_{12} & \cdots & a_{1n} \\ a_{21} & a_{22} & \cdots & a_{2n} \\ \vdots & \vdots & & \vdots \\ a_{m1} & a_{m2} & \cdots & a_{mn} \end{bmatrix}$$

Let the rank of A be r, then there exists at least one r-minor of A which is non-zero. If A_r be a square submatrix of order $r \times r$ such that $|A_r| \neq 0$, then r rows and columns of A_r are linearly independent, it follows that the matrix A has r rows and columns which are linearly independent. As the rank of A is r so that no set of $(r+1)$ rows and columns of A can be linearly independent. Hence the rank of a matrix A is defined to be the maximum number of linearly independent rows and columns of A.

Since on interchanging rows, the rank of A does not change so without loss of generality we may suppose that the first r rows of A are linearly independent. Let $x_1, x_2, x_3, \ldots, x_r$ denote the r independent vectors and let x_t be one of the remaining $(m - r)$ vectors, then the vectors $x_1, x_2, x_3, \ldots, x_r, x_t$ are linearly independent, therefore there exists scalars $\lambda_1, \lambda_2, \lambda_3, \ldots \lambda_r, \lambda_t$, not all zero such that

$$\lambda_1 x_1 + \lambda_2 x_2 + \ldots + \lambda_r x_r + \lambda_t x_t = O$$

Since $x_1, x_2, \ldots, x_r$ are linearly independent, so we take $\lambda_t \neq 0$, thus

$$x_t = \left(-\frac{\lambda_1}{\lambda_t}\right)x_1 + \left(-\frac{\lambda_2}{\lambda_t}\right)x_2 + \ldots + \left(-\frac{\lambda_2}{\lambda_t}\right)x_r$$

It follows that x_t is a linear combination of $x_1, x_2, \ldots, x_r$.

Hence if the rank of a matrix of order $m \times n$ is r, then it has a set of r linearly independent rows (or columns) and $(m - r)$ linearly dependent rows (or columns).

3.3 HOMOGENEOUS LINEAR EQUATIONS

Let us consider a system of linear homogeneous equations as follows

$$\left.\begin{array}{l} a_{11}x_1 + a_{12}x_2 + \ldots + a_{1n}x_n = 0 \\ a_{21}x_1 + a_{22}x_2 + \ldots + a_{2n}x_n = 0 \\ \cdots\cdots\cdots\cdots\cdots\cdots\cdots\cdots\cdots\cdots\cdots\cdots\cdots \\ a_{m1}x_1 + a_{m2}x_2 + \ldots + a_{mn}x_n = 0 \end{array}\right\} \qquad \ldots(1)$$

These equations are m equations in n unknowns. Any set of numbers $x_1, x_2, \ldots, x_n$ that satisfies all the equations (1) is called a solution of (1).

3.3.1 TRIVIAL SOLUTION

The solution $x_1 = 0, x_2 = 0, \ldots x_n = 0$ of the equations (1) is called *trivial* solution.

3.3.2 NON-TRIVIAL SOLUTION

Any other solutions, if exists, is called a *non-trivial* solution of equation (1).

Let the coefficient matrix be

$$A = \begin{bmatrix} a_{11} & a_{12} & \cdots & a_{1n} \\ a_{21} & a_{22} & \cdots & a_{2n} \\ \vdots & \vdots & & \vdots \\ a_{m1} & a_{m2} & \cdots & a_{mn} \end{bmatrix}_{m \times n}$$

and
$$X = \begin{bmatrix} x_1 \\ x_2 \\ x_3 \\ \vdots \\ x_n \end{bmatrix}_{n \times 1}, O = \begin{bmatrix} 0 \\ 0 \\ 0 \\ \vdots \\ 0 \end{bmatrix}_{m \times 1}$$

Then the system of equation (1) can also be written as

$$AX = O \qquad \qquad \ldots (2)$$

This equation (2) is called a *matrix equation*.

THEOREM 1. *If X_1 and X_2 are two non-trivial solutions of $AX = O$, then $k_1X_1 + k_2X_2$ is also a solution of $AX = O$, where k_1 and k_2 are any arbitrary numbers.*

Proof. We have

$$AX = O \text{ and } AX_1 = O, AX_2 = O \text{ are given.}$$

Now consider,

$$A(k_1X_1 + k_2X_2) = k_1(AX_1) + k_2(AX_2) = k_1(O) + k_2(O) = O$$

Hence $k_1X_1 + k_2X_2$ is the solution of $AX = O$.

THEOREM 2. *If the rank of A is r, then the number of linearly independent solutions of the equation $AX = O$ which is a system of m homogeneous linear equations in n unknowns is $(n - r)$.*

Proof. Since the equation is $AX = O$ $\qquad \qquad \ldots (1)$

where
$$A = \begin{bmatrix} a_{11} & a_{12} & \cdots & a_{1n} \\ a_{21} & a_{22} & \cdots & a_{2n} \\ \vdots & \vdots & & \vdots \\ a_{m1} & a_{m2} & \cdots & a_{mn} \end{bmatrix}_{m \times n}$$

and
$$X = \begin{bmatrix} x_1 \\ x_2 \\ x_3 \\ \vdots \\ x_n \end{bmatrix}_{n \times 1}, O = \begin{bmatrix} 0 \\ 0 \\ 0 \\ \vdots \\ 0 \end{bmatrix}_{m \times 1}$$

Since the rank of $A = r$, so A has r linearly independent columns. Suppose the matrix A can be written as

$$A = [c_1 c_2 \ldots c_r \ldots c_n]_{1 \times n}$$

where $c_1, c_2, \ldots c_r, \ldots c_n$ are column vectors of the matrix A. Each $c_1, c_2, \ldots c_n$ has m vectors. Thus the equation (1) can be written as

$$x_1c_1 + x_2c_2 + \ldots + x_rc_r + \ldots + x_nc_n = 0 \qquad \qquad \ldots (2)$$

But each $c_{r+1}, c_{r+2}, \ldots c_n$ is a linear combination of $c_1, c_2, \ldots c_r$. Then

$$\left. \begin{array}{l} c_{r+1} = p_{11}c_1 + p_{12}c_2 + \ldots + P_{1r}c_r \\ c_{r+2} = p_{21}c_1 + p_{22}c_2 + \ldots + P_{2r}c_r \\ \cdots\cdots\cdots\cdots\cdots\cdots\cdots\cdots\cdots\cdots\cdots\cdots\cdots \\ C_n = p_{k1}c_1 + p_{k2}c_2 + \ldots + P_{kr}c_r \end{array} \right\} \qquad \ldots (3)$$

where $$k = (n - r)$$

Now (3) can be written as

$$
\left.\begin{array}{l}
p_{11}c_1 + p_{12}c_2 + \ldots + p_{1r}c_r - 1.c_{r+1} + 0.c_{r+2} + \ldots + 0.c_n = 0 \\
p_{21}C_1 + p_{22}C_2 + \ldots + p_{2r}c_r + 0.c_{r+1} + 1.c_{r+2} + \ldots + 0.c_n = 0 \\
\ldots\ldots\ldots\ldots\ldots\ldots\ldots\ldots\ldots\ldots\ldots\ldots\ldots\ldots\ldots\ldots \\
p_{k1}c_1 + p_{k2}c_2 + \ldots + p_{kr}c_r + 0.c_{r+1} - 0.c_{r+2} - \ldots + -1.c_n = 0
\end{array}\right\} \quad \ldots (4)
$$

Thus equation (2) and (4) are same, so comparing we get

$$
X_1 = \begin{bmatrix} p_{11} \\ p_{12} \\ \vdots \\ p_{1r} \\ -1 \\ 0 \\ \vdots \\ 0 \end{bmatrix}, X_2 = \begin{bmatrix} p_{21} \\ p_{22} \\ \vdots \\ p_{2r} \\ 0 \\ -1 \\ \vdots \\ 0 \end{bmatrix}, \ldots X_{n-r} = \begin{bmatrix} p_{k1} \\ p_{k2} \\ \vdots \\ p_{kr} \\ 0 \\ 0 \\ \vdots \\ -1 \end{bmatrix}
$$

where $$k = (n - r).$$

Hence we obtained $(n - r)$ solutions of the equation $AX = O$. Next we have to show that X_1, X_2, X_{n-r} are linearly independent.

For this let us have

$$l_1 X_1 + l_2 X_2 + \ldots + l_{n-r} X_{n-r} = O \qquad \ldots (5)$$

Now comparing the $(r+1)^{th}$, $(r+2)^{th}$, n^{th} components on both sides of (5), we get

$$l_1 = 0 = l_2 = \ldots = l_{n-r}$$

Hence $X_1, X_2, X_3, \ldots, X_{n-r}$ are linearly independent. Finally we shall have that every solution of the equation $AX = O$ is a linear combination of $X_1, X_2, \ldots, X_{n-r}$. Suppose X is any solution of $AX = O$ with components $x_1, x_2, \ldots x_n$. Then

$$X + x_{r+1}X_1 + x_{r+2}X_2 + \ldots + x_n X_{n-r} \qquad \ldots (6)$$

is also a solution of $AX = O$

Obviously, let $(n - r)$ components of the vector (6) be all equal to zero. Let $z_1, z_2, \ldots z_r$ be the first r components of (6). Then $(z_1, z_2, \ldots, z_r, 0, 0, \ldots 0)$ is a solution of $AX = O$. Therefore from (2), we get

$$z_1 c_1 + z_2 c_2 + \ldots + z_r c_r = O$$

This implies $z_1 = 0 = z_2 = \ldots = z_r$ because $c_1, c_2, \ldots c_r$ are linearly independent, and hence (6) comes out to be zero, then

$$X = -x_{r+1}X_1 - x_{r+2}X_2 - \ldots - x_n X_{n-r}$$

This shows that every solution of $AX = O$ is a linear combination of $X_1, X_2, \ldots X_{n-r}$.

3.4 NATURE OF THE SOLUTION OF THE EQUATION AX = O

Since $AX = O$ is a matrix equation of a system of m homogeneous linear equations in n unknowns and A is a coefficient matrix of order $m \times n$. Let the rank of A be r. Then obviously r cannot be greater than n. So that either r is n or r is less than n. Therefore these are some cases.

Case I. If $r = n$, then the equation $AX = O$, will have no linearly independent solution. So in this case only trivial solution will exist.

Case II. If $r < n$, then there will be $(n - r)$ linearly independent solution of $AX = O$ and thus in this case we shall have infinite solutions.

Case III. Suppose the number of equations is less than number of unknowns, *i.e.*, $m < n$ and since $r \leq m$, then obviously $r < n$. Thus in this case a non-zero solution will exist. Therefore, the equation $AX = O$ will have infinite solution.

WORKING RULE

	In order to determine the solutions of the equation $AX = O$, we proceed to the following steps :
STEP 1.	Reduce the matrix A to Echelon form by applying E-row transformations only. The Echelon form gives the rank of A.
STEP 2.	Let A be matrix of order $m \times n$ and let $\rho(A) = r$. If $r = n$, then $AX = O$ will have zero solution only. If $r < n$, then we will assign $n - r$ arbitrarily chosen values to $n - r$ unknowns.
STEP 3.	Let B be the Echelon form of A, then the equation $AX = O$ is equivalent to the equation $BX = O$. Reduce $BX = O$ to a system of equations and choose $n - r$ unknowns in this system of equations for assigning arbitrary values like c_1, $c_2...$, c_{n-r}.
STEP 4.	By back substitution of $(n - r)$ unknowns to the system of equations reduced from $BX = O$, we finally obtain the solutions. In case of $r < n$, we get infinite solutions.

Solved Examples

Example 1. *Find the non-trivial solutions of the equations:*

$$x + y - 6z = 0$$
$$-3x + y + 2z = 0$$
$$x - y + 2z = 0$$

Solution. The given system of equations can be written as

$$AX = O \qquad \qquad ...(1)$$

where
$$A = \begin{bmatrix} 1 & 1 & -6 \\ -3 & 1 & 2 \\ 1 & -1 & 2 \end{bmatrix}, X = \begin{bmatrix} x \\ y \\ z \end{bmatrix} \text{ and } O = \begin{bmatrix} 0 \\ 0 \\ 0 \end{bmatrix}$$

Reducing the matrix A into Echelon form, we have

Appling $R_2 \rightarrow R_2 + 3R_1, R_3 \rightarrow R_3 - R_1$, we get

$$A \sim \begin{bmatrix} 1 & 1 & -6 \\ 0 & 4 & -16 \\ 0 & -2 & 8 \end{bmatrix}$$

Again applying $R_2 \rightarrow \frac{1}{4}R_2$, we get

$$A \sim \begin{bmatrix} 1 & 1 & -6 \\ 0 & 1 & -4 \\ 0 & -2 & 8 \end{bmatrix}$$

Again applying $R_3 \rightarrow R_3 + 2R_2$ we get

$$A \sim \begin{bmatrix} 1 & 1 & -6 \\ 0 & 1 & -4 \\ 0 & 0 & 0 \end{bmatrix}$$

The last equivalent matrix in Echelon form with two non-zero rows, therefore $\rho(A) = 2$

Thus the given system of equations is equivalent to

$$\begin{bmatrix} 1 & 1 & -6 \\ 0 & 1 & -4 \\ 0 & 0 & 0 \end{bmatrix} \begin{bmatrix} x \\ y \\ z \end{bmatrix} = \begin{bmatrix} 0 \\ 0 \\ 0 \end{bmatrix}$$

$$\Rightarrow \qquad x + y - 6z = 0 \qquad \qquad \dots (2)$$
$$y - 4z = 0 \qquad \qquad \dots (3)$$

Let us put $z = c$ in (3), we get

$$y = 4c$$

Now putting $y = 4c$ and $z = c$ in (2), we get

$$x = 2c$$

Hence the non-trival solutions of the given system of equatons are $x = 2c, y = 4c$, $z = c$, where c is a non-zero arbitrary number.

Example 2. *Show that the only real value of λ for which the following equations have non-zero solutions is 6:*

$$x + 2y + 3z = \lambda x, \ 3x + y + 2z = \lambda y, \ 2x + 3y + z = \lambda z$$

Solution. The given system of equations can be rewritten as

$$(1 - \lambda)x + 2y + 3z = 0 \qquad \qquad \dots (1)$$
$$3x + (1 - \lambda)y + 2z = 0 \qquad \qquad \dots (2)$$
$$2x + 3y + (1 - \lambda)z = 0 \qquad \qquad \dots (3)$$

This system of equations can be written as

$$AX = O \qquad \qquad \dots (4)$$

where

$$A = \begin{bmatrix} 1 - \lambda & 2 & 3 \\ 3 & 1 - \lambda & 2 \\ 2 & 3 & 1 - \lambda \end{bmatrix}, X = \begin{bmatrix} x \\ y \\ z \end{bmatrix} \text{and } O = \begin{bmatrix} 0 \\ 0 \\ 0 \end{bmatrix}$$

For non-zero solutions, we must have $|A| = 0$

i.e., $\qquad \begin{vmatrix} 1 - \lambda & 2 & 3 \\ 3 & 1 - \lambda & 2 \\ 2 & 3 & 1 - \lambda \end{vmatrix} = 0$

$$\Rightarrow \qquad (1 - \lambda)(1 - \lambda)(1 - \lambda) + 8 + 27 - 6(1 - \lambda) - 6(1 - \lambda) - 6(1 - \lambda) = 0$$
$$\Rightarrow \qquad 1 - \lambda^3 - 3\lambda + 3\lambda^2 + 35 - 18(1 - \lambda) = 0$$
$$\Rightarrow \qquad -\lambda^3 + 3\lambda^2 + 15\lambda + 18 = 0$$
$$\Rightarrow \qquad \lambda^3 - 3\lambda^2 - 15\lambda - 18 = 0$$
$$\Rightarrow \qquad (\lambda - 6)(\lambda^2 + 3\lambda + 3) = 0$$

Since $\lambda^2 + 3\lambda + 3 = 0$ given imaginary roots, therefore the only real value of λ for which the system of equations is to have a non-zero solution is 6.

Example 3. *Does the following system of equations possess a common non-zero solution:*

$$x + y + z = 0$$
$$2x - y - 3z = 0$$
$$3x - 5y + 4z = 0$$
$$x + 17y + 4z = 0$$

Solution. The coefficient matrix is

$$A = \begin{bmatrix} 1 & 1 & 1 \\ 2 & -1 & -3 \\ 3 & -5 & 4 \\ 1 & 17 & 4 \end{bmatrix}$$

First reduce A into Echelon form.

Performing $R_2 \rightarrow R_2 - 2R_1, R_3 \rightarrow R_3 - 3R_1, R_4 \rightarrow R_4 - R_1$

$$\sim \begin{bmatrix} 1 & 1 & 1 \\ 0 & -3 & -5 \\ 0 & -8 & 1 \\ 0 & 16 & 3 \end{bmatrix}$$

performing $R_2 \rightarrow -\dfrac{1}{3} R_2$

$$\sim \begin{bmatrix} 1 & 1 & 1 \\ 0 & 1 & \dfrac{5}{3} \\ 0 & -8 & 1 \\ 0 & 16 & 3 \end{bmatrix}$$

performing $R_3 \rightarrow R_3 + 8R_2, R_4 \rightarrow R_4 - 16R_4$

$$\sim \begin{bmatrix} 1 & 1 & 1 \\ 0 & 1 & \dfrac{5}{3} \\ 0 & 0 & \dfrac{43}{3} \\ 0 & 0 & \dfrac{71}{3} \end{bmatrix}$$

performing $R_3 \rightarrow \dfrac{3}{43} R_3$

$$\sim \begin{bmatrix} 1 & 1 & 1 \\ 0 & 1 & \dfrac{5}{3} \\ 0 & 0 & 1 \\ 0 & 0 & -\dfrac{71}{3} \end{bmatrix}$$

performing $R_4 \to R_4 + \dfrac{71}{3}R_3$

$$\sim \begin{bmatrix} 1 & 1 & 1 \\ 0 & 1 & \dfrac{5}{3} \\ 0 & 0 & 1 \\ 0 & 0 & 0 \end{bmatrix}$$

This is an Echleon form and having three non-zero rows so *A* has the rank 3. Since there are 3 number of unknown, hence a trival solution exists here, *i.e.*, $x = 1, y = 0, z = 0$.

Example 4. *Find all the solutions of the following system of linear homogeneous equations.*

$$x - 2y + z - w = 0$$
$$x + y - 2z + 3w = 0$$
$$4x + y - 5z + 8w = 0$$
$$5x - 7y + 2z - w = 0$$

Solution. The coefficient matrix is given by

$$A = \begin{bmatrix} 1 & -2 & 1 & -1 \\ 1 & 1 & -2 & 3 \\ 4 & 1 & -5 & 8 \\ 5 & -7 & 2 & -1 \end{bmatrix}$$

Change this matrix into Echelon form as follows:

performing $R_2 \to R_2 - R_1, R_3 \to R_3 - 4R_1$ and $R_4 \to R_4 - 5R_1$

$$\sim \begin{bmatrix} 1 & -2 & 1 & -1 \\ 0 & 3 & -3 & 4 \\ 0 & 9 & -9 & 12 \\ 0 & 3 & -3 & 4 \end{bmatrix}$$

performing $R_2 \to \dfrac{1}{3}R_2$

$$\sim \begin{bmatrix} 1 & -2 & 1 & -1 \\ 0 & 1 & -1 & \dfrac{4}{3} \\ 0 & 9 & -9 & 12 \\ 0 & 3 & -3 & 4 \end{bmatrix}$$

performing $R_3 \to R_3 - 9R_2, R_4 \to R_4 - 3R_2$

$$\sim \begin{bmatrix} 1 & -2 & 1 & -1 \\ 0 & 1 & -1 & \dfrac{4}{3} \\ 0 & 0 & 0 & 0 \\ 0 & 0 & 0 & 0 \end{bmatrix}$$

This is an Echelon form having two non-zero rows. Hence rank of $A = 2$.

Therefore the given system of equation is equivalent to

$$\begin{bmatrix} 1 & -2 & 1 & -1 \\ 0 & 1 & -1 & \dfrac{4}{3} \\ 0 & 0 & 0 & 0 \\ 0 & 0 & 0 & 0 \end{bmatrix} \begin{bmatrix} x \\ y \\ z \\ w \end{bmatrix} = 0$$

or $\qquad x - 2y + z - w = 0$ ... (1)

$$y - z + \frac{4}{3}w = 0 \qquad \text{... (2)}$$

Let $\qquad z = c_1, w = c_2$

From (2) $\qquad y = c_1 - \dfrac{4}{3}c_2$

and from (1) $\qquad x = c_1 - \dfrac{5}{3}c_2$

Hence solution is $x = c_1 - \dfrac{5}{3}c_2, y - c_1 - \dfrac{4}{3}c_2, z = c_1, w = c_2$

where c_1 and c_2 are arbitrary numbers.

EXERCISE 3.1

Find the solution of the following system of linear homogeneous equations:

1.
$$x + 2y + 3z = 0$$
$$3x + 4y + 4z = 0$$
$$7x + 10y + 12z = 0$$

2.
$$x + y - 3z + 2w = 0$$
$$2x - y + 2z - 3w = 0$$
$$3x - 2y + z - 4w = 0$$
$$-4x + y - 3z + w = 0$$

3.
$$x + y + z = 0$$
$$2x + 5y + 7z = 0$$
$$2x - 5y + 3z = 0$$

4.
$$3x + 4y - z - 6w = 0$$
$$2x + 3y + 2z - 3w = 0$$
$$2x + y + 4z - 9w = 0$$
$$x + 3y + 13z + 3w = 0$$

5.
$$2x - 3y + z = 0$$
$$x + 2y - 3z = 0$$
$$4x - y - 2z = 0$$

6.
$$x + 2y + 3z = 0$$
$$2x + 3y + 4z = 0$$
$$7x + 13y + 19z = 0$$

7.
$$x + 3y - 2z = 0$$
$$2x - y + 4z = 0$$
$$x - 11y + 14z = 0$$

8.
$$2x - 2y + 5z + 3w = 0$$
$$4x - y + z + w = 0$$
$$3x - 2y + 3z + 4w = 0$$
$$x - 3y + 7z + 6w = 0$$

Hint to Selected Problems

1. Performing $R_2 \to R_2 - 3R_1, R_3 \to R_3 - 7R_1$, we get

$$A = \begin{bmatrix} 1 & 2 & 3 \\ 0 & -2 & -5 \\ 0 & -4 & -9 \end{bmatrix} \Rightarrow |A| = 10$$

$\Rightarrow$ Rank of A is 3 which is equal to the number of unknown. Therefore, the only solution is $x = y = z = 0$.

3. Do same as (1).

Answers

1. $x = 0 = y = z$ **2.** $x = 0 = y = z = w$ **3.** $x = 0 = y = z$

4. $x = 11c_1 + 6c_2, y = -8c_1 - 3c_2, z = c_1, w = c_2$ **5.** $x = 0 = y = z$

6. $x = c, y = -2c, z = c$ **7.** $x = -\dfrac{10}{7}c, y = \dfrac{8}{7}c, z = c$ **8.** $x = \dfrac{5}{9}c, y = 4c, z = \dfrac{7}{9}c, w = c$

3.5 NON-HOMOGENEOUS EQUATIONS

Let us consider a system of equations which are non-homogeneous as follows:

$$\left.\begin{array}{l} a_{11}x_1 + a_{12}x_2 + \ldots + a_{1n}x_n = b_1 \\ a_{21}x_1 + a_{22}x_2 + \ldots + a_{2n}x_n = b_2 \\ \cdots\cdots\cdots\cdots\cdots\cdots\cdots\cdots\cdots \\ a_{m1}x_1 + a_{m2}x_2 + \ldots + a_{mn}x_n = b_m \end{array}\right\} \quad \ldots (1)$$

These are m equations in n unknowns. Let

$$A = \begin{bmatrix} a_{11} & a_{12} & \cdots & a_{1n} \\ a_{21} & a_{22} & \cdots & a_{2n} \\ \vdots & \vdots & \vdots & \vdots \\ a_{m1} & a_{m2} & \cdots & a_{mn} \end{bmatrix}_{m \times n}$$

$$X = \begin{bmatrix} x_1 \\ x_2 \\ \vdots \\ x_n \end{bmatrix}_{n \times 1}, B = \begin{bmatrix} b_1 \\ b_1 \\ \vdots \\ b_m \end{bmatrix}_{m \times 1}$$

Then the system of equations (1) can also be written as

$$AX = B \qquad \ldots (2)$$

This equation is called a matrix equation. If $x_1, x_2, \ldots x_n$ simultaneously satisfy the equation (2), then $(x_1, x_2, \ldots x_n)$ is called the solution of (2).

3.5.1 CONSISTENCY AND INCONSISTENCY

When there exist one or more than one solution of the equation $AX = B$, then the equations are said to be consistent otherwise they are said to be inconsistent.

3.5.2 AUGMENTED MATRIX

The matrix of the type

$$[A \mid B] = \begin{bmatrix} a_{11} & a_{12} & \cdots & a_{1n} & b_1 \\ a_{21} & a_{22} & \cdots & a_{2n} & b_2 \\ \vdots & \vdots & \cdots & \cdots & \cdots \\ a_{m1} & a_{m2} & \cdots & a_{mn} & b_m \end{bmatrix}$$

is called the augmented matrix of the equations.

3.6 CONDITION FOR CONSISTENCY

THEOREM **(Rouche's Theorem).** *The equation $AX = B$ is consistent if and only if the rank of A and the rank of the augmented matrix $[A|B]$ are same.*

Proof. Since the equation is $AX = B$

The matrix A can be written as

$$A = [c_1, c_2, \ldots c_n] \qquad \ldots (1)$$

where $c_1, c_2, \ldots c_n$ are column vectors. Then the equation (1) can be written as

$$[c_1, c_2, \ldots, c_n] \cdot \begin{bmatrix} x_1 \\ x_2 \\ \vdots \\ x_n \end{bmatrix} = B$$

or

$$x_1 c_1 + x_2 c_2 + \ldots + x_n c_n = B \qquad \ldots (2)$$

Suppose the rank of A is r, then A has r linearly independent columns. Let these columns be $c_1, c_2, \ldots c_r$ and these $c_1, c_2, \ldots c_r$ are linearly independent and remaining $(n - r)$ columns are linear combination of $c_1, c_2, \ldots c_r$.

Necessary condition. Suppose the equations are consistent, there must exist $k_1, k_2, \ldots, k_n$ such that

$$k_1 c_1 + k_2 c_2 + \ldots + k_n c_n = B \qquad \ldots (3)$$

But $c_{r+1}, c_{r+2}, \ldots c_n$ is a linear combination of $c_1, c_2, \ldots c_r$ then from (2) it is obvious that B is also a linear combination of $c_1, c_2, \ldots c_r$ and thus $[A|B]$ has the rank r. Hence the rank of A is same as the rank of $[A|B]$.

Sufficient condition. Suppose rank A = rank $[A|B] = r$. This implies that $[A|B]$ has r linearly independent columns. But $c_1, c_2, \ldots c_r$ of $[A|B]$ are already linearly independent.

Thus B can be expressed as

$$B = k_1 c_1 + K_2 c_2 + \ldots + k_r c_r \qquad \ldots (4)$$

where $k_1, k_2, \ldots k_r$ are scalars.

Now, equation (4) becomes

$$B = k_1 c_1 + K_2 c_2 + \ldots + k_r c_r + 0.c_{r+1} + \ldots + 0. c_n \qquad \ldots (5)$$

Comparing (2) and (5), we get $x_1 = k_1, x_2 = k_2, \ldots x_r = k_r, x_{r+1} = 0, \ldots = x_n = 0$ and these values of $x_1, x_2, \ldots x_n$ are the solution of $AX = B$. Hence the equations are consistent.

REMARKS

- The n equations in n unknowns have a unique solution.
- If rank of A < rank of $[A|B]$, then there is no solution.
- If $r = n$, then there will be a unique solution.
- If $r < n$, then $(n - r)$ variables can be assigned arbitrary values. Thus there will be infinite solutions and $(n - r + 1)$ solutions will be linearly independent.
- If $m < n$ and $r \le m \le n$, then equations will have infinite solutions.

WORKING RULE

In order to determine the solutions of the equation $AX = B$, we procced the following steps:

STEP 1.	Reduce the augmented matrix $[A|B]$ to Echelon form by applying E-row transformations only. The Echelon form gives the rank of A and augmented matrix $[A|B]$.
STEP 2.	(i) If the rank of A is not equal to the rank of $[A|B]$, then the system of equations has no solution, *i.e.*, equations are inconsistent. (ii) If the rank of A is equal to the rank of $[A|B]$, then the equations are consistent and they will have unique solution if rank of A = rank of $[A|B]$ = number of unknowns and then will have infinite solutions if rank of A = rank of $[A|B]$ = number of unknowns
STEP 3.	Let $[A'|B']$ be the reduced Echelon form of $[A|B]$. Now reduce the equation $A'X = B'$ to a system of equations, after solving these equations we get the required solution.

Solved Examples

Example 1. *Show that the equations*

$$x + 2y - z = 3, 3x - y + 2z = 1, 2x - 2y + 3z = 2, x - y + z = -1$$

are consistent and solve them.

Solution. The given equations can be written as:

$$\begin{bmatrix} 1 & 2 & -1 \\ 3 & -1 & 2 \\ 2 & -2 & 3 \\ 1 & -1 & 1 \end{bmatrix} \begin{bmatrix} x \\ y \\ z \end{bmatrix} = \begin{bmatrix} 3 \\ 1 \\ 2 \\ -1 \end{bmatrix}, i.e., AX = B$$

Therefore, augmented matrix is

$$[A|B] = \begin{bmatrix} 1 & 2 & -1 & : & 3 \\ 3 & -1 & 2 & : & 1 \\ 2 & -2 & 3 & : & 2 \\ 1 & -1 & 1 & : & -1 \end{bmatrix}$$

performing $R_2 \to R_2 - 3R_1$, $R_3 \to R_3 - 2R_1 \to R_4 \to R_4 - R_1$

we get $$[A|B] = \begin{bmatrix} 1 & 2 & -1 & : & 3 \\ 0 & -7 & 5 & : & -8 \\ 0 & -6 & 5 & : & -4 \\ 0 & -3 & 2 & : & -4 \end{bmatrix}$$

performing $R_2 \to R_2 - R_3$

$$\sim \begin{bmatrix} 1 & 2 & -1 & : & 3 \\ 0 & -1 & 0 & : & -4 \\ 0 & -6 & 5 & : & -4 \\ 0 & -3 & 2 & : & -4 \end{bmatrix}$$

performing $R_3 \rightarrow R_3 - 6R_2, R_4 \rightarrow R_4 - 3R_2$

$$\sim \begin{bmatrix} 1 & 2 & -1 & \vdots & 3 \\ 0 & -1 & 0 & \vdots & -4 \\ 0 & 0 & 5 & \vdots & 20 \\ 0 & 0 & 2 & \vdots & 8 \end{bmatrix}$$

performing $R_3 \rightarrow \dfrac{1}{5}R_3, R_4 \rightarrow \dfrac{1}{2}R_4$

$$\sim \begin{bmatrix} 1 & 2 & -1 & \vdots & 3 \\ 0 & -1 & 0 & \vdots & -4 \\ 0 & 0 & 1 & \vdots & 4 \\ 0 & 0 & 1 & \vdots & 4 \end{bmatrix}$$

performing $R_4 \rightarrow R_4 - R_3$

$$\sim \begin{bmatrix} 1 & 2 & -1 & \vdots & 3 \\ 0 & -1 & 0 & \vdots & -4 \\ 0 & 0 & 1 & \vdots & 4 \\ 0 & 0 & 0 & \vdots & 0 \end{bmatrix}$$

This is an Echelon form and having three non-zero rows. Thus rank A = rank of $[A|B]$ = 3. Therefore the equations are consistent.

and
$$\begin{bmatrix} 1 & 2 & -1 \\ 0 & -1 & 0 \\ 0 & 0 & 1 \\ 0 & 0 & 0 \end{bmatrix} \begin{bmatrix} x \\ y \\ z \end{bmatrix} = \begin{bmatrix} 3 \\ -4 \\ 4 \\ 0 \end{bmatrix}$$

$\therefore \qquad x + 2y - z = 3, -y = -4, z = 4$

Hence the solution is $x = -1$, $y = 4$, $z = 4$

Example 2. *Solve the following equations by matrix method*:
$$x - 2y + 3z = 6$$
$$3x + y - 4z = -7$$
$$5x - 3y + 2z = 5$$

Solution. The given equations can be written as
$$\begin{bmatrix} 1 & -2 & 3 \\ 3 & 1 & -4 \\ 5 & -3 & 2 \end{bmatrix} \begin{bmatrix} x \\ y \\ z \end{bmatrix} = \begin{bmatrix} 6 \\ -7 \\ 5 \end{bmatrix}$$

i.e., $\qquad AX = B$

$\therefore$ Argumented matrix is

$$[A \mid B] = \begin{bmatrix} 1 & -2 & 3 & \vdots & 6 \\ 3 & 1 & -4 & \vdots & -7 \\ 5 & -3 & 2 & \vdots & 5 \end{bmatrix}$$

performing $R_2 \rightarrow R_2 - 3R_1, R_3 \rightarrow R_3 - 5R_1$, we get

$$[A \mid B] = \begin{bmatrix} 1 & -2 & 3 & \vdots & 6 \\ 0 & 0 & -13 & \vdots & -25 \\ 0 & 7 & -13 & \vdots & -25 \end{bmatrix}$$

performing $R_3 \to R_3 - R_2$

$$\sim \begin{bmatrix} 1 & -2 & 3 & : & 6 \\ 0 & 7 & -13 & : & -25 \\ 0 & 0 & 0 & : & 0 \end{bmatrix}$$

This is an Echelon form and having two non-zero rows and rank A = rank $[A|B]$ = 2. Thus the equations are consistent.

$$\begin{bmatrix} 1 & -2 & 3 \\ 0 & 7 & -13 \\ 0 & 0 & 0 \end{bmatrix} \begin{bmatrix} x \\ y \\ z \end{bmatrix} = \begin{bmatrix} 6 \\ -25 \\ 5 \end{bmatrix}$$

i.e.,

$$x - 2y + 3z = 6$$
$$7y - 13z = -25$$

Let $z = c$, then

$$y = -\frac{25}{7} + \frac{13}{7}c$$

$$x = -\frac{8}{7} + \frac{5}{7}c$$

Hence the solution is $x = -\frac{8}{7} + \frac{5}{7}c, y = -\frac{25}{7} + \frac{13}{7}c, z = c$

where c is an arbitrary constant.

Example 3. *Investigate for what values of λ, μ the simulataneous equations*

$$x + y + z = 6, x + 2y + 3z = 10, x + 2y + \lambda z = \mu$$

have (i) no solution (ii) a unique solution (iii) infinite solution.

Solution. The given equations can be written as

$$\begin{bmatrix} 1 & 1 & 1 \\ 1 & 2 & 3 \\ 1 & 2 & \lambda \end{bmatrix} \begin{bmatrix} x \\ y \\ z \end{bmatrix} = \begin{bmatrix} 6 \\ 10 \\ \mu \end{bmatrix}$$

i.e., $AX = B$

Therefore, augmented matrix is

$$[A|B] = \begin{bmatrix} 1 & 1 & 1 & : & 6 \\ 1 & 2 & 3 & : & 10 \\ 1 & 2 & \lambda & : & \mu \end{bmatrix}$$

performing $R_2 \to R_2 - R_1, R_3 \to R_3 - R_1$, we get

$$\sim \begin{bmatrix} 1 & 1 & 1 & : & 6 \\ 0 & 1 & 2 & : & 4 \\ 0 & 1 & \lambda-1 & : & \mu-6 \end{bmatrix}$$

performing $R_3 \to R_3 - R_2$

$$\sim \begin{bmatrix} 1 & 1 & 1 & : & 6 \\ 0 & 1 & 2 & : & 4 \\ 0 & 0 & \lambda-3 & : & \mu-10 \end{bmatrix}$$

If $\lambda \neq 3$, then rank A = rank $[A|B]$ = 3. Thus in this case a unique solution exists.
If $\lambda = 3$ and $\mu_0 \neq 10$, then rank A=2, rank $[A|B]$ is 3. Thus rank $A \neq$ rank $[A|B]$.

Hence in this case equations are inconsistent.

If $\lambda = 3$ and $\mu = 10$, then rank A = rank $[A|B]$ = 2. Thus in this case infinite solutions exist.

Example 4. *For what values of η the equations $x+y+z=1$, $x+2y+4z = \eta$, $x+4y+10z = \eta^2$ have a solution? Solve them completely in each case.*

Solution. The given system of equations can be written as

$$AX = B \qquad \qquad \qquad \text{... (1)}$$

where

$$A = \begin{bmatrix} 1 & 1 & 1 \\ 1 & 2 & 4 \\ 1 & 4 & 10 \end{bmatrix}, X = \begin{bmatrix} x \\ y \\ z \end{bmatrix}, B = \begin{bmatrix} 1 \\ \eta \\ \eta^2 \end{bmatrix}$$

Augmented matrix $[A|B]$ is given by

$$\begin{bmatrix} 1 & 1 & 1 & \vdots & 1 \\ 1 & 2 & 4 & \vdots & \eta \\ 1 & 4 & 10 & \vdots & \eta^2 \end{bmatrix}$$

Applying $R_2 \to R_2 - R_1, R_3 \to R_3 - R_1$, we get

$$\sim \begin{bmatrix} 1 & 1 & 1 & \vdots & 1 \\ 0 & 1 & 3 & \vdots & \eta-1 \\ 0 & 3 & 9 & \vdots & \eta^2-1 \end{bmatrix}$$

Applying $R_3 \to R_3 - 3R_2$, we get

$$\sim \begin{bmatrix} 1 & 1 & 1 & \vdots & 1 \\ 0 & 1 & 3 & \vdots & \eta-1 \\ 0 & 0 & 0 & \vdots & \eta^2-3\eta+2 \end{bmatrix}$$

This last equivalent matrix is in Echelon form. The given system of equations will have the solutions if

$$\text{rank of } A = \text{rank of } [A|B]$$

For Echelon form, the rank of A is 2 and the augmented matrix $[A|B]$ will have rank 2 if

$$\eta^2 - 3\eta + 2 = 0$$

i.e., if $\qquad (\eta - 2)(\eta - 1) = 0$

i.e., if $\qquad \qquad \eta = 1, 2$

The last equivalent matrix gives the system of equations as follows:

$$\begin{bmatrix} 1 & 1 & 1 \\ 0 & 1 & 3 \\ 0 & 0 & 0 \end{bmatrix}\begin{bmatrix} x \\ y \\ z \end{bmatrix} = \begin{bmatrix} 1 \\ \eta-1 \\ \eta^2-3\eta+2 \end{bmatrix}$$

$$\Rightarrow \qquad \left. \begin{array}{c} x+y+z = 1 \\ y+3z = \eta-1 \end{array} \right\} \qquad \qquad \text{... (2)}$$

Since rank of A = rank of $[A|B]$ if $\eta = 1$ and $\eta = 2$

Now we have two cases:

Case I: When $\eta = 1$

From (2), we have

$$\left.\begin{array}{r} x + y + z = 1 \\ y + 3z = 0 \end{array}\right\} \qquad \cdots (3)$$

Since rank of $A =$ rank of $[A|B] = 2$ and number of unknowns is 3, therefore we will have $3 - 2 = 1$ unknown to be assigned.

Let us assign z to be c_1, therefore put $z = c_1$ in $y + 3z = 0$, we get $y = -3c_1$.

Again putting $y = -3c_1$ and $z = c_1$ in $x + y + z = 1$, we get $x = 1 + 2c_1$

Thus, in this case the solutions are

$$x = 1 + 2c_1, y = -3c_1, z = c_1$$

where c_1 is an arbitrary number.

Case II : When $\eta = 2$

From (2), we have

$$\left.\begin{array}{r} x + y + z = 1 \\ y + 3z = 1 \end{array}\right\} \qquad \cdots (4)$$

Let us assign z to be c_2, therefore, putting $z = c_2$ in $y + 3z = 1$, we get $y = 1 - 3c_2$.

Again, putting $z = c_2, y = 1 - 3c_2$ in $x + y + z = 1$, we get $x = 2c_2$.

Thus, in this case the solutions are

$$x = 2c_2 , y = 1 - 3c_2, z = c_2$$

Where c_2 is an arbitrary number.

❦ EXERCISE 3.2

1. Use matrix method to solve the equations

   $2x - y + 3z = 9, x + y + z = 6, x - y + z = 2.$

2. Show that the equations $x - 3y - 8z + 10 = 0$, $3x + y - 4z = 0$, $2x + 5y + 6z - 13 = 0$ are consistent and solve them.

3. Examine if the system of equations
   $x + y + 4z = 6, 3x + 2y - 2z = 9$,
   $5x + y + 2z = 13$
   is consistent. Find also the solution if it exists.

4. For what values of λ will the following equtions fail to have a unique solution
   $3x - y + \lambda z = 1, 2x + y + z = 2, x + 2y - \lambda z = -1$

   Will the equations have any solution for these values of λ?

5. Solve $2x + 3y + z = 9, x + 2y + 3z = 6,$
   $3x + y + 2z = 8.$

   Solve the following equations by matrix method:

6. $5x + 3y + 7z = 4, 3x + 26y - 2z = 9, 7x + 2y + 10z = 5.$

7. $5x - 6y + 4z = 15, 7x + 4y - 3z = 19, 2x + y + 6z = 46.$

8. $x - y + 2z = 4, 3x + y + 4z = 6, x + y + z = 1.$

9. $x + y + z = 6, x + 2y + 3z = 4, 3x + y - 4z = 0.$

10. $2x - y + 3z = 8, -x + 2y + z = 4, 3x + y - 4z = 0.$

11. Show that the following equations are inconsistent
    $2x - y + z = 4, 3x - y + z = 6, 4x - y + 2z = 7, -x + y - z = 9.$

12. Show that the equations are inconsistent
    $x - 4y + 7z = 14, 3x + 8y - 2z = 13, 7x - 8y + 26z = 5.$

13. Prove that the following system of equations have a unique solution
    $5x + 3y + 14z = 4, y + 2z = 1, x - y + 2z = 0.$

14. Solve the following equations by matrix mehod:
    $x + y + z = 9, 2x + 5y + 7z = 52, 2x + y - z = 0.$

Hint to Selected Problems

1. Consider the augmented matrix and perform the following opeartions sequentially

$R_1 \leftrightarrow R_2, R_2 \rightarrow R_2 - 2R_1, R_3 \rightarrow R_3 - R_1, R_3 \rightarrow R_3 - \dfrac{2}{3} R_2$.

2. Here, Rank $(A) = 2$, which is less than the number of unknowns, therefore, given system of equations have infinite number of solutions.

3. The rank of augmented matrix is equal to the rank of (A). Therefore, the given system of equation is consistent.

4. The coefficient matrix A is non-singular if $\lambda \neq -\dfrac{7}{2}$. Thus the given system of equations have a unique solution if $\lambda \neq \dfrac{7}{2}$.

11. The rank of augmented matrix = 4

   Rank of A is 3.

   Hence the given system of equations is inconsistent.

Answers

1. $x = 1, y = 2, z = 3$

2. $x = 2c - 1, y = 3 - 2c, z = c$

3. Consistent; $x = 2, y = 2, z = \dfrac{1}{2}$

4. $\lambda \neq -\dfrac{7}{2}$ solution is unique; $\lambda = -\dfrac{7}{2}$, no solution.

5. $x = \dfrac{35}{18}, y = \dfrac{29}{18}, z = \dfrac{5}{18}$

6. $x = \dfrac{7}{11}, y = \dfrac{3}{11}, z = 0$

7. $x = 3, y = 4, z = 6$

8. $x = \dfrac{5}{2} - \dfrac{3}{2}c, y = -\dfrac{3}{2} + \dfrac{1}{2}c, z = c$

9. $x = c - 2, y = 8 - 2c, z = c$

10. $x = 2, y = 2, z = 2$ 14. $x = 1, y = 3, z = 4$

3.7 GAUSS ELIMINATION METHOD

In this method, the variables from the system of linear equations are eliminated successively and the system of equations is therefore reduced to an upper triangular system from which the variable are determined by back substitution. This method is described as follows: Let us consider a system of linear equation

$$AX = B \qquad \qquad \ldots(1)$$

assuming det $A \neq 0$. Equation (1) has the following form:

$$\left. \begin{array}{l} a_{11}x_1 + a_{12}x_2 + \ldots + a_{1n}x_n = b_1 \\ a_{21}x_1 + a_{22}x_2 + \ldots + a_{2n}x_n = b_2 \\ \cdots \quad \cdots \quad \cdots \quad \cdots \quad \cdots \quad \cdots \\ \cdots \quad \cdots \quad \cdots \quad \cdots \quad \cdots \quad \cdots \\ a_{n1}x_1 + a_{n2}x_2 + \ldots + a_{nn}x_n = b_n \end{array} \right\} \qquad \ldots(2)$$

Assuming $a_{11} \neq 0$ and divide the first equation by a_{11} and then we subtract this equation multiplied by $a_{21}, a_{31}, \ldots, a_{n1}$ from second, third ... nth equation of (2), we get

$$\left. \begin{array}{l} x_1 + a'_{12}x_2 + \ldots + a'_{1n}x_n = b'_1 \\ a'_{22}x_2 + \ldots + a'_{2n}x_n = b'_2 \\ \cdots \quad \cdots \quad \cdots \quad \cdots \quad \cdots \\ \cdots \quad \cdots \quad \cdots \quad \cdots \quad \cdots \\ a'_{n2}x_2 + \ldots + a'_{n2}x_n = b'_n \end{array} \right\} \qquad \ldots(3)$$

Next, we divide second equation of (3) by a'_{22} (assuming $a'_{22} \neq 0$) and subtract this equation multiplied by $a'_{32}, a'_{42}, \ldots, a'_{n2}$ from third, fourth ... nth equation of (3), we get

$$\left. \begin{array}{l} x_1 + a'_{12}x_2 + \ldots + a'_{1n}x_n = b'_1 \\ x_2 + a''_{23}x_3 + \ldots + a''_{2n}x_n = b''_2 \\ \quad a''_{33}x_3 + \ldots + a''_{3n}x_n = b''_3 \\ \ldots \quad \ldots \quad \ldots \quad \ldots \quad \ldots \\ \quad a''_{3n}x_3 + \ldots + a''_{nn}x_n = b''_n \end{array} \right\} \quad \ldots(4)$$

Continuing in this way we get a system of equation as follows:

$$\left. \begin{array}{l} x_1 + c_{12}x_2 + c_{13}x_3 + \ldots + c_{1n}x_n = d_1 \\ \quad x_2 + c_{23}x_3 + \ldots + c_{2n}x_n = d_2 \\ \qquad \vdots \\ \qquad \vdots \\ \qquad c_{nn}x_n = d_n \end{array} \right\} \quad \ldots(5)$$

This is a form of upper triangular system. From back substitution we can find the solution of the system of given equations.

WORKING RULE

Let us consider these equations

$$\left. \begin{array}{l} a_{11}x_1 + a_{12}x_2 + a_{13}x_3 = b_1 \\ a_{21}x_1 + a_{22}x_2 + a_{23}x_3 = b_2 \\ a_{31}x_1 + a_{32}x_2 + a_{33}x_3 = b_3 \end{array} \right\} \quad \ldots(6)$$

STEP 1. First, eliminate x_1 from second and third equations. Assuming $a_{11} \neq 0$, now dividing first equation by a_{11} and then subtract from second and third after multiplied by a_{21} and respectively we get

$$\left. \begin{array}{l} x_1 + a'_{12}x_2 + a'_{13}x_3 = b'_1 \\ \quad a'_{22}x_2 + a'_{23}x_3 = b'_2 \\ \quad a'_{32}x_2 + a'_{33}x_3 = b'_3 \end{array} \right\} \quad \ldots(7)$$

where $a'_{12} = \dfrac{a_{12}}{a_{11}}$, $a'_{13} = \dfrac{a_{13}}{a_{11}}$, $a'_{22} = a_{22} - a_{21}a'_{12}, a'_{23} = a_{23} - a_{21}a'_{13}$

$a'_{32} = a_{32} - a_{31}a'_{12}, a'_{33} = a_{33} - a_{31}a'_{13}$,

$b'_1 = \dfrac{b_1}{a_{11}}, b'_2 = b_2 - a_{21}b'_1, b'_3 = b_3 - a_{31}b'_1$

STEP 2. Now eliminating x_2 from third equation in (7).

Again assuming $a_{22}' \neq 0$. Dividing second equation in (7) by a_{22}' and then subtract from third equation after multiplied by a_{32}' we get

$$\left. \begin{array}{l} x_1 + a'_{12}x_2 + a'_{13}x_3 = b'_1 \\ \quad x_2 + a''_{23}x_3 = b'_2 \\ \quad\quad a''_{33}x_3 = b'_3 \end{array} \right\} \quad \ldots(8)$$

where $a''_{23} = \dfrac{a'_{23}}{a'_{22}}, a''_{33} = a'_{33} - a'_{32}a''_{23}$, $b''_2 = \dfrac{b'_2}{a'_{22}}, b''_3 = b_3' - a'_{32}b''_2$.

STEP 3. Evaluating x_1, x_2 and x_3 from (8) by back substitution.

REMARKS

- The coefficient a_{11}, a'_{22} and a''_{33} are called pivots.
- This method will fail if any one of the pivots a_{11}, a'_{22} and a''_{33} becomes zero. In such cases, we rewrite the equations in a different order so that the pivots are non-zero.
- From each of the procedure, the largest coefficient of x is chosen as pivot element.
- This method proposes a systematic astrology for reducing the system of equations to the upper triangular form using the forward elimination approach and then for obtaining values of unknowns using back substitution process.

Solved Examples

Example 1. *Solve the following equations by Gauss's elimination method*
$$6x + 3y + 2z = 6$$
$$6x + 4y + 3z = 0$$
$$20x + 15y + 12z = 0.$$

Solution. Here pivot element is 6. Now Divide first equation by 6, we get
$$x + \frac{1}{2}y + \frac{1}{3}z = 1 \quad\quad\quad ...(1)$$
Now eliminating x from second and third equation with the help of (1). Subtract (1) multiplied by 6 and 20 from second and third equation, respectively we get
$$y + z = -6 \quad\quad\quad ...(2)$$
$$5y + \frac{16}{3}z = -20 \quad\quad\quad ...(3)$$
Now eliminating y from (3) with the help of (2), we get
$$\left(\frac{16}{3} - 5\right)z = -20 + 30$$
$$\frac{1}{3}z = 10 \Rightarrow z = 30$$
Substitute the value of z into (2), we get
$$y = -6 - 30 = -36$$
and again substitute the values of y and z into (1), we get
$$x + \frac{1}{2}(-36) + \frac{1}{3}(30) = 1$$
$$x - 18 + 10 = 1 \Rightarrow x = 9$$
Hence the solution of the equations are
$$x = 9, \ y = -36, \ z = 30.$$

Example 2. *By Gauss's elimination method, solve the following equations*
$$5x - y - 2z = 142$$
$$x - 3y - z = -30$$
$$2x - y - 3z = -50$$

Solution. The largest coefficient in first equation is 5, which is pivot element. So divide first equation by 5, we get
$$x - \frac{1}{5}y - \frac{2}{5}z = \frac{142}{5} \quad\quad\quad ...(1)$$

Now eliminating x from second and third equation with help of (1) we get

$$-\frac{14}{5}y - \frac{3}{5}z = -\frac{292}{5} \qquad \qquad \ldots(2)$$

$$-\frac{3}{5}y - \frac{11}{5}z = -\frac{309}{5} \qquad \qquad \ldots(3)$$

Eliminating y from (2) and (3), we get

$$-\frac{145}{5}z = -\frac{3450}{5}$$

or

$$z = \frac{3450}{145} = 23.79$$

Substitute the value of z into (3) we get

$$-\frac{3}{5}y - \frac{11}{5}(23.79) = -\frac{309}{5}$$

$$-\frac{3}{5}y = -\frac{309}{5} + \frac{11(23.79)}{5}$$

$$-3y = -309 + 11(23.79)$$

$$-3y = -47.31$$

or

$$y = 15.77$$

Substitute the values of y and z into (1), we get

$$x - \frac{1}{5}(15.77) - \frac{2}{5}(23.79) = \frac{142}{5}$$

$$x = \frac{142}{5} + \frac{15.77}{5} + \frac{2(23.79)}{5} = \frac{205.35}{5}$$

or

$$x = 41.07$$

Hence the solution are given by

$$x = 41.07, y = 15.77, z = 23.79.$$

Example 3. *Using Gauss's elimination method solve*

$$2x_1 + 4x_2 + x_3 = 3$$
$$3x_1 + 2x_2 - 2x_3 = 2$$
$$x_1 - x_2 + x_3 = 6$$

Solution. Dividing first equation by 2, we get

$$x_1 + 2x_2 + \frac{1}{2}x_3 = \frac{3}{2} \qquad \qquad \ldots(1)$$

Multiplying (1) by 3 and subtract from second and also subtract (1) from third of the given equation, we get

$$4x_2 + \frac{7}{2}x_3 = \frac{5}{2} \qquad \qquad \ldots(2)$$

$$-3x_2 + \frac{1}{2}x_3 = \frac{9}{2} \qquad \qquad \ldots(3)$$

Now dividing (2) by 4 and subtract after multiplies by –3 from (3), we get

$$25x_3 = 51$$

or
$$x_3 = \frac{51}{25} = 2.04$$

Substitute the value of x_3 into (2), we get

$$+4x_2 + \frac{7}{2}(2.04) = \frac{5}{2}$$

$$4x_2 = \frac{5}{2} - \frac{7(2.04)}{2}$$

$$= \frac{5 - 14.28}{2}$$

$$\Rightarrow \qquad x_2 = -\frac{9.28}{8} = -1.16$$

Now substitute the value of x_2 and x_3 into (1), we get

$$x_1 + 2(-1.16) + \frac{1}{2}(2.04) = \frac{3}{2}$$

$$x_1 = \frac{3}{2} + 2(1.16) - \frac{1}{2}(2.04)$$

$$= \frac{3 + 4.64 - 2.04}{2} = \frac{5.6}{2} = 2.8$$

or
$$x_1 = 2.8$$

Hence the solutions are given by

$$x_1 = 2.8, \; x_2 = -1.16, \; x_3 = 2.04.$$

Example 4. *Solve by the Gauss's elimination method.*

$$2x + y + 4z = 12$$
$$8x - 3y + 2z = 23$$
$$4x + 11y - z = 33$$

Solution. Dividing first equation by 2, we get

$$x + \frac{1}{2}y + 2z = 6 \qquad\qquad \text{...(1)}$$

Now subtract (1) after multiplied by 8 and 4 respectively from second and third equation, we get

$$-7y - 14x = -45 \qquad\qquad \text{...(2)}$$
$$9y - 9z = 9 \qquad\qquad \text{...(3)}$$

Now multiplying (4) by 9 and subtract from (3), we get

$$-27z = 9 - \frac{405}{7}$$

or
$$-27z = -\frac{342}{7} \qquad\qquad \text{...(5)}$$

Hence the system of equations reduces to upper triangular form as follows:

$$\left.\begin{array}{l} x + \dfrac{1}{2}y + 2z = 6 \\[2mm] y + 2z = \dfrac{45}{7} \\[2mm] -27z = -\dfrac{342}{7} \end{array}\right\} \qquad\qquad \text{...(6)}$$

By back substitution , we get

$$z = \frac{342}{189} = 1.81$$

and

$$y + 2(1.81) = \frac{45}{7}$$

$$\Rightarrow \qquad y = \frac{45}{7} - 2(1.81) = 6.43 - 3.62$$

$$= 2.81$$

and

$$x + \frac{1}{2}(2.81) + 2(1.81) = 6$$

$$\therefore \qquad x = 6 - \frac{1}{2}(2.81) - 2(1.81)$$

$$= 0.975$$

Hence the solution is $x = 0.975$, $y = 2.81$, $z = 1.81$.

Example 5. *Apply Gauss's elimination method to solve the equations*

$$x + 4y - z = -5$$
$$x + y - 6z = -12$$
$$3x - y - z = 4$$

Solution. Eliminating x from second and third equation with the help of first equation. Subtract first equation from second and after multiplied by 3 from third respectively, we get the system of equations as follows:

$$x + 4y - z = -5 \qquad \qquad \qquad ...(1)$$
$$-3y - 5z = -7 \qquad \qquad \qquad ...(2)$$
$$-13y + 2z = 19 \qquad \qquad \qquad ...(3)$$

Elimination y from (3) with help of (2). Divide (2) by -3 and then this equation is subtracted after multiplies by -13 from (3), we get

$$\left. \begin{array}{l} x + 4y - z = -5 \\[2mm] y + \dfrac{5}{3}z = \dfrac{7}{3} \\[2mm] \dfrac{71}{3}z = \dfrac{148}{3} \end{array} \right\} \qquad \qquad ...(4)$$

By back substitution from (4), we get

$$z = \frac{148}{71}$$

and

$$y = \frac{7}{3} - \frac{5}{3}z = \frac{7}{3} - \frac{5}{3}\left(\frac{148}{71}\right)$$

$$\Rightarrow \qquad y = -\frac{81}{71}$$

and

$$x = -5 - 4y + z$$

$$= -5 - 4\left(-\frac{81}{71}\right) + \left(\frac{148}{71}\right)$$

$$= -5 + \frac{472}{71} + \frac{117}{71}$$

Hence the solution are $x = \dfrac{117}{71}, y = -\dfrac{81}{71}, z = \dfrac{148}{71}$.

Example 6. *Solve the following system by Gauss's elimination method :*

$$2x + y + z = 10$$
$$3x + 2y + 3z = 18$$
$$x + 4y + 9z = 4$$

Solution. We have

$$2x + y + z = 10 \qquad \qquad \text{...(1)}$$
$$3x + 2y + 3z = 18 \qquad \qquad \text{...(2)}$$
$$x + 4y + 9z = 4 \qquad \qquad \text{...(3)}$$

Divide (1) and 2 and subtract after multiplied by 3 from (2) then subtract from (3), we get

$$x + \frac{1}{2}y + \frac{1}{2}z = 5 \qquad \qquad \text{...(4)}$$

$$\frac{1}{2}y + \frac{3}{2}z = 3 \qquad \qquad \text{...(5)}$$

$$\frac{7}{2}y + \frac{17}{2}z = 11 \qquad \qquad \text{...(6)}$$

Now divide (5) by $\dfrac{1}{2}$ and then subtract after multiplied by $\dfrac{7}{2}$ from (6) we get,

$$x + \frac{1}{2}y + \frac{1}{2}z = 5 \qquad \qquad \text{...(7)}$$
$$y + 3z = 6 \qquad \qquad \text{...(8)}$$
$$-2z = -10 \qquad \qquad \text{...(9)}$$

From back substitution in (9), (8) and (7) we get

$$z = 5$$

and

$$y + 3z = 6$$
$$y + 3(5) = 6$$
$$y = 6 - 15 \Rightarrow y = -9$$
$$y = -9$$

and

$$x + \frac{1}{2}(-9) + \frac{1}{2}(5) = 5$$

$$x = 5 + \frac{9}{2} - \frac{5}{2}$$

Hence the solution is $x = 7, y = -9, z = 5$.

Example 7. *By Gauss's elimination method, solve*

$$4x + 11y - z = 33$$
$$x + y + 4z = 12$$
$$8x - 3y + 2z = 20$$

Solution. Given equation are

$$x + y + 4z = 12 \qquad \qquad \text{...(1)}$$
$$4x + 11y - z = 33 \qquad \qquad \text{...(2)}$$
$$8x - 3y + 2z = 20 \qquad \qquad \text{...(3)}$$

Eliminating x from (2) and (3) so subtract (1) after multiplied by 4 and 8 from (2) and (3), we get

$$x + y + 4z = 12 \qquad \text{...(4)}$$
$$7y - 17z = -15 \qquad \text{...(5)}$$
$$-11y - 30z = -76 \qquad \text{...(6)}$$

Now divide (5) by 7 and then subtract after multiplied by -11 from (6), we get

$$x + y + 4z = 12 \qquad \text{...(7)}$$
$$y - \frac{17}{7}z = 12 \qquad \text{...(8)}$$
$$-\frac{397}{7}z = -\frac{697}{7} \qquad \text{...(9)}$$

By back substitution in (9), (8) and (7) we get

$$z = \frac{697}{397} = 1.756$$

From (8)

$$y = \frac{-15}{7} + \frac{17}{7}z = -\frac{15}{7} + \frac{17}{7}\left(\frac{697}{397}\right)$$
$$= \frac{1}{7}\left(\frac{5894}{397}\right) = \frac{5894}{2779} = 2.121$$

From (7) $\qquad x + y + z = 12$

$$\Rightarrow \qquad x + \frac{5894}{2779} + 4\left(\frac{697}{397}\right) = 12$$

$$\Rightarrow \qquad x = 12 - \frac{5894}{2779} + 4\left(\frac{697}{397}\right) = \frac{7938}{2779} = 2.856$$

Hence the solution is $x = 2.856, y = 2.121, z = 1.756$.

Example 8. *Solve by Gauss's elimination method*

$$x + 2y + z = 3$$
$$2x + 3y + 3z = 10$$
$$3x - y + 2z = 13$$

Solution. Given equation are

$$x + 2y + z = 3 \qquad \text{...(1)}$$
$$2x + 3y + 3z = 10 \qquad \text{...(2)}$$
$$3x - y + 2z = 13 \qquad \text{...(3)}$$

Here pivot element of (1) is 1. Now eliminating x from (2) and (3) by subtracting (1) after multiplied by 2 and 3 respectively from (2) and (3), we get

$$x + 2y + z = 3 \qquad \text{...(4)}$$
$$-y + z = 4 \qquad \text{...(5)}$$
$$-7y - z = 4 \qquad \text{...(6)}$$

Now eliminating y from (6) with the help of (5) by subtracting (5) after multiplied by −7 from (6), we get

$$x + 2y + z = 3 \qquad \qquad \text{...(7)}$$
$$-y + z = 4 \qquad \qquad \text{...(8)}$$
$$6z = 32 \qquad \qquad \text{...(9)}$$

By back substitution from (7), (8) and (9), we get

From (9) $$z = \frac{32}{6} = \frac{16}{3}$$

From (8) $$-y = 4 - z$$

$$-y = 4 - \frac{32}{6} = -\frac{8}{6}$$

∴ $$y = \frac{8}{6} = \frac{4}{3}$$

From (7) $$x + 2y + z = 3$$

$$x + 2\left(\frac{8}{6}\right) + \frac{32}{6} = 3$$

$$x = 3 - \frac{32}{6} - \frac{16}{6}$$

$$= \frac{18 - 48}{6} = -\frac{30}{6}$$

$$x = -5$$

Hence the solution is $x = -5, y = \frac{4}{3}, z = \frac{16}{3}$.

3.8 LU DECOMPOSITION METHOD

This method is based on the fact that every square matrix A can be expressed as the form LU where L is a unit lower triangular matrix while U is upper triangular matrix and provided all the principal minors of A are non-singular.

That is, If $A = [a_{ij}]_{n \times n}$, then

$$a_{11} \neq 0, \begin{vmatrix} a_{11} & a_{12} \\ a_{21} & a_{22} \end{vmatrix} \neq 0, \begin{vmatrix} a_{11} & a_{12} & a_{13} \\ a_{21} & a_{22} & a_{23} \\ a_{31} & a_{32} & a_{33} \end{vmatrix} \neq 0, \text{ and so on.}$$

For simplicity and understanding this method, let us consider a system of three equations

$$a_{11}x_1 + a_{12}x_2 + a_{13}x_3 = b_1$$
$$a_{21}x_1 + a_{22}x_2 + a_{23}x_3 = b_2$$
$$a_{31}x_1 + a_{32}x_2 + a_{33}x_3 = b_3$$

These equations can be written in matrix form as follows:

$$AX = B$$

Where $$A = \begin{bmatrix} a_{11} & a_{12} & a_{13} \\ a_{21} & a_{22} & a_{23} \\ a_{31} & a_{32} & a_{33} \end{bmatrix}, X = \begin{bmatrix} x_1 \\ x_2 \\ x_3 \end{bmatrix}, B = \begin{bmatrix} b_1 \\ b_2 \\ b_3 \end{bmatrix} \qquad \text{... (1)}$$

Let $$A = LU$$

where
$$A = \begin{bmatrix} 1 & 0 & 0 \\ l_{21} & 1 & 0 \\ l_{31} & l_{32} & 1 \end{bmatrix}, U = \begin{bmatrix} u_{11} & u_{12} & u_{13} \\ 0 & u_{22} & u_{23} \\ 0 & 0 & u_{33} \end{bmatrix} \qquad \ldots (2)$$

Then from (2)

$$\begin{bmatrix} a_{11} & a_{12} & a_{13} \\ a_{21} & a_{22} & a_{23} \\ a_{31} & a_{32} & a_{33} \end{bmatrix} = \begin{bmatrix} 1 & 0 & 0 \\ l_{21} & 1 & 0 \\ l_{31} & l_{32} & 1 \end{bmatrix} \begin{bmatrix} u_{11} & u_{12} & u_{13} \\ 0 & u_{22} & u_{23} \\ 0 & 0 & u_{33} \end{bmatrix}$$

or
$$\begin{bmatrix} a_{11} & a_{12} & a_{13} \\ a_{21} & a_{22} & a_{23} \\ a_{31} & a_{32} & a_{33} \end{bmatrix} = \begin{bmatrix} u_{11} & u_{12} & u_{13} \\ l_{21}u_{11} & l_{21}u_{12} + u_{22} & l_{21}u_{13} + u_{23} \\ l_{31}u_{11} & l_{31}u_{12} + l_{32}u_{22} & l_{31}u_{13} + l_{32}u_{23} + u_{33} \end{bmatrix}$$

Comparing the two matrices, we get

(i) $\qquad u_{11} = a_{11}, u_{12} = a_{12}, u_{13} = a_{13}$

(ii) $\qquad l_{21}u_{11} = a_{21}$

or $\qquad l_{21} = \dfrac{a_{21}}{u_{11}} = \dfrac{a_{21}}{a_{11}}$

(iii) $\qquad l_{31}u_{11} = a_{31}$

or $\qquad l_{31} = \dfrac{a_{31}}{u_{11}} = \dfrac{a_{31}}{a_{11}}$

(iv) $\qquad l_{21}u_{12} + u_{22} = a_{22}$

or $\qquad l_{21} = \dfrac{a_{22} - u_{22}}{u_{12}}$

or $\qquad u_{22} = a_{22} - l_{21}u_{12} = a_{22} - \dfrac{a_{21}}{a_{11}}.a_{12}$

(v) $\qquad l_{21}u_{13} + u_{23} = a_{23}$

$$u_{23} = a_{23} - l_{21}u_{13}$$
$$u_{23} = a_{23} - \dfrac{a_{21}}{a_{11}}a_{13}$$

(vi) $\qquad l_{31}u_{12} + l_{32}u_{22} = a_{32}$

or $\qquad l_{32} = \dfrac{1}{u_{22}}\left[a_{32} - \dfrac{a_{31}}{a_{11}}a_{12} \right] = \dfrac{\left[a_{32} - \dfrac{a_{31}a_{12}}{a_{11}} \right]}{\left[a_{22} - \dfrac{a_{21}a_{12}}{a_{11}} \right]}$

(vii) $\qquad l_{31}u_{13} + l_{32}u_{23} + u_{33} = a_{33}$

or $\qquad u_{33} = a_{33} - l_{31}u_{13} - l_{32}u_{23}$

$$= a_{33} - \dfrac{a_{31}}{a_{11}}.a_{13} - \dfrac{1}{u_{22}}\left[a_{32} - \dfrac{a_{31}}{a_{11}}.a_{12} \right]\left[a_{23} - \dfrac{a_{21}}{a_{11}}.a_{13} \right]$$

$$= a_{33} - \dfrac{a_{31}a_{13}}{a_{11}} \dfrac{\left[a_{32} - \dfrac{a_{31}a_{12}}{a_{11}} \right]\left[a_{23} - \dfrac{a_{21}}{a_{11}}.a_{13} \right]}{\left[a_{22} - \dfrac{a_{21}.a_{12}}{a_{11}} \right]}$$

Thus from above we can find the elements of L and U. Now L and U are therefore obtained.

Replacing A by LU in (1), we get

$$LUX = B \qquad \qquad \text{... (3)}$$

Now let

$$UX = Y \qquad \qquad \text{... (4)}$$

where

$$Y = \begin{bmatrix} y_1 \\ y_2 \\ y_3 \end{bmatrix}$$

From (3) and (4), we get

$$LY = B$$

or

$$\begin{bmatrix} 1 & 0 & 0 \\ l_{21} & 1 & 0 \\ l_{31} & l_{32} & 1 \end{bmatrix} \begin{bmatrix} y_1 \\ y_2 \\ y_3 \end{bmatrix} = \begin{bmatrix} b_1 \\ b_2 \\ b_3 \end{bmatrix}$$

From this equation, we get

$$\left. \begin{aligned} y_1 &= b_1 \\ y_2 &= b_2 - l_{21} y_1 \\ y_3 &= b_3 - l_{31} b_1 - l_{32} b_2 \end{aligned} \right\} \qquad \text{... (5)}$$

Now from (4)

$$UX = Y$$

$$\begin{bmatrix} u_{11} & u_{12} & u_{13} \\ 0 & u_{22} & u_{23} \\ 0 & 0 & u_{33} \end{bmatrix} \begin{bmatrix} x_1 \\ x_2 \\ x_3 \end{bmatrix} = \begin{bmatrix} y_1 \\ y_2 \\ y_3 \end{bmatrix}$$

From this equation, we get

$$x_3 = \frac{y_3}{u_{33}},$$

$$x_2 = \frac{y_2 - u_{23} x_3}{u_{22}}$$

and

$$x_1 = \frac{y_1 - u_{12} x_2 - u_{13} x_3}{u_{11}}.$$

With the help of L and U, x_1, x_2, x_3 can be calculated.

REMARKS

- LU decomposition method is superior to Gauss's elimination method.
- This method is applicable if the coefficient matrix can be expressed as the product of lower and upper triangular matrix.
- This method can also be renamed as Method of factorization.

Solved Examples

Example 1. *Solve the following equation by decomposition method.*

$$2x_1 + x_2 + x_3 = 2$$
$$x_1 + 3x_2 + 2x_3 = 2$$
$$3x_1 + x_2 + 2x_3 = 2$$

Solution. Given equation are

$$2x_1 + x_2 + x_3 = 2 \qquad \qquad \ldots (1)$$
$$x_1 + 3x_2 + 2x_3 = 2 \qquad \qquad \ldots (2)$$
$$3x_1 + x_2 + 2x_3 = 2 \qquad \qquad \ldots (3)$$

Here, the coefficient matrix A is given by

$$A = \begin{bmatrix} 2 & 1 & 1 \\ 1 & 3 & 2 \\ 3 & 1 & 2 \end{bmatrix}$$

and

$$X = \begin{bmatrix} x_1 \\ x_2 \\ x_3 \end{bmatrix}, B = \begin{bmatrix} 2 \\ 2 \\ 2 \end{bmatrix}$$

$$\therefore \qquad \qquad AX = B$$

Now

$$A = LU$$

$$\begin{bmatrix} 2 & 1 & 1 \\ 1 & 3 & 2 \\ 3 & 1 & 2 \end{bmatrix} = \begin{bmatrix} 1 & 0 & 0 \\ l_{21} & 1 & 0 \\ l_{31} & l_{32} & 1 \end{bmatrix} \begin{bmatrix} u_{11} & u_{12} & u_{13} \\ 0 & u_{22} & u_{23} \\ 0 & 0 & u_{33} \end{bmatrix} \qquad \ldots (4)$$

On comparing two matrices, we get

$$u_{11} = 2, u_{12} = 1, u_{13} = 1$$

$$l_{21}u_{11} = 1 \text{ or } l_{21} = \frac{1}{u_{11}} = \frac{1}{2}$$

$$l_{31}u_{11} = 3 \text{ or } l_{31} = \frac{3}{u_{11}} = \frac{3}{2}$$

$$l_{21}u_{12} + u_{22} = 3$$

$$u_{22} = 3 - l_{21}u_{12} = 3 - \frac{1}{2}(1) = 3 - \frac{1}{2} = \frac{5}{2}$$

$$l_{21}u_{13} + u_{23} = 2$$

$$u_{23} = 2 - l_{21}u_{13} = 2 - \frac{1}{2}(1) = 2 - \frac{1}{2} = \frac{3}{2}$$

$$l_{31}u_{12} + l_{32}u_{22} = 1$$

$$l_{32} = \frac{1 - l_{31}u_{12}}{u_{22}} = \frac{1 - \frac{3}{2}(1)}{\frac{5}{2}} = -\frac{1}{5}$$

$$l_{31}u_{13} + l_{32}u_{23} + u_{33} = 2$$

$$u_{33} = 2 - l_{31}u_{13} - l_{32}u_{23} = 2 - \frac{3}{2}(1) - \left(-\frac{1}{5}\right)\left(\frac{3}{2}\right)$$

$$= 2 - \frac{3}{2} + \frac{3}{10} = \frac{8}{10} = \frac{4}{5}$$

Thus

$$L = \begin{bmatrix} 1 & 0 & 0 \\ l_{21} & 1 & 0 \\ l_{31} & l_{32} & 1 \end{bmatrix} = \begin{bmatrix} 1 & 0 & 0 \\ \dfrac{1}{2} & 1 & 0 \\ \dfrac{3}{2} & -\dfrac{1}{5} & 1 \end{bmatrix}$$

and

$$U = \begin{bmatrix} u_{11} & u_{12} & u_{13} \\ 0 & u_{22} & u_{23} \\ 0 & 0 & u_{33} \end{bmatrix} = \begin{bmatrix} 2 & 1 & 1 \\ 0 & \dfrac{5}{2} & \dfrac{3}{2} \\ 0 & 0 & \dfrac{4}{5} \end{bmatrix}$$

Since

$$LY = B$$

$$\begin{bmatrix} 1 & 0 & 0 \\ \dfrac{1}{2} & 1 & 0 \\ \dfrac{3}{2} & -\dfrac{1}{5} & 1 \end{bmatrix} \begin{bmatrix} y_1 \\ y_2 \\ y_3 \end{bmatrix} = \begin{bmatrix} 2 \\ 2 \\ 2 \end{bmatrix}$$

From these equation , we get

$$y_1 = 2$$

$$\frac{1}{2}y_1 + y_2 = 2$$

$$\frac{3}{2}y_1 - \frac{1}{5}y_2 + y_3 = 2$$

On solving, we get

$$y_1 = 2, y_2 = 1, y_3 = -\frac{4}{5}$$

and since, we have

$$UX = Y$$

$$\begin{bmatrix} 2 & 1 & 1 \\ 0 & \dfrac{5}{2} & \dfrac{3}{2} \\ 0 & 0 & \dfrac{4}{5} \end{bmatrix} \begin{bmatrix} x_1 \\ x_2 \\ x_3 \end{bmatrix} = \begin{bmatrix} y_1 \\ y_2 \\ y_3 \end{bmatrix} = \begin{bmatrix} 2 \\ 1 \\ -\dfrac{4}{5} \end{bmatrix}$$

$$\therefore \qquad 2x_1 + x_2 + x_3 = 2$$

$$\frac{5}{2}x_2 + \frac{3}{2}x_3 = 1$$

$$\frac{4}{5}x_3 = -\frac{4}{5}$$

By back substitution, we get

$$x_3 = -1, x_2 = 1, x_1 = 1$$

Hence solution is $x_1 = 1, x_2 = 1, x_3 = -1$.

Example 2. *Solve the following equations by LU decomposition method*

$$2x + 3y + z = 9$$
$$x + 2y + 3z = 6$$
$$3x + y + 2z = 8$$

Solution. Above equation can be written as follows.

$$AX = B \qquad \qquad \dots (1)$$

Where

$$A = \begin{bmatrix} 2 & 3 & 1 \\ 1 & 2 & 3 \\ 3 & 1 & 2 \end{bmatrix}, X = \begin{bmatrix} x \\ y \\ z \end{bmatrix}, B = \begin{bmatrix} 9 \\ 6 \\ 8 \end{bmatrix}$$

Let

$$A = LU$$

$$\therefore \qquad \begin{bmatrix} 2 & 3 & 1 \\ 1 & 2 & 3 \\ 3 & 1 & 2 \end{bmatrix} = \begin{bmatrix} 1 & 0 & 0 \\ l_{21} & 1 & 0 \\ l_{31} & l_{32} & 1 \end{bmatrix} \begin{bmatrix} u_{11} & u_{12} & u_{13} \\ 0 & u_{22} & u_{23} \\ 0 & 0 & u_{33} \end{bmatrix}$$

$$\begin{bmatrix} 2 & 3 & 1 \\ 1 & 2 & 3 \\ 3 & 1 & 2 \end{bmatrix} = \begin{bmatrix} u_{11} & u_{12} & u_{13} \\ l_{21}u_{11} & l_{21}u_{12} + u_{22} & l_{21}u_{13} + u_{23} \\ l_{31}u_{11} & l_{31}u_{12} + u_{32}u_{22} & l_{31}u_{13} + l_{32}u_{23} + u_{33} \end{bmatrix}$$

Comparing, we get

$$u_{11} = 2, u_{12} = 3, u_{13} = 1$$

and

$$l_{21}u_{11} = 1 \text{ or } l_{21} = \frac{1}{2}$$

$$l_{21}u_{12} + u_{22} = 2$$

$$u_{22} = 2 - l_{21}u_{12}$$

$$= 2 - \frac{1}{2}(3) = 2 - \frac{3}{2} = \frac{1}{2}$$

$$l_{21}u_{13} + u_{23} = 3$$

or

$$u_{23} = 3 - l_{21}u_{13} = 3 - \frac{1}{2}(1) = \frac{5}{2}$$

$$l_{31}u_{11} = 3$$

$$l_{31} = \frac{3}{u_{11}} = \frac{3}{2}$$

and $\quad l_{31}u_{12} + l_{32}u_{22} = 1$

$$l_{32} = \frac{1 - l_{31}u_{12}}{u_{22}} = \frac{1 - \frac{3}{2}(3)}{1/2} = \frac{2 - 9}{1} = -7$$

Now, $\quad l_{31}u_{13} + l_{32}u_{23} + u_{33} = 2$

$$u_{33} = 2 - l_{31}u_{13} - l_{32}u_{23}$$

$$= 2 - \frac{3}{2}(1) - (-7)\left(\frac{5}{2}\right)$$

$$= 2 - \frac{3}{2} + \frac{35}{2} = \frac{36}{2} = 18$$

Thus L and U are given by

$$L = \begin{bmatrix} 1 & 0 & 0 \\ \frac{1}{2} & 1 & 0 \\ \frac{3}{2} & -7 & 1 \end{bmatrix}, U = \begin{bmatrix} 2 & 3 & 1 \\ 0 & \frac{1}{2} & \frac{5}{2} \\ 0 & 0 & 18 \end{bmatrix}$$

Since, $\quad\quad\quad\quad LY = B$

$$\therefore \quad\quad \begin{bmatrix} 1 & 0 & 0 \\ \frac{1}{2} & 1 & 0 \\ \frac{3}{2} & -7 & 1 \end{bmatrix} \begin{bmatrix} y_1 \\ y_2 \\ y_3 \end{bmatrix} = \begin{bmatrix} 9 \\ 6 \\ 8 \end{bmatrix}$$

From above equation, we get

$$y_1 = 9$$

$$\frac{1}{2}y_1 + y_2 = 6$$

$$\frac{3}{2}y_1 - 7y_2 + y_3 = 8$$

Solving these equation , we get

$$y_1 = 9, y_2 = \frac{3}{2}, y_3 = 5$$

Further since $\quad\quad\quad\quad UX = Y$

$$\therefore \quad\quad \begin{bmatrix} 2 & 3 & 1 \\ 0 & \frac{1}{2} & \frac{5}{2} \\ 0 & 0 & 18 \end{bmatrix} \begin{bmatrix} x \\ y \\ z \end{bmatrix} = \begin{bmatrix} 9 \\ \frac{3}{2} \\ 5 \end{bmatrix}$$

$$\therefore \qquad 2x + 3y + z = 9$$

$$\frac{1}{2}y + \frac{5}{2}z = \frac{3}{2}$$

$$\Rightarrow \qquad 18z = 5$$

By back substitution, we get

$$z = \frac{5}{18}$$

and

$$\frac{1}{2}y + \frac{5}{2}\left(\frac{5}{18}\right) = \frac{3}{2}$$

$$\Rightarrow \qquad y = 3 - \frac{25}{18} = \frac{29}{18}$$

and

$$2x + 3\left(\frac{29}{18}\right) + \left(\frac{5}{18}\right) = 9$$

$$2x = 9 - \frac{29}{6} - \frac{5}{18} = 9 - \frac{92}{18}$$

$$\Rightarrow \qquad 2x = \frac{70}{18} \qquad \Rightarrow \qquad x = \frac{35}{18}$$

Hence the solution is

$$x = \frac{35}{18}, y = \frac{29}{18}, z = \frac{5}{18}.$$

❧ EXERCISE 3.3

1. *Solve the following equations by Gauss's ellimination method :*

(i) $x_1 + x_2 + 2x_3 = 4$

  $3x_1 + x_2 - 3x_3 = -4$

  $2x_1 - 3x_2 - 5x_3 = -5$

(ii) $2x_1 + x_2 + 4x_3 = 12$

  $8x_1 - 3x_2 + 2x_3 = 20$

  $4x_1 + 11x_2 - x_3 = 33$

(iii) $x_1 + x_2 + x_3 = 10$

  $2x_1 + x_2 + 2x_3 = 17$

  $3x_1 + 2x_2 + x_3 = 17$

(iv) $2x + 3y - z = 5$

  $4x + 4y - 3z = 3$

  $2x - 3y + 2z = 2$

(v) $2x + y + z = 10$

  $3x + 2y + 3z = 18$

  $x + 4y + 9z = 16$

(v) $2x_1 + 4x_2 + x_3 = 2$

  $3x_1 + 2x_2 - 2x_3 = -2$

  $x_1 - x_2 + x_3 = 6$

2. *Solve the following equations by LU decomposition method :*

(i) $2x + y + 2z = 2$

  $x + 5y + 3z = 4$

  $x + y - z = 0$

(ii) $x + y + 3z = 10$

  $3x + 2y + 4z = 20$

  $3x + 5y - z = 30$

(iii) $2x - 3y + 10z = 3$

  $-x + 4y + 2z = 20$

  $5x + 2y + z = -12$

Answers

1. (i) $x_1 = 1, x_2 = -1, x_3 = 2$ (ii) $x_1 = 3, x_2 = 2, x_3 = 1$ (iii) $x_1 = 2, x_2 = 3, x_3 = 5$

    (iv) $x = 1, y = 2, z = 3$ (v) $x = 7, y = -9, z = 5$ (vi) $x_1 = 2, x_2 = -1, x_3 = 3$

2. (i) $x = \dfrac{1}{5}, y = \dfrac{2}{5}, z = \dfrac{3}{5}$ (ii) $x = 2, y = 5, z = 1$ (iii) $x = -4, y = 3, z = 2$

CHAPTER REVIEW : A COMPETITIVE APPROACH

Selected Terms and Results

TERMS

- **Vector of Order n :** An ordered set of n numbers $(x_1, x_2, ..., x_n)$ is known as vector of order n.

- **Linear Dependence and Independence of Vectors :** Let $x_1, x_2, ..., x_m$ be m vectors. Then they are said to be linearly independent if

$$\lambda_1 x_1 + \lambda_2 x_2 + ... + \lambda_m x_m = 0 \Rightarrow \lambda_1 = \lambda_2 = ... = \lambda_m = 0$$

  and if none of $\lambda_1, \lambda_2, ..., \lambda_m$ is zero then the vectors $x_1, x_2, ..., x_m$ are called linearly dependent.

- **Linear Combination :** A vector X is said to be linear combination of the vectors $x_1, x_2, ...,$ x_m if there exist scalars $\lambda_1, \lambda_2, ..., \lambda_m$ such that $X = \lambda_1 x_1 + \lambda_2 x_2 + ... + \lambda_m x_m$.

- **Homogeneous Linear Equations :** A system of equations $AX = B$ is said to be homogeneous if $B = 0$.

- **Trivial Solution :** The solution $x_1 = 0 = x_2 = ... = x_n$ is called trivial solution.

- **Non-trivial Solution :** A solution which is not trivial is called non-trivial solution.

- **Consistency and Inconsistency :** When there exist, one or more than one solution of the equation $AX = B$, then the equations are said to be consistent, otherwise they are said to be inconsistent.

RESULTS

- If a set of vectors is linearly dependent then at least one member of the set can be expressed as a linear combinations of the remaining vectors.

- If the rank of a matrix of order $m \times n$ is r then it has a set of r linearly independent rows (or columns) and $m - r$ linearly dependent rows (or columns).

- If the rank of A is r, then the number of linearly independent solutions of the equations $AX = 0$ which is a system of m homogeneous linear equations is n unknowns is $(n-r)$.

- If $AX = 0$ is a system of m homogeneous linear equations in n unknowns. Then

  (i) If $r = n$ then $AX = 0$ will have no linearly independent solutions.

  (ii) If $r < n$ then there will be $(n - r)$ linearly independent solution of $AX = 0$. In this case we have infinite solutions.

  (iii) If the number of equations is less than the number of unknowns, *i.e.*, $m < n$ and since $r \le m$ then $r < n$. In this case a non-zero solution will exist. Therefore, the equation $AX = 0$ will have infinite solution.

- The equation $AX = B$ is consistent if and only if the rank of A and the rank of augmented matrix $[A/B]$ are same.

- For the system $AX = B$, we have the following results :

  (i) The n equations in n unknowns have a unique solution.

  (ii) If rank $(A) <$ rank $[A/B]$, then there is no solution.

  (iii) If $r = n$, there will be a unique solution.

  (iv) If $r < n$ then $(n - r)$ variables can be assigned arbitrary values. Therefore, there will be infinte solutions and $(n - r + 1)$ solutions will be linearly independent.

  (v) If $m < n$ and $r \le m < n$ then equation will have infinite solutions.

- *LU* decomposition method is superior to Gauss's elimination method.

- *LU* decomposition method can also be renamed as method of factorization.

- The Gauss's elimination method will fail if any one of the pivots $a_{11}, a''_{22}, a''_{33}$ become zero. In such cases we rewrite the equations in a different order so that the given pivots are non-zero.

Review Questions and Project Work

1. Show that the following equations are consistent :
$$x_1 - x_2 + x_3 = 2$$
$$3x_1 - x_2 + 2x_3 = -6$$
$$3x_1 + x_2 + x_3 = -18$$

2. Show that the following equations are inconsistent :
$$x_1 + x_2 + x_3 = -3$$
$$3x_1 + x_2 - 2x_3 = -2$$
$$2x_1 + 4x_2 + 7x_3 = 7$$

3. For what value of λ, will the following equations fails to have unique solutions :
$$3x - y + \lambda z = 1$$
$$2x + y + z = 2$$
$$x + 2y - \lambda z = -1 \, (\text{Ans} : \lambda = -\frac{7}{2})$$

4. For what value of λ the following system of equations is consistent and has non-trivial solutions:
$$(\lambda - t)x + (3\lambda + 1)y + 2\lambda z = 0$$

$$(\lambda - 1)x + (4\lambda - 2)y + (\lambda + 3)z = 0$$
$$2x + (3\lambda + 1)y + 3(\lambda - 1)z = 0 \, (\text{Ans} : \lambda = 0, 3)$$

5. Show that the system of equations
$$x + y \cos \gamma + z \cos \beta = 0$$
$$x \cos \gamma + y + z \cos \alpha = 0$$
$$x \cos \beta + y \cos \alpha + z = 0$$
has non-trivial solution if $\alpha + \beta + \gamma = 0$.

6. Without actually solving, prove that the following system of equations has a unique solution :
$$5x + 3y + 14z = 4$$
$$y + 2z = 1$$
$$x - y + 2z = 0$$

7. Show that the three equations $-2x + y + z = a$, $x - 2y + z = b$, $x + y - 2z = c$ have no solution unless $a + b + c = 0$ in which case they have infinitely many solutions. Also, find these solution if $a = 1, b = 1, c = -2$. (Ans. $x = k - 1, y = k - 1, z = k$)

Objective Type Questions

FILL IN THE BLANKS

1. The matrix equation $AX = O$ is a system of linear _____ equations.

2. If the rank of $A = r$, then the number of linearly independent solutions of m homogeneous equations in n variables is _____ .

3. If X_1 and X_2 are the solutions of $AX = O$, then _____ is also a solution of $AX = O$.

4. If the rank of A is equal to the number of unknowns in $AX = O$, then there exists only _____ .

5. If the rank of A is less than the number of unknowns in $AX = O$, then there are _____ .

6. The matrix equation $AX = B$ is a system of linear _____ .

7. The equation $AX = B$ is consistent if the rank $A = $ _____ .

8. The rank $A <$ rank$(A|B)$, then the equation $AX = B$ is _____ .

9. If rank $A = $ rank $(A|B)$ and is also equal to the number of unknowns, then there will be _____ .

10. If rank $A = $ rank $(A|B) = r$ and $r < n$ (No. of unknowns), then there will only _____ linearly independent solutions of $AX = B$.

11. If the number of equations is less than the number of unknowns in $AX = B$, then there will always _____ solutions.

12. The system of equations $2x - 2y + 5z + 3w = 0$, $4x - y + z + w = 0$, $3x - 2y + 3z + 4w = 0$, $x - 3y + 7z + 6w = 0$ has _____ linearly independent solutions.

13. The equations $3x + 4y = 5, 6x + 8y = 10$ have _____ solutions.

14. The matrix $[A : B]$ is known as _____ .

15. The system of m non-homogeneous linear equations in n unknowns $AX = B$, is _____ iff the coefficient matrix A and augmented matrix $[A|B]$ are of the same rank.

TRUE/FALSE

Write 'T' for true and 'F' for false statement.

1. The matrix equation $AX = B$ represents a system of linearly homogeneous equations. **(T/F)**

2. The matrix equations $AX = O$ represents a system of non-linear homogeneous equations. **(T/F)**

3. If $A = [a_{ij}]_{m \times n}$ in $AX = O$ and rank $A = n$ then there is only zero solution. **(T/F)**

4. If X_1 is a solution of $AX = O$, then $2X_1$ is not a solution of $AX = O$. **(T/F)**

5. If the rank A is less than the number of unknowns, then there will be infinite solutions of $AX = O$. **(T/F)**

6. The eqution $AX = B$ is inconsistent if rank A is equal to the number of unknowns, then there will be unique solutions. **(T/F)**

7. If the equation $AX = B$ is consistent and rank A is equal to the number of unknowns, then there will be unique solution. **(T/F)**

8. The number of equations is less than the number of variables, then there will be infinite solutions. **(T/F)**

9. If the rank of $A = r$ and there are n variables in $AX = B$ and $r < n$, then the number of linearly independent solutions is $n - r + 1$. **(T/F)**

10. The system of equations $x + 2y + 3z = 1$, $3x + y + 3z = 2$, $4x + 5y + 9z = 4$ has only one solution. **(T/F)**

MULTIPLE CHOICE QUESTIONS

Choose the most appropriate one.

1. There are m equations in n variables, then the order of coefficient matrix is :
   (a) $n \times m$ (b) $m \times m$
   (c) $n \times n$ (d) $m \times n$

2. The matrix equation $AX = O$ represents :
   (a) non-homogeneous linear equations
   (b) homogeneous linear equations
   (c) homogeneous non-linear equations
   (d) none of these

3. The equation $AX = B$ is linear and :
   (a) homogeneous   (b) non-homogeneous
   (c) none of these

4. If X_1 and X_2 are the solutions of $AX = O$, then which one is also the solution of $AX = O$:
   (a) $X_1^2 + X_2^2$ (b) $(X_1 + X_2)^2$
   (c) $X_1 + X_2$ (d) X_1 / X_2

5. If the equation $AX = B$, is consistent and rank of $(A|B) = 4$, then rank of A is :
   (a) 4   (b) 8
   (c) 3   (d) 2

6. The equations $x + 2y + 3z = \lambda x$, $3x + y + 2z = \lambda y$, $2x + 3y + z = \lambda z$ have non-zero solutions if A equals :
   (a) 2   (b) 6
   (c) 3   (d) 0

7. If the rank of $A = r$ and there are n variables in $AX = B$, then equation $AX = B$ has $(n - r)$ linearly independent solutions if :
   (a) $n = r$ (b) $n > r$
   (c) $n < r$ (d) none of these

8. If the rank of A is equal to the number of variables for $AX = O$, then there will be :
   (a) unique solution   (b) infinite solutions
   (c) finite solution   (d) zero solution

9. The system of equations $x + 2y + 3z = 1$, $2x + y + 3z = 2$, $5x + 5y + 9z = 4$ has :
   (a) only one solution
   (b) infinitely many solutions
   (c) no solution
   (d) none of these

Answers

FILL IN THE BLANKS

1. homogeneous   2. $n - r$ 3. $C_1 X_1 + C_2 X_2$ 4. zero solution   5. infinite solution
6. non-homogeneous equations   7. rank $(A|B)$ 8. inconsistent   9. zero solution
10. $n - r$ 11. infinite solution   12. one   13. no solution
14. augmented matrix   15. consistent.

TRUE/ FALSE

1. F	2. F	3. T	4. F	5. T
6. T	7. T	8. T	9. T	10. T

MULTIPLE CHOICE QUESTIONS

1. (d)	2. (b)	3. (b)	4. (c)	5. (a)	6. (b)	7. (b)	8. (d)	9. (a)

COMPETITION CORNER
for JRF, NET/SET, GATE Aspirants

SOME FASCINATING FACTS

- A system of linear equations is completely determined by its augmental matrix.
- A linear equation is said to be degenerate if all the coefficient are zero.
- Two systems of linear equations have the same solutions if and only if each equation in each system is a linear combination of the equations in the other system.
- If a system M of linear equations is obtained from a system L of linear equations by a finite sequence of elementary operations. Then M and L have the same solution.
- A system $AX = B$ of linear equations has a solution if and only if B is a linear combination of the coefficient matrix A.
- Let A be a non-singular matrix that can be brought into triangular form U using only row operations. Then $A = LU$ where L is the above lower triangular matrix with no 1's on the diagonal and U is an upper triangular matrix with no.'s on the diagonal.
- A system L of linear equations is said to be consistent if no linear combination of its equations is a degenerate equation L with a non-zero constant.

SOME IMPORTANT ILLUSTRATIONS

- If $X = [x_1, x_2, ..., x_n]'$ is n-tuples non-zero vector. Then $n \times n$ matrix $V = XX'$ has rank 1.
- If the matrix equation $PX = Q$, the necessary condition for the existence of at least one solution for the unknown vector is that augmented matrix $[P/Q]$ must have the same rank as P.
- If all the four entries of the 2×2 matrix $P = \begin{bmatrix} p_{11} & p_{12} \\ p_{21} & p_{22} \end{bmatrix}$ are non-zero and one of its eigen value is zero, then $p_{11}p_{22} - p_{12}p_{21} = 0$.
- Multiplication of the matrices E and F is G. Matrices E and G are as follows:
$$E = \begin{bmatrix} \cos\theta & -\sin\theta & 0 \\ \sin\theta & \cos\theta & 0 \\ 0 & 0 & 1 \end{bmatrix} \text{ and } G = \begin{bmatrix} 1 & 0 & 0 \\ 0 & 1 & 0 \\ 0 & 0 & 1 \end{bmatrix}$$
Then the value of the matrix
$$F = \begin{bmatrix} \cos\theta & \sin\theta & 0 \\ -\sin\theta & \cos\theta & 0 \\ 0 & 0 & 1 \end{bmatrix}.$$
- The system of equations $Kx + y + z = 1$, $x + Ky + z = K$ and $x + y + Kz = K^3$ does not have a solution if $K = -2$.
- Let $S_1 = \left\{ \alpha = \begin{bmatrix} 1 & -2 & 4 \\ 3 & 0 & -1 \end{bmatrix}, \beta = \begin{bmatrix} 2 & -4 & 8 \\ 6 & 0 & -2 \end{bmatrix} \right\}$ and $S_2 = \{ f = u^3 + 3u + 4, g = u^3 + 4u + 3 \}$ be the given set, then S_1 is linearly dependent but not S_2.
- The system of equations
$$4x_1 + x_2 - 3x_3 - x_4 = 0$$

$$2x_1 + 3x_2 + x_3 - 5x_4 = 0$$
$$x_1 - 2x_2 - 2x_3 + 3x_4 = 0$$
has infinite number of solutions.

- The system of equations
$$x - y + 3z = 4$$
$$x + z = 2$$
$$x + y - z = 0$$
has infinitely many solutions.

- Consider $AX = B$ where $A = \begin{bmatrix} -1 & 2 \\ 2 & -1 \end{bmatrix}$ and $B = \begin{bmatrix} 3 \\ 1 \end{bmatrix}$ then there exists a non-zero unique solution.

- Let A be an $n \times n$ matrix which is both Hermitian and unitary then $A^2 = I$.

- The system of equations $2x + y = 5$; $x - 3y = -1$; $3x + 4y = k$ is consistant when $k = 10$.

- The value of α for which the system of equations $x + y + z = 0$; $y + 2z = 0$; $\alpha x + z = 0$ has more than one solution is -1.

- The number of different $n \times n$ symmetric matrices with each element being either 0 or 1 where $n = 5$ is 2^{15}.

- Let A be a 3×3 matrix and consider the system of equations $AX = \begin{bmatrix} 1 \\ 0 \\ -1 \end{bmatrix}$ then if the system has a unique solution then A is non-singular.

- A system of linear equations $x + 2y - z = 11$, $3x + y - 2z = 10$, $x - 3y = 5$ has no solution.

Self Assessment Test

1. Let L be a system of linear equations with more unknown than equations. Show that L cannot have a unique solution.

2. Prove that the following three statements about a system of linear equations are equivalent :
   (i) The system is consistent.
   (ii) No linear combinations of the equations is the equation $0x_1 + 0x_2 + \ldots + 0x_n = b \neq 0$.
   (iii) The system is reducible to Echelon form.

3. If u and v are solutions of a non-homogeneous system $AX = B$, then show that difference $w = v - u$ is a solution of the associated homogeneous system $AX = O$.

4. Determine whether the following homogeneous system has a non-zero solution :
   $$x_1 - 2x_2 + 3x_3 - 2x_4 = 0$$
   $$3x_1 - 7x_2 - 2x_3 + 4x_4 = 0$$
   $$4x_1 + 3x_2 + 5x_3 + 2x_4 = 0$$

5. Find a matrix P which transform the matrix $A = \begin{bmatrix} 1 & 0 & -1 \\ 1 & 2 & 1 \\ 2 & 2 & 3 \end{bmatrix}$ to diagonal form. Hence, find A^4.

6. Examine for linear dependence $[1, 0, 2, 1]$, $[3, 1, 2, 1]$, $[4, 6, 2, -4]$, $[-6, 0, -3, -4]$ and find the relation between them if possible.

7. Show that the rank of the following $(n+1) \times (n+1)$ matrix where a is real number
   $$\begin{bmatrix} 1 & a & a^2 & \cdots & a^n \\ 1 & a & a^2 & \cdots & a^n \\ \vdots & \vdots & \vdots & \cdots & \vdots \\ 1 & a & a^2 & \cdots & a^n \end{bmatrix} \text{ is 1.}$$

8. Let A be an $m \times n$ matrix where $m < n$. Consider the system of linear equation $AX = B$ where B is $n \times 1$ column vector and $B \neq 0$ then show that the system of equations has a solution if and only if it has infinitely many solutions.

9. In the matrix equation $AX = B$, show that the necessary condition for the existence of at least one solution for the unknown vector X is that augmented matrix $[A|B]$ must have the same rank as A.

10. Show that for the set of equations
    $$x_1 + 2x_2 + x_3 + 4x_4 = 2$$
    $$3x_1 + 6x_2 + 3x_3 + 12x_4 = 6$$
    multiple non-trivial solution exists.

11. Show that the system of simultaneous equations
    $x + 2y + z = 6$; $2x + y + 2z = 6$; $x + y + z = 5$
    has no solution.

12. Show that if $A_{2 \times 2}$ matrix which satisfy $A^2 - A = 0$ then A must be diagonal.

13. Consider the system of linear equations
    $x + y + z = 3$; $x - y - z = 4$; $x - 5y + kz = 6$.
    Find the value of k for which the system has an infinite number of solutions.
    (Ans. $k = -5$)

14. Find the value of λ for which the equations
    $$(\lambda - 1)x + (3\lambda + 1)y + 2\lambda z = 0$$
    $$(\lambda - 1)x + (4\lambda - 2)y + (\lambda + 3)z = 0$$
    $$2x + (3\lambda + 1)y + 3(\lambda - 1)z = 0$$
    are consistent.

EIGEN VALUES AND EIGEN VECTORS OF A MATRIX

4.1 INTRODUCTION

A polynomial in indeterminate λ of the form

$$f(\lambda) = A_0 + A_1\lambda + A_2\lambda^2 + \ldots + A_n\lambda^n$$

where A_0, A_1, A_2, ..., A_n are all square matrices of the same order, is called a matric polynomial of degree n if $A_n \neq O$ (null matrix).

From above definition it is clear that every square matrix can be expressed as a matric polynomial of zero degree. If A is a square matrix, then we can write

$$A = \lambda^{\circ} A$$

Equality of two Matric Polynomials : Two matric polynomials are said to be equal if and only if the coefficients of like powers of λ are the same.

4.2 THE CHARACTERISTIC EQUATION OF A MATRIX

Let A be a square matrix of order $n \times n$ and let

$$A = \begin{bmatrix} a_{11} & a_{12} & \cdots & a_{1n} \\ a_{21} & a_{22} & \cdots & a_{2n} \\ \vdots & \vdots & & \vdots \\ a_{n1} & a_{n2} & \cdots & a_{nn} \end{bmatrix}$$

If λ is indeterminate, then the matrix $A - \lambda I$ is called the characteristic matrix of A, where I is the unit matrix of order $n \times n$.

The determinant

$$|A - \lambda I| = \begin{bmatrix} a_{11} - \lambda & a_{12} & \cdots & a_{1n} \\ a_{21} & a_{22} - \lambda & \cdots & a_{2n} \\ \vdots & \vdots & & \vdots \\ a_{n1} & a_{n2} & \cdots & a_{nn} - \lambda \end{bmatrix}$$

is an ordinary polynomial in λ which is called the characteristic polynomial of A and the equation

$$|A - \lambda I| = 0$$

i.e.,

$$\begin{vmatrix} a_{11} - \lambda & a_{12} & \cdots & a_{1n} \\ a_{21} & a_{22} - \lambda & \cdots & a_{2n} \\ \vdots & \vdots & & \vdots \\ a_{n1} & a_{n2} & \cdots & a_{nn} - \lambda \end{vmatrix} = 0$$

is known as the characteristic equation of A. The roots of the equation $|A - \lambda I| = 0$ are called characteristic roots or latent roots or eigenvalues of A. The set of all eigenvalues of a matrix A is called spectrum of A.

4.3 CHARACTERISTIC VECTORS OR EIGENVECTORS OF A MATRIX

Let $A = [a_{ij}]$ be a matrix of order $n \times n$ and let

$$X = \begin{bmatrix} x_1 \\ x_2 \\ \vdots \\ x_n \end{bmatrix}$$

be a column vector. Consider a vector equation

$$AX = \lambda X \qquad \qquad ...(1)$$

where λ is a scalar.

It is evident that $X = O$ satisfies the equation (1) for every value of λ, thus $X = O$ is a solution of (1). A value of λ for which a non-zero vector *i.e.*, $X \neq O$ satisfies (1) is called an eigenvalue of the matrix A and the non-zero vector X is called an eigenvector of A corresponding to that eigenvalue λ.

Now equation (1) can be written as

$$AX = \lambda IX$$

or $$(A - \lambda I)X = O \qquad \qquad ...(2)$$

where I is the unit matrix of order $n \times n$. Equation (2) represents a matrix equation of a system of n homogeneous equations. The necessary and sufficient condition for the equation (2) to possess a non-zero solution, *i.e.*, $X \neq O$ is that $|A - \lambda I| = 0$, which is a characteristic equation of matrix A.

REMARKS

- The eigenvector is also known as proper vector.
- If X is an eigenvector of a matrix corresponding eigenvalue λ, then for any non-zero scalar kX is also an eigenvector of A corresponding to the same eigenvalue λ.
- Corresponding to an eigenvalue of a matrix A, there will be different eigenvectors of A.
- For a given eigenvector of a matrix A there corresponds one and only one eigenvalue of A.

4.4 RELATION BETWEEN EIGENVALUES AND EIGENVECTORS

THEOREM 1. λ *is an eigenvalue of a matrix A if and only if there exists a non-zero vector X such that $AX = \lambda X$.*

Proof. Suppose that λ is an eigenvalue of A, then
$$|A - \lambda I| = 0$$
$\Rightarrow$ $A - \lambda I$ is a singular matrix.
$\therefore$ The matrix equation $(A - \lambda I) X = O$ has a non-zero solution, thus there exists a non-zero vector X such that
$$(A - \lambda I) X = O \text{ or } AX = \lambda X$$
Conversely, Suppose that there is a non-zero vector X such that
$$AX = \lambda X$$

$$\Rightarrow \qquad\qquad (A - \lambda I)X = O$$

Since the matrix equation $(A - \lambda I)X = O$ has a non-zero solution, then the coefficient matrix $A - \lambda I$ is singular, therefore

$$|A - \lambda I| = 0$$

Hence A is an eigenvalue of A.

THEOREM 2. *If X is an eigenvector of a matrix A corresponding to an eigenvalue of A, then kX is also an eigenvector of A corresponding to the same eigenvalue A, where k is any non-zero number.*

Proof. Since X is an eigenvector of a matrix A corresponding to an eigenvalue of A, then we have

$$AX = \lambda X \qquad\qquad ...(1)$$

Since $k \ne 0$, then multipling both sides of (1) by k, we get

$$k(AX) = k(\lambda X)$$
$$\Rightarrow \qquad\qquad A(kX) = \lambda(kX) \qquad\qquad ...(2)$$

From equation (2), it follows that kX is also an eigenvector of A corresponding to the same eigenvalue λ.

THEOREM 3. *If X is a non-zero eigenvector of a matrix A, then X cannot correspond to more than one eigenvalue of A.*

Proof. If possible, let X be an eigenvector corresponding to eigenvalues λ_1 and λ_2, then

$$AX = \lambda_1 X \qquad\qquad ...(1)$$
and
$$AX = \lambda_2 X \qquad\qquad ...(2)$$

From (1) and (2) we have

$$\lambda_1 X = \lambda_2 X$$
$$\Rightarrow \qquad (\lambda_1 - \lambda_2)X = O$$
$$\Rightarrow \qquad \lambda_1 - \lambda_2 = 0 \qquad\qquad \because X \ne O$$
$$\Rightarrow \qquad \lambda_1 = \lambda_2$$

THEOREM 4: *If X_1 and X_2 be non-zero eigenvectors of a matrix A corresponding to an eigenvalue λ of A, then $k_1 X_1 + k_2 X_2$ is also an eigenvector of A corresponding to eigenvalue λ, where k_1 and k_2 are non-zero number.*

Proof. Since $X_1 \ne 0$, $X_2 \ne 0$ and $k_1 \ne 0$, $k_2 \ne 0$, then $k_1 X_1 + k_2 X_2 \ne O$. Also X_1 and X_2 are eigenvectors of A corresponding to an eigenvalue λ of A, then we have

$$AX_1 = \lambda X_1 \qquad\qquad ...(1)$$
and
$$AX_2 = \lambda X_2 \qquad\qquad ...(2)$$

Multiplying (1) by k_1 and (2) by k_2 and then adding, we get

$$Ak_1 X_1 + Ak_2 X_2 = \lambda(k_1 X_1 + k_2 X_2)$$
$$\Rightarrow \qquad Ak_1 X_1 + Ak_2 X_2 = \lambda(k_1 X_1 + k_2 X_2) \qquad\qquad ...(3)$$

As $k_1 X_1 + k_2 X_2 \ne O$ then from (3) it follows that $k_1 X_1 + k_2 X_2$ is also an eigenvector of A corresponding to an eigenvalue of A.

THEOREM 5. *Let A be an $n \times n$ matrix. Then the distinct eigenvectors corresponding to distinct eigenvalues of A are linearly independent.*

Proof. Since A is an $n \times n$ matrix so it will have atmost n eigenvalues. Let $\lambda_1, \lambda_2, \lambda_3,..., \lambda_m$ be m distinct eigenvalues of A out of n eigenvalues and let $X_1, X_2, X_3,...,X_m$ be

m distinct eigenvectors corresponding to eigenvalues λ_1, λ_2,..., λ_m respectively. Then we have

$$AX_1 = \lambda_1 X_1, AX_2 = \lambda_2 X_2,..., AX_m = \lambda_m X_m$$

Let S = $\{X_1, X_2,..., X_m\}$. Then we have to show that S is linearly independent. We shall prove it by induction hypothesis on m.

If $m = 1$, then there is only one non-zero vector, which is obviously linearly independent.

Suppose the result is true for $m = k$ i.e., $\{X_1, X_2,..., X_m\}$ is linearly independent. Let this set be denoted by S_1, then

$$S_1 = \{X_1, X_2, ..., X_k\}$$

Finally we shall prove that the set $S_1 \cup \{X_{k+1}\}$ is linearly independent.

For scalars $a_1, a_2, a_3,..., a_k, a_{k+1}$ such that

$$a_1 X_1 + a_2 X_2 +... + a_k X_k + a_{k+1} X_{k+1} = O \qquad ...(1)$$

$\Rightarrow \qquad A(a_1 X_1 + a_2 X_2 + ... + a_k X_k + a_{k+1} X_{k+1}) = AO = O$

$\Rightarrow \qquad a_1 AX_1 + a_2 AX_2 + ... + a_k AX_k + a_{k+1} AX_{k+1} = O$

$\Rightarrow \qquad a_1 \lambda_1 X_1 + a_2 \lambda_2 X_2 + ... + a_k \lambda_k X_k + a_{k+1} \lambda_{k+1} X_{k+1} = O \qquad ...(2)$

Multiplying (1) by λ_{k+1} and then substracting it from (2), we get

$$a_1(\lambda_1 - \lambda_{k+1})X_1 + a_2(\lambda_2 - \lambda_{k+1})X_2 + ... + a_k(\lambda_k - \lambda_{k+1})X_k = O$$

Since S_1 is linearly independent, hence

$\Rightarrow \qquad a_1(\lambda_1 - \lambda_{k+1}) = a_2(\lambda_2 - \lambda_{k+1}) = ... = a_k(\lambda_k - \lambda_{k+1}) = 0$

$\Rightarrow \qquad a_1 = a_2 = ... = a_k = 0 \qquad [\because \lambda_1, \lambda_2,..., \lambda_m$ are all distinct.]

Putting $a_1 = a_2 = ... = a_k = 0$ in (1), we get

$$a_{k+1} X_{k+1} = O$$

$\Rightarrow \qquad a_{k+1} = 0 \qquad [\because X_{k+1} \neq O]$

$\therefore$ The set $S_1 \cup \{X_{k+1}\}$ is linearly independent.

Hence, the result is proved by induction.

4.5 EIGENVALUES OF SPECIAL TYPE OF MATRICES

THEOREM 1. *The eigenvalues of a Hermitian matrix are real.*

Proof. Let A be a Hermitian matrix. Let λ be an eigenvalue of A and let X be its corresponding eigenvector.

Then we have

$$AX = \lambda X \qquad ...(1)$$

Pre-multiplying both sides of (1) by X^θ, we have

$$X^\theta AX = X^\theta \lambda X = \lambda X^\theta X \qquad ...(2)$$

Taking conjugate transpose of both sides of (2), we have

$$(X^\theta AX)^\theta = (\lambda X^\theta X)^\theta$$

$\Rightarrow \qquad X^\theta A^\theta (X^\theta)^\theta = \bar{\lambda} X^\theta (X^\theta)^\theta$

$\Rightarrow \qquad X^\theta A^\theta X = \bar{\lambda} X^\theta X \qquad \qquad \because (X^\theta)^\theta = X$

$\Rightarrow \qquad X^\theta AX = \bar{\lambda} X^\theta X \qquad [\because A$ is Hermitian $\Rightarrow A^\theta = A]$

$$\Rightarrow \qquad X^\theta \lambda X = \bar{\lambda} X^\theta X \qquad\qquad \text{[Using (1)]}$$
$$\text{or} \qquad \lambda X^\theta X = \bar{\lambda} X^\theta X$$

$$\text{or} \qquad (\lambda - \bar{\lambda}) X^\theta X = O$$
$$\text{or} \qquad (\lambda - \bar{\lambda}) = 0 \qquad\qquad [\because X \text{ is non-zero} \Rightarrow X^\theta X \neq O]$$
$$\text{or} \qquad \lambda = \bar{\lambda}$$

Hence λ is real.

THEOREM 2. *The eigenvalues of a real symmetric matrix are all real.*

Proof. Let A be a real symmetric matrix, then
$$A' = A \qquad\qquad\qquad\qquad ...(1)$$
Let λ be any eigenvalue of A and let X be its corresponding eigenvector, then we have
$$AX = \lambda X \qquad\qquad\qquad\qquad ...(2)$$
Pre-multiplying both sides of (2) by X' we get
$$X'AX = \lambda X'X \qquad\qquad\qquad\qquad ...(3)$$
Taking transpose of both sides of (3), we get
$$(X'AX)' = (\lambda X'X)'$$
$$\Rightarrow \qquad X'A'(X')' = \bar{\lambda} X'(X')'$$
$$\Rightarrow \qquad X'A'X = \bar{\lambda} X'X \qquad\qquad\qquad [\because (X')' = X]$$
$$\Rightarrow \qquad X'AX = \bar{\lambda} X'X \qquad\qquad\qquad [\because A' = A]$$
$$\Rightarrow \qquad X'\lambda X = \bar{\lambda} X'X \qquad\qquad\qquad [\because AX = \lambda X]$$
$$\Rightarrow \qquad \lambda X'X = \bar{\lambda} X'X$$
$$\Rightarrow \qquad (\lambda - \bar{\lambda}) X'X = O$$
$$\Rightarrow \qquad (\lambda - \bar{\lambda}) = 0 \qquad\qquad\qquad [\because X \neq O \Rightarrow X'X \neq O]$$
$$\Rightarrow \qquad \lambda = \bar{\lambda}$$
$\Rightarrow \quad \lambda$ is real.

THEOREM 3. *The eigenvalues of a skew-Hermitian matrix are either purely imaginary or zero.*

Proof. Let A be a skew-Hermitian matrix, then
$$A^\theta = -A$$
Now $\qquad\qquad (iA)^\theta = -iA^\theta$
$$\Rightarrow \qquad\qquad (iA)^\theta = iA$$
$\Rightarrow \quad iA$ is a Hermitian matrix

Let λ be an eigenvalue of A and X be its corresponding eigenvector, then
$$AX = \lambda X$$
$$\Rightarrow \qquad\qquad iAX = i\lambda X$$
$\therefore$ $i\lambda$ is an eigenvalue of a Hermitian matrix. By theorem 1, we can say that $i\lambda$ is real. It follows that either λ is purely real or zero.

Corollary. *The eigenvalues of a real skew-symmetric matrix are either purely imaginary or zero.*

Proof. If the elements of a skew-Hermitian matrix are all real, then it is a real skew-symmetric.

Therefore, a real skew-symmetric matrix is skew-Hermitian matrix, hence the result follows from theorem 3.

Theorem 4. *The eigenvalues of a unitary matrix are of unit modulus.*

Proof. Let A be a unitary matrix, then

$$A^\theta A = I \qquad \qquad ...(1)$$

Let λ be an eigenvalue of A and let X be its corresponding eigenvector, then

$$AX = \lambda X \qquad \qquad ...(2)$$

Taking conjugate transpose of both sides of (2), we have

$$(AX)^\theta = (\lambda X)^\theta$$

or $$X^\theta A^\theta = \bar\lambda X^\theta \qquad \qquad ...(3)$$

Now $$(X^\theta A^\theta)(AX) = (\bar\lambda X^\theta)(\lambda X) \qquad \text{[Using (2) and (3)]}$$

$$\Rightarrow \qquad X^\theta(A^\theta A)X = \bar\lambda\lambda X^\theta X$$

$$\Rightarrow \qquad X^\theta I X = \bar\lambda\lambda X^\theta X \qquad \text{[Using (1)]}$$

$$\Rightarrow \qquad X^\theta X = \bar\lambda\lambda X^\theta X$$

$$\Rightarrow \qquad (1 - \bar\lambda\lambda)X^\theta X = O$$

$$\Rightarrow \qquad \bar\lambda\lambda = 1 \qquad \qquad [\because X \neq O \Rightarrow X^\theta X \neq O]$$

$$\Rightarrow \qquad |\lambda|^2 = 1$$

$$\Rightarrow \qquad |\lambda| = 1$$

Theorem *The eigenvalues of an orthogonal matrix are of unit modulus.*

Proof. We know that if the elements of a unitary matrix are all real, then it is an orthogonal matrix, therefore an orthogonal matrix is a unitary matrix hence the result follows from theorem 4.

 **WORKING RULE**

To find the eigen values and eigen vectors

Let A be an $n \times n$ matrix, then it will have n eigenvalues. In order to find the eigenvalues and eigenvectors of A, we use the following steps :

STEP 1.	Find the roots of the characteristic equation $|A - \lambda I| = 0$, the roots of λ give the eigenvalues of A.
STEP 2.	Let $X = \begin{bmatrix} x_1 \\ x_2 \\ \vdots \\ x_n \end{bmatrix} \neq O$ be an eigenvector of A corresponding to an eigenvalue λ_1 (say). Then X can be determined from the equation $$(A - \lambda_1 I)X = O$$ which is a system of n homogeneous equations in $x_1, x_2, ..., x_n$. If the rank of $(A - \lambda_1 I)$ is r, then the number of linearly independent solutions is $n - r$.

Solved Examples

Based on the following Results

➡ $|A - \lambda I| = 0$ is known as characteristic equation and the roots of this equation are called characteristic roots or eigen values.

➡ The value of λ for which a non-zero vector, i.e., $X \neq O$ satisfies $AX = \lambda X$ is called an eigenvalues of the matrix A and the non-zero vector X is called an eigenvector of A corresponding to that eigenvalue λ.

➡ λ is an eigenvalue of a matrix A if and only if there exists a non-zero vector X such that $AX = \lambda X$.

Example 1. If λ is a non-zero eigenvalue of a matrix A, then show that $\dfrac{1}{\lambda}$ is an eigenvalue of A^{-1}.

Solution. Let $X \neq O$ be an eigenvector corresponding to the eigenvalue λ of A, then

$$AX = \lambda X$$
$$\Rightarrow \qquad A^{-1}(AX) = A^{-1}(\lambda X) \qquad\qquad [\because A^{-1} \text{ exists.}]$$
$$\Rightarrow \qquad (A^{-1}A)X = \lambda(A^{-1}X)$$
$$\Rightarrow \qquad IX = \lambda(A^{-1}X) \qquad\qquad [\because A^{-1}A = I]$$
$$\Rightarrow \qquad X = \lambda(A^{-1}X) \qquad\qquad [\because IX = X]$$
$$\Rightarrow \qquad A^{-1}X = \left(\dfrac{1}{\lambda}\right)X$$

Hence, $\dfrac{1}{\lambda}$ is an eigenvalue of A^{-1}.

Example 2. Let A be an $n \times n$ matrix. Then show that zero is an eigenvalue of A iff A is singular.

Solution. Let $X \neq O$ be an eigenvector corresponding to the eigenvalue 0 of A, then

$$AX = 0X = O \qquad\qquad ...(1)$$

Since (1) represents a system of homogeneous equations, it will have non-zero solution if and only if $\rho(A) < n$

i.e., $\qquad\qquad$ iff $|A| = 0$

i.e., iff A is singular.

Example 3. If $\lambda_1, \lambda_2, ..., \lambda_n$ are the eigenvalues of A, then show that $k\lambda_1, k\lambda_2, ..., k\lambda_n$ are eigenvalues of kA, where k is any number.

Solution. If $k = 0$, then $kA = 0A = O$. Since each eigenvalue of a zero matrix is zero, therefore $0\lambda_1, 0\lambda_2, ..., 0\lambda_n$ are the eigenvalues of kA if $\lambda_1, \lambda_2, ..., \lambda_n$ are eigenvalues of A.

Next, suppose that $k \neq 0$, then we have

$$|kA - k\lambda I| = k^n|A - \lambda I|$$

Now $\qquad\qquad |kA - k\lambda I| = 0$ iff $|A - \lambda I| = 0$

It follows that $k\lambda$ is an eigenvalue of kA.

Hence, if $\lambda_1, \lambda_2, ..., \lambda_n$ are the eigenvalues of A then $k\lambda_1, k\lambda_2, ..., k\lambda_n$ are the eigenvalues of kA.

Example 4. If X be a non-zero eigenvector of an $n \times n$ matrix A, then prove that for each positive integer n, X is an eigenvector of A^n corresponding to the eigenvalue λ^n.

Solution. Since $X \neq O$ is an eigenvector corresponding eigenvalue λ of A, then we have

$$AX = \lambda X \qquad\qquad ...(1)$$

Now we have to show that $A^n X = \lambda^n X$.

We shall prove this by induction on n.

If $n = 1$, then the result is true by virtue of (1).

Suppose that the result is true for $n = k$, then we have

$$A^k X = \lambda^k X \qquad \qquad ...(2)$$

Now
$$
\begin{aligned}
A^{k+1} X &= (A^k A) X \\
&= A^k(AX) \\
&= A^k(\lambda X) && \text{[Using (1)\}} \\
&= \lambda(A^k X) \\
&= \lambda(\lambda^k X) && \text{[Using (2)]} \\
&= \lambda^{k+1} X \\
A^{k+1} X &= \lambda^{k+1} X
\end{aligned}
$$

Thus, the result is true for $n = k+1$.

Hence by induction the result is true for all positive integers n.

Example 5. *Show that similar matrices have the same eigenvalues.*

Solution. Two matrices A and B of the same order are said to be similar if there exists a non-singular matrix P such that

$$B = P^{-1}AB$$

Let λ be an eigenvalue of A, then X is a root of $|A - \lambda I| = 0$.

Now
$$
\begin{aligned}
B - \lambda I &= P^{-1}AP - \lambda I \\
&= P^{-1}AP - P^{-1}(\lambda I)P && [\because P^{-1}(\lambda I)P = \lambda P^{-1}P = \lambda I] \\
&= P^{-1}(A - \lambda I)P
\end{aligned}
$$
$$
\begin{aligned}
\Rightarrow \qquad |B - \lambda I| &= |P^{-1}||A - \lambda I||P| \\
\Rightarrow \qquad |B - \lambda I| &= |A - \lambda I||P^{-1}||P| \\
\Rightarrow \qquad |B - \lambda I| &= |A - \lambda I||P^{-1}P| \\
\Rightarrow \qquad |B - \lambda I| &= |A - \lambda I| && [\because |P^{-1}P| = |I| = 1]
\end{aligned}
$$

Since λ is a root of $|A - \lambda I| = 0$, therefore λ is also a root of $|B - \lambda I| = 0$, it follows that λ is an eigenvalue of B.

Hence similar matrices have the same eigenvalues.

Example 6. *Let A and B be two matrices of order $n \times n$. Let $X \neq O$ be an eigenvector of A and B corresponding to the eigenvalues λ_1 and λ_2 respectively, then show that X is an eigenvector of AB corresponding to the eigenvalue $\lambda_1 \lambda_2$ of AB.*

Solution. Since $X \neq O$ is an eigenvector of A and B corresponding to the eigenvalues λ_1 and λ_2 respectively, then we have

$$AX = \lambda_1 X \qquad \qquad ...(1)$$
$$\text{and} \qquad BX = \lambda_2 X \qquad \qquad ...(2)$$

Now
$$
\begin{aligned}
(AB)X &= A(BX) \\
&= A(\lambda_2 X) && \text{[Using (2)]} \\
&= \lambda_2(AX) \\
&= \lambda_2(\lambda_1 X) && \text{[Using (1)]} \\
&= (\lambda_2 \lambda_1)X
\end{aligned}
$$

$$(AB)X = (\lambda_1\lambda_2)X \text{ with } X \neq O$$

It follows that X is an eigenvector of AB corresponding to the eigenvalue $\lambda_1\lambda_2$.

Example 7. *Determine the eigenvalues of the matrix :*

$$A = \begin{bmatrix} 1 & 2 & 3 \\ 0 & -4 & 2 \\ 0 & 0 & 7 \end{bmatrix}$$

Solution. The characteristic equation of A is given by

$$|A - \lambda I| = 0$$

i.e.,
$$\begin{vmatrix} 1-\lambda & 2 & 3 \\ 0 & -4-\lambda & 2 \\ 0 & 0 & 7-\lambda \end{vmatrix} = 0$$

i.e., $\quad (1-\lambda)(-2-\lambda)(7-\lambda) = 0$

The roots of this characteristic equation are given by $\lambda = 1, -4, 7$.

These are the required eigenvalues of A.

REMARK

- It is clear that the given matrix A is an upper triangular matrix so that the principal diagonal elements $1, -4, 7$ will be the eigenvalues of A.

Example 8. *Determine the eigenvalues of the matrix :*

$$A = \begin{bmatrix} 0 & 1 & 2 \\ 1 & 0 & -1 \\ 2 & -1 & 0 \end{bmatrix}.$$

Solution. The characteristic equation of A is given

$$|A - \lambda I| = 0$$

or
$$\begin{vmatrix} 0-\lambda & 1 & 2 \\ 1 & 0-\lambda & -1 \\ 2 & -1 & 0-\lambda \end{vmatrix} = 0$$

or $\quad -\lambda(\lambda^2 - 1) - 1(-\lambda + 2) + 2(-1 + 2\lambda) = 0$

or
$$-\lambda^3 + 6\lambda - 4 = 0$$

Solving this equation, we get

$$(\lambda - 2)(\lambda^2 + 2\lambda - 2) = 0$$

$$\Rightarrow \qquad \lambda = 2 \text{ and } \lambda = -1 \pm \sqrt{3}$$

Hence the eigenvalues of A are $2, -1 \pm \sqrt{3}$.

Example 9. *Determine the eigenvalues and eigenvectors of the matrix*

$$A = \begin{bmatrix} 5 & 4 \\ 1 & 2 \end{bmatrix}.$$

Solution. The characteristic equation of A is given by

$$|A - \lambda I| = 0$$

or
$$\begin{vmatrix} 5 - \lambda & 4 \\ 1 & 2 - \lambda \end{vmatrix} = 0$$

or
$$(5 - \lambda)(2 - \lambda) - 4 = 0$$

or
$$\lambda^2 - 7\lambda + 10 - 4 = 0$$

or
$$\lambda^2 - 7\lambda + 6 = 0$$

The roots of this equation are $\lambda = 6, 1$.

Thus, the eigenvalues of A are 6, 1.

Eigenvector corresponding to $\lambda_1 = 6$:

Let $X_1 = \begin{bmatrix} x_1 \\ x_2 \end{bmatrix} \neq O$ be an eigenvector of A corresponding to $\lambda_1 = 6$, then we have

$$AX_1 = 6X_1$$

or
$$(A - 6I)X_1 = O$$

or
$$\begin{bmatrix} 5 - 6 & 4 \\ 1 & 2 - 6 \end{bmatrix} \begin{bmatrix} x_1 \\ x_2 \end{bmatrix} = \begin{bmatrix} 0 \\ 0 \end{bmatrix}$$

or
$$\begin{bmatrix} -1 & 4 \\ 1 & -4 \end{bmatrix} \begin{bmatrix} x_1 \\ x_2 \end{bmatrix} = \begin{bmatrix} 0 \\ 0 \end{bmatrix} \qquad \ldots(1)$$

The non-zero solution of (1) will give X_1.

Applying $R_2 \rightarrow R_2 + R_1$, we have

$$\begin{bmatrix} -1 & 4 \\ 1 & -4 \end{bmatrix} \begin{bmatrix} x_1 \\ x_2 \end{bmatrix} = \begin{bmatrix} 0 \\ 0 \end{bmatrix} \qquad \ldots(2)$$

The coefficient matrix of equation (1) is of rank 1, *i.e.,* $\rho(A - 6I) = 1$, therefore the system of equations (1) will have $2 - 1 = 1$ linearly independent solution. From (2), we have

$$-x_1 + 4x_2 = 0$$

Clearly, $x_1 = 4$ and $x_2 = 1$ satisfy the above equation.

Hence the eigenvector corresponding to eigenvalue $\lambda_1 = 6$ is

$$X_1 = \begin{bmatrix} 4 \\ 1 \end{bmatrix}$$

Eigenvector corresponding to $\lambda_2 = 1$:

Let $X_2 = \begin{bmatrix} x_1 \\ x_2 \end{bmatrix} \neq O$ be an eigenvector of A corresponding to eigen value $\lambda_2 = 1$, then we have

$$AX_2 = \lambda_2 X_2$$

or
$$AX_2 = IX_2$$

or
$$(A - I)X_2 = O$$

or
$$\begin{bmatrix} 5-1 & 4 \\ 1 & 2-1 \end{bmatrix} \begin{bmatrix} x_1 \\ x_2 \end{bmatrix} = \begin{bmatrix} 0 \\ 0 \end{bmatrix}$$

or
$$\begin{bmatrix} 4 & 4 \\ 1 & 1 \end{bmatrix} \begin{bmatrix} x_1 \\ x_2 \end{bmatrix} = \begin{bmatrix} 0 \\ 0 \end{bmatrix} \qquad \qquad \dots(3)$$

The non-zero solution of (3) will give X_2.

Applying $R_2 \to R_2 - \dfrac{1}{4} R_1$, we get

$$\begin{bmatrix} 4 & 4 \\ 0 & 0 \end{bmatrix} \begin{bmatrix} x_1 \\ x_2 \end{bmatrix} = \begin{bmatrix} 0 \\ 0 \end{bmatrix} \qquad \qquad \dots(4)$$

Clearly, $\rho(A - I) = 1$, therefore the system of equations (3) will have $2 - 1 = 1$ linearly independent solution.

From (4), we get
$$4x_1 + 4x_2 = 0$$

Clearly, $x_1 = 1$ and $x_2 = -1$, satisfy above equation.

Hence the eigenvector corresponding to eigenvalue $\lambda_2 = 1$ is

$$X_2 = \begin{bmatrix} 1 \\ -1 \end{bmatrix}.$$

Example 10. *Determine the eigenvalues and eigenvectors of the matrix*

$$A = \begin{bmatrix} 8 & -6 & 2 \\ -6 & 7 & -4 \\ 2 & -4 & 3 \end{bmatrix}.$$

Solution. The characteristic equation of A is given by

$$|A - \lambda I| = 0$$

or
$$\begin{vmatrix} 8-\lambda & -6 & 2 \\ -6 & 7-\lambda & -4 \\ 2 & -4 & 3-\lambda \end{vmatrix} = 0$$

or $\quad (8-\lambda)\{(7-\lambda)(3-\lambda)-16\}+6\{-18+6\lambda+8\}+2\{24-14+2\lambda\} = 0$

or $\qquad\qquad\qquad\qquad\qquad\qquad\qquad \lambda^3 - 18\lambda^2 + 45\lambda = 0$

or $\qquad\qquad\qquad\qquad\qquad\qquad\qquad \lambda(\lambda-3)(\lambda-15) = 0$

The roots of this equation are $\lambda = 0, 3, 15$.

Thus, the eigenvalues of A are $\lambda_1 = 0$, $\lambda_2 = 3$, $\lambda_3 = 15$.

Eigenvector corresponding to $\lambda_1 = 0$:

Let $X_1 = \begin{bmatrix} x_1 \\ x_2 \\ x_3 \end{bmatrix} \neq O$ be an eigenvector corresponding to the eigen value $\lambda_1 = 0$,

then we have $\qquad\qquad\qquad\qquad AX_1 = \lambda_1 X1.$

or $\qquad\qquad\qquad\qquad\qquad\qquad AX_1 = 0X_1$

or $\qquad\qquad\qquad\qquad\qquad\qquad (A - 0I)\, X_1 = O$

or
$$\begin{bmatrix} 8 & -6 & 2 \\ -6 & 7 & -4 \\ 2 & -4 & 3 \end{bmatrix} \begin{bmatrix} x_1 \\ x_2 \\ x_3 \end{bmatrix} = \begin{bmatrix} 0 \\ 0 \\ 0 \end{bmatrix} \qquad \dots(1)$$

The non-zero solution of (1) will give X_1.

Reducing the coefficient matrix of (1) in Echeleon form by applying elementary row transformations.

Applying $R_1 \leftrightarrow R_3$, we get

$$\begin{bmatrix} 2 & -4 & 3 \\ -6 & 7 & -4 \\ 8 & -6 & 2 \end{bmatrix} \begin{bmatrix} x_1 \\ x_2 \\ x_3 \end{bmatrix} = \begin{bmatrix} 0 \\ 0 \\ 0 \end{bmatrix}$$

Applying $R_2 \rightarrow R_2 + 3R_1$, $R_3 \rightarrow R_3 - 4R_1$, we get

$$\begin{bmatrix} 2 & -4 & 3 \\ 0 & -5 & 5 \\ 0 & 10 & -10 \end{bmatrix} \begin{bmatrix} x_1 \\ x_2 \\ x_3 \end{bmatrix} = \begin{bmatrix} 0 \\ 0 \\ 0 \end{bmatrix}$$

Applying $R_3 \rightarrow R_3 + 2R_2$, we get

$$\begin{bmatrix} 2 & -4 & 3 \\ 0 & -5 & 5 \\ 0 & 0 & 0 \end{bmatrix} \begin{bmatrix} x_1 \\ x_2 \\ x_3 \end{bmatrix} = \begin{bmatrix} 0 \\ 0 \\ 0 \end{bmatrix} \qquad \dots(2)$$

Clearly $\rho(A - 0.I) = 2$, therefore the system of equations (2) will have $3 - 2 = 1$ (unknowns – rank) linearly independent solution.

From (2), we have

$$2x_1 - 4x_2 + 3x_3 = 0$$
$$- 5x_2 + 5x_3 = 0$$

Clearly, $x_1 = \dfrac{1}{2}$, $x_2 = 1$ and $x_3 = 1$ satisfy the above equations.

Hence the eigenvector corresponding to eigenvalue $\lambda_1 = 0$ is

$$X_1 = \begin{bmatrix} 1/2 \\ 1 \\ 1 \end{bmatrix}$$

Eigenvector corresponding to $\lambda_2 = 3$:

Let $X_2 = \begin{bmatrix} x_1 \\ x_2 \\ x_3 \end{bmatrix} \neq O$ be an eigenvector of A corresponding to $\lambda_2 = 3$, then we have

$$AX_2 = \lambda_2 X_2$$

or
$$(A - \lambda_2 I)X_2 = O$$

or
$$(A - 3I)X_2 = O$$

or
$$\begin{bmatrix} 8-3 & -6 & 2 \\ -6 & 7-3 & -4 \\ 2 & -4 & 3-3 \end{bmatrix} \begin{bmatrix} x_1 \\ x_2 \\ x_3 \end{bmatrix} = O$$

or
$$\begin{bmatrix} 5 & -6 & 2 \\ -6 & 4 & -4 \\ 2 & -4 & 0 \end{bmatrix}\begin{bmatrix} x_1 \\ x_2 \\ x_3 \end{bmatrix} = O \qquad \qquad \text{...(3)}$$

The non-zero solution of (3) will give X_2.

Applying $R_1 \rightarrow R_1 + R_2$, we get
$$\begin{bmatrix} -1 & -2 & -2 \\ -6 & 4 & -4 \\ 2 & -4 & 0 \end{bmatrix}\begin{bmatrix} x_1 \\ x_2 \\ x_3 \end{bmatrix} = \begin{bmatrix} 0 \\ 0 \\ 0 \end{bmatrix}$$

Applying $R_2 \rightarrow R_2 - 6R_1,\, R_3 \rightarrow R_3 + 2R_1$, we get
$$\begin{bmatrix} -1 & -2 & -2 \\ 0 & 16 & 8 \\ 0 & -8 & -4 \end{bmatrix}\begin{bmatrix} x_1 \\ x_2 \\ x_3 \end{bmatrix} = \begin{bmatrix} 0 \\ 0 \\ 0 \end{bmatrix}$$

Applying $R_2 \rightarrow \dfrac{1}{8}R_2$, we get
$$\begin{bmatrix} -1 & -2 & -2 \\ 0 & 2 & 1 \\ 0 & -8 & -4 \end{bmatrix}\begin{bmatrix} x_1 \\ x_2 \\ x_3 \end{bmatrix} = \begin{bmatrix} 0 \\ 0 \\ 0 \end{bmatrix}$$

Again applying $R_3 \rightarrow R_3 + 4R_2$, we get
$$\begin{bmatrix} -1 & -2 & -2 \\ 0 & 2 & 1 \\ 0 & 0 & 0 \end{bmatrix}\begin{bmatrix} x_1 \\ x_2 \\ x_3 \end{bmatrix} = \begin{bmatrix} 0 \\ 0 \\ 0 \end{bmatrix} \qquad \text{...(4)}$$

Clearly $\rho(A - 3I) = 2$, therefore the system of equations (3) will have $3 - 2 = 1$ linearly independent solution.

From (4), we have
$$-x_1 - 2x_2 - 2x_3 = 0$$
$$2x_2 + x_3 = 0$$

Clearly, $x_1 = -2$, $x_2 = -1$ and $x_3 = 2$ satisfy the above equations.

Hence the eigenvector corresponding to eigenvalue $\lambda_2 = 3$ is
$$X_2 = \begin{bmatrix} -2 \\ -1 \\ 2 \end{bmatrix}$$

Eigenvector corresponding to $\lambda_3 = 15$:

Let $X_3 = \begin{bmatrix} x_1 \\ x_2 \\ x_3 \end{bmatrix} \neq O$ be an eigenvector of A corresponding to $\lambda_3 = 15$, then we have

$$AX_3 = \lambda_3 X_3$$

or
$$(A - \lambda_3 I)X_3 = O$$

or
$$(A - 15I)X_3 = O$$

or
$$\begin{bmatrix} 8-15 & -6 & 2 \\ -6 & 7-15 & -4 \\ 2 & -4 & 3-15 \end{bmatrix}\begin{bmatrix} x_1 \\ x_2 \\ x_3 \end{bmatrix} = O$$

or
$$\begin{bmatrix} -7 & -6 & 2 \\ -6 & -8 & -4 \\ 2 & -4 & -12 \end{bmatrix}\begin{bmatrix} x_1 \\ x_2 \\ x_3 \end{bmatrix} = O \qquad \ldots(5)$$

The non-zero solution of (5) will give X_3.

Applying $R_1 \leftrightarrow R_3$, we get

$$\begin{bmatrix} 2 & -4 & -12 \\ -6 & -8 & -4 \\ -7 & -6 & 2 \end{bmatrix}\begin{bmatrix} x_1 \\ x_2 \\ x_3 \end{bmatrix} = \begin{bmatrix} 0 \\ 0 \\ 0 \end{bmatrix}$$

Applying $R_1 \to \dfrac{1}{2}R_1$, we get

$$\begin{bmatrix} 1 & -2 & -6 \\ -6 & -8 & -4 \\ -7 & -6 & 2 \end{bmatrix}\begin{bmatrix} x_1 \\ x_2 \\ x_3 \end{bmatrix} = \begin{bmatrix} 0 \\ 0 \\ 0 \end{bmatrix}$$

Applying $R_2 \to R_2 +6R_1, R_3 \to R_3 +7R_1$, we get

$$\begin{bmatrix} 1 & -2 & -6 \\ 0 & -20 & -40 \\ 0 & -20 & -40 \end{bmatrix}\begin{bmatrix} x_1 \\ x_2 \\ x_3 \end{bmatrix} = \begin{bmatrix} 0 \\ 0 \\ 0 \end{bmatrix}$$

Applying $R_3 \to R_3 - R_2$, we get

$$\begin{bmatrix} 1 & -2 & -6 \\ 0 & -20 & -40 \\ 0 & 0 & 0 \end{bmatrix}\begin{bmatrix} x_1 \\ x_2 \\ x_3 \end{bmatrix} = \begin{bmatrix} 0 \\ 0 \\ 0 \end{bmatrix} \qquad \ldots(6)$$

Clearly $\rho(A - 15I) = 2$, therefore the system of equations (5) will have $3 - 2 = 1$ linearly independent solution.

From (6), we have
$$x_1 - 2x_2 - 6x_3 = 0$$
$$- 20x_2 - 40x_3 = 0$$

Clearly, $x_1 = 2, x_2 = - 2$ and $x_3 = 1$ satisfy the above equations.

Hence the eigenvector corresponding to eigenvalue $\lambda_3 = 15$ is

$$X_3 = \begin{bmatrix} 2 \\ -2 \\ 1 \end{bmatrix}$$

Example 11. *Determine the eigenvalues and eigenvectors of the matrix*

$$A = \begin{bmatrix} 2 & 1 & 0 \\ 0 & 1 & -1 \\ 0 & 2 & 4 \end{bmatrix}.$$

Solution. The characteristic equation of A is given by

$$|A - \lambda I| = 0$$

or

$$\begin{vmatrix} 2-\lambda & 1 & 0 \\ 0 & 1-\lambda & -1 \\ 0 & 2 & 4-\lambda \end{vmatrix} = 0$$

or $\quad (2-\lambda)\{(1-\lambda)(4-\lambda)+2\} = 0$

or $\quad (2-\lambda)(\lambda^3 - 5\lambda + 6) = 0$

or $\quad (2-\lambda)(2-\lambda)(3-\lambda) = 0$

The roots of this equation are $\lambda = 2, 2, 3$.

Thus the eigenvalues of A are $\lambda_1 = 2, \lambda_2 = 2, \lambda_3 = 3$.

Eigenvector corresponding to the eigenvalue $\lambda_1 = \lambda_2 = 2$:

Let $X_1 = \begin{bmatrix} x_1 \\ x_2 \\ x_3 \end{bmatrix} \neq O$ be an eigenvector corresponding to the eigenvalue 2, then

we have

$$AX_1 = 2X_1.$$

or $\qquad\qquad (A - 2I)\,X_1 = O$

or

$$\begin{bmatrix} 2-2 & 1 & 0 \\ 0 & 1-2 & -1 \\ 0 & 2 & 4-2 \end{bmatrix} \begin{bmatrix} x_1 \\ x_2 \\ x_3 \end{bmatrix} = \begin{bmatrix} 0 \\ 0 \\ 0 \end{bmatrix}$$

or

$$\begin{bmatrix} 0 & 1 & 0 \\ 0 & -1 & -1 \\ 0 & 2 & 2 \end{bmatrix} \begin{bmatrix} x_1 \\ x_2 \\ x_3 \end{bmatrix} = \begin{bmatrix} 0 \\ 0 \\ 0 \end{bmatrix} \qquad \ldots(1)$$

The non-zero solution of (1) will give X_1.

Applying $R_3 \to R_3 + 2R_2$, we get

$$\begin{bmatrix} 0 & 1 & 0 \\ 0 & -1 & -1 \\ 0 & 0 & 0 \end{bmatrix} \begin{bmatrix} x_1 \\ x_2 \\ x_3 \end{bmatrix} = \begin{bmatrix} 0 \\ 0 \\ 0 \end{bmatrix}$$

Applying $R_2 \to R_2 + R_1$, we get

$$\begin{bmatrix} 0 & 1 & 0 \\ 0 & 0 & -1 \\ 0 & 0 & 0 \end{bmatrix} \begin{bmatrix} x_1 \\ x_2 \\ x_3 \end{bmatrix} = \begin{bmatrix} 0 \\ 0 \\ 0 \end{bmatrix} \qquad \ldots(2)$$

From (2), it is clear that $\rho(A - 2I) = 2$, therefore the system of equations (1) will have $3 - 2 = 1$ linearly independent solution.

From (2), we have

$$x_2 = 0 \text{ and } -x_3 = 0$$

Clearly, $x_1 = 1, x_2 = 0, x_3 = 1$ satisfy the above equations.

Hence the eigenvector corresponding to the eigenvalue $\lambda_1 = \lambda_2 = 2$ is

$$X_1 = \begin{bmatrix} 1 \\ 0 \\ 0 \end{bmatrix}$$

Eigenvector corresponding to the eigenvalue $\lambda_3 = 3$:

Let $X_2 = \begin{bmatrix} x_1 \\ x_2 \\ x_3 \end{bmatrix} \neq O$ be an eigenvector corresponding to $\lambda_3 = 3$, then we have

$$AX_2 = 3X_2$$

or $$(A - 3I)X_3 = O$$

or $$\begin{bmatrix} 2-3 & 1 & 0 \\ 0 & 1-3 & -1 \\ 0 & 2 & 4-3 \end{bmatrix} \begin{bmatrix} x_1 \\ x_2 \\ x_3 \end{bmatrix} = \begin{bmatrix} 0 \\ 0 \\ 0 \end{bmatrix}$$

or $$\begin{bmatrix} -1 & 1 & 0 \\ 0 & -2 & -1 \\ 0 & 2 & 1 \end{bmatrix} \begin{bmatrix} x_1 \\ x_2 \\ x_3 \end{bmatrix} = \begin{bmatrix} 0 \\ 0 \\ 0 \end{bmatrix} \qquad \ldots(3)$$

The non-zero solution of (3) will give X_2.

Applying $R_3 \rightarrow R_3 + R_2$, we get

$$\begin{bmatrix} -1 & 1 & 0 \\ 0 & -2 & -1 \\ 0 & 0 & 0 \end{bmatrix} \begin{bmatrix} x_1 \\ x_2 \\ x_3 \end{bmatrix} = \begin{bmatrix} 0 \\ 0 \\ 0 \end{bmatrix} \qquad \ldots(4)$$

Clearly $\rho(A - 3I) = 2$, therefore the system of equations (3) will have $3 - 2 = 1$ linearly independent solution.

From (4), we have

$$-x_1 + x_2 = 0$$
$$-2x_2 - x_3 = 0$$

Clearly, $x_1 = 1$, $x_2 = 1$ and $x_3 = -2$ satisfy the above equations. Hence the eigenvector corresponding to eigenvalue $\lambda_3 = 3$ is

$$X_2 = \begin{bmatrix} 1 \\ 1 \\ -2 \end{bmatrix}$$

Example 12. *Find the eigenvalues and eigenvectors of the matrix*

$$A = \begin{bmatrix} 5 & 4 & 2 \\ 4 & 5 & 2 \\ 2 & 2 & 2 \end{bmatrix}.$$

Solution. The characteristic equation of A is given by

$$|A - \lambda I| = 0$$

or
$$\begin{vmatrix} 5-\lambda & 4 & 2 \\ 4 & 5-\lambda & 2 \\ 2 & 2 & 2-\lambda \end{vmatrix} = 0$$

or $\quad (5-\lambda)\{(5-\lambda)(2-\lambda)-4\}-4\{4(2-\lambda)-4\}+2\{8-2(8-\lambda)\}=0$

or $\hspace{6cm} -\lambda^3 + 12\lambda - 21\lambda + 10 = 0$

or $\hspace{6cm} -(\lambda-1)^2(\lambda-10) = 0$

The roots of this equation are 1, 1, 10.

Thus the eigenvalues of A are $\lambda_1 = 1, \lambda_2 = 1, \lambda_3 = 10$.

Eigenvector corresponding to the eigenvalue $\lambda_1 = \lambda_2 = 1$:

Let $X = \begin{bmatrix} x_1 \\ x_2 \\ x_3 \end{bmatrix} \neq O$ be an eigenvector of A corresponding to

the eigenvalue $\hspace{3cm} \lambda_1 = \lambda_2 = 1$, then
we have
$$AX_1 = IX$$

or $\hspace{3cm} (A - I)X_1 = O$

or $\hspace{2cm} \begin{bmatrix} 5-1 & 4 & 2 \\ 4 & 5-1 & 2 \\ 2 & 2 & 2-1 \end{bmatrix}\begin{bmatrix} x_1 \\ x_2 \\ x_3 \end{bmatrix} = \begin{bmatrix} 0 \\ 0 \\ 0 \end{bmatrix}$

or $\hspace{2cm} \begin{bmatrix} 4 & 4 & 2 \\ 4 & 4 & 2 \\ 2 & 2 & 2 \end{bmatrix}\begin{bmatrix} x_1 \\ x_2 \\ x_3 \end{bmatrix} = \begin{bmatrix} 0 \\ 0 \\ 0 \end{bmatrix}$ $\hspace{2cm}$ ...(1)

Applying $R_1 \leftrightarrow R_3$, we get
$$\begin{bmatrix} 2 & 2 & 1 \\ 4 & 4 & 2 \\ 4 & 4 & 2 \end{bmatrix}\begin{bmatrix} x_1 \\ x_2 \\ x_3 \end{bmatrix} = \begin{bmatrix} 0 \\ 0 \\ 0 \end{bmatrix}$$

Applying $R_2 \rightarrow R_2 - 2R_1$, we get
$$\begin{bmatrix} 2 & 2 & 1 \\ 0 & 0 & 0 \\ 0 & 0 & 0 \end{bmatrix}\begin{bmatrix} x_1 \\ x_2 \\ x_3 \end{bmatrix} = \begin{bmatrix} 0 \\ 0 \\ 0 \end{bmatrix} \hspace{2cm} \text{...(2)}$$

From (2), it is clear that $\rho(A - I) = 1$, therefore the system of equation (1) will have $3 - 2 = 1$ linearly independent solution.

Now from (2), we have
$$x_2 = 0 \quad \text{and} \quad -x_3 = 0$$

Clearly $x_1 = 0, x_2 = 0, x_3 = 0$ satisfy the above equations.

Hence the eigenvector corresponding to the eigenvalue $\lambda_1 = \lambda_2 = 2$ is

$$X_1 = \begin{bmatrix} 1 \\ 0 \\ 0 \end{bmatrix}$$

Eigenvector corresponding to the eigenvalue $\lambda_1 = \lambda_2 = 1$

Let $X = \begin{bmatrix} x_1 \\ x_2 \\ x_3 \end{bmatrix} \neq 0$ be an eigenvector corresponding to the eigenvalue $\lambda_1 = \lambda_2 = 1$,

then we have

$$AX = IX$$

or $$(A - I)X = O$$

or $$\begin{bmatrix} 5-1 & 4 & 2 \\ 4 & 5-1 & 2 \\ 2 & 2 & 2-1 \end{bmatrix} \begin{bmatrix} x_1 \\ x_2 \\ x_3 \end{bmatrix} = \begin{bmatrix} 0 \\ 0 \\ 0 \end{bmatrix}$$

or $$\begin{bmatrix} 4 & 4 & 2 \\ 4 & 4 & 2 \\ 2 & 2 & 1 \end{bmatrix} \begin{bmatrix} x_1 \\ x_2 \\ x_3 \end{bmatrix} = \begin{bmatrix} 0 \\ 0 \\ 0 \end{bmatrix} \qquad \qquad \qquad ...(1)$$

Applying $R_1 \rightarrow R_3$, we get

$$\begin{bmatrix} 2 & 2 & 1 \\ 4 & 4 & 2 \\ 4 & 4 & 2 \end{bmatrix} \begin{bmatrix} x_1 \\ x_2 \\ x_3 \end{bmatrix} = \begin{bmatrix} 0 \\ 0 \\ 0 \end{bmatrix}$$

Applying $R_2 \rightarrow R_2 - 2R_1$, $R_3 \rightarrow R_3 - 2R_1$, we get

$$\begin{bmatrix} 2 & 2 & 1 \\ 0 & 0 & 0 \\ 0 & 0 & 0 \end{bmatrix} \begin{bmatrix} x_1 \\ x_2 \\ x_3 \end{bmatrix} = \begin{bmatrix} 0 \\ 0 \\ 0 \end{bmatrix} \qquad \qquad \qquad ...(2)$$

From (2), it is clear that $P(A-I) = 1$, therefoe the equations (1) will have $3-1 = 2$ linearly independent solutions.

From (2), we have

$$2x_1 + 2x_2 + x_3 = 0$$

Since this equation has two linearly independent solutions so we take $x_2 = c_1$ and $x_3 = c_2$, where c_1 and c_2 are non-zero scalars, then $x_1 = -c - \dfrac{c_2}{2}$.

Therefore, $$\begin{bmatrix} x_1 \\ x_2 \\ x_3 \end{bmatrix} = \begin{bmatrix} -c_1 - \dfrac{c_2}{2} \\ c_1 \\ c_2 \end{bmatrix} = c_1 \begin{bmatrix} -1 \\ 1 \\ 0 \end{bmatrix} + c_2 \begin{bmatrix} -1/2 \\ 0 \\ 1 \end{bmatrix}$$

Hence the eigenvectors corresponding to the eigenvalue $\lambda_1 = \lambda_2 = 1$ are

$$X_1 = \begin{bmatrix} -1 \\ 0 \\ 1 \end{bmatrix} \text{ and } X_2 = \begin{bmatrix} -1/2 \\ 0 \\ 1 \end{bmatrix}$$

Eigenvector corresponding to the eigenvalue $\lambda_3 = 10$:

Let $X_3 = \begin{bmatrix} x_1 \\ x_2 \\ x_3 \end{bmatrix} \neq O$ be an eigenvector corresponding to $\lambda_3 = 10$, then we have

$$AX_3 = 10X_3$$

or

$$(A - 10I)X_3 = O$$

or

$$\begin{bmatrix} 5-10 & 4 & 2 \\ 4 & 5-10 & 2 \\ 2 & 2 & 2-10 \end{bmatrix}\begin{bmatrix} x_1 \\ x_2 \\ x_3 \end{bmatrix} = \begin{bmatrix} 0 \\ 0 \\ 0 \end{bmatrix}$$

or

$$\begin{bmatrix} -5 & 4 & 2 \\ 4 & -5 & 2 \\ 2 & 2 & -8 \end{bmatrix}\begin{bmatrix} x_1 \\ x_2 \\ x_3 \end{bmatrix} = \begin{bmatrix} 0 \\ 0 \\ 0 \end{bmatrix} \qquad \ldots(3)$$

Applying $R_1 \to R_1 + R_2$, we get

$$\begin{bmatrix} -1 & -1 & 4 \\ 4 & -5 & 2 \\ 2 & 2 & -8 \end{bmatrix}\begin{bmatrix} x_1 \\ x_2 \\ x_3 \end{bmatrix} = \begin{bmatrix} 0 \\ 0 \\ 0 \end{bmatrix}$$

Applying $R_2 \to R_2 + 4R_1, R_3 \to R_3 + 2R_1$, we get

$$\begin{bmatrix} -1 & -1 & 4 \\ 0 & -9 & 18 \\ 0 & 0 & 0 \end{bmatrix}\begin{bmatrix} x_1 \\ x_2 \\ x_3 \end{bmatrix} = \begin{bmatrix} 0 \\ 0 \\ 0 \end{bmatrix} \qquad \ldots(4)$$

From (4), it is clear that $\rho(A - 10I) = 2$, therefore the system of equations (3) will have $3 - 1 = 2$ linearly independent solutions.

From (4), we have

$$-x_1 - x_2 + 4x_3 = 0$$
$$-9x_2 + 18x_3 = 0$$

Let us take $x_3 = c$, then $x_2 = 2c$ and $x_1 = 2c$.

Therefore,

$$\begin{bmatrix} x_1 \\ x_2 \\ x_3 \end{bmatrix} = \begin{bmatrix} 2c \\ 2c \\ c \end{bmatrix} = c\begin{bmatrix} 2 \\ 2 \\ 1 \end{bmatrix}$$

Hence the eigenvector corresponding to eigenvalue $\lambda_3 = 10$ is

$$X_3 = \begin{bmatrix} 2 \\ 2 \\ 1 \end{bmatrix}.$$

Example 13. *Find the eigenvalues and eigenvectors of the matrix*

$$A = \begin{bmatrix} 2 & 0 & 1 & -3 \\ 0 & 2 & 10 & 4 \\ 0 & 0 & 2 & 0 \\ 0 & 0 & 0 & 3 \end{bmatrix}.$$

Solution. The characteristic equation of A is given by
$$|A - \lambda I| = 0$$

or
$$A = \begin{vmatrix} 2-\lambda & 0 & 1 & -3 \\ 0 & 2-\lambda & 10 & 4 \\ 0 & 0 & 2-\lambda & 0 \\ 0 & 0 & 0 & 3-\lambda \end{vmatrix} = 0$$

or
$$(\lambda - 2)^2 (\lambda - 3) = 0$$

The roots of characteristic equations are 2, 2, 2, 3.

Thus, the eigenvalues of A are $\lambda_1 = 2, \lambda_2 = 2, \lambda_3 = 2, \lambda_4 = 3$.

Eigenvector corresponding to $\lambda_1 = \lambda_2 = \lambda_3 = 2$:

Let $X = \begin{bmatrix} x_1 \\ x_2 \\ x_3 \\ x_4 \end{bmatrix} \neq O$ be an eigenvector of A corresponding to the eigenvalue 2, then we have

$$AX = 2X.$$

or
$$(A - 2I) X = O$$

or
$$\begin{bmatrix} 2-2 & 0 & 1 & -3 \\ 0 & 2-2 & 10 & 4 \\ 0 & 0 & 2-2 & Q \\ 0 & 0 & 0 & 3-2 \end{bmatrix} \begin{bmatrix} x_1 \\ x_2 \\ x_3 \\ x_4 \end{bmatrix} = \begin{bmatrix} 0 \\ 0 \\ 0 \\ 0 \end{bmatrix}$$

or
$$\begin{bmatrix} 0 & 0 & 1 & -3 \\ 0 & 0 & 10 & 4 \\ 0 & 0 & 0 & 0 \\ 0 & 0 & 0 & 1 \end{bmatrix} \begin{bmatrix} x_1 \\ x_2 \\ x_3 \\ x_4 \end{bmatrix} = \begin{bmatrix} 0 \\ 0 \\ 0 \\ 0 \end{bmatrix} \qquad \dots(1)$$

Applying $R_3 \leftrightarrow R_4$, we get
$$\begin{bmatrix} 0 & 0 & 1 & -3 \\ 0 & 0 & 10 & 4 \\ 0 & 0 & 0 & 1 \\ 0 & 0 & 0 & 0 \end{bmatrix} \begin{bmatrix} x_1 \\ x_2 \\ x_3 \\ x_4 \end{bmatrix} = \begin{bmatrix} 0 \\ 0 \\ 0 \\ 0 \end{bmatrix}$$

Applying $R_2 \rightarrow R_2 - 10R_1$, we get
$$\begin{bmatrix} 0 & 0 & 1 & -3 \\ 0 & 0 & 0 & 34 \\ 0 & 0 & 0 & 1 \\ 0 & 0 & 0 & 0 \end{bmatrix} \begin{bmatrix} x_1 \\ x_2 \\ x_3 \\ x_4 \end{bmatrix} = \begin{bmatrix} 0 \\ 0 \\ 0 \\ 0 \end{bmatrix}$$

Applying $R_2 \rightarrow \dfrac{1}{34} R_2$, we get
$$\begin{bmatrix} 0 & 0 & 1 & -3 \\ 0 & 0 & 0 & 1 \\ 0 & 0 & 0 & 1 \\ 0 & 0 & 0 & 0 \end{bmatrix} \begin{bmatrix} x_1 \\ x_2 \\ x_3 \\ x_4 \end{bmatrix} = \begin{bmatrix} 0 \\ 0 \\ 0 \\ 0 \end{bmatrix}$$

Applying $R_2 \to R_3 - R_2$, we get

$$\begin{bmatrix} 0 & 0 & 1 & -3 \\ 0 & 0 & 0 & 1 \\ 0 & 0 & 0 & 0 \\ 0 & 0 & 0 & 0 \end{bmatrix}\begin{bmatrix} x_1 \\ x_2 \\ x_3 \\ x_4 \end{bmatrix} = \begin{bmatrix} 0 \\ 0 \\ 0 \\ 0 \end{bmatrix} \qquad \ldots(2)$$

From (2), it is clear that $\rho(A - 2I) = 2$, therefore the equation (1) will have $4 - 2 = 2$ linearly independent solution.

Equation (2) reduces to

$$\left.\begin{array}{c} x_3 - 3x_4 = 0 \\ x_4 = 0 \end{array}\right\} \Rightarrow x_4 = 0, x_3 = 0$$

Let us take $x_1 = c_1, x_2 = c_2$, then we have

$$X = \begin{bmatrix} x_1 \\ x_2 \\ x_3 \\ x_4 \end{bmatrix} = \begin{bmatrix} c_1 \\ c_2 \\ 0 \\ 0 \end{bmatrix} = c_1\begin{bmatrix} 1 \\ 0 \\ 0 \\ 0 \end{bmatrix} + c_2\begin{bmatrix} 0 \\ 1 \\ 0 \\ 0 \end{bmatrix}$$

Hence the eigenvector corresponding to eigenvalue 2 are

$$X_1 = \begin{bmatrix} 1 \\ 0 \\ 0 \\ 0 \end{bmatrix}, X_2 = \begin{bmatrix} 0 \\ 1 \\ 0 \\ 0 \end{bmatrix}$$

Eigenvector corresponding to the eigenvalue $\lambda_4 = 3$:

Let $X = \begin{bmatrix} x_1 \\ x_2 \\ x_3 \\ x_4 \end{bmatrix} \neq O$ be an eigenvector corresponding to the eigenvalue 3, then we have

$$AX = 3X$$

or $$(A - 3I)X = O$$

or $$\begin{bmatrix} 2-3 & 0 & 1 & -3 \\ 0 & 2-3 & 10 & 4 \\ 0 & 0 & 2-3 & 0 \\ 0 & 0 & 0 & 3-3 \end{bmatrix}\begin{bmatrix} x_1 \\ x_2 \\ x_3 \\ x_4 \end{bmatrix} = \begin{bmatrix} 0 \\ 0 \\ 0 \\ 0 \end{bmatrix}$$

or $$\begin{bmatrix} -1 & 0 & 1 & -3 \\ 0 & -1 & 10 & 4 \\ 0 & 0 & -1 & 0 \\ 0 & 0 & 0 & 0 \end{bmatrix}\begin{bmatrix} x_1 \\ x_2 \\ x_3 \\ x_4 \end{bmatrix} = \begin{bmatrix} 0 \\ 0 \\ 0 \\ 0 \end{bmatrix} \qquad \ldots(3)$$

From (3), it is clear that $\rho(A - 3I) = 3$, therefore the equation (3) will have $4 - 3 = 1$ linearly independent solution.

Equation (3) reduces to

$$-x_1 + x_3 - 3x_4 = 0$$
$$-x_2 + 10x_3 + 4x_4 = 0$$
$$-x_3 = 0$$

Let us take $x_4 = c$, then from above equaions, we get

$$x_1 = -3c, \ x_2 = 4c, \ x_3 = 0, \ x_4 = c$$

Therefore,

$$\begin{bmatrix} x_1 \\ x_2 \\ x_3 \\ x_4 \end{bmatrix} = \begin{bmatrix} -3c \\ 4c \\ 0 \\ c \end{bmatrix} = c \begin{bmatrix} -3 \\ 4 \\ 0 \\ 1 \end{bmatrix}$$

Hence the eigenvector corresponding to eigenvalue 3 is

$$X = \begin{bmatrix} -3 \\ 4 \\ 0 \\ 1 \end{bmatrix}.$$

Example 14. *Find the eigenvalues and eigenvectors of the matrix*

$$A = \begin{bmatrix} 2 & -1 \\ 5 & -2 \end{bmatrix}.$$

Solution. The characteristic equation of A is given by

$$|A - \lambda I| = 0$$

or
$$\begin{vmatrix} 2-\lambda & -1 \\ 5 & -2-\lambda \end{vmatrix} = 0$$

or $\quad (2-\lambda)(-2-\lambda) + 5 = 0$

or $\quad \lambda^2 + 1 = 0$

The roots of this equation are $\lambda = \pm i$.
Thus, the eigenvalues of A are $\lambda_1 = -i, \lambda_2 = i$.
[This example confirms that a matrix with real entries may have complex eigenvalues.]
Eigenvector corresponding to the eigenvlaue $\lambda_1 = -i$:

Let $X = \begin{bmatrix} x_1 \\ x_2 \end{bmatrix} \neq O$ be an eigenvector of A corresponding to the eigenvalue

$\lambda_1 = -i$, then we have

$$AX = \lambda_1 X$$

or $\qquad AX = (-i)X$

or $\qquad (A - (-i)I)X_1 = O$

or $\quad \begin{bmatrix} 2-(-i) & -1 \\ 5 & -2-(-i) \end{bmatrix} \begin{bmatrix} x_1 \\ x_2 \end{bmatrix} = \begin{bmatrix} 0 \\ 0 \end{bmatrix}$

or $\quad \begin{bmatrix} 2+i & -1 \\ 5 & -2+i \end{bmatrix} \begin{bmatrix} x_1 \\ x_2 \end{bmatrix} = \begin{bmatrix} 0 \\ 0 \end{bmatrix}$ $\qquad \ldots (1)$

Applying $R_1 \to \left(\dfrac{1}{2+i}\right)R_1$, we get

$$\begin{bmatrix} 1 & -\dfrac{1}{2+i} \\ 5 & -2+i \end{bmatrix}\begin{bmatrix} x_1 \\ x_2 \end{bmatrix} = \begin{bmatrix} 0 \\ 0 \end{bmatrix}$$

Now

$$\dfrac{-1}{2+i} = \dfrac{-1(2-i)}{(2+i)(2-i)} = \dfrac{-2+i}{5} = \dfrac{-2}{5}+\dfrac{i}{5}$$

$\therefore$

$$\begin{bmatrix} 1 & -\dfrac{2}{5}+\dfrac{i}{5} \\ 5 & -2+i \end{bmatrix}\begin{bmatrix} x_1 \\ x_2 \end{bmatrix} = \begin{bmatrix} 0 \\ 0 \end{bmatrix}$$

Applying $R_2 \to R_2 - 5R_1$, we get

$$\begin{bmatrix} 1 & -\dfrac{2}{5}+\dfrac{i}{5} \\ 0 & 0 \end{bmatrix}\begin{bmatrix} x_1 \\ x_2 \end{bmatrix} = \begin{bmatrix} 0 \\ 0 \end{bmatrix} \qquad \ldots(2)$$

From (2), it is clear that $\rho(A - (-i)I) = 1$, therefore the equation (1) will have $2 - 1 = 1$ linearly independent solution.

Equation (2) reduces to

$$x_1 + \left(-\dfrac{2}{5}+\dfrac{i}{5}\right)x_2 = 0$$

Let us take $x_2 = c$, then we get

$$x_1 = -c\left(\dfrac{-2}{5}+\dfrac{i}{5}\right) = c\left(\dfrac{2}{5}-\dfrac{i}{5}\right)$$

$\therefore$

$$\begin{bmatrix} x_1 \\ x_2 \end{bmatrix} = \begin{bmatrix} c\left(\dfrac{2}{5}-\dfrac{i}{5}\right) \\ c \end{bmatrix} = \dfrac{c}{5}\begin{bmatrix} 2-i \\ 5 \end{bmatrix}$$

Hence the eigenvector corresponding to eigenvalue $\lambda_1 = -i$ is

$$X_1 = \begin{bmatrix} 2-i \\ 5 \end{bmatrix}$$

Eigenvector corresponding to the eigenvalue $\lambda_2 = i$:

Let $X = \begin{bmatrix} x_1 \\ x_2 \end{bmatrix} \neq O$ be an eigenvector corresponding to the eigenvalue $\lambda_2 = i$, then we have

$$AX = \lambda_2 X$$
$$AX = i(X)$$

or $\qquad (A - (i)I)X = O$

or

$$\begin{bmatrix} 2-i & -1 \\ 5 & -2-i \end{bmatrix}\begin{bmatrix} x_1 \\ x_2 \end{bmatrix} = \begin{bmatrix} 0 \\ 0 \end{bmatrix} \qquad \ldots(3)$$

Applying $R_1 \to \left(\dfrac{1}{2-i}\right)R_1$, we get

$$\begin{bmatrix} 1 & -\dfrac{1}{2-i} \\ 5 & -2-i \end{bmatrix}\begin{bmatrix} x_1 \\ x_2 \end{bmatrix} = \begin{bmatrix} 0 \\ 0 \end{bmatrix}$$

Now
$$\frac{-1}{2-i} = \frac{-1(2+i)}{(2-i)(2+i)} = \frac{-2-i}{5} = \frac{-2}{5} - \frac{i}{5}$$

$$\therefore \quad \begin{bmatrix} 1 & -\frac{2}{5} - \frac{i}{5} \\ 5 & -2-i \end{bmatrix} \begin{bmatrix} x_1 \\ x_2 \end{bmatrix} = \begin{bmatrix} 0 \\ 0 \end{bmatrix}$$

Applying $R_2 \to R_2 - 5R_1$, we get

$$\begin{bmatrix} 1 & -\frac{2}{5} - \frac{i}{5} \\ 0 & 0 \end{bmatrix} \begin{bmatrix} x_1 \\ x_2 \end{bmatrix} = \begin{bmatrix} 0 \\ 0 \end{bmatrix} \qquad \ldots(4)$$

From (4) it is clear that $\rho(A - (i)I) = 1$, therefore the system of equation (3) will have $2 - 1 = 1$ linearly independent solution.

Equation (4) reduces to

$$x_1 + \left(-\frac{2}{5} - \frac{i}{5} \right) x_2 = 0$$

Let us take $x_2 = c$, then $x_1 = \left(\frac{2+i}{5} \right) c$

$$\therefore \quad \begin{bmatrix} x_1 \\ x_2 \end{bmatrix} = \begin{bmatrix} \left(\frac{2+i}{5} \right) c \\ c \end{bmatrix} = \frac{c}{5} \begin{bmatrix} 2+i \\ 5 \end{bmatrix}$$

Hence the eigenvector corresponding to eigenvalue $\lambda_2 = i$ is

$$X_1 = \begin{bmatrix} 2+i \\ 5 \end{bmatrix}$$

REMARK

• Let A be an $n \times n$ matrix with real entries. If λ is a complex eigenvalue of A with associated eigenvector X, then $\bar{\lambda}$ is also an eigenvalue of A with associated eigenvector $\bar{X}$.

EXERCISE 4.1

1. Prove that a square matrix A and its transpose A' have the same set of eigenvalues.

2. Let A be an $n \times n$ matrix and let $g(x)$ be any polynomial. If λ is an eigenvalue of A, then prove that $g(\lambda)$ is an eigenvalue of $g(A)$.

3. Show that the eigenvalues of a triangular matrix are just the diagonal elements of the matrix.

4. Let $A = $ dig. $(\lambda_1, \lambda_2,..., \lambda_n)$ be a diagonal matrix. Prove that each λ_i $(i = 1, 2, 3,..., n)$ is an eigenvalue of A.

5. Let A be an 3×3 matrix. If $\lambda_1, \lambda_2, \lambda_3$ are the eigenvalues of A, then find the eigenvalues of the matrix $(I + aA)^{-1} (1 + bA)$, where a, b are scalars such that $a\lambda_i \neq -1$ for $i = 1, 2, 3$.

6. Let A and B be two $n \times n$ matrices. Let X be an eigenvector of A and B both. Show that X is also an eigenvector of $aA + bB$, where a, b are scalars.

7. Prove that the eigenvectors of a real symmetric matrix corresponding to two distinct eigenvalues are orthogonal.

8. Prove that the eigenvectors of a Hermitian matrix corresponding to two distinct eigenvalues are orthogonal.

9. (i) If λ is an eigenvalue of a matrix A, then show that $k + \lambda$ is an eigenvalue of $A + kI$.

   (ii) If the matrix A has characteristic roots $\lambda_1, \lambda_2,..., \lambda_n$ show that the matrix A^2 has such roots as $\lambda_1^2, \lambda_2^2,..., \lambda_n^2$.

10. (i) Find the eigenvalues of a matrix $\begin{bmatrix} 1 & 4 \\ 2 & 3 \end{bmatrix}$.

  (v) $\begin{bmatrix} 6 & -2 & 2 \\ -2 & 3 & -1 \\ 2 & -1 & 3 \end{bmatrix}$ (vi) $\begin{bmatrix} -2 & 2 & -3 \\ 2 & 1 & -6 \\ -1 & -2 & 0 \end{bmatrix}$

  (ii) Find the eigenvalues of the matrix
$$A = \begin{bmatrix} a & h & g \\ 0 & b & f \\ 0 & 0 & c \end{bmatrix}.$$

  (vii) $\begin{bmatrix} 1 & 2 & 3 \\ 0 & 2 & 3 \\ 0 & 0 & 2 \end{bmatrix}$ (viii) $\begin{bmatrix} 1 & 1 & 0 \\ 0 & 2 & 2 \\ 0 & 0 & 3 \end{bmatrix}$

11. Find the eigenvalues and eigenvectors of the following matrices :

  (ix) $\begin{bmatrix} 3 & 1 & 1 \\ 2 & 4 & 2 \\ 1 & 1 & 3 \end{bmatrix}$ (x) $\begin{bmatrix} 2 & 1 & 0 \\ 0 & 2 & 1 \\ 0 & 0 & 2 \end{bmatrix}$

  (i) $\begin{bmatrix} 2 & -4 \\ -1 & -1 \end{bmatrix}$ (ii) $\begin{bmatrix} -1 & 0 \\ 0 & 1 \end{bmatrix}$

  (iii) $\begin{bmatrix} 1 & 1 \\ -2 & 4 \end{bmatrix}$ (iv) $\begin{bmatrix} 10 & -18 \\ 6 & -11 \end{bmatrix}$

13. Find the eigenvalues and eigenvectors of the matrix

$$A = \begin{bmatrix} 1 & 1 & 0 & 0 \\ 0 & 2 & 0 & 0 \\ 0 & 0 & 1 & 1 \\ 0 & 0 & -2 & 4 \end{bmatrix}$$

12. Find the eigenvalues and eigenvectors of the following matrices :

  (i) $\begin{bmatrix} 0 & 1 & 0 \\ 0 & 0 & 1 \\ 1 & -3 & 3 \end{bmatrix}$ (ii) $\begin{bmatrix} 5 & 8 & 16 \\ 4 & 1 & 8 \\ -4 & -4 & -11 \end{bmatrix}$

14. Find all the characteristic roots and the corresponding characteristic vectors of the matrix

$$A = \begin{bmatrix} 2 & 1 & -1 \\ 0 & 3 & -2 \\ 2 & 4 & -3 \end{bmatrix}$$

  (iii) $\begin{bmatrix} 1 & -1 & -1 \\ -1 & 1 & -1 \\ -1 & -1 & 1 \end{bmatrix}$ (iv) $\begin{bmatrix} 1 & 2 & 2 \\ 1 & 2 & -1 \\ -1 & 1 & 4 \end{bmatrix}$

Answers

5. $\dfrac{1+b\lambda_1}{1+a\lambda_1}, \dfrac{1+b\lambda_2}{1+a\lambda_2}, \dfrac{1+b\lambda_3}{1+a\lambda_3}$ **10.** (i) $-1, 5$ (ii) a, b, c

11.   (i) $\lambda_1 = -2, X_1 = \begin{bmatrix} 1 \\ 1 \end{bmatrix}; \lambda_2 = 3, X_2 = \begin{bmatrix} -4 \\ 1 \end{bmatrix}$ (ii) $\lambda_1 = 1, X_1 = \begin{bmatrix} 0 \\ 1 \end{bmatrix}; \lambda_2 = -1, X_2 = \begin{bmatrix} 1 \\ 0 \end{bmatrix}$

  (iii) $\lambda_1 = 2, X_1 = \begin{bmatrix} 1 \\ 1 \end{bmatrix}; \lambda_2 = 3, X_2 = \begin{bmatrix} 1 \\ 2 \end{bmatrix}$ (iv) $\lambda_1 = -2, X_1 = \begin{bmatrix} 3 \\ 2 \end{bmatrix}; \lambda_2 = 1, X_2 = \begin{bmatrix} 2 \\ 1 \end{bmatrix}$

12.   (i) $\lambda_1 = \lambda_2 = \lambda_3 = 1, X = \begin{bmatrix} 1 \\ 1 \\ 1 \end{bmatrix}$ (ii) $\lambda_1 = 1, X_1 = \begin{bmatrix} -2 \\ -1 \\ 1 \end{bmatrix}; \lambda_2 = -3, X_2 = \begin{bmatrix} -1 \\ 1 \\ 0 \end{bmatrix}; \lambda_3 = -3, X_3 = \begin{bmatrix} -2 \\ 0 \\ 1 \end{bmatrix}$

  (iii) $\lambda_1 = -1, X_1 = \begin{bmatrix} 1 \\ 1 \\ 1 \end{bmatrix}; \lambda_2 = 2, X_2 = \begin{bmatrix} -1 \\ 1 \\ 0 \end{bmatrix}; \lambda_3 = 2, X_3 = \begin{bmatrix} -1 \\ 0 \\ 1 \end{bmatrix}$

  (iv) $\lambda_1 = 1, X_1 = \begin{bmatrix} 2 \\ -1 \\ 1 \end{bmatrix}; \lambda_2 = 3, X_2 = \begin{bmatrix} 1 \\ 1 \\ 0 \end{bmatrix}; \lambda_3 = 3, X_3 = \begin{bmatrix} 1 \\ 0 \\ 1 \end{bmatrix}$

  (v) $\lambda_1 = 2, X_1 = \begin{bmatrix} -1 \\ 0 \\ 2 \end{bmatrix}; \lambda_2 = 2, X_2 = \begin{bmatrix} 1 \\ 2 \\ 0 \end{bmatrix}; \lambda_3 = 8, X_3 = \begin{bmatrix} 2 \\ -1 \\ 1 \end{bmatrix}$

  (vi) $\lambda_1 = -3, X_1 = \begin{bmatrix} -2 \\ 1 \\ 0 \end{bmatrix}; \lambda_2 = -3, X_2 = \begin{bmatrix} 3 \\ 0 \\ 1 \end{bmatrix}; \lambda_3 = 5, X_3 = \begin{bmatrix} 1 \\ 2 \\ 1 \end{bmatrix}$

(vii) $\lambda_1 = \lambda_2 = 1, X = \begin{bmatrix} 1 \\ 0 \\ 0 \end{bmatrix}; \lambda_3 = 2, X_1 = \begin{bmatrix} 2 \\ 1 \\ 0 \end{bmatrix}$ (viii) $\lambda_1 = 1, X_1 = \begin{bmatrix} 1 \\ 0 \\ 0 \end{bmatrix}; \lambda_2 = 2, X_2 = \begin{bmatrix} 2 \\ 1 \\ 0 \end{bmatrix}; \lambda_3 = 3, X_3 = \begin{bmatrix} 1 \\ 2 \\ 1 \end{bmatrix}$

(ix) $\lambda_1 = 2, X_1 = \begin{bmatrix} -1 \\ 1 \\ 0 \end{bmatrix}; \lambda_2 = 2, X_2 = \begin{bmatrix} -1 \\ 0 \\ 1 \end{bmatrix}; \lambda_3 = 6, X_3 = \begin{bmatrix} 1 \\ 2 \\ 1 \end{bmatrix}$ (x) $\lambda_1 = \lambda_2 = \lambda_3 = 2, X = \begin{bmatrix} 1 \\ 0 \\ 0 \end{bmatrix}$

13. $\lambda_1 = 1, X_1 = \begin{bmatrix} 1 \\ 0 \\ 0 \\ 0 \end{bmatrix}; \lambda_2 = 2, X_2 = \begin{bmatrix} 1 \\ 1 \\ 0 \\ 0 \end{bmatrix}; \lambda_3 = 2, X_3 = \begin{bmatrix} 0 \\ 0 \\ 1 \\ 1 \end{bmatrix}; \lambda_4 = 3, X_4 = \begin{bmatrix} 0 \\ 0 \\ 0 \\ 1 \end{bmatrix}$

4.6 THE CAYLEY-HAMILTON THEOREM

THEOREM 1. *Every square matrix satisfies its characteristic equation.*

or let A be a square matrix of order n and the characteristic equation of A is

$$|A - \lambda I| = (-1^n) [\lambda^n + a_1 \lambda^{n-1} + a_2 \lambda^{n-2} + \ldots + a_{n-1}\lambda + a_n] = 0$$

then its matrix equation

$$X^n + a_1 X^{n-1} + a_2 X^{n-2} + \ldots + a_{n-1}X + a_n I = O$$

is satisfied by the matrix X = A

i.e. $\qquad A^n + a_1 A^{n-1} + a_2 A^{n-2} + \ldots + a_{n-1}A + a_n I = O$

where I is a unit matrix of order n and O is null matrix of order n.

Proof. Since A and I are two square matrices of order n and λ is any characteristic root of A, then the matrix $(A - \lambda I)$ is also a square matrix of order n whose elements are at most of degree one in λ. Therefore Adj. $(A - \lambda I)$ will have its elements a polynomials in λ of degree $n - 1$ or less and thus Adj. $(A - \lambda I)$ can be expressed as a matrix polynomial in λ as follows :

$$\text{Adj. } (A - \lambda I) = B_0 \lambda^{n-1} + B_1 \lambda^{n-2} + \ldots + B_{n-2}\lambda + B_{n-1} \qquad \ldots(1)$$

where $B_0, B_1, \ldots, B_{n-1}$ are the square matrices of order n.

Since we know that $A(\text{Adj } A) = |A| I_n$

$\therefore \qquad (A - \lambda I) \text{ Adj. } (A - \lambda I) = |A - \lambda I| I$

or $\qquad (A - \lambda I) \text{ Adj. } (A - \lambda I) = (-1^n) [\lambda^n + a_1 \lambda^{n-1} + a_2 \lambda^{n-2} + \ldots + a_{n-1}\lambda + a_n] I$
$$\ldots(2)$$

Multiplying both sides of (1) by $(A - \lambda I)$, we get

$$(A - \lambda I) \text{ Adj.}(A - \lambda I) = (A - \lambda I)[B_0 \lambda^{n-1} + B_1 \lambda^{n-2} + \ldots + B_{n-2}\lambda + B_{n-1}]$$
$$\ldots(3)$$

From (2) and (3), we get

$$(A - \lambda I)[B_0 \lambda^{n-1} + B_1 \lambda^{n-2} + \ldots + B_{n-2}\lambda + B_{n-1}]$$

$$= (-1^n) [\lambda^n + a_1 \lambda^{n-1} + a_2 \lambda^{n-2} + \ldots + a_{n-1}\lambda + a_n] I$$

Now comparing the coefficients of like powers of λ, we get

$$\left.\begin{array}{r}
-IB_0 = (-1)^n I \\
AB_0 - IB_1 = (-1)^n a_1 I \\
AB_1 - IB_2 = (-1)^n a_2 I \\
\cdots\cdots\cdots\cdots\cdots\cdots \\
AB_{n-2} - IB_{n-3} = (-1)^n a_{n-1} I \\
AB_{n-1} = (-1)^n a_n I
\end{array}\right\} \qquad \ldots(4)$$

Premultiplying first, second, third etc. equations of (4) by A^n, A^{n-1}, A^{n-2}, etc. respectively and then adding, we get

$-A^n B_0 + A^n B_0 - A^{n-1} B_1 + A^{n-1} B_1 + \ldots$
$$= (-1)^n [A^n + a_1 A^{n-1} + \ldots + a_n I]$$

or $\qquad 0 = (-1)^n [A^n + a_1 A^{n-1} + \ldots + a_n I]$

Hence $\qquad A^n + a_1 A^{n-1} + \ldots + a_n I = O$

Corollary 1. *If A be a non-singular matrix of order $n \times n$ and its characteristic polynomial is*
$$|A - \lambda I| = (-1)^n [\lambda^n + a_1 \lambda^{n-1} + \ldots + a_{n-1} \lambda + a_n]$$
then $\qquad det (A) = (-1)^n a_n.$

Proof. We have
$$|A - \lambda I| = (-1)^n [\lambda^n + a_1 \lambda^{n-1} + \ldots + a_{n-1} A + a_n]$$
Putting $\lambda = 0$, we get
$$|A - 0I| = (-1)^n a_n$$
$\Rightarrow \qquad |A| = (-1)^n a_n$

Corollary 2. *If $\lambda_1, \lambda_2, \ldots, \lambda_n$ are eigenvalues of a square matrix of order $n \times n$, then*
$$det (A) = \lambda_1 \lambda_2 \lambda_3 \ldots \lambda_n.$$

Proof. If λ is an eigenvalue of A, then it is a root of characteristic equation of A. The characteristic equation of A is given by
$$|A - \lambda I| = (-1)^n [\lambda^n + a_1 \lambda^{n-1} + \ldots + a_{n-1} \lambda + a_n] = 0$$
Since $\lambda_1, \lambda_2, \ldots, \lambda_n$ are the eigenvalues of A, hence
$(-1)^n [\lambda^n + a_1 \lambda^{n-2} + \ldots + a_{n-1} \lambda + a_n] = (\lambda_1 - \lambda)(\lambda_2 - \lambda) \ldots (\lambda_n - \lambda) = 0$
Comparing the constant terms of both sides, we get
$$(-1)^n a_n = \lambda_1 \lambda_2 \lambda_3 \ldots \lambda_n$$
From Corollary 1,
$$det (A) = (-1)^n a_n$$
Hence $\qquad det (A) = \lambda_1 \lambda_2 \lambda_3 \ldots \lambda_n$

Corollary 3. *Let A be an $n \times n$ matrix with characteristic polynomial*
$$f(t) = (-1)^n [t^n + a_1 t^{n-1} + \ldots + a_{n-1} t + a_n]$$
Then A is invertible iff $a_n \neq 0$ and its inverse is

$$A^{-1} = \left(\frac{-1}{a_n}\right) [A^{n-1} + a_1 A^{n-2} + \ldots + a_{n-2} A + a_{n-1} I]$$

Proof. By Corollary 1,

$$|A| = (-1)^n a_n$$

$\therefore \qquad |A| \neq 0 \Leftrightarrow a_n \neq 0$

i.e., A is invertible iff $a_n \neq 0$

By Cayley-Hamilton theorem, we have

$$f(A) = O$$

$$\Rightarrow \qquad (-1)^n [A^n + a_1 A^{n-1} + \dots + a_{n-1} A + a_n I] = O$$

$$\Rightarrow \qquad A^n + a_1 A^{n-1} + \dots + a_{n-1} A + a_n I = O$$

$$\Rightarrow \qquad A^{-1} [A^n + a_1 A^{n-1} + \dots + a_{n-1} A + a_n I] = A^{-1} O = O$$

$$\Rightarrow \qquad A^{n-1} + a_1 A^{n-2} + \dots + a_{n-2} A + a_{n-1} I + a_n A^{-1} = O$$

Hence $\qquad A^{-1} = \left(\dfrac{-1}{a_n} \right) [A^{n-1} + a_1 A^{n-2} + \dots + a_{n-2} A + a_{n-1} I]$

Corollary 4. *If $\lambda_1, \lambda_2, \dots, \lambda_n$ are the eigenvalue of a matrix A of order $n \times n$, then*

$$Tr(A) = Trace \ of \ A = \sum_{i=1}^{n} \lambda_i$$

Proof. Let $A = [a_{ij}]_{n \times n}$. Then Tr $(A) = \displaystyle\sum_{i=1}^{n} \lambda_i$

Now $\qquad |A - \lambda I| = \begin{vmatrix} a_{11-\lambda} & a_{12} & \cdots & a_{1n} \\ a_{21} & a_{22-\lambda} & \cdots & a_{2n} \\ \vdots & & & \\ a_{n1} & a_{n2} & \cdots & a_{nn-\lambda} \end{vmatrix}$

$$= (a_{11} - \lambda) A_{11} - a_{12} A_{12} + a_{13} A_{13} + \dots + (-1)^n a_{1n} A_n$$

where $A_{11} = $ Minor of $a_{11-\lambda}$, $A_{12} = $ Minor of a_{12} and so on.

Clearly A_{11} is a polynomial of degree $n - 1$ in λ.

Therefore

$$A_{11} = (a_{22} - \lambda)(a_{33} - \lambda) \dots (a_{nn} - \lambda) + O(\lambda^{n-3})$$

$$\Rightarrow \quad (a_{11} - \lambda) A_{11} = (a_{11} - \lambda)(a_{22} - \lambda)(a_{33} - \lambda) \dots (a_{nn} - \lambda) + O(\lambda^{n-2})$$

Similarly,

$$A_{12} = A \text{ polynomial of degree } n - 2 \text{ in } \lambda$$

$$A_{13} = A \text{ polynomial of degree } n - 2 \text{ in } \lambda$$

$$\dots\dots\dots\dots\dots\dots\dots\dots\dots\dots\dots$$

$$A_{1n} = A \text{ polynomial of degree } n - 2 \text{ in } \lambda$$

$$\therefore \qquad |A - \lambda I| = (-1)^n \left[\lambda^n - \left(\sum_{i=1}^{n} a_{ii} \right) \lambda^{n-1} + O\left(\lambda^{n-2} \right) \right] \qquad \dots(1)$$

Since $\lambda_1, \lambda_2, \dots, \lambda_n$ are eigenvalues of A, hence

$$|A - \lambda I| = (-1)^n \left[\lambda^n - \left(\sum_{i=1}^{n} \lambda_i \right) \lambda^{n-1} + O\left(\lambda^{n-2} \right) \right] \qquad \dots(2)$$

From (1) and (2)

$$(-1)^n \left[\lambda^n - \left(\sum_{i=1}^{n} a_{ii} \right) \lambda^{n-1} + O\left(\lambda^{n-2}\right) \right] = (-1)^n \left[\lambda^n - \left(\sum_{i=1}^{n} \lambda_i \right) \lambda^{n-1} + O\left(\lambda^{n-2}\right) \right]$$

Equating the coefficient of λ^{n-1} of both sides, we get

$$\sum_{i=1}^{n} a_{ii} = \sum_{i=1}^{n} \lambda_i$$

Hence $$\text{Tr}(A) = \sum_{i=1}^{n} a_{ii} = \sum_{i=1}^{n} \lambda_i$$

Corollary 5. *If the characteristic equation of a matrix A of order $n \times n$ is*

$$|A - \lambda I| = (-1)^n \, [\lambda^n + a_1 \lambda^{n-1} + a_2 \lambda^{n-2} + \dots + a_{n-1} \lambda + a_n] = 0$$

then Tr (A) = $-a_1$

Proof.   If $\lambda_1, \lambda_2, \dots, \lambda_n$ are the eigenvalues of a matrix A, then

$$|A - \lambda I| = (-1)^n \, (\lambda - \lambda_1)(\lambda - \lambda_2)\dots(\lambda - \lambda_n) = 0$$

$$= (-1)^n \left[\lambda^n - \left(\sum_{i=1}^{n} \lambda_i \right) \lambda^{n-1} + \dots \right] = 0 \qquad \dots(1)$$

But $|A - \lambda I| = (-1)^n \, [\lambda^n + a_1 \lambda^{n-1} + a_2 \lambda^{n-2} + \dots + a_{n-1} \lambda + a_n] = 0$ ...(2)

From (1) and (2) we get

$$(-1)^n \left[\lambda^n - \left(\sum_{i=1}^{n} \lambda_i \right) \lambda^{n-1} + \dots \right] = (-1)^n [\lambda^n + a_1 \lambda^{n-1} + a_2 \lambda^{n-2} + \dots + a_{n-1} \lambda + a_n]$$

Taking the coefficient of λ^{n-1} on both sides, we get

$$\sum_{i=1}^{n} \lambda_i = -a_1$$

By Corollary 4, we have

$$\text{Tr}(A) = \sum_{i=1}^{n} \lambda_i$$

Hence   Tr (A) = $- a_1$

Corollary 6. *Let A be a matrix of order $n \times n$. If m be a positive integer such that $m \geq n$, then Am is linearly expressible in terms of those of lower order of A.*

Proof.   By Cayley-Hamilton theorem,

$$A^n + a_1 A^{n-1} + a_2 A^{n-2} + \dots + a_{n-1} A + a_n I = O \qquad \dots(1)$$

Multiplying (1) by A^{m-n}, we get

$$A^m + a_1 A^{m-1} + a_2 A^{m-2} + \dots + a_n A^{m-n} = O$$

or $A^m = (-a_1) A^{m-1} + (-a_2) A^{m-2} + \dots + (-a_n) A^{m-n}$

Hence the result.

Solved Examples

Based on the following Results

➡ Every square matrix satisfies its characteristic equation.

➡ If $\lambda_1, \lambda_2, ..., \lambda_n$ are eigenvalues of a square matrix of order $n \times n$ then det $(A) = \lambda_1.\lambda_2...\lambda_n$.

Example 1. *Find the characteristic equation of the matrix*

$$A = \begin{bmatrix} 1 & 0 & 2 \\ 0 & 2 & 1 \\ 2 & 0 & 3 \end{bmatrix}$$

and verify that it is satisfied by A and hence find its inverse.

Solution. The characteristic equation of A is given by

$$|A - \lambda I| = 0$$

or

$$\begin{vmatrix} 1-\lambda & 0 & 2 \\ 0 & 2-\lambda & 1 \\ 2 & 0 & 3-\lambda \end{vmatrix} = 0$$

or $\quad (1-\lambda)\{(2-\lambda)(3-\lambda)-0\}+2\{0-2(2-\lambda)\} = 0$

or $\quad\quad\quad\quad\quad -\lambda^3 + 6\lambda^2 - 7\lambda - 2 = 0$

or $\quad\quad\quad\quad\quad \lambda^3 - 6\lambda^2 + 7\lambda + 2 = 0$

Next we have to show that

$$A^3 - 6A^2 + 7A + 2I = O$$

Now $\quad\quad A^2 = A.A$

$$= \begin{bmatrix} 1 & 0 & 2 \\ 0 & 2 & 1 \\ 2 & 0 & 3 \end{bmatrix}\begin{bmatrix} 1 & 0 & 2 \\ 0 & 2 & 1 \\ 2 & 0 & 3 \end{bmatrix}$$

$$= \begin{bmatrix} 5 & 0 & 8 \\ 2 & 4 & 5 \\ 8 & 0 & 13 \end{bmatrix}$$

and $\quad A^3 = A^2.A = \begin{bmatrix} 5 & 0 & 8 \\ 2 & 4 & 5 \\ 8 & 0 & 13 \end{bmatrix}\begin{bmatrix} 1 & 0 & 2 \\ 0 & 2 & 1 \\ 2 & 0 & 3 \end{bmatrix} = \begin{bmatrix} 21 & 0 & 34 \\ 12 & 8 & 23 \\ 34 & 0 & 55 \end{bmatrix}$

$\therefore \quad A^3 - 6A^2 + 7A + 2I$

$$= \begin{bmatrix} 21 & 0 & 34 \\ 12 & 8 & 23 \\ 34 & 0 & 55 \end{bmatrix} - 6\begin{bmatrix} 5 & 0 & 8 \\ 2 & 4 & 5 \\ 8 & 0 & 13 \end{bmatrix} + 7\begin{bmatrix} 1 & 0 & 2 \\ 0 & 2 & 1 \\ 2 & 0 & 3 \end{bmatrix} + 2\begin{bmatrix} 1 & 0 & 0 \\ 0 & 1 & 0 \\ 0 & 0 & 1 \end{bmatrix}$$

$$= \begin{bmatrix} 21-30+7+2 & 0-0+0+0 & 34-48+14+0 \\ 12-12+0+0 & 8-24+14+2 & 23-30+7+0 \\ 34-48+14+0 & 0-0+0+0 & 55-78+21+2 \end{bmatrix}$$

$$= \begin{bmatrix} 0 & 0 & 0 \\ 0 & 0 & 0 \\ 0 & 0 & 0 \end{bmatrix} = O$$

Hence $A^3 - 6A^2 + 7A + 2I = O$ ...(1)

To find A^{-1} :

Since the characteristic equation of A is

$$\lambda^3 - 6\lambda^2 + 7\lambda + 2 = 0$$

$\therefore$ $|A| = (-1)^3 2 = -2 \neq 0$ $[\because |A| = (-1)^n a_n]$

$\Rightarrow A^{-1}$ exist.

Premultiplying (1) by A^{-1}, we get

$$A^2 - 6A + 7I + 2A^{-1} = O$$

$\Rightarrow$ $A^{-1} = -\dfrac{1}{2}[A^2 - 6A + 7I]$

$\Rightarrow$ $A^{-1} = -\dfrac{1}{2} \left\{ \begin{bmatrix} 5 & 0 & 8 \\ 2 & 4 & 5 \\ 8 & 0 & 13 \end{bmatrix} - 6\begin{bmatrix} 1 & 0 & 2 \\ 0 & 2 & 1 \\ 2 & 0 & 3 \end{bmatrix} + 7\begin{bmatrix} 1 & 0 & 0 \\ 0 & 1 & 0 \\ 0 & 0 & 1 \end{bmatrix} \right\}$

$$= -\dfrac{1}{2} \begin{bmatrix} 6 & 0 & -4 \\ 2 & -1 & -1 \\ -4 & 0 & 2 \end{bmatrix}$$

Hence $A^{-1} = -\dfrac{1}{2} \begin{bmatrix} 6 & 0 & -4 \\ 2 & -1 & -1 \\ -4 & 0 & 2 \end{bmatrix} = \dfrac{1}{2} \begin{bmatrix} -6 & 0 & 4 \\ -2 & 1 & 1 \\ 4 & 0 & -2 \end{bmatrix}$

Example 2. *Find the characteristic equation of the matrix*

$$A = \begin{bmatrix} 2 & -1 & 1 \\ -1 & 2 & -1 \\ 1 & -1 & 2 \end{bmatrix}$$

and verify that it is satisfied by A and hence find A^{-1}.

Solution. The characteristic equation of A is given by

$$|A - \lambda I| = 0$$

or $\begin{vmatrix} 2-\lambda & -1 & 2 \\ -1 & 2-\lambda & -1 \\ 1 & -1 & 2-\lambda \end{vmatrix} = 0$

or $(2-\lambda)\{(2-\lambda)(2-\lambda)-1\} + 1\{-2+\lambda+2\} + 1\{1-2+\lambda\} = 0$

or $-\lambda^3 + 6\lambda^2 - 9\lambda + 4 = 0$

or $\lambda^3 - 6\lambda^2 + 9\lambda - 4 = 0$

Next we have to show that

$$A^3 - 6A^2 + 9A - 4I = O$$

Now $A^2 = A.A$

$$= \begin{bmatrix} 2 & -1 & 1 \\ -1 & 2 & -1 \\ 1 & -1 & 2 \end{bmatrix} \begin{bmatrix} 2 & -1 & 1 \\ -1 & 2 & -1 \\ 1 & -1 & 2 \end{bmatrix} = \begin{bmatrix} 6 & -5 & 5 \\ -5 & 6 & -5 \\ 5 & -5 & 6 \end{bmatrix}$$

and $\quad A^3 = A^2.A = \begin{bmatrix} 6 & -5 & 5 \\ -5 & 6 & -5 \\ 5 & -5 & 6 \end{bmatrix} \begin{bmatrix} 2 & -1 & 1 \\ -1 & 2 & -1 \\ 1 & -1 & 2 \end{bmatrix} = \begin{bmatrix} 22 & -21 & 21 \\ -21 & 22 & -21 \\ 21 & -21 & 22 \end{bmatrix}$

Now $A^3 - 6A^2 + 9A + 2I$

$$= \begin{bmatrix} 22 & -21 & 21 \\ -21 & 22 & -21 \\ 21 & -21 & 22 \end{bmatrix} - 6\begin{bmatrix} 6 & -5 & 5 \\ -5 & 6 & -5 \\ 5 & -5 & 6 \end{bmatrix} + 9\begin{bmatrix} 2 & -1 & 1 \\ -1 & 2 & -1 \\ 1 & -1 & 2 \end{bmatrix} - 4\begin{bmatrix} 1 & 0 & 0 \\ 0 & 1 & 0 \\ 0 & 0 & 1 \end{bmatrix}$$

$$= \begin{bmatrix} 22-36+18-4 & -21+30-9-0 & 21-30+9-0 \\ -21+30-9-0 & 22-36+18-4 & -21+30-9-0 \\ 21-30+9-0 & -21+30-9-0 & 22-36+18-4 \end{bmatrix} = \begin{bmatrix} 0 & 0 & 0 \\ 0 & 0 & 0 \\ 0 & 0 & 0 \end{bmatrix} = O$$

Hence $\qquad A^3 - 6A^2 + 9A + 4I = O$ ...(1)

Since $|A| = 2(4-1) + 1(-2+1) + 1(1-2) = 6 - 1 - 1 = 4 \neq 0$

$\Rightarrow A^{-1}$ exist.

Premultiplying (1) by A^{-1}, we get

$$A^2 - 6A + 9I - 4A^{-1} = O$$

$\Rightarrow \qquad\qquad A^{-1} = +\dfrac{1}{4}[A^2 - 6A + 9I]$

$\Rightarrow \qquad A^{-1} = \dfrac{1}{4}\left\{ \begin{bmatrix} 6 & -5 & 5 \\ -5 & 6 & -5 \\ 5 & -5 & 6 \end{bmatrix} - 6\begin{bmatrix} 2 & -1 & 1 \\ -1 & 2 & -1 \\ 1 & -1 & 2 \end{bmatrix} + 9\begin{bmatrix} 1 & 0 & 0 \\ 0 & 1 & 0 \\ 0 & 0 & 1 \end{bmatrix} \right\}$

$$= \dfrac{1}{4}\begin{bmatrix} 6-12+9 & -5+6+0 & 5-6+0 \\ -5+6+0 & 6-12+9 & -5+6+0 \\ 5-6+0 & -5+6+0 & 6-12+9 \end{bmatrix}$$

$\therefore \qquad\qquad A^{-1} = \dfrac{1}{4}\begin{bmatrix} 3 & 1 & -1 \\ 1 & 3 & 1 \\ -1 & 1 & 3 \end{bmatrix}$

Example 3. *Find the characteristic equation of the matrix*

$$A = \begin{bmatrix} 1 & 2 & 0 \\ 2 & -1 & 0 \\ 0 & 0 & -1 \end{bmatrix}$$

and hence find A^{-1}.

Solution. The characteristic equation of A is given by

$$|A - \lambda I| = 0$$

or $\qquad \begin{vmatrix} 1-\lambda & 2 & 0 \\ 2 & -1-\lambda & 0 \\ 0 & 0 & -1-\lambda \end{vmatrix} = 0$

$$\text{or} (1-\lambda)\{(-1-\lambda)(-1-\lambda)-0\}-2\{2(-1-\lambda)-0\}=0$$

or $$(1-\lambda)(1+\lambda)^2+4(1+\lambda)=0$$

or $$1+\lambda^2+2\lambda-\lambda-\lambda^3-2\lambda^2+4+4\lambda=0$$

or $$-\lambda^3-\lambda^2+5\lambda+5=0$$

or $$\lambda^3+\lambda^2-5\lambda-5=0$$

By Cayley-Hamilton theorem, we have

$$A^3+A^2-5A-5I=O \qquad \qquad \text{...(1)}$$

Since $\quad |A-\lambda I|=-\lambda^3-\lambda^2+5\lambda+5$

$\Rightarrow \quad |A|=5\neq 0 \qquad \qquad \text{(Putting } \lambda=0)$

$\Rightarrow A^{-1}$ exists.

Premultiplying (1) by A^{-1}, we get

$$A^2+A-5I-5A^{-1}=0$$

$\Rightarrow \qquad A^{-1}=\dfrac{1}{5}[A^2+A-5I] \qquad \qquad \text{...(2)}$

Now $\qquad A^2=A.A=\begin{bmatrix}1 & 2 & 0\\ 2 & -1 & 0\\ 0 & 0 & -1\end{bmatrix}\begin{bmatrix}1 & 2 & 0\\ 2 & -1 & 0\\ 0 & 0 & -1\end{bmatrix}=\begin{bmatrix}5 & 0 & 0\\ 0 & 5 & 0\\ 0 & 0 & 5\end{bmatrix}$

So $\qquad A^{-1}=\dfrac{1}{5}\left\{\begin{bmatrix}5 & 0 & 0\\ 0 & 5 & 0\\ 0 & 0 & 1\end{bmatrix}+\begin{bmatrix}1 & 2 & 0\\ 2 & -1 & 0\\ 0 & 0 & -1\end{bmatrix}-5\begin{bmatrix}1 & 0 & 0\\ 0 & 1 & 0\\ 0 & 0 & 1\end{bmatrix}\right\}$

$\therefore \qquad A^{-1}=\dfrac{1}{5}\begin{bmatrix}1 & 2 & 0\\ 2 & -1 & 0\\ 0 & 0 & -5\end{bmatrix}$

Example 4. *Show that the matrix*

$$A=\begin{bmatrix}0 & c & -b\\ -c & 0 & a\\ b & -a & 0\end{bmatrix}$$

satisfies Cayley-Hamilton Theorem.

Solution. The characteristic equation of A is given by

$$|A-\lambda I|=0$$

or $$\begin{vmatrix}-\lambda & c & -b\\ -c & -\lambda & a\\ b & -a & -\lambda\end{vmatrix}=0$$

or $$-\lambda(\lambda^2+a^2)-c(c\lambda-ab)-b(ca+b\lambda)=0$$

or $$-\lambda^3-\lambda(a^2+b^2+c^2)=0$$

or $$\lambda^3+\lambda(a^2+b^2+c^2)=0$$

We have to show that
$$A^3 + A(a^2 + b^2 + c^2) = O$$

Now
$$A^2 = A.A = \begin{bmatrix} 0 & c & -b \\ -c & 0 & a \\ b & -a & 0 \end{bmatrix}\begin{bmatrix} 0 & c & -b \\ -c & 0 & a \\ b & -a & 0 \end{bmatrix}$$

$$= \begin{bmatrix} -(c^2+b^2) & ab & ac \\ ab & -(c^2+a^2) & bc \\ ac & bc & -(a^2+b^2) \end{bmatrix}$$

and
$$A^3 = A^2.A = \begin{bmatrix} -(c^2+b^2) & ab & ac \\ ab & -(c^2+a^2) & bc \\ ac & bc & -(a^2+b^2) \end{bmatrix}\begin{bmatrix} 0 & c & -b \\ -c & 0 & a \\ b & -a & 0 \end{bmatrix}$$

$$= \begin{bmatrix} 0 & -c^3-b^2c-a^2c & bc^2+b^2+a^2b \\ c^3+a^2c+b^2c & 0 & -ab^2-ac^2-a^3 \\ -bc^2-b^3-a^2b & ac^2+ab^2+a^3 & 0 \end{bmatrix}$$

$$= \begin{bmatrix} 0 & -c(a^2+b^2+c^2) & b(a^2+b^2+c^2) \\ c(a^2+b^2+c^2) & 0 & -a(a^2+b^2+c^2) \\ -b(a^2+b^2+c^2) & a(a^2+b^2+c^2) & 0 \end{bmatrix}$$

$$A^3 = -(a^2+b^2+c^2)\begin{bmatrix} 0 & c & -b \\ -c & 0 & a \\ b & -a & 0 \end{bmatrix}$$

$$= -(a^2+b^2+c^2)A$$

Hence, $A^3 + (a^2 + b^2 + c^2)A = O$

Example 5. *Verify Cayley-Hamilton theorem for the matrix*

$$A = \begin{bmatrix} 1 & 1 & 0 & 0 \\ 0 & 2 & 0 & 0 \\ 0 & 0 & -1 & 1 \\ 0 & 0 & -2 & 4 \end{bmatrix}$$

Solution. The characteristic equation of the matrix A is given by

$$|A - \lambda I| = 0$$

or
$$\begin{vmatrix} 1-\lambda & 1 & 0 & 0 \\ 0 & 2-\lambda & 0 & 0 \\ 0 & 0 & 1-\lambda & 1 \\ 0 & 0 & -2 & 4-\lambda \end{vmatrix} = 0$$

or

$$(1-\lambda)\begin{vmatrix} 2-\lambda & 0 & 0 \\ 0 & 1-\lambda & 1 \\ 0 & -2 & 4-\lambda \end{vmatrix} = 0$$

[Expanding along first column]

or $\qquad (1-\lambda)\big[(2-\lambda)\{(1-\lambda)(4-\lambda)+2\}\big]=0$

or $\qquad (1-\lambda)\big[(2-\lambda)(1-\lambda)(4-\lambda)+2(2-\lambda)\big]=0$

or $\qquad (1-\lambda)(2-\lambda)\big(\lambda^2-5\lambda+6\big)=0$

or $\qquad (1-\lambda)(2-\lambda)(2-\lambda)(3-\lambda)=0$

or $\qquad (\lambda-1)(\lambda-2)^2(\lambda-3)=0$

We have to show that

$$(A-I)(A-2I)^2(A-3I) = O$$

Now $\quad A-I = \begin{bmatrix} 1 & 1 & 0 & 0 \\ 0 & 2 & 0 & 0 \\ 0 & 0 & 1 & 1 \\ 0 & 0 & -2 & 4 \end{bmatrix} - \begin{bmatrix} 1 & 0 & 0 & 0 \\ 0 & 1 & 0 & 0 \\ 0 & 0 & 1 & 0 \\ 0 & 0 & 0 & 1 \end{bmatrix}$

$$= \begin{bmatrix} 0 & 1 & 0 & 0 \\ 0 & 1 & 0 & 0 \\ 0 & 0 & 0 & 1 \\ 0 & 0 & -2 & 3 \end{bmatrix}$$

and $\quad A-2I = \begin{bmatrix} 1 & 1 & 0 & 0 \\ 0 & 2 & 0 & 0 \\ 0 & 0 & 1 & 1 \\ 0 & 0 & -2 & 4 \end{bmatrix} - 2\begin{bmatrix} 1 & 0 & 0 & 0 \\ 0 & 1 & 0 & 0 \\ 0 & 0 & 1 & 0 \\ 0 & 0 & 0 & 1 \end{bmatrix}$

$$= \begin{bmatrix} -1 & 1 & 0 & 0 \\ 0 & 0 & 0 & 0 \\ 0 & 0 & -1 & 1 \\ 0 & 0 & -2 & 2 \end{bmatrix}$$

$\therefore \quad (A-2I)^2 = \begin{bmatrix} -1 & 1 & 0 & 0 \\ 0 & 0 & 0 & 0 \\ 0 & 0 & -1 & 1 \\ 0 & 0 & -2 & 2 \end{bmatrix}\begin{bmatrix} -1 & 1 & 0 & 0 \\ 0 & 0 & 0 & 0 \\ 0 & 0 & -1 & 1 \\ 0 & 0 & -2 & 2 \end{bmatrix}$

$$= \begin{bmatrix} 1 & -1 & 0 & 0 \\ 0 & 0 & 0 & 0 \\ 0 & 0 & -1 & 1 \\ 0 & 0 & -2 & 2 \end{bmatrix}$$

$$A - 3I = \begin{bmatrix} 1 & 1 & 0 & 0 \\ 0 & 2 & 0 & 0 \\ 0 & 0 & 1 & 1 \\ 0 & 0 & -2 & 4 \end{bmatrix} - 3 \begin{bmatrix} 1 & 0 & 0 & 0 \\ 0 & 1 & 0 & 0 \\ 0 & 0 & 1 & 0 \\ 0 & 0 & 0 & 1 \end{bmatrix}$$

$$= \begin{bmatrix} -2 & 1 & 0 & 0 \\ 0 & -1 & 0 & 0 \\ 0 & 0 & -2 & 1 \\ 0 & 0 & -2 & 1 \end{bmatrix}$$

Now $(A - I)(A - 2I)^2(A - 3I)$

$$= \begin{bmatrix} 0 & 1 & 0 & 0 \\ 0 & 1 & 0 & 0 \\ 0 & 0 & 0 & 1 \\ 0 & 0 & -2 & 3 \end{bmatrix} \begin{bmatrix} 1 & -1 & 0 & 0 \\ 0 & 0 & 0 & 0 \\ 0 & 0 & -1 & 1 \\ 0 & 0 & -2 & 2 \end{bmatrix} \begin{bmatrix} -2 & 1 & 0 & 0 \\ 0 & -1 & 0 & 0 \\ 0 & 0 & -2 & 1 \\ 0 & 0 & -2 & 1 \end{bmatrix}$$

$$= \begin{bmatrix} 0 & 1 & 0 & 0 \\ 0 & 1 & 0 & 0 \\ 0 & 0 & 0 & 1 \\ 0 & 0 & -2 & 3 \end{bmatrix} \begin{bmatrix} -2 & 2 & 0 & 0 \\ 0 & 0 & 0 & 0 \\ 0 & 0 & 0 & 0 \\ 0 & 0 & 0 & 0 \end{bmatrix}$$

$$= \begin{bmatrix} 0 & 0 & 0 & 0 \\ 0 & 0 & 0 & 0 \\ 0 & 0 & 0 & 0 \\ 0 & 0 & 0 & 0 \end{bmatrix} = O$$

$\therefore \qquad (A - I)(A - 2I)^2(A - 3I) = O$

Hence the Cayley-Hamilton theorem is verified.

Example 6. *Use Cayley-Hamilton theorem to express* $2A^5 - 3A^4 + A^2 - 4I$ *as a linear polynomial in A, where :*

$$A = \begin{bmatrix} 3 & 1 \\ -1 & 2 \end{bmatrix}$$

Solution. The characteristic equation of A is given by

$$|A - \lambda I| = 0$$

or

$$\begin{vmatrix} 3 - \lambda & 1 \\ -1 & 2 - \lambda \end{vmatrix} = 0$$

or

$$(3 - \lambda)(2 - \lambda) + 1 = 0$$

or

$$\lambda^2 - 5\lambda + 7 = 0$$

By Cayley-Hamilton theorem, we have

$$A^2 - 5A + 7I = O \qquad \qquad ...(1)$$

$$\Rightarrow \qquad A^2 = 5A - 7I \qquad \qquad ...(2)$$

Now

$$A^3 = A^2.A$$

$$= (5A - 7I)A = 5A^2 - 7A$$

$$\therefore \qquad A^3 = 5A^2 - 7A \qquad\qquad\qquad ...(3)$$

Again,
$$A^4 = A^3 . A = (5A^2 - 7A)A = 5A^3 - 7A^2$$
$$\Rightarrow \quad A^4 = 5(5A^2 - 7A) - 7(5A - 7I) \qquad \text{[Using (2) and (3)]}$$
$$\Rightarrow \quad A^4 = 25A^2 - 35A - 35A + 49I$$
$$\Rightarrow \quad A^4 = 25(5A - 7I) - 70A + 49I \qquad \text{[Using (2)]}$$
$$\Rightarrow \quad A^4 = 125 A - 175I - 70A + 49I$$
$$\Rightarrow \quad A^4 = 55A - 126I \qquad\qquad\qquad ...(4)$$

Also
$$A^5 = A^4 . A = (55A - 126I)A$$
$$A^5 = 55A^2 - 126A$$
$$\Rightarrow \quad A^5 = 55(5A - 7I) - 126A \qquad \text{[Using (2)]}$$
$$\therefore \quad A^5 = 149A - 385I \qquad\qquad\qquad ...(5)$$

Now
$$2A^5 - 3A^4 + A^2 - 4I$$
$$= 2 (149A - 385I) - 3(55A - 126I) + 5A - 7I - 4I$$
$$\qquad\qquad\qquad \text{[Using (2), (4) and (5)]}$$
$$= 298A - 770I - 165A + 378I + 5A - 11I$$
$$= 138A - 403I$$
$$\therefore \quad 2A^5 - 3A^4 + A^2 - 4I = 138A - 403I \text{ which is a linear polynomial in } A.$$

❧ EXERCISE 4.2

1. Verify Cayley-Hamilton theorem for the matrix
$$A = \begin{bmatrix} 1 & 1 \\ 8 & 1 \end{bmatrix}$$
and use it to find A^{-1}.

2. Use Cayley-Hamilton theorem to find the inverse of the matrix
$$A = \begin{bmatrix} 2 & 1 \\ 5 & 3 \end{bmatrix}$$

3. Verify Cayley-Hamilton theorem for the matrix
$$A = \begin{bmatrix} 0 & 0 & 0 \\ 3 & 1 & 0 \\ -2 & 1 & 4 \end{bmatrix}$$
and hence find A^{-1}.

4. Verify Cayley-Hamilton theorem for the following matrix:
$$A = \begin{bmatrix} 2 & 0 \\ 0 & 1 \end{bmatrix}$$

5. Show that the matrix
$$A = \begin{bmatrix} 1 & 2 \\ 1 & 1 \end{bmatrix}$$
satisfies Cayley-Hamilton theorem.

6. State the Cayley-Hamilton theorem and verify it for the matrix
$$A = \begin{bmatrix} 1 & 0 & -2 \\ 0 & 0 & 0 \\ -2 & 0 & 4 \end{bmatrix}$$

7. Verify Cayley-Hamilton theorem for the matrix
$$A = \begin{bmatrix} 1 & 4 \\ 2 & 3 \end{bmatrix}$$
and hence obtain A^{-1}.

8. Verify Cayley-Hamilton theorem for the matrix
$$A = \begin{bmatrix} 1 & 2 & 1 \\ 0 & 1 & -1 \\ 3 & -1 & 1 \end{bmatrix}$$
and hence find A^{-1}.

9. Verify that the matrix
$$A = \begin{bmatrix} 1 & 2 & 1 \\ -1 & 0 & 3 \\ 2 & -1 & 1 \end{bmatrix}$$
satisfies its characteristic equation.

10. Show that the matrix
$$A = \begin{bmatrix} 2 & 2 & 1 \\ 1 & 3 & 1 \\ 1 & 2 & 2 \end{bmatrix}$$
satisfies Cayley-Hamilton theorem.

11. Verify Cayley-Hamilton theorem for the matrix
$$A = \begin{bmatrix} 1 & \sqrt{2} & 0 \\ \sqrt{2} & -1 & 0 \\ 0 & 0 & 1 \end{bmatrix}$$
and hence find A^{-1}.

12. Verify Cayley-Hamilton theorem for the matrix
$$A = \begin{bmatrix} 1 & 0 & 2 \\ 0 & -1 & 1 \\ 0 & 1 & 0 \end{bmatrix}$$
and hence find A^{-1}.

13. Verify Cayley-Hamilton theorem for the matrix
$$A = \begin{bmatrix} 1 & 3 & 7 \\ 4 & 2 & 3 \\ 0 & 2 & 1 \end{bmatrix}$$
and hence find A^{-1}.

14. Verify Cayley-Hamilton theorem for the matrix
$$A = \begin{bmatrix} 1 & 2 & 3 \\ 3 & -2 & 1 \\ 4 & 2 & 1 \end{bmatrix}$$
and hence find A^{-1}.

15. Verify Cayley-Hamilton theorem for the matrix
$$A = \begin{bmatrix} 1 & 1 & 3 \\ 5 & 2 & 6 \\ -2 & -1 & -3 \end{bmatrix}$$

16. Verify Cayley-Hamilton theorem for the matrix
$$A = \begin{bmatrix} 3 & 2 & 4 \\ 4 & 3 & 2 \\ 2 & 4 & 3 \end{bmatrix}$$

17. Verify Cayley-Hamilton theorem for the matrix
$$A = \begin{bmatrix} 2 & 3 & -2 \\ 0 & 5 & 4 \\ 1 & 0 & 1 \end{bmatrix}$$

18. If $A = \begin{bmatrix} 1 & 2 \\ -1 & 3 \end{bmatrix}$ express $A^6 - 4A^5 + 8A^4 - 12A^3 + 14A^2$ as a linear polynomial in A.

Answers

1. $A^{-1} = -\dfrac{1}{7}\begin{bmatrix} 1 & -1 \\ -8 & 1 \end{bmatrix}$

2. $A^{-1} = \begin{bmatrix} 3 & -1 \\ -5 & 2 \end{bmatrix}$

3. $A^{-1} = \dfrac{1}{5}\begin{bmatrix} 4 & 1 & -1 \\ -12 & 2 & 3 \\ 5 & 0 & 0 \end{bmatrix}$

7. $A^{-1} = -\dfrac{1}{3}\begin{bmatrix} 3 & -4 \\ -2 & 1 \end{bmatrix}$

8. $A^{-1} = \dfrac{1}{9}\begin{bmatrix} 0 & 3 & 3 \\ 3 & 2 & -1 \\ 3 & -7 & -1 \end{bmatrix}$

11. $A^{-1} = -\dfrac{1}{3}\begin{bmatrix} -1 & -\sqrt{2} & 0 \\ -\sqrt{2} & 1 & 0 \\ 0 & 0 & -3 \end{bmatrix}$

13. $A^{-1} = \dfrac{1}{10}\begin{bmatrix} -4 & 11 & -5 \\ -4 & 1 & 25 \\ 8 & -2 & -10 \end{bmatrix}$

14. $A^{-1} = \dfrac{1}{36}\begin{bmatrix} -4 & 4 & 8 \\ 1 & -11 & 0 \\ 14 & 6 & -8 \end{bmatrix}$

18. $-4A + 5I$

4.7 DIAGONALISATION OF A MATRIX

Let A be a square matrix of order n. Then A is said to be diagonalizable iff it is similar to a diagonal matrix.

Therefore, A is diagonalizable if there exists a non-singular matrix P such that
$$P^{-1}AP = D$$
where D is a diagonal matrix and the matrix P is said to transform A to diagonal form.

Theorem 1. *An $n \times n$ matrix is diagonalizable if and only if it possesses n linearly independent eigenvectors.*

Proof. Let A be $n \times n$ matrix. Suppose that A is diagonalizable, then it is similar to a diagonal matrix. Let $D = \text{diag}\,(\lambda_1, \lambda_2, ..., \lambda_n)$ be that diagonal matrix, therefore there exists a non-singular matrix P (say) such that
$$P^{-1}AP = D$$
$$\Rightarrow \qquad AP = PD \qquad\qquad ...(1)$$

The eigenvalues of D are $\lambda_1, \lambda_2, ..., \lambda_n$ and A is similar to D, therefore $\lambda_1, \lambda_2, ..., \lambda_n$ are the only eigenvalues of A.

Suppose that $X_1, X_2, ..., X_n$ are column vectors of P

i.e., $P = [X_1, X_2, ..., X_n]$

Since P is invertible so that $X_1, X_2, ..., X_n$ are n linearly independent vectors. Now from (1) we get

$$A[X_1, X_2, ..., X_n] = [X_1, X_2, ..., X_n]\text{dia}.[\lambda_1, \lambda_2, ..., \lambda_n]$$

or $[AX_1, AX_2, ..., AX_n] = [\lambda_1 X_1, \lambda_2 X_2, ..., \lambda_n X_n]$

$\Rightarrow$ $AX_1 = \lambda_1 X_1, AX_2 = \lambda_2 X_2, ..., AX_n = \lambda_n X_n$

Therefore, $X_1, X_2, ..., X_n$ are the eigenvectors of A corresponding to the eigenvalues of A. Also $X_1, X_2, ..., X_n$ are linearly independent, hence A has n linearly independent eigenvectors.

Conversely. Suppose that A has n linearly independent eigenvectors $X_1, X_2, ..., X_n$ then there are scalars $\lambda_1, \lambda_2, ..., \lambda_n$ (not necessarily distinct) such that

$$AX_1 = \lambda_1 X_1, AX_2 = \lambda_2 X_2, ..., AX_n = \lambda_n X_n$$

Let $P = [X_1, X_2, ..., X_n]$. Since $X_1, X_2, ..., X_n$ are linearly independent eigenvectors, therefore P is invertible.

Let $D = \text{dia}. (\lambda_1, \lambda_2, ..., \lambda_n)$. Then

$$AP = A [X_1, X_2, ..., X_n]$$
$$= [AX_1, AX_2, ..., AX_n]$$
$$= [\lambda_1 X_1, \lambda_2 X_2, ..., \lambda_n X_n]$$
$$= [X_1, X_2, ..., X_n]\text{dia}.[\lambda_1, \lambda_2, ..., \lambda_n]$$

$\Rightarrow$ $AP = PD$

$\Rightarrow$ $P^{-1}AP = D$ $[\because P^{-1} \text{ exists.}]$

$\Rightarrow$ A is similar to a diagonal matrix.

$\Rightarrow$ A is diagonalizable.

Corollary. *If the eigenvalues of an n x n matrix are all distinct then it is essentially diagonalizable.*

Proof. Let A be an $n \times n$ matrix. Since the eigenvectors of A corresponding to the distinct eigenvalues are linearly independent, therefore A has n linearly independent eigenvectors, hence by above theorem, A is diagonalizable.

REMARK

- In view of above theorem, if A is diagonalizable and P diagonalizes A, then

$$P^{-1}AP = \begin{bmatrix} \lambda_1 & 0 & 0 & \cdots & 0 \\ 0 & \lambda_2 & 0 & \cdots & 0 \\ 0 & 0 & \lambda_3 & \cdots & 0 \\ \vdots & & & & \\ 0 & 0 & 0 & \cdots & \lambda_n \end{bmatrix}$$

if and only if the jth column of P is an eigenvector of A corresponding to the eigenvalue λ_i of A for $i = 1, 2, 3, ... n$.

Solved Examples

Based on the following Results

➡ A square matrix A is said to be diagonalizable iff it is similiar to a diagonal matrix.

➡ An $n \times n$ matrix is diagonalizable if and only if it possesses n linearly independent eigenvectors.

➡ If the eigenvalues of an $n \times n$ matrix are all distinct then it is necessarily diagonalizable.

➡ If the eigenvalues and corresponding eigenvectors of a matrix A are given, then we can find the matrix A by the relation

$$A = PDP^{-1}$$

where $D = $ dia. $(\lambda_1, \lambda_2, ..., \lambda_n)$ and P is the matrix containing eigenvector X_i corresponding eigenvalues $\lambda_i : i \in N$.

Example 1. *Consider the matrix*

$$A = \begin{bmatrix} 3 & 2 & 0 \\ 2 & 0 & 0 \\ 1 & 0 & 2 \end{bmatrix}$$

Find an invertible matrix P such that $P^{-1}AP$ is a diagonal matrix. Also find the diagonal matrix.

Solution. The characteristic equation of A is given by

$$|A - \lambda I| = 0$$

or

$$\begin{vmatrix} 3 - \lambda & 2 & 0 \\ 2 & 0 - \lambda & 0 \\ 1 & 0 & 2 - \lambda \end{vmatrix} = 0$$

or $\quad (3 - \lambda)\{-\lambda(2 - \lambda)\} - 2\{2 \ (2 - \lambda)\} = 0$

or $\quad\quad\quad (2 - \lambda)(4 - \lambda)(\lambda + 1) = 0$

∴ $\quad$ −1, 2, 4 are eigenvalues of A.

Since A has distinct eigenvalues so that A is diagonalisable, therefore there exists a non-singular matrix P such that $P^{-1}AP$ is a diagonalisable.

Now to find P we shall find eigenvectors of A.

Eigenvector corresponding to $\lambda_1 = -1$

Let $X_1 = \begin{bmatrix} x_1 \\ x_2 \\ x_3 \end{bmatrix} \neq O$ be an eigenvector corresponding to eigenvalue $\lambda_1 = -1$, then

or $\quad\quad\quad\quad\quad\quad AX_1 = (-1)X_1$

or $\quad\quad\quad\quad\quad\quad (A + I)X_1 = O$

or $\quad \begin{bmatrix} 3 - (-1) & 2 & 0 \\ 2 & 0 - (-1) & 0 \\ 1 & 0 & 2 - (-1) \end{bmatrix} \begin{bmatrix} x_1 \\ x_2 \\ x_3 \end{bmatrix} = \begin{bmatrix} 0 \\ 0 \\ 0 \end{bmatrix}$

or $\quad \begin{bmatrix} 4 & 2 & 0 \\ 2 & 1 & 0 \\ 1 & 0 & 3 \end{bmatrix} \begin{bmatrix} x_1 \\ x_2 \\ x_3 \end{bmatrix} = \begin{bmatrix} 0 \\ 0 \\ 0 \end{bmatrix}$ $\quad\quad\quad\quad$ ...(1)

Applying $R_1 \leftrightarrow R_3$ we get

$$\begin{bmatrix} 1 & 0 & 3 \\ 2 & 1 & 0 \\ 4 & 2 & 0 \end{bmatrix} \begin{bmatrix} x_1 \\ x_2 \\ x_3 \end{bmatrix} = \begin{bmatrix} 0 \\ 0 \\ 0 \end{bmatrix}$$

Applying $R_2 \rightarrow R_2 - 2R_1, R_3 \rightarrow R_3 - 4R_1$, we get

$$\begin{bmatrix} 1 & 0 & 3 \\ 0 & 1 & -6 \\ 0 & 2 & -12 \end{bmatrix} \begin{bmatrix} x_1 \\ x_2 \\ x_3 \end{bmatrix} = \begin{bmatrix} 0 \\ 0 \\ 0 \end{bmatrix}$$

Applying $R_3 \rightarrow R_3 - 2R_2$, we get

$$\begin{bmatrix} 1 & 0 & 3 \\ 0 & 1 & -6 \\ 0 & 0 & 0 \end{bmatrix} \begin{bmatrix} x_1 \\ x_2 \\ x_3 \end{bmatrix} = \begin{bmatrix} 0 \\ 0 \\ 0 \end{bmatrix} \qquad \qquad ...(2)$$

Clearly $\rho(A + I) = 2$, therefore the equation (1) will have $3 - 2 = 1$ linearly independent solution.

From (2), we have
$$x_1 + 3x_3 = 0$$
$$x_2 - 6x_3 = 0$$

Let us put $x_3 = c$ (an arbitrary constant), then
$$x_2 = c, x_1 = -3c$$

Now
$$\begin{bmatrix} x_1 \\ x_2 \\ x_3 \end{bmatrix} = \begin{bmatrix} -3c \\ 6c \\ c \end{bmatrix} = c \begin{bmatrix} -3 \\ 6 \\ 1 \end{bmatrix}$$

$\therefore$
$$X_1 = \begin{bmatrix} -3 \\ 6 \\ 1 \end{bmatrix}$$

Eigenvector corresponding to $\lambda_2 = 2$:

Let $X_2 = \begin{bmatrix} x_1 \\ x_2 \\ x_3 \end{bmatrix} \neq O$ be an eigenvector corresponding to eigenvalue $\lambda_2 = 2$, then

or
$$AX_2 = 2X_2$$

or
$$(A - 2I)X_2 = O$$

or
$$\begin{bmatrix} 3-2 & 2 & 0 \\ 2 & 0-2 & 0 \\ 1 & 0 & 2-2 \end{bmatrix} \begin{bmatrix} x_1 \\ x_2 \\ x_3 \end{bmatrix} = \begin{bmatrix} 0 \\ 0 \\ 0 \end{bmatrix}$$

or
$$\begin{bmatrix} 1 & 2 & 0 \\ 2 & -2 & 0 \\ 1 & 0 & 0 \end{bmatrix} \begin{bmatrix} x_1 \\ x_2 \\ x_3 \end{bmatrix} = \begin{bmatrix} 0 \\ 0 \\ 0 \end{bmatrix} \qquad \qquad ...(3)$$

Applying $R_2 \rightarrow R_2 - 2R_1, R_3 \rightarrow R_3 - R_1$, we get

$$\begin{bmatrix} 1 & 2 & 0 \\ 0 & -6 & 0 \\ 0 & -2 & 0 \end{bmatrix} \begin{bmatrix} x_1 \\ x_2 \\ x_3 \end{bmatrix} = \begin{bmatrix} 0 \\ 0 \\ 0 \end{bmatrix}$$

Applying $R_3 \leftrightarrow R_2$ we get

$$\begin{bmatrix} 1 & 2 & 0 \\ 0 & -2 & 0 \\ 0 & -6 & 0 \end{bmatrix} \begin{bmatrix} x_1 \\ x_2 \\ x_3 \end{bmatrix} = \begin{bmatrix} 0 \\ 0 \\ 0 \end{bmatrix}$$

Applying $R_3 \rightarrow R_3 - 3R_2$, we get

$$\begin{bmatrix} 1 & 2 & 0 \\ 0 & -2 & 0 \\ 0 & 0 & 0 \end{bmatrix} \begin{bmatrix} x_1 \\ x_2 \\ x_3 \end{bmatrix} = \begin{bmatrix} 0 \\ 0 \\ 0 \end{bmatrix} \qquad \ldots(4)$$

Clearly $\rho(A - 2I) = 2$, therefore the equation (3) will have $3 - 2 = 1$ linearly independent solution.

From (4), we have $\qquad x_1 + 2x_2 = 0$
$$-2x_2 = 0$$

Let us put $x_3 = c$, we get $x_2 = 0$, $x_1 = 0$

Now
$$\begin{bmatrix} x_1 \\ x_2 \\ x_3 \end{bmatrix} = \begin{bmatrix} 0 \\ 0 \\ c \end{bmatrix} = c \begin{bmatrix} 0 \\ 0 \\ 1 \end{bmatrix}$$

$\therefore \qquad\qquad X_2 = \begin{bmatrix} 0 \\ 0 \\ 1 \end{bmatrix}$

Eigenvector corresponding to $\lambda_3 = 4$:

Let $X_3 = \begin{bmatrix} x_1 \\ x_2 \\ x_3 \end{bmatrix} \neq O$ be an eigenvector corresponding to eigenvalue $\lambda_3 = 4$, then

or $\qquad\qquad\qquad\qquad AX_3 = 4X_3$

or $\qquad\qquad\qquad\qquad (A - 4I)X_3 = O$

or
$$\begin{bmatrix} 3-4 & 2 & 0 \\ 2 & 0-4 & 0 \\ 1 & 0 & 2-4 \end{bmatrix} \begin{bmatrix} x_1 \\ x_2 \\ x_3 \end{bmatrix} = \begin{bmatrix} 0 \\ 0 \\ 0 \end{bmatrix}$$

or
$$\begin{bmatrix} -1 & 2 & 0 \\ 2 & -4 & 0 \\ 1 & 0 & -2 \end{bmatrix} \begin{bmatrix} x_1 \\ x_2 \\ x_3 \end{bmatrix} = \begin{bmatrix} 0 \\ 0 \\ 0 \end{bmatrix} \qquad \ldots(5)$$

Applying $R_2 \rightarrow R_2 + 2R_1, R_3 \rightarrow R_3 + R_1$, we get

$$\begin{bmatrix} -1 & 2 & 0 \\ 0 & 0 & 0 \\ 0 & 2 & -2 \end{bmatrix} \begin{bmatrix} x_1 \\ x_2 \\ x_3 \end{bmatrix} = \begin{bmatrix} 0 \\ 0 \\ 0 \end{bmatrix}$$

Applying $R_2 \leftrightarrow R_3$, we get

$$\begin{bmatrix} -1 & 2 & 0 \\ 0 & 2 & -2 \\ 0 & 0 & 0 \end{bmatrix} \begin{bmatrix} x_1 \\ x_2 \\ x_3 \end{bmatrix} = \begin{bmatrix} 0 \\ 0 \\ 0 \end{bmatrix} \qquad \ldots(6)$$

Clearly $\rho(A - 4I) = 2$, therefore the equation (5) will have $3 - 2 = 1$ linearly independent solution.

From (6), we have

$$-x_1 + 2x_2 = 0$$
$$2x_2 - 2x_3 = 0$$

Let us put $x_2 = c$, we get $x_2 = 2c$, $x_3 = c$

Now

$$\begin{bmatrix} x_1 \\ x_2 \\ x_3 \end{bmatrix} = \begin{bmatrix} 2c \\ c \\ c \end{bmatrix} = c \begin{bmatrix} 2 \\ 1 \\ 1 \end{bmatrix}$$

$\therefore$

$$X_3 = \begin{bmatrix} 2 \\ 1 \\ 1 \end{bmatrix}$$

Since

$$P = [X_1, X_2, X_3]$$

$\therefore$

$$= \begin{bmatrix} -3 & 0 & 2 \\ 6 & 0 & 1 \\ 1 & 1 & 1 \end{bmatrix}$$

and

$$P^{-1}AP = \text{dia.}(-1, 2, 4) = \begin{bmatrix} -1 & 0 & 0 \\ 0 & 2 & 0 \\ 0 & 0 & 4 \end{bmatrix} = D$$

Example 2. *Show that the matrix*

$$A = \begin{bmatrix} -9 & 4 & 4 \\ -8 & 3 & 4 \\ -16 & 8 & 7 \end{bmatrix}$$

is diagonalizable. Also find the diagonal form and a diagonalizing matrix P.

Solution. The characteristic equation of A is given by

$$|A - \lambda I| = 0$$

or

$$\begin{vmatrix} -9 - \lambda & 4 & 4 \\ -8 & 3 - \lambda & 4 \\ -16 & 8 & 7 - \lambda \end{vmatrix} = 0$$

Applying $C_1 \rightarrow C_1 + C_2 + C_3$, we get

$$\begin{vmatrix} -1 - \lambda & 4 & 4 \\ -1 - \lambda & 3 - \lambda & 4 \\ -1 - \lambda & 8 & 7 - \lambda \end{vmatrix} = 0$$

or

$$-(1 + \lambda) \begin{vmatrix} 1 & 4 & 4 \\ 1 & 3 - \lambda & 4 \\ 1 & 8 & 7 - \lambda \end{vmatrix} = 0$$

Applying $R_1 \rightarrow R_2 - R_1$ and $R_3 \rightarrow R_3 - R_1$, we get

or

$$-(1 + \lambda) \begin{vmatrix} 1 & 4 & 4 \\ 1 & -1 - \lambda & 0 \\ 1 & 4 & 3 - \lambda \end{vmatrix} = 0$$

or

$$-(1 + \lambda)(-1 - \lambda)(3 - \lambda) = 0$$

or

$$(1 + \lambda)^2 (3 + \lambda) = 0$$

$\therefore$ $-1, -1, 3$ are eigenvalues of A.

Eigenvector corresponding to eigenvalue -1 :

Let $X = \begin{bmatrix} x_1 \\ x_2 \\ x_3 \end{bmatrix} \neq O$ be an eigenvector corresponding to eigenvalue -1, then

or $\hspace{5cm} AX = (-1)X$

or $\hspace{5cm} (A + I)X = O$

or $\hspace{2cm} \begin{bmatrix} -9+1 & 4 & 4 \\ -8 & 3+1 & 4 \\ -16 & 8 & 7+1 \end{bmatrix} \begin{bmatrix} x_1 \\ x_2 \\ x_3 \end{bmatrix} = \begin{bmatrix} 0 \\ 0 \\ 0 \end{bmatrix}$

or $\hspace{2cm} \begin{bmatrix} -8 & 4 & 4 \\ -8 & 4 & 4 \\ -16 & 8 & 8 \end{bmatrix} \begin{bmatrix} x_1 \\ x_2 \\ x_3 \end{bmatrix} = \begin{bmatrix} 0 \\ 0 \\ 0 \end{bmatrix}$ $\hspace{2cm}$ ...(1)

Applying $R_1 \to \dfrac{1}{4} R_1$, we get

$$\begin{bmatrix} -2 & 1 & 1 \\ -8 & 4 & 4 \\ -16 & 8 & 8 \end{bmatrix} \begin{bmatrix} x_1 \\ x_2 \\ x_3 \end{bmatrix} = \begin{bmatrix} 0 \\ 0 \\ 0 \end{bmatrix}$$

Applying $R_2 \to R_2 - 4R_1, R_3 \to R_3 - 8R_1$, we get

$$\begin{bmatrix} -2 & 1 & 1 \\ 0 & 0 & 0 \\ 0 & 0 & 0 \end{bmatrix} \begin{bmatrix} x_1 \\ x_2 \\ x_3 \end{bmatrix} = \begin{bmatrix} 0 \\ 0 \\ 0 \end{bmatrix} \hspace{2cm} ...(2)$$

Clearly $\rho(A + I) = 1$, therefore the equation (1) will have $3 - 1 = 2$ linearly independent solution.

From (2), we have

$$-2x_1 + x_2 + x_3 = 0$$

Let us put $x_1 = c_1, x_2 = c_2$ then $x_3 = 2c_1 - c_2$

Now $\hspace{2cm} \begin{bmatrix} x_1 \\ x_2 \\ x_3 \end{bmatrix} = \begin{bmatrix} c_1 \\ c_2 \\ 2c_1 - c_2 \end{bmatrix} = c_1 \begin{bmatrix} 1 \\ 0 \\ 2 \end{bmatrix} + c_2 \begin{bmatrix} 0 \\ 1 \\ -1 \end{bmatrix}$

$\therefore \hspace{3cm} X_1 = \begin{bmatrix} 1 \\ 0 \\ 2 \end{bmatrix}, X_2 = \begin{bmatrix} 0 \\ 1 \\ -1 \end{bmatrix}$

Eigenvector corresponding to eigenvalue 3 :

Let $X_3 = \begin{bmatrix} x_1 \\ x_2 \\ x_3 \end{bmatrix} \neq O$ be an eigenvector corresponding to eigenvalue 3, then

or $\hspace{5cm} AX_3 = 3X_3$

or $\hspace{5cm} (A - 3I)X_2 = O$

or
$$\begin{bmatrix} -9-3 & 4 & 4 \\ -8 & 3-3 & 4 \\ -16 & 8 & 7-3 \end{bmatrix} \begin{bmatrix} x_1 \\ x_2 \\ x_3 \end{bmatrix} = \begin{bmatrix} 0 \\ 0 \\ 0 \end{bmatrix}$$

or
$$\begin{bmatrix} -12 & 4 & 4 \\ -8 & 0 & 4 \\ -16 & 8 & 4 \end{bmatrix} \begin{bmatrix} x_1 \\ x_2 \\ x_3 \end{bmatrix} = \begin{bmatrix} 0 \\ 0 \\ 0 \end{bmatrix} \qquad \ldots(3)$$

Applying $R_2 \leftrightarrow R_1$ we get
$$\begin{bmatrix} -8 & 0 & 4 \\ -12 & 4 & 4 \\ -16 & 8 & 4 \end{bmatrix} \begin{bmatrix} x_1 \\ x_2 \\ x_3 \end{bmatrix} = \begin{bmatrix} 0 \\ 0 \\ 0 \end{bmatrix}$$

Applying $R_1 \rightarrow \dfrac{1}{4}R_1$, we get
$$\begin{bmatrix} -2 & 0 & 1 \\ -12 & 4 & 4 \\ -16 & 8 & 4 \end{bmatrix} \begin{bmatrix} x_1 \\ x_2 \\ x_3 \end{bmatrix} = \begin{bmatrix} 0 \\ 0 \\ 0 \end{bmatrix}$$

Applying $R_2 \rightarrow R_2 - 6R_1$, $R_3 \rightarrow R_3 - 8R_1$, we get
$$\begin{bmatrix} -2 & 0 & 1 \\ 0 & 4 & -2 \\ 0 & 8 & -4 \end{bmatrix} \begin{bmatrix} x_1 \\ x_2 \\ x_3 \end{bmatrix} = \begin{bmatrix} 0 \\ 0 \\ 0 \end{bmatrix} \qquad \ldots(4)$$

Applying $R_3 \rightarrow R_3 - 2R_2$, we get
$$\begin{bmatrix} -2 & 0 & 1 \\ 0 & 4 & -2 \\ 0 & 0 & 0 \end{bmatrix} \begin{bmatrix} x_1 \\ x_2 \\ x_3 \end{bmatrix} = \begin{bmatrix} 0 \\ 0 \\ 0 \end{bmatrix}$$

Clearly $\rho(A - 3I) = 2$, therefore the equation (3) will have $3 - 2 = 1$ linearly independent solution.

From (4), we have
$$-2x_1 + x_3 = 0$$
$$4x_2 - 2x_3 = 0$$

Let us put $x_2 = c$, we get $x_3 = 2c$, $x_1 = c$

Now
$$\begin{bmatrix} x_1 \\ x_2 \\ x_3 \end{bmatrix} = \begin{bmatrix} c \\ c \\ 2c \end{bmatrix} = c\begin{bmatrix} 1 \\ 1 \\ 2 \end{bmatrix}$$

$\therefore$
$$X_3 = \begin{bmatrix} 1 \\ 1 \\ 2 \end{bmatrix}$$

Since
$$P = [X_1, X_2, X_3]$$

$\therefore$
$$P = \begin{bmatrix} 1 & 0 & 1 \\ 0 & 1 & 1 \\ 2 & -1 & 2 \end{bmatrix}$$

Also
$$P^{-1}AP = \text{diag.}(-1, -1, 3) = \begin{bmatrix} -1 & 0 & 0 \\ 0 & -1 & 0 \\ 0 & 0 & 3 \end{bmatrix} = D$$

Example 3. *Show that the matrix* $A = \begin{bmatrix} 8 & -8 & -2 \\ 4 & -3 & -2 \\ 3 & -4 & 1 \end{bmatrix}$ *is diagonalizable.*

Solution. If the matrix A has distinct eigenvalues, then it is essentially diagonalizable, so we shall find its eigenvalues.

The characteristic equation of A is given by

$$|A - \lambda I| = 0$$

or

$$\begin{vmatrix} 8-\lambda & -8 & -2 \\ 4 & -3-\lambda & -2 \\ 3 & -4 & 1-\lambda \end{vmatrix} = 0$$

Applying $R_1 \rightarrow R_1 - (R_2 + R_3)$, we get

$$\begin{vmatrix} 1-\lambda & -1+\lambda & -1+\lambda \\ 4 & -3-\lambda & -2 \\ 3 & -4 & 1-\lambda \end{vmatrix} = 0$$

or

$$(1+\lambda)\begin{vmatrix} 1 & -1 & -1 \\ 4 & -3-\lambda & -2 \\ 3 & -4 & 1-\lambda \end{vmatrix} = 0$$

Applying $C_2 \rightarrow C_2 + C_1, C_3 \rightarrow C_3 + C_1$, we get

or

$$(1-\lambda)\begin{vmatrix} 1 & 0 & 0 \\ 4 & 1-\lambda & 2 \\ 3 & -1 & 4-\lambda \end{vmatrix} = 0$$

or $(1-\lambda)\{(1-\lambda)(4-\lambda)+2\} = 0$

or $(1-\lambda)(\lambda^2 - 5\lambda + 6) = 0$

or $(1-\lambda)(2-\lambda)(3-\lambda) = 0$

∴ A has 1, 2, 3 eigenvalues which are all distinct, hence A is diagonalizable.

REMARK

- If some eigenvalues of a matrix are the same, then it need not be non-diagonalizable.

4.8 ALGEBRAIC AND GEOMETRIC MULTIPLICITY OF AN EIGENVALUE

(i) **Algebraic multiplicity :** Let A be an $n \times n$ matrix and let

$$|A - \lambda I_n| = (\lambda - \lambda_1)^{n_1}(\lambda - \lambda_2)^{n_2}(\lambda - \lambda_3)^{n_3} \ldots (\lambda - \lambda_k)^{n_k}$$

where $n_1 + n_2 + n_3 + \ldots + n_k = n$. Then the numbers $n_1, n_2, n_3, \ldots, n_k$ are the algebraic multiplicities of the eigenvalues $\lambda_1, \lambda_2, \lambda_3, \ldots, \lambda_n$ respectively.

(ii) **Eigenspace of a matrix A corresponding to eigenvalue λ:**

Let A be an $n \times n$ matrix, let λ be an eigenvalue of A.

For eigenvalue λ, we find eigenvectors

$$X = \begin{bmatrix} x_1 \\ x_2 \\ x_3 \\ \vdots \\ x_n \end{bmatrix}$$

by solving the linear system
$$(A - \lambda I)X = O$$
The set of all vectors X satisfying $AX = \lambda X$ is called the eigenspace of A corresponding to eigenvalue A, which is denoted by E_k.

(iii) **Geometric multiplicity :**

Let A be an $n \times n$ matrix and λ be one of its eigenvalues, then the dimension of eigenspace E_λ of A corresponding to λ is called the geometric multiplicity of λ.

REMARKS

- Dimension of E_λ = Dimension of the null space $(A - \lambda I)$.
- Geometric multiplicity of $\lambda = n - \rho(A - \lambda I)$.

Theorem 1. **(Rank-Multiplicity Theorem).**

The geometric multiplicity of an eigenvalue cannot exceed its algebraic multiplicity.

Proof. Let A be an $n \times n$ matrix and λ be one of its eigenvalues. Suppose that the geometric multiplicity of λ is m, then there exists m linearly independent column vectors $X_1, X_2, ..., X_m$ such that
$$AX_i = \lambda X_i (i = 1, 2, ..., m) \qquad ...(1)$$
Since $\{ X_1, X_2, ..., X_m\}$ is linearly independent so it can be extended to a basis
$$\{X_1, X_2, ..., X_m, X_{m+1}, ..., X_n\}$$
for a vector space V_n.
The set $\{X_1, X_2, ..., X_m, X_{m+1}, ..., X_n\}$ is linearly independent since it is a basis for V_n, therefore we have a non-singular matrix P whose column vectors are
$$X_1, X_2, ..., X_m, X_{m+1}, ..., X_n.$$
$$P = [X_1, X_2, ..., X_m, X_{m+1}, ..., X_n]$$
Now $\quad P^{-1}AP = P^{-1}A[X_1, X_2, ..., X_m, X_{m+1}, ..., X_n]$
$$= P^{-1}[AX_1, AX_2, ..., AX_m, AX_{m+1}, ..., AX_n]$$
$$= P^{-1}[\lambda X_1, \lambda X_2, ..., \lambda X_m, AX_{m+1}, ..., AX_n] \qquad \text{[Using (1)]}$$
$$= [\lambda P^{-1}X_1, \lambda P^{-1}X_2, ..., \lambda P^{-1}X_m, P^{-1}AX_{m+1}, ..., P^{-1}AX_n]$$
Clearly, $\lambda P^{-1}X_1, \lambda P^{-1}X_2, ..., \lambda P^{-1}X_m$ are the first, second, ... , mth column of $P^{-1}AP$. But $P^{-1}X_1$ is the first column of $P^{-1}P = I$

$$\therefore \quad \lambda P^{-1}X_1 = \begin{bmatrix} \lambda \\ 0 \\ 0 \\ \vdots \\ 0 \end{bmatrix}_{n \times 1} , \lambda P^{-1}X_2 = \begin{bmatrix} 0 \\ \lambda \\ 0 \\ \vdots \\ 0 \end{bmatrix}_{n \times 1} , ... \lambda P^{-1}X_m = \begin{bmatrix} 0 \\ 0 \\ \vdots \\ \lambda \\ 0 \\ \vdots \\ 0 \end{bmatrix}_{n \times 1}$$

$$\therefore \qquad P^{-1}AP = \begin{bmatrix} \lambda I_m & B \\ 0 & C \end{bmatrix}$$

where B is a matrix of order $m \times (n - m)$ and C is a matrix of order $(n - m) \times (n - m)$ and O is a null matrix of order $(n - m) \times m$.

The characteristic polynomial of $P^{-1}AP$ is

$$|P^{-1}AP - xI| = \begin{vmatrix} (\lambda - x)I_m & B \\ O & C - xI_{n-m} \end{vmatrix}$$

$$= |(\lambda - x)I_m| |C - xI_{n-m}|$$

$$= (\lambda - x)^m |C - xI_{n-m}| \qquad \qquad ...(2)$$

From (2), it is clear that $(\lambda - x)^m$ is a factor of the characteristic polynomial of $P^{-1}AP$, therefore λ is an eigenvalue of $P^{-1}AP$ of the algebraic multiplicity at least m.

Since A and $P^{-1}AP$ have same eigenvalues, so that λ is an eigenvalue of A of algebraic multiplicity at least m.

$\Rightarrow$ Algebraic multiplicity of $\lambda \geq m$

$\Rightarrow$ Algebraic multiplicity of $\lambda \geq$ geometric multiplicity of λ.

Theorem 2. *The necessary and sufficient conditions for a square matrix to be similar to a diagonal matrix is that the geometric multiplicity of each of its eigenvalues coincides with the algebraic multiplicity.*

Proof. Let A be an $n \times n$ matrix.

Necessary Condition : Suppose that A is similar to a diagonal matrix $D = \text{dia.} (\lambda_1, \lambda_2, ..., \lambda_n)$, then there exists a non-singular matrix P such that

$$P^{-1}AP = D = \text{dia.} (\lambda_1, \lambda_2, ..., \lambda_n) \qquad ...(1)$$

$\therefore$ $\lambda_1, \lambda_2, ..., \lambda_n$ are the eigenvalues of A not necessarily distinct.

Let t be an eigenvalue of A of algebraic multiplicity p, then we have

$$\lambda_1 = t, \ \lambda_2 = t , ..., \lambda_p = t \qquad ...(2)$$

If $\rho(A - tI) = m$, then the system of equations

$$(A - tI) X = O$$

will have $n - m$ linearly independent solutions and therefore $n - m$ will be the geometric multiplicity of t.

So, we have to prove that $p = n - m$.

Since the rank of a matrix does not change on premultiplication and post-multiplication by a non-singular matrix.

$\therefore$

$$\rho(A - tI) = \rho\left[P^{-1}(A - tI)P \right]$$

$$= \rho\left[P^{-1}AP - tI \right]$$

$$= \rho[D - tI]$$

$$= \rho\left(\text{dia.} \left[\lambda_1 - t, \lambda_2 - t, ..., \lambda_p - t, \lambda_{p+1} - t, \lambda_n - t\right]\right)$$

$$= \rho\left(\text{dia.} \left[0, 0, ..., 0, \lambda_{p+1} - t, \lambda_n - t\right]\right) \qquad \text{[Using (2)]}$$

$$\rho(A - tI) = n - p$$

$\Rightarrow \qquad\qquad m = n - p \qquad\qquad\qquad \because \rho(A - tI) = m$

$\Rightarrow \qquad\qquad p = m - n$

Sufficient Condition : Suppose that the geometric multiplicity of each eigenvalue of A coincides with its algebraic multiplicity. Then we have to show that A is similar to a diagonal matrix, *i.e.*, A is diagonalizable.

Suppose that A has $\lambda_1, \lambda_2, \dots \lambda_k$ distinct eigenvalues of multiplicity $n_1, n_2, \dots n_k$ respectively, then we have

$$n_1 + n_2 + \dots + n_k = n$$

Let

$$\left.\begin{array}{l} X_{11}, X_{12}, \dots, X_{1n_1} \\ X_{21}, X_{22}, \dots, X_{2n_2} \\ \dots\dots\dots\dots\dots\dots\dots \\ \dots\dots\dots\dots\dots\dots\dots \\ X_{k1}, X_{k2}, \dots, X_{kn_k} \end{array}\right\} \qquad \dots(3)$$

be linearly independent sets of eigenvectors corresponding to the eigenvalues $\lambda_1, \lambda_2, \dots \lambda_k$ respectively.

Now we prove that the n vectors given by (3) are linearly independent.

Let

$$(a_{11}X_{11} + a_{12}X_{12} + \dots + a_{1n_1}X_{1n_1}) + (a_{21}X_{21} + a_{22}X_{22} + \dots + a_{2n_2}X_{2n_2})$$
$$+ \dots + (a_{k1}X_{k1} + a_{k2}X_{k2} + \dots + a_{kn_K}X_{kn_K}) = O \qquad \dots(4)$$

or $\quad X_1 + X_2 + \dots + X_k = O \qquad\qquad\qquad\qquad \dots(5)$

where

$$\left.\begin{array}{l} X_1 = a_{11}X_{11} + a_{12}X_{12} + \dots + a_{1n_1}X_{1n_1} \\ X_2 = a_{21}X_{21} + a_{22}X_{22} + \dots + a_{2n_2}X_{2n_2} \\ \dots\dots\dots\dots\dots\dots\dots\dots\dots\dots\dots\dots\dots\dots\dots \\ X_k = a_{k1}X_{k1} + a_{k2}X_{k2} + \dots + a_{kn_k}X_{kn_k} \end{array}\right\} \qquad \dots(6)$$

From (6), we see that X_1 is a linear combination of $X_{11}, X_{12}, \dots, X_{1n_1}$ which are eigenvectors of A corresponding to the eigenvalue λ_1. If $X_1 \neq O$ then X_1 is also the eigenvector of A corresponding to eigenvalue λ_1, similarly for other eigenvectors $X_2, X_3, \dots, X_k$.

Suppose $X_i \neq O$, then from (3), we see that a system of eigenvectors of A corresponding to distinct eigenvalues of A is linearly dependent which is not possible, hence

$$X_i = O \ \forall \ i = 1, 2, 3, \dots, k$$

$\Rightarrow \qquad\qquad a_{i1}X_{i1} + a_{i2}X_{i2} + \dots + a_{in_i}X_{in_1} = O \qquad\qquad$ [Using (6)]

But $X_{i1}, X_{i2}, \dots, X_{in_1}$ is a set of linearly independent vectors, therefore

$$a_{i1} = a_{i2} = \dots = a_{in_1} = 0 \ \forall \ i = 1, 2, 3, \dots, k$$

It follows that the n vectors given by (3) are linearly independent. Therefore A has n linearly independent eigenvectors, hence A is diagonalizable and hence A is similar to a diagonal matrix.

Solved Examples

Example 1. *Show that the following matrices are not similar to diagonal matrices:*

$$(i) \quad \begin{bmatrix} 2 & 3 & 4 \\ 0 & 2 & -1 \\ 0 & 0 & 1 \end{bmatrix} \qquad (ii) \quad \begin{bmatrix} 2 & -1 & 1 \\ 2 & 2 & -1 \\ 1 & 2 & -1 \end{bmatrix}$$

Solution. (i) Let

$$A = \begin{bmatrix} 2 & 3 & 4 \\ 0 & 2 & -1 \\ 0 & 0 & 1 \end{bmatrix}$$

Clearly, A is an upper triangular matrix therefore its diagonal elements 2, 2, 1 are the eigenvalues of A.

The algebraic multiplicity of eigenvalue is 2.

Next, we find the geometric multiplicity of eigenvalue 2.

Let $X = \begin{bmatrix} x_1 \\ x_2 \\ x_3 \end{bmatrix} \neq O$ be an eigenvector corresponding to eigenvalue 2, then

or $$AX_1 = 2X_1$$
or $$(A - 2I)X_1 = O$$

or $$\begin{bmatrix} 2-2 & 3 & 4 \\ 0 & 2-2 & -1 \\ 0 & 0 & 1-2 \end{bmatrix} \begin{bmatrix} x_1 \\ x_2 \\ x_3 \end{bmatrix} = \begin{bmatrix} 0 \\ 0 \\ 0 \end{bmatrix}$$

or $$\begin{bmatrix} 0 & 3 & 4 \\ 0 & 0 & -1 \\ 0 & 0 & -1 \end{bmatrix} \begin{bmatrix} x_1 \\ x_2 \\ x_3 \end{bmatrix} = \begin{bmatrix} 0 \\ 0 \\ 0 \end{bmatrix}$$

Applying $R_3 \to R_3 - R_2$, we get

$$\begin{bmatrix} 0 & 3 & 4 \\ 0 & 0 & -1 \\ 0 & 0 & 0 \end{bmatrix} \begin{bmatrix} x_1 \\ x_2 \\ x_3 \end{bmatrix} = \begin{bmatrix} 0 \\ 0 \\ 0 \end{bmatrix}$$

Clearly $\rho(A - 4I) = 2$, therefore the geometric multiplicity of 2 is $3 - 2 = 1$, which is not equal to the algebraic multiplicity of eigenvalue 2. Hence A is not similar to the diagonal matrix.

(ii) Let

$$A = \begin{bmatrix} 2 & -1 & 1 \\ 2 & 2 & -1 \\ 1 & 2 & -1 \end{bmatrix}$$

The characteristic equation of A is given by

$$|A - \lambda I| = 0$$

or $$\begin{vmatrix} 2-\lambda & -1 & 1 \\ 2 & 2-\lambda & -1 \\ 1 & 2 & -1-\lambda \end{vmatrix} = 0$$

Applying $C_3 \to C_3 + C_2$, we get

$$\begin{vmatrix} 2-\lambda & -1 & 0 \\ 2 & 2-\lambda & 1-\lambda \\ 1 & 2 & 1-\lambda \end{vmatrix} = 0$$

or $\qquad (1-\lambda)\begin{vmatrix} 2-\lambda & -1 & 0 \\ 2 & 2-\lambda & 1 \\ 1 & 2 & 1 \end{vmatrix} = 0$

Again applying $R_2 \to R_2 - R_3$, we get

or $\qquad (1-\lambda)\begin{vmatrix} 2-\lambda & -1 & 0 \\ 1 & -\lambda & 0 \\ 1 & 2 & 1 \end{vmatrix} = 0$

or $\qquad (1-\lambda)\{-\lambda(2-\lambda)+1\} = 0$

or $\qquad (1-\lambda)(\lambda^2 - 2\lambda + 1) = 0$

or $\qquad (1-\lambda)^3 = 0$

$\therefore$ 1, 1, 1 are the eigenvalues of A, thus the algebraic multiplicity of 1 is 3. Next, we find the geometric multiplicity of eigenvalue 1.

Let $X = \begin{bmatrix} x_1 \\ x_2 \\ x_3 \end{bmatrix} \neq O$ be an eigenvector A corresponding to eigenvalue 1, then

or $\qquad\qquad\qquad AX = (1)X$

or $\qquad\qquad\qquad (A - I)X = O$

or $\qquad \begin{bmatrix} 2-1 & -1 & 1 \\ 2 & 2-1 & -1 \\ 1 & 2 & -1-1 \end{bmatrix}\begin{bmatrix} x_1 \\ x_2 \\ x_3 \end{bmatrix} = \begin{bmatrix} 0 \\ 0 \\ 0 \end{bmatrix}$

or $\qquad \begin{bmatrix} 1 & -1 & 1 \\ 2 & 1 & -1 \\ 1 & 2 & -2 \end{bmatrix}\begin{bmatrix} x_1 \\ x_2 \\ x_3 \end{bmatrix} = \begin{bmatrix} 0 \\ 0 \\ 0 \end{bmatrix}$

Applying $R_2 \to R_2 - 2R_1$, $R_3 \to R_3 - R_1$, we get

$$\begin{bmatrix} 1 & -1 & 1 \\ 0 & 3 & -3 \\ 0 & 3 & -3 \end{bmatrix}\begin{bmatrix} x_1 \\ x_2 \\ x_3 \end{bmatrix} = \begin{bmatrix} 0 \\ 0 \\ 0 \end{bmatrix}$$

Again applying $R_3 \to R_3 - R_2$, we get

$$\begin{bmatrix} 1 & -1 & 1 \\ 0 & 3 & -3 \\ 0 & 0 & 0 \end{bmatrix}\begin{bmatrix} x_1 \\ x_2 \\ x_3 \end{bmatrix} = \begin{bmatrix} 0 \\ 0 \\ 0 \end{bmatrix}$$

Clearly $\rho(A - I) = 2$, therefore the geometric multiplicity of 2 is $3 - 2 \Rightarrow 1$, which is not equal to the algebraic multiplicity of eigenvalue 2. Hence A is not similar to the diagonal matrix.

❧ EXERCISE 4.3

1. Prove that the matrix $A = \begin{bmatrix} 0 & 1 \\ -1 & 0 \end{bmatrix}$ is not diagonalized over R the set of all real numbers, however A is diagonalizable over C the set of all complex numbers. Find an invertible matrix P over C such that $P^{-1}AP$ is a diagonal matrix.

2. Show that the matrix
$$A = \begin{bmatrix} 1 & -6 & -4 \\ 0 & 4 & 2 \\ 0 & -6 & -3 \end{bmatrix}$$
is similar to a diagonal matrix. Also find the transforming matrix and diagonal matrix.

3. Transform the matrix
$$\begin{bmatrix} 8 & -12 & 5 \\ 15 & -25 & 11 \\ 24 & -42 & 19 \end{bmatrix}$$
into diagonal form.

4. Show that each of the following matrices is similar to a diagonal matrix. Also in each case find the diagonal form D and a diagonalizing matrix P :

(i) $\begin{bmatrix} 8 & -6 & 2 \\ -6 & 7 & -4 \\ 2 & -4 & 3 \end{bmatrix}$ (ii) $\begin{bmatrix} 6 & -2 & 2 \\ -2 & 3 & -1 \\ 2 & -1 & 3 \end{bmatrix}$

(iii) $\begin{bmatrix} 17 & 18 & -6 \\ -18 & 19 & -6 \\ -9 & 9 & 2 \end{bmatrix}$ (iv) $\begin{bmatrix} 4 & 2 & -2 \\ -5 & 3 & 2 \\ -2 & 4 & 1 \end{bmatrix}$

5. Find the non-singualar matrix P such that $P^{-1}AP$ is a diagonal matrix, where
$$A = \begin{bmatrix} 1 & -3 & 3 \\ 3 & -5 & 3 \\ 6 & -6 & 6 \end{bmatrix}$$

6. Let $A = \begin{bmatrix} -9 & 4 & 4 \\ -8 & 3 & 4 \\ -16 & 8 & 7 \end{bmatrix}$. Find an invertible matrix P such that $P^{-1}AP$ is a diagonal matrix.

7. Test the matrix $A = \begin{bmatrix} 3 & 1 & 0 \\ 0 & 3 & 0 \\ 0 & 0 & 4 \end{bmatrix}$ for diagonal-izability.

8. Show that the following matrices are not similar to diagonal matrices :

(i) $\begin{bmatrix} 2 & 1 & 0 \\ 0 & 2 & 1 \\ 0 & 0 & 2 \end{bmatrix}$ (ii) $\begin{bmatrix} 3 & 10 & 5 \\ -2 & -3 & -4 \\ 3 & 5 & 7 \end{bmatrix}$

Answers

1. $P = \begin{bmatrix} -i & i \\ 1 & 1 \end{bmatrix}$

2. $P = \begin{bmatrix} 1 & 2 & 2 \\ -2 & -2 & 1 \\ 3 & 3 & -2 \end{bmatrix}, D = \begin{bmatrix} 1 & 0 & 0 \\ 0 & 1 & 0 \\ 0 & 0 & 0 \end{bmatrix}$

4. (i) $D = \begin{bmatrix} 0 & 0 & 0 \\ 0 & 3 & 0 \\ 0 & 0 & 15 \end{bmatrix}, P = \begin{bmatrix} 1 & 2 & 2 \\ 2 & 1 & -2 \\ 2 & -2 & 1 \end{bmatrix}$

4. (ii) $D = \begin{bmatrix} 2 & 0 & 0 \\ 0 & 2 & 0 \\ 0 & 0 & 8 \end{bmatrix}, P = \begin{bmatrix} -1 & 1 & 2 \\ 0 & 2 & -1 \\ 2 & 0 & 1 \end{bmatrix}$

(iii) $D = \begin{bmatrix} -2 & 0 & 0 \\ 0 & 1 & 0 \\ 0 & 0 & 1 \end{bmatrix}, P = \begin{bmatrix} 2 & 1 & -1 \\ 2 & 1 & 0 \\ 1 & 0 & 3 \end{bmatrix}$

(iv) $D = \begin{bmatrix} 1 & 0 & 0 \\ 0 & 2 & 0 \\ 0 & 0 & 5 \end{bmatrix}, P = \begin{bmatrix} 2 & 1 & 0 \\ 1 & 1 & 1 \\ 4 & 2 & 1 \end{bmatrix}$

5. $P = \begin{bmatrix} 1 & -1 & 1 \\ 1 & 0 & 1 \\ 0 & 1 & 2 \end{bmatrix}$

6. $P = \begin{bmatrix} 1 & 1 & 1 \\ 0 & 2 & 1 \\ 2 & 0 & 2 \end{bmatrix}$

7. Not diagonalizable

Selected Terms and Results

TERMS

- **Matric Polynomial :** A polynomial in indeterminate λ of the form

$$f(\lambda) = A_0 + A_1\lambda + A_2\lambda^2 + \ldots + A_n\lambda^n$$

where A_0, A_1, ..., A_n are all square matrices of the same order is called a matric polynomial.

- **Characteristic Equation :** The equation $|A - \lambda I| = 0$ is called characteristic equation.

- **Characteristic Root or Latent Root and Eigenvalue :** The roots of the equation $|A - \lambda I| = 0$ are called characteristic root or latent root of eigenvalue.

- **Characteristic Vector of Eigenvector :** A non-zero vector X is called an eigenvector of A corresponding to the eigenvalue λ.

RESULTS

- Two matric polynomials are said to be equal if and only if the coefficient of like powers of λ are same.

- Corresponding to an eigenvalue of a matric A, there will be different eigenvectors of A.

- For a given eigenvector of a matrix A, there corresponds one and only one eigenvalue of A.

- If λ is an eigenvalue of a matrix A if and only if there exists a non-zero vector X such that $AX = \lambda X$.

- If X is an eigenvector of a matrix A corresponding to an eigenvalue of A, then kX is also an eigenvector of A corresponding to the same eigenvector λ, where k is any non-zero vector.

- If X is a non-zero eigenvector of a matrix A then X cannot correspond to more than one eigenvalue of A.

- Let A be an $n \times n$ matrix. Then the disjoint eigenvectors corresponding to distinct eigenvalue of A are likely independent.

- The eigenvalues of a Hermitian matrix are real.

- The eigenvalues of a real symmetric matrix are all real.

- The eigenvalue of a skew-Hermitian matrix are either pure imaginary or zero.

- The eigenvalues of a real skew-symmetric matrix are either purely imaginary or zero.

- The eigenvalues of a unitary matrix are of unit modulus.

- The eigenvalues of an orthogonal matrix are of unit modulus.

- If λ_1, λ_2, ..., λ_n are the eigenvalues of A, then $k\lambda_1$, $k\lambda_2$, ..., $k\lambda_n$ are eigen values of kA, where K is any number.

- If X be a non-zero eigenvector of an $n \times n$ matrix A then for each positive integer n, X is an eigenvector of A^n corresponding to the eigenvalue λ_n.

- Similar matrices have the same eigenvalues.

- The eigenvectors of real symmetric matrix corresponding to two distinct eigenvalues are orthogonal.

- The eigenvectors of Hermitian matrix corresponding to two distinct eigenvalues are orthogonal.

- Every square matrix satisfies its characteristic equation.

- If $\lambda_1, \lambda_2, ..., \lambda_n$ are eigenvalues of a square matrix of order $n \times n$ then $\det(A) = \lambda_1.\lambda_2.\lambda_3...\lambda_n$.

- An $n \times n$ matrix is diagonalizable if and only if it possesses n linearly independent eigenvectors.

- If the eigenvalues of an $n \times n$ matrix are all distinct then it is essentially diagonalizable.

- The geometric multiplicity of an eigenvalue cannot exceed its algebraic multiplicity.

- The necessary and sufficient conditions for a square matrix to be similar to a diagonal matrix is that the geometric multiplicity of each of its eigenvalues coincide with the algebraic multiplicity.

Review Questions and Project Work

1. Prove that the sum of all eigenvalues of a matrix is equal to the sum of diagonal elements of the matrix.

2. Prove that the scalar λ is an eigenvalues of a matrix A if and only if $(A - \lambda I)$ is singular.

3. Show that the square metrices A and its transpose A' have the same characterisctic roots.

4. Prove that the product of eigenvalues of a square matrix is equal to the determinant of the matrix.

5. If $\lambda_1, \lambda_2,...,\lambda_n$ are the eigenvalues of an n-square matrix A and k is a scalar, prove that the eigen values of $A - kI$ are $\lambda_1 - k, \lambda_2 - k, ..., \lambda_n - k$.

6. If A is non-singular matrix, show that the eigenvalues of A^{-1} are the reciprocals of the eigenvalues of A.

7. Show that if λ is an eigenvalue of a matrix A, then $k + \lambda$ is an eigenvalue of the matrix $A + kI$.

8. If λ is an eigenvalue of a non-singular matrix, prove that $\dfrac{|A|}{\lambda}$ is an eigenvalue of adjoint A.

9. Show that the two matrices $A, C^{-1}AC$ have the same eigenvalues.

10. Show that the matrix $A = \begin{bmatrix} 0 & c & -b \\ -c & 0 & a \\ b & -a & 0 \end{bmatrix}$ satisfies Caylay-Hamilton theorem. Hence, find A^{-1}.

11. If $A = \begin{bmatrix} 1 & 0 & 0 \\ 1 & 0 & 1 \\ 0 & 1 & 0 \end{bmatrix}$, show that for every integer $n \geq A$, $A^n = A^{n-2} + A^3 - A$. Hence, evaluate A^{20}.

12. If $A = \begin{bmatrix} 1 & 2 \\ -1 & 3 \end{bmatrix}$, express $A^6 - 4A^5 + 8A^4 - 12A^3 + 14A^2$ as a linear polynomial in A.

Objective Type Questions

FILL IN THE BLANKS

1. If A is a square matrix, then its characteristic equation is _____ .

2. If the matrix $(A - \lambda I)$ is singular, then λ is an _____ of A.

3. If the matrix $(A - \lambda I)$ is singular, then there exists a non-zero vector X such that $AX =$ _____ .

4. If X is an eigenvector corresponding to the eigenvalue of A, the kX is also eigenvector corresponding to _____ .

5. The characteristic vectors corresponding to distinct characteristic roots of a matrix are linearly _____ .

6. The characteristic roots of a Hermitian matrix are _____ .

7. The characteristic roots of a unitary matrix are of _____ .

8. The characteristic roots of the matrix $\begin{bmatrix} 5 & 4 \\ 1 & 2 \end{bmatrix}$ are _____ .

9. The eigenvalues of the matrix A and A^T are _____ .

10. If 0 is the characteristic root of A, then A is _____ .

11. The characteristic roots of a triangular matrix are just the _____ elements of the same matrix.

12. If $\lambda_1, \lambda_2, ..., \lambda_n$ are the characteristic roots of A, then $a\lambda_1, a\lambda_2, ..., a\lambda_n$ are the characteristic roots of _____ .

13. If λ is an eigenvalues of A, then the eigenvalue of A^{-1} is _____ .

14. If λ is a characteristic root of a matrix A (non-singular) then the characteristic root of Adj. A is _____ .

15. Every square matrix satisfies its _____ .

16. If $A^2 - 5A + 6I = 0$ for the matrix A, then the eigenvalues of A are _____ .

17. If $A^2 - 4A + 5I = 0$ for the matrix A, then its inverse is _____ .

18. If 1, 2, 3 are eigenvalues of A, then the eigenvalues of A^2 are _____ .

TRUE/FALSE

Write 'T' for true and 'F' for false statement.

1. If λ is an eigenvalue of A, then $(A - \lambda I)$ is non-singular. **(T/F)**

2. If $\lambda = 0$ is an eigenvalue of A, then A is singular. **(T/F)**

3. If $X \neq 0$ is an eigenvector corresponding to the eigenvalue 3 of A, then $(A - 3I)X = 0$. **(T/F)**

4. The characteristic roots of a skew Hermitian matrix are always real. **(T/F)**

5. The eigenvalues of $A = \begin{bmatrix} 1 & 0 & 0 \\ 0 & 1 & 0 \\ 0 & 0 & 1 \end{bmatrix}$ are 1, 0, 1. **(T/F)**

6. The eigenvector to the eigenvalue 1 of $\begin{bmatrix} 5 & 4 \\ 1 & 2 \end{bmatrix}$ is $X = [1 - \lambda]^T$. **(T/F)**

7. The characteristic roots of $\begin{bmatrix} 2 & 3 & 4 \\ 0 & -2 & 1 \\ 0 & 0 & 1 \end{bmatrix}$ are 2, -2, 1. **(T/F)**

8. If $|A| \neq 0$ and λ is a characteristic root, then $\dfrac{|A|}{\lambda}$ is a characteristic root of Adj (A). **(T/F)**

9. If A is a square matrix of order 3, then it has 2 characteristic roots. **(T/F)**

10. If the characteristic equation of A is $\lambda^3 - 6\lambda^2 + 11\lambda + a = 0$. The sum of all its eigenvalues is -6. **(T/F)**

11. If the matrix A satisfies the equation $\lambda^2 - 4\lambda + \mu = 0$ and one of the eigenvalue of A is 3, then the value of μ is 3. **(T/F)**

12. If $A^3 + a_1 A^2 + a_2 A + a_3 I = 0$, then the order A is 3. **(T/F)**

13. If $A^2 - 3A + 2I = 0$, then $|A| = 2$. **(T/F)**

14. Every square matrix satisfies its characteristic equation. **(T/F)**

15. If $\lambda_1, \lambda_2, ..., \lambda_n$ are the characteristic roots of A, then the characteristic roots of A^2 are $\lambda_1, \lambda_2, ..., \lambda_n$. **(T/F)**

16. The characteristic equation of $A = \begin{bmatrix} 3 & 1 \\ -1 & 2 \end{bmatrix}$ is $\lambda^2 - 3\lambda + 7 = 0$. **(T/F)**

MULTIPLE CHOICE QUESTIONS

Choose the most appropriate one :

1. If $\lambda = 0$ is an eigenvalues of A, then det (A) is:
   (a) 0   (b) 1
   (c) λ (d) none of these

2. If $|A| \neq 0$ and λ is an eigenvalue of A, then the eigenvalue of Adj. (A) is :
   (a) λ (b) λ^2
   (c) $1/\lambda$ (d) 0

3. If $|A| \neq 0$ and λ is an eigenvalue of A, then the eigenvalue of Adj. (A) is :
   (a) $\lambda |A|$ (b) $\dfrac{1}{\lambda}$
   (c) λ^2 (d) $\dfrac{|A|}{\lambda}$

4. If a matrix A has $(n - 1)$ characteristic roots, then its order is :
   (a) n (b) $n - 1$
   (c) n^2 (d) $n + 1$

5. If 1, -1, 3 are the characteristic roots of A, then characteristics roots of A^2 are :
   (a) 1, 2, 3   (b) 1, -2, 3
   (c) 1, 4, 9   (d) -1, 2, -3

6. If $A^3 - 7A^2 + aA + bI = 0$, then the sum of eigenvalues of A is:
   (a) -7 (b) 7
   (c) a (d) b

7. If $A^3 - 6A^2 + 11A - 6I = 0$, then the det (A) is:
   (a) 11   (b) -6
   (c) 6   (d) -11

8. If $(-1)^n[A^n + a_1 A^{n-1} + ... + a_n I] = 0$, then the order of A is :
   (a) n (b) $n - 1$
   (c) $n + 1$ (d) n^2

9. The eigenvalue of A is λ, then the eigenvalue of A^T is :
   (a) λ^2 (b) $\lambda - 1$
   (c) $\lambda + 1$ (d) λ

10. If λ is a characteristic root of A, then its characteristic equation is :
    (a) $A - \lambda I = 0$ (b) $|A - \lambda I| = 0$
    (c) $|A| = 0$ (d) none of these

11. If 1, 2, 3 are the eigenvalues of A, then 10, 20, 30 are the eigenvalues of the matrix :
    (a) $5A$ (b) A^2
    (c) $10A$ (d) $100A$

12. If λ is a characteristic root of A^{-1}, then the characteristic root of A is :
    (a) λ (b) $\dfrac{1}{\lambda}$
    (c) $\lambda - 1$ (d) λ^2

13. If 3, 5, 6 are the characteristic roots of a matrix A, then –2, 0, 1 are the characteristic roots of :

(a) A (b) $A + 5I$

(c) $A - 5I$ (d) $A - 6I$

14. If $A^2 - 2A + I = 0$ for a square matrix A, then the product of its characteristic roots are :

(a) 2   (b) –1

(c) –2   (d) 1

15. If a matrix is singualar, then one of its characteristic root is :

(a) 1   (b) –1

(c) 0   (d) none of these

16. The characteristic roots of a real symmetric matrix are :

(a) zero or purely imaginary

(b) zero or real

(c) zero

(d) none of these

Answers

FILL IN THE BLANKS

1. $|A - \lambda I| = 0$ **2.** eigenvalue   **3.** λX

4. λ **5.** Independent   **6.** real

7. unit modulus   **8.** 6, 1   **9.** same

10. singular   **11.** diagonal   **12.** aA

13. $\dfrac{1}{\lambda}$ **14.** $\dfrac{|A|}{\lambda}$ **15.** characteristic equation

16. 2, 3   **17.** $-\dfrac{1}{5}[A - 4I]$ **18.** 1, 4, 9.

TRUE/ FALSE

| **1.** F | **2.** T | **3.** T | **4.** F | **5.** F | **6.** T | **7.** T | **8.** T |
| **9.** F | **10.** F | **11.** T | **12.** T | **13.** F | **14.** T | **15.** F | **16.** F |

MULTIPLE CHOICE QUESTIONS

| **1.** (a) | **2.** (c) | **3.** (d) | **4.** (b) | **5.** (c) | **6.** (b) | **7.** (c) | **8.** (a) | **9.** (d) |
| **10.** (b) | **11.** (c) | **12.** (b) | **13.** (c) | **14.** (d) | **15.** (c) | **16.** (a) | | |

COMPETITION CORNER
for JRF, NET/SET, GATE Aspirants

SOME FASCINATING FACTS

- Similar matrices have the same characteristic polynomials.
- Let A be an n-square matrix. Then the characteristic polynomial is
$$\Delta(t) = t^n - S_1 t^{n-1} + S_2 t^{n-2} + \ldots + (-1)^n S_n$$
where S_k is the sum of the principal minor of order k.
- Let A be a square matrix over the complex field C, then A has at least one eigenvalue.
- The geometric multiplicity of an eigenvalue λ of a matrix A does not exceed its algebraic multiplicity.
- Let A be a real symmetric matrix then each root λ of its characteristic polynomial is real.
- The minimal polynomial $m(t)$ of matrix A divides every polynomial that has A as a zero. In particular $m(t)$ divides the characteristic polynomial $\Delta(t)$ of A.
- A scalar λ is an eigenvalue of a linear operator T if and only if λ is a root of the minimal polynomial $m(t)$ of T.
- The matrix A and its transpose A^T have the same characteristic polynomial.
- The scalar 0 is an eigenvalue of T if and only if T is singular.
- Let λ be an eigenvalue of a linear opeartor $T : V \to V$ and let E_λ consist of all eigenvectors belonging to λ, then E_λ is a subspace of V.
- Following are equivalent :
  (i) The scalar λ is an eigenvalue of A.
  (ii) The matrix $\lambda I - A$ is singular.
  (iii) The scalar λ is a root of the characteristic plynomial $\Delta(t)$ A.
- If $V_1, V_2, \ldots, V_n$ are non-zero eigenvectors of T belonging to distinct eigenvalues $\lambda_1, \lambda_2, \ldots, \lambda_n$, then $V_1, V_2, \ldots, V_n$ are linearly independent.
- If $\Delta(t) = (t - a_1)(t - a_2) \ldots (t - a_n)$ is the characteristic polynomial of an n-square matrix A and suppose that the n roots a_i are distinct. Then A is similar to the diagonal matrix $D = \text{dia}.(a_1, a_2, \ldots, a_n)$.
- If $f(t)$ and $g(t)$ be monic polynomials (leading coefficient 1 of minimal degree for which A is a root then $f(t) = g(t)$.

SOME IMPORTANT ILLUSTRATIONS

- If $A = \begin{bmatrix} 4 & 1 & -1 \\ 2 & 5 & -2 \\ 1 & 1 & 2 \end{bmatrix}$ then eigenvalues of A are 3 and 5 and the maximum set of linearly independent vectors of A is $\{(1, -1, 0), (1, 0, 1), (1, 2, 1)\}$.
- The characteristic polynomial of $A = \begin{bmatrix} 3 & -1 & 1 \\ 7 & -5 & 1 \\ 6 & -6 & 2 \end{bmatrix}$ is given by $t^3 - 12t + 16 = 0$.
- Since a linear map $L : U \to V$ may have many matrix representation it is not possible for L to have many characteristic polynomials.
- If $L : R^2 \to R^2$ be the linear opeartor which rotates each vector $V \in R^2$ by an angle $\theta = \dfrac{\pi}{2}$ then L has no eigenvalues and hence no eigenvectors.
- If λ be an eigenvalue of a linear operator S $T : V \to V$. Let E_λ be the eigenspaces of λ, *i.e.*, the set of all eigenvectors of T belonging to λ then E_λ is a subspace of V.
- The algebraic multiplicity of eigenvalue λ is defined to be the multiplicity of λ as a root of the characteristic polynomial of T. The geometric multiplicity of λ is defined by the dimension of its subspace.
- The matrix $A = \begin{bmatrix} 1 & -1 \\ 2 & -1 \end{bmatrix}$ is diagonalizable.
- The matrix $A = \begin{bmatrix} 1 & 1 \\ 0 & 1 \end{bmatrix}$ is not diagonalizable.
- The minimal polynomial of the matrix $A = \begin{bmatrix} 4 & -2 & 2 \\ 6 & -3 & 4 \\ 3 & -2 & 3 \end{bmatrix}$ is given by $t^2 - 3t + 2$.
- If $m(t)$ be the minimal polynomial of an square matrix A then the characteristic polynomial of A divides $(m(t))^n$.
- If V is a non-zero eigenvector of linear map S and T then V is an eigenvector of $S + T$.
- If A be an n-square matrix for which $A^k = 0$ for some $k > n$ then $A^n = 0$.

Self Assessment Test

1. Find the eigenvalue of $A = \begin{bmatrix} 1 & 2 & 3 \\ 0 & -4 & 2 \\ 0 & 0 & 7 \end{bmatrix}$.

2. Show that the characteristic vector corresponding to characteristic root λ of matrix A is also a characteristic vector of every matrix $f(A)$ where $f(x)$ is any scalar polynomial and the corresponding root for $f(A)$ is $f(\lambda)$.

   In general show that if $g(x) = \dfrac{f_1(x)}{f_2(x)}$ and $|f_2(A)| \neq 0$ and $g(\lambda)$ is a characteristic root of $g(A) = \dfrac{f_1(A)}{f_2(A)}$.

3. Show that any two eigenvectors corresponding to two distinct characteristic roots of a
   (i) Hermitian
   (ii) real symmetric
   (iii) unitary
   matrix are orthogonal.

4. Find the characteristic roots and characteristic vectors for the matrix $A = \begin{bmatrix} 2 & \sqrt{2} \\ \sqrt{2} & 1 \end{bmatrix}$ and show that the matrix A satisfies its characteristic equation.

5. Verify Cayley-Hamilton theorem for the matrix $A = \begin{bmatrix} 0 & 0 & 1 \\ 3 & 1 & 0 \\ -2 & 1 & 4 \end{bmatrix}$.
   Hence, evaluate A^{-1}.

6. Prove that the eigenvalues of the matrix $\begin{bmatrix} a_1 & a_2 & a_3 \\ 0 & b_2 & 0 \\ 0 & 0 & c_3 \end{bmatrix}$ are a_1, b_2, c_3.

7. Find all eigenvalues and eigenvectors of the matrix $A = \begin{bmatrix} 3 & 1 & 4 \\ 0 & 2 & 6 \\ 0 & 0 & 5 \end{bmatrix}$.

8. Find a matrix P which transform the matrix $A = \begin{bmatrix} 1 & 0 & -1 \\ 1 & 2 & 1 \\ 2 & 2 & 3 \end{bmatrix}$ to diagonal form. Hence, find A^4.

9. If the characteristic roots of $\begin{bmatrix} 3 & 7 \\ 2 & 5 \end{bmatrix}$ are λ_1 and λ_2, show that the characteristic roots of $\begin{bmatrix} 5 & -7 \\ -2 & 3 \end{bmatrix}$ are $\dfrac{1}{\lambda_1}$ and $\dfrac{1}{\lambda_2}$.

10. If A is any matrix which satisfy $A^3 - A^2 + A - I = 0$ and $A_{3 \times 3}$ then show that $A^4 = A = I$.

11. Let $A = \begin{bmatrix} 2 & 3 \\ x & y \end{bmatrix}$. If the eigenvalues of A are 4 and 8 then show that $x = -4$ and $y = 10$.

12. Let A be an $n \times n$ complex matrices whose characteristic polynomial is given by $f(t) = t^n + c_{n-1} t^{n-1} + \dots + c_1 t + c_0$ then show that $\det(A) = c_0$.

13. If the eigenvalues of a 3×3 matrix A are $1, 2$ and -3 then show that $A^{-1} = \dfrac{1}{6}\left[7I - A^2\right]$.

14. If a square matrix of order 10 has exactly 4 distinct eigenvalues, then show that the degree of its minimal polynomial is at least 4.

15. Show that the matrix $M = \begin{bmatrix} 0 & 1 & 2 & 0 \\ 1 & 0 & 1 & 0 \\ 2 & 1 & 0 & 2 \\ 0 & 0 & 2 & 0 \end{bmatrix}$ has both positive and negative real eigenvalues.

16. Show that the characteristic polynomial of $A = \begin{bmatrix} 5 & -6 & -6 \\ -1 & 4 & 2 \\ 3 & -6 & -4 \end{bmatrix}$ is
    $$\det|xI - A| = (x-2)^2 (x-1).$$

VECTOR SPACES

5.1 INTRODUCTION

We are familiar with the concept of semigroup, a group and a ring. A group was obtained from a semigroup by imposing certain restrictions on the composition of a semigroup. A ring was obtained by defining a certain composition on a group structure and by giving rules that connected the group composition with the new composition. We shall now discuss another algebraic structure called a vector space or linear space which is going to involve a group structure, a ring structure and an operation connecting the elements of these two structures.

In order to discuss a vector space we need two basic things. One of them is the set of vectors and the other is the set of scalars. Therefore, to define a vector space we need a field F. The elements of field F are called the scalars. In addition, we need two binary operations. One of them is internal composition and the other is external composition. Now we distinguish the internal and external compositions as follows:

5.1.1 INTERNAL COMPOSITION

Let R be any set. If $a * b \in R$ for all $a,b \in R$ and $a*b$ is a unique, then '*' is known as the internal composition. That is, a binary operation defined over the vectors is called internal composition (vector addition).

5.1.2 EXTERNAL COMPOSITION

Let V be the set of vectors and F be a field. Then a binary operation defined between the vectors and scalars is called external composition. That is, if $a \, o \, \alpha \in V$ for all $\alpha \in V$ and $a \in F$ and $a \, o \, \alpha$ is unique, then o is called an external composition or scalar multiplication.

5.2 PROLOGUE TO VECTOR SPACE

We have already settled that a non-empty set with a binary composition is a groupoid, an associative groupoid is a semigroup; a semigroup with an identity is a monoid, a monoid in which each element has an inverse is known as a group. Each link in this chain imposes one additional condition on the binary composition defined on the set. Let us put down all these conditions together and look at the *group structure* a fresh.

"Let G be a non-empty set and * be a binary operation defined on it, then the structure $(G, *)$ is said to be a group if the following axioms are satisfied.

(i) Closure property. $a * b \in G \; \forall \, a, b \in G$

(ii) Associativity. The operation * is associative on G, i.e.,

$$a*(b*c) = (a*b)*c \; \forall \, a, b, c \in G$$

(iii) **Existence of identity.** There exists an element $e \in G$, such that

$$a*e = e*a = a \ \forall \ a \in G$$

e is called identity of *in G.

(iv) **Existence of inverse.** For each element $a \in G$, there exists an element $b \in G$ such that

$$a * b = b*a = e$$

The element b is called the inverse of element a with respect to * and we write

$$b = a^{-1}$$

REMARKS

- When we say * is a binary operation defined on a non-empty set G, it implies that G is closed for the binary operation * , *i.e.*,

$$a \in G, b \in G \Rightarrow a * b \in G \ \forall \ a, b \in G$$

- A group is not simply a set, but it is an *algebraic structure*.
- Because of the associativity, the paranthesis can be dropped in products of more than two elements of a group and instead of writing $a*(b*c)$ or $(a*b)*c$ we may simply write $a*b*c$. The associative law can be extended to any finite number of elements.

5.2.1 ABELIAN OR COMMUTATIVE GROUP

A group $(G, *)$ is said to be *abelian* or *commutative* if $a*b = b*a \ \forall \ a, b \in G$. The group which are not abelian, called non-abelian or non-commutative.

REMARKS

- An abelian group under addition is sometimes called a '*module*'.
- The commutative group is also known as *Abelian group* after the name of famous mathematician *Abel*.
- The smallest group for a given composition is the set $\{e\}$, containing identity element.
- A group consisting of the identity element only, is called a trivial group, others are called non-trivial groups.

5.2.2 FINITE AND INFINITE GROUP

If a group contains a finite number of elements, called a finite group.

If the number of elements in a group is infinite, then it is called an *infinite group*.

5.2.3 ORDER OF A GROUP

The number of elements in a finite group is called the order of the group. It is denoted by o (G).

An infinite group is called a group of infinite order.

☞ ILLUSTRATIONS

(1) The set Z of integers is an infinite abelian group with respect to the operation of addition but Z is not a group with respect to the multiplication.

(2) Let $G = [1]$, then G is an abelian group of order 1 with respect to multiplication.

5.2.4 FIELD

Let F be any non-empty set equipped with two binary operations addition $(+)$ and multiplication $(.)$, *i.e.*, for all $a, b \in F$, $a+b \in F$ and $a . b \in F$. Then the algebraic structure $(F, +, .)$ is said to be field if it satisfies the following conditions :

(i) Addition is associative, *i.e.*, $(a+b)+c = a+(b+c) \ \forall \ a, b, c \in F$.

(ii) Addition is commutative, *i.e.*, $a + b = b + a \;\; \forall \, a, \, b \in F$

(iii) $\exists$ an identity element 0 in F such that $a + 0 = 0 + a = a \;\; \forall \, a \in F$.

(iv) To each element $a \in F \; \exists \, -a \in F$ such that $a + (-a) = 0$

(v) Multiplication is commutative, *i.e.*, $a.b = b.a \;\; \forall a, \, b \in F$.

(vi) Multiplication is associative, *i.e.*, $(a. \, b).c = a. \, (b. \, c) \;\; \forall a, b, \; c \in F$.

(vii) There exists a non-zero element 1 (one) in F such that $a.1 = a \;\; \forall a \in F$

(viii) To every non-zero element $a \in F$ there corresponds an element a^{-1} in F such that $a. \, a^{-1} = 1$

(ix) Multiplication is distributive over addition *i.e.*,
$$a.(b+c) = a.b + a.c \;\; \forall a, b, c \in F$$

5.2.5 SUBFIELD

Let F be a field. A non-empty subset K of F is said to be subfield of F if K is closed *w.r.t.* addition and multiplication in F and K itself is a field.

☛ **ILLUSTRATIONS**

(1) The set Q *of* rational numbers is a field.

(2) The set R of real numbers is a field.

(3) Q is a subfield of R.

(4) C is a field.

(5) R is a subfield of C.

5.3 VECTOR SPACES

Let V be a non-empty set of vectors and F be a field. Then an algebraic structure $(V, +, .)$ together with two binary operations vectors addition and scalar multiplication is said to be vector space over F if this structure satisfies the following conditions.

(*i*) $(V, +)$ is an abelian group.

(*ii*) $a(\alpha + \beta) = a\alpha + a\beta, \; \forall \, \alpha, \, \beta \in V \textit{ and } \forall \, a \in F$

(*iii*) $(a + b)\alpha = a\alpha + b\alpha, \; \forall \, \alpha \in V \textit{ and } \forall \, a, b \in F$

(*iv*) $(ab)\alpha = a(b\alpha), \; \forall \, \alpha \in V \textit{ and } \forall \, a, b \in F$

(*v*) $1\alpha = \alpha, \; \forall \, \alpha \in V \textit{ and } 1 \in F$

This vector space V over F is denoted by $V(F)$.

For example, if $F = R$, the field of real numbers, then $V(R)$ is a vector space and it is called a real vector space.

☛ **ILLUSTRATIONS**

(1) Let $R^2 = \{(a_1, a_2) : a_1 \in R, \, a_2 \in R\}$. The set R^2 is a vector space over R with addition and scalar multiplication defined as follows :
$$(a_1, a_2) + (b_1, b_2) = (a_1 + b_1, \, a_2 + b_2)$$
$$c(a_1, a_2) = (ca_1, ca_2), \; \forall \, a_1, a_2, b_1, b_2, c \in R$$

(2) Vector in 3-dimensional space form a vector space over R with respect to addition and scalar multiplication of vectors.

(3) Let R^n be the set of n-tuples of real numbers, *i.e.*,
$$R^n = \{(a_1, a_2, ..., a_n) : a_i \in R$$

Then R^n is a vector space over R with pointwise addition and scalar multiplication as defined in (1).

(4) Let C^n be the set of all ordered n-tuples of complex numbers. Then C^n is a vector space over C with addition and scalar multiplication.

▉ Solved Examples ▉

Example 1. *Show that a field K can be regarded as a vector space over any subfield F of K.*

Solution. Since $F \subset K$, then we have to prove that K is a vector space over F. Now K is a set of vectors and the elements of F are the scalars. Addition of vectors is the addition composition in the field K so that $(K, +)$ forms an abelian group. Also the composition of scalar multiplication is the multiplication composition in the field k, then

$$a\alpha \in K \; \forall \; a \in F \text{ and } \alpha \in K$$

If 1 is the unity element of K, then 1 is also the unity element of subfield F. Now we shall make the following observations :

(i) $a(\alpha+\beta) = a\alpha + a\beta \; \forall \; a \in F \text{ and } \alpha, \beta \in K$ [By left distributive law in K]

(ii) $(a + b)\alpha = a\alpha + b\alpha \; \forall \; a, b \in F \text{ and } \forall \; \alpha \in K$ [By right distributive law in K]

(iii) $(ab)\alpha = a(b\alpha) \; \forall \; a, b \in F \text{ and } \forall \; \alpha \in K$ [By associativity of multiplication in K]

(iv) $1\alpha = \alpha \; \forall \; \alpha \in K, 1 \in F$. Since 1 is also the unity element of the field K, so that $1\alpha = \alpha \; \forall \; \alpha \in K$.

Hence K is a vector space over the field F, which is denoted by $K(F)$.

REMARKS

- As the field F is a subfield of itself so that F (F) is a vector space, therefore, in particular R(R) and C (C) are vector spaces with respect to usual addition and multiplication.

- Since R is a subfield of C so that C(R) is a vector space. But R(C) is not a vector space because R is not closed with respect to scalar multiplication, for example.

$$1+i \in C \text{ and } 3 \in R$$

$$3(1+i) = 3+3i \notin R$$

- Similarly, R(Q) is a vector space but Q(R) is not a vector space.

Example 2. *Show that the set of all ordered n-tuples forms a vector space over a field F.*

Solution. If $a_1, a_2, a_3,..., a_n$ are n elements of a field F, then an ordered set $\alpha=(a_1, a_2,.., a_n)$ is called an n-tuple over F.

Let V be the set of all ordered n-tuples, then

$$V = \{(a_1, a_2,..., a_n) : a_1, a_2,...a_n \in F\}$$

Now we define equality of two n-tuples, addition of two n-tuples and multiplication of an n-tuple with a scalar as follows:

Equality of two n-tuples:

Let $(a_1, a_2, ..., a_n)$ and $(b_1, b_2, ..., b_n)$ be two n-tuple of V. Then they are said to be equal if and only if $a_i = b_i$ for $i=1, 2, 3, ..., n$.

Addition of two n-tuples:

Let $(a_1, a_2, ..., a_n)$ and $(b_1, b_2, ..., b_n)$ be two n-tuple of V, then

$$(a_1, a_2,..., a_n) + (b_1, b_2,..., b_n) = (a_1 + b_1, a_2 + b_2,..., a_n + b_n)$$

Scalar multiplication of an n-tuple : Let $(a_1, a_2, ..., a_n)$ be an n-tuple and $a \in F$, then

$$a(a_1, a_2, ..., a_n) = (aa_1, aa_2,... aa_n).$$

Now we shall show that V is a vector space with respect to addition composition and scalar multiplication.

(i) Closure property: For all $\alpha = (a_1, a_2, ..., a_n) \in V$ and $\beta = (b_1, b_2, ..., b_n) \in V$

$$\alpha + \beta = (a_1 + b_1, \ a_2 + b_2, ..., a_n + b_n)$$

Since $a_1 + b_1, a_2 + b_2, ..., a_n + b_n$ are all elements of F so that

$$\alpha + \beta \in V \ \forall \alpha, \beta \in V$$

Hence V is closed for addition of n-tuples.

(ii) Associativity of addition in V: For all $\alpha = (a_1, a_2, ..., a_n)$, $\beta = (b_1, b_2, ..., b_n)$ and $\gamma = (c_1, c_2, ..., c_n)$ of V

$$\alpha + (\beta + \gamma) = (a_1, a_2, ..., a_n) + [(b_1, b_2, ..., b_n) + (c_1, c_2, ..., c_n)]$$

$$= (a_1, a_2, ..., a_n) + [(b_1 + c_1, b_2 + c_2, ..., b_n + c_n)$$

$$= (a_1 + (b_1 + c_1), a_2 + (b_2 + c_2), ..., a_n + (b_n + c_n))$$

$$= ((a_1 + b_1) + c_1, (a_2 + b_2) + c_2, ..., (a_n + b_n) + c_n)$$

$$[\because F \text{ is associative for addition.}]$$

$$= ((a_1 + b_1), (a_2 + b_2), ..., (a_n + b_n)) + (c_1, c_2, ..., c_n)$$

$$= [(a_1, a_2, ..., a_n) + (b_1, b_2, ..., b_n)] + (c_1, c_2, ..., c_n)$$

$$= (\alpha + \beta) + \gamma$$

(iii) Existence of additive identity in V. Since $0 \in F$ so that $0 = (0, 0, ..., 0) \in V$. Also if $\alpha = (a_1, a_2, ..., a_n)$ is any element of V, then

$$\alpha + 0 = (a_1, a_2, ..., a_n) + (0, 0, ..., 0)$$

$$= (a_1 + 0, a_2 + 0, ..., a_n + 0)$$

$$= (a_1, a_2, ..., a_n) = \alpha$$

$\therefore \qquad\qquad \alpha + 0 = \alpha \ \forall \ \alpha \in V.$

Similarly $\qquad 0 + \alpha = \alpha \forall \ \alpha \in V$

Hence $\qquad\qquad 0 = (0, 0, ..., 0)$ is the additive identity in V.

(iv) Existence of additive inverse in V. If $\alpha = (a_1, a_2, ..., a_n)$ is any element of V, then $-\alpha = (-a_1, -a_2, ..., -a_n) \in V$ because $-a_1, -a_2, ..., -a_n \in F$.

Also, we have
$$-\alpha + \alpha = (-a_1, -a_2, ..., -a_n) + (a_1, a_2, ... a_n)$$

$$= (-a_1 + a_1, -a_2 + a_2, ..., -a_n + a_n)$$

$$= (0, 0, ..., 0) = 0$$

$$-\alpha + \alpha = 0 \forall \alpha \in V$$

Similarly, $\quad \alpha + (-\alpha) = 0 \ \forall \ \alpha \in V$

Hence $(-a_1, -a_2, ..., -a_n)$ is the additive inverse of $(a_1, a_2, ..., a_n)$.

(v) Commutativity of addition in V. For all $\alpha = (a_1, a_2, ..., a_n)$ and $\beta = (b_1, b_2, ..., b_n) \in V$, we have

$$\alpha + \beta = (a_1, a_2, ..., a_n) + (b_1, b_2, ..., b_n)$$

$$= (a_1 + b_1, a_2 + b_2, ..., a_n + b_n)$$

$$= (b_1 + a_1, b_2 + a_2, ..., b_n + a_n)$$

$$= (b_1, b_2, ..., b_n) + (a_1 + a_2, ..., a_n)$$

$$= \beta + \alpha$$

$$\therefore \qquad \alpha + \beta = \beta + \alpha \ \forall \, \alpha, \, \beta \in V$$

Hence, V is an abelian group under addition of n-tuples.

Now we observe that

(i) For all $\alpha = (a_1, a_2, ..., a_n)$ and $\beta = (b_1, b_2, ..., b_n)$ and $a \in F$

$$a \, (\alpha + \beta) = a(a_1 + b_1, a_2 + b_2, ..., a_n + b_n)$$

$$= (a(a_1 + b_1), a(a_2 + b_2), ..., a(a_n + b_n))$$

$$= (aa_1 + ab_1, aa_2 + ab_2, ..., aa_n + ab_n)$$

$$= (aa_1, aa_2, ..., aa_n) + (ab_1, ab_2, ..., ab_n)$$

$$= a(a_1, a_2, ..., a_n) + a(b_1, b_2, ..., b_n)$$

$$= a\alpha + a\beta$$

(ii) For all $a, b \in F$ and $\alpha = (a_1, a_2, ..., a_n) \in V$

$$(a + b) \, \alpha = (a + b) \, (a_1, a_2, ..., a_n)$$

$$= ((a + b) \, a_1, (a + b) \, a_2, ..., (a + b) \, a_n)$$

$$= (aa_1 + ba_1, aa_2 + ba_2, ..., aa_n + ba_n)$$

$$= (aa_1, aa_2, ..., aa_n) + (ba_1, ba_2, ..., ba_n)$$

$$= a(a_1, a_2, ..., a_n) + b(a_1, a_2, ..., a_n)$$

$$= a\alpha + b\beta$$

(iii) For all $a, b \in F$ and $\alpha = (a_1, a_2, ..., a_n) \in V$

$$(ab) \, \alpha = (ab)(a_1, a_2, ..., a_n)$$

$$= ((ab) \, a_1, (ab) \, a_2, ..., (ab) \, a_n)$$

$$= (a \, (ba_1), a(ba_2), ..., a(ba_n))$$

$$= a(ba_1, ba_2, ..., ba_n)$$

$$= a[b(a_1, a_2, ..., ba_n)]$$

$$= a \, (b\alpha)$$

(iv) If 1 is the unity element of F and $\alpha = (a_1, a_2, ..., a_n) \in V$ then

$$1\alpha = 1(a_1, a_2, ..., a_1) = (1a_1, 1a_2, ..., 1a_n) = (a_1, a_2, ..., a_n) = \alpha.$$

Hence, V is a vector space over the field F. This vector space is denoted by $V_n \, (F)$. Sometimes, it is also denoted by F^n or $F^{(n)}$ or $F^n \, (F)$.

REMARK

- V_2 (F) = $\{(a_1, a_2) : a_1, a_2, \in F\}$ is a vector space of all ordered pairs over F. Similarly V_3 (F) = $\{(a_1, a_2, a_3) : a_1, a_2, a_3 \in F\}$ forms a vector space of all ordered triads over F.

Example 3. *Show that the set of all $m \times n$ matrices with their elements as real numbers is a vector space over the field F of real numbers with respect to addition of matrices as addition of vectors and multiplication of a matrix by a scalar as scalar multiplication.*

Solution. Let M_{mn} = $\{A, B, C, ...\}$ be the set of all $m \times n$ matrices. We shall show that M_{mn} (F) will form abelian group under addition.

(i) **Closure property.** For all $A, B \in M_{mn}$, we have $A + B \in M_{mn}$. Hence M_{mn} is closed under addition of matrices.

(ii) **Associativity.** For all $A, B, C \in M_{mn}$, we have
$$A + (B + C) = (A + B) + C$$

(iii) **Existence of identity.** If O be the null matrix of order $m \times n$, then $O \in M_{mn}$. Also for all $A \in M_{mn}$, we have
$$A + O = A = O + A$$
Hence, O is additive identity in M_{mn}.

(iv) **Existence of inverse.** If $A \in M_{mn}$, then $-A \in M_{mn}$. Also for all $A \in M_{mn}$, we have
$$(-A) + A = O = A + (-A)$$
Hence, $-A$ is the additive inverse of A.

(v) **Commutativity.** For all $A, B \in M_{mn}$, we have
$$A + B = B + A$$
Hence, M_{mn} is an abelian group under addition.

If $a \in F$ and A = $[a_{ij}]_{m \times n} \in M_{mn}$ then
$$aA = a[a_{ij}]_{m \times n} = [aa_{ij}]_{m \times n} \in M_{mn}$$
Now we observe that:

(i) For all $A = [a_{ij}]_{m \times n}, B = [b_{ij}]_{m \times n}$ in M_{mn} and $a \in F$

Then, $a(A + B) = a([a_{ij}]_{m \times n} + [b_{ij}]_{m \times n})$ $= a([a_{ij} + b_{ij}]_{m \times n})$

$\qquad = [a(a_{ij} + b_{ij})]_{m \times n}$ $= [(aa_{ij} + ab_{ij})]_{m \times n}$

$\qquad = [aa_{ij}]_{m \times n} + [ab_{ij}]_{m \times n}$ $= a[a_{ij}]_{m \times n} + a[b_{ij}]_{m \times n}$

$\qquad = aA + aB$

(ii) For all $a, b \in F$ and $A = [a_{ij}]_{m \times n}$

$\qquad (a + b)A = (a + b)[a_{ij}]m \times n$ $= [(a + b)a_{ij}]_{m \times n}$

$\qquad = [(aa_{ij} + ba_{ij})]_{m \times n}$ $= [aa_{ij}]_{m \times n} + [ba_{ij}]_{m \times n}$

$\qquad = a[a_{ij}]_{m \times n} + b[a_{ij}]_{m \times n}$

$\qquad = aA + bA$

(iii) For all $a, b \in F$ and $A = [a_{ij}]_{m \times n} \in M_{mn}$

$$(ab)A = [ab][a_{ij}]_{m \times n} = [(ab)\, a_{ij}]_{m \times n}$$

$$= [a(ba_{ij})]_{m \times n} = a[ba_{ij}]_{m \times n}$$

$$= a(b[a_{ij}]_{m \times n}) = a(bA)$$

(iv) Since $1 \in F$ and $A = [a_{ij}]_{m \times n} \in M_{mn}$, then

$$1A = 1[a_{ij}]_{m \times n} = [1a_{ij}]_{m \times n} = [a_{ij}]_{m \times n} = A$$

Hence, M_{mn} is a vector space over F.

REMARK

- If M_{mn} is a set of all $m \times n$ matrices with their elements as rational numbers and F is the field of real numbers, then $M_{mn}(F)$ will not form a vector space because $\sqrt{2} \in F$ and if $A \in M_{mn}(F)$, then $\sqrt{2} A \notin M_{mn}(F)$.

Example 4. *Show that the set of all polynomials over a field F is a vector space.*

Solution. If $(F, +, .)$ be the field, then $F[x]$ denotes the set of all polynomials in the indeterminate x with coefficient from F.

Now we define the addition of polynomials and scalar multiplication of polynomial as follows :

Let $\quad f(x) = a_0 + a_1 x + a_2 x^2 + \dots$

$\qquad g(x) = b_0 + b_1 x + b_2 x^2 + \dots$

be any two polynomials of $F(x)$, where $a_0, a_1, a_2, \dots \in F$ and $b_0, b_1, b_2, \dots \in F'$. *Addition of polynomial in F[x].*

For $f(x)$ and $g(x) \in F'[x]$, we have

$$f(x) + g(x) = (a_0 + a_1 x + a_2 x^2 \dots) + (b_0 + b_1 x + b_2 x^2 + \dots)$$

$$= (a_0 + b_0) + (a_1 + b_1)x + (a_2 + b_2)x^2 + \dots$$

Scalar multiplication in F[x] by an element of F.

If $\quad f(x) = a_0 + a_1 x + a_2 x^2 + \dots \in F[x]$ and $h \in F$ then

$$kf(x) = k(a_0 + a_1 x + a_2 x^2 + \dots)$$

$$= (ka_0) + (ka_1)x + (ka_2)x^2 + \dots$$

Now we shall prove that $F[x]$ is a vector space over F for the above compositions:

(i) Closure property. For all $f(x) = a_0 + a_1 x + a_2 x^2 + \dots \in F[x]$ and

$$g(x) = b_0 + b_1 x + b_2 x^2 + \dots \in F[x], \text{ then we have}$$

$$f(x) + g(x) = (a_0 + b_0) + (a_1 + b_1)x + (a_2 + b_2)x^2 + \dots$$

Since $a_0 + b_0, a_1 + b_1, a_2 + b_2, \dots \in F$, therefore $f(x) + g(x) \in F[x]$.

(ii) Associativity. For all $f(x), g(x), h(x) \in F[x]$ where

$$f(x) = a_0 + a_1 x + a_2 x^2 + \dots$$

$$g(x) = b_0 + b_1 x + b_2 x^2 + \dots$$

$$h(x) = c_0 + c_1 x + c_2 x^2 + \dots$$

we have

$$f(x) + [g(x) + h(x)] = (a_0 + a_1 x + a_2 x^2 + ...)$$

$$+ [(b_0 + c_0) + (b_1 + c_1)x + (b_2 + c_2)x^2 + ...]$$

$$= [a_0 + (b_0 + c_0)] + [a_1 + (b_1 + c_1)]x + [a_2 + (b_2 + c_2)]x^2 + ...$$

$$= [(a_0 + b_0) + c_0] + [(a_1 + b_1) + c_1)]x + [(a_2 + b_2) + c_2)]x^2 + ...$$

$$= [(a_0 + b_0) + (a_1 + b_1)x + (a_2 + b_2)x^2 ...] + [c_0 + c_1 x + c_2 x^2 + ...]$$

$$= [f(x) + g(x)] + h(x)$$

(iii) **Existence of additive identity.**

If $0(x)$ denotes the zero polynomial over F, then

$$0(x) = 0 + 0x + 0x^2 + ...$$

$$\therefore \qquad 0(x) \in F[x]$$

Also, for all $f(x) = a_0 + a_1 x + a_2 x^2 + ...$, we have

$$0(x) + f(x) = (0 + 0x + 0x^2 + ...) + (a_0 + a_1 x + a_2 x^2 + ...)$$

$$= (0 + a_0) + (0 + a_1)x + (0 + a_2)x^2 + ...$$

$$= a_0 + a_1 x + a_2 x^2 + ...$$

$$= f(x)$$

Similarly, $f(x) + 0(x) = f(x)$

Hence $0(x)$ is the additive identity in $F[x]$.

(iv) **Existence of additive inverse.**

If $\quad f(x) = a_0 + a_1 x + a_2 x^2 + ... \in F[x]$, then define

$$-f(x) = -a_0 + (-a_1)x + (-a_2)x^2 + ...$$

Since $-a_0, -a_1, -a_2, ... \in F$, therefore $-f(x) \in F[x]$

Also, we have

$$-f(x) + f(x) = (-a_0 + (-a_1)x + (-a_2)x^2 + ...) + (a_0 + a_1 x + a_2 x^2 + ...)$$

$$= (-a_0 + a_0) + (-a_1 + a_1)x + (-a_2 + a_2)x^2 + ...$$

$$= 0 + 0x + 0x^2 + ...$$

$$= 0(x)$$

Similarly, $f(x) + f(-f(x)) = 0(x)$

Hence, $-f(x)$ is an additive inverse of $f(x)$.

(v) **Commutativity under addition.**

For all $\quad f(x) = a_0 + a_1 x + a_2 x^2 + ...$ and $g(x) = b_0 + b_1 x + b_2 x^2 + ...$ of $F[x]$,

we have

$$f(x) + g(x) = (a_0 + b_0) + (a_1 + b_1)x + (a_2 + b_2)x^2 + ...$$

$$= (b_0 + a_0) + (b_1 + a_1)x + (b_2 + a_2)x^2 + ...$$

$$= g(x) + f(x)$$

Hence $F(x)$ is an abelian group with respect to addition of polynomials. Further, we observe that

(i) If $k \in F$ and $f(x) = a_0 + a_1 x + a_2 x^2 + ...$ and $g(x) = b_0 + b_1 x + b_2 x^2 + ...$ are in $F(x)$, then

$$k[f(x) + g(x)] = k[(a_0 + b_0) + (a_1 + b_1)x + (a_2 + b_2)x^2 + ...]$$

$$= k(a_0 + b_0) + k(a_1 + b_1)x + k(a_2 + b_2)x^2 + ...$$

$$= (ka_0 + kb_0) + (ka_1 + kb_1)x + (ka_2 + kb_2)x^2 + ...$$

$$= [(ka_0) + (ka_1)x + (ka_2)x^2 + ...]$$

$$+ [(kb_0) + (kb_1)x + (kb_2)x^2 + ...]$$

$$= k[a_0 + a_1 x + a_2 x^2 + ...] + k[b_0 + b_1 x + b_2 x^2 + ...]$$

$$= k f(x) + k g(x)$$

(ii) If $k_1, k_2 \in F$ and $f(x) = a_0 + a_1 x + a_2 x^2 + ... \in F[x]$, then

$$(k_1 + k_2)f(x) = (k_1 + k_2)[a_0 + a_1 x + a_2 x^2 + ...]$$

$$= (k_1 + k_2)a_0 + (k_1 + k_2)a_1 x + (k_1 + k_2)a_2 x^2 + ...$$

$$= (k_1 a_0 + k_2 a_0) + (k_1 a_1 + k_2 a_1)x + (k_1 a_2 + k_2 a_2)x^2 + ...$$

$$= [(k_1 a_0) + (k_1 a_1)x + (k_1 a_2)x^2 + ...]$$

$$+ [(k_2 a_0) + (k_2 a_1)x + (k_2 a_2)x^2 + ...]$$

$$= k_1(a_0 + a_1 x + a_2 x^2 + ...) + k_2(a_0 + a_1 x + a_2 x^2 + ...)$$

$$= k_1 f(x) + k_2 f(x)$$

(iii) If 1 is the unity element in F and $f(x) \in F(x)$ where $f(x) = a_0 + a_1 x + a_2 x^2 + ...$, then

$$1. f(x) = (1a_0) + (1a_1)x + (1a_2)x^2 + ...$$

$$= a_0 + a_1 x + a_2 x^2 + ...$$

$$= f(x)$$

Hence, $F[x]$ is a vector space over F with respect to the addition and scalar multiplication of polynomials.

Example 5. *Show that the set of all convergent sequences is a vector space over the field of real numbers.*

Solution. Let V denote the set of all convergent sequences of real numbers.

Let $\qquad \alpha = (\alpha_1, \alpha_2, ..., \alpha_n, ...) = (\alpha_n)$

$$\beta = (\beta_1, \beta_2, ..., \beta_n, ...) = (\beta_n)$$
$$\gamma = (\gamma_1, \gamma_2, ..., \gamma_n, ...) = (\gamma_n) \text{ , etc.}$$

any members of V.

Now we shall show that $(V, +)$ is an abelian group.

(i) Closure property. For all convergent sequences $\alpha = (\alpha_n)$ and $\beta = (\beta_n)$ in V, we have

$$\alpha + \beta = \langle \alpha_n \rangle + \langle \beta_n \rangle = \langle \alpha_n + \beta_n \rangle_{\text{.}}$$

$\therefore$ $\quad$ $\alpha + \beta$ is also convergent.

Hence $\quad$ $\alpha + \beta \in V \; \forall \alpha, \beta \in V$.

(ii) Associativity. For all convergent sequences $\alpha = \langle \alpha_n \rangle, \beta = \langle \beta_n \rangle$ and $\gamma = \langle \gamma_n \rangle$ in V, we have

$$\alpha + (\beta + \gamma) = \langle \alpha_n \rangle + \langle \langle \beta_n \rangle + \langle \gamma_n \rangle \rangle \;\; = \langle \alpha_n \rangle + \langle \langle \beta_n + \gamma_n \rangle \rangle$$

$$= \langle \alpha_n + (\beta_n + \gamma_n) \rangle \;\; = \langle (\alpha_n + \beta_n) + \gamma_n \rangle$$

$$= \langle (\alpha_n + \beta_n) \rangle + \langle \gamma_n \rangle \;\; = (\langle \alpha_n \rangle + \langle \beta_n \rangle) + \langle \gamma_n \rangle$$

$$= (\alpha + \beta) + \gamma$$

(iii) Existence of identity in V. Since $\langle 0 \rangle = \langle 0, 0, 0, ..., 0, ... \rangle$ is a convergent sequence so that $\langle 0 \rangle \in V$.

$$\text{If } \alpha = \langle \alpha_n \rangle \in V, \text{ then } \langle 0 \rangle + \alpha = \langle 0 \rangle + \langle \alpha_n \rangle$$
$$= \langle 0 + \alpha_n \rangle = \langle \alpha_n \rangle = \alpha$$

Similarly, $\quad \alpha + \langle 0 \rangle = \alpha$

Hence $\langle 0 \rangle$ is an identity in V.

(iv) Existence of inverse in V. If $\alpha = \langle \alpha_n \rangle \in V$, then $-\alpha = \langle -\alpha_n \rangle \in V$.

Also, $-\alpha + \alpha = \langle -\alpha_n \rangle + \langle \alpha_n \rangle$
$$= \langle -\alpha_n + \alpha_n \rangle = \langle 0 \rangle$$

Similarly, $\alpha + (-\alpha) = \langle 0 \rangle$

Hence, $-\alpha$ is an additive inverse of α.

(v) Commutativity in V. For all $\alpha = \langle \alpha_n \rangle$ and $\beta = \langle \beta_n \rangle$ of V, we have

$$\alpha + \beta = \langle \alpha_n \rangle + \langle \beta_n \rangle \;\; = \langle \alpha_n + \beta_n \rangle$$

$$= \langle \beta_n + \alpha_n \rangle \;\; = \langle \beta_n \rangle + \langle \alpha_n \rangle$$

$$= \beta + \alpha$$

Hence, $(V, +)$ is an abelian group.

Next, if $a \in R$ (set of all real numbers) and $\alpha = \langle \alpha_n \rangle \in V$, then $a\alpha = a \langle \alpha_n \rangle = \langle a\alpha_n \rangle$ which is also convergent.

$\therefore \quad a\alpha \in V \ \forall a \in R$ and $\alpha \in V.$

Now we observe that

(i) If $\quad \alpha = \langle \alpha_n \rangle, \beta = \langle \beta_n \rangle \in V$ and $a \in R$, then

$$a(\alpha + \beta) = a(\langle \alpha_n \rangle + \langle \beta_n \rangle) = a\langle (\alpha_n + \beta_n) \rangle$$

$$= \langle a(\alpha_n + \beta_n) \rangle = \langle a\alpha_n \rangle + \langle a\beta_n \rangle = a\langle \alpha_n \rangle + a\langle \beta_n \rangle$$

$$= a\alpha + a\beta$$

(ii) If $a, b \in R$ and $\alpha = \langle \alpha_n \rangle \in V$, then

$$(a + b)\alpha = (a + b)\langle \alpha_n \rangle = \langle (a + b)\alpha_n \rangle$$

$$= \langle (a\alpha_n + b\alpha_n) \rangle = \langle a\alpha_n \rangle + \langle b\alpha_n \rangle$$

$$= a\langle \alpha_n \rangle + b\langle \alpha_n \rangle$$

$$= a\alpha + b\alpha$$

(iii) If $a, b \in R$ and $\alpha = \langle \alpha_n \rangle \in V$, then

$$(ab)\alpha = (ab)\langle \alpha_n \rangle$$

$$= \langle (ab)\alpha_n \rangle = \langle a(b\alpha_n) \rangle = a\langle b\alpha_n \rangle$$

$$= a(b\langle \alpha_n \rangle) = a(b\alpha)$$

(iv) Since 1 is the unity of element in R and $\alpha = \langle \alpha_n \rangle \in V$, then

$$1\alpha = 1\langle \alpha_n \rangle = \langle 1\alpha_n \rangle = \langle \alpha_n \rangle = \alpha$$

Hence V is a vector space over the field of real numbers.

Example 6. *Let V be the set of all pairs (x, y) of real numbers, and let F be the field of real numbers. Define*

$$(x, y) + (x_1, y_1) = (x + x_1, 0)$$

and $\qquad c(x, y) = (cx, 0)$

Is V, with these operations, a vector space over the field of real number?

Solution. Clearly, for the operation of addition, the identity element does not exist. For, suppose that the ordered pair $(x_1, y_1) \in V$ is the identity element for addition. Then we must have

$$(x, y) + (x_1, y_1) = (x, y) = (x_1, y_1) + (x, y) \forall x, y \in R \qquad \qquad \dots (1)$$

But by definition of addition, we have

$$(x, y) + (x_1, y_1) = (x, y) = (x + x_1, 0)$$

$\Rightarrow \qquad \qquad (x, y) = (x + x_1, 0) \qquad \qquad$ [Using (1)]

If $y \neq 0$, then we cannot have

$$(x, y) = (x + x_1, 0)$$

Therefore, there does not exist an element $(x_1, y_1) \in V$ such that

$$(x, y) + (x_1, y_1) = (x, y) \forall (x, y) \in V$$

Hence, the additive identity in V does not exist and hence V is not a vector space over the field of real numbers.

Example 7. *Let V be the set of all pairs (x, y) of real numbers and let F be the field of real numbers. Define*

$$(x, y) + (x_1, y_1) = (3y + 3y_1, -x - x_1)$$
$$c(x, y) = (3cy, -cx)$$

Verify that V, with these operations, is not a vector space over the field of real numbers.

Solution. Clearly, V is closed under addition, but it is not associative under addition.

As, $(x, y) + [(x_1, y_1) + (x_2, y_2)]$

$$= (x, y) + [(3y_1 + 3y_2, -x_1 - x_2)]$$

$$= [3y + 9y_1 + 9y_2, -x - (-x_1 - x_2)]$$

$$= (3y + 9y_1 + 9y_2, -x + x_1 + x_2)$$

and $[(x, y) + (x_1, y_1)] + (x_2, y_2)$

$$= [(3y + 3y_1, -x - x_1)] + (x_2, y_2)$$

$$= (9y + 9y_1 + 3y_2, -x - x_1 - x_2)$$

so that $(x, y) + [(x_1, y_1) + (x_2, y_2)] \neq [(x, y) + (x_1, y_l)] + (x_2, y_2)$

Therefore, V is not associative under addition.

Hence, V is not a vector space.

Example 8. *Show that the set of all real valued continuous functions defined on $[0, 1]$ is a vector space over field of reals.*

Solution. Let V be the set of all real valued continuous functions defined on $[0, 1]$. Now we have to show that V is a vector space over R (field of real numbers) under vector addition and scalar multiplication which is defined as follows:

$$(f + g)(x) = f(x) + g(x), \forall f, g \in V$$

and $(af)(x) = af(x), \forall f \in V$ and $a \in R$

First we show that $(V, +)$ is an abelian group.

Let $f, g \in V$, then

$$(f + g)(x) = f(x) + g(x), \forall f, g \in V$$

$\therefore f + g \in V.$ Thus V is closed under vector addition.

Now let $0(x) \in V$, Then we have

$$f(x) + 0(x) = (f + 0)x = f(x), \forall f \in V$$

$\therefore 0(x)$ is the additive identity in V

Let $-f \in V$, then we have

$$-f(x) + f(x) = (-f + f)x = 0(x), \forall f \in V$$

$\therefore -f$ is an additive inverse of f in V.

Since vector addition is always associative as well as commutative, consequently $(V, +)$ is an abelian group.

Further, since V is closed under scalar multiplication therefore, af is a real valued continuous function defined on $[0, 1]$.

(i) If $a \in R$ and $f, g \in V$, then we have

$$a[(f+g) x] = a[f(x) + g(x)]$$

$$= af(x) + ag(x) = (af + ag)(x)$$

$$\therefore \qquad a(f + g) = af + ag$$

(ii) If $a, b \in R$ and $f \in V$, then we have

$$[(a + b)f](x) = (a + b)f(x) = af(x) + bf(x) = (af + bf)(x)$$

$$\therefore \qquad (a + b)f = af + bf$$

(iii) If $a, b \in R$ and $f \in V$, then we have

$$[(ab)f](x) = (ab)f(x) = a(bf(x)) = [a(bf)](x)$$

$$\therefore \qquad (ab)f = a(bf)$$

(iv) If $1 \in R$ and $f \in V$, then we have

$$(1f)(x) = 1f(x) = f(x)$$

$$\therefore \qquad 1f = f, \forall f \in V$$

Hence V is a vector space over R.

Example 10. *Let V be the set of all pairs (x, y) of real numbers, and let F be the field of real numbers. Examine in each of the following cases whether V is a vector space over the field of real numbers or not?*

(i) $(x, y) + (x_1, y_1) = (x + x_1, y + y_1)$; $c(x, y) = (|c| x, |c| y)$

(ii) $(x, y) + (x_1, y_1) = (x + x_1, y + y_1)$; $c(x, y) = (0, cy)$

(iii) $(x, y) + (x_1, y_1) = (x + x_1, y + y_1)$; $c(x, y) = (c^2 x, c^2 y)$

Solution.   (i) In this case, we shall show that

$$(a + b)\alpha \ne a\alpha + b\alpha \, \forall a, b \in F \text{ and } \alpha \in V$$

Let $a, b \in F$ and $\alpha = (x, y) \in V$

$$(a + b)\alpha = (a + b)(x, y)$$

$$= (|a + b| x, |a + b| y) \text{ (By addition of scalar multiplication)}$$

Also, $a\alpha + b\alpha = a(x, y) + b(x, y)$

$$= (|a| x, |a| y) + (|b| x, |b| y)$$

$$= |(|a| + |b|)x, (|a| + |b|)y|$$

$$\therefore \qquad (a + b)\alpha \ne a\alpha + b\alpha \, \forall a, b \in F \text{ and } \alpha \in V.$$

Hence, V is not a vector space.

(ii) In this case we shall show that $1\alpha \ne \alpha \forall \alpha \in V$.

Let $\alpha = (x, y) \in V$ and $1 \in F$

$$1\alpha = 1(x, y) = (0, 1y) = (0, y)$$

If $x \ne 0$, then $\alpha \ne (0, y)$

$$\therefore \qquad 1\alpha \ne \alpha \, \forall \alpha \in V$$

Hence V is not a vector space.

(iii) In this case we shall show that

$$(a + b)\alpha \ne a\alpha + b\alpha \, \forall a, b \in F \text{ and } \forall a \in V$$

Let $\quad\quad\quad \alpha = (x, y) \in V$ and $a, b \in F$

$$(a + b)\alpha = (a + b)(x, y)$$

$$= ((a + b)^2 x, (a + b)^2 y)$$

$\quad\quad\quad\quad\quad\quad\quad$ (By definition of scalar multiplication given by (iii))

Also, $\quad a\alpha + b\alpha = a(x, y) + b(x, y)$

$$= (a^2 x, a^2 y) + (b^2 x, b^2 y)$$

$$= (a^2 x + b^2 x, a^2 y + b^2 y)$$

$$= ((a^2 + b^2)x, (a^2 + b^2)y)$$

Since $\quad (a + b)^2 = a^2 + b^2 + 2ab \neq a^2 + b^2$

$\therefore \quad\quad (a + b)\alpha \neq a\alpha + b\alpha \, \forall \, a, b \in F$ and $\forall \alpha \in V$.

Hence V is not a vector space.

Example 11. *How many elements are there in the vector space of polynomials of degree at most n in which the coefficients are the elements of the field $Z(p)$, the integer modulo p over the field $Z(p)$, p being a prime number?*

Solution. Since $Z(p)$ is a field under addition and multiplication modulo p

i.e., $\quad\quad\quad\quad I(p) = ((0, 1, 2, 3, ..., p - 1) +_p, \times_p)$

Clearly, the number of distinct element in $I(p)$ is p.

Let $f(x)$ be a polynomial of degree at most n over the field $I(p)$.

Then $\quad\quad\quad\quad f(x) = a_0 + a_1 x + a_2 x^2 + ... + a_n x^n$

where $a_0, a_1, a_2, ..., a_n \in Z(p)$

In $f(x)$, there are $n + 1$ terms and each term has a coefficient from $Z(p)$. But $I(p)$ has p distinct elements, therefore we must have $p \times p \times p \times ... \times p$ upto $(n+1)$ times, *i.e.,* p^{n+1} distinct polynomials of degree atmost n over the field $I(p)$.

Hence, if P_n is the vector space of polynomials of degree at most n over the field $Z(p)$, then P_n has p^{n+1} distinct elements.

Example 12. *Let $K = Z_3$, the integers modulo 3. How many elements are in the vector space $V = K^4$?*

Solution. There are three choices 0, 1 or 2 for each of the four components of a vector in V. Hence, V has $3 \times 3 \times 3 \times 3 = 3^4 = 81$ elements.

Example 13. *Is Z_7 a vector space over Z_5?*

Solution. We know that Z_n = set of integers modulo n.

Then, $Z_5 = \{0, 1, 2, 3, 4\}$ and $Z_7 = \{0, 1, 2, 3, 4, 5, 6\}$

Now Z_5 is not a subfield of Z_7 as $2 + 3 = 0$ in Z_6 but $2 + 3 \neq 0$ in Z_7. Hence Z_7 is not a vector space over Z_5.

Example 14. *Show that the set $V = \{a_0 + a_1 x + a_2 x^2 : a_0, a_1, a_2 \in R\}$ of all polynomials of degree 2 over R is a vector space over R w.r.t. the composition*

$$(a_0 + a_1 x + a_2 x^2) + (b_0 + b_1 x + b_2 x^2) = (a_0 + b_0) + (a_1 + b_1)x + (a_2 + b_2)x^2 \quad ...(1)$$

and $\quad\quad\quad\quad a(a_0 + a_1 x + a_2 x^2) = aa_0 + aa_1 x + aa_2 x^2 \quad\quad\quad\quad ...(2)$

Solution. We can easily verify that $(V, +)$ is an abelian group with additive identity $0 = 0 + 0.x + 0.x^2$ and the additive inverse of $a_0 + a_1x + a_2x^2 \in V$ is $(-a_0) + (-a_1)x + (-a_2)x^2 \in V$.

Let $\alpha, \beta \in R, f, g \in V$ by (2) $\alpha f \in V$ and

(i) $\alpha(f + g) = \alpha f + \alpha g$ (ii) $(\alpha + \beta)f = \alpha f + \beta f$

(iii) $\alpha(\beta f) = (\alpha \beta)f$ (iv) $1.f = f$

To verify the first property let $f = a_0 + a_1x + a_2x^2, g = b_0 + b_1x + b_2x^2 \in V$

Then $\alpha(f + g) = \alpha[(a_0 + b_0) + (a_1 + b_1)x + (a_2 + b_2)x^2]$ [By (1)]

$$= \alpha(a_0 + b_0) + \alpha(a_1 + b_1)x + \alpha(a_2 + b_2)x^2 \quad\quad \text{[By (2)]}$$

$$= (\alpha a_0 + \alpha a_1x + \alpha a_2x^2) + (\alpha b_0 + \alpha b_1x + \alpha b_2x^2)$$

$$= \alpha f + \alpha g$$

In a similar way, we can verify the remaining properties. Hence V is a vector space over R.

Hence W is not a subspace of V.

❧ EXERCISE 5.1

1. Show that the complex field **C** is a vector space over the field R of reals.

2. Prove that the set of all vectors in a plane is vector space over the field of real numbers.

3. Let V be the set of all pairs (x, y) of real numbers, and let F be the field of real numbers. Define

$$(x, y) + (x_1, y_1) = (x + x_1, y + y_1)$$

$$c(x, y) = (cx, y)$$

Show that with these operations V is not a vector space over the field of real numbers.

4. Let V be the set of all ordered pairs (x, y) of reals and let F be the field of real numbers, then show that V is not a vector space over F with respect to addition and multiplication defined as follows:

$$(x, y) + (x_1, y_1) = (x + x_1, y + y_1)$$

$$c(x, y) = (cx + cy)$$

$$\forall(x, y), (x_1, y_1) \in V \text{ and } c \in F$$

5. Let V be the set of ordered pairs (z_1, z_2) of complex numbers. Show that V is a vector space over the real field R with addition in V and scalar multiplication on V defined by

$$(z_1, z_2) + (w_1, w_2) = (z_1 + w_1, z_2 + w_2)$$

and $a(z_1, z_2) = (az_1, az_2)$

$$\forall z_1, z_2, w_1, w_2 \in C \text{ and } a \in R$$

6. Let S be a set and V be the set of all subsets of S. Define vector addition and scalar multiplication as follows:

$$A + B = A \cup B \, \forall \, A, B \in V$$

$$cA = A \, \forall \, c \in R$$

Is V a vector space over R with these operations.

7. Find which of the following for a given matrix addition and scalar multiplication over the given field form a vector space:

(i) $V = $ set of all 2×2 matrices of the form $\begin{bmatrix} a & 1 \\ 1 & b \end{bmatrix}$ over R

(ii) $V = $ set of all 2×2 matrices of the form $\begin{bmatrix} a & 0 \\ 0 & a \end{bmatrix}$ over R

(iii) $V = $ set of all 2×2 matrices of the form $\begin{bmatrix} a & a+b \\ a+b & a \end{bmatrix}$ over R

8. Let S be any non-empty set and F be any field. Let V be the set of all functions from S to F, *i.e.*, let

$$V = \{f : f : S \to F\}$$

Let us define sum of two elements f and g in V as follows:

$$(f + g)(x) = f(x) + g(x)\ \forall x \in S$$

Also, let us define scalar multiplication of an element f in V by an element c in F as follows:

$$(cf)(x) = cf(x)\forall x \in S$$

Then $V(F)$ is a vector space.

9. Let U and W be vector spaces over a field F. Let V be the set of ordered pairs, *i.e.*,

$$V = \{(u, w) : u \in U, w \in W\}$$

Show that U is a vector space over F with addition in V and scalar multiplication on V defined by

$$(u, w) + (u_1, w_1) = (u + u_1, w + w_1)$$

and $\qquad k(u, w) = (ku, kw)$

for all $u, u_1 \in U$, $w, w_1 \in W$ and $k \in F$.

10. Let V be the set of ordered pairs (x, y) of real numbers. Show that V is not a vector

space over R with addition in V and scalar multiplication on V defined by:

(i) $(x, y) + (x_1, y_1) = (x, y)$ and $c(x, y) = (cx, cy)$

(ii) $(x,\ y) + (x_1,\ y_1) = (x + x_1,\ y + y_1)$ and

$$c(x, y) = (c^2 x, c^2 y)$$

(iii) $(x, y) + (x_1, y_1) = (0, 0)$ and $c(x, y) = (cx, cy)$

(iv) $(x, y) + (x_1, y_1) = (xx_1, yy_1)$ and

$$c(x, y) = (cx, cy)$$

(v) $(x, y) + (x_1, y_1) = (x + x_1, y + y_1)$ and

$$c(x, y) = c(x, 0)$$

11. Let $V = \{<a_n> : a_n \in R\}$, *i.e.*, V is the set of all real sequences. Prove that V is a vector space over R.

12. Let F be a field and

$$F^n = \{(\alpha_1, \alpha_2, \alpha_3, ..., \alpha_n) : \alpha_i \in F, 1 \le i \le n\}$$

Show that F^n is a vector space over F under the composition

$$\alpha(a_0 + a_1 x + a_2 x^2 + ... + a_m x^m)$$
$$= \alpha a_0 + \alpha a_1 x + ... + \alpha a_m x^m \quad (m < n)$$

Answers

6. V is not vector space.   7. (i) No. (ii) Yes (iii) Yes

5.4 ELEMENTARY PROPERTIES OF VECTOR SPACES

THEOREM 1. *Let $V(F)$ be a vector space over a field F and 0 be the zero (null) vector of V. Then prove that*

(i) $a0 = 0,\ \forall\ a \in F$

(ii) $0\alpha = 0,\ \forall\ \alpha \in V$

(iii) $a(-\alpha) = -(a\alpha),\ \forall\ a \in F, \alpha \in V$

(iv) $(-a)\alpha = -(a\alpha)\ \forall\ a \in F, \alpha \in V$

(v) $a(\alpha - \beta) = a\alpha - a\beta,\ \forall\ a \in F,\ \alpha, \beta \in V$

(vi) $a\alpha = 0 \Rightarrow a = 0\ \ or\ \ \alpha = 0$

(vii) $a\alpha = a\beta \Rightarrow \alpha = \beta,\ \forall\ a \in F, \alpha, \beta \in V, a \ne 0$

(viii) $a\alpha = b\alpha \Rightarrow a = b\ \ \forall\ a, b \in F, \alpha \in V\ and\ \alpha \ne 0$.

Proof. (i) $\qquad\qquad\qquad\qquad a0 = 0$

we have $\qquad\qquad a0 = a(0 + 0)$ $\qquad\qquad\qquad [\because 0 + 0 = 0]$

$\qquad\qquad\qquad\quad = a0 + a0$ $\qquad\qquad$ [By property of vector space]

or $\qquad\qquad 0 + a0 = a0 + a0$ $\qquad\qquad\qquad\quad [\because\ 0 + a0 = a0]$

or $\qquad\qquad\qquad 0 = a0$

(ii) $\qquad\qquad\qquad\qquad 0\alpha = 0$

We have $\qquad\qquad 0\alpha = (0 + 0)\alpha$ $\qquad\qquad\qquad [\because 0 + 0 = 0, 0 \in F]$

$\qquad\qquad\qquad\quad = 0\alpha + 0\alpha$ $\qquad\qquad\qquad$ [By the definition of $V(F)$]

$$\text{or} \qquad 0 + 0\alpha = 0\alpha + 0\alpha \qquad \qquad [\because 0\alpha \in V \therefore 0 + 0\alpha = 0\alpha]$$

$$\text{or} \qquad 0 = 0\alpha$$

(iii) $\qquad a\,(-\alpha) = -(a\alpha)$

We have,

$$a0 = 0 \qquad \qquad [\text{From(i)}]$$

$$\text{or} \qquad a(-\alpha + \alpha) = 0 \qquad \qquad [\because -\alpha + \alpha = 0]$$

$$\text{or} \qquad a(-\alpha) + a\alpha = 0 \qquad \qquad [\text{By the definition of } V(F)]$$

(iv) $\qquad (-a)\,\alpha = -(a\alpha)$

Since, we have

$$0\alpha = 0 \qquad \qquad [\text{From (ii)}]$$

$$\text{or} \qquad (-a + a)\alpha = 0 \qquad \qquad [\because -a + a = 0, a \in F]$$

$$\text{or} \qquad (-a)\alpha + a\alpha = 0 \qquad \qquad [\text{By the definition of } V(F)]$$

$$\text{or} \qquad (-a)\,\alpha = -(a\alpha)$$

(v) $\qquad a(\alpha - \beta) = a\alpha - a\beta$

We have,

$$a(\alpha - \beta) = a\,[\alpha + (-\beta)\,]$$

$$= a\alpha + a(-\beta) \qquad \qquad [\text{By definition of } V(F)]$$

$$\therefore \qquad a(\alpha - \beta) = a\alpha - a\beta \qquad \qquad [\because a(-\beta) = -(a\beta)]$$

(vi) $\qquad a\alpha = 0 \Rightarrow a = 0 \quad \text{or} \quad \alpha = 0$

Suppose $0 \ne a \in F$, then a^{-1} exists in F.

$$\text{Now} \qquad a\alpha = 0 \qquad \qquad [\text{given}]$$

$$\Rightarrow \qquad a^{-1}a\alpha = a^{-1}0$$

$$\Rightarrow \qquad (a^{-1}a)\alpha = 0 \qquad \qquad [\because a^{-1}0 = 0]$$

$$\Rightarrow \qquad 1\alpha = 0 \qquad \qquad [aa^{-1} = 1]$$

$$\Rightarrow \qquad \alpha = 0 \qquad \qquad [\because 1\alpha = \alpha]$$

Suppose $\alpha \ne 0$, then to prove $a = 0$, let us assume that $a \ne 0$, then a^{-1} exists. Since we have

$$a\alpha = 0$$

$$\Rightarrow \qquad a^{-1}(a\alpha) = a^{-1}\,0$$

$$\Rightarrow \qquad (a^{-1}a)\alpha = 0 \qquad \qquad [\because a^{-1}a = 1]$$

$$\Rightarrow \qquad 1\alpha = 0$$

$$\Rightarrow \qquad \alpha = 0 \qquad \qquad [\because 1\alpha = \alpha]$$

$\Rightarrow$ This gives a contradiction because we have taken $\alpha \ne 0$.

Hence, $\qquad a = 0$

(vii) We have,

$$a\alpha = a\beta \Rightarrow a\alpha - a\beta = 0$$

$$\Rightarrow \qquad a(\alpha - \beta) = 0$$

$$\Rightarrow \qquad \alpha - \beta = 0 \qquad \qquad [\because a \ne 0 \text{ and from (vi)}\,]$$

$$\Rightarrow \qquad \alpha = \beta$$

Hence, $\qquad a\alpha = a\beta \Rightarrow$ for all $a \neq 0 \in F$ and $\alpha, \beta \in V$

(viii) We have,

$$a\alpha = b\alpha \Rightarrow a\alpha - b\alpha = 0$$

$$\Rightarrow \qquad (a-b)\alpha = 0$$

$$\Rightarrow \qquad a - b = 0 \qquad\qquad [\because \alpha \neq 0 \text{ and from (vi)}]$$

$$\Rightarrow \qquad a = b$$

Hence, $\qquad a\alpha = b\alpha \Rightarrow a = b.$

5.5 VECTOR SUBSPACES: VECTOR SPACE WITHIN VECTOR SPACE

Just like a subgroup and a subring, we do have the concept of a vector subspace which is generally addresses as a subspace.

Definition. *Let W be a non-empty subset of V, where V is a vector space over a field F. Then W is said to be a vector subspace of V(F) if W is itself a vector space aver F with respect to the same operations as defined on V.*

For Example: The set $W = \{(a, 0, b) : a, b \in R\}$ is a subspace of $R^3(R)$.

5.6 ELEMENTARY PROPERTIES OF VECTOR SUBSPACES

THEOREM 1. *The necessary and sufficient conditions for a non-empty subset W of V(F) to be a subspace are that:*

(i) $\alpha \in W, \beta \in W \Rightarrow \alpha - \beta \in W$

(ii) $a \in F, \alpha \in W \Rightarrow a\alpha \in W$

Proof. Suppose W is a subspace of a vector space $V(F)$. Then if

$$\beta \in W \Rightarrow -\beta \in W$$

$$\therefore \qquad \alpha \in W, -\beta \in W \Rightarrow a + (-\beta) \in W$$

$$[\because W \text{ is closed under vector addition.}]$$

$$\Rightarrow \qquad \alpha - \beta \in W$$

and $\qquad a \in F, \alpha \in W \Rightarrow a\alpha \in W \quad [\because W \text{ is closed under scalar multiplication.}]$

Conversely, Suppose W is a subset of V and

(i) $\alpha \in W, \beta \in W \Rightarrow \alpha - \beta \in W$

(ii) $a \in F, \alpha \in W \Rightarrow a\alpha \in W$

Now we have to show that W is a subspace. For this purpose we proceed as follows:

$$\alpha \in W, \alpha \in W \Rightarrow \alpha - \alpha \in W. \qquad\qquad\qquad \text{[From (i)]}$$

$$\Rightarrow 0 \in W \Rightarrow \text{identity exists.}$$

and $\qquad 0 \in W, \alpha \in W \Rightarrow 0 - \alpha \in W. \qquad\qquad\qquad \text{[From (i)]}$

$$\Rightarrow -\alpha \in W \Rightarrow \text{inverse exists.}$$

Now, $\qquad \alpha \in W, -\beta \in W \Rightarrow \alpha - (-\beta) \in W.$

$$\Rightarrow \alpha + \beta \in W \qquad\qquad\qquad \text{[From (i)]}$$

$\therefore \quad W$ is closed under vector addition.

Also vector addition is always associative and commutative. Thereore $(W, +)$ is an abelian group.

From (ii) it is obvious that W is closed under multiplication. Space V is a vector space over F, therefore remaining properties will also hold in W. Hence W is a vector space and Hence W is itself a subspace.

THEOREM 2. *The necessary and sufficient condition for a non-empty subset of W of a vector space $V(F)$ to be a subspace of V is*

$$a, b \in F, \alpha, \beta \in W \Rightarrow a\alpha + b\beta \in W$$

Proof. Suppose W is a subspace of a vector space $V(F)$. Then W is closed under vector addition and multiplication, therefore we have

$$a \in F, \alpha \in W \Rightarrow a\alpha \in W$$

and $\qquad b \in F, \beta \in W \Rightarrow b\beta \in W$

$\therefore \qquad a\alpha \in W, b\beta \in W \Rightarrow a\alpha + b\beta \in W$

Conversely, Suppose W is a subset of $V(F)$ and

$$a, b \in F, \alpha, \beta \in W \Rightarrow a\alpha + b\beta \in W \text{ is given.}$$

Then we have to show that W is a subset of $V(F)$.

Now taking $a = 1$, $b = 1$, then

$$1 \in F; \alpha, \beta \in W \Rightarrow 1\alpha + 1\beta \in W$$

$\Rightarrow \qquad\qquad \alpha + \beta \in W \qquad\qquad\qquad [1\alpha = \alpha, 1\beta = \beta]$

$\therefore \quad W$ is closed under vector addition.

And taking $a = 0$, $b = -1$, we have

$$0\alpha + (-1)\beta \in W$$

$\Rightarrow \qquad\qquad 0 + (-\beta) \in W \qquad\qquad\qquad [\because (-1)\beta = -\beta]$

$\Rightarrow \qquad\qquad -\beta \in W$

$\therefore \quad$ Additive inverse exists in W.

Again, taking $a = 0, b = 0$, we have

$$0\alpha + 0\beta \in W$$

$\Rightarrow \qquad\qquad 0 \in W \qquad\qquad\qquad [\because 0\alpha = 0, \text{similarly}, 0\beta = 0]$

Since $W \subseteq V$, therefore vector addition is associative and commutative.

Thus, W is an abelian group under vector addition.

Further, taking $\beta = 0$, we have

$$a\alpha + b0 \in W$$

$\Rightarrow \qquad\qquad a\alpha \in W \qquad\qquad\qquad [\because b0 = 0 \text{ and } a\alpha + 0 = a\alpha]$

$\therefore \quad W$ is closed under scalar multiplication.

The rest properties will hold in W because $W \subseteq V$ and these properties hold in V.

Hence W is a vector space and consequently W is a subspace of $V(F)$.

5.7 ALGEBRA OF SUBSPACES

THEOREM 1. *The intersection of any two subspaces of a vector space is a subspace.*

Proof. Let $V(F)$ be a vector space over F and W_1, W_2 be two subspaces of $V(F)$. Then, we have to show that $W_1 \cap W_2$ is a subspace of $V(F)$

Let $\qquad\qquad \alpha, \beta \in W_1 \cap W_2 \Rightarrow \alpha, \beta \in W_1$ and $\alpha, \beta \in W_2$

Since, W_1 and W_2 are subspaces of V, so we have

$$a, b \in W \text{ and } \alpha, \beta \in W_1 \Rightarrow a\alpha + b\beta \in W_1 \qquad \text{...(1)}$$

and, $\quad a, b \in F$ and $\alpha, \beta \in W_2 \Rightarrow a\alpha + b\beta \in W_2 \qquad \text{...(2)}$

From (1) and (2), we get

if $\quad a,b \in F$ and $\alpha,\beta \in W_1 \cap W_2 \Rightarrow a\alpha + b\beta \in W_1 \cap W_2$

Hence, $W_1 \cap W_2$ is a subspace of V.

THEOREM 2. *The intersection of an arbitrary collection of a subspaces of a vector space is also a subspace.*

Proof. Let $\{W_\lambda : \lambda \in \Lambda\}$ be an arbitrary collection of subspaces of a vector space V(say). Then we have to show that $\cap \{W_\lambda : \lambda \in \Lambda\}$ is a subspace of V.

Let $\qquad\qquad \alpha,\beta \in \cap \{W_\lambda : \lambda \in \Lambda\}$

$\Rightarrow \qquad\qquad \alpha,\beta \in W_\lambda$ for each $\lambda \in \Lambda$.

Since, each W_λ is a subspace of V, then for any two scalars $a,b \in F$, we have

$$a\alpha + b\beta \in W_\lambda \text{ for each } \lambda \in \Lambda$$

$\Rightarrow \qquad\qquad a\alpha + b\beta \in \cap \{W_\lambda : \lambda \in \Lambda\}$

Hence, $\cap \{W_\lambda : \lambda \in \Lambda\}$ is a subspace of V.

THEOREM 3. *The union of two subspaces of a vector space is not necessarily a subspace.*

Proof. Let W_1, W_2 be two subspaces of a vector space V and suppose that

$$W_1 = \{(a_1, a_2, 0) : a_1, a_2 \in F\}$$

and $\qquad\qquad W_2 = \{(a_1, 0, a_3) : a_1, a_3 \in F\}$

Obviously, W_1 and W_2 are subspaces of $R^3(R)$. By definition of W_1 and W_2, we have $W_1 \cup W_2$ containing all triads, *i.e.*, 3-tuples of the form $(a_1, a_2, 0)$ and those of the form $(a_1, 0, a_3)$.

Now, if we consider the elements $\alpha = (1,2,0)$ and $\beta = (3,0,5)$ of $W_1 \cup W_2$, then for scalars $a=1$ and $b=2$.

$$a\alpha + b\beta = 1(1,2,0) + 2(3,0,5) = (1,2,0) + (6,0,10)$$

$$= (7,2,10) \notin W_1 \cup W_2$$

Thus, if $\alpha \in W_1 \cup W_2$ and $\beta \in W_1 \cup W_2$, then it is not necessarily implied that $a\alpha + b\beta \in W_1 \cup W_2$ for some $a, b \in F$.

Hence, $W_1 \cup W_2$ is a subspace of $R^3(R)$.

THEOREM 4. *The union of two subspaces of a vector space is a subspace iff one is contained in the other.*

Proof. Let $V(F)$ be a vector space and W_1, W_2 be two subspaces of V.

Suppose $W_1 \subseteq W_2$ or $W_2 \subseteq W_1$. Then we have to show that $W_1 \cup W_2$ is a subspace of V.

Now $\qquad W_1 \cup W_2 = W_2$ if $W_1 \subseteq W_2$ and W_2 is a subspace, therefore $W_1 \cup W_2$ is subspace.

Also, $W_1 \cup W_2 = W_1$ if $W_1 \subseteq W_2$ and since W_1 is a subspace, therefore $W_1 \subseteq W_2$ is a subspace V.

Conversely, Suppose $W_1 \cup W_2$ is a subspace of V. Then we have to show that

$$W_1 \subseteq W_2 \text{ or } W_2 \subseteq W_1.$$

Let us assume that W_1 is not a subset of W_2 and W_2 is not a subset of W_1.

Now, W_1 is not a subset of W_2, this implies that there exists an element α in W_1

which is not in W_2.

Also, W_2 is not a subset of W_1, therefore there exists an element β in W_2 which is not in W_1. But we have $\alpha \in W_1 \cup W_2$ and $\beta \in W_1 \cup W_2$ and since $W_1 \cup W_2$ is a subspace of V, we have.

$$a, b \in F,\ \alpha, \beta \in W_1 \cup W_2 \Rightarrow a\alpha + b\beta \in W_1 \cup W_2$$

Now taking $a = 1$, $b = 1$, we have

$$1\alpha + 1\beta \in W_1 \cup W_2 \qquad [\because 1\alpha \in W_1 \cup W_2 \subseteq V \therefore 1\alpha = \alpha]$$

$$\Rightarrow \qquad \alpha + \beta \in W_1 \cup W_2$$

$$\Rightarrow \qquad \alpha + \beta \in W_1 \text{ or } \alpha + \beta \in W_2$$

Suppose $\alpha + \beta \in W_1$ and $\alpha \in W_1$, then $(\alpha + \beta) - \alpha \in W_1$, because W_1 is a subspace of V. Therefore $\beta \in W_1$ and this gives a contradiction.

Now, suppose $\alpha + \beta \in W_2$ and $\beta \in W_2$, then

$$(\alpha + \beta) - \beta \in W_2$$

$$\Rightarrow \qquad \alpha \in W_2 \qquad\qquad [\because W_2 \text{ is a subspace.}]$$

Hence, either $\qquad W_1 \subseteq W_2 \text{ or } W_2 \subseteq W_1$

Solved Examples

Based on the following Results

➡ The necessary and sufficient conditions for a non-empty subset W of $V(F)$ to be a subspace are that
   (i) $\alpha \in W, \beta \in W \Rightarrow \alpha - \beta \in W$

   (ii) $a \in F, \alpha \in W \Rightarrow a\alpha \in W$
   or equivalently $\qquad a, b \in F,\ \alpha,\ \beta \in W \Rightarrow a\alpha + b\beta \in W.$

➡ The intersection of two subspaces of a vector space is a subspace.

➡ The union of two subspaces of a vector space is a subspace if and only if one is contained in the other.

Example 1. *Show that the set $W = \{(a, b, c): a - 3b + 4c = 0\}$ is a subspace of the 3-tuple space $R^3 (R)$.*

Solution. Let $\alpha = (a_1, b_1, c_1)$ and $\beta = (a_2, b_2, c_2)$ be any two elements of W, such that

$$a_1 - 3b_1 + 4c_1 = 0 \text{ and } a_2 - 3b_2 + 4c_2 = 0$$

For $a, b \in R$, we have

$$a\alpha + b\beta = a(a_1, b_1, c_1) + b(a_2, b_2, c_2)$$

$$= (aa_1, ab_1, ac_1) + (ba_2, bb_2, bc_2)$$

$$= (aa_1 + ba_2, ab_1 + bb_2, ac_1 + bc_2)$$

Now $(aa_1 + ba_2) - 3(ab_1 + bb_2) + 4(ac_1 + bc_2)$

$$= (aa_1 - 3ab_1 + 4ac_1) + (ba_2 - 3bb_2 + 4bc_2)$$

$$= a(a_1 - 3b_1 + 4c_1) + b(a_2 - 3b_2 + 4c_2)$$

$$= a.0 + b.0 = 0$$

So, $\qquad a\alpha + b\beta \in W$

Thus, $\alpha \in W, \beta \in W \Rightarrow a\alpha + b\beta \in W \ \forall a, b \in R.$

Hence, W is a subspace of R^3 (R).

Example 2. *Show that the set $W = \{(a_1, a_2, 0)\}: a_1, a_2 \in F\}$ is a subspace of $V_3(F)$.*

Solution. We have $W = \{(a_1, a_2, 0): a_1, a_2 \in F\}$

Clearly, $(0, 0, 0) \in W \Rightarrow W$ is non-empty.

Let $\alpha, \beta \in W$. Then $\alpha = (a_1, a_2, 0)$ and $\beta = (b_1, b_2, 0)$ for $a_1, a_2, b_1, b_2 \in F$,

Now, for any $a, b \in F$,

$$a\alpha + b\beta = a(a_1, a_2, 0) + b(b_1, b_2, 0)$$

$$= (aa_1, aa_2, 0) + (bb_1, bb_2, 0)$$

$$= (aa_1 + bb_1, aa_2 + bb_2, 0)$$

Since $aa_1 + bb_1, aa_2 + bb_2 \in F$, therefore $a\alpha + b\beta \in W$

Hence W is a subspace of V_3 (F).

Example 3. *Let W be the collection of all elements from the space $M_2(F)$ of the form $\begin{bmatrix} a & b \\ -b & a \end{bmatrix}$. Show that W is a subspace of M_2 (F).*

Solution. We have $W = \left\{ \begin{bmatrix} a & b \\ -b & a \end{bmatrix} : a, b \in F \right\}$

Clearly, $\begin{bmatrix} 0 & 0 \\ 0 & 0 \end{bmatrix} \in W \Rightarrow W$ is non-empty.

Let $\alpha, \beta \in W$. Then $\alpha = \begin{bmatrix} a_1 & b_1 \\ -b_1 & a_1 \end{bmatrix}$ and $\beta = \begin{bmatrix} a_2 & b_2 \\ -b_2 & a_2 \end{bmatrix}$ for all $a_1, b_1, a_2, b_2 \in F$.

If a, b are any element of F, then

$$a\alpha + b\beta = a\begin{bmatrix} a_1 & b_1 \\ -b_1 & a_1 \end{bmatrix} + b\begin{bmatrix} a_2 & b_2 \\ -b_2 & a_2 \end{bmatrix}$$

$$= \begin{bmatrix} aa_1 & ab_1 \\ -ab_1 & aa_1 \end{bmatrix} + \begin{bmatrix} ba_2 & bb_2 \\ -bb_2 & ba_2 \end{bmatrix}$$

$$= \begin{bmatrix} aa_1 + ba_2 & ab_1 + bb_2 \\ -(ab_1 + bb_2) & aa_1 + ba_2 \end{bmatrix}$$

Since $aa_1 + ba_2, ab_1 + bb_2 \in F$, therefore $a\alpha + b\beta \in W$.

Hence W is a subspace of M_2 (F).

Example 4. *Which of the following sets of vectors $\alpha = (a_1, a_2, \ldots a_n) \in R^n$ are subspaces of $R^n (n \geq 3)$?*

(i) all α such that $a_1 \leq 0$

(ii) all α such that a_3 is an integer

(iii) all α such that $a_2 + 4a_3 = 0$

(iv) all α such that $a_1 + a_2 + \ldots + a_n = k$ (a given constant)

Solution.   (i) Let $W = \{(a_1, a_2, ..., a_n) : a_1 \leq 0\}$.

Clearly, $(-1, a_2, a_3, ..., a_n) \in W \Rightarrow W$ is non-empty.

Let $\alpha, \beta \in W$. Then we have

$$\alpha = (a_1, a_2, ..., a_n) \text{ and } \beta = (b_1, b_2, ..., b_n)$$

with $a_1 \leq 0$ and $b_1 \leq 0$.

If a, b be any element of R, then

$$a\alpha + b\beta = a(a_1, a_2, ..., a_n) + b(b_1, b_2, ..., b_n)$$

$$= (aa_1, aa_2, ..., aa_n) + (bb_1, bb_2, ..., bb_n)$$

$$= (aa_1 + bb_1, aa_2 + bb_2, ..., aa_n + bb_n)$$

Since $a_1 \leq 0$ and $b_1 \leq 0$

If $a < 0$ and $b < 0$, then $aa_1 > 0$ and $bb_1 > 0$ so that $aa_1 + bb_1 > 0$. Thus $a\alpha + b\beta \notin W$. Hence W is not a subspace.

(ii) Let $W = \{(a_1, a_2, ..., a_n) : a_3 \text{ is an integer}\}$

Clearly, $(a_1, a_2, 1, a_4, ..., a_n) \in W \Rightarrow W$ is non-empty.

Let $\alpha = (a_1, a_2, 2, a_4, ..., a_n)$ be in W, $a = \dfrac{1}{3}$ an element of R, then

$$a\alpha = \frac{1}{3}(a_1, a_2, 2, a_4, ..., a_n)$$

$$= \left(\frac{a_1}{3}, \frac{a_2}{3}, \frac{2}{3}, \frac{a_4}{3} \cdots \frac{a_n}{3}\right)$$

$$\Rightarrow \qquad a\alpha \notin W$$

Hence W is not a subspace of R^n.

(iii) Let $W = \{(a_1, a_2, a_3, ..., a_n) : a_2 + 4a_3 = 0\}$

Clearly, $(a_1, -4, 1, ..., a_n) \in W \Rightarrow W$ is non-empty.

Let $\alpha = (a_1, a_2, a_3, ... a_n)$ and $\beta = (b_1, b_2, b_3, ... b_n)$ be any two elements in W such that $a_2 + 4a_3 = 0$ and $b_2 + 4b_3 = 0$

Let a, b be any elements of R, then

$$a\alpha + b\beta = a(a_1, a_2, a_3, ..., a_n) + b(b_1, b_2, b_3, ..., b_n)$$

$$= (aa_1 + bb_1, aa_2 + bb_2, aa_3 + bb_3, ..., aa_n + bb_n)$$

Now, $(aa_2 + bb_2) + 4(aa_3 + bb_3) = a(a_2 + 4a_3) + b(b_2 + 4b_3)$

$$= a.0 + b.0 = 0$$

Since $aa_1 + bb_1, aa_2 + bb_2, aa_3 + bb_3, ..., aa_n + bb_n \in R$,

therefore $a\alpha + b\beta \in W$.

Hence W is a subspace of R^n.

(iv) Let $W = \{(a_1, a_2, ..., a_n) : a_1 + a_2 + ... + a_n = k \text{ (given)}\}$

If $k = 0$, then W is a subspace of R^n, but if $k \neq 0$, then W is not a subspace.

Example 5. *If a_1, a_2, a_3 are fixed elements of a field F, then the set W of all ordered triads (x_1, x_2, x_3) of elements of field F, such that $a_1x_1 + a_2x_2 + a_3x_3 = 0$ is a subspace of $V_3(F)$.*

Solution. We have

$$W = \{(x_1, x_2, x_3) : a_1x_1 + a_2x_2 + a_3x_3 = 0, a_1, a_2, a_3 \text{ are fixed}\}.$$

Clearly, $(0, 0, 0) \in W \Rightarrow W$ is non-empty.

Let $\alpha = (x_1, x_2, x_3)$ and $\beta = (y_1, y_2, y_3)$ be any two elements of W, then $x_1, x_2, x_3, y_1, y_2, y_3$ are elements of F such that

$$a_1x_1 + a_2x_2 + a_3x_3 = 0$$
$$a_1y_1 + a_2y_2 + a_3y_3 = 0$$

If a, b be any two elements of F then

$$a\alpha + b\beta = a(x_1, x_2, x_3) + b(y_1, y_2, y_3)$$

$$= (ax_1 + by_1, ax_2 + by_2, ax_3 + by_3)$$

Now $a_1(ax_1 + by_1) + a_2(ax_2 + by_2) + a_3(ax_3 + by_3)$

$$= a(a_1x_1 + a_2x_2 + a_3x_3) + b(a_1y_1 + a_2y_2 + a_3y_3)$$

$$= a.0 + b.0 = 0.$$

Since, $ax_1 + by_1, ax_2 + by_2, ax_3 + by_3 \in F$, therefore, $a\alpha + b\beta \in W$.

Hence W is a subspace of $V_3(F)$.

Example 6. *Let R be the field of real numbers. Which of the following are subspaces of V_3 (R)?*
   (i) $W_1 = \{(x, x, x) : x \in R\}$
   (ii) $W_2 = \{(x, y, z) : x, y, z \text{ are rational numbers}\}$

Solution. (i)   Since $(0, 0, 0) \in W_1 \Rightarrow W_1$ is non-empty.

Let $\alpha = (x, x, x)$ and $\beta = (y, y, y)$ be any two elements of W_1.

If a, b are any two elements of R, then

$$a\alpha + b\beta = a(x, x, x) + b(y, y, y)$$

$$= (ax + by, ax + by, ax + by)$$

Since $ax + by \in R$, therefore $a\alpha + b\beta \in W_1$

Hence W_1 is a subspace of V_3 (R).

(ii)   Since $(0, 0, 0) \in W_2 \Rightarrow W_2$ is non-empty.

Let $\alpha = (x, y, z)$ be an element of W_2, then x, y, z are rational numbers.

If $a = \sqrt{5}$ be an element of R, then

$$a\alpha = \sqrt{5}(x, y, z) = (\sqrt{5}x, \sqrt{5}y, \sqrt{5}z)$$

But $\sqrt{5}x, \sqrt{5}y, \sqrt{5}z$ are not rationals, therefore $a\alpha \notin W_2$.

Hence W_2 is not a subspace of V_3 (R).

Example 7. *Let V be the vector space of all 2×2 matrices over the field R. Show that W is not a subspace of V, where W contains all 2×2 matrices with zero determinant.*

Solution. Clearly, $\begin{bmatrix} 0 & 0 \\ 0 & 0 \end{bmatrix} \in W \Rightarrow W$ is non-empty.

Let $A = \begin{bmatrix} a & 0 \\ 0 & 0 \end{bmatrix}, B = \begin{bmatrix} 0 & 0 \\ 0 & b \end{bmatrix}$, where $a, b \in R$ and $a \neq 0, b \neq 0$. Then

$$|A| = 0 = |B| \Rightarrow A, B \in W$$

Now
$$A + B = \begin{bmatrix} a & 0 \\ 0 & 0 \end{bmatrix} + \begin{bmatrix} 0 & 0 \\ 0 & b \end{bmatrix} = \begin{bmatrix} a & 0 \\ 0 & b \end{bmatrix}$$

and $\qquad |A + B| \neq 0 \Rightarrow A + B \notin W$

Hence, W is not a subspace of V.

Example 8. *Let $M_n (F)$ be the vector space of all $n \times n$ matrices over the field F. Let W be the subset of $M_n(F)$ consisting of all symmetric matrices. Show that W is a subspace of $M_n(F)$.*

Solution. Let
$$W = \{[a_{ij}]_{n \times n} : a_{ij} = a_{ji}\}.$$

Clearly, $O \in W$, where O is the null matrix of order $n \times n$. Thus W is non-empty.

Let $A = [a_{ij}]_{n \times n}$ and $A = [b_{ij}]_{n \times n}$ be any two elements of W, then

$$a_{ij} = a_{ji} \text{ and } b_{ij} = b_{ji}$$

If a, b be any two elements of F, then
$$aA + bB = a[a_{ij}]_{n \times n} + b[b_{ij}]_{n \times n}$$

$$= [aa_{ij} + bb_{ij}]_{n \times n}$$

$$= [aa_{ji} + bb_{ji}]_{n \times n} \qquad [\because a_{ij} = a_{ji} \text{ and } b_{ij} = b_{ji}]$$

Therefore, $aA + bB$ is a symmetric matrix so that $aA + bB \in W$.
Hence, W is a subspace of $M_n(F)$.

Example 9. *Let $V(F)$ be the vector space of all $n \times 1$ matrices over the field F. Let A be an $m \times n$ matrix over F. Then the set W of all $n \times 1$ matrices X over F such that $AX = O$ is a subspace of V, here O is a null matrix of the type $m \times 1$.*

Solution. Clearly, $O_{n \times 1} \in W \Rightarrow W$ is non-empty.
Let X, Y be any elements of W. Then X and Y are of order $n \times 1$ matrices over F such that $\qquad\qquad AX = O, AY = O$
If $a \in F$, then $aX + Y$ is also an $n \times 1$ matrix over F.

Now $\qquad A(aX + Y) = A(aX) + AY = a(AX) + AY$

$$= a.O + O = O + O = O$$

$\Rightarrow \qquad A(aX + Y) \in W$
Hence, W is a subspace of V.

Example 10. *Let V be the vector space of all polynomials in an indeterminate x over a field F, i.e., $V = F[x]$. Let W be a subset of V consisting of all polynomials of degree $\leq n$. Then W is a subspace of V.*

Solution. Let α, β be any two elements of W.

Let $$\alpha = a_0 + a_1 x + \ldots + a_r x^r, r \leq n$$

and $$\beta = b_0 + b_1 x + \ldots + b_s x^s, s \leq n$$

where $$a_0, a_1, \ldots, a_r, b_0, b_1, b_2, \ldots b_s \in F$$

If $a, b \in F$, then

$$a\alpha + b\beta = a(a_0 + a_1 x + \ldots + a_r x^r) + b(b_0 + b_1 x + \ldots + b_s x^s)$$
$$= (aa_0 + aa_1 x + \ldots + aa_r x^r) + (bb_0 + bb_1 + \ldots + bb_s x^s).$$

If $s > r$ and setting $a_k = 0$ for all $r + 1 \leq k \leq s$, then

$$a\alpha + b\beta = (aa_0 + bb_0) + (aa_1 + bb_1)x \ldots + (aa_s + bb_s)x^s$$

Since $aa_0 + bb_0, aa_1 + bb_1, \ldots, aa_s + bb_s \in F$, then $a\alpha + b\beta$ is a polynomial of degree $\leq n$, therefore $a\alpha + b\beta \in W$.

Hence W is a subspace of V.

Example 11. *Prove that the set of all solutions (a, b, c) of the equation $a + b + 2c = 0$ is a subspace of vector space $V_3(R)$.*

Solution. Let $$W = \{(a, b, c): a + b + 2c = 0\}$$

Clearly, $(0, 0, 0) \in W \Rightarrow W$ is non-empty.

Let $\alpha = (a_1, b_1, c_1)$ and $\beta = (a_2, b_2, c_2)$ be any two elements of W, then

$$a_1 + b_1 + 2c_1 = 0$$
and $$a_2 + b_2 + 2c_2 = 0$$

If a, b be any two elements of R, then

$$a\alpha + b\beta = a(a_1, b_1, c_1) + b(a_2, b_2, c_2)$$
$$= (aa_1, ab_1, ac_1) + (ba_2, bb_2, bc_2)$$
$$= (aa_1 + ba_2, ab_1 + bb_2, ac_1 + bc_2)$$

Now, $aa_1 + ba_2 + ab_1 + bb_2 + 2(ac_1 + bc_2)$
$$= a(a_1 + b_1 + 2c_1) + b(a_2 + b_2 + 2c_2)$$
$$= a.0 + b.0$$
$$= 0 + 0 = 0$$

Since, $aa_1 + ba_2, ab_1 + bb_2, ac_1 + bc_2 \in R$ then $a\alpha + b\beta \in W$

Hence W is a subspace of $V_3(R)$.

Example 12. *Let $V = R^3(R)$ the real vector space and let $W_1 = \{(0, y, z): y, z \in R\}$, $W_2 = \{(x, y, 0): x, y \in R\}$. What is $W_1 \cap W_2$? Is it subspace of V? Is $W_1 \cup W_2$ subspace of V?*

Solution. Clearly, W_1 and W_2 are subspaces of V.

From the definition of W_1 and W_2, we define $W_1 \cap W_2$ as follows:

$$W_1 \cap W_2 = \{(0, y, 0) : y \in R\}.$$

Let $\alpha = (0, y_1, 0)$ and $\beta = (0, y_2, 0)$ be any two elements of $W_1 \cap W_2$.
If a, b be any elements of R, then

$$a\alpha + b\beta = a(0, y_1, 0) + b(0, y_2, 0)$$
$$= (0, ay_1, 0) + (0, by_2, 0)$$
$$= (0, ay_1 + by_2, 0)$$

Since, $ay_1 + by_2 \in R$ therefore $a\alpha + b\beta \in W_1 \cap W_2$

Hence $W_1 \cap W_2$ is a subspace of V.

Next, let $\alpha = (0, y_1, z_1) \in W_1$ and $\beta = (x_1', y_1', 0) \in W_2$ where $y_1, z_1, x_1', y_1' \in R$.

Now α and β are both elements of $W_1 \cup W_2$, then for $a, b \in R$, we have
$$a\alpha + b\beta = a(0, y_1, z_1) + b(x_1', y_1', 0)$$
$$= (bx_1', ay_1 + by_1', az_1)$$

Since, $(bx_1', ay_1 + by_1', az_1)$ belongs neither to W_1 nor to W_2, therefore $(bx_1', ay_1 + by_1', az_1) \notin W_1 \cup W_2$ or $a\alpha + b\beta \notin W_1 \cup W_2$.

Hence $W_1 \cup W_2$ is not a subspace of V.

Example 13. *Let V be the (real) vector space of all functions f from R into R. Which of the following sets of functions are subspaces of V?*
  (i) all f such that $f(x^2) = [f(x)]^2$
  (ii) all f which are continuous.

Solution. (i) Let $W = \{f : f(x^2) = [f(x)]^2\}$.

Clearly, $O(x) \in W \Rightarrow W$ is non-empty.

Let f and g be any two elements of W, then
$f(x)^2 = [f(x)]^2$ and $g(x)^2 = [g(x)]^2$.

If a and b are any two elements of R, then

$$(af + bg)(x^2) = (af)(x^2) + (bg)(x^2) = af(x^2) + bg(x^2)$$
$$= a[f(x)]^2 + b[g(x)]^2$$

Now $[(af + bg)(x)]^2 = [af(x) + bg(x)]^2$
$$= a^2[f(x)]^2 + b^2[g(x)]^2 + 2abf(x)g(x)$$

∴ $(af + bg)(x^2) \neq [(af + bg)(x)]^2$

⇒ $af + bg \notin W$

Hence W is not a subspace of V.

(ii) Let $W = \{f : f \text{ is continuous}\}$

Since f is continuous so that af is also continuous for any $a \in R$.
Clearly, $O(x) \in W \Rightarrow W$ is non-empty.
Let f and g be any two elements of W, then f and g are continuous.
If a, b are any two elements of W, then $af + bg$ is also continuous, therefore $af + bg \in W$.
Hence W is a subspace of V.

🐚 EXERCISE 5.2

1. Show that the set $W = \{(a, 0,0): a \in R\}$ is a subspace of V_3 (R).

2. Show that the subset $W = \{(a,b,c): a+b+c=0\}$ of R^3 is a subspace of R^3.

3. Show that the set W of the elements of the vector space V_3 (R) of the form $(x+2y, y, -x +3y)$ where $x, y \in R$, is a subspace of V_3(R).

4. Let $V = R^3$. Show that the set $W = \{(a, b, c): a, b, c \in Q\}$ of V is not a subspace of V.

5. Let F be the field of integers modulo 2, V be the set of all 2×2 matrices over F. Show that $V(F)$ is a finite vector space. Give two non-trivial subspaces of this vector space.

6. Let V be a vector space of all $n \times n$ matrices. Prove that the set W consisting of all $n \times n$ real matrices which commute with a given matrix T of V forms a subspace of V.

7. Let C be the field of complex numbers and let n be a positive integer ($n \geq 2$). Let V be the vector space of all $n \times n$ matrices over C. Which of the following sets of matrices A in V are subspaces of V?

   (i) All invertible A;

   (ii) All non-invertible A;

   (iii) All A such that $AB=BA$, where B is some fixed matrix in V.

8. Which of the following sets of vectors $\alpha = (a_1, a_2,...,a_n)$ in R^n are subspaces of R^n ($n \geq 3$)?

   (i) All α such that $a_1 \geq 0$

   (ii) All α such that $a_1 + 3a_2 = a_3$

   (iii) All α such that $a_2 = a_1^2$

   (iv) all α such that $a_1 a_2 = 0$

   (v) all α such that a_2 is rational.

9. Determine whether or not W is a subspace of R^3 if W consists of those vectors $(a, b, c) \in R^3$ for which:

   (i) $a=2b$ (ii) $ab = 0$

   (iii) $a \leq b \leq c$ (iv) $a = b^2$

   (v) $k_1 a + k_2 b + k_3 c = 0, k_1, k_2, k_3 \in R$

10. Let V be the vector space of all polynomials over the field R. Determine whether or not W is a subspace of V where

    (i) W contains all polynomials with integral coefficients .

    (ii) W consists of all polynomials $a_0 + a_1 x^2 + a_2 x^4 + ... + a_n x^{2n}$, i.e., polynomials with only even powers of x.

11. Let V be the (real vector) space of all functions f from R into R. Which of the following sets of functions are subspaces of V?

    (i) all f such that $f(0) = f(1)$

    (ii) all f such that $f(3) = 1+f(-5)$

    (iii) all f such that $f(-1) = 0$

    (iv) all f which are bounded.

12. Let $AX= B$ be a non-homogeneous system of linear equations in n unknowns over a field F. Show that the solution set W of the system is not a subspace of F^n.

13. Let V be the vector space of all $n \times n$ square matrices over a field F. Show that W is a subspace of V if W consists of all matrices which are:

    (i) skew-symmetric

    (ii) upper triangular

    (iii) diagonal

    (iv) scalar

14. Let V be the vector space of infinite sequences $(a_1, a_2, ..., a_n, ...)$ in a field F. Show that W is a subspace of V if :

    (i) W consists of all sequences with 0 as the first component;

    (ii) W consists of all sequences with only a finite number of non-zero components.

15. Let W be the set of all vectors of the form $(x, 2x, -3x, x) x \in R$, then prove that W is subspace of V_4 (R).

16. Let W be the set of all five-tuples of real numbers of the form $(x, 2x, -3x, 5x, x), x \in R$. Show that W is a subspace of V_5 (R).

17. If R is the field of real numbers and
$$W_1 = \{(x,0,0): x \in R\}$$
$$W_2 = \{(0,y,0): y \in R\}$$

    are two subspaces of $V_3(R)$. What is $W_1 \cap W_2$: Is it a subspace of V? Is $W_1 \cup W_2$ a subspace of V?

18. Give an example of a subset W of a vector space $V=R^2$ which is not a subspace of V but for which :

    (i) $W + W = W$

    (ii) $W + W \subset W$

19. Let V be the vector space of $n \times n$ matrices over a field F. Let W_1 and W_2 be the subspaces of upper triangular matrices and lower triangular matrices respectively, find:

   (i) $W_1 + W_2$

   (ii) $W_1 \cap W_2$

Answers

5. $W_1 = \left\{ \begin{bmatrix} a & 0 \\ 0 & b \end{bmatrix} : a, b \in F \right\}$ and $W_2 = \left\{ \begin{bmatrix} 0 & a \\ b & 0 \end{bmatrix} : a, b \in F \right\}$ are two non-trivral subspaces.

7. (i) not a subspace (ii) not a subspace   (iii) a subspace

8. (i) not a subspace (ii) a subspace   (iii) not a subspace
   (iv) not a subspace (v) not a subspace

9. (i) a subspace   (ii) not a subspace   (iii) not a subspace
   (iv) not a subspace (v) a subspace

10. (i) not a subspace (ii) a subspace

11. (i) a subspace   (ii) not a subspace   (iii) a subspace
    (iv) a subspace

17. (i) $W_1 \cup W_2 = \{(0, 0, 0)\}$ (ii) $W_1 \cup W_2 = \{(x, y, 0) : x, y \in R\}$

18. (i) $W = \{(0, 0) \ (0, 1) \ (0, 2) \ (0, 3), \ldots\}$ (ii) $W = \{(0, 8), (0, 6), (0, 7), \ldots\}$

19. (i) $W_1 + W_2 = V$ (ii) $W_1 \cap W_2$ = set of all diagonal matrices.

5.8 LINEAR SUM OF TWO SUBSPACES

Let W_1 and W_2 be two subspaces of a vector $V(F)$. Then the linear sum of W_1 and W_2 is the set of all those elements each one of which is expressible as the sum of an element of W_1 and an element of W_2. The linear sum of W_1 and W_2 can be written as $W_1 + W_2$. That is

$$W_1 + W_2 = \{\alpha + \beta : \alpha \in W_1, \beta \in W_2\}$$

REMARK

• If $a \in W_1$, then $a = a + 0$ with $0 \in W_1$ and $0 \in W_2$ so $W_1 \subseteq W_1 + W_2$, Similarly, $W_2 \subseteq W_1 + W_2$.

THEOREM 1. *The linear sum of two subspaces of a vector space is also a subspace.*

Proof. Let W_1 and W_2 be two subspaces of a vector space $V(F)$. Then we have to show that $W_1 + W_2$ is a subspace of $V(F)$.

Let α, β be any two arbitrary elements of $W_1 + W_2$,

Then, $\alpha, \beta \in W_1 + W_2$

$\Rightarrow$ $\alpha = \alpha_1 + \alpha_2$ and $\beta = \beta_1 + \beta_2$, where $\alpha_1, \beta_1 \in W_1$ and $\alpha_2, \beta_2 \in W_2$.

Since, $\alpha_1, \alpha_2, \beta_1, \beta_2 \in V$

$\therefore$ $W_1 + W_2 \subseteq V$.

Since, W_1 and W_2 are subspaces of V. Then

      $\alpha_1, \beta_1 \in W_1$ $\Rightarrow a\alpha_1 + b\beta_1 \in W_1$ for some $a, b \in F$

and $\alpha_2, \beta_2 \in W_2$ $\Rightarrow a\alpha_2 + b\beta_2 \in W_2$

Now, $a\alpha_1 + b\beta_1 \in W_1$ and $a\alpha_2 + b\beta_2 \in W_2$

$$\Rightarrow \quad (a\alpha_1 + b\beta_1) + (a\alpha_2 + b\beta_2) \in W_1 + W_2$$

$$\Rightarrow \quad a(\alpha_1 + \alpha_2) + b(\beta_1 + \beta_2) \in W_1 + W_2$$

$$\Rightarrow \quad a\alpha + b\beta \in W_1 + W_2$$

$$\therefore \qquad \alpha, \beta \in W_1 + W_2 \, , a, b \in F \Rightarrow a\alpha + b\beta \in W_1 + W_2$$

Hence, $W_1 + W_2$ is a subspace.

5.9 DIRECT SUM OF VECTOR SUBSPACES

Let W_1 and W_2 be two subspaces of a vector space V. Then V is said to be the direct sum of W_1 and W_2 if each element of V can be uniquely expressed as the sum of an element of W_1 and an element of W_2. If V is direct sum of W_1 and W_2, then it can be written as $V = W_1 \oplus W_2$.

In general, if V is the direct sum of $W_1, W_2, \ldots \ldots W_n$, then

$$V = W_1 \oplus W_2 \oplus \ldots \oplus W_n$$

Here $W_1, W_2, \ldots, W_n$ *are called* complementary spaces.

THEOREM 1. *The necessary and sufficient condition for a vector space V to be the direct sum of two of its subspaces W_1 and W_2 are :*

    (i) $V = W_1 + W_2$

    (ii) $W_1 \cap W_2 = \{0\}$

Proof.

Condition is necessary:

Suppose V is the direct sum of W_1 and W_2, then each element of V can be uniquely expressed as sum of an element of W_1 and an element of W_2, so in particular each element of V is expressible as the sum of an element of W_1 and element of W_2, this concludes that

$$V = W_1 + W_2$$

Next, we shall show that $W_1 \cap W_2 = \{0\}$, for this let, if possible, there be a non-zero vector in $W_1 \cap W_2$ and let it be $\alpha \in W_1 \cap W_2$. Then we may write

$$\alpha = \alpha + 0 \text{ with } \alpha \in W_1 \text{ and } 0 \in W_2.$$

and $\qquad\qquad\quad \alpha = 0 + \alpha$ with $0 \in W_1$ and $\alpha \in W_2.$

Since $W_1 + W_2 \in V$, so $\alpha \in V$ and $V = W_1 \oplus W_2$ therefore, α can be uniquely expressed as the sum of an element of W_1 and an element of W_2. Thus contains only zero vector. This implies $W_1 \cap W_2 = \{0\}$.

Condition is Sufficient:

Suppose the conditions:

    (i) $\quad V = W_1 + W_2$ (ii) $W_1 \cap W_2 = \{0\}$

hold then we shall show that $V = W_1 \oplus W_2$.

From (i) we conclude that each element of V can be expressed as the sum of an element of W_1 and an element of W_2. Therefore, we shall only show that this representation is unique.

Let, if possible, an element $\alpha \in V$ has two representations, that is,

$$\alpha = \alpha_1 + \alpha_2 \text{ with } \alpha_1 \in W_1 \text{ and } \alpha_2 \in W_2.$$

and $\qquad\qquad\qquad \alpha' = \alpha'_1 + \alpha'_2$ with $\alpha'_1 \in W_1$ and $\alpha'_2 \in W_2.$

$$\Rightarrow \qquad\qquad \alpha_1 + \alpha_2 = \alpha'_1 + \alpha'_2$$

$$\Rightarrow \qquad \alpha_1 - \alpha'_1 = \alpha'_2 - \alpha_2$$
$$\alpha_1, \alpha'_1 \in W_1 \Rightarrow \alpha_1 - \alpha'_1 \in W_1$$

and $\qquad\qquad \alpha_2, \alpha'_2 \in W_2 \Rightarrow \alpha'_2 - \alpha_2 \in W_2 \qquad$ [$\because W_1$ and W_2 are subspaces.]

$$\therefore \qquad \alpha_1 - \alpha'_1 = \alpha'_2 - \alpha_2 \in W_1 \cap W_2.$$

But $\qquad\qquad W_1 \cap W_2 = \{0\}$, this implies

$$\alpha_1 - \alpha'_1 = 0 = \alpha'_2 - \alpha_2$$

$$\Rightarrow \qquad\qquad \alpha_1 = \alpha'_1 \text{ and } \alpha'_2 = \alpha_2.$$

This shows that each element of V can be uniquely expressed as the sum of an element of W_1 and an element of W_2. Hence, $V = W_1 \oplus W_2$, *i.e.*, V is the direct sum of W_1 and W_2.

Solved Examples

Based on the following Results

➡ The necessary and sufficient conditions for a non-empty subset W of $V(F)$ to be a subspace is that
$$a, b \in F, \alpha, \beta \in W \Rightarrow a\alpha + b\beta \in W$$

➡ $W_1 + W_2 = \{\alpha + \beta : \alpha \in W_1, \beta \in W_2\}$

➡ The necessary and sufficient condition for a vector space V to be the direct sum of two of its subspaces W_1 and W_2 are
  (i) $V = W_1 + W_2$ (ii) $W_1 \cap W_2 = \{0\}$

Example 1. *In $V = R^3$. Let W_1 be the xy-plane and let W_2 be the z-plane given by*
$$W_1 = \{(x, y, 0) : x, y \in R\} \quad and \quad W_2 = \{(0, y, z) : y, z \in R\}$$
Show that $\qquad V = \{W_1 \oplus W_2\}$

Solution. Let $(x, y, z) \in V$, then this element can be written as the sum of an element of W_1 and an element of W_2 on one and only one way *i.e.*,
$$(x, y, z) = (x, y, 0) + (0, 0, z)$$
Accordingly, V is the direct sum of W_1 and W_2, that is $V = \{W_1 \oplus W_2\}$

Example 2. *In $V = R^3$ and W_1 be the xy-plane and let W_2 be the yz-plane:*
$$W_1 = \{(x, y, 0) : x, y \in R\} \quad and \quad W_2 = \{(0, y, z) : y, z \in R\}$$
then show that V is not the direct sum of W_1 and W_2.

Solution. Let (a, b, c) be any element of R^3, then
$$(a, b, c) = (x, y, 0) + (0, y, z) = (x, 2y, z)$$
$$\Rightarrow \qquad a = x, b = 2y, c = z$$
$$\Rightarrow \qquad a, b, c \in R \qquad\qquad [\because x, y, z \in R]$$
$\Rightarrow$ every element of V can be written as the sum of an element of W_1 and an element of W_2.
But such sums are not unique, for example,
Let $(3, 5, 7) \in R^3$, then
$$(3, 5, 7) = (3, 2, 0) + (0, 3, 7), (3, 2, 0) \in W_1 \text{ and } (0, 3, 7) \in W_2.$$
Also $\qquad (3, 5, 7) = (3, 1, 0) + (0, 4, 7), (3, 1, 0) \in W_1 \text{ and } (0, 4, 7) \in W_2.$

Hence, V is not the direct sum of W_1 and W_2.

Example 3. *If $V_3(R)$ is a vector space and $W_1 = \{(a, 0, c) : a, c \in R\}$ and $W_2 = \{(0, b, c) : b, c \in R\} $ are two subspaces of $V_3(R)$, then show that $V = W_1 + W_2$ and $V \neq W_1 \oplus W_2$.*

Solution. Let (x, y, z) be an arbitrary element of $V_3(R)$, then

$$(x, y, z) = (a, 0, c) + (0, b, c) = (a, b, 2c)$$

$\Rightarrow \qquad\qquad x = a, y = b, z = 2c$

$\Rightarrow \qquad\qquad x, y, z \in R \qquad\qquad\qquad\qquad [\because a, b, c \in R]$

$\Rightarrow$ every element of $V_3(R)$ can be written as the sum of an element of W_1 and an element of W_2.

$\Rightarrow \qquad\qquad V = W_1 + W_2$

But such representations is not unique, for example,

Let $(3, 5, 6) \in V_3(R)$, then

$(3, 5, 6) = (3, 0, 2) + (0, 5, 4); (3, 0, 2) \in W_1$ and $(0, 5, 4) \in W_2$.

Also $\quad (3, 5, 6) = (3, 0, 5) + (0, 5, 1); (3, 0, 5) \in W_1$ and $(0, 5, 1) \in W_2$.

So here $(3, 5, 6)$ can be written as the sum of elements of W_1 and W_2 in two ways. Hence $V \neq W_1 \oplus W_2$.

Example 4. *Let V be a vector space of all functions from R into R; let V_e be the subset of even functions, such that $f(-x) = f(x)$; let V_0 be the subset of odd functions $f(-x) = -f(x)$. Prove that*

(a) V_e and V_0 are subspaces of V.

(b) $V_e + V_0 = V$

(c) $V_e \cap V_0 = \{0\}$

Solution. (a) Since, V is a vector space of all functions, therefore,

$$V_e \subseteq V, V_0 \subseteq V.$$

Let $\quad f(x), g(x) \in V_e$, so that $f(-x) = f(x)$ and $g(-x) = g(x)$ and let $a, b \in R$, then consider,

$$h(x) = af(x) + bg(x)$$

Now $\quad h(-x) = af(-x) + bg(-x)$

or $\quad h(-x) = af(x) + bg(x) \qquad\qquad [f(-x) = f(x), g(-x) = g(x)]$

$$= h(x)$$

$\Rightarrow \qquad\qquad h(x) \in V_e$

$\Rightarrow af(x) + bg(x) \in V_e$

Consequently, if $f(x), g(x) \in V_e$, then $af(x) + bg(x) \in V_e$. Hence V_e is a subspace of V.

Similarly, we can prove that V_0 is a subspace of V.

(b) Let $f(x)$ be any element of V.

Consider $\quad f(x) = \dfrac{1}{2} [f(x) + f(-x)] + \dfrac{1}{2} [f(x) - f(-x)]$

Let $\quad \alpha(x) = \dfrac{1}{2} [f(x) + f(-x)]$ and $\beta(x) = \dfrac{1}{2} [f(x) - f(-x)]$

$\therefore \qquad\qquad f(x) = \alpha(x) + \beta(x).$

Also $\quad \alpha(-x) = \dfrac{1}{2} [f(-x) + f(x)] = \alpha(x)$

$$\therefore \qquad \alpha(x) = V_e.$$

and $\qquad \beta(-x) = \dfrac{1}{2}[f(-x) - f(x)] = \dfrac{1}{2}[f(x) - f(-x)] = \beta(x)$

$$\therefore \qquad \beta(x) = V_0.$$

Consequently, $f(x) = \alpha(x) + \beta(x)$ where, $\alpha(x) = V_e$ and $\beta(x) = V_0$. Hence every element of V can be expressed as the sum of an element of V_e and an element of V_0. That is,

$$V = V_e + V_0.$$

(c) $V_e \cap V_0 = \{0\}$

Let if possible, there exists a non-zero function $f(x)$, which belongs to $V_e \cap V_0$ Therefore, if $f(x) \in V_e$, then $f(-x) = f(x)$ and $f(x) \in V_0$, then $f(-x) = -f(x)$. So that,

$$f(x) = -f(x)$$
$$\Rightarrow \qquad 2\, f(x) = 0 \Rightarrow \ f(x) = 0$$

This gives a contradiction, because $f(x)$ is assumed to be non-zero function. Hence every function of $V_e \cap V_0$ is a zero function. Consequently, $V_e \cap V_0 = \{0\}$.

Example 5. *If W_1 and W_2 are subspaces of a vector space $V(F)$, then show that $W_1 + W_2$ is also a subspace of $V(F)$.*

Solution. Let us define

$$W_1 + W_2 = \{\alpha_1 + \alpha_2; \alpha_1 \in W_1, \alpha_2 \in W_2\}.$$

We have to show that $W_1 + W_2$ is a subspace of V.

Let $\quad \alpha \in W_1 + W_2 \Rightarrow \alpha = \alpha_1 + \alpha_2$ for some $\alpha_1 \in W_1$ and $\alpha_2 \in W_2$

$$\Rightarrow \alpha = \alpha_1 + \alpha_2 \text{ for some } \alpha_1, \alpha_2 \in V \Rightarrow \alpha \in V$$

$$\therefore \qquad W_1 + W_2 \subseteq V$$

Now, let $\alpha_1 + \alpha_2 \in W_1 + W_2$ and $\beta_1 + \beta_2 \in W_1 + W_2$, where $\alpha_1 + \alpha_2 \in W_1$ and $\beta_1 + \beta_2 \in W_2$.

Since, W_1 and W_2 are subspaces of V, then

$$a, b \in F, \alpha_1, \beta_1 \in W_1 \Rightarrow a\alpha_1 + b\beta_1 \in W_1$$
$$a, b \in F, \alpha_2, \beta_2 \in W_2 \Rightarrow a\alpha_2 + b\beta_2 \in W_2$$
$$a\alpha_1 + b\beta_1 \in W_1, a\alpha_2 + b\beta_2 \in W_2$$

$$\Rightarrow (a\alpha_1 + b\beta_1) + (a\alpha_2 + b\beta_2) \in W_1 + W_2$$
$$\Rightarrow a(\alpha_1 + \alpha_2) + b(\beta_1 + \beta_2) \in W_1 + W_2$$

Since, $\alpha_1 + \alpha_2 \in W_1 + W_2$ and $\beta_1 + \beta_2 \in W_1 + W_2$.

Thus we have $\qquad a, b \in F, \alpha_1 + \alpha_2, \beta_1 + \beta_2 \in W_1 + W_2$

$$\Rightarrow a(\alpha_1 + \alpha_2) + b(\beta_1 + \beta_2) \in W_1 + W_2$$

Hence, $W_1 + W_2$ is subspace of $V(F)$.

Example 6. *Let R be the field of real numbers, show that the set $W=\{(x,2y,3z): x,y, z \in R\}$ is a subspace of $V_3(R)$.*

Solution. Since, $W=\{(x, 2y, 3z) : x, y, z \in R\}$.

Let $\alpha, \beta \in W$, where $\alpha=\{x_1, 2y_1, 3z_1\}$ and $\beta=\{x_2, 2y_2, 3z_2\}$ and $x_1, y_1, z_1, x_2, y_2, z_2 \in R$,

If $a, b \in R$, then

$$a\alpha+b\beta=a(x_1, 2y_1, 3z_1)+ b(x_2, 2y_2, 3z_2)$$
$$=(ax_1, 2ay_1, 3az_1)+ (bx_2, 2by_2, 3bz_2)$$
$$=(ax_1+bx_2), 2(ay_1+by_2), 3(az_1+bz_2)$$

$\therefore \qquad a\alpha+b\beta \in W,$

Because $ax_1+bx_2, ay_1+by_2, az_1+bz_2 \in R$.

Hence, W is a subspace of $V_3(R)$.

Example 7. *Show that the set of all real valued continuous functions defined on $[0,1]$ is a vector space over field of reals.*

Solution. Let V be the set of all real valued continuous functions defined on $[0,1]$. Now we have to show that V is a vector space over R (field of real numbers) under vector addition and scalar multiplication which is defined as follows:

$$(f+g)(x)= f(x)+g(x), \forall f, g \in V$$

and $\qquad (af)(x)= af(x), \forall f \in V$ and $a \in R.$

First we shall show that $(V,+)$ is an abelian group.

Let $f, g \in V$, then

$$(f+g)(x)= f(x)+g(x), \forall f, g \in V$$

$\therefore \qquad f+g \in V.$ Thus V is closed under vector addition.

Now let $0(x) \in V$, we have

$$f(x)+0(x)= (f+0)(x)= f(x), \forall f \in V$$

$\therefore \qquad 0(x)$ is the additive identity in V.

Let $-f \in V$, then we have

$$-f(x)+f(x)= (-f+f)(x)= 0(x), \forall f \in V$$

$\therefore \qquad -f$ is the additive inverse of V.

Since vector addition is always associative as well as commutative, consequently $(V,+)$ is an abelian group.

Further, since V is closed under scalar multiplication therefore af is a real valued continuous function defined on $[0,1]$.

(i) If $a \in R$ and $f, g \in V$, then we have

$$a[(f+g)x]=a[f(x)+g(x)]$$
$$=af(x)+ag(x)=(af+ag)(x)$$

$\therefore \qquad a(f+g)=af+ag.$

(ii) If $a, b \in R$ and $f \in V$, then we have

$$[(a+b)f](x) = (a+b)f(x) = af(x) + bf(x) = (af + bf)(x)$$

∴ $\qquad (a+b)f = af + bf.$

(iii) If $a, b \in R$ and $f \in V$, then we have

$$[(ab)f](x) = (ab)f(x) = a[bf(x)] = [a(bf)](x).$$

∴ $\qquad (ab)f = a(bf).$

(iv) If $1 \in R$ and $f \in V$, then we have

$$(1f)(x) = 1f(x) = f(x)$$

∴ $\qquad 1f = f, \ \forall f \in V$

Hence V is a vector space over R.

Example 8. *Show that $R^2(R)$ is not a vector space when addition and scalar multiplication composition are defined by*

$$(a_1, a_2) + (b_1, b_2) = (a_1 + b_1, a_2 + b_2)$$

and $\qquad a(a_1, a_2) = (aa_1, a_2), \ \forall \ a, a_1, a_2, b_1, b_2 \in R.$

Solution. Suppose $a = 1$ and $b = 2$ and $(a_1, a_2) = (3, 4)$, then using the given compositions, we have

$$(a+b)(a_1, a_2) = (1+2) . (3+4)$$
$$= 3 . (3, 4) = (3.3, 4) = (9, 4)$$

and $a . (a_1, a_2) + b . (a_1, a_2) = 1 . (3, 4) + 2 . (3, 4)$
$$= (3.1, 4) + (3.2, 4)$$
$$= (3, 4) + (6, 4)$$
$$= (3 + 6, 4 + 4) = (9, 8)$$

∴ $\qquad (a+b) . (a_1, a_2) \neq a . (a_1, a_2) + b . (a_1, a_2).$

Hence, R^2 (R) is not a vector space.

Example 9. *Prove the solution set W of the differential equation*

$$2 \frac{d^2 y}{dx^2} - 9 \frac{dy}{dx} + 2y = 0$$

is a subspace of vector space of all real valued functions of R.

Solution. Let $W = \left\{ y : 2 \frac{d^2 y}{dx^2} - 9 \frac{dy}{dx} + 2y = 0 \right\}$, be the set of all solutions of the given

differential equation where, $y = f(x)$.

Now if we define a real valued function denoted by 0 on R by $0(x) = 0, \ \forall \ x \in R$, then $0(x)$ satisfies the given differential equation, so that $0(x) \in W$.

Let $y_1 = f(x)$ and $y_2 = g(x)$ be any two elements of W, then we have

$$2 \frac{d^2 f(x)}{dx^2} - \frac{df(x)}{dx} + 2f(x) = 0 \qquad \text{.... (1)}$$

and $\qquad 2 \frac{d^2 g(x)}{dx^2} - \frac{dg(x)}{dx} + 2g(x) = 0 \qquad \text{.... (2)}$

Let a, b be any two scalars.

Now, multiplying (1) by a and (2) by b and then adding, we get

$$2\frac{d^2}{dx^2}[af(x) + bg(x)] - 9\frac{d}{dx}[af(x) + bg(x)] + 2[af(x) + bg(x)] = 0$$

which shows that $af(x) + bg(x)$ is also the solution of the given differential equation.

So that $\qquad\qquad [af(x) + bg(x)] \in W.$

$\therefore \qquad\qquad f(x) \in W, g(x) \in W \Rightarrow af(x) + bg(x) \in W \; \forall \, a, b \in R.$

Hence W is a subspace of a vector space of all real valued functions of R.

Example 10. *Let V be the vector space of all functions from the real field* R *into* R. *Show that the set $W = \{f : f(7) = 2 + f(1)\}$ is not a subspace of V.*

Solution. Let f and g be any two elements of W, i.e.,

$$f(7) = 2 + f(1) \text{ and } g(7) = 2 + g(1).$$

Then $\qquad (f + g)(7) = f(7) + g(7)$

$$= 2 + f(1) + 2 + g(1)$$
$$= 4 + f(1) + g(1)$$
$$= 4 + (f + g)(1) \neq 2 + (f + g)(1).$$

Hence $f + g \neq W$, and so W is not a subspace of V.

Example 11. *Let V be the vector space of all square $n \times n$ matrices over a field of reals* R. *Show that W is a subspace of V, where*

  (i) W consists of the symmetric matrices.

  (ii) W consists of all matrices which commute with a given matrix M, i.e.,

$$W = \{A \in V : AM = MA\}.$$

Solution. (i) The null matrix $0 \in W$, as all its entries being zero and it is symmetric.

Let $\quad A = [a_{ij}]$ and $B = [b_{ij}]$ be any two elements of W.

Then $a_{ij} = a_{ji}$ and $b_{ij} = b_{ji}$.

For any scalars $a, b \in R$, we have

$$aA + bB = a[a_{ij}] + b\,[b_{ij}] = [aa_{ij} + bb_{ij}] = [c_{ij}] = C$$

where, $\qquad\qquad c_{ij} = aa_{ij} + bb_{ij}$

Now, $\qquad\qquad c_{ij} = aa_{ij} + bb_{ij} = aa_{ji} + bb_{ji} = c_{ji}$

Thus $aA + bB$ is symmetric so it belongs to W. Hence W is subspace of V.

(ii) The null matrix $0 \in W$ as $OM = MO$.

Now suppose A and B be any two elements of W then, $AM = MA$ and $BM = MB$.

For $a, b \in R$, we have

$$(aA + bB)M = (aA)M + (bB)M = a(AM) + b(BM)$$
$$= a(MA) + b(MB) = M(aA + bB)$$

$\therefore \qquad\qquad aA + bB \in W, \; A, B \in W$ and $\forall \, a, b \in R.$

Hence W is a subspace of V.

Example 12. *let $V = R^3$. Show the set $W = \{(a, b, c) : a^2 + b^2 + c^2 \leq 1\}$ is not a subspace of V.*

Solution. Let $\alpha = (1, 0, 0) \in W, \quad \beta = (0, 1, 0) \in W.$ But we have

$$\alpha + \beta = (1, 0, 0) + (0, 1, 0)$$
$$= (1, 1, 0) \notin W \text{ as } 1^2 + 1^2 + 0^0 = 2 > 1.$$

Hence W is not a subspace of V.

Example 13. *Let V be the vector space of all 2 × 2 matrices over the real field R. Show that W is not a subspace of V, where*

    (i) *W consists of all matrices with zero determinant,*

    (ii) *W consists of all matrices A from which $A^2 = A$.*

Solution.   (i) Let $A = \begin{pmatrix} 1 & 0 \\ 0 & 0 \end{pmatrix}$ and $B = \begin{pmatrix} 0 & 0 \\ 0 & 1 \end{pmatrix}$ be two elements of W. But,

$$A + B = \begin{pmatrix} 1 & 0 \\ 0 & 0 \end{pmatrix} + \begin{pmatrix} 0 & 0 \\ 0 & 1 \end{pmatrix}$$

$$= \begin{pmatrix} 1 & 0 \\ 0 & 1 \end{pmatrix} \notin W \text{ as } |A + B| = 1 \neq 0$$

∴ *W* is not a subspace of *V*.

    (ii) The unit matrix $I = \begin{pmatrix} 1 & 0 \\ 0 & 1 \end{pmatrix} \notin W$ as $I^2 = I$.

    But $2I = \begin{pmatrix} 2 & 0 \\ 0 & 2 \end{pmatrix} \notin W$ as

$$(2I)^2 = \begin{pmatrix} 2 & 0 \\ 0 & 2 \end{pmatrix}\begin{pmatrix} 2 & 0 \\ 0 & 2 \end{pmatrix} = \begin{pmatrix} 4 & 0 \\ 0 & 4 \end{pmatrix} \neq 2I.$$

Hence *W* is not a subspace of *V*.

🕮 EXERCISE 5.3

1. Show that the complex field C is a vector space over the field R of reals.

2. Let *V* be the set of all pairs of real numbers and let *F* be the field of real numbers and define

    $(x, y) + (x_1, y_1) = (3y + 3y_1, -x - x_1)$ and $c(x, y) = (3cy, -cx)$.

    Show that *V* is a vector space over *F*.

3. Show that the set $W = \{(a_1, a_2, 0) : a_1, a_2 \in F\}$ is a subspace of $V_3(F)$.

4. Show that the set *W* of the elements of the vector space $V_3(R)$ of the form

    $(x + 2y, y, -x+3y)$

    where $x, y \in R$ is a subspace of $V_3(R)$.

5. Prove that the set of all solutions (a, b, c) of the equation $a+b+2c = 0$ is a subspace of vector space $V_3(R)$.

6. Prove that the arbitrary intersection of subspaces of a vector space is a subspace.

7. Let $V = R^3$. Show that the set $W = \{(a,b,c) : a,b,c \in Q]$ is not a subspace of R^3.

Hint to Selected Problems

1. Let C be the set of vectors and R the set of scalars, since C is a field so that $a\alpha \in C$, this show that C is closed under scalar multiplication. $1 \in R$ so $1 \in C$.

    Also for $a, b \in C$ and $a,b \in R$,

         $a, b \in R \Rightarrow a,b \in C$ $(\because R \subseteq C)$

  ∴ $a(\alpha + \beta\} = a\alpha + a\beta$ and $(a + b)\alpha = a\alpha + b\alpha$.

    also, $(ab)\alpha = a(b\alpha)$, $1.\alpha = \alpha, \forall \alpha \in C$.   Hence C is a vector space over R.

4. $W = ((x+2y, y, -x+3y) : x, y \in R)$, then show that $a\alpha + b\beta \in W \forall \alpha, \beta \in W$ and $a, b \in R$.

    $a\alpha + b\beta = a(x_1 + 2y_1, y_1, -x_1 + 3y_1] + b(x_2 + 2y_2, y_2, -x_2 + 3y_2]$

$$= (ax_1 + 2ay_1, ay_1, -ax_1+3ay_1)+(bx_2 +2by_2, by_2, -bx_2+3by_2)$$
$$= (ax_1 + bx_2 +2(ay_1+by_2), ay_1+, by_2-(ax_1+ bx_2)+3(ay_1+by_2))\in W$$

$\therefore a\alpha+b\beta \in W \Rightarrow W$ is a subspace.

7. $(a, b, c) \in W, (a, b, -c)\in W$ as $-c\in Q$.

   Now $(a, b, c) + (a, b, -c) = (2a, 2b, 0)\in W$. Hence W is not a subspace.

5.10 LINEAR COMBINATION OF VECTORS

Definition. *Let V be a vector space over a field F and $\alpha_1, \alpha_2, \ldots ,\alpha_n \in V$, then any vector $\alpha \in V$ can be expressed as below:*

$$\alpha = a_1\alpha_1+ a_2\alpha_2+\ldots+ a_n\alpha_n$$

where $a_1, a_2,\ldots, a_n\in F$, is said to be the linear combination of vectors $\alpha_1, \alpha_2,\ldots, \alpha_n$.

Definition. *Let V(F) be a vector space aver F and let S be any non-empty subset of V, then the set of all linear combination of finite elements of S, is called the linear span of S. It is denoted by L(S). Therefore, we have*

$$L(S) = \{a_1\alpha_1+ a_2\alpha_2+\ldots+ a_n\alpha_n: a_1, a_2,\ldots, a_n\in F$$

and $\alpha_1, \alpha_2,\ldots, \alpha_n$ are a finite elements of S}

THEOREM 1. *The linear span L(S) of a non-empty subset S of a vector space V(F) is the smallest subspace of V containing S.*

Proof. By definition of $L(S)$, we have

$$L(S) = \{a_1\alpha_1 +a_2\alpha_2 + \ldots + a_n\alpha_n : a_i \in V\}$$

Let $\alpha \in S$, then $\alpha = 1.\alpha, 1\in F$, so $\alpha\in L(S)$

$\therefore \qquad S\subseteq L(S)$

Now, we shall show that $L(S)$ is a subspace.

Let α, β be any two arbitrary elements of $L(S)$, then

$$\alpha = a_1\alpha_1+a_2\alpha_2+\ldots+ a_n\alpha_n; \text{ for } \alpha_1, \alpha_2,\ldots, \alpha_n \in S \text{ and for } a_1, a_2,\ldots,a_n \in F$$

Also, $\quad \beta = b_1\beta_1+b_2\beta_2+ \ldots + b_m\beta_m$ for all $a, b \in F$, we have

$$a\alpha + b\beta = a(a_1\alpha_1+a_2\alpha_2+ \ldots + a_n\alpha_n)+ b(b_1\beta_1+b_2\beta_2 + \ldots + b_m\beta_m)$$
$$= (aa_1)\alpha_1+(aa_2)\alpha_2+ \ldots + (aa_n)\alpha_n+ (bb_1)\beta_1+(bb_2)\beta_2+ \ldots + (bb_m)\beta_m$$

This implies that $a\alpha + b\beta$ is a linear combination of finite number of elements of S, so $a\alpha + b\beta \in L(S)$. Hence $L(S)$ is a subspace of V.

Next, we shall show that $L(S)$ is the smallest subspace containing S.

For this there is a subspace W of V containing S. Let $\alpha_1, \alpha_2,\ldots,\alpha_t \in S \subset W$ and W being a subspace, then

$$a_1\alpha_1+ a_2\alpha_2 +\ldots + a_t\alpha_t \in W; \text{ for all } a_i \in F$$

This implies that W contains all linear combinations of finite elements of S, therefore $L(S)\in W$.

Hence $L(S)$ is the smallest subspace of V containing S.

THEOREM 2. *If S, T are two subsets of a vector space V, then*

   (i) $S\subseteq T \Rightarrow L(S) \subseteq L(T)$.

   (ii) $L(S \cup T) = L(S) + L(T)$

   (iii) $L[L(S)] = L(S)$

Proof.   Let α be an arbitrary element of $L(S)$, then

$$\alpha \in L(S) \Rightarrow \alpha = a_1\alpha_1 + a_2\alpha_2 + \dots + a_n\alpha_n$$

where, $\alpha_1, \alpha_2,\dots,\alpha_n \in S$ and $a_1, a_2,\dots,a_n \in F$.

(i) Since $S \subseteq T$, so that $\alpha_1, \alpha_2, \dots, \alpha_n \in T$, therefore α is also the linear combination of finite elements of T. This implies $\alpha \in L(T)$.

Thus $\qquad\qquad \alpha \in L(S) \Rightarrow \alpha \in L(T)$

Hence $\qquad\qquad L(S) \subseteq L(T)$ if $S \subseteq T$.

(ii) Since $S \subseteq S \cup T$ and $T \subseteq S \cup T$, then from (i), we have

$$L(S) \subseteq L(S \cup T)$$

and $\qquad\qquad L(T) \subseteq L(S \cup T)$

$\Rightarrow \qquad\qquad L(S) + L(T) \subseteq L(S \cup T) \qquad\qquad\dots(1)$

Next, let α be an arbitrary element of $L(S \cup T)$, then α is a linear combination of finite elements of $S \cup T$. This implies that some of $\alpha_i \in S$ or some of $\alpha_j \in T$. This shows that α is a linear combination of finite elements of S and finite elements of T, therefore $\alpha \in L(S) + L(T)$. Thus,

$$L(S \cup T) \subseteq L(S) + L(T) \qquad\qquad\dots(2)$$

From (1) and (2), we get

$$L(S \cup T) = L(S) + L(T) \qquad\qquad\dots(3)$$

(iii) Since $S \subseteq T(S)$, then from (i), we have

$$L(S) \subseteq L[L(S)] \qquad\qquad\dots(4)$$

Next, let α be any arbitrary element of $L[L(S)]$, then α is a linear combination of finite elements of $L(S)$. Suppose, we have

$$\alpha = b_1\beta_1 + b_2\beta_2 + \dots + b_n\beta_n = \sum_{i=1}^{n} b_i\beta_i$$

where, each $\beta_i \in L(S)$ for all $b_1, b_2, \dots, b_n \in F$. Also each β_i is a linear combination of finite elements of S, so that

$$\beta_1 = a_{11}\alpha_1 + a_{12}\alpha_2 + \dots + a_{1m}\alpha_m$$

$$\beta_1 = a_{21}\alpha_1 + a_{22}\alpha_2 + \dots + a_{2t}\alpha_t$$

$$\dots\quad\dots\quad\dots\quad\dots\quad\dots\quad\dots\quad\dots$$

On putting the values of $\beta_1, \beta_2, \dots$, etc. in (2), we see that α is a linear combination of finite elements of S. Thus $\alpha \in L(S)$

$\therefore \qquad\qquad L[L(S)] \subseteq L(S) \qquad\qquad\dots(5)$

From (4) and (5), we get $L(S) = L[L(S)]$.

THEOREM 3.   *The linear sum of two subspaces W_1 and W_2 of a vector space $V(F)$ is generated by their union. That is, $W_1 + W_2 = L(W_1 \cup W_2)$.*

Proof.   We have already proved that the linear sum of two subspaces is also a subspace and linear span of a subset of a vector space is also a subspace.

Therefore, $W_1 + W_2$ and $L(W_1 \cup W_2)$ are subspaces of $V(F)$.

Let α be any arbitrary element of $W_1 + W_2$, then

$$\alpha \in W_1 + W_2$$

$\Rightarrow$ $\qquad\qquad\qquad \alpha = \alpha_1 + \alpha_2$ for some $\alpha_1 \in W_1$ and $\alpha_2 \in W_2$

Since, $\alpha_1 \in W_1$ and $\alpha_2 \in W_2$, so, $\alpha_1 + \alpha_2 \in W_1 + W_2$

Also, we may write $\alpha = \alpha_1 + \alpha_2 = 1. \alpha_1 + 1.\alpha_2$. This implies that α is a linear combination of finite elements namely α_1 and α_2 of $W_1 \cup W_2$, so that $\alpha \in L\ (W_1 + W_2)$

$\therefore$ $\qquad\qquad\qquad \alpha \in W_1 + W_2 \Rightarrow a \in L\ (W_1 \cup W_2)$

Thus, $\qquad W_1 + W_2 \subseteq L\ (W_1 \cup W_2)$. $\qquad\qquad\qquad\qquad\qquad$ ... (1)

But $L\ (W_1 \cup W_2)$ being the smallest subspace containing $W_1 \cup W_2$, and since $W_1 + W_2$ is a subspace containing $W_1 \cup W_2$, therefore

$$L(W_1 \cup W_2) \subseteq W_1 \cup W_2 \qquad\qquad\qquad\qquad\qquad\text{... (2)}$$

From (1) and (2), we get

$$W_1 + W_2 = L\ (W_1 \cup W_2)$$

5.11 LINEAR DEPENDENCE AND INDEPENDENCE OF VECTORS

In this section, we shall discuss the concept of linear dependence or linear independence which lays the foundations for the key notions (viz., dimensions) of the theory of vector spaces.

Definition 1. *Let $V(F)$ be a vector space over a field F. Then a finite set $\{\alpha_1, \alpha_2, ..., \alpha_n\}$ of vectors of V is said to be linearly dependent if there exists scalars $a_1, a_2, ..., a_n$ not all of them equal to zero such that*

$$a_1 \alpha_1 + a_2 \alpha_2, ...; + a_n \alpha_n = 0$$

Definition 2. *Let $V(F)$ be a vector space over F. Then a finite set of vectors $\{\alpha_1, \alpha_2, ..., \alpha_n\}$ of V is said to linearly independent if for every expressions of the type.*

$$a_1 \alpha_1 + a_2 \alpha_2 + ... + a_n \alpha_n = 0$$

where $a_1, a_2, ..., a_n \in F$ implies $a_1 = 0, a_2 = ... = a_n$.

REMARKS

- S is linearly independent
  $\Leftrightarrow$ S is not linearly dependent $\Leftrightarrow$ no finite subset of S is linearly dependent.
  $\Leftrightarrow$ each finite subset of S is linearly independent.
  $\Leftrightarrow$ whenever $\sum\limits_{i=1}^{n} a_i\alpha_i = 0, a_i \in F, \alpha_i \in S, i = 1, 2, ..., n$ and $n \in N$, then each $a_i = 0$.

- Any infinite set is linearly independent if it's every finite subset is linearly independent otherwise it is linearly dependent.

THEOREM 1. *If $\alpha_1, \alpha_2, ..., \alpha_n \in V$ are linearly independent, then every element in their linear span has a unique representation in the form*

$$a_1 \alpha_1 + a_2 \alpha_2, ..., + a_n\alpha_n \text{ with } a_i \in F.$$

Proof. By definition of linear span, we know that every element in the linear span is of the form $a_1 \alpha_1 + a_2 \alpha_2 + ... + a_n\alpha_n$. Therefore, we only show the uniqueness of the representation.

Let if possible,

$a_1 \alpha_1 + a_2 \alpha_2 + ... + a_n\alpha_n$ and $b_1 \alpha_1 + b_2 \alpha_2 + ... + b_n\alpha_n$ be two forms of an element in linear span of $\alpha_1, \alpha_2, ..., \alpha_n$, then

$$a_1 \alpha_1 + a_2 \alpha_2 + ... + a_n\alpha_n = b_1 \alpha_1 + b_2 \alpha_2 + ... + b_n\alpha_n$$

$\Rightarrow \qquad (a_1 - b_1)\alpha_1 + (a_2 - b_2)\alpha_2 + \ldots + (a_n - b_n)\alpha_n = 0$

Since, $\alpha_1, \alpha_2, \ldots, \alpha_n$ are linearly independent so, we have

$$a_1 - b_1 = 0, \quad a_2 - b_2 = 0, \ldots, a_n - b_n = 0$$

$\Rightarrow \qquad a_1 = b_1, \qquad a_2 = b_2, \ldots, \qquad a_n = b_n .$

Hence, every element in the linear span of linearly independent vectors $\alpha_1, \alpha_2, \ldots, \alpha_n$ has a unique form of $a_1 \alpha_1 + a_2 \alpha_2 + \ldots + a_n \alpha_n$.

THEOREM 2. *If $\alpha_1, \alpha_2, \ldots, \alpha_n \in V$, then either they are linearly independent or some α_k is a linear combination of preceding ones $\alpha_1, \alpha_2, \ldots, \alpha_{k-1}$.*

Proof. If $\alpha_1, \alpha_2, \ldots, \alpha_n \in V$ are linearly independent, then nothing is to prove. So assume that $\alpha_1, \alpha_2, \ldots, \alpha_n$ are not linearly independent. Then, there are some $a_i \in F$ which are non-zero such that

$$a_1 \alpha_1 + a_2 \alpha_2 + \ldots + a_n \alpha_n = 0$$

Let k be the largest integer for which $a_k \neq 0$. Since $a_i = 0$ for $i > k$, and

$$a_1 \alpha_1 + a_2 \alpha_2 + \ldots + a_k \alpha_k = 0$$

$\Rightarrow \qquad a_k = a_k^{-1}(-a_1\alpha_1 - a_2\alpha_2 - \ldots - a_{k-1}\alpha_{k-1}) \qquad [\because a_k \neq 0]$

$\Rightarrow \qquad a_k = (-a_k^{-1}a_1)\alpha_1 + (-a_k^{-1}a_2)\alpha_2 + \ldots + (-a_k^{-1}a_{k-1})\alpha_{k-1}$

$\Rightarrow \qquad a_k$ is a linear combination of preceding ones $\alpha_1, \alpha_2, \ldots, \alpha_{k-1}$.

Solved Examples

Example 1. *Is the vector $(2, -5, 3)$ in the subspace of R^3 spanned by the vectors $(1, -3, 2)$, $(2, -4, -1)$, $(1, -5, 7)$?*

Solution. Let $\alpha = (2, -5, 3)$ and $\alpha_1 = (1, -3, 2)$, $\alpha_2 = (2, -4, -1)$ and $\alpha_3 = (1, -5, 7)$

Again let $\qquad \alpha = a_1\alpha_1 + a_2\alpha_2 + a_3\alpha_3 \quad$ where $\quad a_1, a_2, a_3 \in R$

Then $\qquad (2, -5, 3) = a_1(1, -3, 2) + a_2(2, -4, -1) + a_3(1, -5, 7)$

$\Rightarrow \qquad (2, -5, 3) = (a_1 + 2a_2 + a_3, -3a_1 - 4a_2 - 5a_3, 2a_1 - a_2 + 7a_3)$

$\therefore \qquad a_1 + 2a_2 + a_3 = 2$ ... (1)

$\qquad -3a_1 - 4a_2 - 5a_3 = -5$ ... (2)

$\qquad 2a_1 - a_2 + 7a_3 = 3$ ... (3)

Eliminating a_2 between (1) and (2), we get

$$-a_1 - 3a_3 = -1$$

or $\qquad a_1 = 1 - 3a_3$ ... (4)

Again, eliminating a_2 between (2) and (3), we get

$$11a_1 + 33a_3 = 17$$ ... (5)

Here no values of a_3 and a_1 will satisfy both equations (4) and (5). Thus, equations (1), (2) and (3) have no solution.

Hence α cannot be expressed as a linear combination of α_1, α_2 and α_3.

Hence, the vector $(2, -5, 3)$ is not spanned by the vectors $(1, -3, 2)$, $(2, -4, -1)$ and $(1, -5, 7)$.

Example 2. *In the vector space* R^3 *express the vector* $(1, -2, 5)$ *as a linear combination of the vectors* $(1, 1, 1)$, $(1, 2, 3)$ *and* $(2, -1, 1)$.

Solution. Let $\alpha = (1, -2, 5), \alpha_1 = (1,1,1), \alpha_2 = (1,2,3)$ and $\alpha_3 = (2, -1, 1)$.

Let
$$\alpha = a_1\alpha_1 + a_2\alpha_2 + a_3\alpha_3$$

where $a_1, a_2, a_3 \in R$. Then
$$(1, -2, 5) = a_1(1,1,1) + a_2(1,2,3) + a_3(2, -1, 1)$$

$$\Rightarrow \quad (1, -2, 5) = (a_1 + a_2 + 2a_3, a_1 + 2a_2 - a_3, a_1 + 3a_2 + a_3)$$

$$\therefore \qquad\qquad a_1 + a_2 + 2a_3 = 1 \qquad\qquad \ldots (1)$$
$$a_1 + 2a_2 - a_3 = -2 \qquad\qquad \ldots (2)$$
$$a_1 + 3a_2 + a_3 = 5 \qquad\qquad \ldots (3)$$

Eliminating a_1 between (1) and (2), we get
$$-a_2 + 3a_3 = 3 \qquad\qquad \ldots (4)$$

Eliminating a_1 between (2) and (3), we get
$$-a_2 - 2a_3 = -7$$

or
$$a_2 + 2a_3 = 7 \qquad\qquad \ldots (5)$$

Solving (4) and (5), we get
$$a_3 = 2, a_2 = 3$$

Putting the values of a_2 and a_3 in (1), we get
$$a_1 = -6$$

Hence, $(1, -2, 5) = -6(1, 1, 1) + 3(1, 2, 3) + 2(2, -1, 1)$

Example 3. *For what values of m, the vector* $(m, 3, 1)$ *is a linear combination of the vectors* $(3, 2, 1)$ *and* $(2, 1, 0)$?

Solution. Let $\alpha = (m, 3, 1), \alpha_1 = (3, 2, 1)$ and $\alpha_2 = (2, 1, 0)$

Let $\qquad\qquad \alpha = a_1\alpha_1 + a_2\alpha_2$ where $a_1, a_2 \in R$

Then $\qquad\qquad (m, 3, 1) = a_1(3, 2, 1) + a_2(2, 1, 0)$

$$\Rightarrow \qquad\qquad (m, 3, 1) = (3a_1 + 2a_2, 2a_1 + a_2, a_1)$$
$$\therefore \qquad\qquad 3a_1 + 2a_2 = m \qquad\qquad \ldots (1)$$
$$2a_1 + a_2 = 3 \qquad\qquad \ldots (2)$$
$$a_1 = 1 \qquad\qquad \ldots (3)$$

Solving (1), (2) and (3), we get
$$a_1 = 1, a_2 = 1, m = 5$$

Hence, the reaquired value of m is 5.

Example 4. *In the vector space* R^4 *determine whether or not the vector* $(3, 9, -4, -2)$ *is a linear combination of the vectors* $(1, -2, -0, 3)$, $(2, 3, 0, -1)$ *and* $(2, -1, 2, 1)$.

Solution. Let $\alpha = (3, 9, -4, -2), \alpha_1 = (1, -2, 0, 3), \alpha_2 = (2, 3, 0, -1)$ and $\alpha_3 = (2, -1, 2, 1)$.

Let
$$\alpha = a_1\alpha_1 + a_2\alpha_2 + a_3\alpha_3$$

where $\alpha_1, \alpha_2, \alpha_3 \in R$.

Then, $(3,9,-4,-2) = a_1(1,-2,0,3) + a_2(2,3,0,-1) + a_3(2,-1,2,1)$

$\therefore$ $a_1 + 2a_2 + 2a_3 = 3$ ...(1)

$-2a_1 + 3a_2 - a_3 = 9$ ... (2)

$2a_3 = -4$ ... (3)

$3a_1 - a_2 + a_3 = -2$ ... (4)

Solving (1), (2) and (3), we get

$$a_1 = 1, a_2 = 3, a_3 = -2$$

These values satisfy the equation (4). Thus the system of equations (1), (2), (3) and (4) is consistent and has a solution.

Hence, α can be written as a linear combination of vectors α_1, α_2 and α_3.

Example 5. *Write the polynomial $f(x) = x^2 + 4x - 3$ over R as a linear combination of the polynomials*

$$f_1(x) = x^2 - 2x + 5, f_2(x) = 2x^2 - 3x \text{ and } f_3(x) = x + 3$$

Solution. Let $f(x) = a_1 f_1(x) + a_2 f_2(x) + a_3 f_3(x)$ where $a_1, a_2, a_3 \in R$

Then, $x^2 + 4x - 3 = a_1(x^2 - 2x + 5) + a_2(2x^2 - 3x) + a_3(x + 3)$

$\Rightarrow$ $x^2 + 4x - 3 = (a_1 + 2a_2)x^2 + (-2a_1 - 3a_2 + a_3)x + 5a_1 + 3a_3$

$\therefore$ $a_1 + 2a_2 = 1$ ... (1)

$-2a_1 - 3a_2 + a_3 = 4$ ... (2)

$5a_1 + 3a_3 = -3$ ... (3)

Eliminating a_2 between (1) and (2), we get

$-a_1 + 2a_3 = 11$ ... (4)

Solving (3) and (4), we get

$$a_1 = -3, a_3 = 4$$

Putting these values in (1), we get

$$a_2 = 2$$

Hence $x^2 + 4x - 3 = -3(x^2 - 2x + 5) + 2(2x^2 - 3x) + 4(x + 3)$

Example 6. *In the vector space R^3, let $\alpha = (1,2,1), \beta = (3,1,5), \gamma = (3,-4,7)$ Show that the subspace spanned by $S = \{\alpha, \beta\}$ and $T = \{\alpha, \beta, \gamma\}$ are the same.*

Solution. We have to show that $L(S) = L(T)$

From given sets S and T we have

$$S \subseteq T \quad \Rightarrow \quad L(S) \subseteq L(T)$$

Now we show that γ can be expressed as a linear combination of α and β.

Let $\gamma = a\alpha + b\beta$

$\Rightarrow$ $(3,-4,7) = a(1,2,1) + b(3,1,5)$

$\Rightarrow$ $(3,-4,7) = (a + 3b, 2a + b, a + 5b)$

$\therefore$ $a + 3b = 3$ ... (1)

$$2a + b = -4 \qquad \qquad \dots (2)$$

$$a + 5b = 7 \qquad \qquad \dots (3)$$

Solving (1) and (2), we get

$$a = -3, b = 2$$

These values satisfy the equation (3).

Thus $\qquad \qquad \gamma = -3\alpha + 2\beta$

Now let $\delta \in L(T)$, then δ can be expressed as a linear combination of α, β and γ but γ can be replaced by $-3\alpha + 2\beta$, therefore, δ can be expressed as a linear combination of α and β.

Thus $\qquad \qquad \delta \in L(S)$

$\therefore \qquad \qquad L(T) \subseteq L(S)$

Hence $\qquad \qquad L(S) = L(T)$

Example 7. *Find a condition on a, b, c such that* $\alpha = (a, b, c)$ *is a linear combination of vectors* $(1, -3, 2)$ *and* $(2, -1, 1)$.

Solution. Let $\alpha = a_1(1, -3, 2) + a_2(2, -1, 1)$

$\Rightarrow \qquad \qquad (a, b, c) = a_1(1, -3, 2) + a_2(2, -1, 1)$

$\Rightarrow \qquad \qquad (a, b, c) = (a_1 + 2a_2, -3a_1 - a_2, 2a_1 + a_2)$

$\therefore \qquad \qquad a_1 + 2a_2 = a \qquad \qquad \dots (1)$

$$-3a_1 - a_2 = b \qquad \qquad \dots (2)$$

$$2a_1 + a_2 = c \qquad \qquad \dots (3)$$

Solving (1) and (2), we get

$$a_1 = -\frac{1}{5}(a + 2b), a_2 = \frac{1}{5}(3a + b)$$

Putting these values in (3), we get

$$a - 3b - 5c = 0$$

Thus, the system of equations (1), (2) and (3) is consistent iff $a - 3b - 5c = 0$. Hence, α is a linear combination of $(1, -3, 2)$ and $(2, -1, 1)$ if and only if

$$a - 3b - 5c = 0 \cdot$$

Example 8. *Show that* $(1, 1, 1)$, $(0, 1, 1)$ *and* $(0, 1, -1)$ *generate* R^3.

Solution. In order to show that $(1, 1, 1)$, $(0, 1, 1)$ and $(0, 1, -1)$ generate R^3, we have to show that any vector of R^3 is a linear combination of $(1, 1, 1)$, $(0, 1, 1)$ and $(0, 1, -1)$.

Let $\quad \alpha = (a, b, c) \in R^3$ and let

$$(a, b, c) = a_1(1, 1, 1) + a_2(0, 1, 1) + a_3(0, 1, -1)$$

where $a_1, a_2, a_3 \in R$

Then, $\qquad (a, b, c) = (a_1, a_1 + a_2 + a_3, a_1 + a_2 - a_3)$

$\therefore \qquad \qquad a_1 = a \qquad \qquad \dots (1)$

$$a_1 + a_2 + a_3 = b \qquad \qquad \dots (2)$$

$$a_1 + a_2 - a_3 = c \qquad \qquad \dots (3)$$

Solving (1), (2) and (3), we get

$$a_1 = a, a_2 = \frac{b+c-2a}{2}, a_3 = \frac{b-c}{2}.$$

Thus, the system of equations (1), (2) and (3) is consistent and has a solution. Hence (1, 1, 1), (0, 1, 1) and (0, 1, –1) generate R^3.

Example 9. *Write the matrix* $E = \begin{bmatrix} 3 & 1 \\ 1 & -1 \end{bmatrix}$ *as a linear combination of the matrices*

$A = \begin{bmatrix} 1 & 1 \\ 1 & 0 \end{bmatrix}, B = \begin{bmatrix} 0 & 0 \\ 1 & 1 \end{bmatrix}$ *and* $C = \begin{bmatrix} 0 & 2 \\ 0 & -1 \end{bmatrix}.$

Solution. Let $E = xA + yB + zC$ where $x, y, z, \in R$

Then, $\begin{bmatrix} 3 & 1 \\ 1 & -1 \end{bmatrix} = x\begin{bmatrix} 1 & 1 \\ 1 & 0 \end{bmatrix} + y\begin{bmatrix} 0 & 0 \\ 1 & 1 \end{bmatrix} + z\begin{bmatrix} 0 & 2 \\ 0 & -1 \end{bmatrix}$

$\Rightarrow \quad \begin{bmatrix} 3 & 1 \\ 1 & -1 \end{bmatrix} = \begin{bmatrix} x & x \\ x & 0 \end{bmatrix} + \begin{bmatrix} 0 & 0 \\ y & y \end{bmatrix} + \begin{bmatrix} 0 & 2z \\ 0 & -z \end{bmatrix}$

$\Rightarrow \quad \begin{bmatrix} 3 & 1 \\ 1 & -1 \end{bmatrix} = \begin{bmatrix} x & x+2z \\ x+y & y-z \end{bmatrix}$

$\therefore \quad x = 3, x+2z = 1, x+y = 1, y-z = -1$

Solving these equations, we get

$$x = 3, y = -2, z = -1.$$

Since, these values also satisfy the last equation, they form a solution of the system.

Hence $\qquad\qquad E = 3A - 2B - C$

Example 10. *Show that in the vector space $V_n(F)$, the system of n vectors $e_1 = (1, 0, 0,...,0)$, $e_2 = (0,1,0,...,0)$, ..., $e_n = (0,0,0,...,1)$ is linearly independent where 1 denotes the unity of the field F.*

Solution. Let $a_1e_1 + a_2e_2 + ... + a_ne_n = 0$ for $a_i \in F$

Then $a_1(1, 0, 0,...,0) + a_2(0,1,0,...,0) + ... + a_n(0,0,0,...,1) = (0,0,0,...,0)$

$\Rightarrow \qquad (a_1, a_2,..., a_n) = (0, 0, 0,..., 0)$

$\Rightarrow \qquad a_1 = 0, a_2 = 0, ..., a_n = 0$

Hence, the set of n vectors $e_1, e_2, ..., e_n$ is linearly independent.

REMARK

- In particular $\{(1, 0, 0), (0, 1, 0), (0, 0, 1)\}$ is a linearly subset of $V_3 (F)$

Example 11. *Prove that every superset of a linearly dependent set of vectors is linearly dependent.*

Solution. Let $S = \{\alpha_1, \alpha_2,..., \alpha_n\}$ be a linearly dependent set of vectors. Then there exists $a_1, a_2, ... a_n$ not all zero such that

$$a_1\alpha_1 + a_2\alpha_2 + ... + a_n\alpha_n = 0 \qquad\qquad ... (1)$$

Let $S' = \{\alpha_1, \alpha_2,...\alpha_n, \beta_1, \beta_2,..., \beta_m\}$ be a superset of S.

Then, from (1), we have

$$a_1\alpha_1 + a_1\alpha_2 + ... + a_n\alpha_n + 0.\beta_1 + 0.\beta_2 + ... + 0.\beta_m = 0 \qquad\qquad ... (2)$$

Therefore, in this relation the scalar coefficients are not all zero. Hence S' is linearly dependent.

Example 12. *Show that the system of three vectors* (1, 3, 2), (1, –7, –8), (2, 1, –1) *of* V_3 (R) *is linearly dependent.*

Solution. Let $a, b, c \in R$ such that
$$a(1,3,2) + b(1,-7,-8) + c(2,1,-1) = (0,0,0)$$

$$\Rightarrow \qquad (a+b+2c, 3a-7b+c, 2a-8b-c) = (0,0,0)$$

$$\therefore \qquad \left.\begin{array}{r} a+b+2c = 0 \\ 3a-7b+c = 0 \\ 2a-8b-c = 0 \end{array}\right\} \qquad \qquad \dots(1)$$

The coefficient matrix is given by
$$A = \begin{bmatrix} 1 & 1 & 2 \\ 3 & -7 & 1 \\ 2 & -8 & -1 \end{bmatrix}$$

$$\Rightarrow \qquad |A| = \begin{vmatrix} 1 & 1 & 2 \\ 3 & -7 & 1 \\ 2 & -8 & -1 \end{vmatrix}$$

$$= 1(7+8) - 1(-3-2) + 2(-24+14)$$

$$= 15+5-20 = 20-20 = 0$$

$\Rightarrow$ Rank of A is less than 3 which is less than the number of variables a, b, and c, therefore, the system (1) of homogeneous equations has a non-zero solution. Thus, a, b, c are not all zero. Hence the system of three given vectors (1, 3, 2), (1, –7, –8), (2, 1, –1) is linearly dependent.

Example 13. *Show that* $S = \{(1,2,4), (1, 0, 0), (0, 1, 0), (0, 0, 1)\}$ *is a linearly dependent subset of the vector space* V_3 (R) *where* R *is the field of real numbers.*

Solution. We know that the set $\{(1, 0, 0), (0, 1, 0), (0, 1, 0), (0, 0, 1)\}$ is linearly independent of V_3 (R).

Now, $\qquad (1, 2, 4) = 1(1, 0, 0) + 2(0, 1, 0) + 4(0, 0, 1)$

$\Rightarrow \quad$ (1, 2, 4) is a linear combination of vectors (1, 0, 0), (0, 1, 0) and (0, 0, 1).

Hence, $\{(1, 2, 4), (1, 0, 0), (0, 1, 0), (0, 0, 1)\}$ is linearly dependent.

Example 14. *If* α, β, γ *are linearly independent vectors of a vector space* V(F) *where* F *is any field of complex numbers, then so also are* $\alpha+\beta$, $\beta+\gamma$, $\gamma+\alpha$.

Solution. Let a, b, c be scalars such that
$$a(\alpha + \beta) + b(\beta + \gamma) + c(\gamma + \alpha) = 0$$

$$\Rightarrow \quad (a+c)\alpha + (a+b)\beta + (b+c)\gamma = 0 \qquad \qquad \dots(1)$$

Since α, β, γ are linearly independent, then all the coefficients in the relation (1) must be zero.

$$\therefore \qquad \left.\begin{array}{r} a+c = 0 \\ a+b = 0 \\ b+c = 0 \end{array}\right\} \qquad \qquad \dots(2)$$

Thus, equation (2) represents a system of homogeneous equations. The coefficient matrix of this system is given by
$$A = \begin{bmatrix} 1 & 0 & 1 \\ 1 & 1 & 0 \\ 0 & 1 & 1 \end{bmatrix}$$

$$\Rightarrow \qquad |A| = \begin{bmatrix} 1 & 0 & 1 \\ 1 & 1 & 0 \\ 0 & 1 & 1 \end{bmatrix}$$

$$\Rightarrow \qquad |A| = 1(1-0) + 1(1-0) = 2 \neq 0$$

Therefore, the rank of $A = 3$ which is equal to the number of unknowns a, b, c which implies that the system of equations (2) has only zero solution, *i.e.*, $a = 0, b = 0$ and $c = 0$.

Hence $\alpha + \beta, \; \beta + \gamma, \; \gamma + \alpha$ are also linearly independent.

Example 15. *Show that three vectors* $(1, 1, -1)$, $(2, -3, 5)$ *and* $(-2, 1, 4)$ *of* R^3 *are linearly independent.*

Solution. Let $a, b, c \in R$ such that
$$a(1,1,-1) + b(2,-3,5) + c(-2,1,4) = (0,0,0)$$

$$\Rightarrow \quad (a+2b-2c, \; a-3b+c, \; -a+5b+4c) = (0, 0, 0)$$

$$\therefore \qquad \left. \begin{array}{r} a + 2b - 2c = 0 \\ a - 3b + c = 0 \\ -a + 5b + 4c = 0 \end{array} \right\} \qquad \qquad ...(1)$$

Here, equation (2) represents a system of homogeneous equations. The coefficient matrix of this system is given by

$$A = \begin{bmatrix} 1 & 2 & -2 \\ 1 & -3 & 1 \\ -1 & 5 & 4 \end{bmatrix}$$

$$\Rightarrow \qquad |A| = \begin{vmatrix} 1 & 2 & -2 \\ 1 & -3 & 1 \\ -1 & 5 & 4 \end{vmatrix}$$

$$\Rightarrow \qquad |A| = 1(-12-5) - 2(4+1) - 2(5-3)$$

$$\Rightarrow \qquad |A| = -17 - 10 - 4 = -31 \neq 0$$

$$\Rightarrow \qquad \text{Rank } (A) = 3$$

which is equal to the number of unknowns a, b, c. Therefore, the system (1) has only zero solution, *i.e.*, $a = 0, b = 0, c = 0$.

Hence, the vectors $(1, 1, -1)$, $(2, -3, 5)$ and $(-2, 1, 4)$ of R^3 are linearly independent.

Example 16. *In* $V_3(R)$, *where R is the field of real numbers, examine each of the following sets of vectors for linear dependence:*

   (i) $\{(1, 3, 2), (1, -7, -8), (2, 1, -1)\}$

   (ii) $\{(0, 2, -4), (1, -2, -1), (1, -4, 3)\}$

  (iii) $\{(1, 2, 0), (0, 3, 1), (-1, 0, 1)\}$

  (iv) $\{(-1, 2, 1), (3, 0, -1), (-5, 4, 3)\}$

   (v) $\{(2, 3, 5), (4, 9, 25)\}$

  (vi) $\{2, 1, 2), (8, 4, 8)\}$

Solution. (i) Let $a, b, c \in R$ such that
$$a(1, 3, 2) + b(1, -7, -8) + c\,(2, 1, -1) = (0, 0, 0)$$

$$\Rightarrow \quad (a + b + 2c, 3a - 7b + c, 2a - 8b - c) = (0,0,0)$$

$$\therefore \quad \begin{aligned} a+b+2c &= 0 \\ 3a-7b+c &= 0 \\ 2a-8b-c &= 0 \end{aligned} \Bigg\} \qquad \qquad \dots (1)$$

The coefficient matrix of the system of equations is given by

$$A = \begin{bmatrix} 1 & 1 & 2 \\ 3 & -7 & 1 \\ 2 & -8 & -1 \end{bmatrix}$$

$$\Rightarrow \quad |A| = \begin{vmatrix} 1 & 1 & 2 \\ 3 & -7 & 1 \\ 2 & -8 & -1 \end{vmatrix}$$

$$\Rightarrow \quad |A| = 1(7+8) - (-3-2) + 2(-24+14)$$
$$\Rightarrow \quad |A| = 15 + 5 - 20 = 0$$
$$\Rightarrow \quad \text{Rank } (A) < 3.$$

Therefore, system (1) has non-zero solution.

Hence, the set $\{(1, 3, 2), (1, -7, -8), (2, 1, -1)\}$ is linearly dependent.

(ii) Let $a, b, c \in R$ such that
$$a(0, 2, -4) + b(1, -2, -1) + c(1, -4, 3) = (0, 0, 0)$$
$$\Rightarrow \quad (b + c, 2a - 2b - 4c, -4a - b + 3c) = (0, 0, 0)$$

$$\therefore \quad \begin{aligned} b + c &= 0 \\ 2a - 2b - 4c &= 0 \\ -4a - b + 3c &= 0 \end{aligned} \Bigg\} \qquad \qquad \dots(1)$$

The coefficient matrix of system (1) is given by

$$A = \begin{bmatrix} 0 & 1 & 1 \\ 2 & -2 & -4 \\ -4 & -1 & 3 \end{bmatrix}$$

$$\Rightarrow \quad |A| = \begin{vmatrix} 0 & 1 & 1 \\ 2 & -2 & -4 \\ -4 & -1 & 3 \end{vmatrix}$$

$$\Rightarrow \quad |A| = -1(6-18) + 1(-2-8)$$
$$\Rightarrow \quad |A| = 10 - 10 = 0$$

Therefore, system (1) has non-zero solutions.

Hence $\{(0, 2, -4), (1, -2, -1), (1, -4, 3)\}$ is linearly dependent.

(iii) Let $a, b, c \in R$ such that
$$a(1, 2, 0) + b(0, 3, 1) + c(-1, 0, 1) = (0, 0, 0)$$
$$\Rightarrow \quad (a - c, 2a + 3b, b + c) = (0, 0, 0)$$

$$\therefore \quad \begin{aligned} a - c &= 0 \\ 2a + 3b &= 0 \\ b + c &= 0 \end{aligned} \Bigg\} \qquad \qquad \dots (1)$$

The coefficient matrix of system (1) is given by

$$A = \begin{bmatrix} 1 & 0 & -1 \\ 2 & 3 & 0 \\ 0 & 1 & 1 \end{bmatrix}$$

$$\Rightarrow \qquad |A| = \begin{vmatrix} 1 & 0 & -1 \\ 2 & 3 & 0 \\ 0 & 1 & 1 \end{vmatrix}$$

$$\Rightarrow \qquad |A| = 1(3-0) - 1(2-0)$$

$$\Rightarrow \qquad |A| = 3 - 2 = 1 \neq 0$$

$$\Rightarrow \qquad \text{Rank } (A) = 3$$

which is equal to the number of unknowns a, b, c. Therefore, the system has only zero solution, *i.e.*, $a = 0$, $b = 0$, $c = 0$.

Hence, $\{(1, 2, 0), (0, 3, 1), (-1, 0, 1)\}$ is linearly independent.

(iv) Let $a, b, c \in R$ such that

$$a(-1, 2, 1) + b(3, 0, -1) + c(-5, 4, 3) = (0, 0, 0)$$

$$\Rightarrow \quad (-a + 3b - 5c, 2a + 4c, a - b + 3c) = (0, 0, 0)$$

$$\therefore \qquad \begin{aligned} -a + 3b - 5c &= 0 \\ 2a + 4c &= 0 \\ a - b + 3c &= 0 \end{aligned} \qquad \qquad \dots (1)$$

The coefficient matrix of the system (1) is given by

$$A = \begin{bmatrix} -1 & 3 & -5 \\ 2 & 0 & 4 \\ 1 & -1 & 3 \end{bmatrix}$$

$$\Rightarrow \qquad |A| = \begin{vmatrix} -1 & 3 & -5 \\ 2 & 0 & 4 \\ 1 & -1 & 3 \end{vmatrix}$$

$$\Rightarrow \qquad |A| = -1(0+4) - 3(6-4) - 5(-2-0)$$

$$\Rightarrow \qquad |A| = -4 - 6 + 10 = 0$$

$$\Rightarrow \qquad \text{Rank } (A) < 3.$$

Therefore, the system (1) has non-zero solution.

Hence, the set $\{(-1, 2, 1), (3, 0, -1), (-5, 4, 3)\}$ is linearly dependent.

(v) Let $a, b \in R$ such that

$$a(2, 3, 5) + b(4, 9, 25) = (0, 0, 0)$$

$$\Rightarrow \quad (2a + 4b, 3a + 9b, 5a + 25b) = (0, 0, 0)$$

$$\therefore \qquad \left. \begin{aligned} 2a + 4b &= 0 \\ 3a + 9b &= 0 \\ 5a + 25b &= 0 \end{aligned} \right\} \qquad \qquad \dots (1)$$

The coefficient matrix of the system (1) is given by

$$A = \begin{bmatrix} 2 & 4 \\ 3 & 9 \\ 5 & 25 \end{bmatrix}$$

Clearly, Rank $A = 2$ which is equal to the number of unknowns a, b. Therefore, the system (1) has only zero solution, *i.e.*, $a = 0$, $b = 0$.

Hence, the set $\{(2, 3, 5), (4, 9, 25)\}$ is linearly independent.

(vi) Let a, $b \in$ R, such that $a(2, 1, 2) + b(8, 4, 8) = (0, 0, 0)$

$\Rightarrow \quad$ $(2a + 8b, a + 4b, 2a + 8b) = (0, 0, 0)$

$$\therefore \qquad \left. \begin{array}{r} 2a + 8b = 0 \\ a + 4b = 0 \\ 2a + 8b = 0 \end{array} \right\}$$

The coefficient matrix of the system (1) is given by

$$A = \begin{bmatrix} 2 & 8 \\ 1 & 4 \\ 2 & 8 \end{bmatrix}$$

Clearly, Rank $A = 2$ which is equal to the number of unknowns a, b.

Therefore, the system has non-zero solution.

Hence, the set $\{(2, 1, 2), (8, 4, 8)\}$ is linearly dependent.

Example 17. *If α_1, α_2 are vectors of $V(F)$ and a, $b \in F$, show that the set $\{\alpha_1, \alpha_2, a\alpha_1 + b\alpha_2\}$ is linearly dependent.*

Solution. We have

$$(-a)\alpha_1 + (-b)\alpha_2 + 1(a\alpha_1 + b\alpha_2) = (-a + a)\alpha_1 + (-b + b)\alpha_2$$

$$= 0\alpha_1 + 0\alpha_2$$

$$= 0 \text{ (zero vector)}$$

$$\therefore \quad (-a)\alpha_1 + (-b)\alpha_2 + 1(a\alpha_1 + b\alpha_2) = 0$$

Therefore, in this relation the scalar coefficients are $-a$, $-b$ and 1. As $1 \neq 0$, therefore whatever may be the scalars $-a$ and $-b$, the set $\{\alpha_1, \alpha_2, a\alpha_1 + b\alpha_2\}$ is linearly dependent.

Example 18. *In the vector space $F[x]$ of all polynomials over the field F the infinite set $S = \{1, x, x^2, x^3, ...\}$ is linearly independent.*

Solution. Let $S_n = \{x^{m_1}, x^{m_2}, x^{m_3}, ..., x^{m_n}\}$ be any finite subset of S having n vectors, where $m_1, m_2, m_3, ..., m_n$ are some non-negative integers.

Let $a_1, a_2, a_3, ..., a_n$ be scalars such that

$$a_1 x^{m_1} + a_2 x^{m_2} + a_3 x^{m_3} + ... + a_n x^{m_n} = 0 \qquad \text{(Zero polynomial)}$$

$\Rightarrow \quad a_1 x^{m_1} + a_2 x^{m_2} + a_3 x^{m_3} + ... + a_n x^{m_n} = 0 + 0.x + 0.x^2 + ...$

$\Rightarrow \quad a_1 = 0, a_2 = 0, a_3 = 0, a_n = 0$

(By the definition of equality of two polynomials)

Thus every finite subset of S is linearly independent.

Hence, S is linearly independent.

Example 19. *Prove that in R [x], the vector space of all polynomials in x over R, the system of* $p(x) = 1 + x + 2x^2$, $q(x) = 2 - x + x^2$, $r(x) = -4 + 5x + x^2$ *is linearly dependent.*

Solution. Let a, b, $c \in R$ such that

$$a\, p(x) + bq(x) + cr(x) = 0 \qquad \text{(zero polynomial)}$$

$$\Rightarrow \quad a(1 + x + 2x^2) + b(2 - x + x^2) + c(-4 + 5x + x^2) = 0 + 0.x + 0.x^2$$

$$\Rightarrow \quad (a + 2b - 4c) + (a - b + 5c)x + (2a + b + c)x^2 = 0 + 0.x + 0.x^2$$

$$\therefore \quad \left. \begin{array}{r} a + 2b - 4c = 0 \\ a - b + 5c = 0 \\ 2a + b + c = 0 \end{array} \right\} \qquad \ldots (1)$$

The coefficient matrix of the system of equivalent is given by

$$A = \begin{bmatrix} 1 & 2 & -4 \\ 1 & -1 & 5 \\ 2 & 1 & 1 \end{bmatrix}$$

$$\Rightarrow \quad |A| = \begin{vmatrix} 1 & 2 & -4 \\ 1 & -1 & 5 \\ 2 & 1 & 1 \end{vmatrix}$$

$$\Rightarrow \quad |A| = 1(-1 - 5) - 2(1 - 10) - 4(1 + 2)$$

$$\Rightarrow \quad |A| = -6 + 18 - 12 = 0$$

$\Rightarrow \quad$ Rank $A < 3$ which is the number of unknowns a, b, c.

Therefore, the system (1) has non-zero solution, *i.e*, not all a, b, c are zero.

Hence $p(x)$, $q(x)$, $r(x)$ are linearly dependent.

Example 20. *Show that the set* $\{1, x, 1 + x + x^2\}$ *is a linearly independent set of vectors in the vector space of all polynomials over the field of real numbers.*

Solution. Let a, b, $c \in R$ such that

$$a.1 + bx + c(1 + x + x^2) = 0 \qquad \text{(zero polynomial)}$$

$$\Rightarrow \quad a + bx + c(1 + x + x^2) = 0 + 0x + 0x^2$$

$$\Rightarrow \quad (a + c) + (b + c)x + cx^2 = 0 + 0x + 0x^2$$

$$\therefore \quad \left. \begin{array}{r} a + c = 0 \\ b + c = 0 \\ c = 0 \end{array} \right\} \qquad \ldots (1)$$

Solving the system (1) of equations, we get

$$a = 0, b = 0, c = 0$$

Hence, $\{1, x, 1 + x + x^2\}$ is linearly independent.

Example 21. *Let* $V = R^3$ *(R) . Find a set of linearly independent vector of V which contains vector* (1, 1, 1).

Solution. We know that in R^3(R) a set $\{(1, 0, 0), (0, 1, 0), (0, 0, 1)\}$ is linearly independent.

Now, we have $\quad (1, 1, 1) = (0, 0, 0) + (0, 1, 0) + (0, 0, 1)$

Thus, we consider a set $S = \{(1,0,0), (0,1,0), (1,1,1)\}$

Now we show that S is linearly independent.

Let $a, b, c \in R$ such that

$$a(1,0,0) + b(0,1,0) + c(1,1,1) = (0,0,0)$$

$$\Rightarrow \qquad (a+c, b+c, c) = (0,0,0)$$

$$\therefore \qquad\qquad a+c = 0$$
$$b+c = 0$$
$$c = 0$$

Solving these equations, we get

$$a = 0, b = 0, c = 0$$

Hence, the set S is linearly independent which is a required set.

REMARK

- In this question, we may replace any one of the vectors (1, 0, 0), (0, 1, 0) and (0,0,1) by (1, 1, 1).

Example 22. *Find a maximal linearly independent subsystem of the system of vectors* $\alpha_1 = (2, -2, -4)$, $\alpha_2 = (1, 9, 3)$, $\alpha_3 = (-2, -4, 1)$ *and* $\alpha_4 = (3, 7, -1)$.

Solution. We know that A set S of linearly independent vectors is a maximal linearly independent set if every set of vectors which contains S as a proper subset is linearly dependent.

Let A denote the matrix whose rows are the vectors $\alpha_1, \alpha_2, \alpha_3$ and α_4,

i.e., $\qquad A = \begin{bmatrix} \alpha_1 \\ \alpha_2 \\ \alpha_3 \\ \alpha_4 \end{bmatrix}$

or $\qquad A = \begin{bmatrix} 2 & -2 & -4 \\ 1 & 9 & 3 \\ -2 & -4 & 1 \\ 3 & 7 & -1 \end{bmatrix}$

Now we shall reduce this matrix to Echelon form by using the row transformations.

Applying $R_1 \leftrightarrow R_2$, we have

$$A \sim \begin{bmatrix} 1 & 9 & 3 \\ 2 & -2 & -4 \\ -2 & -4 & 1 \\ 3 & 7 & -1 \end{bmatrix}$$

Applying $R_2 \rightarrow R_2 - 2R_1, R_3 \rightarrow R_3 + 2R_1, R_4 \rightarrow R_4 - 3R_1$

$$A \sim \begin{bmatrix} 1 & 9 & 3 \\ 0 & -20 & -10 \\ 0 & 14 & 7 \\ 0 & -20 & -10 \end{bmatrix}$$

Applying $R_2 \rightarrow \dfrac{1}{20} R_2$

$$A \sim \begin{bmatrix} 1 & 9 & 3 \\ 0 & -1 & -1/2 \\ 0 & 14 & 7 \\ 0 & -20 & -10 \end{bmatrix}$$

Applying $R_3 \rightarrow R_3 - 14R_2, R_4 \rightarrow R_4 + 20R_2$

$$A \sim \begin{bmatrix} 1 & 9 & 3 \\ 0 & 1 & 1/2 \\ 0 & 0 & 0 \\ 0 & 0 & 0 \end{bmatrix}$$

which is in Echelon form having two non-zero rows.

$\Rightarrow$ Rank $A = 2$

Therefore, the maximum number of linearly independent row vectors in the matrix $A =$ Rank $A = 2$.

Clearly, the vectors α_1 and α_2 are linearly independent.

Hence, $\{\alpha_1, \alpha_2\}$ is a maximal linearly independent subsystem of the given system of vectors.

REMARK

- Here we observe that none of α_1, α_2, α_3, and α_4 is a scalar multiple of any of remaining three vectors, so that any two of the given four vectors form a maximal linearly independent subsystem of the given system of vectors.

📚 EXERCISE 5.4

1. If S is a subspace of V, then prove that $L(S) = S$.
2. Show that $L(S)$ is the intersection of all the subspaces of a vector space V containing S.
3. Find k so that the vector $(1, k, 5)$ is a linear combination of $(1, -3, 2)$ and $(2, -1, 1)$.
4. For which value of k will the vector $(1, -2, k)$ in R^3 be a linear combination of the vectors $(3, 0, -2)$ and $(2, -1, -5)$?
5. Express the vector $(3, 4)$ in R^3 as a linear combination of the vectors $(1, 3)$ and $(-1, 1)$.
6. Is the vector $(3, -1, 0, -1)$ in the subspace of R^4 spanned by the vectors $(2, -1, 3, 2)$, $(-1, 1, 1, -3)$ and $(1, 1, 9, -5)$?
7. Write $(1, 7, -4)$ as a linear combination of the vectors $(1, -3, 2)$ and $(2, -1, 1)$ in R^3.
8. Write $(2, -5, 4)$ as a linear combination of the vectors $(1, -3, 2)$ and $(2, -1, 1)$ in R^3.
9. Write E as a linear combination of

$$A = \begin{bmatrix} 1 & 1 \\ 0 & -1 \end{bmatrix}, B = \begin{bmatrix} 1 & 1 \\ -1 & 0 \end{bmatrix} \text{ and } C = \begin{bmatrix} 1 & -1 \\ 0 & 0 \end{bmatrix}$$

where (i) $E = \begin{bmatrix} 3 & -1 \\ 1 & -2 \end{bmatrix}$ (ii) $E = \begin{bmatrix} 2 & 1 \\ -1 & -2 \end{bmatrix}$.

10. Express $f(x)$ as a linear combination of the polynomials

$$f_1(x) = 2x^2 + 3x - 4, f_2(x) = x^2 - 2x - 3$$

where (i) $f(x) = 3x^2 + 8x - 5$

(ii) $f(x) = 4x^2 - 6x - 1$

11. Find conditions on a, b, $c \in R$ so that $(a, b, c) \in R^3$ belongs to the space spanned by the vectors $(2, 1, 0)$, $(1, -1, 2)$ and $(0, 3, -4)$.

12. Show that the vectors $(1, 2, 3)$, $(0, 1, 2)$ and $(0, 0, 1)$ generate R^3.

13. Show that the vectors $(1, 2, 5)$, $(1, 3, 7)$ and $(1, -1, -1)$ do not generate R^3.

14. Let V be the vector space of all polynomials in x, then show that the polynomials $1, x, x^2, x^3$, ... generate V.

15. In a vector space R^4, let $\alpha_1 = (1,2,-1,3)$, $\alpha_2 = (2,4,1,-2)$, $\alpha_3 = (3,6,3,-7)$, $\beta_1 = (1,2,-4,11)$, $\beta_2 = (2,4,-5,14)$. Show that the subspace spanned by $S = (\alpha_1, \alpha_2, \alpha_3)$ and $T = (\beta_1, \beta_2)$ are the same.

16. In a vector space R^3, let $\alpha_1 = (1,1,-1)$, $\alpha_2 = (2,3,-1)$, $\alpha_3 = (3,1,-5), \beta_1 = (1,-1,-3)$, $\beta_2 = (3,-2,-8)$ and $\beta_3 = (2,1,-3)$. Show that the subspace spanned by $S = \{\alpha_1, \alpha_2, \alpha_3\}$ and $T = \{\beta_1, \beta_2, \beta_3\}$ are the same.

17. Find one vector in R^3 which generates the intersection of W_1 and W_2 where $W_1 = \{(a,b,0) : a,b \in R\}$ and W_2 is the space generated by the vectors $(1, 2, 3)$ and $(1, -1, 1)$.

18. If α, β and γ are vectors such that $\alpha + \beta + \gamma = 0$, then show that α and β span the same subspace as β and γ.

19. Show that the vectors $(1, 0, -2)$, $(0, 2, 1)$, $(-1, 2, 3)$ are linearly dependent in R^3.

20. Show that the vectors $(1, 2, -1)$, $(-1, 1, 0)$, $(1, 3, -1)$ are linearly independent in R^3.

21. Show that the set $\{(1, 2, 1, 0), (3, -4, 5, 6), (2, -1, 3, 3), (-2, 6, -4, -6)\}$ of $V_4(R)$ is linearly dependent.

22. Examine whether the set of vectors $(2, 3, -1)$, $(-1, 4, -2)$ and $(1, 18, -4)$ is linearly dependent or not in $V_3(R)$.

23. Show that the vectors $(1, 1, 0, 0)$, $(0, 1, -1, 0)$ and $(0, 0, 0, 3)$ in R^4 are linearly independent.

24. Determine whether the set $\{(-1, 2, 1), (3, 1, -2)\}$ of vectors in V_3 (Q) is linearly dependent or independent, Q being the field of rational numbers.

25. Show that the vectors $(0, 2, -4)$, $(1, -2, -1)$, $(1, -4, 3)$ in R^3 are linearly dependent. Also express $(0, 2, -4)$ as a linear combination of $(1, -2, -1)$ and $(1, -4, 3)$.

26. Show that the vectors $(1, -2, 1)$, $(2, 1, -1)$, $(7, -4, 1)$ in R^3 are linearly dependent. Also express $(1, -2, 1)$ as a linear combination of $(2, 1, -1)$ and $(7, -4, 1)$.

27. Prove that the four vectors $(1, 0, 0)$, $(0, 1, 0)$, $(0, 0, 1)$ and $(1, 1, 1)$ in V_3 (C) form a linearly dependent set but any of three of them are linearly independent.

28. Find a linearly independent subset T of the set $S = \{\alpha_1, \alpha_2, \alpha_3, \alpha_4\}$ where $\alpha_1 = (1, 2, -1)$, $\alpha_2 = (-3, -6, 3)$, $\alpha_3 = (2, 1, 3)$, $\alpha_4 = (8, 7, 7)$ in R^3 which spans the same space as S.

29. Let α_1, α_2, α_3, be the vectors of $V(F)$, a, $b \in F$. Show that the set $\{\alpha_1, \alpha_2, \alpha_3\}$ is linearly dependent if the set $\{\alpha_1 + a\alpha_2 + b\alpha_3, \alpha_2, \alpha_3\}$ is linearly dependent.

30. If α, β, γ are linearly independent vectors of $V(F)$ where F is the field of complex numbers, then prove that:

(i) $\alpha + \beta, \alpha - \beta, \alpha - 2\beta + \gamma$

(ii) $\alpha + 3\beta - 2\gamma, 2\alpha + \beta - \gamma, 3\alpha + \beta + \gamma$

are also linearly independent.

31. Let $S = \{\alpha_1, \alpha_2, \alpha_3\}$ be a linearly independent set of vectors in R^n:

(i) Prove that the set $\{\alpha_1 - \alpha_2, \alpha_2 - \alpha_3, \alpha_1 + \alpha_3\}$ is linearly independent.

(ii) Prove that the set $\{\alpha_1 - \alpha_2, \alpha_2 - \alpha_3, \alpha_1 - \alpha_3\}$ is linearly dependent.

32. Let $\{\alpha_1, \alpha_2, \ldots \alpha_{r-1}, \alpha_r, \alpha_{r+1}, \ldots, \alpha_k\}$ be a linearly independent set of k vectors in R^n

and let $\beta_r = \sum_{j=1}^{k} a_j \alpha_j$ with $a_r \neq 0$. Prove that $\{\alpha_1, \alpha_2, \ldots \alpha_{r-1}, \beta_r, \alpha_{r+1}, \ldots, \alpha_k\}$ is linearly independent.

33. Show that there is one vector in the set $\{(1, 1, 0), (0, 1, 1), (1, 0, -1), (1, 0, 1)\}$ which cannot be written as a linear combination of the other vectors in the set.

34. If $V(F)$ is a vector space and $\alpha_1, \alpha_2, \ldots \alpha_n \in V$ and $\lambda_2, \lambda_3, \ldots \lambda_n \in F$ such that $\{\alpha_2 + \lambda_2\alpha_1, \alpha_3 + \lambda_3\alpha_1, \ldots, \alpha_n + \lambda_n\alpha_1\}$ is linearly dependent then show that $\{\alpha_1, \alpha_2, \ldots \alpha_n\}$ is linearly dependent.

35. Show that the set $\{1, x, x(1-x)\}$ is a linearly independent set of vectors in the space of all polynomials over the real field.

36. Find whether the vectors $2x^3 + x^2 + x + 1$, $x^3 + 3x^2 + x - 2$ and $x^3 + 2x^2 - x + 3$ of R[x], the vector space of all polynomials over the real numer field, and linearly independent or not.

37. Show the set of functions $\{x, |x|\}$ is linearly independent in vector space of continuous functions defined in $(-1, 1)$.

38. Let V be the vector space of functions from R into R. Show that f, g, $h \in V$ are linearly independent where:

(i) $f(x) = e^{2x}, g(x) = x^2, h(x) = x$

(ii) $f(x) = \sin x, g(x) = \cos x, h(x) = x$.

39. Prove that a set of vectors which contains zero vector is linearly dependent.

40. Prove that a set consisting of exactly one non-zero vector is linearly independent.

41. Prove that the three non-coplanar vectors are linearly independent.

42. Show that any three non-zero non-coplanar vectors are linearly dependent in R^3.

43. Let V be the vector space of 2×2 matrices over R. Determine whether that matrices

$$A = \begin{bmatrix} 1 & 2 \\ 3 & 1 \end{bmatrix}, B = \begin{bmatrix} 3 & -1 \\ 2 & 2 \end{bmatrix}, C = \begin{bmatrix} 1 & -5 \\ -4 & 0 \end{bmatrix}$$

of V are linearly dependent.

44. In each case, determine whether or not the given vector α is linearly dependent on the given set S:

(i) $\alpha = (-2,1,4), S = \{(1,1,1),(0,1,1),(0,0,1)\}$

(ii) $\alpha = (1,2,1), S = \{(1,0,-2),(0,2,1),$
$(1, 2, -1), (-1, 2, 3)\}$

(iii) $\alpha = (-4,4,2), S = \{(2,1,-3),(1,-1,3)\}$

(iv) $\alpha = (1,2,3), S = \{(1,1,1),(0,1,1),(0,0,1)\}$

(v) $\alpha = (2,13,-5), S = \{(1,2,-1),(3,6,-3),$
$(-1,1,0),(0,6,-2),(2,4,-2)\}$

45. Show that every vector in R^3 is dependent vector on the set $\{\alpha_1, \alpha_2, \alpha_3\}$, where $\alpha_1 = (1, 0, 0)$, $\alpha_2 = (1, 1, 0)$ and $\alpha_3 = (1, 1, 1)$.

46. Find all subsets of maximal linear independent vectors from the following set of vectors $\{(1, 0, 1, 0), (0, 1, 0, 2), (1, 1, 0, 1), (-1, 0, 2, 0)\}$.

47. Test the vectors $(0, 1, 0, 1, 1)$, $(1, 0, 1, 0, 1)$, $(0, 1, 0, 1, 1)$ and $(1, 1, 1, 1, 1)$ in V_5 over the field of rational numbers for linear independence.

Answers

3. $k = -8$ **4.** $k = -8$ **5.** $(3,4) = \dfrac{7}{2}(1,3) + \dfrac{1}{2}(-1,1)$ **6.** No **7.** $(1, 7, -4) = -3(1, -3, 2) + 2(2, -1, 1)$

8. $(2, -5, 4)$ is not a linear combination of $(1, -3, 2)$ and $(2, -1, 1)$ **9. (i)** $E = 2A - B + 2C$

(ii) Not possible **10. (i)** $f(x) = 2f_1(x) - f_2(x)$ **(ii)** Not possible **11.** $2a - 4b - 3c = 0$ **17.** $(-2,5,0)$

22. linearly independent **24.** linearly independent **28.** $\{\alpha_1, \alpha_3\}$ **36.** linearly independent

44. (i) linearly dependent (ii) linearly independent (iii) linearly independent (iv) linearly dependent

(v) linearly dependent   **46.** all four vectors   **47.** linearly dependent

5.12 BASIS OF A VECTOR SPACE

Definition. *Let V be a vector space over a field F and let S be any non-empty subset of V. Then S is said to be a basis of V if*

(i) *S is linearly independent.*

(ii) *$L(S) = V$, i.e., every element of V is a linear combination of finite elements of S.*

For Example: The set $S = \{(1, 0, 0), (0, 1,0), (0, 0, 1)\}$ forms a basis of $V_3(R)$, and is called usual basis.

REMARKS

- The zero space has no basis.
- Every finitely generated vector space has a basis.
- Every non-zero vector space has a basis.
- A vector space many have more than one basis.

5.13 FINITE DIMENSIONAL VECTOR SPACE

Let $V(F)$ be a vector space over a field F and let S be any non-empty subset of V, then $V(F)$ is said to be finite dimensional if S is finite subset of V such that $L(S) = V$. If this set contains n elements, then the dimension of V is n.

THEOREM 1. *If $S = \{\alpha_1, \alpha_2, ..., \alpha_n\}$ is the basis of a vector space $V(F)$, then each element of V is uniquely expressible as a linear combination of elements of S.*

Proof. Since S is the basis of a vector space $V(F)$, then by the definition of basis, each element of V is a linear combination of elements of S. Thus, we only show the uniqueness. Let there be two different sets $\{a_1, a_2,...,a_n\}$ and $\{b_1, b_2,...,b_n\}$ of scalars corresponding to an element $\alpha \in V$ such that

$$\alpha = a_1\alpha_1 + a_2\alpha_2 + ... + a_n\alpha_n$$

and $\qquad \alpha = b_1\alpha_1 + b_2\alpha_2 + ... + b_n\alpha_n \qquad\qquad ....(1)$

$\Rightarrow \quad a_1\alpha_1 + a_2\alpha_2 + ... + a_n\alpha_n = b_1\alpha_1 + b_2\alpha_2 + ... + b_n\alpha_n$

$\Rightarrow \quad a_1\alpha_1 - b_1\alpha_1 + a_2\alpha_2 - b_2\alpha_2 + ... + a_n\alpha_n - b_n\alpha_n = 0$

$\Rightarrow \quad (a_1 - b_1)\alpha_1 + (a_2 - b_2)\alpha_2 + ... + (a_n - b_n)\alpha_n = 0$

Since the set $S = \{\alpha_1, \alpha_2,..., \alpha_n\}$ is linearly independent so that

$$a_1 - b_1 = 0, \ a_2 - b_2 = 0,..., a_n - b_n = 0$$

$\Rightarrow \qquad a_1 = b_1, \qquad a_2 = b_2, ..., \qquad a_n = b_n.$

Hence, the expression (1) is unique.

THEOREM 2. **(Existence Theorem).** *Every finitely generated vector space has a finite basis.*

Step Outlines : To make the proof easier, use the following steps :

Step 1. *If S is linearly independent, that theorem is obvious.*

Step 2. *Assume S is linearly dependent and obtained the set $S_1 = (\alpha_1, \alpha_2, ... \alpha_{k-1}, \alpha_{k+1}, ... \alpha_n)$ by eliminating α_k from S.*

Step 3. *Again if S_1 is linearly independent, result is obvious. And if S_1 is linearly dependent, proceed same as in step-2.*

Proof. Let V be a vector space over F which is generated by a finite set (say) $S = \{\alpha_1, \alpha_2,..,\alpha_n\}$ of vectors of V. Without loss of any generality we may assume that all the elements in S are non-zero, because zero vector in the linear combination of elements of S is zero.

If S is linearly independent, then S forms a finite basis for V and in this case theorem is proved. So, we assume that S is linearly dependent, then there exists some α_k $(2 \leq k \leq n)$ in S such that α_k is a linear combination of preceding vectors $\alpha_1, \alpha_2, ... \alpha_{k-1}$. Therefore,

$$\alpha_k = a_1\alpha_1 + a_2\alpha_2 + ... + a_{k-1}\alpha_{k-1} = \sum_{i=1}^{k-1} a_i\alpha_i \qquad\qquad ...(1)$$

for some scalars $a_i's \in F$.

But S generates V so that an arbitrary element $\alpha \in V$ is expressible as a linear combination of elements of S.

$\therefore \qquad \alpha = b_1\alpha_1 + b_2\alpha_2 + ... + b_k\alpha_k + ... + b_n\alpha_n$

$\qquad\qquad = \sum_{i \neq k} b_i\alpha_i + b_k\alpha_k, \text{for some} \, b_i's \in F$

$$= \sum_{\substack{i \neq k}} b_i \alpha_i + b_k \sum_{i=1}^{k-1} a_i \alpha_i$$

$$= (b_1 + b_k a_1)\alpha_1 + (b_2 + b_k a_2)\alpha_2 + ...$$
$$+ (b_{k-1} + b_k a_{k-1})\alpha_{k-1} + (b_{k+1}) \alpha_{k+1} + ... + b_n \alpha_n$$

$\Rightarrow$ α is a linear combination of $a_1 \alpha_1 + a_2 \alpha_2 + ... + a_{k-1} \alpha_{k-1}$.

Thus the set,

$$S_1 = \{\alpha_1, \alpha_2, ..., \alpha_{k+1}..., \alpha_k\}.$$

Obtained by eliminating some α_k from S also generates V.

If S_1 in linearly independent, then S_1 will form a basis of V and the theorem is proved in this case. If S_1 is linearly dependent, then by above process, we obtain a new set

$$S_2 = \{\alpha_1, \alpha_2, ..., \alpha_{k-1}, \alpha_{k+1}, ..., \alpha_{i-1}, \alpha_{i+1}, ..., \alpha_n\}$$

by eliminating some α_i $(i > k)$ from S_1, which generates V. If S_2 is linearly independent, then S_2 will form a basis. If S_2 is linearly dependent, then we continue the above process, till after a finite number of steps we obtain a linearly independent set which generates V. At the most by repeating the above process we may obtain a singleton set which is always linearly independent and it generating V and will form a basis of V. Hence, in every finitely generated vector space there exists a finite basis.

THEOREM 3. *If V is a finite-dimensional vector space and if $a_1, a_2, ..., a_m$ span V, then some subset of $\alpha_1, \alpha_2, ..., \alpha_m$ forms a basis of V.*

Proof. Since a finite-dimensional vector space has a basis of V containing a finite number of elements. Let these vectors be $\alpha_1, \alpha_2, ..., \alpha_n$. Thus every element in V has a unique representation of the form

$$a_1 \alpha_1 + a_2 \alpha_2 + a_n \alpha_n; \text{ for } a_1, a_2, ..., a_n \in F.$$

If $\alpha \in V$, then

$$\alpha = a_1 \alpha_1 + a_2 \alpha_2 + ... + a_n \alpha_n$$

Now define a map ϕ from V into $F^{(n)}$ by

$$\phi (a_1 \alpha_1 + a_2 \alpha_2 + ... + a_n \alpha_n) = (a_1, a_2, ..., a_n)$$

Since $a_1 \alpha_1 + a_2 \alpha_2 + ... + a_n \alpha_n$ is a unique representation so that ϕ is well defined, one-to-one and onto and also preserves the composition.

Thus V is isomorphic to $F^{(n)}$ for some n, where n is the number of elements in some basis of V over F. If some other basis of V has m elements, then V would be isomorphic to $F^{(m)}$. Since both $F^{(n)}$ and $F^{(m)}$ are isomorphic to V, therefore, $F^{(n)}$ and $F^{(m)}$ are isomorphic to each other. This implies $n = m$. Hence the theorem.

THEOREM 4. *If $\{\alpha_1, \alpha_2, ..., \alpha_n\}$ is a basis of $V(F)$ and if $\beta_1, \beta_2, ..., \beta_m \in V$, are linearly independent over F, then $m \leq n$.*

Proof. Since the set $\{\alpha_1, \alpha_2, ..., \alpha_n\}$ is a basis of $V(F)$, then every element in V is a linear combination of the elements of basis, so in particular, β_m is a linear combination of $\alpha_1, \alpha_2, ..., \alpha_n$. Therefore, the set $\{\beta_m, \alpha_1, \alpha_2, ..., \alpha_n\}$ is linearly dependent. Also, this set spans V because $\alpha_1, \alpha_2, ..., \alpha_n$ span V. Thus proper subset of the set

$\{\beta_m, \alpha_1, \alpha_2,..., \alpha_n\}$ forms a basis of V. Let this proper set be $\{\beta_m, \alpha_1, \alpha_2, ..., \alpha_n\}$ with $k \le n - 1$. In forming this new basis at least one α_i is replaced by some one β_j. Repeat this procedure with the set $\{\beta_{m-1}, \alpha_{i_1}, \alpha_{i_2}, ..., \alpha_{i_k}\}$ which is obviously linearly dependent, so we can extract a basis of the form $\{\beta_{m-1}, \beta_m, \alpha_{j_1}, \alpha_{j_2}, ..., \alpha_{j_s}\}$ with $s \le n - 2$. Continuing this procedure with the set $\{\beta_2, \beta_3, ..., \beta_{m-1}, \beta_m, \alpha_x, \alpha_y\}$. Since β_1 is not a linear combination of $\{\beta_2, \beta_3, ... , \beta_{m-1}, \beta_m, \alpha_x, \alpha_y\}$, therefore, the basis $\{\beta_2, \beta_3, ... , \beta_{m-1}, \beta_m, \alpha_x, \alpha_y\}$ must contain some α's. To get this basis. We have introduced $m-1$ β's and each such introduction costs at least one α's and yet there is an 'α' left. Thus $m - 1 \le n - 1$ implying $m \le n$.

THEOREM 5. *If V is a finite-dimensional vector space over F, then any two bases of V have the same number of elements.*

Proof. Let S_1 and S_2 be any two bases of $V(F)$ and let
$$S_1 = \{\alpha_1, \alpha_2,..,\alpha_m\} \text{ and } S_2 = \{\beta_1, \beta_2,..., \beta_n\}.$$
Then we shall have to show that $m = n$.

Since S_1 forms a basis of V, so that every element of V is uniquely expressible as a linear combination of the elements of S_1. In particular β_1 is uniquely expressible as a linear combination of S_1. Thus the set $S_3 = \{\beta_1, \alpha_1, \alpha_2,..., \alpha_{k-1}\}$ is now linearly dependent.

Therefore, there exists an element α_k in S_s which is linear combination of proceeding ones, $\beta_1, \alpha_1, \alpha_2,..., \alpha_{k-1}$. But every element of V can be expressed as a linear combination of $\alpha_1, \alpha_2,..., \alpha_k, ..., \alpha_m$. Also, α_k is a linear combination of $\beta_1, \alpha_1, \alpha_2,..., \alpha_{k-1}$. This implies that each element of V is expressible as a linear combination of $\{\beta_1, \alpha_1, \alpha_2,..., \alpha_{k-1}, ..., \alpha_m\}$. Thus the set
$$S_4 = \{\beta_1, \alpha_1, \alpha_2,..., \alpha_{k-1}, \alpha_{k+1},..., \alpha_m\}$$
generates V. This set is obtained by adjoining β_1 to S_3 and eliminating α_k and S_3. Since $\beta_2 \in V$ so it is the linear combination of elements of S_4, therefore, the set is obtained by adjoining β_2 to S_4 and eliminating α_1 as before. Thus, the set $S_5 = \{\beta_2, \beta_1, \alpha_1, \alpha_2,..., \alpha_{k-1}, \alpha_{k+1},..., \alpha_{l-1}, \alpha_{l+1}, ... , \alpha_m\}$ generates V.

Continuing the above manner, we observe that each step consists of an inclusion of one β's and the exclusion of an α's and the resulting set generates V. Since all the α's cannot be exhausted before the β's. If it is so then a proper subset of S_2 generates V which is a contradiction of linear independence of S_2. Hence $m < n$. Similarly, if we change the role of S_1 and S_2, we obtain $n < m$. Hence $m = n$.

THEOREM 6. **(Extension Theorem).** *If $V(F)$ is a finite dimensional vector space, then every linearly independent subset of V is either a basis of V or can be extended to form a basis of V.*

Proof. Suppose the dimension of $V = n$. Let $S = \{a_1, a_2, ..., a_n\}$ be a basis of V and let $S_1 = \{\beta_1, \beta_2, ..., \beta_m\}$ be a linearly independent subset of V.

Since S is a basis of V so that every element of V is expressible as a linear combination of elements of S, in particular, β_m is a linear combination of elements of S, therefore, the set obtained by adjoining β_m to S is linearly dependent. Since, the superset of a linearly dependent set is linearly dependent, it follows that the set,

$$S_2 = \{\beta_1, \beta_2, ..., \beta_n, \; \alpha_1, \alpha_2, ..., \alpha_n\}$$

is linearly dependent.

Since the set S_1 is linearly independent, then there exists an element α_k in S_2 which can be expressed as a linear combination of preceding ones $\beta_1, \beta_2, ..., \beta_m, \alpha_1, \alpha_2$,...., α_{k-1}, therefore each element in V can be expressed as a linear combination of $\beta_1, \beta_2, ..., \beta_m, \alpha_1, \alpha_2, ..., \alpha_{k-1}, \alpha_{k+1}, ... , \alpha_n$. Thus the set

$$S_3 = \{\beta_1, \beta_2, ..., \beta_m, \alpha_1, \alpha_2, ..., \alpha_{k-1}, \alpha_{k+1}, ... , \alpha_n\}$$

is obtained be eliminating α_k from S_2 which generates V. If S_3 is linearly independent, then S_3 forms a basis of V containing S_1. Thus, in this case the theorem is proved.

If S_3 is linearly dependent, then repeat the above process of adjoining and eliminating, till after a finite number of steps we obtain a linearly independent set which generates V and contains S_1. At the most by repeating the above process we may obtain the set S_1 itself which generates V and being linearly independent will form a basis of V.

Hence, either S_1 is a basis of V or can be extended to form a basis of V.

THEOREM 7. *Let V be a finite-dimensional vector space and let dim. $V = n$. Then*

    (i) any subset of V which contains more than n vectors is linearly dependent.

    (ii) no subset of V which contains less than n vectors can span V.

Proof.

    (i) Since $V(F)$ is n-dimensional, every basis of V will contain n vectors. Let S be any subset of V which contains more than n vectors. Let, if possible, S be linearly independent, then by previous theorem either S is a basis of V or can be extended to form a basis of V. But in both cases the basis of V contains more than n elements which is contradictory to the fact that V is n-dimensional. Hence S is linearly dependent.

    (ii) Let S be a subset of V which contains less than n elements and which can span V. Then every element of V can be expressed as a linear combination of elements of S. If S is linearly independent, then S forms a basis of V which show that dim. $V < n$ which is contradictory to the fact that dim. $V = n$. On the other hand it S is linearly dependent, then S cannot span V which is again a contradiction because we have assumed that S spans V. Hence in both cases such subset S cannot exist. Consequently no subset of V containing less than n vectors can span V.

5.14 DIMENSION OF SUBSPACE OF A VECTOR SPACE

THEOREM 1. *Let S be a linearly independent subset of a vector space V. Suppose β is a vector in V which is not in the subspace spanned by S. Then the set obtained by adjoining β to S is linearly independent.*

Proof. Let $S = \{\alpha_1, \alpha_2, ... \alpha_n\}$ be a linearly independent subset of V. Then we shall show that the set

$$S_1 = \{\beta, \alpha_1, \alpha_2, ,...\alpha_n\}$$

obtained by adjoining β to S is also linearly independent where $\beta \in V$, but not in the subspace of V which is spanned by S.

Since $\alpha_1, \alpha_2, ... , \alpha_n$ are distinct vectors in S such that

$$a_1 \alpha_1 + a_2 \alpha_2 + ... + a_n \alpha_n + b\beta = 0 \qquad ...(1)$$

where, all a's. are zero.

We actually show that $b = 0$. Let, if possible, $b \neq 0$. Then from (1), we have

$$\beta = \left(-\frac{a_1}{b}\right)\alpha_1 + \left(-\frac{a_2}{b}\right)\alpha_2 + ... + \left(-\frac{a_n}{b}\right)\alpha_n$$

$\Rightarrow$ β is a linear combination of $\alpha_1, \alpha_2, ..., \alpha_n$.

$\Rightarrow$ β is in the subspace of V spanned by $\alpha_1, \alpha_2, ..., \alpha_n$.

But it is contradictory to the hypothesis that β is not in the subspace spanned by S. Hence $b = 0$. Consequently, the set S_1 is linearly independent.

THEOREM 2. *If W is a subspace of a finite-dimensional vector space V, every linearly independent subset of W is finite and is a part of a (finite) basis for W.*

Proof. Let S be a linear independent subset of W and let S_1 be a linearly independent subset of W containing S. Then S_1 contains not more than dim. V elements.

If S spans W, then S is a basis for W and in this case, theorem is proved. If S does not span W, then we find a vector β_1 in W such that the set

$$S_2 = S \cup \{\beta_1\}$$

is linearly independent (By theorem 1). If S_2 spans W, it will form a basis for W, If S_2 does not span W, then again we find a vector β_2 in W such that the set

$$S_3 = S \cup \{\beta_1, \beta_2\}$$

is linearly independent.

Continuing above process up to finite number of steps less than dim. V, we get a set ,

$$S_m = S \cup \{\beta_1, \beta_2,.., \beta_{m-1}\}$$

which is linearly independent and is a basis for W or a part of basis for W.

THEOREM 3. *If V(F) is a finite-dimensional vector space and W is a subspace of V, then W is finite dimensional and dim. W ≤ dim V. In particulars, if W is a proper subspace of V, then dim. W< dim. V. Also V = W if and only if dim. V = dim. W.*

Proof. Let dim. $V = n$. Then every basis of V will contain n vectors of V, therefore, every subset having vectors more than n will be linearly dependent. Thus, a linearly independent set of vectors in W contains at most n-elements. Let $S = \{\alpha_1, \alpha_2, ,...,\alpha_m\}$, with $m \geq n$ being a maximal linearly independent set in W. If α is an arbitrary element in W, then the set

$$S_1 = \{\alpha_1, \alpha_2, ,...,\alpha_m\}$$

will be linearly dependent because S being a maximal linearly independent set. Therefore, α is a linear combination of $\alpha_1, \alpha_2, ,...,\alpha_m$ which shows that S spans V. Hence W is finite-dimensional. Also, dim. $W = m \leq n = $ dim. V.

$\therefore$ \qquad\qquad\qquad dim. $W < $ dim. V.

Next, if $V = W$, then every basis of V is also the basis of W which shows that dim. V = dim W. On the other hand, if dim. V = dim. W, then every basis of W will contain the vectors equal to dim. V so it will also generate V. Thus each one of V and W is generated by some basis. Hence $V = W$.

THEOREM 4. **(Existence of Complementary Subspace).** *Every subspace of a finite dimensional vector space has a complement.*

Proof. Let $V(F)$ be a finite dimensional vector space and let W_1 be its subspace.

Then our aim is to find out a subspace W_2 of V such that $V = W_1 \oplus W_2$. Since $V(F)$ is a finite dimensional vector space so that W_1 will be finite dimensional.

Let $S_1 = \{\alpha_1, \alpha_2, ..., \alpha_n\}$ be the basis of W_1, then by extension theorem, we have, S_1 can be extended to form a basis of V. Let this extended set be

$$S_2 = \{\alpha_1, \alpha_2, ,...,\alpha_n, \beta_1, \beta_2, ,...,\beta_m\} \text{ be the basis of } V.$$

Let us suppose that the set $\{\beta_1, \beta_2, ,...,\beta_m\}$ generates a subspace and let this subspace be W_2.

Now we shall show that $V = W_1 \oplus W_2$ or equivalently $V = W_1 + W_2$ and $W_1 \cap W_2 = \{0\}$.

Let γ be an arbitrary element of V and S_2 being the basis for V, then there exists scalars $a_1, a_2, ..., a_n, b_1, b_2, ..., b_m$ such that

$$\gamma = a_1\alpha_1 + a_2\alpha_2 + ... + a_n\alpha_n + b_1\beta_1 + b_2\beta_2 + ... + b_m\beta_m = \alpha + \beta$$

Where
$$\alpha = \sum_{i=1}^{n} a_i\alpha_i \text{ and } \beta = \sum_{j=1}^{m} b_j\beta_j$$

Since $\{\alpha_1, \alpha_2, ..., \alpha_n\}$ generates W_1 and $\{\beta_1, \beta_2, ,...,\beta_m\}$ generates W_2, therefore $\alpha \in W_1$ and $\beta \in W_2$.

Thus, every element of V is expressible as the sum of an element of W_1 and an element of W_2.

$$\therefore \qquad V = W_1 + W_2$$

Again,
$$\alpha = \sum_{i=1}^{n} a_i\alpha_i \in W_1 \text{ and } \beta = \sum_{j=1}^{m} b_j\beta_j \in W_2$$

Let if possible $W_1 \cap W_2 \neq \{0\}$. Then there exists a non-zero element which belong to both W_1 and W_2. Let it be x. Then $x \in W_1$ and $x \in W_2$.

$$\therefore \qquad x = \sum_{i=1}^{n} a_i'.\alpha_i \text{ and } x = \sum_{j=1}^{m} b_j'.\beta_j$$

$$\Rightarrow \qquad \sum_{i=1}^{n} a_i'.\alpha_i = \sum_{j=1}^{m} b_j'.\beta_j$$

$$\Rightarrow \qquad \sum_{i=1}^{n} a_i'.\alpha_i + \sum_{j=1}^{m} (-b_j').\beta_j = 0$$

Since the set S_2 is linearly independent, so that

$$a_i' = 0, \text{ for each } i = 1, 2 ,.., n$$

and
$$b_j' = 0,, \text{ for each } i = 1, 2 ,.., m$$

$$\Rightarrow \qquad x = 0$$

which is a contradiction, because we have taken $x \neq 0$.

Thus, the contradiction arises by assuming that $W_1 \cap W_2 \neq \{0\}$

Hence, $W_1 \cap W_2 = \{0\}$.

Consequently

$$V = W_1 \oplus W_2.$$

REMARK

- Here W_2 is the subspace complementary to the subspace W_2 of finite dimensional vector space V.

THEOREM 5. *If W_1 and W_2 are two finite dimensional subspaces of a vector space $V(F)$, then $W_1 + W_2$ is finite dimensional and $dim. W_1 + dim. W_2 = dim. (W_1 \cap W_2) + dim. (W_1 + W_2)$.*

Proof. Since W_1 and W_2 are subspaces of V so that $W_1 \cap W_2$ will be a subspace of V and its dimension is finite. Let dim. $W_1 = m$, dim. $W_2 = n$ and dim. $(W_1 \cap W_2) = r$.

Let $\{\alpha_1, \alpha_2, ..., \alpha_r\}$ be a basis of $W_1 \cap W_2$. Therefore, we can extend this basis to a basis of W_1 and also to a basis of W_2.

Let, $S_1 = \{\alpha_1, \alpha_2, ... \alpha_r, \beta_1, \beta_2, ... \beta_{m-r}\}$

and $S_2 = \{\alpha_1, \alpha_2, ... \alpha_r, \gamma_1, \gamma_2, ... \gamma_{n-r}\}$

be the basis of W_1 and W_2 respectively. Consider the set

$$S = \{\alpha_1, \alpha_2, ... \alpha_r, \beta_1, \beta_2, ... \beta_{m-r}, \gamma_1, \gamma_2, ... \gamma_{n-r}\}$$

Now we have to show that S will form a basis for $W_1 + W_2$. For this, we shall show that S is linearly independent and spans $W_1 + W_2$. For this, suppose

$$\sum a_i \alpha_i + \sum b_j \beta_j + \sum c_k \gamma_k = 0 \text{ for } a_i' s, b_j' s, c_k' s \in F.$$

Then $\sum c_k \gamma_k = \sum a_i \alpha_i + \sum b_j \beta_j \Rightarrow \sum c_k \gamma_k \in W_1$

Also, $\sum c_k \gamma_k \in W_2$. It follows that $\sum c_k \gamma_k \in W_1 \cap W_2$ and we have $\sum c_k \gamma_k \in d_i \alpha_i$ for some scalars $d_1, d_2, ..., d_r$. Since the set $\{\alpha_1, \alpha_2, ... \alpha_r, \gamma_1, \gamma_2, ... \gamma_{n-r}\}$ is linearly independent hence all the scalars $c_1 = 0 = c_2 = ... = c_{n-r}$.

Thus $\sum a_i \alpha_i + \sum b_j \beta_j = 0$

and since the set $\{\alpha_1, \alpha_2, ... \alpha_r, \beta_1, \beta_2, ... \beta_{m-r}\}$ is also linearly independent,

then

$$a_1 = 0 = a_2 = ... = a_r$$

and $$b_1 = 0 = b_2 = ... = b_{m-r}$$

Thus the set $S = \{\alpha_1, \alpha_2, ... \alpha_r, \beta_1, \beta_2, ... \beta_{m-r}, \gamma_1, \gamma_2, ... \gamma_{n-r}\}$ is linearly independent.

Now we shall show that S spans $W_1 + W_2$.

Let α be an arbitrary element of $W_1 + W_2$ then it can be written as $\alpha = \beta + \gamma$ with $\beta \in W_1$ and $\gamma \in W_2$. Now S_1 and S_2 being the basis of W_1 and W_2 respectively, β and γ can be expressed uniquely in the form

$$\beta = \sum_{i=1}^{r} a_i \alpha_i + \sum_{j=1}^{n-r} b_j \beta_j, \text{ for some } a_i's \text{ and } b_j's.$$

and
$$\gamma = \sum_{i=1}^{r} e_i \alpha_i + \sum_{j=1}^{n-r} c_j \gamma_j \text{ , for some } e_i's \text{ and } c_j's.$$

$\therefore$
$$\alpha = \beta + \gamma = \sum_{i=1}^{r} (a_i + e_i)\alpha_i + \sum_{j=1}^{n-r} b_j \beta_j + \sum_{j=1}^{n-r} c_j \gamma_j .$$

$\Rightarrow$ α is a linear combination of elements of S.

$\Rightarrow$ S spans $W_1 + W_2$.

Hence, S is basis of $W_1 + W_2$ so that $W_1 + W_2$ is finite-dimensional with dimensional $(m + n - r)$.

Finally,
$$\dim W_1 + \dim W_2 = m + n = r + (m + n - r)$$
$$= \dim .(W_1 \cap W_2) + \dim .(W_1 + W_2).$$

THEOREM 6. *If a finite dimensional vector space $V(F)$ be the direct sum of its two subspaces W_1 and W_2, then $\dim.V = \dim. W_1 + \dim.W_2$.*

Proof. Since V is finite dimensional, therefore W_1 and W_2 are also finite-dimensional. Let
$$\dim.W_1 = m, \dim.W_2 = n$$

Also $V = W_1 \oplus W_2$, implying that

(i) $V = W_1 + W_2$

(ii) $W_1 \cap W_2 = \{0\}$.

Let $S_1 = \{\alpha_1, \alpha_2, ..., \alpha_m\}$ be a basis of W_1 and the set $S_2 = \{\beta_1, \beta_2, ..., \beta_n\}$ is a basis of W_2.

Now consider a set
$$S_3 = \{\alpha_1, \alpha_2, ..., \alpha_m, \beta_1, \beta_2, ..., \beta_n\}$$

We claim that S_3 forms a basis of V.

For some scalars $a_1, a_2, ..., a_m, b_1, b_2, ..., b_n \in F$, we have
$$a_1 \alpha_1 + a_2 \alpha_2 + ... + a_m \alpha_m + b_1 \beta_1 + b_2 \beta_2 + ... + b_n \beta_n = 0$$

$\Rightarrow$ $a_1 \alpha_1 + a_2 \alpha_2 + ... + a_m \alpha_m = -(b_1 \beta_1 + b_2 \beta_2 + ... + b_n \beta_n)$

$\Rightarrow$ $a_1 \alpha_1 + a_2 \alpha_2 + ... + a_m \alpha_m \in W_1$ as $b_1 \beta_1 + b_2 \beta_2 + ... + b_n \beta_n \in W_2$

and $b_1 \beta_1 + b_2 \beta_2 + ... + b_n \beta_n \in W_2$ as $a_1 \alpha_1 + a_2 \alpha_2 + ... + a_m \alpha_m \in W_1$

$\Rightarrow$ $a_1 \alpha_1 + a_2 \alpha_2 + ... + a_m \alpha_m \in W_1 \cap W_2$

$\Rightarrow$ $b_1 \beta_1 + b_2 \beta_2 + ... + b_n \beta_n \in W_1 \cap W_2$

But from (i) $W_1 \cap W_2 = \{0\}$.

$\Rightarrow$ $a_1 \alpha_1 + a_2 \alpha_2 + ... + a_m \alpha_m = 0$

and $b_1 \beta_1 + b_2 \beta_2 + ... + b_n \beta_n = 0$

Since S_1 and S_2 both are linearly independent, therefore
$$a_1 = 0 = a_2 = ... = a_m , b_1 = 0 = b_2 = ... = b_n.$$

$\Rightarrow$ S_1 is linearly independent.

Next, let γ be an arbitrary element of V, then
$$\gamma = \alpha + \beta, \ \alpha \in W_1, \ \beta \in W_2 \qquad\qquad [\because V = W_1 + W_2]$$

Since $\alpha \in W_1 \Rightarrow \alpha \in a_1 \alpha_1 + a_2 \alpha_2 + ... + a_m \alpha_m$ for some $a_i's \in F$.

and $\beta \in W_2 \Rightarrow \beta \in b_1 \beta_1 + b_2 \beta_2 + ... + b_m \beta_m$ for some $b_i's \in F$.

$$\therefore \qquad \gamma = \alpha + \beta = a_1\alpha_1 + a_2\alpha_2 + \ldots\ldots + a_m\alpha_m + b_1\beta_1 + b_2\beta_2 + \ldots.. + b_n\beta_n$$

$\Rightarrow \quad S_3$ generates V, thus S_3 forms a basis of V.

Accordingly, dim.$V = m+n = $ dim.$W_1 + $ dim.W_2.

REMARK

- It should be noted that in case of finite dimensional spaces, since a basis is a maximal linearly independent subset of the vector space, so the dimension of a finite dimensional vector spaces may be regarded as the maximum of numbers of elements in all linearly independent subsets. If this is adopted as definition of the dimension of a vector space, is n iff it has a basis consisting of n elements.

Solved Examples

Example 1. *Let V be the vector space of ordered pairs of complex numbers over the real field R, i.e., let V be the vector space C^2(R). Show that the set S={(1, 0), (i, 0), (0, 1), (0, i)} is a basis for V.*

Solution. First, we shall show that S is linearly independent.

Let $a, b, c, d \in$ R, such that

$$a(1,0) + b(i,0) + c(0,1) + d(0,i) = (0,0)$$

$$\Rightarrow \qquad (a + ib, c + id) = (0,0)$$

$$\therefore \qquad \left. \begin{array}{l} a + ib = 0 \\ c + id = 0 \end{array} \right\} \qquad \ldots (1)$$

Solving the system (1), we get

$$a = 0, b = 0, c = 0, d = 0$$

Therefore, S in linearly independent.

Now, we shall show that $L(S) = V$.

Let $(a+ib, c+id)$ be any element of V where $a, b, c, d \in$ R.

Then, $\quad (a+ib, c+id) = a(1, 0) + b(i, 0) + c(0, 1) + d(0, i)$.

Therefore, every element of V can be expressible as a linear combination of the elements of S.

Thus, $\qquad L(S) = V$

Hence , S is a basis of V.

Example 2. *Show that the set S={(1, 2), (3, 4)} forms a basis for R^2.*

Solution. Since the dimension of R^2 is 2 and S contains 2 elements.

Now we shall show that S is linearly independent.

Let $a, b \in$ R such that

$$a(1,2) + b(3,4) = (0,0)$$

$$\Rightarrow \qquad (a + 3b, 2a + 4b) = (0,0)$$

$$\therefore \qquad \left. \begin{array}{l} a + 3b = 0 \\ 2a + 4b = 0 \end{array} \right\} \qquad \ldots (1)$$

The coefficient matrix of the system (1) is

$$A = \begin{bmatrix} 1 & 3 \\ 2 & 4 \end{bmatrix}$$

$$\Rightarrow \qquad |A| = \begin{vmatrix} 1 & 3 \\ 2 & 4 \end{vmatrix}$$

$$\Rightarrow \qquad |A| = 4 - 6 = -2 \neq 0$$

$\Rightarrow$ Rank $A = 2$ which is equal to number of unknowns.

Therefore, the system (1) has only zero solution, *i.e.*, $a = 0$, $b = 0$ so that S is linearly independent.

Hence S forms a basis of R^2.

Example 3. *Let V be the vector space of all 2×2 matrices over the field F. Prove that V has dimension 4 by exhibiting a basis for V which has 4 elements.*

Solution. Let $S = \{\alpha_1, \alpha_2, \alpha_3, \alpha_4\}$, where

$$\alpha_1 = \begin{bmatrix} 1 & 0 \\ 0 & 0 \end{bmatrix}, \alpha_2 = \begin{bmatrix} 0 & 1 \\ 0 & 0 \end{bmatrix}, \alpha_3 = \begin{bmatrix} 0 & 0 \\ 1 & 0 \end{bmatrix}, \text{and } \alpha_4 = \begin{bmatrix} 0 & 0 \\ 0 & 1 \end{bmatrix}$$

are the four elements of V.

Now we shall show that S forms a basis of V.

Let a, b, c, $d \in F$ such that

$$a\alpha_1 + b\alpha_2 + c\alpha_3 + d\alpha_4 = 0$$

$$\Rightarrow \quad a\begin{bmatrix} 1 & 0 \\ 0 & 0 \end{bmatrix} + b\begin{bmatrix} 0 & 1 \\ 0 & 0 \end{bmatrix} + c\begin{bmatrix} 0 & 0 \\ 1 & 0 \end{bmatrix} + d\begin{bmatrix} 0 & 0 \\ 0 & 1 \end{bmatrix} = \begin{bmatrix} 0 & 0 \\ 0 & 0 \end{bmatrix}$$

$$\Rightarrow \qquad\qquad \begin{bmatrix} a & b \\ c & d \end{bmatrix} = \begin{bmatrix} 0 & 0 \\ 0 & 0 \end{bmatrix}$$

$$\Rightarrow \quad a = 0, b = 0, c = 0, d = 0$$

Therefore, S is linearly independent.

Next, we shall show that $L(S) = V$.

Let $\begin{bmatrix} a & b \\ c & d \end{bmatrix}$ be any element of V.

Then $\qquad \begin{bmatrix} a & b \\ c & d \end{bmatrix} = a\alpha_1 + b\alpha_2 + c\alpha_3 + d\alpha_4$

$$\Rightarrow \qquad L(S) = V.$$

Hence S forms a basis of V which has four elements, therefore

$$\dim.V = 4$$

Example 4. *Let $\alpha = (1,2,1), \beta = (2,9,0)$ and $\gamma = (3,3,4)$. Show that the set $S = \{\alpha, \beta, \gamma\}$ is a basis of R^3.*

Solution. The dimension of $R^3 = 3$

Now we shall show that S is linearly independent.

Let a, b, $c \in R$ such that

$$a\alpha + b\beta + c\gamma = 0$$

$$\Rightarrow \qquad a(1,2,1) + b(2,9,0) + c(3,3,4) = (0,0,0)$$

$$\Rightarrow \qquad (a + 2b + 3c, 2a + 9b + 3c, a + 4c) = (0,0,0)$$

$\therefore$
$$\begin{aligned} a + 2b + 3c &= 0 \\ 2a + 9b + 3c &= 0 \\ a + 4c &= 0 \end{aligned} \right\} \qquad \text{... (1)}$$

The coefficient matrix of the system (1) is

$$A = \begin{bmatrix} 1 & 2 & 3 \\ 2 & 9 & 3 \\ 1 & 0 & 4 \end{bmatrix}$$

$\Rightarrow$
$$|A| = \begin{vmatrix} 1 & 2 & 3 \\ 2 & 9 & 3 \\ 1 & 0 & 4 \end{vmatrix}$$

$\Rightarrow \qquad |A| = 1(36 - 0) - 2(8 - 3) + 3(0 - 9)$

$\Rightarrow \qquad |A| = 36 - 10 - 27 = -1 \neq 0$

$\Rightarrow$ Rank $(A) = 3$ which is equal to the number of unknowns. Therefore the system (1) has only zero solution, *i.e.*, $a = 0$, $b = 0$, $c = 0$.

$\therefore$ S is a linearly independent which has 3 elements.

Hence, S forms a basis of R^3.

Example 5. *Consider the basis $S = \{\alpha_1, \alpha_2, \alpha_3\}$ of R^3 where $\alpha_1 = (1, 1, 1)$, $\alpha_2 = (1, 1, 0)$, $\alpha_3 = (1, 0, 0)$. Express $(2, -3, 5)$ in terms of the basis elements $\alpha_1, \alpha_2, \alpha_3$.*

Solution. Since $S = \{\alpha_1, \alpha_2, \alpha_3\}$ forms a basis of R^3.

Then every elements of R^3 can be expressed as the linear combination α_1, α_2 and α_3.

Now for $a, b, c \in R$, we have

$$(2, -3, 5) = a\alpha_1 + b\alpha_2 + c\alpha_3 \qquad \text{...(1)}$$

$\Rightarrow \qquad (2, -3, 5) = a(1,1,1) + b(1,1,0) + c(1,0,0)$

$\Rightarrow \qquad (2, -3, 5) = (a + b + c, a + b, a)$

$\therefore$
$$\begin{aligned} a + b + c &= 2 \\ a + b &= -3 \\ a &= 5 \end{aligned}$$

Solving these equations, we get

$$a = 5, \ b = -8, \ c = 5$$

Putting the values of a, b and c in (1), we have

$$(2, -3, 5) = 5\alpha_1 - 8\alpha_2 + 5\alpha_3$$

Example 6. *Show that the vectors $\alpha_1 = (1, 0, -1), \alpha_2 = (1, 2, 1), a_3 = (0, -3, 2)$ form a basis of R^3. Express each of the standard basis vectors as a linear combination of $\alpha_1, \alpha_2, \alpha_3$.*

Solution. The dimension of $R^3 = 3$.

Let $S = \{\alpha_1, \alpha_2, \alpha_3\}$. Now we shall show that S is linearly independent.

Let $a, b, c \in R$ such that

$$a_1\alpha_1 + a_2\alpha_2 + a_3a_3 = 0$$

$$\Rightarrow \qquad a(1,0,-1) + b(1,2,1) + c(0,-3,\ 2) = (0,0,0)$$

$$\Rightarrow \qquad (a+b, 2b-3c, -a+b+2c) = (0,0,0)$$

$$\therefore \qquad\qquad a+b = 0$$

$$2b - 3c = 0$$

$$-a + b + 2c = 0$$

The coefficient matrix of the above equations is

$$A = \begin{bmatrix} 1 & 1 & 0 \\ 0 & 2 & -3 \\ -1 & 1 & 2 \end{bmatrix}$$

$$\Rightarrow \qquad |A| = \begin{vmatrix} 1 & 1 & 0 \\ 0 & 2 & -3 \\ -1 & 1 & 2 \end{vmatrix}$$

$$\Rightarrow \qquad |A| = 1(4+3) - 1(0-3)$$

$$\Rightarrow \qquad |A| = 7 + 3 = 10 \neq 0$$

$\Rightarrow \qquad$ Rank $(A) = 3$ which is equal to the number of unknowns a, b, c.

Thus, above equation has only zero solution, *i.e.*, $a = 0, b = 0, c = 0$.

Therefore, S is linearly independent.

Also dim $R^3 = 3$. Hence S forms a basis of R^3.

The standard basis of R^3 is $\{e_1, e_2, e_3\}$, where $e_1 = (1, 0, 0), e_2 = (0, 1, 0), \ e_3 = (0, 0, 1)$

Now let $\alpha = (p, q, r)$ be any element of R^3. Since S forms a basis of R^3. Then there exists $x, y, z \in R$ such that

$$\alpha = x\alpha_1 + y\alpha_2 + z\alpha_3 \qquad\qquad \dots (1)$$

$$\Rightarrow \qquad (p,q,r) = x(1,0,-1) + y(1,2,1) + z(0,-3,2)$$

$$\Rightarrow \qquad (p,q,r) = (x+y, 2y-3z, -x+y+2z)$$

$$\therefore \qquad\qquad p = x + y \qquad\qquad \dots (2)$$

$$q = 2y - 3z \qquad\qquad \dots (3)$$

$$r = -x + y + 2z \qquad\qquad \dots (4)$$

Adding (2) and (4), we get

$$2y + 2z = p + r \qquad\qquad \dots (5)$$

Subtracting (3) from (5), we get

$$5z = p + r - q$$

$$\therefore \qquad\qquad z = \frac{1}{5}(p - q + r)$$

Putting the value of z in (3), we get

$$2y = q + 3z = q + \frac{3}{5}(p - q + r)$$

or $$2y = \frac{3}{5}p + \frac{2}{5}q + \frac{3}{5}r$$

∴ $$y = \frac{1}{10}(3p + 2q + 3r)$$

Putting the value of y in (2), we get
$$x = p - y$$

or $$x = p - \frac{1}{10}(3p + 2q + 3r)$$

∴ $$x = \frac{1}{10}(7p - 2q - 3r)$$

Now, we shall express e_1, e_2, e_3 in terms of $\alpha_1, \alpha_2, \alpha_3$.

So, for $$e_1 = (1,0,0) = (p,q,r)$$

⇒ $$p = 1, q = 0, r = 0$$

Then $$x = \frac{1}{10}(7) = \frac{7}{10}, y = \frac{3}{10}, z = \frac{1}{5}$$

Thus from (1), we get
$$e_1 = \frac{7}{10}\alpha_1 + \frac{3}{10}\alpha_2 + \frac{1}{5}\alpha_3$$

For $$e_2 = (0,1,0) = (p,q,r)$$

⇒ $$p = 0, q = 1, r = 0$$

Then, $$x = -\frac{1}{5}, y = \frac{1}{5}, z = -\frac{1}{5}$$

∴ From (1), we get
$$e_2 = -\frac{1}{5}\alpha_1 + \frac{1}{5}\alpha_2 - \frac{1}{5}\alpha_3$$

For $$e_3 = (0,0,1) = (p,q,r)$$

⇒ $$p = 0, q = 0, r = 1$$

Then, $$x = \frac{-3}{10}, y = \frac{3}{10}, z = \frac{1}{5}$$

∴ From (1), we get
$$e_3 = \frac{-3}{10}\alpha_1 + \frac{3}{10}\alpha_2 + \frac{1}{5}\alpha_3 .$$

Example 7. *Show that the set $S = \{1, x, x^2, ..., x^n\}$ of $n+1$ polynomials in x is a basis of the vectors space $P_n(R)$ of all polynomials in x (of degree at most n) over the field of real numbers.*

Solution. Here $S = \{1, x, x^2, ..., x^n\}$ is a subset of $P_n(R)$.

First, we shall show that S is linearly independent.

Let $a_0, a_1, a_2, ..., a_n \in R$ such that
$$a_0.1 + a_1 x + a_2 x^2 + ... + a_n x^n = 0(x), i.e., \text{ zero polynomial}$$

⇒ $$a_0 + a_1 x + a_2 x^2 + ... + a_n x^n = 0 + 0x + 0x^2 + ... + 0x^n$$

$\Rightarrow$ $\qquad a_0 = 0, a_1 = 0, a_2 = 0, ..., a_n = 0$

Therefore, S is linearly independent.

Now we shall show that S spans $P_n(\text{R})$.

Let $p(x) = a_0 + a_1x + a_2x^2 + ... + a_nx^n$ be any polynomial of $P_n(\text{R})$ where $a_0, a_1, a_2, ... a_n \in \text{R}$.

Therefore, $p(x)$ is a linear combination of the elements of S. Thus S spans $P_n(\text{R})$

Hence S forms a basis of $P_n(\text{R})$.

Example 8. *Given that each set S below spans* R^3, *find a basis of* R^3 *which is contained in S:*

(*i*) $\{(1, 0, 2), (0, 1, 1), (2, 1, 5), (1, 1, 3), (1, 2, 1)\}$

(*ii*) $\{(2, 6, -3), (5, 15, -8), (3, 9, -5), (1, 3, -2), (5, 3, -2)\}$

Solution. (i) Let $S = \{(1, 0, 2), (0, 1, 1), (2, 1, 5), (1, 1, 3), (1, 2, 1)\}$

Since dim. $\text{R}^3 = 3$ and S spans R^3.

Now we shall find maximal linearly independent set containing S.

Let A be a matrix whose rows are the elements of S. Then

$$A = \begin{bmatrix} 1 & 0 & 2 \\ 0 & 1 & 1 \\ 2 & 1 & 5 \\ 1 & 1 & 3 \\ 1 & 2 & 1 \end{bmatrix}$$

We shall reduce this matrix to Echelon form by using row transformation.

Applying $R_3 \to R_3 - 2R_1, R_4 \to R_4 - R_1, R_5 \to R_5 - R_1$, we get

$$A \sim \begin{bmatrix} 1 & 0 & 2 \\ 0 & 1 & 1 \\ 0 & 1 & 1 \\ 0 & 1 & 1 \\ 0 & 2 & -1 \end{bmatrix}$$

Applying $R_3 \to R_3 - R_2, R_4 \to R_4 - R_2$ and, $R_5 \to R_5 - 2R_2$, we get

$$A \sim \begin{bmatrix} 1 & 0 & 2 \\ 0 & 1 & 1 \\ 0 & 0 & 0 \\ 0 & 0 & 0 \\ 0 & 0 & -2 \end{bmatrix}$$

This is an Echelon form which has three non-zero rows. Thus Rank $(A) = 3$. Here non-zero rows are corresponding to the vectors $(1, 0, 2), (0, 1, 1)$ and $(1, 2, 1)$ of S.

Therefore, S contains maximal linearly independent subset $\{(1, 0, 2), (0, 1, 1), (1, 2, 1)\}$ which has 3 elements which is equal to the dimension of R^3. Hence, the set $\{(1, 0, 2), (0, 1, 1), (1, 2, 1)$ forms a basis of R^3.

(ii) Let $S = \{(2, 6, -3), (5, 15, -8), (3, 9, -5), (1, 3, -2), (5, 3, -2)\}$.

Since dim. $R^3 = 3$ and S spans R^3.

Now we shall find maximal linearly independent set contained in S.

Let A be a matrix whose rows and the elements of S.

Then
$$A = \begin{bmatrix} 2 & 6 & -3 \\ 5 & 15 & -8 \\ 3 & 9 & -5 \\ 1 & 3 & -2 \\ 5 & 3 & -2 \end{bmatrix}$$

Now we shall reduce this matrix to echelon form by using row transformations.
Applying $R_4 \leftrightarrow R_1$, we get
$$A \sim \begin{bmatrix} 1 & 3 & -3 \\ 5 & 15 & -8 \\ 3 & 9 & -5 \\ 2 & 6 & -3 \\ 5 & 3 & -2 \end{bmatrix}$$

Applying $R_2 \rightarrow R_2 - 5R_1, R_3 \rightarrow R_3 - 3R_1, R_4 \rightarrow R_4 - 2R_1$ and $R_5 \rightarrow R_5 - 5R_1$,

we get
$$A \sim \begin{bmatrix} 1 & 3 & -2 \\ 0 & 0 & 2 \\ 0 & 0 & 1 \\ 0 & 0 & 1 \\ 0 & -12 & 8 \end{bmatrix}$$
Applying $R_5 \leftrightarrow R_2$, we get
$$A \sim \begin{bmatrix} 1 & 3 & -2 \\ 0 & -12 & 8 \\ 0 & 0 & 1 \\ 0 & 0 & 1 \\ 0 & 0 & 2 \end{bmatrix}$$

Applying $R_4 \rightarrow R_4 - R_3, R_5 \rightarrow R_5 - 2R_3$, we get
$$A \sim \begin{bmatrix} 1 & 3 & -2 \\ 0 & -12 & 8 \\ 0 & 0 & 1 \\ 0 & 0 & 0 \\ 0 & 0 & 0 \end{bmatrix}$$

This is an Echelon form which has three non-zero rows. Thus Rank $(A) = 3$. Here non-zero rows are corresponding to the vectors $(1, 3, -2)$, $(5, 3, -2)$ and $(3, 9, -5)$ of S.

Therefore, S contains maximal linearly independent subset $\{(3, 9, -5), (1, 3, -2), (5, 3, -2)\}$ which has 3 elements which is equal to the dimension of R^3. Hence, the set $\{(3, 9, -5), (1, 3, -2), (5, 3, -2)\}$ forms a basis of R^3.

Example 9. *Given that the set S is a basis of* R^4 *and that T is linearly independent. Extend T to a basis of* R^4, *where*

$$S = \{(1,0,0,0),(0,0,1,0),(5,1,11,0),(-4,0,-6,1)\}$$

and $T = \{(1,0,1,0),(0,2,0,3)\}$

Solution. Since dim. $R^4 = 4$ and T contains two linearly independent elements so in order to extend it to form a basis of R^4 we shall include two elements from S to T till T becomes linearly independent.

First include an element $(1, 0, 0, 0)$ to T then, we have

$$T_1 = \{(1, 0, 1, 0), (0, 2, 0, 3), (1, 0, 0, 0)\}$$

Now, we shall check whether T_1 is linearly independent or not.

Let A be a matrix whose rows are the elements of T_1. Then

$$A = \begin{bmatrix} 1 & 0 & 1 & 0 \\ 0 & 2 & 0 & 3 \\ 1 & 0 & 0 & 0 \end{bmatrix}$$

Applying $R_3 \rightarrow R_3 - R_1$

$$A \sim \begin{bmatrix} 1 & 0 & 1 & 0 \\ 0 & 2 & 0 & 3 \\ 0 & 0 & -1 & 0 \end{bmatrix}$$

which is an Echelon form having 3 non-zero rows so that Rank (A) =3. Thus, T_1 is linearly independent but T_1 contains 3 elements and dim. $R^4 = 4$, so it is not a basis of R^4.

We again include an element $(0, 0, 1, 0)$ to T_1, we have

$$T_2 = \{(1,0,1,0),(0,2,0,3),(1,0,0,0)(0,0,1,0)\}.$$

Again we shall test the linear dependence.

Let A be a matrix whose rows are the elements of T_1. Then

$$A = \begin{bmatrix} 1 & 0 & 1 & 0 \\ 0 & 2 & 0 & 3 \\ 1 & 0 & 0 & 0 \\ 0 & 0 & 1 & 0 \end{bmatrix}$$

Applying $R_3 \rightarrow R_3 \rightarrow R_1$, we get

$$A \sim \begin{bmatrix} 1 & 0 & 1 & 0 \\ 0 & 2 & 0 & 3 \\ 0 & 0 & -1 & 0 \\ 0 & 0 & 1 & 0 \end{bmatrix}$$

Applying $R_4 \rightarrow R_4 + R_3$

$$A \sim \begin{bmatrix} 1 & 0 & 1 & 0 \\ 2 & 2 & 0 & 3 \\ 0 & 0 & -1 & 0 \\ 0 & 0 & 0 & 0 \end{bmatrix}$$

which is an Echelon form having 3 non-zero rows, so Rank $(A) = 3$. Thus T_2 is linearly dependent.

Now we include an element $(5, 1, 11, 0)$ in place of $(0, 0, 1, 0)$ to T_2, then we have

$$T_3 = \{(1, 0, 1, 0), (0, 2, 0, 3), (1, 0, 0, 0), (5, 1, 11, 0)\}$$

Again we shall test the linear dependence.

Let A be a matrix whose rows are the elements of T_3. Then

$$A = \begin{bmatrix} 1 & 0 & 1 & 0 \\ 0 & 2 & 0 & 3 \\ 1 & 0 & 0 & 0 \\ 5 & 1 & 11 & 0 \end{bmatrix}$$

Applying $R_3 \to R_3 - R_1$ and $R_4 \to R_4 - 5R_1$, we get

$$A \sim \begin{bmatrix} 1 & 0 & 1 & 0 \\ 0 & 2 & 0 & 3 \\ 0 & 0 & -1 & 0 \\ 0 & 1 & 6 & 0 \end{bmatrix}$$

Applying $R_4 \to R_4 - \dfrac{1}{2} R_2$, we get

$$A \sim \begin{bmatrix} 1 & 0 & 1 & 0 \\ 0 & 2 & 0 & 3 \\ 0 & 0 & -1 & 0 \\ 0 & 0 & 6 & -3/2 \end{bmatrix}$$

Applying $R_4 \to R_4 + 6R_3$, we get

$$A \sim \begin{bmatrix} 1 & 0 & 1 & 0 \\ 0 & 2 & 0 & 3 \\ 0 & 0 & -1 & 0 \\ 0 & 0 & 0 & -3/2 \end{bmatrix}$$

which is an Echelon form having 4 non-zero rows so Rank $(A) = 4$. Thus T_3 is linearly independent which has 4 elements. Hence the set

$\{(1, 0, 1, 0), (0, 2, 0, 3), (1, 0, 0, 0), (5, 1, 11, 0)\}$ is a basis.

Example 10. *Let W be a subspace of a vector space V(F) of dimension r. Then show that a set of r vectors in W is a basis of W if and only if it is linearly independent.*

Solution. If a set of r vectors in W is a basis, then by definition of a basis, it is linearly independent.

Let $S = \{\alpha_1, \alpha_2, ..., \alpha_r\}$ be a set of r linearly independent vectors in W. Then S can be extended to a basis of W. Now S has r elements and dimension of W is r and hence, all bases of W have r elements. In particular, the basis to which S is extended has r elements therefore, S is a basis of W.

Example 11. *Let W be a subspace of a vector space V(F) of dimension r. Then show that a set of r vectors in W is a basis if and only if it spans W.*

Solution. If a set of r vectors in W is a basis if and onl if it spans W(By definition).

Let $S = \{\alpha_1, \alpha_2, ..., \alpha_r\}$ be a set of r vectors which spans W. Then S contains a basis of W.

But any basis of W contains r vectors. Therefore, the basis contained in S is not a proper subset of S, and S is a basis of W.

Example 12. *If n vectors span a vector space V containing r linearly independent vectors, then show that $n \geq r$.*

Solution. Let S be a subset of V containing n vectors and S spans V. Then there exists a linearly independent subset T and S which also spans V. Therefore, T will form a basis of V. If T contains m elements then dim. $V = m$ and $m \geq n$.

Since dim. $V = m$, then any subset of V containing more than m elements will be linealry dependent. Therefore, if there is a linearly independent subset of V containing r vectors, then $r \leq m$ but $m \leq n$, hence $n \leq r$.

Example 13. *Extend the linearly independent subset $\{(1, 0, 1), (0, -1, 1)\}$ of $V_3(R)$ to form a basis of V_3 (R).*

Solution. Since the dim. V_3 (R)$=3$ and the subset $\{(1, 0, 1), (0, -1, 1)\}$ has two linearly independent vectors, therefore, in order to extend this subset to form a basis of V_3 (R) we shall include one vector of $V_3(R)$ such that they are linearly independent.

Let $\qquad\qquad S = \{(1,0,1),(0,-1,1)\}$

Then, $\qquad L(S) = \{a(1,0,1) + b(0,-1,1) : a, b \in R\}$
$$= \{a, -b, a + b) : a, b \in R\}$$

Clearly, in this span, the third component is the sum of first and negative of second component. Therefore, we shall include a vector of $V_3(R)$ which does not have this property.

Now there are many such vectors in V_3 (R), one of which is (1, 0, 0). So we include this vector to S. Thus the set $\{(1, 0, 1), (0, -1, 1), (1, 0, 0)\}$ is linearly independent and has 3 elements.

Hence, this set is a required basis of $V_3(R)$ which is the extension of S.

Example 14. *Extend the linearly dependent subset $\{(1, -1, 0, 0), (1, 1, 1, 0)\}$ of V_4 (R) to form a basis of $V_4(R)$.*

Solution. Let $S = \{(1,-1,0,0),(1,1,1,0)\}$. Then
$$L(S) = \{a(1,-1,0,0) + b(1,1,1,0) : a, b \in R\}$$
$$= \{(a + b, -a + b, b, 0) : a, b \in R\}$$
Clearly, in this span, the fourth coordinate is zero.

Then (0, 0, 0, 1) is clearly not in this span, so that we include this vector to S, we get an enlarged linearly independent set
$$S' = \{(1,-1,0,0),(1,1,1,0),(0,0,0,1)\}$$

Again we extend this set

Now, $\quad L(S') = \{a(1,-1,0,0) + b(1,1,1,0) + c(0,0,0,1) : a,b,c \in R\}$
$\quad\quad\quad = \{(a+b,-a+b,b,c) : a,b,c \in R\}$

In this span, we observe that for given a, b and c the second coordinate is always $-a + b$, so we try to find such vector which does not follow this hypothesis.

Clearly, $(1, -2, 0, 0)$ is not in $L(S')$. Now we include this vector to S', we get a set
$$S'' = \{(1,-1,0,0),(1,1,1,0),(0,0,0,1),(1,-2,0,0)\}$$
which is linearly independent and has 4 elements.

Hence, S'' is a required basis of $V_4(R)$ which is the extension of S.

Example 15. *Let W be the subspace of $V_4(R)$ generated by the vectors $(1, -2, 5, -3)$, $(2, 3, 1, -4)$, $(3, 8, -3, -5)$, then*

   (i) *Find a basis and dimension of W.*

   (ii) *Extend the basis of W to a basis of $V_4(R)$.*

Solution.   (i) Let $S = \{(1,-2,5,-3),(2,3,1,-4),(3,8,-3,-5)\}$ and $L(S)=W$.

Now we shall find maximal linearly independent subset of S.

Let A be a matrix whose rows are the elements of S, then

$$A = \begin{bmatrix} 1 & -2 & 5 & -3 \\ 2 & 3 & 1 & -4 \\ 3 & 8 & -3 & -5 \end{bmatrix}$$

We shall have to reduce A to an Echelon form by using row transformations.

Applying $R_2 \to R_2 - 2R_1, R_3 \to R_3 - 3R_1$, we get

$$A \sim \begin{bmatrix} 1 & -2 & 5 & -3 \\ 0 & 7 & -9 & 2 \\ 0 & 14 & -18 & 4 \end{bmatrix}$$

Again applying $R_3 \to R_3 - 2R_2$, we get

$$A \sim \begin{bmatrix} 1 & -2 & 5 & -3 \\ 0 & 7 & -9 & 2 \\ 0 & 0 & 0 & 0 \end{bmatrix}$$

which is an Echelon form and has 2 non-zero rows representing the co-ordinate vectors $(1, -2, 5, -3)$ and $(0, 7, -9, 2)$ that form a basis of rows space i.e., $T = \{(1,-2,5,-3),(0,7,-9,2)\}$ is a basis of W. Thus dim. $W = 2$.

(ii) Since dim. $V_4(R) = 4$, so in order to form a basis of $V_4(R)$ we shall extend the set T by including two vectors.

Let us take these two vectors as $(0, 0, 1, 0)$ and $(0, 0, 0, 1)$, so including both in T such that the set
$$T' = \{(1,-2,5,-3),(0,7,-9,2),(0,0,1,0),(0,0,0,1)\}$$
is linearly independent because a matrix.

$$\begin{bmatrix} 1 & -2 & 5 & -3 \\ 0 & 7 & -9 & 2 \\ 0 & 0 & 1 & 0 \\ 0 & 0 & 0 & 1 \end{bmatrix}$$

whose rows are the elements of T' is in Echelon form with 4 non-zero rows showing that T' is linearly independent.

Hence T' is a basis of $V_4(R)$ which is obtained by extending a basis of W.

Example 16. Let W be the subspace of $V_4(R)$ generated by the set of vectors

$$S = \{(1,1,0,-1),(1,2,3,0),(2,3,3,-1)\}$$

and W_2 the subspace of $V_4(R)$ generated by the set of vectors

$$T = \{(1,2,2,-2),(2,3,2,-3),(1,3,4,-3)\}$$

Find:

(*i*) $\dim.(W_1 + W_2)$ \hspace{2cm} (ii) $\dim.(W_1 \cap W_2)$

Solution. (i) We know that

$$L(W_1 \cup W_2) = W_1 + W_2$$

Then W_1+W_2 is a subspace generated by the set of vectors of $S \cup T$ where $S \cup T = \{(1,1,0,-1),(1,2,3,0),(2,3,3,-1),(1,2,2,-2)(2,3,2,-3),(1,3,4,-3)\}$

Let A be a matrix whose rows are the elements of $S \cup T$. Then

$$A = \begin{bmatrix} 1 & 1 & 0 & -1 \\ 1 & 2 & 3 & 0 \\ 2 & 3 & 3 & -1 \\ 1 & 2 & 2 & -2 \\ 2 & 3 & 2 & -3 \\ 1 & 3 & 4 & -3 \end{bmatrix}$$

Now we shall reduce A to an Echelon form as follows:

Applying $R_2 \to R_2 - R_1, R_3 \to R_3 - 2R_1, R_4 \to R_4 - R_1, R_5 \to R_5 - 2R_1$ and $R_6 \to R_6 - R_1$

$$A \sim \begin{bmatrix} 1 & 1 & 0 & -1 \\ 0 & 1 & 3 & 1 \\ 0 & 1 & 3 & 1 \\ 0 & 1 & 2 & -1 \\ 0 & 1 & 2 & -1 \\ 0 & 2 & 4 & -2 \end{bmatrix}$$

Again Applying $R_3 \to R_3 - R_2, R_4 \to R_4 - R_2, R_5 \to R_5 - R_2$ and $R_6 \to R_6 - 2R_2$, we get

$$A \sim \begin{bmatrix} 1 & 1 & 0 & -1 \\ 0 & 1 & 3 & 1 \\ 0 & 0 & 0 & 0 \\ 0 & 0 & -1 & -2 \\ 0 & 0 & -1 & -2 \\ 0 & 0 & -2 & -4 \end{bmatrix}$$

Applying $R_4 \leftrightarrow R_3$, we get

$$A \sim \begin{bmatrix} 1 & 1 & 0 & -1 \\ 0 & 1 & 3 & 1 \\ 0 & 0 & -1 & -2 \\ 0 & 0 & 0 & 0 \\ 0 & 0 & -1 & -2 \\ 0 & 0 & -2 & -4 \end{bmatrix}$$

Applying $R_5 \to R_5 - R_3, R_6 \to R_6 - 2R_3$

$$A \sim \begin{bmatrix} 1 & 1 & 0 & -1 \\ 0 & 1 & 3 & 1 \\ 0 & 0 & -1 & -2 \\ 0 & 0 & 0 & 0 \\ 0 & 0 & 0 & 0 \\ 0 & 0 & 0 & 0 \end{bmatrix}$$

which is an Echleon form and has 3 non-zero rows.

$$\therefore \qquad \dim.(W_1 + W_2) = 3$$

(ii) First we find the dim. W_1 and W_2.

Let A_1 be a matrix whose rows are the elements of S, then

$$A_1 = \begin{bmatrix} 1 & 1 & 0 & -1 \\ 1 & 2 & 3 & 0 \\ 2 & 3 & 3 & -1 \end{bmatrix}$$

Applying $R_2 \to R_2 - R_1$ and $R_3 \to R_3 - 2R_1$, we get

$$A \sim \begin{bmatrix} 1 & 1 & 0 & -1 \\ 0 & 1 & 3 & 1 \\ 0 & 1 & 3 & 1 \end{bmatrix}$$

Again applying $R_3 \to R_3 - R_2$ we get

$$A \sim \begin{bmatrix} 1 & 1 & 0 & -1 \\ 0 & 1 & 3 & 1 \\ 0 & 0 & 0 & 0 \end{bmatrix}$$

which is an Echelon form and has 2 non-zero rows.

$$\therefore \qquad \dim. W_1 = 2$$

Let A_2 be a matrix whose rows are the elements of T, then

$$A_2 = \begin{bmatrix} 1 & 2 & 2 & -2 \\ 2 & 3 & 2 & -3 \\ 1 & 3 & 4 & -3 \end{bmatrix}$$

Applying $R_2 \to R_2 - 2R_1, R_3 \to R_3 - R_1$, we get

$$A_2 \sim \begin{bmatrix} 1 & 2 & 2 & -2 \\ 0 & -1 & -2 & 1 \\ 0 & 1 & 2 & -1 \end{bmatrix}$$

Again applying $R_3 \rightarrow R_3 + R_2$, we get

$$A_2 \sim \begin{bmatrix} 1 & 2 & 2 & -2 \\ 0 & -1 & -2 & 1 \\ 0 & 1 & 0 & 0 \end{bmatrix}$$

which is an Echelon form and has 2 non-zero rows.

$\therefore \qquad$ dim. $W_2 = 2$

We know that

$$\text{dim}.(W_1 + W_2) = \text{dim}.W_1 + \text{dim}.W_2 - \text{dim}.(W_1 \cap W_2)$$

$\Rightarrow \qquad 3 = 2 + 2 - \text{dim}.(W_1 \cap W_2)$

$\therefore \qquad \text{dim}.(W_1 \cap W_2) = 1$

Example 17. *Let V be the vector space of 2×2 symmetric matrix over R. Show that dim. $V = 3$*

Solution. An arbitrary 2×2 symmetric matrix is of the form $\begin{bmatrix} a & b \\ b & c \end{bmatrix}$ where a, b, $c \in$ R.

Now setting (i) $a = 1, b = 0, c = 0$, (ii) $a = 0, b = 1, c = 0$, (iii) $a = 0, b = 0, c = 1$, thus we obtain the following matrices

$$A = \begin{bmatrix} 1 & 0 \\ 0 & 0 \end{bmatrix}, B = \begin{bmatrix} 0 & 1 \\ 1 & 0 \end{bmatrix}, C = \begin{bmatrix} 0 & 0 \\ 0 & 1 \end{bmatrix}$$

We shall show that the set $S = \{A, B, C\}$ is a basis of V.

First we show that S is linearly independent.

Let $x, y, z \in$ R such that

$$xA + yB + zC = 0$$

$\Rightarrow \quad x\begin{bmatrix} 1 & 0 \\ 0 & 0 \end{bmatrix} + y\begin{bmatrix} 0 & 1 \\ 1 & 0 \end{bmatrix} + z\begin{bmatrix} 0 & 0 \\ 0 & 1 \end{bmatrix} = \begin{bmatrix} 0 & 0 \\ 0 & 0 \end{bmatrix}$

$\Rightarrow \qquad\qquad \begin{bmatrix} x & y \\ y & z \end{bmatrix} = \begin{bmatrix} 0 & 0 \\ 0 & 0 \end{bmatrix}$

$\Rightarrow \qquad\qquad x = 0, y = 0, z = 0$

$\therefore \quad S = \{A, B, C\}$ is linearly independent.

Now, we show that $L(S) = V$.

Let $\begin{bmatrix} a & b \\ b & c \end{bmatrix}$ be any element of V, then we have

$$\begin{bmatrix} a & b \\ b & c \end{bmatrix} = \begin{bmatrix} a & 0 \\ 0 & 0 \end{bmatrix} + \begin{bmatrix} 0 & b \\ b & c \end{bmatrix} + \begin{bmatrix} 0 & 0 \\ 0 & c \end{bmatrix}$$

$\Rightarrow \qquad \begin{bmatrix} a & b \\ b & c \end{bmatrix} = a\begin{bmatrix} 1 & 0 \\ 0 & 0 \end{bmatrix} + b\begin{bmatrix} 0 & 1 \\ 1 & 0 \end{bmatrix} + c\begin{bmatrix} 0 & 0 \\ 0 & 1 \end{bmatrix}$

$\Rightarrow \qquad \begin{bmatrix} a & b \\ b & c \end{bmatrix} = aA + bB + cC$

$\Rightarrow \quad$ Every element of V is linear combination of elements of S

$\Rightarrow$ $\qquad$ $L(S) = V$

Therefore, S forms a basis of V which has 3 elements.

Hence, dim. $V = 3$.

Example 18. *Let W be the subspace of R^3 defined by $W = \{(a, b, c): a+b+c = 0\}$. Find a basis and dimension of W.*

Solution. Since $(1, 2, 3) \in R^3$ but $(1, 2, 3) \notin W$, therefore $W \neq R^3$. Thus dim. $W < 3$.

Now $\alpha_1 = (1, 0, -1)$ and $\alpha_2 = (0, 1, -1)$ are two vectors of W. Also α_1 cannot be expressed as a multiple of α_2 so that α_1, α_2 are linearly independent.

Therefore, $\{(1, 0, -1), (0, 1, -1)\}$ is a basis of W and dim. $W = 2$.

Example 19. *Let W be the subspace of R^3 defined by $W = \{(a, b, c) : a = b = c\}$. Find a basis and dimension of W.*

Solution. Since $(1, 1, 1) \in W$. Let $\alpha = (1, 1, 1)$. Then any vector β of W is of the form $\beta = (a, a, a)$ for all $a \in R$. Thus $\beta = a\alpha$. Therefore α spans W.

Hence $\{(1, 1, 1)\}$ is a basis of W and dim. $W = 1$

Example 20. *Let W_1 and W_2 be distinct subspaces of V and dim. $W_1 = 4$, dim. $W_2 = 4$ and dim. $V = 6$. Find the possible dimensions of $W_1 \cap W_2$.*

Solution. Since W_1 and W_2 be distinct subspaces of V, then $W_1 \subseteq W_1 + W_2$ and $W_2 \subseteq W_1 + W_2$.

$\therefore$ $\qquad$ dim. $(W_1 + W_2) >$ dim. W_1 or dim W_2.

$\Rightarrow$ $\qquad$ dim. $(W_1 + W_2) > 4$ $\qquad\qquad\qquad\qquad$ ... (1)

Also $\qquad\qquad$ dim. $V = 6$

$\Rightarrow$ $\qquad$ dim. $(W_1 + W_2) \leq 6$ $\qquad\qquad\qquad\qquad$ ...(2)

From (1) and (2), we conclude that dim. $(W_1 + W_2)$ is either 5 or 6.

When dim. $(W_1 + W_2) = 5$:

We know that

$\qquad\qquad$ dim. $(W_1 \cap W_2) =$ dim. $W_1 +$ dim. $W_2 -$ dim. $(W_1 + W_2)$

$\Rightarrow$ $\qquad$ dim. $(W_1 \cap W_2) = 4 + 4 - 5$

$\Rightarrow$ $\qquad$ dim. $(W_1 \cap W_2) = 3$

When $\qquad$ dim. $(W_1 + W_2) = 6$:

$\qquad\qquad$ dim. $(W_1 \cap W_2) = 4 + 4 - 6$

$\Rightarrow$ $\qquad$ dim. $(W_1 \cap W_2) = 2$

Example 21. *Let W_1 and W_2 be subspaces of R^3 for which dim. $W_1 = 1$, dim. $W_2 = 2$ and $W_1 \not\subseteq W_2$. Show that $R^3 = W_1 \oplus W_2$.*

Solution. Since $W_1 \not\subseteq W_2$ then $(W_1 \cap W_2) \subseteq W_1$ and $W_2 \cap W_1 \subseteq W_2$

$\therefore$ $\qquad$ dim. $(W_1 \cap W_2) <$ dim. W_1

$\Rightarrow$ $\qquad$ dim. $(W_1 \cap W_2) < 1$

$\Rightarrow \qquad\qquad \text{dim. } (W_1 \cap W_2) = 0$

$\Rightarrow \qquad\qquad W_1 \cap W_2 = \{0\}$

Also, $\qquad \text{dim. } (W_1 + W_2) > \text{dim. } W_2$

$\Rightarrow \qquad\qquad \text{dim. } (W_1 + W_2) > 2 \qquad\qquad\qquad\qquad\qquad \text{... (1)}$

But dim. $R^3 = 3$ and $W_1 + W_2 \subseteq R^3$

$\therefore \qquad\qquad \text{dim. } (W_1 + W_2) \le 3 \qquad\qquad\qquad\qquad\qquad \text{... (2)}$

From (1) and (2), we conclude that

$\qquad\qquad \text{dim. } (W_1 + W_2) = 3$

$\Rightarrow \qquad\qquad W_1 + W_2 = R^3$

Thus, $R^3 = W_1 + W_2$ and $W_1 \cap W_2 = \{0\}$, hence $R^3 = W_1 \oplus W_2$

Example 22. *Show that if $S = \{\alpha, \beta, \gamma\}$ is a basis of $C^3(C)$, then the set $S' = \{\alpha+\beta, \beta+\gamma, \gamma+\alpha\}$ is also a basis of $C^3(C)$.*

Solution. Since dim. $C^3 = 3$, therefore any subset of C^3 having 3 linearly independent vectors will form a basis of C^3.

Further, since $S = \{\alpha, \beta, \gamma\}$ is a basis of C^3, so that α, β, γ are linearly independent.

If $S' = \{\alpha+\beta, \beta+\gamma, \gamma+\alpha\}$ is linearly independent then it will form a basis of C^3, so we shall test the dependence of S'.

Let $a, b, c \in C$ such that

$\qquad\qquad a(\alpha + \beta) + b(\beta + \gamma) + c(\gamma + \alpha) = 0$

$\Rightarrow \qquad\qquad (a+c)\alpha + (a+b)\beta + (b+c)\gamma = 0$

$\Rightarrow \qquad\qquad a+c = 0, a+b = 0, b+c = 0 \qquad [\because \alpha, \beta, \gamma \text{ are linearly independent.}]$

$\Rightarrow \qquad\qquad a = 0, b = 0, c = 0$

$\Rightarrow \quad S'$ is linearly independent.

Hence $S' = \{\alpha+\beta, \beta+\gamma, \gamma+\alpha\}$ is also a basis of $C^3(C)$.

Example 23. *Give a basis for each of the following vector space over the indicated fields:*

(i) $R(\sqrt{2})$ over R \qquad (ii) $Q(2^{1/4})$ over Q

where Q, R are field of rational and real numbers.

Solution. (i) We have $\qquad R(\sqrt{2}) = \{a + \sqrt{2}b : a, b \in R\}$

Now zero element of $R(\sqrt{2})$ can be written as $0 = 0 + 0 . \sqrt{2}$

Let $S = \{1, \sqrt{2}\}$, then $S \subseteq R(\sqrt{2})$. Now we shall show that S forms a basis of $R(\sqrt{2})$.

Let $a, b \in R$, such that

$\qquad\qquad a.1 + \sqrt{2}\,b = 0$

$\Rightarrow \qquad\qquad a + \sqrt{2}b = 0 + 0 . \sqrt{2}$

$\Rightarrow \qquad\qquad a = 0, b = 0$

$\Rightarrow$ S is linearly independent.

Let $x + \sqrt{2}y$ be any element of $R(\sqrt{2})$. Then $x + \sqrt{2}y = x.1 + \sqrt{2}.y$

$\Rightarrow$ every element of $R(\sqrt{2})$ is expressible as the linear combination of elements of S.

$\Rightarrow$ $L(S) = R(\sqrt{2})$

Hence, $S = \{1, \sqrt{2}\}$ is a basis of $R(\sqrt{2})$ having two elements so that dim. $R(\sqrt{2})=2$

(ii)   We have

$$Q(2^{1/4}) = \{a + (2^{1/4})b : a, b \in Q\}$$

The zero element of $Q(2^{1/4})$ is $0 = 0 + (2^{1/4}).0$

Let $S = \{1, 2^{1/4}\}$, then $S \subseteq Q(2^{1/4})$. Now we shall see that S forms a basis of $Q(2^{1/4})$.

Let $a, b \in Q$ such that $a.1 + b.(2^{1/4}) = 0$

$\Rightarrow$ $a + b(2^{1/4}) = 0 + 0.(2^{1/4})$

$\Rightarrow$ $a = 0, b = 0$

$\Rightarrow$ S is linearly independent.

Let $x + (2^{1/4})y$ be any element of $Q(2^{1/4})$

Then, $x + (2^{1/4})y = x.1 + (2^{1/4}).y$

$\Rightarrow$ every element of $Q(2^{1/4})$ is expressible as a linear combination of elements of S.

$\Rightarrow$ $L(S) = Q(2^{1/4})$

Hence, $S = \{1, 2^{1/4}\}$ is a basis of $Q(2^{1/4})$ and dim. $Q(2^{1/4})=2$.

Example 24. *Determine dim V/W where V= C(R) and W = R(R).*

Solution. We know that $\{1, i\}$ is a basis of $C(R)$ and $\{1\}$ is a basis of $R(R)$. Therefore, dim $V = 2$ and dim $W = 1$.

Hence,   dim V/W = dim V – dim W = 2– 1 = 1

Example 25. *Let $V = C^2(R)$ and $W = R^2(R)$, find dim (V/W).*

Solution. We know that $\{(1, 0), (0, 1), (1, 0), (i, 0), (0, i)\}$ is a basis of $C^2(R)$ and $\{(1, 0), (0, 1)\}$ is a basis of $R^2(R)$.

So,   dim $V = 4$ and dim $W = 2$

Hence,   dim V/W = dim V – dim W

                    $= 4 – 2 = 2$

EXERCISE 5.5

1. Show that the set $\{(1, i, 0), (2i, 1, 1), (0, 1+i, 1-i)\}$ is a basis of V_3 (C).

2. Show that the vectors $(2, -1, 0)$, $(3, 5, 1)$, $(1, 1, 2)$ forms a basis of V_3 (R).

3. Determine whether the set $\{(-1, 1, 2), (2, -3, 1), (10, -14, 0)\}$ of vectors is a basis of V_3(R).

4. In the vector space R^3, let $\alpha = (1, 2, 1)$, $\beta = (3, 1, 5)$, $\gamma = (3, -4, 7)$. Show that there exists more than one basis for the subspace spanned by the set $S = \{\alpha, \beta, \gamma\}$.

5. Select a basis, if any, of R^3 (R) from the set $\{\alpha_1, \alpha_2, \alpha_3, \alpha_4\}$, where $\alpha_1 = (1, -3, 2)$, $\alpha_2 = (2, 4, 1)$, $\alpha_3 = (3, 1, 3)$, $\alpha_4 = (1, 1, 1)$.

6. Determine whether the following vectors form a basis of R^3 or not: $(1, 1, 2)$, $(1, 2, 5)$, $(5, 3, 4)$.

7. Show that the vectors $\alpha_1 = (1, 0, -1)$, $\alpha_2 = (1, 2, 1)$ and $\alpha_3 = (0, -3, 2)$ form a basis of V_3(R).

8. Determine a basis of the subspace spanned by the vectors $\alpha_1 = (1, 2, 3)$, $\alpha_2 = (2, 1, -1)$, $\alpha_3 = (1, -1, -4)$ and $\alpha_4 = (4, 2, -2)$.

9. Show that the set
   $S = \{(1,0,0),(1,1,0),(1,1,1),(0,1,0)\}$ space the vector space V_3(R) but is not a basis set.

10. Find three vectors in R^3 which are linearly dependent and are such that any two of them are linearly independent.

11. Let V be the vector space of all 3×3 symmetric matrices over F. Show that dim. $V=6$.

12. Prove that the space of all $m \times n$ matrices over the field F has dimension mn, by exhibiting a basis for this space.

13. What is the dimension of a vector space V of $m \times n$ symmetric matrices over a field F?

14. If $\{\alpha_1, \alpha_2, \alpha_3\}$ is a basis of V_3(R) show that $\{\alpha_1 + \alpha_2, \alpha_2 + \alpha_3, \alpha_3 + \alpha_1\}$ is also a basis of V_3(R).

15. Show that the vectors $(1, 1, 1, 1)$, $(0, 1, 1, 1)$, $(0, 0, 1, 1)$, $(0, 0, 0, 1)$ form a basis of R^4.

16. Let V be the vector of all polynomials in x of degree at most n. Show that $\{1, x-1, (x-1)^2, ..., (x-1)^n\}$ is a basis of V.

17. Show that the set of matrices
    $$S = \left\{ \begin{bmatrix} 1 & 1 \\ 1 & 0 \end{bmatrix}, \begin{bmatrix} 0 & 0 \\ 1 & 1 \end{bmatrix}, \begin{bmatrix} 1 & 0 \\ 0 & 1 \end{bmatrix}, \begin{bmatrix} 0 & 1 \\ 1 & 1 \end{bmatrix} \right\}$$
    forms a basis for the vector space $M_{2\times 2}$ of all 2×2 matrices over R.

18. Let V be a finite dimensional vector space over a field F, and let $\alpha_1, \alpha_2, ... \alpha_n$ be a basis of V. If $\beta_1, \beta_2, ... \beta_m$ in V are linearly independent over F, then prove that $m \le n$.

19. Extend the linearly independent subset $\{(1, 0, 1, 0), (0, 0, 0, 1)\}$ of V_4 to form a basis of V_4.

20. Given that the set S space R^3, find a basis of R^3 which is contained in S, where
    $S = \{(1,1,0),(2,2,0),(2,4,1),(5,9,2),$
    $(7,13,3),(1,2,1)\}$

21. Given that the set S is a basis of R^4 and the set T is linearly independent, then extend T to a basis of R^4, where
    $S = \{(1, 1, 0, 0), (0, 1, 1, 0), (0, 0, 0, 1), (0, 1, 0, 1)\}$ and $T = \{(1, 0, 2, 3), (0, 1-2, -3)\}$.

22. If W_1 and W_2 are finite dimensional subspaces with the same dimension, and if $W_1 \subseteq W_2$ then $W_1 = W_2$.

23. If W is a subspace of a finite dimensional vector space V, prove that any basis of W can be extended to form a basis of V.

24. Let V be a vector space. Let W be a subspace of V generated by the vectors $\alpha_1, \alpha_2, ... \alpha_n$. Prove that W is spanned by a linearly independent subset of $\alpha_1, \alpha_2, ... \alpha_n$.

25. In a vector space V over the field F, let $B = \{\alpha_1, \alpha_2, ... \alpha_n\}$ span V. Prove that the following two statement are equivalent:
    (i) B is linearly independent.
    (ii) If $\alpha \in V$, then the expression $\alpha = \sum_{i=1}^{n} a_i \alpha_i$ with $a_i \in F$ is unique.

26. Let W be the subspace of R^4 generated by the vectors $(1, -2, 5, -3)$, $(2, 3, 1, -4)$ and $(3, 8, -3, -5)$. Find a basis and the dimension of W. Extend the basis of W to a basis of the whole space R^4.

27. Find a basis and the dimension of the subspace W of R^4 spanned by $(1, 4, -1, 3)$, $(2, 1, -3, -1)$ and $(0, 2, 1, -5)$.

28. Find a basis and the dimension of the subspace W of R^4 spanned W of R^4 spanned by $(1,-4,-2,1)$, $(1,-3,-1,2)$ and $(3,-8,-2,7)$.

29. Let W be the subspace of R^3 defined by $W = \{(a, b, c) : c = 3a\}$. Find a basis and dimension of W.

30. Find bases for the subspaces $W_1 = \{(a, b, 0): a, b \in R\}$, $W_2 = \{(0, b, c): b, c \in R\}$ of R^3. Find the dimensions of W_1, W_2 and $W_1 \cap W_2$.

31. Let W_1 and W_2 be the subspaces of R^4.
$W_1 = \{(a,b,c,d) : b+c+d=0\}$
$W_2 = \{(a,b,c,d) : a+b = 0, c = 2d\}$.
Find the dimension of $W_1 \cap W_2$.

32. Let W_1 and W_2 be subspaces of V and dim. $W_1 = 4$, dim. $W_2 = 5$ and dim. $V = 7$. Find the possible dimensions of $W_1 \cap W_2$.

33. Let E be a subfield of a field F a subfield of a field K, i.e., $E \subset F \subset K$. Let K be of dimension n over F and that F be of dimension m over E, show that K is of dimension mn over E.

34. Find a basis of C cover R.

35. Find the dimension of $Q(\sqrt{2}, \sqrt{3})$ over Q.

36. Show that a system X consisting of the vectors $\alpha_1 = (1, 0, 0, 0)$, $\alpha_2 = (0, 1, 0, 0)$, $\alpha_3 = (0, 0, 1, 0)$ and $\alpha_4 = (0, 0, 0, 1)$ is a basis set of $R^4(R)$.

Answers

3. No

5. $\{\alpha_1, \alpha_2, \alpha_3\}$

6. No

8. $\{\alpha_1, \alpha_2\}$

10. $(1, 0, 0), (0, 1, 0), (1, 1, 0)$

13. $\dfrac{n(n+1)}{2}$

19. $\{(1, 0, 1, 0), (0, 0, 0, 1), (1, 1, 2, 0), (1, 1, 1, 1)\}$

20. $\{(1, 1, 0), (2, 4, 1)(1, 2, 1)\}$

21. $\{(1, 0, 2, 3), (0, 1, -2, -3), (0, 0, 0, 1), (0, 1, 0, 1)\}$

26. A basis of $W = \{(1, -2, 5, -3), (0, 7, -9, 2)\}$, dim. $W = 2$
A basis of $R^4 = \{(1, -2, 5, -3), (0, 7, -9, 2), (0, 0, 1, 0), (0, 0, 0, 1)\}$

27. A basis of $W = \{(1, 4, -1, 3), (0, -7, -1, -7), (0, 0, 5, -49)\}$ and dim. $W = 3$

28. A basis of $W = \{(1, -4, -2, 1), (0, 1, 1, 1)$ and dim. $W = 2$

29. A basis of $W = \{(1, -4, -2, 1), (0, 1, 1, 1)\}$, dim. $W = 2$

30. A basis of $W_1 = \{(1, 0, 0), (0, 1, 0)\}$, dim. $W_1 = 2$
A basis of $W_2 = \{(0, 1, 0), (0, 0, 1)\}$, dim. $W_2 = 2$, dim. $(W_1 \cap W_2) = 1$.

31. dim. $(W_1 \cap W_2) = 1$.

32. The possible dimensions of $(W_1 \cap W_2)$ and 2, 3 and 4.

34. $\{1, i\}$

35. dim. $Q(\sqrt{2}, \sqrt{3}) = 4$

5.15 COSETS

Let V be a vector space and W be its any subspace, then the sets,

$$\alpha + W = \{\alpha + \beta : \forall \beta \in W\}$$

and $\quad W + \alpha = \{\beta + \alpha : \forall \beta \in W\}$,

where α is any arbitrary element of V, are called left and right cosets respectively.

Since addition is commutative in V, so

$$\alpha + W = W + \alpha \quad \forall \alpha \in V$$

Therefore, there is no matter we take right coset or left coset.

REMARKS

- If $W + \alpha = W + \beta$, then $\alpha - \beta \in W$ and so conversely.
- Any two cosets are either disjoint or identical.
- Since W is an additive subgroup of V, so we take additive cosets.

15.1 ADDITION AND MULTIPLICATION OF TWO COSETS

Let V be a vector space and W be its any subspace and let α, β be any two arbitrary elements of V, and let $\alpha + W$ and $\beta + W$ are two cosets of W in V, then

$$(\alpha + W) + (\beta + W) = (\alpha + \beta) + W.$$

and
$$(\alpha + W)(\beta + W) = (\alpha \beta) + W.$$

REMARKS

* Zero elements of the set $\alpha + W = \{\alpha + \beta : \beta \in W\}$ is W.
* Unit elements of this set $1 + W$.

5.16 QUOTIENT SPACES

Let V be a vector space over a field F and let W be any subspace of V. Let $\alpha \in V$.

Then, $\qquad W + \alpha = \{w + \alpha : w \in W\}$ is called the right coset of W in V.

Similarly $\qquad \alpha + W = \{\alpha + w : w \in W\}$ is called the left coset of W in V.

Clearly, the sets of $W + \alpha$ and $\alpha + W$ are both subsets of the vector space $V(F)$. Now since $(V, +)$ is an abelian group, therefore $W + \alpha$ and $\alpha + W$ are equal.

Thus, we can say that $W + \alpha$ is a coset of W in V generated by α. Let V/W represents the set of all cosets of W in V, i.e.,

$$V/W = \{W + \alpha : \alpha \in V\}$$

Further, we define vector addition and scalar multiplication of V/W as follows :

$$(W + \alpha) + (W + \beta) = W + (\alpha + \beta)$$

and
$$a(W + \alpha) = W + a\alpha \ \forall \ \alpha, \beta \in V, a \in F$$

Then, V/W is a vector space over F for these compositions. This vector space is known as *quotient space or factor space.*

THEOREM 1. *If V is a vector space over a field F and if W is a subspace of V, then the set*

$$V/W = \{\alpha + W : \forall \alpha \in V\}$$

is a vector space over the field F with respect to the linear compositions.

(i) $(\alpha + W) + (\beta + W) = (\alpha + \beta) + W, \ \forall \ \alpha, \beta \in V$

(ii) $a(\alpha + W) = a\alpha + W, \ \forall \ \alpha \in V$ *and* $a \in F$

Proof. First of all, we shall show that the above compositions are well defined. For this, let $\qquad \alpha + W = \alpha' + W$ and $\beta + W = \beta' + W$, then we have

$$\alpha + W = \alpha' + W \quad \text{and} \quad \beta + W = \beta' + W$$

$\Rightarrow \qquad \alpha - \alpha' \in W \qquad$ and $\quad \beta - \beta' \in W$

$\Rightarrow \qquad (\alpha - \alpha') + (\beta - \beta') \in W$

$\Rightarrow \qquad (\alpha + \beta) - (\alpha' + \beta') \in W$

$\Rightarrow \qquad (\alpha + \beta) + W = (\alpha' + \beta') + W$

$\Rightarrow \qquad (\alpha + W) + (\beta + W) = (\alpha' + W) + (\beta' + W)$

$\Rightarrow \quad$ The composition given in (i) is well defined.

Also, $\qquad\qquad\qquad\qquad \alpha + W = \alpha' + W$

$$\Rightarrow \qquad \alpha - \alpha' \in W$$

$$\Rightarrow \qquad a(\alpha - \alpha') \in W \text{ for some } a \in F.$$

$$\Rightarrow \qquad (a\alpha - a\alpha') \in W$$

$$\Rightarrow \qquad (a\alpha + W) = (a\alpha' + W).$$

$\Rightarrow$ The composition given in (ii) is well defined.

(i) Next, $\alpha + W$, $\beta + W$ and $\gamma + W$ be any three elements of V/W, then

$$[(\alpha + W) + (\beta + W)] + (\gamma + W) = [(\alpha + \beta) + W] + (\gamma + W)$$
$$= [(\alpha + \beta) + \gamma] + (W)$$
$$= [\alpha + (\beta + \gamma)] + (W)$$
$$= (\alpha + W) + [(\beta + \gamma) + (W)]$$
$$= (\alpha + W) + [(\beta + W) + (\gamma + W)]$$

$\therefore$ The first composition given in (i) is associative in V/W.

(ii) Let $\alpha + W$ be an element of V/W, then

$$(\alpha + W) + (0 + W) = (\alpha + 0) + W = \alpha + W$$

and $\qquad (0 + W) + (\alpha + W) = (0 + \alpha) + W = \alpha + W$

Thus, $0 + W$ is an additive identity in V/W.

(iii) Since, if $\alpha \in V$ and $-\alpha \in V$, so $\alpha + W \in V/W$ and $(-\alpha) + W \in V/W$. Then

$$(\alpha + W) + [(-\alpha) + W] = [\alpha + (-\alpha)] + W = 0 + W = W$$

and $\qquad (0 + W) + (\alpha + W) = (0 + \alpha) + W = \alpha + W$

Thus, $(-\alpha) + W$ is an additive inverse of $\alpha + W$ in V/W.

(iv) Let $\alpha + W$, $\beta + W$ be any two elements of V/W, then

$$(\alpha + W) + (\beta + W) = (\alpha + \beta) + W = (\beta + \alpha) + W$$
$$[\because \text{ addition is commutative in } V.]$$
$$= (\beta + W) + (\alpha + W)$$

Thus addition is commutative in V/W.

(v) Let $\alpha + W$, $\beta + W$ be any elements of V/W and $a \in F$, then

$$a.[(\alpha + W) + (\beta + W)] = a.[(\alpha + \beta) + W]$$
$$= a.(\alpha + \beta) + W$$
$$= (a.\alpha + a.\beta) + W$$
$$= (a\alpha + W) + (a\beta + W)$$
$$= a(\alpha + W) + a(\beta + W)$$

(vi) Let $\alpha + W$ be an arbitrary element of V/W and $a, b \in F$, then

$$(a + b)(\alpha + W) = (a + b)\alpha + W = (a\alpha + b\alpha) + W$$
$$= (a\alpha + W) + (b\alpha + W)$$
$$= a(\alpha + W) + b(\alpha + W)$$

(vii) Let $\alpha + W$ be an arbitrary element of V/W and $a, b \in F$, then

$$(ab)(\alpha + W) = (ab)\alpha + W$$
$$= a(b\alpha) + W \quad \text{[By elementary property of (v)]}$$
$$= a(b\alpha + W) = a(b(\alpha + W)).$$

(viii) Let $\alpha + W$ be an arbitrary element of V/W and $1 \in F$, then

$$1 . (\alpha + W) = (1. \alpha) + W = \alpha + W. \qquad [\because 1. \alpha = 1]$$

Hence V/W is a vector space and this vector space is known as *Quotient space.*

THEOREM 2. *If V is finite dimensional and if W is a subspace of V, then W is finite-dimensional and dim. (V/W) = dim.V – dim. W.*

Proof. Let dim. $V = n$, then any $n+1$ elements in V are linearly dependent, in particular any $n+1$ elements in W are also linearly dependent. Thus we can find a maximal linearly independent subset of W.

Let, $\qquad S = \{\alpha_1, \alpha_2,..., \alpha_m\}$ with $m \le n$.

If $\alpha \in W$, then the set $\{\alpha_1, \alpha_2,..., \alpha_m\}$ is linearly dependent such that

$$a\alpha + a_1\alpha_1 + a_2\alpha_2 +... + a_m\alpha_m = 0$$

where, not all of the $a_i's$ are zero, If $a = 0$, then we get each $a_i = 0$ as S being linearly independent, which contradicts that $a = 0$. Thus $a \ne 0$ so that

$$\alpha = - a^{-1} (a_1\alpha_1 + a_2\alpha_2 +... + a_m\alpha_m)$$

$\Rightarrow \qquad \alpha_1, \alpha_2,...,\alpha_m$ span W

$\Rightarrow W$ is finite dimensional with dim . $W = m$.

Since S forms a basis of W so it can be extended to form a basis of V. Let the extended set $S_1 = (\alpha_1, \alpha_2,..., \alpha_m, \beta_1, \beta_2,..., \beta_{n-m}\}$ be the basis of V, as n being the dim. V.

Consider a set $S_2 = \{ W + \beta_1, W + \beta_2,...,W+\beta_{n-m}\}$ of cosets in V/W. Now, we have to prove that S_2 forms a basis of V/W.

For some scalars $b_1, b_2,...,b_{n-m}$, we have

$$b_1 (W + \beta_1) + b_2(W + \beta_2)+...+ b_{n-m}(W+\beta_{n-m}\} = W + 0$$

where, $W + 0$ being the zero element of V/W.

$\Rightarrow \qquad (W + b_1\beta_1) + (W + b_2 \beta_2)+...+ (W+ b_{n-m} \beta_{n-m}\} = W + 0$

$\Rightarrow \qquad W + (b_1\beta_1+ b_2 \beta_2+...+ b_{n-m} \beta_{n-m}) = W + 0$

$\Rightarrow \qquad b_1\beta_1+ b_2 \beta_2+...+ b_{n-m} \beta_{n-m} \in W$

Since S generates W so that for some scalars $c_1, c_2,...,c_m$ we have

$$b_1\beta_1+ b_2 \beta_2+...+ b_{n-m} \beta_{n-m} = c_1\alpha_1+ c_2 \alpha_2+...+ c_m \alpha_m$$

$\Rightarrow \quad c_1\alpha_1+ c_2 \alpha_2+...+ c_m \alpha_m +(-b_1)\beta_1+ (-b_2)\beta_2+...+ (-b_{n-m})\beta_{n-m}= 0$

$\Rightarrow \qquad c_1 = 0 = c_2 = ... = c_m , b_1 = 0 = b_2 = ... = b_{n-m} ,$

$$[\because S_1 \text{ is linearly independent.}]$$

$\Rightarrow \qquad$ each $b_i = 0$ for $i = 1, 2, ..., n - m$

$\Rightarrow \qquad S_2$ is linearly independent.

Also, if $W + \beta$ is an arbitrary element of V/W, then $\beta \in V$ and S_1 being the basis of V, therefore, β can be expressible as a linear combination of elements of S_1 . That is, for some scalars $d_1, d_2,...,d_m, e_1, e_2, ..., e_{n - m}$

$$\beta = d_1\alpha_1 + d_2\alpha_2 + \dots + d_m\alpha_m + e_1\beta_1 + e_2\beta_2 + \dots + e_{n-m}\beta_{n-m}$$

$$= \sum_{i=1}^{m} d_i\alpha_i + \sum_{j=1}^{n-m} e_j\beta_j$$

$$\therefore \qquad W + \beta = W + \left(\sum_{i=1}^{m} d_i\alpha_i + \sum_{j=1}^{n-m} e_j\beta_j \right)$$

$$W + \beta = \left(W + \sum_{i=1}^{m} d_i\alpha_i \right) + \left(W + \sum_{j=1}^{n-m} e_j\beta_j \right) \qquad \left[\because \sum_{i=1}^{m} d_i\alpha_i \in W \right]$$

$$= W + \sum_{i=1}^{n-m} e_i\beta_i = W + (e_1\beta_1 + e_2\beta_2 + \dots + e_{n-m}\beta_{n-m})$$

$$= (W + e_1\beta_1) + (W + e_2\beta_2) + \dots + (W + e_{n-m}\beta_{n-m})$$

$$= e_1(W + \beta_1) + e_2(W + \beta_2) + \dots + e_{n-m}(W + \beta_{n-m})$$

$\Rightarrow S_2$ generates V/W.

Thus, S_2 forms a basis of V/W and having $(n - m)$ elements so that

dim. $V/W = n - m$

Accordingly,

dim .$V/W = n - m = $ dim. $V - $ dim. W.

5.17 ISOMORPHISM

Let U and V be two vector spaces over the same field F. Then a mapping $T : U \rightarrow V$
which associates to each element $\alpha \in U$ to a unique element $T(\alpha) \in V$ such that
$$T(a\alpha + b\beta) = aT(\alpha) + bT(\beta), \ \forall \ \alpha, \beta \in U, a, b \in F$$

is called a linear transformation of U into V.

A linear transformation T of U onto V is called an isomorphism if it is one-one. Therefore,
we can say that U is isomorphic to V and write this as $U \cong V$.

THEOREM 1. *Every n-dimensional vector space V(F) is isomorphic to F^n (F).*

Proof.   Since the vector space V is n-dimensional. Let the set

$$S = \{\alpha_1, \alpha_2, \dots, \alpha_n\} \text{ form the basis for } V.$$

Then every vector V can be expressed as the linear combination of the elements of S.
Let a be any arbitrary vector in V, then there exists a unique order set $\{a_1, a_2, \dots, a_n\}$
of scalars such that

$$\alpha = \sum_{i=1}^{n} a_i\alpha_i .$$

Consider a mapping

$$T : V \rightarrow F^n \text{ by } T(\alpha) = (a_1, a_2, \dots, a_n) \ \forall \ \alpha \in V.$$

(i) T is linear.

Let $\alpha = \sum\limits_{i=1}^{n} a_i\alpha_i$ and $\beta = \sum\limits_{i=1}^{n} b_i\alpha_i$ be any two elements of V,

then for all $a, b \in F$, we have

$$T(a\alpha + b\beta) = T\left(a\sum_{i=1}^{n} a_i\alpha_i + b\sum_{i=1}^{n} b_i\alpha_i\right) = T\left(\sum_{i=1}^{n}(aa_i + bb_i)\alpha_i\right)$$

$$= (aa_1 + bb_1, aa_2 + bb_2, ..., aa_n + bb_n)$$
$$= (aa_1, aa_2,..., aa_n) + (bb_1, bb_2, ..., bb_n)$$
$$= a(a_1, a_2,...,a_n) + b(b_1, b_2,...,b_n) = a\,T(\alpha) + bT(\beta).$$

$\therefore$ T is linear.

(ii) T is one-one.

Suppose that

$$T(\alpha) = T(\beta)$$

$\Rightarrow \qquad T\left(\sum\limits_{i=1}^{n} a_i\alpha_i\right) = T\left(\sum\limits_{i=1}^{n} b_i\alpha_i\right)$

$\Rightarrow \qquad (a_1, a_2,...,a_n) = (b_1, b_2, ..., b_n)$

$\Rightarrow \qquad\qquad a_i = b_i,$ for each $i = 1, 2, 3, ..., n$

$\Rightarrow \qquad\qquad \sum\limits_{i=1}^{n} a_i\alpha_i = \sum\limits_{i=1}^{n} b_i\alpha_i \qquad \Rightarrow \alpha = \beta$

$\therefore$ T is one-one.

(iii) T is onto

Since, corresponding to each element $(aa_1, aa_2,...,aa_n) \in F^n$, there exists a vector

$$\sum_{i=1}^{n} a_i\alpha_i \in V \text{ such that } T\left(\sum_{i=1}^{n} a_i\alpha_i\right) = (a_1, a_2,.....,a_n)$$

$\Rightarrow \qquad T$ is onto.

Hence, V is isomorphic to F^n, i.e., $V \cong F^n$.

THEOREM 2. *If W_1 and W_2 are the complementary subspaces of a vector space $V(F)$, then the correspondence that assigns to each other $\alpha \in W_2$, the coset $W_1 + \alpha$ is an isomorphism between W_2 and V/W_1.*

Proof. Since W_1 and W_2 are the complementary subspaces of V, then $V = W_1 \oplus W_2$, or we have

$$V = W_1 + W_2 \text{ and } W_1 \cap W_2 = \{0\}$$

Consider a mapping

$$T : W_2 \to V/W_1$$

defined by $\qquad T(\alpha) = W_1 + \alpha, \forall\, \alpha \in W_2$

First to show that T is linear.

Let α, β be any two elements in W_2, then for all $a, b \in F$, we have

$$T(a\alpha + b\beta) = W_1 + (a\alpha + b\beta) = (W_1 + a\alpha) + (W_1 + b\beta)$$
$$= a(W_1 + \alpha) + b(W_1 + \beta) = aT(\alpha) + bT(\beta)$$

$\therefore$ T is linear.

Second, to show that T is one-one.

Suppose that
$$T(\alpha) = T(\beta)$$
$\Rightarrow$ $W_1 + \alpha = W_2 + \beta \Rightarrow \alpha - \beta \in W_1$
$\Rightarrow$ $\alpha - \beta \in W_1 \cap W_2$ as $\alpha - \beta \in W_2$
$\Rightarrow$ $\alpha - \beta = 0$ $[\because W_1 \cap W_2 = \{0\}]$
$\Rightarrow$ $\alpha = \beta$

$\therefore$ T is one-one.

Finally, to show that T is onto.

Let $W_1 + \gamma$ be an arbitrary element of V/W_1 so that $\gamma \in V$. But V is the direct sum of W_1 and W_2, therefore, γ can be uniquely expressed as the sum of an element of W_1 and an element of W_2. So there exists $\alpha \in W_1$ and $\beta \in W_2$ such that
$$\gamma = \alpha + \beta$$
$\Rightarrow$ $W_1 + \gamma = W_1 + (\alpha + \beta)$
$$= (W_1 + \alpha) + (W_1 + \beta)$$
$$= (W_1 + 0) + (W_1 + \beta) \quad [\because \alpha \in W_1 \Rightarrow W_1 + \alpha = W_1 = W_1 + 0]$$
$$= (W_1 + \beta) = T(\beta)$$

which shows that to each element $W_1 + \gamma$ in V/W_1, there exists a unique element $\beta \in W$ such that
$$T(\beta) = W_1 + \gamma$$

$\therefore$ T is onto.

Hence, W_2 is isomorphic to V/W_1.

THEOREM 3. *Any two finite dimensional vector spaces over the same field are isomorphic if and only if they are of same dimension.*

Proof. Let $U(F)$ and $V(F)$ be two spaces over same field F. Suppose that both are of same dimension n (say).

Let $S_1 = \{\alpha_1, \alpha_2, \ldots, \alpha_n\}$ and $S_2 = \{\beta_1, \beta_2, \ldots, \beta_n\}$ be the basis for U and V respectively. So, corresponding to each element $\alpha \in U$, there exists unique scalars $a_1, a_2, \ldots, a_n \in F$ such that

$$a = a_1\alpha_1 + a_2\alpha_2 + \ldots + a_n\alpha_n = \sum_{i=1}^{n} a_i\alpha_i$$

Consider a mapping $T : U \to V$.

by $$T(\alpha) = \sum_{i=1}^{n} a_i\beta_i \;,\forall\; a_i \in F \text{ and for all } \alpha \in U$$

First we show that T is linear.

Let α, β be any two elements of U, then $\alpha = \sum_{i=1}^{n} a_i \alpha_i$ and $\beta = \sum_{i=1}^{n} b_i \alpha_i$.

For $a, b \in F$, we have

$$T(a\alpha + b\beta) = T\left(a\sum_{i=1}^{n} a_i \alpha_i + b\sum_{i=1}^{n} b_i \alpha_i\right)$$

$$= T\left(\sum_{i=1}^{n} (aa_i + bb_i)\alpha_i\right) = \sum_{i=1}^{n} (aa_i + bb_i)\beta_i$$

$$= \sum_{i=1}^{n} aa_i\beta_i + \sum_{i=1}^{n} bb_i\beta_i$$

$$= a\sum_{i=1}^{n} a_i\beta_i + b\sum_{i=1}^{n} b_i\beta_i$$

$$= aT(\alpha) + bT(\beta)$$

$\therefore$ T is a linear.

Now, we show that T is one-one.

Suppose   for $\alpha, \beta \in U$

$\Rightarrow$ $\sum_{i=1}^{n} a_i\beta_i = \sum_{i=1}^{n} b_i\beta_i$

$\Rightarrow$ $\sum_{i=1}^{n} (a_i - b_i)\beta_i = 0$ for each $i = 1, 2, 3, ..., n$

                                 $[\because S_2$ is linearly indepenmdent.$]$

$\Rightarrow$ $\sum_{i=1}^{n} a_i\beta_i = \sum_{i=1}^{n} b_i\beta_i \Rightarrow \alpha = \beta$.

$\therefore$ T is one-one.

Finally, we show that T is onto.

Let γ be any element of V, then there exists scalars $c_1, c_2, ..., c_n$ of F such that

$$g = \sum_{i=1}^{n} c_i\beta_i \; ,$$

Now, $T\left(\sum_{i=1}^{n} c_i\alpha_i\right) = \sum_{i=1}^{n} c_i\beta_i = \gamma$

$\Rightarrow$ T-image of $\sum_{i=1}^{n} c_i\alpha_i \in U$ is $\gamma = \sum_{i=1}^{n} c_i\beta_i$

$\therefore$ T is a onto.

Hence T is an isomorphism, *i.e.*, $U \cong V$

Conversely, suppose that $U \cong V$ and let T be the corresponding isomorphism.

Let $S_1 = \{\alpha_1, \alpha_2, ..., \alpha_n\}$ be the basis of U.

We claim that the set

$$S_2 = \{T(\alpha_1), T(\alpha_2), ..., T(\alpha_n)\}$$

is the basis for V.

First we show that S_2 is linearly independent.

For some scalars $a_1, a_2, \ldots, a_n \in F$, the relation

$$a_1 T(\alpha_1) + a_2 T(\alpha_2) + \ldots a_n T(\alpha_n) = 0$$

$$\Rightarrow \quad T(a_1\alpha_1 + a_2\alpha_2 + \ldots a_n\alpha_n) = 0 \qquad [\because T \text{ is a linear.}]$$

$$T\left(\sum_{i=1}^{n} a_i\alpha_i\right) = T(0) \qquad [\because T(0) = (0)]$$

$$\sum_{i=1}^{n} a_i\alpha_i = 0 \qquad [\because T \text{ is one-one.}]$$

$$\Rightarrow \quad a_i = 0 \text{ for each } i = 1, 2, 3, \ldots, n. \qquad [\because S_1 \text{ is linearly independent.}]$$

$\therefore \quad S_2$ is linearly independent.

Secondly, We show that S_2 spans V.

Let γ be any element of V. Since T is one-one onto mapping, then there exists a unique vector $\alpha = \sum_{i=1}^{n} a_i\alpha_i \in U$ such that

$$\gamma = T(\alpha) = T\left(\sum_{i=1}^{n} a_i\alpha_i\right) = \sum_{i=1}^{n} a_i T(\alpha_i) \qquad [\because T \text{ is a linear.}]$$

Thus, $\gamma \in V$ is expressible as a linear combinations of elements of S_2, therefore S_2 spans V.

Consequently $\dim V = n = \dim U$.

Hence, both the vectors $U(F)$ and $V(F)$ are finite dimensional of dimension n.

REMARK

- The dimension of the solution space W of the homogeneous system of linear equations $AX = O$ is $(n-r)$ where, n is the number of unknowns and r is the rank of the coefficient matrix A.

Miscellaneous Solved Examples

Example 1. *Prove that if two vectors are linearly dependent, one of them is a scalar multiple of the other.*

Solution. Let V be a vector space over F and α, β be two vectors of V. These vectors are linearly dependent, then there exists scalars $a, b \in F$ not both equal to zero such that

$$a\alpha + b\beta = 0$$

If $a \neq 0$, then a^{-1} exists in F, we have

$$a^{-1}(a\alpha) + a^{-1}(b\beta) = a^{-1}0$$

$$\Rightarrow \quad (a^{-1}a)\alpha + (a^{-1}b)\beta = 0 \qquad [\because a^{-1}0 = 0]$$

$$\Rightarrow \quad 1\alpha + a^{-1}b\beta = 0 \qquad [\because aa^{-1} = 1]$$

$$\Rightarrow \quad \alpha + (a^{-1}b)\beta = 0 \qquad [\because 1\alpha = \alpha]$$

$$\Rightarrow \quad \alpha = -(a^{-1}b)\beta \qquad [\because 1\alpha = \alpha]$$

$$\Rightarrow \quad \alpha \text{ is a scalar multiple of } \beta.$$

Similarly, if $b \neq 0$, then we obtain β as a scalar multiple of α. Hence, one of α or β is scalar multiple of the other.

Example 2. *Show that the set $S=\{(1,2,1),(3,1,5),(3,-4,7)\}$ is linearly dependent where $S \subseteq V_3(R)$.*

Solution. Let $a, b, c \in R$ such that

$$a(1,2,1)+b(3,1,5)+c(3,-4,7)= (0,0,0)$$

or $\qquad (a+3b+3c,2a+b-4c,a+5b+7c)= (0,0,0)$

or $\qquad \left.\begin{array}{l} a + 3b + 3c = 0 \\ 2a + b - 4c = 0 \\ a + 5b + 7c = 0 \end{array}\right\}$ ...(1)

Equation (1) is a system of linear homogeneous equations.

∴ Coefficient matrix is given by

$$A=\begin{bmatrix} 1 & 3 & 3 \\ 2 & 1 & -4 \\ 1 & 5 & 7 \end{bmatrix}$$

Now, $\qquad \det(A)= 1(7+20)-3(14+4)+3(10-1)$

$$= 27-54+27=0$$

∴ Rank of A is less than 3, *i.e.*, rank of A is less than the number of variables a,b and c so there exists a non-zero solutions. Thus, a,b,c are all not equal to zero. Hence, the given set S is linearly dependent.

Example 3. *Show that the vectors $(1,1,2,4)$, $(2,-1,-5,2)$, $(1,-1,-4,0)$ and $(2,1,1,6)$ are linearly dependent in R^4.*

Solution. Let $a, b, c, d \in R$ such that

$$a(1,1,2,4)+b(2,-1,-5,2)+c(1,-1,-4,0)+d(2,1,1,6)= (0,0,0,0)$$

or $\qquad (a+2b+c+2d,a-b-c+d,2a-5b-4c+d,4a+2b+6d)= (0,0,0,0)$

or $\qquad \left.\begin{array}{l} a + 2b + c + 2d=0 \\ a - b - c - d=0 \\ 2a - 5b - 4c + d = 0 \\ 4a + 2b + 6d = 0 \end{array}\right\}$ ...(1)

Therefore, equation (1) represents a system of linear homogeneous equation. Now the coefficient matrix of these equation is

$$A=\begin{bmatrix} 1 & 2 & 1 & 2 \\ 1 & -1 & -1 & 1 \\ 2 & -5 & -4 & 1 \\ 4 & 2 & 0 & 6 \end{bmatrix}$$

performing $R_2 \rightarrow R_2-R_1$, $R_3 \rightarrow R_3-2R_1$, $R_4 \rightarrow R_4-4R_1$, we get

$$\sim \begin{bmatrix} 1 & 2 & 1 & 2 \\ 0 & -3 & -2 & -1 \\ 0 & -9 & -6 & -3 \\ 0 & -6 & -4 & -2 \end{bmatrix}$$

performing $R_3 \rightarrow R_3 - 3R_2$, $R_4 \rightarrow R_4 - 2R_2$

$$\sim \begin{bmatrix} 1 & 2 & 1 & 2 \\ 0 & -3 & -2 & -1 \\ 0 & 0 & 0 & 0 \\ 0 & 0 & 0 & 0 \end{bmatrix}$$

Thus, this matrix is in Echelon form and having two non-zero rows so the rank of A is 2 which is less than the number of unknown. Therefore, the system of equation has non-zero solution. That is a, b, c, d are not all zero, Hence the given vectors are linearly dependent.

Example 4. *Show that the vectors* $(1,1,2)$, $(1,2,5)$, $(5,3,4)$ *do not form a basis of* R^3.

Solution. Since we know that the set $(1,0,0)$, $(0,1,0)$, $(0,0,1)$ forms a basis of R^3 so dimension of $R^3 = 3$.

Let $a, b, c \in R$ such that

$$a(1,1,2) + b(1,2,5) + c(5,3,4) = (0,0,0)$$

or $\qquad (a+b+5c, a+2b+3c, 2a+5b+4c) = (0,0,0)$

or $\qquad \left. \begin{array}{r} a + b + 5c = 0 \\ a + 2b + 3c = 0 \\ 2a + 5b + 4c = 0 \end{array} \right\}$ ...(1)

The equation (1) represents the system of linear homogeneous equations. Then the coefficient matrix of these equation is given by

$$A = \begin{bmatrix} 1 & 1 & 5 \\ 1 & 2 & 3 \\ 2 & 5 & 4 \end{bmatrix}$$

$\therefore \qquad |A| = 1(8-15) - 1(4-6) + 5(5-4) = -7 + 2 + 5 = 0$

Thus A is a singular matrix and therefore the rank of A is less than of 3, *i.e.*, less than the number of unknown a, b, c. Then the scalars a, b, c are all not equal to zero. Hence, the given set of vectors is not linearly independent, and hence this given set does not form a basis of R^3.

Example 5. *Show that the vectors* $(2,1,4)$, $(1,-1,2)$, $(3,1,-2)$ *form a basis of* R^3.

Solution. Since we know that the set $\{(1,0,0), (0,1,0), (0,0,1)\}$ forms a basis of R^3. Then dim of $R^3 = 3$. Therefore, if the given set $\{(2,1,4), (1,-1,2), (3,1,-2)\}$ is linearly independent, then it will form the basis of R^3.

Let $a, b, c \in R$ such that

$$a(2,1,4) + b(1,-1,2) + c(3,1,-2) = (0,0,0)$$

or $\qquad (2a+b+3c, a-b+c, 4a+2b-2c) = (0,0,0)$

or $\qquad \left. \begin{array}{r} 2a + b + 3c = 0 \\ a - b + c = 0 \\ 4a + 2b - 2c = 0 \end{array} \right\}$ ...(1)

The equation (1) represents a system of three linear homogeneous equation. Then the coefficient matrix of these equation is given by

$$A = \begin{bmatrix} 2 & 1 & 3 \\ 1 & -1 & 1 \\ 4 & 2 & -2 \end{bmatrix}$$

Now $\qquad | A | = 2(2-2) - 1(-2-4) + 3(2+4)$

$$= 0 + 6 + 18 = 24 \neq 0$$

$\therefore \qquad$ The matrix A is a non-singular matrix and thus the rank of A is 3 which is equal to the number of unknowns a, b, c. Hence, the system of equations has only zero solution, *i.e.*, $a=0$, $b=0$, $c=0$. Consequently the given set is linearly independent and hence forms a basis of R^3.

Example 6. *Find the dimension of the solution W of the system of linear equations.*
$$x+2y-4z+3r-s=0; \quad x+2y-2z+4r+s=0; \quad 2x+4y-2z+3r+4s=0$$

Solution. Above equation can be written as $AX=O$

where $A=\begin{bmatrix} 1 & 2 & -4 & 3 & -1 \\ 1 & 2 & -2 & 4 & 1 \\ 2 & 4 & -2 & 3 & 4 \end{bmatrix}$ and $X=\begin{bmatrix} x \\ y \\ z \\ r \\ s \end{bmatrix}$, $O=\begin{bmatrix} 0 \\ 0 \\ 0 \end{bmatrix}$.

Now reduce A to Echelon form as follows:

$$A \sim \begin{bmatrix} 1 & 2 & -4 & 3 & -1 \\ 0 & 0 & 2 & 1 & 2 \\ 0 & 0 & 6 & 3 & 6 \end{bmatrix} \qquad \begin{bmatrix} R_2 \to R_2 - R_1 \\ R_3 \to R_3 - 2R_1 \end{bmatrix}$$

$$\sim \begin{bmatrix} 1 & 2 & -4 & 3 & -1 \\ 0 & 0 & 2 & 1 & 2 \\ 0 & 0 & 0 & 0 & 0 \end{bmatrix} \qquad \begin{bmatrix} R_3 \to R_3 - 3R_2 \end{bmatrix}$$

Echelon form of A has two non-zero rows, so that rank of $A=2$ and number of unknowns$= 5$. Hence dim.$W = 5 - 2 = 3$.

Example 7. *Prove that a set of non-zero vectors $\{x_1, x_2, ..., x_n\}$ is linearly dependent if some of these vectors say x_i is a linear combination of the preceding vectors $x_1, x_2, ..., x_{i-1}$ and conversely.*

Solution. Let us first assume that x_i can be expressed as a linear combination of $x_1, x_2, ..., x_{i-1}$. Then we have
$$x_i = a_1 x_1 + a_2 x_2 + ... + a_{i-1} x_{i-1} \qquad \qquad ...(1)$$
For some scalars $a_1, a_2, ..., a_{i-1} \in F.$

Now equation (1) can be written as:
$$a_1 x_1 + a_2 x_2 + ... + a_{i-1} x_{i-1} + (-1) x_i = 0$$

or $\qquad a_1 x_1 + a_2 x_2 + ... + a_{i-1} x_{i-1} + (-1)x_i + 0\, x_{i+1} + 0\, x_{i+2} + ... + 0\, x_n = 0$

This equation shows that this relation has at least one non-zero coefficient (scalar) of x_i which is -1. It shows that the set $\{x_1, x_2, ..., x_n\}$ is linearly dependent. Then we have a relation.
$$a_1 x_1 + a_2 x_2 + ... + a_n x_n = 0$$

in which not all a_i's are zero.

If i is the greatest positive integer less than or equal to n such that $a_i \neq 0$, so that
$$a_{i+1} = 0 = a_{i+2} = ... = a_n, \text{ then we have}$$
$$a_1 x_1 + a_2 x_2 + ... + a_{i-1} x_{i-1} + a_i x_i + 0.\, x_{i+1} + 0.\, x_{i+2} + ... + 0.\, x_n = 0$$

$\Rightarrow \qquad x_i = - a_i^{-1}(a_1 x_1 + a_2 x_2 + ... + a_{i-1} x_{i-1}).$

$\Rightarrow x_i$ is the linear combination of preceding vectors $x_1, x_2, ..., x_{i-1}$.

Example 8. *Prove that the vector (α_1, α_2), $(\beta_1, \beta_2) \in R \times R$ are linearly dependent iff $\alpha_1 \beta_1 - \alpha_2 \beta_2 = 0$*

Solution. The vectors (α_1, α_2) and (β_1, β_2) are linearly dependent if and only if there exists scalar $a, b \in R$ such that

$$a(\alpha_1, \alpha_2) + b(\beta_1, \beta_2) = (0,0)$$

where, $(0,0)$ is the vector in R $\times$ R.

$\Leftrightarrow \qquad (a\alpha_1, a\alpha_2) + (b\beta_1, b\beta_2) = (0,0)$

$\Leftrightarrow \qquad (a\alpha_1 + b\beta_1, a\alpha_2 + b\beta_2) = (0,0)$

$\Leftrightarrow \qquad a\alpha_1 + b\beta_1 = 0, \ a\alpha_2 + b\beta_2 = 0$

The equations $a\alpha_1 + b\beta_1 = 0$ and $a\alpha_2 + b\beta_2 = 0$ have non-trival (non-zero) solutions iff the determinant of coefficients.

$$\begin{vmatrix} \alpha_1 & \beta_1 \\ \alpha_2 & \beta_2 \end{vmatrix} = 0, \ i.e., \ \alpha_1\beta_2 - \beta_1\alpha_2 = 0.$$

Hence, the vectors (α_1, α_2) and (β_1, β_2) are linearly dependent if $\alpha_1\beta_2 - \beta_1\alpha_2 = 0$.

Example 9. *Prove that the vectors $(2, i, -i)$, $(2i, -1, 1)$, $(1, 2, 3)$ are linearly independent in $V_3(R)$ but linearly dependent in $V_3(C)$.*

Solution. Let $\alpha_1 = (2, i, -i)$, $\alpha_2 = (2i, -1, 1)$ and $\alpha_3 = (1, 2, 3)$ and let $\alpha_1, \alpha_2, \alpha_3$ be three scalars, Then

$\Rightarrow \qquad a_1\alpha_1 + a_2\alpha_2 + a_3\alpha_3 = 0$

$\Rightarrow \quad a_1 (2, i, -i) + \alpha_2(2i, -1, 1) + \alpha_3(1, 2, 3) = (0, 0, 0)$

$\Rightarrow \quad (2a_1 + 2ia_2 + a_3, ia_1 - a_2 + 2a_3, -ia_1 + a_2 + 3a_3) = (0, 0, 0)$

$\therefore \qquad\qquad 2a_1 + 2ia_2 + a_3 = 0 \qquad\qquad\qquad …(1)$

$\qquad\qquad\qquad ia_1 - a_2 + 2a_3 = 0 \qquad\qquad\qquad …(2)$

$\qquad\qquad\qquad -ia_1 + a_2 + 3a_3 = 0 \qquad\qquad\qquad …(3)$

Now adding (2) and (3), we get

$$5a_3 = 0 \Rightarrow a_3 = 0$$

Putting $a_3 = 0$ in any one of above equations, we get

$$ia_2 - a_2 = 0 \text{ or } a_2 = ia_1. \qquad\qquad …(4)$$

Case I: In $V_3(R)$, equation (4) is satisfied when $a_1 = 0$, $a_2 = 0$.

Thus in case of $V_3(R)$, equation (4) is satisfied when $a_1 = 1$, $a_2 = i$.

$\therefore \qquad\qquad 1(2, i, -i) + i(2i, -1, 1) + 0(1, 2, 3) = (0, 0, 0)$

Thus in case of $V_3(C)$, the vectors α_1, α_2 and α_3 are linearly dependent.

Example 10. *Prove that the vectors $(1, 1, 0)$ $(3, 1, 3)$ and $(5, 3, 3)$ are linearly dependent.*

Solution. Let $\alpha_1 = (1, 1, 0), \alpha_2 = (3, 1, 3)$ and $\alpha_3 = (5, 3, 3)$ and let α_1, α_2 and α_3 be some scalars such that

$$a_1\alpha_1 + a_2\alpha_2 + a_3\alpha_3 = 0$$

$\Rightarrow \qquad a_1(1, 1, 0) + a_2(3, 1, 3) + a_3(5, 3, 3) = (0, 0, 0)$

$\Rightarrow \quad (a_1 + 3a_2 + 5a_3, a_1 + a_2 + 3a_3, 3a_2 + 3a_3) = (0, 0, 0)$

$\Rightarrow \qquad\qquad a_1 + 3a_2 + 5a_3 = 0 \qquad\qquad\qquad …(1)$

$\qquad\qquad\qquad a_1 + a_2 + 3a_3 = 0 \qquad\qquad\qquad …(2)$

$\qquad\qquad\qquad 3a_2 + 3a_3 = 0 \qquad\qquad\qquad\qquad …(3)$

Coefficient matrix of above equation is $A = \begin{bmatrix} 1 & 3 & 5 \\ 1 & 1 & 3 \\ 0 & 3 & 3 \end{bmatrix}$

Reduce this matrix into Echelon from as follows:

$$A \sim \begin{bmatrix} 1 & 3 & 5 \\ 0 & -2 & -2 \\ 0 & 3 & 3 \end{bmatrix} \qquad\qquad [R_2 \rightarrow R_2 - R_1]$$

$$\sim \begin{bmatrix} 1 & 3 & 5 \\ 0 & -2 & -2 \\ 0 & 0 & 0 \end{bmatrix} \qquad\qquad \left[R_3 \rightarrow R_3 + \frac{3}{2} R_2\right]$$

This is the Echelon from and having two non-zero rows so that rank of $A=2$. Thus the equation (1), (2) and (3) have non-zero solutions. Hence the vectors (1, 1, 0), (3, 1, 3) and (5, 3, 3) are linearly dependent.

Example 11. *Under what conditions on the scalar 'a' are the vectors (a, 1, 0), (1, a, 1) and (0, 1, a) in* R^3 *linearly dependent?*

Solution. Let $\alpha_1 = (a, 1, 0)$, $\alpha_2 = (1, a, 1)$ and $\alpha_3 = (0, 1, a)$ and let a_1, a_2 and a_3 be three scalars such that

$$a_1\alpha_1 + a_2\alpha_2 + a_3\alpha_3 = 0$$

$\Rightarrow \quad a_1(a,1,0) + a_2(1,a,1) + a_3(0,1,a) = (0,0,0)$

$\Rightarrow \quad (a_1 a + a_2, a_1 + a_2 a + a_3, a_2 + a_3 a) = (0,0,0)$

$\Rightarrow \qquad\qquad\begin{aligned} a_1 a + a_2 &= 0 & \text{......(1)} \\ a_1 + a_2 a + a_3 &= 0 & \text{......(2)} \\ a_2 + a_3 a &= 0 & \text{......(3)} \end{aligned}$

Let A be the coefficient matrix of the equations (1), (2), (3)

$$\therefore \qquad\qquad A = \begin{bmatrix} a & 1 & 0 \\ 1 & a & 1 \\ 0 & 1 & a \end{bmatrix}$$

For non-trivial solution of (1), (2) and (3), we must have

$$\begin{vmatrix} a & 1 & 0 \\ 1 & a & 1 \\ 0 & 1 & a \end{vmatrix} = 0$$

$\Rightarrow \qquad a[a^2 - 1] - 1[a - 0] = 0 \Rightarrow a(a^2 - 2) = 0$

$\Rightarrow \qquad\qquad a = 0, a = \pm \sqrt{2}$

Hence the vectors (a, 1, 0), (1, a, 1) and (0, 1, a) are linearly dependent when $a = 0; \pm \sqrt{2}$.

Example 12. *For what value of m, the vector (m, 3, 1) a linear combination of* $e_1 = (3, 2, 1)$ *and* $e_2 = (2, 1, 0)$.

Solution. For some scalars a_1 and a_2 such that

$$(m, 3, 1) = a_1 e_1 + a_2 e_2 = a_1(3,2,1) + a_2(2,1,0)$$
$$= (3a_1 + 2a_2, 2a_1 + a_2, a_1)$$

$\Rightarrow \qquad\qquad\begin{aligned} 3a_1 + 2a_2 &= m & \text{... (i)} \\ 2a_1 + a_2 &= 3 & \text{... (ii)} \\ a_1 &= 1 & \text{... (iii)} \end{aligned}$

Putting the value of $a_1 = 1$ in (2), we get $2(1) + a_2 = 3$, $a_2 = 3 - 2 = 1$.

Now putting the values of a_1 and a_2 in (1), we get $m = 3(1) + 2(1) = 3 + 2 = 5$

Example 13. *Show that the set $\{(1,2,1), (2,1,0), (I, -1,2)\}$ forms a basis for $V_3(R)$.*

Solution. Since we know that the set $\{(1, 0, 0), (0, 1, 0),(1, -1,2)\}$ is linearly independent, then it will form the basis of $V_3(R)$.

Let $a, b, c \in R$ such that $a(1,2,1) + b(2,1,0) + c(1,-1,2)=(0,0,0)$

$\Rightarrow \qquad (a + 2b + c, 2a + b - c, a + 2c) = (0, 0, 0)$

$$a + 2b + c = 0 \qquad\qquad\qquad ...(1)$$
$$\Rightarrow \qquad 2a + b - c = 0 \qquad\qquad\qquad ...(2)$$
$$a + 2c = 0 \qquad\qquad\qquad ...(3)$$

The equation (1), (2), (3) represent a system of three linear homogeneous equations. Then the coefficient matrix of these equations is given by :

$$A = \begin{bmatrix} 1 & 2 & 1 \\ 2 & 1 & -1 \\ 1 & 0 & 2 \end{bmatrix}$$

Now, $\qquad |A| = 1(2-0) - 2(4+1) + 1(0-1) = 2 - 10 - 1 = -9 \neq 0$.

Therefore, matrix A is non-singular and thus the rank of A is 3 which is equal to the number of unknowns a, b, c. Hence the system of equations has only zero solution, i.e., $a = 0, b = 0, c = 0$. Consequently, the given set is linearly independent and hence forms a basis of $V_3(R)$.

Example 14. *Under what conditions on the scalar 'a', do the vectors $(1, 1, 1)$ and $(1,a, a^2)$ form a basis of $C^3(C)$.*

Solution. Since dim. $C^3(C) = 3$ so that every basis of $C^3(C)$ must have three vectors. Therefore, a must have two values. Let a_1 and a_2 be the two values of a such that $a_1^2 = a_2^2$. In other words, we are to find the condition on the scalar 'a' such that the vectors $(1,1,1)$, $(1, a, a^2)$, $(1,- a, a^2)$ form a basis of $C^3(C)$. Since dim. $C^3(C) = 3$. therefore, we find the condition such that $(1,1,1)$, $(1, a, a^2)$ and $(1, - a, a^2)$ are linearly independent.

For some scalar $a_1, a_2, a_3 \in C$, we have

$$a_1 (1, 1, 1) + a_2(1,a,a^2) + a_3(1,-a, a^2) = (0, 0, 0)$$

$\Rightarrow \quad (a_1 + a_2 + a_3, a_1 +aa_2 - aa_3, a_1 + a^2a_2 +a^2a_3) = (0, 0, 0)$

$$a_1 + a_2 + a_3 = 0 \qquad\qquad ... (1)$$
$$\Rightarrow \qquad a_1 + aa_2 - aa_3 = 0 \qquad\qquad ... (2)$$
$$a_1 + a^2a_2 + a^2a_3 = 0 \qquad\qquad ... (3)$$

This system of homogeneous equations has a trivial solution if and only if

$$\begin{vmatrix} 1 & 1 & 1 \\ 1 & a & -a \\ 1 & a^2 & a^2 \end{vmatrix} \neq 0$$

$\Rightarrow \quad 1(a^3+a^3) - 1(a^2 + a) + 1(a^2 - a) \neq 0$

$\Rightarrow \qquad 2a^3 - 2a \neq 0 \Rightarrow 2a(a^2 - 1) \neq 0 \Rightarrow a \neq 0, a \neq \pm 1$.

Hence the vectors $(1,1,1)$ and $(1, a, a^2)$ form a basis of $C^3(C)$ if any only if $a \neq 0, \ a \neq \pm 1$.

Example 15. *Let V(R) be the vector space of all complex numbers a + ib over the field of reals R and let T be a mapping V(R) to V_2(R) defined as*

$$T(a+ib) = (a, b)$$

Show that T is an isomorphism.

Solution. If T is one-one, onto and linear, then T is an isomorphism. Therefore, we first show that T is one-one.

T is one-one:

Let $\alpha = a + ib$, $\beta = c + id$ be any two elements of V(R), where a, b, c, d∈R.

We have $\qquad\qquad T(\alpha) = T(\beta)$

$\Rightarrow \qquad\qquad T(a + ib) = T(c + id) \qquad\qquad$ (By definition of T)

$\Rightarrow \qquad\qquad\quad (a,b) = (c,d)$

$\Rightarrow \qquad\qquad\qquad a = c, b = d$

$\Rightarrow \qquad\qquad\quad a + ib = c + id$

$\Rightarrow \qquad\qquad\qquad\quad \alpha = \beta$

∴ $\quad$ T is one-one.

Now we show that T is onto.

T is onto:

Let (a, b) be an arbitrary element of V_2(R) where a, b∈R. Then there exists element $a+ib$ in V(R) such that

$$T(a + ib) = (a,b)$$

∴ $\quad$ Finally, we show that T is linear.

T is linear:

Let $\alpha = a + ib, \beta = c + id$ be any two elements of V(R), where a, b, c, d∈ R and k_1 and k_2 be any two elements of R, then

$$T(k_1\alpha + k_2\beta) = T(k_1(a + ib) + k_2(c + id))$$

$\Rightarrow \qquad T(k_1\alpha + k_2\beta) = T((k_1a \dotplus k_2c) + i(k_1b + k_2d))$

$\Rightarrow \qquad T(k_1\alpha + k_2\beta) = (k_1a + k_2c, k_1b + k_2d) \qquad$ (By definition of T)

$\Rightarrow \qquad T(k_1\alpha + k_2\beta) = (k_1a, k_1b) + (k_2c, k_2d)$

$\Rightarrow \qquad T(k_1\alpha + k_2\beta) = k_1(a,b) + k_2(c,d)$

$\Rightarrow \qquad T(k_1\alpha + k_2\beta) = k_1T(a + ib) + k_2T(c + id)$

$\Rightarrow \qquad T(k_1\alpha + k_2\beta) = k_1T(\alpha) + k_2T(\beta)$

∴ $\quad$ T is linear.

Hence, T is an isomorphism.

Example 16. *If f : U → V is an isomorphism of the vector space U into the vector V, then a set of vectors $\{f(\alpha_1), f(\alpha_2),..., f(\alpha_r)\}$ is linearly independent if and only if the set $\{\alpha_1, \alpha_2, ..., \alpha_r\}$ is linearly independent.*

Solution. Let $\{\alpha_1, \alpha_2, ..., \alpha_r\}$ be a set of linearly independent vectors of U. Then, we show that the set $\{f(\alpha_1), f(\alpha_2),..., f(\alpha_r)\}$ is linearly independent.

Let $\alpha_1, \alpha_2, ..., \alpha_r \in F$ such that

$$a_1f(\alpha_1) + a_2f(\alpha_2) + ... + a_r f(\alpha_r) = 0'$$

$\Rightarrow \qquad f(a_1\alpha_1 + a_2\alpha_2 + ... + a_r\alpha_r) = 0' \qquad\qquad (\because f \text{ is linear.})$

$\Rightarrow \qquad f(a_1\alpha_1 + a_2\alpha_2 + ... + a_r\alpha_r) = f(0)$ $\qquad\qquad (\because f(0) = 0')$

$\Rightarrow \qquad a_1\alpha_1 + a_2\alpha_2 + ... + a_r\alpha_r = 0$ $\qquad\qquad (\because f \text{ is one-one.})$

$\Rightarrow \qquad a_1 = 0, a_2 = 0, ..., a_r = 0$

$\qquad\qquad\qquad\qquad\qquad (\because \alpha_1, \alpha_2, ..., \alpha_r \text{ are linearly independent.})$

$\therefore \quad f(\alpha_1), f(\alpha_2), ..., f(\alpha_r)$ are linearly independent.

Conversely, suppose $f(\alpha_1), f(\alpha_2), ..., f(\alpha_r)$ are linearly independent, then we show that $\alpha_1, \alpha_2, ..., \alpha_r$ are linearly independent.

Let $a_1, a_2, ..., a_r \in F$ such that

$$a_1\alpha_1 + a_2\alpha_2 + ... + a_r\alpha_r = 0$$

$\Rightarrow \qquad f(a_1\alpha_1 + a_2\alpha_2 + ... + a_r\alpha_r) = f(0)$

$\Rightarrow \qquad a_1 f(\alpha_1) + a_2 f(\alpha_2) + ... + a_r f(\alpha_r) = 0'$ $\qquad (\because f \text{ is linear.})$

$\Rightarrow \qquad a_1 = 0, a_2 = 0, ..., a_r = 0$

$\qquad\qquad\qquad\qquad (\because f(\alpha_1), f(\alpha_2), ..., f(\alpha_r) \text{ are linearly independent.})$

$\therefore \quad \alpha_1, \alpha_2, ..., \alpha_r$ are linearly independent.

Example 17. *If V is a finite dimensional vector space and f is an isomorphism of V into V, prove that f must map V onto V.*

Solution. Let dim. $V = n$ and f is an isomorphism of V into V, therefore, f is a homomorphism and one-one.

Now we shall prove that f is onto V.

Let $S = \{\alpha_1, \alpha_2, ..., \alpha_n\}$ be a basis of V. Then, we claim $S' = \{f(\alpha_1), f(\alpha_2), ..., f(\alpha_n)\}$ is a basis of V.

Let $a_1, a_2, ..., a_n \in F$ such that

$$a_1 f(\alpha_1) + a_2 f(\alpha_2) + ... + a_n f(\alpha_n) = 0'$$

$\Rightarrow \qquad f(a_1\alpha_1 + a_2\alpha_2 + ... + a_n\alpha_n) = f(0)$ $\qquad (\because f \text{ is homomorphism.})$

$\Rightarrow \qquad a_1\alpha_1 + a_2\alpha_2 + ... + a_n\alpha_n = 0$ $\qquad (\because f \text{ is one-one.})$

$\Rightarrow \qquad a_1 = 0, a_2 = 0, ..., a_n = 0$ $\qquad (\because S \text{ is linearly independent.})$

$\therefore \quad S'$ is linearly independent.

Since dim. $V = n$ and S' is a linearly independent subset of V containing n elements, therefore S' is a basis of V.

Now we shall show that f is onto V.

Let α be any element of V, then there exists scalars $c_1, c_2, ... c_n \in F$ such that

$$\alpha = c_1 f(\alpha_1) + c_2 f(\alpha_2) + ... + c_n f(\alpha_n) \qquad (\because L(S') = V)$$

$\Rightarrow \qquad \alpha = f(c_1\alpha_1 + c_2\alpha_2 + ... + c_n\alpha_n)$ $\qquad (\because f \text{ is homomorphism.})$

$\Rightarrow \quad \alpha$ is f-image of an element $c_1\alpha_1 + c_2\alpha_2 + ... + c_n\alpha_n \in V$.

$\therefore \quad f$ is onto V.

Example 18. *If V is finite dimensional vector space and f is a homomorphism of V onto V, prove that f must be one-one and so an isomorphism.*

Solution. Let dim. $V = n$, and f be a homomorphism of V onto V, then we shall see that f is one-one.

Let $S = \{\alpha_1, \alpha_2, ..., \alpha_n\}$ be a basis of V, then we claim that

$S' = \{f(\alpha_1), f(\alpha_2), \ldots f(\alpha_n)\}$ is also a basis of V.

Since S is a basis of V, then every vector of V can be expressed as a linear combination of vectors of S.

Therefore, if $\alpha \in V$, then there exists scalars $a_1, a_2, \ldots, a_n$ such that
$$\alpha = a_1\alpha_1 + a_2\alpha_2 + \ldots + a_n\alpha_n$$

Since f is onto V. Then for every $\alpha \in V$ there exists $\alpha\beta \in V$ such that
$$\beta = f(\alpha)$$
$$\Rightarrow \quad \beta = f(a_1\alpha_1 + a_2\alpha_2 + \ldots + a_n\alpha_n)$$
$$\Rightarrow \quad \beta = a_1 f(\alpha_1) + a_2 f(\alpha_2) + \ldots + a_n f(\alpha_n) \quad (\because f \text{ is homomorphism.})$$
$$\Rightarrow \quad \beta \text{ is expressible in a linear combination of elements of } S'.$$
$$\Rightarrow \quad L(S') = V$$

Since dim. $V = n$ and S' is a subset of V containing n elements such that $L(S') = V$, therefore S' is a basis of V, so that S' is linearly independent.

Now we shall show that f is one-one.

Let γ and δ be two elements of V, then there exists scalars $c_1, c_2, \ldots, c_n$; $d_1, d_2, \ldots, d_n$ such that
$$\gamma = c_1\alpha_1 + c_2\alpha_2 + \ldots + c_n\alpha_n$$
and
$$\delta = d_1\alpha_1 + d_2\alpha_2 + \ldots + d_n\alpha_n$$

We assume that
$$f(\gamma) = f(\delta)$$
$$\Rightarrow \quad f(c_1\alpha_1 + c_2\alpha_2 + \ldots + c_n\alpha_n) = f(d_1\alpha_1 + d_2\alpha_2 + \ldots + d_n\alpha_n)$$
$$\Rightarrow \quad c_1 f(\alpha_1) + c_2 f(\alpha_2) + \ldots + c_n f(\alpha_n) = d_1 f(\alpha_1) + d_2 f(\alpha_2) + \ldots + d_n f(\alpha_n)$$
$$(\because f \text{ is homomorphism.})$$
$$\Rightarrow \quad (c_1 - d_1)f(\alpha_1) + (c_2 - d_2)f(\alpha_2) + \ldots + (c_n - d_n)f(\alpha_n) = 0'$$
$$\Rightarrow \quad c_1 - d_1 = 0, c_2 - d_2 = 0, \ldots, c_n - d_n = 0$$
$$[\because f(\alpha_1), f(\alpha_2), \ldots, f(\alpha_n)] \text{ are linearly independent.}]$$
$$\Rightarrow \quad c_1 = d_1, c_2 = d_2, \ldots, c_n = d_n$$
$$\Rightarrow \quad c_1\alpha_1 + c_2\alpha_2 + \ldots + c_n\alpha_n = d_1\alpha_1 + d_2\alpha_2 + \ldots + d_n\alpha_n$$
$$\Rightarrow \quad \gamma = \delta$$
$$f(\gamma) = f(\delta) \Rightarrow \gamma = \delta$$

Hence, f is one-one and f is an isomorphism of V onto V.

Example 19. *Let $V(F)$ and $W(F)$ be two finite dimensional vector spaces such that dim. V=dim. W. If f is an isomorphism of V into W, prove that f must map V onto W.*

Solution. Let dim.$V = $ dim.$W = n$ and f be an isomorphism of V into W, then we shall prove that f is onto W.

Let $S = \{\alpha_1, \alpha_2, \ldots, \alpha_n\}$ be a basis of V, then we claim that $S' = \{f(\alpha_1), f(\alpha_2), \ldots, f(\alpha_n)\}$ is a basis of W.

Let $a_1, a_2, \ldots, a_n \in F$, such that
$$a_1 f(\alpha_1) + a_2 f(\alpha_1) + \ldots + a_n f(\alpha_n) = 0'$$

$\Rightarrow \quad f(a_1\alpha_1 + a_2\alpha_2 + ... + a_n\alpha_n) = f(0) \qquad (\because f \text{ is homomorphic and } f(0)=0')$

$\Rightarrow \quad a_1\alpha_1 + a_2\alpha_2 + ... + a_n\alpha_n = 0 \qquad\qquad (\because f \text{ is one-one.})$

$\Rightarrow \quad a_1 = 0, a_2 = 0, ..., a_n = 0 \qquad (\because \alpha_1, \alpha_2, ..., \alpha_n \text{ are linearly independent.})$

$\Rightarrow \quad S'$ is linearly independent.

Since dim. $W = n$ and S' is a linearly independent subset of W containing n elements, therefore S' is a basis of W, thus every element of W can be expressed as a linear combination of elements of W.

Now we shall show that f is onto W.

Since $S = \{\alpha_1, \alpha_2, ... \alpha_n\}$ is a basis of V, then every element of V can be expressed as a linear combination of elements of S. Therefore, if $\alpha \in V$, then there exists scalar $c_1, c_2, ..., c_n \in F$ such that

$$\alpha = c_1\alpha_1 + c_2\alpha_2 + ... + c_n\alpha_n$$

$\Rightarrow \qquad\qquad f(\alpha) = f(c_1\alpha_1 + c_2\alpha_2 + ... + c_n\alpha_n)$

$\Rightarrow \qquad\qquad f(\alpha) = c_1 f(\alpha_1) + c_2 f(\alpha_2) + ... + c_n f(\alpha_n)$

$\Rightarrow \quad f(\alpha)$ is a linear combination of elements of S'.

$\Rightarrow \qquad\qquad f(\alpha) \in W$

$\therefore \quad$ For every $\alpha \in V$, we have $f(\alpha) \in W$.

Hence, f is onto W.

Example 20. *Show that the mapping $f : V_3 (F) \to V_2 (F)$ defined by $f(a_1, a_2, a_3) = (a_1, a_2)$ is a homomorphism of $V_3 (F)$ onto $V_2 (F)$.*

Solution. Let $\alpha = (a_1, a_2, a_3), \beta = (b_1, b_2, b_3)$ be any elements of $V_3 (F)$ and $a, b \in F$, then

$$a\alpha + b\beta = a(a_1, a_2, a_3) + b(b_1, b_2, b_3)$$

$\Rightarrow \qquad\qquad a\alpha + b\beta = (aa_1 + bb_1, aa_2 + bb_2, aa_3 + bb_3)$

Also, $\qquad\quad f(\alpha) = f(a_1, a_2, a_3) = (a_1, a_2)$

$\qquad\qquad\qquad f(\beta) = f(b_1, b_2, b_3) = (b_1, b_2)$

Now $\qquad\quad f(a\alpha + b\beta) = f(aa_1 + bb_1, aa_2 + bb_2, aa_3 + bb_3)$

$\Rightarrow \qquad\quad f(a\alpha + b\beta) = (aa_1 + bb_1, aa_2 + bb_2) \qquad\qquad$ (By definition of f)

$\Rightarrow \qquad\quad f(a\alpha + b\beta) = (aa_1, aa_2) + (bb_1, bb_2)$

$\Rightarrow \qquad\quad f(a\alpha + b\beta) = a(a_1, a_2) + b(b_1, b_2)$

$\Rightarrow \qquad\quad f(a\alpha + b\beta) = af(\alpha) + bf(\beta)$

$\Rightarrow f$ is a homomorphism of $V_3(F)$ into $V_2(F)$.

Now we shall show that f is onto.

For any $a_1, a_2, a_3 \in F$, we have $(a_1, a_2, a_3) \in V_3(F)$ and $(a_1, a_2) \in V_3(F)$. Thus for any element $(a_1, a_2, a_3) \in V_3(F)$ there exists an element $(a_1, a_2) \in V_2(F)$ such that $f(a_1, a_2, a_3) = (a_1, a_2)$. Hence, f is onto $V_2(F)$.

Hence, f is a homomophism of $V_3(F)$ onto $V_2(F)$.

EXERCISE5.6

1. Show that the set $S = \{(1, 0, 0), (0, 1, 0), (0, 0, 1)\}$ spans the vector space $V_3(\text{R})$.

2. Show that the set $S = \{(1, 2, 4), (1, 0, 0), (0, 1, 0), (0, 0, 1)\}$ is linearly dependent subset of the vector space $V_3(\text{R})$.

3. Show that the set $\{1, x, 1 + x + x^2\}$ is linearly independent set of vectors in the vector space of all polynomials over real number field.

4. Show that the vectors $(1, 1, -1)$, $(2, -3, 5)$, $(-2, 1, 4)$ of R^3 are linearly independent.

5. Show that the vectors $(1, 3, 2)$, $(1, -7, -8)$, $(2, 1, -1)$ of $V_3(\text{R})$ are linearly dependent.

6. Is the vector $(3, -1, 0, 1)$ in the subspace of R^4 spanned by the vectors $(2, -1, 3, 2)$, $(-1, 1, 2, -3)$, $(1, 1, 9, -5)$?

7. Is the vector $(2, -5, 3)$ in the subspace of R^3 spanned by the vectors $(1, -3, 2)$, $(2, -4, -1)$, $(1, -5, 7)$?

8. Show that every subset of a linearly independent set of vectors is linearly independent.

9. Examine each of the following sets of vectors for linear dependent in the vector space $V_3(\text{R})$

   (i) $\{(1, 2, 0), (0, 3, 1), (-1, 0, 1)\}$

   (ii) $\{(-1, 2, 1), \{3, 0, -1\}, (-5, 4, 3)\}$

10. Let $f : V_2(\text{R}) \to V_2(\text{R})$ be defined as $f(a_1, b_1) = (b_1, a_1)$. Show that f is an isomorphism.

11. Define a homomorphism of a vector space $V(F)$ into a vector space $W(F)$. Show that the mapping
    $$f : (a, b) \to (a + 2, b + 3)$$
    of $V_2(\text{R})$ into itself is not a homomorphism.

12. Give an example of a one-one homomorphism of an infinite dimensional vector space which is not an isomorphism.

13. Let f be a homomorphism from a vector space U into a vector space V. If S is a subspace of U, prove that $f(S)$ will be a subspace of V.

14. If V is finite dimensional and f is a homomorphism of V into itsefl which is not onto, prove that there is some $\alpha \neq 0$ in V such that $f(\alpha) = 0$.

15. If f is an isomorphism of a vector space V onto a vector space W, prove that f maps a basis of V onto a basis of W.

CHAPTER REVIEW : A COMPETITIVE APPROACH

Selected Terms and Results

TERMS

- **Vector space:** An algebraic structure $(V, +, .)$ is said to be vector space over a field F if
  - (i) $(V, +)$ is an abelian group.
  - (ii) $a(\alpha+\beta) = a\alpha+a\beta \ \forall \alpha, \beta \in V, a \in F$
  - (iii) $(a+b)\alpha = a\alpha +b\alpha \ \forall \alpha \in V, a, b \in F$
  - (iv) $(ab)\alpha = a(b\alpha) \ \forall \alpha \in V, a, b \in F$
  - (v) $1. \alpha = \alpha \forall \ \alpha \in V$

- **Vector Subspace:** A non-empty subset W of a vector space $V(F)$ which itself is a vector space is called vector subspace of $V(F)$.

- **Linear sum:** The linear sum of two subspaces W_1 and W_2 is the set of all those elements each one of which is expressible as the sum of an element of W_1 and an element of W_2.

- **Direct Sum:** V is said to be direct sum of W_1 and W_2 if each element of V can be uniquely expressed as the sum of an element of W_1 and an element of W_2.

- **Linear Combination of Vectors:** Let $V(F)$ be a vector space and $\alpha_1, \alpha_2, ...\alpha_n \in V$. Then any vector $\alpha \in V$ can be expressed as α $= a_1\alpha_1+a_2\alpha_2+...+a_n\alpha_n$, where $a'_i s \in F$ is said to be linear combination of vectors $\alpha_1, \alpha_2, ..., \alpha_n$.

- **Linear Span:** Let $V(F)$ be a vector space and S be any non-empty subset of V, then set of all linear combination of finite elements of S is called the linear span of S.

- **Linearly Dependent Vectors:** A finite set $\{\alpha_1, \alpha_2, ..., \alpha_n\}$ of vectors of V is said to be linearly dependent if there exists scalars $\alpha_1, \alpha_2, ..., \alpha_n$ not all of them equal to zero such that
$$a_1\alpha_1+a_2\alpha_2+ ...+ a_n\alpha_n = 0$$

- **Linearly Independant Vectors:** A finite set of vectors $\{\alpha_1, \alpha_2,..., \alpha_n\}$ of the vector space $V(F)$ is said to be linearly independent if for every expression of the type $a_1\alpha_1+a_2\alpha_2+...+ a_n\alpha_n = 0$, $a_i's \in F$ implies $a_1 = a_2 = ... = a_n = 0$.

- **Basis of a Vector Space:** A non-empty subset S of a vector space $V(F)$ is said to be basis if
  - (i) S is linearly independent.
  - (ii) $L(S) = V$

- **Finite Dimensional Vector Space:** Let S be a non-empty subset of a vector space $V(F)$, then $V(F)$ is said to be finite dimensional if S is finite subset of V such that $L(S) = V$.

- **Cosets:** Let W be a subspace of a vector space $V(F)$ then the set $\alpha+W = \{\alpha+\beta, \forall\beta \in W\}$ and $W+\alpha = \{\beta + \alpha \ \forall\beta \in W\}$ are called left and right cosets respectively.

- **Quotient space:** Let $V/W = \{ W+\alpha : \alpha \in V\}$ be the set of all cosets of W in V such that $(W+\alpha)+(W+\beta)=W+(\alpha+\beta)$ and $a(W+\alpha) = W+a\alpha$. Then vector space V/W is called quotient space.

RESULTS

- The intersection of any two subspaces of a vector space is a subspace.

- The union of two subspaces of a vector space is a subspace iff one is contained in the other.

- The linear span $L(S)$ of a non-empty subset S of a vector space $V(F)$ is the smallest subspace of V containing S.

- The zero space has no basis.

- Every finitely generated vector space has a finite basis **(Existence theorem)**.

- Every non-zero vector space has a basis. A vector space may have more than one basis.

- Every linearly independent subset of a vector space V is either a basis of V or can be extended to form a basis of V (Extension theorem).

- Every subspace of a finite dimensional vector space has a compliment.

- Any two cosets are either disjoint or identical.

- Every n-dimensional vector space $V(F)$ is isomorphic to $F^n(F)$.

- Any two finite dimensional vector space over the same field are isomorphic if and only if they are of same dimension.

- If two vectors are linearly dependent, then one of them is a scalar multiple of other.

- A set of non-zero vector $[x_1, x_2,..., x_n]$ is linearly dependent if some of these vectors, say x_i is a linear combination of the preceding vectors $x_1, x_2,..., x_{i-1}$ and conversely.
- If W_1 and W_2 are subspaces of a vector space $V(F)$, then $W_1 + W_2$ is also a subspace of $V(F)$.
- The set of all real valued continuous function defined in [0, 1] is a vector space over field of reals.
- The complex field C is a vector space over the field of reals.
- Arbitrary intersection of subspaces of a vector space is a subspace.
- Linear transformation is also known as vector space homomorphism.
- Let V be an m-dimensional vector space over the field F and let V be an n-dimensional vector space over F. Then the vector space $L(U, V)$ is finite dimensional and has dimension mn.

Review Questions and Project Work

1. If V is a vector space over an infinite field F then show that it is not possible to write V as union of a finite number of proper subspaces.

2. Show that $L(S)$ is the smallest subspace of V containing S.

3. Show that following vectors are linearly dependent :
   (i) $(1, 1, 2)$, $(-3, 1, 0)$, $(1, -1, 1)$ $(1, 2, -3)$ in $R^3(R)$.
   (ii) $(1, -1, 2, 0)$, $(3, 0, 0, 1)$, $(2, 1, -1, 0)$, $(1, -1, 2, 0)$ in $R^4(R)$.

4. Show that following vectors are linearly independent :
   (i) $(1, 1, 0)$, $(1, 0, 1)$, $(0, 1, 1)$ in $R^3(R)$.
   (ii) $(1, 0, 0)$, $(1, 1, 1)$, $(1, 2, 3)$ in $R^3(R)$.

5. Show that the vectors (v_1, v_2) and (w_1, w_2) in **C** are linearly dependent if $v_1 w_2 = v_2 w_1$.

6. Let S be a finite subset of a vector space such that S is linearly independent and every proper superset of S in V is linearly dependent show that S is a basis of V.

7. Verify that following is an inner product on R^2.
   $$(u, v) = x_1 y_1 - 2x_1 y_2 - 2x_2 y_1 + 5x_2 y_2$$

8. If W is a subspace of V and $v \in V$ satisfies
   $$(v, w) + (w, v) \le (w, w) \forall \ w \in W$$
   prove that $(v, w) = 0 \ \forall w \in W$, where V is an inner product space over F.

9. Let T be a linear operator on V and let Rank T^2 = Rank T then show that

   Range $(T) \subseteq$ Ker $(T) = \{0\}$

10. Show that a necessary and sufficient condition for the map $T : F^2 \to F^2$ such that $T(x_1, x_2) = (\alpha x_1 + \beta \ x_2)$, $(\alpha, \beta, \gamma, \delta$, are some fixed element of $F)$ to be an isomorphism is that $\begin{vmatrix} \alpha & \beta \\ \gamma & \delta \end{vmatrix} \ne 0$

11. Let A be $n \times n$ matrix over F. Show that A is invertible if and only if rows of A are linearly independent over F.

12. Let $u, v \in V$ and that $f(u) = 0 \Rightarrow f(v) = 0$ for all $f \in V^*$. Show that $v = \alpha u$ for some scalar α.

13. Let T be a linear operator on R^2 which is represented in the standard ordered basis by the matrix $A = \begin{bmatrix} 0 & -1 \\ 1 & 0 \end{bmatrix}$, show that T has no eigen values in R.

14. Let V be the vector space of all real valued continuous functions. Define $T : V \to V$ by $Tf(x) = \int_0^x f(t)dt$ show that T has no eigen values.

15. Let a, b, c be elements of the field F and
    $$A = \begin{bmatrix} 0 & 0 & c \\ 1 & 0 & b \\ 0 & 1 & a \end{bmatrix}$$
    Prove that the characteristic polynomial of A is same as that of its minimal polynomial.

Objective Type Questions

FILL IN THE BLANKS

1. In a vector space $(V, \ '+', \ '.')$ the external composition is also known as _____ .

2. In a vector space $V(F)$, the vector addition is also known as_____ .

3. The elements of a field F for $V(F)$ are called_____ .

4. In a vector space $(V, \ '+', \ '.')$, $(V, +)$ must be _____ .

5. The additive identity for the vector space $V(F)$ is _____.

6. If $a\,\alpha \in V$ for $a \in F$ and $\alpha \in V$, then V is closed under _____.

7. If $a \in F$ and $\alpha \in V$, then $(ab)\alpha =$ _____.

8. If α, β, $\gamma \in V$ and $\alpha + \beta = \gamma$, then $\alpha + \beta - \gamma =$ _____.

9. If F is any field, then F is a vector space over _____.

10. If $V(F)$ is a vector space and $0 \in V$, then _____ $= 0 \; \forall \; \alpha \in V$.

11. If $V(F)$ is a vector space and $a \in F$, $\alpha \in V$, then $a\alpha = 0 \Rightarrow a =$ _____ or $\alpha =$ _____.

12. If W be a subset of a vector space $V(F)$ and $a\alpha + b\beta \in W$, for all a, $b \in F$ and α, $\beta \in W$, then W is a _____.

13. If $W = \{(a_1, a_2, 0): a_1, a_2 \in F\}$, then W is a _____ of vector space.

14. $R(C)$ is _____.

15. For any non-empty subset W of $V(F)$, if $\alpha - \beta \in W$ and $a\alpha \in W$ for all $a \in F$ and α, $\beta \in W$, then W is _____.

16. If W_1 and W_2 are two subspaces of $V(F)$, then $W_1 \cap W_2$ is a _____.

17. If W_1 and W_2 are two subspaces of $V(F)$ and either $W_1 \subseteq W_2$ or $W_2 \subseteq W_1$, then _____ is a subspace of $V(F)$.

18. $L(\phi) =$ _____.

19. If $S = \{(1,0,0), (0,1,0), (0,0,1)\}$ is a subset of $V_3(F)$, then $L(S) =$ _____.

20. If $a \neq 0$, $b \neq 0 \in F$ and $a\alpha + b\beta = 0$ for α, $\beta \in V$, then α, β are _____.

21. The vectors $(1, 0, 0)$, $(0, 1, 0)$, $(0, 0, 1)$ are linearly _____.

22. For any subset S of $V(F)$, $L(S) = V$, then S is a basis of $V(F)$ if S is linearly _____.

23. Every superset of a linearly dependent set of vectors is linearly _____.

24. The vectors in a basis are linearly _____.

25. Every vector space has a _____.

26. Let $S = \{\alpha\}$ and $a \neq 0 \in V(F)$, then S is always linearly _____.

27. Any infinite set of vectors of V is linearly independent if its every finite subset is linearly _____.

28. If a basis of vector space has 4 elements, then dimension of the vector space is _____.

29. The set $\{(1, 0, 0), (0, 1, 0), (0, 0, 1)\}$ forms a .... of $V_3(F)$.

30. The set $S = \{(1, 0), (0, 1)\}$ is a basis of $V_n(F)$ for $n =$ _____.

31. Any subset containing $(n+1)$ vectors of an n-dimensional vector space is linearly _____.

32. M and N are two subspaces of a vector space V, then $V = M \oplus N$ if
(i) _____ and (ii) _____

TRUE/FALSE

Write 'T' for true and 'F' for false statement.

1. In a vector space $V(F)$, the vector addition is also called an internal composition. **(T/F)**

2. The elements of V are scalars. **(T/F)**

3. The elements of F are vectors. **(T/F)**

4. Let $V(F)$ be a vector space. Then the zero space $\{0\}$ is called a trivial subspace. **(T/F)**

5. The set $W = \{(a, 0, b) : a, b \in R\}$ is not a subspace of $R^3(R)$. **(T/F)**

6. Let $V(F)$ be a vector space and $\alpha \in V$, $a \in F$, then $a0 = 0$ and $0\alpha = 0$. **(T/F)**

7. If K is a field and $F \subseteq K$, then $K(F)$ is a vector space. **(T/F)**

8. The field of complex numbers is not a vector space over a field of real number. **(T/F)**

9. The field of real number is a vector space over a field of complex numbers. **(T/F)**

10. If any subset W of V is closed under addition and scalar multiplication in V, then W is a subspace of V. **(T/F)**

11. The intersection of two subspaces of a vector space is also a subspace. **(T/F)**

12. If $\{(1, 0), (0, 1)\} \subseteq V_2(F)$, then $L\{1, 0), (0, 1)\} = F^2$. **(T/F)**

13. $L(\phi) = \{0\}$ **(T/F)**

14. The single non-zero vector is always linearly dependent. **(T/F)**

15. If W_1 and W_2 are two subspaces of $V(F)$, then $W_1 + W_2$ is also a subspace of $V(F)$. **(T/F)**

16. In a vector space, every subset of a linearly independent set is linearly dependent. **(T/F)**

17. The set containing the zero vector is linearly dependent. **(T/F)**

18. The vectors $(1 + i, 2i)$, $(1,1 + i)$ in $C^2(\mathbf{C})$ are linearly dependent but in $C^2(R)$ are linearly independent. **(T/F)**

19. For a non-empty subset S of $V(F)$, $L(S) = V$ and S is linearly independent, then S is a basis of $V(F)$. **(T/F)**

20. Every vector space has a finite basis. **(T/F)**

21. The vectors in a basis are linearly independent. **(T/F)**

22. Every vector space has a basis. **(T/F)**

MULTIPLE CHOICE QUESTIONS

Choose the most appropriate one :

Problem Set-1

1. In a vector space $V(F)$, a 0 equals :
(a) 0
(b) **0**
(c) a
(d) 1

2. In a vector space $V(F)$, $\alpha \in V$ and $a, b \in F$, then $(ab)\alpha$ equals :
(a) $a(b\alpha)$
(b) ab
(c) a
(d) $\alpha(ab)$

3. If W_1 and W_2 are two subspaces of $V(F)$, then $W_1 \cup W_2$ is a subspace if:
(a) $W_1 - W_2$
(b) $W_1 \subseteq W_2$
(c) $W_1 \cap W_2$
(d) none of these

4. $L(\phi)$ equals :
(a) 0
(b) ϕ
(c) $\{0\}$
(d) none of these

5. If $S = \{(1, 0), (0, 1) \subseteq V_2(R)\}$, then $L(S)$ equals :
(a) R
(b) S
(c) R^3
(d) R^2

6. If $\alpha = k\beta$ for α, $\beta \in V$ and $k \in F$, then $\{\alpha, \beta\}$ is linearly :
(a) dependent
(b) independent
(c) None of these

7. The set $S = \{(1, 0, 0), (0, 1, 0), (0, 0, 1)\}$ forms a basis for $V_n(R)$ if 'n' equals :
(a) 2
(b) 3
(c) 4
(d) 1

8. The dimension of a vector space $R^3(R)$ is :
(a) 2
(b) 4
(c) 1
(d) 3

9. Which of the sets is linearly dependent ?
(a) $\{0\}$
(b) $\{\phi\}$
(c) $\{1\}$
(d) $\{\alpha\}$

10. A subset S of $V(F)$ forms a basis of $V(F)$ if S is linearly independent and :
(a) $L(S) = S$
(b) $L(S) = V$
(c) $L(S) = F$
(d) none of these

11. Which of the following is not a vector space ?
(a) R(R)
(b) C(C)
(c) R(C)
(d) C(R)

12. Condition that vectors (a_1, a_2) and (b_1, b_2) are linearly dependent is :
(a) $a_1 b_1 + a_2 b_2 = 0$
(b) $a_1 b_2 + a_2 b_1 = 0$
(c) $a_1 b_2 - a_2 b_1 = 0$
(d) $a_1 b_1 - a_2 b_2 = 0$

Problem Set-2

1. Let $f(x)$ be the ring of polynomials in one variable x over a field F with the relation $x^n = 0$ for fixed $n \in N$. Then dimension of $f(x)$ over F is:
(a) 1
(b) n
(c) $n-1$
(d) none of these

2. Let $x = (3, 2, -1), y = (2, 4, 1), z = (4, 0, -3)$ and $w = (10, 4, -5)$ be vectors in R^3, which one of the following is correct?
(a) $2x - y = z$, $y + 2z = w$
(b) $2x + y = z$, $y - 2z = w$
(c) both (a) and (b) are true
(d) None of the above

3. The set $s = [a+ib, c+id]$ is a basis for the vector space C over R iff :
(a) $ad - bc = 0$
(b) $ad - bc \neq 0$
(c) $ad + bc \neq 0$
(d) none of these

4. Let V be a vector space over the field F of dimension n. Consider the statements:
  (i) Every subset of V containing n elements is a basis of V.
  (ii) No linearly independent subset of V contains more than n elements. Which of the above statment is/are true
(a) (i) only
(b) (ii) only
(c) both (i) and (ii)
(d) none of these

5. Let V be the vector space over the field of real numbers spanned by the set
$S = \{(0,1,0,0),(1,1,0,0),(1,0,1,0),$
$\qquad (0,0,1,0),(1,1,1,0),(1,0,0,0)\}$

Then dimension of V is :
(a) 1
(b) 2
(c) 3
(d) none of these

6. If $S = \{(1, i, 0), (2i, 1, 1), (0, 1+i, 1-i)\}$ be a subset of the complex vector space C^3 and $T = \{(1,1,1),(1,1,0),(1,0,0)\}$ be a subset of the real vector space R^3. Which of the following statement is correct ?
   (a) S and T both are basis
   (b) S is a basis but T not
   (c) S is not a basis
   (d) none of these

7. Consider the real vector space R^3. The subspace $\{(x, y, z) \in R^3 : y = x\}$ of R^3 is generated by which one of the following ?
   (a) $\{(1, 1, 0), (0, 0, 1)\}$
   (b) $\{(1, 1, 0), (1, 0, 0)\}$
   (c) Both (a) and (b) are true
   (d) None of the above

8. The dimension of the vector space formed by the solution of the system of equations
   $$x_1 + x_2 + x_3 = 0$$
   $$x_1 + 2x_2 = 0$$
   $$x_2 - x_3 = 0$$
   is :
   (a) 1   (b) 2
   (c) 3   (d) none of these

9. The dimension of the vector space defined in the above question over the field R is :
   (a) 1   (b) 2
   (c) 3   (d) none of these

10. Let $\alpha = (1, 2, 3)$, $\beta = (3, 1, 0)$, $\gamma = (2, 1, 3)$ and $\delta = (-1, 3, 6)$ then :
   (a) γ is a linear combination of α and β
   (b) δ is a linear combination of α and β
   (c) Both (a) and (b) are true
   (d) None of the above

11. If W_1, W_2, W_3 are the subspaces of a vector space V such that $W_3 \subseteq W_1$. Then which one of the following is true ?
   (a) $W_1 \cup (W_2 + W_3) = W_2 + W_1 \cup W_3$
   (b) $W_1 \cap (W_2 + W_3) = W_2 + W_1 \cap W_3$
   (c) Both (a) and (b) are true.
   (d) None of the above

12. Let V be a vector space over an infinite field F such that dim. $V = 2$, then the number of distinct subspace V is :
   (a) 1   (b) 2
   (c) 3   (d) none of these

13. The dimension of the subspace of R^3 spanned by $(-3, 0, 1)$, $(1, 2, 1)$ and $(3, 0, -1)$ is :
   (a) 1   (b) 2
   (c) 3   (d) none of these

14. Choose the correct one .
   (a) R is a vector space over C
   (b) R is a vector space over N
   (c) R is a vector space over Z
   (d) None of the above

15. The dimension of R(Q) is :
   (a) 1   (b) finite
   (c) infinite   (d) none of these

16. Let V be a vector space of ordered pairs of complex numbers over the real field R. Then, dimension of V is :
   (a) 2   (b) 1
   (c) 4   (d) none of these

17. The zero vector in the vector space R^4 is :
   (a) $(0, 0)$ (b) $(0, 0, 0, 0)$
   (c) $(0, 0, 0)$ (d) none of these

18. The dimension of the vector space spanned by $(1, -2, 3, 1)$ and $(1, 1, -2, 3)$ is :
   (a) 1   (b) 2
   (c) 4   (d) none of these

19. Let V be the vector space of all 2×2 matrices over the field F. Then dimension of V is :
   (a) 2   (b) 4
   (c) 3   (d) none of these

20. The dimension of the vector space $V_3(R)$ is :
   (a) 1   (b) 2
   (c) 3   (d) none of these

21. The dimension of the vector space $C(R)$ is :
   (a) 1   (b) 2
   (c) 4   (d) none of these

22. The dimension of the vector space $C^2(R)$ is :
   (a) 1   (b) 2
   (c) 4   (d) none of these

23. The dimension of the vector space spanned by $(1, -2, 3, -1)$ and $(1, 1, -2, 3)$ is :
   (a) 1   (b) 2
   (c) 3   (d) none of these

24. Let W be a subspace of a finite dimensional vector space $V(F)$ then dim.(V/W) is equal to :
   (a) dim. V – dim. W (b) dim. V + dim. W
   (c) dim. V (d) none of these

25. The number of elements in the vector space of polynomials of degree at most n in which the coefficients are the elements of the field Z(P) over the field Z(P), P being a prime number is :

(a) P^n

(b) P^{n+1}

(c) P^{n-1}

(d) none of these

26. The set V$= [(x, y) \in R^2 : xy \geq 0]$ is :

(a) a vector space over R^2

(b) not a vector space over R^2

(c) a vector space over R

(d) none of the above

27. If V be the vector space of $n \times n$ matrices over R and let W be the subspace of matrices with entries in each row adding up to zero, then the dimension of W is :

(a) n

(b) $n(n-1)$

(c) $\dfrac{n}{n-1}$

(d) $n(n+1)$

28. Let $V_1 = \{(a, b, c, d) : b - 2c + d = 0\}$ and $V_2 = \{(a, b, c, d) : a = d, b = 2c\}$ are subspaces of R^2 then :

(a) dimension of V_1 is 3

(b) dimension of V_2 is 2

(c) both (a) and (b) are true

(d) none of the above

29. Which of following is/are true for the vector space C(R) ?

(a) $\{1, i\}$ form a basis for C.

(b) 1 and i are linearly independent.

(c) $\{1, i\}$ span C(R).

(d) All are true.

30. If V is a vector space of $n \times n$ matrix over the field of reals and V_1, V_2 are the subspaces of symmetric and antisymmetric matrices respectively then :

(a) $V = V_1 + V_2$

(b) $V = V_1 \oplus V_2$

(c) $V_1 \cap V_2 = \{0\}$

(d) all are true

31. If $V = R^3$ also W $= \{(x, y, 0) : x, y \in R\}$ then:

(a) W is a subspace of V

(b) $\alpha, \beta \in W \Rightarrow \alpha + \beta \in W$

(c) both (a) and (b) are true

(d) none of the above

32. Let V be the vector space of all functions from R to R and $W = \{f : 2f(3) = f(1)\}$ then :

(a) $\alpha, \beta \in W \Rightarrow \alpha + \beta \in W$

(b) $\alpha, \beta \in W \Rightarrow \alpha + 3\beta \in W$

(c) W is a subspace of V

(d) all are true

33. Let L be a finite dimensional linear space and let M_1, M_2 be subspaces of L such that $L = M_1 + M_2$ and dim $L =$ dim $M_1 +$ dim M_2 then :

(a) $L = M_1 \oplus M_2$

(b) $M_1 \cap M_2 = \{0\}$

(c) both (a) and (b) are true

(d) none of the above

34. If $S = [v_1, v_2, ..., v_n]$ be a basis of a finite dimensional vector space $V(F)$ and dim.$V(F) = n$ then :

(a) S is linearly independent

(b) $L(S) = V$

(c) both (a) and (b) are true

(d) none of the above

35. If a finite dimensional vector space $V(F)$ be the direct sum of its subspace V and W, i.e., $V = V \oplus W$ then :

(a) dim. $(U \cap W) = 0$

(b) dim. V = dim. $U +$ dim. W

(c) $U \cap W = \{0\}$

(d) all are true

Problem Set-3

1. Let dimension of a vector space V be n. If any set $S \subset V$ and S have m elements, $m > n$. Then :

(a) S is linearly independent

(b) S is linearly dependent

(c) S is zero space

(d) none of the above

2. The vector space which has only the additive identity element zero is :

(a) Complex space

(b) Real space

(c) Null space

(d) none of these

3. If V and W are finite dimensional vector spaces over the same field. Then V and W are isomorphic iff:

(a) dim. $V >$ dim. W

(b) dim. $V =$ dim. W

(c) dim. $V <$ dim. W

(d) none of these

4. The vectors $a_1, a_2, ..., a_n$ ar linearly dependent if for scalars $c_1, c_2, ..., c_n$.

$$c_1 a_1 + c_2 a_2 + ... + c_n a_n = 0$$

implies:

(a) $c_1, c_2, ..., c_n$ are not zero

(b) $c_1 = c_2 = ... = c_n = 0$

(c) $c_1 = c_2 = c_3 = ... = c_n$

(d) none of these

5. If B_1 and B_2 are two basis of the vector space V. Then :

(a) B_1 and B_2 have same numbers of elements

(b) B_1 and B_2 have distinct numbers of elements

(c) $B_1 = B_2$

(d) none of these

6. If V is finite dimensional vector space, and dim. $V = n$. Then any subset of V which contains m vectors is linearly dependent if:

(a) $m > n$ (b) $m < n$

(c) $m = n$ (d) none of these

7. (A). The intersection of any number of subspaces of a vector space $V(F)$ is a subspace of $V(F)$.

(B). The union of any number of subspaces of a vector space $V(F)$ is a subspace of $V(F)$. :

(a) (A) is false, (B) is true

(b) (A) is true, (B) is false

(c) (A) and (B) both are true

(d) (A) and (B) both are false.

8. If α_1 and α_2 are linearly dependent vectors, then :

(a) $\alpha_1 = c\alpha_2$ for some scalar c

(b) $\alpha_1 \neq c\alpha_2$

(c) $\alpha_1 > \alpha_2$

(d) $\alpha_1 < \alpha_2$

9. If $V = R^3$ is a vector space and $W = \{(x, y, 1)\} : x, y \in R\}$ is a subset of V, then :

(a) W is a subspace of V

(b) W is not a subspace of V

(c) W is not a vector space

(d) none of these

10. If V is a finite dimensional vector space, if W is a subspace of V, then:

(a) W is finite dimensional

(b) W is infinite dimensional

(c) The dimension of W is greater than V

(d) none of these

11. If V is finite dimensional vector space, dim. $V = m$ and W_1 and W_2 are two subspaces of V and dim. $W_1 = n_1$ and dim. $W_2 = n_2$, then :

(a) $n_1 + n_2 = m$ (b) $n_1 + n_2 \leq m$

(c) $n_1 + n_2 \geq m$ (d) none of these

12. If A is a matrix, then:

(a) Row rank (A) = column rank (B)

(b) Row rank $(A) \neq$ column rank (B)

(c) Row rank $(A) >$ column rank (B)

(d) none of these

13. A non-empty subset U of $V(F)$: $\alpha, \beta \in U$, $c\alpha + \beta \in U$, c any scalar, then :

(a) U is a subspace of V

(b) U is a superspace of V

(c) V is a subspace of U

(d) none of these

14. If a system contains a single non-zero vector, it is always:

(a) Linearly indpendent

(b) Linearly dependent

(c) Subspace

(d) none of these

15. Any set containing linearly dependent set is:

(a) Linearly independent

(b) Linearly dependent

(c) Null set

(d) none of these

16. If W_1 and W_2 are subspaces of V, then following is false:

(a) $W_1 \cup W_2$ is a subspaces of V

(b) $W_1 \cap W_2$ is a subspaces of V

(c) $W_1 + W_2$ is a subspaces of V

(d) $W_1 \cup W_2$ is not a subspaces of V

17. A set of single non-zero vector is:

(a) Linearly dependent

(b) Linearly independent

(c) Basis

(d) None of these

18. If V is vector space and V is finite dimensional then its basis have:

(a) Infinite number of elements

(b) Finite number of elements

(c) Number of elements are not defined

(d) none of these

19. Let 0 be a zero vector in vector space V, then $\{0\}$ is:

(a) Zero subspace of V

(b) Null space of V

(c) Identity space of V

(d) none of these

20. The dimension of the vector space is:

(a) Number of elements in vector space

(b) Number of elements in basis of the vector space

(c) Subspace of vector space

(d) none of these

21. If W_1 and W_2 are finite dimensional subspaces of vector space V, then:

(a) $\dim.(W_1+W_2) = \dim. W_1 + \dim. W_2$

(b) $\dim.(W_1+W_2) = \dim. W_1 + \dim. W_2$
$+ \dim. (W_1 \cup W_2)$

(c) $\dim.(W_1+W_2) = \dim. W_1 + \dim. W_2$
$- \dim. (W_1 \cap W_2)$

(d) $\dim.(W_1+W_2) = \dim. W_1 + \dim. W_2$
$+ \dim. (W_1 \cap W_2)$

22. The singleton set $\{\alpha\}$ is linearly independent iff :

(a) $\alpha = 0$ (b) $\alpha \neq 0$

(c) α is scalar   (d) none of these

23. Let V and W be finite dimensional vector spaces such that $\dim. V = n$ and $\dim. W = m$, then $\dim. L (V, W)$ is :

(a) mn (b) $m + n$

(c) m/n (d) $m - n$

24. If V is a vector space with $\dim. V = n$, then dimension of hyperspace of V is:

(a) n (b) $n - 1$

(c) $n + 1$ (d) 0

25. If $W_1 + W_2 + ... + W_n$ are subspaces of vector space, then :

(a) $W_1 + W_2 + ... + W_n$ is not a vector space

(b) $W_1 + W_2 + ... + W_n$ is a vector space

(c) $W_1 \cup W_2 \cup ... \cup W_n$ is not a vector space

(d) none of these

26. If S is a linearly dependent set and set W contains S, then:

(a) W is linearly independnet

(b) W is linearly dependent

(c) $W = S$

(d) none of these

27. If a set contains a zero vector, then:

(a) set is linearly dependent

(b) set is linearly independent

(c) set is null set

(d) none of these

28. If W_1 and W_2 are the subspaces of a vector space V and $W_1 \cup W_2$ is also a subspace of V, then :

(a) either $W_1 \subset W_2$ or $W_2 \subset W_1$

(b) $W_1 \cap W_2 = \phi$

(c) $W_1 = W_2$

(d) None of these

29. If W_1 and W_2 are subspaces of vector space and W_1 is the subspace spanned by vectors α_1, α_2, ..., α_n, then if $\alpha_1, \alpha_2, ..., \alpha_n \in W_2$, then :

(a) $W_1 = W_2$

(b) $W_1 \subset W_2$

(c) $W_1 \supset W_2$

(d) $W_1 \cap W_2 = \phi$

30. Let V be a finite dimensional vector space. I is an identify transformation, then range of I is :

(a) $\{0\}$ (b) V

(c) ϕ (d) none of these

31. If V is finite dimensional and $\dim. V = n$, $S \subset V$ is any subset with $m < n$ elements, then:

(a) subset S can span V

(b) subset S can not span V

(c) subset S is equal to V

(d) none of these

32. If $Q(\sqrt{2}, \sqrt{3})$ is the smallest field containing Q, $\sqrt{2}$ and $\sqrt{3}$, then dimension of $Q(\sqrt{2}, \sqrt{3})$ is :

(a) 4   (b) 2

(c) 3   (d) 1

Answers

FILL IN THE BLANKS

1. scalar multiplication **2.** internal composition **3.** scalars **4.** abelian group **5.** zero vectors **6.** scalar multiplication. **7.** $a(b\alpha)$ **8.** zero vector *i.e.*, 0 **9.** F **10.** 0α **11.** 0, **0** **12.** subspace **13.** subspace **14.** not a vector space **15.** subspace **16.** subspace **17.** $W_1 \cup W_2$ **18.** $\{0\}$ **19.** V_3 **20.** linearly dependent **21.** independent **22.** independent **23.** dependent **24.** independent **25.** basis **26.** independent **27.** independent **28.** 4 **29.** basis **30.** 2 **31.** dependent **32.** $V = M+N, M \cup N = \{0\}$

TRUE/ FALSE

1. T	**2.** F	**3.** F	**4.** T	**5.** F	**6.** F	**7.** T	**8.** F	**9.** F
10. T	**11.** T	**12.** T	**13.** T	**14.** T	**15.** T	**16.** F	**17.** F	**18.** T
19. T	**20.** F	**21.** T	**22.** T					

MULTIPLE CHOICE QUESTIONS

Problem set-1

1. (b)	**2.** (a)	**3.** (b)	**4.** (c)	**5.** (d)	**6.** (a)	**7.** (b)	**8.** (d)	**9.** (a)
10. (b)	**11.** (c)	**12.** (c)						

Problem set-2

1. (b)	**2.** (a)	**3.** (b)	**4.** (b)	**5.** (c)	**6.** (b)	**7.** (a)	**8.** (c)	**9.** (b)
10. (b)	**11.** (a)	**12.** (c)	**13.** (b)	**14.** (a)	**15.** (c)	**16.** (c)	**17.** (b)	**18.** (b)
19. (b)	**20.** (c)	**21.** (b)	**22.** (c)	**23.** (a)	**24.** (a)	**25.** (b)	**26.** (b)	**27.** (b)
28. (c)	**29.** (d)	**30.** (d)	**31.** (c)	**32.** (d)	**33.** (c)	**34.** (c)	**35.** (d)	

Problem set-3

1. (b)	**2.** (c)	**3.** (b)	**4.** (b)	**5.** (a)	**6.** (a)	**7.** (b)	**8.** (a)	**9.** (a)
10. (a)	**11.** (b)	**12.** (a)	**13.** (a)	**14.** (a)	**15.** (b)	**16.** (a)	**17.** (b)	**18.** (b)
19. (a)	**20.** (b)	**21.** (c)	**22.** (b)	**23.** (a)	**24.** (b)	**25.** (b)	**26.** (b)	**27.** (a)
28. (a)	**29.** (b)	**30.** (b)	**31.** (b)	**32.** (a)				

COMPETITION CORNER
for JRF, NET/SET, GATE Aspirants

SOME FASCINATING FACTS

1. In the definition of a vector space V over a field F, the condition $f(x)=x \ \forall \ x \in V$ can be replaced by the conditon '$\lambda x=0$' holds only if $\lambda=0$ or $x=0$, $\lambda \in F, x \in V$.

2. If R is considered as a vector space over Q, then the necessary and sufficient condition that the vector 1 and x in R be linearly independent is that the real number x be irrational.

3. In a vector space every subset of a linearly independent set is linearly independent and every superset of a linearly dependent set is linearly dependent.

4. If W is a subspace of a vector space V then there is a one-one correspondence between subspaces of V which contain W and subspaces of V/W.

5. Let W be a subspace of a vector space $V(F)$ for $a,b \in W$ we define $a \equiv b(\mathrm{mod}\ W)$ if and only if $a-b \in W$ and for all $\alpha \in F$, $\alpha(a-b) \in W$. Also congruence modulo W is an equivalence relation on V.

6. Every subspaces of a finite dimensional vector space has a complement.

7. If U is a subspace of a finite dimensional vector space V then every complement of U is equal to dim. V–dim. U.

8. All complements of a subspace are of same dimension.

9. Complement of a subspace is not unique.

10. If $V(F)$ be a vector space of dimension n. If V_1 and V_2 are subspaces each of dimensions strictly greater than $\dfrac{n}{2}$ then $V_1 \cap V_2 \neq \{0\}$.

11. Every inner product space is a metric space.

12. If x,y are vectors in a Euclidean space such that $\|x\| = \|y\|$ then $x+y$ is orthogonal to $x–y$.

13. Schwarz inequality implies that cosine of an angle is of absolute value at most 1.

14. A linear transformation $T:U \to V$ is one-one if and only if it maps every linearly independent subset of U into a linearly independent subset of V.

SOME IMPORTANT ILLUSTRATIONS

1. A field F itself is a vector space over the field F.

2. C is a vector space over R.

3. The set V of all $m \times n$ matrices with their elements as real numbers is a vector space over the field F of real numbers w.r.t. addition of matrices as addition of vectors and multiplication of a matrix by a scalar as scalar multiplication.

4. R is not a vector space over C.

5. Q is not a vector space over R.

6. R is a vector space over Q.

7. The set of all polynomials in x of degree less than equal to 2 is a vector space.

8. Let R^n be the set of all n-types of real numbers,*i.e.*,
$R^n = \{a_1, a_2, \ldots \ldots, a_n : a_i\text{'s} \in R\}$
Then R^n is a vector space over R with additon and scalar multiplication defined as follows:

$(a_1, a_2, \ldots, a_n) + (b_1, b_2, \ldots, b_n) = (a_1+b_1, a_2+b_2, \ldots, a_n+b_n), c(a_1, a_2, \ldots, a_n) = (ca_1, ca_2, \ldots, ca_n)$

9. The set V of all real valued functions of $(0,1)$ is a vector space over R with respect to addition and scalar multiplication of functions.

10. The set V of all those polynomials functions over R with coefficeint in R which are of degree $\leq n$ together with the zero function in R is a vector space over R.

11. The set of all Hermitian matrices of order n is a vector space over R with respect to matrix addition and multiplication of a matrix by a scalar.

12. Let S be the set of all matrix of the form
$\begin{pmatrix} a & b \\ -b & a \end{pmatrix}, a,b \in C$

Then S is a vector space over C.

13. The set of all odd functions from R to itself is a vector space w.r.t. addition and scalar multiplication of functions.

14. The set of all 4×4 complex matrices *w.r.t.* matrix addition and multiplication of a matrix by a scalar is a vector space over C.

15. The set of all $n \times n$ matrices over Q is a vector space over Q *w.r.t.* matrix addition and multiplication of a matrix by a scalar.

16. The set of all n-rowed real skew-symmetric matrices over Q *w.r.t.* matrix addition and multiplication by a scalar is a vector space over R.

17. The set of all skew-Hermitian matrices of order n is a vector space over R *w.r.t.* matrix addition and multiplication of a matrix by a scalar.

18. Each of the following set of matrices (form given below) is a vector space over C *w.r.t.* matrix addition and matrix multiplication by a scalar.

(i) $\begin{pmatrix} x & y \\ z & 0 \end{pmatrix} : x, y, z \in C$

(ii) $\begin{pmatrix} x & 0 \\ 0 & y \end{pmatrix} : x, y \in C$

(iii) $\begin{pmatrix} x & 0 \\ 0 & 0 \end{pmatrix} : x \in C$

Self Assessment Test

1. In $V_3(R)$, examine each of the following set of vectors for linear dependence :
   (i) $\{(1,2,0), (0,3,1), (-1,0,1)\}$
   (ii) $\{(-1,2,1), (3,0,-1), (-5,4,3)\}$
   (iii) $\{(1,3,2), (1,-7,-8), (2,1,-1)\}$
   (iv) $\{(1,1,-1), (2,-3,5), (-2,1,4)\}$

2. Prove that the four vectors $(1,0,0)$, $(0,1,0)$, $(0,0,1)$, $(1,1,1)$ in $V_3(C)$ form a linearly dependent set but any of them are linearly independent.

3. If α, β and γ are vectors such that $\alpha + \beta + \gamma = 0$ then show that α and β span the same subspaces as β and γ.

4. In the vector space R^3, let $\alpha = (1,2,1)$, $\beta = (3,1,5)$, $\gamma = (3,-4,7)$. Show that the subspace spanned by $S = \{\alpha, \beta\}$ and $T = \{\alpha, \beta, \gamma\}$ are the same.

5. Let f be a linear transformation from a vector space U into a vector space V. If S is a subspaces of U, show that $f(S)$ will be a subspaces of V.

6. Show that the space of all real functions is the direct sum of the subspaces of odd functions and even functions.

7. Show that the vectors S and α of the vector space R over Q are linearly independent iff α is an irrational number but that the same is not true in the real vector space of R.

8. Show that $\{a+ib, c+id\}$ forms a basis of the vector space of complex numbers over the field of real numbers.

9. Let $W = (1,2,3)$ be a vector in Euclidean space R^3. Find an orthonormal basis of $W^\perp$.

10. Let P be orthogonal, prove that $\|Pu\| = \|u\|$ for every $u \in V$.

11. If A is orthogonally equivalent to B. Show that B is orthogonally equivalent to A.

12. Find an orthonormal basis for the subspaces U of R^4 spanned by $V_1 = (1,1,1,1)$, $V_2 = (1,2,4,5), V_3 = (1,-3,-4,-2)$.

13. Let V be a vector space of polynomials $f(t)$ with inner product $(f, g) = \int\limits_{-1}^{1} f(t)g(t)d(t)$. Apply the Gram-schmidt algorithm to the set $\{1, t, t^2, t^3\}$ to obtain an orthonormal set $\{f_0, f_1, f_2, f_3\}$.

14. Find the matrix A which represents the given inner product on R^2 with respect to usual basis $\{(1,0), (0,1)\}$ of R^2.

15. Find the matrix relative to the basis $\{1+i, 1+2i\}$.

16. Let $F : R^2 \to R^2$ be defined by $F(1,0) = (2,4)$ and $F(0,1) = (5,8)$.
    Find the matrix A representing F with respect to the usual basis for R^2.

17. If a vector space has one basis that contains infinitely many elements, prove that every basis contains infinitely many element.

18. If $\{u,v,w\}$ is a linearly independent subset of a vector space, show that $\{u, u+v, u+v+w\}$ is also linearly independent.

19. Let $V(F)$ be a vector space and a subset $S = \{v_1, v_2,...,v_n\}$ of $V(F)$ be a linearly independent set. If $v \notin V(F)$ and $v \notin L(S)$, then show that $S_1 = \{v, v_1, v_2,...,v_n\}$ is a linearly independent set.

20. Let V be a vector space of all polynomial function of degree less than or equal to two from the field of real numbers R into itself. For a fixed $t \in R$, let
    $g_1(x) = 1, g_2(x) = x+t, g_3(x) = (x+t)^2$
    Prove that $\{g_1, g_2, g_3\}$ is a basis for V.

■ ■ ■

6 LINEAR TRANSFORMATIONS

6.1 INTRODUCTION

The concept of homomorphism is easily carried to vector spaces. To begin with linear transformation (vector space homomorphism) is a function from one vector space to another. Like a ring homomorphism, it is supposed to preserve both the vector space operations. The process of taking functional values and performing the vector space operation should be commutative. This requires that the scalar field in case of either space should be the same.

Definition. *Let U and V be two vector spaces over the same field F. A mapping $T:U \rightarrow V$ is said to be a linear transformation from U into V which associates to each element α of U to a unique element $T(\alpha)$ of V such that*

$$T(a\alpha+b\beta) = aT(\alpha) + bT(\beta)$$

for all α and β in U and all scalars a, b in F.

REMARKS

- Linear transformation is also known as vector space homomorphism.
- If a linear transformation is onto, then it is known as isomorphism.

☛ ILLUSTRATIONS

(1) If V is any vector space over F, then the identity transformation I, defined by $I(\alpha)=\alpha$, $\forall\ \alpha\in V$ is a linear transformation from V into V. Also the zero transformation 0 denoted by $0(\alpha) = 0$, is a linear transformation.

(2) Let F be a field of real numbers and let V be the vector space of all polynomials, then a mapping

$$D : V \rightarrow V$$

given by $D[f(x)] = \dfrac{d}{dx}[f(x)]$, $\forall\ f(x) \in V$ is a linear transformation.

Since for any $f(x)$ and $g(x) \in V$ and $a, b \in F$

$$D[af(x)+bg(x)] = \frac{d}{dx}[af(x)+bg(x)] = \frac{d}{dx}[af(x)] + \frac{d}{dx}[bg(x)]$$

$$= a\frac{d}{dx}[f(x)] + b\frac{d}{dx}[g(x)] = aD[f(x)] + bD[g(x)]$$

(3) Let R be the field of real numbers and let V be the vector space of all functions from R into R which are continuous.

Then a mapping $T : V \rightarrow V$ given by

$$T[f(x)] = \int_0^x f(t)\,dt \text{ is a linear transformation.}$$

For any $f(x)$, $g(x) \in V$ and $a, b \in R$

$$\therefore \quad T[af(x) + bg(x)] = \int_0^x [af(t) + \int_0^x bg(t)] dt$$

$$= \int_0^x af(t) dt + \int_0^x bg(t) dt$$

$$= a\int_0^x f(t) dt + b\int_0^x g(t) dt = aT[f(x)] + bT[g(x)]$$

(4) Let V be the vector space of all $m \times n$ matrices over a field F and let P be a fixed $m \times n$ matrix and Q be a fixed matrix of order $n \times n$.

Then a mapping $T : V \to V$ given by $T(A) = PAQ$, $\forall A \in V$ is a linear transformation. For any two matrices, $A, B \in V$ and $a, b \in F$

$$T(aA + bB) = P(aA + bB)Q = (aPA + bPB)Q$$
$$= aPAQ + bPBQ = aT(A) + bT(B).$$

6.2 SOME DEFINITIONS

(i) Linear operator : Let $V(F)$ be a vector space. Then a linear transformation from V into V is called a linear operator.

(ii) Zero transformation : Let U and V be two vector spaces over the same field F. Then the zero transformation of U into V is a mapping T defined by

$$T(\alpha) = 0 \ \forall \ \alpha \in U$$

where 0 is the zero vector of V.

(iii) Identity transformation : Let $V(F)$ be a vector space then a linear transformation $I : V \to V$ is said to be identity transformation defined by

$$I(\alpha) = \alpha \ \forall \ \alpha \in V$$

(iv) Negative of a linear transformation : Let U and V be two vector spaces over the same field F. Let T be a linear transformation of U into V. Then a linear transformation $-T$ of U into V defined by $\quad (-T)(\alpha) = -[T(\alpha)] \ \forall \ \alpha \in V$

is called the negative of a linear transformation T.

6.3 PROPERTIES OF LINEAR TRANSFORMATIONS

THEOREM 1. *Let $U(F)$ and $V(F)$ be two vector spaces and T be a linear transformation of U into V. Then*

(i) *$T(0) = 0$, where 0 on LHS is the zero vector of U and 0 on RHS is the zero vector of V.*

(ii) *$T(-\alpha) = -T(\alpha) \ \forall \ \alpha \in U$*

(iii) *$T(\alpha - \beta) = T(\alpha) - T(\beta) \ \forall \ \alpha, \beta \in U$*

(iv) *$T(\alpha_1 a_1 + \alpha_2 a_2 + \dots + \alpha_n a_n) = a_1 T(\alpha_1) + a_2 T(\alpha_2) + \dots + a_n T(\alpha_n)$*

$$\forall \ \alpha_1, \alpha_2, \dots, \alpha_n \in U \text{ and } a_1, a_2, \dots, a_n \in F.$$

Proof. (i) If $\alpha \in U$, then $T(\alpha) \in V$. Since V is a vector space, then we have

$$T(\alpha) + 0 = T(\alpha)$$

$$\Rightarrow \qquad T(\alpha) + 0 = T(\alpha + 0) \qquad\qquad [\because \alpha + 0 = \alpha \text{ in } U]$$

$$\Rightarrow \qquad T(\alpha) + 0 = T(\alpha) + T(0) \qquad\qquad [\because T \text{ is linear.}]$$

$$\Rightarrow \qquad\qquad\qquad 0 = T(0)$$

[By left cancellation for addition in V]

(ii)   We have, $\forall\ \alpha \in U$

$$T[\alpha + (-\alpha)] = T(\alpha) + T(-\alpha) \qquad [\because T \text{ is linear.}]$$
$$\Rightarrow \qquad T(0) = T(\alpha) + T(-\alpha)$$
$$\Rightarrow \qquad 0 = T(\alpha) + T(-\alpha) \qquad [\because T(0) = 0]$$
$$\Rightarrow \qquad T(-\alpha) = -T(\alpha)$$

(iii)   We have, $\forall\ \alpha,\ \beta \in U$

$$T[\alpha + (-\beta)] = T(\alpha) + T(-\beta) \qquad [\because T \text{ is linear.}]$$
$$\Rightarrow \qquad T(\alpha-\beta) = T(\alpha) - T(\beta) \qquad [\text{Using part (ii)}]$$

(iv)   Since $\alpha_1 a_1 + \alpha_2 a_2 + \ldots + \alpha_n a_n$ is a linear combination of vectors of U. Now we shall prove the result by induction on n.

For $n = 1$, $\qquad T(\alpha_1 a_1) = \alpha_1 T(a_1) \qquad [\because T \text{ is linear.}]$

For $n = 2$, $\qquad T(\alpha_1 a_1 + \alpha_2 a_2) = a_1 T(\alpha_1) + a_2 T(\alpha_2) \qquad [\because T \text{ is linear.}]$

Suppose the result is true for $n - 1$ values, *i.e.*

$$T(\alpha_1 a_1 + \alpha_2 a_2 + \ldots + \alpha_{n-1} a_{n-1}) = a_1 T(\alpha_1) + a_2 T(\alpha_2) + \ldots + a_{n-1} T(\alpha_{n-1}) \qquad \ldots(1)$$

Now, $T(\alpha_1 a_1 + \alpha_2 a_2 + \ldots + \alpha_{n-1} a_{n-1} + \alpha_n a_n)$

$$= T[(\alpha_1 a_1 + \alpha_2 a_2 + \ldots + \alpha_{n-1} a_{n-1})] + a_n T(\alpha_n) \qquad [\because T \text{ is linear.}]$$
$$= a_1 T(\alpha_1) + a_2 T(\alpha_2) + \ldots + a_{n-1} T(\alpha_{n-1}) + a_n T(\alpha_n) \qquad [\text{Using (1)}]$$

Hence, the result is proved by induction.

THEOREM 2. *Let U and V be two finite-dimensional vector spaces over the same field F and let $\{\alpha_1, \alpha_2, \ldots, \alpha_n\}$ be an ordered basis for U and let $\{\beta_1, \beta_2, \ldots, \beta_n\}$ be an ordered set in V. Then there is precisely one linear transformation T from U into V such that $T(\alpha_j) = \beta_j$, $j = 1, 2, 3, \ldots n$.*

Proof. Since the set $\{\alpha_1, \alpha_2, \ldots, \alpha_n\}$ is a basis of $U(F)$, then for each $\alpha \in U$, there are some scalars $a_1, a_2, \ldots, a_n$ such that

$$\alpha = a_1 \alpha_1 + a_2 \alpha_2 + \ldots + a_n \alpha_n = \sum_{i=1}^{n} a_i \alpha_i$$

For this vector α we define $T : U \to V$ given by

$$T(\alpha) = a_1 \beta_1 + a_2 \beta_2 + \ldots + a_n \beta_n = \sum_{i=1}^{n} a_i \beta_i$$

Then T is well defined for each vector α in U and a vector $T(\alpha) \in V$. From the definition it is clear that $T(\alpha_j) = \beta_j$ for each j.

Now we shall show that T is linear. For this if $\alpha = \sum_{i=1}^{n} a_i \alpha_i$ and $\beta = \sum_{i=1}^{n} b_i \alpha_i$ are any two vectors in U, then for all $a, b \in F$, we have

$$T(a\alpha + b\beta) = \left[a \sum_{i=1}^{n} a_i \alpha_i + b \sum_{i=1}^{n} b_i \alpha_i \right] = T\left(\sum_{i=1}^{n} a a_i \alpha_i + \sum_{i=1}^{n} b b_i \alpha_i \right)$$

$$= T\left(\sum_{i=1}^{n} a a_i + \sum_{i=1}^{n} b b_i \right) \alpha_i = \sum_{i=1}^{n} (a a_i + b_i b_i) \beta_i$$

$$= a \sum_{i=1}^{n} a_i \beta_i + b \sum_{i=1}^{n} b_i \beta_i$$

$$= aT \left(\sum_{i=1}^{n} a_i \alpha_i \right) + bT \left(\sum_{i=1}^{n} b_i \alpha_i \right) = aT(\alpha) + bT(\beta)$$

Now, we shall show the uniqueness of T.

Let if possible, T_1 be another linear transformation from U into V such that $T_1(\alpha_j) = \beta_j, j = 1, 2, ..., n$.

Then for any vector $\alpha = \sum_{i=1}^{n} a_i \alpha_i$, we have

$$T_1(\alpha) = T_1 \left(\sum_{i=1}^{n} a_i \alpha_i \right) = \sum_{i=1}^{n} a_i T_1(\alpha_i) \qquad [\because T_1 \text{ is linear.}]$$

$$= \sum_{i=1}^{n} a_i \beta_i \qquad [\because T_1(\alpha_i) = \beta_i]$$

$$= T \left(\sum_{i=1}^{n} a_i \alpha_i \right) = T(\alpha)$$

$$\Rightarrow \qquad T_1 = T \qquad [\because \alpha \text{ is an arbitrary vector.}]$$

Hence T is unique.

6.4 ALGEBRA OF LINEAR TRANSFORMATIONS

THEOREM 1. *Let U and V be two vector spaces over the field F. Let T_1 and T_2 be two linear transformations from U into V then the function $(T_1 + T_2)$ defined by*

$$(T_1 + T_2)(\alpha) = T_1(\alpha) + T_2(\alpha), \ \forall \alpha \in U$$

is a linear transformation from U into V. If c is any element of F, then the function (cT) defined by

$$(cT)(\alpha) = cT(\alpha)$$

is a linear transformation from U into V.

The set of all transformations $L(U, V)$ from U into V, together with the addition and scalar multiplication defined above, is a vector space over the field F.

Proof. For $\alpha, \beta \in U$ and $a, b \in F$, we have

$$(T_1 + T_2) (a\alpha + b\beta) = T_1(a\alpha + b\beta) + T_2(a\alpha + b\beta) \qquad [\text{By definition}]$$

$$= [aT_1(\alpha) + bT_1(\beta)] + [aT_2(\alpha) + bT_2(\beta)]$$

$$[\because T_1 \text{ and } T_2 \text{ are linear transformations.}]$$

$$= [aT_1(\alpha) + aT_2(\alpha)] + [bT_1(\beta) + bT_2(\beta)]$$

$$= a(T_1 + T_2)(\alpha) + b (T_1 + T_2)(\beta)$$

$\therefore \ T_1 + T_2$ is a linear transformation.

Again, T is linear transformation and c is any scalar, then for $\alpha, \beta \in U$ and $a, b \in F$, We have

$$(cT) (a\alpha + b\beta) = c[T(a\alpha + b\beta)] \qquad [\text{By definition}]$$

$$= c[aT(\alpha) + bT(\beta)] \qquad [\because T \text{ is linear transformation.}]$$
$$= c[aT(\alpha)] + c[bT(\beta)] = (ca)T(\alpha) + (cb)T(\beta)$$
$$= (ac)T(\alpha) + (bc)T(\beta)]$$
$$\qquad\qquad\qquad [\because \text{ Multiplication is commutative in } F.]$$
$$= a(cT)(\alpha) + b(cT)(\beta)$$

$\therefore$ cT is a linear transformation.

Now we shall show that the set of all linear transformations $L(U, V)$ from U into V forms a vector space with respect to the above defined compositions. First we show that $\{L(U, V), +\}$ is an abelian group :

(i) Closure Property.

If $T_1, T_2 \in L(U, V)$, then we have already proved that $T_1 + T_2$ is linear transformation, so that $T_1, T_2 \in L(U,V)$.

(ii) Associative Property.

For all $T_1, T_2, T_3 \in L(U,V)$ and for all $\alpha \in U$, we have
$$[(T_1 + T_2) + T_3](\alpha) = (T_1 + T_2)(\alpha) + T_3(\alpha)$$
$$= [T_1(\alpha) + T_2(\alpha)] + T_3(\alpha)$$
$$= T_1(\alpha) + [T_2(\alpha) + T_3(\alpha)]$$
$$\qquad\qquad\qquad [\because \text{ Addition is associative in } V.]$$
$$= T_1(\alpha) + (T_2 + T_3)(\alpha) = [T_1 + (T_2 + T_3)](\alpha)$$
$$\therefore \qquad (T_1 + T_2) + T_3 = T_1 + (T_2 + T_3)$$

(iii) Commutative Property.

For all $T_1, T_2, \in L(U,V)$ and $\alpha \in U$, we have
$$(T_1 + T_2)(\alpha) = T_1(\alpha) + T_2(\alpha)$$
$$= T_2(\alpha) + T_1(\alpha) \qquad [\because \text{ Addition is commutative in } V.]$$
$$= (T_2 + T_1)(\alpha)$$
$$\therefore \qquad T_1 + T_2 = T_2 + T_1$$

(iv) Existence of Identity .

The zero transformation, denoted by 0 and defined by $0(\alpha) = 0$, $\forall \alpha \in U$ is a linear transformation.

Also if $T \in L(U, V)$, then
$$(T + 0) = 0 + T = T, \text{ for all } T.$$
$\therefore$ $0 \in L(U, V)$ and is identity transformation.

(v) Existence of Inverse.

For each $T \in L(U, V)$, there exists $(-T) \in L(U,V)$, defined by
$$(-T)(\alpha) = -T(\alpha), \forall \alpha \in U, (-T) \text{ is linear and } T + (-T) = (-T) + T = 0.$$
$\therefore$ $(-T)$ is the additive inverse of T.

(vi) Distributive Property.

For all $T_1, T_2 \in L(U, V)$, $\alpha \in U$ and $a \in F$,
$$a[(T_1 + T_2)](\alpha) = a(T_1 + T_2)(\alpha) = a[T_1(\alpha) + T_2(\alpha)]$$
$$= aT_1(\alpha) + aT_2(\alpha) = (aT_1 + aT_2)(\alpha)$$

$$\therefore \qquad a(T_1 + T_2) = aT_1 + aT_2$$

Also, for all $T \in L(U, V)$ and $\alpha \in U$, $a, b \in F$,

$$[(a+b)T](\alpha) = (a+b)T(\alpha) = aT(\alpha) + bT(\alpha) = (aT + bT)(\alpha)$$

$$\Rightarrow \qquad (a+b)T = aT + bT$$

(vii) For all $T \in L(U, V)$, $\alpha \in U$ and $a, b \in F$,

$$[(ab)T](\alpha) = (ab)T(\alpha) = a[bT(\alpha)] = a(bT)(\alpha)$$

$$(ab).T = a.(b\,T)$$

(viii) If 1 is the unity in F, then for all $T \in L(U, V)$ and $\alpha \in U$

$$(1.T)(\alpha) = 1.T(\alpha) = T(\alpha)$$

$\therefore$ Hence $L(U, V)$ is a vector space.

REMARK

- The vector space $L(U, V)$ is also denoted by Hom. (U, V), *i.e.* (the set of all homomorphism from U into V).

THEOREM 2. *Let U be an m-dimensional vector space over the field F, and let V be an n-dimensional vector space over F. Then the vector space L(U, V) is finite dimensional and has dimension mn.*

Proof. Since U and V both are finite dimensional vector spaces of dimensions m and n respectively, therefore, let

$$\beta = \{\alpha_1, \alpha_2, \dots \alpha_m\} \quad \text{and} \quad \beta' = \{\beta_1, \beta_2, \dots \beta_n\}$$

be the ordered basis of U and V respectively.

For each pair of integers (i, j) with $1 \le i \le m$ and $1 \le j \le n$, we define a linear transformation T_{ij} from U into V by

$$T_{ij}(\alpha_k) = \begin{cases} 0; & \text{if } k \ne j \\ \beta_i; & \text{if } k = j \end{cases}$$

The existence and uniqueness of above linear transformations follows from preceding theorem. It is obvious that there are mn linear transformations of the type T_{ij}, so we claim that these mn transformations form a basis of $L(U, V)$.

(i) For mn scalars a_{ij}, we have

$$\sum_{i=1}^{n} \sum_{j=1}^{m} a_{ij} T_{ij} = \mathbf{0} \quad \text{[Zero transformation]}$$

$$\Rightarrow \qquad \sum_{i=1}^{n} \sum_{j=1}^{m} a_{ij} T_{ij}(\alpha_k) = \mathbf{0}(\alpha_k), \forall \alpha_k \in U, 1 \le k \le n$$

$$\Rightarrow \qquad \sum_{i=1}^{n} \sum_{j=1}^{m} a_{ij} T_{ij}(\alpha_k) = \mathbf{0}$$

$$\Rightarrow \qquad \sum_{i=1}^{n} \left(\sum_{j=1}^{m} a_{ij} T_{ij}(\alpha_k) \right) = \mathbf{0}$$

$$\Rightarrow \sum_{i=1}^{n} [a_{i1} T_{i1}(\alpha_k) + a_{i2} T_{i2}(\alpha_k) + \dots + a_{im} T_{im}(\alpha_k)] = \mathbf{0}$$

$$\Rightarrow \sum_{i=1}^{n} a_{i1}T_{i1}(\alpha_k) + \sum_{i=1}^{n} a_{i2}T_{i2}(\alpha_k) + ... + \sum_{i=1}^{n} a_{im}T_{im}(\alpha_k) = 0$$

$$\Rightarrow \quad a_{11}T_{11}(\alpha_k) + a_{21}T_{21}(\alpha_k) + ... + a_{n1}T_{n1}(\alpha_k)$$
$$+ a_{12}T_{12}(\alpha_k) + a_{22}T_{22}(\alpha_k) + ... + a_{n2}T_{n2}(\alpha_k)$$
$$+ \text{..}$$
$$+ a_{1m}T_{1m}(\alpha_k) + a_{2m}T_{2m}(\alpha_k) + ... + a_{nm}T_{nm}(\alpha_k) = 0$$
$$= a_{11}\beta_1 + a_{21}\beta_2 + ... + a_{n1}\beta_n + a_{12}\beta_1 + a_{22}\beta_2 + ... + a_{n2}\beta_n$$
$$\text{..}$$
$$+ \alpha_{1m}\beta_1 + \alpha_{2m}\beta_2 + ... + a_{nm}\beta_n = 0$$

$$\left(\begin{array}{l} \because T_{ij}(\alpha_k) = 0, \, (j \neq k) \\ T_{ij}(\alpha_k) = \beta_i, \, (j = k) \end{array} \right)$$

Since $\beta' = \{\beta_1, \beta_2, ..., \beta_n\}$ is a basis of V, therefore it is linearly independent so that

$$a_{11} = 0 = a_{21} = ... = a_{n1}$$
$$a_{12} = 0 = a_{22} = ... = a_{n2}$$
$$\text{...}$$
$$\text{...}$$
$$a_{1m} = 0 = a_{2m} = ... = a_{nm}$$

Thus, $\{T_{ij} : 1 \leq i \leq m, \, i \leq j \leq n\}$ is linearly independent.

(ii) Now we show that $\{T_{ij} : 1 \leq i \leq m, \, i \leq j \leq n\}$ spans $L(U, V)$. For this, let T be an arbitrary linear transformation from U into V, i.e. $T \in L(U, V)$.

For $\alpha_j \in U$, $T(\alpha_j) \in V$ and $\beta' = \{\beta_1, \beta_2, ..., \beta_n\}$ is a basis of V so that
$$T(\alpha_j) = a_{1j}\beta_1 + a_{2j}\beta_2 + ... + a_{nj}\beta_n; \, 1 < j < n$$
where $a_{1j}, a_{2j}, ..., a_{nj}$ are the coordinates of vector $T(\alpha_j)$ in β'.

$$T(\alpha_j) = \sum_{i=1}^{n} a_{ij}\beta_i = \sum_{i=1}^{n}\sum_{j=1}^{m} a_{ij}T_{ij}(\alpha_j)$$

$$\Rightarrow \quad T = \sum_{i=1}^{n}\sum_{j=1}^{m} a_{ij}T_{ij}$$

$\Rightarrow$ $\{T_{ij} : 1 \leq i \leq m, \, 1 \leq j \leq n\}$ generates $L(U, V)$.

$\Rightarrow$ $\{T_{ij} : 1 \leq i \leq m, \, 1 \leq j \leq n\}$ is a basis of $L(U, V)$.

Hence $L(U, V)$ is finite dimensional and dim. $L(U, V) = mn$.

THEOREM 3. *Let U, V and W be vector spaces aver the field F. Let T_1 be a linear transformation from U into V and T_2 be a linear transformations from V into W, then the composed function T_2T_1 is defined by*

$$(T_2T_1)(\alpha) = T_2[T_1(\alpha)], \text{ for all } \alpha \in U$$

is a linear transformation from U into W.

Proof. For $\alpha, \beta \in U$ and $a, b \in F$, we have
$$(T_2T_1(a\alpha + b\beta)) = T_2[T_1(a\alpha + b\beta)] = T_2[aT_1(\alpha) + bT_1(\beta))] \, [\because T_1 \text{ is linear.}]$$
$$= a(T_2T_1)(\alpha) + b(T_2T_1)(\beta) \qquad [\because T_2 \text{ is linear.}]$$
$\therefore$ T_2T_1 is a linear transformation from U into W.

6.5 LINEAR OPERATOR

Definition. *If V is a vector space over the field F, then a linear transformation from V into V is called a linear operator.*

In case of above theorem if U, V and W are replaced by V, then T_1 and T_2 are linear operators on the space V and $T_2 T_1$ is also a linear operator on V. Thus the vector space $L(V, V)$ has a 'multiplication' defined on it by composition. In this case the operator $T_1 T_2$ is also defined but in general $T_2 T_1 \neq T_1 T_2$. Therefore, if T is a linear operator on V, then we can compose T with T as follows :

$$T^2 = T\,T$$
$$T^3 = T\,T\,T$$

in general, $\qquad T^n = T\,T...T$ (n times) for $n = 1, 2, 3, ...$

REMARK

- If $T \neq 0$, then we define $T^0 = 1$ (Identity transformation).

6.6 ALGEBRA OF LINEAR OPERATORS

THEOREM 1. *Let V be a vector space over the field F and let T, T_1, , T_2 and T_3 be linear operators on V and let c be an element in F, then*

    (*i*) *$IT = TI = T$, I being an identity operator.*

    (*ii*) *$T_1 (T_2 + T_3) = T_1 T_2 + T_1 T_3$; $(T_2 + T_3) T_1 = T_2 T_1 + T_3 T_1$*

    (*iii*) *$T_1 (T_2 T_3) = (T_1 T_2) T_3$*

    (*iv*) *$c(T_1 T_2) = (cT_1) T_2 = T_1 (cT_2)$*

    (*v*) *$T0 = 0T = 0$, 0 being zero linear operator.*

Proof.   (i) For $\alpha \in V$

$$(IT)(\alpha) = I[T(\alpha)] = T(\alpha) \qquad\qquad [\because I(\alpha) = \alpha]$$
$$IT = T$$

Also, $\qquad (TI)(\alpha) = T[I(\alpha)] = T(\alpha)$

$\Rightarrow \qquad\qquad\quad TI = T$

Thus, $\qquad\quad IT = TI = T$

(ii) For any $\alpha \in V$

$$[\,T_1(T_2 + T_3)\,](\alpha) = T_1[(T_2 + T_3)(\alpha)] = T_1\,[T_2\,(\alpha) + T_3\,(\alpha)]$$
$$= (T_1 T_2)(\alpha) + (T_1 T_3)(\alpha) = (T_1 T_2 + T_1 T_3)(\alpha)$$

$\therefore \qquad T_1(T_2 + T_3) = T_1 T_2 + T_1 T_3$

Similarly,

$$(T_2 + T_3)T_1 = T_2 T_1 + T_3 T_1$$

(iii) For any $\alpha \in V$

$$[T_1(T_2 T_3)](\alpha) = T_1[(T_2 T_3)(\alpha)] = T_1[T_2(T_3(\alpha))]$$
$$= (T_1 T_2)[T_3(\alpha)] = [(T_1 T_2)T_3](\alpha)$$
$$T_1(T_2 T_3) = (T_1 T_2)T_3$$

(iv) For any $\alpha \in V$, $c \in F$

$$[c(T_1T_2)](\alpha) = c[(T_1T_2)(\alpha)] = c\,[T_1(T_2(\alpha))]$$
$$= (cT_1)[T_2(\alpha)] = [(cT_1)T_2](\alpha)$$

$\therefore \qquad c(T_1T_2) = (cT_1)T_2$

Also, $[c(T_1T_2)](\alpha) = (cT_1)[T_2(\alpha)] = T_1(cT_2(\alpha)) = T_1[(cT_2)](\alpha)$

$\therefore \qquad c(T_1T_2) = T_1(cT_2)$

Thus, $\qquad c(T_1T_2) = (cT_1)T_2 = T_1(cT_2)$

(v) For any $\alpha \in V$,

$$(T0)(\alpha) = T[0(\alpha)] = T(0) \qquad\qquad [\because 0(\alpha) = 0]$$
$$= 0$$

Similarly, $\qquad 0T = 0$.

6.7 RANGE AND NULL SPACE OF A LINEAR TRANSFORMATION

(i) Range space of a linear transformation. If T is a linear transformation from U into V, then the range of T is a subspace of V. Let R_T be the range of T, that is, the set of all vectors β in V such that $T(\alpha) = \beta$ for some $\alpha \in U$,

i.e. $R_T = \{\beta \in V : T(\alpha) = \beta,$ for some $\alpha \in U\}$.

If U is finite dimensional, then the dimension of range of T is called rank of T and is denoted by $\rho(T)$.

(ii) Null space of a linear transformation. If T is a linear transformation from a vector space U into a vector space V, then the null space of T denoted by $N(T)$ is the set of all vectors α in U such that $T(\alpha) = 0$, where 0 is the zero vector in V, i.e.,

$N(T) = \{\alpha \in U : T(\alpha) = 0\}$.

If U is finite-dimensional, then the dimension of null space $N(T)$ is called nullity of T and is denoted by $n(T)$.

REMARK

- Kernel of T is also known as null space of T.

THEOREM 1. *Let U and V be vector spaces over the field F and let T be a linear transformation from U into V. Suppose U is finite-dimensional. Then*

$$rank\ (T) + nullity\ (T) = dim.\ U$$

i.e., $\qquad\qquad \rho(T) + n(T) = dim.\ U$

Proof. Let $\{\alpha_1, \alpha_2, ..., \alpha_k\}$ be the basis of N_T, the null space of T. Let the dimension of U be n, so that $\alpha_{k+1}, \alpha_{k+2}, ..., \alpha_n \in U$ such that $\{\alpha_1, \alpha_2, ..., \alpha_n\}$ forms a basis of U. Therefore, dim. N_T, $= k$ and dim. $U = n$.

We claim that $[T(\alpha_{k+1}), T(\alpha_{k+2}),..., T(\alpha_n)]$ is a basis of range of T.

For scalars $a_i \in F$ we have

$$a_{k+1}T(\alpha_{k+1}) + a_{k+2}\,T(\alpha_{k+2}) + ... + a_n\,T(\alpha_n) = 0$$

$$\Rightarrow \quad \sum_{i=k+1}^{n} a_i T(\alpha_i) = 0 \qquad \Rightarrow \quad T\left(\sum_{i=k+1}^{n} a_i\alpha_i\right) = 0$$

$$\Rightarrow \quad \sum_{i=k+1}^{n} a_i\alpha_i \in N_T$$

Since $\{\alpha_1, \alpha_1, ..., \alpha_k\}$ is the basis of N_T, so that for some scalars $b_1, b_2,,..., b_k$, we have

$$\sum_{i=k+1}^{n} a_i\alpha_i = b_1\alpha_1 + b_2\alpha_2 + ... + b_k\alpha_k$$

$$\Rightarrow \quad b_1\alpha_1 + b_2\alpha_2 + ... + b_k\alpha_k - \sum_{i=k+1}^{n} a_i\alpha_i = 0$$

$$\Rightarrow \quad b_1\alpha_1 + b_2\alpha_2 + ... + b_k\alpha_k + (-a_{k+1})\alpha_{k+1} + ... + (-a_n)\alpha_n = 0$$

Since $\alpha_1, \alpha_2, ..., \alpha_n$ are linearly independent, we must have

$$b_1 = 0 = b_2 = ... = b_k = a_{k+1} = a_n.$$

$\therefore$ $[T(\alpha_{k+1}), T(\alpha_{k+2}), ... , T(\alpha_n)]$ is linearly independent.

Now, we shall show that $T(\alpha_{k+1}), T(\alpha_{k+2}), ... , T(\alpha_n)]$ spans range of T.

For this, let $T(\alpha) \in R_T$ (range of T) for some $\alpha \in U$.

Since $\{\alpha_1, \alpha_2, ..., \alpha_n\}$ spans U so that

$$\alpha = a_1\alpha_1 + a_2\alpha_2 + ... + a_n\alpha_n$$

For $a_i' s \in F$, we have

$$\begin{aligned} T(\alpha) &= T(a_1\alpha_1 + a_2\alpha_2 + ... + a_n\alpha_n) \\ &= T(a_1\alpha_1) + T(a_2\alpha_2) + ... + T(a_n\alpha_n) \quad [\because T \text{ is linear.}] \\ &= a_1T(\alpha_1) + a_2T(\alpha_2) + ... + a_kT(\alpha_k) + a_{k+1}T(\alpha_{k+1}) + ... + a_nT(\alpha_n) \\ &= a_{k+1}T(\alpha_{k+1}) + a_{k+2}T(\alpha_{k+2}) + ... + a_nT(\alpha_n) \\ &\qquad\qquad\qquad\qquad\qquad\qquad\qquad [\because T(\alpha_i) = 0, 1 \leq i \leq k] \end{aligned}$$

Thus, $T(\alpha_{k+1})...T(\alpha_n)$ spans R_T.

Hence $[T(\alpha_{k+1}),..., T(\alpha_n)]$ is a basis of R_T.

Accordingly, $\dim. R_T = n - k = \dim .U - \dim. N_T$

$\therefore$ $\dim.R_T + \dim.N_T = \dim. U$

Hence rank(T) + nullity(T) = dim. U

6.8 PRODUCT OF LINEAR TRANSFORMATIONS

THEOREM 1. *Let $U(F)$, $V(F)$ and $W(F)$ be the vector spaces. Let T be a linear transformation from U into V and S a linear transformation from V into W. Then the composite function ST, called the product of linear transformations, defined by*

$$(ST)(\alpha) = S[T(a)] \,\forall\, \alpha \in U$$

is a linear transformation from U into W.

Proof. Since $T : U \to V$ is a linear transformation, hence $T(\alpha) \in V$ for $\alpha \in U$.

Also, $S : V \to W$ is a linear transformation, hence $S(\beta) \in W$ for $\beta \in V$.

$\therefore$ For $\beta = T(\alpha) \in V$, $S(T(\alpha)) \in W$

Thus, $(ST)(\alpha) \in W$, therefore, ST is a function from U into W. Now we shall show that ST is a linear transformation from U into W. Let $\alpha_1, \alpha_2 \in U$ and $a, b \in F$, then

$$\begin{aligned} (ST)(a\alpha_1 + b\alpha_2) &= S[T(a\alpha_1 + b\alpha_2)] \\ &= S[aT(\alpha_1) + bT(\alpha_2)] \quad [\because T \text{ is linear.}] \\ &= aS(T(\alpha_1)) + bS(T(\alpha_2)) \quad [\because S \text{ is linear.}] \\ &= a(ST)(\alpha_1) + b(ST)(\alpha_2) \end{aligned}$$

Hence ST is a linear transformation.

REMARK

- In above theorem, if U, V and W are replaced by V, then T and S are linear operators on the vector space V and ST is also a linear operator on V. Also TS exists and is a linear operator on V. However, in general $TS \neq ST$.

Solved Examples

Example 1. *Let T_1 and T_2 be linear operators on R^2 defined as follows:*

$$T_1(x_1, x_2) = (x_2, x_1)$$

and $\qquad T_2(x_1, x_2) = (x_1, 0)$

show that $\qquad T_1T_2 \neq T_2T_1.$

Solution. Let $\alpha = (x_1, x_2) \in R^2$. Then

$$(T_1T_2)(\alpha) = T_1(T_2(\alpha))$$
$$= T_1[T_2(x_1, x_2)]$$
$$= T_1(x_1, 0)$$
$$= (0, x_1)$$

and $\qquad (T_2T_1)(\alpha) = T_2(T_1(\alpha))$
$$= T_2[T_1(x_1, x_2)]$$
$$= T_2(x_2, x_1)$$
$$= (x_2, 0)$$

Clearly, $\qquad (T_1T_2)(\alpha) \neq (T_2T_1)(\alpha) \ \forall \ \alpha \in R^2$

Hence, $\qquad T_1T_2 \neq T_2T_1.$

Example 2. *Let $V(R)$ be the vector space of all polynomial functions in x with coefficients in the field R of real numbers. Let D and T be two linear operators on V defined by*

$$D(f(x)) = \frac{d}{dx} f(x)$$

and $\qquad T(f(x)) = \int_0^x f(x)\,dx$

for every $f(x) \in V$. Then show that $DT = I$ (Identity opearator) and $TD \neq I$.

Solution. Let $f(x) = a_0 + a_1 x + a_2 x^2 + \ldots \in V$, where $a_1, a_2, \ldots, \in R$. Then

$$(DT)(x) = D[T(f(x))]$$

$$= \left[\int_0^x f(x)\,dx \right] = D\left[\int_0^x (a_0 + a_1 x + a_2 x^2 + \ldots)\,dx \right]$$

$$= D\left[a_0 x + \frac{a_1 x^2}{2} + \frac{a_2 x^3}{3} + \ldots \right]$$

$$= \frac{d}{dx}\left[a_0 x + \frac{a_1 x^2}{2} + \frac{a_2 x^3}{3} + \ldots \right]$$

$$= a_0 + a_1 x + a_2 x^2 + \ldots = f(x) = I(f(x)).$$

$\therefore \qquad (DT)(f(x)) = I(f(x)) \ \forall \ f(x) \in V.$

Thus, $\qquad DT = I$

Now, $(TD)(f(x)) = T[D(f(x))]$

$$= T\left[\frac{d}{dx}f(x)\right] = T\left[\frac{d}{dx}(a_0 + a_1 x + a_2 x^2 + \ldots)\right]$$

$$= T[a_1 + 2a_2 x + \ldots] = \int_0^x (a_1 + 2a_2 x + \ldots)dx$$

$$= a_1 x + a_2 x^2 + \ldots \neq f(x) = I(f(x))$$

$\therefore \qquad (TD)(f(x)) \neq I(f(x)) \ \forall \ f(x) \in V.$

Thus, $\qquad (TD) \neq I$

Hence in general $DT \neq TD$.

Example 3. *Let V(R) be the vector space of all polynomials in x with coefficients in the field R. Let D and let T be two linear transformations on V defined by*

$$D(f(x)) = \frac{d}{dx}f(x) \ \forall \ f(x) \in V$$

and $\qquad T(f(x)) = xf(x) \ \forall \ f(x) \in V$

then show that DT ≠ TD. Also, show that DT − TD = I.

Solution. Let $f(x) \in V$. Then

$$(DT)\ (f(x)) = D[T(f(x))] = D[xf(x)]$$

$$= \frac{d}{dx}[xf(x)] = f(x) + x\frac{d}{dx}f(x) \qquad \ldots(1)$$

Also, $\qquad (TD)(f(x)) = T(D(f(x)))$

$$= T\left(\frac{d}{dx}f(x)\right) = x.\frac{d}{dx}f(x) \qquad \ldots(2)$$

Therefore, from (1) and (2), we can say that there exists $f(x) \in V$ such that

$$(DT)(f(x)) \neq (TD)(f(x))$$

Hence, $\qquad DT \neq TD.$

Also, $(DT)(f(x)) - (TD)(f(x)) = f(x) = I(f(x))$

$\therefore \qquad DT - TD = I$

6.9 POLYNOMIALS IN A LINEAR OPERATOR

Let T be a linear operator on a vector space V(F). Then TT is a linear operator on V. Since the product of linear operators is an associative operation, therefore, if n is a positive integer, then we define

$$T^2 = TT$$

$$T^3 = TTT$$

in general, $\qquad T^n = TT\ldots T \ (n \text{ times})$

Clearly, T^n is also a linear operators on V. Also, $T^0 = I$, which is defined as identity operator.

If m and n are non-negative integers, then we see that

$$T^m T^n = T^{m+n}$$

and $\qquad (T^m)^n = T^{mn}$

The set $L(V, V)$ of linear operators on V is a vector space over the field. If $a_0, a_1, a_2, ..., a_n \in F$, then

$$p(T) = a_0 I + a_1 T + a_2 T_2 + ... + a_n T^n \in L(V, V).$$

Thus, $p(T)$ is also a linear operator on V, we call it as a polynomial in linear operator T.

6.10 INVERTIBLE LINEAR TRANSFORMATION

A linear transformation T from a vector space $U(F)$ into $V(F)$ is called invertible or regular if there exists a unique linear transformation T^{-1} (called the inverse of T) from $V(F)$ into $U(F)$ such that (T^{-1}) is the identity linear transformation on U and (TT^{-1}) is the identity transformation on V.

Furthermore, T is invertible iff

        (i) T is one-to-one.

        (ii) T is onto, i.e., $R(T) = V$

THEOREM 1. *Let U and V be vector spaces over the same field F and let T be a linear transformation from U into V. If T is invertible, then T^{-1} is a linear transformation from V into U.*

Proof. Since T is invertible, hence for each $\beta \in V$, there is a unique $\alpha \in U$ such that

$$T(\alpha) = \beta \Leftrightarrow T^{-1}(\beta) = \alpha$$

Now, we shall show that T^{-1} is linear.

For $\alpha_1, \alpha_2 \in U$ and $a, b \in F$

$$T(a\alpha_1 + b\alpha_2) = aT(\alpha_1) + bT(\alpha_2) \qquad [\because T \text{ is linear.}]$$

But for β_1 and β_2 in V, there are unique $\alpha_1, \alpha_2 \in U$ respectively, such that

$$T(\alpha_1) = \beta_1 \Leftrightarrow T^{-1}(\beta_1) = \alpha_1$$

and $\qquad\qquad T(\alpha_2) = \beta_2 \Leftrightarrow T^{-1}(\beta_2) = \alpha_2$

Thus, we have

$$T(a\alpha_1 + b\alpha_2) = a\beta_1 + b\beta_2$$

$\Rightarrow \qquad\qquad a\alpha_1 + b\alpha_2 = T^{-1}(a\beta_1 + b\beta_2) \quad [\because a\alpha_1 + b\alpha_2 \text{ is unique in } V.]$

$\Rightarrow \quad aT^{-1}(\beta_1) + b\,T^{-1}(\beta_2) = T^{-1}(a\beta_1 + b\beta_2)$

Hence T^{-1} is a linear transformation.

THEOREM 2. *Let T_1 be an invertible linear transformation from $U(F)$ into $V(F)$ and T_2 an invertible linear transformation from $V(F)$ into $W(F)$. Then $T_1 T_2$ is invertible and $(T_2 T_1)^{-1} = T_1^{-1} T_2^{-1}$.*

Proof. To show $T_2 T_1$ is invertible, we shall show that it is one-one and onto.

If $\alpha_1, \alpha_2 \in U$ such that $(T_2 T_1)(\alpha_1) = (T_2 T_1)(\alpha_2)$, then

$$(T_2 T_1)(\alpha_1) = (T_2 T_1)(\alpha_2) \Rightarrow T_2[T_1(\alpha_1)] = T_2[T_1(\alpha_2)]$$

$$\Rightarrow T_1(\alpha_1) = T_1(\alpha_2) \qquad\qquad [\because T_2 \text{ is one-one.}]$$

$$\Rightarrow \alpha_1 = \alpha_2 \qquad\qquad\qquad [\because T_1 \text{ is one-one.}]$$

Thus, $T_2 T_1$ is one-one.

Also, T_1 and T_2 being onto, then for each $\beta \in V$, there exists a unique $\alpha \in U$ such that

$$T_1(\alpha) = \beta$$

and for each $\gamma \in W$, there exists a unique $\beta \in V$ such that $T_2(\beta) = \gamma$.

Thus, $\gamma \in W \Rightarrow$ there exists $\beta \in V : \gamma = T_2(\beta)$.

$\Rightarrow$ there exists $\alpha \in U : \gamma = T_2 (T_1 (\alpha))$ $[\because T_1 (\alpha) = \beta]$

$\Rightarrow$ there exists $\alpha \in U : \gamma = (T_2 T_1)(\alpha)$.

Therefore $(T_2 T_1)$ is onto. Hence $(T_2 T_1)$ is invertible.

Also, $(T_2 T_1)(T_1^{-1} T_2^{-1}) = T_2(T_1 T_1^{-1})T_2^{-1} = (T_2 I)T_2^{-1} = T_2 T_2^{-1} = I$

Similarly $(T_1^{-1} T_2^{-1})(T_2 T_1) = T_1^{-1}(T_2^{-1} T_2)T_1 = T_2^{-1}(I T_1) = T_1^{-1} T_1 = I$

Hence $(T_2 T_1^{-1}) = T_1^{-1} T_2^{-1}$

6.11 NON-SINGULAR LINEAR TRANSFORMATIONS

Let U and V be vector spaces over the field F. Then a linear transformation T from U into V is called non-singular if the null space of T is (0).

Thus, if T is non-singular, then

$$T(\alpha) = 0 \qquad \Rightarrow \alpha = 0$$

Also, when T is non-singular and $\alpha, \beta \in U$

$$T(\alpha) = T(\beta) \qquad \Rightarrow T(\alpha) - T(\beta) = 0$$

$\Rightarrow$ $T(\alpha-\beta) = 0$ $[\because T$ is linear.$]$

$\Rightarrow$ $\alpha-\beta = 0$ $[\because T$ is non-singular.$]$

$\Rightarrow$ $\alpha = \beta$

Hence T is non-singular, implies that T is one-one.

THEOREM 1. *Let T be a linear transformation from $U(F)$ into $V(F)$. Then T is non-singular if and only if T carries each linearly independent subset of U onto a linearly independent subset of V.*

Proof. Let us first suppose that T is non-singular. Now, let

$$S = \{\alpha_1, \alpha_2,,...,\alpha_k)$$

be an arbitrary linearly independent subset of U. Then we have to show that the set

$$S_1 = \{T(\alpha_1), T(\alpha_2),...,T(\alpha_k)\}$$

is linearly independent subset of V.

For scalars $a_1, a_2,... a_k \in F$ we have

$$a_1 T(\alpha_1) + a_2 T(\alpha_2) + ... + a_k T(\alpha_k)\} = 0$$

$\Rightarrow$ $T(a_1\alpha_1) + T(a_2\alpha_2) + ... + T(a_k\alpha_k)\} = 0$ $[\because T$ is linear.$]$

$\Rightarrow$ $T(a_1\alpha_1 + a_2\alpha_2 + ... + a_k\alpha_k) = 0$ $[\because T$ is linear.$]$

$\Rightarrow$ $a_1\alpha_1 + a_2\alpha_2 + ... + a_k\alpha_k = 0$ $[\because T$ is non-singular.$]$

$\Rightarrow$ $a_1 = a_2 = ... = a_k = 0$ $[\because S$ is linear independent.$]$

Hence S_1 is linearly independent.

Conversely, suppose that T carries each linearly independent subset of U into a linearly independent subset of V. Let α be a non-zero vector in U, then $\{\alpha\}$ is linearly independent so is $\{T(\alpha)\}$. Consequently, $T(\alpha) \neq 0$ because the set consisting of the zero vector alone is dependent. Therefore, the null space of T is the zero space and hence T is non-singular.

THEOREM 2. *Let U and V be finite dimensional vector spaces over the field F such that dim. U = dim. V. If T is a linear transformation from U into V, then the following are equivalent:*

(i) *T is invertible.*

(ii) *T is non-singular.*

(iii) *T is onto, that is, the range of T is V.*

(iv) *If $\{\alpha_1, \alpha_2,..., \alpha_n\}$ is a basis of U, then $\{T(\alpha_1), T(\alpha_2),..., T(\alpha_2)\}$ is a basis of V.*

Proof. **(i)** $\Rightarrow$ **(ii)** : Since T is invertible, so it is one-one and onto, therefore, T is non-singular.

(ii) $\Rightarrow$ **(iii):** Let T be non-singular and let $\{\alpha_1, \alpha_2 , ..., \alpha_n\}$ be the basis of U, then the set $\{T(\alpha_1),T(\alpha_2),...,T(\alpha_n)\}$ is linearly independent subset of V, but dim. U = dim. V, therefore $\{T(\alpha_1), ..., T(\alpha_n)\}$ is a basis for V.

For any $\beta \in V, a_1, a_2, ..., a_n \in F$, we have

$$\beta = a_1 T(\alpha_1) + a_2 T(\alpha_2) + ...+ a_n T(\alpha_n)$$
$$\beta = T(a_1 \alpha_1 + a_2 \alpha_2 + ...+ a_n \alpha_n) \qquad [\because T \text{ is linear.}]$$
$$\Rightarrow \qquad \beta \in R_T$$

Thus, $\qquad V \subseteq R_T$, but $R_T \subseteq V$

$$\therefore \qquad R_T = V$$

i.e. the range of $T = V$

(iii) $\Rightarrow$ **(iv)** : Suppose range of $T = V$. Let the set $\{\alpha_1, \alpha_2 , ..., \alpha_n\}$ be a basis of U so that an arbitrary element $\alpha \in U$ is expressible as linear combination of $\alpha_1, \alpha_2 , ..., \alpha_n$.

$$\therefore \qquad \alpha = b_1 \alpha_1 + b_2 \alpha_2 + ...+ b_n \alpha_n \text{ for some scalars, } b_1, b_2, ..., b_n \in F$$
$$\Rightarrow \qquad T(\alpha) = T(b_1 \alpha_1 + b_2 \alpha_2 + ...+ b_n \alpha_n)$$
$$= b_1 T(\alpha_1) + b_2 T(\alpha_2) +, ...,+ b_n T(\alpha_n) \qquad [\because T \text{ is linear.}]$$

This shows that each element of range of T is expressible as a linear combination of $\{T(\alpha_1) + T(\alpha_2), ..., T(\alpha_n)\}$.Thus, the set

$$\{T(\alpha_1) + T(\alpha_2), ..., T(\alpha_n)\}.$$

spans R_T . Since $R_T = V$. Also, dim. U = dim. $V = n$.

Hence $\{T(\alpha_1), T(\alpha_2), ..., T(\alpha_n)\}$ forms a basis of V.

(iv) $\Rightarrow$ **(i):** Let $\{\alpha_1, \alpha_2 ,..., \alpha_n\}$ be a basis of U such that $\{T(\alpha_1), T(\alpha_2), ..., T(\alpha_n)\}$ is a basis of V.

Let α be an arbitrary element of U, then for $b_1, b_2 , ..., b_n \in F$, we have

$$\alpha = b_1 \alpha_1 + b_2 \alpha_2 + ...+ b_n \alpha_n$$

Now, $\qquad T(\alpha) = 0$

$$\Rightarrow \qquad T(b_1 \alpha_1 + b_2 \alpha_2 + ...+ b_n \alpha_n) = 0$$
$$\Rightarrow \qquad b_1 T(\alpha_1) + b_2 T(\alpha_2) + ...+ b_n T(\alpha_n) = 0 \qquad [\because T \text{ is linear.}]$$
$$\Rightarrow \qquad b_1 = b_2 = ... = b_n = 0$$
$$[\because \{T(\alpha_1), T(\alpha_2), ..., T(\alpha_n)\} \text{ is linearly independent.}]$$
$$\Rightarrow b_1 \alpha_1 + b_2 \alpha_2 + ... + b_n \alpha_n = 0 \Rightarrow \alpha = 0$$

Hence, T is non-singular and therefore T is one-one.

Also, $\{T(\alpha_1), T(\alpha_2), ..., T(\alpha_n)\}$ spans V and range of T is V. Consequently, T is one-one and hence T is invertible.

THEOREM 3. *A linear transformation T on a finite dimensional vector space is invertible iff T is non-singular.*

Proof. Let $V(F)$ be a vector space and let dim. $V = n$ and T be a linear transformation on V. If T is invertible, then it is one-one and hence T is non-singular.

Conversely, if T is non-singular, then T is one-one. Now in order to prove that T is invertible we will show that T is onto. Let $S = \{\alpha_1, \alpha_2, ..., \alpha_n\}$ be a basis of V, then we shall show that $S' = \{T(\alpha_1), T(\alpha_2), ..., T(\alpha_n)\}$ is a basis of V.

S' is linearly independent : Let $a_1, a_2, ..., a_n \in F$ and let

$$a_1 T(\alpha_1) + a_2 T(\alpha_2) + ... + a_n T(\alpha_n) = 0$$
$$\Rightarrow \qquad T(a_1\alpha_1 + a_2\alpha_2 + ... + a_n\alpha_n) = 0$$
$$\Rightarrow \qquad T(a_1\alpha_1 + a_2\alpha_2 + ... + a_n\alpha_n) = T(0) \qquad\qquad [\because T(0) = 0]$$
$$\Rightarrow \qquad a_1\alpha_1 + a_2\alpha_2 + ... + a_n\alpha_n = 0 \qquad\qquad [\because T \text{ is one-one.}]$$
$$\Rightarrow \qquad a_1 = 0, a_2 = 0, ..., a_n = 0 \qquad [\because S \text{ is linearly independent.}]$$

$\therefore$ S' is linearly independent.

Since dim. $V = n$ and S' contains n linearly independent vectors. Therefore, S' must be a basis of V. Thus each vector of V can be expressed as a linear combination of vectors of S'.

Let $\alpha \in V$. Then there exists $c_1, c_2, ..., c_n \in F$ such that

$$\alpha = c_1 T(\alpha_1) + c_2 T(\alpha_2) + ... + c_n T(\alpha_n)$$
$$\Rightarrow \qquad \alpha = T(c_1\alpha_1 + c_2\alpha_2 + ... + c_n\alpha_n)$$

Now $c_1\alpha_1 + c_2\alpha_2 + ... + c_n\alpha_n \in V$ and α is the T-image. Hence T is onto.

THEOREM 4. *A linear transformation T on a finite dimensional vector space is invertible iff T is onto.*

Proof. Let $V(F)$ be a finite dimensional vector space and let T be a linear transformation on V. If T is invertible, then T is onto.

Conversely, Let T be onto. Now, in order to prove that T is invertible we shall prove that T is one-one. Let dim. $V = n$.

Let $S = \{a_1, a_2, ..., a_n\}$ be a basis of V, then we claim that $S' = \{T(\alpha_1), T(\alpha_2), ..., T(\alpha_n)\}$ is also a basis of V.

Let α be any element of V and T is onto V, then there exists $\beta \in V$ such that $T(\beta) = \alpha$.

Also, $\qquad \beta = a_1\alpha_1 + a_2\alpha_2 + ... + a_n\alpha_n$ for $a_1, a_2, ..., a_n \in F$

Then, $\qquad \alpha = T(\beta) = T(a_1\alpha_1 + a_2\alpha_2 + ... + a_n\alpha_n)$

$$\Rightarrow \qquad \alpha = a_1 T(\alpha_1) + a_2 T(\alpha_2) + ... + a_n T(\alpha_n)$$
$$\Rightarrow \qquad L(S') = V.$$

Since dim. $V = n$ and S' is a subset of V containing n vectors with $L(S') = V$, then S' must be a basis of V. Thus S' is linearly independent.

Let $\gamma = c_1\alpha_1 + c_2\alpha_2 + ... + c_n\alpha_n$ and $\delta = d_1\alpha_1 + d_2\alpha_2 + ... + d_n\alpha_n$ be any element of V. We have

$$T(\gamma) = T(\delta)$$

$\Rightarrow \qquad T(c_1\alpha_1 + c_2\alpha_2 + ... + c_n\alpha_n) = T(d_1\alpha_1 + d_2\alpha_2 + ... + d_n\alpha_n)$

$\Rightarrow c_1T(\alpha_1) + c_2T(\alpha_2) + ... + c_nT(\alpha_n) = d_1T(\alpha_1) + d_2T(\alpha_2) + ... + d_nT(\alpha_n)$

$\Rightarrow \quad (c_1 - d_1)T(\alpha_1) + (c_2 - d_2)T(\alpha_2) + ... + (c_n - d_n)T(\alpha_n) = 0$

$\Rightarrow \quad c_1 - d_1 = 0, c_2 - d_2 = 0, ..., c_n - d_n = 0 \qquad [\because S' \text{ is linearly independent.}]$

$\Rightarrow \quad c_1 = d_1, c_2 = d_2, ..., c_n = d_n$

$\Rightarrow \qquad\qquad\qquad \gamma = \delta.$

Hence T is one-one.

Solved Examples

Example 1. *Describe explicitly a linear transformation from $V_3(R)$ into $V_3(R)$ which has its range the subspace spanned by $(1, 0, -1)$ and $(1, 2, 2)$.*

Solution. We know that the set $\{(1, 0, 0), (0, 1, 0), (0, 0, 1)\}$ is a basis of $V_3(R)$.

Also $\{(1, 0, 0), (0, 1, 0), (0, 0, 1)\}$ is a subset of $V_3(R)$ which has same number of vectors as the above basis set has. Then there exists a unique linear transformation T from $V_3(R)$ into $V_3(R)$ such that

$$T(1, 0, 0) = (1, 0, -1)$$
$$T(0, 1, 0) = (1, 2, 2)$$
$$T(0, 0, 1) = (0, 0, 0)$$

Now the vectors $T(1, 0, 0)$, $T(0, 1, 0)$ and $T(0, 0, 1)$ span the range of T i.e. the vectors $(1, 0, -1)$, $(1, 2, 2)$ and $(0, 0, 0)$ span the range of T. Thus the range of T is the subspace of $V_3(R)$ spanned by the set $\{(1, 0, -1), (1, 2, 2)\}$ because the vector $(0, 0, 0)$ can be omitted from the spanning set.

Let (x, y, z) be any element of $V_3(R)$, then

$$(x, y, z) = x(1, 0, 0) + y(0, 1, 0) + z(0, 0, 1)$$
$$\Rightarrow \qquad T(x, y, z) = xT(1, 0, 0) + yT(0, 1, 0) + zT(0, 0, 1)$$
$$\Rightarrow \qquad T(x, y, z) = x(1, 0, -1) + y(1, 2, 2) + z(0, 0, 0)$$
$$\therefore \qquad T(x, y, z) = (x + y, \ 2y, \ -x + 2y)$$

which is the required linear transformation.

Example 2. *Let T be a linear operator on a vector space $V(F)$. If $T^2 = 0$, what can you say about the relation of the range of T to the null space of T? Give an example of a linear operator on $V_2(R)$ such that $T^2 = 0$ but $T \neq 0$.*

Solution. Since $T^2 = 0$, then for $\alpha \in V$

$$T^2(\alpha) = 0(\alpha) \Rightarrow T[T(\alpha)] = 0$$
$$T(\alpha) \in N(T) \qquad\qquad \text{[By definition of null space]}$$

But $\qquad\qquad\qquad\qquad T(\alpha) \in R(T) \ \forall \ \alpha \in V$

$\therefore \qquad\qquad\qquad\qquad R(T) \subset N(T)$

Hence when $T^2 = 0$, the range of T is contained in null space of T.

Next, let T be a linear map from $V_2(R)$ into $V_2(R)$ such that

$$T(a, b) = (0, a) \ \forall \ (a, b) \in V_2(R)$$

Obviously, $\qquad\qquad\qquad T \neq 0.$

Also, $\qquad\qquad T^2(a, b) = T[T(a, b)] = T[(0, a)] = (0, 0) = 0(a, b)$

$\Rightarrow \qquad\qquad\qquad T^2 = 0.$

Example 3. *If* $T : R^3 \rightarrow R^3$ *is a linear operator defined by* $T(x, y, z) = (x + z, x - z, y)$. *Show that T is invertible and find* $T^{-1}(2, 4, 6)$.

Solution. Let $T(x, y, z) = (0, 0, 0)$

$\Rightarrow \quad (x + z, x - z, y) = (0, 0, 0)$

$\Rightarrow \quad x + z = 0, x - z = 0, y = 0$

Solving these equations, we get $x = 0, y = 0, z = 0$. Therefore, for $\alpha \in R^3$, $T(\alpha) = 0 \Rightarrow \alpha = 0$. Thus T is non-singular. Hence T is invertible.

Now $\quad\quad T(x, y, z) = (p, q, r)$

$\Rightarrow \quad\quad x + z, x - z, y) = (p, q, r)$

$\Rightarrow \quad\quad\quad x + z = p, x - z = q, y = r$

$\Rightarrow \quad\quad\quad\quad x = \dfrac{p+q}{2}, y = r, z = \dfrac{p-q}{2}$

$\Rightarrow \quad\quad T^{-1}(p, q, r) = (x, y, z)$

$\Rightarrow \quad\quad T^{-1}(p, q, r) = \left(\dfrac{p+q}{2}, r, \dfrac{p-q}{2}\right)$

$\therefore \quad\quad T^{-1}(2, 4, 6) = (3, 6, -1)$

Example 4. *A linear transformation T is defined on $V_2(C)$ by* $T(a, b) = (\alpha a + \beta b, \gamma a + \delta b)$, *where* $\alpha, \beta, \gamma, \delta$ *are fixed elements of C. Prove that T is invertible if and only if* $\alpha\delta - \beta\gamma \neq 0$.

Solution. Since dim. $V_2(C) = 2$. Therefore, T is a linear transformation on a finite dimensional vector space, so that T will be invertible if and only if the null space of T contains only zero vector. The zero vector of $V_2(C)$ is $(0, 0)$.

Thus, T is invertible iff $\quad T(x, y) = (0, 0)$

i.e. iff $\quad\quad (\alpha x + \beta y, \gamma x + \delta y) = (0, 0)$

iff $\quad\quad\quad\quad \alpha x + \beta y = 0$ and $\gamma x + \delta y = 0$

Now these equations have only zero solution, *i.e.* $x = 0, y = 0$ if and only if

$$\begin{vmatrix} \alpha & \beta \\ \gamma & \delta \end{vmatrix} \neq 0 \Rightarrow \alpha\delta - \gamma\beta \neq 0$$

Hence T is invertible iff $\alpha\delta - \beta\gamma \neq 0$.

Example 5. *Find two linear operators T and S on $V_2(R)$ such that* $TS = 0$ *but* $ST \neq 0$.

Solution. Consider two linear operators T and S on $V_2(R)$ defined as

$$T(a, b) = (a, 0) \; \forall \; (a, b) \in V_2(R) \text{ and } a, b \in R$$

and $\quad\quad S(a, b) = (0, a) \; \forall \; (a, b) \in V_2(R)$ and $a, b \in R$

Now $\quad\quad (TS)(a, b) = T[S(a, b)]$

$= T(0, a)$

$= (0, 0) = 0 \; \forall (a, b) \in V_2(R)$

$TS = 0$

and $\quad\quad (ST)(a, b) = S[T(a, b)]$

$= S(a, 0) = (0, a) \; \forall \; (a, b) \in V_2(R)$

$\therefore \quad\quad ST \neq 0$

Example 6. *Let T be a linear transformation on a vector space V(F). Prove that the set of all linear transformation S on V for which TS = 0 is a subspace of the vector space of all linear transformations.*

Solution. Let $L(V, V)$ denote the set of all linear transformations on vector space $V(F)$.

Let $W = \{S : S$ is a linear transformation on V and $TS = 0\}$

Now we shall prove that W is a subspace of $L(V, V)$.

Let $S_1, S_2 \in W$, then $TS_1 = 0$, $TS_2 = 0$. If $a, b \in F$ and let $\alpha \in V$, then

$$[T(aS_1 + bS_2)](\alpha) = T[(aS_1 + bS_2)(\alpha)]$$

$$\text{[By the product of linear transformations]}$$

$$= T[(aS_1)(\alpha) + (bS_2)(\alpha)]$$

$$= T[aS_1(\alpha) + bS_2(\alpha)]$$

$$= aT[S_1(\alpha)] + bT[S_2(\alpha)] \qquad [\because T \text{ is linear}]$$

$$= a(TS_1)(\alpha) + b(TS_2)(\alpha)$$

$$= a0(\alpha) + b0(\alpha) = a0 + b0 = 0 = 0(\alpha)$$

$\therefore \quad [T(aS_1 + bS_2)](\alpha) = 0\ (\alpha)\ \forall\ \alpha \in V.$

Thus, $T(aS_1 + bS_2) = 0$, therefore $aS_1 + bS_2 \in W$.

Hence, W is a subspace of $L(V, V)$.

Example 7. *If $T : U \to V$ is a linear transformation and U is finite dimensional, show that U and range of T have the same dimension iff T is non-singular. Determine all non-singular linear transformations*

$$T : V_4(R) \to V_3(R)$$

Solution. We know that

$$\text{dim. } U = \text{rank of } T + \text{nullity of } T$$

$$\Rightarrow \qquad \text{dim. } U = \text{dim. of range of } T + \text{dim. of null space}$$

$\therefore$ dim. $U =$ dim. of range of T iff dim. of null space of T is zero, *i.e.*, iff T is non-singular.

Let T be a linear transformation from $V_4(R)$ into $V_3(R)$. Then T will be non-singular iff dim. of $V_4(R) =$ dim. of range of T.

Since dim. $V_4(R) = 4$ and dim. $V_3(R) = 3$, therefore, the dim. of range of $T = 3$ because range of T $\leq V_3(R)$. Thus dim. $V_4(R)$ cannot be equal to the dim. of range of T.

Hence, T cannot be non-singular. Consequently, there can be no non-singular linear transformation from $V_4(R)$ into $V_3(R)$.

Example 8. *Let V be a finite dimensional vector space and T be a linear operator on V. Suppose that rank (T^2) = rank (T). Prove that the range and null space of T are disjoint, i.e. have only the zero vector in common.*

Solution. We know that, dim. $V = $ rank $(T) + $ nullity (T) ...(1)

Now, T_2 is also a linear operator on V, then

$$\text{dim. } V = \text{rank } (T^2) + \text{nullity } (T^2) \qquad \text{...(2)}$$

From (1) and (2), we have

rank $(T) + $ nullity $(T) = $ rank $(T^2) + $ nullity (T^2)

$\Rightarrow$ $\qquad$ nullity (T) = nullity (T^2) $\qquad\qquad$ [$\because$ rank (T) = rank (T^2)]

$\Rightarrow$dim. of null space of T = dim. of null space of T^2.

If $\alpha \in$ null space of T, then

$$T(\alpha) = 0$$

$\Rightarrow$ $\qquad\qquad$ $T[T(\alpha)] = T(0)$

$\Rightarrow$ $\qquad\qquad$ $T^2(\alpha) = 0$ $\qquad\qquad\qquad\qquad$ [$\because$ $T(0) = 0$]

$\therefore$ $\qquad\qquad$ $\alpha \in$ null space of T^2

$\therefore$ null space of $T \subseteq$ null space of T^2.

But null space of T and null space of T^2 are both subspaces of V and have the same dimension.

Then, $\quad$ null space of T = null space of T^2

$\Rightarrow$ $\qquad$ null space of $T^2 \subseteq$ null space of T

$\Rightarrow$ $\qquad\qquad$ $T^2(\alpha) = 0$

$\Rightarrow$ $\qquad\qquad$ $T(\alpha) = 0$ $\qquad\qquad\qquad\qquad\qquad\qquad$ ...(1)

Let $\beta \neq 0$ and $\beta \in R(T) \cap N(T)$, then $\beta \in R(T)$ and $\beta \in N(T)$.

Now $\qquad\qquad\qquad$ $\beta \in N(T) \Rightarrow T(\beta) = 0$

Also $\qquad\qquad\qquad$ $\beta \in R(T) \Rightarrow \exists\ \alpha \in V$ such that $T(\alpha) = \beta$.

Now $\qquad\qquad$ $T(\alpha) = \beta$

$\Rightarrow$ $\qquad\qquad$ $T[T(\alpha)] = T(\beta) = 0$.

Thus there exists $\alpha \in V$ such that $T[T(\alpha)] = 0$ but $T(\alpha) = \beta \neq 0$ which is against to the equation (1). Therefore, there exists no $\beta \in R(T) \cap N(T)$ such that $\beta \neq 0$. Hence $R(T) \cap N(T) = \{0\}$.

Example 9. *If A and B are linear transformations on vector space $V(F)$, then show that a necessary and sufficient condition that both A and B be invertible is that both AB and BA be invertible.*

Solution. $\qquad$ **Condition is necessary.** Let A and B be two invertible linear transformations on a vector space V. Then,

$$AA^{-1} = I = A^{-1}A$$

and $\qquad\qquad$ $BB^{-1} = I = B^{-1}B$

Now $\qquad$ $(AB)(B^{-1}A^{-1}) = A(BB^{-1})A^{-1}$

$\qquad\qquad\qquad\qquad$ $= (AI)A^{-1}$

$\qquad\qquad\qquad\qquad$ $= AA^{-1} = I$

and $\qquad$ $(B^{-1}A^{-1})(AB) = B^{-1}(A^{-1}A)B$

$\qquad\qquad\qquad\qquad$ $= B^{-1}(IB) = B^{-1}B = I$

$\therefore$ $\qquad$ $(AB)(B^{-1}A^{-1}) = I = (B^{-1}A^{-1})(AB)$

Thus AB is invertible.

Also, we have

$\qquad$ $(BA)(A^{-1}B^{-1}) = I = (A^{-1}B^{-1})(BA)$

Thus, BA is invertible.

Condition is sufficient. Let AB and BA be both invertible. Then AB and BA are both one-one and onto.

Now we shall show that A and B are invertible. First, we shall show that A is invertible.

A is one-one. Let α_1, $\alpha_2 \in V$. Then

$$A(\alpha_1) = A(\alpha_2)$$

$$\Rightarrow \qquad B[A(\alpha_1)] = B[A(\alpha_2)]$$

$$\Rightarrow \qquad (BA)(\alpha_1) = (BA)(\alpha_2)$$

$$\Rightarrow \qquad \alpha_1 = \alpha_2 \qquad\qquad\qquad [\because BA \text{ is one-one.}]$$

$\therefore$ A is one-one.

A is onto. Let $\beta \in V$. Since AB is onto, then there exists $\alpha \in V$ such that

$$(AB)(\alpha) = \beta.$$

$$\Rightarrow \qquad A[B(\alpha)] = \beta.$$

Thus $\beta \in V$, then there exists $B(\alpha) \in V$ such that $A[B(\alpha)] = \beta$.

$\therefore$ A is onto.

Hence A is invertible.

Similarly, if we interchange the role of AB and BA, we find that B is invertible.

❧ EXERCISE 6.1

1. Describe explicitly the linear transformation T from F^2 to F^2 such that $T(e_1) = (a, b)$, $T(e_2) = (c, d)$, where $e_1 = (1, 0)$, $e_2 = (0, 1)$.

2. If $T : R^2 \to R^2$ is the linear transformation for which $T(1, 1) = 3$ and $T(0, 1) = -2$, find $T(a, b)$.

3. Describe explicitly a linear transformation from $V_3(R)$ into $V_4(R)$ which has its range the subspace spanned by the vectors $(1, 2, 0, 4)$, $(2, 0, -1, -3)$.

4. Find a linear mapping $T : R^3 \to R^4$ whose image is generated by $(1, -1, 2, 3)$ and $(2, 3, -1, 0)$.

5. Let $T : F^2 \to F^2$ be a linear operator defined by $T(x, y) = (x + y, x) \;\forall\; (x, y) \in F^2, x, y \in F$. Show that T is invertible and find a rule for a T^{-1} like the one which defines T.

6. If $T : R^3 \to R^2$ is a linear operator defined by $T(x, y, z) = (2x, 4x - y, 2x + 3y - z)$, show that T is invertible and find T^{-1} $(2, 4, 6)$.

7. Let $T : R^3 \to R^3$ be a linear operator defined by $T(x, y, z) = (x - 3y - 2z, y - 4z, z) \forall (x, y, z) \in R^3$. Show that T is non-singular and find a formula for T^{-1} and hence find $T^{-1}(1, 2, 3)$.

8. Let T be the (unique) linear operator on C^3 for which $T(1, 0, 0) = (1, 0, i)$, $T(0, 1, 0) = (0, 1, 1)$, $T(0, 0, 1) = (i, 1, 0)$. Show that T is not invertible.

9. Let S and T be the linear operators on R^2 defined by S $(a, b) = (b, a)$ and $T(a, b) = (a, 0)$. Give rules like the one defining S and T for each of the linear transformations $(T + S)$, TS, ST, S^2, T^2.

10. Let V be of finite dimension and T be a linear operator on V for which $TS = I$, for some operator S on V. Show that T is invertible. Also show that $S = T^{-1}$.

11. Show that if two linear transformations of a finite dimensional vector space coincide on a basis of that vector space, then they are identical.

12. If T is a linear transformation on a finite dimensional vector space V such that range of T is a proper subset of V, show that there exists a non-zero element α in V with $T(\alpha) = 0$.

13. Let $T : R^3 \to R^3$ be defined as $T(x, y, z) = (0, x, y)$, show that $T \neq 0$, $T^2 \neq 0$ and $T^3 = 0$.

14. Let T be a linear transformation from a vector space U into a vector V with kernel of $T \neq 0$. Show that there exists vectors α_1 and α_2 in U such that $\alpha_1 \neq \alpha_2$ and $T\alpha_1 = T\alpha_2$.

15. Let T be a linear transformation from $V_3(R)$ into $V_2(R)$ and let S be a linear transformation from $V_2(R)$ into $V_3(R)$. Prove that the transformation ST is not invertible.

16. Prove that the set of invertible linear operators on a vector space V with the operation of composition forms a group. Check if this group is commutative.

17. If $V(F)$ be the vector space of all polynomials in x and D and T be the two linear operators on V defined by

$$D[f(x)] = \frac{df(x)}{dx}, T[f(x)] = xf(x)$$

for each $f(x) \in V$. Then show that the product of these operators is not commutative, *i.e.* $DT \neq TD$ and $(TD)^2 = TD + T^2D^2$.

18. If $\{\alpha_1, \alpha_2,..., \alpha_n\}$ and $\{\beta_1, \beta_2,..., \beta_n\}$ are linearly independent sets of vectors in a finite dimensional vector space V, then there exists an invertible linear transformation T on V such that

$$T(\alpha_i) = \beta_i, i = 1, 2, ..., n$$

19. Let U and V be vector spaces over the same field F and S be an isomorphism of U onto V. Prove that $T \to STS^{-1}$ is an isomorphism of $L(U, U)$ and onto $L(V, V)$.

20. If V is the space of all polynomials of degree $\leq n$ over a field F, prove that the differentiation operator on V is nilpotent.

21. Let $T_1 : \mathrm{R}^3 \to \mathrm{R}^3$ be a linear operator defined by $T(x, y, z) = (2x, 2x - 5y, 2y + z)$. Find T^{-1}.

22. Let $T : \mathrm{R}^3 \to \mathrm{R}^3$ be a linear operator defined by $T(x, y, z) = (x \cos \theta - y \sin \theta, x \sin \theta + y \cos \theta, z)$. Is T singular or non-singular ?

23. Let $T : \mathrm{R}^3 \to \mathrm{R}^2$ be defined as

$$T(x, y) = (2x - 4y, 3x - 6y).$$

Is T non-singular ? If not, find $\alpha \neq 0$ in R^2 such that $T(\alpha) = 0$.

24. Let $T : \mathrm{R}^3 \to \mathrm{R}^3$ be defined by $T(x, y, z) = (x + y - 2z, x + 2y + z, 2x + 2y - 3z)$. Is T non-singular ?

25. Let $T : \mathrm{R}^3 \to \mathrm{R}^3$ be defined by $T(x, y, z) = (x + y + z, x + 2y - z, 3x + 5y - z)$. Is T non-singular? If not, find $\alpha \neq 0$ in R^3 such that $T(\alpha) = 0$.

26. Let V be a vector space of all real polynomials in t. Let $T : V \to V$ be the linear operator defined by $T[f(t)] = tf(t) \ \forall \ f(t) \in V$. Is T singular or non-singular ?

Answers

1. $T(x, y) = (xa + yc, xb + yd)$
2. $T(x, y) = 5x - 2y$
3. $T(x, y, z) = (x + 2y, 2x - y, - 4x - 3y)$
4. $T(x, y, z) = (x + 2y, -x + 3y, 2x - y, 3x)$
5. $T^{-1}(p, q) = (q, p - q)$
6. $T^{-1}(2, 4, 6) = (1, 0, - 4)$
7. $T^{-1}(p, q, r) = (p + 3q + 14r, q + 4r, r)$; $T^{-1}(1, 2, 3) = (49, 14, 3)$
9. $(T + S)(x, y) = (x + y, x)$; $(TS)(x, y) = (y, 0)$; $(ST)(x, y) = (0, x)$; $(S^2)(x, y) = (x, y)$; $(T^2)(x, y) = (x, 0)$.

16. Non-commutative
21. $T^{-1}(p, q, r) = \left(\dfrac{p}{2}, \dfrac{p-q}{5}, \dfrac{-2p + 2q + 5r}{5}\right)$

22. Non-singular
23. T is singular and $\alpha = (-2, 1)$ such that $T(\alpha) = 0$
24. T is non-singular.
25. T is singular and $\alpha = (-3, 2, 1)$ such that $T(\alpha) = 0$
26. T is non-singular.

6.12 COORDINATE VECTOR

Let V be a finite dimensional vector space over a field F and let dim. $V = n$, then $B = \{\alpha_1, \alpha_2, ..., \alpha_n\}$ is a basis of V and for $\alpha \in V$, suppose that

$$\alpha = a_1 \alpha_1 + a_2 \alpha_2 + ... + a_n \alpha_n$$

for $a_i's \in F$. Then the coordinate vector of α relative to β, which we write as a column vector unless otherwise specified or implied, is

$$[\alpha]_B = \begin{bmatrix} a_1 \\ a_2 \\ \vdots \\ a_n \end{bmatrix}$$

6.13 MATRIX REPRESENTATION OF A LINEAR TRANSFORMATION

Let U be an m-dimensional vector space over a field F and let V be an n-dimensional vector space over the field F. Let $B = \{\alpha_1, \alpha_2, ..., \alpha_m\}$ and $B' = \{\beta_1, \beta_2, .., \beta_n\}$ be the basis of U and

V respectively. If T is a linear transformation from U into V, then $T(\alpha_1)$, $T(\alpha_2)$, ...,$T(\alpha_m)$ are vectors in V. Since $B' = \{\beta_1, \beta_2,..., \beta_n\}$ is a basis of V so that each $T(\alpha_i)$ is a linear combination of the elements of B'. For $a_{ij} \in F$, $1 \le i \le m$, $1 \le j \le n$, we have

$$T(\alpha_1) = \{a_{11}\beta_1 + a_{12}\beta_2 + ,..., + a_{1n}\beta_n)$$
$$T(\alpha_2) = \{a_{21}\beta_1 + a_{22}\beta_2 + ,..., + a_{2n}\beta_n)$$
$$..$$
$$T(\alpha_m) = \{a_{m1}\beta_1 + a_{m2}\beta_2 + ,..., + a_{mn}\beta_n)$$

Definition. *The transpose of the above matrix of coefficients, denoted by $[T]_B$ is called the matrix representation of T relative to the ordered basis B.*

Thus,
$$[T]_B = \begin{bmatrix} a_{11} & a_{21} & \cdots & a_{m1} \\ a_{12} & a_{22} & \cdots & a_{m2} \\ \cdots & \cdots & \cdots & \cdots \\ a_{1n} & a_{2n} & \cdots & a_{mn} \end{bmatrix}_{n \times m}$$

For Example : Let V be the vector space of polynomials in 't' over the field of reals R, of degree ≤ 3, and let

$$D : V \to V$$

be the differential operator defined by $D\,[p(t)] = \dfrac{d}{dt}[p(t)]$

We compute the matrix of D in the basis $B = [1, t, t^2, t^3]$ as follows

$$D(1) = 0 = 0 + 0t + 0t^2 + 0t^3$$
$$D(t) = 1 = 1 + 0t + 0t^2 + 0t^3$$
$$D(t^2) = 2t = 0 + 2t + 0t^2 + 0t^3$$
$$D(t^3) = 3t^2 = 0 + 0t + 3t^2 + 0t^3$$

Thus, the matrix of D relative to B is given by

$$[D]_B = \begin{bmatrix} 0 & 1 & 0 & 0 \\ 0 & 0 & 2 & 0 \\ 0 & 0 & 0 & 3 \\ 0 & 0 & 0 & 0 \end{bmatrix}$$

THEOREM 1. *Let U be an m-dimensional vector space aver the field F and V an n-dimensional vector space over the field F. Let B be an ordered basis for U and B' an ordered basis for V. Let T be any linear transformation from U into V. Then for any vector $\alpha \in U$.*

$$[T]_B\,[\alpha]_B = [T(\alpha)]_{B'}$$

Proof. Let $B = \{\alpha_1, \alpha_2,..., \alpha_m\}$ be an ordered basis for U and $B' = \{\beta_1, \beta_2,..., \beta_n\}$ an ordered basis for V. T is linear transformation from U into V, then T is determined by its action on the vectors α_j, $1 \le i \le m$. Each of m vectors $T(\alpha_i)$ is uniquely expressible as a linear combination of elements of B' :

$$T(\alpha_i) = \sum_{j=1}^{n} a_{ij}\beta_j \qquad ...(1)$$

where $a_{i1}, a_{i2},..., a_{in}$ are the coordinates of $T(\alpha_i)$ in the ordered basis B'.

If α be any vector in U, then

$$\alpha = a_1\alpha_1 + a_2\alpha_2 + \ldots + a_m\alpha_n$$

$$\therefore \quad [\alpha]_B = \begin{bmatrix} a_1 \\ a_2 \\ \vdots \\ a_m \end{bmatrix}$$

Now, $T(\alpha) = T(a_1\alpha_1 + a_2\alpha_2 + \ldots + a_m\alpha_m)$

$$= a_1 T(\alpha_1) + a_2 T(\alpha_2) + \ldots + a_m T(\alpha_m)$$

$$= a_1 \sum_{j=1}^{n} a_{1j}\beta_j + a_2 \sum_{j=1}^{n} a_{2j}\beta_j + \ldots + a_m \sum_{j=1}^{n} a_{mj}\beta_j \qquad \text{[using (1)]}$$

$$= aTa\left(\sum_{i=1}^{n} a_i\alpha_i\right) + bT\left(\sum_{i=1}^{n} b_i\alpha_i\right) = aT(\alpha) + bt(\beta)$$

$$= a_1 a_{11}\beta_1 + a_1 a_{12}\beta_2 + \ldots + a_1 a_{1n}\beta_n + a_2 a_{21}\beta_1 + a_2 a_{22}\beta_2 + \ldots + a_2 a_{2n}\beta_n$$

$$+ a_m a_{m1}\beta_1 + a_m a_{m2}\beta_2 + \ldots + a_m a_{mn}\beta_n$$

$$\therefore \quad [T(\alpha)]_{B'} = \begin{bmatrix} a_1 a_{11} + a_2 a_{21} + \ldots + a_m a_{m1} \\ a_1 a_{12} + a_2 a_{22} + \ldots + a_m a_{m2} \\ \ldots \ldots \ldots \ldots \ldots \ldots \ldots \ldots \ldots \\ a_1 a_{1n} + a_2 a_{2n} + \ldots + a_m a_{mn} \end{bmatrix}_{n \times m}$$

$$= \begin{bmatrix} a_{11} & a_{21} & \cdots & a_{m1} \\ a_{12} & a_{22} & \cdots & a_{m2} \\ \cdots & \cdots & \cdots & \cdots \\ a_{1n} & a_{2n} & \cdots & a_{mn} \end{bmatrix}_{n \times m} \cdot \begin{bmatrix} a_1 \\ a_2 \\ \vdots \\ a_m \end{bmatrix}_{m \times 1}$$

$$[T(\alpha)]_{B'} = [T]_B [\alpha]_{B'}.$$

THEOREM 2. *Let U, V and W be vector spaces over the field F of respective dimensions n, m and p. Let T_1 be a linear transformation from U into V and T_2 a linear transformation from V into W. If B, B' and B'' are the ordered bases for the spaces U, V and W respectively, if A is the matrix of T_1, relative to the pair B, B' and B is the matrix of T_2 relative to the pair B' and B'', then the matrix of $(T_2 T_1)$ relative to the pair B, B'' is the product matrix C = BA.*

Proof. Let $B = \{\alpha_1, \alpha_2, \ldots, \alpha_n\}$, $B' = \{\beta_1, \beta_2, \ldots, \beta_m\}$ and $B'' = \{\gamma_1, \gamma_2, \ldots, \gamma_p]$ be the bases of U, V and W respectively. If α is any vector in U, then

$$[T_1\{\alpha\}]_{B'} = [T_1]_B [\alpha]_B \qquad \text{[By above theorem]}$$

$$= A[\alpha]_B \qquad [\because A = [T_1]_B]$$

and $$[T_2(T_1(\alpha))]_{B''} = [T_2]_{B'} [T_1(\alpha)]_{B'} = B[T_1(\alpha)]_{B'} \qquad [\because B = [T_2]B']$$

$$\therefore \quad [(T_2 T_1)(\alpha)]_{B''} = BA[\alpha]_B.$$

Hence by the definition and uniqueness of the representing matrix, we must have $C = BA$ as the matrix of $(T_2 T_1)$ relative to B, B''.

THEOREM 3. *Let V be an n-dimensional vector space over the field F and B be an ordered basis of V. If T_1 and T_2 are linear operators from V into V, then*

(i) $[T_1+T_2]_B = [T_1]_B+[T_2]_B$

(ii) $[cT_1]_B = c[T_1]_B$, *for* $c \in F$

(iii) $[T_2 T_1]_B = [T_1]_B[T_2]_B.$

Proof. Let $\{a_1, a_2, ..., a_n\}$ be the basis of V, then for $a_{ij} \in F$ and $b_{ij} \in F$, $1 \le i \le n$, $1 \le j \le n$, we have

$$T_1(\alpha_1) = a_{11}\alpha_1 + a_{12}\alpha_2 + ... + a_{1n}\alpha_n$$
$$T_1(\alpha_2) = a_{21}\alpha_1 + a_{22}\alpha_2 + ... + a_{2n}\alpha_n$$

$$\cdots$$

$$T_1(\alpha_n) = a_{n1}\alpha_1 + a_{n2}\alpha_2 + ... + a_{nn}\alpha_n$$

$$\therefore \quad [T_1]_B = \begin{bmatrix} a_{11} & a_{21} & \cdots & a_{n1} \\ a_{12} & a_{22} & \cdots & a_{n2} \\ \cdots & \cdots & \cdots & \cdots \\ a_{1n} & a_{2n} & \cdots & a_{nn} \end{bmatrix}$$

Also,

$$T_2(\alpha_2) = b_{11}\alpha_1 + b_{12}\alpha_2 + ... + b_{1n}\alpha_n$$
$$T_2(\alpha_2) = b_{21}\alpha_1 + b_{22}\alpha_2 + ... + b_{2n}\alpha_n$$

$$\cdots$$

$$T_2(\alpha_n) = b_{n1}\alpha_1 + b_{n2}\alpha_2 + ... + b_{nn}\alpha_n$$

$$\therefore \quad [T_2]_B = \begin{bmatrix} b_{11} & b_{21} & \cdots & b_{n1} \\ b_{12} & b_{22} & \cdots & b_{n2} \\ \cdots & \cdots & \cdots & \cdots \\ b_{1n} & b_{2n} & \cdots & b_{nn} \end{bmatrix}$$

(i) $(T_1 + T_2)(\alpha_1) = (T_1)(\alpha_1) + (T_2)(\alpha_1)$

$$= (a_{11} + b_{11})\alpha_1 + (a_{12} + b_{12})\alpha_2 + ... + (a_{1n} + b_{1n})\alpha_n$$

$(T_1 + T_2)(\alpha_2) = (T_1)(\alpha_2) + (T_2)(\alpha_2)$

$$= (a_{21} + b_{21})\alpha_1 + (a_{22} + b_{22})\alpha_2 + ... + (a_{2n} + b_{2n})\alpha_n$$

$(T_1 + T_2)(\alpha_n) = (T_1)(\alpha_n) + (T_2)(\alpha_n)$

$$= (a_{n1} + b_{n1})\alpha_1 + (a_{n2} + b_{n2})\alpha_2 + ... + (a_{nn} + b_{nn})\alpha_n$$

$$[T_1 + T_2]_B = \begin{bmatrix} (a_{11}+b_{11}) & (a_{21}+b_{21}) & \cdots & (a_{n1}+b_{n1}) \\ (a_{12}+b_{12}) & (a_{22}+b_{22}) & \cdots & (a_{2n}+b_{2n}) \\ \cdots & \cdots & \cdots & \cdots \\ (a_{1n}+b_{1n}) & (a_{2n}+b_{2n}) & \cdots & (a_{nn}+b_{nn}) \end{bmatrix}$$

$$= \begin{bmatrix} a_{11} & a_{21} & \cdots & a_{n1} \\ a_{12} & a_{22} & \cdots & a_{n2} \\ \cdots & \cdots & \cdots & \cdots \\ a_{1n} & a_{2n} & \cdots & a_{nn} \end{bmatrix} + \begin{bmatrix} b_{11} & b_{21} & \cdots & b_{n1} \\ b_{12} & b_{22} & \cdots & b_{2n} \\ \cdots & \cdots & \cdots & \cdots \\ b_{1n} & b_{2n} & \cdots & b_{nn} \end{bmatrix}$$

$$= [T_1]_B + [T_2]_B$$

(ii) $\quad (cT_1)(\alpha_1) = cT_1(\alpha_1) = ca_{11}\alpha_1 + ca_{12}\alpha_2 + \ldots + ca_{1n}\alpha_n$

$\quad\quad (cT_1)(\alpha_2) = cT_1(\alpha_2) = ca_{21}\alpha_1 + ca_{22}\alpha_2 + \ldots + ca_{2n}\alpha_n$

$$\ldots\ldots\ldots\ldots\ldots\ldots\ldots$$

$\quad\quad (cT_1)(\alpha_n) = cT_1(\alpha_n) = ca_{n1}\alpha_1 + ca_{n2}\alpha_2 + \ldots + ca_{nn}\alpha_n$

$$\therefore \quad (cT_1)_B = \begin{bmatrix} ca_{11} & ca_{21} & \cdots & ca_{n1} \\ ca_{12} & ca_{22} & \cdots & ca_{n2} \\ \cdots & \cdots & \cdots & \cdots \\ ca_{1n} & ca_{2n} & \cdots & ca_{nn} \end{bmatrix} = c\begin{bmatrix} a_{11} & a_{21} & \cdots & a_{n1} \\ a_{12} & a_{22} & \cdots & a_{n2} \\ \cdots & \cdots & \cdots & \cdots \\ a_{1n} & a_{2n} & \cdots & a_{nn} \end{bmatrix} = c[T_1]_B$$

(iii) $\quad (T_2 T_1)(\alpha_1) = T_2(T_1(\alpha_1))$

$\quad\quad\quad\quad\quad\quad\quad = T_2(a_{11}\alpha_1 + a_{12}\alpha_2 + \ldots + a_{1n}\alpha_n)$

$\quad\quad\quad\quad\quad\quad\quad = a_{11}T_2(\alpha_1) + a_{12}T_2(\alpha_2) + \ldots + a_{1n}T_2(\alpha_n) \quad\quad [\because T_2 \text{ is linear.}]$

$\quad\quad\quad\quad\quad\quad\quad = a_{11}(b_{11}\alpha_1 + b_{12}\alpha_2 + \ldots + b_{1n}\alpha_n)$

$\quad\quad\quad\quad\quad\quad\quad\quad\quad + a_{12}(b_{21}\alpha_1 + b_{22}\alpha_2 + \ldots + b_{2n}\alpha_n)$

$$\ldots\ldots\ldots\ldots\ldots\ldots\ldots$$

$\quad\quad\quad\quad\quad\quad\quad\quad\quad + a_{1n}(b_{n1}\alpha_1 + b_{n2}\alpha_2 + \ldots + b_{nn}\alpha_n)$

$\quad\quad\quad\quad\quad\quad\quad = (a_{11}b_{11} + a_{12}b_{21} + \ldots + a_{1n}b_{n1})\alpha_1$

$\quad\quad\quad\quad\quad\quad\quad\quad\quad + (a_{11}b_{12} + a_{12}b_{22} + \ldots + a_{1n}b_{n2})\alpha_2$

$$\ldots\ldots\ldots\ldots\ldots\ldots\ldots$$

$\quad\quad\quad\quad\quad\quad\quad\quad\quad + (a_{11}b_{1n} + a_{12}b_{2n} + \ldots + a_{1n}b_{nn})\alpha_n$

$(T_2 T_1)(\alpha_2) = T_2(T_1(\alpha_2)) = T_2(a_{21}\alpha_1 + a_{22}\alpha_2 + \ldots + a_{2n}\alpha_n)$

$\quad\quad\quad\quad\quad\quad\quad = a_{21}T_2(\alpha_1) + a_{22}T_2(\alpha_2) + \ldots + a_{2n}T_2(\alpha_n)$

$\quad\quad\quad\quad\quad\quad\quad = a_{21}(b_{11}\alpha_1) + b_{12}\alpha_2 + \ldots + b_{1n}\alpha_n)$

$\quad\quad\quad\quad\quad\quad\quad\quad\quad + a_{22}(b_{21}\alpha_1 + b_{22}\alpha_2 + \ldots + b_{2n}\alpha_n)$

$$\ldots\ldots\ldots\ldots\ldots\ldots\ldots$$

$\quad\quad\quad\quad\quad\quad\quad\quad\quad + a_{2n}(b_{n1}\alpha_1 + b_{n2}\alpha_2 + \ldots + b_{nn}\alpha_n)$

$\quad\quad\quad\quad\quad\quad\quad = a_{21}b_{11} + a_{22}b_{21} + \ldots + a_{2n}b_{n1})\alpha_1$

$\quad\quad\quad\quad\quad\quad\quad\quad\quad + (a_{21}b_{12} + a_{22}b_{22} + \ldots + a_{2n}b_{n2})\alpha_2$

$$\ldots\ldots\ldots\ldots\ldots\ldots\ldots$$

$\quad\quad\quad\quad\quad\quad\quad\quad\quad + (a_{21}b_{1n} + a_{22}b_{2n} + \ldots + a_{2n}b_{nn})\alpha_n$

Similarly,

$\quad (T_2 T_1)(\alpha_n) = T_2(T_1(\alpha_n)) = T_2(a_{n1}\alpha_1 + a_{n2}\alpha_2 + \ldots + a_{nn}\alpha_n)$

$\quad\quad\quad\quad\quad\quad\quad = a_{n1}T_2(\alpha_1) + a_{n2}T_2(\alpha_2) + \ldots + a_{nn}T_2(\alpha_n)$

$\quad\quad\quad\quad\quad\quad\quad = a_{n1}(b_{11}\alpha_1 + b_{12}\alpha_2 + \ldots + b_{1n}\alpha_n)$

$\quad\quad\quad\quad\quad\quad\quad\quad\quad + a_{n2}(b_{21}\alpha_1 + b_{22}\alpha_2 + \ldots + b_{2n}\alpha_n)$

$$\ldots\ldots\ldots\ldots\ldots\ldots\ldots$$

$\quad\quad\quad\quad\quad\quad\quad\quad\quad + a_{nn}(b_{n1}\alpha_1 + b_{n2}\alpha_2 + \ldots + b_{nn}\alpha_n)$

$\quad\quad\quad\quad\quad\quad\quad = (a_{n1}b_{11} + a_{n2}b_{21} + \ldots + a_{nn}b_{n1})\alpha_1$

$\quad\quad\quad\quad\quad\quad\quad\quad\quad + (a_{n1}b_{12} + a_{n2}b_{22} + \ldots + a_{nn}b_{n2})\alpha_2$

$$\ldots\ldots\ldots\ldots\ldots\ldots\ldots$$

$\quad\quad\quad\quad\quad\quad\quad\quad\quad + (a_{n1}b_{1n} + a_{n2}b_{2n} + \ldots + a_{nn}b_{nn})\alpha_n$

$$[T_2T_1]_B = \begin{bmatrix} a_{11}b_{11} + a_{12}b_{21} + ... + a_{1n}b_{n1} \\ a_{11}b_{12} + a_{12}b_{22} + ... + a_{1n}b_{n2} \\ .... \\ a_{11}b_{1n} + a_{12}b_{2n} + ... + a_{1n}b_{nn} \end{bmatrix}$$

$$\begin{bmatrix} a_{21}b_{11} + ... + a_{2n}b_{n1}, ... a_{n1}b_{11} + ... + a_{nn}b_{n1} \\ a_{21}b_{12} + ... + a_{2n}b_{n2}, ... a_{n1}b_{12} + ... + a_{nn}b_{n2} \\ .... \\ a_{21}b_{1n} + ... + a_{2n}b_{nn} ... a_{n1}b_{1n} + ... + a_{nn}b_{nn} \end{bmatrix}$$

$$= \begin{bmatrix} b_{11} & b_{21} & \cdots & b_{n1} \\ b_{12} & b_{22} & \cdots & b_{n2} \\ ... & ... & ... & ... \\ b_{1n} & b_{2n} & \cdots & b_{nn} \end{bmatrix} \begin{bmatrix} a_{11} & a_{21} & \cdots & a_{n1} \\ a_{12} & a_{22} & \cdots & a_{n2} \\ ... & ... & ... & ... \\ a_{1n} & a_{2n} & \cdots & a_{nn} \end{bmatrix} = [T_2]_B[T_1]_B$$

6.14 CHANGE OF BASIS

It has been shown that we can represent vectors by tuples (column vectors) and linear operators by matrix once we have selected a basis.

In this section we will see how the representation of matrix of linear transformation changes if we take another basis.

Let $\{\alpha_1, \alpha_2, ... , \alpha_n\}$ be a basis of V and let $\{\beta_1, \beta_2, ... , \beta_n\}$ be another basis of V and suppose

$$\beta_1 = a_{11}\alpha_1 + a_{12}\alpha_2 + ... + a_{1n}\alpha_n$$
$$\beta_2 = a_{21}\alpha_1 + a_{22}\alpha_2 + ... + a_{2n}\alpha_n$$
$$...$$
$$\beta_n = a_{n1}\alpha_1 + a_{n2}\alpha_2 + ... + a_{nn}\alpha_n$$

Then the transpose of the coefficient matrix of above equation is called the *transition matrix* from the basis $\{\alpha_1, \alpha_2, ..., \alpha_n\}$ to the basis $\{\beta_1, \beta_2, ..., \beta_n\}$

$$P = \begin{bmatrix} a_{11} & a_{21} & \cdots & a_{n1} \\ a_{12} & a_{22} & \cdots & a_{n2} \\ ... & ... & ... & ... \\ a_{1n} & a_{2n} & \cdots & a_{nn} \end{bmatrix}$$

REMARK

- P is invertible and its P^{-1} is the transition matrix from new basis to old basis.

  For Example: Let $\{(1,0),(0,1)\}$ and $\{(1,1),(-1,0)\}$ be two bases of R^2, then $(1,1) = 1.(0,1) + 1(0,1)$ and $(-1,0) = -1(1,0) + 0.(0,1)$

  $$P = \begin{bmatrix} 1 & -1 \\ 1 & 0 \end{bmatrix}$$

THEOREM 1. *Let P be the transition matrix from a basis B to a basis B' in a vector space V. Then for any vector $a \in V$, $P[\alpha]_{B'} = [\alpha]_B$ and $[\alpha]_{B'} = P^{-1}[a]_B$.*

Proof. Let V be an n-dimensional vector space and let,
$$B=\{\alpha_1,\alpha_2,......,\alpha_n\} \text{ and } B'=\{\beta_1,\beta_2,......,\beta_n\}$$
be two bases of V and let P be the transition matrix from B to B'. Then we have,
$$\beta_1=a_{11}\alpha_1+a_{12}\alpha_2+......+a_{1n}\alpha_n$$
$$\beta_2=a_{21}\alpha_1+a_{22}\alpha_2+......+a_{2n}\alpha_n$$

$$\beta_n=a_{n1}\alpha_1+a_{n2}\alpha_2+......+a_{nn}\alpha_n; \text{ for } \alpha_{ij} \in F$$

$$\therefore \quad P=\begin{bmatrix} a_{11} & a_{21} & ... & a_{n1} \\ a_{12} & a_{22} & ... & a_{n2} \\ ... & ... & ... & ... \\ a_{1n} & a_{2n} & ... & a_{nn} \end{bmatrix}.$$

Now suppose $\alpha \in V$ such that
$$\alpha=b_1\beta_1+b_2\beta_2+......+b_n\beta_n.$$
Substituting β's from above, we obtain,
$$\alpha=b_1(a_{11}\alpha_1+a_{12}\alpha_2+......+a_{1n}\alpha_n)+b_2(a_{21}\alpha_1+a_{22}\alpha_2+......+a_{2n}\alpha_n)+$$
$$+b_n(a_{n1}\alpha_1+a_{n2}\alpha_2+......+a_{2n}\alpha_n)$$
$$= (b_1a_{11}+b_2a_{12}+......+b_na_{1n})\alpha_1+ (b_1a_{12}+b_2a_{22}+......+b_na_{2n})\alpha_2$$
$$+(b_1a_{1n}+b_2a_{2n}+......+b_na_{2n})\alpha_n$$

Thus, $[\alpha]_{B'}=\begin{bmatrix} b_1 \\ b_2 \\ ... \\ b_n \end{bmatrix}$ and $[\alpha]_B=\begin{bmatrix} b_1a_{11} + b_2a_{12} +......+ b_na_{1n} \\ b_1a_{12} + b_2a_{22} +......+ b_na_{2n} \\ .... \\ b_1a_{1n} + b_2a_{2n} +......+ b_na_{nn} \end{bmatrix}$

Accordingly,

$$P[\alpha]_{B'}=\begin{bmatrix} a_{11} & a_{21} & ... & a_{n1} \\ a_{12} & a_{22} & ... & a_{n2} \\ ... & ... & ... & ... \\ a_{1n} & a_{2n} & ... & a_{nn} \end{bmatrix} \cdot \begin{bmatrix} b_1 \\ b_2 \\ ... \\ b_n \end{bmatrix} = \begin{bmatrix} b_1a_{11} + b_2a_{12} +......+ b_na_{1n} \\ b_1a_{12} + b_2a_{22} +......+ b_na_{2n} \\ .... \\ b_1a_{1n} + b_2a_{2n} +......+ b_na_{nn} \end{bmatrix}$$

$$=[\alpha]_B$$

Furthermore, since P is invertible , hence
$$P[\alpha]_{B'}=[\alpha]_B \qquad \Rightarrow \quad P^{-1}P[\alpha]_{B'}= P^{-1}[\alpha]_B$$
$$\Rightarrow \quad I[\alpha]_{B'}= P^{-1}[\alpha]_B \qquad \Rightarrow \qquad [\alpha]_{B'}= P^{-1}[\alpha]_B$$

THEOREM 2. *Let P be the transition matrix from a basis to a basis B' in a vector space V. Then for any linear operator T on V,*
$$[T]_{B'}=P^{-1}[T]_BP.$$

Proof. Let α be any vector in V. then we have
$$[T]_B[\alpha]_B=[T(\alpha)]_B \qquad\qquad ...(1)$$
and
$$P[\alpha]_{B'}=(\alpha)_B \qquad\qquad ...(2)$$
$$\Rightarrow \quad [T]_BP[\alpha]_{B'}= [T]_B[\alpha]_B=[T(\alpha)]_B \qquad\qquad [\text{Using (1)}]$$
$$\Rightarrow \quad P^{-1}[T]_BP[\alpha]_{B'}= P^{-1}[T(\alpha)]_B=[T(\alpha)]_{B'} \qquad\qquad [\text{By theorem (1)}]$$
$$= [T]_{B'}[\alpha]_{B'} \qquad\qquad [\text{Using (1)}]$$
$$\Rightarrow \quad P^{-1}[T]_BP = [T]_{B'} \qquad\qquad [\because [\alpha]_{B'}\in F \text{ are arbitrary.}]$$

6.15 SIMILARITY OF MATRICES

Let A and B be two square matrices each of order n over the field F. Then B is similar to A if there exists an invertible matrix C of order n over the field F such that

$$B = C^{-1}AC \text{ or } A = CBC^{-1}$$

THEOREM 1. *The relation of similarity is an equivalence relation in the set of all $n \times n$ matrices over the field F.*

Proof. Let M_n be the set of all $n \times n$ matrices over the field F.

If $A, B \in M_n$, then B is similar to A if there exists an invertible matrix C in M_n such that

$$B = C^{-1}AC.$$

Now, in order to prove equivalence relation, we shall prove that the relation is reflexive, symmetric and transitive.

Reflexive. Let $A \in M_n$, then there exists an $n \times n$ invertible matrix I_n such that

$$A = I_n^{-1}AI_n$$

where I_n is the unit matrix over F.

Thus, A is similar to A itself.

∴   The relation of similarity on M_n is reflexive.

Symmetric. Let $A, B \in M_n$ such that A is similar to B, then there exists an invertible matrix $C \in M_n$ such that

$$A = C^{-1}BC$$

$\Rightarrow \qquad CAC^{-1} = C(C^{-1}BC)C^{-1}$

$\Rightarrow \qquad CAC^{-1} = (CC^{-1})B(CC^{-1})$

$\Rightarrow \qquad CAC^{-1} = B$

$\Rightarrow \qquad B = CAC^{-1}$

$\Rightarrow \qquad B = (C^{-1})^{-1}AC^{-1}$ \hfill $[\because C^{-1} \text{ is invertible.}]$

$\Rightarrow \quad B$ is similiar to A.

∴   The relation of similarity on M_n is symmetric.

Transitive. Let $A, B, C \in M_n$ such that A is similar to B and B is similar to C.

Now A is similar to B, then there exists an invertible matrix $P \in M_n$ such that

$$A = P^{-1}BP$$

Also B is similar to C, then there exists an invertible matrix $Q \in M_n$ such that

$$B = Q^{-1}CQ$$

Now $\qquad A = P^{-1}BP$

$\Rightarrow \qquad A = P^{-1}(Q^{-1}CQ)P$

$\Rightarrow \qquad A = (P^{-1}Q^{-1})C(QP)$

$\Rightarrow \qquad A = (QP)^{-1}C(QP)$ \hfill $[\because (QP)^{-1} = P^{-1}Q^{-1}]$

$\Rightarrow \quad A$ is similar to C.

∴   The relation of similarity on M_n is transitive.

Hence, the relation of similarity of matrices is an equivalence relation on the set of $n \times n$ matrices over the field F.

THEOREM 2. *Similar matrices have the same determinant.*

Proof. Let A and B be two square matrices of order $n \times n$ over a field F such that B is similar to A. Then there exists an invertible matrix C of order $n \times n$ over the field F such that

$$B = C^{-1}AC$$

$$\Rightarrow \qquad \det B = \det(C^{-1}AC)$$

$$\Rightarrow \qquad \det B = (\det C^{-1})(\det A)(\det C)$$

$$\Rightarrow \qquad \det B = (\det C^{-1})(\det C)(\det A)$$

$$\Rightarrow \qquad \det B = (\det C^{-1}C)(\det A)$$

$$\Rightarrow \qquad \det B = (\det I_n)(\det A) \qquad\qquad [\because C^{-1}C = I_n]$$

$$\Rightarrow \qquad \det B = \det A \qquad\qquad [\because \det I_n = 1]$$

Hence A and B have the same determinant.

6.16 SIMILARITY OF LINEAR TRANSFORMATIONS

Let S and T be two linear transformations on a vector space $V(F)$. Then T is similar to S if there exists an invertible linear transformation P on V such that

$$T = PSP^{-1}$$

THEOREM 1. *The relation of similarity is an equivalence relation in the set of all linear transformations on a vector space $V(F)$.*

Proof. Let $L(V, V)$ be the set of all linear transformations on V over a field F. Let S and T be two elements of $L(V, V)$. Then T is similar to S if there exists an invertible linear transformation $P \in L(V, V)$ such that

$$T = PSP^{-1}$$

Now we shall show that the relation of similarity is an equivalence relation.

Reflexive. Let $T \in L(V, V)$, Then there exists an invertible linear transformation $I \in L(V, V)$ such that

$$T = ITI^{-1}$$

where I is an identity transformation.

$\therefore \quad T$ is similar to T.

Symmetric. Let $T, S \in L(V, V)$ such that T is similar to S, then there exists an invertible linear transformation $P \in L(V, V)$ such that

$$T = PSP^{-1}$$

$$\Rightarrow \qquad P^{-1}TP = P^{-1}(PSP^{-1})P$$

$$\Rightarrow \qquad P^{-1}TP = (P^{-1}P)S(P^{-1}P)$$

$$\Rightarrow \qquad P^{-1}TP = ISI \qquad\qquad [\because P^{-1}P = I = PP^{-1}]$$

$$\Rightarrow \qquad P^{-1}TP = S$$

$$\Rightarrow \qquad S = P^{-1}TP$$

$$\Rightarrow \qquad S = P^{-1}T(P^{-1})^{-1} \qquad\qquad [(P^{-1})^{-1} = P]$$

$$\Rightarrow \quad S \text{ is similar to } T.$$

$\therefore \quad$ If T is similar to S, then S is similar to T.

Transitive. Let $T_1, T_2, T_3 \in L(V, V)$ such that T_1 is similar to T_2 and T_2 is similar to T_3. Now T_1 is similar to T_2, then there exists an invertible linear transformation $P \in L(V, V)$ such that

$$T = PT_2P^{-1}$$

Also T_2 is similar to T_3, then there exists an invertible linear transformation $Q \in L(V, V)$ such that

$$T_2 = QT_3Q^{-1}$$

Now $$T_1 = PT_2P^{-1}$$
$$\Rightarrow \qquad T_1 = P(QT_3Q^{-1})P^{-1}$$
$$\Rightarrow \qquad T_1 = (PQ)T_3(Q^{-1}P^{-1})$$
$$\Rightarrow \qquad T_1 = (PQ)T_3(PQ)^{-1} \qquad \qquad [\because (PQ)^{-1} = Q^{-1}P^{-1}]$$
$\Rightarrow \quad T_1$ is similar and T_3.

$\Rightarrow \quad T_1$ is similar to T_2 and T_2 is similar to T_3 then T_1 is similar to T_3.

Hence the relation of similarity of linear transformations on a vector space $V(F)$ is an equivalence relation.

THEOREM 2. *Let T be a linear operator on an n-dimensional vector space V(F) and let B and B′ be two ordered bases for V. Then the matrix of T relative to B′ is similar to the matrix of T relative to B.*

Proof. Since dim. $V = n$. Let $B = \{\alpha_1, \alpha_2 ..., \alpha_n\}$ and $B' = (\beta_1, \beta_2,..., \beta_n)$ any two ordered bases of V.

Let $[T]_B = [a_{ij}]_{m \times n}$ be the matrix of T relative to B.

and $[T]_{B'} = [b_{ij}]_{n \times n}$ be the matrix of T relative to B'.

Then we have

$$T(\alpha_j) = \sum_{i=1}^{n} a_{ij}\alpha_i, j = 1,2,...n \qquad \qquad ...(1)$$

and

$$T(\beta_j) = \sum_{i=1}^{n} b_{ij}\beta_i, j = 1,2,...n \qquad \qquad ...(2)$$

Let S be the linear operator on V defined by

$$S(\alpha_j) = \beta_j, j = 1, 2,..., n \qquad \qquad ...(3)$$

Clearly, S maps a basis B onto a basis B', therefore S is necessary invertible. Let $[S]_B$ be the matrix of S relative to B, then $[S]_B$ is also invertible.

Now if $[S]_B = [p_{ij}]_{n \times n}$, then

$$S(\alpha_j) = \sum_{i=1}^{n} p_{ij}\alpha_i, j = 1,2,...n \qquad \qquad ...(4)$$

We have, $T(\beta_j) = T[S(\alpha_j)]$ [using (3)]

$$\Rightarrow \qquad T(\beta_j) = T\left[\sum_{i=1}^{n} p_{ij}\alpha_i\right] \qquad \qquad \text{[Using (4)]}$$

$$\Rightarrow \qquad T(\beta_j) = T\left[\sum_{k=1}^{n} p_{kj}\alpha_k\right] \qquad \text{[Replacing } i \text{ by } k \text{ which is immaterial]}$$

$$\Rightarrow \qquad T(\beta_j) = \sum_{i=1}^{n} p_{kj}T(\alpha_k) \qquad \qquad [\because T \text{ is linear]}$$

$$\Rightarrow \qquad T(\beta_j) = \sum_{k=1}^{n} p_{kj}\sum_{i=1}^{n}\alpha_{ik}\alpha_i \qquad \qquad \text{[Using (1)]}$$

$$T(\beta_j) = \sum_{i=1}^{n}\left(\sum_{i=1}^{n} a_{ik}p_{kj}\right)\alpha_i \qquad \qquad ...(5)$$

From (2), on replacing i by k, we have

$$T(\beta_j) = \sum_{k=1}^{n} b_{ki}\beta_k$$

$$\Rightarrow \qquad T(\beta_j) = \sum_{k=1}^{n} b_{ki}S(\alpha_k) \qquad \text{[Using (3)]}$$

$$\Rightarrow \qquad T(\beta_j) = \sum_{k=1}^{n} b_{ki}\sum_{i=1}^{n} p_{ki}\alpha_i \qquad \text{[Using (4) on replacing } j \text{ by } k]$$

$$\Rightarrow \qquad T(\beta_j) = \sum_{i=1}^{n}\left(\sum_{k=1}^{n} p_{ik}b_{kj}\right)\alpha_i \qquad \text{...(6)}$$

Now from (5) and (6), we have

$$\sum_{i=1}^{n}\left(\sum_{k=1}^{n} a_{ik}p_{kj}\right)\alpha_i = \sum_{i=1}^{n}\left(\sum_{i=1}^{n} p_{ik}b_{kj}\right)\alpha_i$$

$$\Rightarrow \qquad \sum_{i=1}^{n} a_{ik}p_{kj} = \sum_{i=1}^{n} p_{ik}b_{kj}$$

$$\Rightarrow \qquad [a_{ik}]_{n\times n}[p_{kj}]_{n\times n} = [p_{ik}]_{n\times n}[b_{kj}]_{n\times n} \qquad \text{[By matrix multiplication]}$$

$$\Rightarrow \qquad [T]_B[S]_B = [S]_B[T]_B$$

$$\Rightarrow \qquad [S]_B^{-1}[T]_B[S]_B = [T]_B$$

Hence $[T]_B$ is similar to $[T]_{B'}$.

THEOREM 3. *Let V be an n-dimensional vector space over the field F and T_1 and T_2 be two linear operators on V. If there exists two ordered bases B and B' for V such that $[T_1]_B = [T_2]_{B'}$, then T_2 is similar to T_1.*

Proof. dim. $V = n$. Let $B = \{\alpha_1, \alpha_2, ..., \alpha_n\}$ and $B' = \{\beta_1, \beta_2, ... \beta_n\}$ be two ordered bases for V.

Let $$[T_1]_B = [T_2]_{B'} = [a_{ij}]_{n\times n}$$

Then we have $$T_1(\alpha_j) = \sum_{i=1}^{n} a_{ij}\alpha_i, j = 1,2,...,n \qquad \text{...(1)}$$

and $$T_2(\beta_j) = \sum_{i=1}^{n} a_{ij}\beta_i, i = 1,2,...,n \qquad \text{...(2)}$$

Let S be the linear operator on V defined by

$$S(\alpha_j) = \beta_j, j = 1, 2, ..., n \qquad \text{...(3)}$$

Clearly, S maps a basis B onto a basis B' of V, therefore, S is invertible.

Now we have

$$T_2(\beta_j) = T_2[S(\alpha_j)] \qquad \text{[using (3)]}$$

$$\Rightarrow \qquad T_2(\beta_j) = (T_2S)(\alpha_j). \qquad \text{...(4)}$$

From (2), $$T_2(\beta_j) = \sum_{i=1}^{n} a_{ij}\beta_i$$

$$\Rightarrow \qquad T_2(\beta_j) = \sum_{i=1}^{n} a_{ij}S(\alpha_i) \qquad \text{[Using (3)]}$$

$$\Rightarrow \qquad T_2(\beta_j) = S\left(\sum_{i=1}^{n} a_{ij}\alpha_i\right) \qquad\qquad [\because S \text{ is linear.}]$$

$$\Rightarrow \qquad T_2(\beta_j) = S[T_1((\alpha_j))]$$

$$\Rightarrow \qquad T_2(\beta_j) = (ST_1)(\alpha_j) \qquad\qquad\qquad \dots(5)$$

From (4) and (5), we have

$$(T_2 S)(\alpha_j) = (ST_1)(\alpha_j), j = 1, 2, \dots, n$$

Since $T_2 S$ and ST_1, agree on a basis B on V, then we have

$$T_2 S = ST_1$$

$$\Rightarrow \qquad T_2 SS^{-1} = ST_1 S^{-1} \qquad\qquad [\because S^{-1} \text{ exists.}]$$

$$\Rightarrow \qquad T_2 = ST_1 S^{-1}$$

Hence, T_2 is similar to T_1.

6.17 DETERMINANT OF A LINEAR TRANSFORMATION ON A FINITE DIMENSIONAL VECTOR SPACE

Let $V(F)$ be an n-dimensional vector space and T be a linear operator on V. If B and B' be two ordered bases of V, then the matrices $[T]_B$ and $[T]_{B'}$ are similar and similar matrices have the same determinant. Then the determinant of T is the determinant of the matrix of T relative to any ordered basis for V.

6.18 SCALAR TRANSFORMATION

Let $V(F)$ be a vector space and T be a linear transformation on V. Then for a fixed scalar $c \in F$, the linear transformation T on V is said to be a scalar transformation of V if

$$T(\alpha) = c\alpha \ \forall \ \alpha \in V$$

Also, we may write $T = cI$, where I is the identity transformation on V.

6.19 TRACE OF A MATRIX

Let there be a square matrix of order n over a field F. Then the trace of A, denoted by trace A or $tr\ A$, is the sum of the elements of A lying along the principal diagonal.

If $A = [a_{ij}]_{n \times n}$, then $\qquad tr\ A = a_{11} + a_{22} + \dots + a_{nn} = \displaystyle\sum_{i=1}^{n} a_{ii}.$

THEOREM 1. *Let A and B be two square matrices of order n over a field F and $\lambda \in F$. Then*

    (i) $tr\ (\lambda A) = \lambda\ tr\ A$

    (ii) $tr\ (A + B) = tr\ A + tr\ B$

    (iii) $tr\ (AB) = tr\ (BA)$

Proof. Let $A = [a_{ij}]_{n \times n}$ and $B = [b_{ij}]_{n \times n}$ and $\lambda \in F$. Then

$$\lambda A = [\lambda a_{ij}]_{n \times n}.$$

(i) $\qquad tr\ (\lambda A) = \displaystyle\sum_{i=1}^{n} \lambda a_{ii}$

$$= \lambda \sum_{i=1}^{n} a_{ij} = \lambda\ tr\ A$$

(ii) $$A + B = [a_{ij} + b_{ij}]_{n \times n}, \text{ then}$$

$$\text{tr}\,(A + B) = \sum_{i=1}^{n} (a_{ii} + b_{ii})$$

$$= \sum_{i=1}^{n} a_{ii} + \sum_{i=1}^{n} b_{ii} = \text{tr}\,A + \text{tr}\,B$$

(iii) By the multiplication of matrices, we have

$$AB = [c_{ij}]_{n \times n}$$

where $$c_{ij} = \sum_{k=1}^{n} a_{ik} b_{kj} \qquad \qquad ...(1)$$

and $$BA = [d_{ij}]_{n \times n} \text{ where } d_{ij} = \sum_{k=1}^{n} b_{ik} a_{kj} \qquad \qquad ...(2)$$

Now, $$\text{tr}\,(AB) = \sum_{i=1}^{n} c_{ii}$$

$$= \sum_{i=1}^{n} \left(\sum_{k=1}^{n} a_{ik} b_{ki} \right) \qquad \qquad \text{[Using (1)]}$$

$$= \sum_{k=1}^{n} \left(\sum_{i=1}^{n} a_{ik} b_{ki} \right)$$

(Interchanging the order of summation)

$$= \sum_{k=1}^{n} \left(\sum_{i=1}^{n} b_{ki} a_{ik} \right)$$

$$= \sum_{i=1}^{n} d_{kk} = \text{tr}(BA).$$

THEOREM 2. *Similar matrices have the same trace.*

Proof. Let A and B be two similar matrices. Then there exist an invertible matrix C such that

$$B = C^{-1}AC$$

Let $D = C^{-1}A$, then

$$B = DC$$

$\Rightarrow$ $\quad$ $\text{tr}\,B = \text{tr}\,(DC)$

$\Rightarrow$ • $\quad$ $\text{tr}\,B = \text{tr}\,(CD)$ $\qquad \qquad [\because \text{tr}\,(AB) = \text{tr}\,(BA)]$

$\Rightarrow$ $\quad$ $\text{tr}\,B = \text{tr}\,(CC^{-1}A)$

$\Rightarrow$ $\quad$ $\text{tr}\,B = \text{tr}\,(IA)$

$\Rightarrow$ $\quad$ $\text{tr}\,B = \text{tr}\,A$

6.20 TRACE OF A LINEAR TRANSFORMATION ON A FINITE DIMENSIONAL VECTOR SPACE

Let V be an n-dimensional vector space over a field F and T be a linear operator on V. If B and B' are two ordered bases of V, then the matrices $[T]_B$ and $[T]_{B'}$ are similar and the similar matrices have the same trace. Thus, the *trace of a linear transformation T* is the trace of the matrix of T relative to any ordered basis for V.

Solved Examples

Based on the following Results

➡ For any matrix A, the trace of A is the sum of the elements of A lying along the principal diagonal

➡ $tr(\lambda A) = \lambda\, tr(A)$

➡ $(A^{-1})^{-1} = A$

➡ $tr(A+B) = tr(A) + tr(B)$

➡ $tr(AB) = tr(BA)$

➡ Similar matrices have the same trace.

➡ The trace of a linear transformation T is the trace of the matrix of T relative to any ordered basis for V.

Example 1. *Let $T: R^2(R) \to R^2(R)$, where for any $(x, y) \in R^2$, $T(x, y) = \left(2x, \dfrac{1}{2}y\right)$. Find the matrix associated with T with respect to the ordered basis $\{(1, 0), (0, 1)\}$.*

Solution. Let $B = \{(1, 0), (0,1)\}$ be an ordered basis of $R^2(R)$ and $T(x, y) = \left(2x, \dfrac{1}{2}y\right)$, then

$$T(1, 0) = (2, 0) \text{ and } T(0,1) = \left(0, \frac{1}{2}\right)$$

Now, $$T(1, 0) = (2, 0) = 2(1, 0) + 0.(0, 1)$$

and $$T(0, 1) = \left(0, \frac{1}{2}\right) = 0(1, 0) + \frac{1}{2}(0, 1).$$

Thus, the matrix associated with T w.r.t. B is

$$[T]_B = \begin{bmatrix} 2 & 0 \\ 0 & \dfrac{1}{2} \end{bmatrix}.$$

Example 2. *Find the matrix of the linear transformation T on $V_3(R)$ defined as $T(a, b, c) = (2b + c, a - 4b, 3a)$ with respect to the ordered basis B and also with respect to the ordered basis B' where*

(i) $B = \{(1, 0, 0),(0, 1, 0),(0, 0, 1)\}$

(ii) $B' = \{(1, 1, 1),(1, 1, 0),(1, 0, 0)\}$

Solution. (i)   We have

$$T(a, b, c) = (2b + c, a - 4b, 3a)$$

Then $$T(1, 0, 0) = (0, 1, 3)$$
$$T(0, 1, 0) = (2, -4, 0)$$
$$T(0, 0, 1) = (1, 0, 0)$$

Now $$T(1, 0, 0) = 0(1, 0, 0) + 1(0, 1, 0) + 3(0, 0, 1) \qquad ...(1)$$
$$T(0, 1, 0) = 2(1, 0, 0) - 4(0, 1, 0) + 0(0, 0, 1) \qquad ...(2)$$
$$T(0, 0, 1) = 1(1, 0, 0) + 0(0, 1, 0) + 0(0, 0, 1) \qquad ...(3)$$

The matrix of T relative to B is given by

$$[T]_B = \begin{bmatrix} 0 & 2 & 1 \\ 1 & -4 & 0 \\ 3 & 0 & 0 \end{bmatrix}$$

(ii) Let (a, b, c) be any element of $V_3(\mathrm{R})$, then there exists $x, y, z \in \mathrm{R}$ such that

$$(a, b, c) = x\,(1, 1, 1) + y\,(1, 1, 0) + z\,(1, 0, 0)$$

$$\Rightarrow \quad (a, b, c) = (x + y + z, x + y, x)$$

$$\Rightarrow \quad x + y + z = a, x + y = b, x = c$$

$$\therefore \quad x = c, y = b - c, z = a - b$$

$$\therefore \quad (a, b, c) = c\,(1, 1, 1) + (b - c)\,(1, 1, 0) + (a - b)\,(1, 0, 0) \quad ...(1)$$

Now $T(1, 1, 1) = (3, -3, 3)$, $T(1, 1, 0) = (2, -3, 3)$ and $T(1, 0, 0) = (0, 1, 3)$. Then from (1), we get

$$T(1, 1, 1) = (3, -3, 3) = 3(1, 1, 1) - 6(1, 1, 0) + 6(1, 0, 0)$$

$$T(1, 1, 0) = (2, -3, 3) = 3(1, 1, 1) - 6(1, 1, 0) + 5(1, 0, 0)$$

$$T(1, 0, 0) = (0, 1, 3) = 3\,(1, 1, 1) - 2\,(1, 1, 0) - 1(\,1, 0, 0)$$

Therefore, the matrix of T relative to B' is given by

$$[T]_B = \begin{bmatrix} 3 & 3 & 3 \\ -6 & -6 & -2 \\ 6 & 5 & -1 \end{bmatrix}$$

Example 3. *Let T be a linear operator on R^2 defined by $T(x, y) = (2y, 3x - y)$. Find the matrix representation of T relative to the basis $\{(1, 3), (2, 5)\}$.*

Solution. Let (x, y) be any element of R^2. Then there exist $a, b \in \mathrm{R}$ such that

$$(x, y) = a\,(1, 3) + b\,(2, 5)$$

$$\Rightarrow \quad (x, y) = (a + 2b, 3a + 5b)$$

$$\Rightarrow \quad a + 2b = x, 3a + 5b = y$$

Solving these equations, we get

$$a = 2y - 5x, b = 3x - y,$$

$$\therefore \quad (x, y) = (2y - 5x)(1, 3) + (3x - y)(2, 5) \quad ...(1)$$

Since, $\quad T(x, y) = (2y, 3x - y)$

Then, $\quad T(1, 3) = (6, 0), T(2, 5) = (10, 1)$

Now from (1) $\quad T(1, 3) = (6, 0) = -30(1, 3) + 18(2, 5)$

$$T(2, 5) = (10, 1) = -48\,(1, 3) + 29\,(2, 5)$$

Therefore, the matrix of T relative to the given basis is $\begin{bmatrix} -30 & -48 \\ 18 & 29 \end{bmatrix}$.

Example 4. *Show that the vector $\alpha_1 = (1, 0, -1)$, $\alpha_2 = (1, 2, 1)$, $\alpha_3 = (0, -3, 2)$ form a basis for R^3. Express each of the standard basis vectors as a linear combination of $\alpha_1, \alpha_2, \alpha_3$.*

Solution. Let $B' = \{\alpha_1, \alpha_2, \alpha_3\}$. First, we shall show that B' is linearly independent.

Let $a, b, c \in \mathrm{R}$ such that $\quad a\alpha_1 + b\alpha_2 + c\alpha_3 = 0$

$$\Rightarrow \quad a(1, 0, -1) + b\,(1, 2, 1) + c\,(0, -3, 2) = (0, 0, 0)$$

$$\Rightarrow \quad (a + b, 2b - 3c, -a + b + 2c) = (0, 0, 0) \quad ...(1)$$

$$\Rightarrow \quad a + b = 0, 2b - 3c = 0, -a + b + 2c = 0$$

The coefficient matrix of these equations is

$$A = \begin{bmatrix} 1 & 1 & 0 \\ 0 & 2 & -3 \\ -1 & 1 & 2 \end{bmatrix}$$

$$\Rightarrow \quad |A| = \begin{vmatrix} 1 & 1 & 0 \\ 0 & 2 & -3 \\ -1 & 1 & 2 \end{vmatrix} = 1(4 + 3) - 1(0 - 3) = 10 \neq 0$$

$\Rightarrow$ rank of $A = 3$, which is the number of variables a, b, c. Hence, the system of equation (1) has only zero solution, i.e. $a = 0$, $b = 0$, $c = 0$.

Therefore, B' is linearly independent containing 3 elements since dim. $R_3 = 3$, hence B' forms a basis for R^3.

Let $B = \{e_1, e_2, e_3\}$ be the standard basis for R^3, where $e_1 = (1, 0, 0)$, $e_2 = (0, 1, 0)$, $e_3 = (0, 0, 1)$.

Now, we have
$$\alpha_1 = (1, 0, -1) = 1e_1 + 0e_2 - 1e_3$$
$$\alpha_2 = (1, 2, 1) = 1e_1 + 2e_2 + 1e_3$$
$$\alpha_3 = (0, -3, 2) = 0e_1 - 3e_2 + 2e_3$$

Let P be the transition matrix from the basis B to B', then

$$P = \begin{bmatrix} 1 & 1 & 0 \\ 0 & 2 & -3 \\ -1 & 1 & 2 \end{bmatrix} \qquad \therefore |P| = 10$$

Now we shall find P^{-1} :

The cofactors of the elements of the first row of P are

$$\begin{vmatrix} 2 & -3 \\ 1 & 2 \end{vmatrix}, -\begin{vmatrix} 0 & -3 \\ -1 & 2 \end{vmatrix}, \begin{vmatrix} 0 & 2 \\ -1 & 1 \end{vmatrix}, i.e. 7, 3, 2$$

The cofactors of the elements of the second row of P are

$$-\begin{vmatrix} 1 & 0 \\ 1 & 2 \end{vmatrix}, \begin{vmatrix} 1 & 0 \\ -1 & 2 \end{vmatrix}, -\begin{vmatrix} 1 & 1 \\ -1 & 1 \end{vmatrix}, i.e. -2, 2, -2$$

The cofactors of the elements of the third row of P are

$$\begin{vmatrix} 1 & 0 \\ 2 & -3 \end{vmatrix}, -\begin{vmatrix} 1 & 0 \\ 0 & -3 \end{vmatrix}, \begin{vmatrix} 1 & 1 \\ 0 & 2 \end{vmatrix}, i.e. -3, 3, -2$$

$$\text{adj. } P = \text{transpose of the matrix} \begin{bmatrix} 7 & 3 & 2 \\ -2 & 2 & -2 \\ -3 & 3 & 2 \end{bmatrix}$$

$$= \begin{bmatrix} 7 & -2 & -3 \\ 3 & 2 & 2 \\ 2 & -2 & 2 \end{bmatrix}$$

$$P^{-1} = \frac{\text{adj.} A}{|P|} = \frac{1}{10} \begin{bmatrix} 7 & -2 & -3 \\ 3 & 2 & 2 \\ 2 & -2 & 2 \end{bmatrix}$$

Now $e_1 = 1e_1 + 0e_2 + 0e_3$.

Therefore, the coordinate matrix of e_1 relative to B is $[e_1]_B = \begin{bmatrix} 1 \\ 0 \\ 0 \end{bmatrix}$

So that the coordinate matrix of e_1 relative to B' is

$$[e_1]_{B'} = P^{-1}[e_1]_B = \frac{1}{10} \begin{bmatrix} 7 & -2 & -3 \\ 3 & 2 & 2 \\ 2 & -2 & 2 \end{bmatrix} \begin{bmatrix} 1 \\ 0 \\ 0 \end{bmatrix} = \frac{1}{10} \begin{bmatrix} 7 \\ 3 \\ 2 \end{bmatrix} = \begin{bmatrix} 7/10 \\ 3/10 \\ 1/5 \end{bmatrix}$$

$\therefore$ $$e_1 = \frac{7}{10}\alpha_1 + \frac{3}{10}\alpha_2 + \frac{1}{5}\alpha_3.$$

Similarly, $[e_2]_B = \begin{bmatrix} 0 \\ 1 \\ 0 \end{bmatrix}, [e_3]_B = \begin{bmatrix} 0 \\ 0 \\ 1 \end{bmatrix}$

$$\therefore \quad \left[e_2\right]_{B'} = P^{-1}\begin{bmatrix} 0 \\ 1 \\ 0 \end{bmatrix} = \frac{1}{10}\begin{bmatrix} -2 \\ 2 \\ -2 \end{bmatrix} = \begin{bmatrix} -1/5 \\ 1/5 \\ -1/5 \end{bmatrix}$$

$$\left[e_3\right]_{B'} = P^{-1}\begin{bmatrix} 0 \\ 0 \\ 1 \end{bmatrix} = \frac{1}{10}\begin{bmatrix} -3 \\ 3 \\ 2 \end{bmatrix} = \begin{bmatrix} -3/15 \\ 3/10 \\ 1/5 \end{bmatrix}$$

$$\therefore \qquad e_2 = -\frac{1}{5}\alpha_1 + \frac{1}{5}\alpha_2 - \frac{1}{5}\alpha_3$$

$$e_3 = -\frac{3}{10}\alpha_1 + \frac{3}{10}\alpha_2 + \frac{1}{5}\alpha_3.$$

Example 5. *Let T be a linear operator on* R^3 *defined by*

$$T(x, y, z) = (3x + z, -2x + y, -x + 2y + 4z)$$

Prove that T is invertible and find a formula for T^{-1}.

Solution. Let $B = \{(1, 0, 0), (0, 1, 0), (0, 0, 1)\}$ be the standard basis for R^3.
Let A be the matrix of T relative to B, then

$$A = [T]_B$$

Now, $T(1, 0, 0) = (3, -2, 1) = 3(1, 0, 0) - 2(0, 1, 0) - 1(0, 0, 1)$

$$T(0, 1, 0) = (0, 1, 2) = 0(1, 0, 0) + 1(0, 1, 0) + 2(0, 0, 1)$$

and $T(0, 1, 1) = (1, 0, 4) = 1(1, 0, 0) + 0(0, 1, 0) + 4(0, 0, 1)$

$$\therefore \quad A = [T]_B = \begin{bmatrix} 3 & 0 & 1 \\ -2 & 1 & 0 \\ -1 & 2 & 4 \end{bmatrix}$$

Now $\quad |A| = \begin{vmatrix} 3 & 0 & 1 \\ -2 & 1 & 0 \\ -1 & 2 & 4 \end{vmatrix} = 3(4 - 0) + 1(-4 + 1) = 9 \neq 0$

Since $|A| \neq 0$, therefore A is invertible and hence T is invertible.
Now we shall find A^{-1}. For this we find adj. A.
The cofactors of the first row of A are

$$\begin{vmatrix} 1 & 0 \\ 2 & 4 \end{vmatrix}, \begin{vmatrix} -2 & 0 \\ -1 & 4 \end{vmatrix}, \begin{vmatrix} -2 & 1 \\ -1 & 4 \end{vmatrix}, i.e.\, 4, 8, -3$$

The cofactors of the second row of A are

$$-\begin{vmatrix} 0 & 1 \\ 2 & 4 \end{vmatrix}, \begin{vmatrix} 3 & 1 \\ -1 & 4 \end{vmatrix}, \begin{vmatrix} 3 & 0 \\ -1 & 2 \end{vmatrix}, i.e.\, 2, 13, -6$$

The cofactors of the third row of A are

$$\begin{vmatrix} 0 & 1 \\ 1 & 0 \end{vmatrix}, -\begin{vmatrix} 3 & 1 \\ -2 & 0 \end{vmatrix}, \begin{vmatrix} 3 & 0 \\ -2 & 1 \end{vmatrix}, i.e. -1, -2, 3$$

adj. A = transpose of the cofactors matrix

$$= \begin{bmatrix} 4 & 2 & -1 \\ 8 & 13 & -2 \\ -3 & -6 & 3 \end{bmatrix}$$

$$A^{-1} = \frac{adj.A}{|A|} = \frac{1}{9}\begin{bmatrix} 4 & 2 & -1 \\ 8 & 13 & -2 \\ -3 & -6 & 3 \end{bmatrix}$$

Since we know that
$$[T^{-1}]_B = [T]_B^{-1} = A^{-1}$$
Now we shall find the formula for T^{-1}.

Let $\alpha = (p, q, r)$ be any element of R^3 and B is a standard basis for R^3. Then

$$[\alpha]_B = \begin{bmatrix} p \\ q \\ r \end{bmatrix}$$

$$[T^{-1}(\alpha)]_B = [T^{-1}]_B[\alpha]_B = A^{-1}[\alpha]_B = \frac{1}{9}\begin{bmatrix} 4 & 2 & -1 \\ 8 & 13 & -2 \\ -3 & -6 & 3 \end{bmatrix}\begin{bmatrix} p \\ q \\ r \end{bmatrix}$$

$$[T^{-1}(\alpha)]_B = \frac{1}{9}\begin{bmatrix} 4p + 2q - r \\ 8p + 13q - 2r \\ -3p - 6q + 3r \end{bmatrix}$$

$$T^{-1}(\alpha) = T^{-1}(p,q,r) = \left(\frac{4p + 2q - r}{9}, \frac{8p + 13q - 2r}{9}, \frac{-3p - 6q + 3r}{9} \right)$$

Example 6. *Consider the vector space $V(R)$ of all 2×2 matrices over the field R of real numbers. Let T be the linear transformation on V sending each matrix X onto AX, where $A = \begin{bmatrix} 1 & 1 \\ 1 & 1 \end{bmatrix}$. Find the matrix of T with respect to the ordered basis $B = \{E_1, E_2, E_3, E_4\}$, for V where*

$$E_1 = \begin{bmatrix} 1 & 0 \\ 0 & 0 \end{bmatrix}, E_2 = \begin{bmatrix} 0 & 1 \\ 0 & 0 \end{bmatrix}, E_3 = \begin{bmatrix} 0 & 0 \\ 1 & 0 \end{bmatrix}, E_4 = \begin{bmatrix} 0 & 0 \\ 0 & 1 \end{bmatrix}$$

Solution. We have
$$T(X) = AX$$

Then
$$T(E_1) = AE_1 = \begin{bmatrix} 1 & 1 \\ 1 & 1 \end{bmatrix}\begin{bmatrix} 1 & 0 \\ 0 & 0 \end{bmatrix} = \begin{bmatrix} 1 & 0 \\ 0 & 0 \end{bmatrix}$$

$$= 1\begin{bmatrix} 1 & 0 \\ 0 & 0 \end{bmatrix} + 0\begin{bmatrix} 0 & 1 \\ 0 & 0 \end{bmatrix} + 1\begin{bmatrix} 0 & 0 \\ 1 & 0 \end{bmatrix} + 0\begin{bmatrix} 0 & 0 \\ 0 & 1 \end{bmatrix}$$

$\therefore$
$$T(E_1) = 1E_1 + 0E_2 + 1E_3 + 0E_4$$

$$T(E_2) = AE_2 = \begin{bmatrix} 1 & 1 \\ 1 & 1 \end{bmatrix}\begin{bmatrix} 0 & 1 \\ 0 & 0 \end{bmatrix} = \begin{bmatrix} 0 & 1 \\ 0 & 0 \end{bmatrix}$$

$$= 0\begin{bmatrix} 1 & 0 \\ 0 & 0 \end{bmatrix} + 1\begin{bmatrix} 0 & 1 \\ 0 & 0 \end{bmatrix} + 0\begin{bmatrix} 0 & 0 \\ 1 & 0 \end{bmatrix} + 1\begin{bmatrix} 0 & 0 \\ 0 & 1 \end{bmatrix}$$

$\therefore$
$$T(E_2) = 0E_1 + 1E_2 + 0E_3 + 1E_4$$

$$T(E_3) = AE_3 = \begin{bmatrix} 1 & 1 \\ 1 & 1 \end{bmatrix}\begin{bmatrix} 0 & 0 \\ 1 & 0 \end{bmatrix} = \begin{bmatrix} 0 & 0 \\ 1 & 0 \end{bmatrix}$$

$$= 1\begin{bmatrix} 1 & 0 \\ 0 & 0 \end{bmatrix} + 0\begin{bmatrix} 0 & 1 \\ 0 & 0 \end{bmatrix} + 1\begin{bmatrix} 0 & 0 \\ 1 & 0 \end{bmatrix} + 0\begin{bmatrix} 0 & 0 \\ 0 & 1 \end{bmatrix}$$

$\therefore$
$$T(E_3) = 1E_1 + 0E_2 + 1E_3 + 0E_4$$

and
$$T(E_4) = AE_4 = \begin{bmatrix} 1 & 1 \\ 1 & 1 \end{bmatrix}\begin{bmatrix} 0 & 0 \\ 0 & 1 \end{bmatrix} = \begin{bmatrix} 0 & 0 \\ 0 & 1 \end{bmatrix}$$

$$= 0\begin{bmatrix} 1 & 0 \\ 0 & 0 \end{bmatrix} + 1\begin{bmatrix} 0 & 1 \\ 0 & 0 \end{bmatrix} + 0\begin{bmatrix} 0 & 0 \\ 1 & 0 \end{bmatrix} + 1\begin{bmatrix} 0 & 0 \\ 0 & 1 \end{bmatrix}$$

$$\therefore \qquad T(E_4) = 0E_1 + 1E_2 + 0E_3 + 1E_4$$

The matrix of T relative to B is $[T]_B = \begin{bmatrix} 1 & 0 & 1 & 0 \\ 0 & 1 & 0 & 1 \\ 1 & 0 & 1 & 0 \\ 0 & 1 & 0 & 1 \end{bmatrix}$.

Example 7. *Let a linear map $T : P_3 \to P_2$ be defined by*

$$T(a_0 + a_1 x + a_2 x^2 + a_3 x^3) = a_3 + (a_2 + a_3)x + (a_0 + a_1)x^2$$

where $P_n[x]$ = set of all polynomials of degree $\leq n$. Find the matrix of T with respect to the ordered bases $B = \{1, (x-1), (x-1)^2, (x-1)^3\}$ and $B' = \{1, x, x^2\}$.

Solution. Since B and B' are the bases of P_3 and P_2, respectively, hence we shall express the images of each element of B in terms of an element of B'.

Now
$$T(1) = T(1 + 0 . x + 0 . x^2 + 0 . x^3)$$
$$= 0 + (0 + 0)x + (1 + 0)x^2 = x^2$$

$\therefore \qquad T(1) = 0.1 + 0 . x + 1 . x^2$

$\therefore \qquad T(x - 1) = T(-1 + 1 . x + 0 . x^2 + 0 . x^3)$
$$= 0 + (0 + 0)x + (-1 + 1)x^2 = 0$$

$\Rightarrow \qquad T(x - 1) = 0 . 1 + 0 . x + 0 . x^2$

$$T[(x - 1)^2] = T(1 - 2x + x^2)$$
$$= T(1 + (-2)x + 1 . x^2 + 0 . x^3)$$
$$= 0 + (1 + 0)x + (1 - 2)x^2$$
$$= x - x^2.$$

$\Rightarrow \qquad T[(x - 1)^2] = 0.1 + 1.x + (-1) x^2$

$\therefore \qquad T[(x - 1)^3] = T(-1 + 3x - 3x^2 + x^3)$
$$= 1 + (-3 + 1)x + (-1 + 3) x^2$$
$$= 1 - 2x + 2x^2$$

$\Rightarrow \qquad T[(x - 1)^3] = 1.1 (-2)x + 2x^2$

Thus, the matrix of T relative to the ordered bases B and B' is

$$_B[T]_{B'} = \begin{bmatrix} 0 & 0 & 0 & 1 \\ 0 & 0 & 1 & -2 \\ 1 & 0 & -1 & 2 \end{bmatrix}$$

Example 8. *Let A be an $m \times n$ matrix of real entries. Prove that $A = 0$ (null matrix) if and only if trace $(A^T A) = 0$.*

Solution. Let $A = [a_{ij}]_{m \times n}$, then $A^T = [b_{ij}]_{n \times m}$, where $b_{ij} = a_{ij}$

Also, $A^T A$ is a matrix of order $n \times n$.

Let
$$A^T A = [b_{ij}]_{n \times m} [a_{ij}]_{m \times n} = [c_{ij}]_{n \times n}$$

where
$$c_{ij} = \sum_{k=1}^{n} b_{ik} a_{kj}$$

$\therefore \qquad \text{tr}(A^T A) = \sum_{i=1}^{n} c_{ij} = \sum_{i=1}^{n} \left(\sum_{k=1}^{m} b_{ik} a_{ki} \right)$

$$= \sum_{i=1}^{n}\left(\sum_{k=1}^{m} a_{ki}a_{ki}\right) \qquad [\because b_{ik} = a_{ki}]$$

$$= \sum_{i=1}^{n}\left(\sum_{k=1}^{m} a_{ki}^2\right)$$

$$\therefore \qquad \text{tr}(A^T A) = \sum_{i=1}^{n}(a_{1i}^2 + a_{2i}^2 + \dots + a_{mi}^2)$$

If tr $(A^TA) = 0$, then from (1), we have

$$\sum_{i=1}^{n}(a_{1i}^2 + a_{2i}^2 + \dots + a_{mi}^2) = 0$$

$\Rightarrow$ the sum of the squares of all the elements of $A = 0$

$\Rightarrow$ each element of $A = 0$

$\Rightarrow$ A is a null matrix.

$\Rightarrow$ $A = 0$.

Conversely, If A is null matrix, then A^TA is also a null matrix.

$$\text{tr}(A^TA) = 0.$$

Example 9. *Let T and S be linear operators on the finite dimensional vector space $V(F)$, prove that*

(i) det $(TS) =$ det (T) det (S)

(ii) T is invertible iff det $T \neq 0$.

Solution.   (i)   Let B be any ordered basis of V then we have

$$[TS]_B = [T]_B[S]_B$$

$\Rightarrow$ det $([TS]_B) =$ det $([T]_B[S]_B)$

$\Rightarrow$ det $([TS]_B) =$ det $([T]_B)$ det $([S]_B)$.

Since the determinant of a linear transformation is equal to the determinant of its matrix with respect to any ordered basis, therefore

det $(TS) =$ det (T) det (S)

(ii)   If T is invertible, then there exists a linear transformation T^{-1} on V such that

$$T^{-1}T = I = TT^{-1}$$

$\Rightarrow$ det $(T^{-1}T) =$ det $(I) =$ det $([I]_B)$ [For any ordered basis B]

$\Rightarrow$ det (T^{-1}) det $(T) = 1$ $[\because [I]_B$ is a unit matrix.]

Now det (T) and det $(T^{-1}) \in F$ and F is a field and in a field the product of elements can be zero iff at least one of them is zero.

$\therefore$ det(T^{-1}) det$(T) = 1$

$\Rightarrow$ det $(T) \neq 0$.

Conversely, Suppose that det $(T) \neq 0$.

Then for any ordered basis B of V, we have

det$([T]_B) \neq 0$

$\Rightarrow$ $[T]_B$ is invertible.

$\Rightarrow$ T is invertible.

Example 10. If T and S are similar linear transformation on a finite dimensional vector space $V(F)$, then $\det T = \det S$.

Solution. Since T and S are similar linear transformations, then there exists an invertible linear transformation P on V such that

$$T = PS\,P^{-1}$$

$\Rightarrow \qquad \det T = \det (PS\,P^{-1})$

$\Rightarrow \qquad \det T = (\det P)\,(\det S)\,(\det P^{-1})$

$\Rightarrow \qquad \det T = (\det P)\,(\det P^{-1})\,(\det S)$

$\Rightarrow \qquad \det T = (\det P\,P^{-1})(\det S)$

$\Rightarrow \qquad \det T = (\det I)\,(\det S)$

$\Rightarrow \qquad \det T = \det S \qquad\qquad\qquad [\because \det I = 1]$

EXERCISE 6.2

1. Let T be the linear operator on R^2 defined by $T(x, y) = (4x - 2y, 2x + y)$. Compute the matrix of T relative to the basis $\{\alpha_1, \alpha_2\}$ where $\alpha_1 = (1, 1)$, $\alpha_2 = (-1, 0)$.

2. Find the matrix of the following linear maps with respect to the standard basis of R^n:

(i) $T : R \to R^2$ defined by $T(x) = (3x, 5x)$

(ii) $T : R^3 \to R^2$ defined by $T(x, y, z) = (2x - 4y + 9z, 5x + 3y - 2z)$.

3. Let T be the linear operator on R^2 defined by $T(x, y) = (4x - 2y, 2x + y)$.

(i) What is the matrix of T relative to the standard basis B of R^2 ?

(ii) Find the transition matrix P from the ordered basis B to the ordered basis $B' = \{\alpha_1, \alpha_2\}$, where $\alpha_1 = (1, 1)$, $\alpha_2 = (-1, 0)$. Hence find the matrix of T relative to the ordered basis B'.

4. Let T be the linear operator on R^2 defined by $T(a, b) = (a, 0)$. Write the matrix of T in the standard ordered basis $B = \{(1, 0), (0, 1)\}$. If $B' = \{(1, 1), (1, 2)\}$ is another ordered basis for R^2, find the transition matrix P from the basis B to the basis B'. Hence find the matrix of T relative to the basis B'

5. Find the matrix representation of the linear mappings relative to the usual bases for R^n.

(i) $T : R^3 \to R^2$ defined by $T(x, y, z) = (x, y, 0)$

(ii) $T : R^3 \to R^2$ defined by $T(x, y, z) = (z, y + z, x + y + z)$.

6. Let $V = R^3$ and $T : V \to V$ be a linear mapping defined by $T(x, y, z) = (x + z, -2x + y, -x + 2y + z)$. What is the matrix of T relative to the basis $B = \{(1, 0, 1), (-1, 1, 1), (0, 1, 1)\}$?

7. Let $T : V_2(R) \to V_3(R)$ be a linear transformation defined by $T(x, y) = (x + y, 2x - y, 7y)$. If B and B' are the standard bases of $V_2(R)$ and $V_3(R)$ respectively, then find the matrix of T with respect to these basis.

8. Let T be the linear operator on R^3 defined by $T(x, y, z) = (x + y + z, -x - y - 4z, 2x - z)$. What is the matrix of T in the ordered basis $B = \{(1, 1, 1), (0, 1, 1), (1, 0, 1)\}$?

9. If the matrix of a linear transformation $V_3(C)$ with respect to the basis $B = \{(1,0,0), (0,1,0), (0,0,1)\}$ is $\begin{bmatrix} 0 & 1 & 1 \\ 1 & 0 & -1 \\ -1 & -1 & 0 \end{bmatrix}$.

what is the matrix of T relative to the basis $B' = \{(0, 1, -1), (1, -1, 1), (-1, 0, 1)\}$.

10. Find the matrix relative to the basis $\{\alpha_1, \alpha_2, \alpha_3\}$ where, $\alpha_1 = \left(\dfrac{2}{3}, \dfrac{2}{3}, -\dfrac{1}{3}\right)$, $\alpha_2 = \left(\dfrac{1}{3}, -\dfrac{2}{3}, -\dfrac{2}{3}\right)$, $\alpha_3 = \left(\dfrac{2}{3}, -\dfrac{1}{3}, \dfrac{2}{3}\right)$ of R^3, of the linear transformation $T : R^3 \to R^3$ whose matrix relative to the standard basis is $\begin{bmatrix} 2 & 0 & 0 \\ 0 & 4 & 0 \\ 0 & 0 & 3 \end{bmatrix}$.

11. Let a linear map $T : V_3(R) \to V_3(R)$ be defined by

$$T(x, y, z) = (x - y + z, 2x + 3y + \dfrac{1}{2} z, x + y - 2z).$$

If B is the standard basis and another basis B' is $\{(1, 1, 0), (1, 2, 3), (-1, 0, 1)\}$, then find the matrix of T relative to the bases B and B'.

12. Let a linear map $T : V_3(\mathbb{R}) \to V_2(\mathbb{R})$ be defined by $T(e_1) = 2e_1{}' - e_2{}'$, $T(e_2) = e_1{}' + 2e_2{}'$ and $T(e_3) = 0e_1{}' + 0e_2{}'$, where $B = \{e_1, e_2, e_3\}$ and $B' = \{e_1{}', e_2{}'\}$ are the standard bases in $V_3(\mathbb{R})$ and $V_2(\mathbb{R})$ respectively. Find the matrix of T relative to these bases. If the bases are
$$B_1 = \{(1, 1, 0), (1, 0, 1), (0, 1, 1)\}$$
and $\quad B_2 = \{(1, 1), (1, -1)\}$
then find the matrix of T relative to the bases B_1 and B_2.

13. Let $D : P_2 \to P_1$ be the differential map defined by $D(p(x)) = p'(x)$. If $B = \{x^2, x, 1\}$ and $B' = \{x, 1\}$ be the ordered bases for P_3 and P_2, respectively, then find the matrix of D relative to the bases B and B', where P_n is a vector space of all polynomials of degree $\leq n$.

14. Explain what is meant by the matrix of a linear transformation on V relative to a basis of V. Let F be a field and V, the set of all polynomials in x over F of degree ≤ 5. If $D : V \to V$ is defined by $D[f(x)] = f'(x)$ where $f'(x)$ is the derivative of $f(x)$, show that D is a linear transformation on V. Find the matrix of D relative to the above bases.

15. The set $\{1, x, e^x, xe^x\}$ is a basis of a vector space V of functions $f : \mathbb{R} \to \mathbb{R}$. Let D be the differential operator on V defined by $D [f(t)] = f'(t)$. Find the matrix of D relative to the above bases.

16. The set $\{e^{3x}, xe^{3x}, x^2e^{3x}\}$ is a basis of a vector space V of function $f : \mathbb{R} \to \mathbb{R}$. Let D be the differential operator on V defined by $D[f(t)] = f'(t)$. Find the matrix of D relative to the above basis.

17. Let V be the vector space of those polynomial functions from reals into itself which have degree ≤ 3. Let $B = \{f_1, f_2, f_3, f_4\}$ where $f_i(x) = x^{i-1}$ $\{1 \leq i \leq 4\}$. Show that B forms basis for V. For any real numbers t let $g_i(x) = (x + t)^{i-1}$. Show that $B' = \{g_1, g_2, g_3, g_4\}$ is also a basis for V. If D is the differential operator on V, write the matrices of D in the ordered bases B and B'.

18. Let T be the linear operator on $\mathbb{R}^3$ defined by
$$T(x_1, x_2, x_3) = (3x_1 + x_3, -2x_1 + x_2, -x_1 + 2x_2 + 4x_3)$$
  (i) What is the matrix of T in the standard basis for $\mathbb{R}^3$?
  (ii) Find the transition matrix P from the ordered basis B to the ordered basis $B' = \{\alpha_1, \alpha_2, \alpha_3\}$ where $\alpha_1 = (1, 0, 1)$, $\alpha_2 = (-1, 2, 1)$, $\alpha_3 = (2, 1, 1)$. Hence find the matrix of T relative to the ordered basis B'.

19. Let $A : \mathbb{R}^2 \to \mathbb{R}^2$ be a linear operator by the matrix $A = \begin{bmatrix} 5 & -7 \\ 2 & 3 \end{bmatrix}$. Let B be the matrix representation of A relative to the basis $\{(1, 4), (3, 10)\}$. Find B'.

20. Let V be a vector space of all 2×2 matrices over $\mathbb{R}$ and let $T : V \to V$ be a linear operator defined by $T(A) = MA$ where $M = \begin{bmatrix} 1 & 2 \\ 3 & 4 \end{bmatrix}$. Find the matrix of T with respect to the ordered basis $B = \{E_1, E_2, E_3, E_4\}$ for V where
$$E_1 = \begin{bmatrix} 1 & 0 \\ 0 & 0 \end{bmatrix}, E_2 = \begin{bmatrix} 0 & 1 \\ 0 & 0 \end{bmatrix},$$
$$E_3 = \begin{bmatrix} 0 & 0 \\ 1 & 0 \end{bmatrix}, E_4 = \begin{bmatrix} 0 & 0 \\ 0 & 1 \end{bmatrix}$$

21. Let $T : \mathbb{R}^3 \to \mathbb{R}^3$ be a linear operator defined by
$$T(x, y, z) = (2x - z, x + 2y - 4z, 3x - 3y + z)$$
Find the trace of T.

22. Find the trace of the following operator on $\mathbb{R}^3$:
$$T(x, y, z) = (a_1x + a_2y + a_3z, b_1x + b_2y + b_3z, c_1x + c_2y + c_3z).$$

23. Consider the complex field $\mathbb{C}$ as a vector space over the real field $\mathbb{R}$. Let T be the conjugate operator on $\mathbb{C}$ defined by $T(z) = z$. Find det. T and tr (T).

24. Let V be the vector space of all 2×2 matrices over the field F and let P be a fixed 2×2 matrix over F. Let T be the linear operator on V defined by $T(A) = PA \;\forall\; A \in V$. Prove that trace $(T) = 2$ trace (P).

25. Let V be the vector space of 2×2 matrices over $\mathbb{R}$ and let $M = \begin{bmatrix} 1 & 2 \\ 3 & 4 \end{bmatrix}$. Let T be the linear operator on V defined by $T(A) = MA \;\forall\; A \in V$. Find the trace of T.

26. Show that the only matrix similar to the identity matrix I is I itself.

27. Show that the only matrix similar to the zero matrix is the zero matrix itself.

28. If two linear transformations A and B on $V(F)$ are similar, then show that A^2 and B^2 are also similar and if A, B are invertible, then A^{-1}, B^{-1} are also similar.

29. If A and B are linear transformations on $V(F)$ and if at least one of them is invertible, then AB and BA are similar.

30. If T and S are linear transformations on a finite dimensional vector space V such that $TS = 0$, $T \neq 0$, $S \neq 0$ then det $T = $ det S.

31. If $\{\alpha_1, \alpha_2, ..., \alpha_n\}$ and $\{\beta_1, \beta_2, ..., \beta_n\}$ are bases in the same finite dimensional vector space $V(F)$ and if T is a linear transformation on V such that
$$T(\alpha_i) = \beta_i, i = 1, 2, ..., n$$
then det $T \neq 0$.

32. Let T be the linear operator on $\mathbb{R}^3$ defined by
$$T(x, y, z) = (2y + z, x - 4y, 3x)$$
(i) Find the matrix of T in the basis $B = \{(1, 1, 1), (1, 1, 0), (1, 0, 0)\}$.

(ii) Verify that $[T]_B[\alpha]_B = [T(\alpha)]_B \ \forall \ \alpha \in \mathbb{R}^3$.

33. Let $B = \{(1, 0, 0), (0, 1, 0), (0, 0, 1)\}$ and $B' = \{(1, 1, 1), (1, 1, 0), (1, 0, 0)\}$ be two bases for $\mathbb{R}^3$.

(i) Find the transition matrix P from B to B'.

(ii) Show that $P[\alpha]_{B'} = [\alpha]_B \ \forall \ \alpha \in \mathbb{R}^3$.

(iii) Show that $[T]_{B'} = P^{-1}[T]_B P$, where $T(x, y, z) = (2y + z, x - 4y, 3x)$.

Answers

1. Matrix of T $= \begin{bmatrix} 3 & -2 \\ 1 & 2 \end{bmatrix}$

2. (i) $\begin{bmatrix} 3 \\ 5 \end{bmatrix}$ (ii) $\begin{bmatrix} 2 & -4 & 9 \\ 5 & 3 & -2 \end{bmatrix}$

3. (i) $\begin{bmatrix} 4 & -2 \\ 2 & 1 \end{bmatrix}$ (ii) $\begin{bmatrix} 1 & -1 \\ 1 & 0 \end{bmatrix}, \begin{bmatrix} 3 & -2 \\ 1 & 2 \end{bmatrix}$

4. $[T]_B = \begin{bmatrix} 1 & 0 \\ 0 & 0 \end{bmatrix}; P = \begin{bmatrix} 1 & 2 \\ 1 & 1 \end{bmatrix}; [T]_B = \begin{bmatrix} -1 & -2 \\ 1 & 2 \end{bmatrix}$

5. (i) $\begin{bmatrix} 1 & 0 & 0 \\ 0 & 1 & 0 \\ 0 & 0 & 0 \end{bmatrix}$ (ii) $\begin{bmatrix} 0 & 0 & 1 \\ 0 & 1 & 1 \\ 1 & 1 & 1 \end{bmatrix}$

6. $[T]_B = \begin{bmatrix} 2 & 1 & 2 \\ 0 & 1 & 1 \\ -2 & 2 & 0 \end{bmatrix}$

7. $_B[T]_{B'} = \begin{bmatrix} 1 & 1 \\ 2 & -7 \\ 0 & 7 \end{bmatrix}$

8. $[T]_B = \begin{bmatrix} -4 & -2 & -4 \\ -2 & -3 & -6 \\ 7 & 4 & 1 \end{bmatrix}$

9. $[T]_B = \begin{bmatrix} 1 & 0 & 0 \\ 0 & 0 & 0 \\ 0 & 0 & -1 \end{bmatrix}$

10. $\begin{bmatrix} 3 & -2/3 & -2/3 \\ -2/3 & 10/3 & 0 \\ -2/3 & 0 & 8/3 \end{bmatrix}$

11. $_B[T]_{B'} = \begin{bmatrix} 2 & 6 & 0 \\ 0 & -3/2 & -1/4 \\ 1 & 11/2 & -5/4 \end{bmatrix}$

12. $_B[T]_{B'} = \begin{bmatrix} 2 & 1 & 0 \\ -1 & 2 & 0 \end{bmatrix}; _{B_1}[T]_{B_2} = \begin{bmatrix} 2 & 1/2 & 3/2 \\ 1 & 3/2 & 1/2 \end{bmatrix}$

13. $_B[D]_{B'} = \begin{bmatrix} 2 & 0 & 0 \\ 0 & 1 & 0 \end{bmatrix}$

14. $\begin{bmatrix} 0 & 0 & 2 & 0 & 0 \\ 0 & 0 & 0 & 6 & 0 \\ 0 & 0 & 0 & 0 & 12 \\ 0 & 0 & 0 & 0 & 0 \\ 0 & 0 & 0 & 0 & 0 \end{bmatrix}$

15. $[D] = \begin{bmatrix} 0 & 1 & 0 & 0 \\ 0 & 0 & 0 & 0 \\ 0 & 0 & 1 & 1 \\ 0 & 0 & 0 & 1 \end{bmatrix}$

16. $[D] = \begin{bmatrix} 3 & 1 & 0 \\ 0 & 3 & 2 \\ 0 & 0 & 3 \end{bmatrix}$

17. $[D]_B = \begin{bmatrix} 0 & 1 & 0 & 0 \\ 0 & 0 & 2 & 0 \\ 0 & 0 & 0 & 3 \\ 0 & 0 & 0 & 0 \end{bmatrix}; [D]_{B'} = \begin{bmatrix} 0 & 1 & 0 & 0 \\ 0 & 0 & 2 & 0 \\ 0 & 0 & 0 & 3 \\ 0 & 0 & 0 & 0 \end{bmatrix}$

18. (i) $[T]_B = \begin{bmatrix} 3 & 0 & 1 \\ -2 & 1 & 0 \\ -1 & 2 & 4 \end{bmatrix}$ (ii) $[T]_{B'} = \begin{bmatrix} 17/4 & 35/4 & 11/2 \\ -3/4 & 15/4 & -3/2 \\ -1/2 & -7/2 & 0 \end{bmatrix}$

19. $B = \begin{bmatrix} 136 & 329 \\ -53 & -128 \end{bmatrix}$

20. $[T]_B = \begin{bmatrix} 1 & 0 & 2 & 0 \\ 0 & 1 & 0 & 2 \\ 3 & 0 & 4 & 0 \\ 0 & 3 & 0 & 4 \end{bmatrix}$

21. tr $(T) = 5$

22. tr $(T) = a_1 + b_2 + c_3$

23. det $(T) = -1$, tr $(T) = 0$

25. tr $(T) = 10$

32. $[T]_{B'} = \begin{bmatrix} 3 & 3 & 3 \\ -6 & -6 & -2 \\ 6 & 5 & -1 \end{bmatrix}$

33. $P = \begin{bmatrix} 1 & 1 & 1 \\ 1 & 1 & 0 \\ 1 & 0 & 0 \end{bmatrix}$

Miscellaneous Solved Examples

Example 1. *Show that the mapping T defined by*

$T(a, b) = (\alpha+\beta, \alpha-\beta, \beta)$, $\forall (\alpha,\beta) \in V_2(R)$ is a linear transformation.

Solution. Obviously, T is a mapping from $V_2(R)$ into $V_3(R)$, because

$(\alpha+\beta, \alpha-\beta, \beta) \in V_3(R)$ $\forall (\alpha, \beta)$

For each $a, b \in F$ and $(\alpha_1, \beta_1), (\alpha_2, \beta_2) \in V_2(R)$

$T[a(\alpha_1,\beta_1) + b(\alpha_2,\beta_2)] = T(a\alpha_1 + b\alpha_2, a\beta_1 + b\beta_2)$
$= [(a\alpha_1 + b\alpha_2) + (a\beta_1 + b\beta_2), (a\alpha_2 + b\alpha_2) - (a\beta_1 + b\beta_2), (a\beta_1 + b\beta_2)]$
$= a(\alpha_1 + \beta_1, \alpha_1 - \beta_1, \beta_1) + b(\alpha_2 + \beta_2, \alpha_2 - \beta_2, \beta_2)$
$= aT(\alpha_1, \beta_1) + bT(\alpha_2, \beta_2)$

Hence T is linear.

Example 2. *Which of the following function T from R^2 into R^2 are linear transformation?*

(i) $T(x_1, x_2) = (x_1^2, x_2)$

(ii) $T(x_1, x_2) = (\sin x_1, x_2)$

(iii) $T(x_1, x_2) = (x_1^2 - x_2, 0)$.

Solution. (i) Let (x_1, x_2) and (y_1, y_2) be any two vectors in R^2 and $a, b \in R$, then

$T[a(x_1,x_2) + b(y_1,y_2)] = T[(ax_1 + by_1), (ax_2 + by_2)]$
$= [(ax_1 + by_1)^2, (ax_2 + by_2)]$ $\quad [\because T(x_1, x_2) = (x_1^2, x_2)]$
$= [a^2x_1^2 + b_2y_1^2 + 2abx_1y_1, ax_2 + by_2]$
$\neq aT(x_1, x_2) + bT(y_1 + y_2)$

$\therefore$ T is not a linear transformation.

(ii) Let $(x_1, x_2), (y_1, y_2) \in R^2$ be an arbitrary vector and let $a, b \in R$, then

$T[a(x_1, x_2) + b(y_1, y_2)] = T(ax_1 + by_1), (ax_2 + by_2)]$
$= [\sin(ax_1 + by_1), ax_2 + by_2]\{\because T\{x_1, x_2) = (\sin x_1, x_2)\}$

$\therefore$ T is not a linear transformation.

(iii) Let $(x_1, x_2), (y_1, y_2) \in R^2$ be an arbitrary vector and let $a, b \in R$, then

$T[a(x_1,x_2) + b(y_1, y_2)] = T[(ax_1 + by_1), (ax_2 + by_2)]$
$= (ax_1 + by_1) - (ax_2 + by_2), 0]$ $\quad [\because T(x_1,x_2) = (x_1 - x_2, 0)]$
$= [a(x_1 - x_2) + b(y_1 - y_2), 0]$
$= aT(x_1,x_2) + bT(y_1, y_2)$

Hence T is a linear transformation.

Example 3. *Show that the mapping $T : R^3 \to R^2$, defined by*

$$T(\alpha, \beta, \gamma) = (\alpha, \beta), \ \forall (\alpha, \beta, \gamma) \in R^3$$

is a homomorphism (linear transformation) of the vector space $R^3(R)$ onto $R^2(R)$.

Solution. Let $(\alpha_1, \beta_1, \gamma_1), (\alpha_2, \beta_2, \gamma_2) \in R^3(R)$ be an arbitrary vector and let $a, b \in R$, then

$$T[a(\alpha_1, \beta_1, \gamma_1) + b(\alpha_2, \beta_2, \gamma_2)]$$
$$= T[(a\alpha_1 + b\alpha_2, a\beta_1 + b\beta_2, a\gamma_1 + b\gamma_2)]$$
$$= (a\alpha_1 + b\alpha_2, a\beta_1 + b\beta_2) \qquad\qquad [\because T(\alpha, \beta, \gamma) = (\alpha, \beta)]$$
$$= a(\alpha_1, \beta_1) + b(\alpha_2, \beta_2) = aT(\alpha_1, \beta_1, \gamma_1) + bT(\alpha_2, \beta_2, \gamma_2)$$

Thus, T is a homomorphism.

Further, if $(\alpha, \beta) \in R^2$, then $(\alpha, \beta, \gamma) \in R^3$ and we have $T(\alpha, \beta, \gamma) = (\alpha, \beta)$.

Hence, T is onto.

Example 4. *Let F be the field of complex numbers and let T be the function from R^3 onto R^3 defined by $T(a_1, a_2, a_3) = (a_1 - a_2 + 2a_3, 2a_1 + a_2 - a_3, -a_1 - 2a_2)$. Verify that T is a linear transformation. Describe the null space of T.*

Solution. Let $\alpha = (a_1, a_2, a_3)$ and $\beta = (b_1, b_2, b_3)$ be any two vectors in R^3 and $a, b \in R$, then

$$a\alpha + b\beta = a(a_1, a_2, a_3) + b(b_1, b_2, b_3)$$
$$= (aa_1, aa_2, aa_3) + (bb_1, bb_2, bb_3)$$
$$= (aa_1 + bb_1, aa_2 + bb_2, aa_3 + bb_3)$$

Now, $T(a\alpha + b\beta) = T(aa_1 + bb_1, aa_2 + bb_2, aa_3 + bb_3)$

$$= [(aa_1 + bb_1) - (aa_2 + bb_2) + 2(aa_3 + bb_3), 2(aa_1 + bb_1) + (aa_2 + bb_2)$$
$$- (aa_3 + bb_3), -(aa_1 + bb_1) - 2(aa_2 + bb_2)]$$
$$= [a(a_1 - a_2 + 2a_3) + b(b_1 - b_2 + 2b_3), a(2a_1 + a_2 - a_3)$$
$$+ b(2b_1 + b_2 - b_3), a(-a_1 - 2a_2) + b(-b_1 - 2b_2)]$$
$$= [a(a_1 - a_2 + 2a_3, 2a_1 + a_2 - a_3, -a_1 - 2a_2)$$
$$+ b(b_1 - b_2 + 2b_3, 2b_1 + b_2 - b_3, -b_1 - 2b_2)]$$

Thus, T is a linear transformation.

Next, By the definition of null space of T, we have

$$N_T = \{\alpha \in R^3 : T(\alpha) = 0 = (0, 0, 0)\}$$

Let $\qquad\qquad \alpha = (a_1, a_2, a_3) \in R^3$

$\therefore \qquad\qquad T(\alpha) = T(a_1, a_2, a_3) = (0, 0, 0)$

$\Rightarrow \quad (a_1 - a_2 + 2a_3, 2a_1 + a_2 - a_3, -a_1 - 2a_2) = (0, 0, 0)$

$\Rightarrow \qquad \left. \begin{array}{r} a_1 - a_2 + 2a_3 = 0 \\ 2a_1 + a_2 - a_3 = 0 \\ -a_1 - 2a_2 = 0 \end{array} \right\}$ ...(1)

Now, we find the solution of system of equation (1).

Let A be the coefficient matrx of (1), then we have,

$$A = \begin{bmatrix} 1 & -1 & 2 \\ 2 & 1 & -1 \\ -1 & -2 & 0 \end{bmatrix}$$

Performing $R_2 \to R_2 - 2R_1$, $R_3 \to R_3 + R_1$, we get

$$A = \begin{bmatrix} 1 & -1 & 2 \\ 0 & 3 & -5 \\ 0 & -3 & 2 \end{bmatrix}$$

Performing $R_3 \rightarrow R_3 + R_2$, we get

$$\sim \begin{bmatrix} 1 & -1 & 2 \\ 0 & 3 & -5 \\ 0 & 0 & -3 \end{bmatrix}$$

This matrix is in Echelon form and having three non-zero rows, thus its rank=3, which is equal to the number of unknowns. Hence the system of equations (1) has only trival solution, i.e. $a_1=0$, $a_2=0$, $a_3 = 0$. Consequently, $N_T = \{(0, 0, 0)\}$.

Example 5. *Describe explicitly the linear transformation $T : R^2 \rightarrow R^3$ such that $T(2, 3) = (4, 5)$ and $T(1, 0) = (0, 0)$.*

Solution. Let $\alpha = (2, 3)$, $\beta = (1, 0)$ and let $a, b \in R$, then

$$a\alpha + b\beta = 0(0, 0) \qquad\qquad \Rightarrow a(2, 3) + b(1, 0) = (0, 0)$$

$\Rightarrow \qquad\qquad (2a + b, 3a) = (0, 0)$

$\Rightarrow \qquad\qquad 2a + b = 0$

$\qquad\qquad\qquad 3a = 0$

$\Rightarrow \qquad\qquad a = 0, b = 0$

$\Rightarrow$ the set $\{\alpha, \beta\} = \{(2, 3), (1, 0)\}$ is linearly independent.

Also, dim. $R^2 = 2$, thus $\{(2, 3), (1, 0)\}$ forms a basis of R^2.

Let (x, y) be any element of R^2 and for some scalars p and q in R, we have

$$(x, y) = p\alpha + q\beta = p(2, 3) + q(1, 0) = (2p + q, 3p)$$

$\Rightarrow \qquad\qquad x = 2p + q, y = 3p \Rightarrow p = \dfrac{y}{3}, q = \dfrac{3x - 2y}{3}$

$\therefore \qquad\qquad (x, y) = \dfrac{y}{3}(2, 3) + \dfrac{3x - 2y}{3}(1, 0)$

Now, $\qquad\qquad T(x, y) = T\left[\dfrac{y}{3}(2, 3) + \dfrac{3x - 2y}{3}(1, 0)\right]$

$$= \dfrac{y}{3}T(2, 3) + \dfrac{3x - 2y}{3}T(1, 0) \qquad\qquad [\because T \text{ is linear.}]$$

$$= \dfrac{y}{3}(4, 5) + \dfrac{3x - 2y}{3}(0, 0) = \left(\dfrac{4y}{3}, \dfrac{5y}{3}\right).$$

Example 6. *If a map $T : V_2(R) \rightarrow V_3(R)$ defined by $T(a, b) = (a+b, a-b, b)$ is a linear transformation. Find the range, rank, null-space and nullity of T.*

Solution. *Determination of range of T, i.e., R_T and rank :*

Since the ordered set $\{(1, 0), (0, 1)\}$ forms a basis of $V_2(R)$. Then by definition of T, we have,

$$T(1, 0) = (1+0, 1-0, 0) = (1, 1, 0)$$

and $\qquad\qquad T(0, 1) = (0+1, 0-1, 1) = (1, -1, 1)$.

Since $(1, 0), (0, 1)$ generates $V_2(R)$. Therefore

$\qquad T(1, 0), T(0, 1)$ will generate $T(V_2(R)) = R_T$

$\Rightarrow \qquad (1, 1, 0), (1, -1, 1)$ generates R_T.

Also, for some scalars $a, b \in R$, such that

$$a (1, 1, 0) + b (1, -1, 1) = (0, 0, 0)$$

$$\Rightarrow \qquad (a+b, a-b, b) = (0, 0, 0)$$

$$\Rightarrow \qquad a+b = 0, a-b = 0, b = 0$$

$$\Rightarrow \qquad a = 0, b = 0.$$

$\therefore$ $\{(1,1,0), (1, -1, 1)\}$ is linearly independent and spans R_T, so it forms a basis of R_T. Hence dim. $R_T = 2$.

Determination of null space and nullity of T.

Since T is a linear transformation from $V_2(R)$ into $V_3(R)$. Therefore,

$$\text{dim. } R_T + \text{dim. } N_T = \text{dim. } V_2(R)$$

$$2 + \text{dim. } N_T = 2 \Rightarrow \text{dim. } N_T = 0$$

Thus, nullity of $T = 0$.

Since dim. $N_T = 0$

$\Rightarrow$ Null space of T, *i.e.* N_T is a zero space.

$\Rightarrow \qquad N_T = \{(0, 0)\}.$

Example 7. *Let T be a linear operator on a vector space V(F). If $T_2 = 0$, what can you say about the relation of the range of T to the null space of T? Give an example of a linear operator on $V_2(R)$ such that $T^2 = 0$ but $T \neq 0$.*

Solution. Since $T^2 = 0$, then for $\alpha \in V$

$$T^2(\alpha) = 0(\alpha)$$

$$\Rightarrow \qquad T[T(\alpha)] = 0$$

$$\Rightarrow \qquad T(\alpha) \in N_T \qquad \qquad \text{[By the definition of null space]}$$

But $T(\alpha) \in R_T \ \forall \ \alpha \in V$

$\therefore \qquad\qquad R_T \subset N_T$

Hence when $T^2 = 0$, the range of T is contained in null space of T.

Next, let T be a linear map from $V_2(R)$ into $V_2(R)$ such that

$$T(a, b) = (0, a), \forall \ (a, b) \in V_2(R)$$

Obviously, $T \neq 0$.

Also, $T^2(a, b) = T[T (a, b)] = T[(0, a)] = \{0, 0\} = 0 (a, b)$

$$\Rightarrow \qquad T^2 = 0.$$

Example 8. *Find a linear transformation $T : R^2 \rightarrow R^2$ such that $T(1, 0) = (1,1)$ and $T(0, 1) = (-1, 2)$. Prove that T maps the square with vertices $(0, 0), (1, 0), (1, 1)$ and $(0,1)$ into a parallelogram.*

Solution. Since the ordered set $\{\{1, 0), (0, 1)\}$ forms a basis of R^2, so that for some scalars p and q in R and $(x, y) \in R^2$ such that

$$(x, y) = p(1, 0) + q(0, 1) \qquad\qquad\qquad ...(1)$$

$$\Rightarrow \qquad T(x, y) = T[p(1, 0) + q(0, 1)]$$

$$= pT(1, 0) + qT(0, 1)$$

$$= p(1, 1) + q(-1, 2) = (p - q, p + 2q) \qquad ...(2)$$

From (1), we have

$$(x, y) = (p, q) \quad \Rightarrow \quad p = x, q = y.$$

From (2), we get
$$T(x, y) = (x-y, x + 2y).$$ ...(3)
This is the required linear transformation.

Next, let A, B, C and D be the vertices of a square with A $(0, 0)$, B $(1, 0)$, C $(1, 1)$ and D $(0, 1)$ and let P, Q, R and S be the T-images of A, B, C and D respectively. Then we have,

$$P = T(A) = T (0, 0) = (0, 0) \qquad \text{[Using (3)]}$$
$$Q = T(B) = T(1,0) = (1,1) \qquad \text{[Using (3)]}$$
$$R = T(C) = T(1,1) = (0,3) \qquad \text{[Using (3)]}$$
and
$$S = T(D) = T(0,1) = (-1, 2) \qquad \text{[Using (3)]}$$

Now, PQ = Distance between $(0, 0)$ and $(1, 1)$ = $\sqrt{(1-0)^2 + (1-0)^2} = \sqrt{2}$

PS = Distance between $(0,0)$ and $(-1,2)$ = $\sqrt{(-1-0)^2 + (2-0)^2} = \sqrt{1+4} = \sqrt{5}$

RS = Distance between $(0,3)$ and $(-1,2)$ = $\sqrt{(-1-0)^2 + (2-3)^2} = \sqrt{1+1} = \sqrt{2}$
and

QR = Distance between $(1, 1)$ and $(0, 3)$ = $\sqrt{(0-1)^2 + (3-1)^2} = \sqrt{1+4} = \sqrt{5}$

Hence $PQRS$ is a parallelogram.

Example 9. *Let $T : V_3(R) \rightarrow V_3(R)$ be a linear transformation defined by*
$$T(a, b, c) = (3a, a-b, 2a + b + c), \forall\ a, b, c \in R.$$
Prove that T is invertible and find T^{-1}. Also prove that $(T^2 - I)(T - 3I) = 0$.

Solution. Let $\alpha = (a_1, b_1, c_1)$ and $\beta = (a_2, b_2, c_2)$ be two vectors in $V_3(R)$. Suppose that
$$T(\alpha) = T(\beta) \Rightarrow T(a_1, b_1, c_1) = T(a_2, b_2, c_2)$$
$$\Rightarrow \quad (3a_1, a_1 - b_1, 2a_1 + b_1 + c_1) = (3a_2, a_2 - b_2, 2a_2 + b_2 + c_2)$$
$$\Rightarrow \qquad \left.\begin{array}{r} 3a_1 = 3a_2 \\ a_1 - b_1 = a_2 - b_2 \\ 2a_1 + b_1 + c_1 = 2a_2 + b_2 + c_2 \end{array}\right\}$$
$$\Rightarrow \qquad a_1 = a_2,\ b_1 = b_2 \text{ and } c_1 = c_2 \quad \Rightarrow \quad \alpha = \beta.$$

Thus, T is one-one.

Since, $V_3 (R)$ is a 3-dimensional vector space, then T is onto also. Thus, T is one-one onto mapping. Hence T is invertible.

Determination of T^{-1}.

Let $\qquad\qquad T(a, b, c) = (p, q, r)$, then $T^{-1} (p, q, r) = (a, b, c)$
$$T (a, b, c) = (p, q, r) \Rightarrow (3a, a - b, 2a + b + c) = (p, q, r)$$
$$\Rightarrow \qquad \begin{cases} p = 3a \\ q = a - b \\ r = 2a + b + c \end{cases}$$
$$\Rightarrow \qquad a = \frac{p}{3},\ b = \frac{p}{3} - q, c = r - p + q$$
$$\therefore \qquad T^{-1}(p, q, r) = \left(\frac{p}{3}, \frac{p}{3} - q, r - p + q\right)$$

To prove that $(T^2 - I)(T - 3I) = 0$

$$(T - 3I) (a, b, c) = T (a, b, c) - 3I (a, b, c)$$
$$= T(a,b,c) - 3(a,b,c) [\because I \text{ is identity transformation.}]$$
$$= (3a, a - b, 2a + b + c) - 3(a, b, c)$$
$$= (3a, a - b, 2a + b + c) - (3a, 3b, 3c)$$
$$= (0, a - 4b, 2a + b - 2c)$$
$$[(T^2 - I) (T - 3I)] (a, b, c) = (T^2 - I) [(T - 3I) (a, b, c)]$$
$$= (T^2 - I)(0, a - 4b, 2a + b - 2c)$$
$$= T^2(0, a - 4b, 2a + b - 2c) - I(0, a - 4b, 2a + b - 2c)$$
$$= T[T(0, a - 4b, 2a + b - 2c)] - (0, a - 4b, 2a + b - 2c)$$
$$= T[T(0, -a + 4b, 0 + a - 4b + 2a + b - 2c)]$$
$$\qquad\qquad -(0, a - 4b, 2a + b - 2c)$$
$$= T[0, -a + 4b, 3a - 3b - 2c)] - (0, a - 4b, 2a + b - 2c)$$
$$= (0, 0 - (-a + 4b), 0 + (-a + 4b) + (3a - 3b - 2c)$$
$$\qquad\qquad -(0, a - 4b, 2a + b - 2c)$$
$$= (0, a - 4b, 2a + b - 2c) - (0, a - 4b, 2a + b - 2c)$$
$$= (0, 0, 0) = 0 = 0 (a, b, c)$$
$$\Rightarrow \qquad (T^2 - I) (T - 3I) = 0.$$

Example 10. *If T is a linear transformation on a vector space V such that $T^2 - T + I = 0$, then show that T is invertible.*

Solution. Since $\qquad T^2 - T + I = 0$, then
$$T^2 = T - I$$

For every $\alpha_i \in V$, we have
$$T^2(\alpha_i) = (T - I)(\alpha_i) \Rightarrow T[T(\alpha_i)] = T(\alpha_i) - I(\alpha_i)$$
$$\Rightarrow \qquad T[T(\alpha_i)] = T(\alpha_i) - \alpha_i.$$

Now, for some $\beta_i \in V$, such that $T(\alpha_i) = \beta_i$
$$\Rightarrow \qquad T(\beta_i) = \beta_i - \alpha_1 \qquad\qquad\qquad ...(1)$$

To show that T is one-one :

For $\beta_1, \beta_2 \in V$, suppose that
$$T(\beta_1) = T(\beta_2) \Rightarrow \beta_1 - \alpha_1 = \beta_2 - \alpha_1 \qquad \text{[Using (1)]}$$
$$\Rightarrow \qquad\qquad \beta_1 = \beta_2$$
$\therefore \quad T$ is one-one.

To show T is onto :

For every $\beta_i \in V$, there exists $\beta_i - \alpha_i \in V$ such that $T(\beta_i) = \beta_i - \alpha_i$.

Thus T is also onto. Hence T is one-one and onto. Hence T is invertible.

Example 11. *Let $T : R^2(R) \to R^2(R)$, where for any $(x,y) \in R^2, T(x, y) = \left(2x, \dfrac{1}{2}y\right)$. Find the matrix associated with T w.r.t. the ordered basis $\{(1,0), (0,1)\}$.*

Solution. Let $B = \{(1,0), (0,1)\}$ be an ordered basis of $R^2(R)$ and $T(x,y) = \left(2x, \dfrac{1}{2}y\right)$, then
$$T(1, 0) = (2, 0) \text{ and } T(0, 1) = \left(0, \dfrac{1}{2}\right)$$

Now, $T(1, 0) = (2, 0) = 2(1, 0) + 0(0, 1)$ and $T(0, 1) = \left(0, \dfrac{1}{2}\right) = 0(1,0) + \dfrac{1}{2}(0,1)$

Hence the matrix associated with T w.r.t., B is $[T]_B = \begin{bmatrix} 2 & 0 \\ 0 & \dfrac{1}{2} \end{bmatrix}$

Example 12. *Find the matrix representation of a linear map $T : R^3 \to R^3$ defined $T(x, y, z) = (z, y+z, x+y+z)$ relative to the basis $\{(1,0,1),(-1,2,1),(2,1,1)\}$.*

Solution. Suppose $B=\{(1, 0, 1),(-1, 2, 1),(2, 1, 1\}$ is the basis of R^3 and $T(x, y, z)=(z, y+z, x+y+z)$ then

$$\left. \begin{aligned} T(1,0,1) &= (1,1,2) \\ T(-1,2,1) &= (1,3,0) \\ T(2,1,1) &= (1,2,4) \end{aligned} \right\} \qquad \ldots(1)$$

For some $a, b, c \in R$ and $(x, y, z) \in R^3$ we have

$$(x, y, z) = a(1,0,1)+b(-1,2,1)+c(2,1,1) = (a - b + 2c,\ 2b+c,\ a+b+c)$$

$\Rightarrow \qquad\qquad x = a - b + 2c,\ y = 2b + c,\ z = a + b + c$

$\Rightarrow \qquad\qquad a = \dfrac{1}{4}(-x - 3y + 5z),\ b = \dfrac{1}{4}(-x + y + z),\ c = \dfrac{1}{4}(2x + 2y - 2z)$

So $\qquad (x, y, z) = \dfrac{1}{4}(-x - 3y + 5z)(1,0,1) + \dfrac{1}{4}(-x + y + z)(-1,2,1)$

$$+ \dfrac{1}{4}(2x + 2y - 2z)(2,1,1) \ \ldots(2)$$

Putting $x = 1, y = 1, z = 1$ in (2) and using (1), we get

$$T(1,0,1) = (1,1,2) = \dfrac{3}{2}(1,0,1) + \dfrac{1}{2}(-1,2,1) + 0(2,1,1) \qquad\qquad \ldots(3)$$

Putting $x = 1, y = 3, z = 0$ in (2) and using (1), we get

$$T(-1,2,1) = (1,3,0) = 0 \cdot(1,0,1)+1 \cdot(-1,2,1)+1 \cdot(2,1,1) \qquad \ldots(4)$$

Putting $x = 1, y = 2, z = 4$ in (2) and using (1), we get

$$T(2, 1, 1) = (1, 2, 4) = \dfrac{13}{4}(1,0,1)+ \dfrac{5}{4}(-1,2,1) - \dfrac{1}{2}(2,1,1) \quad \ldots(5)$$

Now, the matrix of coefficients of equations (3), (4) and (5) is,

$$\begin{bmatrix} \dfrac{3}{2} & \dfrac{1}{2} & 0 \\ 0 & 1 & 1 \\ \dfrac{13}{4} & \dfrac{5}{4} & -\dfrac{1}{2} \end{bmatrix}$$

Thus, the matrix representation of T relative to B is the transpose of above matrix:

$$[T]_B = \begin{bmatrix} \dfrac{3}{2} & 0 & \dfrac{13}{4} \\ \dfrac{1}{2} & 1 & \dfrac{5}{4} \\ 0 & 1 & -\dfrac{1}{2} \end{bmatrix}$$

Example 13. *Let T be linear operator in* R^3 *defined by*

$$T(x_1, x_2, x_3) = (3x_1 + x_3, -2x_1 + x_2, -x_1 + 2x_2 + 4x_3)$$

Find the matrix of T in the ordered basis $\{\alpha_1, \alpha_2, \alpha_3\}$, *where*

$$\alpha_1 = (1, 0, 1), \ \alpha_2 = (-1, 2, 1) \ and \ \alpha_3 = (2, 1, 1)$$

Solution. Suppose $B = \{\alpha_1, \alpha_2, \alpha_3\}$ is the basis of R^3 where, $\alpha_1 = (1,0,1)$, $\alpha_2 = (-1,2,1)$ and $\alpha_3 = (2,1,1)$ also $T : R^3 \to R^3$ defined by

$$T(x_1, x_2, x_3) = (3x_1 + x_3, -2x_1 + x_2, -x_1 + 2x_2 + 4x_3)$$

Then, we get

$$\left. \begin{array}{l} T(\alpha_1) = T(1,0,1) = (4,-2,3) \\ T(\alpha_2) = T(-1,2,1) = (-2,4,9) \\ T(\alpha_3) = T(2,1,1) = (7,-3,4) \end{array} \right\} \qquad \ldots(1)$$

Let $(x, y, z) = a\alpha_1 + b\alpha_2 + c\alpha_3$ for some $a, b, c \in R$. Then

$$(x, y, z) = a(1, 0, 1) + b(-1, 2,1) + c(2,1,1) = (a - b + 2c, \ 2b + c, \ a + b + c)$$

$$\Rightarrow \qquad x = a - b + 2c, y = 2b + c, z = a + b + c$$

$$\Rightarrow \qquad a = \frac{1}{4}(-x - 3y + 5z), b = \frac{1}{4}(-x + y + z), c = \frac{1}{4}(2x + 2y - 2z)$$

$$\therefore \quad (x, y, z) = \frac{1}{4}(-x - 3y + 5z)\alpha_1 + \frac{1}{4}(-x + y + z)\alpha_2 + \frac{1}{4}(2x + 2y - 2z)\alpha_3 \quad \ldots(2)$$

Putting $x = 4, y = -2, z = 3$ in (2) and using (1), we get

$$T(\alpha_1) = (4, -2, 3) = \frac{17}{4}\alpha_1 - \frac{3}{4}\alpha_2 - \frac{1}{2}\alpha_3 \qquad \ldots(3)$$

Putting $x = -2, y = 4, z = 9$ in using (1), we get

$$T(\alpha_2) = (-2, 4, 9) = \frac{35}{4}\alpha_1 + \frac{15}{4}\alpha_2 - \frac{7}{2}\alpha_3 \qquad \ldots(4)$$

Putting $x = 7, y = -3, z = 4$ in (2) and using (1), we get

$$T(\alpha_3) = (7, -3, 4) = \frac{11}{2}\alpha_1 - \frac{3}{2}\alpha_2 + 0.\alpha_3 \qquad \ldots(5)$$

Now, the coefficient matrix of the system of equation (3),(4) and (5) is

$$\begin{bmatrix} 17/4 & -3/4 & 11/2 \\ 35/4 & 15/4 & -7/2 \\ 11/2 & -3/2 & 0 \end{bmatrix}$$

Thus, the matrix of T relative to B is obtained by taking the transpose of above coefficeint matrix :

$$[T]_B = \begin{bmatrix} 17/4 & 35/4 & 11/2 \\ -3/4 & 15/4 & -3/2 \\ -1/2 & -7/2 & 0 \end{bmatrix}.$$

Example 14. *If the matrix of a linear transformation T on a vector space* $V_2(C)$ *w.r.t. the ordered basis* $B = \{(1, 0, (0, -1)\}$ *is* $\begin{bmatrix} 1 & 1 \\ 1 & 1 \end{bmatrix}$ *what is the matrix w.r.t. the ordered basis* $B' = \{(1, 1), (1, -1)\}$.

Solution. Since $B = \{(1, 0, (0, -1)\}$ and $B' = \{(1,1), (1, -1)\}$ and

$$[T]_B = \begin{bmatrix} 1 & 1 \\ 1 & 1 \end{bmatrix}$$

Determination of $[T]_{B'}$.

Let $\qquad\qquad \alpha_1 = (1,0), \alpha_2 = (0,-1)$

Since, $\qquad [T]_B = \begin{bmatrix} 1 & 1 \\ 1 & 1 \end{bmatrix}$ is the matrix of T w.r.t. B, then

$$T(\alpha_1) = T(1,0) = 1.\alpha_1 + 1.\alpha_2$$
$$= 1.\,(1, 0) + 1.\,(0, -1) = (1, 0) + (0, -1) = (1, -1)$$
$$T(\alpha_2) = T(0, -1) = 1.(\alpha_1) + 1.\alpha_2$$
$$= 1.(1,0) + 1.(0,-1) = (1,-1)$$

If $(a, b) \in V_2(C)$, then we can write, for some $p, q \in C$

$\Rightarrow \qquad\qquad (a,b) = p(1, 0) + q(0,-1) = (p,-q)$

$\Rightarrow \qquad\qquad p = a, q = -b$

Now, $\qquad T(a,b) = T(p\alpha_1 + q\alpha_2) = pT(\alpha_1) + qT(\alpha_2)$

$\therefore \qquad\qquad T(a,b) = (a-b, a+b).$ ...(1)

Further, since $\quad B' = \{(1,1), (1, -1)\}$ is another basis of $V_2(C)$, let

$$\beta_1 = (1,1), \beta_2 = (1,-1)$$

$$\begin{aligned} T(\beta_1) &= T(1,1) = (0,0) \\ T(\beta_2) &= T(1,-1) = (0,0) \end{aligned} \qquad \begin{aligned} &[\text{Using}(1)] \\ &[\text{Using}(2)] \end{aligned} \qquad ...(2)$$

Let $(x,y) \in V_2(C)$ such that

$$(x, y) = p_1\beta_1 + q_1\beta_2, \text{ for some } p_1, q_1 \in C$$
$$= p_1(1, 1) + q_1(1, -1) = (p_1 + q_1, p_1 - q_1)$$

$\Rightarrow \qquad x = p_1 + q_1, y = p_1 - q_1 \Rightarrow \quad p_1 = \dfrac{x+y}{2}, q_1 = \dfrac{x-y}{2}$

$\therefore \quad (x, y) = \left(\dfrac{x+y}{2}\right)\beta_1 + \begin{bmatrix} 0 & 0 \\ 0 & 2 \end{bmatrix}\beta_2 = \dfrac{x+y}{2}(1,1) + \dfrac{x-y}{2}(1,-1) \qquad ...(3)$

Putting $x = 0, y = 0$ in (3) and using (2), we get

$$T(1,1) = (0, 0) = 0.(1,1) + 0.(1,-1) \qquad\qquad ...(4)$$

Putting $x = 2, y = -2$ in (3) and using (2), we get

$$T(1, -1) = (2, -2) = 0.(1, 1) + 2(1, -1) \qquad\qquad ...(5)$$

Now, coefficient matrix of the system of equations (4) and (5) is

$$\begin{bmatrix} 0 & 0 \\ 0 & 2 \end{bmatrix}$$

Thus, the matrix of T relative to B' is the transpose of above coefficient matrix :

$$[T]_B = \begin{bmatrix} 0 & 0 \\ 0 & 2 \end{bmatrix}$$

Example 15. *Let $B = \{(1,0), (0,1)\}$ and $B' = \{(1,2), (2,3)\}$ be any two bases of R^2*

(i) Find the transition matrices P from B to B'

(ii) Verify that $[\alpha]_B = P^{-1}[\alpha]_{B'}, \forall \alpha \in R^2$

(iii) Verify that $P^{-1}[T]_B P = [T]_{B'}$, where $T(x, y) = (2x - 3y, x + y)$.

Solution. Let $\alpha_1 = (1, 0)$, $\alpha_2 = (0, 1)$, $\alpha'_1 = (1, 2)$ *and* $\alpha'_2 = (2, 3)$.

Since $\quad T(x, y) = (2x - 3y, x + y)$

(i) $\qquad \alpha'_1 = (1, 2) = 1.(1,0) + 2(0, 1) = 1.\alpha_1 + 2\alpha_2$ ...(1)

and $\qquad \alpha'_2 = (2, 3) = 2.(1, 0) + 3(0,1) = 2.\alpha_1 + 3\alpha_2$ ...(2)

The coefficient matrix of the system of equation (1) and (2) is given by

$$\begin{bmatrix} 1 & 2 \\ 2 & 3 \end{bmatrix}.$$

∴ The transition matrix P is the transpose of the coefficient matrix is given by

$$P = \begin{bmatrix} 1 & 2 \\ 2 & 3 \end{bmatrix}.$$

Let α be an arbitrary element of R^2, then

$$\alpha = a_1\alpha_1 + a_2\alpha_2, \text{ for some } a_1, a_2 \in R.$$
$$= a_1(1, 0) + a_2(0, 1) = (a_1, a_2)$$

If $\alpha = \{x, y\}$; $x,y \in R$, then

$$(x, y) = (a_1, a_2) \Rightarrow \quad x = a_1, y = a_2$$

∴ $\qquad (x, y) = x\alpha_1 + y\alpha_2$

Thus, $[\alpha]_B = \begin{bmatrix} x \\ y \end{bmatrix}$, which is the co-ordinate vector of a w.r.t. B.

Again, $\qquad (x, y) = a'_1\alpha'_1 + a'_2\alpha'_2$

$$= a'_1(1, 2) + a'_2(2, 3) = (a'_1 + 2a'_2, 2a'_1 + 3a'_2)$$

$\Rightarrow \qquad x = a'_1 + 2a'_2, y = 2a'_1 + 3a'_2$

$\Rightarrow \qquad a'_1 = 2y - 3x, a'_2 = 2x - y.$

∴ $\qquad (x, y) = (2y - 3x)\alpha'_1 + (2x - y)\alpha'_2$ ...(3)

Thus, $\qquad [\alpha]_{B'} = \begin{bmatrix} 2y - 3x \\ 2x - y \end{bmatrix}$

∴ $\qquad P[\alpha]_{B'} = \begin{bmatrix} 1 & 2 \\ 2 & 3 \end{bmatrix} \begin{bmatrix} 2y - 3x \\ 2x - y \end{bmatrix}$

$$= \begin{bmatrix} 1(2y - 3x) + 2(2x - y) \\ 2(2y - 3x) + 3(2x - y) \end{bmatrix} = \begin{bmatrix} x \\ y \end{bmatrix} = [\alpha]_B$$

(ii) Since $\qquad P = \begin{bmatrix} 1 & 2 \\ 2 & 3 \end{bmatrix}$

$$|P| = (1 \times 3 - 2 \times 2) = -1.$$

Matrix of co-factors of element of P is

$\begin{bmatrix} 3 & -2 \\ -2 & 1 \end{bmatrix}$ so, adj. $P = \begin{bmatrix} 3 & -2 \\ -2 & 1 \end{bmatrix}$

∴ $\qquad P^{-1} = \frac{1}{p} \begin{bmatrix} 3 & -2 \\ -2 & 1 \end{bmatrix}$

$$= \frac{1}{p} \begin{bmatrix} 3 & -2 \\ -2 & 1 \end{bmatrix} = -1 \begin{bmatrix} 3 & -2 \\ -2 & 1 \end{bmatrix} = \begin{bmatrix} -3 & 2 \\ 2 & -1 \end{bmatrix}$$

Since $\qquad T(x, y) = (2x - 3y, x + y)$

$$\therefore \qquad T(\alpha_1) = T(1, 0) = (2, 1) = 2(1, 0)+1.(0, 1) = 2\alpha_1+\alpha_2 \qquad ...(4)$$
$$T(\alpha_2) = T(0, 1) = (-3, 1) = -3(1, 0)+1.(0, 1) = -3\alpha_1+\alpha_2 \ ...(5)$$

$\therefore$ $[T]_B$ = Transpose of the coefficient matrix of the system of equations (4) and (5)

$$= \begin{bmatrix} 2 & -3 \\ 1 & 1 \end{bmatrix}$$

Also,
$$\left. \begin{array}{l} T(\alpha'_1) = T(1,2) = (-4,3) \\ T(\alpha'_2) = T(2,3) = (-5,5) \end{array} \right\} \qquad\qquad\qquad ...(6)$$

From (3), we have
$$(x, y) = (2x - 3x)\,\alpha'_1 + (2x - y)\,\alpha'_2 \qquad\qquad ...(7)$$

Putting $x = -4, y = 3$ in (7) and using (6), we get
$$T(\alpha'_1) = (-4, 3) = 18\alpha'_1 - 11\alpha'_2 \qquad\qquad ...(8)$$

Putting $x = -5, y = 5$ in (7) and using (6), we get
$$T(\alpha'_2) = (-5, 5) = 25\alpha'_1 - 15\alpha'_2 \qquad\qquad ...(9)$$

$\therefore$ $[T]_{B'}$ = Transpose of coefficient matrix of the system of equations (8) and (9)

$$= \begin{bmatrix} 18 & -11 \\ 25 & -15 \end{bmatrix} = \begin{bmatrix} 18 & 25 \\ -11 & -15 \end{bmatrix}$$

So,
$$P^{-1}[T]_B P = \begin{bmatrix} -3 & 2 \\ 2 & -1 \end{bmatrix} \begin{bmatrix} 2 & -3 \\ 1 & 1 \end{bmatrix} \begin{bmatrix} 1 & 2 \\ 2 & 3 \end{bmatrix}$$
$$= \begin{bmatrix} -4 & 11 \\ 3 & -7 \end{bmatrix} \begin{bmatrix} 1 & 2 \\ 2 & 3 \end{bmatrix} = \begin{bmatrix} 18 & 25 \\ -11 & -15 \end{bmatrix} = [T]_{B'}$$

Example 16. *If T_1, T_2 and T_3 are linear transformations on a vector space $V(F)$ such that $T_1T_2 = T_3T_1 = I$, then T_1 is invertible and $T_1^{-1} = T_2 = T_3$.*

Solution. First we shall show that T_1 is one-one :

For $\alpha, \beta \in V$, suppose that
$$T_1(\alpha) = T_1(\beta) \Rightarrow T_3(T_1(\alpha)) = T_3(T_1(\beta))$$
$$\Rightarrow \qquad (T_3T_1)(\alpha) = (T_3T_1)(\beta) \Rightarrow \qquad I(\alpha) = I(\beta) \qquad [\because T_3T_1 = I]$$
$$\Rightarrow \qquad \alpha = \beta$$

Thus, T_1 is one-one.

Secondly, we shall show that T_1 is onto : Let $\beta \in V$ be an arbitrary vector.

Since $T_2 : V \to V,$ then $T_2(\beta) \in V$.

Let us take $\qquad T_2(\beta) = \alpha \in V$

Now, $\qquad\qquad T_2(\beta) = \alpha \qquad\qquad \Rightarrow \qquad T_1(T_2(\beta)) = T_1(\alpha)$
$$\Rightarrow \qquad (T_1T_2)(\beta)) = T_1(\alpha) \qquad \Rightarrow \qquad\qquad I(\beta) = T_1(\alpha) \qquad [\because T_1T_2 = 1]$$

$\therefore$ For any β, there exists $\alpha \in V$ such that $T_1(\alpha) = \beta$. Hence T_1 is onto .

Thus, T_1 is one-one and onto and hence T_1 is invertible.

Next, since $\qquad T_1T_2 = T_3T_1 = I$
$$\therefore \qquad\qquad T_1T_2 = I \qquad\qquad \Rightarrow \qquad T_1^{-1}(T_1)T_2 = T_1^{-1}I$$
$$\Rightarrow \qquad (T_1^{-1}T_1)T_2 = T_1^{-1} \qquad\qquad \Rightarrow \qquad IT_2 = T_1^{-1}I$$
$$\Rightarrow \qquad\qquad T_2 = T_1^{-1}$$

Also, $\qquad T_3T_1 = I \qquad\qquad \Rightarrow (T_3T_1)T_1^{-1} = I\,T_1^{-1}$

$\Rightarrow \qquad\qquad T_3(T_1T_1^{-1}) = T_1^{-1} \qquad\qquad \Rightarrow \qquad T_3 I = T_1^{-1} \Rightarrow T_3 = T_1^{-1}.$

Hence $\qquad\qquad T_1^{-1} = T_2 = T_3.$

Example 17. *Let T be an invertible linear operator on a vector space V(F). Then show that*

    (i) *aT is also an invertible linear operator, when $a \neq 0$ and $a \in F$*

    (ii) $(aT)^{-1} = \left(\dfrac{1}{a}\right)T^{-1}$, *where $a \neq 0$ and $a \in F$.*

    (iii) T^{-1} *is invertible and $(T^{-1})^{-1} = T$.*

Solution. Since T is invertible so that it is one-one and onto.

    (i)   Let $\alpha, \beta \in V$ and $0 \neq a \in F$ and suppose that

$$(aT)(\alpha) = (aT)(\beta) \qquad \Rightarrow a(T(\alpha)) = a(T(\beta))$$

$\Rightarrow \qquad\qquad T(\alpha) = T(\beta) \qquad \Rightarrow \alpha = \beta \qquad\qquad [\because T \text{ is one-one.}]$

$\therefore \qquad\qquad aT$ is one-one.

Let β be any vector in V, then there exists a vector α in V such that

$$T(\alpha) = \beta \qquad\qquad\qquad\qquad [\because T \text{ is onto.}]$$

$\Rightarrow \qquad\qquad a(T)(\alpha) = a(\beta) \qquad \Rightarrow (aT)(\alpha) = a\beta$

$\Rightarrow \qquad\qquad (aT)(\alpha) = \gamma \quad \text{for some } \gamma = a\beta \in V.$

$\therefore \qquad\qquad aT$ is onto.

Thus, aT is one-one and onto and hence aT is invertible.

    (ii) Consider,

$$(aT)\left(\frac{1}{a}T^{-1}\right) = a.\frac{1}{a}\left(TT^{-1}\right)$$

$$= 1.(TT^{-1}) = I \qquad\qquad [\because T \text{ is invertible.}]$$

Also, $\qquad \left(\dfrac{1}{a}T^{-1}\right)(aT) = I$

Hence, $\qquad (aT)^{-1} = \left(\dfrac{1}{a}\right)T^{-1}$

    (iii) Since $\qquad T^{-1}T = TT^{-1} = I$, this implies that inverse of T^{-1} is T,

i.e. $\qquad\qquad (T^{-1})^{-1} = T.$

$\therefore \qquad\qquad (T^{-1})^{-1}T^{-1} = T^{-1}(T^{-1})^{-1} = I$

$\Rightarrow \qquad T^{-1}$ exists.

Hence, T^{-1} exists and $(T^{-1})^{-1} = T$.

Example 18. *Let V be the vector spaces of all $n \times n$ matrices over the field F, and let B be a fixed $n \times n$ matrix, if*

$$T(A) = AB - BA \; \forall \, A \in V$$

verify that T is a linear transformation from V into V.

Solution. Since A and B are two $n \times n$ matrices, then AB and BA are $n \times n$ matrices and so $AB - BA$.

$\therefore \qquad\qquad T(A) = AB - BA \in V$

Thus T is a function from V into V.

Let $A_1, A_2 \in V$ and $a, b \in F$. Then $aA_1 + bA_2$ is a $n \times n$ matrix so $aA_1 + bA_2 \in V$.

Now $\qquad\qquad T(aA_1 + bA_2) = (aA_1 + bA_2)B - B(aA_1 + bA_2)$

$$= aA_1B + bA_2B - aBA_1 - bBA_2$$
$$= a(A_1B - BA_1) + b(A_2B - BA_2)$$
$$= aT(A_1) + bT(A_2).$$

∴ T is a linear transformation from V into V.

Example 19. *Let $T : R^3 \to R^2$ be defined as*

$$T(\alpha_1) = (1, 0), T(\alpha_2) = (2, -1), T(\alpha_3) = (4, 3)$$

Find $T(2, -3, 5)$, where $\alpha_1 = (1, 1, 1), \alpha_2 = (1, 1, 0), \alpha_3 = (1, 0, 0)$ and $\{\alpha_1, \alpha_2, \alpha_3\}$ forms a basis of R^3.

Solution.   Since $\{\alpha_1, \alpha_2, \alpha_3\}$ forms a basis of R^3. Hence

$$(2, -3, 5) = a\alpha_1 + b\alpha_2 + c\alpha_3 \text{ for } a, b, c \in R$$
$$\Rightarrow \qquad (2, -3, 5) = a(1, 1, 1) + b(1, 1, 0) + c(1, 0, 0)$$
$$\Rightarrow \qquad (2, -3, 5) = (a + b + c, a + b, a)$$
$$\therefore \qquad a + b + c = 2, a + b = -3, a = 5$$

Solving these equations, we get

$$a = 5, b = -8, c = 5$$
$$\therefore \qquad (2, -3, 5) = 5\alpha_1 - 8\alpha_2 + 5\alpha_3$$

Then, $$T(2, -3, 5) = 5T(\alpha_1) - 8T(\alpha_2) + 5T(\alpha_3)$$
$$\Rightarrow \qquad T(2, -3, 5) = 5(1, 0) - 8(2, -1) + 5(4, 3)$$
$$\therefore \qquad T(2, -3, 5) = (9, 2, 3).$$

Example 20. *Let V be an n-dimensional vector space over the field F and let T be a linear tranformation from V into V such that the range and null space of T are identical. Prove that n is even. Give an example of such a linear transformation.*

Solution.   Since dim. $V = n$ and $R(T) = N(T)$ therefore, dim. $R(T) = $ dim. $N(T)$

Now $$\rho(T) + v(T) = \text{dim. } V$$
$$\Rightarrow \qquad \rho(T) + v(T) = n$$
$$\Rightarrow \qquad 2\rho(T) = n \qquad\qquad\qquad [\because \rho(T) = v(T)]$$
$$\Rightarrow \quad n \text{ is even.}$$

For Example : Let $T : V_2(R) \to V_2(R)$ be a transformation defined by $T(a, b) = (b, 0) \; \forall \; a, b \in R$.

Let $\alpha = (a_1, b_1), \beta = (a_2, b_2)$ be the elements of $V_2(R)$ and $x, y \in R$, then

$$T(x\alpha + y\beta) = T[x(a_1, b_1) + y(a_2, b_2)]$$
$$= T(xa_1 + ya_2, xb_1 + yb_2)$$
$$= (xb_1 + yb_2, 0)$$
$$= xT(a_1, b_1) + yT(a_2, b_2)$$
$$= xT(\alpha) + yT(\beta)$$

∴ T is a linear transformation from $V_2(R)$ into $V_2(R)$.

Now, dim. $V_2(R) = 2$ then $\{(1, 0), (0, 1)\}$ is a basis of $V_2(R)$.

Then, we have $T(1, 0) = (0, 0)$ and $T(0, 1) = (1, 0)$

∴   Range of $T = \{T(1, 0), T(0, 1)\} = \{(0, 0), (1, 0)\}$

Thus the range of T is a subspace of $V_2(R)$ spanned by the vector $(0, 0)$ and $(1, 0)$. Now the vector $(0, 0)$ is zero vector so we can omit this vector from the range of T. Therefore, the range of T is the subspace of $V_2(R)$ spanned by the vector $(1, 0)$.

∴　　Range of $T = \{a(1, 0) : a \in R\} = \{(a, 0) : a \in R\}$

Now let (a, b) be any element of null space of T.

Then　　　　　　　　　$(a, b) \in N(T) \Rightarrow T(a, b) = (0, 0)$

$\Rightarrow$　　　　　　　　　$(b, 0) = (0, 0)$

$\Rightarrow$　　　　　　　　　$b = 0.$

∴　　　　　Null space of $T = \{(a, 0) : a \in R\}$

Thus,　　　　　range of T = null space of T

Also, dim. $V_2(R) = 2$, which is even.

Example 21. *Let V be a vector space and T a linear transformation from V into V. Prove that the following two statements about T are equivalent : (i) the intersection of the range space of T and the null space of T is the zero subspace, i.e.* $R(T) \cap N(T) = \{0\}$ *and (ii)* $T[T(\alpha)] = 0 \Rightarrow T(\alpha) = 0.$

Solution. We shall show that (i) $\Rightarrow$ (ii) and (ii) $\Rightarrow$ (i)

(i) $\Rightarrow$ (ii) : Since $R(T) \cap N(T) = (0)$

Now　　　　　　　　　$T[T(\alpha)] = 0.$

$\Rightarrow$　　　　　　　　　$T(\alpha) \in N(T)$

But for every $a \in V$, $T(\alpha) \in R(T)$.

∴　　　　　　　　　$T(\alpha) \in R(T) \cap N(T)$

$\Rightarrow$　　　　　　　　　$T(\alpha) = 0$　　　　　　　　　$[\because R(T) \cap N(T) = \{0\}]$

(ii) $\Rightarrow$ (i) : Let $\alpha \neq 0$ and $\alpha \in R(T) \cap N(T)$ then $\alpha \in R(T)$ and $\alpha \in N(T)$

Now　　　　　　　　　$\alpha \in N(T) \Rightarrow T(\alpha) = 0.$

Also $\alpha \in R(T) \Rightarrow$ there exists $\beta \in V$ such that $T(\beta) = \alpha$

$$T[T(\beta)] = T(\alpha) = 0$$

Thus, there exists $\beta \in V$ such that $T[T(\beta)] = 0$ but $T(\beta) = \alpha \neq 0$. But it gives contradiction to the given hypothesis (ii).

Therefore, there exists no $\alpha \in R(T) \cap N(T)$ such that $\alpha \neq 0$.

Hence　$R(T) \cap N(T) = \{0\}$

☙ EXERCISE 6.3

1. Show that the following mappings are linear:
   (i) $T : R^2 \to R^2$ defined by $T(x, y) = (2x-y, x)$.
   (ii) $T : R^3 \to R^2$ defined by
   $T(x, y, z) = (z, x+y)$.
   (iii) $T : R \to R^2$ defined by $T(x) = (2x, 3x)$.
   (iv) $T : R^2 \to R^2$ defined by
   $T(x, y) = (ax+by, cx+dy)$
   where $a, b, c, d \in R$.

2. Show that the following mappings are linear:
   (i) $T : R^2 \to R^2$ defined by $T(x, y) = (x^2, y^2)$.
   (ii) $T : R^3 \to R^2$ defined by
   $T(x, y, z) = (x+1, y+z)$.
   (iii) $T : R^2 \to R^2$ defined by $T(x, y) = |x-y|$.

3. Show that the map $T : R^2 \to R^3$ defined by $T(a, b) = (a-b, b-a, -a)$ is linear transformation. Find the range, rank, null space and nullity.

4. Let $F : R^3 \to R^2$ be a map given by $F(a, b, c) = (a, b)$, $\forall (a, b, c) \in R^3$. Prove that F is a (homomorphism) linear tranformation. Also, find the kernel of F (null space of F).

5. Give an exmple of a linear tranformation T on $V_3(R)$ such that $T \neq 0$, $T^2 \neq 0$ but $T^3 = 0$ where 0 is the zero tranformation.

6. Let there be a linear operator on R^3 given by $T(x, y, z) = (2x, 2x-5y, 2y + z)$. Find T^{-1}.

7. Let $T : R^3 \to R^2$ be a linear map defind by $T(x, y, z) = (x+ 2y-z, y+z, x+y-2z)$. Find basis and dimension of (i) range of T (ii) null space of T.

8. Prove that a linear transformation T on a finite-dimensional vector space $V(F)$ is invertible if and only if T is nonsingular.

9. Describe explicitly a linear transformation from $V_3(R)$ into $V_3(R)$ which has its range subspace spanned by $(1, 0, -1)$ and $(1, 2, 2)$.

10. Let T be the linear oeprator on R^2 defined by $T (x, y) = (4x-2y, 2x + y)$. Compute the matrix of T relative to the basis $\{\alpha_1, \alpha_2\}$, where $\alpha_1=(1, 1), \alpha_2=(-1, 0)$.

11. Let $V = R^3$ and $T : V \to V$ be a linear map defined by $T (x, y, z) = (x+z, -2x + y, -x + 2y + z)$. What is matrix of T w.r.t. basis $B =\{(1, 0, 1), (-1, 1, 1), (0, 1, 1)\}$?

12. If the matrix of a linear transformation on $V_3(C)$ with respect to the basis $B= \{(1, 0\ 0),$ $(0, 1, 0),\ (0, 0, 1)\}$ is $\begin{bmatrix} 0 & 1 & 1 \\ 1 & 0 & -1 \\ -1 & -1 & 0 \end{bmatrix}.$

What is the matrix of T w.r.t. basis $B'=\{(0,1,-1),\ (1, -1, 1),\ (-1, 0, 1)$ and $B' = (1, 1, -1), (-1, 0, 1) (1, 2, 1)\}$

13. Find the matirx representation of the linear mappings relative to the usual basis

$B = \{(1, 0, \dots 0), (0, 1, \dots,1), (0, 0,\dots, 1)\}.$
For R^n

(i) $T : R^3 \to R^2$ defined by
$\qquad T (x, y, z) = (2x -4y + 9z,\ 5x + 3y-2z)$

(ii) $T : R \to R^2$ defined by $T (x) = (3x, 5x).$

14. Let T be a linear operator on R^3 defined by $T(x, y) = (2y, 3x -y)$. Find the matrix representation of T relative to the basis $B = \{(1, 3), (2, 5)\}$

15. Let F be a linear operator on R^3 defined by $F (x, y, z)= (2y+z, x-4y, 3x)$

(i) Find the matrix of F in the basis
$\qquad B' = \{(1, 1, 1), (1, 1, 0), (1, 0, 0)\}$

(ii) Verify that
$\qquad [F]_B = [F(\alpha)]_{B'} = [F(\alpha)]_{B'},\ \forall\ \alpha \in R^3.$

16. Let $B= \{(1, 0, 0), (0, 1, 0), (0, 0, 1)\}$ and $B' =\{(1, 1, 1), (1, 1, 0), (1, 0, 0)\}$ be two basis of R^3

(i) Show that $P [\alpha]_{B'} = [\alpha]_{B'}\ \forall \alpha \in R^3.$

(ii) Show that $[T]_{B'} = P^{-1}[T]_B P$, where
$\qquad T(x, y, z) = (2y+z, x-4y, 3x).$

17. Find two linear tranformations T and S on a vector space $R^2(R)$ such that $TS=0$ but $ST\neq 0$.

Hints to Selected Problems

1. (i) Let $\alpha=(x_1, y_1, z_1),\ \beta= (x_2, y_2, z_2) \in R^3$ and $a, b \in R$.

Then $a\alpha + b\beta = a(x_1, y_1, z_1)+ b(x_2, y_2, z_2)$
$= (ax_1+ bx_2, ay_1 + by_2, az_1 + bz_2)$
$\therefore\ T(a\alpha + b\beta)$
$= T (ax_1+ bx_2, ay_1 + by_2, az_1 + bz_2)$
$= (az_1+ bz_2, ax_1 + bx_2+ ay_1 + by_2)$
$= \{az_1+ bz_2, a(x_1 + y_1) + b(x_2 + y_2)]$
$= \{az_1, a(x_1+y_1)\} + \{bz_2, b(x_2+y_2)\}$
$= a(z_1, x_1+y_1)+b(z_2, x_2+y_2)$
$= aT (\alpha) + bT(\beta)$
Hence T is linear.

2. (iii) Let $\alpha = (x_1, y_1,),\ \beta= (x_2, y_2) \in R^2$ and $a, b \in R$.

Then $a\alpha +b\beta =a(x_1, y_1) + b(x_2, y_2)$
$= (ax_1+bx_2, ay_1+by_2)$
$\therefore\ T(a\alpha +b\ \beta) =T(ax_1+ bx_2, ay_1 + by_2)$
$= |(ax_1+bx_2)- ay_1+ by_2)|$
$= |a (x_1-y_1)+b(x_2-y_2)|$
$\leq |a(x_1-y_1)| + |b(x_2-y_2)|$
$\qquad$ (by triangular inequality)
$\leq aT(\alpha+bT(\beta).$
$\therefore\ T(a\alpha +b\ \beta) \neq aT (\alpha)+bT(\beta).$
Hence T is not linear.

3. $B = \{(1, 0)\ (0, 1)\}$ be the basis of R^2. Then $T(1, 0)=(1, -1, -1)$ and $T (0, 1) = (-1,1, 0)$. Range space is generated by the set $B_1 = \{(1, -1, -1), (-1, 1, 0)\}$.
Clearly B_1 is linearly independent.
Thus rank $(T) = 2$.
For Null space.
$N (T) =\{\alpha \in R^2 : T(\alpha)=0\}$
$\qquad = \{(x, y)\in R^2 : T(x, y)=0\}$
Now, $T(x, y)= (x - y, y- x, -x)= (0, 0, 0)$
$\Rightarrow \qquad x-y =0, y-x = 0, -x = 0$
$\Rightarrow \qquad\quad x = 0,\quad y = 0, -x = 0$
$\therefore\quad$ Null space of $T = \{0\}$ and nullity $(T) = 1$.

4. Let $\alpha= (a_1, b_1, c_1),\ \beta=(a_2, b_2, c_2)$ and $a, b, c \in R$. Then
$\qquad a\alpha + b\beta = (aa_1+ ba_2, ab_1+ bb_2, ac_1+ bc_2)$
$\therefore\ F(a\alpha+\beta) = (aa_1 + ba_2, ab_1+bb_2)$
$\qquad\qquad = (aa_1 + ab_1) + (ba_1+bb_2)$
$\qquad\qquad = a(a_1 +b_1) + b(a_2 +b_2)$
$\qquad\qquad = aF(a_1, b_1, c_1) + bF(a_2, b_2, c_2)$
$\qquad\qquad = aF(\alpha) + bF(\beta)$
$\therefore\ F$ is linear.
Now Ker $F = \{\alpha \in R^3 : F (\alpha) = 0\}\ \{(a, b, c)\in R^3 : F(a, b, c) = 0\}$

$= \{(a, b, c)\} \in R^3 : (a, b) = (0, 0)\}$

$= \{(a, b, c)\} \in R^3 : a = 0, b = 0\}$

$= \{(0, 0, c) : c \in R\}.$

6. Since $T(x, y, z,) = (2x, 2x - 5y, 2y + z)$

$(x, y, z) = T^{-1}(2x, 2x - 5y, 2y + z)$

Let $p = 2x, q = 2x - 5y, r = 2y + z$

so that $x = \dfrac{p}{2}, y = \dfrac{p-q}{5}$ and $z = \dfrac{5r - 2p + 2q}{5}$

Then $T^{-1}(p, q, r) = \left(\dfrac{p}{2}, \dfrac{p-q}{5}, \dfrac{5r - 2p + 2q}{5}\right)$

9. Since range space is generated by $(1, 0, -1)$ and $(1, 2, 2)$.

So, we take $T(1, 0, 0) = (1, 0, -1)$,

$T(0, 1, 0) = (1, 2, 2)$ and $T(0, 0, 1) = (0, 0, 0)$.

Let $(x, y, z) \in V_3(R)$, then

$(x, y, z) = x(1, 0, 0) + y(0, 1, 0) + z(0,0,1)$

$T(x,y,z) = xT(1,0,0) + yT(0, 1, 0) + zT(0,0,1)$

$= x(1, 0, -1) + y(1, 2, 2) + z(0, 0, 0)$

$= (x+y, 2y, -x+2y).$

10. $T : R^2 \to R^2$ given by $T(x, y) = (4x-2y, 2x+y)$

$T(\alpha_1) = T(1,1) = (2,3) = a(1,1) + b(-1, 0);$

$T(\alpha_2) = T(-1,0) = (-4,-2) = c(1, 1) + d(-1, 0).$

$\therefore a - b = 2, a = 3$ and $c - d = -4. c = -2,$

i.e. $a = 3, b = 1$ and $c = -2, d = 2$

Hence, $[T]_{(\alpha_1, \alpha_2)} = \begin{bmatrix} 3 & -2 \\ 1 & 2 \end{bmatrix}$

12. $B = (1, 0, 0), (0, 1, 0), (0, 0, 1)$

and $[T]_B = \begin{bmatrix} 0 & 1 & 1 \\ 1 & 0 & -1 \\ -1 & -1 & 0 \end{bmatrix}$

$\therefore T(1, 0, 0) = (0, 1, -1); T(0,1,0) = (1,0, -1);$

$T(0, 0, 1) = (1, -1, 0).$

Let $(x, y, z) \in V_3(C)$ Then

$(x, y, z) = x(1, 0, 0) + y(0, 1, 0) + z(0, 0, 1).$

$\therefore T(x, y, z) = x T(1, 0, 0) + yT(0, 1, 0) + zT(0, 0, 1)$

$= x(0, 1, -1) + yT(1, 0, -1) + z(1, -1, 0).$

$= (y + z, x-z, -x -y).$

Now for $B_1 = \{(0, 1, -1), (1, -1, 1), (-1, 0, 1)\}$

$T(0, 1, -1) = (0. 1, -1) = a_1(0, 1, -1)$

$\qquad + b_1(1, -1, 1) + c_1(-1, 0, 1)$

$T(1, -1, 1) = (0, 0, 0) = a_2(0, 1, -1)$

$\qquad + b_2(1, -1, 1) + c_2(-1, 0, 1)$

$\therefore \qquad b_1 - c_1 = 0, a_1 - b_1 = 1, -a_1 + b_1 + c_1 = -1$

$\qquad b_2 - c_2 = 0, a_2 - b_2 = 0, -a_2 + b_2 + c_2 = 0$

and $\qquad b_3 - c_3 = 1, a_3 - b_3 = -2, -a_3 + b_3 + c_3 = 1.$

Solving these equations, we get $a_1 = 1, b_1 = 0,$

$c_1 = 0; a_2 = 1, b_2 = 0, c_2 = 0; a_3 = 0, b_3 = 0,$

$c_3 = -1.$

$\therefore \quad [T]_B = \begin{bmatrix} 1 & 0 & 0 \\ 0 & 0 & 0 \\ 0 & 0 & -1 \end{bmatrix}$

15 (ii) From (i)

$[F]_{B'} = \begin{bmatrix} 3 & 3 & 3 \\ -6 & -6 & -2 \\ 6 & 5 & -1 \end{bmatrix}$

Let $a \in R^3$ so taking $\alpha = (a, b, c)$

$\therefore (a, b, c) = p_1(1, 1, 1) + q_1(1, 1, 0)$

$\qquad + r_1(1, 0, 0)$

$\Rightarrow \qquad p_1 = c, q_1 = b - c, r_1 = a - b.$

$\therefore \qquad [\alpha]_{B'} = \begin{bmatrix} p_1 \\ q_1 \\ r_1 \end{bmatrix} = \begin{bmatrix} c \\ b - c \\ a - b \end{bmatrix}$

$F(\alpha) = F(a, b, c) = (2b+c, a - 4b, 3a)$

$= x(1, 1, 1) + y(1, 1, 0) + z(1, 0, 0)$

$\therefore \quad x = 3a, y = -2a - 4b, z = -a + 6b + c$

So that $F[\alpha]_{B'} = \begin{bmatrix} x \\ y \\ z \end{bmatrix} = \begin{bmatrix} 3a \\ -2a - 4b \\ -a + 6b + c \end{bmatrix}$

Now

$[F]_{B'}[\alpha]_{B'} = \begin{bmatrix} 3 & 3 & 3 \\ -6 & -6 & -2 \\ 6 & 5 & -1 \end{bmatrix}\begin{bmatrix} c \\ b - c \\ a - b \end{bmatrix}$

$= \begin{bmatrix} 3a \\ -2a - 4b \\ -a + 6b + c \end{bmatrix} = [F(\alpha)]_{B'}$

Answers

5. Kernel of $F = \{(0, 0, 0) \in R^3 : T(0, 0, c) = (0, 0, 0)\}$

5. $T : V_3(R) \to V_3(R)$ defined by $T(a, b, c) = (0, a, b)$ such that $T \neq 0, T^2 \neq 0$ and $T^3 \neq 0.$

6. $T^{-1}(x,y,z) = \left[\dfrac{x}{2}, \dfrac{x - y}{5}, \dfrac{-2x + 2y + 5z}{5}\right]$

7. (i) $\{(1, 0, 1), (2, 1, 1)\}$ is a basis of R_T and dim. $R_T = 2$

(ii) $\{(3, -1, 1\}$ is a basis of N_T and dim. $N_T = 1.$

9. $T(a,b,c) = (a+b, 2b, 2b-a)$

10. Matrix of T relative to $\{\alpha_1, \alpha_2\}$ is $\begin{bmatrix} 3 & -2 \\ 1 & 2 \end{bmatrix}$ **11.** $[T]_B = \begin{bmatrix} 2 & 1 & 2 \\ 0 & 1 & 1 \\ -2 & 2 & 0 \end{bmatrix}$

12. $[T]_B = \begin{bmatrix} 1 & 0 & 0 \\ 0 & 0 & 0 \\ 0 & 0 & -1 \end{bmatrix}$ **13.** (i) $[T]_B = \begin{bmatrix} 2 & -4 & 9 \\ 5 & 3 & -2 \end{bmatrix}$ (ii) $[T]_B = \begin{bmatrix} 3 \\ 5 \end{bmatrix}$

14. $[T]_B = \begin{bmatrix} -30 & -48 \\ 18 & 29 \end{bmatrix}$ **15.** $[T]_{B'} = \begin{bmatrix} 3 & 3 & 3 \\ -6 & -6 & -2 \\ 6 & 5 & -1 \end{bmatrix}$ **16.** (i) $P = \begin{bmatrix} 1 & 1 & 1 \\ 1 & 1 & 0 \\ 1 & 0 & 0 \end{bmatrix}$

17. $T(a, b) = (2a, 0)$ and $S(a, b) = (0, 2a)$

CHAPTER REVIEW : A COMPETITIVE APPROACH

Selected Terms and Results

TERMS

- **Linear Transformation :** Let $U(F)$ and $V(F)$ be two vector spaces. Then mapping $T : U \to V$ is said to be a linear transformation from U into V which associates to each element $\alpha \in U$ to a unique element $T(\alpha)$ of V such that

  $T(a\alpha + b\beta) = aT(\alpha) + bT(\beta) \ \forall \ \alpha, \ \beta \in U$, $a, b \in F$.

- **Linear Operator :** Let $V(F)$ be a vector space then a linear transformation from V into V is called a linear operator.

- **Zero Transformation :** $T(\alpha) = 0 \ \forall \ \alpha \in U$

- **Identity Transformation :** $I(\alpha) = \alpha \ \forall \ \alpha \in V$

- **Range of a Linear Transformation :** $R(T) = \{T(\alpha) \in V : \alpha \ \forall \ \alpha \in U\}$

- **Null space of a Linear Transformation :** $N(T) = \{\alpha \in U : T(\alpha) = 0\}$

- **Rank of a Linear Transformation :** $\rho(T) = \dim. \ R(T)$

- **Nullity of a linear transformation :** $v(T) = \dim. \ N(T)$

- **Invertible Transformation :** T is invertible iff T is one to one and T is onto.

- **Non-singular Transformation :** A linear transformation $T : U \to V$ is said to be non-singular if the null space of T is $\{0\}$, *i.e.* if $\alpha \in U$ and $T(\alpha) = 0 \Rightarrow \alpha = 0$

- **Singular Transformation :** T is singular if there exists a vector $\alpha \neq 0$ of U such that $T(\alpha) = 0$.

- **Similarity :** Let S and T be two linear transformations on a vector space $V(F)$ then T is similiar to S if there exists an invertible linear transformation P on V such that $T = PSP^{-1}$.

- **Trace of a linear transformation :** The trace of a linear transformation T is the trace of the matrix of T relative to any ordered basis for V.

RESULTS

- Linear transformation is also known as homomorphism.

- If T is a linear transformation from U to V then range of T is a subspace of V and null space of T is a subspace of U.

- Let $T : U \to V$ be a linear transformation. If U is finite dimensional then the range of T is a finite dimensional subspace of V.

- Rank(T) + nullity(T) = dim U

- If $T : U \to V$ is a linear transformation and T is invertible then T^{-1} is a linear transformation from V into U.

- $TT^{-1} = I = T^{-1}T$

- If T_1, T_2 and T_3 are linear transformations on a vector space $V(F)$ such that $T_1T_2 = T_3T_1 = I$ then T_1 invertible and $T_1^{-1} = T_2 = T_3$.

- The necessary and sufficient condition for a linear transformation T on a vector space $V(F)$ to be invertible is that there exists a linear transformation S on V such that $TS = I = ST$.

- $(T_1T_2)^{-1} = T_2^{-1}.T_1^{-1}$ and $(T^{-1})^{-1} = T$

- A linear transformation T from $U(F)$ into $V(F)$ is non-singular iff T is one-one.

- A linear transformation T on a finite dimensional vector space is invertible iff T is non-singular.

- A linear transformation T on a finite dimensional vector space is invertible iff T is onto.

- If T is a linear transformation on a vector space V such that $T^2 - T + I = 0$ then T is invertible.

- If T_1 and T_2 are linear transformations on vector space $V(F)$. Then a necessary and sufficient condition that both T_1 and T_2 be invertible is that both T_1T_2 and T_2T_1 are invertible.

- The relation of similarity is an equivalence relation on the set of all linear transformation on a vector space $V(F)$.

- If T and S are similar linear transformations on a finite dimensional vector space $V(F)$ then det. T = det. S.

- If dim $U = m$ and dim $V = n$ then

  $\dim [T : U \to V] = m.n$

Review Questions and Project Work

1. Let $T : R^2 \to R^2$ be defined by $T(x, y) = (x+y, x)$. Show that T is linear.

2. Show that the mapping $T : R^2 \to R^3$ defined by $T(x, y) = (x+3, 2y, x+y)$ is not linear.

3. Find a linear map $T : R^3 \to R^4$ whose image is spanned by $(1, 2, 0, -4)$ and $(2, 0, -1, -3)$.

4. Show that the mapping $T : R^2 \to R^2$ defined by $T(x, y) = (x-y, x-2y)$ is non-singular.

5. Let $T : U \to V$ be a non-singular linear transformation, then show that image of any linearly independent set is linearly independent.

6. Let $T : U \to V$ be a linear transformation such that U has finite dimension and dim $U = $ dim V. Then show that T is an isomorphism if and only if T is non-singular.

7. Let T and S be linear operator on R^2 defined by $T(x, y) = (0, x)$ and $S(x, y) = (x, 0)$, show that
   (i) $TS = 0$ but $ST \neq 0$
   (ii) $S^2 = S$

8. Find a linear transformation $T : R^3 \to R^3$ whose image is spanned by $(1, 2, 3)$ and $(4, 5, 6)$.

9. Prove that if $T : U \to V$ is onto then dim $V \leq$ dim U.

10. Let T and S be a linear operators on V and T is non-singular. If V is finite dimensional, show that rank $(ST) = $ rank $(TS) = $ rank (S).

11. Let V be finite dimensional vector space and T is a linear operator on V such that rank(T^2)=rank (T), show that Ker $T \cap$ Im$(T) = \{0\}$.

12. Show that the mapping defined by $T_1(x, y) = (x, 2y)$, $T_2(x, y) = (y, x+y)$ and $T_3(x, y) = (0, x)$ are linearly independent.

Objective Type Questions

FILL IN THE BLANKS

1. A linear transformation T on a vector space $V(F)$ is an isomorphism if and only if Ker(T) = _____ .

2. Let $T : R^3 \to R^3$ be the projection mapping into the xy-plane defined by $T(x, y, z) = (x, y, 0)$. Then T is singular since non-zero vector on the z-axis map into _____ .

3. A linear transformation T is non-sigular if $T(X) = 0 \Rightarrow$ _____ .

4. Since the identity transformation $I : U \to U$ defined by $I(\alpha) = \alpha$ for all $\alpha \in U$ is one-one and onto therefore I is _____ .

5. The matrix of the identity linear transformation $T : U \to U$ relative to any pair of bases is the _____ matrix.

6. If T is linear transformation from $R^2(R)$ to $R^3(R)$. Then order of matrix A associated with T is _____ .

7. If a mapping $T : U \to V$ satisfies the condition $T(a\alpha+b\beta) = aT(\alpha) + bT(\beta) \; \forall \, a, b \in F, \alpha, \beta \in U$. Then T is called a _____ .

8. If $T : V_3(R) \to V_2(R)$ be defined by the rules $T(x_1, x_2, x_3) = (x_1 - x_2, x_1 + x_2)$. Then T is _____ .

9. Let f be a homomorphism of $V(F)$ into $V(F)$ then $f(-\alpha) = $ _____ $\forall \, \alpha \in U$.

10. The kernel of a homomorphism $f : U \to V$ is a _____ of $U(F)$.

TRUE/FALSE

Write 'T' for true and 'F' for false statement.

1. If $T : U(F) \to V(F)$. Then the kernel of T and range of T are subspaces of U and V respectively.   **(T/F)**

2. A linear trasformation $T : U \to V$ over F is one-one mapping if and only if Ker.$(T) = \{0\}$.   **(T/F)**

3. The null space of a linear transformation T is always linearly independent if it always forms a basis.   **(T/F)**

4. The inverse of an isomorphism is again an isomorphism.   **(T/F)**

5. The rank of a linear transformtion is not changed by multiplication by an isomorphism on either side.   **(T/F)**

6. The sum of the dimensions of the image and kernel of a linear transformation is equal to the dimension of its domain.   **(T/F)**

7. T is an epimorphism if and only if $T_1 T_2 = 0 \Rightarrow T_2 = 0$ and T_1 is monomorphism if and only if $T_1 T_2 = 0 \Rightarrow T = 0$.   **(T/F)**

8. $(T^{-1})^{-1} = T$ if T is singular.   **(T/F)**

9. To every linear transformation there correspond a unique matrix. **(T/F)**

10. A matrix of a linear transformation $T : U \to V$ is square if and only if dim U = dim V. **(T/F)**

11. No standard basis exists for vector spaces other than $V_n(F)$ or R^n or C^n or Q^n. **(T/F)**

12. Two linear operator are similar if and only if they have the same matrices in two basis. **(T/F)**

13. If $B = PAP^{-1}$ and $A = P^{-1}BP$ then A and B are similar. **(T/F)**

14. Similar matrices have the same minimal polynomials. **(T/F)**

15. If A is similar to B, B is similar to C then A is similar to C. **(T/F)**

MULTIPLE CHOICE QUESTIONS

Choose the most appropriate one :

1. If $T : V \to V$ defined by $T(x_1, x_2) = (x_1, -x_2)$ then $T^{-1}(y_1, y_2)$ is :
   (a) (y_1, y_2)　　　　(b) $y_1 - y_2$
   (c) $y_1 + y_2$　　　　(d) none of these

2. Let $U(F)$ and $V(F)$ be m and n dimensional vector spaces and $T : U \to V$ be a linear transformation then the matrix T will be of order :
   (a) $m \times n$　　　　(b) $n \times m$
   (c) $n \times n$　　　　(d) none of these

3. Let $T : R \to R^2$ defined by $T(x) = T(3x, 5x)$ then the matrix representing T relative to the usual basis for R^n is :
   (a) $\begin{bmatrix} 3 \\ 5 \end{bmatrix}$　　　　(b) $\begin{bmatrix} 3,5 \end{bmatrix}$
   (c) $\begin{bmatrix} 5 \\ 3 \end{bmatrix}$　　　　(d) none of the above

4. If T be a linear transformation on R^2 defined by $T(x, y) = \{(4x-2y), (2x+y)\}$. The matrix of T relative to the basis $\{\alpha_1, \alpha_2\}$ where $\alpha_1 = (1, 1)$, $\alpha_2 = (-1, 0)$ is :
   (a) $\begin{pmatrix} 2 & -2 \\ 1 & 3 \end{pmatrix}$
   (b) $\begin{bmatrix} 3 & -2 \\ 1 & 2 \end{bmatrix}$
   (c) both (a) and (b) are true
   (d) none of the above

5. If T_1 and T_2 are similar linear transformation on a finite dimensional vector space $V(F)$, then :
   (a) $\det(T_1) \neq \det(T_2)$
   (b) $\det(T_1) = \det(T_2)$
   (c) $\det(T_1) - \det(T_2) > 0$
   (d) none of the above

6. Which of the following mapping $T : V_2(R) \to V_2(R)$ is a linear transformation?
   (a) $T(x, y) = (x+y, x)$
   (b) $T(x, y) = (1+x, y)$

   (c) both (a) and (b) are true
   (d) none of the above

7. If $T : U \to V$ be a linear transformation then $R(T)$ is :
   (a) a subspace of U
   (b) a subspace of V
   (c) a subspace of U and V
   (d) none of the above

8. If $T : U \to V$ be a linear transformation then $N(T)$ is :
   (a) a subspace of U
   (b) a subspace of V
   (c) a subspace of U and V
   (d) none of the above

9. Let the mapping $f : U \to V$ be one-one and onto linear mapping then $f^{-1} : V \to U$ is :
   (a) one-one but not linear
   (b) onto but not linear
   (c) linear
   (d) none of the above

10. Let $T : U \to V$ be a linear transformation and if U is finite dimensional vector space then :
    (a) dim $R(T) <$ dim U
    (b) dim $R(T) >$ dim U
    (c) dim $R(T) \leq$ dim U
    (d) none of the above

11. The mapping $f : V_2(R) \to V_2(R)$ defined by $f(a_1, b_1) = (b_1, a_1)$ is:
    (a) one-one　　　　(b) onto
    (c) isomorphic　　　(d) none of the above

12. The mapping $f : U(F) \to V(F)$ is an isomorphism if :
    (a) f is one-one
    (b) f is one-one, onto
    (c) f is one-one, onto, linear
    (d) none of the above

13. The kernel of the linear transformation $T : U \to V$ is :
    (a) subspace of U　　(b) subgroup of U
    (c) subset of U　　　(d) none of the above

14. If U and V are two finite dimensional vector spaces over the same field F and T is a linear transformation from U into V then which of the following statement is true ?
(a) Rank (T) = dim U
(b) Rank (T) = dim R(T)
(c) Rank (T) = dim V
(d) None of the above

15. Let $T : U(F) \to V(F)$ be a linear transformation then which is true?
(a) Rank(T) + dim(V) = nullity(V)
(b) Rank(T) + nullity(T) = dim V
(c) Both (a) and (b) are true
(d) None of the above

16. Let $T : U(F) \to V(F)$ be a linear transformation then which of the following statement is true?
(a) Nullity (T) = dim (Ker(T))
(b) Nullity of T = dim V
(c) Nullity (T) = dim U
(d) None of the above

17. Let $T : U \to V$ be a linear and onto map where U is n-dimensional vector space, then which of the following statement is true :
(a) T is one-one $\Rightarrow$ dim $U = n$
(b) T is one-one $\Rightarrow$ dim $V = n$
(c) T is one-one $\Rightarrow$ dim $V < n$
(d) None of the above

18. Which of the following is a linear transformation :
(a) $T(x, y) = (x, y^2)$
(b) $T(x, y) = (x+y, x-y)$
(c) $T(x, y) = (\sin x, y)$
(d) None of the above

19. If $T : R^2 \to R^2$ such that $T(1, 0) = (1, 1)$ and $T(0, 1) = (-1, 2)$, then T maps the square with vertices $(0, 0)$, $(1, 0)$, $(1, 1)$ and $(0, 1)$ into a :
(a) square
(b) parallelogram
(c) rectangle
(d) none of the above

20. The intersection of range of T and the null space of T is the :
(a) dim(T)
(b) rank(T)
(c) zero space of V
(d) none of the above

21. If V is a finite dimensional vector space and T be a linear operator on V such that rank (T^2) = rank(T) then range and null space of T are:
(a) zero space
(b) disjoint
(c) equal
(d) none of the above

22. If $T(T(\alpha)) = 0$ then $T(\alpha) = :$

(a) α
(b) 0
(c) α^2
(d) none of the above

23. A linear transformation T on a finite dimensional vector space is invertible iff :
(a) T is onto
(b) T is non-singular
(c) both (a) and (b) are true
(d) none of the above

24. If T is a linear transformation on a finite dimensional vector space $V(F)$ and if $T_1 T_2 = I$. Then :
(a) T_1 is invertible
(b) T_2 is invertible
(c) both T_1, T_2 are invertible
(d) none of the above

25. If T is a linear transformation on V such that $T^2 - T + I = 0$. Then T is :
(a) singular
(b) invertible
(c) non-zero
(d) none of the above

26. If T_1, T_2 be linear operators on a vector space $V(F)$, then both T_1 and T_2 are invertible if and only if :
(a) only $T_1 T_2$ is invertible
(b) only $T_2 T_1$ is invertible
(c) both $T_1 T_2$ and $T_1 T_2$ are invertible
(d) none of the above

27. Let T be a linear operator on a finite dimensional vector space such that $N(T) = 0$ then :
(a) T is non-singular (b) T is one-one
(c) T is onto (d) all are true

28. Let T be a linear operator on a finite dimensional vector space of dim. n such that rank $T = n$ then :
(a) T is non-singular (b) T is one-one
(c) T is onto (d) All are true

29. If T_1 and T_2 are two linear transformations from $V_3(R)$ to $V_2(R)$ defined by $T_1(a, b, c) = (2a, b+c)$, $T_2(a, b) = (b, a)$ then :
(a) $T_1 T_2$ is not defined
(b) $T_1 T_2$ is defined
(c) both (a) and (b) are true
(d) none of the above

30. Let T be a linear transformation on R^2 into itself such that $T(1, 0) = (1, 2)$ and $T(1, 1) = (0, 2)$ then $T(a, b) =$
(a) $(a, 2b)$
(b) $(a-b, 2a)$
(c) $(a+b, 2b)$
(d) none of the above

31. Let $T : V \to V$ be a linear transformation of vector space V and $ToT = 0$, then :
    (a) T is non-singular
    (b) T is zero transformation
    (c) T is identity transformation
    (d) none of the above

32. Which of the following is not a linear transformation ?
    (a) $(x_1, x_2) \to (x_2, x_1)$
    (b) $(x_1, x_2) \to (x_1 + x_2, x_2)$
    (c) $(x_1, x_2) \to (x_1 + 1, x_2)$
    (d) none of the above

33. Let T be a linear transformation from a 3-dimensional vector space U into a 2-dimensional vector space V, then T:
    (a) can be surjective
    (b) can be surjective but not injective
    (c) both (a) and (b) are true
    (d) none of the above

34. Let $T : R^2 \to R^3$ defined by
    $T(x, y) = (-x-y, 3x+8y, 9x-11y)$
    be a linear transformation then rank and nullity of T are respectively given by :
    (a) 2 and 0   (b) 0 and 2
    (c) 1 and 2   (d) none of the above

35. A linear transformation $T : R^2 \to R^2$ such that $T(3, 1) = (2, -4)$ and $T(1, 1) = (0, 2)$, then value of $T(7, 8) =$
    (a) $(1, 3)$ (b) $(3, -1)$
    (c) $(3, 1)$ (d) none of the above

36. Let S be any subset of a vector space $V(F)$, then S is a subspace of :
    (a) V (b) V'
    (c) V'' (d) none of the above

37. If $T : V_2(R) \to V_3(R)$ defined by $T(a, b) = (a+b, a-b, b)$ is a linear transformation, then nullity of T is :
    (a) 1   (b) 2
    (c) 0   (d) none of the above

38. For a linear transformation $T : R^3 \to R^2$ defined by $T(x, y, z) = (x+y, y+z)$ is :
    (a) linear
    (b) has a proper subspace as kernel
    (c) both (a) and (b) are true
    (d) none of the above

39. If T_1 and T_2 are linear operator on V, then :
    (a) $(T_1 T_2)' = T_1' T_2'$
    (b) $(T_1 T_2)' = T_2' T_1'$

(c) both (a) and (b) are true
(d) none of the above

40. For a linear transformation $T : R^{12} \to R^6$, the Kernel is having dimension 7, then the dimension of the range of T is :
    (a) 4   (b) 5
    (c) 6   (d) 17

41. Let V be a finite dimensional vector space over F. The minimal polynomial for the zero operator is:
    (a) x (b) $x - 1$
    (c) $x + 1$ (d) none of the above

42. Let V be a subspace and T a linear operator on V. If W is a subspace of V then W is invariant under T if and only if $\alpha \in T \Rightarrow$
    (a) $T(\alpha) = 0$ (b) $T(\alpha) = \alpha$
    (c) $T(\alpha) \in W$ (d) none of the above

43. If $T : V \to V$ is a linear operator for dim $V = n$ and T has n distinct eigenvalues, then :
    (a) T is invertible
    (b) T is diagonalizable
    (c) both (a) and (b) are true
    (d) none of the above

44. T is non-singular if and only if :
    (a) nullity of $T = 0$ (b) nullity of $T \neq 0$
    (c) rank $T = 0$ (d) none of the above

45. If $T : R^4 \to R^3$ be the linear transformation defined by
    $T(x, y, z, w) = (x-y+z+w, x+2z-w, x+y+3z-3w)$
    then dimension of its range is :
    (a) 1   (b) 2
    (c) 3   (d) 0

46. For a linear transformation $T : R^{10} \to R^6$, then Kernel is having dimension 5. Then dimension of the range of T is :
    (a) 3   (b) 4
    (c) 5   (d) 6

47. If the linear transformation $T : R^2 \to R^3$ is such that $T(0, 1) = (2, 3, 1)$ and $T(1, 1) = (3, 0, 2)$ then $T(x, y) =$
    (a) $(2x+y, 3x-3y, x+y)$
    (b) $(2x+y, 3x+3y, x+y)$
    (c) $(2x+y, 3x-3y, x-y)$
    (d) none of the above

48. Let $T : R^7 \to R^7$ be a linear transformation such that $T^2 = 0$ then rank(T) :
    (a) $= 3$ (b) < 3
    (c) > 3 (d) none of the above

49. The numbner of all non-singular linear transformation $T : R^4 \to R^3$ is :

(a) 0   (b) 1

(c) 2   (d) none of the above

50. If U and V be a vector space of dimension 5 and 6 respectively. Then dim[Hom(U, V)] :

(a) 6   (b) 5

(c) 30   (d) none of the above

Answers

FILL IN THE BLANKS

1. $\{0\}$ **2.** 0   **3.** $x = 0$ **4.** invertible   **5.** identity   **6.** 2×3

7. linear transformation   **8.** linear transformation   **9.** $-f(\alpha)$ **10.** subspace

TRUE/ FALSE

1. T   **2.** T   **3.** T   **4.** T   **5.** T   **6.** T   **7.** T   **8.** F   **9.** F

10. T   **11.** T   **12.** T   **13.** T   **14.** T   **15.** T

MULTIPLE CHOICE QUESTIONS

1. (b)   **2.** (a)   **3.** (a)   **4.** (b)   **5.** (b)   **6.** (a)   **7.** (b)   **8.** (a)   **9.** (c)

10. (c)   **11.** (c)   **12.** (c)   **13.** (a)   **14.** (b)   **15.** (b)   **16.** (a)   **17.** (b)   **18.** (b)

19. (b)   **20.** (c)   **21.** (b)   **22.** (b)   **23.** (c)   **24.** (c)   **25.** (b)   **26.** (c)   **27.** (d)

28. (d)   **29.** (c)   **30.** (b)   **31.** (a)   **32.** (c)   **33.** (c)   **34.** (a)   **35.** (b)   **36.** (a)

37. (c)   **38.** (c)   **39.** (b)   **40.** (b)   **41.** (a)   **42.** (c)   **43.** (c)   **44.** (a)   **45.** (b)

46. (c)   **47.** (a)   **48.** (c)   **49.** (a)   **50.** (c)

 **COMPETITION CORNER**

for **JRF, NET/SET, GATE Aspirants**

SOME FASCINATING FACTS

- If U and V are two subspaces of a vector space W then

  $(U + V)|U \cong V|(V \cap U)$

- If T is a linear operator on a finite dimensional vector space over F and minimal polynomial $p(x)$ of T is a product of distinct linear factors then T is a diagonalizable.

- A linear transformation is nothing but a vector space homomorphism.

- Every linear transformation is a group homomorphism.

- Elements of quotient space $V|W$ are actually subsets of V and therefore $V|W \subset P(V)$.

- A linear transformation from U into V forms a vector space with vector addition and scalar multiplication.

- A linear transformation $T : U \rightarrow V$ is an isomorphism iff the null space of T consists of a single element.

- If $T : U \rightarrow V$ is linear then $T(0) = 0$.

- If U and V are vector spaces over a field K. Let $\{u_1, u_2, ..., u_n\}$ be a basis of U and $v_1, v_2, ..., v_n$ be any vector in V then there exists a unique linear mapping $T : U \rightarrow V$ such that $T(u_1) = v_1$; $T(u_2) = v_2 ... T(u_n) = v_n$.

- If $T : U \rightarrow V$ be a linear transformation. Then the Kernel of T is a subspace of U and the image of T is a subspace of V.

- If $u_1, u_2, ..., u_n$ span a vector space U and $T : U \rightarrow V$ is linear then $T(u_1), T(u_2), ..., T(u_n)$ span image(T).

- Let A be any $m \times n$ matrix over a field F viewed as a linear transformation $A : F^n \rightarrow F^n$ then

  Ker (A) = null sp(A)

  Image (A) = column space (A)

- If dim. $U = n$, dim. $V = m$, then dim.(Hom(U, V)) $= mn$

SOME IMPORTANT ILLUSTRATIONS

- The mapping $T : R^2 \rightarrow R^2$ defined by

  $T(x, y) = (x+y, x)$ is linear transformation.

- The mapping $T : R^3 \rightarrow R$ defined by $T(x, y, z) = 2x-3y+4z$ then T is linear transformation.

- The mapping $T : R^2 \rightarrow R^3$ defined by $T(x, y) = (x+1, 2y, x+y)$ is not linear.

- The mapping $T : R^3 \rightarrow R^2$ defined by $T(x, y, z) = (|x|, 0)$ is not linear.

- The mapping $T : R^2 \rightarrow R^2$ defined by $T(x, y) = (x^2, y^2)$ is not linear.

- The mapping $T : R^3 \rightarrow R^2$ defined by $T(x, y, z) = (x+1, y+z)$ is not linear.

- The mapping $T : R^2 \rightarrow R$ defined by $T(x) = (2x, -3x)$ is linear.

- The mapping $T : R^2 \rightarrow R^2$ defined by $T(x, y) = (2x-y, x)$ is linear.

- Let $T : V \rightarrow V$ be defined by $T(A) = M+A, A \in V$, then T is linear if and only if $M = 0$.

- There is a unique linear map $T : R^2 \rightarrow R^2$ for which $T(1, 2) = (2, 3)$ and $T(0, 1) = (1, 4)$.

- T is not linear if C is viewed as a vector space over itself.

- Let $T : R^3 \rightarrow R^3$ be the linear mapping which rotates a vector about the z-axis through an angle θ :

  $T(x, y, z) = (x \cos \theta - y \cos \theta, x \sin \theta + y \cos \theta, z)$

  then Ker.$(T) = \{0\}$.

- If U has finite dimension and $T : U \rightarrow V$ is linear, then Im.T has finite dimension and dim.(Im.T) $\leq$ dim. U.

- The sum of the dimensions of the image and kernel of a linear mapping is equal to the dimensions of its domain.

- Let $T_1 : U \rightarrow V$ and $T_2 : V \rightarrow W$ be linear then rank$(T_2 \circ T_1) \leq$ rank(T_1).

- If $T : R^2 \rightarrow R^2$ be defined by $T(x, y) = (2x-4y, 3x-6y)$. Then T is singular.

- If $T : V \rightarrow V$ be the linear transformation which multiplies a polynomial by t $i.e.\, T[f(t)] = t.f(t)$, then T is non-singular.

- If $T : U \rightarrow V$ is linear then image $T(L)$ of line segment L in U is a line segment in V.

- If $T : U \rightarrow V$ is a linear transformation and X is a convex subset of U then image $T(X)$ is a convex subset of V.

- If the mapping $T : U \rightarrow V$ is linear, then there exists a basis of U and basis of V such that the matrix representation A of T has the form $T = \begin{pmatrix} I & 0 \\ 0 & 0 \end{pmatrix}$ where I is the r-square identity matrix and r is the rank of T.

- If A is a matrix representation of a linear operator T. Then B is also a matrix representation of T if and only if B is similar to A.

- If B is similar to A then B^{-1} is similar to A^{-1}.

Self Assessment Test

1. Let T and S be linear transformation on a finite dimensional vector space, where S is non-singular. Then show that
   $$\text{rank}(ST) = \text{rank}(TS) = \text{rank } T.$$

2. Let U and V be vector space of the same dimension n and $T : U \to V$ be a linear trasformation, then show that T is injective iff T is surjective.

3. Let U and V be vector spaces of dimension m and n respectively and $T : U \to V$ be a linear transformation. Then show that
   (i) $m > n \Rightarrow T$ is not injective.
   (ii) $m < n \Rightarrow T$ is not surjective.
   (iii) $m = n \Rightarrow T$ is injective iff T is surjective.

4. Let $T : U \to V$ be a linear transformation then show that T is non-singular iff image of a linearly independent set is linearly independent.

5. Let $T : C^2 \to C^2$ be a linear transformation defined by $T(z_1, z_2) = (\alpha z_1 + \beta z_2; \gamma z_1 + \delta z_2)$, $\alpha, \beta, \gamma, \delta \in R$.
   Show that T is non-singular if $\alpha\delta - \beta\gamma \neq 0$.

6. Let $T : R^3 \to R^2$ be a linear transformation given by
   $$T(x, y, z) = (x, y) \ \forall \ (x, y, z) \in R^3$$
   with respect to the standard basis of R^3 and the basis $\{(1, 0), (1, 1)\}$ of R^2. Show that matrix representation of T is given by
   $$\begin{bmatrix} 1 & -1 & 0 \\ 0 & 1 & 0 \end{bmatrix}.$$

7. Let $T : R^3 \to R^3$ be a linear transformation given by $T(x, y, z) = (x, y, 0)$, show that the null space is generated by $(0, 0, 1)$.

8. Show that nullity of $T : R^2 \to R^2 : T(x, y) = (x+y, x-y)$ is zero.

9. If the linear transformation $T : R^2 \to R^3$ is such that $T(1, 0) = (2, 3, 1)$ and $T(1, 1) = (3, 0, 2)$ then show that
   $$T(x, y) = (2x+y; 3x-3y; x+y)$$

10. Let $T : R^n \to R^n$ be a linear transformation and A be the standard matrix for T. Show that T is one-one if and only if the columns of A are linearly dependent.

11. Let $T : R^n \to R^m$ be a linear transformation and $A_{m \times n}$ be its matrix representation. Show that columns of A span R^n implies T is onto.

12. Show that two transforamtions $T : R^3 \to R^2$ and $S : R^3 \to R^2$ defined as $T(x, y, z) = (x+1, y+z)$ and $S(x, y, z) = (|x|, 0)$ are not linear.

13. If the minimal polynomial of a linear map $T : R^5 \to R^5$ is $x^2(x^3 - 1)$. Show that $\det(T) = 0$.

14. Let C be the vector space of all complex numbers over C and $T : C \to C$ be defined by $T(z) = \bar{z}$, $z \in$ C. Show that T is not linear map but a well defined map.

15. Let $T : U \to V$ be a linear transformation such that U has finite dimension and $\dim U = \dim V$. Show that T is an isomorphism if and only if T is one-one, onto and non-singular.

7 LINEAR FUNCTIONALS

7.1 INTRODUCTION

We shall study the linear mappings from a vector space V into its field F of scalars. Naturally all the theorems and results for arbitrary linear mappings on V hold for this special case. However, such type of mappings are treated separately because of their fundamental importance and because of the relationship between V and F, but all the theorems and results do not apply in the general case.

Definition. *Let V be a vector space over a field F. Then a linear transformation $\phi : V \to F$ is called a linear functional (or linear form) if for every α, $\beta \in V$ and every a, $b \in F$.*

$$\phi(a\alpha + b\beta) = a\phi(\alpha) + b\phi(\beta).$$

☛ ILLUSTRATIONS

(1) Let V be the vector space of polynomials in t over R. Then, the integral operator $\phi : V \to$ R defined by

$$\phi[p(t)] = \int_0^1 p(t)\,dt \text{ is a linear functional.}$$

(2) Let V be the vector space of n-square matrices over the field F. Let $T : V \to F$ be the trace mapping defined by $T(A) = a_{11} + a_{12} + \dots + a_{1n}$ where $A = [a_{ij}]$. This mapping is a linear functional.

(3) If V is a vector over the field F, then a mapping $T : V \to F$ defined by

$$T(\alpha) = 0, \ \forall \ a \in V$$

is a linear functional. This functional is also known as zero functional.

Definition. *Let V be a vector space over the field F. Let $\phi : V \to F$ be a linear functional. Then the negative of a linear functional f, defined by $-f$ is a linear functional from V into F defined by*

$$(-\phi)(\alpha) = -\phi(\alpha) \ \forall \ \alpha \in V$$

THEOREM 1. *Let V be a vector space over the field F. Let $\phi : V \to F$ be a linear functional. Then*

(i) *$\phi(\mathbf{0}) = 0$, where $\mathbf{0}$ is zero vector of V and 0 is zero element of F.*

(ii) *$\phi(-\alpha) = -\phi(\alpha) \ \forall \ \alpha \in V$.*

Proof. (i) Let $\alpha \in V$. Then

$$\alpha + \mathbf{0} = \alpha \qquad\qquad [\because \mathbf{0} \in V]$$
$$\Rightarrow \qquad \phi(\alpha + \mathbf{0}) = \phi(\alpha)$$
$$\Rightarrow \qquad \phi(\alpha) + \phi(\mathbf{0}) = \phi(\alpha) \qquad\qquad [\because \phi \text{ is linear.}]$$
$$\Rightarrow \qquad \phi(\alpha) + \phi(\mathbf{0}) = \phi(\alpha) + 0 \qquad\qquad [\because 0 \in F]$$
$$\Rightarrow \qquad \phi(\mathbf{0}) = 0 \qquad\qquad [\text{By left cancellation law for addition in } F]$$

(ii) We have

$$\alpha + (-\alpha) = \mathbf{0} \;\forall\; \alpha \in V$$

$$\Rightarrow \quad \phi[\alpha + (-\alpha)] = \phi(\mathbf{0})$$

$$\Rightarrow \quad \phi(\alpha) + \phi(-\alpha) = \phi(\mathbf{0}) \qquad\qquad [\because \phi \text{ is linear.}]$$

$$\Rightarrow \quad \phi(\alpha) + \phi(-\alpha) = 0 \qquad\qquad [\because \phi(\mathbf{0}) = 0]$$

$$\Rightarrow \quad \phi(-\alpha) = -\phi(\alpha)$$

7.2 DUAL SPACES

The set of linear functionals on a vector space V over the field F is also a vector space over F with addition and scalar multiplication defined by,

(i) $(\phi_1 + \phi_2)(\alpha) = \phi_1(\alpha) + \phi_2(\alpha), \;\forall\; \alpha \in V$

(ii) $(c\phi)(\alpha) = c\phi(\alpha), \; c \in F, \alpha \in V$

where ϕ_1, ϕ_2, ϕ are linear functionals on V.

This space is called the dual space (conjugate space) of V and is denoted by V^*. We also write $V^* = L(V, F)$.

Theorem 1. *Let V be a vector space over the field F. Let V^* be the set of all linear functionals on V with addition and scalar multiplication defined by*

(i) $(\phi_1 + \phi_2)(\alpha) = \phi_1(\alpha) + \phi_2(\alpha) \;\forall\; \alpha \in F$

(ii) $(c\phi)(\alpha) = c\phi(\alpha) \;\forall\; \alpha \in F, c \in F$

where $\phi_1, \phi_2 \in V^*$ forms a vector space over the field F.

Proof. **(i) Closure of addition in V^* :** Let ϕ_1 and ϕ_2 be linear functionals on V such that

$$(\phi_1 + \phi_2)(\alpha) = \phi_1(\alpha) + \phi_2(\alpha) \;\forall\; \alpha \in F$$

Since $\phi_1(\alpha) + \phi_2(\alpha) \in F$, then $\phi_1 + \phi_2$ is a function from V into F.

Let $a, b \in F$ and $\alpha, \beta \in V$. Then

$$(\phi_1 + \phi_2)(a\alpha + b\beta) = \phi_1(a\alpha + b\beta) + \phi_2(a\alpha + b\beta)$$

$$= [a\phi_1(\alpha) + b\phi_1(\beta)] + [a\phi_2(\alpha) + b\phi_2(\beta)]$$

$$[\because \phi_1, \phi_2 \text{ are linear.}]$$

$$= a[\phi_1(\alpha) + \phi_2(\alpha)] + b[\phi_1(\beta) + \phi_2(\beta)]$$

$$= a(\phi_1 + \phi_2)(\alpha) + b[\phi_1 + \phi_2)(\beta)$$

$$\therefore \quad (\phi_1 + \phi_2)(a\alpha + b\beta) = a(\phi_1 + \phi_2)(\alpha) + b(\phi_1 + \phi_2)(\beta) \;\forall\; a, b \in F, \;\forall\; \alpha, \beta \in V.$$

Thus, V^* is closed with respect to addition defined on it.

(ii) Associativity of addition in V^* : Let $\phi_1, \phi_2, \phi_3 \in V^*$. Then $\forall\; \alpha \in V$,

$$[\phi_1 + (\phi_2 + \phi_3)](\alpha) = \phi_1(\alpha) + (\phi_2 + \phi_3)(\alpha) \qquad\qquad [\text{By (i)}]$$

$$= \phi_1(\alpha) + [\phi_2(\alpha) + \phi_3(\alpha)] \qquad\qquad [\text{By (i)}]$$

$$= [\phi_1(\alpha) + \phi_2(\alpha)] + \phi_3(\alpha)$$

$$[\because \text{Addition in } F \text{ is associative}]$$

$$= [(\phi_1 + \phi_2)(\alpha)] + \phi_3(\alpha) \qquad\qquad [\text{By (i)}]$$

$$= [(\phi_1 + \phi_2) + \phi_3](\alpha) \qquad\qquad [\text{By (i)}]$$

$$\therefore \quad [\phi_1 + (\phi_2 + \phi_3)](\alpha) = [(\phi_1 + \phi_2) + \phi_3](\alpha) \;\forall\; \alpha \in V$$

$$\therefore \quad \phi_1 + (\phi_2 + \phi_3) = (\phi_1 + \phi_2) + \phi_3 \qquad [\text{By equality of two functional}]$$

(iii) Existence of additive identity :

Let **0** be the zero linear functional in V. Then
$$\mathbf{0}(\alpha) = 0 \; \forall \; \alpha \in V$$

$\Rightarrow$ $\mathbf{0} \in V^*$ and $\alpha \in V$, then

$$(\mathbf{0} + \phi)(\alpha) = 0(\alpha) + \phi(\alpha) \hspace{3cm} \text{[By (i)]}$$

$\Rightarrow$ $(\mathbf{0} + \phi)(\alpha) = 0 + \phi(\alpha)$

$\Rightarrow$ $(\mathbf{0} + \phi)(\alpha) = \phi(\alpha) \hspace{2.5cm} [\because 0 \text{ is an additive identity in } F.]$

$\therefore$ $(\mathbf{0} + \phi)(\alpha) = \phi(\alpha) \; \forall \; \alpha \in V$

$\therefore$ $\mathbf{0} + \phi = \phi \; \forall \; \phi \in V^*.$

Therefore, **0** is the additive identity in V^*.

(iv) Existence of additive inverse of each element in V^* :

Let $\phi \in V^*$, then $-\phi \in V^*$ such that
$$(-\phi)(\alpha) = -\phi(\alpha) \; \forall \; \alpha \in V$$

Now $(-\phi + \phi)(\alpha) = (-\phi)(\alpha) + \phi(\alpha) \hspace{2cm} \text{[By (i)]}$

$\Rightarrow$ $(-\phi + \phi)(\alpha) = -\phi(\alpha) + \phi(\alpha)$

$\Rightarrow$ $(-\phi + \phi)(\alpha) = 0 = \mathbf{0}(\alpha)$

$\therefore$ $-\phi + \phi = \mathbf{0} \; \forall \; \phi \in V^*$

Thus each element of V^* possesses additive inverse.

(v) Commutativity of addition in V^* :

Let $\phi_1, \phi_2 \in V^*$ and α be any element in V, then

$$[\phi_1 + \phi_2](\alpha) = \phi_1(\alpha) + \phi_2(\alpha) \hspace{3cm} \text{[By (i)]}$$

$\Rightarrow$ $(\phi_1 + \phi_2)(\alpha) = \phi_2(\alpha) + \phi_1(\alpha) \hspace{1cm} [\because \text{ addition in } F \text{ is commutative.}]$

$\Rightarrow$ $(\phi_1 + \phi_2)(\alpha) = (\phi_2 + \phi_1)(\alpha) \hspace{2.5cm} \text{[By (i)]}$

$\therefore$ $(\phi_1 + \phi_2) = (\phi_2 + \phi_1) \; \forall \; \phi_1, \phi_2 \in V^*.$

Hence V^* is an abelian group under addition defined on it.

Further, we make the following observations :

(vi) Closure under scalar multiplication in V^*:

Let $\phi \in V^*$ and $c \in F$. Then we define $c\phi$ as follows :
$$(c\phi)(\alpha) = c\phi(\alpha) \; \forall \; \alpha.$$

Since $c\phi(\alpha) \in F$, therefore, $c\phi$ is a functional from V into F.

Let $a, b \in F$ and $\alpha, \beta \in V$. Then

$$(c\phi)\,(a\alpha + b\beta) = c\phi(a\alpha + b\beta) \hspace{3cm} \text{[By (ii)]}$$

$\Rightarrow$ $(c\phi)(a\alpha + b\beta) = c[a\,\phi(\alpha) + b\phi(\beta)] \hspace{1.5cm} [\because \phi \text{ is linear.}]$

$\Rightarrow$ $(c\phi)(a\alpha + b\beta) = (ca)\phi(\alpha) + (cb)\phi(\beta) \hspace{1.5cm} [\because F \text{ is a field.}]$

$\Rightarrow$ $(c\phi)(a\alpha + b\beta) = (ac)\phi(\alpha) + (bc)\,\phi(\beta)$

$\Rightarrow$ $(c\phi)(a\alpha + b\beta) = a[c\phi(\alpha)] + b[c\phi(\beta)]$

$\Rightarrow$ $(c\phi)(a\alpha + b\beta) = a[(c\phi)(\alpha)] + b[(c\phi)(\beta)]$

$\therefore$ $c\phi$ is a linear functional on V.

$\therefore$ $c\phi \in V^* \; \forall \; \phi \in V^*, c \in F.$

Therefore, V^* is closed with respect to scalar multiplication defined on it.

(1) Let $c \in F$ and $\phi_1, \phi_2 \in V^*$. If α be any element in V, then

$$[c(\phi_1 + \phi_2)](\alpha) = c[(\phi_1 + \phi_2)(\alpha)] \qquad \text{[By(ii)]}$$
$$\Rightarrow \quad [c(\phi_1 + \phi_2)](\alpha) = c[\phi_1(\alpha) + (\phi_2)(\alpha)] \qquad \text{[By (i)]}$$
$$\Rightarrow \quad [c(\phi_1 + \phi_2)](\alpha) = c\phi_1(\alpha) + c\phi_2(\alpha)$$
$$\Rightarrow \quad [c(\phi_1 + \phi_2)](\alpha) = c\phi_1(\alpha) + c\phi_2(\alpha) \qquad \text{[By (ii)]}$$
$$\Rightarrow \quad [c(\phi_1 + \phi_2)](\alpha) = (c\phi_1 + c\phi_2)(\alpha) \qquad \text{[By (i)]}$$
$$\therefore \quad c(\phi_1 + \phi_2) = c\phi_1 + c\phi_2$$

(2) Let $a, b \in F$ and $\phi \in V^*$. If α be any element in V, then

$$[(a + b)\phi](\alpha) = (a + b) \phi(\alpha) \qquad \text{[By (ii)]}$$
$$\Rightarrow \quad [(a + b)\phi](\alpha) = a\phi(\alpha) + b\phi(\alpha) \qquad [\because F \text{ is a field.}]$$
$$\Rightarrow \quad [(a + b)\phi](\alpha) = (a\phi)(\alpha) + (b\phi)(\alpha) \qquad \text{[By (ii)]}$$
$$\Rightarrow \quad [(a + b)\phi](\alpha) = (a\phi + b\phi)(\alpha) \qquad \text{[By (i)]}$$
$$\therefore \quad (a + b)\phi(\alpha) = a\phi (\alpha) + b\phi(\alpha)$$

(3) Let $a, b \in F$ and $\phi \in V^*$. If α be any element in V, then

$$[(ab)\phi](\alpha) = (ab) \phi(\alpha) \qquad \text{[By (ii)]}$$
$$\Rightarrow \quad [(ab)\phi](\alpha) = a[b\phi(\alpha)] \qquad [\because F \text{ is a field.}]$$
$$\Rightarrow \quad [(ab)\phi](\alpha) = a[(b\phi)(\alpha)] \qquad \text{[By (ii)]}$$
$$\Rightarrow \quad [(ab)\phi](\alpha) = [a(b\phi)](\alpha) \qquad \text{[By (ii)]}$$
$$\therefore \quad (ab)\phi = a(b\phi) \; \forall \; \phi \in V^*$$

(4) Let 1 be the multiplication identity in F and $\phi \in V^*$.

If α be any element in V, then

$$(1\alpha)(\alpha) = 1 \; \phi(\alpha) \qquad \text{[By (ii)]}$$
$$\Rightarrow \quad (1\alpha)(\alpha) = \phi(\alpha) \qquad [\because F \text{ is a field.}]$$
$$\therefore \quad 1\phi = \phi \; \forall \; \phi \in V^*$$

Hence V^* forms a vector space over the field F.

THEOREM 2. *Let V be a finite-dimensional vector space over the field F. Then, dim. $V^* = $ dim. V.*

Proof. Let V be n-dimensional vector space. Let V^* be its dual space, so that

$$\text{dim. } V = n, L(V, F) = V^*, \text{dim. } F = 1.$$

Since we know that

$$\text{dim. } L(U, V) = (\text{dim. } U) (\text{dim. } V)$$
$$\Rightarrow \quad \text{dim. } L(V, F) = (\text{dim.} V) (\text{dim. } F)$$
$$\Rightarrow \quad \text{dim. } V^* = (\text{dim.} V) .1 \; \Rightarrow \; \text{dim. } V^* = \text{dim. } V$$

THEOREM 3. *Let V be a finite dimensional vector space and $\alpha \neq 0$ in V, then there is an element $\phi \in V^*$ such that $\phi(\alpha) \neq 0$.*

Proof. Let V be n-dimensional vector space over the field F and let $\alpha \neq 0$ be arbitrary non-zero vector of V and let $\{\alpha_1, \alpha_2, .., \alpha_n\}$ be the basis of V. Then there exists unique scalars $a_i \in F$ such that

$$\alpha = a_1\alpha_1 + a_2\alpha_2 + ... + a_n\alpha_n = \sum_{j=1}^{n} a_j\alpha_j$$

Suppose $\{\phi_1, \phi_2, ..., \phi_n\}$ is dual basis of V^*, then

$$\phi_i(\alpha_j) = \begin{cases} 1, & i = j \\ 0, & i \neq j \end{cases} = \delta_{ij}.$$

Now,
$$\phi_i(\alpha) = \phi_i\left(\sum_{j=1}^{n} a_j\alpha_j\right)$$

$$= \sum_{j=1}^{n} a_j\phi_i(\alpha_j) = \sum_{j=1}^{n} a_j\delta_{ij} = a_i$$

$\Rightarrow \qquad \phi_1(\alpha) = a_1, \phi_2(\alpha) = a_2, ..., \phi_n(\alpha) = a_n$

Since all the scalars a_i are not zero, this implies that there exists a linear functional $\phi \in V^*$ such that $\phi(\alpha) \neq 0$.

7.3 DUAL BASIS

Let $B = \{\alpha_1, \alpha_2, ..., \alpha_n\}$ be a basis for V, then there exists a unique linear functionals f on V for each i such that

$$\phi_i(\alpha_j) = \begin{cases} 1, & \text{if } i = j \\ 0, & \text{if } i \neq j \end{cases} = \delta_{ij} \text{ (Kronecker delta)}$$

Thus we obtain from B, a set of n distinct linear functional $\phi_1, \phi_2, ..., \phi_n$ on V. These functionals are also linearly independent and generate V^*.

Therefore, the set $B^* = \{\phi_1, \phi_2, ..., \phi_n\}$ forms a basis for V^*. This basis is called *dual basis* of B.

THEOREM 1. *Let V be an n-dimensional vector space over the field F and let $B = \{\alpha_1, \alpha_2, ..., \alpha_n\}$ be an ordered basis of V. If $\{\alpha_1, \alpha_2, ..., \alpha_n\}$ is an ordered set of n scalars, then there exists a unique functional ϕ on V such that*

$$\phi(\alpha_i) = a_i, \, i = 1, 2, ... n$$

Proof.

(i) Existence of ϕ

Since $B = \{\alpha_1, \alpha_2, ..., \alpha_n\}$ is a basis for V, then every element of V can be expressed as a linear combination of vectors of B.

Therefore, if $\alpha \in V$, then there exist scalars $x_1, x_2, ..., x_n \in F$ such that

$$f(\alpha) = x_1a_1 + x_2a_2 + ... + x_na_n$$

Now for this $\alpha \in V$, we define

$$\phi(\alpha) = x_1\alpha_1 + x_2\alpha_2 + ... + x_n\alpha_n$$

which is a unique element in F, therefore ϕ is a well-defined rule for associating with each $\alpha \in V$ in a unique scalar $\phi(\alpha) \in F$.

Thus, ϕ is a function from V into V.

Since each $\alpha_i \in V$ is uniquely expressed as a linear combination of vectors of B.

$$\therefore \qquad \alpha_i = 0\alpha_1 + 0\alpha_2 + ... + 1\alpha_i + 0.\alpha_{i+1} + ... + 0\alpha_n.$$

Then, by definition of ϕ, we have

$$\phi(\alpha_i) = \phi(0\alpha_1 + 0\alpha_2 + ... + 1\alpha_i + 0.\alpha_{i+1} + ... + 0\alpha_n).$$

$$= 0a_1 + 0a_2 + ... + 1a_i + 0\alpha_{i+1} + ... + 0\alpha_n$$

$$\therefore \qquad \phi(\alpha_i) = a_i, \, i = 1, 2, 3, ..., n$$

(ii) ϕ **is a linear functional.**

Let $a, b \in F$ and $\alpha, \beta \in V$. Also, let

$$\alpha = x_1\alpha_1 + x_2\alpha_2 + \dots + x_n\alpha_n$$

and $\quad \beta = y_1\alpha_1 + y_2\alpha_2 + \dots + y_n\alpha_n$ for each $x_i, y_i \in F$.

Then

$$\begin{aligned}
\phi(a\alpha + b\beta) &= \phi[a(x_1\alpha_1 + x_2\alpha_2 + \dots + x_n\alpha_n) + b(y_1\alpha_1 + y_2\alpha_2 + \dots + y_n\alpha_n)] \\
&= \phi[(ax_1 + by_1)\alpha_1 + (ax_2 + by_2)\alpha_2 \dots + (ax_n + by_n)\alpha_n] \\
&= (ax_1 + by_1)a_1 + (ax_2 + by_2)a_2 \dots + (ax_n + by_n)a_n \\
&= a(x_1a_1 + x_2a_2 + \dots + x_na_n) + b(y_1a_1 + y_2a_2 + \dots + y_na_n)
\end{aligned}$$

$\therefore \qquad \phi(a\alpha + b\beta) = a\phi(\alpha) + b\phi(\beta)$

$\therefore \qquad \phi$ is linear functional on V.

Hence there exists a linear functional ϕ on V such that $\phi(\alpha_i) = a_i, i = 1, 2, \dots, n$

(iii) ϕ **is unique.**

Suppose ϕ_1 is a linear functional on V such that

$$\phi_1(\alpha_i) = a_i, i = 1, 2, \dots, n$$

Let $\alpha = x_1\alpha_1 + x_2\alpha_2 + \dots + x_n\alpha_n$ be any vector of V. Then

$$\begin{aligned}
\phi_1(\alpha_i) &= \phi_1(x_1\alpha_1 + x_2\alpha_2 + \dots + x_n\alpha_n) \\
&= x_1\phi_1(\alpha_1) + x_2\phi_2(\alpha_2) + \dots + x_n\phi_n(\alpha_n) \qquad [\because \phi_1 \text{ is linear}] \\
&= x_1a_1 + x_2a_2 + \dots + x_na_n \qquad\qquad\quad [\text{By the definition of } \phi] \\
&= \phi(\alpha) \qquad\qquad\qquad\qquad\qquad\qquad [\text{By the definition of } \phi]
\end{aligned}$$

$\therefore \qquad \phi_1(\alpha_i) = \phi(\alpha) \; \forall \; \alpha \in V.$

Hence $\quad \phi_1 = \phi$ which shows that ϕ is unique.

THEOREM 2. *If $\{\alpha_1, \alpha_2, \dots, \alpha_n\}$ is a basis of a vector space V over the field F. Let $\phi_1, \phi_2, \dots, \phi_n \in V^*$ be the linear functionals defined by*

$$\phi_i(\alpha_j) = \delta_{ij} = \begin{cases} 1, & \text{if } i = j \\ 0, & \text{if } i \neq j \end{cases}$$

Then $\{\phi_1, \phi_2, \dots, \phi_n\}$ is a basis of V^*.

Proof. First we show that $\{\phi_1, \phi_2, \dots, \phi_n\}$ is linarly independent.

For $a_i \in F$ such that $a_1\phi_1 + a_2\phi_2, \dots, a_n\phi_n = 0$.

Applying both sides to α_1, we get

$$(a_1\phi_1 + a_2\phi_2, \dots, a_n\phi_n)(\alpha_1) = 0(\alpha_1) = 0$$

$\Rightarrow \qquad (a_1\phi_1(\alpha_1) + a_2\phi_2(\alpha_1), \dots, a_n\phi_n(\alpha_1) = 0$

$\Rightarrow \qquad a_1.1 + a_2.0 + \dots + a_n.0 \qquad\qquad \Rightarrow \quad a_1 = 0$

Similarly, for $i = 2, 3, \dots, n$ we have

$a_1\phi_1(\alpha_i) + a_2\phi_2(\alpha_i) + \dots + a_i\phi_i(\alpha_i) + \dots + a_n\phi_n(\alpha_i) = 0 \Rightarrow a_i = 0$. Thus

$a_1 = 0, a_2 = 0, \dots a_n = 0$. Hence $\{\phi_1, \phi_2, \dots, \phi_n\}$ is linearly independent.

Now we show that $\{\phi_i\}$ spans V^*.

For this, let ϕ be an arbitrary element of V^* and suppose that

$$\phi(\alpha_1) = c_1, \phi(\alpha_2) = c_2, \dots, \phi(\alpha_n) = c_n$$

and set $\qquad\qquad \psi = c_1\phi_1 + c_2\phi_2 + \dots + c_n\phi_n$, then

$$\psi(\alpha_1) = c_1\phi_1(\alpha_1) + c_2\phi_2(\alpha_1) + \dots + c_n\phi_n(\alpha_1) = c_1$$

Similarly, for $i = 2, 3, ..., n$. We have

$$\psi(\alpha_i) = c_i$$

Thus, $\phi(\alpha_i) = \psi(\alpha_i)$ for $i = 1, 2, 3, ..., n$. Since ϕ and ψ both agree on the basis vectors of V.

$\therefore$

$$\phi = \psi = c_1\phi_1 + c_2\phi_2 + ... + c_n\phi_n$$

$\Rightarrow$ $\{\phi_1, \phi_2, ..., \phi_n\}$ spans V^*.

Hence $\{\phi_1, \phi_2, ..., \phi_n\}$ forms a basis of V^*.

THEOREM 3. *Let $B = \{\alpha_1, \alpha_2, ..., \alpha_n\}$ be a basis of V and let $B^* = \{\phi_1, \phi_2, ..., \phi_n\}$ be the dual basis of V^*. Then, for any vector $\alpha \in V$, such that*

$$\alpha = \phi(\alpha)\alpha_1 + \phi(\alpha)\alpha_2 + ... + \phi_n(\alpha)\alpha_n$$

and for any linear functional $\phi \in V^*$, $\phi = \phi(\alpha)\phi_1 + \phi(\alpha)\phi_2 + ... + \phi_n(\alpha)\phi_n$.

Proof. Since $B = \{\alpha_1, \alpha_2, ..., \alpha_n\}$ is a basis for V and $B^* = \{\phi_1, \phi_2, ..., \phi_n\}$ is a basis for V^*, then there is a unique ϕ_j for each j such that

$$\phi_j(\alpha_i) = \delta_{ij} \qquad \qquad ...(1)$$

Since $\{\phi_1, \phi_2, ..., \phi_n\}$ generates V^*, then for some scalar $a_i \in F$ and for $\phi \in V^*$ such that

$$\phi = a_1\phi_1 + a_2\phi_2 + ... + a_n\phi_n = \sum_{j=1}^{n} a_j\phi_j \qquad \qquad ...(2)$$

Then

$$\phi(\alpha_i) = \sum_{j=1}^{n} a_j\phi_j(\alpha_i) = \sum_{j=1}^{n} a_j\delta_{ij} \qquad \text{[Using (1)]}$$

$$\phi(\alpha_i) = a_i, i = 1, 2, ..., n.$$

Thus (2) becomes

$$\phi = \phi(\alpha_1)\phi_1 + \phi(\alpha_2)\phi_2 + ... + \phi(\alpha_n)\phi_n.$$

Similarly, $B = (\alpha_1, \alpha_2, ..., \alpha_n)$ generates V, then for $\alpha \in V$ some scalars $b_i \in F$ such that

$$\alpha = b_1\alpha_1 + b_2\alpha_2 + ... + b_n\alpha_n = \sum_{i=1}^{n} b_i\alpha_j \qquad \qquad .... (3)$$

Then,

$$\phi_j(\alpha) = \sum_{i=1}^{n} b_i\alpha_j(\alpha_i) = \sum_{i=1}^{n} b_i\delta_{ij} \qquad \text{[Using (1)]}$$

$$= b_j \qquad \text{for } j = 1, 2, ..., n.$$

Hence, (3) becomes

$$\alpha = \phi_1(\alpha)\alpha_1 + \phi_2(\alpha)\alpha_2 + ... + \phi_n(\alpha)\alpha_n.$$

THEOREM 4. *Let V be a finite dimensional vector space over the field F. Then dim $V^* =$ dim. V.*

Proof. Let V be n-dimensional vector space. Let V^* be its dual space, so that

$$\text{dim. } V = n, L(V, F) = V^*, \text{dim. } F = 1$$

Since we know that

$$\text{dim. } L(U, V) = (\text{dim. } U)(\text{dim. } V)$$

$\Rightarrow$

$$\text{dim. } L(V, F) = (\text{dim. } V)(\text{dim. } F)$$

$\Rightarrow$

$$\text{dim. } V = (\text{dim. } V) . 1 \Rightarrow \text{dim. } V^* = \text{dim. } V.$$

THEOREM 5. *Let V be a finite dimensional vector space and a ≠ 0 in V, then there is an element $\phi \in V^*$ such that $\phi(\alpha) \neq 0$.*

Proof. Let V be n-dimensional vector space over the field F and let $\alpha \neq 0$ be arbitrary non-zero vector of V and let $\{\alpha_1, \alpha_2, .., \alpha_n\}$ be the basis of V. Then there exists unique scalars $a_i \in F$ such that

$$\alpha = a_1\alpha_1 + a_2\alpha_2 + ... + a_n\alpha_n = \sum_{j=1}^{n} a_j\alpha_j$$

Suppose $\{\phi_1, \phi_2, ..., \phi_n\}$ is dual basis of V^*, then

$$\phi_i(\alpha_j) = \begin{cases} 1, & i = j \\ 0, & i \neq j \end{cases} = \delta_{ij}.$$

Now,

$$\phi_i(\alpha) = \phi_i\left(\sum_{j=1}^{n} a_j\alpha_j\right)$$

$$= \sum_{j=1}^{n} a_j\phi_i(\alpha_j) = \sum_{j=1}^{n} a_j\delta_{ij} = a_i$$

$$\Rightarrow \quad \phi_1(\alpha) = a_1, \phi_2(\alpha) = a_2, .., \phi_n(\alpha) = a_n$$

Since all the scalars a_i are not zero, this implies that there exists a linear functional $\phi \in V^*$ such that $\phi(\alpha) \neq 0$.

THEOREM 6. *Let V be an n-dimensional vector space over the field F. If α, β are any two different vectors in V, there exists a linear functional ϕ on V such that $\phi(\alpha) \neq \phi(\beta)$.*

Proof. Since $\alpha \neq \beta$

$$\Rightarrow \quad \alpha - \beta \neq 0.$$

As $\alpha - \beta \neq 0$, exists a linear functional ϕ on V such that

$$\phi(\alpha - \beta) \neq 0$$

$$\Rightarrow \quad \phi(\alpha) - \phi(\beta) \neq 0$$

$$\Rightarrow \quad \phi(\alpha) \neq \phi(\beta)$$

7.4 SECOND DUAL SPACE : BIDUAL SPACE

We have observed that every vector space V has a dual space V^* which consists of all linear functionals on V. Therefore V^* will have a dual space denoted by V^{**} which is called the *second dual of V*. The second dual is also denoted by V''.

If V is finite dimensional, then

$$\text{dim. } V = \text{dim. } V^* = \text{dim. } V^{**} \Rightarrow V \equiv V^* \equiv V^{**}.$$

THEOREM 1. *Let V be an n-dimensional vector space over the field F. If $\alpha \in V$, then the function L_α on V^* defined by*

$$L_\alpha(\phi) = \phi(\alpha) \ \forall \ \alpha \in V^*.$$

is a linear functional on V^*, i.e. $L_\alpha \in V^{**}$.

Also the mapping $\alpha \to L_\alpha$ is an isomorphism of V onto V^{**}.

Proof. For $\alpha \in V$ and $\phi \in V^*$, $\phi(\alpha)$ is a unique element of F.

Therefore a mapping L_α defined by

$$L_\alpha(\phi) = \phi(\alpha) \ \forall \ \alpha \in V.$$

is a function from V^* into F.

Now, we shall show that L_α is a linear functional on V.

Let $a, b \in F$ and $\phi_1, \phi_2 \in V^*$. Then

$$L_\alpha (a\phi_1 + b\phi_2) = (a\phi_1 + b\phi_2)(\alpha) \qquad \text{[By the definitions of } L_\alpha]$$
$$= (a\phi_1)(\alpha) + (b\phi_2)(\alpha)$$
$$= a\phi_1(\alpha) + b\phi_2(\alpha)$$

$$\text{[By scalar multiplication of linear functional]}$$
$$= aL_\alpha(\phi_1) + bL_\alpha(\phi_2)$$

$\therefore$ L_α is a linear functional on V^*. Thus $L_\alpha \in V^{**}$.

Let $f : V \to V^{**}$ be a function defined by

$$f(\alpha) = L_\alpha \ \forall \ \alpha \in V.$$

Now we shall show that f is an isomorphism.

f is one-one : Let $\alpha, \beta \in V$ and assume that

$$f(\alpha) = f(\beta)$$
$$\Rightarrow \qquad L_\alpha = L_\beta$$
$$\Rightarrow \qquad L_\alpha(\phi) = L_\beta(\phi) \ \forall \ \phi \in V^*$$
$$\Rightarrow \qquad \phi(\alpha) = \phi(\beta) \qquad \text{[By the definition of } L_\alpha, L_\beta]$$
$$\Rightarrow \qquad \phi(\alpha) - \phi(\beta) = 0 \ \forall \ \phi \in V^*$$
$$\Rightarrow \qquad \phi(\alpha - \beta) = 0 \ \forall \ \phi \in V^*$$

Since, we know that if $\alpha - \beta \neq 0$, then there exists a linear functional ϕ on V such that $\phi(\alpha - \beta) \neq 0$. But here

$$\phi(\alpha - \beta) = 0 \ \forall \ \phi \in V^*$$
$$\Rightarrow \qquad \alpha - \beta = 0$$
$$\Rightarrow \qquad \alpha = \beta.$$

$\therefore$ f is one-one.

f is a linear transformation :

Let $a, b \in F$ and $\alpha, \beta \in V$. Then

$$f(a\alpha + b\beta) = L_{a\alpha + b\beta}$$

Now every $\phi \in V^*$, we have

$$L_{a\alpha + b\beta} (\phi) = (a\alpha + b\beta)$$
$$= a\phi(\alpha) + b\phi(\beta) \qquad \text{[}\because \phi \text{ is linear.]}$$
$$= aL_\alpha(\phi) + bL_\beta(\phi)$$
$$= (aL_\alpha)(\phi) + (bL_\beta)(\phi)$$
$$= (aL_\alpha + bL_\beta)(\phi)$$
$$\Rightarrow \qquad L_{a\alpha + b\beta} = aL_\alpha + bL_\beta$$
$$= af(\alpha) + bf(\beta)$$
$$\Rightarrow \qquad f(a\alpha + b\beta) = af(\alpha) + bf(\beta).$$

$\therefore$ f is linear transformation from V into V^{**}.

Since, dim. V = dim. V^{**}, therefore f is one-one implies that f must be onto.

Hence f is an isomorphism from V onto V^{**}.

REMARKS

- The mapping $\alpha \to L_\alpha$ is called natural mapping.
- For finite dimensional vector space, this natural mapping is an isomorphism. Thus this property of vector space is called reflexivity.
- In future we shall identify V^{**} with V through the natural isomorphism $\alpha \leftrightarrow L_\alpha$. Thus, we shall say that the element L of V^{**} is the same as the element α of V iff $L = L_\alpha$, i.e. iff $L(\phi) = \phi(\alpha) \; \forall \; \phi \in V^*$.

THEOREM 2. *Let V be a finite-dimensional vector space over the field F. If L is a linear functional on the dual space V^*, then there is a unique vector $\alpha \in V$ such that*

$$L(\phi) = \phi(\alpha), \; \phi \in V^*$$

Proof. This theorem is an immediate consequence of theorem 1. Therefore, in order to prove this theorem we should first prove theorem 1 and then proceed as follows: The mapping $\alpha \to L_\alpha$ is one-to-one correspondence between V and V^{**}. Therefore, if $L \in V^{**}$, then there exists a unique vector $\alpha \in V$ such that $L = L_\alpha$, i.e. $L(\phi) = \phi(\alpha) \; \phi \in V^*$

THEOREM 3. *Let V be a finite dimensional vector space over the field F. Each basis for V^* is the dual of some basis for V.*

Proof. Let dim. $V = n = $ dim. V^*

Let $B' = \{\phi_1, \phi_2, \dots \phi_n\}$ be a basis for V^*. Then there exists a dual basis
$$(B^*)^* = \{L_1, L_2, \dots, L_n\}$$
for V^{**} such that $L_i(\phi_j) = \delta_{ij}$ $\qquad \qquad \dots(1)$

Since V is n-dimensional, then for each i there is a unique vector α_i in V such that
$$L_i = L_{\alpha_i}$$
where $\qquad L_{\alpha_i}(\phi) = \phi(\alpha_i) \forall \phi \in V^*$ $\qquad \qquad \dots(2)$

Further since, the correspondence $\alpha \leftrightarrow L_\alpha$ is an isomorphism of V onto V^{**}. Therefore, under this isomorphism a basis is mapped onto a basis.

Therefore, $B = \{\alpha_1, \alpha_2 \dots, \alpha_n\}$ is a basis for V because it is the image set of a basis V under above isomorphism.

Putting $\phi = \phi_j$ in (2) we get

$$\phi_j(\alpha_i) = L_{\alpha_i}(\phi_j) = L_i\left(\phi_j\right) = \delta_{ij} \qquad \qquad \text{[From (1)]}$$

$\therefore \quad B^* = \{\phi_1, \phi_2, \dots, \phi_n\}$ is the dual of the basis B.

THEOREM 4. *Let V be a finite dimensional vector space over the field F. Let B be a basis for V and let B^* be the dual basis of B. Then*

$$B^{**} = (B^*)^* = B.$$

Proof. Let dim. $V = n$. Let $B = \{\alpha_1, \alpha_2, \dots, \alpha_n\}$ be a basis for V, $B^* = \{\phi_1, \phi_2, \dots, \phi_n\}$ be the dual basis of B in V^* and $B^{**} = \{L_1, L_2, \dots, L_n\}$ be the dual basis of B in V^{**}. Then

$$\phi_i(\alpha_j) = \delta_{ij} \text{ and } L_i(\phi_j) = \delta_{ij} \text{ ; for } i = 1, 2, \dots, n; j = 1, 2, \dots, n.$$

If α be any element of V, then there exists unique $L_\alpha \in V^{**}$ such that

$$L_\alpha(\phi) = \phi(\alpha) \; \forall \; \phi \in V^*$$

Now putting $\alpha = \alpha_i$ and $\phi = \phi_j$, we get
$$L_{\alpha_i}(\phi_j) = \phi_j(\alpha_i) = \delta_{ij} = L_i(\phi_j).$$
Thus L_{α_i} and L_i agree on a basis for V^*.
$$\therefore \qquad\qquad L_{\alpha_i} = L_i$$
If V^{**} is identified with V through the natural isomorphism $\alpha \leftrightarrow L_\alpha$, then L_α is considered as the same element as α.
$$\therefore \qquad\qquad L_i = L_\alpha = \alpha_i \text{ for } i = 1, 2, ..., n$$
Hence, $\qquad\qquad B^{**} = B.$

7.5 NATURAL MAPPING

Let V be an n-dimensional vector space, V^* its dual and V^{**} the dual of V^*. Then a mapping $\upsilon \to \bar{\upsilon}, \upsilon \in V, \bar{\upsilon} \in V^{**}$, where, υ is a linear functional on V^* defined by $\bar{\upsilon}(\phi) = \phi(\upsilon), \phi \in V^*$ is an isomorphism. This mapping is called a natural mapping.

REMARK

- If V is not finite dimensional, then the natural mapping can never be onto V^{**}. However, it is always linear and one-one.

7.6 ANNIHILATOR

Let V be a vector space over the field F and V^* its dual. Let W be a subset of V which is not necessarily a subspace. Then a linear functional $\phi \in V^*$ is called an annihilator of W if $\phi(\alpha) = 0$ for every $\alpha \in W$, which is denoted by W^0.

That is, the set of all linear functional ϕ on V such that $\phi(\alpha) = 0$, $\forall \alpha \in V$, i.e., $\phi(W) = \{0\}$ is called *annihilator* of W.

Also, $\qquad\qquad W^0 = \{\phi \in V^*: \phi(\alpha) = 0, \forall \alpha \in V\}$

REMARKS

- Annihilator of V is the zero functional on V.
- $\{0\}^0 = V^*$.

THEOREM 1. W^0 is a subspace of V^*.

Proof. By definition of W^0, it is clear that $0 \in W^\circ$ and $W^\circ \subseteq V^*$.
Now suppose $\phi_1, \phi_2 \in W^0$ and for any scalars $a, b \in F$ and for any $a \in W$.
$$(a\phi_1 + b\phi_2)(\alpha) = a\phi_1(\alpha) + b\phi_2(\alpha) = a.0 + b.0 \qquad [\because \phi_1, \phi_2 \in W^0]$$
$$= 0$$
$$\Rightarrow \qquad a\phi_1 + b\phi_2 \in W^0$$
Hence W^0 is a subspace of V^*.

THEOREM 2. *Let V be a finite-dimensional vector space over the field F and let W be a subspace of V. Then $dim.W + dim.W^0 = dim.V.$*

Proof. Let V be n-dimensional vector space over the field F. Let Dim. $W = m$.
Let W be a subspace of V. Then W^0 is a subspace of V^*(dual of V).
Since W is a subspace of V so that
$$\dim . W \le \dim . V, \text{ i.e. } m \le n.$$
Let $\{a_1, a_2, ..., a_n\}$ be a basis of W so it can be extended to form a basis of V,

therefore choose vectors $\{\alpha_{m+1}, \alpha_{m+2},\dots, \alpha_n\}$ in V such that
$$B=\{\alpha_1, \alpha_2,\dots, \alpha_m, \alpha_{m+1},\dots, \alpha_n)$$
is a basis of V. Let $\{\phi_1, \phi_2,\dots, \phi_n)$ be the basis of V^* which is the dual of B.
Now we claim that $\{\phi_{m+1}, \phi_{m+2},\dots,\phi_n)$ is a basis of W^0. Obviously, $\phi_i \in W^0$ for $i \geq m +1$, because

$$\phi_i(\alpha_j) = \delta_{ij} = \begin{cases} 1, & if \ i \neq j \\ 0, & if \ i = j \end{cases}$$

and $\qquad\qquad \delta_{ij}=0$ if $i \geq m+1$ and $j \leq m$.

Since $\{\phi_{m+1}, \phi_{m+2},\dots, \phi_n\}$ is a subset of linearly independent set $\{\phi_1, \phi_2,\dots, \phi_n\}$ hence $\{\phi_{m+1}, \phi_{m+2},\dots, \phi_n\}$ is linearly independent. Now we shall show that $\{\phi_{m+1}, \phi_{m+2},\dots, \phi_n\}$ spans W^0.

Let $\phi \in W^0$ be an arbitrary linear functional, so that
$$\phi(\alpha_i)= 0, \text{ for } 1 \leq i \leq m \qquad\qquad \dots(1)$$
Since $W^0 \subseteq V^*$, then $\phi \in V^*$

But $\{\phi_1, \phi_2,\dots, \phi_n\}$ generates V^*, therefore, we have

$$\phi = \sum_{i=1}^{n} \phi(\alpha_i)\phi_i \qquad\qquad \text{[By theorem]}$$

$$= \phi(\alpha_1)\phi_1 + \phi(\alpha_2)\phi_2 + \dots + \phi(\alpha_m)\phi_m + \phi(\alpha_{m+1})\phi_{m+1}$$
$$+ \phi(\alpha_{m+2})\phi_{m+2} + \dots + \phi(\alpha_n)\phi_n$$

$$= \phi(\alpha_{m+1})\phi_{m+1} + \phi(\alpha_{m+2})\phi_{m+2} + \dots + \phi(\alpha_n)\phi_n = \sum_{i=1}^{n} \phi(\alpha_i)\phi_i$$

This shows that $\{\phi_{m+1}, \phi_{m+2},\dots, \phi_n\}$ spans W^0.

Thus $\{\phi_{m+1}, \phi_{m+2},\dots, \phi_n\}$ forms a basis of W^0.

Accordingly, $\qquad\qquad$ dim. $W^0 = n - m =$ dim. $V -$ dim. W

Hence $\qquad\qquad$ dim. $W +$ dim. $W^0 =$ dim. V.

Corollary. *If W and W_1 are two subspaces of a vector space V which are annihilated by the subspace W^0, then dim. $W =$ dim. W_1.*

Proof. Since W and W_1 are both annihilated by the W^0, and both are subspaces of V, then by above theorem we have
$$\text{dim.}W + \text{dim.}W^0 = \text{dim. } V \qquad\qquad \dots(1)$$
and $\qquad\qquad$ dim.$W_1 +$ dim.$W^0 =$ dim. V $\qquad\qquad \dots(2)$
On using (1) and (2), we get dim. $W =$ dim.W_1

THEOREM 3. *Let W_1 and W_2 be subspaces of a finite dimensional vector space over the field F, then $W_1 = W_2$ if and only if $W_1^0 = W_2^0$.*

Proof. If $W_1 = W_2$, then obviously, $W_1^0 = W_2^0$. Conversely, if $W_1^0 = W_2^0$, then we can show that $W_1 = W_2$.

Let if possible, $W_1 \neq W_2$, then there is at least one vector in W_1 which is not in W_2. Suppose $\alpha \in W_2$ and $\alpha \neq W_1$. then there is a linear functional ϕ such that $\phi(\beta) = 0 \ \forall \ \beta \in W$ but $\phi(\alpha) \neq 0$. This implies that $\phi \in W_1^0$ but $\phi \in W_2^0$ and thus $W_1^0 \neq W_2^0$. Hence if $W_1^0 = W_2^0$, then $W_1 = W_2$.

7.7 ANNIHILATOR OF AN ANNIHILATOR

Definition. Let V be a vector space over the field F, V^* its dual space and let V^{**} be the dual space of V^*.

Let W be a subset of a vector space $V(F)$. Then W^0 is a subspace of V^*. Therefore,

$$(W^0)^0 = W^{00} = \{\psi \in V^{**} : \psi(\phi) = 0, \forall\, \phi \in W^0\}$$

Thus W^{00} is called an *annihilator of* W^0.

Since V^* is natural isomorphism to V^{**}, then we can write W^{00} as follows :

$$W^{00} = \{\alpha \in V : \phi(\alpha) = 0 \,\forall\, \phi \in W^0\}$$

THEOREM 1. *If W is a subspace of a finite-dimensional vector space V over the field F, then $W = W^{00}$.*

Proof. In order to prove $W = W^{00}$, we prove that

(i) $W \subset W^{00}$ (ii) $W^{00} \subset W$

By definition, we have

$$W^0 = \{\phi \in V^* : \phi(\alpha) = 0, \forall \phi \in W^0\} \qquad \ldots (1)$$

and

$$W^{00} = \{\psi \in V^{**} : \psi(\phi) = 0, \forall \phi \in W^0\}$$

or

$$W^{00} = \{\alpha \in V : \phi(\alpha) = 0, \text{ for all } \phi \in W^0\} \qquad \ldots(2)$$

Let $\alpha \in W$ and $\phi \in W^0$ be arbitrary. Then

$$\alpha \in W \Rightarrow \phi(\alpha) = 0 \,\forall\, \phi \in W^0 \Rightarrow \alpha \in W^{00} \qquad \text{[From (2)]}$$

$$\therefore \qquad W \subset W^{00}$$

Further, we know that

$$\text{dim.}W + \text{dim. } W^0 = \text{dim. } V \qquad \ldots(3)$$

and

$$\text{dim. } W^0 + \text{dim. } W^{00} = \text{dim. } V^*. \qquad \ldots (4)$$

From (3) and (4), we get dim. W = dim. W^{00} $[\because \text{dim. } V = \text{dim.} V^*]$

As $W \subset W^{00} \Rightarrow W$ is a subspace of W^{00} with the property that dim. W = dim. W^{00}.

Hence $W = W^{00}$

Solved Examples

Based on the following Results

→ A linear transformation $\phi : V \to F$ is called linear functional if for every α, $\beta \in V$ and a, $b \in F$,
$\phi(a\alpha + b\beta) = a\phi(\alpha) + b\phi(\beta)$

→ dim.(W) + dim (W^0) = dim .V

→ The set of linear functionals on a vector space V over the field F is also a vector space such that $(\phi_1 + \phi_2) = \phi_1(\alpha) + \phi_2(\alpha) \forall \alpha \in V$ and $\phi[c\alpha] = c\phi(\alpha)$, $c \in F$, where ϕ_1, ϕ_2 are linear functional on V. This space is called the dual space or conjugate space.

Example 1. *If $B = \{(-1, 1, 1), (1, -1, 1), (1, 1, -1)\}$ is a basis of $V_3(\mathrm{R})$, then find the dual basis of B.*

Solution. Let $\alpha_1 = (-1, 1, 1), \alpha_2 = (1, -1, 1), \alpha_3 = (1, 1, -1)$.

$\therefore \qquad B = \{\alpha_1, \alpha_2, \alpha_3\}$ is a basis of $V_3(\mathrm{R})$.

Let $\qquad B^* = (\phi_1, \phi_2, \phi_3)$ be the dual of B such that

$$\phi_i(\alpha_j) = \begin{cases} 1 \text{ if } & i = j \\ 0 \text{ if } & i \neq j \end{cases} \qquad \ldots(1)$$

Now write,

$$\phi_1(x, y, z) = a_1x + b_1y + c_1z$$
$$\phi_2(x, y, z) = a_2x + b_2y + c_2z$$
$$\phi_3(x, y, z) = a_3x + b_3y + c_3z$$

Determination of ϕ_1.

$$\phi_1(\alpha_1) = \phi_1(-1, 1, 1) = -a_1 + b_1 + c_1$$
$$\phi_1(\alpha_2) = \phi_1(1, -1, 1) = a_1 - b_1 + c_1$$
$$\phi_1(\alpha_3) = \phi_1(-1, 1, -1) = a_1 + b_1 - c_1$$

But using (1), we have

$$\phi_1(\alpha_1) = 1, \phi_1(\alpha_2) = 0, \phi_1(\alpha_3) = 0,$$

$$\therefore \qquad \left.\begin{array}{l} -a_1 + b_1 + c_1 = 1 \\ a_1 - b_1 + c_1 = 1 \\ a_1 + b_1 - c_1 = 1 \end{array}\right\} \qquad \dots (2)$$

From last two equations of (2), we have

$$\frac{a_1}{(-1)(-1) - (1)(1)} = \frac{b_1}{(1)(1) - (1)(-1)} = \frac{c_1}{(1)(1) - (-1)(1)} = k$$

$$\Rightarrow \qquad \frac{a_1}{1-1} = \frac{b_1}{1+1} = \frac{c_1}{1+1} = k$$

$$a_1 = 0, \; b_1 = 2k, \; c_1 = 2k.$$

Putting the values of a_1, b_1 and c_1 in the first equation of (2), we get

$$0 + 2k + 2k = 1 \Rightarrow 4k = 1 \Rightarrow k = \frac{1}{4}$$

$$\therefore \qquad a_1 = 0, b_1 = \frac{1}{2}, c_1 = \frac{1}{2}$$

Thus

$$\phi_1(x, y, z) = \frac{1}{2}(y + z)$$

Determination of ϕ_2.

$$\phi_2(\alpha_1) = \phi_2(-1, 1, 1) = -a_2 + b_2 + c_2$$
$$\phi_2(\alpha_2) = \phi_2(1, -1, 1) = a_2 - b_2 + c_2$$
$$\phi_2(\alpha_3) = \phi_2(1, 1, -1) = a_2 + b_2 - c_2$$

Again, $\qquad \phi_2(\alpha_1) = 0, \phi_2(\alpha_2) = 1, \phi_2(\alpha_3) = 0.$

So that above equations become:

$$-a_2 + b_2 + c_2 = 0$$
$$a_2 - b_2 + c_2 = 1$$
$$a_2 + b_2 - c_2 = 0$$

Solving these equations, we get

$$a_2 = \frac{1}{2}, b_2 = 0, c_2 = \frac{1}{2}$$

$$\therefore \qquad \phi_2(x, y, z) = \frac{1}{2}(x + z)$$

Determination of ϕ_3.

$$\phi_3(\alpha_1) = \phi_3(-1, 1, 1) = -a_3 + b_3 + c_3$$
$$\phi_3(\alpha_2) = \phi_3(1, -1, 1) = a_3 - b_3 + c_3$$
$$\phi_3(\alpha_3) = \phi_3(-1, 1, -1) = a_3 + b_3 - c_3$$

Again, $\phi_3(\alpha_1) = 0, \phi_3(\alpha_2) = 1, \phi_3(\alpha_3) = 1.$

So that above equations become:

$$-a_3 + b_3 + c_3 = 0$$
$$a_3 - b_3 + c_3 = 0$$
$$a_3 + b_3 - c_3 = 1$$

Solving these equations, we get $a_3 = \dfrac{1}{2}, b_3 = \dfrac{1}{2}, c_3 = 0$

$\therefore$ $$\phi_3(x, y, z) = \frac{1}{2}(x + y)$$

Hence $\{\phi_1, \phi_2, \phi_3\}$ is the dual basis,

Where $\phi_1(x, y, z) = \dfrac{1}{2}(y + z), \phi_2(x, y, z) = \dfrac{1}{2}(x + z), \phi_3(x, y, z) = \dfrac{1}{2}(x + y).$

Example 2. *If $\{\alpha_1 = (1, -2, 3), \alpha_2 = (1, -1, 1), \alpha_3 = (2, -4, 7)\}$ is a basis of R^3, then find the dual basis $\{\phi_i\}$.*

Solution. Let $\{\phi_1, \phi_2, \phi_3\}$ be the dual to the basis $\{\alpha_1, \alpha_2, \alpha_3\}$ such that

Let $$\phi_i(\alpha_j) = \begin{cases} 1 \text{ if } & i = 1 \\ 0 \text{ if } & i \neq j \end{cases}$$ ...(1)

$$\phi_1(x, y, z) = a_1 x + b_1 y + c_1 z$$
$$\phi_2(x, y, z) = a_2 x + b_2 y + c_2 z$$
$$\phi_3(x, y, z) = a_3 x + b_3 y + c_3 z$$

Determination of ϕ_1.

$$\phi_1(\alpha_1) = \phi_1(1, -2, 3) = a_1 - 2b_1 + 3c_1$$
$$\phi_2(\alpha_2) = \phi_1(1, -1, 1) = a_1 - b_1 + c_1$$
$$\phi_3(\alpha_3) = \phi_1(2, -4, 7) = 2a_1 - 4b_1 - 7c_1 \qquad \text{[Using (1)]}$$

But $\phi_1(\alpha_1) = 1, \phi_1(\alpha_2) = 0, \phi_1(\alpha_3) = 0,$

So that above equations become:

$$a_1 - 2b_1 + 3c_1 = 1$$
$$a_1 - b_1 + c_1 = 0$$
$$2a_1 - 4b_1 - 7c_1 = 0$$

Solving these equations, we get

$$a_1 = -3, b_1 = -5, c_1 = 2.$$

$\therefore$ $\phi_1(x, y, z) = -3x - 5y - 2z.$

Determination of ϕ_2.

$$\phi_2(\alpha_1) = \phi_2(1, -2, 3) = a_2 - 2b_2 + 3c_2$$
$$\phi_2(\alpha_2) = \phi_2(1, -1, 1) = a_2 - b_2 + c_2$$
$$\phi_1(\alpha_3) = \phi_3(2, -4, 7) = 2a_2 - 4b_2 + 7c_2$$

Since, $\phi_2(\alpha_1) = 0, \phi_2(\alpha_2) = 1, \phi_2(\alpha_3) = 0$, the above equation become

$$a_2 - 2b_2 + 3c_2 = 0$$
$$a_2 - b_2 + c_2 = 1$$

$$2a_2 - 4b_2 + 7c_2 = 0$$

Solving these equations, we get

$$a_2 = 2, b_2 = 1, c_2 = 0$$

$$\therefore \qquad \phi_2(x, y, z) = 2x + y$$

Determination of ϕ_3.

$$\phi_3(\alpha_1) = \phi_3(1, -2, 3) = a_3 - 2b_3 + 3c_3$$
$$\phi_3(\alpha_2) = \phi_3(1, -1, 1) = a_3 - b_3 + c_3$$
$$\phi_3(\alpha_3) = \phi_3(2, -4, 7) = 2a_3 - 4b_3 + 7c_3$$

Again, $\phi_3(\alpha_1) = 0$, $\phi_3(\alpha_2) = 0$, $\phi_3(\alpha_3) = 1$, the above equations become

$$a_3 - 2b_3 + 3c_3 = 0$$

$$a_3 - b_3 + c_3 = 0$$

$$2a_3 - 4b_3 + 7c_3 = 0$$

Solving the equations, we get

$$a_3 = 1, b_3 = 2, c_3 = 1$$

$$\therefore \qquad \phi_3(x, y, z) = x + 2y + z.$$

Hence, $\{\phi_1, \phi_2, \phi_3\}$ is the dual basis, where

$$\phi_1(x, y, z) = -(3x + 5y + 2z), \phi_2(x, y, z) = 2x + y, \phi_3(x, y, z) = 2x + 2y + 2z$$

Example 3. *Find the dual basis of the set*

$$B = \{(1, -1, 3), (0, 1, -1), (0, 3, -2)\} \text{ for } V_3(R)$$

Solution. Let us suppose $\alpha_1 = (1, -1, 3)$, $\alpha_2 = (0, 1, -1)$ and $\alpha_3 = (0, 3, -2)$ and let $B^* = \{(\phi_1, \phi_2, \phi_3)\}$ be the dual to the basis B such that

$$\phi_i(\alpha_j) = \begin{cases} 1 \text{ if } & i = j \\ 0 \text{ if } & i \neq j \end{cases} \qquad \qquad \text{...(1)}$$

Suppose

$$\phi_1(x, y, z) = a_1 x + b_1 y + c_1 z$$
$$\phi_2(x, y, z) = a_2 x + b_2 y + c_2 z$$
$$\phi_3(x, y, z) = a_3 x + b_3 y + c_3 z$$

Determination of ϕ_1.

$$\phi_1(\alpha_1) = \phi_1(1, -1, 3) = a_1 - b_1 + 3c_1$$
$$\phi_1(\alpha_2) = \phi_1(0, 1, -1) = b_1 - c_1$$
$$\phi_1(\alpha_3) = \phi_1(0, 3, -2) = 3b_1 - 2c_1$$

Since $\phi_1(\alpha_1) = 1$, $\phi_1(\alpha_2) = 0$, $\phi_1(\alpha_3) = 0$, [Using (1)]

then above equations become:

$$a_1 - b_1 + 3c_1 = 1$$
$$b_1 - c_1 = 0$$
$$3b_1 - 2c_1 = 0$$

Solving these equations, we get

$$a_1 = 1, b_1 = 0, c_1 = 0.$$

$$\therefore \qquad \phi_1(x, y, z) = x$$

Determination of ϕ_2.

$$\phi_2(\alpha_1) = \phi_2(1, -1, 3) = a_2 - b_2 + 3c_2$$

$$\phi_2 (\alpha_2) = \phi_2 (0, 1, -1) = b_2 - c_2$$
$$\phi_2 (\alpha_3) = \phi_3 (0, 3, -2) = 3b_2 - 2c_2$$

Since $\qquad \phi_2 (\alpha_1) = 0, \phi_2 (\alpha_2) = 1, \phi_2 (\alpha_3) = 0,$

the above equation become

$$a_2 - b_2 + 3c_2 = 0$$
$$b_2 - c_2 = 1$$
$$3b_2 - 2c_2 = 0$$

Solving these equations, we get

$$a_2 = 7, b_2 = -2, c_2 = -3$$

$\therefore \qquad\qquad \phi_2(x, y, z) = 7x - 2y - 3z$

Determination of ϕ_3.

$$\phi_3 (\alpha_1) = \phi_3 (1, -1, 3) = a_3 - b_3 + 3c_3$$
$$\phi_3 (\alpha_2) = \phi_3 (0, 1, -1) = b_3 - c_3$$
$$\phi_3 (\alpha_3) = \phi_3 (0, 3, -2) = 3b_3 - 2c_3$$

Since, $\qquad \phi_3 (\alpha_1) = 0, \phi_3 (\alpha_2) = 0, \phi_3 (\alpha_3) = 1,$

the above equations become

$$a_3 - b_3 + 3c_3 = 0$$
$$b_3 - c_3 = 0$$
$$3b_3 - 2c_3 = 0$$

Solving the equations, we get

$$a_3 = -2, b_3 = 1, c_3 = 1$$

$\therefore \qquad\qquad \phi_3 (x, y, z) = -2x + y + z.$

Hence, $\{\phi_1, \phi_2, \phi_3\}$ is the dual basis, where

$$\phi_1 (x, y, z) = x, \phi_2 (x, y, z) = 7x - 2y - 3z, \phi_3 (x, y, z) = -2x + y + z$$

Example 4. *A basis of the vector space $R^3(R)$ is*

$$B = \{\alpha_1 = (1,1,0), \alpha_2 = (1,0,1), \alpha_3 = (0,1,1)\}$$

and f is a linear functional on R^3 such that $f(\alpha_1) = 1, f(\alpha_2) = -1, f(\alpha_3) = 3$, then find f, when $\alpha = (1, -1, 3)$.

Solution. Since $B = \{\alpha_1 = (1, 1, 0), \alpha_2 = (1, 0, 1), \alpha_3 = (0, 1, 1)\}$ is a basis of $R^3(R)$.

Let $B^* = \{\phi_1, \phi_2, \phi_3\}$ be the basis. Then

$$\phi_i(\alpha_j) = \begin{cases} 1 & \text{if } i = j \\ 0 & \text{if } i \neq j \end{cases} \qquad\qquad \text{...(1)}$$

Now write

$$\phi_1(x, y, z) = a_1x + b_1y + c_1z$$
$$\phi_2(x, y, z) = a_2x + b_2y + c_2z$$
$$\phi_3(x, y, z) = a_3x + b_3y + c_3z$$

Determination of ϕ_1 .

$$\phi_1(\alpha_1) = \phi_1(1,1,0) = a_1 + b_1$$
$$\phi_1(\alpha_2) = \phi_1(1,0,1) = a_1 + c_1$$
$$\phi_1(\alpha_3) = \phi_1(0,1,1) = b_1 + c_1$$

Since $\qquad \phi_1(\alpha_1) = 1, \phi_1(\alpha_2) = 0, \phi_1(\alpha_3) = 0 \qquad\qquad$ [Using (1)]

Then, the above equations become
$$a_1 + b_1 = 1$$
$$a_1 + c_1 = 0$$
$$b_1 + c_1 = 0$$

Solving these equations, we get
$$a_1 = \frac{1}{2}, b_1 = \frac{1}{2}, c_1 = -\frac{1}{2}$$

$\therefore$
$$\phi_1(x, y, z) = \frac{1}{2}(x + y - z)$$

Determination of ϕ_2 .
$$\phi_2(\alpha_1) = \phi_2(1,1,0) = a_2 + b_2$$
$$\phi_2(\alpha_2) = \phi_2(1,0,1) = a_2 + c_2$$
$$\phi_2(\alpha_3) = \phi_3(0,1,1) = b_2 + c_2.$$

Using (1) we have $\phi_2(\alpha_1) = 1, \phi_2(\alpha_2) = 1, \phi_2(\alpha_3) = 0$,
the above equations become
$$a_2 + b_2 = 0$$
$$a_2 + c_2 = 1$$
$$b_2 + c_2 = 0.$$

Solving these equations, we get
$$a_2 = \frac{1}{2}, b_2 = -\frac{1}{2}, c_2 = \frac{1}{2}.$$

$\therefore$
$$\phi_2(x, y, z) = \frac{1}{2}(x - y + z).$$

Determination of ϕ_3 .
$$\phi_3(\alpha_1) = \phi_3(1,1,0) = a_3 + b_3$$
$$\phi_3(\alpha_2) = \phi_3(1,0,1) = a_3 + c_3$$
$$\phi_3(\alpha_3) = \phi_3(0,1,1) = b_3 + c_3$$

Since
$$\phi_3(\alpha_1) = 0, \phi_3(\alpha_2) = 0, \phi_3(\alpha_3) = 1.$$

Then, the above equations become
$$a_3 + b_3 = 0$$
$$a_3 + c_3 = 0$$
$$b_3 + c_3 = 1$$

Solving these equations, we get
$$a_3 = -\frac{1}{2}, b_3 = \frac{1}{2}, c_3 = \frac{1}{2}$$

$$\phi_3(x,y,z) = \frac{1}{2}(-x + y + z)$$

Since f is a linear functional R^3 such that $f(\alpha_1) = 1, f(\alpha_2) = -1, f(\alpha_3) = 3$.
Then by theorem, we have
$$f = f(\alpha_1)\phi_1 + f(\alpha_2)\phi_2 + f(\alpha_3)\phi_3$$
$$f = \phi_1 - \phi_2 + 3\phi_3.$$

$\therefore$
$$\phi(x, y, z) = \phi_1(x, y, z) - \phi_2(x, y, z) + 3\phi_3(x, y, z).$$

$$= \frac{1}{2}(x + y - z) - \frac{1}{2}(x - y + z) + \frac{3}{2}(-x + y + z)$$

$$= -\frac{3}{2}x + \frac{5}{2}y + \frac{1}{2}z.$$

Determination of f when $\alpha = (1, -1, 3)$

$$f(\alpha) = f(1, -1, 3) = -\frac{3}{2}(1) + \frac{5}{2}(-1) + \frac{1}{2}(3).$$

$$= -\frac{3}{2} - \frac{5}{2} + \frac{3}{2} = -\frac{5}{2}.$$

Example 5. *If W_1 and W_2 are subspaces of a vector space V over a field F and $W_1 \subset W_2$, show that $W_2^0 \subset W_1^0$.*

Solution. Since W_1 and W_2 are subspace of V with the condition that $W_1 \subset W_2$. Then we shall show that $W_2^0 \subset W_1^0$.

Let $\phi \in W_2^0$ be an arbitrary linear functional.

Then, we have $\phi \in W_2^0 \Rightarrow \phi(\alpha) = 0, \forall \alpha \in W_1$

$\Rightarrow \phi(\alpha) = 0, \forall \alpha \in W_2$ $[\because W_1 \subset W_2]$

$\Rightarrow \phi \in W_1^0$

Hence $W_2^0 \subset W_1^0$.

Example 6. *If W is a subset of a vector space $V(F)$, then*

     (i) $W^0 = [L(W)]^0$ *(ii) $W^{00} = L(W)$*

Solution. (i) Let $W = \{\alpha_1, \alpha_2, \dots \alpha_n\}$ be a subset of a vector space $V(F)$. We shall show that $W^0 = [L(W)]^0$

By the definition of W^0, we have

$$W^0 = \{\phi \in V^* : \phi(\alpha) = 0, \alpha \in W\}$$

Since $W \subseteq L(W)$, therefore

$$[L(W)]^0 \subset W^0 \qquad \dots(1)$$

Now, $\phi \in W^0 \Rightarrow \phi(\alpha) = 0, \forall \alpha \in W$

$\Rightarrow \phi(\alpha) = 0, \forall \alpha \in L(W)$ $\because W \subset L(W)]$

$\Rightarrow \phi \in [L(W)]^0$

$\therefore$ $W^{00} \subset [L(W)]^0 \qquad \dots(2)$

From (1) and (2), we get

$$W^0 = [L(W)]^0$$

(ii) Since we know that

$$W^{00} = W \qquad \dots(3)$$

$\therefore$ $W^{00} = (W^0)^0 = [[L(W)]^0]^0$ $[\because W^0 = [L(W)]^0]$

             $= L(W)$ [Using (3)]

Hence $W^{00} = L(W)$.

Example 7. *Let W_1 and W_2 be subspaces of a finite dimensional vector space V over the field F. Prove that*

     (i) $(W_1 + W_2)^0 = W_1^0 \cap W_2^0$

     (ii) $W_1^0 + W_2^0 = (W_1 \cap W_2)^0$

Solution. (i) To prove that : $(W_1 + W_2)^0 = W_1^0 \cap W_2^0$.

Since we know that : $W_1 \subset W_2 \Rightarrow W_2^0 \subset W_1^0$.

Obviously, $W_1 \subset W_1 + W_2$ and $W_2 \subset W_1 + W_2$

$\therefore$ $(W_1 + W_2)^0 \subset W_1^0$ and $(W_1 + W_2)^0 \subset W_2^0$

$\Rightarrow$ $(W_1 + W_2)^0 \subset W_1^0 \cap W_2^0$ ...(1)

Now, let $\phi \in W_1^0 \cap W_2^0$ be an arbitrary linear functional then

$\phi \in W_1^0 \cap W_2^0 \Rightarrow \phi \in W_1^0$ and $\phi \in W_2^0$

$\Rightarrow \phi(\alpha) = 0, \forall \alpha \in W_1$ and $\phi(\beta) = 0, \forall \beta \in W_2$

Let $\gamma \in W_1 + W_2$ be an arbitrary vector so that

$y = \alpha + \beta$, for some $\alpha \in W_1$, $\beta \in W_2$

$\therefore$ $\phi(\gamma) = \phi(\alpha + \beta) = \phi(\alpha) + \phi(\beta)$ [$\because \phi$ is linear.]

$= 0 + 0 = 0$

$\Rightarrow$ $\phi(\gamma) = 0$, for all $\gamma \in W_1 + W_2$

$\Rightarrow$ $\phi \in (W_1 + W_2)^0$

Thus, $W_1^0 \cap W_2^0 \subset (W_1 + W_2)^0$ ... (2)

From (1) and (2), we get

$(W_1 + W_2)^0 = W_1^0 \cap W_2^0$

(ii) Above result (i) is taken for the vector space $V^*(F)$ in place of $V(F)$, then we have $(W_1^0 + W_2^0)^0 = W_1^{00} \cap W_2^{00}$

$\Rightarrow$ $(W_1^0 + W_2^0)^0 = W_1 \cap W_2$ [$\because W^{00} = W$]

$\Rightarrow$ $(W_1^0 + W_2^0)^{00} = (W_1 \cap W_2)^0$

$\Rightarrow$ $W_1^0 + W_2^0 = (W_1 \cap W_2)^0$ [$\because W^{00} = W$]

Example 8. *Let vectors $\alpha_1 = (1, 1, 1)$, $\alpha_2 = (1, 1, -1)$ and $\alpha_3 = (1, -1, -1)$ form a basis of $V_3(C)$. If $\{\phi_1, \phi_2, \phi_3\}$ is the dual basis and if $\alpha = (0, 1, 0)$, find $\phi_1(\alpha)$, $\phi_2(\alpha)$ and $\phi_3(\alpha)$.*

Solution. Since $\alpha_1, \alpha_2, \alpha_3$ form a basis of $V_3(C)$; if $\alpha \in V_3(C)$, then

$\alpha = a_1\alpha_1 + a_2\alpha_2 + a_3\alpha_3, a_1, a_2, a_3 \in C$ then

and $\phi_1(\alpha) = a_1, \phi_2(\alpha) = a_2, \phi_3(\alpha) = a_3$

Now $\alpha = a_1\alpha_1 + a_2\alpha_2 + a_3\alpha_3$

$\Rightarrow$ $(0, 1, 0) = a_1(1, 1, 1) + a_2(1, 1, -1) + a_3(1, -1, -1)$

$\Rightarrow$ $(0, 1, 0) = (a_1 + a_2 + a_3, a_1 + a_2 - a_3, a_1 - a_2 - a_3)$

$\therefore$ $a_1 + a_2 + a_3 = 0$

$a_1 + a_2 - a_3 = 1$

$a_1 - a_2 - a_3 = 0.$

Solving these equations, we get

$$a_1 = 0, a_2 = \frac{1}{2}, a_3 = -\frac{1}{2}.$$

$$\phi_1(\alpha) = 0, \phi_2(\alpha) = \frac{1}{2}, \phi_3(\alpha) = -\frac{1}{2}.$$

Example 9. *Prove that if f is a linear functional on an n-dimensional vector space V(F) then the set of all those vectors α for which f(α) = 0 is a subspace of V. What is the dimension of that subspace ?*

Solution. Let $N = \{\alpha \in V : f(\alpha) = 0\}$.

Since we know that $f(0) = 0$, therefore, $0 \in N$.

Thus, N is a non-empty set.

Let $a, b \in F$ and $\alpha, \beta \in N$. Then

$$f(\alpha) = 0, f(\beta) = 0.$$

Now $f(a\alpha + b\beta) = af(\alpha) + bf(\beta)$ [∵ f is linear.]

$$= a0 + b0$$

$$= 0 + 0 = 0.$$

∴ $a\alpha + b\beta \in N.$

Thus, for all $a, b \in F$ and $\alpha, \beta \in N$, we have $a\alpha + b\beta \in N$.

Therefore, N is a subspace of V. This subspace N is the null space of f.

We know that

$$\dim. V = \dim. N + \dim. (\text{range of } f)$$

Now we have two cases :

Case I.   If f is a zero linear functional on V, then the range of f has only zero element of F.

Therefore,   dim. (range of f) = 0.

Then,   dim. N = dim. V = n.

Case II.   If f is a non-zero linear functional on V then f is onto F, therefore, dim. (range of f) is equal to the dimension of F^1 is 1.

Then,   dim. N = dim. $V - 1 = n - 1$.

Example 10. *If f is a non-zero linear functional on a vector space V and if x is an arbitrary scalar, does there necessarily exist a vector α in V such that f(α) = x ?*

Solution. Since f is a non-zero linear functional on V, then there must exist some non-zero element β in V such that

$$f(\beta) = y \neq 0 \text{ in } F.$$

If x is any element in F, then

$$x = x.1 = xy^{-1}y \qquad\qquad [\because y \neq 0, y \in F]$$

$$= (xy^{-1})y$$

$$= (xy^{-1})f(\beta)$$

$$= f[(xy^{-1})\beta] \qquad\qquad [\because f \text{ is linear functional.}]$$

Therefore there exists $\alpha = (xy^{-1})\beta \in V$ such that $f(\alpha) = x$.

Example 11. *Let V be a vector space over the field F. Let f be a non-zero linear functional on V and let N be the null space of f. Fix a vector α_0 in V which is not in N. Prove that for each α in V there is a scalar c and a vector β in N such that $\alpha = c\alpha_0 + \beta$. Prove that c and β are unique.*

Solution. Since f is a non-zero linear functional on V, then there exists a non-zero vector α_0 in V such that $f(\alpha_0) \neq 0$.

Therefore, by the definition of null space of f, we have $\alpha_0 \notin N$.

Let $f(\alpha_0) = y \neq 0$. Let α be any element of V such that

$$f(\alpha) = x$$
$$\Rightarrow \qquad f(\alpha) = (xy^{-1})y \qquad\qquad [\because y \in F \text{ and } y \neq 0 \Rightarrow y^{-1} \text{ exists}]$$
$$\Rightarrow \qquad f(\alpha) = cy \text{ where } c = xy^{-1} \in F$$
$$\Rightarrow \qquad f(\alpha) = cf(\alpha_0) \qquad\qquad [\because y = f(a_0)]$$
$$\Rightarrow \qquad f(\alpha) = f(c\alpha_0) \qquad\qquad [\because f \text{ is linear functional}]$$
$$\Rightarrow \qquad f(\alpha) - f(c\alpha_0) = 0$$
$$\Rightarrow \qquad f(\alpha - c\alpha_0) = 0$$
$$\Rightarrow \qquad \alpha - c\alpha_0 \in N$$
$$\Rightarrow \qquad \alpha - c\alpha_0 = \beta \text{ for some } \beta \in N$$
$$\Rightarrow \qquad \alpha = c\alpha_0 + \beta.$$

Now we shall show that c and β are unique.

Let, if possible, c' and β' be such that $a = c'\alpha_0 + \beta'$

Then, $\qquad\qquad c\alpha_0 + \beta = c'\alpha_0 + \beta' \qquad\qquad\qquad …(1)$
$$\Rightarrow \quad (c - c')\alpha_0 + (\beta - \beta') = 0$$
$$\Rightarrow \quad f[(c - c')\alpha_0 + (\beta - \beta')] = f(0)$$
$$\Rightarrow \quad (c{-}c')f(\alpha_0) + f(\beta - \beta')] = f(0)$$
$$\Rightarrow \qquad (c - c')f(\alpha_0) = 0 \qquad [\because \beta, \beta' \in N \Rightarrow \beta - \beta' \in N \Rightarrow f(\beta - \beta') = 0]$$
$$\Rightarrow \qquad\qquad c - c' = 0 \qquad\qquad\qquad\qquad\qquad [\because f(\alpha_0) \neq 0]$$
$$\Rightarrow \qquad\qquad c = c'$$

Putting $c = c'$ in (1), we get

$$\beta = \beta'$$

Hence c and β are unique.

Example 12. *If f and g are in V^* such that $f(\alpha) = 0 \Rightarrow g(\alpha) = 0$, prove that $g = kf$ for some $k \in F$.*

Solution. Since $f(\alpha) = 0 \Rightarrow g(\alpha) = 0$. Therefore, if α belongs to the null space of f, then α will also belong to the null space of g,

$\therefore$ Null space of $f \subseteq$ Null space of g.

Now, we have two cases :

Case I. If f is zero linear functional, then Null space of $f = V$
$$\Rightarrow \qquad\qquad V \subseteq \text{Null space of } g.$$
But $\qquad$ Null space of $g \subseteq V$.
$\therefore \qquad$ Null space of $g = V$.

Therefore, g is also zero linear functional.

Hence, $\qquad\qquad g = kf \ \forall \ k \in F.$

Case II. If f is a non-zero linear functional on V, then there exists a non-zero vector α_0 in V such that

$$f(\alpha_0) = y \neq 0, y \in F$$

If $\alpha \in V$, then we may write

$$\alpha = c\alpha_0 + \beta$$

where $c \in F$ and $\beta \in$ Null space of f and both are unique.

We have $\qquad\qquad g(\alpha) = g(c\alpha_0 + \beta)$

$$= cg(\alpha 0) + g(\beta)$$

$$= cg(\alpha 0)$$
$$[\because \beta \in \text{Null space of } f \Rightarrow f(\beta) = 0 \Rightarrow g(\beta) = 0]$$

Also
$$(kf)(\alpha) = kf(\alpha)$$
$$= kf(c\alpha_0 + \beta)$$
$$= kf(c\alpha_0) + kf(\beta)$$
$$= kcf(\alpha_0) \qquad\qquad [\because f(\beta) = 0]$$

Now, let $k = \dfrac{g(\alpha_0)}{f(\alpha_0)}$, then $k \in F$ and

$$(kf)(\alpha) = c\frac{g(\alpha_0)}{f(\alpha_0)}f(\alpha_0) = cg(\alpha_0)$$

$$\therefore \qquad\qquad (kf)(\alpha) = g(\alpha) \ \forall \ \alpha \in V$$

Hence, $\qquad\qquad kf = g$ for $k \in F$.

Example 13. *Let V be the vector space of polynomials over R of degree ≤ 1, i.e., $V = \{a + bx;$
$a, b \in R\}$. Let ϕ_1 and ϕ_2 be linear functionals on V defined by*

$$\phi_1\big[f(x)\big] = \int_0^1 f(x)dx \text{ and } \phi_2\big[f(x)\big] = \int_0^2 f(x)dx.$$

Find the basis $\{f_1, f_2\}$ of V which is dual to $\{\phi_1, \phi_2\}$.

Solution. Let
$$f_1(x) = a_1 + b_1 x, a_1, b_1 \in R$$
$$f_2(x) = a_2 + b_2 x, a_2, b_2 \in R$$

Since $\{f_1, f_2\}$ is dual to $\{\phi_1, \phi_2\}$. Then
$$\phi_1(f_1(x)) = 1, \phi_1(f_2(x)) = 0$$
and
$$\phi_2(f_1(x)) = 0, \phi_2(f_2(x)) = 1$$

Now
$$\phi_1(f_1(x)) = 1 \Rightarrow \qquad \int_0^0 (f_1(x)) = 1$$

$$\Rightarrow \qquad \int_0^1 (a_1 + b_1 x)dx = 1 \Rightarrow \left[a_1 x + b_1 \frac{x^2}{2}\right]_0^1 = 1$$

$$\Rightarrow \qquad\qquad a_1 + \frac{b_1}{2} = 1 \qquad\qquad \ldots(1)$$

$$\phi_2(f_1(x)) = 0 \Rightarrow \qquad \int_0^2 (f_1(x)) = 0$$

$$\Rightarrow \qquad \int_0^1 (a_1 + b_1 x)dx = 0 \Rightarrow \left[a_1 x + b_1 \frac{x^2}{2}\right]_0^2 = 0$$

$$\Rightarrow \qquad\qquad 2a_1 + 2b_1 = 0$$
$$\Rightarrow \qquad\qquad a_1 + b_1 = 0 \qquad\qquad \ldots(2)$$

Solving (1) and (2), we get
$$a_1 = 2, b_1 = -2$$

Putting these values of a_1 and b_1 in $f_1(x) = a_1 + b_1 x$, we get
$$f_1(x) = 2 - 2x.$$

Again $\qquad \phi_2(f_2(x)) = 0$

$\Rightarrow \qquad \int_0^1 (f_1(x)) = 0$

$\Rightarrow \qquad \int_0^1 (a_2 + b_2 x)\, dx = 0 \quad \Rightarrow \quad \left[a_2 x + b_2 \dfrac{x^2}{2} \right]_0^1 = 0$

$\Rightarrow \qquad a_2 + \dfrac{b_2}{2} = 0 \qquad\qquad\qquad …(3)$

$\phi_2(f_2(x)) = 1 \quad \Rightarrow \quad \int_0^2 (f_2(x)) = 1$

$\Rightarrow \qquad \int_0^2 (a_2 + b_2 x)\, dx = 1 \quad \Rightarrow \quad \left[a_2 x + b_2 \dfrac{x^2}{2} \right]_0^2 = 0$

$\Rightarrow \qquad 2a_2 + 2b_2 = 1 \qquad\qquad\qquad …(4)$

Solving (3) and (4), we get

$$a_2 = -\dfrac{1}{2},\, b_2 = 1$$

Putting these values of a_2 and b_2 in $f_2(x) = a_2 + b_2 x$, we get

$$f_2(x) = -\dfrac{1}{2} + x.$$

Hence, $\qquad f_1(x) = 2 - 2x,\, f_2(x) = -\dfrac{1}{2} + x$

Example 14. *If W_1 and W_2 are subspaces of a vector space V, and if $V = W_1 \oplus W_2$, then $V^* = W_1^\circ \oplus W_2^\circ$.*

Solution. Since $V = W_1 \oplus W_2$, then we have

$$V = W_1 + W_2 \text{ and } W_1 \cap W_2 = \{0\}.$$

Therefore, in order to prove that $V^* = W_1^\circ \oplus W_2^\circ$ we shall prove that

(i) $V^* = W_1^\circ + W_2^\circ$ and (ii) $W_1^\circ \cap W_2^\circ = \{0\}$

First we prove that $W_1^\circ \cap W_2^\circ = \{0\}$.

Let $\phi \in W_1^\circ \cap W_2^\circ$. Then $\phi \in W_1^\circ$ and $\phi \in W_2^\circ$.

Now, since $V = W_1 + W_2$ so if $\alpha \in V$, then

$$\alpha = \alpha_1 + \alpha_2 \text{ where } \alpha_1 \in W_1 \text{ and } \alpha_2 \in W_2.$$

Then $\qquad \phi(\alpha) = \phi(\alpha_1 + \alpha_2)$

$\qquad\qquad = \phi(\alpha_1) + \phi(\alpha_2) \qquad\qquad [\because \phi \text{ is linear functional.}]$

$\qquad\qquad = 0 + 0 \qquad\qquad [\because \phi \in W_1^\circ \text{ and } \alpha_1 \in W_1 \Rightarrow \phi(\alpha_1) = 0$

$\qquad\qquad\qquad\qquad\qquad \text{similarly } \phi(\alpha_2) = 0]$

$\qquad\qquad = 0$

$\therefore \qquad \phi(\alpha) = 0\ \forall\ \alpha \in V$

$\Rightarrow \qquad \phi = 0 \qquad\qquad\qquad \text{(Zero functional on } V)$

$\therefore \quad W_1^\circ \cap W_2^\circ = \{0\}$

Secondly, we shall prove that $V^* = W_1^\circ + W_2^\circ$.

Since, $V = W_1 \oplus W_2$, if $\alpha \in V$, then α can be uniquely written as

$$\alpha = \alpha_1 + \alpha_2 \text{ where } \alpha_1 \in W_1 \text{ and } \alpha_2 \in W_2.$$

Let $\phi \in V^*$. Now define two functions ϕ_1 and ϕ_2 from V into f for each of ϕ such that

$$\phi_1(\alpha) = \phi_1(\alpha_1 + \alpha_2) = \phi(\alpha_2) \qquad \qquad \text{...(1)}$$

and $\qquad \qquad \phi_2(\alpha) = \phi_2(\alpha_1 + \alpha_2) = \phi(\alpha_1) \qquad \qquad \text{...(2)}$

Now to show that ϕ_1 is a linear function.

Let $a, b \in F$ and $\alpha, \beta \in V$, and $\alpha = \alpha_1 + \alpha_2$, $\beta = \beta_1 + \beta_2$ where $\alpha_1, \beta_1 \in W_1$ and $\alpha_2, \beta_2 \in W_2$. Then

$$\therefore \qquad \phi_1(a\alpha + b\beta) = \phi_1[a(\alpha_1 + \alpha_2) + b(\beta_1 + \beta_2)]$$
$$= \phi_1[(a\alpha_1 + b\beta_1) + (a\alpha_2 + b\beta_2)]$$
$$= \phi[(a\alpha_2 + b\beta_2)] \qquad \qquad \text{[From (1)]}$$
$$\text{[As } a\alpha_1 + b\beta_1 \in W_1 \text{ and } a\alpha_2 + b\beta_2 \in W_2]$$
$$= a\phi(\alpha_2) + b\phi(\beta_2) \qquad \qquad [\because \phi \text{ is linear]}$$
$$= a\phi_1(\alpha) + b\phi_1(\beta) \qquad \qquad \text{[From (1)]}$$

$\therefore \quad \phi_1$ is a linear functional on $V \Rightarrow \phi_1 \in V^*$.

Now to show that $\phi_1 \in W_1^\circ$.

Let α_1 be any vector in W_1. Then $\alpha_1 \in V = W_1 \oplus W_2$.

$\therefore \qquad \alpha_1 = \alpha_1 + 0$, where $\alpha_1 \in W_1$, $0 \in W_2$.

Then, from (1), $\phi_1(\alpha_1) = \phi_1(\alpha_1 + 0) = \phi(0) = 0$.

$\therefore \qquad \qquad \phi_1(\alpha_1) = 0 \ \forall \ \alpha_1 \in W_1$

$\Rightarrow \qquad \qquad \phi_1 \in W_1^\circ.$

Similarly, we may prove that ϕ_2 is a linear functional on V and $\phi_2 \in W_2^\circ$.

Finally, we shall show that $\phi = \phi_1 + \phi_2$.

Let $\alpha \in V$ and $\alpha = \alpha_1 + \alpha_2$, where $\alpha_1 \in W_1$, $\alpha_2 \in W_2$.

Then, $\qquad (\phi_1 + \phi_2)(\alpha) = \phi_1(\alpha) + \phi_2(\alpha)$

$$= \phi(\alpha_2) + \phi(\alpha_1) \qquad \qquad \text{[From (1) and (2)]}$$
$$= \phi(\alpha_1) + \phi(\alpha_2)$$
$$= \phi(\alpha_1 + \alpha_2) \qquad \qquad [\because \phi \text{ is linear.]}$$
$$= \phi(\alpha).$$

$\therefore \qquad (\phi_1 + \phi_2)(\alpha) = \phi(\alpha) \ \forall \ \alpha \in V$

$\therefore \qquad \qquad \phi = \phi_1 + \phi_2$

Thus, $\qquad \phi \in V^* \Rightarrow \phi = \phi_1 + \phi_2$, where $\phi_1 \in W_1^\circ, \phi_2 \in W_2^\circ$.

$\therefore \qquad \qquad V^* = W_1^\circ + W_2^\circ.$

Therefore, $\qquad V^* = W_1^\circ + W_2^\circ$ and $W_1^\circ \cap W_2^\circ = \{0\}$

Hence, $\qquad \qquad V^* = W_1^\circ \oplus W_2^\circ.$

Example 15. *Let W be the subspace of* R^4 *spanned by the vectors* $\alpha_1 = (1, 2, -3, 4)$ *and* $\alpha_2 = (0, 1, 4, -1)$. *Find a basis of annihilator of W.*

Solution. By the definition of annihilator of W, we have

$$\phi \in W^\circ \text{ if } \phi(\alpha) = 0 \ \forall \ \alpha \in W.$$

Therefore, it is sufficient to find a basis set of linear functionals $\phi(x, y, z, t) = a_1x + a_2y + a_3z + a_4t$ for which $\phi(\alpha_1) = 0$, $\phi(\alpha_2) = 0$.

Then, $\phi(\alpha_1) = 0$

$\Rightarrow \qquad \phi(1, 2, -3, 4) = a_1 + 2a_2 - 3a_3 + 4a_4 = 0$ ...(1)

and $\phi(\alpha_2) = 0$

$\Rightarrow \qquad \phi(0, 1, 4, -1) = a_2 + 4a_3 - a_4 = 0$ ...(2)

There are two equations in four unknowns a_1, a_2, a_3, a_4.

Therefore, in order to find their solution we take two unknowns as free variables.

Let us taking a_3 and a_4 as free variables.

Set $a_3 = 1$, $a_4 = 0$, we get

$$a_1 + 2a_2 = 3$$
$$a_2 = -4.$$

Solving these equations, we get $a_1 = 11$, $a_2 = -4$.

$\therefore$ We obtain the solution $a_1 = 11$, $a_2 = -4$, $a_3 = 1$, $a_4 = 0$

and hence the linear functional $\phi_1(x, y, z, t) = 11x - 4y + z$

Again, set $a_3 = 0$, $a_4 = -1$, we get

$$a_1 + 2a_2 = 4$$
$$a_2 = -1.$$

Solving these equations, we get $a_1 = 6$, $a_2 = -1$.

$\therefore$ We obtain the solution $a_1 = 6$, $a_2 = -1$, $a_3 = 0$, $a_4 = 1$.

and hence the linear functional $\phi_2(x, y, z, t) = 6x - y - t$

Hence the set of linear functionals $\{\phi_1, \phi_2\}$ is a basis of W°.

🐝 EXERCISE 7.1

1. Let $\phi : R^2 \to R$ and $\psi : R^2 \to R$ be the linear functional defined by $\phi(x, y) = x + 2y$ and $\psi(x, y) = 3x - y$.
   Find (i) $\phi + \psi$ (ii) 4ϕ (iii) $2\phi - 5\psi$

2. Let $\phi : R^3 \to R$ and $\psi : R^3 \to R$ be the linear functionals defined by $\phi(x, y, z) = 2x - 3y + z$ and $\psi(x, y, z) = 4x - 2y + 3z$.
   Find (i) $\phi + \psi$ (ii) 3ϕ (iii) $2\phi - 2\psi$

3. Find the dual basis of the basis set $\{(2, 1), (3, 1)\}$ for R^2.

4. Let ϕ be the linear functional on R^2 defined by $\phi(2, 1) = 15$ and $\phi(1, -2) = -10$. Find $\phi(x, y)$ and in particular, find $\phi(-2, 7)$.

5. Find the dual basis of the basis set $B = \{(1, 0, 0), (0, 1, 0), (0, 0, 1)\}$ for $V_3(R)$.

6. Let V be the vector space of polynomials over R of degree ≤ 2. Let ϕ_1, ϕ_2 and ϕ_3 be the linear

functionals on V defined by

$$\phi_1[f(t)] = \int_0^1 f(t)dt, \phi_2[f(t)] = f'(t),$$
$$\phi_3[f(t)] = f(0)$$

Here, $f(t) = a + bt + ct' \in V$ and $f'(t)$ denotes the derivatives of $f(t)$. Find the basis $\{f_1(t), f_2(t), f_3(t)\}$ of V which is dual to $\{\phi_1, \phi_2, \phi_3\}$.

7. Prove that there are infinitely many linear functionals ϕ for which $\phi(0, 1, 2) = 1$ and $\phi(1, 1, -1) = 2$.

8. Prove that there is no linear functional ϕ on $V_3(R)$ for which
   (i) $\phi(1, 1, 2) = 1$, $\phi(2, 2, 4) = 3$
   (ii) $\phi(1, 1, 2) = -1$, $\phi(1, 0, 1) = 2$, $\phi(2, 1, 3) = 2$.
   [Hint : (i) $\{(1, 1, 2), (2, 2, 4)\}$ is not linearly independent.**]**

9. Define a non-zero linear functional ϕ in C^3 such that $\phi(\alpha) = 0 = \phi(\beta)$ where $\alpha = (1, 1, 1)$ and $\beta = (1, 1, -1)$.

10. Prove that every finite dimensional vector space V is isomorphic to its second dual

space V^{**} under an isomorphism which is independent of the choice of a basis in V.

11. Define the dual space V^* of the vector space V, If V is finite-dimensional vector space over the field F and V^* be the dual of V. Then, prove that dim. V = dim. V^*.

Answers

1. (i) $4x + y$ (ii) $4x + 8y$ (iii) $-13x + 9y$

2. (i) $6x - 5y + 4z$ (ii) $6x - 9y + 3z$ (iii) $-16x + 4y - 13z$

3. $B^* = \{\phi_1, \phi_2\}$, where $\phi_1(x, y) = 3y - x$, $\phi_2(x, y) = x - 2y$.

4. $\phi(x, y) = 4x + 7y$; $\phi(-2, 7) = 41$

5. $B^* = \{\phi_1, \phi_2, \phi_3\}$ where $\phi_1 = (x, y, z) = x$, $\phi_2 = (x, y, z) = y$, $\phi_3 = (x, y, z) = z$

6. $f_1(t) = 3t - \dfrac{3}{2}t^2, f_2(t) = -\dfrac{1}{2}t + \dfrac{3}{4}t^2, f_3(t) = 1 - 3t + \dfrac{3}{2}t^2$

7. $\phi(x, y, z) = c(x - y)$, where c is any non-zero scalars.

7.8 EIGENVALUES AND EIGENVECTORS OF A LINEAR TRANSFORMATION

Let V be a finite dimensional vector space over a field F and let T be a linear operator on V, then a scalar $\lambda \in F$ is called an eigenvalue of T if there exists a non-zero vector $\alpha \in V$ such that

$$T(\alpha) = \lambda\alpha.$$

Also, each such non-zero vector $\alpha \in V$ is called an eigenvector of T corresponding to λ. The set E_λ of all such vectors is a subspace of V called the eigenspace of λ.

THEOREM 1. *Non-zero eigenvectors belonging to distinct eigenvalues are linearly independent.*

Proof. Let $T : V \to V$ be a linear operator and let $\alpha_1, \alpha_2, \ldots, \alpha_n$ be non-zero eigenvectors of T corresponding to distinct eigenvalues $\lambda_1, \lambda_2, \ldots, \lambda_n$. Then we have to show that $\alpha_1, \alpha_2, \ldots, \alpha_n$ are linearly independent.

We shall prove by induction on n. If $n = 1$ then α_1 is linearly independent as $\alpha_1 \neq 0$, so assume that for $n > 1$, we have $\alpha_1, \alpha_2, \ldots, \alpha_{n-1}$ are linearly independent. Suppose

$$a_1\alpha_1 + a_2\alpha_2 + \ldots + a_n\alpha_n = 0 \qquad \ldots(1)$$

for $a_1, a_2, \ldots, \in F$.

Applying T to (1), we get

$$T(a_1\alpha_1 + a_2\alpha_2 + + \ldots + a_n\alpha_n) = T(0) = 0$$

$$\Rightarrow \quad a_1 T(\alpha_1) + a_2 T(\alpha_2) + \ldots + a_n T(\alpha_n) = 0 \qquad \{\because T \text{ is linear.}\} \qquad \ldots(2)$$

But by given hypothesis

$$T(\alpha_i) = \lambda_i\alpha_i, \forall\ i.$$

Therefore, (2) becomes

$$a_1\lambda_1\alpha_1 + a_2\lambda_2\alpha_2 + \ldots + a_n\lambda_n\alpha_n = 0 \qquad \ldots(3)$$

Now multiplying (1) by λ_n, we get

$$a_1\lambda_n\alpha_1 + a_2\lambda_n\alpha_2 + \ldots + a_n\lambda_n\alpha_n = 0 \qquad \ldots(4)$$

Subtracting (4) from (3), we get

$$a_1(\lambda_1 - \lambda_n)\,\alpha_1 + a_2(\lambda_2 - \lambda_n)\,\alpha_2 + ... + a_{n-1}(\lambda_{n-1} - \lambda_n)\,\alpha_{n-1} = 0 \qquad ...(5)$$

Since $\alpha_1, \alpha_2, ..., \alpha_{n-1}$ are linearly independent, also λ_i are distinct, therefore from (5) we obtain

$$a_1 = a_2 = ... = a_{n-1} = 0.$$

Substituting these values of $a_1, a_2...a_{n-1}$ into (1), we get

$$a_n\alpha_n = 0 \Rightarrow a_n = 0 \qquad [\because \alpha_n \neq 0]$$

So, $a_1 = a_2 = ... = a_n = 0$, hence $\alpha_1, \alpha_2, .., \alpha_n$ are linearly independent.

THEOREM 2. *Let $T : V \to V$ be a linear operator on a finite dimensional vector space over a field F. Then $\lambda \in F$ is an eignvalue of T if and only if the operator $(\lambda I - T)$ is singular. The eigenspace of X is then the kernel of $(\lambda I - T)$.*

Proof. Suppose $\lambda \in F$ is an eigenvalue of T, then we shall show that $(\lambda I - T)$ is singular. Let α be any non-zero vector of V, then by definition of eigenvalue of T, we have

$$T(\alpha) = \lambda\alpha \quad \text{or} \quad \lambda\alpha - T(\alpha) = 0$$
$$\Rightarrow \qquad (\lambda I)(\alpha) - T(\alpha) = 0 \Rightarrow (\lambda I\alpha - T)(\alpha) = 0$$
$$\Rightarrow \qquad |(\lambda I - T)| = 0$$
$$\Rightarrow \qquad \lambda I - T = 0 \Rightarrow \lambda I - T \text{ is singular.}$$

Conversely, suppose $(\lambda I - T)$ is singular, then we shall show that λ is an eigenvalue of T.

Let $\alpha \neq 0$ be an element of V so it is linearly independent.

Since $(\lambda I - T)$ is singular so that

$$|(\lambda I - T)| = 0 \Rightarrow (\lambda I - T)(\alpha) = 0 [\because \alpha \text{ is linearly independent.}]$$
$$\Rightarrow \qquad (\lambda I)(\alpha) - T(\alpha) = 0 \Rightarrow \lambda I - T(\alpha) = 0 \qquad [\because I(\alpha) = \alpha]$$
$$\Rightarrow \qquad T(\alpha) = \lambda\alpha$$
$$\Rightarrow \qquad \lambda \text{ is an eigenvalue of } T.$$

Also, $\qquad (\lambda I - T)(\alpha) = 0 \Rightarrow \alpha \in$ kernel of $(\lambda I - T)$.

7.8.1 Monic Polynomials

Let T be a linear operator on a n-dimensional vector space V over a field F. If B is an ordered basis of V, then $|(xI - T) = (xI - [T]_B)|$ is called a monic polynomial of degree n in $F[x]$. The monic polynomial $|(xI - T)|$ in the variable x is also called characteristic polynomial of T, it is denoted by $\Delta(x)$.

REMARKS

- By above theorem, λ is an eigenvalue of T if and only if λ is a root of this polynomial in F.
- The degree of characteristic polynomial of T is exactly equal to n, then n-dimension of V, T cannot have more than n eigenvalues, counted with multiplicity.

THEOREM 3. **(Cayley Hamilton Theorem).** *If T be a linear transformation on n-dimensional vector space and $\Delta(\lambda)$ be its characteristic polynomial then $\Delta(T) = 0$.*

Proof. Let $T : V(F) \to V(F)$ be a linear operator and let B be an ordered basis of V and the matrix of T corresponding to B be V.

$$[T]_B = A = [a_{ij}]_{n \times n} \text{ for } a_{ij}{'s} \in F$$

where, $$T(\alpha_j) = \sum_{i=1}^{n} a_{ij}\alpha_i.$$

If I is an identity matrix of order $n \times n$ and λ an indeterminate scalar, then the characteristic polynomial of T is given by

$$\Delta(\lambda) = |T - \lambda I| = 0 = |[T]_B - \lambda I| = 0 = |A - \lambda I| = 0$$

$$\therefore \qquad \Delta(\lambda) = \begin{vmatrix} a_{11} - \lambda & a_{12} & \dots & a_{1n} \\ a_{21} & a_{21} - \lambda & \dots & a_{2n} \\ \dots & \dots & \dots & \dots \\ a_{n1} & a_{n2} & \dots & a_{nn} - \lambda \end{vmatrix}$$

$$= b_n\lambda^n + b_{n-1}\lambda^{n-1} + \dots b_1\lambda + b_0 \text{ (say)}.$$

Thus the characteristic polynomial of A is

$$\Delta(\lambda) = 0, \text{ i.e., } b_n\lambda^n + b_{n-1}\lambda^{n-1} + \dots b_1\lambda + b_0 = 0 \qquad \dots(1)$$

Since the elements of $A - \lambda I$ are polynomials of degree at most one in λ so that each element of adj.$(A - \lambda I)$ is the polynomial at most of degree $n-1$, in λ. Therefore,

$$\text{adj.}(A - \lambda I) = B_0 + B_1\lambda + \dots + B_{n-1}\lambda^{n-1} \qquad \dots(3)$$

where B_i's are square matrices of degree $n \times n$.

Since $(A - \lambda I_n) \text{ adj } (A - \lambda I_n) = |A - \lambda I_n| I_n$

or $\qquad (A - \lambda I)(B_0 + B_1\lambda + B_2\lambda^2 + \dots + B_{n-1}\lambda^{n-1}) = (b_n\lambda^n + b_{n-1}\lambda^{n-1} + \dots + b_0) I.$

Comparing the coefficients of like powers in λ, on both sides, we have

$$AB_0 = b_0 I$$
$$AB_1 - B_0 = b_1 I$$
$$AB_2 - B_1 = b_2 I$$
$$\dots \qquad \dots \qquad \dots \qquad \dots$$
$$AB_{n-1} - B_{n-1} = b_{n-1}I - B_{n-1} = b_n I.$$

Multiplying above equation by $I, A, A^2, \dots, A^n$ respectively and then adding, we get

$$0 = b_0 I + b_1 A + b_2 A^2 + \dots + b^n A^n$$

or $\qquad \Delta(A) = 0$ $\qquad\qquad$ [Using (1)]

or $\qquad \Delta(T) = 0.$

THEOREM 4. *If $\lambda \in F$ is a characteristic root (eigenvalue) of T, then for any polynomial $p(x) \in F[x]$, $p(\lambda)$ is a characteristic root of $p(T)$.*

Proof. Since $\lambda \in F$ is a characteristic root of T, there exists a non-zero vector a in V such that

$$T(\alpha) = \lambda\alpha. \qquad \dots(1)$$

Now, $\qquad T^2(\alpha) = T[T(\alpha)] = T(\lambda\alpha)$ $\qquad\qquad$ [Using (1)]

$\qquad\qquad\qquad = \lambda T(\alpha)$ $\qquad\qquad\qquad\qquad$ [$\because$ T is linear.]

$\qquad\qquad\qquad = \lambda^2\alpha.$ $\qquad\qquad\qquad\qquad$ [Using (1)]

Continuing in this way, we get

$$T^k(\alpha) = \lambda^k\alpha \qquad \dots(2)$$

for all positive integer k.

Let $\quad p(x) = a_0x^n + a_1x^{n-1} + \ldots + a_{n-1}x + a_n, \, a_i \in F$

Then $\quad p(T) = a_0T^n + a_1T^{n-1} + \ldots + a_{n-1}T + a_nI.$

Now
$$
\begin{aligned}
[p(T)](\alpha) &= (a_0T^n + a_1T^{n-1} + \ldots + a_{n-1}T + a_n I)(\alpha) \\
&= a_0T^n(\alpha) + a_1T^{n-1}(\alpha) + \ldots + a_{n-1}T(\alpha) + a_n I(\alpha) \\
&= a_0\lambda^n\alpha + a_1\lambda^{n-1}\alpha + \ldots + a_{n-1}\lambda\,\alpha + a_n\,\alpha \\
&= (a_0\lambda^n + a_1\lambda^{n-1} + \ldots + a_{n-1}\lambda + a_n)(\alpha) \\
&= [p(\lambda)](\alpha)
\end{aligned}
$$

$\Rightarrow \quad [p(\lambda)I - p(T)](\alpha) = 0$

$\Rightarrow \quad [p(\lambda)I - p(T)] = 0$

$\Rightarrow \quad p(\lambda)$ is a characteristic root of $p(T)$.

7.9 MINIMAL POLYNOMIAL

A monic polynomial $m_T(x) \in F[x]$ of least degree, is said to be a minimal polynomial of T if $m_T(T) = 0$.

THEOREM 1. *If $\lambda \in F$ is a characteristic root of T then λ is a root of the minimal polynomial of T. In particular, T only has a finite number of characteristic roots in F.*

Proof. Let $m(x) \in F[x]$ be a minimal polynomial of T, then we shall show that λ is a root of $m(x) = 0$.

If $\lambda \in F$ is a characteristic root of T, then there exists a non-zero vector α in V such that
$$ T(\alpha) = \lambda\alpha. $$

By above theorem,
$$ [m(T)](\alpha) = [m(\lambda)](\alpha) = m(\lambda)\,\alpha. $$

But $m(T) = 0$ as $m(x)$ is a minimal polynomial of T.

$\therefore \qquad 0\,\alpha = m(\lambda)\alpha$

$\Rightarrow \qquad m(\lambda)\,\alpha = 0 \qquad$ [By elementary property of vector space]

Since $\alpha \neq 0$ so that $m(\lambda) = 0$.

Thus λ is a root of $m(x)$.

Also, V is a finite dimensional vector space; if dim $V = n$ then degree of $m(x) \leq n^2$, therefore $m(x)$ has atmost n^2 roots implying that $m(x)$ has only a finite number of roots in F, thus there can only be finite number of characteristic roots of T in F.

THEOREM 2. *If T and S are two linear operators with S invertible, then T and STS^{-1} have the same minimal polynomial.*

Proof. Now, $\quad (STS^{-1})^2 = (STS^{-1})((STS^{-1}) = ST(S^{-1}S)TS^{-1}$

$\qquad\qquad\qquad = ST^2S^{-1} \qquad\qquad\qquad\qquad\qquad [S^{-1}S = I\,]$

Continuing in this way, we get

$\qquad\qquad (STS^{-1})^k = ST^kS^{-1} \qquad\qquad\qquad\qquad\qquad \ldots(1)$

Let $m_T(x)$ be a minimal polynomial of T, thus $m_T(T) = 0$ and let
$$ m_T(x) = a_0x^n + a_1x^{n-1} + \ldots + a_{n-1}x + a_n. $$

Then, $\qquad a_0T^n + a_1T^{n-1} + \ldots + a_{n-1}T + a_nI = 0.$

Now, $\quad m_T(STS^{-1}) = a_0(STS^{-1})^n + a_1(STS^{-1})^{n-1} + \ldots + a_n(STS^{-1})$

$\qquad\qquad\qquad = a_0ST^n S^{-1} + a_1ST^{n-1}S^{-1} + \ldots + a_nSTS^{-1} \qquad$ [Using (1)]

$$= S(a_0 T^n + a_1 ST^{n-1} + ... + a_n I)S^{-1}$$
$$m_T(STS^{-1}) = S . 0 . S^{-1}. \qquad \text{[Using (2)]}$$
$$= 0$$

Hence $m_T(x)$ is also the minimal polynomial of STS^{-1}.

THEOREM 3. *Let V be a finite dimensional vector space over a field F and let T be a linear operator on V. Then there exists a vector α_j in V such that $m_T^{\alpha_j}(x) = m_T(x)$.*

Proof. For any vector $\beta \in V$, the polynomial $m_T^{\beta}(x)$ divides $m_T(x)$. Thus each $m_T^{\beta}(x)$ is a factor of $m_T(x) \in F[X]$. Since V is finite-dimensional so for all $\beta \in V$, we get only finite number of polynomials. Let these polynomials be $p_1(x), p_2(x), ..., p_k(x)$ corresponding to the vectors $\alpha_1, \alpha_2, ..., \alpha_k$, that is

$$m_T^{\alpha_i}(x) = p_i(x) \qquad \text{for } i = 1, 2, ..., k.$$

Let $V_i = \ker p_i(T)$.

Then $V \subseteq \overset{k}{\underset{i=1}{\cup}}$ and so $V = V_j$ for some j, therefore,

$$(p_j(T))_\beta = 0, \forall \ \beta \in V$$

$$\Rightarrow \qquad\qquad p_j(T) = m_T^{\alpha_i}(x) = m_T(x)$$

7.10 INVARIANCE OF LINEAR OPERATOR

Let $T : V \to V$ be a linear operator. A subspace W of V is said to be T-invariant or invariant under T if T maps W into itself, that is, if $a \in W$ implies $T(\alpha) \in W$. In this case T restricted to W defines a linear operator on W; that is T induces a linear operator $\hat{T}(\alpha) = T(\alpha), \forall \ \alpha \in W$.

THEOREM 1. *Let $T : V \to V$ be a linear operator and let $p(x)$ be any polynomial. Then the kernel of $p(T)$ is T-invariant*

Proof. Let $\alpha \in \ker p(T)$, then $(p(T))(\alpha) = 0$.

Now we shall show that $T(\alpha) \in \ker p(T)$.

Since $\qquad\qquad\qquad p(x)x = x \, p(x)$ so that $p(T)T = Tp(T)$.

$\therefore \qquad\qquad\qquad (p(T)T)(\alpha) = T(p(T)(\alpha)) = T(0) = 0$

$\Rightarrow \qquad\qquad\qquad p(T)(T(\alpha)) = 0.$

Thus, $\qquad\qquad\qquad T(\alpha) \in \ker p(T).$

THEOREM 2. *Let $T : V \to V$ be a linear operator, and suppose that $p(x) = f(x) \, g(x)$ are polynomials such that $p(T) = 0$ and $f(x)$ and $h(x)$ are relatively prime. Then V is the direct sum of the T-invariant subspace W_1 and W_2 where, $W_1 = \ker f(T)$ and $W_2 = \ker g(T)$.*

Proof. Since $f(x)$ and $g(x)$ are relatively prime, then there exist two polynomials $r(x)$ and $s(x)$ such that

$$r(x)f(x) + s(x)g(x) = I.$$

So, for the operator T, we get

$$r(T)f(T) + s(T)g(T) = I. \qquad ...(1)$$

Let $a \in V$, then form (1), we have

$$\alpha = [r(T)f(T)](\alpha) + [s(T)g(T)](\alpha) \qquad ...(2)$$

Since $\quad [g(T)(r(T)f(T))](\alpha) = r(T)[f(T)g(T)(\alpha)]$
$$= r(T)[p(T)(\alpha)]f = r(T)0. \alpha \qquad [\because p(T) = 0]$$
$$= 0.$$
$\therefore \qquad [r(T)f(T)](\alpha) \in W_2 = \ker g(T).$

Similarly, $[s(T)g(T)](\alpha) \in W_1 = \ker f(T).$

Thus, α is the sum of an element of W_1 and an element of W_2. Hence
$$V = W_1 + W_2.$$

Now, to show that $V = W_1 \oplus W_2$, we must show that a sum $\alpha = \beta + \gamma$ with $\beta \in W_1$ and $\gamma \in W_2$, is uniquely determined by α.

Applying the operator $r(T)f(T)$ to $\alpha = \beta + \gamma$ and using $f(T)(\beta) = 0$, we get
$$[r(T)f(T)](\alpha) = [r(T)f(T)](\beta) + [r(T)f(T)](\gamma) = [r(T)f(T)](\gamma) ...(3)$$

Also, from (1), we get
$$[r(T)f(T)](\gamma) + [s(T)g(T)]\gamma = \gamma$$
$$\Rightarrow \qquad\qquad\qquad \gamma = [r(T)f(T)](\gamma) \qquad [\because (g(T))(\gamma) = 0] \qquad ...(4)$$

From (3) and (4), we get
$$\gamma = [r(T)f(T)](\alpha)$$

which shows that γ is uniquely determined by α.

Similarly, β is uniquely determined by α. Hence $V = W_1 \oplus W_2$.

THEOREM 3. *If T_1 is the restriction of T to W_1 and T_2 is the restriction of T to W_2, where $T : V \to V$ is a linear operator and $p(x) = f(x)g(x)$ such that $p(x)$ is the minimal polynomial of T and $f(x)$ and $g(x)$ are monic. Then $f(x)$ and $g(x)$ are minimal polynomials of T_1 and T_2 respectively.*

Proof. Let $m_1(x)$ and $m_2(x)$ be the minimal polynomials of T_1 and T_2 respectively, then we shall show that $m_1(x) = f(x)$ and $m_2(x) = g(x)$

Since $\qquad\qquad W_1 = \ker f(x)$ and $W_2 = \ker g(x)$

$\therefore \qquad\qquad f(T_1) = 0$ and $g(T_2) = 0$

Thus, $m_1(x)$ divides $f(x)$ and $m_2(x)$ divides $g(x)$.

Let $p(x)$ be the least common multiple of $m_1(x)$ and $m_2(x)$. But $m_1(x)$ and $m_2(x)$ are relatively prime. Accordingly, we have
$$p(x) = m_1(x) m_2(x).$$

But we have,
$$p(x) = f(x)g(x)$$

Since $m_1(x)$ is monic and divides $f(x)$ and $m_2(x)$ is also monic and divides $g(x)$.

Thus $\qquad\qquad f(x) = m_1(x)$ and $g(x) = m_2(x).$

THEOREM 4. **(Primary Decomposition Theorem).** *Let $T : V \to V$ be a linear operator with minimal polynomial*
$$m(x) = p_1(x)^{n_1} p_2(x)^{n_2} ...p_r(x)^{n_r}$$

where the $p_i(x)$ are distinct monic irreducible polynomials. Then V is the direct sum of T-invariant subspaces $W_1, W_2,...,W_r$, where $W_i = \ker p_i(T)^{n_i}$. Moreover, $p_i(x)^{n_i}$ is the minimal polynomial of the restriction of T to W_i.

Proof. We shall prove the theorem by induction on r. If $r = 1$, then the theorem is trivially true.

Suppose, that the theorem is true for $r - 1$.

By theorem 1, we may write V as the direct sum of T- invariant subspaces W_1 and V_1 where $W_1 = \ker p_1(T)^{n_i}$ and $V_1 = \ker(p_2(T)^{n_2}...p_r(T)^{n_r})$.

Now by theorem 1, the minimal polynomials of the restrictions of T to W_1 and V_1 are respectively $p_1(x)^{n_i}$ and $p_2(x)^{n_2}...p_r(x)^{n_r}$.

Let T_1 be the restriction of T to V_1. Then by induction hypothesis, V_1 is the direct sum of subspaces $W_2, W_3, .. ., W_r$ such that

$$W_i = \ker p_i(T)^{n_i}$$

where $p_i(x)^{n_i}$ is the minimal polynomial for the restriction of T_1 to W_i.

But $\qquad \ker p_i(T)^{n_i} \subseteq V_1$ for $i = 2, 3,..., r$.

Since $\qquad p_i(T_1)^{n_i} \,|\, p_2(x)^{n_2}...p_r(x)^{n_r}$. Thus the $\ker p_i(T)^{n_i}$ is the same as the $\ker p_i(T_1)^{n_i} = W_1$. Also the restriction of T to W_i is the same as the restriction of T_1 to $W_i (i = 2,3,...,r)$. Hence $p_i(x)^{n_i}$ is also the minimal polynomial for the restriction of T to W_i.

Hence $\qquad V = W_1 \oplus W_2 \oplus ... \oplus W_r$.

Solved Examples

Based on the following Results

➡ A scalar $\lambda \in F$ is called an eigenvalue of a linear operator T if their exists a non-zero vector $\alpha \in V$ such that $T(\alpha) = \lambda \cdot \alpha$.

➡ λ is an eigenvalue of T if and only if λ is the root of characteristic polynomial.

➡ Let T be a linear transformation on n-dimensional vector space and $\Delta(\lambda)$ be its characteristic equation then $\Delta(T) = 0$ [Cayley-Hamilton theorem]

Example 1. *Let $I : V \to V$ be the identity mapping on any non-zero vector space V. Show that $\lambda = 1$ is an eigenvalue of I. What is the eigenspace E_1 of $\lambda = 1$?*

Solution. Let $\alpha \neq 0$ be any vector of V, then
$$I(\alpha) = \alpha = 1\alpha.$$
$\Rightarrow \quad \lambda = 1$ is an eigenvalue of I.

Also, every vector in V is an eigenvector corresponding to 1, therefore $E_1 = V$.

Example 2. *Show that 0 is an eigenvalue of T if and only if T is singular.*

Solution. Let α be a non-zero vector of V.

0 is an eigenvalue of $T \Leftrightarrow$ there exists a non-zero vector such that
$$T(\alpha) = 0 . \alpha$$
$\Leftrightarrow \qquad T(\alpha) = 0$
$\Leftrightarrow \qquad T$ is singular.

Example 3. *Let λ be an eigenvalue of a linear operator $T : V \to V$. Let E_λ be the eigenspace of A. Show that E_λ is a subspace of V.*

Solution. In order to show E_λ to be subspace of V, we shall show that

(i) if $\alpha \in E_\lambda$ then $k\alpha \in E_\lambda$ for any scalar $k \in F$,

(ii) if $\alpha, \beta \in E_\lambda$, then $\alpha + \beta \in E_\lambda$

(i)   Since $\alpha \in E_\lambda$, so we have
$$T(\alpha) = \lambda\alpha$$
then $\quad\quad T(k\alpha) = kT(\alpha) = k(\lambda x) = \lambda(k\alpha).$
$$k\alpha \in E_\lambda.$$

(ii) Since $\alpha, \beta \in E_\lambda$, we have
$$T(\alpha) = \lambda\alpha, \quad T(\beta) = \lambda\beta.$$
Then, $\quad T(\alpha+\beta) = T(\alpha) + T(\beta) = \lambda\alpha + \lambda\beta = \lambda(\alpha+\beta).$
$\therefore \quad\quad\quad \alpha + \beta \in E_\lambda.$

Hence, E_λ is a subspace of V.

Example 4. *Let V be the vector space of differentiable functions on* R *and D:V → V be the differential operator. Show that functions* $e^{a_1 t}, e^{a_2 t},...,e^{a_n t}$ *, where* $a_1, a_2,...,a_n$ *are distinct non-zero scalars are eigenvectors of D, to which eigenvalue* λ_i *does* $e^{a_i t}$ *belong ?*

Solution. Since $e^{a_i t} \neq 0$ for all $a_i \neq 0.$
Then, we have
$$D(e^{a_i t}) = a_i e^{a_i t}, \quad \forall\, i = 1, 2, 3,...,n.$$
This equation shows that $e^{a_i t}$ is the eigenvector of D corresponding to $\lambda_i = a_i$.

Example 5. *Suppose λ is an eigenvalue of an invertible operator T. Show that* λ^{-1} *is an eigenvalue of* T^{-1} *.*

Solution. Since T is invertible so that it is non-singular, therefore, $\lambda \neq 0$. Also λ is an eigenvalue of T, then there exists a non-zero vector $\alpha \in V$ such that
$$T(\alpha) = \lambda\alpha.$$
Multiplying both sides by T^{-1} , we get
$$T^{-1}[T(\alpha)] = T^{-1}(\lambda\alpha)$$
$\Rightarrow \quad\quad (T^{-1} T)(\alpha) = T^{-1}(\lambda\alpha) \Rightarrow I(\alpha) = \lambda[T^{-1}(\alpha)]$
$\Rightarrow \quad\quad\quad \alpha = \lambda\, T^{-1}(\alpha) \Rightarrow T^{-1}(\alpha) = \dfrac{1}{\lambda}\,\alpha = \lambda^{-1}\alpha.$

Hence, λ^{-1} is an eigenvalue of T^{-1}.

Example 6. *Suppose α is a non-zero eigenvector of linear maps S and T . Show that α is an eigenvector of S + T .*

Solution. Since α is an eigenvector of S and T . Let λ_1 and λ_2 be eigenvalues of S and T corresponding to α respectively.
Then, $\quad\quad S(\alpha) = \lambda_1\alpha$ and $T(\alpha) = \lambda_2\alpha$
Now, $\quad (S + T)(\alpha) = S(\alpha) + T(\alpha) = \lambda_1\alpha + \lambda_2\alpha$
$$(S + T)(\alpha) = (\lambda_1 + \lambda_2)\alpha$$
Thus, α is an eigenvector $S + T$ corresponding to the eigenvalue $\lambda_1 + \lambda_2$.

Example 7. *Let A be a square matrix of order n×n with its elements belong to F. The left multiplication by A defines a linear operator*
$$T_A : F^{n \times m} \to F^{n \times m}$$

such that $T_A(B) = AB.$

A scalar $\lambda \in F$ is an eigenvale of T_A if and only if λ is an eigenvalue of A.

Solution. Suppose λ is an eigenvalue of A, then there exists a non-zero vector $\alpha \in F^n$ such that

$$A(\alpha) = \lambda\alpha.$$

Let B be an $n \times m$ matrix whose first column is α and all other columns zero, then B is a non-zero matrix and

$$T_A(B) = AB = \lambda B \qquad\qquad [\because A(\alpha) = \lambda\alpha]$$

Thus, λ is an eigenvalue of T_A.

Conversely, if λ, is eigenvalue of T_A, then there exists a non-zero matrix $B \in F^{n \times m}$ such that

$$T_A(B) = \lambda B = AB$$

Let α be one of the non-zero columns of B, then

$$AB = \lambda B.$$

$\therefore$ $A(\alpha) = \lambda\alpha.$

Thus, λ is an eigenvalue of A.

7.11 DIAGONALIZATION

Let V be a finite dimensional vector space over a field F. Then a linear operator $T : V \to V$ is said to be diagonalizable if V has a basis consisting of eigenvectors of T only. Equivalently, T has n linearly independent eigenvectors, if V is n-dimensional vector space.

For Example: A scalar multiple of identity operator is diagonalizable.

THEOREM 1. *A linear operator $T : V \to V$ can be represented by a diagonal matrix A iff V has a basis consisting of eigenvectors of T. In this case the diagonal elements of A are the corresponding eigenvalues.*

Proof. Let V be n-dimensional vector space. Let $B = \{\alpha_1, \alpha_2,...,\alpha_n\}$ be an ordered basis of V such that each α_j is the eigenvector of T.

If $\lambda_1, \lambda_2,..., \lambda_n$ are the corresponding eigenvalue of T, then

$$\left.\begin{array}{l} T(\alpha_1) = \lambda_1\alpha_1 \\ T(\alpha_2) = \lambda_2\alpha_2 \\ \cdots\ \cdots\ \cdots\ \cdots \\ T(\alpha_n) = \lambda_n\alpha_n \end{array}\right\} \qquad\qquad ...(1)$$

System (1) can also be rewritten as

$$T(\alpha_1) = \lambda_1\alpha_1 + 0\alpha_2 + 0\alpha_3 +...+ 0\alpha_n$$
$$T(\alpha_2) = 0\alpha_1 + \lambda_2\alpha_2 + 0\alpha_3 +...+ 0\alpha_n$$
$$... \qquad ... \qquad ... \qquad .... \qquad ....$$
$$T(\alpha_n) = 0\alpha_1 + 0\alpha_2 + 0\alpha_3 +...+ \lambda_n\alpha_n$$

Thus, the matrix representation of T, with respect to B is given by

$$A = [T]_B = \begin{vmatrix} \lambda_1 & 0 & 0 \\ 0 & \lambda_2 & 0 \\ \vdots & \vdots & \vdots \\ 0 & 0 & \lambda_n \end{vmatrix}.$$

Hence, T can be represented by a diagonal matrix A whose elements are the eigenvalues of T.

Conversely, suppose T can be represented by a diagonal matrix A given by

$$A = \begin{vmatrix} \lambda_1 & 0 & \cdots & 0 \\ 0 & \lambda_2 & \cdots & 0 \\ \vdots & & & \\ 0 & 0 & \cdots & \lambda_n \end{vmatrix}.$$

Since V is n-dimensional, there exists a basis $\{\alpha_1, \alpha_2,...,\alpha_n\}$ of V for which

$$T(\alpha_i) = \lambda_i \alpha_i, \qquad \forall\ i = 1, 2,...,n.$$

This equation shows that each α_i in the basis is an eigenvector of T corresponding to each λ_i.

REMARK

- Let T be a linear operator on an n-dmiensional vector space V over a field F. If T has n distinct eigenvalues, then T is diagonalizable.

Definition. *Let T be a linear operator on V and let λ be an eigenvalue of T. Then $E_\lambda = ker(T - \lambda I)$ is called the eigenspace of T corresponding to λ.*

The dimension of E_λ is called the *geometric multiplicity* of λ and the multiplicity of λ as a root of the characteristic polynomial is called the *algebraic multiplicity* of λ.

THEOREM 2. *Let λ be an eigenvalue of a linear operator $T : V \to V$. Then the geometric multiplicity of λ does not exceed its algebraic multiplicity.*

Proof. Suppose the geometric multiplicity of λ be r. Then there are r linearly independent eigenvector of T corresponding to λ. Let these be $\alpha_1, \alpha_2,...,\alpha_r$.

If $r = \dim. V$, then $B = \{\alpha_1, \alpha_2,...,\alpha_r\}$ is an ordered basis of V and

$$[T]_B = \lambda I_r.$$

Thus $|xI_r - [T]_B| = (x - \lambda)^r$ so that the algebraic multiplicity of λ is r.

If $r < \dim.V = n$, then the set $B = \{\alpha_1, \alpha_2,...,\alpha_r\}$ can be extended to form a basis of V. Let the extension of B is given by

$$B_1 = \{\alpha_1, \alpha_2,...,\alpha_r, \beta_{r+1}, \beta_{r+2},...,\beta_n\}$$

We have

$$T(\alpha_1) = \lambda \alpha_1$$
$$T(\alpha_2) = \lambda \alpha_2$$
$$\cdots \quad \cdots \quad \cdots$$
$$T(\alpha_r) = \lambda \alpha_r$$
$$T(\beta_{r+1}) = a_{11}\alpha_1 +... + a_{1r}\alpha_r + a_{1(r+1)}\beta_{r+1} +...+ a_{1n}\beta_n$$
$$T(\beta_{r+2}) = a_{21}\alpha_1 +... + a_{2r}\alpha_r + a_{2(r+1)}\beta_{r+1} +...+ a_{2n}\beta_n$$
$$\cdots \quad \cdots \quad \cdots \quad \cdots \quad \cdots \quad \cdots \quad \cdots \quad \cdots$$
$$T(\beta_n) = a_{n1}\alpha_1 +... + a_{nr}\alpha_r + a_{n(r+1)}\beta_{r+1} +...+ a_{nn}\beta_n.$$

Thus the matrix of T in the basis B_1, is given by

$$[T]_{\beta_1} = \begin{vmatrix} \lambda & 0 & \cdots & 0 & a_{11} & a_{21} & \cdots & a_{n1} \\ 0 & \lambda & \cdots & 0 & a_{12} & a_{22} & \cdots & a_{n2} \\ \cdots & \cdots & \cdots & \cdots & \cdots & \cdots & \cdots & \cdots \\ 0 & 0 & \cdots & \lambda & a_{1r} & a_{2r} & \cdots & a_{nr} \\ 0 & 0 & \cdots & 0 & a_{1(r+1)} & a_{2(r+1)} & \cdots & a_{n(r+1)} \\ 0 & 0 & \cdots & 0 & a_{1(r+2)} & a_{2(r+2)} & \cdots & a_{n(r+2)} \\ \cdots & \cdots & \cdots & \cdots & \cdots & \cdots & \cdots & \cdots \\ 0 & 0 & \cdots & 0 & a_{1n} & a_{2n} & \cdots & a_{nn} \end{vmatrix}$$

$$= \begin{bmatrix} \lambda I_r & \vdots & A \\ \cdots & \cdots & \cdots \\ O & \vdots & B \end{bmatrix}$$

Since $[T]_{\beta_1}$ is a block triangular matrix, the characteristic polynomial of λI_r is $(x-\lambda)^r$, which must divide the characteristic polynomial of $[T]_{\beta_1}$ and hence T. Hence the algebraic multiplicity of λ for the operator T is at least r.

THEOREM 3. *Let T be a linear operator on an n- dimensional vector space V over a field F and let $\lambda_1, \lambda_2,...,\lambda_k$ be all the distinct eigenvalues of T. Then the following statements are equivalent :*

 (i) *T is diagonalizable*

 (ii) *$V = E_{\lambda_1} \oplus E_{\lambda_2} \oplus ... \oplus E_{\lambda_k}$, $E_{\lambda_i} = \ker(T - \lambda_i I)$*

 (iii) *the characteristic polynomial of T splits over F, and the algebraic multiplicity of each eigenvalue equals its geometric multiplicity*

 (iv) *the minimal polynomial of T is $(x - \lambda_1)(x - \lambda_2)...(x - \lambda_k)$.*

Proof. **(i)** $\Rightarrow$ **(ii).** If T is diagonalizable, the V has an ordered basis

$$B = \{\alpha_1, \alpha_{12},..., \alpha_{1n}, \alpha_{21}, \alpha_{22},..., \alpha_{2n_2},..., \alpha_{k1},..., \alpha_{kn_k}\}$$

Such that each α_{i_m} is an eigenvector of T corresponding to eigevalue λ_{i_m}. Obviously $n_1 + n_2 + ... + n_k = n$. Thus it is enough to show that for each m, the set $\{\alpha_{m1}, \alpha_{m2},..., \alpha_{mn_m}\}$ is a basis of E_{λ_m}.

Let if possible $\{\alpha_{m1}, \alpha_{m2},...., \alpha_{mn_m}\}$ is not a basis of $E_{\lambda m}$ for some $m = 1$, 2, ..., k. Then there exists vector $\beta_m \in E_{\lambda m}$ such that $\{\alpha_{m1}, \alpha_{m2},...., \alpha_{mn_m}$ $\beta_m\}$ is a linearly independent subset of $E_{\lambda m}$. But $B \cup \{\beta_m\}$ is a linearly independent subset of V, which gives a contradiction as B spans V, therefore, $\{\alpha_{m1}, \alpha_{m2},...., \alpha_{mn_m}\}$ is a basis of $E_{\lambda m}$.

(ii) $\Rightarrow$ **(i).** It is obvious.

(i) $\Rightarrow$ **(iii).** If T is diagonalizable then there is an order basis B on V such that $[T]_B$ is a diagonal matrix with its diagonal elements as eigenvalues of T, so that characteristic ploynomial split over F. Since.

$$V = E_{\lambda_i} \oplus E_{\lambda_2} .... \oplus E_{\lambda_k}$$

so, if B_i is a basis of E_{λ_i} then

$$B = \bigcup_{i=1}^{k} B_i$$

is a basis of V and

$$[T]_B = \text{diag.} \ ([T_1]_{B_1} \dots [T_k]_{B_k})$$

where T_i is the linear operator of E_{λ_i} induced by T.

Let dim. $E_{\lambda_i} = n_i$, since $T_i(\alpha) = T(\alpha) = \lambda_i$ for all $\alpha \in E_{\lambda_i}$) so that

$$[T_i]B_i = (\lambda_i I_{n_i}).$$

Hence the characteristic polynomial is

$$(x - \lambda_1)^{n_1}, (x - \lambda_2)^{n_2}, \dots, (x - \lambda_k)^{n_k}$$

which shows that the algebraic multiplicity of each λ_i equals geometric multiplicity.

(iii) $\Rightarrow$ **(iv).** Since characteristic polynomial of T splits over F and it is given by

$$(x - \lambda_1)^{n_1}, (x - \lambda_2)^{n_2}, \dots, (x - \lambda_k)^{n_k}$$

Also the geometric multiplicity and the algebraic multiplicity of each eigevalues are equal, so for each λ_i, there are n_i linearly independent eignvalues of T which are $\alpha_{i1}, \alpha_{i2}, \dots \alpha_{in_i}$

Therefore, $B = (\alpha_{11}, \dots \alpha_{1n_1}, \alpha_{21}, \dots \alpha_{2n_2}, \dots, \alpha_{n1}, \dots \alpha_{kn_k})$ is a basis of V and the linear operator $(T - \lambda_1 I)(T - \lambda_2 I) \dots (T - \lambda_k I)$ maps all the basis elements of B to the zero vector.

Thus $(T - \lambda_1 I)(T - \lambda_2 I) \dots (T - \lambda_k I) = 0$, hence $(x - \lambda_1)(x - \lambda_2) \dots (x - \lambda_k)$ is minimal polynomial of T.

THEOREM 4. *A linear operator $T : V \rightarrow V$ has a diagonal matrix respresentation if and only if its minimal polynomial $m(x)$ is product of distinct linear polynomials.*

Proof. Suppose $m(x)$ is a product of distinct linear polynomials, let

$$m(x) = (x - \lambda_1)(x - \lambda_2) \dots (x - \lambda_n) \qquad \dots(1)$$

where $\lambda_1, \lambda_2, \dots, \lambda_n$, are distinct scalars.

By primary decomposition theorem V is direct sum of subspaces $W_1, W_2, \dots, W_n$, where,

$$W_1 = \ker(T - \lambda_1 I), \ W_2 = \ker(T - \lambda_2 I), \ W_n = \ker(T - \lambda_n I).$$

If $\alpha \in W_i$, then

$$(T - \lambda_1 I)(\alpha) = 0 \Rightarrow T(\alpha) = \lambda_i \alpha$$

$\Rightarrow$ everyvector in W_i is an eignevector corresponding to the eigenvalue λ_i.

Thus, the union of bases for $W_1, W_2 \dots W_n$, form a basis of V. This basis consists of eigen vectors and so T is a diagonlizable

Conversely, suppose, T has a diagonal matrix representation, implying that V has a basis consisting of eigevectors of T.

Let $\lambda_1, \lambda_2, \dots \lambda_r$ be the distinct eigenvalus of T, then

$$P(T) = (T - \lambda_1 I)(T - \lambda_2 I) \dots (T - \lambda_r I).$$

Since $(T - \lambda_1 I)$ maps each bsais vector to 0 so that $p(T) = 0$.

Thus, the minimal polynomial $m(x)$ of T divides the polynomial $P(x) = (x - \lambda_1)(x - \lambda_2) \dots (x - \lambda_r)$. According $m(x)$ is the product of distinct linear polynomials.

Solved Examples

Example 1. *If T is a linear operator on R^3 which is represented in the standard ordered basis by*

$$A = \begin{bmatrix} 5 & -6 & -6 \\ -1 & 4 & 2 \\ 3 & -6 & -4 \end{bmatrix}$$

then show that T is diagonalizable.

Solution. The characteristic polynomial of A is

$$\Delta(x) = |xI - A| = \begin{vmatrix} x-5 & 6 & 6 \\ 1 & x-4 & -2 \\ -3 & 6 & x+4 \end{vmatrix}$$

$$= (x-5)\{(x-4)(x+4) + 12\} - 6\{x + 4 - 6\} + 6\{6 + 3(x-4)\}$$
$$= (x-5)(x^2 - 4) - 6x + 12 + 18x - 36$$
$$= x^3 - 5x^2 - 4x + 20 - 6x + 12 + 18 - 36$$
$$= x^3 - 5x^2 + 8x - 4$$
$$\Delta(x) = (x-1)(x-2)^2$$

Now we find the minimal polynomial of T.

The possible form of the minimal polynomial $m(x)$ of T are :

(i) $m_1(x) = (x-1)(x-2)$

(ii) $m_2(x) = (x-1)(x-2)^2$

Since $m_2(A) = 0$ may be the minimal polyonmial of T of $m_1(A) \neq 0$, but if $m_1(A) = 0$, then the minimal polynomial will be $m_1(x)$ but not the $m_2(x)$. So we test whether $m_1(A) = 0$, or $m_1(A) \neq 0$.

Since, $m_1(A) = (A - I)(A - 2I)$.

Now $A - I = \begin{bmatrix} 5-1 & -6 & -6 \\ -1 & 4-1 & 2 \\ 3 & -6 & -4-1 \end{bmatrix} = \begin{bmatrix} 4 & -6 & -6 \\ -2 & 3 & 2 \\ 3 & -6 & -5 \end{bmatrix}$

and $A - 2I = \begin{bmatrix} 5-2 & -6 & -6 \\ -1 & 4-2 & 2 \\ 3 & -6 & -4-2 \end{bmatrix} = \begin{bmatrix} 3 & -6 & -6 \\ -1 & 2 & 2 \\ 3 & -6 & -6 \end{bmatrix}$

So $(A - I)(A - 2I) = \begin{bmatrix} 4 & -6 & -6 \\ -2 & 3 & 2 \\ 3 & -6 & -5 \end{bmatrix} = \begin{bmatrix} 3 & -6 & -6 \\ -1 & 2 & 2 \\ 3 & -6 & -6 \end{bmatrix} = \begin{bmatrix} 0 & 0 & 0 \\ 0 & 0 & 0 \\ 0 & 0 & 0 \end{bmatrix} = 0$

$m_1(A) = 0$

Thus the minimal polynomial of T is $m_1(x) = (x-1)(x-2)$
which is the product of monic linear polynomial over F. Hence T is diagonalizable.

Example 2. *Let $T : R^3 \to R^3$ be definded by $T(x, y, z) = (2x + 3y - 2z, 5y + 4z, x - z)$. Find the characteristic polynomial $\Delta(t)$ of T.*

Solution. Let $B = \{(1, 0, 0), (0, 1, 0), (0, 0, 1)\}$ be the standard basis of R^3. So that

$$T(1, 0, 0) = (2, 0, 1)$$

$$T\,(0,\,1,\,0) = (3,\,5,\,0)$$
$$T\,(0,\,0,\,1) = (-2,\,4,\,-1)$$

Thus, $\quad [T]_B = \begin{bmatrix} 2 & 3 & -2 \\ 0 & 5 & 4 \\ 1 & 0 & -1 \end{bmatrix}$

The characteristic polynomial of T is given by

$$\Delta(t) = \begin{vmatrix} t-2 & -3 & 2 \\ 0 & t-5 & -4 \\ -1 & 0 & t+1 \end{vmatrix}$$

$$= (t-2)\,\{(t-5)\,(t+1) - 0\} + 3\,(0-4)\} + 2\,\{0 + t - 5\}$$
$$= (t-2)\,(t^2 - 4\,t - 5) - 12 + 2\,t - 10$$
$$= t^3 - 2t^2 + 4t^2 + 8t - 5t + 10 - 12 + 2t - 10$$
$$= t^3 - 6t^2 + 5t - 12$$

Example 3. *If A be a square matrix given by* $A = \begin{bmatrix} 3 & 0 & 0 \\ 0 & 2 & -5 \\ 0 & 1 & -2 \end{bmatrix}$

then find all the eigenvalues of A viewed as matrics over (i) Real field R (ii) Complex field C. Also, find in which case the matrix A is diagonalizable.

Solution. The characteristic polynomial of A is given by

$$\Delta(t) = |\,tI - A\,| = \begin{vmatrix} t-3 & 0 & 0 \\ 0 & t-2 & 5 \\ 0 & -1 & t+2 \end{vmatrix}$$

$$= (t-3)\{(t-2)\,(t+2) + 5\}$$
$$\Delta\,(t) = (t-3)\,(t^2 + 1) = (t-3)\,(t-i)\,(t+i)$$

The roots of $\Delta t = 0$ are $t = 3,\,-i,\,i$.

(i) If A is a matrix over the field of reals, then A has only one eigenvalue which is 3. Thus, A cannot be diagonalized.

(ii) If A is a matrix over the field of complex numbers, then A has three distinct eigenvalues, so that the minimal polynomial of A is equal to $\Delta\,(t)$ which is the product of linear polynomials. Hence A is digonalizable.

Example 4. *If T be a linear tranformation on V (F). Then the following are equivalent :*
(i) λ *is the characteristic value of T*
(ii) *the transformation* $T - \lambda\,I$ *is singular.*
(iii) $|\,(T - \lambda\,I)\,| = 0$

Solution. **(i)** $\Rightarrow$ **(ii).** If λ is a characteristic value of T, then there exists a non-zero vector $\alpha \in V$ such that

$$T\,(\alpha) = \lambda\alpha \Rightarrow T(\alpha) = \lambda I(\alpha) \qquad\qquad [\because I(\alpha) = \alpha]$$
$$\Rightarrow \quad T\,(\alpha) - \lambda\,I\,(\alpha) = 0 \Rightarrow (T - \lambda I)\,(\alpha) = 0 \text{ with } \alpha \neq 0.$$
$$\therefore \quad T - \lambda I \text{ is singular.}$$

(ii) $\Rightarrow$ **(iii).** Since $(T - \lambda I)$ is singular, *i.e.* it is not invertible so that $|\,(T - \lambda I)\,| = 0$

(iii) $\Rightarrow$ **(i).** Since $|(T - \lambda I)| = 0$, therefore $(T - \lambda I)$ is not invertible, *i.e.*

$$(T-\lambda I)\alpha = 0 \Rightarrow T(\alpha) - \lambda I(\alpha) = 0$$

$$\Rightarrow \qquad T(\alpha) - \lambda\alpha = 0 \Rightarrow T(\alpha) = \lambda\alpha$$

$$\therefore \qquad \lambda \text{ is an eigenvalue of } T.$$

Example 5. *Show that the matrix* $A = \begin{bmatrix} 1 & 2 \\ 0 & 1 \end{bmatrix}$ *is diagonalizable over the field of complex numbers.*

Solution. The characteristic polynomial of A is

$$\Delta(t) = |tI - A| = \begin{vmatrix} t-1 & -2 \\ 0 & t-1 \end{vmatrix} = (t-1)^2$$

Thus A has only one eigenvalue which is I.

The eigenvectors of A of $\lambda = 1$ are given by the solution of the homogeneous system

$$(I - A)\begin{bmatrix} x_1 \\ x_2 \end{bmatrix} = \begin{bmatrix} 0 \\ 0 \end{bmatrix} \text{ or } \begin{bmatrix} 0 & -2 \\ 0 & 0 \end{bmatrix}\begin{bmatrix} x_1 \\ x_2 \end{bmatrix} = \begin{bmatrix} 0 \\ 0 \end{bmatrix} \text{ or } -2x_2 = 0$$

Above equation gives $x_2 = 0$ and x_1 is assigned let $x_2 = 1$, thus A has only one linearly independent eigenvector $\begin{bmatrix} 1 \\ 0 \end{bmatrix}$. Hence A is not diagonlizable.

Example 6. *Let V be the vector space of functions which have B = {sinθ , cosθ} as a basis and let D be the differential operator on V. Find the characterstic polynomial $\Delta(t)$ of D.*

Solution. First, we find the matrix A which represents D in the basis B :

$$D(\sin\theta) = \cos\theta = 0(\sin\theta) + 1.(\cos\theta)$$

$$D(\cos\theta) = -\sin\theta = (-1)\sin\theta + 0(\cos\theta)$$

Thus, $\qquad [D]_B = A\begin{bmatrix} 0 & 1 \\ -1 & 0 \end{bmatrix}$

Now $\qquad \Delta(t) = \begin{bmatrix} t & -1 \\ 1 & t \end{bmatrix} = t^2 + 1$

This is the required characteristic polynomial of A.

Example 7. *What is the algebraic and geometric multiplicity of $\lambda = -2$, where $\lambda = -2$ is one of the eigenvalue of the matrix*

$$A = \begin{bmatrix} -3 & 1 & -1 \\ -7 & 5 & -1 \\ -6 & 6 & -2 \end{bmatrix}$$

Solution. The characteristic polynomial of A is given by.

$$\Delta(t) = |tI - A| = \begin{vmatrix} t+3 & -1 & 1 \\ 7 & t-5 & 1 \\ 6 & -6 & t+2 \end{vmatrix}$$

$$= (t + 3)\{(t - 5)(t + 2) + 6\} + 1\{7(t + 2) - 6\} + 1\{-42 - 6(t-5)\}$$

$$= (t + 3)\{t^2 - 3t - 10 + 6\} + 7t + 8 - 42 - 6t + 30$$

$$= (t + 3)(t^2 - 3t - 4) + t - 4\}$$

$$= t^3 + 3t^2 - 3t^2 - 9t - 4 - 12 + t - 4$$

$\Delta(t) = t^3 - 12t - 16 = (t + 2)^2 - (t - 4)$.

Since the factor $(t + 2)$ occurs twice in $\Delta(t)$, so that the algebaric multiplicity of $\lambda = -2$ is two.

Now we find a basis of the eigenspace of $\lambda = -2$.

Let $X = \begin{bmatrix} x_1 \\ x_2 \\ x_3 \end{bmatrix}$ be non-zero eigenvector corresponding to $\lambda = -2$, then

$$(\lambda I - A)X = 0 \text{ or } (-2I - A)X = 0$$

or $\begin{bmatrix} 1 & -1 & 1 \\ 7 & -7 & 1 \\ 6 & -6 & 0 \end{bmatrix}\begin{bmatrix} x_1 \\ x_2 \\ x_3 \end{bmatrix} = \begin{bmatrix} 0 \\ 0 \\ 0 \end{bmatrix}$

or $\begin{aligned} x_1 - x_2 + x_3 &= 0 \\ 7x_1 - 7x_2 + x_3 &= 0 \\ 6x_1 - 6x_2 &= 0 \end{aligned}$

or $\begin{aligned} x_1 - x_2 + x_3 &= 0 \\ 7x_1 - 7x_2 + x_3 &= 0 \\ x_1 - x_2 &= 0 \end{aligned}$

This system has only one independent solution, *i.e.* $x_1 = 1$, $x_2 = 1$ and $x_3 = 0$. Thus $\alpha = (1, 1, 0)$ forms, a basis of the eigenspace E_2. Hence the geometirc multiplicity of $\lambda = -2$ is one as dim. $E_2 = 1$.

📖 EXERCISE 7.2

1. Let α and β be eigenvectors of T corresponding to two eigenvalues λ and μ respectively. Show that for non-zero scalars a and b, $a\alpha + b\beta$ is not an eigenvector of T.

2. Suppose $T : V \rightarrow V$ is a linear operator on a vector space V with dimension n. Define the characteristic polynomial $\Delta(t)$ of T.

3. Suppose a linear map $T : V \rightarrow V$ may have matrix representations, is it possible for T to have many characteristic polynomials?

4. Let $T: R^2 \rightarrow R^2$ be the linear operator which rotates each vector $\alpha \in R^2$ by an angle $\theta = \pi/2$. Show geometrically that T has no eigenvalues and hence no eigenvectors.

5. Let λ be an eigenvalue of a linear operator $T:V \rightarrow V$. Let E_λ be the set of all eigenvectors of T belonging to λ. Show that E_λ is a subspace of V.

6. Let λ be an eigenvalue of a linear operator $T: V \rightarrow V$. Define the algebraic multiplicity and the geometric multiplicity of λ.

7. Let A and B be n-square matrices. Show that AB and BA have the same eigenvalues.

8. Show that if λ is an eigenvalue of T and $p(x) \in F[x]$, then $p(\lambda)$ is an eigenvalue of $p(T)$.

9. Find eigenvalues and eigenvectors for the following matrices over C, the field of complex numbers:

(i) $A = \begin{bmatrix} 6 & -1 & 2 \\ 4 & 1 & 2 \\ -10 & 0 & 3 \end{bmatrix}$ (ii) $A = \begin{bmatrix} 4 & 2 & 2 \\ 3 & 3 & 2 \\ -3 & -1 & 0 \end{bmatrix}$

(iii) $A = \begin{bmatrix} 0 & 0 & 1 \\ 1 & 0 & -1 \\ 0 & 1 & 1 \end{bmatrix}$.

10. Prove that if every non-zero is an eigenvector of T, then T is a scalar multiple of the identity operator.

11. Prove the if char $F \neq 2$, then T can be expressed as the sum of two invertible linear operators.

12. If $V_2(R)$ is a vector space and T is a linear transformation on $V_2(R)$ whose matrix relative to the basis of $V_2(R)$ is $A = \begin{bmatrix} 1 & 2 \\ 3 & 2 \end{bmatrix}$. Find the dimension of the eigenspace of T.

13. Find the minimal polynomial for real matrix
$$A = \begin{bmatrix} 7 & 4 & -1 \\ 4 & 7 & -1 \\ -4 & -4 & 4 \end{bmatrix}.$$

14. Find the minimal polynomial for real matrix
$$A = \begin{bmatrix} 1 & 1 & 1 \\ 1 & 1 & 1 \\ 1 & 1 & 1 \end{bmatrix}.$$

15. Find the invertible matrix P such that $P^{-1}AP$ is a diagonal matrix, where $A = \begin{bmatrix} 1 & 4 \\ 2 & 3 \end{bmatrix}$.

16. If T be the linear transformation on $V_3(R)$ which is represented in the standard ordered basis by the matrix
$$\begin{bmatrix} -9 & 4 & 4 \\ -8 & 3 & 4 \\ -16 & 8 & 7 \end{bmatrix}.$$
Prove that T is diagonalizable.

17. Show that the matrix $\begin{bmatrix} 1 & 1 \\ 0 & 1 \end{bmatrix}$ is not diagonalizable.

18. If T is linear operator on $V_3(R)$ which is represented in the standard ordered basis by the matrix:
$$A = \begin{bmatrix} 5 & -6 & -6 \\ -1 & 4 & 2 \\ -3 & -6 & -4 \end{bmatrix}.$$ Find the eigenvalues of A and prove that T is diagonalizable.

19. Let λ be an eigenvalue of $T \in L(V)$ with algebraic multiplicity m. Prove that T is not diagonalizable if rank $(T-\lambda I)>n-m$, where $n=\dim. V$.

20. A linear operator $T : R^3 \rightarrow R^3$ defined by $T(x,y,z) = \{2x + y,y-z,2y+4z\}$.
   (i) Find the characteristic polynomial $\Delta(t)$ of T.
   (ii) Find the eigenvalues of T.
   (iii) Is T diagonalizable ?

21. Suppose α is a non-zero eigenvector of T. Show that, for any $k \in F$, is an eigenvector of kT.

22. Suppose λ is an eigenvalue of a linear operator T :
   (i) Show that λ^2 is an eigenvalue of T^2
   (ii) Show that λ^n is an eigenvalue of T^n for $n \geq 1$.

=== Hint to Selected Problems ===

1. Since α, β are the eigenvectors of T corresponding to two distinct eigenvalues λ and μ. Then we have

$T(\alpha) = \lambda\alpha$ and $T(\beta) = \mu\beta$. Now for a, b (non-zero) scalars,

$T(a\alpha + b\beta) = aT(\alpha) + bT(\beta) = a\lambda\alpha + b\mu\beta \neq v(a\alpha + b\beta)$, for some scalar v.

$\therefore$ $a\alpha + b\beta$ is not an eigenvector of T.

5. $E_\lambda = \{\alpha \in V : T(\alpha) = \lambda\alpha\}$. Let $\alpha,\beta \in E_\lambda$ and $a,b \in F$, then

$T(a\alpha + b\beta) = aT(\alpha) + bT(\beta) = a\lambda\alpha + b\mu\beta$
$\qquad [\because T(\alpha) = \lambda\alpha, T(\beta) = \lambda\beta]$
$\qquad = \lambda(a\alpha + b\beta)$.

$\therefore$ $a\alpha + b\beta$ is also an eigenvector of T corresponding to λ.

$\therefore$ $a\alpha + b\beta \in E_\lambda$. Hence E_λ is subspace of V.

8. Since λ is an eigenvalue of T so that for a non-zero vector $\alpha \in V, T(\alpha) = \lambda\alpha$. For $p(x) = F[x]$.
We have $(p(T))(\alpha) = p(T)(\alpha) = p(\lambda\alpha)$
$\qquad\qquad\qquad\qquad [\because T(\alpha)=\lambda\alpha]$
$\qquad\qquad = p(\lambda)\alpha$.
This show that $p(\lambda)$ is an eigenvalue of $p(T)$.

10. Let α be any non-zero vector and if α is an eigenvector of T, then $T(\alpha) = \lambda\alpha$, for some scalar λ

$(T(\alpha)) = \lambda\alpha = \lambda I(\alpha)$ $\qquad [\because I(\alpha)=\alpha]$

$(T - \lambda I)(\alpha) = 0, \forall\, \alpha \in V$
$\Rightarrow \qquad\qquad T = \lambda I$

12. Since matrix representation of T is given by
$$A = \begin{bmatrix} 1 & 2 \\ 3 & 2 \end{bmatrix}$$
Eigenvalues of A are the root of the equation
$\Delta(t) = |A-tI| = 0$ or $\begin{vmatrix} 1-t & 2 \\ 3 & 2-t \end{vmatrix} = 0$

or $(1-t)(2-t) - 6 = 0$ or $t^2-3t-4 = 0$
or $\qquad (t-4)(t+1) = 0$ or $t = -1, 4$.

Let $X_1 = \begin{bmatrix} x_1 \\ x_2 \end{bmatrix}$ be an eigenvector corresponding to $\lambda = -1$

$\therefore$ $(A-\lambda I)X_1 = 0$ or $\begin{bmatrix} 2 & 2 \\ 3 & 3 \end{bmatrix}\begin{bmatrix} x_1 \\ x_2 \end{bmatrix} = \begin{bmatrix} 0 \\ 0 \end{bmatrix}$

or $2x_1 + 2x_2 = 0, 3x_1 + 3x_2 = 0$, i.e., $x_1 + x_2 = 0$. So we have one equation in two variables, then set $x_2 = -1, x_1 = 1$.

$\therefore X_1 = \begin{bmatrix} 1 \\ -1 \end{bmatrix}$. Again $X_2 = \begin{bmatrix} x_1 \\ x_2 \end{bmatrix}$ be an eigenvector corresponding to $\lambda = 4$.

$\therefore$ $(A-\lambda I)X_1 = 0$ or $\begin{bmatrix} -3 & 2 \\ 3 & -2 \end{bmatrix}\begin{bmatrix} x_1 \\ x_2 \end{bmatrix} = \begin{bmatrix} 0 \\ 0 \end{bmatrix}$.

or $-3x_1 + 2x_2 = 0, 3x_1 - 2x_2 = 0$

∴ We have one equation in two variables so set $x_2 = 3$, and $x_1 = 2$.

∴ $$X_2 = \begin{bmatrix} 2 \\ 3 \end{bmatrix}.$$

Hence eigenspace of T is generated by two vectors so the dimension of eigenspace of T is 2.

16. The characteristic polynomial of the matrix of T

is given by $\Delta(t) = \begin{vmatrix} -9-t & 4 & 4 \\ -8 & 3-t & 4 \\ -16 & 8 & 7-t \end{vmatrix}$

$= (-9-t)\{(3-t)(7-t) - 32\} + 4\{-64$

$\qquad + (-8(7-t)) + 4\{-64 + 16(3-t)\}$

$= -(9+t)\{21 - 10t + t^2 - 32\} + 4\{-8 - 8t\}$

$\qquad\qquad\qquad + 4\{-16 - 16t\}$

$= -(9+t)(t^2 - 10t - 11) - 32 - 32t - 64 - 64t$

$= -(t^3 - 10t^2 - 11t + 9t^2 - 90t - 99) - 96 - 96t$

$= -t^3 + t^2 + 101t + 99 - 96 - 96t$

∴ $\Delta(t) = -t^3 + t^2 + 5t + 3 = (t+1)^2(3-t)$.

There are two polynomials $m_1(t) = (t + 1)^2$ $(3-1)$ and $m_2(t) = (t + 1)(3 - t)$ one of them may be minimal polynomial.

Now

$$(A + I)(3I-A) = \begin{bmatrix} -8 & 4 & 4 \\ -8 & 4 & 4 \\ -16 & 8 & 8 \end{bmatrix} \begin{bmatrix} 12 & -4 & -4 \\ 8 & 0 & -4 \\ 16 & -8 & -4 \end{bmatrix}$$

$$= \begin{bmatrix} 0 & 0 & 0 \\ 0 & 0 & 0 \\ 0 & 0 & 0 \end{bmatrix} = O.$$

∴ $m_2(A) = O$

Thus $m_2(t) = (t+ 1)(3 - t)$ is a minimal polynomial of A which is a product of linear polynomials.

Hence the matrix of T is diagonalizable,

21. Let λ be an eigenvalue of T corresponding to which α is its eigenvector. Thus $T(\alpha) = \lambda\alpha$.

For $k \in F.$ $(kT)(\alpha) = k(T(\alpha)) = k\lambda\alpha = (k\lambda)\alpha$. Hence α is also an eigenvector of T corresponding to eigenvalue $k\lambda$.

22. Let $\alpha \neq 0$ be an eigenvector of T corresponding to an eigenvalue k. Then $T(\alpha) = \lambda\alpha$.

(i) $(T^2)(\alpha) = T(T(\alpha)) = T(\lambda\alpha) = \lambda(T(\alpha))$

$\qquad\qquad = \lambda(\lambda\alpha) = \lambda^2\alpha.$

∴ λ^2 is an eigenvalue of T^2. Similarly we may prove that λ^n is an eigenvalue of T^n.

Answers

3. No

9. *(ii)* 1,2,4;[0, 1, –1]′,[1, 1, –2]′[1, 1, –1]′   *(iii)* [1,0,1]′;[1–(1+i)i]′,[1–(1–t)–1]′

12. Two

13. $m(x) = x^2 - 15 + 36$

14. 0,3;[10–1]′,[aaa]′, $a \in F$

15. $p = \begin{bmatrix} 1 & 2 \\ 1 & -1 \end{bmatrix}$

18. 1,2,2

20. *(i)* $\Delta(t) = t^3 - 7t^2 + 16t - 12$ *(ii)* 2,2,3   *(iii)* No

CHAPTER REVIEW : A COMPETITIVE APPROACH

Selected Terms and Results

TERMS

- **Linear Functionals :** Let $V(F)$ be a vector space. A linear transformation $f : V \rightarrow F$ is called a linear functional on V,

  i.e. a function $f : V \rightarrow F$ is said to be linear functional on V, if $f(a\alpha + b\beta) = af(\alpha) + bf(\beta) \quad \forall \, a, b \in F, \alpha, \beta \in V$.

- **Zero Functional :** Let $V(F)$ be a vector space. The linear functional $f : V \rightarrow F$ defined by $f(\alpha) = 0 \; \forall \; \alpha \in V$ is said to be zero functional.

- **Dual Space :** Let $V(F)$ be a vector space. Then the set V' of all linear functional on V is also a vector space over F. The vector space V' is called the dual space of V.

- **Dual Basis :** Let V be n-dimensional vector space over the field F and $B = \{\alpha_1, \alpha_2,..., \alpha_n\}$ be a basis for V. Then uniquely determined basis $B' = (f_1, f_2, ..., f_n)$ for V' such that $f_i(\alpha_j) = \delta_{ij}$ is called dual basis of B.

- **Second Dual Basis :** Let V' be dual space of a vector space V. Then the set $V'' \{= (V')'\}$ consisting of all linear functional on V' is called second dual space of V.

- **Annihilator :** Let $V(F)$ be a vector space and S is a subset of V. The annihilator of S is the set S° of all linear functional f on V such that

  $f(\alpha) = 0 \; \forall \; \alpha \in S.$

- **Annihilator of an Annihilator :** Let $V(F)$ be a vector space, S any subset of V. Then S° is a subset of V'. Then annihilator of an annihilator is given by

  $S^{\circ\circ} = \{L \in V'' : L(f) = 0 \; \forall f \in S^\circ\}$

- **Invariance :** Let V be a vector space and T a linear operator on V. If W is a subspace of V. Then W is said to be invariant under T if $\alpha \in W \Rightarrow T(\alpha) \in W$.

- **Reducibility :** Let W_1 and W_2 be two subspaces of a vector space V and T be a linear operator on V, then T is said to be reduced by the pair (W_1, W_2) if

  (i) $V = W_1 \oplus W_2$

  (ii) W_1 and W_2 both are invariant under T.

- **Diagonalizable Operators :** Let T be a linear operator on the finite dimensional vector space V. Then T is said to be diagonalizable if there is a basis B for V each vector of which is a characteristic vector of T.

- **Diagonalizable Matrix :** A matrix A over a field F is said to be diagonalizable if it is similar to a diagonal matrix over the field F.

RESULTS

- Linear functional is a scalar valued function.
- Trace function is a linear functonal on the space of all $n \times n$ matrices over the field F.
- The dual space of V is also called the conjugate spaces.
- Two linear functionals of V are equal if they agree on a basis of V.
- If V is an n-dimensional vector space over the field F. Then V is isomorphic to its dual space V'.
- Let V be an n-dimensional vector space over the field F. If α is a non-zero vector in V, there exists a linear functionals f on V such that $f(\alpha) \neq 0$.
- Let V be an n-dimensional vector space over the field F. If $f(\alpha) = 0 \; \forall f \in V'$ then $\alpha = 0$.
- If V is finite dimensional then
  $$\dim. V = \dim. V' = \dim. V''.$$

- Let V be a finite dimensional vector space over the field F. Then each basis for V' is the dual of some basis for V.
- If f is a linear functional on an n-dimensional vector space $V(F)$ then the set of all those vectors α for which $f(\alpha) = 0$ is a subspace of V of dimension $n - 1$.
- Every finite dimensional vector space V is isomorphic to its second conjugate space V^{**} under an isomorphism which is independent of the choice of a basis in V.
- $\dim. W + \dim. W^\circ = \dim. V$
- If S is any subset of a vector space $V(F)$, then S° is a subspace of V'.
- If V is finite dimensional and W is a subspace of V then W is isomorphic to V'/W.

- If $V(F)$ be finite dimensional and W be a subspace of U then $W^{\circ\circ} = W$.
- If S_1, S_2 are two subsets of a vector space V such that $S_1 \subseteq S_2$, then $S_2^{\circ} \subseteq S_1^{\circ}$.
- Let V be a vector space over F and S, any subset of U then $S^{\circ} = [L(S)]^{\circ}$ and $S^{\circ\circ} = L(S)$.
- If W_1 and W_2 are two subspaces of a finite dimensional vector space V then $(W_1 + W_2)^{\circ} = W_1^{\circ} \cap W_2^{\circ}$ and $(W_1 \cap W_2)^{\circ} = W_1^{\circ} + W_2^{\circ}$. Also, $W_1^{\circ} = W_2^{\circ} \Leftrightarrow W_1 = W_2$.
- If T is a linear operator on a vector space V then the range of T and null space of T are both invariant under T.
- The subspace spanned by two subspaces each of which is invariant under same linear operator T, is itself invariant under T.
- If α is a characteristic vector of T, then α cannot correspond to more than one characteristic value of T.
- Distinct characteristic vectors of T corresponding to distinct characteristic value of T are linearly independent.
- If T is a linear operator on n-dimensional vector space V then T cannot have more than n distinct characteristic values.
- Every square matrix satisfies its characteristic equation.
- A necessary and sufficient condition that an $n \times n$ matrix A over a field F be diagonalizable is that A has n linearly independent characteristic vectors in $V_n(F)$.
- Let T be a linear operator on a finite dimensional vector space V then 0 is a characteristic value of T iff T is not invertible.
- If C is a characterstic value of an invertible transformation T then C^{-1} is a characteristic value of T^{-1}.
- If A and B are similar linear operator on a finite dimensional vector space V then A and B have the same characteristic polynomial.
- If an $n \times n$ matrix A has n distinct eigenvalues then A is diagonalizable.
- The minimal polynomial of a matrix or of a linear opertor is unique.
- The minimal polynomial of a matrix is a divisor of the characteristic polynomial of that matrix.
- Similar matrix have the same minimal polynomials.

Review Questions and Project Work

1. Let V be a finite dimensional vector space, then show that each basis for V^* is the dual of some basis for V.

2. If V is an n-dimensional vector space over the field F, show that V is isomorphic to its dual space V^*.

3. Show that there are infinitely many linear functional f for which $f(0, 1, 2) = 1$; $f(1, 1, -1) = 2$.

4. Show that the dual of an n-dimensional vector space is also n-dimensional.

5. Let $[a, b]$ be a closed interval on the real line and $C[a, b]$ be the space of all real valued continuous functions defined on $[a, b]$, show that the relation
$$L(g) = \int_a^b g(t)dt$$
defines linear functional L on $C[a, b]$.

6. Show that if $f \in V^*$ annihilates a subset S of V, then f annihilates linear space $L(S)$ of S.

7. If W is an m-dimensional suspace of an n-dimesional vector space V, show that $W^{\circ} = n - m$.

8. Let V be a finite dimensional vector space over the field F and let W_1, W_2 be two subspaces of V, then show that
$$W_1^{\circ} + W_2^{\circ} \Rightarrow W_1 = W_2.$$

9. Let T be a linear transformation on R^3 defined by $T(x_1, x_2, x_3) = (3x_1 + x_3, -2x_1 + x_2, -x_1 + 2x_2 + 4x_3)$ Show that T is invertible.

10. Show that the matrix $\begin{bmatrix} 1 & 0 \\ 0 & 1 \\ 1 & 1 \end{bmatrix}$ determines a linear transormation from V_3 to V_2 defined by $T(e_1) = l_1$, $T(e_2) = l_2$, $T(e_3) = l_1 + l_2$ relative to the standard basis (e_1, e_2, e_3) and (l_1, l_2).

11. Show that each linear functional is a linear transformation but not conversely.

12. If V is finite dimensional vector space and V is a subspace of V and $A(W)$ is annihilator of V. Prove that
$$\dim. A(W) = \dim. W - \dim. V$$

Objective type Questions

FILL IN THE BLANKS

1. If V^* is a dual space of V, then dim. $V =$ _____ .

2. Let V be a vector space and $\overline{V}$ be its dual space. Let W be a non-empty subset of V, then W° is a subspace of _____ .

3. If W is a subspace of a finite dimensional vector space $V(F)$, then, $W^{\circ\circ} =$ _____ .

4. M is an m-dimensional subspace of a n-dimensional vector space V and M° is an annihilator of M, then dim. $M^\circ =$ _____ .

5. If W is a subset of a vector space $V(F)$, then $[L(W)]^\circ =$ _____ .

6. If W_1 and W_2 are subsets of a vector space with $W_1 \subset W_2$, then _____ .

7. V is a finite-dimensional vector space, V^* is its dual space, $x, y \in V$, $x \neq y$, then there is an $f \in V^*$ such that _____ .

8. Let W_1 and W_2 be subspaces of a vector space over a field F, then $W_1^\circ + W_2^\circ$ _____ .

TRUE/FALSE

Write 'T' for true and 'F' for false statement.

1. Let V^* be the dual space of V and V^{**} be the dual of V^*, then dim. $V^* <$ dim. V^{**}. **(T/F)**

2. If W is any subset of a finite dimensional vector space V, then $W^{\circ\circ} = L(W)$. **(T/F)**

3. If W_1 and W_2 are subspaces of $V(F)$ with $W_1 \subset W_2$, then $W_1^\circ \subset W_2^\circ$. **(T/F)**

4. Every linear functional is a linear transformation. **(T/F)**

5. If W is a subset of V with $W = V$, then $W^{\circ\circ} = \{0\}$. **(T/F)**

6. If W is a subset of V with $W = \{0\}$, then $W^\circ = V$. **(T/F)**

7. If W_1 and W_2 are two subspaces of $V(F)$, then $W_1 = W_2$ iff $W_1^\circ = W_2^\circ$. **(T/F)**

8. If W is a subspace of a finite dimensional vector space $V(F)$, then $W = W^{\circ\circ}$. **(T/F)**

MULTIPLE CHOICE QUESTIONS

Choose the most appropriate one :

1. If W_1 and W_2 are subsets of a vector space V such that $W_2 \subset W_1$, then :
   (a) $W_1^\circ \subset W_2^\circ$ (b) $W_1^\circ = W_2^\circ$
   (c) $W_2^\circ \subset W_1^\circ$ (d) none of these

2. If V is a finite dimensional vector space and V^* is its dual and V^{**} is the dual of V^*, then :
   (a) V is isomorphic to V^* and but not to V^{**}
   (b) V is isomorphic to V^{**} but not to V^*
   (c) V is not isomorphic to both V^* and V^{**}
   (d) none of these

3. If W is a subspace of a finite dimensional vector space $V(F)$, then dim. $V +$ dim. W° is equal to :
   (a) dim. V (b) 0
   (c) 1   (d) none of these

4. If W_1 and W_2 are subspaces of a vector space $V(F)$ then $(W_1 \cap W_2)^\circ$ is equal to :
   (a) $W_1^\circ + W_2^\circ$ (b) $W_1^\circ \cup W_2^\circ$
   (c) $W_1^\circ \cap W_2^\circ$ (d) none of these

5. If W_1 and W_2 are subspaces of a vector space $V(F)$, then $(W_1 + W_2)^\circ$ is equal to :
   (a) $W_1^\circ \cup W_2^\circ$ (b) $W_1^\circ \cap W_2^\circ$

   (c) $W_1^\circ + W_2^\circ$ (d) none of these

6. If W_1 and W_2 are subspaces of a vector space $V(F)$ such that $V = W_1 \oplus W_2$ then $W_1^\circ \oplus W_2^\circ$ is equal to :
   (a) V^* (b) V
   (c) V^{**} (d) none of these

7. If W_1 and W_2 are subspaces of a vector space $V(F)$ which are annihilated by the subspace W°, then dim. $W_1 +$ dim. W_2 is equal to :
   (a) $2($dim. $V -$ dim. $W^\circ)$
   (b) dim. $V -$ dim. W°
   (c) $\dfrac{1}{2}($dim. $V -$ dim. $W^\circ)$
   (d) $2 ($dim. $W^\circ -$ dim. $V)$.

8. If T be a linear transformation on R^2 defined by
   $$T(x, y) = \{(4x-2y), (2x+y)\}$$
   The matrix of T is relative to the basis $[\alpha_1, \alpha_2]$ where $\alpha_1 = (1, 1)$, $\alpha_2 = (-1, 0)$ is:
   (a) $\begin{bmatrix} 3 & -2 \\ 1 & 2 \end{bmatrix}$ (b) $\begin{bmatrix} 3 & 1 \\ 2 & 2 \end{bmatrix}$
   (c) $\begin{bmatrix} 3 & 2 \\ 1 & 2 \end{bmatrix}$ (d) none of these

9. Let $f : \mathrm{R}^2 \to \mathrm{R}$ be a linear functional defined by $f(x, y) = x+2y$ then $4f(x, y)$ is :

(a) $4x-8y$ (b) $4x+8y$

(c) $2x+4y$ (d) none of these

10. Let f be a linear functional on $V(F)$. Let $T : V \to F$ and $T(-\alpha) = -T(\alpha) \; \forall \; \alpha \in V$. Then $T(-\alpha)$ is the :

(a) negative of $T(\alpha)$

(b) additive inverse of $T(\alpha)$ in F

(c) zero linear functional

(d) none of the above

11. If V is a vector space over the field F and let V^* of all linear functionals on V is also a vector space over F, then the vector space V^* is called:

(a) subset of V (b) subfield of V

(c) dual space of V (d) none of these

12. The property that a finite dimensional vector space V is isomorphic to the second dual space V^{**} is called :

(a) reflexivity   (b) symmetry

(c) transitivity   (d) none of these

13. Let V be a vector space over F, then a linear transformation $f : V \to F$ is called :

(a) dual space   (b) linear functional

(c) linear space   (d) none of these

14. If W_1 and W_2 are subspaces of a vector space $V(F)$ which are annihilated by the subspace W°, then dim. W_1 + dim. W_2 is equal to :

(a) $2(\text{dim. } V - \text{dim. } W^\circ)$

(b) dim. V – dim. W°

(c) $\frac{1}{2}(\text{dim. } V - \text{dim. } W^\circ)$

(d) none of the above

15. If W_1 and W_2 are subspaces of a vector space $V(F)$ such that $V = W_1 \oplus W_2$ then $W_1^\circ \oplus W_2^\circ$ is equal to :

(a) V (b) V^*

(c) V^{**} (d) None of these

16. If V is a finite dimensional vector space and V^* is its dual and V^{**} is the dual of V^*, then :

(a) V is isomorphic to V^*

(b) V is isomorphic to V^{**}

(c) V is not isomorphic to V^* and V^{**}

(d) None of the above

17. Let V be a finite dimensional vector space $V(F)$. If S is any subset of V then $S^{\circ\circ}$ is equal to:

(a) $L(S)$ (b) $L(S^\circ)$

(c) S (d) none of these

18. Consider the basis $\{u_1, u_2, u_3\}$ of R^3 where

$u_1 = (1, 0, 0)$, $u_2 = (1, 1, 0)$, $u_3 = (1, 1, 1)$. Let $\{f_1, f_2, f_3\}$ be the dual basis of $\{u_1, u_2, u_3\}$ and f be a linear functional defined by $f(a,b,c) = a+b+c$, $(a, b, c) \in R^3$. If $f = \alpha_1 f_1 + \alpha_2 f_2 + \alpha_3 f_3$ then $(\alpha_1, \alpha_2, \alpha_3)$ is equal to :

(a) $(1, 3, 2)$ (b) $(1, 2, 3)$

(c) $(3, 2, 1)$ (d) none of these

19. The linear operator T for which the minimal polynomial $m(t) = t^3 - 5t^2 + 6t + 8$ is :

(a) $\begin{bmatrix} 0 & 0 & -8 \\ 1 & 0 & -6 \\ 0 & 1 & 5 \end{bmatrix}$ (b) $\begin{bmatrix} 0 & 0 & 8 \\ 1 & 0 & 6 \\ 0 & 1 & 5 \end{bmatrix}$

(c) $\begin{bmatrix} 0 & 0 & -8 \\ 1 & 0 & 6 \\ 1 & 0 & 5 \end{bmatrix}$ (d) none of these

20. Let T be a linear operator on a finite dimensional vector space V. If

$$m(\lambda) = \lambda^r + a_{r-1}\lambda^{r-1} + a_{r-2}\lambda^{r-2} + \ldots + a_1\lambda + a_0$$

$(a_0 \neq 0)$

be a minimal polynomial of T, then :

(a) T is invertible

(b) T is singular

(c) both (a) and (b) are true

(d) none of the above

21. Let T be a linear operator on a finite dimensional vector space V.

If $m(\lambda) = \lambda^r + a_{r-2}\lambda^{r-1} + a_{r-3}\lambda^{r-2} + \ldots + a_0\lambda$

be a minimal polynomial of T, then :

(a) T is invertible   (b) T is non-singular

(c) T is singular   (d) none of these

22. Let $T : V_2(R) \to V_2(R)$ where T be reflection of the points through the line $y = -x$ then (with respect to standard basis) T must be :

(a) diagonalizable

(b) invertible

(c) both (a) and (b) are true

(d) none of the above

23. Let the minimal polynomial of a linear transformation T from R^4 to R^4 be x^2+x+1, then T has :

(a) no real eigenvalues

(b) all real eigenvalues

(c) exactly two real eigenvalues

(d) none of the above

24. Let A be a real 4×4 matrix with characteristic polynomial $(T^2 + 1)^2$ which of the following is true ?

(a) A is nilpotent.

(b) A is invertible.

(c) Both (a) and (b) are true.

(d) None of the above

25. Let V be an n-dimensional vector space over the field F. Then dimension of the dual space of V is :

(a) n

(b) n^2

(c) $n(n+1)$

(d) none of these

Answers

FILL IN THE BLANKS

1. dim. V^*	**2.** V^*	**3.** W	**4.** $(n-m)$
5. W^*	**6.** $W_1^{\circ} \subset W_2^{\circ}$	**7.** $f(x) \neq f(y)$	**8.** $(W_1 \cap W_2)^{\circ}$.

TRUE/ FALSE

1. F	**2.** T	**3.** F	**4.** T	**5.** T	**6.** F	**7.** T	**8.** T

MULTIPLE CHOICE QUESTIONS

1. (a)	**2.** (c)	**3.** (a)	**4.** (a)	**5.** (b)	**6.** (a)	**7.** (a)	**8.** (a)	**9.** (b)
10. (b)	**11.** (c)	**12.** (a)	**13.** (b)	**14.** (a)	**15.** (b)	**16.** (c)	**17.** (a)	**18.** (a)
19. (a)	**20.** (a)	**21.** (a)	**22.** (b)	**23.** (a)	**24.** (b)	**25.** (a)		

COMPETITION CORNER
for JRF, NET/SET, GATE Aspirants

SOME FASCINATING FACTS

- Let V be an n-dimensional vector space over the field F. If α, $\beta \in V$ such that $\alpha \neq \beta$ then there exists a linear functional f on V such that $f(\alpha) \neq f(\beta)$.

- If V is finite dimensional, then
  $$\text{dim. } V = \text{dim. } V^* = \text{dim. } V^{**}.$$

- If V is not finite dimensional space then correspondence $\alpha \rightarrow L_\alpha$ would never onto, however it is always one-one and linear.

- If $V(F)$ be a finite dimensional vector spaces and $S \subseteq V$. Let α be non-zero vector in S. Then, $\exists$ a linear functional f on V such that $f(\alpha) \neq 0$. Therefore, there is $f \in V^*$ such that $f \notin S^\circ$, which implies $S^\circ \neq V^*$, i.e. S° is a proper subset of V^*.

- $S^{\circ\circ}$ is a subspace of S^{**}.

- If V is finite dimensional then V^{**} is identical with V through the natural isomorphism $\alpha \leftrightarrow L_\alpha$. Therefore, we may regard $S^{\circ\circ}$ is a subspace of V.

- Let V be a finite dimensional vector space and T be a linear transformation from V into U then
  (i) $[R(T)]^\circ = N(T^*)$
  (ii) $[N(T)]^\circ = R(T^*)$

- Only matrix similar to identity matrix I is I itself.

- Every finite dimensional vector space V is isomorphic to its second conjugate V^{**} under an isomorphism which is independent of the choice of a basis in V.

SOME IMPORTANT ILLUSTRATIONS

- Let A be an n-square matrix then its characteristic polynomial is
  $$\Delta(t) = t^n - S_1 t^{n-1} + S_2 t^{n-2} + \ldots + (-1)^n S_n.$$
  where S_k is the sum of the principal minors of order k.

- If A is a square matrix over the complex field C, then A has atleast one eigenvalue.

- A linear operator T can be represented by a diagonal matrix D if and only if there exists a basis S of V consisting of eigenvectors of T. In this case, the diagonal element of D are the corresponding eigenvalues.

- The minimal polynomial of $A = \begin{bmatrix} 2 & 2 & -5 \\ 3 & 7 & -15 \\ 1 & 2 & -4 \end{bmatrix}$ is given by $m(t) = t^2 - 4t + 3$.

- The characteristic polynomial of the matrix given by
  $$A = \begin{bmatrix} 1 & 2 & 3 \\ 3 & 0 & 4 \\ 6 & 4 & 5 \end{bmatrix} \text{ is given by}$$
  $\Delta(t) = t^3 - 6t^2 - 35t - 38.$

- If $A = \begin{bmatrix} 3 & -4 \\ 2 & -6 \end{bmatrix}$. Then it has $(4, 1)$ as a non-zero solution.

Therefore, $(4, 1)$ is eigenvector belonging to $\lambda = 2$.

- The eigenvalues of the matrix $\begin{bmatrix} 11 & -8 & 4 \\ -8 & -1 & -2 \\ 4 & -2 & -4 \end{bmatrix}$ are given by $\lambda = -5, -5, 16$.

- The minimal polynomial of the matrix $A = \begin{bmatrix} 4 & -2 & 2 \\ 6 & -3 & 4 \\ 3 & -2 & 3 \end{bmatrix}$ is given by $f(t) = (t-2)(t-1)$ or $g(t) = (t-2)(t-1)^2$.

- Let $m(t)$ be the minimal polynomial of n-square matrix A, then characteristic polynomial $\Delta(t)$ of A divides $[m(t)]^n$.

- The characteristic polynomial $\Delta(t)$ and the minimal polynomial $m(t)$ of a matrix A have the same irreducible factors.

- If V is an eigenvectors of linear operator T_1 and T_2. Then V is also an eigenvector of the linear operator $k_1 T_1 + k_2 T_2$ where k_1, k_2 are scalars.

- If $\lambda \neq 0$ is an eigenvalue of the composition $T_1 \circ T_2$ of linear operators T_1 and T_2, then λ is also an eigenvalue of the composition $T_2 \circ T_1$.

- A is a scalar matrix kI if and only if the minimal polynomial of A is $m(T) = t - k$.

- Let A be an n-square matrix for which $A^k = 0$ for some $k > n$, then $A^n = 0$

- If V is vector space of all polynomials in x defined on $[a, b]$ with coefficients are real numbers. Then the mapping

  $f : V \to R$ defined by $f(V(x)) = \int_a^b V(x)dx$ $\forall\, v(x) \in V(x)$ is a linear functional on V.

- If $f_1 = V_2(R) \to R$ and $f_2 = V_2(R) \to R$ be the linear functional defined by $f_1(a, b) = a + 2b$, $f_2(a, b) = 3a - b$ $\forall\, a, b \in R$ then

- $(f_1 + f_2)(a, b) = 4a + b$; $4f_1(a, b) = 4a + 8b$ and $(2f_1 - 5f_2)(a, b) = -13a + 9b$.

- Let V be a vector space of real continuous function on the interval $a \le t \le b$. Let $J : V \to R$ be the integral operator defined by

  $J(F) = \int_a^b f(t)dt$

  then J is a linear functional on V.

- Let ϕ be the linear functional on R^2 defined by $\phi(2, 1) = 15$ and $\phi(1, -2) = -10$ then $\phi = \phi(x, y) = 4x + 7y$.

- The dual basis for the basis of $R^2 : (v_1 = (2, 1),$ $v_2 = (3, 1))$ is given by

  $\{\phi_1(x, y) = -x + 3y, \phi_2(x, y) = x - 2y\}$

Self Assessment Test

1. Let V be the vector space of n-square matrices over the field K. Also, let $D : V \to K$ be the determinant function, *i.e.* $D(A) = \det(A)$. Show that D is not a linear functional on V.

2. If V is the vector spoace of all real polynomials and D, the derivative operator on V, *i.e.* $D[f(t)] = \dfrac{df}{dt}$. Show that D is not a linear functional on V.

3. Find the dual basis of the basis of R^3 : $\{(1, -1, 3), (0, 1, -1), (0, 3, -2)\}$.

4. Find the basis which is dual to the usual basis (e_1, e_2, e_3) of R^3.

5. If U and W are subspaces of V, prove that $(U+W)° = U° \cap W°$.

6. Show that the natural mapping from V to V^{**} is linear.

7. Find a maximum set of linearly independent eigenvectors of $A = \begin{bmatrix} 4 & 1 & -1 \\ 2 & 5 & -2 \\ 1 & 1 & 2 \end{bmatrix}$.

8. Show that the matrix $A = \begin{bmatrix} 1 & 1 \\ 0 & 1 \end{bmatrix}$ is not diagonalizable.

9. Let $A = \begin{bmatrix} -3 & 1 & -1 \\ -7 & 5 & -1 \\ -6 & 6 & -2 \end{bmatrix}$, find the characteristic polynomial $\Delta(t)$ and eigenvalues of A.

10. Find all eigenvalues and a maximum set of linearly independent eigenvectors of the matrix $A = \begin{bmatrix} 5 & -1 \\ 1 & 3 \end{bmatrix}$.

11. Show that 0 is an eigenvalue of T if and only if T is singular.

12. If $B = \{(1, 1, 1), (1, 1, -1), (1, -1, -1)\}$ be a basis of $V_3(R)$ then find its dual basis $B^* = \{f_1, f_2, f_3\}$ and hence for any vector $\alpha = (0, 1, 0) \in V_3(R)$. Find $f_1(\alpha), f_2(\alpha)$ and $f_3(\alpha)$.

13. If f is a non-zero linear functional on a vector space V and if x is an arbitrary scalar, does there necessrily exist a vector $\alpha \in V$ such that $f(\alpha) = x$?

14. Prove that if W is a subset of the vector space $V(F)$, then $W° = [L(W)]°$ and $W°° = L(W)$.

15. If $V = M \oplus N$, then prove the following:

   (i) $M^* \cong N°$ (ii) $N^* \cong M°$

(iii) $V^* = M° \oplus N°$

   where $*$ denote the dual and $°$ denote the annihilators.

16. Show that if W contains a non-zero vector then $W° \neq V^*$.

17. Show that the annihilator of the vector space $V(F)$ is the zero subspace of V^* consisting of zero functional.

18. Find the dual basis of the basis set $B = \{(1, -2, 3), (1, -1, 1), (2, -4, 7)\}$ of $V_3(R)$.

19. If $B = \{(1, -1, 3), (0, 1, -1), (0, 3, -2)\}$ be a basis of $V_3(R)$. Find its dual basis B^*.

20. If $\{e_1, e_2, e_3\}$ be a basis for vector space $R^3(R)$, find its dual basis where $e_1 = (1, 0, 0); e_2 = (0, 1, 0), e_3 = (0, 0, 1)$.

8

INNER PRODUCT SPACES

Let V be a vector space over $F(\text{R } or \text{ C})$. An inner product on V is a function $(.,.): V \times V \to F$ which satisfies the following properties:

    (i) $(\alpha, \alpha) > 0$ *for all non-zero vectors α in V.*

    (ii) $(\alpha,\beta) = \overline{(\beta,\alpha)}, \forall \alpha, \beta \in V$

    (iii) $(a\alpha + b\beta, \gamma) = a(\alpha, \gamma) + b(\beta, \gamma) \; \forall \; \alpha, \beta, \gamma \in V$ and $a, b \in F$

For Example:

    (1)   On F^n, we may defined an inner product, known as standard inner product as follows:

         Let $\alpha = (a_1, a_2, ..., a_n)$ and $\beta = (b_1, b_2, ..., b_n)$ be two elements in F^n. Then standard inner product is defined by

$$(\alpha,\beta) = \sum_{i=1}^{n} a_i \overline{b_i} \text{ , if } a_i, b_i \in F = C$$

       and $$(\alpha,\beta) = \sum_{i=1}^{n} a_i b_i \text{ , if } a_i, b_i \in F = R.$$

    (2)   Let V be the vector space of all continuous complex-valued functions on the unit interval $0 \le t \le 1$.

       Let $$(f,g) = \int_0^1 f(t)\overline{g(t)} d(t).$$

       Then $(., .)$ is an inner product on V.

       For Example: Let V and W be vector spaces over F and suppose that $(., .)$ is an inner product on W. If T is a non-singular linear transformation from V into W, then $p_T(\alpha,\beta) = (T(\alpha), T(\beta))$ defines an inner product p_T on V.

REMARKS

- In case of reals, the standard inner product is often called the dot or scalar product and it is denoted by (α, β).

- $(\alpha, a\beta + b\gamma) = \overline{(a\beta + b\gamma, \alpha)} = \overline{a}\,\overline{(\beta, \alpha)} + \overline{b}\,\overline{(\gamma, \alpha)} = \overline{a}(\alpha,\beta) + \overline{b}(\alpha, \gamma)$.

Definition 1. *A vector space V together with an inner product is called an inner product space.*

Definition 2. *The norm or length of a vector $\alpha \in V$ is defined by $\|\alpha\| = \sqrt{\alpha, \alpha}$.*

Definition 3. *A vector $\alpha \in V$ is called a unit vector, if $\|\alpha\| = 1$.*

Definition 4. *A finite-dimensional real inner product space is often called an Euclidean space.*

Definition 5. *A complex inner product space is often called a unitary space.*

REMARKS

- The distance between two vectors α and β in an inner product space V is $d(\alpha,\beta) = \|\alpha - \beta\|$.

- If θ be the angle between two vectors α and β in an inner product space V, then $\cos\theta = \dfrac{(\alpha,\beta)}{\|\alpha\|\|\beta\|}$.

THEOREM 1. *Let V be an inner product space over F and α, $\beta \in V$. Then*

(i) $\|\alpha \pm \beta\|^2 = \|\alpha\|^2 + 2\mathrm{Re}.(\alpha, \beta) + \|\beta\|^2$

where Re .(α, β) denotes the real parts of (α, β).

(ii) $\|\alpha + \beta\|^2 + \|\alpha - \beta\|^2 = 2\|\alpha\|^2 + 2\|\beta\|^2$ [Parallelogram law]

(iii) $\|k\alpha\| = |k|\,\|\alpha\| \;\forall\; k \in F$

(iv) $4(\alpha,\beta) = \begin{cases} \|\alpha + \beta\|^2 - \|\alpha - \beta\|^2 & , \text{if } F = R \\ \|\alpha + \beta\|^2 - \|\alpha - \beta\|^2 + i\|\alpha + i\beta\|^2 - i\|\alpha - i\beta\|^2 & , \text{if } F = C \end{cases}$

[Polarization identities]

(v) $|(\alpha,\beta)| \le \|\alpha\|\,\|\beta\|$ [Cauchy-Schwarz Inequality]

(vi) $\|\alpha \pm \beta\| \le \|\alpha\| + \|\beta\|$ [Triangle inequality]

(vii) $|\,\|\alpha\| - \|\beta\|\,| \le \|\alpha - \beta\|$

Proof.

(i) $\|\alpha \pm \beta\|^2 = (\alpha \pm \beta, \alpha \pm \beta) = (\alpha, \alpha \pm \beta) \pm (\beta, \alpha \pm \beta)$

$= (\alpha,\alpha) \pm (\alpha,\beta) \pm (\beta,\alpha) + (\beta,\beta)$

$= \|\alpha\|^2 \pm (\alpha,\beta) + \overline{(\alpha,\beta)} + \|\beta\|^2$

$= \|\alpha\|^2 + 2\mathrm{Re}.(\alpha,\beta) + \|\beta\|^2.$

(ii) From (i), we have

$\|\alpha + \beta\|^2 = \|\alpha\|^2 + 2\mathrm{Re}.(\alpha,\beta) + \|\beta\|^2$

and $\|\alpha - \beta\|^2 = \|\alpha\|^2 - 2\mathrm{Re}.(\alpha,\beta) + \|\beta\|^2$

Adding these equations, we get

$\|\alpha + \beta\|^2 + \|\alpha - \beta\|^2 = 2\|\alpha\|^2 + 2\|\beta\|^2$

(iii) $\|k\alpha\|^2 = (k\alpha,k\alpha) = k(\alpha,k\alpha) = k\,\bar{k}\,(\alpha,\alpha) = |k|^2\|\alpha\|^2.$

$\therefore$ $\|k\alpha\| = |k|\,\|\alpha\|.$

(iv) From (i), we have

$\|\alpha + \beta\|^2 = \|\alpha\|^2 + 2\mathrm{Re}.(\alpha,\beta) + \|\beta\|^2$

and $\|\alpha - \beta\|^2 = \|\alpha\|^2 - 2\mathrm{Re}.(\alpha,\beta) + \|\beta\|^2$

On subtracting

$$\|\alpha+\beta\|^2 - \|\alpha-\beta\|^2 = 4\mathrm{Re}.(\alpha,\beta)$$

If $F = R$, then

$$\mathrm{Re}.(\alpha,\beta) = (\alpha,\beta)$$

$$\therefore \quad \|\alpha+\beta\|^2 - \|\alpha-\beta\|^2 = 4(\alpha,\beta)$$

This proves the first result.

Also, if $\quad\quad\quad F = C$.

From (i), we have

$$\|\alpha+\beta\|^2 - \|\alpha-\beta\|^2 = 2(\alpha,\beta) + 2(\beta,\alpha) \qquad \text{...(1)}$$

Now, $\quad\quad \|\alpha+i\beta\|^2 = (\alpha+i\beta, \alpha+i\beta) = (\alpha, \alpha+i\beta) + (i\beta, \alpha+\beta)$

$$= (\alpha,\alpha) + i(\alpha,\beta) + (i\beta,\alpha) - i(i\beta,\beta)$$

$$= \|\alpha\|^2 - i(\alpha,\beta) + i(\beta,\alpha) - i.i(\beta,\beta)$$

$$= \|\alpha\|^2 - i(\alpha,\beta) + i(\beta,\alpha) + \|\beta\|^2$$

$$\therefore \quad\quad\quad i\|\alpha+i\beta\|^2 = i\|\alpha\|^2 + (\alpha,\beta) - (\beta,\alpha) + i\|\beta\|^2 \qquad \text{...(2)}$$

Similarly, $\;\; -i\|\alpha-i\beta\|^2 = -i\|\alpha\|^2 + (\alpha,\beta) - (\beta,\alpha) - i\|\beta\|^2 \qquad \text{...(3)}$

Adding (1), (2) and (3), we get

$$\|\alpha+\beta\|^2 - \|\alpha-\beta\|^2 + i\|\alpha+i\beta\|^2 - \|\alpha-i\beta\|^2 = 4(\alpha,\beta)$$

This proves the second result.

(v) If $\beta = 0$, then the statement is trivially true. If $\beta \neq 0$, then for $a, b \in F$, using (i) and (iii), we have

$$\|a\alpha + b\beta\|^2 \geq 0$$

$$\Rightarrow \quad |a|^2\|\alpha\|^2 + 2\,\mathrm{Re}.a\bar{b}\,(\alpha,\beta) + |b|^2\|\beta\|^2 \geq 0$$

in particular, $a = \|\beta\|^2$ and $b = -(\alpha,\beta)$, then

$$\Rightarrow \quad \|\alpha\|^2\|\beta\|^4 + 2\mathrm{Re}.\|\beta\|^2\{-\overline{(\alpha,\beta)}\,(\alpha,\beta)\} + |(\alpha,\beta)|^2\,\|\beta\|^2 \geq 0$$

$$\Rightarrow \quad \|\alpha\|^2\|\beta\|^4 - 2|(\alpha,\beta)^2|\,\|\beta\|^2 + |(\alpha,\beta)|^2\|\beta\|^2 \geq 0$$

$$\Rightarrow \quad \|\alpha\|^2\|\beta\|^4 - |(\alpha,\beta)|^2\|\beta\|^2 \geq 0$$

$$\Rightarrow \quad \|\alpha\|^2\|\beta\|^2 - |(\alpha,\beta)|^2 \geq 0 \qquad\qquad [\because \beta \neq 0]$$

$$\therefore \quad\quad\quad\quad |(\alpha,\beta)| \leq \|\alpha\| + \|\beta\|$$

(vi) From (i), we have

$$\|\alpha \pm \beta\|^2 = \|\alpha\|^2 \pm 2\mathrm{Re}.(\alpha,\beta) + \|\beta\|^2$$

$$\leq \|\alpha\|^2 \pm 2|\,\mathrm{Re}.(\alpha,\beta)| + \|\beta\|^2$$

$$\leq \|\alpha\|^2 \pm 2|\alpha, \beta| + \|\beta\|^2$$

$$\leq \|\alpha\|^2 \pm 2\|\alpha\|\,\|\beta\| + \|\beta\|^2 \qquad\qquad \text{[Using v]}$$

$$\leq (\|\alpha\| + \|\beta\|)^2$$

$$\therefore \quad\quad\quad \|\alpha \pm \beta\| \leq \|\alpha\| + \|\beta\|$$

(vii) From (vi), we have

$$\|\alpha\| = \|(\alpha - \beta) + \beta\| \le \|\alpha - \beta\| + \|\beta\|$$

so that $\|\alpha\| - \|\beta\| \le \|\alpha - \beta\|$

Now interchanging α and β, we get

$$\|\beta\| - \|\alpha\| \le \|\beta - \alpha\| = \|\alpha - \beta\|$$

$$\therefore \quad \pm(\|\alpha\| - \|\beta\|) \le \|\alpha - \beta\| \text{ or } |\, \|\alpha\| - \|\beta\| \,| \le \|\alpha - \beta\|$$

8.2 ORTHOGONALITY AND ORTHONORMALITY

Definition 1. *Let V be an inner product space. Then any two vectors α, $\beta \in V$ are said to be orthogonal to each other if $(\alpha, \beta) = 0$.*

For Example:

(1) The zero vector is orthogonal to every vector of V.
As $(0, \alpha) = (0\alpha, \alpha)$ $[\because 0\alpha = 0]$
                                 $= 0 \, (\alpha, \alpha) = 0 .$

(2) The vector (a, b) in R^2 is orthogonal to $(-b, a)$ with respect to the standard inner product, for $((a, b), (-b, a)) = -ab + ba = 0$.

(3) The standard basis of either R^n or C^n is an orthogonal with respect to the standard inner product.

Definition 2. *Let V be an inner product space. If S and T are the subsets of V, then S is said to be orthogonal to T, denoted by $S \perp T$, if for each vector $s \in S$ and $t \in T$, $(s, t) = 0$.*

Definition 3. *Let V be an inner product space, and S is a subset of V. Then a set of all those vectors of V which are orthogonal to each vector of S, is called the orthogonal complement of S, denoted by $S^{\perp}$,*
   i.e., $S^{\perp} = \{\alpha \in V : (\alpha, s) = 0, \forall s \in S\}.$

It can easily be verified that $S^{\perp}$ is a subspace of V, and if S is a subspace of V, then

$$S \cap S^{\perp} = \{0\}.$$

Also, if $S \perp T$, then $T \subseteq S^{\perp}.$

Definition 4. *The set of vectors in V is called an orthonormal set if*

(i) $\|\alpha\| = 1, \forall \alpha \in V$

(ii) $(\alpha, \beta) = 0, \forall \alpha, \beta \in V.$

For Example : The standard basis of either R^n or C^n is an orthonormal set with respect to the standard inner product.

Definition 5. *A maximal orthonormal set of vectors in an inner product space V is called a complete orthonormal set.*

Definition 6. *If V be a finite dimensional inner product space, then a basis of V which is also orthonormal set (orthogonal set) is called an orthonormal basis (orthogonal basis).*

THEOREM 1.   *An orthogonal set of non-zero vectors is linearly independent.*

Proof.   Let S be a finite or infinite orthogonal set of non-zero vectors of an inner product space $V(F)$. Suppose $\alpha_1, \alpha_2, ..., \alpha_n$ are distinct vectors in S.

Let $\displaystyle\sum_{i=1}^{n} a_i \alpha_i = 0$ for $a_i \in F.$

Then, for any $k = 1, 2, 3, \ldots, n$ suppose $\alpha_k = 0.$

So, $\left(\sum_{i=1}^{n} a_i \alpha_i, \alpha_k\right) = (0, \alpha_k) = 0$

or $(a_1\alpha_1 + a_2\alpha_2 + ... + a_k\alpha_k + ... + a_n\alpha_n, \alpha_k) = 0$

or $a_1(\alpha_1, \alpha_k) + a_2(\alpha_2, \alpha_k) + ... + a_k(\alpha_k, \alpha_k) + ... + a_n(\alpha_n, \alpha_k) = 0$

or $a_k(\alpha_k, \alpha_k) = 0$ $\qquad [\because (\alpha_i, \alpha_j) = 0 \text{ for } i \neq j]$

or $a_k = 0$ as $(\alpha_k, \alpha_k) = 1$

$\therefore \qquad a_k = 0$ for $k = 1, 2, 3, ..., n$

Hence, S is a linearly independent subset of V.

THEOREM 2. *In a finite dimensional inner product space a complete orthonormal set is a basis.*

Proof. Let S be a complete orthonormal subset of a finite dimensional inner product space V. Since V is finite dimensional so that S is linearly independent.

Also, S is finite set.

Let $\qquad S = \{\alpha_1, \alpha_2, ..., \alpha_m\}$.

Then, for some $\alpha \in V$, we have

$$\beta = \alpha - \sum_{i=1}^{m} (\alpha, \alpha_i)\alpha_i .$$

Let $\alpha_k \neq 0$ for $1 \leq k \leq m$. Then

$$(\beta, \alpha_k) = \left(\alpha - \sum_{i=1}^{m} (\alpha, \alpha_i)\alpha_i, \alpha_k\right) = (\alpha, \alpha_k) - \sum_{i=1}^{m} (\alpha, \alpha_k)(\alpha_i, \alpha_k)$$

$$= (\alpha, \alpha_k) - (\alpha, \alpha_k)(\alpha_k, \alpha_k) \quad [\because (\alpha_i, \alpha_k) = 0 \text{ for } i \neq k]$$

$$= (\alpha, \alpha_k) - (\alpha, \alpha_k) \qquad [\because (\alpha_k, \alpha_k) = 1]$$

$\therefore \qquad (\beta, \alpha_k) = 0, \quad \forall k = 1, 2, 3, ..., m.$

Thus, $\beta \perp S$, Since $\alpha_k \neq 0$, so that

$$\beta = 0 \Rightarrow \alpha = \sum_{i=1}^{m} (\alpha, \alpha_i)\alpha_i.$$

Hence, S forms a basis of V.

THEOREM 3. **(Gram-Schmidt Orthogonalization).** *Let V be an inner product space and let $\beta_1, \beta_2, .., \beta_n$ be any linearly independent vectors in V. Then one may construct orthogonal vectors $\alpha_1, \alpha_2, ..., \alpha_n$ in V such that for each $k = 1, 2, 3, ..., n$, the set $\{\alpha_1, \alpha_2, ..., \alpha_k\}$ forms a basis for the subspace spanned by $\beta_1, \beta_2 ..... \beta_k$.*

Proof. Let $S_k = \{\beta_1, \beta_2 ..... \beta_k\}$, for $k = 1, 2, 3, ..., n$. We inductively construct orthogonal vectors $\{\alpha_1, \alpha_2, ..., \alpha_k\}$ such that $\{\alpha_1, \alpha_2, ..., \alpha_k\}$ is a basis of the subspace spanned by S_k.

First let $\alpha_1 = \beta_1$. The other vectors are then defined inductively as follows :

Suppose $\alpha_1, \alpha_2, ..., \alpha_m$ for $m = 1, 2, 3, ..., n$ have been constructed so that for each $k = 1, 2, 3, ..., m$, $\{\alpha_1, \alpha_2, ..., \alpha_k\}$ is an orthogonal basis of the subspace of V which is spanned by $\{\beta_1, \beta_2, ..., \beta_k\}$.

Now, we construct α_{m+1}. Let,

$$\alpha_{m+1} = \beta_{m+1} - \sum_{k=1}^{m} \frac{(\beta_{m+1}, \alpha_k)}{\|\alpha_k\|^2} \alpha_k. \qquad \ldots(1)$$

Then, $\alpha_{m+1} \neq 0$, for otherwise β_{m+1} will be a linear combination of $\alpha_1, \alpha_2, ..., \alpha_m$ and so a linear combination of $\{\beta_1, \beta_2, ..., \beta_n\}$, which will give a contradiction because $\{\beta_1, \beta_2, ..., \beta_n\}$ is linear independent.

Also, if $1 \leq j \leq m$, then

$$(\alpha_{m+1}, \alpha_j) = \left(\beta_{m+1} - \sum_{k=1}^{m} \frac{(\beta_{m+1}, \alpha_k)}{\|\alpha_k\|^2} (\alpha_k, \alpha_j) \right)$$

$$= (\beta_{m+1}, \alpha_j) - \sum_{k=1}^{m} \frac{(\beta_{m+1}, \alpha_k)}{\|\alpha_k\|^2} (\alpha_k, \alpha_j)$$

$$= (\beta_{m+1}, \alpha_j) - (\beta_{m+1}, \alpha_j) = 0$$

Thus, $\{\alpha_1, \alpha_2, ..., \alpha_{m+1}\}$ is an orthogonal set consisting $m+1$ non-zero vectors in a subspace spanned by $\{\beta_1, \beta_2, ..., \beta_{m+1}\}$. Hence it is basis for this subspace.

Corollary. *Every finite dimensional inner product space has an orthonormal basis.*

Proof. First, we obtain a orthogonal basis by Gram-Schmidt process as $\{\alpha_1, \alpha_2, ..., \alpha_n\}$ and then replace each vector α_k by $\alpha_k / \|\alpha_k\|$ to obtain orthonormal basis.

THEOREM 4. *Let $S = \{\alpha_1, \alpha_2, ..., \alpha_n\}$ be an orthogonal set of non-zero vectors in an inner product space V. If a vector $\beta \in V$ is the linear combination of elements of S, then*

$$\beta = \sum_{k=1}^{m} \frac{(\beta, \alpha_k)}{\|\alpha_k\|^2} \alpha_k.$$

Proof. Since $\beta \in L(S)$, then there exists scalars $a_1, a_2, ..., a_m$ such that

$$\beta = a_1 \alpha_1 + a_2 \alpha_2 + ... + a_m \alpha_m \qquad \ldots(1)$$

Now

$$(\beta, \alpha_k) = \left(\sum_{i=1}^{m} a_i \alpha_i, \alpha_k \right)$$

$$= \sum_{i=1}^{m} a_i (\alpha_i, \alpha_k) \qquad \text{[By linearity]}$$

$$= a_k (\alpha_k, \alpha_k) \qquad [\because (a_i, \alpha_k) = 0 \text{ for } i \neq k]$$

$$= a_k \|\alpha_k\|^2 \qquad [\because \text{Each } \alpha_k \neq 0 \text{ so } \|\alpha_k\|^2 = (\alpha_k, \alpha_k) \neq 0]$$

$$\therefore \qquad a_k = \frac{(\beta, \alpha_k)}{\|\alpha_k\|^2}, \ k = 1, 2, 3, ..., m.$$

Putting the values of $a_1, a_2, ..., a_m$ in (1), we get

$$\beta = \sum_{k=1}^{m} \frac{(\beta, \alpha_k)}{\|\alpha_k\|^2} \alpha_k.$$

THEOREM 5. *Let $S = \{\alpha_1, \alpha_2, ..., \alpha_m\}$ be an orthonormal set of vectors in an inner product space V. If a vector $\beta \in L(S)$, then*

$$\beta = \sum_{k=1}^{m} (\beta, \alpha_k) \alpha_k.$$

Proof. Since $\beta \in L(S)$, then there exist scalars $a_1, a_2, ..., a_m$ such that

$$\beta = a_1\alpha_1 + a_2\alpha_2 + ... + a_m\alpha_m = \sum_{k=1}^{m} a_i\,\alpha_k. \qquad ...(1)$$

Now $\quad (\beta, \alpha_k) = \left(\sum_{i=1}^{m} a_i\,\alpha_i, \alpha_k \right)$

$$= \sum_{i=1}^{m} a_i(\alpha_i, \alpha_k) \qquad\qquad \text{[By Linearity]}$$

$$= a_k \qquad [\because (a_i, \alpha_k)=0 \text{ for } i \neq k \text{ and } (\alpha_i, \alpha_k)=1 \text{ for } i = k]$$

$\therefore \qquad\qquad a_k = (\beta, \alpha_k), k = 1, 2, 3, ..., m.$

Putting the values of $a_1, a_2, ..., a_m$ in (1), we get

$$\beta = \sum_{k=1}^{m} (\beta, \alpha_k)\alpha_k.$$

THEOREM 6. *Any orthonormal set of vectors in an inner product space is linearly independent.*

Proof. Let S be any orthonormal set of vectors in an inner product space V. Let $S_1 = \{\alpha_1, \alpha_2, ..., \alpha_m\}$ be a finite subset of S containing m distinct vectors. Then we have

$$(\alpha_i, \alpha_j) = \begin{cases} 1, & \text{if } i = j \\ 0, & \text{if } i \neq j \end{cases} \qquad ...(1)$$

Suppose for scalars, $a_1, a_2, ..., a_m \in F$, we have

$$a_1\alpha_1 + a_2\alpha_2 + ... + a_m\alpha_m = 0$$

or $\qquad\qquad \sum_{k=1}^{m} a_i\alpha_i = 0$

Now $\left(\sum_{i=1}^{m} a_i\alpha_i, \alpha_k \right) = (0, \alpha_k) = 0$

$\Rightarrow \qquad\qquad \sum_{i=1}^{m} a_i(\alpha_i, \alpha_k) = 0$

$\Rightarrow \qquad\qquad\qquad a_k = 0 \qquad\qquad\qquad\qquad \text{[Using (1)]}$

$\therefore \qquad\qquad\qquad a_k = 0, k = 1, 2, 3, ..., m.$

$\therefore$ For $a_k = 0, k = 1, 2, 3, ..., m$ we have $\sum_{i=1}^{m} a_i\alpha_i = 0.$

S_1 is a linearly independent subset which shows that every finite subset of S is linearly independent, hence S is linearly independent.

THEOREM 7. *Let $S = \{\alpha_1, \alpha_2, ..., \alpha_m\}$ be an orthonormal set in an inner product space V and let $\beta \in V$, then*

$$\gamma = \beta - \sum_{i=1}^{m} (\beta, \alpha_i)\alpha_i$$

is orthogonal to each of $\alpha_1, \alpha_2, ..., \alpha_m$ and consequently to subspace spanned by S.

Proof.   Now

$$(\gamma, \alpha_k) = \left(\beta - \sum_{i=1}^{m} (\beta, \alpha_i) \alpha_i, \alpha_k \right)$$

$$= (\beta, \alpha_k) - \left(\sum_{i=1}^{m} (\beta, \alpha_i) \alpha_i, \alpha_k \right) \qquad \text{[By linearity]}$$

$$= (\beta, \alpha_k) - \sum_{i=1}^{m} (\beta, \alpha_i)(\alpha_i, \alpha_k) \qquad \text{[By linearity]}$$

$$= (\beta, \alpha_k) - (\beta, \alpha_k)$$

$$\qquad\qquad [\because (\alpha_i, \alpha_k) = 1 \text{ for } i = k \text{ and otherwise zero}]$$

$$= 0.$$

$\therefore (\gamma, \alpha_k) = 0$ for all $k = 1, 2, 3, ..., m$

which shows that γ is orthogonal to each α_i.

Again, let α be any vector in the subspace spanned by S, then $\alpha \in L(S)$.

Since, $\alpha \in L(S)$, then there exist scalars $a_1, a_2, ..., a_m$ such that

$$\alpha = a_1 \alpha_1 + a_2 \alpha_2 + ... + a_m \alpha_m = \sum_{i=1}^{m} a_i \alpha_i$$

Now

$$(\gamma, \alpha) = \left(\gamma, \sum_{i=1}^{m} a_i \alpha_i \right)$$

$$= \sum_{i=1}^{m} \bar{a}_i (\gamma, \alpha_i)$$

$$= \sum_{i=1}^{m} \bar{a}_i \cdot 0 \qquad [\because (\gamma, \alpha_i) = 0 \; \forall \; i = 1, 2, ..., m]$$

$$= 0$$

which shows that γ is orthogonal to the subspace spanned by S.

THEOREM 8.   **(Bessel's inequality).** *If* $S = \{\alpha_1, \alpha_2, ..., \alpha_m\}$ *is any finite orthonormal set in an inner product space* $V(F)$ *and if for any vector* $\alpha \in V.$

$$x_i = (\alpha, \alpha_i) \text{ for } i = 1, 2, ..., m$$

then

$$\sum_{i=1}^{m} |x_i|^2 \leq ||\alpha||^2 \text{ or } \sum_{i=1}^{m} |(\alpha, \alpha_i)|^2 \leq ||\alpha||^2.$$

Also show that $\beta = \alpha - \sum_{i=1}^{m} x_i \alpha_i$ *is orthogonal to each* $\alpha_i \in S.$

Proof.   Since V is an inner product space and $S \subseteq V$ so that

$$\beta = \alpha - \sum_{i=1}^{m} x_i \alpha_i \in V.$$

Now

$$||\beta||^2 = (\beta, \beta)$$

$$= \left(\alpha - \sum_{i=1}^{m} x_i \alpha_i, \alpha - \sum_{i=1}^{m} x_i \alpha_i \right)$$

$$= \left(\alpha, \alpha - \sum_{i=1}^{m} x_i \alpha_i \right) - \left(\sum_{i=1}^{m} x_i \alpha_i, \alpha - \sum_{i=1}^{m} x_i \alpha_i \right) \qquad \text{[By linearity]}$$

$$= \left(\alpha - \sum_{i=1}^{m} x_i \alpha_i, \alpha \right) - \sum_{i=1}^{m} x_i \left(\alpha_i, \alpha - \sum_{i=1}^{m} x_i \alpha_i \right) \qquad \text{[By linearity]}$$

$$= \overline{(\alpha, \alpha)} - \overline{\left(\sum_{i=1}^{m} x_i \alpha_i, \alpha \right)} - \sum_{i=1}^{m} x_i (\alpha_i, \alpha) - \sum_{i=1}^{m} x_i \left(\alpha_i - \sum_{i=1}^{m} x_i \alpha_i \right)$$

$$= \| \alpha \|^2 - \sum_{i=1}^{m} \overline{x_i} \overline{(\alpha_i, \alpha)} - \sum_{i=1}^{m} x_i \overline{(\alpha, \alpha_i)} - \sum_{i=1}^{m} x_i \left\{ -\sum_{i=1}^{m} \overline{x_i} (\alpha_i, \alpha_i) \right\}$$

$$= \| \alpha \|^2 - \sum_{i=1}^{m} \overline{x_i} (\alpha, \alpha_i) - \sum_{i=1}^{m} x_i \overline{(\alpha, \alpha_i)} + \sum_{i=1}^{m} x_i \overline{x_i} \qquad [\because (\alpha_i, \alpha_i) = 1]$$

$$= \| \alpha \|^2 - \sum_{i=1}^{m} \overline{x_i} x_i - \sum_{i=1}^{m} x_i \overline{x_i} + \sum_{i=1}^{m} x_i \overline{x_i} \qquad [\because x_i = (\alpha, \alpha_i)]$$

$$= \| \alpha \|^2 - \sum_{i=1}^{m} | x_i |^2$$

$$\therefore \qquad \| \beta \|^2 = \| \alpha \|^2 - \sum_{i=1}^{m} | x_i |^2$$

But $\qquad 0 \le \| \beta \|^2$

$\Rightarrow \qquad 0 \le \| \alpha \|^2 - \sum_{i=1}^{m} | x_i |^2$

$\therefore \qquad \sum_{i=1}^{m} | x_i |^2 \le \| \alpha \|^2$

or $\qquad \sum_{i=1}^{m} | (\alpha, \alpha_i) |^2 \le \| \alpha \|^2.$

Also, $\qquad (\beta, \alpha_j) = \left(\alpha - \sum_{i=1}^{m} x_i \alpha_i, \alpha_j \right) \qquad \text{for } j = 1, 2, ..., m$

$$= (\alpha, \alpha_j) - \sum_{i=1}^{m} x_i (\alpha_i, \alpha_j) \ \text{ for } j = 1, 2, ..., m$$

$$= (\alpha, \alpha_j) - x_j (\alpha_j, \alpha_j) \ \text{ for } j = 1, 2, ..., m \qquad [\because (\alpha_i, \alpha_j) = 0, i \ne j]$$

$$= (\alpha, \alpha_j) - x_j \qquad\qquad\qquad\qquad\qquad [\because (\alpha_j, \alpha_j) = 1]$$

$$= (\alpha, \alpha_j) - (\alpha, \alpha_j) \qquad\qquad\qquad\qquad [\because x_j = (\alpha, \alpha_j)]$$

$$= 0$$

$\therefore \qquad (\beta, \alpha_j) = 0 \ \forall \ j = 1, 2, ..., m.$

Hence, β is orthogonal to each $\alpha_j \in S$.

8.3 ORTHOGONAL EXPANSION

Let V be a finite dimensional inner product space and let dim. $V = n$.

Let $S = \{\alpha_1, \alpha_2, ..., \alpha_n\}$ be an orthonormal basis for V. Then for any vector $\alpha \in V$,

$$\alpha = (\alpha, \alpha_1)\alpha_1 + (\alpha, \alpha_2)\alpha_2 + ... + (\alpha, \alpha_n)\alpha_n \qquad ...(1)$$

The coefficients (α, α_i), $i = 1, 2, ..., n$ are called the fourier coefficients of α with respect to S and the expression (1) is called the fourier expansion of α with respect to S. This expansion is also known as *orthogonal expansion* if the vectors $\alpha_1, \alpha_2, ..., \alpha_n$ are orthogonal.

Solved Examples

Example 1. *Decide which of the following functions define an inner product on* R². *For* $\alpha = [a_1, b_1]^t$, $\beta = [a_2, b_2]^t$

(i) $(\alpha, \beta) = a_1 b_2 + a_2 b_1$
(ii) $(\alpha, \beta) = a_1 b_1 + a_2 b_2$
(iii) $(\alpha, \beta) = a_1 a_2 b_1 b_2$
(iv) $(\alpha, \beta) = a_1 a_2 - a_1 b_2 - a_2 b_1 + 3b_1 b_2$

Solution.
(i) Since $(\alpha, \alpha) = a_1 b_1 + a_1 b_1 = 2a_1 b_1$.

∴ (α, α) may or may not be positive.

So that the function, defined by $(\alpha, \beta) = a_1 b_2 + a_2 b_1$ is not an inner product.

(ii) Similarly, the function defined by $(\alpha, \beta) = a_1 b_1 + a_2 b_2$ is not an inner product.

(iii) (a) $(\alpha, \alpha) = a_1^2 b_1^2 > 0, \forall \alpha \in R^2$

(b) $(\beta, \alpha) = b_1 b_2 a_1 a_2$ \hfill [By definition]

$= a_1 a_2 b_1 b_2 = \bar{a}_1 \bar{a}_2 \bar{b}_1 \bar{b}_2$ \hfill [∵ *a's* and *b's* are real.]

$= \overline{a_1 a_2 b_1 b_2} = \overline{(\alpha, \beta)}$

(c) Let $\lambda, \mu \in R$ and let $\alpha = [a_1, b_1]^t$, $\beta = [a_2, b_2]^t$ and $\gamma = [a_3, b_3]^t$, then

$$\lambda\alpha + \mu\beta = \lambda[a_1 b_1]^t + \mu[a_2 b_2]^t = [\lambda a_1 + \mu a_2, \lambda b_1 + \mu b_2]^t$$

Now, $(\lambda\alpha + \mu\beta, \gamma) = (\lambda a_1 + \mu a_2)a_3 + (\lambda b_1 + \mu b_2)b_3$

and, $\lambda(\alpha, \gamma) + \mu(\beta, \gamma) = \lambda(a_1 a_3 b_1 b_3) + \mu(a_2 a_3 b_2 b_3)$.

∴ $(\lambda\alpha + \mu\beta, \gamma) \neq \lambda(\alpha, \gamma) + \mu(\beta, \gamma)$.

Thus the function defined by

$$(\alpha, \beta) = a_1 a_2 b_1 b_2$$

is not an inner product.

(iv) We verify the three axioms of an inner product:

(a) When $a_1 \neq 0, b_1 \neq 0$,

$$(\alpha, \alpha) = a_1^2 - 2a_1 b_1 + 3b_1^2 = a_1^2 - 2a_1 b_1 + b_1^2 + 2b_1^2$$

$$= (a_1 - b_1)^2 + 2b_1^2 > 0$$

∴ $(\alpha, \alpha) > 0, \forall 0 \neq \alpha \in R^2$.

(b) $(\beta, \alpha) = a_2 a_1 - a_2 b_1 - b_2 a_1 + 3b_2 b_1 = a_1 a_2 - a_1 b_2 - a_2 b_1 + 3b_1 b_2$

$= (\alpha, \beta) = \overline{(\alpha, \beta)}$ \hfill [∵ $\alpha, \beta \in R^2$]

∴ $(\beta, \alpha) = \overline{(\alpha, \beta)}, \forall \alpha, \beta \in R^2$.

(c)   Let $\gamma = [a_3, b_3]^t$, and $\lambda, \mu \in R$, then

$$\lambda\alpha + \mu\beta = \lambda[a_1, b_1]^t + \mu[a_2, b_2]^t$$

$$(\lambda\alpha + \mu\beta) = [\lambda a_1 + \mu a_2, \lambda b_1 + \mu b_2]^t$$

Now,

$$(\lambda\alpha + \mu\beta, \gamma) = ([\lambda a_1 + \mu a_2, \lambda b_1 + \mu b_2]^t, [a_3, b_3]^t)$$

$$= (\lambda a_1 + \mu a_2)a_3 - (\lambda a_1 + \mu a_2)b_3 - (\lambda b_1 + \mu b_2)a_3 + 3(\lambda b_1 + \mu b_2)b_3$$

$$= \lambda(a_1 a_3 - a_1 b_3 - a_3 b_1 + 3b_1 b_3) + \mu(a_2 a_3 - a_2 b_3 - a_3 b_2 + 3b_2 b_3)$$

$$= \lambda(\alpha, \gamma) + \mu(\beta + \gamma)$$

Hence, the function defined in (iv) is an inner product.

Example 2.   *Let V be a vector space of real continuous functions on the interval $0 \le t \le 1$ with inner product defined by $(f, g) = \int_0^1 f(t)g(t)\,dt$ and the polynomials $f(t) = t+2$, $g(t) = 3t - 2$ and $h(t) = t^2 - 2t - 3$.*

Find (f, g), (f, h), $\|f\|$ and $\|g\|$.

Solution.   (i) $(f, g) = \int_0^1 f(t)g(t)dt = \int_0^1 (t + 2)(3t - 2)dt$

$$= \int_0^1 (3t^2 - 4t - 4)dt = [t^3 + 2t^2 - 4t]_0^1 = (1 + 2 - 4) = -1$$

(ii) $(f, h) = \int_0^1 f(t)h(t)dt = \int_0^1 (t + 2)(t^2 - 2t - 3)dt$

$$= \int_0^1 (t^3 - 7t - 6)dt = \left[\frac{t^4}{4} - \frac{7}{2}t^2 - 6t\right]_0^1$$

$$= \left(\frac{1}{4} - \frac{7}{2} - 6\right) = -\frac{37}{4}$$

(iii) $(f, f) = \int_0^1 f(t)f(t)dt = \int_0^1 (t + 2)^2 dt$

$$= \int_0^1 (t^2 + 4t + 4)dt = \left[\frac{t^3}{3} + 2t^2 + 4t\right]_0^1$$

$$= \left(\frac{1}{3} + 2 + 4\right) = \frac{19}{3}$$

$$\therefore \quad \|f\| = \sqrt{(f, f)} = \sqrt{\frac{19}{3}} = \frac{1}{3}\sqrt{57}$$

(iv) $(g, g) = \int_0^1 g(t)g(t)dt = \int_0^1 (3t - 2)^2 dt$

$$= \int_0^1 (9t^2 - 12t + 4)dt = [3t^2 - 6t^2 + 4t]_0^1$$

$$= (3 - 6 + 4) = 1$$

$$\therefore \quad \|g\| = \sqrt{g, g} = \sqrt{1} = 1.$$

Example 3.   *Apply Gram-Schmidt orthogonalization process to the vectors $\beta_1 = (1, 0, 1)$, $\beta_2 = (1, 0, -1)$ and $\beta_3 = (0, 3, 4)$ to obtain an orthonormal basis $(\alpha_1, \alpha_2, \alpha_3)$ for R^3 with standard inner product.*

Solution. Let
$$\alpha_1 = \frac{\beta_1}{\|\beta_1\|} = \frac{(1,0,1)}{\sqrt{1^2+0^2+1^2}} = \frac{(1,0,1)}{\sqrt{2}} = \left(\frac{1}{\sqrt{2}},0,\frac{1}{\sqrt{2}}\right)$$

and
$$\gamma_2 = \beta_2 - (\beta_2,\alpha_1)\alpha_1.$$

Now, $(\beta_2,\alpha_1) = (1,0,-1).\left(\frac{1}{\sqrt{2}},0,\frac{1}{\sqrt{2}}\right) = \frac{1}{\sqrt{2}}+0-\frac{1}{\sqrt{2}} = 0$

$\therefore \quad \gamma_2 = (1,0,-1) - 0\left(\frac{1}{\sqrt{2}},0,\frac{1}{\sqrt{2}}\right) = (1,0,-1).$

Now
$$\alpha_2 = \frac{\gamma_2}{\|\gamma_2\|} = \frac{(1,0,-1)}{\sqrt{1^2+0^2+(-1)^2}} = \frac{(1,0,-1)}{\sqrt{2}} = \left(\frac{1}{\sqrt{2}},0,-\frac{1}{\sqrt{2}}\right)$$

Again, let $\gamma_3 = \beta_3 - (\beta_3,\alpha_2)\alpha_2 - (\beta_3,\alpha_1)\alpha_1.$

Now $(\beta_3,\alpha_2) = (0,3,4).\left(\frac{1}{\sqrt{2}},0,-\frac{1}{\sqrt{2}}\right) = (0+0-2\sqrt{2}) = -2\sqrt{2}$

and $(\beta_3,\alpha_1) = (0,3,4).\left(\frac{1}{\sqrt{2}},0,\frac{1}{\sqrt{2}}\right) = (0+0+2\sqrt{2}) = 2\sqrt{2}$

$\therefore$
$$\gamma_3 = (0,3,4) - (-2\sqrt{2})\left(\frac{1}{\sqrt{2}},0,-\frac{1}{\sqrt{2}}\right) - 2\sqrt{2}\left(\frac{1}{\sqrt{2}},0,\frac{1}{\sqrt{2}}\right)$$

$$= (0,3,4) + (2,0,-2) - (2,0,2)$$

$$= (0,3,4) + (0,0,-4) = (0,3,0)$$

So,
$$\alpha_3 = \frac{\gamma_3}{\|\gamma_3\|} = \frac{(0,3,0)}{\sqrt{0^2+3^2+0^2}} = \frac{(0,3,0)}{3} = (0,1,0)$$

Thus the required orthonormal basis is
$$\left\{\left(\frac{1}{\sqrt{2}},0,\frac{1}{\sqrt{2}}\right),\left(\frac{1}{\sqrt{2}},0,-\frac{1}{\sqrt{2}}\right),(0,1,0)\right\}.$$

Example 4. *Let* $\beta_1 = (3,0,4)$, $\beta_2 = (-1,0,7)$, $\beta_3 = (2,9,11)$ *be vectors in* R^3 *equipped with the standard inner product. Obtain an orthogonal basis.*

Solution. Let $\alpha_1 = \beta_1 = (3,0,4)$, then
$$\|\alpha_1\|^2 = 3^2+0^2+4^2 = 25$$

Now $\alpha_2 = \beta_2 - \frac{(\beta_2,\alpha_1)}{\|\alpha_1\|^2}\alpha_1 = (-1,0,7) - \frac{\{(-1,0,7).(3,0,4)\}}{25}(3,0,4)$

$$= (-1,0,7) - \frac{(-3+0+28)}{25}(3,0,4)$$

$$= (-1,0,7) - \frac{25}{25}(3,0,4) = (-1,0,7) - (3,0,4)$$

$$\alpha_2 = (-4,0,3) \text{ and } \|\alpha_2\|^2 = 16+0+9 = 25$$

Also, $\alpha_3 = \beta_3 - \dfrac{(\beta_3,\alpha_1)}{\|\alpha_1\|^2}\alpha_1 - \dfrac{(\beta_3,\alpha_2)}{\|\alpha_2\|^2}\alpha_2$

$$= (2,9,11) - \frac{\{(2,9,11).(3,0,4)\}}{25}(3,0,4) - \frac{\{(2,9,11).(-4,0,3)\}}{25}(-4,0,3)$$

$$= (2,9,11) - \frac{(6+0+44)}{25}(3,0,4) - \frac{(-8+0+33)}{25}(-4,0,3)$$

$$= (2, 9, 11) - 2(3, 0, 4) - (-4, 0, 3)$$

$$= (2, 9, 11) - (6, 0, 8) - (-4, 0, 3) = (0, 9, 0).$$

Thus the required orthogonal basis is {(3, 0, 4), (-4, 0, 3), (0, 9, 0)}.

Example 5. *Let W be a subspace of the inner product space V spanned by {(0,1,1,0), (0,5,-3,-2), (-3, -3, 5, -7)}. Find orthonormal basis for W.*

Solution. Given that β_1 = (0, 1, 1, 0), β_2 = (0, 5, -3, -2), β_3 = (-3, -3, 5, -7).

Let $\alpha_1 = \beta_1 = (0, 1, 1, 0)$.

Then $\|\alpha_1\|^2 = (\beta_1,\beta_1) = 0^2 + 1^2 + 1^2 + 0^2 = 2$

Now $\alpha_2 = \beta_2 - \dfrac{(\beta_2,\alpha_1)}{\|\alpha_1\|^2}\alpha_1$

$$= (0,5,-3,-2) - \frac{\{(0,5,-3,-2),(0,1,1,0)\}}{2}(0,1,1,0)$$

$$= (0,5,-3,-2) - \frac{(0+5-3+0)}{2}(0,1,1,0)$$

$$= (0, 5, -3, -2) - (0, 1, 1, 0)$$

$$= (0, 4, -4, -2)$$

Then $\|\alpha_2\|^2 = 0+16+16+4 = 36$

Again $\alpha_3 = \beta_3 - \dfrac{(\beta_3,\alpha_1)}{\|\alpha_1\|^2}\alpha_1 - \dfrac{(\beta_3,\alpha_2)}{\|\alpha_2\|^2}\alpha_2$

$$= (-3,-3,5,-7) - \frac{\{(-3,-3,5,-7),(0,1,1,0)\}}{2}(0,1,1,0)$$
$$- \frac{\{(-3,-3,5,-7),(0,4,-4,-2)\}}{36}(0,4,-4,-2)$$

$$= (-3,-3,5,-7) - \frac{(0-3+5+0)}{3}(0,1,1,0)$$
$$- \frac{(0-12-20+14)}{36}(0,4,-4,-2)$$

$$= (-3,-3,5,-7) - (0,1,1,0) + \frac{1}{2}(0,4,-4,-2)$$

$$= (-3, -3, 5, -7) - (0, 1, 1, 0) + (0, 2, -2, -1)$$

$$= (-3, -2, 2, -8)$$

Then, $\|\alpha_3\|^2 = 9 + 4 + 4 + 64 = 81$

Thus, the required orthonormal basis of W is

$$B = \left\{ \frac{\alpha_1}{\|\alpha_1\|}, \frac{\alpha_2}{\|\alpha_2\|}, \frac{\alpha_3}{\|\alpha_3\|} \right\}$$

$$= \left\{ \frac{(0,1,1,0)}{\sqrt{2}}, \frac{(0,4,-4,-2)}{\sqrt{36}}, \frac{(-3,-2,2,-8)}{\sqrt{81}} \right\}$$

$$= \left\{ \frac{1}{\sqrt{2}}(0,1,1,0), \frac{1}{3}(0,2,-2,-1), \frac{1}{9}(-3,-2,2,-8) \right\}.$$

Example 6. *In an inner product space*

$$\| x + y \| = \| x \| + \| y \|$$

then prove that x, y are linearly dependent but converse is not true.

Solution. Given that

$$\| x + y \| = \| x \| + \| y \|$$

$\Rightarrow$ $\| x + y \|^2 = (\| x \| + \| y \|)^2$

$\Rightarrow$ $\| x + y \| = \| x \|^2 + \| y \|^2 + 2 \| x \| \| y \|$ ...(1)

But $\| x + y \|^2 = \| x \|^2 + 2\,\mathrm{Real}(x,y) + \| y \|^2$ ...(2)

From (1) and (2), we get

$$\mathrm{Real}(x,y) = \| x \| \| y \|$$

$\Rightarrow$ $|(x,y)| \geq \| x \| \| y \|$ ...(3)

But by Schwarz's inequality, we have

$$|(x,y)| \leq \| x \| \| y \|$$ ...(4)

From (3) and (4), we get

$$|(x,y)| = \| x \| \| y \|$$ $\Rightarrow$ $x = ky$ for $k \in F.$

Hence, x and y are linearly dependent.

But converse is not true for this, we have an example.

Let $x = (-1, 0, 1), y = (2, 0, -2) \in V_3(\mathrm{R})$. Then $y = -2x$ which means that x and y are linearly dependent.

But $\| x \| = \sqrt{(-1)^2 + 0^2 + 1^2} = \sqrt{2}$

$$\| y \| = \sqrt{2^2 + 0^2 + (-2)^2} = 2\sqrt{2}$$

then $\| x \| + \| y \| = 3\sqrt{2}$

Also $x + y = (-1, 0, 1) + (2, 0, -2) = (1, 0, -1)$

Then $\| x + y \| = \sqrt{1^2 + 0^2 + (-1)^2} = \sqrt{2}$

Clearly $\| x + y \| \neq \| x \| + \| y \|$

Example 7. *Show that two vectors x, y in a real inner product space are orthogonal if and only if*

$$\| x + y \|^2 = \| x \|^2 + \| y \|^2 .$$

Solution. Let $V(F)$ be a real inner product space and $x, y \in V(F)$ are orthogonal, then $(x, y) = 0$.

Now, $\qquad \| x + y \|^2 = (x + y, x + y)$

$$= (x + y, x) + (x + y, y)$$

$$= (x, x) + (y, x) + (x, y) + (y, y)$$

$$= \| x \|^2 + (y, x) + (x, y) + \| y \|^2$$

$$= \| x \|^2 + (x, y) + (x, y) + \| y \|^2$$

$$[\because \text{ Inner product is real so } (x, y) = (y, x)]$$

$$= \| x \|^2 + 2(x, y) + \| y \|^2$$

$$= \| x \|^2 + \| y \|^2 \qquad\qquad [\because (x, y) = 0]$$

Conversely, for $x, y \in V(F)$, we have

$$\| x + y \|^2 = \| x \|^2 + \| y \|^2$$

$$\Rightarrow \quad \| x \|^2 + 2(x, y) + \| y \|^2 = \| x \|^2 + \| y \|^2$$

$$\Rightarrow \qquad\qquad 2(x, y) = 0 \quad \Rightarrow \quad (x, y) = 0$$

Thus, x, y are othogonal.

Example 8. *If x and y are orthogonal unit vectors, then find the distance between x and y.*

Solution. Given that $\qquad (x, y) = 0, (x, x) = 1, (y, y) = 1$

Then the distance between x and y

$$\| x - y \| = (x - y, x - y) \;\; = (x - y, x) - (x - y, y)$$

$$= (x, x) - (y, x) - (x, y) + (y, y)$$

$$= 1 - 0 - 0 + 1 = 2.$$

Example 9. *Show that two vectors x and y in a complex inner product space are orthogonal if and only if*

$$\| ax + by \|^2 = \| ax \|^2 + \| by \|^2$$

for all pairs of scalars a and b.

Solution. Let $x, y \in V(F)$ be orthogonal, where $V(F)$ is a complex inner product space.

$\therefore \qquad\qquad (x, y) = 0.$

Then $\qquad\qquad \| ax + by \|^2 = (ax + by, ax + by)$

$$= (ax, ax + by) + (by, ax + by)$$

$$= \overline{(ax + by, ax)} + \overline{(ax + by, by)}$$

$$= \overline{(ax, ax)} + \overline{(by, ax)} + \overline{(ax, by)} + \overline{(by, by)}$$

$$= \| ax \|^2 + (ax, by) + \overline{(ax, by)} + \| by \|^2$$

$$= \| ax \|^2 + 2\mathrm{Real}(ax, by) + \| by \|^2$$

$$= \| ax \|^2 + 2\mathrm{Real}\{a\bar{b}(x, y)\} + \| by \|^2$$

$$= \| ax \|^2 + \| by \|^2 \qquad\qquad [\because (x, y) = 0]$$

$$\therefore \qquad \| ax + by \|^2 = \| ax \|^2 + \| by \|^2$$

Conversely, for $x, y \in V(F)$, we have

$$\| ax + by \|^2 = \| ax \|^2 + \| by \|^2$$

$$\Rightarrow \| ax \|^2 + 2\mathrm{Real}(ax, by) + \| by \|^2 = \| ax \|^2 + \| by \|^2$$

$$\Rightarrow \qquad\qquad 2\,\mathrm{Real}\,(ax, by) = 0$$

$$\Rightarrow \qquad\qquad 2\,\mathrm{Real}\,\{a\bar{b}(x, y)\} = 0$$

$$\Rightarrow \qquad\qquad (x, y) = 0$$

$\therefore$ x and y are orthogonal.

Example 10. *If x and y are vectors in a real inner product space and if $\| x \| = \| y \|$, then x–y and x+y are orthogonal.*

Solution. We have $(x - y, x + y) = (x, x + y) - (y, x + y)$

$$= \overline{(x + y, x)} - \overline{(x + y, y)}$$

$$= \overline{(x, x)} + \overline{(y, x)} - \overline{(x, y)} - \overline{(y, y)}$$

$$= \| x \|^2 + (x, y) - \overline{(x, y)} - \| y \|^2 = \| x \|^2 - \| y \|^2$$

$$[\because \text{ In real inner product space } (x, y) = \overline{(x, y)}\,]$$

$$= 0 \qquad\qquad\qquad [\because \| x \| = \| y \|]$$

$$\therefore \qquad (x - y, x + y) = 0$$

which shows that $x - y$ is orthogonal to $x + y$.

Example 11. *Find the cosine of the angle between x and y if*
   (i) $x = (1, -3, 2)$, $y = (2, 1, 5)$ in V_3 (R)
   (ii) $x = 2t - 1, y = t^2$ in the space of polynomial in t and the inner product defined as

$$(p, q) = \int_0^1 p(t)\,q(t)\,dt \quad for\ p, q \in V_3(\mathrm{R}).$$

Solution.   (i) $x = (1, -3, 2)$, $y = (2, 1, 5)$, then

$$\| x \| = \sqrt{1^2 + (-3)^2 + 2^2} = \sqrt{14}$$

$$\| y \| = \sqrt{2^2 + 1^2 + 5^2} = \sqrt{30}$$

and $(x, y) = ((1, -3, 2), (2, 1, 5))$

$$= 1 \times 2 + (-3) \times 1 + 2 \times 5$$

$$= 2 - 3 + 10 = 9$$

If θ be the angle between x and y, then

$$\cos\theta = \frac{(x,y)}{\|x\|\,\|y\|} = \frac{9}{\sqrt{14}\,\sqrt{30}} = \frac{9}{\sqrt{420}}$$

(ii)

$$(x,y) = \int_0^1 (2t-1)(t^2)\,dt$$

$$= 2\int_0^1 t^3 dt - \int_0^1 t^2 dt$$

$$= 2\left[\frac{t^4}{4}\right]_0^1 - \left[\frac{t^3}{3}\right]_0^1$$

$$= \frac{1}{2} - \frac{1}{3} = \frac{1}{6}$$

$$\|x\|^2 = (x,x) = \int_0^1 (2t-1)(2t-1)\,dt$$

$$= \int_0^1 (4t^2 + 1 - 4t)\,dt$$

$$= \left[4\frac{t^3}{3} + t - 2t^2\right]_0^1 = \frac{4}{3} + 1 - 2 = \frac{1}{3}$$

$\therefore$

$$\|x\| = \sqrt{\frac{1}{3}} = \frac{1}{\sqrt{3}}$$

$$\|y\|^2 = (y,y) = \int_0^1 t^2 \cdot t^2 dt = \int_0^1 t^4 dt = \left[\frac{t^5}{5}\right]_0^1 = \frac{1}{5}$$

$\therefore$

$$\|y\| = \sqrt{\frac{1}{5}} = \frac{1}{\sqrt{5}}.$$

If θ be the angle between x and y then

$$\cos\theta = \frac{(x,y)}{\|x\|\,\|y\|} = \frac{1/6}{\dfrac{1}{\sqrt{3}} \cdot \dfrac{1}{\sqrt{5}}} = \frac{\sqrt{15}}{6}.$$

Example 12. *Find the vector of unit length which is orthogonal to the vector* $\alpha = (2, 1, -6)$ *of* R^3 *with respect to standard inner product.*

Solution. Let $\beta = (x, y, z)$ be a vector in R^3 which is orthogonal to α.

Since α and β are orthogonal, then

$$(\alpha, \beta) = 0$$

$\Rightarrow \qquad ((2, 1, 6), (x, y, z)) = 0$

$\Rightarrow \qquad 2x - y + 6z = 0$

Clearly, $x = 2, y = -2, z = -1$ satisfy above equation.

Thus, $\beta = (x, y, z) = (2, -2, -1)$ is a vector in R^3 which is orthogonal to α.

Now, $$\|\beta\| = \sqrt{2^2 + (-2)^2 + (-1)^2} = \sqrt{4+4+1} = \sqrt{9} = 3$$

Thus, the required vector $= \dfrac{\beta}{\|\beta\|}$

$$= \frac{1}{3}(2,-2,-1) = \left(\frac{2}{3}, -\frac{2}{3}, -\frac{1}{3}\right).$$

⮞ EXERCISE 8.1

1. Show that $(\alpha, \beta) = x_1 y_1 + x_2 y_2 - x_3 y_3$ is not an inner product on R^3 where $\alpha = (x_1, x_2, x_3)$ and $\beta = (y_1, y_2, y_3)$.

2. Decide which of the following functions define an inner product on R^2 for $\alpha = [x_1. \ x_2]^t$, $\beta = [y_1. \ y_2]^t$:
   (i) $(\alpha, \beta) = x_1 y_2 - x_2 y_1$
   (ii) $(\alpha, \beta) = 2x_1 y_1 + x_1 y_2 + x_2 y_1 + 2x_2 y_2$.

3. Decide which of the following functions define an inner product on C^2. For

   $$\alpha = [x_1 x_2]^t, \ \beta = [y_1 y_2]^t$$

   (i) $(\alpha, \beta) = x_1 \bar{y}_2$ (ii) $(\alpha, \beta) = x_1 \bar{y}_1 + x_2 \bar{y}_2$
   (iii) $(\alpha, \beta) = 2x_1 \bar{y}_1 + i(x_2 \bar{y}_1 - x_1 \bar{y}_2) + 2x_2 \bar{y}_2$

4. Let α and β be vectors in an inner product space such that $\|\alpha + \beta\| = 8, \|\alpha - \beta\| = 6, \|\alpha\| = 7$, find $\|\beta\|$.

5. Prove that for α, β and γ in an inner product space V.

$$\|\alpha - \beta\| \ \|\gamma\| \le \|\beta - \gamma\| \ \|\alpha\| + \|\gamma - \alpha\| \ \|\beta\|.$$

6. Expand $(3\alpha_1 + 2\alpha_2, 5\beta_1 - 6\beta_2 + 4\beta_3)$.

7. Let $S = \{\alpha_1, \alpha_2, \alpha_3, \alpha_4\}$ be an orthonormal basis of R^4 and let $\alpha, \beta \in V$ be represented by $(1, 2, 3, -1)$ and $(2, 4, -1, 1)$ respectively. Compute inner product (α, β).

8. Find the orthonormal basis of $V_3(R)$ with standard inner product using Gram-Schmidt orthogonalization process to the vectors $\beta_1 = (1, 0, 1), \beta_2 = (1, 2, -2)$ and $\beta_3 = (2, -1, 1)$.

9. Orthonormalize the set of linearly independent vectors $\{(1,0,1,1), (-1,0,-1,1), (0,-1,1,1)\}$ of $V_4(R)$.

10. Find an orthonormal basis of the subspace of $V_2(C)$ with standard inner product spanned by $\alpha_1 = (1,0,i), \alpha_2 = (2, 1, 1+i)$.

11. Let W be a subspace of the inner product space V spanned by $\{(0, 1, 1, 0), (0, 5, -3, -2), (-3, -3, 5, -7)\}$. Find an orthonormal basis for W.

Hints to Selected Problems

1. Let $\alpha = (3, 4, 5)$, then $(\alpha, \alpha) = 3 \times 3 + 4 \times 4 - 5 \times 5 = 9 + 16 - 25 = 0$
   $\therefore \qquad (\alpha, \alpha) = 0$ where $\alpha = 0$.
   Hence $(\alpha, \beta) = x_1 y_1 + x_2 y_2 - x_3 y_3$ is not an inner product on R^3.

4. Since we know that $\|\alpha + \beta\|^2 + \|\alpha - \beta\|^2 = 2\|\alpha\|^2 + 2\|\beta\|^2$.

   $\therefore \qquad\qquad 8^2 + 6^2 = 2(7)^2 + 2\|\beta\|^2$

   $\therefore \qquad\qquad\qquad \|\beta\| = 1$

6. $(3\alpha_1 + 2\alpha_2, 5\beta_1 - 6\beta_2 + 4\beta_3) = 3(\alpha_1, 5\beta_1 - 6\beta_2 + 4\beta_3) + 2(\alpha_2, 5\beta_1 - 6\beta_2 + 4\beta_3)$
   $\qquad\qquad\qquad\qquad = 15(\alpha_1, \beta_1) - 18(\alpha_1, \beta_2) + 12(\alpha_1, \beta_3) + 10(\alpha_2, \beta_1) - 12(\alpha_2, \beta_2) + 8(\alpha_2, \beta_3)$

7. $(\alpha, \beta) = ((1, 2, 3, -1), (2, 4, -1, 1)) = 1 \times 2 + 2 \times 4 + 3 \times (-1) + (-1) \times 1 = 2 + 8 - 3 - 1 = 6$

8. Let $\gamma_1 = \beta_1$, and let $\alpha_1 = \dfrac{\gamma_1}{\|\gamma_1\|} = \dfrac{\beta_1}{\|\beta_1\|} = \dfrac{(1,0,1)}{\sqrt{1^2 + 0^2 + 1^2}} = \dfrac{1}{\sqrt{2}}(1,0,1)$.

   Now $\qquad\qquad \gamma_2 = \beta_2 - (\beta_2, \alpha_1)\alpha_1$

   $$= (1, 2, -2) - \frac{1}{\sqrt{2}}(1 \times 1 + 2 \times 0 + (-2) \times 1)\left(\frac{1}{\sqrt{2}}, 0, \frac{1}{\sqrt{2}}\right)$$

$$= (1,2,-2) + \frac{1}{\sqrt{2}}\left(\frac{1}{\sqrt{2}},0,\frac{1}{\sqrt{2}}\right) = (1,2,-2) + \left(\frac{1}{2},0,\frac{1}{2}\right) = \left(\frac{3}{2},2,-\frac{3}{2}\right)$$

So,
$$\alpha_2 = \frac{\gamma_2}{\|\gamma_2\|} = \frac{\left(\frac{3}{2},2,\frac{3}{2}\right)}{\sqrt{\frac{9}{4}+4+\frac{9}{4}}} = \left(\frac{3}{\sqrt{34}},\frac{4}{\sqrt{34}},-\frac{3}{\sqrt{34}}\right)$$

Now
$$\gamma_3 = \beta_3 - (\beta_3,\alpha_2)\,\alpha_2 - (\beta_3,\alpha_1)\alpha_1$$

$$(\beta_3,\alpha_2) = \frac{1}{\sqrt{34}}(2\times3+(-1)\times4+1\times(-3)) = \frac{1}{\sqrt{34}}(6-4-3) = -\frac{1}{\sqrt{34}}$$

$$(\beta_3,\alpha_1) = \frac{1}{\sqrt{2}}(2\times1+(-1)\times0+1) = \frac{1}{\sqrt{2}}(2+1) = \frac{3}{\sqrt{2}}$$

$$\therefore \quad \gamma_3 = (2,-1,1) + \frac{1}{\sqrt{34}}\left(\frac{3}{\sqrt{34}},\frac{4}{\sqrt{34}},-\frac{3}{\sqrt{34}}\right) - \frac{3}{\sqrt{2}}\left(\frac{1}{\sqrt{2}},0,\frac{1}{\sqrt{2}}\right)$$

$$= (2,-1,1) + \frac{1}{34}(3,4,-3) - \frac{3}{2}(1,0,1) = \left(\frac{10}{17},\frac{15}{17},\frac{10}{17}\right) = \frac{5}{17}(2,-3,-2)$$

$$\therefore \quad \alpha_3 = \frac{\gamma_3}{\|\gamma_3\|} = \frac{\frac{5}{17}(2,-3,-2)}{\frac{5}{17}\sqrt{4+9+4}} = \left(\frac{2}{\sqrt{17}},\frac{-3}{\sqrt{17}},\frac{-2}{\sqrt{17}}\right).$$

===== **Answers** =====

2. (i) Not inner product (ii) Inner product
3. (i) Not inner product (ii) Inner product   (iii) Not inner product
4. $\|\beta\| = 1$ **6.** $(\alpha_1,\beta_2) -18(\alpha_1,\beta_2)+12(\alpha_1,\beta_3)+10(\alpha_2,\beta_1)-12(\alpha_2,\beta_2)+8\,(\alpha_2,\beta_3)$

7. 6   **8.** $\left\{\left(\frac{1}{\sqrt{2}},0,\frac{1}{\sqrt{2}}\right),\left(\frac{3}{\sqrt{34}},\frac{4}{\sqrt{34}},\frac{-3}{\sqrt{34}}\right),\left(\frac{2}{\sqrt{17}},\frac{-3}{\sqrt{17}},\frac{-2}{\sqrt{17}}\right)\right\}$

9. $\left\{\left(\frac{1}{\sqrt{3}},0,\frac{1}{\sqrt{3}},\frac{1}{\sqrt{3}}\right),\left(-\frac{1}{\sqrt{6}},0,-\frac{1}{\sqrt{6}},\frac{2}{\sqrt{6}}\right),\left(-\frac{1}{\sqrt{6}},-\frac{2}{\sqrt{6}},\frac{1}{\sqrt{6}},0\right)\right\}$

10. $\left\{\frac{1}{\sqrt{2}}(1,0,i),\frac{1}{\sqrt{2}}\left(\frac{1+i}{2},1,\frac{1-i}{2}\right)\right\}$ **11.** $\left\{\frac{1}{\sqrt{2}}(0,1,1,0),\frac{1}{3}(0,2,-2,-1),\frac{1}{9}(-3,-2,2,-8)\right\}$

8.4 THE ADJOINT OF A LINEAR TRANSFORMATION

Definition 1. *Let U and V be finite-dimensional inner product space and T : U → V be a linear transformation. Then there exists a unique linear mapping T*: V→ U such that for all α∈V and β∈U*

$$(T(\alpha),\,\beta) = (\alpha,\,T^*(\beta)) \text{ or } (T\alpha,\,\beta) = (\alpha,\,T^*\beta)$$

The mapping T* is called the adjoint of T.

Definition 2. *If A∈F $^{m\times n}$, then A* denotes the conjugate transpose of the matrix A. If we take standard inner product spaces F^n and F^m, and consider A as a linear transformation from F^n to F^m, as α → A(α), then for α∈F^n and β∈F^m*

$$(A(\alpha),\,\beta) = \beta*(A(\alpha)) = (A^*(\alpha))^* \beta = (\beta,\,A^*(\alpha)).$$

Thus A* = $(\overline{A})^t$ is called adjoint of linear transformation A between standard inner product space.

8.5 PROPERTIES OF THE ADJOINT

THEOREM 1. *Let U and V be finite-dimensional inner product spaces over the same field F, then*

(i) $(T_1+T_2)^* = T_1^* + T_2^*$, *for* $T_1, T_2 \in L(U, V)$

(ii) $(\lambda T)^* = \bar{\lambda} T^*$, *for* $T \in L(U, V)$ *and* $\lambda \in F$

(iii) $(T_1 T_2)^* = T_2^* T_1^*$, *for* $T_1, T_2 \in L(U, V)$

(iv) $(T^*)^* = T, T \in L(U, V)$

(v) *If* $T \in L(U, V)$ *and T is invertible, then* $(T^*)^{-1} = (T^{-1})^*$.

Proof. (i) Let $\alpha \in U$ and $\beta \in V$

$$(\alpha, T_1 + T_2)^*(\beta)) = ((T_1 + T_2)(\alpha), (\beta))$$
$$= (T_1(\alpha) + T_2(\alpha), \beta) = (T_1(\alpha), \beta) + (T_2(\alpha), \beta)$$
$$= (\alpha, T_1^*(\beta)) + (\alpha, T_2^*(\beta)) = (\alpha, (T_1^* + T_2^*)(\beta))$$

Therefore, by uniqueness of adjoint mapping, $(T_1 + T_2)^* = T_1^* + T_2^*$.

(ii) Let $\alpha \in U$ and $\beta \in V$, then

$$(\alpha, (\lambda T)^*(\beta)) = (\lambda T(\alpha), \beta) = \lambda(T(\alpha), \beta)$$
$$= \lambda(\alpha, T^*(\beta)) = (\alpha, (\bar{\lambda} T^*)(\beta)).$$

$$\Rightarrow \qquad (\lambda T)^* = \bar{\lambda} T^*$$

(iii) Let $\alpha \in U$ and $\beta \in V$ then

$$(\alpha, (T_1 T_2)^*(\beta)) = ((T_1 T_2)(\alpha), \beta) = (T_1(T_2(\alpha)), \beta)$$
$$= (T_2(\alpha), T_1^*(\beta)) = (\alpha, T_2^* T_1^*(\beta)).$$

$$\therefore \qquad (T_1 T_2)^* = T_2^* T_1^*$$

(iv) Let $\alpha \in U$ and $\beta \in V$, then

$$(\alpha, (T^*)^*(\beta)) = (T^*(\alpha), \beta) = \overline{(\beta, T^*(\alpha))}$$

$$= \overline{(T(\beta), \alpha)} = (\alpha, T(\beta)).$$

$$\therefore \qquad (T^*)^* = T$$

(v) Let $\alpha \in U$ and $\beta \in V$, then as

$$(\alpha, \beta) = (I(\alpha), \beta) [\because I(\alpha) = \alpha \text{ as } I \text{ an identity operator}]$$
$$= (\alpha, I^*(\beta))$$

$$\Rightarrow \qquad I^* = I$$

Since T is invertible, then

$$T T^{-1} = I \Rightarrow (T T^{-1})^* = I^*$$

$$\Rightarrow \qquad (T^{-1})^* T^* = I \Rightarrow (T^*)^{-1} = (T^{-1})^*$$

THEOREM 2. *Let U and V be finite dimensional inner product spaces over F and let* $T \in L(U, V)$. *If B and B' are ordered orthonormal bases of U and V respectively, then the matrix representation of T* with respect to these bases is the conjugate transpose of the matrix representation of T with respect to these bases, that is,*

$$[T^*]_{B'} = [T]^*_B$$

Proof. Let $B = \{\alpha_1, \alpha_2, ..., \alpha_n\}$ and $B' = \{\beta_1, \beta_2, ...\beta_m\}$ be ordered orthonormal bases of U and V respectively. Let $[T]_B = A \in F^{m \times n}$ and let $[T]_{B'} = C \in F^{n \times m}$.

Then $$T(\alpha_j) = \sum_{k=1}^{m} a_{kj} \beta_k \qquad \qquad ... (1)$$

for $j = 1, 2, 3, \ldots, n$ and so

$$\left(T(\alpha_j),(\beta_i)\right) = \sum_{k=1}^{m} a_{kj}(\beta_k,\beta_i) = a_{ij} \qquad [\because B' \text{ is orthonormal}]$$

$\therefore$ $(T(\alpha_j), \beta_i)$ is (i, j)th entry of $m \times n$ matrix $[T]_B$.

It follows that

$$[C]_{ij} = [T^*(\beta_j),\alpha_i] = \overline{[\alpha_i, T^*(\beta_j)]}$$

$$= \overline{[T(\alpha_i),\beta_j]} = \overline{[A]_{ji}} = [A^*]_{ij}$$

$$\Rightarrow \qquad C = A^* \Rightarrow [T^*]_{B'} = [T]_B^*$$

REMARK

- Let T be a linear operator on a finite dimensional inner product space V over F. Then $|T^*| = |\overline{T}|$ and $tr\, T^* = tr.\overline{T}$ where, $tr.$ stands for trace of a matrix.

8.6 SELF-ADJOINT TRANSFORMATION

REMARKS

- A linear transformation T on an inner product space $V(F)$ is said to be self-adjoint if $T^* = T$ and T is said to be skew-adjoint if $T^* = -T$.
- In an Euclidean space, the self-adjoint transformation is called symmetric and in an unitary space it is called Hermitian.

THEOREM 1. *Let T_1 and T_2 be self-adjoint transformations on an inner product space $V(F)$. Then*

(i) *$T_1 + T_2$ is self-adjoint*

(ii) *$T_1 T_2$ is self-adjoint if and only if $T_1 T_2 = T_2 T_1$*

(iii) *T^{-1} is self-adjoint if T is invertible*

(iv) *λT is self-adjoint iff λ is real for $T \neq 0$ and $\lambda \neq 0$.*

Proof. It is given that $T_1^* = T_1$ and $T_2^* = T_2$

(i) $((T_1 + T_2)(\alpha), (\beta)) = (T_1(\alpha) + T_2(\alpha), \beta) = (T_1(\alpha), \beta) + (T_2(\alpha),\beta)$

$$= (\alpha, T_1^*(\beta)) + (\alpha, T_2^*(\beta)) = (\alpha, T_1(\beta)) + (\alpha, T_2(\beta))$$

$$= (\alpha, (T_1+T_2)(\beta)) \qquad \ldots (1)$$

Since, $((T_1 + T_2)(\alpha), (\beta)) = (\alpha, (T_1+T_2)^*(\beta))$ $\ldots (2)$

From (1) and (2), we get

$$(\alpha, (T_1 + T_2)^*(\beta)) = (\alpha,(T_1+T_2)(\beta))$$

$$\Rightarrow \qquad (T_1+T_2)^* = (T_1+T_2)$$

Thus T_1+T_2 is self-adjoint.

(ii) Suppose $T_1 T_2$ is self-adjoint, then

$$(T_1 T_2)^* = T_1 T_2 \qquad \Rightarrow T_2^* T_1^* = T_1 T_2 \quad [\text{By reversal rule}]$$

$$\Rightarrow \qquad T_2 T_1 = T_1 T_2 \qquad\qquad [\because T_1^* = T_1 \text{ and } T_2^* = T_2\,]$$

Conversely, suppose $T_2 T_1 = T_1 T_2$, then we have to show that $T_1 T_2$ is self-adjoint.

Now, $\qquad (T_1 T_2)^* = T_2^* T_1^* = T_2 T_1 = T_1 T_2$ $[\because T_2 T_1 = T_1 T_2]$

Thus $T_1 T_2$ is self-adjoint.

(iii) Since T is invertible, so

$$TT^{-1} = T^{-1}T = I \Rightarrow (TT^{-1})^* = (T^{-1}T)^* = I^*$$

$\Rightarrow \qquad (T^{-1})^*T^* = T^*(T^{-1})^* = I \qquad\qquad [\because I^* = I]$

$\Rightarrow \qquad (T^{-1})^*T = T(T^{-1})^* = I \qquad\qquad [\because T^* = T]$

$\therefore \qquad (T^{-1})^* = T^{-1}.$ Thus T^{-1} is self- adjoint.

(iv) For each $\alpha, \beta \in V$, we have

$$((\lambda T)(\alpha), \beta) = (\alpha, (\lambda T)^*(\beta)) \qquad\qquad ...(3)$$

Also, $\qquad ((\lambda T)(\alpha), \beta) = (\lambda T(\alpha), \beta) = \lambda\, (T(\alpha). \beta)$

$$= \lambda(\alpha, T^*(\beta)) = \lambda(\alpha, T(\beta)) \qquad [\because T^* = T]$$

$$= (\alpha, (\bar\lambda\, T)(\beta)) \qquad\qquad ... (4)$$

From (3) and (4) , we get

$$(\alpha, (\lambda T)^*(\beta)) = (\alpha, (\bar\lambda\, T)(\beta)) \Rightarrow (\lambda T)^* = \bar\lambda\, T$$

$\Rightarrow \qquad (\lambda T)^* = \lambda T$ iff $\bar\lambda = \lambda.$

Hence, λT is self-adjoint iff λ is real.

THEOREM 2. *If T is a self-adjoint transformation on an inner product space $V(F)$, then S^*TS is self-adjoint for all S, also if S is invertible and $S^* T S$ is self-adjoint, then T is self-adjoint.*

Proof. Let T be self-adjoint, so that

$$T^* = T.$$

Now, $\qquad (S^*TS)^* = S^*T^*(S^*)^* \qquad\qquad$ [By reversal rule]

$$= S^* TS \qquad\qquad [\because T^* T \text{ and } (S^*)^* = S]$$

$\therefore\ S^* T S$ is self-adjoint

Secondly, let S be invertible, so that

$$SS^{-1} = S^{-1}S = I$$

Also S^* is invertible , then

$$S^* (S^*)^{-1} = (S^*)^{-1} S^* = I. \qquad\qquad ... (2)$$

Since S^*TS is self-adjoint so.

$$(S^*TS)^* = S^*TS \Rightarrow S^* T^*(S^*)^* = S^*TS$$

$\Rightarrow \qquad S^*T^*S = S^*TS$

$\Rightarrow \qquad (S^*)^{-1} S^*T^*S = (S^*)^{-1} S^*TS \qquad$ [By left inverse of S^*]

$\Rightarrow \qquad IT^*S = ITS \qquad\qquad$ [Using (2)]

$\Rightarrow \qquad T^*S = TS \ \Rightarrow (T^*S)S^{-1} = (TS)S^{-1} \quad$ [By right inverse of T]

$\Rightarrow \qquad T^*(SS^{-1}) = T(SS^{-1}) \Rightarrow T^*I = T I \qquad$ [Using (1)]

$\Rightarrow \qquad T^* = T.$

Thus T is self-adjoint.

THEOREM 3. *If T is a linear transformation on an inner product space V(F). Then T^* is self-adjoint$\Leftrightarrow [T] = [T^*]$ with respect to the orthonormal basis B of V(F).*

Proof. Let the matrix of T be

$$[T]_B = [a_{ij}] \qquad \qquad ...(1)$$

relative to B.

Then the matrix of T^* relative to basis B^* the dual of B, is

$$[T^*]_B = [\bar{a}_{ij}] \qquad \qquad ...(2)$$

First, suppose that T is self-adjoint, so $T^* = T$.

Since orthonormal basis is self dual, i.e., $B^* = B$, then $a_{ij} = \bar{a}_{ij} \Rightarrow [T] = [T^*]$.

Conversely, suppose that

$$[T] = [T^*], i.e., \ a_{ij} = \bar{a}_{ij} \qquad \qquad ...(3)$$

Let $B = \{\alpha_1, \alpha_2,...,\alpha_n\}$ be an ordered orthonormal basis of V. Then we have

$$T(\alpha_j) = \sum_{i=1}^{n} a_{ij}\alpha_i \qquad \qquad ...(4)$$

where $$[T] = [a_{ij}]$$

Also from (2), we have $[T^*] = [\bar{a}_{ji}]$

where, $$T^*(\alpha_j) = \sum_{i=1}^{n} \bar{a}_{ji}\alpha_i \quad \text{for } j = 1, 2,..., n. \qquad \qquad ...(5)$$

Let $\alpha \in V$ be any vector. Then for $b_j \in F$, we have $\alpha = \sum_{j=1}^{n} b_j\alpha_i \qquad [\because B \text{ spans } V]$

so, $$T(\alpha) = T\left(\sum_{j=1}^{n} b_j\alpha_j\right) = \sum_{j=1}^{n} b_j T(\alpha_j) = \sum_{j=1}^{n} b_j \sum_{j=1}^{n} a_{ij}\alpha_i \qquad [\text{Using (4)}]$$

$$= \sum_{i=1}^{n}\left(\sum_{i=1}^{n} \bar{a}_{ji}b_j\right)\alpha_i = \sum_{i=1}^{n}\left(\sum_{j=1}^{n} a_{ji}b_j\right)\alpha_i \qquad [\text{Using (3)}]$$

$$= \sum_{j=1}^{n} b_j\left(\sum_{i=1}^{n} \bar{a}_{ji}\alpha_i\right) = \sum_{j=1}^{n} b_j \cdot T^*(\alpha_j) \qquad [\text{Using (5)}]$$

$$= T^*\left(\sum_{j=1}^{n} b_j\alpha_j\right)$$

$\therefore \qquad \qquad T(\alpha) = T^*(\alpha).$

Since α is an arbitrary vector of V. Thus, $T^* = T$ and hence T is self-adjoint.

THEOREM 4. *If T be a self-adjoint linear transformation on an inner product space V(F), then $T = 0 \Leftrightarrow (T(\alpha),\alpha) = 0, \forall \alpha \in V$.*

Proof. Let $T = 0$, then for $\alpha \in V$

$$(T(\alpha),\alpha) = (0(\alpha),\alpha) = (0, \alpha) = 0.$$

Conversely, let $(T(\alpha),\alpha) = 0 \ \forall \ \alpha \in V.$

Now consider the identity

$$(T(\alpha),\beta) + (T(\beta), \alpha) = (T(\alpha+\beta),(\alpha+\beta)) - (T(\alpha),\alpha) - (T(\beta),\beta), \text{ for } \forall \ \alpha, \beta \in V.$$

Then, $(T(\alpha), \beta) + (T(\beta), \alpha) = 0 - 0 - 0 = 0$ or $(T(\alpha), \beta) + (\beta, T^*(\alpha)) = 0$

or $(T(\alpha), \beta) + \overline{(T^*(\alpha), \beta)} = 0$ or $(T(\alpha), \beta) + \overline{(T(\alpha), \beta)} = 0$ [$\because T$ self-adjoint]

$\Rightarrow \qquad 2\,\text{Real}\,\{(T(\alpha), \beta)\} = 0 \Rightarrow \qquad \text{Real}\,(T(\alpha), \beta) = 0 \qquad \qquad ...(1)$

There arises two cases :

Case I. If $V(F)$ is an Euclidean space, then $(T(\alpha), \beta)$ is real, so from (1)

$$(T(\alpha), \beta) = 0, \forall\, \alpha, \beta \in V \quad \Rightarrow \quad T = 0.$$

Case II. If $V(F)$ is an inner product over a field of complex numbers, then

$$(T(i\alpha), \beta) = \text{Real}\, i\ (T(\alpha), \beta) = 0 \Rightarrow |(T(\alpha), \beta)| = 0$$

$$\Rightarrow \qquad |(T(\alpha), \beta)| = 0 \ \Rightarrow\ T = 0$$

Solved Examples

Based on the following Results

➠ Let $T : U \to V$ be a linear transformation. Then $T^* : V \to U$ is said to be adjoint of T if $(T(\alpha), \beta) = (\alpha, T^*(\beta)); \alpha \in U, \beta \in V.$

➠ $T^* = T \Leftrightarrow T$ is self- adjoint.

➠ $T^* = -T \Leftrightarrow T$ is skew-adjoint.

➠ If T_1 and T_2 are self-adjoint then $T_1 + T_2$ is self-adjoint and $T_1.T_2$ is self-adjoint iff $T_1 T_2 = T_2 T_1$.

Example 1. *If T is a linear transformation on $V_3(F)$, so that*

$$T(a, b, c) = (a + b, b, a + b + c)$$

for arbitrary $(a, b, c) \in V_3$. Find $T^*(a_1, b_1, c_1)$.

Solution. Let $(a_1, b_1, c_1) \in V_3(F)$, then by definition of T^*

$((a, b, c), T^*(a_1, b_1, c_1)) = (T(a, b, c), (a_1, b_1, c_1))$

$\qquad\qquad = ((a + b, b, a + b + c), (a_1, b_1, c_1))$

$\qquad\qquad = (a + b)\bar{a}_1 + b\bar{b}_1 + (a + b + c)\bar{c}_1$

$\qquad\qquad = a(\bar{a}_1 + \bar{c}_1) + b(\bar{b}_1 + \bar{c}_1 + \bar{a}_1) + c\bar{c}_1$

$\qquad\qquad = a\overline{(a_1 + c_1)} + b\overline{(b_1 + c_1 + a_1)} + c\bar{c}_1$

$\qquad\qquad = ((a, b, c), (a_1 + c_1, b_1 + c_1 + a_1, c_1))$

Since (a, b, c) and (a_1, b_1, c_1) are arbitrary elements in $V_3(F)$, so we have

$\qquad T^*(a_1, b_1, c_1) = (a_1 + c_1, b_1 + c_1 + a_1, c_1)$

Example 2. *Let T be the linear operator on $\mathbb{C}^3$ defined by*

$$T(x, y, z) = (2x + (1 - i)y, (3 + 2i)x - 4iz, 2ix + (4 - 3i)y - 3z).$$

Find $\qquad T^*(x, y, z)$.

Solution. Let $B = \{\{1, 0, 0), (0, 1, 0), (0, 0, 1)\}$ be the standard basis of $\mathbb{C}^3$. Then the matrix of T relative to B is

$$[T] = \begin{bmatrix} 2 & 1 - i & 0 \\ 3 + 2i & 0 & -4i \\ 2i & 4 - 3i & -3 \end{bmatrix}.$$

Now, the conjugate transpose of $[T]$ is

$$[T^*] = \begin{bmatrix} 2 & 3-2i & -2i \\ 1+i & 0 & 4+3i \\ 0 & 4i & -3 \end{bmatrix}.$$

Hence, $\quad T^*(x, y, z) = (2x + (3-2i)y - 2iz,\ (1+i)x + (4+3i)z,\ 4iy - 3z).$

Example 3. Let T be the linear operator on $\mathbf{C}^3$ defined by

$$T(x, y, z) = (2x + iy,\ y - 5iz,\ x + (1-i)y + 3z)$$

Find $T^*(x, y, z)$.

Solution. Let $B = \{(1,0,0),(0,1,0),(0,0,1)\}$ be the standard basis of $\mathbf{C}^3$. Then

$$T(1,0,0) = (2,0,1)$$
$$T(0,1,0) = (i,\ 1,1-i)$$
$$T(0,0,1) = (0,\ -5i,\ 3).$$

The matrix of T is

$$[T] = \begin{bmatrix} 2 & i & 0 \\ 0 & 1 & -5i \\ 1 & 1-i & 3 \end{bmatrix}.$$

Now the conjugate transpose of $[T]$ is

$$[T^*] = \begin{bmatrix} 2 & 0 & 1 \\ -i & 1 & 1+i \\ 0 & 5i & 3 \end{bmatrix}.$$

Hence, $\quad T^*(x, y, z) = (2x + z,\ -ix + y + (1+i)z,\ 5iy + 3z).$

Example 4. Let T be a linear operator on V, let W be a T-invariant subspace of V. Show that $W^\perp$ is invariant under T^*.

Solution. Let $\alpha \in W^\perp$. Since W is T-invariant subspace of V, then $T(\beta) \in W$ for $\beta \in W$.

Now $\quad\quad\quad \alpha \in W^\perp \Rightarrow (T(\beta), \alpha) = 0 \quad \forall\ T(\beta) \in W$

$$\Rightarrow (\beta, T^*(\alpha)) = 0 \quad \Rightarrow T^*(\alpha) \perp \beta \ \forall\ \beta \in W$$

$$\Rightarrow T^*(\alpha) \in W^\perp$$

Hence, $W^\perp$ is invariant under T^*.

Example 5. Use the definition of adjoint to show that $0^* = 0$.

Solution. For every $\alpha, \beta \in V$,

$$(0(\alpha), \beta) = (0, \beta) = 0 = (\alpha, 0) = (\alpha, 0(\beta))$$

$$\Rightarrow \quad\quad (\alpha, 0^*(\beta)) = (\alpha, 0(\beta)) \quad \Rightarrow \quad 0^* = 0.$$

Example 6. If T be a linear transformation on an inner product space $V(F)$ satisfying $T^2 = T$, then show that T is self-adjoint $\Leftrightarrow$ $T^*T = TT^*$.

Solution. It is given that $T^2 = T$. Suppose that T is self-adjoint, i.e., $T^* = T$.

Then, for $\alpha, \beta \in V$, we have

$$(T(\alpha), \beta) = (\alpha, T^*(\beta)) = (\alpha, T(\beta)) \quad\quad\quad\quad [\because T^* = T]$$

$$= (\alpha, T^2(\beta)) \qquad [\because T = T^2]$$
$$= (\alpha, TT(\beta)) = (\alpha, T^*T(\beta)) \qquad [\because T = T^*]$$

$$\therefore \qquad (T(\alpha), \beta) = (\alpha, T^*T(\beta))$$

$$\Rightarrow \qquad T^* = T^*T \Rightarrow T = TT^* \qquad \qquad ...(1)$$

Also, $\qquad (T(\alpha), \beta) = (\alpha, TT^*(\beta))$

$$\Rightarrow \qquad T^* = TT^* \Rightarrow T = T^*T \qquad \qquad ...(2)$$

From (1) and (2), $TT^* = T^*T$

Conversely, suppose that

$$T^*T = TT^* = T^2T^* \qquad \qquad [\because T^2 = T]$$
$$= T(TT^*) = T(T^*T)$$
$$T^*T = (TT^*)T$$

$$\Rightarrow \qquad T^* = TT^* \Rightarrow T = (TT^*)^* = T^*T = TT^* = T^*$$

$$\therefore \qquad T^* = T.$$

Hence, T is self-adjoint.

Example 7. *If $V_2(C)$ be an inner product space having standard basis $B = \{(1, 0), (0,1)\}$ and T be a linear transformation on $V_2(C)$ defined by*

$$T(0,1) = (1, -2), \ T(0, 1) = (i, -1).$$

Then, find $T(x, y)$ and $T^*(x, y)$.

Solution. Since T is linear, so that

$$T(x, y) = T[x(1, 0) + y(0, 1)] = xT(1, 0) + yT(0,1)$$
$$= x(1, -2) + y(i, -1) = (x + iy, -2x - y).$$

The matrix of T relative to B is

$$[T] = \begin{bmatrix} 1 & i \\ -2 & -1 \end{bmatrix}.$$

If T^* is the adjoint of T, then the matrix of T^* relative to B^* (the dual basis of B)is

$$[T^*] = \begin{bmatrix} 1 & -2 \\ -i & -1 \end{bmatrix}.$$

Thus, $\qquad T^*(x, y) = (x - 2y, -ix, -y)$

Example 8. *If T be a linear transformation on an inner product space $V_2(C)$ having $B = \{(1,0), (0,1)\}$ as basis and if T is defined by*

$$T(1, 0) = (1 + i, 2), \ T(0, 1) = (i, i)$$

then find the matrix of T^* relative to B. Does T^* commute with T?

Solution. Let $(x, y) \in V_2(C)$. Then

$$(x, y) = x(1, 0) + y(0, 1).$$

$$\therefore \qquad T(x, y) = xT(1, 0) + yT(0,1) = x(1 + i, 2) + y(i, i)$$
$$= ((1 + i) x + iy, 2x + iy). \qquad \qquad ...(1)$$

The matrix of T relative to B is

$$[T] = \begin{bmatrix} 1+i & i \\ 2 & i \end{bmatrix}.$$

$\therefore$ The matrix of T^* relative to B is

$$[T^*] = [\bar{T}]^t = \begin{bmatrix} 1-i & 2 \\ -i & -i \end{bmatrix}.$$

Thus, we have

$$T^*(x, y) = ((1-i)x + 2y, -ix - iy). \qquad \qquad ...(2)$$

Now, $\quad TT^*(x, y) = T(T^*(x,y)) = T((1-i)x + 2y, -ix-iy) \qquad$ [Using (2)]

$$= ((1+i)\{(1-i)x+2y\}+i(-ix-iy), 2\{(1-i)x+2y\}+i(-ix-iy)\}$$

$$= ((1-i^2x + 2(1+i)y + x + y, 2(1-i)x+4y+x+y)$$

$$TT^*(x, y) = (3x+(3+2i)y, (3-2i)x + 5y) \qquad \qquad ...(3)$$

Also, $\quad T^*T(x, y) = T^*[(x, y)] = T^*((1+i)x + iy, 2x + iy) \qquad$ [Using (1)]

$$= ((1-i)\{(1+i)x+iy\} + 2(2x+iy), -i(1+i)x+y-2ix + y)$$

$$= (2x+i(1-i)y + 4x+2iy, (1-3i)x+2y) \qquad \qquad ...(4)$$

$$T^*T(x, y) = (6x + (1+3i)y, (1-3i)x + 2y)$$

From (3) and (4), we get $TT^* \neq T^*T$.

Hence, T^* does not commute with T.

Example 9. *If T is a self-adjoint operator on V and λ be an eigenvalue of T show that λ is real.*

Solution. Let α be a non-zero eigenvector of T belonging to λ, then

$$T(\alpha) = \lambda\alpha \qquad \qquad ...(1)$$

Since $\alpha \neq 0$, so (α, α) is positive.

Now, $\quad \lambda(\alpha,\alpha) = (\lambda\alpha, \alpha) = (T(\alpha), \alpha) \qquad$ [Using (1)]

$$= (\alpha, T^*(\alpha)) \qquad$$ [By the definition of adjoint]

$$= (\alpha, T(\alpha)) \qquad\qquad [\because T^* = T]$$

$$= (\alpha, \lambda\alpha)$$

$$\lambda(\alpha,\alpha) = \bar{\lambda}(\alpha, \alpha)$$

Since $(\alpha, \alpha) \neq 0$ so that $\lambda = \bar{\lambda}$, this shows that λ is real.

Example 10. *If T is a self-adjoint operator on V and $T^2 = 0$, show that $T = 0$.*

Solution. For any $\alpha \in V$, we have

$$\|T(\alpha)\|^2 = (T(\alpha), T(\alpha)) = (\alpha, T^*T(\alpha)) = (\alpha, T^2(\alpha)) \qquad [\because T^* = T]$$

$$= (\alpha, 0(\alpha)) \qquad\qquad\qquad [\because T^2 = 0]$$

$$= (\alpha, 0) = 0$$

$\Rightarrow \quad \|T(\alpha)\| = 0 \quad \Rightarrow \quad T(\alpha) = 0 \ \forall \ \alpha \in V$

$\Rightarrow \qquad\qquad T = 0$

Example 11. *If T is skew-symmetric transformation on an Euclidean space V, then $(T(\alpha), \alpha) = 0$ for $\alpha \in V$. Is converse also true?*

Solution. Since T is skew-symmetric so that
$$T^* = -T \qquad \qquad \dots(1)$$
then for $\alpha \in V$, we have
$$(T(\alpha), \alpha) = (\alpha, T^*(\alpha)) = (\alpha, (-T)(\alpha)) \qquad \text{[Using (1)]}$$
$$= -(\alpha, T(\alpha)) = -\overline{(T(\alpha), \alpha)}$$
$$= -(T(\alpha), \alpha) \qquad \qquad [\because V \text{ is Euclidean space.}]$$
$$\Rightarrow \quad 2(T(\alpha), \alpha) = 0 \quad \Rightarrow \quad (T(\alpha), \alpha) = 0.$$
Conversely, if $(T(\alpha), \alpha) = 0 \quad \forall \, \alpha \in V$
$$\Rightarrow \qquad \qquad T = 0$$
Thus, converse is not true.

8.7 CONGRUENT OPERATORS

Two linear operators T_1 and T_2 on an inner product vector space are said to be congruent if there exists an invertible linear operator P such that $T_2 = P^*T_1P$.

THEOREM 1. *If T_1 and T_2 be two congruent operators then T_1^* and T_2^* are also congruent.*

Proof. Given
$$T_2 = P^*T_1P$$
$$T_2^* = (P^*T_1P)^* = P^*T_1^*P^{**} = P^*T_1^*P.$$
Hence T_1^* and T_2^* are also congruent.

THEOREM 2. *If T_1, T_2 are invertible and congruent then their inverses are also congruent.*

Proof. Given $T_2 = P^*T_1P$ where P is invertible operator and hence P^* is also invertible and
$$(P^*)^{-1} = (P^{-1})^* \qquad \qquad \dots(1)$$
From
$$T_2 = P^*T_1P \text{ we get } T_2^{-1} = (P^*T_1P)^{-1}$$
or
$$T_2^{-1} = P^{-1}T_1^{-1}(P^*)^{-1} = [(P^{-1})^*]T_1^{-1}(P^{-1})^* \text{ by 1.}$$
Put $(P^{-1})^* = Q$, and since P^{-1} is invertible and hence $(P^{-1})^*$, *i.e.*, Q is also invertible.
$$\therefore \qquad \qquad T_2^{-1} = Q^*T_1^{-1}Q, \text{ where } Q \text{ is invertible.}$$
Hence T_1^{-1} and T_2^{-1} are also congruent.

THEOREM 3. *The relation of congruency of linear operators is an equivalence relation.*

Proof. Define $T_2 = P^*T_1P$, where P is invertible operator.

A relation is said to be an equivalence relation if it is reflexive, symmetric and transitive.

(i) **Reflexive.** $\qquad T_1 = IT_1I = I^*T_1I$
$$I^* = I \text{ and also } I \text{ is invertible.}$$
Hence T_2 is congruent to T_1.

(ii) **Symmetric.** $\qquad T_2 = P^*T_1P$
$$\Rightarrow \qquad (P^*)^{-1}T_2 = (P^*)^{-1}P^*T_1P = T_1P$$
$$\Rightarrow \qquad (P^{-1})^*T_2P^{-1} = T_1PP^{-1} = T_1.$$
or $\qquad \qquad T_1 = (P^{-1})^*T_2^*P^{-1} = Q^*T_2Q \text{ where } Q = P^{-1}$
which is invertible as P is invertible. Hence symmetric.

(iii) **Transitive.** Let T_1, T_2, T_3 be three invertible operators where

$$T_1 \text{ and } T_2 \text{ are congruent} \Rightarrow T_1 = P^* T_2 P$$
$$T_2 \text{ and } T_3 \text{ are congruent} \Rightarrow T_2 = Q^* T_3 Q$$

So, $\quad T_1 = P^* T_2 P = P^* Q^* T_3 Q P = (QP)^* T_3 Q P = R^* T_3 R.$

where $R = QP$ is also invertible as Q and P are invertible.

$\therefore$ T_1 and T_3 are also congruent. Hence transitive.

THEOREM 4. *If T_1, T_2 be congruent and T_1 being skew, i.e., $T^* = -T$ then T_2 is also skew.*

Proof. We have $\quad T_2 = P^* T_1 P$

$\therefore \qquad\qquad T_2{}^* = (P^* T_1 P)^* = P^* T_1{}^* P^{**} = P^* (-T_1) P$
$\qquad\qquad\qquad = -(P^* T_1 P) = -T_2$

$\therefore$ T_2 is also skew.

8.8 INNER PRODUCT VECTOR SPACE ISOMORPHISM

Definition 1. *Let V and W be two vector spaces over the same field F, then an isomorphism of V onto W is a linear transformation $T : V \to W$, which is one-one, onto (i.e., invertible) and preserves the vector space compositions.*

Definition 2. *If V and W be inner product vector spaces then an isomorphism of V into W is a linear transformation $T = V \to W$ which is one-one onto and preserves vector space compositions and in addition preserves the inner products also which means that*

$$(T\alpha, T\beta) = (\alpha, \beta) \; \forall \; \alpha, \beta \in V$$

Hence an inner product vector space isomorphism of V onto W is a vector space isomorphism of V onto W with the additional property that it preserves inner products.

i.e., $\qquad\qquad (T\alpha, T\beta) = (\alpha, \beta) \; \forall \; \alpha, \beta \in V.$

Interpretation (i) *'T preserves Inner products'.*

$$(T\alpha, T\beta) = (\alpha, \beta) \; \forall \; \alpha, \beta \in V$$

Choosing $\beta = \alpha$ we get

$$(T\alpha, T\alpha) = (\alpha, \alpha) \; \forall \; \alpha \in V.$$

or $\qquad\qquad \|T\alpha\|^2 = \|\alpha\|^2 \text{ or } \|T\alpha\| = \|\alpha\|.$

i.e., T preserves norms also.

Also if $\alpha \neq 0$ then $\|\alpha\| \neq 0$ and hence $\|T\alpha\| \neq 0 \Rightarrow T\alpha \neq 0$ when $\alpha \neq 0$. It means that T is non-singular which implies that T is injective.

(ii) *T preserves distances, i.e., T is an isometry.*

Now $\qquad\qquad \|T\alpha - T\beta\| = \|T(\alpha - \beta)\| = \|\alpha - \beta\|$ as T preserves norms.

Now $\qquad\qquad d(\alpha, \beta) = \|\alpha - \beta\|$

$\qquad\qquad d(T\alpha, T\beta) = \|T\alpha - T\beta\| = \|\alpha - \beta\| = d(\alpha, \beta).$

Hence T preserves distances.

THEOREM 1. *Let V and W be inner product spaces over the same field and let T be a linear transformation from V to W, then T preserves inner product if and only if*

$$\|T\alpha\| = \|\alpha\| \; \forall \; \alpha \in V$$

Proof. Given that $(T\alpha, T\beta) = (\alpha, \beta)$

i.e., T preserves inner products.

Choosing $\beta = \alpha$ we get $(T\alpha, T\alpha) = (\alpha, \alpha)$

$$\Rightarrow \qquad \|T\alpha\|^2 = \|\alpha\|^2 \Rightarrow \|T\alpha\| = \|\alpha\|,$$

i.e., T preserves norms also.

Conversly, let $\|T\alpha\| = \|\alpha\|$, *i.e.* T preserves norms.

To prove that $(T\alpha, T\beta) = (\alpha, \beta)$

Now we know from polarization identity, that

$$4(\alpha,\beta) = \|\alpha+\beta\|^2 + \|\alpha-\beta\|^2 + i\|\alpha+i\beta\|^2$$
$$- i\|\alpha-i\beta\|^2 \text{ for } V$$

$$4(T\alpha,T\beta) = \|T\alpha+T\beta\|^2 + \|T\alpha-T\beta\|^2 + i\|T\alpha+iT\beta\|^2$$
$$- i\|T\alpha-iT\beta\|^2 \text{ for } W$$

Since T is linear

$$\therefore \qquad T\alpha + T\beta = T(\alpha+\beta)$$

$$4(T\alpha,T\beta) = \|T(\alpha+\beta)\|^2 + \|T(\alpha-\beta)\|^2 + i\|T(\alpha+i\beta)\|^2$$
$$- i\|T(\alpha-i\beta)\|^2$$

Hence we get $(T\alpha, T\beta) = (\alpha, \beta)$.

i.e., T preserves inner products also.

THEOREM 2. *Let V and W be inner product vector spaces over the same field and let T be a linear transformation from V into W, then the following three conditions on T are equivalent.*   *--:*

(1) *T preserves inner products.*

(2) *T preserves norms.*

(3) *T is an isometry.*

Proof. **(1) $\Rightarrow$ (2).** Let T preserves inner products, *i.e*, $(T\alpha, T\beta) = (\alpha, \beta)$ given

Choosing $\beta = \alpha$ we get $(T\alpha, T\alpha) = (\alpha, \alpha)$.

or $\qquad \|T\alpha\|^2 = \|\alpha\|^2$ or $\|T\alpha\| = \|\alpha\|$, *i.e.,* T preserves norms.

(2) $\Rightarrow$ (3). Let T preserves norm, *i.e.,* $\|T\alpha\| = \|\alpha\|$ given.

Applying the above result on vector $\alpha - \beta$ we get

$$\|T(\alpha-\beta)\| = \|\alpha-\beta\|$$

or $\qquad \|T\alpha - T\beta\| = \|\alpha-\beta\|$ as T is linear.

i.e., T is an isometry.

(3) $\Rightarrow$ (1). Let T be an isometry, *i.e.,* $\|T\alpha - T\beta\| = \|\alpha-\beta\|$ given.

Choosing $\beta = 0$ we get $\|T\alpha\| = \|\alpha\|$

or $\qquad (T\alpha, T\alpha) = (\alpha, \alpha)$ ...(1)

We have now to prove that $(T\alpha, T\beta) = (\alpha, \beta)$, i.e., T preserves inner products also.

Now by (1) we have

$$(T(a + \beta), T(\alpha+\beta)) = (\alpha + \beta, \alpha+ \beta)$$

or $\quad (T\alpha + T\beta, T\alpha + T\beta) = (\alpha + \beta, \alpha+ \beta)$

or $\quad\quad (T\alpha, T\alpha)+(T\alpha, T\beta) + (T\beta, T\alpha) + (T\beta, T\beta)$

$$= (\alpha, \alpha) + (\alpha, \beta) + (\beta, \alpha) + (\beta, \beta)$$

or $\quad (T\alpha, T\beta) +(T\beta, T\alpha) = (\alpha, \beta) + (\beta, \alpha) \quad\quad$ by (1) $\quad\quad$ ...(2)

Now if the field F is that of reals then we know that

$$(T\beta, T\alpha) = (T\alpha, T\beta) \text{ and } (\beta, \alpha) = (\alpha, \beta)$$

Hence from (2) we get

$$2(T\alpha, T\beta) = 2(\alpha, \beta) \text{ or } (T\alpha, T\beta) = (\alpha, \beta).$$

But if the field F is that of complex numbers then putting $i\beta$ for β in (2) we get

$$(T\alpha, Ti\beta) + (Ti\beta, T\alpha) = (\alpha, i\beta) + (i\beta, \alpha)$$

or $\quad (T\alpha, iT\beta) + (iT\beta, T\alpha) = (\alpha, i\beta) + (i\beta, \alpha)$

or $\quad i(T\alpha, T\beta) + i(T\beta, T\alpha) = i(\alpha, \beta)+ i(\beta, \alpha)$

or $\quad -i(T\alpha, T\beta) + i(T\beta, T\alpha) = -i(\alpha, \beta)+i(\beta, \alpha)$

or $\quad -(T\alpha, T\beta) + (T\beta, T\alpha) = -(\alpha, \beta)+(\beta, \alpha) \quad\quad$ ...(3)

Subtracting (2) and (3) we get

$$2(T\alpha, T\beta) = 2(\alpha, \beta) \text{ or } (T\alpha, T\beta) = (\alpha, \beta)$$

Hence it preserves inner products.

THEOREM 3. *The relation of Isomorphism in the set of inner product vector spaces is an equivalence relation.*

Proof. Let $V \cong W \Rightarrow$ there exists a linear transformation T which is invertible and hence T^{-1} exists which is also invertible.

Also if $\quad V \cong W \Rightarrow \exists\, T_1$ which is invertible.

$\quad\quad\quad\quad W \cong U \Rightarrow \exists\, T_2$ which is invertible.

then $\quad T_2 T_1 : V \to U$ which is also invertible.

Also $(T_2T_1\, \alpha, T_2\, T_1\, \beta) = (T_2(T_1\alpha), T_2\,(T_1\beta))$

$\quad\quad\quad\quad = (T_1\alpha, T_1\beta)$ because $T_1\alpha, T_1\beta \in W$ and $W \cong U$ through T_2.

$\quad\quad\quad\quad = (\alpha, \beta) \quad \alpha, \beta \in V$ and $V \cong W$ through T_1.

Hence T_1T_2 preserves inner products.

(i) Reflexive. $V \cong V$ through identity transformation and we know that

$$(I\alpha, I\beta) = (\alpha, \beta)$$

(ii) Symmetric. $V \cong W$ through T then $W \cong V$ through T^{-1} which is invertible and also

$$[T^{-1}(T\alpha), T^{-1}(T\beta)] = (T^{-1}T\alpha, T^{-1}T\beta) = (I\alpha, I\beta) = (\alpha, \beta) = (T\alpha, T\beta)$$

(iii) Transitive.

$V \cong W$ through T_1 and $W \cong U$ through T_2 then $V \cong U$ through T_2T_1 which is also invertible and preserves inner products also.

Thus this relation of isomorphism is an equivalence relation.

THEOREM 4. *Let V be unitary vector space of dimension n with inner product (,). Then V is isomorphic to $V_n(C)$ with standard inner product.*

Proof. We know that every finite dimensional inner product vector space has an orthonormal basis set and choose

$B = \{\alpha_1, \alpha_2, ..., \alpha_n\}$ to be an orthonormal basis for V

$$(\alpha_i, \alpha_j) = \delta_{ij} \qquad ...(1)$$

Let β_1 be a vector in V so that it is expressible as a linear combination of the basis set B

$$\therefore \qquad \beta_1 = x_1\alpha_1 + x_2\alpha_2 + ... + x_n\alpha_n = \sum_{i=1}^{n} x_i\alpha_i$$

Now if we define a mapping $T : V \to V_n (C)$ such that $T(\beta_1) = (x_1, x_2,..., x_n)$, *i.e.*, n tuple of the co-ordinates of β_1 relative to basis B.

We have already proved earlier that T defined as above is a vector space isomorphism from V into $V_n(C)$. In order to prove it to be inner product vector space isomorphism we have only to establish that T preserves inner products also.

i.e., $\qquad (T\beta_1, T\beta_2) = (\beta_1, \beta_2)$

Where $\qquad \beta_2 = y_1\alpha_1 + y_2\alpha_2 + ...+ y_n\alpha_n = \sum_{j=1}^{n} y_j\alpha_j$

$\therefore \qquad T\beta_2 = (y_1, y_2, ..., y_n)$

$\therefore \qquad (T\beta_1, T\beta_2) = x_1\bar{y}_1 + x_2\bar{y}_2 +...+ x_n\bar{y}_n$

$$= \sum_{i=1}^{n} x_i\bar{y}_i \qquad ...(3)$$

by definition of standard inner product in $V_n (C)$.

Again $\qquad (\beta_1, \beta_2) = (\sum_i x_i\alpha_i, \sum_i y_j\alpha_j)$

$$= \sum_i \sum_j x_i\bar{y}_j(\alpha_i\alpha_j)$$

Summing *w.r.t. j* all the terms is R.H.S. will vanish except when $j = i$ as $(\alpha_i, \alpha_j) = \delta_{ij}$.

$\therefore \qquad (\beta_1, \beta_2) = \Sigma x_i\bar{y}_i (\alpha_i, \alpha_i) = \Sigma x_i\bar{y}_i, 1 \qquad ...(4)$

From (3) and (4) we get

$$(T\beta_1, T\beta_2) = (\beta_1, \beta_2)$$

i.e., T is an isomorphism preserves inner products also and as such it is an inner product vector space isomorphism.

8.9 UNITARY OPERATORS

If z be any complex number then z will be unimodular ,

i.e., $\qquad |z| = 1$ if $\bar{z}z = z\bar{z} = 1$

From this concept we consider the type of operators for which $TT^* = T^*T = I$. Such operators T will be called unitary operators for complex inner product vector space and orthogonal for real inner product vector space.

Definition 1. *An isomorphism of an inner product vector space V onto itself is called a unitary operator.*

Definition 2. *A linear operator on an inner product vector space which is one-one-onto (invertible), preserves vector space compositions and also preserves inner product (i.e., preserves norm) is called a unitary operator.*

Thus $T : V \to V$ is unitary if

(1) T is invertible and linear

(2) T preserves inner products, *i.e.* preserves norms.

THEOREM 1. *Let T be a linear operator on an inner product vector space V then T is unitary if and only if the adjoint of T, i.e. T^* exists and $TT^* = T^*T = I$.*

Proof. Given T is unitary $\Rightarrow$ T is invertible,

i.e. $\qquad\qquad TT^{-1} = T^{-1}T = I$ $\qquad\qquad\qquad$ ... (1)

and also $\qquad (T\alpha, T\beta) = (\alpha\ \beta)$ $\qquad\qquad\qquad\qquad$ ...(2)

To prove that $\qquad TT^* = T^*T = I$

i.e. inverse of T is adjoint of T, *i.e.* T^*

Now $\qquad\qquad (T\alpha, \beta) = (T\alpha, TT^{-1}\beta) = (\alpha, T^{-1}\beta) \ \forall\ \alpha, \beta$ by (2).

But $\qquad\qquad (T\alpha, \beta) = (\alpha, T^*\beta) = (\alpha, T^{-1}\beta)$

$\therefore \qquad\qquad\qquad T^{-1} = T^* =$ adjoint of T.

Putting in $\quad TT^{-1} = T^{-1}T = I =$ we get $TT^* = T^*T=I$

Conversely , Given $\quad TT^* =T^*T = I$

i.e., T is invertible and $T^{-1} = T^*$.

We have to prove that T is unitary.

T is invertible and also $(T\alpha, T\beta) = (\alpha, T^*T\beta) = (\alpha, I\beta) = (\alpha, \beta)$

$\therefore$ T preserves inner products also.

Hence T is unitary.

Theorem 2. *The set of all unitary operators on an inner product vector space forms a group.*

Proof. **(1) Closure property.**

Let T_1 and T_2 be unitary then they are invertible and preserve inner products or norm, then T_2T_1 or T_1T_2 is also invertible.

Also $\qquad \|T_1T_2\alpha\| = \|T_1(T_2\alpha)\| = \|T_1\beta\|$ where $\beta = T_2(\alpha)$

$\qquad\qquad\qquad = \|\beta\| = \|T_2\alpha\| = \|\alpha\| \ \forall\ \alpha$

Hence $T_2 T_1$ is also unitary, thus the closure property is satisfied.

(2) Associative.

Multiplication of operators is always associative.

(3) Existance of Inverse.

If T be unitary then T being invertible $\Rightarrow T^{-1}$ is also invertible. Let us now show that

$$\left\|T^{-1}\alpha\right\| = \|\alpha\| \forall \alpha \in V$$

Let $T^{-1}\alpha = \beta$ so that $\alpha = T\beta$

$\therefore \qquad\qquad\qquad \|T\beta\| = \|\alpha\|^* \qquad\qquad\qquad\qquad$ ...(1)

But T is unitary

$$\therefore \qquad \|T\beta\| = \|\beta\| \text{ or } \|T^{-1}\alpha\| = \|\alpha\| \text{ by (1)}$$

Hence T^{-1} is also unitary.

(4) Existence of Identity.

Since identity operator is invertible and also $\|I\alpha\| = \|\alpha\|$

i.e., it preserves norms and hence it is also unitary.

Thus the set of all unitary operators form a group.

THEOREM 3. *If V is any inner product vector space and T is a linear operator on V then the following conditions are equivalent.*

(1) $T^*T = I$

(2) $(T\alpha,\ T\beta) = (\alpha,\ \beta)\ \forall\ \alpha,\ \beta \in V.$

(3) $\|T\alpha\| = \|\alpha\|\ \forall \alpha \in V$

Proof. **(1)** $\Rightarrow$ **(2).** Let $\qquad T^*T = I$

Now $\qquad (T\alpha, T\beta) = (T^*T\alpha, \beta) = (I\alpha, \beta) = (\alpha,\ \beta)$

(2) $\Rightarrow$ **(3).** $\qquad (T\alpha, T\beta) = (\alpha,\ \bar{\beta})$ putting $\beta = \alpha$

$$(T\alpha, T\alpha) = (\alpha, \alpha) \Rightarrow \|T\alpha\|^2 = \|\alpha\|^2$$

$$\Rightarrow \|T\alpha\| = \|\alpha\|$$

(3) $\Rightarrow$ **(1).** $\qquad \|T\alpha\| = \|\alpha\| \Rightarrow \|T\alpha\|^2 = \|\alpha\|^2$

$\Rightarrow \qquad (T\alpha, T\alpha) = (\alpha, \alpha) \Rightarrow (T^*T\alpha, \alpha) = (I\alpha, \alpha)$

$\Rightarrow \qquad ((T^*T - I)\alpha, \alpha) = 0 \forall \alpha$

$\Rightarrow \qquad T^*T - I = 0 \Rightarrow T^*T = I$

THEOREM 4. *A linear operator T on finite dimensional inner product vector space (f. I. P. V. S) is unitary if and only if T preserves inner products.*

Proof. Since T is unitary therefore by definition T preserves inner products.

Conversely. T preserves inner products, then T will be unitary if we are able to show that T is invertible also.

Now $\qquad (T\alpha, T\beta) = (\alpha, \beta) \Rightarrow (T\alpha, T\alpha) = (\alpha, \alpha)$

$\Rightarrow \qquad \|T\alpha\| = \|\alpha\|$

if $\alpha \neq 0$, then $\|\alpha\| \neq 0$ hence $\|T\alpha\| \neq 0 \Rightarrow T\alpha \neq 0$

Hence if $\alpha \neq 0$, $T\alpha \neq 0$. It means that T is non-singular which implies that T is one-one.

Now as V is finite dimensional therefore T is one-one implies that T is onto as well. Hence T is invertible.

Finally T is invertible and preserves inner products also and therefore by definition T is unitary.

Now, to prove T is unitary.

If T be unitary then $\|T\alpha\| = \|\alpha\|$

$$\therefore \qquad \|T\alpha - T\beta\| = \|T(\alpha - \beta)\| = \|\alpha - \beta\|$$

i.e., distance between vectors α, β and their T images are preserved under unitary operator. On account of this T is called isometry.

THEOREM 5. *An operator T on an inner product vector space T is unitary if and only if it is an isometric isomorphism of V onto itself.*

Proof. **Case 1. T is Unitary.** By definition T is an isomorphism (inner product) which implies that it preserves norm and as shown above it preserves distance. Hence it is an isometric isomorphism.

Case 2. T is isometric isomorphism.

Since T is isomorphism and as such T is invertible. Again it being isometric therefore it preserves distances or norm or inner products which in turn means that $T^*T = I$ and hence $(T * T) T^{-1} = IT^{-1}$

or $\qquad T * = T^{-1}$, *i.e.* inverse of T is adjoint of T.

Also $\qquad TT^* = TT^{-1} = I$

Thus $\qquad TT^* = T * T = I$ and hence T is unitary.

THEOREM 6. *Let V be finite dimensional inner product vector space and T be a linear operator on V then T is unitary if and only if matrix A of T in some (or every) ordered orthonormal basis is a unitary matrix.*

Proof. Given T is unitary $\Rightarrow T^{-1} = T^*$ and $(T\alpha, T\beta) = (\alpha, \beta)$

Also V is finite dimensional inner product vector space and hence it has an orthonormal basis say

$\{\alpha_1, \alpha_2, \dots \alpha_n\}$ where $(\alpha_i, \alpha_j) = \delta_{ij}$

$\therefore \qquad T\alpha_i = \Sigma a_{ij}\alpha_i, , j = 1, 2\dots n.$

Now $\qquad \delta_{ij} = (\alpha_i, \alpha_j) = (T\alpha_i, T\alpha_j)$

$$= (\sum_k a_{ki}\alpha_k, \sum_p a_{pj}\alpha_p) = \sum_k a_{ki}(\alpha_k, \sum_p a_{pj}\alpha_p)$$

Summing w.r.t. p all the terms will vanish except one when $p = k$.

$$\therefore \qquad \delta_{ij} = \sum_k a_{ki}(\alpha_k, a_{kj}a_k) = \sum_k a_{ki}.\overline{a}_{kj}.1 = \overline{\Sigma a_{kj}a_{kj}}$$

$$= j - i \text{ the element of matrix } A * A$$

Thus $\qquad \delta_{ij} = [A * A]_{j-i}$ where $\delta_{ij} = 0, i \neq j, \delta_{ij} = 1, i = j$

Above shows that $[A^*A]$ is unitary matrix

THEOREM 7. *An operator T on a finite dimensional inner product vector space is unitary if and only if $\{T(\alpha_i)\}$ is a complete orthonormal basis set whenever $\{\alpha_i\}$ is.*

Proof. T is unitary $\Rightarrow T$ is an isometric isomorphism $\Rightarrow T$ is invertible $\Rightarrow T$ is non-singular $\Rightarrow T$ carries a linearly independent set k_{ij} to $\{T(\alpha_i)\}$ which is also linear independent and in case $\{\alpha_i\}$ is a basis then $\{T(\alpha_i)\}$ is also a basis as a linearly independent set consisting of n elements of a finite diminsional vector space of dim. n constitutes a basis for the same.

Further if $\{\alpha_i\}$ is orthonormal then $(\alpha_i, \alpha_j) = \delta_{ij}$

Also $\qquad (T\alpha_i, T\alpha_j) = (\alpha_i, \alpha_j) = \delta_{ij}$

Hence the set $\{T \alpha_i\}$ is also orthonormal.

Again if $\{\alpha_i\}$ is complete then we have to show that $\{T\alpha_i\}$ is also complete. If it be not so then $\exists$ a non zero vector α, s.t.$\{\alpha, T\alpha_1, T\alpha_2, \dots, T\alpha_n\}$ is orthonormal.

$$\Rightarrow \qquad\qquad (\alpha, T\,\alpha_i) = 0 \;\Rightarrow (T^*\,\alpha, \alpha_i) = 0 \;\forall i$$

But $\{\alpha_i\}$ is complete and hence if $(T^*\,\alpha, \alpha_i) = 0 \;\forall\; i$ then it implies that $T^*\alpha = 0$

$$\Rightarrow \quad \alpha = 0 \text{ and hence contradiction as } \alpha \neq 0.$$

Therefore $\{T\alpha_i\}$ constitutes a complete orthonormal set.

Conversely $\{\alpha_i\}$ and $\{T\alpha_i\}$ are complete orthonormal basis so that

$$(\alpha_i, \alpha_j) = \delta_{ij}, \; (T\alpha_i\, T\alpha_j) = \delta_{ij} \qquad\qquad ...(1)$$

We have to prove that T is unitary, *i.e.*, $(T\alpha, T\beta) = (\alpha, \beta)$

Now let $\qquad \alpha = \displaystyle\sum_{i=1}^{n} x_i\alpha_i, \; \beta = \sum_{j=1}^{n} y_j\alpha_j.$

$$\therefore \qquad\qquad (\alpha, \beta) = \left(\sum_i x_i\alpha_i, \sum_j y_j\alpha_j \right)$$

$$= \sum_i \sum_j x_i \bar{y}_j (\alpha_i, \alpha_j)$$

Summing *w.r.t.* j all the terms will vanish because $(\alpha_i, \alpha_j) = \delta_{ij}$ except one when $j = i$ and in that case $(\alpha_i, \alpha_j) = 1$

$$\therefore \qquad\qquad (\alpha, \beta) = \sum_i x_i \bar{y}_i .1 \qquad\qquad\qquad ... (2)$$

$$(T\alpha, T\beta) = \left(T \sum_i x_i\alpha_i, T \sum_j y_j\alpha_j \right)$$

$$= \left(\sum_i x_i T\alpha_i, \sum_j y_j T\alpha_j \right)$$

$$= \sum_i \sum_j x_i y_j (T\alpha_i, T\alpha_j)$$

Summing *w.r.t.* j all the terms will vanish because $(T\alpha_i, T\alpha_j) = \delta_{ij}$ except one when $j = i$ and in that case $(T\alpha_i, T\alpha_j) = 1$ ...(3)

$$\therefore \qquad (T\alpha, T\beta) = \Sigma x_i \bar{y}_i$$

From (2) and (3) we conclude that $(T\alpha, T\beta) = (\alpha, \beta)$ and hence T is unitary

REMARKS

- The above theorem can be restated as: we know that
- The linear operator T on a finite dimensional inner product vector space is unitary if and only if it takes an orthonormal basis into an orthonormal basis of V.
- Let V and W be finite dimensional inner product vector space, over the field F and having the same dimensions. If T is a linear transformation from V to W then the following are equivalent :
  1. T preserves inner products.
  2. T is an inner product vector space isomorphism.
  3. T carries every orthonormal basis of V onto an orthonormal basis for W.
  4. There is at least one orthonormal basis of V which is transferred to orthonormal basis W.

8.10 NORMAL OPERATORS

Let V be a finite dimensional inner product vector space and T be a linear operator on V which commutes with its adjoint, *i.e.*, $TT^* = T^*T$ then T is said to be a normal operator.

☛ **ILLUSTRATIONS**

1. Every self-adjoint operator is normal, *i.e.*, $T = T^* \Rightarrow TT^* = T^*T$.

2. Every unitary operator is normal.

   T is unitary $= TT^* = T^*T = I$ and hence normal.

3. Zero and identity operators being self-adjoint are normal.

4. Every scalar multiple of a self-adjoint or normal operator is a normal operator. Given T is self adjoint or normal

   $$\therefore \qquad\qquad T = T^* \text{ or } TT^* = T^*T \qquad\qquad\qquad ...(1)$$

   Now αT will be normal if

   $$(\alpha T)(\alpha T)^* = (\alpha T)^*(\alpha T)$$

   L.H.S. $(\alpha T)(\bar{\alpha} T^*) = \alpha \bar{\alpha} TT^*$

   R.H.S. $(\bar{\alpha} T^*)(\alpha T) = \bar{\alpha} \alpha T^*T$

   By virtue of condition (1) L.H.S. = R.H.S. and hence αT is normal.

THEOREM 1. *If T_1 and T_2 be normal operators on an inner product vector space with the property that either commutes with the adjoint of the other, then prove that $T_1 + T_2$ and $T_1 T_2$ are normal operators. Also $T - \lambda I$ is normal if T be normal, where $\lambda \in F$.*

Proof.   Given that $T_1 T_1^* = T_1^* T_1$ and $T_2 T_2^* = T_2^* T_2$ ... (1)

                              $T_1 T_2^* = T_2^* T_1$ and $T_2 T_1^* = T_1^* T_2$ ... (2)

To prove that $(T_1 + T_2)$ is normal.

$(T_1 + T_2)$ will be normal if $(T_1 + T_2)(T_1 + T_2)^* = (T_1 + T_2)^*(T_1 + T_2)$

L.H.S. $= (T_1 + T_2)(T_1^* + T_2^*)$

                          $= T_1 T_1^* + T_1 T_2^* + T_2 T_1^* + T_2 T_2^*$

R.H.S. $= (T_1^* + T_2^*)(T_1 + T_2)$

                          $= T_1^* T_1 + T_1^* T_2 + T_2^* T_1 + T_2^* T_2$

                          $= T_1 T_1^* + T_2 T_1^* + T_1 T_2^* + T_2 T_2^*$ by 1 and 2.

L.H.S. = R.H.S. Hence $T_1 + T_2$ is normal.

Now, to prove that $T_1 T_2$ is normal.

We know that $T_1 T_2$ will be normal if

         $(T_1 T_2)(T_1 T_2)^* = (T_1 T_2)^*(T_1 T_2)$

L.H.S. $T_1 T_2 T_2^* T_1^* = T_1(T_2 T_2^*) T_1^*$

                     $= T_1(T_2^* T_2) T_1^*$ by (1)

                     $= (T_1 T_2^*)(T_2 T_1^*)$

                     $= T_2^* T_1 T_1^* T_2$ by (2)

                     $= T_2^* T_1^* T_1 T_2$ by (1)

                     $= (T_1 T_2)^*(T_1 T_2) =$ R.H.S.

Hence $T_1 T_2$ is also normal under given conditions.

Finally prove that $T - \lambda I$ is normal.

Consider, $(T - \lambda I)^* = T^* - (\lambda I)^* = T^* - \bar{\lambda} I I^* = I$

$\therefore \quad (T - \lambda I)^* (T - \lambda I) = (T^* - \bar{\lambda} I)(T - \lambda I)$

$\qquad\qquad\qquad = T^* T - \bar{\lambda} IT - \lambda T^* I + \lambda \bar{\lambda} I$

$\qquad\qquad\qquad = T^* T - \bar{\lambda} T - \lambda T^* + \lambda \bar{\lambda}$ ... (3)

$(T - \lambda I)(T - \lambda I)^* = (T - \lambda I)(T^* - \bar{\lambda} I)$

$\qquad\qquad\qquad = TT^* - \lambda IT^* - \bar{\lambda} TI + \bar{\lambda} I \lambda$

$\qquad T^* T - \lambda T^* - \bar{\lambda} T + \lambda \bar{\lambda} I$ ... (4)

$\therefore \qquad\qquad TT^* = T^*T$ as T^* is normal

Hence from (3) and (4) we get that $T - \lambda I$ is normal.

THEOREM 2. *An operator T on an inner product vector space is normal*

$\qquad\qquad \Leftrightarrow \| T^* x \| = \| Tx \| \; \forall x$

and hence $Tx = 0 \Leftrightarrow T^* x = 0$ when T is normal.

Proof. **Case I.** Given $\| T^* x \| = \| Tx \| \; \forall x$

Now $\qquad\qquad \| T^* x \| = \| Tx \| \Rightarrow \| T^* x \|^2 = \| Tx \|^2$

$\Rightarrow \qquad (T^* x, T^* x) = (Tx, Tx) \Rightarrow (TT^* x, x) = (T^* Tx, x)$

$\Rightarrow \quad ((TT^* - T^* T)x, x) = 0 \, \forall x \Rightarrow TT^* - T^* T = O$

$\Rightarrow \qquad\qquad\qquad TT^* = T^*T$

and hence T is normal.

Therefore, if $\quad (Tx, x) = 0 \; \forall \; x$ then $T = 0$

Case II. Given that T is normal, i.e. $TT^* = T^* T$.

$\| Tx \|^2 = (Tx, Tx) = (x, T^* Tx) = (x, TT^* x) = (T^* x, T^* x)$

$\qquad\qquad = \| T^* x \|^2$

$\| Tx \| = \| T^* x \|$

Now $\qquad\qquad Tx = 0 \Leftrightarrow \| Tx \| = 0 \Leftrightarrow \| T^* x \| = 0 \Leftrightarrow T^* x = 0$

THEOREM 3. *If T is an operator on an inner product vector space then T is normal if and only if its real and imaginary parts commute.*

Proof. We have already proved that any transformation can be written as $T = A + iB$ where A and B are self-adjoint are called the real and imaginary parts of T so $T^* = (A + iB)^* = A^* + (iB)^* = A^* + iB^* = A^* - iB^* = A - iB$, as A and B are self-adjoint.

$TT^* = (A + iB)(A - iB) = A^2 + B^2 + i(BA - AB)$ ...(1)

$T^* T = (A - iB)(A + iB) = A^2 + B^2 + i(AB - BA)$ ...(2)

If $AB = BA$, i.e. real and imaginary parts commute then (1) and (2) are equal, i.e. $TT^* = T^* T$ and hence T is normal.

Conversely if T is normal, i.e., $TT^* = T^* T$ then from (1) and (2) we get

$\qquad\qquad BA - AB = AB - BA = 2BA = 2AB$

$\Rightarrow \qquad\qquad\qquad BA = AB$

i.e. real and imaginary parts commute.

THEOREM 4. *Let T be a normal operator and α be a vector such that*

$$T^2\alpha = 0 \text{ then } T\alpha = 0$$

Hence the range and null space of a normal operator are disjoint.

Proof. Given T is normal

$$\Leftrightarrow \qquad \|Tx\| = \|T*x\| \forall x \qquad \qquad \dots (1)$$

Also $\qquad T^2\alpha = 0 \qquad \qquad \dots (2)$

To prove that $\quad T\alpha = 0$

Now $\qquad T^2\alpha = 0 \Rightarrow T(T\alpha) = 0$

$\Rightarrow \qquad T(\beta) = 0$ where $\beta = T(\alpha) \in V$ as T is a linear operator.

Since $\qquad T(\beta) = 0$ is follows from (1) that

$$\|T*\beta\| = \|T\beta\| = 0$$

$\therefore \qquad T*\beta = 0$ as $\|x\| = 0 \Leftrightarrow x = 0$

$\therefore \qquad (T*\beta, \alpha) = 0$ or $(\beta, T\alpha) = 0$ or $(\beta, \beta) = 0$ as $\beta = T\alpha$

$$\|\beta\|^2 = 0 \text{ or } \beta = 0 \text{ or } T\alpha = 0$$

Let $\beta \neq 0$ belong to null space of T

$\therefore \qquad T\beta = 0.$

If β belongs to range space of T then $\beta = T\alpha$ for some α

$\therefore \qquad T\beta = T(T\alpha) = T^2\alpha$

$\therefore \qquad T\beta = 0 \Rightarrow \qquad T^2\alpha = 0 \Rightarrow T\alpha = 0 \Rightarrow \beta = 0$

$\therefore \quad \beta = 0$ which is a contradiction.

Hence no non-zero vector can belong to both range and null space of T under given condition, *i.e.*, they are disjoint.

8.11 POSITIVE OPERATORS

A linear operator T on an inner product vector space is called a positive operator, *i.e*, $T \geq O$ if

(i) T is self-adjoint.

(ii) $(Tx, x) \geq 0 \ \forall \ x \in V$

For example.

(1) $O, I, TT*$ and T are all self-adjoint operators.

As $\qquad (Ox, x) = 0, (Ix, x) = (x, x) = \|x\|^2 \geq 0$

$$(TT*x, x) = (T*x, T*x) = \|T*x\|^2 \geq 0$$

$$(T*Tx, x) = (Tx, Tx) = \|Tx\|^2 \geq 0$$

and as such they satisfy the second condition also and therefore all the above operators are positive operators.

THEOREM 1. *T is a linear operator on an inner product vector space p is a functional which sends ordered pairs of V to the field, i.e., $p : V \times V \to F$ defined as $p(\alpha, \beta) = (T\alpha, \beta)$. The necessary and sufficient conditions which T must satisfy so that p be an inner product on V are $(T\alpha, \alpha) \geq$ and $T* = T$.*

Proof. 1st Case. Let p be an inner product on V and as such it must satisfy the following axioms of inner product.

1. $p\,(\alpha, \alpha) \geq 0 \Rightarrow (T\alpha, \alpha) \geq 0$

2. $p(\alpha, \beta) = p\,\overline{(\beta, \alpha)}$

$\Rightarrow \qquad (T\alpha, \beta) = (T\overline{\beta, \alpha}) = (\alpha, T\beta)$

$\Rightarrow \qquad T$ is self adjoint, *i.e.*, $T^* = T$

3. $\qquad p\,(c\alpha + \gamma, \beta) = cp(\alpha, \beta) + p(\gamma, \beta)$

$\qquad p\,(c\alpha + \gamma, \beta) = (T(c\alpha + \gamma), \beta) = (cT\alpha + T\gamma, \beta)$

$\qquad\qquad\qquad\qquad = c(T\alpha, \beta) + (T\gamma, \beta)$

$\qquad\qquad\qquad\qquad = cp(\alpha, \beta) + p(\gamma, \beta)$

Hence the third property holds good but the first two impose restrictions on T which are that $(T\alpha, \alpha) \geq 0$ and $T^* = T$ and both these imply that T should be a positive operator in accordance with the definition of $+ive$ operators.

2nd Case. Suppose T is a linear operator on V which satisfies the conditions $T = T^*$ and $(T\alpha, \alpha) \geq 0 \;\forall\; \alpha$, then the function p defined as $p(\alpha, \beta) = (T\alpha, \beta) \;\forall\; \alpha, \beta \in V$ defines an inner product on V.

1. $p(\alpha, \alpha) = (T\alpha, \alpha) \geq 0$ (given) $\Rightarrow p\,(\alpha, \alpha) \geq 0 \;\forall\; \alpha$

2. $p(\alpha, \beta) = (T\alpha, \beta) = (\alpha, T^*\overline{\beta}) = (\alpha, T\beta); \; T^* = T$

$\qquad = \overline{(T\beta, \alpha)} = p\overline{(\beta, \alpha)}$

3. $p(c\alpha + \gamma, \beta) = (T\,(c\alpha + \gamma), \beta) = (c\,T\alpha + T\gamma, \beta)$

$\qquad\qquad = c(T\alpha, \beta) + (T\gamma, \beta) = cp\,(\alpha, \beta) + p\,(\gamma, \beta)$

Hence the function p defined as above defines an inner product on V under given conditions.

Therefore we can say that the following conditions are equivalent :

(a) $(T\alpha, \alpha) \geq 0, T^* = T$

(b) $p(\alpha, \beta) = (T\alpha, \beta)$ defines an inner product on V.

THEOREM 2. *Let V be a finite dimensional inner product vector space with inner product $(,)$. If p is any inner product on V; there is a unique positive linear operator T on V such that $p\,(\alpha, \beta) = (T\alpha, \beta) \;\forall\; \alpha, \beta \in V$.*

Proof. Let us fix up a vector β in V then $p(\alpha, \beta)$ gives a linear functional on V then there exists a unique vector β' in V s.t. $p(\alpha, \beta) = (\alpha, \beta') \;\forall\alpha$.

Now we define a function T from V into V by $T\beta = \beta'$

Hence $\qquad\qquad p(\alpha, \beta) = (\alpha, \beta') = (\alpha, T\beta)$ ...(1)

Now $\qquad\qquad\qquad \overline{p(\alpha, \beta)} = p(\beta, \alpha)$

$\therefore \qquad\qquad\qquad \overline{(\alpha, T\beta)} = (\beta, T\alpha) \text{ by (1)}$

or $\qquad\qquad\qquad (T\beta, \alpha) = (\beta, T\alpha)\,.$

Take conjugates $\quad (\alpha, T\beta) = (T\alpha, \beta)$

$\qquad\qquad\qquad p(\alpha, \beta) = (T\alpha, \beta) \text{ by (1)},$

When this condition is satisfied then by last theorem T is a $+ive$ operator.

Linearity of T.

$\qquad\qquad (T\,(c\alpha_1 + \alpha_2)\beta) = pc\,((c\alpha_1 + \alpha_2), \beta)$

$\qquad\qquad\qquad\qquad\qquad = c\,p(\alpha_1, \beta) + p(\alpha_2, \beta)$

$\qquad\qquad\qquad\qquad\qquad = c(T\alpha_1, \beta) + (T\alpha_2, \beta)$

Uniqueness of T

Suppose that there are two such operators T and u, so that $p(\alpha, \beta) = (T\alpha, \beta) = (u\alpha, \beta)\ \forall \alpha, \beta, \in V$.

$\Rightarrow\qquad (T\alpha - u\alpha, \beta) = 0 \Rightarrow ((T - u)\alpha, \beta) = 0\forall\ \alpha, \beta \in V$

$\Rightarrow\qquad\qquad T - u = 0 \Rightarrow T = u$

THEOREM 3. *Let V be a finite dimensional inner product vector space and T be a linear operator on V. Then T is positive if and only if there is an invertible linear operator u on V such that $T = u^*u$*

Proof. Let us first suppose $T = u^*u$

To prove that T is positive, *i.e.* T is self-adjoint and $(T\alpha, \alpha) \geq 0$.

Now $\qquad\qquad\qquad T = u^*u \Rightarrow T^* = (u^*u)^* = u^*u^{**} = u^*u = T$

Hence T is self-adjoint.

Now $\qquad\qquad (T\alpha, \alpha) = (u^*u\alpha, \alpha) = (u\alpha, u^{**}\alpha) = (u\alpha, u\alpha) \geq 0$

Since u is invertible $\Rightarrow u$ is non-singular $\Rightarrow u\alpha \neq 0$, whenever $\alpha \neq 0$.

$\therefore\qquad\qquad (T\alpha, \alpha) = (u\alpha, u\alpha) > 0,\ u\alpha \neq 0$ when $\alpha \neq 0$

Since T is self Adjoint and $(T\alpha, \alpha) > 0$, hence T is positive.

Conversely, T is positive operator which implies that there exists an inner product on V defined as $p(\alpha, \beta) = (T\alpha, \beta)\forall\ \alpha, \beta$. Again V is finite dimensional and hence $\exists$ an orthonormal basis $\{\alpha_1, \alpha_2, \ldots \alpha_n\}$ w.r.t. inner product and let $\{\beta_1, \beta_2, \ldots \beta_n\}$ be an orthonormal basis w.r.t p, then

$$(a_i, a_j) = \delta_{ij},\ p(\beta_i, \beta_j) = \delta_{ij}$$

We have to prove that $T = u^*u$, where u is an invertible linear operator. Now define a linear operator. u on V such that $u\beta_j = \alpha_j, j = 1, 2 \ldots n$.

Since u carries a basis onto a basis therefore u is invertible and we have to show that $T = u * u$.

Now $\qquad\qquad p(\alpha, \beta) = p(\Sigma x_i\beta_i, \Sigma y_j\beta_j)$

$\{\beta_1, \beta_2, \ldots \beta_n\}$ is an orthonormal basis w.r.t. inner product p.

$\therefore\qquad\qquad\qquad p(\alpha,\beta) = \sum_i\sum_j x_i\overline{y}_j, p(\beta_i.\beta_j)\qquad\qquad\qquad\qquad \ldots(1)$

But $\qquad\qquad p(\beta_i, \beta_j) = \delta_{ij} = (\alpha_i, \alpha_j) = (u\beta_i, u\alpha_i)$ by def. of u.

$\therefore\qquad\qquad p(\alpha,\beta) = \sum_i\sum_j x_i\overline{y}_j, (u\beta_i.u\beta_j)$

$\qquad\qquad\qquad\qquad = (\Sigma x_i u\beta_i, \Sigma y_j u\beta_j)$

$\qquad\qquad\qquad\qquad = (u\Sigma x_i\beta_i, u\Sigma y_j\beta_j)$

$\qquad\qquad\qquad\qquad = (u\alpha, u\beta) = (u*u\alpha, \beta)$

$\therefore\qquad\qquad p(\alpha,\beta) = (u*u\alpha, \beta)$

But by definition $p(\alpha, \beta) = (T\alpha, \beta)\forall\alpha, \beta$ as T is +ive.

$\therefore\qquad\qquad (u^*u\ \alpha, \beta) = (T\alpha, \beta)$

or $\qquad ((u*u - T)\ \alpha, \beta) = 0\forall \beta$

$\therefore\qquad\qquad (u^*u - T) = O\quad$ or $\quad T = u * u$

Solved Examples

Based on the following Results

➥ An operator T is said to be unitary if $T*T = TT^* = I$

➥ An operator T is said to be normal if $T*T = TT^*$

➥ Every adjoint operator is normal.

➥ Every unitary operator is normal.

➥ Zero and identity operators are normal.

➥ T is normal $\Leftrightarrow \|T^* x\| = \|Tx\| \ \forall \ x$

➥ T is normal $\Leftrightarrow$ real and imaginary parts commutes.

➥ Range and null space of a normal operator are disjoint.

➥ An operator T is said to be positive if (i) T is self-adjoint (ii) $(Tx, x) \geq 0$

Example 1. If T_1, T_2 are positive linear operators on an inner product vector space then prove that $T_1 + T_2$ is also positive.

Solution. Given $T_1^* = T_1$, $(T_1\alpha, \alpha) > 0$, $T_2^* = T_2$, $(T_2\alpha; \alpha) > 0$

Now $(T_1 + T_2)^* = T_1^* + T_2^* = T_1 + T_2$

$\therefore \quad T_1 + T_2$ is self-adjoint.

Again $((T_1 + T_2)\alpha, \alpha) = (T_1\alpha + T_2\alpha, \alpha)$

$\qquad\qquad = (T_1\alpha, \alpha + T_2\alpha, \alpha)$ by given conditions.

Hence $T_1 + T_2$ is also positive.

Example 2. If T is an arbitrary operator on V and if α and β are scalars such that $|\alpha| = |\beta|$, prove that $\alpha T + \beta T^*$ is normal.

Solution. We have $\quad |\alpha| = |\beta| \Rightarrow |\alpha|^2 \Rightarrow |\beta|^2 \Rightarrow \alpha\bar{\alpha} = \beta\bar{\beta}$ ... (1)

Now $(\alpha T + \beta T^*)(\alpha T + \beta T^*)^*$

$\qquad\qquad = (\alpha T + \beta T^*)(\bar{\alpha}T^* + \bar{\beta}T) \ T^{**} = T, \text{and} (\alpha T)^* = \bar{\alpha}T^*$

$\qquad\qquad = \alpha\bar{\alpha}TT^* + \alpha\bar{\beta}T^2 + \beta\bar{\alpha}T^{*2} + \beta\bar{\beta}T^*T$

Again $(\alpha T + \beta T^*)^*(\alpha T + \beta T^*)$

$\qquad\qquad = (\bar{\alpha}T^* + \bar{\beta}T)(\alpha T + \beta T^*)$

$\qquad\qquad = \bar{\alpha}\alpha T^*T + \bar{\beta}\alpha T^2 + \alpha\bar{\beta}T^{*2} + \beta\bar{\beta}TT^*$

$\qquad\qquad = \beta\bar{\beta}T^*T + \alpha\bar{\beta}T^2 + \beta\bar{\alpha}T^{*2} + \bar{\alpha}\alpha TT^*$ by (1) ... (3)

Comparing (2) and (3) we find that all the terms are same and hence the operator $\alpha T + \beta T^*$ is such that it commutes with its adjoint under given condition and therefore it is a normal operator.

Example 3. Let V be the set of complex numbers regarded as a real vector space.

Prove that $(\alpha, \beta) = Re (\alpha\bar{\beta})$ defines an inner product on V.

Solution. In order to prove that $(\alpha, \beta) = Re (\alpha\bar{\beta})$ defines an inner product we have to verify the axioms of inner product

1. $(\alpha + \gamma, \beta) = \text{Re } (\alpha + \gamma)\bar{\beta} = \text{Real } (\alpha\bar{\beta} + \gamma\bar{\beta})$

$$= \text{Real } \alpha\bar{\beta} + \text{Real } (\gamma\bar{\beta}) = (\alpha, \beta) + (\gamma, \beta)$$

2. $(\beta, \alpha) = \text{Re } (\beta, \bar{\alpha}) = (\text{Re } (\overline{\beta\alpha}))$ Real parts of two conjugates are equal.

$\therefore$ $(\beta, \alpha) = \text{Re } (\overline{\beta\alpha}) = \text{Re}\overline{(\beta\alpha)} = \text{Re}(\alpha\bar{\beta}) = (\alpha,\beta)$

The property $(\overline{\beta, \alpha}) = (\alpha,\beta)$ is changed to $(\beta, \alpha) = (\alpha,\beta)$ when the field of vector space is real, *i.e.* for $V(R)$.

3. $(c_j, \beta) = \text{Re } (c\alpha)\bar{\beta} = c \text{ Re } (\alpha\bar{\beta}) = c(\alpha, \beta)$

4. $(\alpha, \alpha) = \text{Re}(\alpha\bar{\alpha}) = \text{Re } |\alpha|^2 \geq 0$. It will be zero if and only if $\alpha = 0$.

Hence $(\alpha, \beta) = \text{Re } (\alpha, \bar{\beta})$ defines an inner praduct on $V(R)$.

Example 4. *V is finite dimensional inner product vector space and $V = W + W^{\perp}$, i.e.,*

$$\alpha \quad = \quad \beta \quad + \quad \gamma$$
$$\in V \quad \in W \quad \in W^{\perp}$$

Define a linear operator u on V by $u\alpha = \beta - \gamma$. Show that u is both self-adjoint and unitary.

Solution. Let $\alpha_1 = \beta_1 + \gamma_1, \alpha_2 = \beta_2 + \gamma_2$

$\therefore$ $u\alpha_1 = \beta_1 - \gamma_1, u\alpha_2 = \beta_2 - \gamma_2$

$\therefore$ $(u\alpha_1, \alpha_2) = (\beta_1 - \gamma_1, \beta_2 + \gamma_2)$

$$= (\beta_1,\beta_2) + (\beta_1,\gamma_2) - (\gamma_1,\beta_2) - (\gamma_1,\gamma_2)$$

$$= (\beta_1,\beta_2) - (\gamma_1,\gamma_2) + 0 + 0 \text{ as } \beta^s \in W \text{ and } \gamma^s \in W^{\perp}$$

$(\alpha_1, u\alpha_2) = (\beta_1 + \gamma_1, \beta_2 - \gamma_2) = (\beta_1,\beta_2) - (\gamma_1,\gamma_2)$ as above

$\therefore$ $(u\alpha_1, \alpha_2) = (\alpha_1, u\alpha_2)$ $\therefore$ u is self-adjoint.

u will be unitary if $u^*u = I$.

Now $(u^* u\alpha_1, \alpha_2) = (u^* (\beta_1 - \gamma_1), \alpha_2)$

But u is self-adjoint therefore $u^* = u$

$$= (u(\beta_1 - \gamma_1),\alpha_2) = (u\beta_1 - u\gamma_1,\alpha_2)$$

Now $\beta_1 = \beta_1 + 0$ $\therefore$ $u\beta_1 = \beta_1 - 0 = \beta_1$

$$\gamma_1 = 0 + \gamma_1 \quad \therefore \quad u\gamma_1 = 0 - \gamma_1 = -\gamma_1$$

$\therefore$ $u\beta_1 - u\gamma_1 = \beta_1 + \gamma_1 = \alpha_1$

$\therefore$ $(u^* u\alpha_1,\alpha_2) = (\beta_1 + \gamma_1,\alpha_2) = (\alpha_1,\alpha_2) = (I\alpha_1,\alpha_2)$

$\Rightarrow$ $u^* u = I$ $((u^* u - I)\alpha_1,\alpha_2) = 0$

$\Rightarrow$ $u^* u - I = O$ or $u^* u = I$

Example 5. *V is finite dimensional inner product vector spaces and T is a positive linear operator on V. Let P_T be an inner product on V defined by $p_T(\alpha, \beta) = (T\alpha, \beta)$. u is a linear operator on V and u^* its adjoint w.r.t.(,). Prove that u is unitary w.r.t. inner product p_T if and only if $T = u^*Tu$.*

Solution. u will be unitary w.r.t. inner product p_T if it preserves inner product, *i.e.*

$$p_T (u\alpha, u\beta) = p_T(\alpha, \beta) \text{ under given condition that}$$

$$T = u^*Tu$$

Now $P_T(\alpha,\beta) = (T\alpha,\beta)$ by definition

$$= ((u * Tu)\alpha, \beta) = (u * (Tu\alpha), \beta)$$
$$= (Tu\alpha, u\beta) = p_T(u\alpha, u\beta)$$

Hence u is unitary w.r.t. inner product p_T.

Example 6. *If T be a normal operator and f be a polynomial with complex coefficients then show that the operator f(T) is also normal.*

Solution. Let
$$f(x) = a_0 x^n + a_1 x^{n-1} + \ldots + a_n$$
then
$$f(T) = a_0 T^n + a_1 T^{n-1} + \ldots + a_n I \qquad \ldots(1)$$
Now
$$(cT)^* = \bar{c} T^*$$
$$\therefore \qquad [f(T)]^* = \bar{a_0} T^{*n} + \bar{a_1} T^{*n-1} + \_ \ldots + \bar{a_n} I \qquad \ldots(2)$$

We have to prove that $f(T)[f(T)]^* = [f(T)]^* f(T)$

In order to show that $f(T)$ is normal, we have to prove T is normal

$$\therefore \qquad T^*T = TT^* \qquad \ldots(3)$$
$$\therefore \qquad T^{*2}T = T^*(TT^*) = (T^*T)T^*$$
or
$$T^{*2}T = (TT^*)T = TT^{*2}$$

Now we shall prove that $T^{*p}T = TT^{*p} \qquad \ldots(4)$

Above relation holds for $p = 1, 2$ and hence by induction let us assume that it holds for $p = k$, i.e. $T^{*k}T = TT^{*k}$ and we shall show that it holds for $p = k+1$.

Now $T^{*k+1}T = T^*(T^{*k}T) = T^*(TT^{*k})$ as the relation holds for $p = k$.

$$= (T*T)T^{*k} = (TT^*)T^{*k} = TT^{*k+1}$$

Therefore the relation holds for $p = k+1$ and hence it holds for all values of p, i.e.,
$$T^{*p}T = TT^{*p} \qquad \ldots(5)$$

Now we shall show that $T^{*p}T^{*q} = T^qT^{*p}$ with p fixed $\qquad \ldots(6)$

Above relation holds for $q = 1$ by virtue of (5) and let us assume that it holds for $q = m$, i.e., $T^{*p}T^m = T^m T^{*p}$ and we will show that it holds for $q = m+1$

$$T^{*p}T^{m+1} = (T^{*p}T^m)T = (T^m T^{*p})T = T^m(T^{*p}T)$$

$$= T^m(TT^{*p}) = T^{m+1}T^{*p}$$

Hence the result is true for all values of q with p fixed.

$$\therefore \qquad T^{*p}[f(T)] = T^{*p}[a_0 T^n + a_1 T^{n-1} + \ldots + a_n I]$$

$$= a_0 T^{*p}T^n + a_1 T^{*p}T^{n-1} + \ldots + a_n T^{*p}I$$

$$= a_0 T^n T^{*p} + a_1 T^{n-1}T^{*p} + \ldots + a_n IT^{*p} \quad \text{by (5)}$$

$$= [f(T)]T^{*p}$$

or
$$T^{*p}[f(T)] = [f(T)]T^{*p}$$

Above is going to be true for each p.

$$\therefore \qquad [f(T)]^* = f(T)[f(T)]^*$$

Hence $f(T)$ is going to be a normal operator when T is normal.

8.12 PERPENDICULAR PROJECTION

A projection E on an inner product vector space is called a perpendicular projection if its range and null space are orthogonal.

Alternative we may say that a linear operator E on an inner product vector space is called a perpendicular projection if $E^2 = E$ (criteria for projection (ii) range and null spaces of E are orthogonal (criteria for perpendicular projection).

Again whenever we talk of a projection we always specify its range and null space and speak as E is a projection on M along N. In the case of a perpendicular projection as we know that range and null space are orthogonal so it will be sufficient to specify only the range space and speak as E is a perpendicular projection on M, (it being understood along $M^\perp$).

Similarly we can say that if E be a perpendicular projection on M, (it being understood that it is along $M^\perp$) then $I - E$ is a perpendicular projection of $M^\perp$ (it being understood that it is along $(M^\perp)^\perp$, i.e., $M^{\perp\perp} = M$)

THEOREM 1. *A linear operator E is a perpendicular projection on an inner product space V if and only if $E = E^2 = E^*$.*

i.e., E is idempotent and self-adjoint or

$\Rightarrow$ *E is perpendicular projection $\Leftrightarrow E = E^2 = E^*$.*

Proof. Given E is perpendicular projection and to prove that $E = E^2 = E^*$. Since E is a projection so it must be idempotent i.e. $E^2 = E$.   ...(1)

Again because E is a perpendicular projection therefore its range and null space are orthogonal by definition.

i.e., $V = M \oplus M^\perp$

i.e., $z = x + y\, ; z \in V, x \in M, y \in M^\perp$

where $E(z) = x, E(y) = 0$ and $(x, y) = 0$ ... (2)

Now $(Ez, z) = (x, z) = (x, x + y) = (x, x) + (x, y)$

$= (x, x) + 0 \,\text{by}\,(2)$

$(E^*z, z) = (z, Ez) = (z, x) = (x + y, x) = (x, x) + (y, x)$

$= (x, x) + 0 \,\text{by}\,(2)$

Hence $(Ez, z) = (E^*z, z) \Rightarrow ((E - E^*)z, z) = 0 \,\forall\, z$

$\Rightarrow$ $EE^* = 0 \Rightarrow E = E^*, i.e., E$ is self-adjoint.

Here we know that if E be an arbitrary operator s.t. $(Tx, x) = 0 \,\forall\, x$ then $T = O$. Conversely, given $E = E^2 = E^*$ and we have to prove that E is a perpendicular projection. Since $E^2 = E$ and hence E is a projection and say it is a projection on M along N.

so that $z = x + y$ s.t. $E(x) = x, Ey = 0; z \in V, x \in M, y \in N$

It will be a perpendicular projection if the range and null spaces of E are orthogonal which in other words means that if $x \in M, y \in N$ then we should be able to establish that $(x, y) = 0$ under given conditions.

i.e. $E = E^2 = E^*$

$(x, y) = (Ex, y) = (x, E^*y) = (x, Ey) = (x, 0) = 0$

$Ex = x, x \in M$ and $Ey = 0\,;\, y \in N.$

8.13 PROPERTIES OF A PERPENDICULAR PROJECTION

THEOREM 1. *The perpendicular projections are positive linear operators such that $O \le E \le I$ and have the property that*

$$\|Ez\| \le z \,\forall z \in V$$

Proof. We know that a linear operator T is said to be $+ive$ if T is self adjoint and $(Tx, x) \ge 0$, $\forall x \in V$.

Since E is a perpendicular projection and hence E is self adjoint

and $$E^2 = E$$

Now $$(Ez, z) = (E^2 z, z) = (EEz, z) = (Ez, E * z)$$

$$= (Ez, Ez) = \|Ez\|^2 \ge 0 \qquad \dots (1)$$

Thus E is self-adjoint and also $(Ez, z) \ge 0 \,\forall z$ and hence E is a positive linear operator.

i.e. $$E \ge O \text{ or } O \le E$$

Again when E is a perpendicular projection then as shown earlier $I - E$ is a perpendicular projection so that

$$I - E \ge O \text{ or } E \le I$$

and hence $$O \le E \le I.$$

2nd Part

We have proved in (1) that

$$(Ez, z) = \|Ez\|^2 \ge 0. \qquad \dots(2)$$

We have to prove that

$$\|Ez\| \le \|z\| \,\forall z.$$

But since E is a perpendicular projection therefore $I - E$ is also a perpendicular projection and hence by (1)

$$((I - E)z, z) \ge 0$$

or $$(Iz, z) - (Ez, z) \ge 0$$

or $$(z, z) - (Ez, z) \ge 0$$

or $$\|z\|^2 \ge (Ez, z)$$

or $$(Ez, z) \le \|z\|^2 \,\forall z \in V$$

or $$\|Ez\|^2 \le \|z\|^2 \text{ by (2)}$$

$\therefore$ $$\|Ez\| \le \|z\| \,\forall z \in V$$

Alternative Method

New if E be a perpendicular projection on M, then

$$z = Ez + (I - E) z \in V \qquad \dots (1)$$

where $$x = Ez \in M \qquad E(Ez) = E^2 z = Ez$$

and $$y = (I - E) z \in M^\perp$$

$$E(I - E) z = (E - E^2) z = (E - E) z = 0,$$

i.e., $z = x + y$ where $x \in M$ and $y \in M^\perp$, *i.e.* $(x, y) = 0$

$\therefore$ $$\|z\|^2 = \|x + y\|^2 = (x + y, x + y) = (x, x) + (x, y) + (y, x) + (y, y)$$

or $$\|z\|^2 = \|x + y\|^2 = \|x\|^2 + \|y\|^2$$

$\therefore$ x and y are orthogonal.

$$\therefore \qquad \|z\|^2 = \|Ez\|^2 + \|(I-E)z\|^2 \qquad \qquad \qquad \dots(3)$$

From above it is clear that

$$\|Ez\| \leq \|z\| \quad \forall z \in V$$

Particular Case.

If E is a perpendicular projection on M, then

$$x \in M \Leftrightarrow Ex = x \Leftrightarrow \|Ex\| = \|x\|$$

1st Case. $\qquad\qquad x \in M = x + 0$ where $0 \in N$

$$\therefore \qquad\qquad E(x) = E(x+0) = E(z) = x \text{ by def.}$$

Conversely $\qquad\qquad Ex = x \quad \therefore \quad E(Ex) = Ex \Rightarrow Ex \in M$

$$\Rightarrow x \in M \qquad Ex = x$$

2nd Case. $\qquad\qquad Ex = x \quad \Rightarrow (Ex, Ex) = (x, x)$

$$\Rightarrow \qquad \|Ex\|^2 = \|x\|^2 \Rightarrow \qquad \|Ex\| = \|x\|.$$

Conversely. $\qquad \|Ex\| = \|x\|$

and we have to show that $Ex = x$

We have already proved that

$$\therefore \qquad\qquad \|x\|^2 = \|Ez\|^2 + \|(I-E)z\|^2 \ \forall z \in V.$$

Replace z by x where $x \in M$.

$$\therefore \qquad\qquad \|x\|^2 = \|Ex\|^2 + \|(I-E)x\|^2$$

But $\qquad\qquad \|x\|^2 = \|Ex\|^2$ given

$$\Rightarrow \qquad\qquad \|(I-E)x\|^2 = \|x-Ex\|^2 = 0$$

$$\Rightarrow \qquad\qquad x - Ex = 0 \Rightarrow Ex = x$$

THEOREM 2. *If a linear transformation E satisfies*

(i) $E^2 = E$ and   (ii) $\|Ez\| \leq \|z\| \forall z$

then E is self-adjoint i.e., $E^* = E$.

Proof. Since $E^2 = E$ we conclude that E is a projection say on M along N. Further we know that if E is a perpendicular projection then

$$E^2 = E = E^* \text{ and } \|Ez\| \leq \|z\| \forall z.$$

Here we are given two conditions which E satisfies,

i.e. $\qquad\qquad E^2 = E$ and $\|Ez\| \leq \|z\| \forall z$

and we have to prove that $E^* = E$. In other words we have to show that E is a perpendicular projection which in turn means that we have to establish that the range of E, *i.e.*, M and null space of E, *i.e.*, N are orthogonal. This amounts to proving that $M = N^\perp$. We shall prove this equality by establishing that $N^\perp \subset M$ and $M \subset N^\perp$ under given conditions.

$$N^\perp \subset M$$

Let $\alpha \in N^\perp$ and we will show that this implies that

$$\alpha \in M \text{ so that } N^\perp \subset M$$

Choose $\beta = E\alpha - \alpha$, so that $E\beta = E(E\alpha - \alpha) = E^2\alpha - E\alpha$

or $\qquad\qquad E\beta = E\alpha - E\alpha = 0$

and hence $\beta \in N$ the null space of E

$\therefore$ $\qquad (\alpha, \beta) = 0 \qquad \alpha \in N^\perp$ and $\beta \in N$

Now from $\beta = E\alpha - \alpha$ we get $E\alpha = \alpha + \beta$ where $(\alpha, \beta) = 0$

$\therefore \qquad \|E\alpha\|^2 = \|\alpha\|^2 + \|\beta\|^2 \qquad (\alpha, \beta) = 0 \qquad \qquad \dots(1)$

Also $\|\alpha\|^2 \geq \|E\alpha\|^2$ by given condition

$\therefore \qquad \|\alpha\|^2 \geq \|E\alpha\|^2$, *i.e.* $\geq \|\alpha\|^2 + \|\beta\|^2$ by (1)

But $\qquad \|\alpha\|^2 + \|\beta\|^2 \geq \|\alpha\|^2; \|\beta\|^2 \geq 0$

Hence we have from above

$$\|\alpha\|^2 \geq \|\alpha\|^2 + \|\beta\|^2 \geq \|\alpha\|^2$$

Above relation will hold good only if $\|\beta\|^2 = 0$, *i.e.*, $\beta = 0$ and in this case from $\beta = E\alpha - \alpha$, we get $E\alpha = \alpha$ showing thereby that α belongs to range space of E, *i.e.*, M.

Hence $\qquad \alpha \in N^\perp \Rightarrow \quad \alpha \in M \qquad \therefore \quad N^\perp \subset M. \qquad \dots(2)$

Now we will show that $M \subset N^\perp$.

Let $\gamma \in M$ the range space then $E\gamma = \gamma$ by definition as E is a projection on M and we have to show that $\quad \gamma \in N^\perp$

Again we know that $\qquad V = N^\perp \oplus N$.

and hence any γ can be expressed uniquely as

$$\gamma = \alpha + \beta \text{ with } \alpha \in N^\perp \text{ and } \beta \in N \text{ so that } E\beta = 0$$

as N is the null space of E which is a projection on M along N.

$$\gamma = E\gamma = E(\alpha + \beta) = E\alpha + E\beta = E\alpha \qquad \dots(3)$$

Now $\alpha \in N^\perp$ and we have proved above in first case that

$$N^\perp \subset M$$

$\therefore \quad \alpha \in N^\perp \Rightarrow \alpha \in M$ and hence $E\alpha = \alpha$ as E is a projection on M

$\therefore \quad \gamma = \alpha$ by (3) and hence $y \in M \Rightarrow \gamma \in N^\perp$

$\therefore \qquad M \subset N^\perp \qquad \dots(4)$

Relations (2) and (4) prove that $M = N^\perp$, *i.e.*, range and null spaces are orthogonal so that E is a perpendicular projection and consequently it is self-Adjoint.

8.14 INVARIANCE AND REDUCIBILITY IN INNER PRODUCT SPACE

Let M be a subspace of a vector space V and T be a linear operator on V. M is said to be invariant under T if $x \in M = T(x)$ also belongs to M. In other words $T(M) \subset M$.

Clearly M is invariant under zero operator.

REMARKS

- Every closed linear subspace is invariant under identity operator I.
- It should be clearly understood here that

$$x \in M \Rightarrow T(x) \in M$$

is one way implication. It does not mean that for y in M we can write it as $y = Tx$ with x in M.

Reducibility

A linear operator T on a vector space V is said to be reduced by a pair of subspaces M and N of V if

    (i)   *V = M ⊕ N, i.e., V is the direct sum of subspaces M and N.*

    (ii) *Both M and N are invariant under T.*

REMARKS

- We may also say that the pair (M, N) reduces T.
- If V be an inner product vector space then we know that

$$V = M \oplus M^\perp$$

and if both M and $M^\perp$ are invariant under T, then we simply say M reduces E instead of saying that the pair $(M, M^\perp)$ reduce T or T is reduced by M (and $M^\perp$ being understood).

THEOREM 1.   *A subspace M of an inner product space V is invariant under an operator T if and only if $M^\perp$ is invariant under T *.*

Proof.   Let us first suppose M is invariant under, T i.e., $x \in M \Rightarrow T(x) \in M$.

To prove that $M^\perp$ is invariant under T^*, i.e., $y \in M^\perp \Rightarrow T^*(y) \in M^\perp$.

Let $x \in M$ and $y \in M^\perp$ so that $(x, y) = 0$.

Now $(x, T^*y) = (Tx, y)$

But M is invariant under T ∴ $Tx \in M$ and $y \in M^\perp$

∴ $(Tx, y) = 0$

∴ $(x, T^* y) = 0.$

Above is true for every $x \in M$ and hence T^*y must belong to $M^\perp$ where $y \in M^\perp$

∴ $M^\perp$ is invariant under T^*.

Conversely $M^\perp$ is invariant under T^* (given) and hence $(M^\perp)^\perp$ is invariant under $(T^*)^*$ by Ist case.

But $M^{\perp\perp} = M$ and $T^{**} = T.$

∴ M is invariant under T.

THEOREM 2.   *A subspace M of an inner product space V reduces an operator $T \Leftrightarrow M$ is invariant under both T and T^*.*

Proof.   Let us first suppose M reduces an operator T means that both M and $M^\perp$ are invariant under T.

Since $M^\perp$ is invariant under T hence $(M^\perp)^\perp$ is invariant under T^*.

or, M is invariant under T^*.

Thus M is invariant under both T and T^*.

Conversely M is invariant under both T and T^*.

i.e., M is invariant under T.

M is invariant under $T^* = M^\perp$ is invariant under $(T^*)^* = T$.

Thus M and $M^\perp$ both are invarient under T. In other words it means that M reduces an operator T.

THEOREM 3.   *If E is a perpendicular projection on a subspace M of an inner product space V and T a linear operator on V. Then*

M is invariant under $T \Leftrightarrow TE = ETE$.

$$V = M \oplus M^\perp \text{ so that } z = x + y ; z \in V, x \in M, y \in M^\perp$$

Proof.   *E is a projection on V so that $Ex = x$ and $Ey = 0$.*

Proof. Let us first suppose that M is invariant under T, *i.e.*
$$x \in M \Rightarrow Tx \in M$$
To prove that $TE = ETE$ and we shall establish that
$$TE(z) = ETE(z) \forall z \in V.$$
Now $\qquad TE(z) = TE(x+y) = T(Ex+Ey) = T(x+0) = Tx \qquad$ ... (1)

where $T(x)$ belongs to M as M is invariant under T.

Again $\qquad ETE(z) = E(TE(z)) = E(T(x))$ by (1)

But $Tx \in M$ by (1) and E is a projection on M.

$\therefore \qquad\qquad E(T(x)) = T(x)$.

$\therefore \qquad\qquad ETE(z) = Tx \qquad\qquad$ ... (2)

Hence from (1) and (2) we get
$$ET(z) = ETE(z) \forall z \qquad\qquad \therefore \quad TE = ETE$$

Conversely given $TE = ETE$.

To prove that M is invariant under T.

Let x be any vector $\in M$ and E is a projection on M so that $Ex = x$.

Also T is any operator on V so that $Tx \in V$.

Now $\qquad\qquad Tx = TE(x) \qquad\qquad\qquad Ex = x$

or $\qquad\qquad Tx = ETEx \qquad\qquad \therefore \quad TE = ETE$

$\therefore \qquad\qquad Tx = E(TEx) = E(T(x)) \in M \qquad\qquad$ ...(3)

because $T(x) \in V$ and E is projection on M.

$\therefore \quad E(T(x))$ must belong to M.

$\therefore \quad Tx \in M$ when $x \in M$ by (3) and hence M is invariant under T.

THEOREM 4. *If E is a perpendicular projection on a subspace M of an inner product space V and T a linear operator on V then M reduces $T \Leftrightarrow TE = ET$.*

Proof. Let us suppose that M reduces $T \Rightarrow M$ is invariant under both T and T^*.

$\therefore \qquad\qquad TE = ETE$ as M is invariant under T

$\qquad\qquad T^*E = ET^*E$ as M is invariant under T^*

To prove $\qquad TE = ET$

Since E is a perpendicular projection

$\therefore \qquad\qquad E^* = E$ and hence $ET = E^*T$

$\therefore \qquad\qquad ET = E^*T^{**} = (T^*E)^* = (ET^* E)^*$

$\qquad\qquad T^*E = ET^*E$

$\qquad\qquad\qquad = E^*T^{**}E^* = ETE = TE \Rightarrow ETE = TE$

$\therefore \qquad\qquad ET = TE$

Conversely let $TE = ET$.

To prove that M reduces T, *i.e.*, M is invariant under both T and T^*, *i.e.*, we have to show that
$$TE = ETE \text{ and } T^*E = ET^*E \text{ by Theorem 3}$$
Now $\qquad\qquad ETE = (ET)E = TEE = TE^2 = TE, \qquad E^2 = E$

Hence M is invariant under T.

Again $\qquad\qquad TE = ET \Rightarrow (TE)^* = (ET)^* \Rightarrow E^*T^* = T^*E^*$

$$\Rightarrow \qquad ET^* = T^*E \qquad\qquad E^* = E$$

Now $\qquad ET^* E = (ET^*)E = (T^*E)E = T^*EE$

$$= T^*E^2 = T^*E$$

$\therefore$ M is invariant under T^*.

8.15 ORTHOGONAL PROJECTIONS

THEOREM 1. *E_1 and E_2 are perpendicular projections on M_1 and M_2 respectively where M_1 and M_2 are subspaces of an inner product space then M_1 and M_2 are orthogonal*

$$\Leftrightarrow E_1 E_2 = O \Leftrightarrow E_2 E_1 = O$$

Proof.

We know that

$E_1 E_2 = O$ then $(E_1 E_2)^* = O^* \Rightarrow E_2^* E_1^* = O \Rightarrow E_2 E_1 = O$

and hence it is sufficient to prove that $E_1 E_2 = O$.

Given $M_1 \perp M_2$. To prove that $E_1 E_2 = O$.

Let $z = x_1 + y_1$ corresponding to E_1 a projection on M_1 so that $E_1 z = x_1$

$\qquad z = x_2 + y_2$ corresponding to E_2 a projection on M_2 so that $E_2 z = x$

Now $\qquad (E_1 E_2 z, z) = (E_2 z, E_1^* z) = (E_2 z, E_1 z) E_1 . E_1^*$

$$= x^2 (x_2, x_1) \text{ where } x_2 \in M_2 \text{ and } x_1 \in M_1$$

$$= 0 \qquad M_1 \text{ and } M_2 \text{ are orthogonal}$$

$\therefore \qquad (E_1 E_2 z, z) = 0 \,\forall\, z$

$\therefore \qquad E_1 E_2 = 0 \,(Tx, x) = 0 \,\forall\, x \Rightarrow T = 0$

Conversely, let $E_1 E_2 = O$. To prove that M_1 and M_2 are orthogonal.

Let $x_1 \in M_1$ and $x_2 \in M_2$ so that $E_1 x_1 = x_1$ and $E_2 x_2 = x_2$

$\therefore \qquad (x_1, x_2) = (E_1 x_1, E_2 x_2) = (x_1, E_1^* E_2 x_2)$

$$= (x_1, E_1 E_2 x)$$

$$= (x_1, O x_2) = (x_1, 0) = 0 , E_2 E_2 = O$$

Hence $M_1 \perp M_2$.

Definition (1). *A projection is said to be perpendicular if its ranges and null spaces are orthogonal.*

Definition (2). *Two perpendicular projections E_1 and E_2 are said to be orthogonal if $E_1 E_2 = E_2 E_1 = 0$ which in turn means that M_1 and M_2 their ranges are orthogonal.*

THEOREM 2. *If $E_1, E_2 \dots E_n$ are perpendicular projections on subspaces $M_1, M_2 \dots M_n$ of the inner product vector space V then their sum $E = E_1 + E_2 + \dots + E_n$ is perpendicular projection if and only if E_i^S are pairwise orthogonal i.e. if and only if $E_i E_j = O, i \neq j$ and in this case E is a projection on M where*

$$M = M_1 + M_2 + \dots + M_n.$$

Proof.

Given that E_i^S are perpendicular projection which are pairwise othogonal,

i.e. $\qquad E_i^S = E_i = E_i^* \text{ and } \qquad E_i E_j = 0, i \neq j,$

To prove that $\quad E = E_1 + E_2 + \dots E_n$ is a perpendicular projection

$$E^* = (E_1 + E_2 + \dots + E_n)^* = E_1^* + E_2^* + \dots + E_n^*$$

or $\qquad E^* = E_1 + E_2 + \dots + E_n = E , E_i^* = E_i \text{ for each } i$

Therefore E is self-adjoint.

Again
$$E^2 = (E_1 + E_2 + \ldots + E_n)^2$$
$$= (E_1^2 + E_2^2 + \ldots + E_n^2 + 2\Sigma E_i E_j)$$
$$= E_1 + E_2 + \ldots + E_n + O \text{ by given condition}$$
$$\therefore \qquad E^2 = E$$

Hence E is idempotent and it is already shown to be self-adjoint.

Therefore E is a perpendicular projection.

Conversely let E_i^s be perpendicular projections and $E = E_1 + \ldots + E_n$ is also a perpendicular projection.

To prove that E_i^s are pairwise orthogonal, *i.e.,*

$$E_i E_j = 0, \, i \neq j.$$

We shall make use of the following results :

(1) $\|Ez\| \leq \|z\| \; \forall z$

(2) $\|Ex\| = \|x\|$ if x belongs to range of E

(3) If E_1 and E_2 be perpendicular projections on M_1 and M_2 then
$$M_1 \perp M_2 \Leftrightarrow E_1 E_2 = O = E_2 E_1$$

(4) $(Ez, z) = \|Ez\|^2 \; \forall \; z$

$(Ez, z) = (E^2 z, z) = (Ez, E^* z) = (Ez, Ez)$
$$= \|Ez\|^2$$

Let x be any vector belonging to range M_i of E_i

$$\therefore \qquad \|x\|^2 = \|E_i x\|^2$$

Now $\qquad \|x\|^2 = \|E_i x\|^2$

$$\leq \sum_{j=1}^{n} \left\| E_j x \right\|^2 = \sum_{j=1}^{n} (E_j x, x) \text{ by (4)}$$

$$= (E_1 x, x) + (E_2 x, x) + \ldots + (E_n x, x)$$
$$= ((E_1 + E_2 + \ldots)x, x) = (Ex, x) = \|Ex\|^2 \text{ by (4).}$$
$$\therefore \qquad \|x\|^2 \leq \|Ex\|^2$$

Above will be true if and only if the sign of equality holds throughout the above computation and this would mean that

$$\left\| E_i x \right\|^2 = \sum_{j=1}^{n} \left\| E_j x \right\|^2$$

Above will hold good if and only if

$\|E_j x\| = 0$ for $j \neq i$

This implies that $\qquad E_j x = 0, \, j \neq i$

$\Rightarrow x$ belongs to null sapce of E_j whose range space is M_j and therefore null space is $M_j^\perp$.

Hence $x \in M_i \Rightarrow x \in M_j^\perp, \, j \neq i$

or $\qquad x \in M_i \Rightarrow x$ is perpendicular to M_j
$$\Rightarrow M_i \perp M_j \; \forall \; j \neq i$$

Now E_i and E_j are perpendicular projections whose range M_i and M_j are orthogonal and hence these projections are also orthogonal. Therefore by (3) $E_i E_j = O, i \neq j$, i.e., they are pairwise orthogonal.

2nd Part. To prove that range of E, i.e., $R(E)$ is M where
$$M = M_1 + M_2 + \dots + M_n$$
Let us prove that $R(E) \subseteq M$.

Let $\qquad\qquad\qquad x \in R(E) \qquad \therefore \qquad x = Ex$ by (2)

$\Rightarrow \qquad\qquad\qquad x = (E_1 + E_2 + \dots E_n) x = E_1 x + E_2 x + \dots + E_n x.$

Now by definition $E_j x \in$ Range of E_i, i.e., M_i

$\Rightarrow \qquad\qquad\qquad x \in M_1 + M_2 + \dots + M_n \qquad \Rightarrow \qquad x \in M$

$\qquad\qquad\qquad R(E) \subseteq M \qquad\qquad\qquad\qquad\qquad\qquad\qquad \dots (1)$

Secondly we should prove that $M \subseteq R(E)$.

Let $x \in$ Range of E_i, i.e., M_i then $E_i x = x$ by (2)

or $\qquad\qquad\qquad \|E_i x\| = \|x\|$

Also $E_j = 0, j \neq i$ because $x \in M_j \Rightarrow x \in M_i^\perp$

$M_i \perp M_j$ and $M_i^\perp$ is null space of E_j. Therefore $\quad E_j x = 0$

$\therefore \quad Ex = (E_1 + E_2 + \dots E_n) x = E_1 x + E_2 x + \dots + E_n x = E_i x$

$\therefore \qquad\qquad\qquad \|E_i x\| = \|x\|$ or $\|Ex\| = \|x\| \ \forall \ x \in M_i$

Above shows that $x \in$ Range of E.

Hence $x \in M_i \Rightarrow x \in$ Range of E.

$\therefore \qquad\qquad\qquad M_i \subseteq$ Range of $E \ \forall \ i$

Hence their sum, i.e., $\quad M \subseteq$ Range E.

From (1) and (2) we prove that $M =$ Range of E. $\qquad\qquad\qquad\qquad \dots (2)$

Solved Examples

Based on the following Results

➠ A projection E on an inner product vector space is called a perpendicular projection if its range and null space are orthogonal.

➠ A linear operator E is a perpendicular projection on an inner product space V if and only if
$$E = E^2 = E^*$$

➠ If a linear operator E satisfies (i) $E^2 = E$ and (ii) $\|Ez\| \leq \|z\| \forall z$ then E is self-adjoint.

➠ Two projections E_1 and E_2 are said to be orthogonal iff $E_1 E_2 = E_2 E_1 = 0$

Example 1. *If E and F are perpendicular projections on subspaces M and N of an inner product vector space then prove that EF is a perpendicular projection if and only if $EF = FE$.*

In this case show that EF is a projection on $M \cap N$.

Solution. E is a perpendicular projection on M.

$\therefore \qquad\qquad\qquad \bar{V} = M \oplus M^\perp$

F is a perpendicular projection on N.

$\therefore \qquad\qquad\qquad V = N \oplus N^\perp$

Also $\qquad\qquad\qquad E^2 = E, E^* = E, F^2 = F, F^* = F \qquad\qquad \dots (1)$

Now if EF is a perpendicular projection then it is self-adjoint.

$$(EF)^* = EF \text{ or } F*E* = EF \text{ or } FE = EF \text{ by (1)}$$

Converse given $EF = FE$. To show that EF is a perpendicular projection, *i.e.*, it is idempotent and self-adjoint.

(i) $(EF)^2 = (EF)(EF) = (EF)(FE)$
$$= EF^2E = EFE = E(EF) = E^2F = EF.$$

(ii) $(EF)^* = F^*E^* = FE = EF.$

Therefore EF is a perpendicular projection.

2nd Part. In order to show that EF is a projection on $M \cap N$ we have to show that range space of EF is $M \cap N$. In other words it means that we should show that for any
$$x \in M \cap N, EF(x) = x$$
Now since $x \in M \cap N \Rightarrow x \in M$ and $x \in N$.

$\therefore$ $Ex = x$ and $Fx = x$ by definition of E and F

$\therefore$ $EF(x) = E(F(x)) = E(x) = x$

$\therefore$ EF is a perpendicular projection on $M \cap N$.

REMARK

- We conclude that if E and F be two commutative $(EF = FE)$ perpendicular projections then EF is also a perpendicular projection.

Example 2. *E and F are perpendicular projections with ranges M and N respectively which are subspaces of an inner product vector space V. Prove that the following statements are equivalent to one another.*

(1) $E \leq F$ (2) $\|Ex\| \leq \|Fx\| \ \forall x$

(3) $M \subseteq N$ (4) $EF = E, FE = E$

Solution. **(1)** $\Rightarrow$ **(2)** Given $E \leq F$ and hence by definition
$$(Ex, x) \leq (Fx, x) \ \forall x$$
$$\Rightarrow \qquad \|Ex\|^2 \leq \|Fx\|^2 \Rightarrow \|Ex\| \leq \|Fx\| \forall x.$$
Because we have proved that
$$(Ex, x) = \|Ex\|^2 \ \forall x.$$

(2) $\Rightarrow$ **(3)** Given $\|Ex\| \leq \|Fx\|$.

To Prove that $M \subseteq N$.

Let $x \in M$ then $\|Ex\| = \|x\|$ where E is a projection on M by particular case P.

Again we know that $\|Ez\| \leq \|z\| \ \forall z$

Hence $\|Fx\| \leq \|x\|$, x does not belong to range of F

$\therefore \qquad \|x\| = \|Ex\| \leq \|Fx\| \leq \|x\|$

Therefore we must have the sign of equality all through in the above relation and consequently.
$$\|Fx\| = \|x\| \Rightarrow x \in \text{ range of } F, \text{ i.e., } N$$
Thus $\qquad x \in M \ \Rightarrow x \in N \ \therefore M \subseteq N$

(3) $\Rightarrow$ **(4)** Given $M \subseteq N$ i.e, $x \in M \Rightarrow x \in N$

To prove that $EF = E$ and $FE = E$

Now $\qquad x \in M \qquad\qquad \Rightarrow x \in N$

$\therefore \qquad Ex \in M \qquad\qquad \Rightarrow Ex \in N \quad Ex = x$

where $x \in$ range of E.

Now F is a perpendicular projection on N and $Ex \in N$.

$\therefore \qquad\qquad F(Ex) = Ex$ or $\quad FEx = Ex \Rightarrow FE = E$.

$\therefore \qquad\qquad (FE)^* = E^*$ or $\quad E^*F^* = E^*$ or $EF = E$.

Hence $\qquad\qquad EF = FE = E$

(4) $\Rightarrow$ (1) Given. $EF = FE = E$. To prove $E \le F$.

Now $\qquad\qquad (Ex, x) = \|Ex\|^2 = \|EFx\|^2 = \|E(Fx)\|^2$

$\qquad\qquad\qquad\qquad \le \|Fx\|^2$ or $\le (Fx, x)$.

$\therefore \qquad\qquad \|Ex\|^2 \le \|z\|^2 \,\forall z$ Here $z = Fx$

Since $\qquad\qquad (Ex, x) \le (Fx, x)$

$\therefore \qquad\qquad\qquad E \le F$.

Example 3. *E and F are perpendicular projections with ranges M and N respectively which are sub spaces of an inner product vector space V. Prove that F − E is a projection if and only if E ≤ F.*

In this case show that F − E is a projection on $N \cap M^\perp$.

Solution. Given $F - E$ is a perpendicular projection.

To prove that $E \le F$.

Clearly $\qquad (F - E)^* = F^* - E^* = F - E$.

Also $\qquad (F - E)^2 = F - E \Rightarrow (F - E)(F - E) = F - E$.

$\Rightarrow F^2 - EF - FE + E^2 = F - E$.

$\Rightarrow \quad F - EF - FE + E = F - E$. Above will hold good if

$\qquad\qquad\qquad EF = FE = E$.

Now $\qquad\qquad (Ex, x) = \|Ex\|^2 = \|E\,Fx\|^2 = \|E(Fx)\|^2$

$\qquad\qquad\qquad\qquad \le \|Fx\|^2$ or $\le (Fx, x)$

$\qquad\qquad\qquad \|Ez\|^2 \le \|z\|^2 \,\forall\, z$ and here $z = Fx$

Since $\qquad\qquad (Ex, x) \le (Fx, x) \quad \therefore \quad E \le F$.

Conversely, let $E \le F$. To prove that $F - E$ is a projection.

By a chain of equivalence as proved in (Ex. 2) we prove that $E \le F \Rightarrow EF = FE = E$ and with this as proved above, we can show that $(F - E)^* = F - E$.

and $(F - E)^2 = F - E$ and hence $F - E$ is a perpendicular projection.

Now, in order to show that $F - E$ is a projection on $N \cap M^\perp$ we have to show that range space of $F - E$ is $N \cap M^\perp$. In other words it means that we should show that for any $\qquad\qquad x \in N \cap M^\perp;\ (F - E)x = x$

Now $\qquad\qquad x \in N \cap M^\perp \Rightarrow x \in N$ and $x \in M^\perp$

$\therefore \qquad\qquad Fx = x$ *as F is a projection on N.*

and $\qquad\qquad Ex = 0$ *as $M^\perp$ is null space of E.*

$\therefore \qquad\qquad (F - E)x = Fx - Ex = F(x) - 0 = x$.

Therefore $F - E$ is perpendicular projection on $N \cap M^\perp$

Example 4. *If T is a linear transformation on an inner product vector space V(F) and E be a projection then show that T is an involuntary isometry if and only if E is perpendicular projection.*

Solution.   Given T is an involution, *i.e.*, $T^2 = I$ and in this case if E be a projection, then $T = 2E - I$. Also T is an isometry.

$$\therefore \qquad\qquad TT^* = T^*T = I.$$

To prove x is a perpendicular projection.

i.e. $\qquad\qquad\qquad E = E^2 = E^*$

Since $\qquad\qquad T^2 = I$ and $T = 2E - I \qquad \therefore \quad (2E - I)^2 = I$

$\Rightarrow \qquad\qquad 4E^2 - 4E + I = I \Rightarrow E^2 = E.$ ... (1)

Again $\qquad\qquad T\,T^* = I \qquad \Rightarrow \quad (2E - I)\,(2E - I)^* = I$

$\Rightarrow \qquad (2E - I)(2\bar{E}^* - I^*) = I \Rightarrow (2E - I)\,(2E^* - I) = I$

$\Rightarrow \qquad 4E\,E^* - 2E - 2E^* + I = I \qquad\qquad \bar{E} = E,\, I^* = I$

$\Rightarrow \qquad 2EE^* - E^2 - E^{*2} = O \Rightarrow (E - E^*)^2 = O$

$\Rightarrow \qquad\qquad\qquad E = E^*$ ... (2)

Results (1) and (2) give $E = E^2 = E^*$ and hence E is a perpendicular projection.

Conversely, given $E = E^2 = E^*$. To prove that T is an involution isometry,

$$T^2 = (2E - I)^2 = 4E^2 - 4E + I = 4E - 4E + I = I$$

Hence T is an involution.

Also, $\qquad T\,T^* = (2E - I)\,(2E - I)^* = (2E - I)(\bar{2}E^* - I^*)$

$\qquad\qquad\qquad = (2E - I)\,(2E^* - I) = (2E - I)\,(2E - I) \qquad\qquad E^* = E$

$\qquad\qquad\qquad = (2E - I)^2 = I$

Similarly $\qquad T^*T = I$

8.16 CHARACTERIZATION OF SPECTRA

We know that the set of all characteristic (proper) values of a linear operator T is called spectrum of T. We shall now classify the spectrum of T where T is a linear operator on any inner product vector space $V(F)$ according to various classification of operator T.

THEOREM 1. *If T be a self-adjoint operator on an inner product vector space then the characteristic values of T are real and the characteristic vectors of T corresponding to distinct characteristic values are orthogonal.*

further if T be positive, or strictly positive then the characteristic values are positive or strictly positive respectively.

Proof.   Let λ be the proper values of T so that $\exists\ x \neq 0$ such that $Tx = \lambda x$.

Also T is self-adjoint $\Leftrightarrow (Tx, x)$ is real.

Now $\qquad (Tx, x) = (\lambda x, x)$ by (1) $= \lambda(x, x) = \lambda \|x\|^2$

$$\therefore \qquad\qquad \lambda = \frac{(Tx, x)}{\|x\|^2} \qquad\qquad\qquad\qquad ...(2)$$

Since $\|x\|^2$ is always positive real and (Tx, x) is also real it follows that λ is real. Again we know that T is positive or strictly positive according as $(Tx, x) \geq 0$ or > 0 and hence from (2) it follows that λ is positive or strictly positive.

Now, the characteristic vectors corresponding to distinct characteristic values of a self-adjoint operator are orthogonal.

Let λ_1, λ_2 be two distinct characteristic values of a self-adjoint operator T and x_1, x_2 be the corresponding characteristic vectors so that $Tx_1 = \lambda_1 x_1$, $Tx_2 = \lambda_2 x_2$, where λ_1, λ_2 are real.

We have to prove that x_1, x_2 are orthogonal, i.e., $(x_1, x_2) = 0$

$$\lambda_1(x_1, x_2) = (\lambda_1 x_1, x_2) = (Tx_1, x_2) = (x_1, T * x_2)$$

$$= (x_1, Tx_2)(\text{as } T^* = T) = (x_1, \lambda_2 x_2)$$

$$= \bar{\lambda}_2(x_1, x_2) = \lambda_2(x_1, x_2) \text{ as } \lambda_2 \text{ is real.}$$

$\therefore$ $\qquad\qquad \lambda_1(x_1, x_2) = \lambda_2(x_1, x_2)$

or $\qquad (\lambda_1 - \lambda_2)(x_1, x_2) = 0$

Since λ_1, λ_2 are distinct therefore $(x_1, x_2) = 0$

Hence x_1, x_2 are orthogonal.

THEOREM 2. *If T be an unitary operator (or an isometry) on an inner product vector space $V(F)$ then the characteristic values of T are unimodular and the corresponding distinct characteristic vectors are orthogonal.*

Proof. Let λ be a characteristic value of T so that $\exists x \neq 0$ such that

$$Tx = \lambda x \qquad\qquad\qquad ...(1)$$

Also T is unitary or isometry $\Leftrightarrow (Tx, Tx) = (x, x)$

i.e. T preserves inner products.

or $\qquad\qquad (\lambda x, \lambda x) = (x, x) \text{ by (1) or } \lambda\bar{\lambda}(x, x) = (x, x)$

or $\qquad\qquad |\lambda|^2 \|x\|^2 = \|x\|^2$

Hence $\qquad\qquad |\lambda|^2 = 1 \text{ or } |\lambda| = 1$

i.e., characteristic values are unimodular.

Now, let λ_1, λ_2 be two distinct characteristic values of a unitary or isometric operator T and x_1, x_2 be the corresponding characteristic vectors so that.

$$Tx_1 = \lambda_1 x_1, Tx_2 = \lambda_2 x_2, \text{ where } \lambda_1, \lambda_2 \text{ are unimodular.}$$

Now $\qquad (Tx_1, Tx_2) = (x_1, x_2)$ as T is unitary or isometric

or $\qquad (\lambda_1 x_1, \lambda_2 x_2) = (x_1, x_2)$

or $\qquad \lambda_1\bar{\lambda}_2(x_1, x_2) = (x_1, x_2)$

Since $\qquad\qquad |\lambda_2|^2 = 1$

$\therefore$ $\qquad\qquad \lambda_2\bar{\lambda}_2 = 1 \quad \text{or} \quad \bar{\lambda}_2 = \dfrac{1}{\lambda_2}$

$\therefore$ $\qquad \dfrac{\lambda_1}{\lambda_2}(x_1, x_2) = (x_1, x_2) \quad \text{or} \quad (\lambda_1 - \lambda_2)(x_1, x_2) = 0$

Since λ_1 and λ_2 are distinct, it follows from above that $(x_1, x_2) = 0$, i.e., corresponding characteristic vectors are orthogonal.

THEOREM 3. *Every characteristic vector of a normal operator T is also a characteristic vector for T^*. If T be a normal operator and x is a characteristic vector of T corresponding to characteristic value λ then x is a characteristic vector of $T *$ corresponding to eigenvalue $\bar{\lambda}$. Further the eigenspaces of a normal operator T are pairwise orthogonal.*

Proof. Given that T is normal $\Rightarrow TT^* = T^*T$ $\qquad\qquad ... (1)$

Again $\qquad \|Tx\|^2 = (Tx, Tx) = (x, T * Tx) = (x, TT * x) \text{ by (1)}$

$$= (T^*x, T^*x) = \| T * x\|^2$$

$\therefore$ $\qquad\qquad \|Tx\| = \|T^* x\|$

Now we shall show that $T - \lambda I$ is also normal.   ... (2)

$$(T - \lambda I)^* = T^* - (\lambda I)^* = T^* - \bar{\lambda} I \qquad I^* = I$$

$$\therefore \quad (T - \lambda I)(T - \lambda I)^* = (T - \lambda I)(T^* - \bar{\lambda} I)$$

$$= TT^* - \lambda T^* - \bar{\lambda} T + \bar{\lambda} \lambda \qquad \qquad ... (3)$$

$$(T - \lambda I)^*(T - \lambda I) = (T^* - \bar{\lambda} I)(T - \lambda I)$$

$$= T^*T - \lambda T^* - \bar{\lambda} T + \lambda \bar{\lambda} \qquad \qquad ... (4)$$

From (3) and (4) by the help of (1) we get that $T - \lambda I$ commutes with its adjoint and as such $T - \lambda I$ is a normal operator. Since $T - \lambda I$ is a normal operator therefore for any $x \in V.$

$$\|(T - \lambda I)x\| = \|(T - \lambda I)^* x\| \qquad \text{by (2)}$$

or $\qquad \|(Tx - \lambda x)\| = \|(T^* - \bar{\lambda} T)x\| = \|(T^* x - \bar{\lambda} x)$

Since x is characteristic vector of T corresponding to characteristic value λ

$$\therefore \qquad\qquad Tx = \lambda x$$

$$\therefore \qquad\qquad 0 = \| T^* x - \bar{\lambda} x \|$$

Hence $\qquad\qquad T^* x = \bar{\lambda} x$

Above relation shows that x is also characterstic vector of T^* and corresponding characteristic value is $\bar{\lambda}$. Thus any characteristic vector of T is also a characteristic vector of T^* if T be a normal operator.

2nd part. *Proper spaces of a normal operator T are pairwise orthogonal.*

Let x_i and x_j belong to M_i and M_j the proper (eigen) spaces of a normal operator T and the corresponding characteristic values be λ_i and λ_j respectively so that

$$Tx_i = \lambda_i x_i, \; Tx_j = \lambda_j x_j$$

and $\qquad\qquad T^* x_j = \bar{\lambda}_j x_j$ as T is normal.

Now $\qquad\qquad \lambda_i(x_i, x_j) = (\lambda_i x_i, x_j) = (Tx_i, x_j)$

$$= (x_i, T^* x_j) = (x_i, \bar{\lambda}_j x_j) = \lambda_j(x_i, x_j)$$

$$\therefore \qquad (\lambda_i - \lambda_j)(x_i, x_j) = 0 \qquad \Rightarrow \quad (x_i, x_j) = 0 \text{ as } \lambda_i \neq \lambda_j$$

Now $x_i \in M_i$ and $x_j \in M_j$ and $(x_i, x_j) = 0$ and hence $M_i \perp M_j$, i.e., M_i, M_j are pairwise orthogonal.

THEOREM 4. *If T be a normal operator then each eigen space M_i reduces T.*

Proof. Let $x_i \in M_i$ so that $Tx_i = \lambda_i x_i$ ...(1)

$\Rightarrow \quad T^* x = \bar{\lambda}_i x_i$ because in the case of T being normal a characteristic vector x corresponding to characteristic value λ is also a characteristic vector of T^* with corresponding characteristic value $\bar{\lambda}$.

Since M_i is a sub space $\therefore \qquad \bar{\lambda}_i x_i \in M_i$

Hence $T^* x_i = \bar{\lambda}_i x_i \in M_i$

Above relation shows that M_i is invariant under T^*.

Since M_i is invariant under both T and T^* we say M_i reduces T when T is normal.

THEOREM 5. *Let V be a finite dimensional inner product vector space and T be a self-adjoint linear operator on V. Then there exists an orthonormal basis for V each vector of which is a characteristic vector of T.*

Proof. The self-adjoint operator T on V has a characteristic vector α.

Now if we choose $\alpha_1 = \dfrac{\alpha}{\|\alpha\|}$ so that $\|\alpha_1\| = 1$ then since α is characteristic vector then each scalar multiple of α is also a characteristic vector for the same characteristic value.

$$Tx = \lambda x \quad\Rightarrow\quad T(cx) = cT(x) = c\lambda x = \lambda(c\,x).$$

i.e. if x is characteristic vector then cx is also a characteristic vector and corresponding characteristic value is same λ.

Now if dim. $V = 1$ then $\{\alpha_1\}$ is a basis of V which is orthonormal and consists of α_1 which is a characteristic vector of T.

Assume that the above holds for all spaces whose dimension is less than n the dimension of V.   ... (1)

Let W be a one dimensional subspace spanned by vector α_1. Now α_1 is a characteristic vector of T and W is spanned by α_1 so that. W_1 is invariant under T.

Also we know that if W is invariant under T then $W^\perp$ is invariant under T^*.

But T being self-adjoint it follows that $T^* = T$.

$\therefore$ $W^\perp$ is invariant under T.

Since $V = W \oplus W^\perp$ and dim. $V = n$, dim. $W = 1$.

$\therefore$ dim. $W^\perp = n - 1$, i.e., less than dim V

Therefore $W^\perp$ with the inner product from V is an inner product space of dimension one less than dimension of V.

Let u be a linear operator induced on $W^\perp$ by T, i.e. the restriction of T to $W^\perp$, then u is self-adjoint and hence by induction hypothesis (1) $W^\perp$ has an orthonormal basis $\{\alpha_2, \alpha_3, ..., \alpha_n\}$ consisting of $n-1$ vectors each of which is a characteristic vectors of u. Each of the above vectors is also a characteristic vector of T. Now because $V = W \oplus W^\perp$ it follows that the union of these basis, i.e., $\{\alpha_1, \alpha_2, ... \alpha_n\}$ is an orthonormal basis for V.

THEOREM 6. *Let V be a finite dimensional complex inner product vector space and T be a normal operator on V. Then V has an orthonormal basis each vector of which is a characteristic vector for T.*

Proof. The normal operator T has a characteristic vector α.

Now if we choose $\alpha_1 = \dfrac{\alpha}{\|\alpha\|}$ so that then $\|\alpha\| = 1$, since α is a characteristic vector of T each scalar multiple of α is also a characteristic vector of T for the same characteristic value.

i.e., $Tx = \lambda x \Rightarrow T(cx) = cT(x) = c\lambda x = \lambda(cx)$

i.e., if x is a characteristic vector then cx is also a characteristic vector and corresponding characteristic value is same.

Now if dim. $V = 1$ then $\{\alpha_1\}$ is a basis for V which is orthonormal and consists of α_1 which is a characteristic vector of T.

Assume that the above holds for all spaces whose dimension is less than n the dimension of V.   ...(1)

Let W be a one dimensional subspace generated by α_1. Now α_1 is a characteristic vector of T and W is spanned by α_1.

∴ W_1 is invariant under T.

But by theorem (3) which states that

If T be a normal operator and x is characteristic vector of T then x is also a characteristic vector of T^*.

Therefore α_1 is also a characteristic vector of T^*.

∴ W is invariant under T^*.

⇒ $W^\perp$ is invariant under $(T^*)^*$.

⇒ $W^\perp$ is invariant under T.

Since $V = W \oplus W^\perp$, dim. $V = n$ and dim. $W = 1$

∴   dim. $W^\perp = n-1$.

Therefore $W^\perp$ with the inner product from V is an inner product vector space of dimension one less than dimension of V

Let u be a linear operator induced on $W^\perp$ by T, *i.e.*, restriction of T to $W^\perp$ then u is normal and hence by induction hypothesis (1) $W^\perp$ has an orthonormal basis $\{\alpha_2, \alpha_3 ... \alpha_n\}$ consisting of $n-1$ vectors each of which is a characteristic vector of u. Each of these is also a characteristic vector of T. Now because $V = W \oplus W^\perp$, it follows that the union of the two basis $\{\alpha_1, \alpha_2,... \alpha_n\}$ is an orthonormal basis for V.

8.17 THE SPECTRAL THEOREM FOR NORMAL OPERATORS

THEOREM 1.   *Let T be an arbitrary linear operator on a complex finite dimensional inner product vector space $V(F)$ and $\lambda_1, \lambda_2, ... \lambda_n$ be the set of complex characteristic (eigen) values of T with characteristic (eigen) spaces $M_1, M_2 ... M_n$. Further $E_1, E_2, ... E_n$ are perpendicular projections on the spaces $M_1, M_2,... M_n$ respectively. Then following statements are equivalent :*

1. The subspaces M_i^s are pair wise orthogonal and span V.

2. E_i^s are pairwise orthogonal and

   (a) $E_1 + E_2 + ... + E_n = T$

   (b) $\lambda_1 E_1 + \lambda_2 E_2 + ...+ \lambda_n E_n = T$

3. T is normal operator.

Proof.   We shall prove the above theorem by proving that

$$1 \Rightarrow 2, \ 2 \Rightarrow 3 \text{ and } 3 \Rightarrow 1.$$

(1) ⇒ (2). Given E_i^s are perpendicular projections on M_i^s which are pairwise orthogonal and which span V.

i.e., $V = M_1 \oplus M_2 \oplus M_3 \oplus ... \oplus M_n$

Hence any vector $x \in V$ can be uniquely expressed as

$$x \in V = x_1 + x_2 + ... + x_n, \ x_i \in M_i$$

E_i^s are pairwaise orthogonal.

We know that, M_1, M_2 are subspaces of inner product vector space and E_1, E_2 are perpendicular projections on M_1, M_2 respectively then M_1 and M_2 are orthogonal ⇔ $E_1 E_2 = O \Leftrightarrow E_2 E_1 = O$.

Therefore E_i^s are pairwise orthogonal

i.e. $\qquad\qquad\qquad E_i\,E_j \;=\; O, i \neq j$

$\qquad\qquad E_1 + E_2 + \ldots + E_n = I$

E_i is a projection on M_i and $\quad x = x_1 + x_2 + x_3 + \ldots + x_n.$

$\qquad x_i \in M_i$

$\therefore \qquad\qquad\qquad E_i\,x_i = x_i \;\forall\; i \text{ and } E_i\,x_j = 0$

$\therefore \qquad\qquad\qquad\quad E_i x = x_i$

$\therefore \qquad\qquad\qquad x = x_1 + x_2 + x_3 + \ldots + x_n$

or $\qquad\qquad\qquad x = E_1 x_1 + E_2 x_2 + \ldots + E_n x_n$

$\therefore \qquad\qquad (E_1 + E_2 + \ldots + E_n)x \qquad\qquad\qquad\qquad \ldots (1)$

$\qquad\qquad\quad = (E_1 + E_2 + \ldots + E_n)\,(E_1 x_1 + E_2 x_2 + \ldots + E_n x_n)$

$\qquad\qquad\quad = E_1^2 x_1 + E_2^2 x_2 + \ldots + E_n^2 x_n + O$

$\qquad\qquad E_i E_j = O, i \neq j \text{ and } E_i^2 = E_i$

$\qquad\qquad\quad = E_1 x_1 + E_2 x_2 + \ldots + E_n x_n$

$\qquad\qquad\quad = x \text{ by } (1) = I\,x$

Since above is true for all x. It follows that

$\qquad E_1 + E_2 + \ldots + E_n = I$

$\qquad\qquad\qquad T = \lambda_1 E_1 + \lambda_2 E_2 + \ldots + \lambda_n E_n$

Now $x_i \in M_i$ the eigen space of T corresponding to eigenvalue λ_i.

$\therefore \qquad\qquad\qquad Tx_i = \lambda_i x_i$

$\therefore \qquad\qquad\qquad Tx = T(x_1 + x_2 + \ldots + x_n)$

$\qquad\qquad\qquad\quad = Tx_1 + Tx_2 + \ldots + Tx_n$

$\qquad\qquad\qquad\quad = \lambda_1 x_1 + \lambda_2 x_2 + \ldots + \lambda_n x_n \text{ by } (2)$

$\qquad\qquad\qquad\quad = \lambda_1 E_1 x + \lambda_2 E_2 x + \ldots + \lambda_n E_n x \qquad E_i x = x_i$

$\qquad\qquad\qquad\quad = (\lambda_1 E_1 + \lambda_2 E_2 + \ldots + \lambda_n E_n)x$

Since above holds for all x we have

$\qquad\qquad\qquad T = \lambda_1 E_1 + \lambda_2 E_2 + \ldots + \lambda_n E_n$

Above completes the proof $(1) \Rightarrow (2)$.

(2) $\Rightarrow$ (3).

Given. (a) E_i^s are pairwise orthogonal, i.e. $E_i\,E_j = O, i \neq j$

(b) $\quad I = E_1 + E_2 + \ldots + E_n$

(c) $\quad T = \lambda_1 E_1 + \lambda_2 E_2 + \ldots + \lambda_n E_n$

To prove that T is normal operator, i.e. $TT^* = T^*T$

By (b) $\qquad\qquad T = \lambda_1 E_1 + \lambda_2 E_2 + \ldots + \lambda_n E_n$

$\therefore \qquad\qquad T^* = (\lambda_1 E_1 + \lambda_2 E_2 + \ldots + \lambda_n E_n)^*$

$\qquad\qquad\quad = (\lambda_1 E_1)^* + (\lambda_2 E_2)^* + \ldots + (\lambda_n E_n)^*$

$$= \bar{\lambda}_1 E_1{}^* + \bar{\lambda}_2 E_2{}^* + \dots + \bar{\lambda}_n E_n{}^* \qquad (cT)^* = \bar{c}\, T^*$$

$$= \bar{\lambda}_1 E_1 + \bar{\lambda}_2 E_2 + \dots + \bar{\lambda}_n E_n \qquad E_i^* = \bar{E}_i = E_i$$

$$\therefore \qquad TT^* = (\lambda_1 E_1 + \lambda_2 E_2 + \dots + \lambda_n E_n) \cdot (\bar{\lambda}_1 E_1 + \bar{\lambda}_2 E_2 + \dots + \bar{\lambda}_n E_n)$$

$$= \sum_{i=1}^{n} \lambda_i \bar{\lambda}_i E_i^2 + \sum_{j \neq i} \lambda_i \bar{\lambda}_j E_i E_j$$

$$= \sum_{i=1}^{n} |\lambda_i|^2 E_i + O \text{ by (a) and } E_i^2 = E_i$$

In a similar manner we can show that

$$T^*T = \sum_{i=1}^{n} |\lambda_i|^2 E_i$$

Hence $TT^* = T^*T$ and as such T is a normal operator.

(3) $\Rightarrow$ **(1)**

Given that T is normal, *i.e.*, $TT^* = T^*T$

To prove that

(a) M_i^s are pairwise orthogonal.

(b) M_i^s span V.

We know that if T be a normal operator and x is an eigenvector of T corresponding to eigenvalue λ then x is also an eigen vector of T^* corresponding to eigenvalue $\bar{\lambda}$ and further the eigenspaces of a normal operator T are pairwise orthogonal. Hence we prove that M_i^s are pairwise orthogonal.

M_i^s span V.

If T be a normal operator then each eigenspace M_i reduces T which is other words mean that M_i is invariant under both T and T^*. We have proved above that M_i^s are pairwise orthogonal and E_i^s are perpendicular projections on M_i^s and hence E_i^s are pairwise orthogonal [as the ranges M_i^s are orthogonal] and hence

$$E = E_1 + E_2 + \dots + E_n \text{ is a perpendicular projection on} \qquad \dots (1)$$

$$M = M_1 + M_2 + \dots M_n \qquad \dots (2)$$

where each M_i reduces T as T is a normal operator.

Again we know that, If E is a perpendicular projection on a subspace M of an inner product vector space and T a linear operator on V then

M reduces $\qquad T \Leftrightarrow TE = ET$

Here as shown above each M_i reduces T.

$\Leftrightarrow \qquad TE_i = E_i T$ where E_i is a perpendicular projection on M_i.

$\Leftrightarrow \qquad \Sigma TE_i x = \Sigma E_i Tx. \forall x.$

$\Leftrightarrow \qquad T\Sigma E_i x = (\Sigma E_i)Tx \Leftrightarrow TEx = ETx$ by (1)

$\Leftrightarrow$　　　　　　　$TE = ET$ where E is a projection on $M = \Sigma M_i$ by (1) & (2).

Hence M also reduces, T i.e. M is invariant under both T and T^*.

Since M is invariant under $T^* \Rightarrow M^{\perp}$ is invariant under $(T^*)^*$

$\therefore$　$M^{\perp}$ is invariant under T.

Now let us suppose that $M^{\perp}$ is non-empty and all the eigenvectors of T are members of M. Now $M^{\perp}$ is invariant under T where T is an operator on a finite dimensional non-empty inner product space $M^{\perp}$ which has no eigenvectors and hence no eigenvalues in $M^{\perp}$. (All the eigenvectors are in (M). This is not possible.

Therefore $M^{\perp}$ must be empty.

Again since $V = M \oplus M^{\perp}$ by projection theorem and $M^{\perp}$ being empty we conclude that $V = M$, i.e., M_i^s span V.

8.17.1 Spectral form of T (Normal)

The representation of a normal operator T on an inner product vector space $V(F)$ in the form.

$$T = \lambda_1 E_1 + \lambda_2 E_2 + ... + \lambda_n E_n$$

where $E_1 + E_2 + ... + E_n = I$ and E_i^s are pairwise orthogonal projections on M_i^s (the eigenspaces) which are also orthogonal is called spectral form of T.

Solved Examples

Example 1. *If M_i^s be subspaces of an inner product vector space and E_i^s be perpendicular projections on M_i^s then show that the following two statements are equivalent.*

(a) $V = M_1 \oplus M_2 \oplus ... \oplus M_n$ and this is an orthogonal direct sum, i.e., M_i^s are pairwise orthogonal.

(b) $I = E_1 + E_2 + ... + E_n$ and $E_i E_j = 0, i \neq j$.

Prove further that if B_i is an orthogonal basis for M_i, $i = 0, 1, 2 ... n$ then $B = \sum\limits_{i=1}^{n} B_i$ is an orthogonal basis for V.

Solution. **(a) $\Rightarrow$ (b).** We have proved it in $1 \Rightarrow 2$ of last Theorem.

Now we shall prove that $b \Rightarrow a$.

Let $x \in V$ then $x = Ix = (E_1 + E_2 + ... + E_n)x$

or　　　　　　　$x = E_1 x + E_2 x + ... + E_n x$

But $E_i x \in$ range of E_i which is M_i.

Hence (1) expresses a vector x as the sum of vectors one from each M_i.

$\therefore$　　　　　　　$V = M_1 + M_2 + ... + M_n$

In order to strengthen this relation to

$$V = M_1 \oplus M_2 \oplus ... \oplus M_n$$

we have to show that expression (1) is unique.

Let if possible

$$x = x_1 + x_2 + ... + x_n \text{ with } x_i \in M_i \qquad ...(2)$$

then $\qquad E_i x_i = x_i$

$\therefore \qquad\qquad x = E_1 x_1 + E_2 x_2 + \dots + E_n x_n$

$\therefore \qquad\qquad E_i x = E_i(E_1 x_1 + E_2 x_2 + \dots + E_n x_n) = E_i^2 x_i$

$\therefore \qquad\qquad E_{ij} = O$

or $\qquad\qquad E_i x = E_i x_i$ or $E_i x = x_i \qquad\qquad E_i^2 = E_i \qquad\qquad ..\ (3)$

Hence from (2) and (3) we get

$$x = E_1 x + E_2 x + \dots + E_n x$$

Thus the expression is unique.

$\therefore \qquad\qquad V = M_1 \oplus M_1 + \dots \oplus M_n$

Thus V is direct sum. In order to prove that it is orthogonal direct sum we have to show that M_i and M_j are orthogonal, *i.e.* $(\alpha_i, \alpha_j) = 0$, $\alpha_i \in M_i$, $\alpha_j \in M_j$

$$(\alpha_i, \alpha_j) = (E_i \alpha_i, E_j \alpha_j) = (\alpha_i, E_i * E_j \alpha_j) = (\alpha_i, E_i E_j \alpha_j)$$

$$= (\alpha_i, O\alpha_j) = (\alpha_i, 0) = 0 \qquad E_i E_j = O, i \ne j$$

Hence M_i is orthogonal to M_j, $i \ne j$.

Hence $V = M_1 \oplus M_2 \dots \oplus M_n$ is orthogonal direct sum.

Further, since V is orthogonal direct sum of M_i^s therefore any $x \in V$ is expressible as

$$x = x_1 + x_2 + \dots + x_n \text{ with } x_i \in M_i \text{ and } (x_i, x_j) = 0; i \ne j$$

Now B_i is an orthonormal basis for M_i then each x_i is expressible as linear combination of elements of B_i and this is true for each i.

Hence $x \in V$ is expressible as linear combination of elements of $B_1, B_2 \dots B_n$ the basis of M_i^s or $x \in V$ is expressible as linear combination of elements of $B = \sum\limits_{i=1}^{n} B_i$.

$\therefore \qquad B$ spans V.

Again B is orthonormal set because each B_i is an orthonormal basis set and then any vector in B_i is orthogonal to any vector in B_j as B_i, B_j are basis sets for M_i and M_j which are orthogonal. Hence the set $B = \cup B_i$ is also orthonormal.

Now we know that every orthonormal set is linearly independent as such B is linearly independent and it spans V. Hence $B = \cup B_i$ is a basis set for V.

Example 2. *Let T be a normal operator on a finite dimensional complex inner product space V, then show that V has an orthonormal basis B consisting of characteristic vectors of T. Consequently the matrix of T relative to B is a diagonal matrix.*

Solution. Given, since T is a normal operator therefore by spectral theorem for normal operators we have

$$T = \lambda_1 E_1 + \lambda_2 E_2 + \dots + \lambda_n E_n$$

where λ_i^s are all distinct characteristic vectors of T with characteristic spaces M_i^s E_i^s are perpendicular projections on M_i^s, such that

$$E_1 + E_2 + \dots + E_n = I$$

Proof. Since T is a normal operator therefore the eigen spaces of normal operator are orthogonal, *i.e.*, M_i^s are pairwise orthogonal.

Now let $B_1, B_2 ... B_n$ be orthonormal basis for the eigen spaces $M_1, M_2 ... M_n$ then we shall establish that $B = \cup B_i$ is an orthonormal basis for V and for this we will have to show the following :

(1) B is orthonormal. (2) B is linearly independent. (3) B spans V.

(1) Now B_i and B_j are orthonormal basis sets for M_i and M_j. But M_i and M_j are orthogonal and hence each vector in B_i is orthogonal to each vector in B_j. Hence the set $B = \cup B_j$ is also an orthonormal set.

(2) Further we know that in a finite dimensional inner product vector space an orthonormal set is always a linearly independent set.

(3) Let $x \in V$ then $x = Ix = (E_1 + E_2 + ... + E_n)x$

or $x = E_1 x + E_2 x + ... + E_n x = x_1 + x_2 + ... + x_n$

Now $E_i x \in$ range of E_i which is M_i and hence $x_i \in M_i$

and B_i is an orthonormal basis for M therefore each $x_i \in M_i$ is expressible as a linear combination of elements of B_i, and this is true for each i.

Hence $x \in V$ is expressible as a linear combination of elements of B_i^s which are basis for M_i^s or $x \in V$ is expressible is a linear combination of elements of $B = \cup B_i$

$\therefore$ B spans V.

Hence B is an orthonormal basis set of V.

Now each non-zero vector in M_i, is a characteristic vector of T. Therefore each vector in B_i, which is a basis for M_i is also a characteristic vector of T. Hence each vector in $B = \cup B_i$ is a characteristic vector of T. Hence there exists an orthonormal basis of B consisting of characteristic vectors of T and hence by Th. the matrix of T relative to B is diagonal matrix.

8.18 SPECTRAL THEOREM FOR SELF- ADJOINT OPERATORS

THEOREM 1. *Let T be a self-adjoint operator on a finite dimeneionsl inner product vector space $V(F)$ then there exists n real numbers $\lambda_1, \lambda_2 ... \lambda_n$ and perpendicular projections $E_1 + E_2 + ... + E_n$ (where n is strictly +ive, not greater than the dimension of the vector space) so that*

(i) $\lambda_i, i = 1, 2, ,..., n$ are pairwise distinct.

(ii) E_i^s are pairwise orthogonal and different from zero.

(iii) $E_1 + E_2 + ... + E_n = I$

(iv) $\lambda_1 E_1 + \lambda_2 E_2 + ... + \lambda_n E_n = T$

Proof.   **1.** $\lambda_1, \lambda_2, ..., \lambda_n$ are distinct and real.

We know from f, that the proper values of an operator T on finite dimensional inner product vector space $V(F)$ are the roots of the characteristic equation $|A - \lambda I| = 0$ where A is the matrix of the operator T with respect to basis B.

The above being an equation of nth degree in λ will have n distinct values in case some be repeated.

Thus $\lambda_1, \lambda_2, ... \lambda_n$ where $n \leq$ the dimension of the vector space are distinct

Also which states that if T be a self-adjoint linear operator on and inner product vector space then the characteristic values of T are real and the characteristic vectors of T corresponding to distinct characteristic values are orthogonal.

Thus all the λ_i^s are real and distinct.

Let M_1, M_2, ..., M_n be the proper (eigen) spaces of T corresponding to eigen values λ_1, λ_2, ..., λ_n, i.e., M_i is the subspace consisting of all the characteristic vectors of T corresponding to characteristic value λ_i i.e. consisting of all solutions of $Tx = \lambda_i x$, $x \neq 0 \in V$

Also by above theorem M_i^s are pairwise orthogonal as T is a self-adjoint operator.

2. *E_j^s are pairwise orthogonal and different from zero.*

Let E_1, E_2 ... E_n be perpendicular projections on eigenspaces M_1, M_2,... M_n respectively.

Now $\qquad M_i = \{x : Tx = \lambda_i x\}$, $x \neq 0$

and E_i is a perpendicular projection on M_i which must be different from 0 as $E_i x = x \in M_i$, and if E_i were 0 then $0x = 0$ which would mean that all vectors in M_i are zero vectors. Hence E_i^s are different from zero.

Again since the range spaces M_i^s are pairwise orthogonal therefore the perpendicular projections E_i^s are also pairwise orthogonal, since we know that M_1, M_2 are subspaces of inner product vector space and E_1, E_2 are perpendicular projections on M_1, M_2 respectively then

M_1, M_2 are orthogonal $\Leftrightarrow E_1 E_2 = 0 \Leftrightarrow E_2 E_1 = 0$

Hence E_i^s are pairwise orthogonal and different from zero.

3. $E_1 + E_2 + ... + E_n = 1$

Now since E_i^s are perpendicular projections on M_i^s which are pairwise orthogonal and hence their sum $E = E_1 + E_2 + ... + E_n$ is a perpendicular projection on $M = M_1 + M_2 + ... + M_n$ where each M_i is invariant under T by definition and hence also invariant under T^* because $T^* = T$ as T is self-adjoint.

Now we shall prove that $V = M_1 \oplus M_2 \oplus ... \oplus M_n$.

We know that If E is a perpendicular projection on a sub space M of an inner product vector space and T a linear operator on V then M reduces $T \Leftrightarrow TE = ET$. As shown above each M_i reduces T.

$\Leftrightarrow \qquad TE_i = E_i T$ as E_i is a perpendicular projection on M_i.

$\Leftrightarrow \qquad \Sigma TE_i x = \Sigma E_i Tx \forall x$

$\Leftrightarrow \qquad T\Sigma E_i x = (\Sigma E_i)\, Tx \Leftrightarrow TEx = ETx$

$\Leftrightarrow \qquad TE = ET$ where E is a projection on $M = \Sigma M_i$

Hence M also reduces T, i.e., M is invariant under both T and T^*, i.e., T.

Since M is invariant under $T \Rightarrow M^\perp$ is invariant under T^*, i.e., T.

Now let us suppose that $M^\perp$ is non empty and all the eigenvectors of T are members of M. Now $M^\perp$ is invariant under T where T is an operator on a finite dimensional non-empty inner product space $M^\perp$ which has no eigenvectors and hence no eigenvalues. This is not possible. Hence $M^\perp$ must be empty.

Again since $V = M \oplus M^{\perp}$ by projection theorem and $M^{\perp}$ being empty we conclude that

$$V = M = \Sigma M_i, \text{ i.e., } V \text{ is spanned by } M_i^s. \text{ Since } i^s \text{ are pairwise orthogonal.}$$

Therefore $i = M_1 \oplus M_2 \oplus ... \oplus M_n$

Hence any $x \in V$ can be uniquely expressed as

$$x = x_1 + x_2 + ... + x_n \text{ with } x_i \in M_i \text{ and } (x_i, x_i) = 0 \qquad ... (1)$$

Now E_1 is a projection on M_i

$$\therefore \qquad\qquad E_i x = E_i (x_1 + x_2 + ... + x_n) = x_i$$

$\therefore \quad E_i x_j = 0$ as E_i is a perpendicular projection on M_i which is orthogonal to M_j and $x_j \in M_j$

Now $\qquad\qquad Ix = x_1 + x_2 + ... + x_n$

$$= E_1 x + E_2 x + ... + E_n x \text{ by (1).}$$
$$= (E_1 + E_2 + ... + E_n)x$$

Since above is true for all x

$$\therefore \qquad\qquad E_1 + E_2 + ... + E_n = I$$

4. $\lambda_1 E_1 + \lambda_2 E_2 + ... + \lambda_n E_n = T$

Now $x_i \in M_i$ the eigenspace of T corresponding to eigenvalue λ_i

such that $\qquad\qquad Tx_i = \lambda_i x_i \qquad\qquad\qquad\qquad ... (2)$

$$\therefore \qquad\qquad Tx = T(x_1 + x_2 + ... + x_n) \qquad\qquad\qquad ... (3)$$
$$= Tx_1 + Tx_2 + ... + Tx_n$$
$$= \lambda_1 x_1 + \lambda_2 x_2 + ... + \lambda_n x_n$$
$$= \lambda_1 E_1 x + \lambda_2 E_2 x + ... + \lambda_n E_n x \qquad E_i x = x_i \text{ by } I$$
$$= (\lambda_1 E_1 + \lambda_2 E_2 + ... + \lambda_n E_n)x$$

Since above holds for all $x \in V$

$$\therefore \qquad\qquad T = \lambda_1 E_1 + \lambda_2 E_2 + ... + \lambda_n E_n$$

8.18.1 Spectral form of T (Self-adjoint)

The representation of a linear operator T in the form

$$T = \Sigma \lambda_i E_i = \lambda_1 E_1 + \lambda_2 E_2 + ... + \lambda_n E_n$$

where all λ_i^s are distinct and real, E_i^s are pairwise orthogonal and non-zero, such that $E_1 + E_2 + ... + E_n = I$ is called spectral form of T.

THEOREM 2. *If* $\sum\limits_{i=1}^{n} \lambda_i E_i$ *is the spectral form of a self-adjoint operator T on a finite dimensional inner product vector space then* λ_i^s *are all the distinct proper values of T. If moreover* $1 \le k \le n$ *then there exists polynomials P_k with real coefficients such that $P_k(\lambda_i) = 0$ whenever $i \ne k$ and $P_k(\lambda_k) = 1$. For all such Polynomials $P_k(T) = E_k$, i.e. each E_k is a polynomial in T.*

Proof. Since $T = \sum\limits_{i=1}^{n} \lambda_i E_i$ is the spectral form of T, it follows that E_i^s are perpendicular projections which are pairwise orthogonal and different from zero and also $\Sigma E_i = I$.

Now $E_i \ne O$ therefore $\exists$ a vector x in the range of E_i

such that $E_i x = x$ are $E_j x = 0$ if $i \neq j$.

Also $\qquad\qquad T = \sum_{i=1}^{n} \lambda_i E_i$ $\qquad\qquad$ As given $\qquad$ ... (2)

$\therefore \qquad\qquad Tx = \sum_{i=1}^{n} \lambda_i E_i(x)$

$\qquad\qquad\qquad = \lambda_1 E_1 x + \lambda_2 E_2 x + ... + \lambda_i E_i x + ... + \lambda_n E_n x, \qquad x \in$ Range of E_i

$\qquad\qquad\qquad = \lambda_i E_i x = \lambda_i x$ by (1)

since $\qquad\qquad Tx = \lambda_i x$

Above relation implies that each λ_i is an eigenvalue of T. λ_i^s are distinct.

Let λ be any eigenvalue of T so that $Tx = \lambda x$. $(x \neq 0)$ $\qquad\qquad$ ... (4)

Put $\qquad\qquad T = \sum \lambda_i E_i$ and $x = Ix = (\sum E_i)_x$ in (4)

$\therefore \qquad\qquad \sum(\lambda_i E_i)x = \lambda(\sum E_i) x$

or $\quad \lambda_1 E_1 x + \lambda_2 E_2 x + ... + \lambda_n E_n x = \lambda E_1 x + \lambda E_2 x + ... + \lambda E_n x$.

or $\quad (\lambda - \lambda_1)E_1 x + ... + (\lambda - \lambda_n) E_n (x) = 0$ $\qquad\qquad$ ...(5)

Now $E_1 x, E_2 x, ..., E_n(x)$ are vectors in the ranges of E_i^s which are pairwise orthogonal *i.e.,* ranges are orthogonal and hence the above vectors are pairwise orthogonal.

Operate the sum (5) by $E_1, E_2, ..., E_n$ successively and since $E_i E_j = O$ and $E_i^2 = E_j$ we have

$\qquad (\lambda - \lambda_1)E_1 x = 0, (\lambda - \lambda_1)E_2 x = 0, ... (\lambda - \lambda_n)E_n(x) = 0$

Now all the vectors $E_1 x$ ... $E_n x$ cannot be zero because in that case $(E_1 + E_2 + ... + E_n)x = 0$

or $\qquad\qquad Ix = 0$ or $x = 0$ which is not possible.

Hence at least one of them must be non-zero.

Let $E_i x \neq 0$, $\lambda - \lambda_i = 0$ or $\lambda = \lambda_i$, *i.e.,* λ is one of the λ_i^s .

So, non-zero vectors x_i^s form a linearly independent set since, we know that Every orthogonal set (or orthonormal set) of non-zero vectors in an inner product vector space is linearly independent set. Therefore in relation (6) all non-zero vectors x_i^s will form a linearly independent set and as such all the scalars should be zero.

$\therefore \qquad\qquad \lambda - \lambda_i = 0$ for each i, *i.e.,* λ is equal to one of $\lambda_1, \lambda_2, ... \lambda_n$.

Also each λ_i is real as eigenvalues of self-adjoint operator are real.

3rd part.

Since E_i^s are pairwise orthogonal. Threfore $E_i E_j = O$, $i \neq j$

Also $\qquad\qquad E_i^s = E_i \ \forall i$

$\qquad\qquad\qquad T^0 = I = E_1 + E_2 + ... + E_n$

$\qquad\qquad\qquad T = \sum \lambda_i E_i$

$\therefore \qquad\qquad T^2 = (\sum \lambda_i E_i)(\sum \lambda_j E_j)$

$$= \sum_i \sum_j \lambda_i \lambda_j E_i E_j = \sum_i \lambda_i^2 E_i^2 \text{ by above relation.}$$

or
$$T^2 = \sum \lambda_i^2 E_1 \text{ as } E_i^2 = E_i$$

$$T^3 = \sum \lambda_i^3 E_i \text{ and in general}$$

$$T^n = \sum \lambda_i^n E_i$$

and in case when $n = 0$ we get $I = \sum E_i$ where I is identity operator.
Hence for every +positive integer n we have

$$T^n = \sum \lambda_i^n E_i$$

$$\therefore \qquad P(T) = \sum_{i=1}^{n} P(\lambda_i) E_i \quad \text{for every polynomial } P.$$

Consider the Lagranges polynomial corresponding to scalar (real) $\lambda_1, \lambda_2, \dots \lambda_n$

$$P_k(x) = \frac{(x - \lambda_1)(x - \lambda_2)\dots(x - \lambda_{k-1})(x - \lambda_{k+1})\dots(x - \lambda_n)}{(\lambda_k - \lambda_1)(\lambda_k - \lambda_2)\dots(\lambda_k - \lambda_{k-1})(\lambda_k - \lambda_{k+1})\dots(\lambda_k - \lambda_n)}$$

In the numerator of polynomial P_k the factor $x - \lambda_k$ is missing.
Clearly $p_k(\lambda_i) = 0$, $i \neq k$ as the factor $x - \lambda_i = 0$ when $x = \lambda_i$.
Also $p_k(\lambda_k) = 1$ as N^r and D^r of $p_k(x)$ become identical for $x = \lambda_k$.
In other words $p_k(\lambda_i) = \delta_{ki}$ or $p_i(\lambda_j) = \delta_{ij}$.

$$p_k(T) = E_k$$

$$P(T) = \sum_{i=1}^{n} P(\lambda_i) E_i$$

$$\therefore \qquad p_k(T) = \sum_{i=1}^{n} p_k(\lambda_i) E_i$$

Summing w.r.t. i all the terms will vanish except one when
$$i = k.$$

$$\therefore \qquad p_k(T) = p_k(\lambda_k) E_k = 1. E_k = E_k$$

Solved Examples

Example 1. *Let T be a normal operator on a finite dimensional complex inner product vector space and $\lambda_1, \lambda_2, \dots \lambda_n$ be distinct complex numbers and $E_1, E_2, \dots, E_n$ be non-zero linear operators on V such that*

(a) $T = \lambda_1 E_1 + \lambda_2 E_2 + \dots + \lambda_n E_n$

(b) $I = E_1 + E_2 + \dots + E_n$

(c) $E_i E_j = O$ for $i \neq j$

Then prove that following :

(1) Each E_i is a projection

(2) λ_i^s are precisely the distinct characteristic values of T.

(3) E_i is a projection on M_i where W_i is characteristic space corresponding to characteristic value λ_i.

(4) *Each E_i is a polynomial in T.*

(5) *Each E_i is a perpendicular projection.*

Solution.   **(1)** E_i is a projection. Therefore $I = (E_1 + E_2 + ... + E_n)$

$\therefore \qquad\qquad E_i I = E_i (E_1 + E_2 + ... + E_n)$

$\therefore \qquad\qquad E_i = E_i^2 \quad E_i E_j = O \text{ if } i \neq j$

Since $\qquad\qquad E_i^2 = E_i$ therefore E_i is a projection.

(2) *To show that λ_i is a characteristic value of T.*

i.e., for any non zero vector x, $Tx = \lambda_i x$.

Now E_i is given to be non-zero operator which has been shown to be a projection and hence $\exists$ non-zero vector x in the range of E_i, such that

$$E_i x = x.$$

Now $\qquad\qquad Tx = (\lambda_1 E_1 + \lambda_2 E_2 + ... + \lambda_n E_n)x$

$$= (\lambda_1 E_1 + \lambda_2 E_2 + ... + \lambda_n E_n)E_i x$$

$$= \lambda_i E_i^2 x = \lambda_i E_i x = \lambda_i x$$

Since $Tx = \lambda_i x$ where x is non-zero vector therefore λ_i is a characteristic value of T.

Now we shall show that λ_i^s are distinct characteristic values of T.

Let λ be any characteristic value of T so that $\exists$ a non-zero vector x such that

$$Tx = \lambda x = \lambda I x$$

$\therefore \qquad (\lambda_1 E_1 + \lambda_2 E_2 + ... + \lambda_n E_n)x = \lambda (E_1 + E_2 + ... + E_n)x$

$\therefore \qquad (\lambda_1 - \lambda)E_1 x + ... + (\lambda_1 - \lambda)E_n x = 0$

Operating both sides of above by E_i all the terms will cancel because

$$E_i E_j = O \text{ if } i \neq j$$

$\therefore \qquad (\lambda_i - \lambda)E_i^2 x = 0 \text{ or } (\lambda_i - \lambda)E_i x = 0$

$$E_i^2 = E_i$$

Now $E_i x \neq O$ for each i because if it is true then

$$E_1 x + E_2 x + ... + E_n x = 0$$

or $\qquad\qquad (E_1 + E_2 + ... + E_n)x = 0$

or $\qquad\qquad\qquad Ix = 0 \text{ or } x = 0$

which is not true.

Hence we must have $\lambda_i - \lambda = 0$ or λ must be equal to λ_i for some i.

(3) *E_i is a projection on M_i where W_i is characteristic space corresponding to characteristic value λ_i.*

If M_i is a characteristic space corresponding to characteristic value λ_i then $\exists$ x in M_i such that

$$Tx = \lambda_i x = \lambda_i I x$$

$$(\lambda_1 E_1 + \lambda_2 E_2 + ... + \lambda_n E_n)x = \lambda_i (E_1 + E_2 + ... + E_n)x$$

or $(\lambda_1 - \lambda_i)E_1 x + ... + (\lambda_n - \lambda_i)E_n x = 0$

Operating both sides by E_i we get, all the terms except one will vanish because

$$E_i E_j = O \text{ if } i \neq j$$

∴ $(\lambda_j - \lambda_i)E_j^2 x = 0 \Rightarrow (\lambda_j - \lambda_i)E_j x = 0$

But $\lambda_j - \lambda_i \neq 0$ as λ_i^s are distinct as shown in (2)

∴ $E_j x = 0 \text{ if } j \neq i$

∴ $x \in M_i = Ix = (E_1 + E_2 + ... + E_n)x$

                    $= E_1 x + E_2 x + ... + E_n x = E_i x \in \text{ range of } E_i$

            $E_i x = x \Rightarrow x \in \text{ range of } E_i$

∴ $x \in M_i \Rightarrow x \in \text{ range of } E_i$

∴ $M_i \subset R(E_i)$ ...(1)

Again if $x \in R(E_i)$ then

                  $E_i x = x$

∴ $Tx = (\lambda_1 E_1 + \lambda_2 E_2 + ... + \lambda_n E_n)x$

                    $= (\lambda_1 E_1 + \lambda_2 E_2 + ... + \lambda_n E_n)E_i x$

                    $= \lambda_i E_i^2 x = \lambda_i E_i x = \lambda_i x$

∴ $x \in M_i$ the characteristic space corresponding to characteristic value λ_i

∴ $R(E_i) \subset M_i$ ... (2)

∴   Range of $E_i = M_i$ by (1) and (2)

or E_i is a projection on M_i

(4) *Each E_i is a polynomial in T.*

We have already proved this result.

(5) *E_i is a perpendicular projection.*

E_i is a polynomial in T where T is a normal operator. We know that if T is normal then $f(T)$ is also normal and hence E_i is also normal operator.

i.e., $E_i E_i * = E_i * E_i$

Also E_i being a projection is idempotent, *i.e.,* $E_i^2 = E_i$

Hence E_i is self-adjoint $E_i * = E_i$

Because if $T^2 = T$ then T is Self Adjoint $\Leftrightarrow TT^* = T^*T$

Therefore E_i is a perpendicular projection as $E_i^2 = E_i = E_i *$

When T is self adjoint. The proofs for 1, 2, 3 and 4 will be same.

∴ $E_i^* = [p_i(T)]^* = p_i(T^*)$

                    $(cT)^* = cT^*$ when c is real

∴ $E_i^* = p_i(T)$ $T^* = T,$

                    $= E_i$

Since $E_i^* = E_i$ ∴ E_i is a perpendicular projection .

Example 2. *If T be a normal operator on a finite dimensional complex inner product space V then prove the following :*

(a) each characteristic value of T is real $\Rightarrow$ T is self-adjoint.

(b) each characteristic values T is positive $\Rightarrow$ T is positive.

(c) each characteristic value of T has absolute value I $\Rightarrow$ T is unitary.

Solution. Since T is a normal operator . Therefore,

$$T = \lambda_1 E_1 + \lambda_2 E_2 + \ldots + \lambda_n E_n$$

is the spectral resolution of T where $E_1 + E_2 + \ldots + E_n = I$ and E_i^s are perpendicular projections which are pairwise orthogonal

i.e., $\qquad E_i^* = E_i$ and $E_i E_j = O.\ i \ne j$

on M_i^* (eigenspaces) which are also orthogonal. Also λ_i^s are distinct characteristic values of T.

(a) Each characteristic value of T is real, *i.e.,* $\ \overline{\lambda}_i = \lambda_i \ \forall\ i$

and we have to prove that T is self-adjoint, *i.e.,* $T^* = T$.

$$T = \lambda_1 E_1 + \lambda_2 E_2 + \ldots + \lambda_n E_n$$
$$\therefore \qquad T^* = (\lambda_1 E_1 + \lambda_2 E_2 + \ldots + \lambda_n E_n)^*$$
$$= \overline{\lambda}_1 E_1{}^* + \overline{\lambda}_2 E_2{}^* + \ldots + \overline{\lambda}_n E_n{}^*$$
$$= \lambda_1 E_1 + \lambda_2 E_2 + \ldots + \lambda_n E_n = T$$

Because $\overline{\lambda}_i = \lambda_i$ and $E_i^* = E_i$ is a perpendicular projection.

Since $T^* = T$.

$\therefore$ T is self-adjoint.

(b) Each characteristic values of T is positive *i.e.* $\lambda_i > 0 \ \forall\ i$. To prove that T is positive, *i.e.* ,T is self adjoint and

$$(Tx,\ x) \ge 0 \ \forall\ x$$

Since each characteristic value of T is positive and hence it is real so that by case (a) it follows that T is self-adjoint.

Now $\qquad (T\alpha,\ \alpha) = (T\alpha,\ I\alpha)$

$$= \left(\sum_{i=j}^{n} \lambda_i E_i \right) \alpha, \left(\sum_{j=1}^{n} E_j \right) \alpha$$

$$= \sum_{i=1}^{n} \sum_{j=1}^{n} \lambda_i (E_i \alpha, E_j \alpha)$$

$$= \sum_{i=1}^{n} \sum_{j=1}^{n} \lambda_i (\alpha, E_i{}^* E_j \alpha) = \sum_{i=1}^{n} \sum_{j=1}^{n} \lambda_i (\alpha, E_i E_j \alpha)$$

Summing w.r.t. j all the terms will cancel as $E_i E_j = O\ j \ne i$ except one term when $j = i$.

$$\therefore \qquad (T\alpha, \alpha) = \sum_{i=1}^{n} \lambda_i (\alpha, E_i E_i \alpha) = \sum_{i=1}^{n} \lambda_i (E_i{}^* \alpha, E_i \alpha)$$

$$= \sum_{i=1}^{n} \lambda_1 (E_i \alpha, E_i \alpha) = \sum_{i=1}^{n} \lambda_i \|E_i \alpha\|^2 \qquad \therefore$$

Now $\lambda_i > 0$ and $\|E_i\alpha\|^2 \geq 0$

$\therefore \qquad (T\alpha, \alpha) \geq 0.$

Since T is self-adjoint and also $(T\alpha, \alpha) \geq 0.$

$\therefore \quad T$ is positive

(c)   Each characteristic value of T has absolute value 1,

i.e. $\qquad |\lambda_i| = 1$ or $|\lambda_i|^2 = 1$ or $\lambda_i \bar{\lambda}_i = 1$

To prove that T is unitary.

$$TT^* = (\sum_i \lambda_i E_i)(\sum_j \lambda_j E_j)^*$$

$$= (\Sigma\lambda_i E_i)(\Sigma\bar{\lambda}_j E_j^*)$$

$$= (\Sigma\lambda_i E_i)(\Sigma\bar{\lambda}_i E_j) \qquad\qquad E_i^* = E_j$$

$$= (\lambda_1 E_1 + \lambda_2 E_2 + \ldots + \lambda_n E_n)(\bar{\lambda}_1 E_1 + \bar{\lambda}_2 E_2 + \ldots + \bar{\lambda}_n E_n)$$

$$= (\lambda_1\bar{\lambda}_1 E_1^2 + \lambda_2\bar{\lambda}_2 E_2^2 + \ldots + \lambda_n\bar{\lambda}_n E_n^2)$$

$$= E_1 + E_2 + \ldots + E_n = I \quad \lambda_i\bar{\lambda}_i = 1, E_i^2 = E_i$$

$TT^* = I$ and V is finite dimensional.

$\therefore \qquad\qquad TT^* = I \quad \Leftrightarrow \quad T^*T = 1.$

Hence T is unitary.

Example 3.   *Let T be a normal operator on a finite dimensional complex inner product vector space V, then prove the following :*

(a)   each characteristic value of T is non-negative $\Rightarrow T$ is non-negative.

(b)   each characteristic value of T is different from zero $\Rightarrow T$ is invertible.

(c)   each characteristic value of T is zero or one $\Rightarrow T$ is idempotent.

Solution.   (a)   Let $T = \lambda_1 E_1 + \lambda_2 E_2 + \ldots + \lambda_n E_n$ be the spectral resolution of T then as proved in (a) Ex. 2, $T^* = T$, $T(\alpha, \alpha) \geq 0$ hence T is positive, *i.e.*, T is non-negative.

(b)   Since $\lambda_i \neq 0 \forall i$ then let us define a linear operator S such that

$$S = \frac{1}{\lambda_1}E_1 + \frac{1}{\lambda_2}E_2 + \ldots + \frac{1}{\lambda_n}E_n$$

then $\qquad TS = (\sum_i \lambda_i E_i)\left(\sum_j \frac{1}{\lambda_j}E_j\right) = \sum_i \sum_j \lambda_i E_i \left(\frac{1}{\lambda_i}E_j\right)$

$$E_i E_j = O$$

$\therefore \qquad\qquad TS = \Sigma\lambda_i \frac{1}{\lambda_i}E_i^2 = \Sigma E_i = I$

Similarly $\qquad ST = I.$

Since $\qquad ST = TS = I$ therefore T is invertible.

(c) Given $\quad\lambda_i = 0$ or 1 then in either case $\lambda_i^2 = \lambda_i \forall i$

Now $\quad T^2 = (\Sigma\lambda_i E_i)(\Sigma\lambda_j E_j) = \sum\sum\lambda_i\lambda_j E_i E_j$

$$= \sum_i (\lambda_i E_i)(\lambda_i E_i) \qquad\qquad E_i E_j = 0, i \ne j$$

$$= E\lambda_i^2 E_i^2 = \Sigma\lambda_i E_i \qquad\qquad \lambda_i^2 = \lambda_i,\ E_i^2 = E_i$$

Hence T is idempotent.

Example 4. *If T is a normal operator on a finite dimensional inner product vector space and if W is a subspace of V invariant under T, then show that the restriction of T to W is also normal.*

Solution. As W is a invariant under T then W is also invariant under T^*. Let u be the restriction of T to W and S be restriction of T^* to W so that by definition

$$u\alpha = T\alpha \forall\ \alpha \in W \text{ and } S\alpha = T^*\alpha\ \forall\ \alpha \in W \qquad\qquad ...(1)$$

We will show that $S = u^*$.

Now $\quad (u\alpha, \beta) = (T\alpha, \beta) = (\alpha, T^*\beta) = (\alpha, S\beta)$ by (1)

Also $\quad (u\ \alpha, \beta) = (\alpha, u^*\beta) = (\alpha, S\beta)\ \therefore\ S = u^* \qquad\qquad ...(2)$

Now we have to prove that u is normal. If α be any vector in W then

$$uu^*\alpha = uS\alpha \text{ by (2)} = uT^*\alpha \text{ by (1)}$$

$$= TT^*u = T^*T\alpha \text{ as T is normal.}$$

$$= T^*u\alpha \text{ by (1)} = Su\alpha = u^*u\alpha \text{ by (1) and (2)}$$

Since $\quad uu^*\alpha = u^*u\alpha\ \forall\ \alpha \in W$

$\therefore \qquad uu^* = u^*u$, *i.e.*, u the restriction of T to W is normal.

📚 EXERCISE 8.2

1. Let $T : R^3 \to R^3$ defined by $T(x, y, z) = (3x + 4y - 5z, 2x - 6y + 7z, 5x - 9y + z)$. Find $T^*(x, y, z)$.

2. Let V be an inner product space. For each $\alpha \in V$, there is a mapping $\phi : V \to F$ defined by $\phi(\beta) = (\beta, \alpha), \beta \in V$. Show that ϕ is linear.

3. Show that the product of two self-adjoint operators is self-adjoint if and only if they commute each other.

4. Let λ be an eigenvalue of a linear operator T on V,
   (i) If $T^* = T^{-1}$, then $|\lambda| = 1$.
   (ii) If $T^* = -T$, then λ is purely imaginary.
   (iii) If $T = S^*S$ with S non-singular, then λ is real and positive.

5. If T is a self-adjoint operator on V, then eigenvectors of T belonging to distinct eigenvalues are orthogonal.

6. Show that $T + T^*$ is self-adjoint for any operator T on V.

7. Show that $T - T^*$ is self-adjoint for any operator T on V.

8. Show that any operator T on V can be expressed as the sum of a self-adjoint operator and a skew-adjoint operator.

9. Show that $T^*T - I$ is self-adjoint for any operator T on V.

10. Let T be the linear operator on R^2 defined by $T(x, y) = (y, -x)$. Then $(T(\alpha), \alpha) = 0$ for every $\alpha \in V$, but $T \ne 0$.

Hint to Selected Problems

6. $(T + T^*) = T^* + (T^*)^* = T^* + T = T + T^*$.

$\therefore\ T + T^*$ is self-adjoint. Similarly we can prove that $T - T^*$ is skew.

8. Suppose $T = P + Q$, where $P = \dfrac{T+T^*}{2}$, $Q = \dfrac{T-T^*}{2}$.

Now $P^* = \left(\dfrac{T+T^*}{2}\right)^* = \dfrac{1}{2}\left(T+T^*\right) = \dfrac{1}{2}\left(T+T^*\right) = P \quad \Rightarrow \quad P$ is self-adjoint.

and $Q^* = \dfrac{1}{2}\left(T-T^*\right)^* = \dfrac{1}{2}\left(T^*-T\right) = -\dfrac{1}{2}\left(T-T^*\right) = -Q \Rightarrow \quad Q$ is skew-adjoint.

Hence T can be expressed as the sum of a self-adjoint and a skew-adjoint operators.

11. The matrix of T relative to B is

$$[T] = \begin{bmatrix} i & 1 & -i & 1 \\ 1 & i & 1 & -i \\ -1 & 1 & i & 1 \\ 1 & -i & 1 & i \end{bmatrix}$$

$$[T^*] = [\overline{T}]^t = \begin{bmatrix} -i & 1 & i & 1 \\ 1 & -i & 1 & i \\ -1 & 1 & -i & 1 \\ 1 & i & 1 & -i \end{bmatrix}^t = \begin{bmatrix} -i & 1 & -1 & 1 \\ 1 & -i & 1 & i \\ i & 1 & -i & 1 \\ 1 & i & 1 & -i \end{bmatrix}$$

13. Since $T^* = -T$.

Now $(T^2)^* = (TT)^* = (T^*T) = ((-T)(-T)) = T^2$

$\Rightarrow \qquad T^2$ is self-adjoint and $(T^3)^* = (T^2T)^* = T^*(T^2)^* = (-T)(T^2) = -T^3$

$\Rightarrow \qquad T^3$ is skew-adjoint.

Answers
---

1. $T^*(x,y,z) = (3x+2y+5z,\ 4x-6y-9z,\ -5x+7y+z)$

2. $[T^*] = \begin{bmatrix} -i & 1 & -1 & 1 \\ 1 & -i & 1 & i \\ i & 1 & -i & 1 \\ 1 & i & 1 & -i \end{bmatrix}$ **10.** $[T^*] = \begin{bmatrix} 1 & 1 \\ 2 & -1 \end{bmatrix}$ **13.** T^2 is self-adjoint, T^3 is skew.

CHAPTER REVIEW : A COMPETITIVE APPROACH

Selected Terms and Results

TERMS

- **Inner Product:** Let V be a vector space over R or C. An inner product on V is a function $(\cdot, \cdot)$: $V \times V \to F$ Satisfying the following conditions:

  (i) $(\alpha, \alpha) > 0 \ \forall \ \alpha \in V, \ \alpha \neq 0$

  (ii) $(\alpha, \beta) = \overline{(\beta, \alpha)} \ \forall \ \alpha, \beta \in V$

  (iii) $(a\alpha + b\beta, \gamma) = a(\alpha, \gamma) + b(\beta, \gamma) \ \forall \ \alpha, \beta, \gamma \in V,$
  $a \ b \in F.$

- **Inner Product Space:** A vector space V together with an inner product is called an inner product space.

- **Norm:** The norm or length of a vector $\alpha \in V$ is defined by $\|\alpha\| = \sqrt{(\alpha, \alpha)}$.

- **Euclidean Space:** A finite dimensional real inner product space is often called an Euclidean space.

- **Unitary Space:** A complex inner product space is called a unitary space.

- **Orthogonal Vectors:** Two vectors α, β of an inner product space V are said to be orthogonal to each other if $(\alpha, \beta) = 0$

- **Orthogonal Compliment:**
  $$S^{\perp} = \{\alpha \in V : (\alpha, s) = 0 \ \forall s \in S\}$$

- **Orthonormal Set:** The set of vectors in V is called an orthonormal set if (i) $\|\alpha\| = 1 \forall \alpha \in V$ and (ii) $(\alpha, \beta) = 0 \ \forall \ \alpha, \beta, \in V$

- **Complete Orthonormal Set:** A maximal orthonormal set of vectors in an inner product space V is called complete orthonormal set.

- **Orthogonal and Orthonormal Basis:** Let V be a finite dimensional inner product space then a basis V which is also orthonormal (orthogonal) is called an orthonormal (orthogonal) basis.

- **Adjoint of a Linear Transformation:** Let

$T : U \to V$ be a linear transformation then there exists a unique linear mapping $T^* : V \to U$ such that for all $\alpha \in V, \beta \in U$

$(T(\alpha), \beta) = (\alpha, T^*(\beta))$ or $(T\alpha, \beta) = (\alpha, T^*\beta)$

Then T^* is called the adjoint of T.

- **Self-Adjoint Transformation:** If $T^* = T$

- **Skew Adjoint:** If $T^* = -T$

- **Congruent Operators:** Two linear operators T_1 and T_2 on an inner product vector space are said to be congruent if there exists an invertible linear operator P such that $T_2 = P^* T_1 P$

- **Unitary Operator:** An isomorphism of an inner product space V onto itself is called a unitary operator.

- **Normal Operator:** If $TT^* = T^*T$

- **Positive Operator:** A linear operator T on an inner product vector space is called a positive operator, *i.e.*, $T \geq 0$ if

  (i) T is self-adjoint

  (ii) $(Tx, x) \geq 0 \ \forall \ x \in V$

- **Projection:** If V is the direct sum of two subspaces M and N, *i.e.*, $V = M \oplus N$ then a projection E on M along N is a linear operator such that

  (i) E is idempotent, *i.e.*, $E^2 = E$

  (ii) Range of $E = M$.

- **Perpendicular Projection:** A projection E on an inner product vector space is called a perpendicular projection if its range and null space are orthogonol.

- **Orthogonal projections:** Two perpendicular projection E_1 and E_2 are said to be orthogonal if $E_1 E_2 = E_2 E_1 = 0$

RESULTS

- **Parallelogram Law:**

$$\|\alpha + \beta\|^2 + \|\alpha - \beta\|^2 = 2\|\alpha\|^2 + 2\|\beta\|^2$$

- **Polarization Identity:**

$$4(\alpha, \beta) = \begin{cases} \|\alpha + \beta\|^2 - \|\alpha - \beta\|^2 \\ \|\alpha + \beta\|^2 - \|\alpha + \beta\|^2 + i\|\alpha + i\beta\|^2 - i\|\alpha - i\beta\|^2 \\ \qquad\qquad\qquad\qquad , \text{if } F = C \end{cases}$$

- **Triangle Inequality:** $\|\alpha + \beta\| \le \|\alpha\| + \|\beta\|$
- **Schwarz Inequality:** $(\alpha, \beta) \le \|\alpha\| . \|\beta\|$
- An orthogonal set of non-zero vectors is linearly independent.
- In a finite dimensional inner product space, a complete orthonormal set is a basis.
- Every finite dimensional inner product space has an orthonormal basis
- Any orthonormal set of vectors in an inner product space is linearly independent.
- In an Euclidean space the self-adjoint transformation is called symmetric and in an unitary space it is called Hermition.
- If T is self-adjoint in an inner product space $V(F)$ then $S*TS$ is self-adjoint for all S, also if S is invertible and $S*TS$ is self-adjoint, then T is self-adjoint.
- If T be a self-adjoint transformation on an inner product space $V(F)$ then $T = 0 \Leftrightarrow (T(\alpha), \alpha) = 0$ $\forall \alpha \in V$.
- The largest number of vectors which an orthonormal set in a finite dimensional inner product space contains is called the orthogonal dimension of the space.
- Every complete orthonormal set in a finite dimensional inner product vector space forms a basis for V.
- The correspondance $y \to f_y$ such that $f_y(x) = (x, y) \forall x$ is norm preserving.
- If T_1 and T_2 be two congruent operators then T_1* and T_2* are also congruent.
- The relation of congruency of linear operator is equivalence relation.
- If T_1, T_2 are congruent and T_1 is skew then T_2 is also skew.
- If T be an invertible operator then either both or none of T and T^{-1} are self-adjoint.
- Every linear transformation T on a complex inner product space can be uniquely expressed as $T = A + iB$ where A and B are self-adjoint.
- T is self-adjoint if and only if (Tx, x) is real $\forall x \in V$.
- Let V be a finite dimensional inner product vector space and T be an idempotent operator on V then T is self-adjoint iff $TT* = T*T$.
- If T is self-adjoint or skew and $T^2x = 0$ then $Tx = 0$
- If T_1, T_2 both are self adjoint or skew, then $T_1T_2 + T_2T_1$ is self-adjoint and $T_1T_2 - T_2T_1$ is

skew. Also, if one of T_1, T_2 is self-adjoint and other is skew then both $T_1T_2 + T_2T_1$, $T_1T_2 - T_2T_1$ are skew.
- $T = 0 \Rightarrow (Tx, y) = 0 \forall x, y \in V$.
- T is a linear operator on a complex inner product vector space then $(Tx, x) = 0 \Rightarrow T = 0$ $\forall x \in V$.
- A linear operator T on a complex inner product vector space is self-adjoint if and only if (Tx, x) is real $\forall x \in V$.
- The relation of isomorphism is the set of inner product vector space is an equivalence relation.
- An operator T on an inner product vector space T is unitary iff it is an isometric isomorphism of V onto itself.
- T is normal $\Leftrightarrow$ its real and imaginary parts commute.
- Every identity, adjoint, unitary and zero operator is normal.
- Range and null space of a normal operator are disjoint.
- The sum of two positive operators is again positive.
- If T is an arbitrary operator on V and α, β are scalars such that $|\alpha| = |\beta|$ then $\alpha T + \beta T*$ is normal.
- A linear opertor E is perpendicular projection on an inner product vector space V if and only if $E = E^2 = E*$.
- If E is a perpendicular projection on a subspace M of an inner product space V and T, a linear operator on V, then

  M is invariant under $T \Leftrightarrow TE = ETE$
- Two perpendicular projections E_1 and E_2 are orthogonal if and only if $E_1E_2 = 0$ or $E_2E_1 = 0$
- If a linear operator E satisfies (i) $E^2 = E$ and (ii) $\|Ez\| \le \|z\| \forall z$ then E is self-adjoint.
- If T is a linear operator on an n-dimensional vector space V then T cannot have more than n distinct characteristic values.
- The characteristric vectors corresponding to district characteristic values of a self adjoint operator are orthogonal.
- Eigen spaces of a normal operator T are pairwise orthogonal.
- If T is normal operator then each eigenspace M_i reduces T.

Review Questions and Project Work

1. If $V = M \oplus N$ and E is a projection on M along N then show that (i) Range of $E = M$ (ii) null space of $E = N$.

2. Prove that if E is a projection on M along N if and only if $I-E$ is a projection on N along M.

3. Show that we can always define an inner product on a finite dimensional vector space over real or complex.

4. Show that a positive multiple of an inner product is an inner product.

5. If in an inner product space $\|\alpha + \beta\| = \|\alpha\| + \|\beta\|$ then prove that the vectors are linearly dependent.

6. If α, β are two orthogonal unit vectors, *i.e.*, if $\{\alpha, \beta\}$ is an orthogonal set in an inner product space, show that the distance between α and β is $\sqrt{2}$.

7. If α and β are vectors in a real inner product space and if $\|\alpha\| = \|\beta\|$ then show that $\alpha - \beta$ and $\alpha + \beta$ are orthogonal.

8. Show that an orthogonal set of non-zero vectors in an inner product space V is linearly independent.

9. If T is skew, show that T^2 is not skew but T^3 is skew.

10. Show that the determinant of a unitary operator has absolute value 1.

11. Show that the range and null space of a normal operator are disjoint.

12. Show that the minimal polynomial of a normal operator on a finite dimensional inner product space has distinct roots.

13. Let T be a linear operator on a finite dimensional complex inner product space V. Show that T is normal iff T^* is a polynomial in T.

14. Let T be a normal operator and α a vector such that $T^2\alpha = 0$ then show that $T\alpha = 0$.

Objective Type Questions

FILL IN THE BLANKS

1. The vertical bars in $|(\alpha, \beta)|$ denote the _____ of a complex number.

2. If α, β are vectors in an inner product space then
$$\|\alpha + \beta\|^2 + \ldots\ldots\ldots = \|\alpha\|^2 + \|\beta\|^2$$

3. An orthonormal set of vectors is linearly _____.

4. An orthonormal set of non-zero vectors is linearly _____.

5. Every finite dimensional vector space over the complex field is an _____.

6. A positive multiple of an inner product is an _____.

7. An orthonormal set is said to be _____ orthonormal if it is not a subset of a larger or the normal set.

8. The _____ vector is orthogonal to every vector $\alpha \in V$.

9. If $|(\alpha, \beta)| = \|\alpha\| \cdot \|\beta\|$ then α and β are said to be linearly _____.

10. An inner product space is a normal space and converse is _____.

TRUE/FALSE

Write 'T' for true and 'F' for false statement.

1. If F is the field of reals, then inner product space $V(F)$ is called Euclidean space. **(T/F)**

2. If F is the field of complex numbers, then inner product space $V(F)$ is called unitary space. **(T/F)**

3. We can always define an inner product on a finite dimensional vector space, real or complex. **(T/F)**

4. If α, β are linearly independent, then $|(\alpha, \beta)| = \|\alpha\| \cdot \|\beta\|$ **(T/F)**

5. If α is orthogonal to β, then every scalar multiple of α is orthogonal to β. **(T/F)**

6. The zero vector is the only vector which is orthogonal to itself. **(T/F)**

7. Any orthogonal set of non-zero vectors in an inner product space is linearly independent. **(T/F)**

8. The largest number of vectors contains an orthonormal set is called orthogonal dimension. **(T/F)**

9. The operation of adjoint behaves like the operations of conjuction on complex numbers. **(T/F)**

10. A linear operator T is unitary if and only if the adjoint T^* of T exists and $T^*T = TT^* = I$ **(T/F)**

MULTIPLE CHOICE QUESTIONS

Choose the most appropriate one :

1. A linear operator T on a finite dimensional inner product space V is such that $(T*x, y) = (x, T*y)$ then $T*$ is called :
   - (a) adjoint of T
   - (b) self- adjoint of T
   - (c) conjugate of T
   - (d) none of these

2. The value of $(T_1 + T_2)*$ is :
   - (a) $T_1 + T_2$
   - (b) $T_1* + T_2*$
   - (c) $T_1 T_2*$
   - (d) None of these

3. The value of $(T_1 T_2)*$ is :
   - (a) $T_1* T_2*$
   - (b) $T_2* T_1*$
   - (c) $T_1 T_2*$
   - (d) none of these

4. The value of $(T*)*$ is :
   - (a) T
   - (b) $T**$
   - (c) both (a) and (b) are true
   - (d) none of these

5. The value of $(T*)^{-1}$ is :
   - (a) $T*$
   - (b) T^{-1}
   - (c) $(T^{-1})*$
   - (d) none of these

6. Two linear operators T_1 and T_2 are said to be congruent if there exists an invertible linear operator P such that :
   - (a) $T_1 = P* T_2 P$
   - (b) $T_2 = P* T_1 P$
   - (c) both (a) and (b) are true
   - (d) none of the above

7. The relation of congruency of linear operator is :
   - (a) reflexive
   - (b) symmetric
   - (c) transitive
   - (d) all are true

8. An operator T is said to be self-adjoint if :
   - (a) $T* = T$
   - (b) $T* = -T$
   - (c) $T^2 = T$
   - (d) none of these.

9. If T_1 and T_2 both are self-adjoint then $T_1 + T_2$:
   - (a) is self-adjoint
   - (b) not self-adjoint
   - (c) may or may not be self-adjoint
   - (d) none of these

10. If T be an invertible operator then :
    - (a) T and T^{-1} both are self-adjoint
    - (b) T and T^{-1} both are not self-adjoint
    - (c) either (a) or (b) is true
    - (d) none of the above

11. If T_1 and T_2 be self-adjoint then $T_1 T_2$ is :
    - (a) self-adjoint
    - (b) not self-adjoint
    - (c) self-adjoint iff $T_1 T_2 = T_2 T_1$
    - (d) none of the above

12. Every linear transformation T on a complex inner product space can be uniquely written as $T = A + iB$ where :
    - (a) A is self-adjoint
    - (b) B is self-adjoint
    - (c) both A and B are self-adjoint
    - (d) none of the above

13. A linear operator T on a complex inner product vector space is self-adjoint if and only if :
    - (a) (Tx, x) is real
    - (b) (Tx, x) is complex
    - (c) $(Tx, x) = 0$
    - (d) none of these

14. The relation of isomorphism in the set of inner product space is :
    - (a) reflexive
    - (b) transitive
    - (c) symmetric
    - (d) all are true

15. An operator T is said to be unitary if and only if :
    - (a) $TT* = T*T$
    - (b) $TT* = T*T = I$
    - (c) $TT* \neq T*T$
    - (d) none of these

16. An isomorphism of an inner product vector space V onto itself is called :
    - (a) unitary operator
    - (b) normal operator
    - (c) invertible operator
    - (d) none of the above

17. If T is a linear operator on an inner product vector space V then which is/are true?
    - (a) $T*T = I$
    - (b) $\|T\alpha\| = \|\alpha\|$
    - (c) $(T\alpha, T\beta) = (\alpha, \beta)$
    - (d) All are true.

18. An operator T is said to be self-adjoint if :
    - (a) $TT* = T*T$
    - (b) $TT* \neq T*T$
    - (c) $TT* = T$
    - (d) none of these

19. For a normal operator T, which is/are true?
    - (a) Every self-adjoint operator is normal.
    - (b) Every unitary operator is normal.
    - (c) Identity operator is normal.
    - (d) All are true.

20. If the real and imaginary parts of a linear operator T commutes then T is :
(a) normal
(b) self-adjoint
(c) unitary
(d) none of these.

21. If the range and null space of a projection are orthogonal then it is called :
(a) perpendicular projection
(b) normal
(c) unitary
(d) none of the above

22. A linear operator E is called perpendicular projection if :
(a) $E = E^*$
(b) $E = E^2$
(c) $E = E^* = E^2$
(d) none of these

23. Two perpendicular projections E_1 and E_2 are said to be orthogonal projection if :
(a) $E_1 E_2 = 0$
(b) $E_2 E_1 = 0$
(c) either (a) or (b) is true
(d) none of the above

24. If E and F are perpendicular projections then EF is perpendicular if :
(a) $EF = FE$
(b) $EF \neq FE$
(c) $E^2 = F^2$
(d) none of these

25. A projection is said to be perpendicular if its :
(a) range space is orthogonal
(b) null space is orthogonal
(c) both (a) and (b)
(d) none of the above

26. T be a self-adjoint operator then the characteristic value of T are :
(a) real
(b) zero
(c) complex
(d) none of these

27. The characteristic vectors corresponding to distinct characteristic values of a self-adjoint operator are :
(a) real
(b) orthogonal
(c) zero
(d) none of these

28. Characteristic vectors of a normal operator T and its adjoint T^* are :
(a) equal
(b) unified
(c) can't say
(d) none of these

29. Proper spaces of a normal operator T are :
(a) orthogonal
(b) pairwise orthogonal

(c) not orthogonal
(d) none of the above

30. Spectral theorem states that :
(a) the subspace $M_i^{'s}$ are pairwise orthogonal
(b) $E_i^{'s}$ are pairwise orthogonal
(c) r is normal
(d) all the above

31. In an innerproduct space V, any orthogonal set of non-zero vector is :
(a) linearly dependent
(b) linearly independent
(c) can't say
(d) none of the above

32. Two vectors α, $\beta \in V(F)$ are said to be orthogonal if :
(a) $(\alpha, \beta) \neq 0$
(b) $(\alpha, \beta) = 0$
(c) $(\alpha, \beta) > 1$
(d) none of these

33. Vector $\alpha \in V(F)$ is said to be orthonormal if :
(a) $(\alpha, \alpha) = 0$
(b) $(\alpha, \alpha) \neq 0$
(c) $(\alpha, \alpha) = 1$
(d) none of these

34. In an inner product space, the set of orthonormal vectors is :
(a) linearly dependent
(b) linearly independent
(c) can't say
(d) none of these

35. For α, $\beta \in V(F)$, we have :
(a) $|(\alpha, \beta)| \leq \|\alpha\|.\|\beta\|$
(b) $|(\alpha, \beta)| = \|\alpha\|.\|\beta\|$
(c) $|(\alpha, \beta)| \geq \|\alpha\|.\|\beta\|$
(d) None of these

36. If α, β are vectors in real inner product space such that $\|\alpha\| = \|\beta\|$ then which of the following is true ?
(a) $(\alpha + \beta, \alpha) = 0$
(b) $(\alpha + \beta, \alpha - \beta) = 0$
(c) $(\alpha + \beta, \beta) = 0$
(d) none of these

37. In an inner product $V(F)$ each vector is orthogonal to
(a) zero vector
(b) itself
(c) every vector
(d) none of these

38. If α, β are two orthogonal vectors in an inner product space such that $\|\alpha\| = \|\beta\| = 1$ then the value of $\|\alpha - \beta\| =$
(a) 2
(b) 0
(c) $\sqrt{2}$
(d) 1

39. If $\alpha, \beta \in V$ be such that $\alpha \perp \beta$ then :

(a) $(\alpha, \beta) = 0$ (b) $(\beta, \alpha) = 0$

(c) $(k\alpha, \beta) = 0$ (d) all are true

40. If α, and β are orthonormal set in an inner product space then the distance between α and β is :

(a) 1   (b) 2

(c) $\sqrt{2}$ (d) none of these

41. If $\alpha = (a_1, a_2, a_3)$ and $\beta = (b_1, b_2, b_3) \in V_3(R)$ then which of the following defines an inner product on $V_3(R)$?

(a) $(\alpha, \beta) = a_1 b_1 + a_2 b_2$

(b) $(\alpha, \beta) = 2a_1 b_1 + a_2 b_2 + 4a_3 b_3$

(c) Both (a) and (b) are true

(d) none of the above

42. The angle between two vectors α and β is given by

(a) $\cos\theta = \dfrac{(\alpha, \beta)}{\|\alpha\| \cdot \|\beta\|}$

(b) $\sin\theta = \dfrac{(\alpha, \beta)}{\|\alpha\| \cdot \|\beta\|}$

(c) both (a) and (b) are true

(d) none of the above

43. Let $V(F)$ be a vector space, then an inner product in $V(F)$ is a function from :

(a) $V \times V$ to F (b) $V \times V$ to V

(c) V into F (d) none of these

44. Let $u = (5, 5, 8, 8)$ and $v = (1, 2, 3, 4)$ then $\|u - v\| =$

(a) $\sqrt{6}$ (b) 66

(c) $\sqrt{66}$ (d) none of these

45. Two vectors α and β in a real inner product space are orthogonal if and only if :

(a) $\|\alpha + \beta\| = \|\alpha\| + \|\beta\|$

(b) $\|\alpha + \beta\|^2 = \|\alpha\|^2 + \|\beta\|^2$

(c) $\|\alpha + \beta\| \le \|\alpha\| + \|\beta\|$

(d) none of the above

Answers

FILL IN THE BLANKS

1. modulus **2.** $\|\alpha - \beta\|^2$ **3.** independent **4.** independent **5.** inner product **6.** inner product **7.** complete **8.** zero vector **9.** Independent **10.** not true

TRUE/ FALSE

1. T	**2.** T	**3.** T	**4.** F	**5.** T	**6.** T	**7.** T	**8.** T	**9.** T
10. T								

MULTIPLE CHOICE QUESTIONS

1. (a)	**2.** (b)	**3.** (b)	**4.** (c)	**5.** (c)	**6.** (c)	**7.** (d)	**8.** (a)	**9.** (a)
10. (c)	**11.** (c)	**12.** (c)	**13.** (a)	**14.** (d)	**15.** (b)	**16.** (a)	**17.** (d)	**18.** (a)
19. (d)	**20.** (a)	**21.** (a)	**22.** (c)	**23.** (c)	**24.** (a)	**25.** (c)	**26.** (a)	**27.** (b)
28. (a)	**29.** (b)	**30.** (d)	**31.** (b)	**32.** (b)	**33.** (c)	**34.** (b)	**35.** (a)	**36.** (b)
37. (a)	**38.** (c)	**39.** (d)	**40.** (c)	**41.** (b)	**42.** (a)	**43.** (a)	**44.** (c)	**45.** (b)

COMPETITION CORNER
for JRF, NET/SET, GATE Aspirants

SOME FASCINATING FACTS

1. For inner product space, we consider the map from $V \times V$ to F.

2. In an inner product space, the distributive laws holds with the modification that a scalar with the second member of inner product will be replaced by its conjugate.

3. A real inner product space which is finite dimensional is called Euclidean space and the corresponding complex one is called unitary.

4. Every inner product space is a normal vector space.

5. Every finite dimensional vector space over the field of real or complex is an inner product space and hence, every finite dimensional vector space is a normed vector space.

6. The geometrical interpretation of the parallelogram law is that sum of the squares of the sides of a parallelogram is equal to the sum of the squares of its diagonals.

7. Schwarz's inequality simply emphasizes that cosine of a real angle is less or equal to 1.

8. $|x, y| = \| x \| \cdot \| y \| \Leftrightarrow x$ and y are linearly dependent.

9. If $\| x + y \| = \| x \| + \| y \|$ then x and y are linearly dependent.

10. An orthonormal set being essentially orthogonal is therefore also, linearly independent.

11. Every orthogonal (or orthonormal) set of non-zero vectors in an inner product vector space is linearly independent.

12. Every finite subset of an orthogonal (orthonormal) set of an inner product vector space is linearly independent.

13. A largest number of vectors which an orthonormal set in a finite dimensional inner product vector space contains is called the orthogonal dimension of the space.

14. If x and y are vectors in a real inner product space and if $\| x \| = \| y \|$ then $x - y$ and $x + y$

are orthogonal. Geometrically, it can be interpreted as follows:

Let x, y repesents the sides AB, BC of a parallelogram $ABCD$ which will be a rhombus as $\|x\| = \|y\|$, *i.e.* $AB = BC$. Clearly the vectors $x + y$ and $x - y$ represnets the diagonals of this rhombus which are prependicular to each other under given condition. Hence we estabilish that the diagonals of a rhombus interset at right angles. Conversely, the diagonal $AC = x + y$ and $DB = x - y$ of parallelogram $ABCD$ are at right angles and then $\|x\| = \|y\|$. *i.e.*, $AB = BC$ and B parallellogram is rhombus. Hence we estabilish that if the diagonals of a parallelogram are at right angles, then it is a rhombus.

15. Let V and W be finite dimensional inner product vector space over the field F and having the same dimension. If T is a linear transformation from V to W, then following are equivalent:

   (i) T preserves inner product.

   (ii) T is an inner product vector space isomorphism.

   (iii) T carries every orthonormal basis of V onto an orthonormal basis for W.

   (iv) There is at least one orthonormal basis of V which is transfered to orthonormal basis of W.

16. The range and null space of a normed operator are disjoint.

17. $x \in M \Rightarrow T(x) \in M$ is one way implication. It does not mean that y in M. We can write it as $y = Tx$ with x in M.

18. The product of two commutative projection is again commutative.

19. The set of all characteristic (proper) values of a linear operator T is called spectrum of T.

20. Eigenspaces of a normal operator are pairwise orthogonal.

SOME IMPORTANT ILLUSTRATIONS

1. If T is a linear transformation on V such that $T^2(I - T) = T(I - T)^2 = 0$, then T is a projection on V.

2. The set of all $n \times n$ matrices forms an inner product vector space over F, where F is either real or complex, the inner product being

defined as
$$(A, B) = \text{trace } (AB^*)$$
where B^* is transpose conjugate of matrix B.

3. If $x = (a_1, a_2)$, $y = (b_1, b_2)$, then $(x, y) = a_1 b_1 - a_2 b_1 - a_1 b_2 + 4 a_2 b_2$ defines an inner product on V_2 (R).

4. If $x = (a_1, a_2)$, $y = (b_1, b_2)$, then $(x, y) = 2a_1 \overline{b}_1 + \overline{a}_1 \overline{b}_2 + a_2 \overline{b}_1 + a_2 \overline{b}_2$ defines an inner product on $V_2(F)$ and $(x, y) = 2a_1 b_1 + a_2 b_2 + a_2 b_1 + a_2 b_2$ defines an inner product on $V_2(\mathbf{R})$.

5. If $x = (a_1, a_2)$, $y = (b_1, b_2)$, then $(x, y) = a_1 + a_2 + b_1 + b_2$ does not define an inner product on V_2 (R).

6. $(x, y) = a_1 \overline{b}_2 + a_2$ does not define an inner product in $V_2(F)$.

7. Let P be the space of all polynomials over the complex field C. If $p(t), q(t) \in P$ then $(p, q) = \int_0^1 p(t) \overline{q(t)} dt$ defines an inner product in P.

8. Let V be the vector space of complex $n \times n$ metrices with inner product $(x, y) = \text{Trace } (y^*x)$. If A is fixed $n \times n$ matrix over C then adjoint of the operator which premultiplies each matrix with A is the operator which premultiplies each matrix A^*.

9. Let V be finite dimensional inner product vector space and T, a positive linear operator on V. Let p_T be an inner product on V defined by $p_T(\alpha, \beta) = (T\alpha, \beta)$. Also, u is a linear operator on V and u^* its adjoint with respect to $(\cdot)$, then u is unitary with respect to inner product p_T if and only if $T = u^*Tu$.

10. Let E and F are perpendicular projections with ranges M and N respectively which are subspaces of an inner product vector space V then $F - E$ is a projection on $N \cap M^\perp$ if and only if $E \leq F$.

11. The vector space $C^n = \{(\alpha_1, \alpha_2, ... \alpha_n) : \alpha_i \in C\}$ is an inner product space with respect to the inner product
$$(u, v) = \alpha_1 \overline{\beta}_1 + \alpha_2 \overline{\beta}_2 + ... + \alpha_n \overline{\beta}_n$$
where $u = (\alpha_1, \alpha_2, ... \alpha_n), v = (\beta_1, \beta_2, ... \beta_n) \in C^n$

12. The vector space $R^n = \{(\alpha_1, \alpha_2, ... \alpha_n) : \alpha_i \in R\}$ is an inner product space with respect to the inner product
$$(u, v) = \alpha_1 \beta_1 + \alpha_2 \beta_2 + ... + \alpha_n \beta_n$$
where
$$u = (\alpha_1, \alpha_2, ... \alpha_n), v = (\beta_1, \beta_2, ... \beta_n) \in R^n$$

13. For any $\alpha \in R^2$, $\alpha = (\alpha, e_1) e_1 + 1(\alpha, e_2) e_2$ where $e_1 = (1, 0)$, $e_2 = (0, 1)$

14. Let V be the vector space of all real polynomials of degree ≤ 2, then
$$(f(x), g(x)) = \int_{-1}^{1} f(x) g(x) \, dx \ \forall \ f(x), g(x) \in V$$
is an inner product on V.

15. If W is a subspace of an inner product space V and $v \in V$ satisfies
$$(v, w) + (w, v) \leq (w, w) \ \forall w \in W$$
then $(v, w) = 0$ for all $w \in W$

Self Assessment Test

1. Let V be a vector space, show that the sum of two inner products on V is an inner product on V. Is the difference of two inner products an inner product? show that a positive multiple of an inner product is an inner product.

2. If $u = (-2, -3, 5)$, $v = (4, 6, -10) \in R^2$ verify that
$$|(u, v)| = \|u\| \cdot \|v\|$$

3. Prove that
$$|x_1y_1 - x_2y_1 - x_1y_2 + 4x_2y_2| \le \{(x_1 - x_2)^2 + 3x_2^2\}^{1/2}\{(y_1 - y_2)^2 + 3y_2^2\}^{1/2}$$

4. If W is a subspace of V and $v \in V$ satisfies
$(v, w) + (w, v) \le (w, w) \ \forall \ w \in W$
Prove that $(v, w) = 0 \ \forall w \in W$ where V is an inner product space.

5. Let V be the set of all real functions $y = f(x)$ satisfying
$$\frac{d^3y}{dx^3} + 6\frac{d^2y}{dx^2} + 11\frac{dy}{dx} + 6y = 0$$
Prove that $V(R)$ is a 3-dimensional real inner product space under the inner product
$$(u, v) = \int_0^\infty u \cdot v \, dx \ \forall \ u, v \in V(R)$$

6. Let W be a finite dimensional proper subspace of an inner product space V. Let $\alpha \in V$ and $\alpha \notin W$. Show that there is a vector $\beta \in W$ such that $\alpha - \beta$ is orthogonal to W.

7. Let W be a subspace of an inner product space V. If $\{w_1, w_2, \dots w_n\}$ is basis for w, show that $w \in W^\perp$ if and only if $(w, w_i) = 0$ for $i = 1, 2, \dots n$.

8. Let $\alpha = (a_1, a_2)$, $\beta = (b_1, b_2) \in V_2(C)$, prove that
$$(\alpha, \beta) = a_1\bar{b}_1 + (a_1 + a_2)(\bar{b}_1 + \bar{b}_2)$$
defines an inner product on $V_2(C)$. Show that the norm of the vector $(3, 4)$ in this inner product space is $\sqrt{58}$.

9. If $\alpha_1, \alpha_2, \dots \alpha_n$ are pairwise orthogonal vectors in an inner product space V, then show that
$$\|\alpha_1 + \alpha_2 + \dots + \alpha_n\|^2 = \|\alpha_1\|^2 + \|\alpha_2\|^2 + \dots + \|\alpha_n\|^2$$

10. Let $C(0, \pi)$ have the inner product $(f, g) = \int_0^\pi f(x)g(x)dx$ and let $f_n = \cos nx$, $n = 0$, 1, 2, ... show that if $k \ne l$ then f_k and f_l are orthogonal with respect to the given inner product.

11. Let V be a finite dimensional inner product space of dimension n. If $\{\alpha_1, \alpha_2, \dots, \alpha_m\}$ is an orthogonal set in V, prove that there exists vectors $\alpha_{m+1} \dots \alpha_n$ such that
$$\{\alpha_1, \alpha_2, \dots, \alpha_m, \alpha_{m+1} \dots, \alpha_n\}$$
is an orthonormal basis for V.

12. Let W be a finite dimensional proper subspace of an inner product space V. Let $\alpha \in V$ and $\alpha \notin W$. Show that there is a vector $\beta \in W$ such that $\alpha - \beta$ is orthogonal to W.

13. Show that $(u, v) = x_1y_1x_2y_2$ is not an inner product on R^2 where $u = (x_1, x_2)$, $v = (y_1, y_2)$.

14. Show that $\langle u+v, u-v \rangle = \|u\|^2 + \|v\|^2$

15. Verify the Pythagorean theorem for the following orthogonal set in R^4,
$u = (1, 2, -3, 4)$, $v = (3, 4, 1, -2)$, $w = (3, -2, 1, 1)$

16. Let P be orthogonal projection, show that $\|Pu\| = \|u\|$ for every $u \in V$.

17. Show that any matrix A is orthogonally equivalent to A.

18. Consider the basis $B = \{V_1 = (1, 3), V_2 = (2, 5)\}$ of R^2. Find the matrix A which represents the usual inner product on R^2 with respect to the basis B.

19. If B is a real non-singular matrix, show that B^TB is positive definite.

20. Show that if $u \ne v$, then $d(u, v) > 0$ and $d(u, u) = 0$.

■■■

9 BILINEAR, QUADRATIC AND HERMITIAN FORMS

9.1 INTRODUCTION

Let V be a finite dimensional inner product space over a field F and if T be a linear operator on V, then a function f defined on $V \times V$ by

$$f(\alpha,\beta) = (T(\alpha), \beta), \text{ for } \alpha, \beta \in V$$

may be regarded as a kind of substitute for T and it can be easily seen that f determines the linear mapping T.

For Example. If $B = \{\alpha_1, \alpha_2, ..., \alpha_n\}$ be an orthonormal basis of V, then the elements of the matrix of T relative to B are given by

$$a_{ij} = f(\alpha_j, \alpha_i).$$

In this chapter we extend the notion of inner product to vector space over arbitrary field F and we shall discuss three forms namely Bilinear, Quadratic and Hermitian forms and their properties.

9.2 BILINEAR FORMS

Definition 1. *A bilinear form on a vector space $V(F)$ is a mapping $f : V \times V \to F$ which satisfies the following properties :*

    (i) $f(a\alpha + b\beta, \gamma) = af(\alpha, \gamma) + bf(\beta, \gamma)$

    (ii) $f(\alpha, a\beta + \beta\gamma) = af(\alpha, \beta) + bf(\alpha, \gamma)$ *for all $a, b \in F$ and all $\alpha, \beta, \gamma \in V$.*

We express condition (i) by saying f is linear in its first variable (co-ordinate) and condition (ii) by saying f is linear in its second variable (co-ordinate). Therefore such mapping f is also known as Sesqui-linear form.

    For Example :

    (1)   Let f be the dot product on R^n, then f is bilinear form of R^n.

    (2)   Let f be the dot product on C^n, that is

        $f(\alpha, \beta) = a_1 b_1 + a_2 b_2 + ... + a_n b_n$

        where $\alpha = (a_1, a_2, ..., a_n)$, $\beta = (b_1, b_2, ..., b_n)$, then f is bilinear.

Definition 2. *A vector space V together with a bilinear form f defined above is called a bilinear space which is denoted by (V, f) or $B(V, f)$.*

Definition 3. *Let f be a bilinear form on V over F. Then a vector $\alpha \in V$ is said to be orthogonal to $\beta \in V$ with respect to f if $f(\alpha, \beta) = 0$.*

Definition 4. *A bilinear form f is called reflexive if orthogonality relation with respect to f is symmetric. That is for $\alpha, \beta \in V$, $f(\beta, \alpha) = 0$ whenever $f(\alpha, \beta) = 0$.*

Definition 5. *A bilinear form f is called symmetric if $f(\alpha, \beta) = f(\beta, \alpha)$ for all $\alpha, \beta \in V$.*

Definition 6. *A bilinear form f is called skew-symmetric if $f(\alpha, \beta) = -f(\beta,\alpha)$ for all $\alpha,\beta \in V$.*

Definition 7. *A bilinear form f is called alternating if $f(\alpha, \alpha) = 0$ for all $\alpha \in V$.*

THEOREM 1. *A bilinear form is reflexive if and only if it is either symmetric or alternating.*

Proof. Let f be a reflexive bilinear form on a vector space V over F. Then for $\alpha,\beta,\gamma \in V$, we have

$$f(\alpha, f(\alpha,\beta)\gamma - f(\alpha, \gamma)\beta) = 0$$

$$\Rightarrow \qquad f(f(\alpha,\beta)\gamma - f(\alpha, \gamma)\beta,\alpha) = 0$$

$$\Rightarrow \qquad f(\alpha,\beta)f(\gamma,\alpha) - f(\alpha,\gamma)f(\beta,\alpha) = 0 \qquad \qquad ...(1)$$

In particular for $\gamma = \alpha$, we get

$$f(\alpha,\beta)f(\alpha,\alpha) - f(\alpha,\alpha)f(\beta,\alpha) = 0$$

or $\qquad f(\alpha,\alpha).(f(\alpha,\beta) - f(\beta,\alpha)) = 0 \qquad \qquad ...(2)$

Now assume that f is not symmetric, then we shall prove that f is alternating.
If for $u \in V$, there exists $v \in V$ such that $f(u, v) \neq f(v, u)$
Then from (2), we get

$$f(u, u) = 0$$

Let $u \in V$ be such that $f(\alpha,u)$ for all $\alpha \in V$. Now, choose $v, w \in V$ such that

$$f(v, w) \neq f(w, v).$$

Then from (1), we have

$$f(u, v)\, f(w, u) - f(u, w)\, f(v, u) = 0$$

$$\Rightarrow \qquad f(u, v)\, f(w, u) - f(u, w)\, f(u, v) = 0 \qquad \qquad [\because f(u, v) = f(v, u)]$$

$$\Rightarrow \qquad f(u, v)(\, f(w, u) - f(u, w)) = 0$$

so that $\qquad f(u, v) = 0 = f(v, w)$

Similarly,

$$f(u,w) = 0 = f(w,u).$$

Also, $\qquad f(v, u+w) = f(v,u) + f(v,w).$

$$= f(v,w) + f(w,v) = f(u+w,v)$$

so that $f(u +w, u+w) = 0$. Hence, $f(u, u) = 0, \forall\, u \in V$.
Thus f is alternating.
Conversely, let f be symmetric, then $f(u, v) = f(v, u) \ \forall\, u,v \in V$.
When $f(u, v) = 0$, then $f(v, u) = 0$, this shows that f is reflexive.

REMARK

- A bilinear form is reflexive iff it is either symmetric or skew-symmetric.

THEOREM 2. *Let V be a vector space of dimension n over F. Let $\{f_1,f_2,...,f_n\}$ be a basis of the dual space V^*. Then $\{f_{ij} : i,j = 1,2,..,n\}$ is a basis of $B(V, f)$ where f_{ij} is defined by $f_{ij}(\alpha,\beta) = f_i(\alpha)\, f_j(\beta)$. Thus in particular dim $B(V, f) = n^2$.*

Proof. Let $\{\alpha_1,\alpha_2...,\alpha_n\}$ be the basis of V dual to $\{f_i\}$. We first show that $\{f_{ij}\}$ spans $B(V,f)$. Let $f \in B(V, f)$ and suppose that $f(\alpha_i, \alpha_j) = a_{ij}$. Then we shall show that $f = \sum a_{ij} f_{ij}$.
It is sufficient to show that $f(\alpha_k, \alpha_m) = (\sum a_{ij} f_{ij})(\alpha_k, \alpha_m)$ for $k, m = 1, 2, ..., n$.
We have $(\sum a_{ij} f_{ij})(\alpha_k, \alpha_m) = \sum a_{ij} f_{ij}(\alpha_k, \alpha_m)$

$$= \sum a_{ij} f_i(\alpha_k) f_j(\alpha_m) \qquad \text{[By the definition of } f_{ij}]$$
$$= \sum a_{ij} \cdot \delta_{ik} \delta_{jm} \qquad [\because f_i(\alpha_j) = \delta_{ij}]$$
$$= a_{km} \qquad \left[\because \delta_{ij} = \begin{cases} 1, i = j \\ 0, i \neq j \end{cases} \right]$$
$$= f(\alpha_k, \alpha_m)$$

Hence, $\{f_{ij}\}$ spans $B(V, f)$.

Now, we will show that $\{f_{ij}\}$ is linearly independent.

Let $\sum a_{ij} f_{ij} = 0$ for $i, j = 1, 2, ..., n$.

Then for $k, m = 1, 2, ..., n$, we have

$$\Rightarrow \quad (\sum a_{ij} f_{ij})(\alpha_k, \alpha_m) = 0(\alpha_k, \alpha_m)$$
$$\Rightarrow \quad \sum a_{ij} f_{ij}(\alpha_k, \alpha_m) = 0 \Rightarrow a_{km} = 0$$

Thus, $\{f_{ij}\}$ is linearly independent.

Hence $\{f_{ij}\}$ forms a basis of $B(V, f)$. Further, i and j take all the values from 1 to n therefore there are n^2 elements in $\{f_{ij}\}$, which shows that

dim. $B(V, f) = n^2$.

9.3 BILINEAR FORMS AND MATRICES

Definition 1. *Let f be a bilinear form on V(F) and let S = {$\alpha_1, \alpha_2, ..., \alpha_n$} be a basis of V. Then a matrix A={a_{ij}}, where $a_{ij}=f(\alpha_i, \alpha_j)$ is called the matix representation of f with respect to the basis S, this matrix A is denoted by $[f]_s$. The above matrix A represents f in another way as follows :*

Let $\alpha, \beta \in V$, then for $a_1, a_2, ... a_n, b_1, b_2, ..., b_n \in F$, we have

$$\alpha = a_1 \alpha_1 + a_2 \alpha_2 + ... a_n \alpha_n$$
$$\beta = b_1 \alpha_1 + b_2 \alpha_2 + ... b_n \alpha_n$$

Then, $\quad f(\alpha, \beta) = f(a_1\alpha_1 + a_2\alpha_2 + ... a_n\alpha_n, b_1\alpha_1 + b_2\alpha_2 + ... b_n\alpha_n)$

$$= a_1 b_1 f(\alpha_1, \alpha_1) + a_1 b_2 f(\alpha_1, \alpha_2) + ... + a_n b_n f(\alpha_n, \alpha_n)$$

$$= [a_1, a_2, ..., a_n] A \begin{bmatrix} b_1 \\ b_2 \\ \vdots \\ b_3 \end{bmatrix} = [\alpha]^t_n A [\beta]_s$$

Definition 2. *A matix B is said to be congruent to A if there exists an invertible matix P such that*

$$B = P'AP.$$

THEOREM 1. *Let P be the transition matrix from one basis of V to another. Let A be the matrix of bilinear form f in the original basis. Then B = P'AP is the matrix of f in the new basis.*

Proof. Let $S = \{\alpha_1, \alpha_2, ..., \alpha_n\}$ and $S' = \{\beta_1, \beta_2, ..., \beta_n\}$ be two ordered basis of V. Let T be a linear operator on V defined by

$$T(\alpha_i) = \beta_i \text{ for } i = 1, 2, ..., n. \qquad ...(1)$$

Obviously T is a mapping from S to S' so it is invertible.

Let $P = [p_{ij}]_{n \times n}$ be the matrix of T relative to the basis S so P is also invertible.

Also, $\qquad T(\alpha_j) = \sum p_{ij} \alpha_i = \beta_i \qquad ...(2)$

For any $\alpha, \beta \in V$, we have

$$[\alpha]_s = P[\alpha]_s \qquad \ldots(3)$$

and

$$[\beta]_s = P[\beta]_s \qquad \ldots(4)$$

$$\therefore \qquad [\alpha]^t_s = (P[[\alpha]_{s'})^t = [\alpha]^t_{s'} . P^t. \qquad \ldots(5)$$

Now, from definition of t, we have

$$f(\alpha, \beta) = [\alpha]^t_s A[\beta]_s = [\alpha]^t_{s'} P^t AP = [\beta]_{s'} \, [\text{Using (4) and (5)}]$$

Since α and β are arbitrary elements of V. Thus $P^t AP$ is the matrix of f in the basis S'. If it be B, then $B = P^t AP$ or $[f]_{s'} = P^t [f]_s , P$

REMARKS

- The matrix P defined in above theorem is called transition matrix from ordered basis S to S'.
- The above theorem indicates one main difference between bilinear forms and linear forms operator, both of which can be represented by square matrix, namely, if B and A represent the same linear opearator, then B is similar to A, that is $B = P^{-1} AP$, where P is the change of basis matrix but if B and A represent the same linear form then $B = P^t AP$.

Definition 3. *Let f be a bilinear form on V, then rank of a bilinear form f, denoted by* rank(f) *is defined to be the rank of any matrix representation of f.*

Definition 4. *A bilinear form f on V is said to be degenerate if rank $(f) <$ dim. V and it is said to be non-degenerate if rank $(f) = $ dim. V.*

Definition 5. *Let f and g be two bilinear forms on V and on W respectively over the field F. Then a linear mapping $T : V \to W$ is called an isometry if it preserves the bilinear forms, that is, for all $\alpha, \beta \in V$.*

$$f(\alpha, \beta) = g(T(\alpha), T(\beta)).$$

Also, if T is isomorphism then it is called an isometric isomorphism. If such a mapping exists, then we can say that the bilinear spaces $B(V, f)$ and $B(W, g)$ are isomorphic.

Definition 6. *A mapping $q : V \to F$ is called a quadratic mapping if $q(\alpha) = f(\alpha, \alpha)$, where f is a symmetric bilinear form on V.*

Definition 7. *If in a field F, we have $i + j \neq 0$, then f is obtainable from q from the following identity $f(\alpha, \beta) = \dfrac{1}{2} (q(\alpha + \beta) - q(\alpha) - q(\beta))$.*

This form is called polar form of f.

THEOREM 2. *Let $B(V, f)$ be a finite dimensional bilinear space over F and let $T : B(V, f) \to B(W, g)$ be an isometric isomorphism, where $B(W, g)$ is other bilinear space over F. Then $[f]_s$ and $[g]_{s'}$ are congruent matrices, where S and S' are ordered bases of V and W respectively.*

Proof. For any $\alpha, \beta \in V$, we have

$$f(\alpha, \beta) = g(T(\alpha), T(\beta)) \qquad \ldots(1)$$

Also, $$f(\alpha, \beta) = [\alpha]^t_s [f]_s [\beta]_{s'} \qquad \ldots(2)$$

Similarly, $$g(T(\alpha), T(\beta)) = [T(\alpha)]^t_{s'} [g]_{s'} [T(\beta)]_{s'}$$

$$= ([T]_s [\alpha]_s)^t [g]_{s'} ([T]_s [\beta]_s)$$

$$= ([\alpha]^t_s [T]^t_s) [g]_{s'} ([T]_s [\beta]_s)$$

$$= [\alpha]^t_s ([T]^t_s [g]_{s'} ([T]_s) [\beta]_{s'} \qquad \ldots(3)$$

From (1) and (2), we get

$$g(T(\alpha), T(\beta)) = [\alpha]^t_s [f]_s [\beta]_s \qquad \qquad ...(4)$$

Now from (3) and (4), we have

$$[f]_s = [T]^t_s [g]_{s'} [T]_{s'}$$

Hence, $[f]_s$ and $[g]_{s'}$ are congruent.

THEOREM 3. *Finite dimensional bilinear spaces B(V, f) and B(W, g) over F are isomorphic if and only if f and g admit the same matrix representation with respect to a suitably chosen ordered bases of V and W.*

Proof. Let $T : B(V, f) \to B(W, g)$ be an isometric isomorphism and let

$S = \{\alpha_1, \alpha_2 ...\alpha_n\}$ be an ordered basis of V.

Then $S' = \{T(\alpha_1), T(\alpha_2),..,T(\alpha_n)\}$ is an ordered basis of W and $[T]_s = I_n$,

where $I_{n'}$ is a unit matrix of order n. Therefore from above theorem, we have

$$[f]_s = [g]_{s'}.$$

Thus, f and g admit the same matrix representation.

Conversely, suppose f and g have the same matrix representation, with respect to some ordered bases S and S' of V and W respectively, and dim. V = dim. W.

Let $S = \{\alpha_1, \alpha_2,....., \alpha_n\}$ and $S' = \{\beta_1, \beta_2,....., \beta_n\}$ be ordered bases of V and W such that

$$[f]_s = [g]_{s'}.$$

Let $T : V \in W$ be a mapping defined by $T(a_i) = b_i$, i = 1, 2, ...,n.

Then, T is obviously isomorphism.

Now, for $\alpha, \beta \in V$, we have

$$\alpha = a_1\alpha_1 + a_2\alpha_2 +..... a_n\alpha_n; \quad \text{for } a's \in F$$

and

$$\beta = b_1\alpha_1 + b_2\alpha_2 +..... b_n\alpha_n; \quad \text{for } b's \in F$$

Then,

$$f(\alpha, \beta) = f\left(\sum_i a_i\alpha_i, \sum_j b_j\alpha_j\right) = \left(\sum_i \sum_j a_i b_j f(\alpha_i, \alpha_j)\right)$$

$$= \left(\sum_i \sum_j a_i b_j g(\beta_i, \beta_j)\right) = \left(\sum_i \sum_j a_i b_j g(T(\alpha_i),(T(\alpha_j))\right)$$

$$= g\left(\sum_i a_i T(\alpha_i), \sum_j b_j T(\alpha_j)\right) = g(T(\alpha).T(\beta))$$

Hence T is an isometry.

THEOREM 4. *Let f be an alternating bilinear form on V. Then there exists a basis of V in which f is represented by a matrix of the form.*

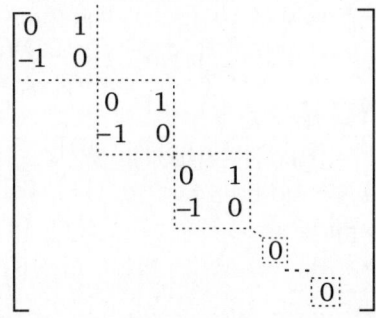

Moreover, the number of $\begin{bmatrix} 0 & 1 \\ -1 & 0 \end{bmatrix}$ is uniquely determined by f (because it is equal to $\dfrac{1}{2}$ rank (f)).

Proof. Suppose $f = 0$, then the theorem is obvious. Also, if dim. $V = 1$, then for $\alpha \in V$ and $a,b \in F$, we have

$$f(a\alpha, b\alpha) = abf(\alpha,\alpha) = 0 \text{ and so } f = 0.$$

Therefore, we may assume that dim. $V > 1$ and $f \neq 0$.

Since $f \neq 0$, so there exists non-zero vectors $\alpha_1, \alpha_2 \in V$ such that $f(\alpha_1, \alpha_2) \neq 0$. In fact multiplying α_1 by an appropriate factor, we may assume that $f(\alpha_1, \alpha_2) = 1$ and so $f(\alpha_2, \alpha_1) = -1$.

Now, α_1 and α_2 are linearly independent. Let if possible α_1 and α_2 are dependent, i.e., $\alpha_2 = k\alpha_1$, then $f(\alpha_1, \alpha_2) = f(\alpha_1, k\alpha_1) = kf(\alpha_1, \alpha_1) = 0$, which is not possible, thus α_1, α_2 must be linearly independent.

Let U be a subspace spanned by α_1 and α_2 that is $U = \text{span}(\alpha_1, \alpha_2)$. Thus, we have

(*i*) The matrix representation of the restriction of f to U in the basis $\{\alpha_1, \alpha_2\}$ is

$$\begin{bmatrix} 0 & 1 \\ -1 & 0 \end{bmatrix}.$$

(ii) If $u \in U$, so $u = a\alpha_1 + b\alpha_2$, then

$$f(u, \alpha_1) = f(a\alpha_1 + b\alpha_2, \alpha_1) = af(\alpha_1,\alpha_1) + bf(\alpha_2, \alpha_1)$$
$$= a0 + b(-1) = -b$$
$$f(u, \alpha_2) = f(a\alpha_1 + b\alpha_2, \alpha_2) = af(\alpha_1,\alpha_2) + bf(\alpha_2, \alpha_2)$$
$$= a(1) + b0 = a$$

Let W consists of those vector $w \in V$ such that

$$f(w,\alpha_1) = 0 \text{ and } f(w,\alpha_2) = 0.$$

We now claim that $V = U \oplus W$. From the definition of U and W, it is clear that $U \cap W = \{0\}$. Therefore, we only have to show that $V = U + W$. Let $v \in V$ and setting

$$\left. \begin{aligned} u &= f(v,\alpha_2)\alpha_1 - f(v,\alpha_1)\alpha_2 \\ w &= v - u \end{aligned} \right\} \qquad \ldots(1)$$

Since u is a linear combination of α and β, then we show that $w \in W$.

Now, $\quad f(u, \alpha_1) = f(f(v, \alpha_2)\alpha_1 - f(v, \alpha_1)\alpha_2, \alpha_1)$
$$= f(v,\alpha_2)f(\alpha_1, \alpha_1) - f(v, \alpha_1) f(\alpha_2, \alpha_1)$$
$$= 0 - (-1) f(v, \alpha_1) = f(v, \alpha_1).$$

So that $\quad f(w, \alpha_1) = f(v - u, \alpha_1)$ [Using (1)]
$$= f(u,\alpha_1) - f(u,\alpha_1) = 0.$$

This implies that $w \in W$ and then by (1),

$$v = u + w$$

where, $u \in U$ and $w \in W$.

Thus, every element of V can be expressed as the linear sum of an element of U and an element of W, *i.e.*, $V = U + W$. Hence, $V = U \oplus W$.

Now, the restriction of f to W is an alternating bilinear form on W. By induction hypothesis, there exists a basis $\alpha_3, \alpha_4, \ldots, \alpha_n \in W$ in which the matrix representation of f restricted to W has the above desired form. Hence, $\alpha_1, \alpha_2, \ldots, \alpha_n$, is a basis of V in which the matrix representing f has the desired form.

THEOREM 5. *Let f be a symmetric bilinear form on V over F. Then V has a basis $\{\alpha_1, \alpha_2, \ldots, \alpha_n\}$ in which f is represented by a diagonal matix i.e., $f(\alpha_i, \alpha_j) = 0$ for $i \neq j$.*

Proof. If $f = 0$ or if dim.$V = 1$, then the theorem is evidently true. Thus, we may assume that $f \neq 0$ and dim.$V = n > 1$. If for every $\alpha \in V$, $q(\alpha) = f(\alpha, \alpha)$, then the polar form of f gives that $f = 0$.

Hence we may assume that there is a vector $\alpha_1 \in V$ such that $f(\alpha_1, \alpha_1) \neq 0$.

Let W_1 be a subspace of V spanned by α_1 and let W_2 consist of those vectors $\alpha \in V$ such that $f(\alpha_1, \alpha) = 0$. We now claim that $V = W_1 \oplus W_2$.

(i) We first show that $W_1 \cap W_2 = \{0\}$.

If possible $W_1 \cap W_2 \neq \{0\}$, then there exists a non-zero vector β such that $\beta \in W_1$ and $\beta \in W_2$.

Now, $\beta \in W_1 \Rightarrow \beta = k\alpha_1$ for some $k \in F$.

Also, $\beta \in W_2 \Rightarrow f(\beta, \beta) = 0 \Rightarrow f(k\alpha_1, k\alpha_1) = 0$

$\Rightarrow k^2 f(\alpha_1, \alpha_1) = 0 \Rightarrow k = 0$ $[\because f(\alpha_1, \alpha_1) \neq 0]$

$\Rightarrow \beta = 0$.

Thus, $W_1 \cap W_2 = \{0\}$.

(ii) Now, to show that $V = W_1 + W_2$.

Let $\beta \in V$ and set

$$y = \beta - \frac{f(\alpha_1, \beta)}{f(\alpha_1, \alpha_1)} \alpha_1 \qquad \ldots(1)$$

Then, $f(\alpha_1, g) = f\left(\alpha_1, \beta - \frac{f(\alpha_1, \beta)}{f(\alpha_1, \alpha_1)} \alpha_1\right) = f(\alpha_1, \beta) - \frac{f(\alpha_1, \beta)}{f(\alpha_1, \alpha_1)} f(\alpha_1, \alpha_1)$

$$= f(\alpha_1, \beta) - f(\alpha_1, \beta) = 0$$

Thus, $\gamma \in W_2$, then from(1), we have

$$\beta = \gamma + \frac{f(\alpha_1, \beta)}{f(\alpha_1, \alpha_1)} \alpha_1$$

This shows that $\beta \in V$ is the sum of an element of W_1, and an element of W_2, thus

$$V = W_1 + W_2 \qquad \ldots(2)$$

Hence, from (1) and (2), we have

$$V = W_1 \oplus W_2$$

Now, f restricted to W_2, is a symmetric bilinear form on W_1. Also,

dim.W_2 = dim.V – dim.W_1 = $n-1$

Thus, by induction, we may assume that there is a basis $\{\alpha_2, \alpha_3, \ldots, \alpha_n\}$ of W_2 such that $f(\alpha_i, \alpha_j) = 0$ for $i \neq j$ and $2 \leq i, i \leq n$. But by definition of W_2, $f(\alpha_i, \alpha_j) = 0$ for $i = 2, 3, \ldots, n$. Therefore, we obtain the basis $\{\alpha_1, \alpha_2, \ldots, \alpha_n\}$ of V such that $f(\alpha_i, \alpha_j) = 0$, for all $i \neq j$.

THEOREM 6. *If f is a non-degenerate bilinear form on V(F). Then the set G of all linear transformation T on V which preserves f, is a group under the product of transformations.*

Proof. It is given that $G = \{T: f(T(\alpha), T(\beta)) = f(\alpha, \beta), \forall \alpha, \beta \in V\}$. Now to show that G is a group under the product of transformations.

(i) Closure property. For each $T_1, T_2 \in G$, we have
$$f((T_1 T_2)(\alpha), (T_1 T_2)(\beta)) = f(T_1(T_2(\alpha)), T_1(T_2(\beta)))$$
$$= f(T_2(\alpha), T_2(\beta)) \qquad [\because T_1 \text{ preserves } f.]$$
$$= f(\alpha, \beta). \qquad [\because T_2 \text{ preserves } f.]$$
Thus, $T_1 T_2$ preserves f so that $T_1 T_2 \in G$.

(ii) Associativity. Multiplication of linear transformations is always associative.

(iii) Existence of Identity. Since $I(\alpha) = \alpha$, $I(\beta) = \beta$.

[By definition of identity mapping]

Now, $\quad f(I(\alpha), I(\beta)) = f(\alpha, \beta), \forall \alpha, \beta \in V$.

This implies that $I \in G$ which is the identity toansformation.

(iv) Existence of inverse. For each $T \in G$. Let α be a vector in the null space of T, so that
$$T(\alpha) = 0, \forall \alpha \in V.$$
Then, $\quad f(T(\alpha), T(\beta)) = f(0, T(\beta)) = 0$.

Since T preserves f, i.e., $f(\alpha, \beta) = f(T(\alpha), T(\beta))$.
$$f(\alpha, \beta) = 0, \quad \forall \beta \in V.$$
Also, f is non-degenerate, therefore $f(\alpha, \beta) = 0, \forall \beta \in V$ is only possibility when $\alpha = 0$.
$$\therefore \qquad T(\alpha) = T(0) = 0.$$
This implies T is non-singular. Now, as V is finite dimensional so T is invertible.

Also, $f(T^{-1}(\alpha), T^{-1}(\beta)) = f(T(T^{-1}(\alpha)), T(T^{-1}(\beta)))$
$$= f((TT^{-1})(\alpha), (TT^{-1})(\beta))$$
$$= f(I(\alpha), I(\beta)) = f(\alpha, \beta)$$
$\therefore \qquad T^{-1}$ preserves f.

Thus, $T^{-1} \in G$. This shows that every element in G has its inverse in G. Hence G is a group.

9.3.1 Isotropic Subspace

Let f be a reflexive bilinear form on $V(F)$ and S be a subspace of V. Define a set $S^{\perp} = \{\beta \in V : f(\alpha, \beta) = 0, \forall \alpha \in S\}$. Clearly, $S^{\perp}$ is a subspace of V. The subspace S is called isotropic if $S \subseteq S^{\perp} \neq \{0\}$, otherwise, anisotropic. Also, S is called totally isotropic if $S \subseteq S^{\perp}$.

The subspace $V^{\perp}$ is called the radical of f, denoted by rad (f). Thus
$$rad\ (f) = \{\beta \in V : f(\alpha, \beta) = 0, \forall \alpha \in V\}$$
If rad (f) = {0}, then f is non-degenerate, otherwise degenerate.

THEOREM 7. *Let f be a reflexive bilinear form on a finite dimensional vector space V over F. Then f is non-degenerate if and only if the matrix of f with respect to an ordered basis of V is invertible.*

Proof. Let B be an ordered basis of V. Let if possible $[f]_B$ is not invertible $i.e.$, det.$[f]_B = 0$, then the system of equations (matrix form)

$$[f]_B X = 0$$

has a non-zero solution. Let this solution be $X_0 \in F$, where $n = \dim . V$ and let $\beta \in V$ such that $\quad X_0 = [b]_B$.

Then for any $\alpha \in V$, we have

$$f(\alpha, \beta) = [\alpha]_B^t \ [f]_B \ [\beta]_B = [\alpha]_B \ [f]_B \ X_0 = [\alpha]_B \ 0 = 0.$$

$\Rightarrow \qquad\qquad \beta \in \text{rad}(f) - \{0\}.$

Thus, $\text{rad}(f) \neq \{0\}$. Hence f is degenerate, therefore if $[f]_B$ is invertible, then f is non-degenerate.

Conversely, let us suppose that f is degenerate, then there exists a non-zero vector $\beta \in V$ such that

$$f(\alpha, \beta) = 0, \ \forall \ \alpha \in V.$$

Let $B = \{\alpha_1, \alpha_2, ..., \alpha_n\}$ be an ordered basis of V, so

$$\beta = a_1 \alpha_1 + a_2 \alpha_2 + ... + a_n \alpha_n, \text{ for } a_i \in F.$$

Then $[\beta]_B = [a_1, a_1, ..., a_n]^t$ is a non-zero element of F^n.

Now, for $i = 1, 2, ..., n$, then i-th entry of $[f]_B \ [\beta]_B$ is equal to

$$[\alpha_i]^t{}_B [f]_B [\beta]_B = f(\alpha_i, \beta) = 0.$$

Hence, $[f]_B$ is not invertible. Therefore, if f is non-degenerate, then $[f]_B$ is invertible.

THEOREM 8. *Let f be a reflexive bilinear form on a vector space $V(F)$. Let W_1 and W_2 be subspaces of V, then*

(i) If $W_1 \subseteq W_2$, then $W_2^\perp \subseteq W_1^\perp$ *(ii) $(W_1 + W_2)^\perp = W_1^\perp \cap W_2^\perp$*

(iii) $W_1 \subseteq W_1^{\perp\perp}$ *(iv) $W_1^\perp \subseteq W_1^{\perp\perp\perp}$.*

Proof.   (i)   Let $\beta \in W_2^\perp$, then $\beta \in V$ such that

$$f(\alpha, \beta) = 0, \ \forall \ \alpha \in W_2.$$

Since every element of W_1 is also the element of W_2, therefore

$$f(\alpha, \beta) = 0 \ \forall \ \alpha \in W_1 \qquad \Rightarrow \beta \in W_1^\perp.$$

$\therefore \qquad\qquad W_2^\perp \subseteq W_1^\perp.$

(ii)   Since W_1 and W_2 are subspaces of V, then $W_1 + W_2$ is also subspace of V.

Let $\alpha \in W_1 + W_2$, then

$\alpha = \alpha_1 + \alpha_2$ where $\alpha_1 \in W_1$ and $\alpha_2 \in W_2$.

Now, $\beta = (W_1 + W_2)^\perp \Leftrightarrow f(\alpha, \beta) = 0 \ \forall \ \alpha \in W_1 + W_2$

$\Leftrightarrow f(\alpha_1 + \alpha_2, \beta) = 0 \ \forall \ \alpha_1 \in W_1$ and $\alpha_2 \in W_2$

$\Leftrightarrow f(\alpha_1, \beta) + f(\alpha_2, \beta) = 0$

$\Leftrightarrow f(\alpha_1, \beta) = 0 \ \forall \ \alpha_1 \in W_1$ and $f(\alpha_2, \beta) = 0 \ \forall \ \alpha_2 \in W_2$

$\Leftrightarrow \beta \in W_1^\perp$ and $\beta \in W_2^\perp \Leftrightarrow \beta \in W_1^\perp \cap W_2^\perp$

$\therefore \qquad (W_1 + W_2)^\perp = W_1^\perp \cap W_2^\perp.$

In a similar way, we may prove result (iii) and (iv).

THEOREM 9. **(Riesz representation theorem).** *Let f be a non-degenerate reflexive bilinear form on a finite dimensional vector space V over F. If ϕ is a linear form on V, then there exists a unique vector $\beta \in V$ such that $\phi(\alpha) = f(\alpha, \beta)$ for all $\alpha \in V$.*

Proof. Let $\psi: V \to V^*$ be a mapping defined by $\psi(y) = \phi(\alpha, \gamma)$, where V^* is dual to V.

Obviously, ψ is well defined homomorphism of vector space. If $\gamma \in \ker.\psi$, then $f(\alpha, \gamma) = 0$, $\forall \alpha \in V$. Since f is non-degenerate so that $\gamma = 0$.

Therefore, ψ is one to one. Also, V is finite dimensional so that dim.V_1 = dim.V^*. Thus, the mapping ψ is surjective.

Hence if $\phi \in V^*$, then there exists a unique vector $\beta \in V$ such that
$$\phi(\beta) = f(\alpha, \beta), \ \forall \alpha \in V.$$

THEOREM 10. *Let f be a non-degenerate reflexive bilinear form on a finite dimensional vector space V over F. Then for a subspace W of V, $W^{\perp\perp} = W$.*

Proof. It is known that
$$W \subseteq W^{\perp\perp} \qquad \qquad \dots(1)$$

Suppose that the inclusion is proper, then there exists a linear map ϕ on V such that $\phi = 0$, for all $\alpha \in W$ and $\phi \neq 0$, for all $\beta \in W^{\perp\perp}$.

Let $\gamma \in V$ such that $\phi(\alpha) = f(\alpha, \gamma)$ for all $\alpha \in V$.

Since $f(\alpha, \gamma) = 0$, for all $\alpha \in W$ and $f(\alpha, \gamma) \neq 0$, for some $\alpha \in W^{\perp\perp}$, this implies $\gamma \in W^{\perp} \sim W^{\perp\perp}$, which gives a contradiction as $W^{\perp} = W^{\perp\perp}$.

Hence $W = W^{\perp\perp}$.

THEOREM 11. *Let f be a reflexive bilinear form on a vector space V over F. If S is a finite dimensional anisotropic subspace of V, then $V = S \oplus S^{\perp}$.*

Proof. Let $\beta \in V$. The mapping $\phi: S \to F$ defined by $\phi(\alpha) = f(\alpha, \beta)$, for all $\alpha \in S$. Obviously

ϕ is linear on S. The bilinear form $\hat{f} = f|_{S \times S}$ is non-degenerate on S. Since if $\gamma \in S$ such that $f(\alpha, \gamma) = 0$, for all $\alpha \in S$, then $\gamma \in S \cap S^{\perp} = \{0\}$. Therefore, by Riesz

representation theorem there exists $\gamma \in S$ such that $\phi(\alpha) = \hat{f}(\alpha, \gamma)$ for all $\alpha \in S$. Thus, for all $\alpha \in S$
$$f(\alpha, \beta) = \hat{f}(\alpha, \gamma) = f(\alpha, \gamma) \ \Rightarrow \ f(\alpha, \beta - \gamma) = 0, \text{ for all } \alpha \in S$$
$$\Rightarrow \quad \beta - \gamma \in S^{\perp} \quad \Rightarrow \quad \beta = \gamma + \beta - \gamma \in S + S^{\perp}.$$
Since β is an arbitrary element of V. Hence $V = S \oplus S^{\perp}$.

THEOREM 12. *A symmetric bilinear form on a finite dimensional vector space V over a field F of characteristic not equal to 2 is diagonolizable.*

Proof. We shall prove the theorem by induction on dim. V.

If dim. $V = 1$, then every non-zero vector of V is an orthogonal basis. Now assume that induction hypothesis holds for all symmetric bilinear forms on vector spaces of dimension less than dim.V.

Let f be a symmetric bilinear form on V. If $f(\alpha, \alpha) = 0$, $\forall \alpha \in V$. Then for $\beta, \gamma \in V$.
$$0 = f(\beta + \gamma, \beta + \gamma) = f(\beta, \beta) + 2f(\beta, \gamma) + f(\gamma, \gamma) \Rightarrow f = 0 \text{ as characteristic of } F \neq 2.$$
This shows that every basis of V is orthogonal.

Assume that for some $\alpha_1 \in V, f(\alpha_1, \alpha_1) \neq 0$.

Let W be a subspace spanned by α_1, then $V = W \oplus W^{\perp}$. If $f_1 = f|_{W^{\perp} \times W^{\perp}}$, then f_1 is a symmetric bilinear form on $W^\perp$.

Since dim. $W^\perp$ = dim.V–1. By induction hypothesis $W^\perp$ has an orthogonal basis B_1. Hence, $B = \{\alpha_1\} \cup B_1$ is an orthogonal basis of V relative to f such that f is of the form

$$[f]_B = \text{diag} \left(\begin{bmatrix} 0 & 1 \\ -1 & 0 \end{bmatrix}, ..., \begin{bmatrix} 0 & 1 \\ -1 & 0 \end{bmatrix}, 0, ..., 0 \right)$$

THEOREM 13. *Let f be a symmetric and non-alternating bilinear form on a finite dimensional vector space V over a field F of characteristic 2. Then f is diagonalizable.*

Proof. Let V_1 be a subspace of V complementary to rad(f). Then $V = V_1 \oplus$ rad f and

$\hat{f} = f|_{V_1 \times V_1}$ is a non-degenerate, non-alternating and symmetric bilinear form on V_1. If V_1 has an orthogonal basis B_1 with respect to $\hat{f}$ and B_2 any basis of rad(f), then $B = B_1 \cup B_2$ is an orthogonal basis of V. Therefore, we can assume that f is also non-degenerate.

Now, we shall prove the main theorem by induction on dim.V.

If dim.$V = 1$, obviously the theorem is true in this case. It is also assumed that the hypothesis hold for all non-degenerate, non-alternating and symmetric bilinear forms on vector spaces of dimension less than dim. V.

Since f is non-alternating, then there exists a vector $\alpha \in V$ such that $f(\alpha,\alpha) = k \neq 0$. Let W be a subspace of V spanned by the vector α. Then $V = W \oplus W^\perp$. Let $f_0 = f|_{W^\perp \times W^\perp}$. If f_0 is non-altenating, then by induction hypothesis, $W^\perp$ has an orthogonal basis B_0 relative to f_0 and so $B = \{\alpha\} \cup B_0$ is an orthogonal basis of V relative to f.

Since f_0 is non-degenerate and if f_0 is alternating then there is a basis
$$B' = \{\alpha_2, \alpha_3, ..., \alpha_n\} \text{ such that}$$

$$[f_0]_{B'} = \text{diag} \left(\begin{bmatrix} 0 & 1 \\ 1 & 0 \end{bmatrix}, ..., \begin{bmatrix} 0 & 1 \\ 1 & 0 \end{bmatrix} \right)$$

Let $W_1 = (\alpha, \alpha_2, \alpha_3)$ and let $f_1 = f|_{W_1 \times W_1}$. Then clearly f_1 is a non-degenerate bilinear form on W_1. Also, the vectors $\beta_1 = \alpha + \alpha_2 + \alpha_3$, $\beta_2 = \alpha + k\alpha_2$, $\beta_3 = \alpha + (1+k) \alpha_2 + \alpha_3$, are mutually orthogonal and the matrix of f_1 relative to the ordered basis $\{\beta_1, \beta_2, \beta_3\}$ is equal to dig.(k, k, k).

Now, let W_2 be a subspace spanned by the vectors $\beta_3, \alpha_4, ..., \alpha_n$.

Since β_3 is a linear combination of vectors α, α_2 and α_3 and $f(\beta_3, \beta_3) = k$, the bilinear form $f_2 = f|_{W_2 \times W_2}$ is non-degenerate, non-alternating and symmetric.

Also dim. $W_2 < \dim V$, then by induction hypothesis, W_2 has an orthogonal basis B''. Hence $B = \{\beta_1, \beta_2\} \cup B''$ is an orthogonal basis of V.

THEOREM 14. *Let V be a finite-dimensional vector space over the field of complex numbers. Let f be a symmetric bilinear form on V which has rank r. Then there is an ordered basis $B = \{\beta_1, \beta_2, ..., \beta_n\}$ for V such that*

(i) the matrix of f in the ordered basis B is diagonal.

(ii) $f(\beta_j, \beta_j) = \begin{cases} 1, j = 1, 2, ..., r \\ 0, j > r \end{cases}$

Proof. By theorem 4, there is an ordered basis $\{\alpha_1, \alpha_2, ..., \alpha_n\}$ of V such that $f(\alpha_i, \alpha_j) = 0$ for $i \neq j$.

Since f has a rank r, therefore the matrix of f is relative to the ordered basis $\{\alpha_1, \alpha_2, ..., \alpha_n\}$.

Hence, $f(\alpha_i, \alpha_j) \neq 0$ for exactly r values of j. Now, by reordering the vectors α_j, we may assume that $f(\alpha_j, \alpha_j) \neq 0$, $j = 1, 2, ..., r$.

Also, the field is the field of complex numbers, and if $\sqrt{f(\alpha_j, \alpha_j)}$ denotes complex square root of $f(\alpha_j, \alpha_j)$ and if we put

$$\beta_j = \begin{cases} \dfrac{1}{\sqrt{f(\alpha_j, \alpha_j)}} \alpha_j, & 1, 2, ..., r \\ \alpha_j, & j > r \end{cases}$$

Then the basis $\{\beta_1, \beta_2, ..., \beta_n\}$ satisfies the given hypothesis (*i*) and (*ii*).

9.3.2 Positive Definite

A symmetric bilinear form f is said to be positive definite if $f(\alpha, \alpha) > 0$ for $0 \neq \alpha \in V$.

THEOREM 15. *Let V be an n-dimensional vector space over the field of real numbers and let f be a symmetric bilinear form on V which has rank r. Then there is an ordered basis $\{\beta_1, \beta_2, ..., \beta_n\}$ for V in which the matrix of f is diagonal and such that*

$$f(\beta_j, \beta_j) = \pm 1, j = 1, 2, ..., r$$

Furthermore, the number of basis vectors β_j for which $f(\beta_j, \beta_j) = 1$ is independent of the choice of basis.

Proof. By theorem 4, there is an ordered basis $\{\alpha_1, \alpha_2, ..., \alpha_n\}$ such that $f(\alpha_i, \alpha_j) = 0, i \neq j$.

Therefore, $f(\alpha_j, \alpha_j) \neq 0$, $1 \leq j \leq r$

and $f(\alpha_j, \alpha_j) = 0$, $j > r$

Let us assume that

$$\beta_j = \begin{cases} \dfrac{1}{\sqrt{|f(\alpha_j, \alpha_j)|}} \alpha_j, & 1 \leq j \leq r \\ \alpha_j, & j > r \end{cases}$$

Then, $\{\beta_1, \beta_2, ..., \beta_n\}$ forms a basis with the given properties.

Now, assume that there are k number of basis vectors β_j for which $f(\beta_j, \beta_j) = 1$. Then we show that the number k is independent of this basis.

Let V^+ be the subspace of V spanned by the basis vectors β_j for which $f(\beta_j, \beta_j) = 1$. and let V^- be the subspace of V spanned by the basis vectors β_j. For which $f(\beta_j, \beta_j) = -1$. Now $k = \dim V^+$, so it is the uniqueness of the dimension of V^+ which we must show. Clearly, if α is a non-zero vector in V^+, then $f(\alpha, \alpha) < 0$ implying f is negative definite on V^-.

Now, let $V^\perp$ be the subspace spanned by the vector β_j for which $f(\beta_j, \beta_j) = 0$, so if $\alpha \in V^\perp$, then $f(\alpha, \beta) = 0$, $\forall \beta \in V$.

Since $\{\beta_1, \beta_2, ..., \beta_n\}$ is a basis of V, then we have
$$V = V^+ \oplus V^- \oplus V^\perp.$$

Next, we claim that if W is any subspace of V on which f is positive definite, then the subspaces W, V^- and $V^\perp$ are independent.

For, suppose $\alpha \in W$, $\beta \in V^-$ and $\gamma \in V^\perp$ and $\alpha + \beta + \gamma = 0$, then

$$\left. \begin{array}{l} 0 = f(\alpha, \alpha + \beta + \gamma) = f(\alpha, \alpha) + f(\alpha, \beta) + f(\alpha, \gamma) \\ 0 = f(\beta, \alpha + \beta + \gamma) = f(\beta, \alpha) + f(\beta, \beta) + f(\beta, \gamma) \end{array} \right\} \qquad ...(1)$$

Since, $\gamma \in V^\perp$ so $f(\alpha, \gamma) = f(\beta, \gamma) = 0$, and since f is symmetric, then (1) reduces to
$$0 = f(\alpha, \alpha) + f(\alpha, \beta) \text{ and } 0 = f(\beta, \beta) + f(\alpha, \beta)$$

From these equations, we obtain that $f(\alpha, \alpha) = f(\beta, \beta)$.

But $f(\alpha, \alpha) \geq 0$ and $f(\beta, \beta) \leq 0$, therefore it follows that
$$f(\alpha, \alpha) = f(\beta, \beta) = 0.$$

Also, f is positive definite on W and negative definite on V^-, then we obtain
$$\alpha = \beta = 0$$
and hence
$$\gamma = 0$$

Since $V = V^+ \oplus V^- \oplus V^\perp$ and $W, V^-, V^\perp$ are independent so that
$$\dim.W \leq \dim V^+.$$

If β_1 is another ordered basis of V which satisfies the conditions of the theorem, then we shall have corresponding subspaces V_1^+, V_1^- and $V_1^\perp$ and using above process, we have $\dim V_1^+ \leq \dim V^+$. Now, interchanging the role of V_1^+ and V^+, we have
$$\dim.V^+ \leq \dim V_1^+.$$

Hence,
$$\dim V^+ = \dim V_1^+.$$

9.3.3 Rank and Signature

Let f be a symmetric bilinear form on V and let V^+, V^- be the subspaces of V such that $f(\beta_j, \beta_j) = 1$ on V^+ and $f(\beta_j, \beta_j) = -1$. On V^-, where $\{\beta_1, \beta_2, \dots, \beta_n\}$ is an ordered basis of V, then the number $(dim V^+ + dim V^-)$ is called the rank of f and the number $(dim\ V^+ - dim\ V^-)$ is called the signature of f.

THEOREM 16. *Let V be an n-dimensional vector space over a subfield of the complex numbers, and let f be a skew-synmetric bilinear form on V. Then the rank $r = 2k$, there is an ordered basis for V in which ihe matix of f is the direct sum of the $(n-r) \times (n-r)$ zero matrix and k copies of the 2×2 matrix $\begin{bmatrix} 0 & 1 \\ -1 & 0 \end{bmatrix}$.*

Proof. Let $\alpha_1, \beta_1, \alpha_2, \beta_2, \dots, \alpha_k, \beta_k$ be the vectors having the following properties:
  (i) $f(\alpha_j, \beta_j) = 1$ for $j = 1, 2, \dots, k$.
  (ii) $f(\alpha_i, \beta_j) = f(\beta_i, \beta_j) = f(\alpha_i, \beta_j) = 0, i \neq j$.
  (iii) If W_j is the two dimensional subspace spanned by α_j and β_j, then
$$V = W_1 \oplus W_2 \oplus, \dots, \oplus W_k \oplus W_0$$
where every vector in W_0 is orthogonal to all α_j and β_j and the restriction of f to W_0 is the zero form.

Assuming that $[\gamma_1, \gamma_2, \dots, \gamma_m]$ be any ordered basis for the subspace W_0, then
$$B = \{\alpha_1, \beta_1, \dots, \alpha_k, \beta_k, \gamma_1, \gamma_2, \dots, \gamma_n\}$$
forms an ordered basis for V.

From conditions (i), (ii) and (iii), it is clear that the matrix of f in the ordered basis B is the direct sum of the $(n - 2k) \times (n - 2k)$ zero matrix and k copies of the 2×2 matrix $\begin{bmatrix} 0 & 1 \\ -1 & 0 \end{bmatrix}$

It is also clear that the rank of this matrix of f is $2k$.

Solved Examples

Based on the following Results

→ A mapping $f : V \times V \to F$ is said to be bilinear form on a vector space if

(i) $f(a\alpha+b\beta,\gamma)=af(\alpha,\gamma)+ bf(\beta,\gamma)$

(ii) $f(\alpha,a\beta+b\gamma)=af(\alpha,\beta)+ bf(\alpha,\gamma) \ \forall \ a,b \in F, \ \alpha,\beta,\gamma \in V$

→ Let f be a bilinear form on V over F. Then a vector $\alpha \in V$ is said to be orthogonal to $\beta \in V$ with respect to f if $f(\alpha,\beta)=0$.

→ Let f be a bilinear form on $V(F)$ and let $S=\{\alpha_1,\alpha_2,.....,\alpha_n\}$ be a basis of V. Then a matrix $A=\{a_{ij}\}$ where $a_{ij}=f(\alpha_i, \alpha_j)$ is called the matrix representation of f w.r.t. the basis S.

Example 1. *Let A be any square matix of order $n \times n$ over F. Then the map $f(X,Y) = X^TAY$ is a bilinear form on F^n.*

Solution. For any $a, b \in F$ and any $X_i, Y_i \in F^n$

$$f(aX_1 +bX_2,Y)=(aX_1 +bX_2)^TAY = (aX_1{}^T +bX_2{}^T) A \in V$$
$$= aX_1^T AY + bX_2^T AY = af(X_1, Y) + bf(X_2, Y).$$

This shows that f is linear in first variable.

Also, $\quad f(X, aY_1 + bY_2) = X^TA(aY_1 + bY_2)$
$$= aX^TAY_1 + bX^TAY_2 = af(X, Y_1) + bf(X, Y_2)$$

This shows that f is linear in second variable.

Hence f is bilinear form on F^n.

Example 2. *Let ϕ and ψ be any linear functionals on a vector space V. Let $f : V \times V \to F$ be defined by $f(\alpha, \beta) = \phi(\alpha)\psi(\beta)$. Show that f is bilinear form.*

Solution. For any $a, b \in F$ and $\alpha_i,\beta_i \in V$, we have

$$f(a\alpha_1 + b\alpha_2,\beta) = \phi(a\alpha_1 + b\alpha_2)\psi(\beta)=(a\phi(\alpha_1)+b\phi(\alpha_2))\psi(\beta)$$
$$= a\phi(\alpha_1)\psi(\beta) +b\phi(\alpha_2)\psi(\beta) = af(\alpha_1,\beta)+bf(\alpha_2,\beta)$$

Also, $\quad f(\alpha,a\beta_1 + b\beta_2)=\phi(\alpha) + \psi(a\beta_1+b\beta_2)=\phi(\alpha)(a\psi(\beta_1)+b\psi(\beta_2))$
$$= a\phi(\alpha)\psi(\beta_1)+b\phi(\alpha)\psi(\beta_2)= af(\alpha,\beta_1)+bf(\alpha,\beta_2)$$

Hence, f is bilinear form.

Example 3. *Let f be a bilinear form on R^2 defined by $f((x_1, x_2),(y_1,y_2))=2x_1y_1 -3x_1y_2+ x_2y_2$. Find the matrix of f in the basis $\{\alpha_1=(1,0),\alpha_2=(1,1)\}$.*

Solution. Let $A = \{a_{ij}\}$ be the matrix of f in the given basis, where $a_{ij}=f(\alpha_i\alpha_j)$, $1 \le i, j \le 2$.

Now, $\qquad a_{11} = f(\alpha_1,\alpha_1) = f(1,0),(1,0)$
$$= 2(1)(1) - 3(1)(0) + (0) (0) = 2$$

$\qquad a_{12} = f(\alpha_1,\alpha_2) = f(1,0),(1,1)$
$$= 2(1)(1) - 3(1)(1) - 0(1) = 2 - 3 = -1$$

$\qquad a_{21} = f(\alpha_1,\alpha_1) = f(1,0),(1,0)$
$$= 2(1)(1) - 3(1)(0) + (0)(0) = 2$$

$\qquad a_{22} = f(\alpha_2,\alpha_2) = f((1,1),(1,1))$
$$= 2(1)(1) - 3(1)(1) + (1)(1) = 2 - 3 + 1 = 0$$

Thus, $\quad A= \begin{bmatrix} 2 & -1 \\ 2 & 0 \end{bmatrix}$ is the matrix of f in the basis $\{\alpha_1,\alpha_2\}$.

Example 4. *Let [f] denote the matrix representation of a bilinear form f on V relative to a basis $S=\{\alpha_1, \alpha_2, ..., \alpha_n\}$ of V. Show that the mapping $\phi: f \to [f]$ is an isomorphism of $B(V, f)$ on the vector space of n-square matrices.*

Solution. Since f is completely determined by the scalars $f(\alpha_i, \alpha_j)$, therefore the mapping ϕ is one-one and onto. Now to show that ϕ is homomorphism.

For any $a, b \in F, f_1, f_2 \in B(V, f)$,

$$(af_1 + bf_2)(\alpha_i, \alpha_j) = af_1(\alpha_i, \alpha_j) + bf_2(\alpha_i, \alpha_j)$$
$$\Rightarrow \qquad [af_1 + bf_2] = a[f_1] + b[f_2].$$

Hence, the result.

Example 5. *Which of the following mappings $f : R^2 \times R^2 \to R$ are bilinear form : if for $\alpha = (x_1, x_2), \beta = (y_1, y_2)$*

(i) $f(\alpha, \beta) = x_1 x_2 + y_1 y_2 - x_2 y_1 - x_1 y_2$

(ii) $f(\alpha, \beta) = (x_1 - y_1)^2 + x_2 y_2$

(iii) $f(\alpha, \beta) = x_1 y_2 \pm x_2 y_1$

(iv) $f(\alpha, \beta) = x_1 y_1 + 1$

Solution. Let $\alpha = (x_1, x_2), \beta = (y_1, y_2)$ and $\gamma = (z_1, z_2)$.

(i) For $a, b \in R$

$$f(a\alpha + b\beta, \gamma) = f(a(x_1, x_2) + b(y_1, y_2), (z_1, z_2))$$
$$= f(ax_1 + by_1, ax_2 + by_2), (z_1, z_2))$$
$$= (ax_1 + by_1)(ax_2 + by_2) + z_1 z_2 - (ax_2 + by_2)z_1 - (ax_1 + by_1)z_2$$
$$= a^2 x_1 x_2 + abx_1 y_2 + aby_1 x_2 + b^2 y_1 y_2 + z_1 z_2$$
$$\qquad\qquad - ax_2 z_1 - by_2 z_1 - ax_1 z_2 - by_1 z_2$$

and $af(\alpha, \gamma) + bf(\beta, \gamma) = af((x_1, x_2), (z_1, z_2)) + bf((y_1, y_2), (z_1, z_2))$
$$= ax_1 x_2 + az_1 z_2 - ax_2 z_1 - ax_1 z_2 + by_1 y_2 + bz_1 z_2 - by_2 z_1 + by_1 z_2$$

Clearly, $f(a\alpha + b\beta, \gamma) \neq af(\alpha, \gamma) + bf(\beta, \gamma)$.

Thus, f is not bilinear form.

(ii) For $a, b \in R$;
$$f(a\alpha + b\beta, \gamma) = f((ax_1 + by_1, ax_2 + by_2), (z_1, z_2))$$
$$= (ax_1 + by_1 - z_1)^2 + (ax_2 + by_2) z_2$$

and $af(\alpha, \gamma) + bf(\beta, \gamma) = af((x_1, x_2), (z_1, z_2)) + bf((y_1, y_2), (z_1, z_2))$
$$= a[(x_1 - z_1)^2 + x_2 z_2] + b[(y_1 - z_1]^2 + y_2 z_2]$$

Clearly $f(a\alpha + b\beta, \gamma) \neq af(\alpha, \gamma) + bf(\beta, \gamma)$.

Thus, f is not bilinear form.

(iii) For $a, b \in R$
$$f(a\alpha + g\beta, \gamma) = f((ax_1 + by_1, ax_2 + by_2), (z_1, z_2)) = (ax_1 + by_1)z_2 \pm (ax_2 + by_2)z_1$$
$$= a(x_1 z_2 + x_2 z_1) + b(y_1 z_2 + y_2 z_1)$$
$$= af((x_1, x_2), (z_1, z_2)) + bf((y_1, y_2), (z_1, z_2)) = af(\alpha, \gamma) + bf(\beta, \gamma)$$

Also, $f(\alpha, a\beta + b\gamma) = f((x_1, x_2)(ay_1 + bz_1, ay_2 + bz_2))$
$$= x_1(ay_2 + bz_2) \pm x_2(ay_1 + bz_1)$$
$$= a(x_1 y_2 \pm x_2 y_1) + b(x_1 z_2 \pm x_2 z_1)$$
$$= af((x_1, x_2), (y_1, y_2)) + bf((x_1, x_2), (z_1, z_2))$$
$$= af(\alpha, \beta) + bf(\alpha, \gamma)$$

Hence f is bilinear form.

(iv) For $a, b \in R$

$$f(a\alpha+b\beta,\gamma) = f((ax_1+by_1,ax_2+by_2),(z_1,z_2))$$
$$= (ax_1+by_1)z_1+1 = a(x_1z_1)+b(y_1z_1)+1$$

and $af(\alpha,\gamma)+bf(\beta,\gamma) = af((x_1,x_2),(z_1,z_2))+bf((y_1,y_2),(z_1,z_2))$
$$= a[x_1z_1+1]+b[y_1z_1+1] = a(x_1z_1)+b(y_1z_1)+a+b$$

Clearly, $f(a\alpha+b\beta,\gamma) \neq af(\alpha,\gamma)+bf(\beta,\gamma)$.

Thus, f is not bilinear form.

Example 6. *Prove that if f is a bilinear form on $V(F)$, then*

(i) $f(-\alpha,\beta) = -f(\alpha,\beta) = f(\alpha,-\beta), \forall \alpha,\beta \in V$

(ii) $f(\alpha,0) = 0 = f(0,\alpha), \forall \alpha \in V$

(i) For any $a \in F$ and any $\alpha,\beta \in V$.

$$f(a\alpha,\beta) = af(\alpha,\beta)$$

taking $a=-1$, we get

$$f(-\alpha,\beta) = -f(\alpha,\beta)$$

Similarly, $f(\alpha,a\beta) = af(\alpha,\beta)$

again taking $a=-1$, we get

$$f(\alpha,-\beta) = -f(\alpha,\beta).$$
$$f(-\alpha,\beta) = -f(\alpha,\beta) = f(\alpha,-\beta).$$

(ii) $\qquad f(\alpha,0) = f(\alpha,0.0)$ $\qquad\qquad\qquad\qquad [\because 0.0=0]$

$$= 0f(\alpha,0) = 0.$$

Similarly, $f(0,\alpha) = 0.$

$\therefore \qquad f(\alpha,0) = 0 = f(0,\alpha), \forall \alpha \in V.$

Example 7. *Let F be a field of characteristic 2. Verify that the mapping $\phi: F \times F \to F$ defined by $\phi(x,y) = xy$ is a skew-symmetric bilinear form which is not alternating.*

Solution. First we show that ϕ is bilinear.

For $a,b \in F$ and $x,y,z \in F$, we have $(ax+by,z) \in F \times F$ so,

$$\phi(ax+by,z) = (ax+by)z$$
$$= axz+byz = a\phi(x,z)+b\phi(x,z)$$

Also, $\phi(x,ay+bz) = x(ay+bz) = axy+bxz = a\phi(x,y)+b\phi(x,z)$

Thus, Q is bilinear form.

Since F is of characteristic 2 so that $2xy = 0, \forall x,y \in F$.

$\Rightarrow \qquad\qquad xy+xy=0 \quad \Rightarrow \quad xy=-xy$

$\Rightarrow \qquad\qquad \phi(x,y) = -\phi(x,y), \forall x,y \in F$. Thus, ϕ is skew-symmetric.

Now, ϕ is skew-symmetric so that

$$\phi(x,x) = -\phi(x,x), \forall x \in F \text{ or } 2\phi(x,x)=0$$

$\therefore \qquad\qquad \phi(x,x) \neq 0$ as characteristic of $F=2$.

This shows that ϕ is not alternating form.

Example 8. *If f is a bilinear form on $V(F)$ and S be a subspace of V, then the set*

$$S^\perp = \{\beta \in V: f(\alpha,\beta) = 0, \forall \alpha \in S\} \text{ is a subspace of } V.$$

Solution. Let $\alpha \in S$ so $f(\alpha,0)=0$. Therefore $0 \in S^\perp$. Suppose $\alpha_1,\alpha_2 \in S^\perp$ and $k \in F$. Then

$f(\alpha,\alpha_1)=0$ and $f(\alpha,\alpha_2)=0$.

Thus, $\qquad f(\alpha, \alpha_1+\alpha_2) = f(\alpha,\alpha_1)+f(\alpha,\alpha_2)=0+0=0$

and $\qquad\qquad\qquad f(\alpha, k\alpha_1)= kf(\alpha,\alpha_1)=k.0=0$.

This implies $\alpha_1+\alpha_2 \in S^\perp$ and $k\alpha_1 \in S^\perp$.

Hence $S^\perp$ is a subspace of V.

REMARKS

- The subspace S is called isotropic if $S \cap S^\perp \neq \{0\}$ otherwise anisotropic.
- S is called totally isotropic if $S \subseteq S^\perp$.
- The subspace $V^\perp$ is called the radical of f, i.e., rad $f = \{\beta \in V : f(\alpha,\beta)=0 \ \forall \alpha \in V\}$

Example 9. *Show that a bilinear form f is symmetric if and only if any matrix A representing f is symmetric.*

Solution. Suppose f is symmetric and A represents f. For X, X^TAY, is a scalar therefore, it is equal to its transpose. Then

$$f(X,Y) = X^TAY = (X^TAY)^T = Y^TA^TX$$

Since f is symmetric, i.e., $f(X, Y) = f(Y, X)$

So, $\qquad Y^TA^TX=Y^TAX, \ \forall \ X,Y$

$\Rightarrow \qquad\qquad A^T = A \qquad \Rightarrow \ A$ is symmetric.

Conversely, suppose A is symmetric. Then

$$f(X, Y)=X^TAY= (X^TAY)^T = Y^TA^TX = Y^TAX \qquad\qquad [\because A^T = A]$$
$$= f(Y, X)$$

$\therefore f$ is symmetric.

Example 10. *Let V be a finite-dimensional vector space and L_1,L_2 linear functionals on V. Show that the equation $f(\alpha, \beta)= L_1(\alpha)L_2(\beta) - L_1(\beta)L_2(\alpha)$ defnes a skew-symmetric bilinear form on V. Show that f =0 if and only if L_1,L_2 are linearly dependent.*

Solution. First we show f is linear form on V.

Let $\alpha,\beta,\gamma \in V$ and $a,b \in F$, we have

$$f(\alpha,a\beta+b\gamma)=L_1(\alpha)L_2(a\beta+b\gamma) - L_1(a\beta+b\gamma)L_2(\alpha)$$
$$= L_1(\alpha)\{aL_2(\beta)+ bL_2(\gamma)\} - \{aL_1(\beta)+bL_1(\gamma)\}L_2(\alpha)$$
$$[\because L_1, L_2 \text{ are linear.}]$$
$$= aL_1(\alpha)L_2(\beta)+bL_1(\alpha)L_2(\gamma) - aL_1(\beta)L_2(\alpha)-bL_1(\gamma)L_2(\alpha)$$
$$= a\{L_1(\alpha)L_2(\beta)-L_1(\beta)L_2(\alpha)\}+b\{L_1(\alpha)L_2(\gamma)-L_1(\gamma)L_2(\alpha)\}$$
$$= af(\alpha, \beta)+bf(\alpha, \gamma).$$

$\therefore \quad f$ is also linear in second coordinate.

Thus f is a bilinear form on V.

Secondly, we show that f is skew-symmetric.

For any $\alpha,\beta \in V$, we have

$$f(\alpha,\beta)=L_1(\alpha)L_2(\beta)-L_1(\beta)L_2(\alpha)$$
$$= -\{L_1(\beta)L_2(\alpha)-L_1(\alpha)L_2(\beta)\}$$
$$= -f(\beta,\alpha)$$

$\therefore \qquad\qquad f(\alpha,\beta)=-f(\beta, \alpha) \ \forall \ \alpha, \beta \in V$

Hence f is skew-symmetric.

Next, we show that $f = 0$ iff L_1, L_2 are linearly dependent.

Suppose L_1, L_2 are linearly dependent then

$$
\begin{aligned}
f(\alpha, \beta) &= L_1(\alpha)\,L_2(\beta) - L_1(\beta)L_2(\alpha) \\
&= (kL_2)(\alpha)L_2(\beta) - (kL_2)(\beta)L_2(\alpha) \\
&= kL_2(\alpha)L_2(\beta) - kL_2(\beta)L_2(\alpha) \\
&= k(L_2(\alpha)L_2(\beta) - L_2(\beta)L_2(\alpha)) \\
&= k.0 \\
&= 0
\end{aligned}
$$

Since α, β are arbitrary element of V so that $f = 0$.

Conversely, suppose $f = 0 \ \forall \ \alpha < \beta \in V$, then

$\Rightarrow \qquad L_1(\alpha)\,L_2(\beta) - L_1(\beta)L_2(\alpha) = 0$.

This is possible only if either $L_1 = kL_2$ or $L_2 = kL_1$.

Hence L_1, L_2 are linearly dependent.

Example 11. *Let n be a positive integer, and let V be the space of all $n \times n$ matrices over the field of complex numbers. Show that the equation*

$$f(A, B) = n\,\mathrm{tr}(AB) - \mathrm{tr}(A)\mathrm{tr}(B)$$

defines a bilinear form f on V. Is it true that $f(A, B) = f(B, A)$ for all A, B?

Solution. Let $A, B, D \in V$ and $a, b \in C$ (complex field).

Then, $\qquad$
$$
\begin{aligned}
f(A, aB+bD) &= n\,\mathrm{tr}[A(aB+bD)] - \mathrm{tr}(A)\mathrm{tr}(aB+bD) \\
&= n\,\mathrm{tr}(aAB+bBD) - \mathrm{tr}(A)\{\mathrm{tr}(aB) + \mathrm{tr}(bD)\} \\
&= na\,\mathrm{tr}(AB) + nb\,\mathrm{tr}(BD) - a\,\mathrm{tr}(A)\,\mathrm{tr}(B) - \mathrm{tr}(A)\mathrm{tr}(bD) \\
&= na\,\mathrm{tr}(AB) + nb\,\mathrm{tr}(BD) - a\,\mathrm{tr}(A)\,\mathrm{tr}(B) - b\,\mathrm{tr}(A)\,\mathrm{tr}(D) \\
&= a[n\,\mathrm{tr}(AB) - \mathrm{tr}(A)\,\mathrm{tr}(B)] + b[n\,\mathrm{tr}(BD) - \mathrm{tr}(A)\,\mathrm{tr}(D)] \\
&= af(A,B) + bf(A,D)
\end{aligned}
$$

$\therefore f$ is linear in one coordinate.

Also, $\qquad$
$$
\begin{aligned}
f(aB+bD, A) &= n\,\mathrm{tr}((aB+bD)A) - \mathrm{tr}(aB+bD)\mathrm{tr}(A) \\
&= n\,\mathrm{tr}(aBA+bDA) - \mathrm{tr}(aB)\,\mathrm{tr}(A) - \mathrm{tr}(bD)\,\mathrm{tr}(A) \\
&= na\,\mathrm{tr}(BA) + nb\,\mathrm{tr}(DA) - a\,\mathrm{tr}(B)\,\mathrm{tr}(A) - b\,\mathrm{tr}(D)\,\mathrm{tr}(A) \\
&= a[n\,\mathrm{tr}(DA) - \mathrm{tr}(B)\,\mathrm{tr}(A)] + b[n\,\mathrm{tr}(DA) - \mathrm{tr}(D)\,\mathrm{tr}(A)] \\
&= af(B,A) + bf(D,A).
\end{aligned}
$$

$\therefore f$ is linear in second coordinate.

Hence, f is a bilinear form on V.

Further, since $\mathrm{tr}(AB) = \mathrm{tr}(BA)$ for all A and B, then

$\qquad n\,\mathrm{tr}(AB) - \mathrm{tr}(A)\,\mathrm{tr}(B) = n\,\mathrm{tr}(BA) - \mathrm{tr}(B)\mathrm{tr}(A)$

$\Rightarrow \qquad\qquad f(A, B) = f(B, A)$ for all A and B.

☙ EXERCISE 9.1

1. Decide which of the following mappings $f : R^2 \times R^2 \to R$ are bilinear, if for $\alpha = [x_1, x_2]^t, \beta = [y_1, y_2]^t$ in R^2:

(i) $f(\alpha, \beta) = (x_1 + x_2)(y_1 + y_2)$

(ii) $f(\alpha, \beta) = (x_1 + x_2)/(y_1 + y_2)$

(iii) $f(\alpha, \beta) = x_1 x_2 y_1 y_2$

(iv) $f(\alpha, \beta) = (3x_1 + 5)y_1 + (3y_2 + 5)x_1$

(v) $f(\alpha, \beta) = x_1 y_2 + x_2 y_1^2$

(vi) $f(\alpha, \beta) = (x_1 + y_1)^2 - (x_1 - y_1)^2$

2. Let f be the bilinear form on R^2 defined by $f((x_1,y_1),(x_2,y_2))=x_1y_1+x_2y_2$.
   Find the matrix of f in each of the following bases:

   (i) $\{(1,0),(0,1)\}$ (ii) $\{(1,-1),(1,1)\}$

   (iii) $\{(1,2),(3,4)\}$

3. Let V be the vector space of all 2 x 3 matrices over R, and let f be the bilinear form on V defined by $f(X,Y)=\text{trace}(X^tAY)$ where

   $$A=\begin{bmatrix} 1 & 2 \\ 3 & 4 \end{bmatrix}.$$

   Find the matrix of f in the ordered basis $\{E^{11}, E^{12}, E^{13}, E^{21}, E^{22}, E^{23}\}$, where E^{ij} is the matrix whose non-zero entry is 1 in $(i,j)^{th}$ position.

4. Let $\alpha=(x_1, x_2, x_3)$ and $\beta=(y_1, y_2, y_3)$ and let
   $f(\alpha,\beta)=3x_1y_1-2x_1y_2+5x_2y_1+7x_2y_2-8x_3y_3+4x_3y_2-x_3y_3$.
   Express f in matrix notation.

5. Let f and g be bilinear forms on V. Show that the sum $f+g$ defined by
   $(f+g)(\alpha,\beta)=f(\alpha,\beta)+g(\alpha,\beta)$ is bilinear.

6. Let f be a bilinear form on V and let $k \in F$. show that the map kf, defined by
   $(kf)(\alpha,\beta)=kf(\alpha,\beta)$ is bilinear.

7. Let $B(V, f)$ denote the collection of all bilinear forms on V. Show that $B(V, f)$ is a vector space with respect to addition $f+g$ and scalar multiplication kf defined by
   $(f + g)(\alpha,\beta)=f(\alpha,\beta)+g(\alpha,\beta)$
   $(kf)(\alpha,\beta)=kf(\alpha,\beta)\ \forall \alpha,\beta \in V$

8. Let f be the bilinear form on R^2 defined by
   $f(\alpha,\beta)=3x_1y_1-2x_1y_2+4x_2y_1-x_2y_2$
   where $\alpha=(x_1,x_2), \beta=(y_1, y_2)$.

   (i) Express f in matrix form with respect to an ordered basis $B = \{(1, 0), (0, 1)\}$.

   (ii) Express f in matrix form with respect to an ordered basis $B = \{(1, 1), (1,2)\}$.

(iii) Find the matrix P such that $[f]_{B'}=P^T[f]_B P$.

9. Prove that if char. $F \neq 2$, then a bilinear form on V over F is skew-symmetric if and only if it is altenating.

10. Prove that if char. $F \neq 2$ and f is a symmetric and altenating bilinear form on $V(F)$, then $f=0$.

11. Prove that in a reflexive bilinear space if $\alpha_1, \alpha_2, ..., \alpha_n$ are mutually orthogonal anisotropic vectors, then $\{\alpha_1, \alpha_2, ..., \alpha_n\}$ are linearly independent.

12. Prove that a diagonalizable bilinear form is symmetric.

13. Let $B(V, f)$ and $B(W, g)$ be isomorphic bilinear spaces. Prove that f is reflexive if and only if g is reflexive.

14. If f is reflexive, non-degenerate bilinear form on a vector space $V(F)$, then prove that for $\alpha, \beta \in V$

   (i) if $f(\alpha,\gamma)=f(\beta,\gamma)$ for all $\gamma \in V$, then $\alpha=\beta$.

   (ii) if $f(\gamma,\alpha)=f(\gamma,\beta)$ for all $\gamma \in V$, then $\alpha=\beta$.

15. Let f be any bilinear form on a finite dimensional vector space V. Let W be the subspace of all β such that $f(\alpha, \beta)= 0$ for every $\alpha \in V$. Show that
   $\text{rank}(f)=\dim.V - \dim.W$

16. Let f, g be bilinear forms on a finite dimensional vector space $V(F)$. Suppose g is non-singular. Show that there exists unique operators T_1, T_2 on V such that
   $f(\alpha,\beta)=g(T_1(\alpha),\beta)=g(\alpha,T_2(\beta))$
   for all $\alpha,\beta \in V$.

17. Let f be a symmetric bilinear form on $\mathbf{C}^n$ and g a skew-symmetric bilinear form on $\mathbf{C}^n$. Suppose $f+g=g$. Show that $f = g = 0$.

18. Let m and n be positive integers and F a field. Let V be the vector space of all $m \times n$ matrices over F. Let A be a fixed $m \times m$ matrix over F. Define $f_A(X,Y)=\text{trace}(X^tAY)$. Then show that f_A is bilinear.

Hint to Selected Problems

1. Since $\alpha=(x_1,x_2), \beta=(y_1,y_2)$, $\gamma=(z_1,z_2)$.
   $\therefore$ For $a,b \in F$, $a\alpha+b\beta = a(x_1, x_2) +b(y_1,y_2)$
   $= (ax_1 +by_1, ax_2+by_2)$
   $f(a\alpha + b\beta, \gamma)$
   $= f((ax_1 + by_1, ax_2 + by_2),(z_1, z_2))$
   $=(ax_1 +by_1 +ax_2+by_2)(z_1+z_2)$
   $\quad\quad [\because f(\alpha,\beta)=(x_1 +x_2)(y_1 +y_2)]$
   $=(ax_1+ax_2)(z_1+z_2)+(by_1+by_2)(z_1+z_2)$

   $=a(x_1 + x_2)(z_1+ z_2)+b(y_1+ y_2)(z_1+ z_2)$
   $= af(\alpha,\beta)+bf(\alpha, \gamma)$
   Now, $f(\gamma, a\alpha+b\beta)$
   $= f((z_1,z_2),(ax_1 +by_1, ax_2 +by_2))$
   $=(z_1 +z_2)(ax_1 + by_1 + ax_2 + by_2)$
   $= a(z_1 + z_2)(x_1 + x_2) + b(z_1 + z_2)(y_1 + y_2)$
   $= af(\gamma, \alpha) + bf(\gamma,\beta)$
   Hence f is bilinear.

2. (i) Since $f(x_1, y_1), (x_2, y_2)) = x_1 y_1 + x_2 y_2$
$$B = \{(1,0), (0,1)\}.$$

Let $\alpha_1 = (1,0), \alpha_2 = (0,1)$

$\alpha_{11} = f(\alpha_1, \alpha_1) = f((1, 0, (1, 0))$
$$= 1 \times 0 + 1 \times 0 = 0$$

$\alpha_{12} = f(\alpha_1, \alpha_2) = f((1, 0), (0, 1))$
$$= 1 \times 0 + 0 \times 1 = 0$$

$\alpha_{21} = f(\alpha_2, \alpha_1) = f(0, 1), (1, 0))$
$$= 0 \times 1 + 1 \times 0 = 0$$

$\alpha_{22} = f(\alpha_2, \alpha_2) = f((0, 1), (0, 1))$
$$= 0 \times 1 + 0 \times 1 = 0$$

∴ The matix of f relative to $B = \begin{bmatrix} 0 & 0 \\ 0 & 0 \end{bmatrix}$.
Other parts are similar to (i).

3. $E^{11} = \begin{bmatrix} 1 & 0 & 0 \\ 0 & 0 & 0 \end{bmatrix}, E^{12} = \begin{bmatrix} 0 & 1 & 0 \\ 0 & 0 & 0 \end{bmatrix},$

$E^{13} = \begin{bmatrix} 0 & 0 & 1 \\ 0 & 0 & 0 \end{bmatrix}, E^{21} = \begin{bmatrix} 0 & 0 & 0 \\ 1 & 0 & 0 \end{bmatrix},$

$E^{22} = \begin{bmatrix} 0 & 0 & 0 \\ 0 & 1 & 0 \end{bmatrix}, E^{23} = \begin{bmatrix} 0 & 0 & 0 \\ 0 & 0 & 1 \end{bmatrix}$

Now,

$(E^{11})^t A(E^{11}) = \begin{bmatrix} 1 & 0 \\ 0 & 0 \\ 0 & 0 \end{bmatrix} \begin{bmatrix} 1 & 2 \\ 3 & 4 \end{bmatrix} \begin{bmatrix} 1 & 0 & 0 \\ 0 & 0 & 0 \end{bmatrix}$

$= \begin{bmatrix} 1 & 2 \\ 0 & 0 \\ 0 & 0 \end{bmatrix} \begin{bmatrix} 1 & 0 & 0 \\ 0 & 0 & 0 \end{bmatrix} = \begin{bmatrix} 1 & 0 & 0 \\ 0 & 0 & 0 \\ 0 & 0 & 0 \end{bmatrix}$

∴ trace $(E^{11})^t A(E^{11}) = 1$

∴ $a^{11} = f(E^{11}, E^{11}) = 1$

and $(E_{11})^t A(E^{12}) = \begin{bmatrix} 1 & 0 \\ 0 & 0 \\ 0 & 0 \end{bmatrix} \begin{bmatrix} 1 & 2 \\ 3 & 4 \end{bmatrix} \begin{bmatrix} 0 & 1 & 0 \\ 0 & 0 & 0 \end{bmatrix}$

$= \begin{bmatrix} 1 & 2 \\ 0 & 0 \\ 0 & 0 \end{bmatrix} \begin{bmatrix} 0 & 1 & 0 \\ 0 & 0 & 0 \end{bmatrix} = \begin{bmatrix} 0 & 1 & 0 \\ 0 & 0 & 0 \\ 0 & 0 & 0 \end{bmatrix}$

∴ trace$[(E^{11})^t AE^{12}] = 0$

∴ $a_{12} = f(E^{11}, E^{12}) = 0$
and

$(E_{11})^t A(E^{13}) = \begin{bmatrix} 1 & 2 \\ 0 & 0 \\ 0 & 0 \end{bmatrix} \begin{bmatrix} 0 & 0 & 1 \\ 0 & 0 & 0 \end{bmatrix} = \begin{bmatrix} 0 & 0 & 1 \\ 0 & 0 & 0 \\ 0 & 0 & 0 \end{bmatrix}$

∴ trace$[(E^{11})^t AE^{13}] = 0$, so $a_{13} = 0$

and

$(E^{11})^t A(E^{21}) = \begin{bmatrix} 1 & 2 \\ 0 & 0 \\ 0 & 0 \end{bmatrix} \begin{bmatrix} 0 & 0 & 0 \\ 1 & 0 & 0 \end{bmatrix} = \begin{bmatrix} 2 & 0 & 0 \\ 0 & 0 & 0 \\ 0 & 0 & 0 \end{bmatrix}$

∴ trace$[(E^{11})^t AE^{21}] = 0$, so $a_{14} = 0$
and

$(E^{11})^t A(E^{22}) = \begin{bmatrix} 1 & 2 \\ 0 & 0 \\ 0 & 0 \end{bmatrix} \begin{bmatrix} 0 & 0 & 0 \\ 0 & 1 & 0 \end{bmatrix} = \begin{bmatrix} 0 & 2 & 0 \\ 0 & 0 & 0 \\ 0 & 0 & 0 \end{bmatrix}$

∴ trace$[(E^{11})^t AE^{22}] = 0$, so $a_{15} = 0$
and

$(E^{11})^t A(E^{23}) = \begin{bmatrix} 1 & 2 \\ 0 & 0 \\ 0 & 0 \end{bmatrix} \begin{bmatrix} 0 & 0 & 0 \\ 0 & 0 & 1 \end{bmatrix} = \begin{bmatrix} 0 & 0 & 2 \\ 0 & 0 & 0 \\ 0 & 0 & 0 \end{bmatrix}$

∴ trace$[(E^{11})^t AE^{23}] = 0$, so $a_{16} = 0$

Hence first row of the matrix of f relative to the given basis is given by $a_{11}, a_{12}, a_{13}, a_{14}, a_{15}$ and a_{16} i.e., 1, 0, 0, 2, 0, 0.

In the similar manner, we can also find other rows.

6. For $\alpha, \beta, \gamma \in V, a, b \in F$,
$$(kf)\ (a\alpha + b\beta, \gamma) = k[f(a\alpha + b\beta, \gamma]$$
$$= k[af(\alpha, \gamma) + bf(\alpha, \gamma)]$$
$$= a(kf)(\alpha, \gamma) + b(kf)(\alpha, \gamma).$$

Similarly,
$$(kf)(\gamma, a\alpha + b\beta) = a(kf)(\gamma, \alpha) + b(kf)(\gamma, \beta).$$

Hence kf is bilinear.

8. $f(\alpha, \beta) = 3x_1 y_1 - 2x_1 y_2 + 4x_2 y_1 - x_2 y_2$

Since, $\alpha = (x_1, x_2), \beta = (y_1, y_2)$

For $B = ((1, 0), (0, 1))$.

Let $\alpha_1 = (1, 0), \alpha_2 = (0, 1)$
$a_{11} = f(\alpha_1, \alpha_1) = f((1, 0)\ (1, 0)) = 3$
$a_{12} = f(\alpha_1, \alpha_2) = f((1, 0)\ (0, 1)) = -2$
$a_{13} = f(\alpha_2, \alpha_1) = f((0, 1)\ (1, 0)) = 4$
$a_{14} = f(\alpha_2, \alpha_2) = f((0, 1)\ (0, 1)) = -1$

∴ $[f]_B = \begin{bmatrix} 3 & -2 \\ 4 & -1 \end{bmatrix}$

Similarly, for $B' = \{(1, 1), (1, 2)\}$

$[f]_{B'} = \begin{bmatrix} 4 & 1 \\ 7 & 3 \end{bmatrix}$

Let P be the matrix whose columns are the basis vectors in B'.

∴ $P = \begin{bmatrix} 1 & 1 \\ 1 & 2 \end{bmatrix}$

Then, $P^t [f]_B P = \begin{bmatrix} 1 & 1 \\ 1 & 2 \end{bmatrix} \begin{bmatrix} 3 & -2 \\ 4 & -1 \end{bmatrix} \begin{bmatrix} 1 & 1 \\ 1 & 2 \end{bmatrix} = \begin{bmatrix} 4 & 1 \\ 7 & 3 \end{bmatrix}$

$= [f]_{B'}$

9. Since char. $F \neq 2$, *i.e.*, $1 + 1 \neq 0$. Let f be a bilinear form.

Suppose f is alternating.

Let $\alpha, \beta \in V$, then

$$0 = f(\alpha+\beta, \alpha+\beta)$$

$$[\because f \text{ is alternating.}]$$

$$= f(\alpha, \alpha) + f(\alpha, \beta) + f(\beta, \alpha) + f(\beta, \beta)$$

$$= f(\alpha, \beta) + f(\beta, \alpha)$$

$$[\because f(\alpha, \alpha)=0, \ f(\beta, \beta)=0]$$

$$\Rightarrow \quad f(\alpha, \beta) = -f(\beta, \alpha)$$

$\Rightarrow$ f is skew-symmetric.

Conversely, suppose f is skew-symmetric, then

$$f(\alpha, \alpha) = -f(\alpha, \alpha) \ \forall \ \alpha \in V$$

$$\Rightarrow 2f(\alpha, \alpha) = 0$$

$$\Rightarrow \quad f(\alpha, \alpha) = 0 \ \forall \alpha \in V$$

$$[\text{as } 1+1 \neq 0 \ i.e., 2 \neq 0]$$

$\Rightarrow$ f is alternating.

12. Let f be a diagonalizable bilinear form. Then $f(\alpha_i, \alpha_j)=0$ for $i \neq j$.

$\therefore$ $B = \{\alpha_1, \alpha_2, \ldots, \alpha_n\}$ forms an orthogonal basis with respect to f.

Let $\alpha, \beta \in V$, then

$$\alpha = \sum_i^n a_i \alpha_i, \ \beta = \sum_{j=1}^n b_j \beta_j \text{ for } a_i, b_j \in F$$

$$f(\alpha, \beta) = f\left(\sum_{i=1}^n a_i \alpha_i, \sum_{j=1}^n b_j \alpha_j\right)$$

$$= \sum_i \sum_j a_i b_j f(\alpha_i \alpha_j)$$

$$= \sum_i^n a_i b_i f(\alpha_i^j, \alpha_i) \quad [\because f(\alpha_i, \alpha_j) = 0, i \neq j]$$

and $f(\beta, \alpha)$

$$= f\left(\sum_{j=1}^n b_j \alpha_j, \sum_{i=1}^n a_i \alpha_i\right) = \sum_i \sum_j b_j a_i f(\alpha_j \alpha_i)$$

$$= \sum_{i=1}^n b_i a_i f(\alpha_i, \alpha_i)$$

Since $a_i b_i = b_i a_i$.

$\therefore \quad f(\alpha, \beta) = f(\beta, \alpha) \ \forall \alpha, \beta \in V.$

Hence, f is symmetric.

14. Since f is reflexive and non-degenerate. Then for $\alpha, \beta, \gamma \in V$

$$f(\alpha, \gamma) = 0, \text{ whenever } f(\gamma, \alpha) = 0$$

and $f(0, \gamma) = 0 \ \forall \ \gamma \in V.$

(i) $f(\alpha, \gamma) = f(\beta, \gamma)$.

$\therefore f(\alpha-\beta, \gamma) = f(\alpha, \gamma) - f(\beta, \gamma) = 0 \ \forall \ \gamma \in V$

$$\Rightarrow \quad \alpha - \beta = 0 \quad \therefore \quad \alpha = \beta.$$

(ii) $f(\gamma, \alpha-\beta) = f(\gamma, \alpha) - f(\gamma, \beta), \ \forall \ \gamma \in V$

$$= 0 \ \forall \ \gamma \in V$$

$$\Rightarrow f(\alpha-\beta, \gamma) = 0$$

$$\therefore \quad \alpha - \beta = 0, f \text{ is non degenerate.}$$

$$\Rightarrow \quad \alpha = \beta$$

Answers

1. No one is bilinear.   **2.**   (i) $\begin{bmatrix} 0 & 0 \\ 0 & 0 \end{bmatrix}$ (ii) $\begin{bmatrix} -2 & 0 \\ 0 & 2 \end{bmatrix}$ (iii) $\begin{bmatrix} 4 & 14 \\ 14 & 24 \end{bmatrix}$

3. $\begin{bmatrix} I_3 & 2I_3 \\ 2I_3 & I_3 \end{bmatrix}$, where I_3 is a unit matrix of order 3×3.

4. $\begin{bmatrix} 3 & -2 & 0 \\ 5 & 7 & -8 \\ 0 & 4 & -1 \end{bmatrix}$ **8.** (i) $\begin{bmatrix} 3 & -2 \\ 4 & -1 \end{bmatrix}$ (ii) $\begin{bmatrix} 4 & 1 \\ 7 & 3 \end{bmatrix}$ (iii) $P = \begin{bmatrix} 1 & 1 \\ 1 & 2 \end{bmatrix}$

9.4   QUADRATIC FORMS

Definition 1. *A mapping $q: V \to F$ is called a quadratic form if $q(\alpha) = f(\alpha, \alpha)$ for some bilinear form f on V.*

Alternatively, a quadratic form is a polynomial $q(X) = X^t A X$, where

$X^t = (x_1, x_2, \ldots, x_n)$ and A is a symmetric matrix. That is

$$q(x)=(x_1,x_2,.....,x_n)\begin{bmatrix} a_{11} & a_{12} & \cdots & a_{1n} \\ a_{21} & a_{22} & \cdots & a_{2n} \\ \cdots & \cdots & \cdots & \cdots \\ a_{n1} & a_{n2} & \cdots & a_{nn} \end{bmatrix}\begin{bmatrix} x_1 \\ x_2 \\ \vdots \\ x_n \end{bmatrix}$$

$$= \sum_i \sum_j a_{ij} x_i x_j \quad \text{for } 1 \le i \le n,\ 1 \le j \le n$$

$$q(x) = a_{11}x_1^2 + a_{12}x_2^2 + \ldots + a_{nm}x_n^2 + 2\sum_{i<j} a_{ij}\, x_i\, x_j$$

Clearly, $q(X)$ is a polynomial in which every term has degree two.

Definition 2. *Let V be a real inner-product space and suppose that A is a real symmetric matrix linear transformation on V. Then a real valued function Q(α) defined by Q(α) = (A(α),α), is called the quadratic form associated with A.*

For Example :

(1) $5x^2 - 6xy + 8y^2$ *is a real quadratic form in two variables x and y.*

(2) $2x^2 - y^2 + 2z^2 - 2yz - 4zx + 6xy$ *is a real quadratic form in three variables x,y and z.*

Definition3. *If the matrix A is diagonal, then the corresponding quadratic form q has diagonal representation as*

$$q(X) = X^t A X = a_{11}x_1^2 + a_{22}x_2^2 + \ldots + a_{nn}x_n^2.$$

9.5 REAL SYMMETRIC BILINEAR AND QUADRATIC FORMS: LAW OF INERTIA

In this section, we shall discuss symmetric bilinear forms and quadratic forms on vector spaces over the field of reals. These forms appear in many branches of mathematics and physics.

Definition 1. *If q(X) = X^t AX is a real quadratic form, then the rank of q is defined as the rank of the matix A.*

Definition 2. *Let q be a real quadratic form and f be a bilinear form, then the signature of f and of q are defned by sig. (f) = sig. (q) = p−N, where p is the number of positive entries and N the number of negative entries in any diagonal representation of f and q.*

THEOREM 1. **(Sylvester's Theorem or Law of Inertia).** *Let f be a symmetric bilinear form on V over R. Then there is a basis of V in which f is represented by a diagonal matrix, every other diagonal representation has the same number p of positive entries and the same number N of negative entries.*

Proof. We know that there is a basis $\{\alpha_1, \alpha_2,...,\alpha_n\}$ of V such that $f(\alpha_i, \alpha_j) = 0$ for $i \ne j$. This shows that f is represented by a diagonal matrix, say, with p positive and N negative entries.

Now assume that $\{\beta_1, \beta_2,...,\beta_n\}$ is another basis of V in which f is represented by a diagonal matrix say, with p_1 positive and N_1 negative entries. Without any loss of generality, we may assume that the positive entries in each matrix representation appear first.

Since rank(f) = $p + N = p_1 + N_1$. Now we only prove that $p = p_1$.

Let U be a subspace of V spanned by the vectors $\alpha_1, \alpha_2,...,\alpha_p$ and let W be a subspace of V spanned by $\beta_{p_1+1} + \beta_{p_1+2} + ... + \beta_n$. Then $f(\alpha,\alpha) > 0$ for every non-zero $\alpha \in U$ and $f(\alpha,\alpha) \leq 0$ for every non-zero vector $\alpha \in W$.

Obviously, $U \cap W = \{0\}$. Also, dim.$U = p$ and dim.$W = n{-}p_1$. Thus

$$\text{dim.}(U{+}W) = \text{dim. } U {+} \text{dim. } W - \text{dim. } (U \cap W)$$
$$= p{+}n{-}p_1{-}0 = p{-}p_1{+}n.$$

But   dim.$(U{+}W) \leq$ dim. $V = n$.

$\therefore$ dim $(U{+}W) \leq n$ or $p{-}p_1{+}n \leq n$ or $p \leq p_1$.

Similarly, $p_1 \leq p$ and therefore $p = p_1$.

THEOREM 2. *If $q(x) = X^t A X$ be a real quadratic form of rank r in n variables, then there exists a real orthogonal transformation $X = PY$ which transform $q(X)$ to the form*

$$\lambda_1 y_1{}^2 + \lambda_2 y_2{}^2 + ... + \lambda_r y_r{}^2$$

where $\lambda_1, \lambda_2,...\lambda_r$ are the r non-zero eigenvalues of A and $n{-}r$ eigenvalues of A being equal to zero and $Y = (y_1, y_2, ..., y_r)^t$.

Proof. Since A is a real symmetric matrix of rank r, therefore there exists a real orthogonal matrix P such that

$$P^{-1}AP = \text{diag.}[\lambda_1, \lambda_2,...\lambda_r, 0, 0, ...0].$$

Also, P is orthogonal so $P^{-1} = P^t$.

$\therefore$ $P^{-1}AP = P^t AP = \text{diag.}[\lambda_1, \lambda_2,...\lambda_r, 0, 0, ...0].$

Now, consider the real orthogonal transformation $X = PY$, then we have

$$q(X) = X^t A X = (PY)^t A(PY)$$
$$= Y^t P^t APY = Y^t \text{ diag.}[\lambda_1, \lambda_2,...\lambda_r, 0, 0, ...0]Y$$
$$= \lambda_1 y_1{}^2 + \lambda_2 y_2{}^2 + ... + \lambda_r y_r{}^2.$$

THEOREM 3. *Every real quadratic form*

$$q(x_1,x_2,..,x_n) = X^t A X = y_1{}^2 + y_2{}^2 + y^2{}_p - y^2{}_{p+1} - ... - y_r{}^2.$$

where r is the rank of A and p is the number of positive eigenvalues of A.

Proof. From above theorem, we have

$$Q^{-1}AQ = Q^t AQ = \text{diag.}[\lambda_1, \lambda_2,...\lambda_r, 0, 0, ...0].$$

where Q is a real orthogonal matrix.

Let $\lambda_1, \lambda_2,...\lambda_p$ be positive eigenvalues and $\lambda_{p+1}, \lambda_{p+2},...,\lambda_r$ be negative. Let D be $n \times n$ real matrix diagonal elements as

$$\frac{1}{\sqrt{\lambda_1}}, \frac{1}{\sqrt{\lambda_2}}, ..., \frac{1}{\sqrt{\lambda_p}}, \frac{1}{\sqrt{(-\lambda_{p+1})}}, \frac{1}{\sqrt{(-\lambda_{p+2})}}, ..., \frac{1}{\sqrt{(-\lambda_p)}}, 1, 1, ..., 1$$

Then D is a non-singular matrix and also $D' = D$.

If we take $P = QD$, then P is also a real non-singular matrix, so that

$$P'AP = (QD)'A(QD) = D'Q'AQD$$
$$= D' \text{ diag.}[\lambda_1, \lambda_2,...\lambda_r, 0, 0, ...0]D$$
$$= \text{diag.}[1, 1, ...1, -1, -1, ..., -1, 0, 0, ..., 0].$$

It is noted that here 1 and -1 appear p times and $r - p$ times respectively.

Now consider a real non-singular linear transformation $X = PY$ which reduces $X'AX$ to the form $Y'P'APY$ as

$$q(X) = y_1{}^2 + y_2{}^2 + ... + y^2{}_p - y^2{}_{p+1} - y^2{}_{p+1} - ... - y_r{}^2.$$

9.6 ORTHOGONAL DIAGONALIZATION OF THE QUADRATIC FORM

Let q be a quadratic form on a vector space V over F determined by a symmetric bilinear form f. Let $B = \{\alpha_1, \alpha_2, \ldots \alpha_n\}$ be an ordered orthogonal basis of V with respect to f. Then $[f]_B = \text{diag.}(\lambda_1, \lambda_2, \ldots \lambda_n)\lambda_i = f(\alpha_i, \alpha_i)$.

For $\beta \in V$, we may write

$$\beta = a_1\alpha_1 + a_2\alpha_2 + \ldots + a_n\alpha_n, a_i \in F$$

$$= \sum_{i=1}^{n} a_i\alpha_i$$

so that $\quad q(\alpha) = f(\alpha, \alpha)$ [By definition]

$$= f\left(\sum_{i=1}^{n} a_i\alpha_i, \sum_{j=1}^{n} a_j\alpha_j\right)$$

$$= a_1^2 f(\alpha_1, \alpha_1) + a_2^2 f(\alpha_2, \alpha_2) + \ldots + a_n^2 f(\alpha_n, \alpha_n) \quad [\because f(\alpha_i, \alpha_j) = 0, i \neq j]$$

$$= \lambda_1 a_1^2 + \lambda_2 a_2^2 + \ldots + \lambda_n y_n^2$$

Since, $\quad [\beta]_B = [a_1 a_2 \ldots a_n]'$ [By definition of co-ordinate of a vector]

$\therefore \quad q(\alpha) = [\beta]_B^t \text{ diag.}(\lambda_1, \lambda_2, \ldots \lambda_n)[\beta]_B$.

This is called the diagonalization.

Definition 1. *Let V be an n-dimensional vector space over R. Then a quadratic form q on V is called positive semidefinite if $q(\alpha) \geq 0$ for all $\alpha \in V$.*

Definition 2. *A quadratic form q on n-dimensional vector space V is said to be positive definite if $q(\alpha) > 0$ for all $\alpha \in V - \{0\}$.*

Definition 3. *A quadratic form q is said to be negative semidefinite if $q(\alpha) \leq 0$ for all $\alpha \in V$.*

Definition 4. *A quadratic form q is said to be negative definite if $q(\alpha) < 0$ for all $\alpha \in V - \{0\}$.*

If the quadratic form is none of above, then it is called *indefinite*.

 **WORKING RULE**

To describe the algorithm which diagonalizes quadratic form $q(X)$ on R^n by means of an orthogonal change of co-ordinates $X = PY$.	
STEP 1.	Find the symmetric matrix A which represents q and find its characteristic polynomial $\Delta(t)$.
STEP 2.	Find the eigenvalues of A which are the roots of the equation $\Delta(t) = 0$.
STEP 3.	For each eigenvalue λ of A in step 2. Find an orthogonal basis of its eigenspace.
STEP 4.	Normalize all eigenvectors which are obtained in step 3 which then form an orthogonal basis of R^n.
STEP 5.	Find the matrix P whose columns are the normalized eigenvectors obtained in step 4.

Then $X = PY$ is the required orthogonal change of co-ordinates and the diagonal entries in $P'AP$ will be the eigenvalues of A which correspond to the columns of P.

To descibe the algorithm which gives the matrix P such that $P'AP$ is diagonal.

(i) First form the matrix $M = (A : I)$

(ii) Then apply the row and column operations to M in such a way that the row operation will change both halves of M, but the column operations will only change the left half of M. Then algorithm will finally transform M into the form $M' = (D:Q)$ where D is diagonal matrix. Then $P=Q'$, and $P'AP=D$.

Solved Examples

Example 1. *Find the quadratic form $q(x, y)$ corresponding to the symmetric matrix*

$$A = \begin{bmatrix} 5 & -3 \\ -3 & 8 \end{bmatrix}$$

Solution. We have $q(x, y) = (x, y) \begin{bmatrix} 5 & -3 \\ -3 & 8 \end{bmatrix} \begin{bmatrix} x \\ y \end{bmatrix} = (5x - 3y, -3x + 8y) \begin{bmatrix} x \\ y \end{bmatrix}$

$$= x(5x - 3y) + y(-3x + 8y) = 5x^2 - 3xy - 3yx + 8y^2$$

$$= 5x^2 - 6xy + 8y^2$$

Example 2. *Find the quadratic form $q(x_1, x_2, x_3)$ corresponding to the symmetric matrix*

$$A = \begin{bmatrix} 1 & 2 & -4 \\ 2 & 3 & 5 \\ -4 & 5 & -7 \end{bmatrix}.$$

Solution. Let $X = \begin{bmatrix} x_1 \\ x_2 \\ x_3 \end{bmatrix}$,

Then $q(X) = X'AX = [x_1 x_2 x_3] \begin{bmatrix} 1 & 2 & -4 \\ 2 & 3 & 5 \\ -4 & 5 & -7 \end{bmatrix} \begin{bmatrix} x_1 \\ x_2 \\ x_3 \end{bmatrix}$

$$= [x_1 + 2x_2 - 4x_3, 2x_1 + 3x_2 + 5x_3, -4x_1 + 5x_2 - 7x_3] \begin{bmatrix} x_1 \\ x_2 \\ x_3 \end{bmatrix}$$

$$= x_1(x_1 + 2x_2 - 4x_3) + x_2(2x_1 + 3x_2 + 5x_3) + x_3(-4x_1 + 5x_2 - 7x_3)$$

$$= x_1^2 + 2x_1x_2 - 4x_1x_3 + 2x_2x_1 + 3x_2^2 + 5x_2x_3 - 4x_3x_1 + 5x_3x_2 - 7x_3^2$$

$$= x_1^2 + 3x_2^2 - 7x_3^2 + 4x_1x_2 - 8x_1x_3 + 10x_2x_3$$

Example 3. *The following expressions define quadratic forms q on R^2. Find the symmetric bilinear form f corresponding to each q.*

(i) ax_1^2 (ii) bx_1x_2 (iii) $x_1^2 + 9x_2^2$

(iv) $2x_1^2 - \dfrac{1}{3} x_1x_2$ (v) $3x_1x_2 - x_2^2$ (vi) $4x_1^2 + 6x_1x_2 - 3x_2^2$.

Solution. Let $X = \begin{bmatrix} x_1 \\ x_2 \end{bmatrix}$. By definition of quadratic form, we have

(i) $f((x_1, x_2), (x_1, x_2)) = q(x_1, x_2) = ax_1^2$

(ii) $f((x_1, x_2), (x_1, x_2)) = bx_1x_2$

(iii) $f((x_1, x_2), (x_1, x_2)) = x_1^2 + 9x_2^2$

(iv) $f((x_1, x_2),(x_1, x_2)) = 2x_1{}^2 - \dfrac{1}{3}\, x_1 x_2$

(v) $f((x_1, x_2),(x_1, x_2)) = 3x_1 x_2 - x_2{}^2$

(vi) $f((x_1, x_2),(x_1, x_2)) = 4x_1{}^2 + 6x_1 x_2 - 3x_2{}^2$

Example 4. *Find the symmetric matrix A corresponding to the quadratic form*

$$q(x, y, z) = 3x^2 + 4xy - y^2 + 8xz - 6yz + z^2.$$

Solution. Since the symmetric matrix $A = [a_{ij}]$ representing $q(x_1, x_2, ..., x_n)$ has the diagonal entry a_{ij} equal to the coefficient of $x_i{}^2$, and has the entries a_{ij} and a_{ji} each equal to half the coefficient of $x_i x_j$.

Thus $\qquad A = \begin{bmatrix} 3 & 2 & 4 \\ 2 & -1 & -3 \\ 4 & -3 & 1 \end{bmatrix}$

Aliter. $q(x, y, z)$ may be rewritten as

$$q(x, y, z) = (3x^2 + 2xy + 4xz) + (2yx - y^2 - 3yz) + (4zx - 3zy + z^2)$$
$$= x(3x + 2y + 4z) + y(2x - y - 3z) + z(4x - 3y + z)$$

$$q(x, y, z) = (x, y, z) \begin{bmatrix} 3 & 2 & 4 \\ 2 & -1 & -3 \\ 4 & -3 & 1 \end{bmatrix} \begin{bmatrix} x \\ y \\ z \end{bmatrix}$$

Thus $\qquad A = \begin{bmatrix} 3 & 2 & 4 \\ 2 & -1 & -3 \\ 4 & -3 & 1 \end{bmatrix}$

Example 5. *Find the symmetric matrix A which corresponds to*

$$q(x, y, z) = 4xy + 5y^2.$$

Solution. Since $q(x, y, z)$ indicates that there are three variables, so $q(x, y, z)$ may be written as

$$q(x, y, z) = 0.\, x^2 + 4xy + 5y^2 + 0\,.xz + 0.\, yz + 0.z^2$$
$$= (0.\, x^2 + 2xy + 0\,.xz) + (2yx + 5y^2 + 0\,.\, yz) + (0.\, 2z + 0.\, zy + 0.z^2)$$
$$= x\,(0.\, x + 2y + 0\,.\, z) + y\,(2x + 5y + 0\,.z) + z(0.\, x + 0.\, y + 0.z)$$

$$= (x, y, z) \begin{bmatrix} 0 & 0 & 0 \\ 2 & 5 & 0 \\ 0 & 0 & 0 \end{bmatrix} \begin{bmatrix} x \\ y \\ z \end{bmatrix}$$

Thus, $\qquad A = \begin{bmatrix} 0 & 2 & 0 \\ 2 & 5 & 0 \\ 0 & 0 & 0 \end{bmatrix}$

Example 6. *Consider the quadratic form*

$$q(x, y, z) = x^2 + 4xy + 3y^2 - 6xz + 10yz + 7z^2.$$

Find a non-singular linear substitution expressing the variables x, y, z in terms of variables r, s, t such that q(r, s, t) is diagonal.

Solution.

$$A = \begin{bmatrix} 1 & 2 & -3 \\ 2 & 3 & 5 \\ -3 & 5 & 7 \end{bmatrix}$$

Now, $M = (A \mid I)$.

$$\therefore \quad M \begin{pmatrix} 1 & 2 & -3 & \vdots & 1 & 0 & 0 \\ 2 & 3 & 5 & \vdots & 0 & 1 & 0 \\ -3 & 5 & 7 & \vdots & 0 & 0 & 1 \end{pmatrix}$$

Apply the row operations $R_2 \to R_2 - 2R_1$ and $R_3 \to R_3 + 3R_1$ to M and then the corresponding column operations $C_2 \to C_2 - 2C_1$ and $C_3 \to C_3 + 3C_1$ to A, we get

$$\begin{pmatrix} 1 & 2 & -3 & \vdots & 1 & 0 & 0 \\ 0 & -1 & 11 & \vdots & -2 & 1 & 0 \\ 0 & 11 & -2 & \vdots & 3 & 0 & 1 \end{pmatrix}$$

and then

$$\begin{pmatrix} 1 & 0 & 0 & \vdots & 1 & 0 & 0 \\ 0 & -1 & 11 & \vdots & -2 & 1 & 0 \\ 0 & 11 & -2 & \vdots & 3 & 0 & 1 \end{pmatrix}$$

Next apply the row operations $R_3 \to R_3 + 11R_2$ and then the corresponding column operation $C_3 \to C_3 + 11C_2$, we finally obtain,

$$\begin{pmatrix} 1 & 0 & 0 & \vdots & 1 & 1 & 0 \\ 0 & -1 & 0 & \vdots & -2 & 1 & 0 \\ 0 & 0 & 119 & \vdots & -19 & 11 & 1 \end{pmatrix}$$

Thus

$$P = \begin{pmatrix} 1 & -2 & -19 \\ 0 & 1 & 11 \\ 0 & 0 & 1 \end{pmatrix} \text{ and } P'AP = \begin{pmatrix} 1 & 0 & 0 \\ 0 & -1 & 0 \\ 0 & 0 & 119 \end{pmatrix}$$

Thus the linear substitutions are given by

$$\begin{bmatrix} x \\ y \\ z \end{bmatrix} = P \begin{bmatrix} r \\ s \\ t \end{bmatrix}$$

$$\therefore \quad x = r - 2s - 19t, \ y = s + 11t, \ z = t$$

Hence $q(r, s, t) = (r, s, t) P'AP \begin{bmatrix} r \\ s \\ t \end{bmatrix}$

$$= (r, s, t) \begin{bmatrix} 1 & 0 & 0 \\ 0 & -1 & 0 \\ 0 & 0 & 119 \end{bmatrix} \begin{bmatrix} r \\ s \\ t \end{bmatrix} = r^2 - s^2 + 119r^2$$

Example 7. *Give an example of a quadratic form q on R^2 such that $q(\alpha) = 0$ and $q(\beta) = 0$ for some $\alpha, \beta \in R^2$, but $q(\alpha + \beta) \neq 0$.*

Solution.

Let $q(x, y) = x^2 - y^2$ and $\alpha = (1, 1), \beta = (1, -1)$

Then $q(\alpha) = q(1, 1) = 1^2 - 1^2 = 0$

And $q(\beta) = q(1, -1) = 1^2 - (-1^2) = 0$

But $q(\alpha + \beta) = q((1, 1) + (1, -1)) = q(2, 0) = 2^2 - 0^2 = 4 \neq 0$.

Example 8. *Find the signature of the quadratic form $q(x, y, z)$*

$$q(r, y, z) = x^2 + 4xy + 3y^2 - 6xz + 10yz + 7z^2.$$

Solution. From example 6, the equivalent diagonal form of $q(x, y, z)$ is

$$q(r, s, t) = r^2 - s^2 + 119t^2.$$

Obviously, this quadratic form has $p = 2$ positive entries on the diagonal and $N = 1$ negative entry on the diagonal. Thus sig.$(q) = p - N = 2 - 1 = 1$.

Example 9. *Let $q(x, y, z) = x^2 + 2y^2 - 4xz - 4yz + 6z^2$. Is q positive definite?*

Solution. The symmetric matrix corresponding to the quadratic form is

$$A = \begin{bmatrix} 1 & 0 & -2 \\ 0 & 2 & -2 \\ -2 & -2 & 7 \end{bmatrix}$$

Now convert A into diagonal form by applying $R_3 \to R_3 + 2R_1$, and $C_3 \to C_3 + 2C_1$ and then $R_3 \to R_3 + R_2$ and $C_3 \to C_3 + C_2$,

$$A = \begin{bmatrix} 1 & 0 & -2 \\ 0 & 2 & -2 \\ -2 & -2 & 7 \end{bmatrix} - \begin{bmatrix} 1 & 0 & 0 \\ 0 & 2 & -2 \\ 0 & -2 & 3 \end{bmatrix} - \begin{bmatrix} 1 & 0 & 0 \\ 0 & 2 & 0 \\ 0 & 0 & 1 \end{bmatrix}.$$

The diagonal representation of q only contains positive entries, 1,2 and 1 on the diagonal. Hence q is positive definite.

Example 10. *Show that $q(x, y)' = ax^2 + byx + cy^2$ is positive definite if and only if $b^2 - 4ac < 0$.*

Solution. Suppose $\alpha = (x, y) \neq 0$ say $y \neq 0$. Let $t = \dfrac{x}{y}$. Then,

$$q(\alpha) = y^2 \left[a\left(\frac{x}{y}\right)^2 + b\left(\frac{x}{y}\right) + c \right] = y^2(ar^2 + bt + c)$$

Let $s = ar^2 + bt + c$, thus s lies above the t-axis *i.e.*, s is positive for every value of t if and only if $b^2 - 4ac < 0$. Hence q is positive definite if and only if $b^2 - 4ac < 0$.

Example 11. *Find an orthogonal change of co-ordinates which diagonalizes the real quadratic form $q(x, y) = 2x^2 - 4xy + 5y^2$.*

Solution. The matrix A corresponding to the given quadratic form is

$$A = \begin{bmatrix} 2 & -2 \\ -2 & 5 \end{bmatrix}$$

Then its characteristic polynomial $\Delta(t)$ is given by

$$\Delta(t) = |tI - A| = \begin{bmatrix} t - 2 & -2 \\ 2 & t - 5 \end{bmatrix} = (t - 6)(t - 1)$$

$\therefore$ The eigenvalues of A are 6 and 1.

Let $X_1 = \begin{bmatrix} x_1 \\ x_2 \end{bmatrix} \neq 0$ be an eigenvector corresponding to $t = 6$, then

$$(tI - A)X_1 = 0 \Rightarrow \quad (6I - A)X_1 = 0$$

$$\Rightarrow \quad \begin{bmatrix} 4 & 2 \\ 2 & 1 \end{bmatrix}\begin{bmatrix} x_1 \\ x_2 \end{bmatrix} = \begin{bmatrix} 0 \\ 0 \end{bmatrix} \quad \Rightarrow 4x_1 + 2x_2 = 0 \text{ and } 2x_1 + x_2 = 0$$

A non-zero solution of these equation is

$$X_1 = \begin{bmatrix} 1 \\ -2 \end{bmatrix}$$

Now, let $\quad X_2 = \begin{bmatrix} x_1 \\ x_2 \end{bmatrix} \neq 0$ be an eigenvector corresponding to $t = 1$, then

$$(tI - A)X_2 = 0 \text{ or } (I - A)X_2 = 0 \text{ or } \begin{bmatrix} -1 & 2 \\ 2 & -4 \end{bmatrix}\begin{bmatrix} x_1 \\ x_2 \end{bmatrix} = \begin{bmatrix} 0 \\ 0 \end{bmatrix}$$

$$\Rightarrow \quad -x_1 + 2x_2 = 0 \text{ and } 2x_1 - 4x_2 = 0$$

The non-zero solution of these equations is

$$X_2 = \begin{bmatrix} 2 \\ 1 \end{bmatrix}$$

Normalizer X_1 and X_2 to obtain the orthogonal basis

$$\{\alpha_1 = (1/\sqrt{5}, -2/\sqrt{5}), \alpha_2 = (2/\sqrt{5}, 1/\sqrt{5})\}$$

Then $\quad P = \begin{bmatrix} 1/\sqrt{5} & 2/\sqrt{5} \\ -2/\sqrt{5} & 1/\sqrt{5} \end{bmatrix}$ and $P'AP = \begin{bmatrix} 6 & 0 \\ 0 & 1 \end{bmatrix}$

Thus the required orthogonal change of co-ordinates is

$$\begin{bmatrix} x \\ y \end{bmatrix} = P\begin{bmatrix} x' \\ y' \end{bmatrix} \text{ i.e., } x = \frac{x'}{\sqrt{5}} + \frac{2y'}{\sqrt{5}}, y = \frac{2x'}{\sqrt{5}} + \frac{y'}{\sqrt{5}}$$

Under this change of co-ordinates q is transformed into the diagonal form

$$q(x', y') = 6x'^2 + y'^2$$

Example 12. *Let q be the quadratic form on R^2 given by*

$$q(x_1 + x_2) = ax_1^2 + 2bx_1x_2 + cx_2^2, a \neq 0$$

Find an invertible linear operator U on R^2 such that

$$(U^* q)(x_1, x_2) = ax_1^2 + \left(c + \frac{b^2}{a}\right)x^2.$$

Solution. Diagonalize q by the method known as "completing the square".

$$\therefore \qquad q(x_1, x_2) = a\left(x_1^2 + \frac{2b}{a}x_1x_2\right) + cx_2^2$$

$$= a\left(x_1^2 + \frac{2b}{a}x_1x_2 + \frac{b^2}{a^2}x_2^2\right) - \frac{b^2}{a}x_2^2 + cx_2^2$$

$$= a\left(x_1 + \frac{b}{a}x_2\right)^2 + \left(c - \frac{b^2}{a}\right)x_2^2.$$

Let $s = x_1 + \dfrac{b}{a}x_2, t = a_2$, then this linear substitution yields the quadratic form as

$$q'(s,t) = as^2 + \left(c - \frac{b^2}{a}\right)t^2$$

or $\quad (U^* q)(s,t) = as^2 + \left(c - \dfrac{b^2}{a}\right)t^2$

But $\quad q(U(s, t)) = (U^* q)\ (s,t) = as^2 + \left(c - \dfrac{b^2}{a}\right)t^2$

Now if we put $x_1 = s - \dfrac{b}{a}t, x_2 = t$ in given quadratic, then

$$q\left(s - \frac{b}{a}t, t\right) = a\left(s - \frac{b}{a}t\right)^2 + 2b\left(s - \frac{b}{a}t\right)t + ct^2$$

$$= as^2 + \left(c - \frac{b^2}{a}\right)t^2$$

$\therefore \quad q(U(s,t)) = q\left(s - \dfrac{b}{a}t, t\right)$

$\therefore \quad U(s,t) = \left(s - \dfrac{b}{a}t, t\right)$

or $\quad U^{-1}(s,t) = \left(s + \dfrac{b}{a}t, t\right)$

This is the required invertible linear operator.

Example 13. *Let q be the quadratic form on* R^2 *given by* $q(x_1, x_2) = 2bx_1x_2$. *Find an invertible linear operator U on* R^2 *such that*

$$(U^* q)\ (x_1, x_2) = \frac{b}{2}x_1^2 - \frac{b}{2}x_2^2.$$

Solution. Making the complete square in q.

$$q(x_1, x_2) = 2bx_1x_2$$

$$q(x_1, x_2) = \frac{b}{2}[(x_1 + x_2)^2 - (x_1 - x_2)^2]$$

Let $s = x_1 + x_2$, $t = x_1 - x_2$, then this linear substitution yields the quadratic form as

$$q'(s, t) = bs^2 - bt^2$$

or $\quad (U^* q)(s,t) = \dfrac{b}{2}s^2 - \dfrac{b}{2}t^2$

But we know that

$$(U^* q)(s,t) = q(U(s,t)) = \frac{b}{2}s^2 - \frac{b}{2}t^2.$$

Now if we put $x_1 = \dfrac{s+t}{2}, x_2 = \dfrac{s-t}{2}$ in the given quadratic then.

$$q\left(\frac{s+t}{2}, \frac{s-t}{2}\right) = 2b\left(\frac{s+t}{2}\right)\left(\frac{s-t}{2}\right)$$

$$= \frac{b}{2}(s^2 - t^2) = \frac{b}{2}s^2 - \frac{b}{2}t^2$$

$\therefore \qquad q(U(s,t)) = q\left(\dfrac{s+t}{2}, \dfrac{s-t}{2}\right)$

$\therefore \qquad U(s,t)) = \left(\dfrac{s+t}{2}, \dfrac{s-t}{2}\right)$

This is the required invertible linear operator.

9.7 QUADRATIC FORMS AND MATRICES

An expression of the form :

$$\sum_{i=1}^{n} \sum_{j=1}^{n} a_{ij} x_i x_j$$

where $a_{ij} \in F$ (a field), is called a quadratic form in n variables $x_1, x_2, ..., x_n$ over a field F, which is denoted by $Q(x_1, x_2, ..., x_n)$ or by Q.

9.7.1 Real Quadratic Form

The quadratic form

$$Q = \sum_{i=1}^{n} \sum_{j=1}^{n} a_{ij} x_i x_j$$

is called a real quadratic form if a_{ij} are all real numbers $i.e., a_{ij} \in R$ (the field of all real numbers).

For Example :

(1) $ax^2 + 2hxy + by^2$ is a real quadratic form in two variables x and y.

(2) $x^2 + y^2 + z^2 + 2yz + 2zx + 2xy$ is a real quadratic form in three variables x, y and z.

(3) $x_1x_2 + x_2x_3 + x_3x_4$ is a real quadratic form in four variables x_1, x_2, x_3 and x_4.

Theorem 1. *Every quadratic form in n variables $x_1, x_2,..., x_n$ over a field F can be expressed in the form X'BX, where $X = [x_1, x_2,..., x_n]'$ is a column vector and B is a symmetric matrix of order n over the field F.*

Proof. Let $\qquad Q = \sum_{i=1}^{n} \sum_{j=1}^{n} a_{ij} x_i x_j$ ...(1)

be a quadratic form in n variables $x_1, x_2, ..., x_n$ over a field F.

Writing the equation (1) such that the terms $a_{ij}x_ix_j$ and $a_{ji}x_jx_i$ are taken together, we get

$$Q = a_{11}x_1^2 + (a_{12} + a_{21})x_1x_2 + (a_{13} + a_{31})x_1x_3 + ... + (a_{1n} + a_{n1})x_1x_n$$
$$+ a_{22}x_2^2 + (a_{23} + a_{32})x_2x_3 + ... + (a_{2n} + a_{n2})x_2x_n$$
$$+ a_{33}x_3^2 + (a_{34} + a_{43})x_3x_4 + ... + (a_{3n} + a_{3n})x_3x_n$$
$$+ .... + a_{nn}x_n^2$$

$$...(2)$$

Set $b_{ij} = \dfrac{1}{2}(a_{ij} + a_{ji})$, then $bij = b_{ji}$ and $b_{ij} + b_{ji} = a_{ij} + a_{ji}$, using these relations, equations (2) can be written as

$$Q = b_{11}x_1^2 + b_{12}x_1x_2 + \ldots + b_{1n}x_1x_n + b_{21}x_2x_1 + b_{22}x_2^2 + \ldots$$
$$\ldots + b_{2n}x_2x_n + \ldots + b_{n1}x_nx_1 + b_{n2}x_nx_2 + \ldots + b_{nn}x_n^2$$

$$\therefore \quad Q = \sum_{i=1}^{n}\sum_{j=1}^{n} b_{ij}x_ix_j \qquad\qquad \ldots(3)$$

Let $B = [b_{ij}]$ be a matrix of order $n \times n$. Clearly B is a symmetric matrix.

Let $X = \begin{bmatrix} x_1 \\ x_2 \\ \vdots \\ x_n \end{bmatrix}$, then $\qquad X' = [x_1, x_2, \ldots, x_n]$

Therefore $X'BX$ is a matrix of order 1×1 *i.e.*, $X'BX$ has a single element and this single element is $\displaystyle\sum_{i=1}^{n}\sum_{j=1}^{n} b_{ij}x_ix_j{}'$.

If we regard a matrix of order 1×1 equal to its single element, then we have

$$X'BX = \sum_{i=1}^{n}\sum_{j=1}^{n} b_{ij}x_ix_j$$

But we have $b_{ii} = a_{ii}$ and $b_{ij} = b_{ji}, = \dfrac{1}{2}(a_{ij} + a_{ji})$, then we have

$$\sum_{i=1}^{n}\sum_{j=1}^{n} a_{ij}x_ix_j = \sum_{i=1}^{n}\sum_{j=1}^{n} b_{ij}x_ix_j$$

Hence, $\qquad X'BX = \displaystyle\sum_{i=1}^{n}\sum_{j=1}^{n} a_{ij}x_ix_j$.

9.8   MATRIX OF QUADRATIC FORM $\displaystyle\sum_{i=1}^{n}\sum_{j=1}^{n} a_{ij}x_ix_j$

If $Q = \displaystyle\sum_{i=1}^{n}\sum_{j=1}^{n} a_{ij}x_ix_j$ is a quadratic form in n variables $x_1, x_2, \ldots, x_n$ over a field, then there exists a unique symmetric matrix B such that

$$X'BX = \sum_{i=1}^{n}\sum_{j=1}^{n} a_{ij}x_ix_j$$

The symmetric matrix B is called the matrix of the quadratic form.

In order to find the matrix of the quadratic form $\displaystyle\sum_{i=1}^{n}\sum_{j=1}^{n} a_{ij}x_ix_j$ we shall adjust the coefficients, a_{ij} in such a way that its coefficients form a symmetric matrix.

9.9 CONVERSION OF A SYMMETRIC MATRIX INTO QUADRATIC FORM

Let $A = [a_{ij}]$ be an $n \times n$ symmetric matrix over a field F and let $X = \begin{bmatrix} x_1 \\ x_2 \\ \vdots \\ x_n \end{bmatrix}$, then the quadratic form of A is given by

$$Q = X'AX = [x_1, x_2, ..., x_n]A\begin{bmatrix} x_1 \\ x_2 \\ \vdots \\ x_n \end{bmatrix}$$

$$= \sum_{i=1}^{n} \sum_{j=1}^{n} a_{ij}x_i x_j$$

Solved Examples

Example 1. *Find the matrix corresponding to the following quadratic form :*
$$Q = 2x_1^2 - 7x_3^2 + 4x_1x_2 - 6x_2x_3$$

Solution. The given quadratic form can be written as
$$Q = 2x_1^2 + (2 + 2)x_1x_2 + (0 + 0)x_1x_3 + 0.x_2^2$$
$$+ (-3 - 3)x_2x_3 + (0 + 0)x_3x_1 + (-7)x_3^2$$
$$= 2x_1^2 + 2x_1x_2 + 0.x_1x_3 + 2x_2x_1 + 0.x_2^2 + (-3)x_2x_3$$
$$+ 0x_3x_1 + (-3)x_3x_2 + (-7)x_3^2$$

Therefore, the coefficients of quadratic form will form a matrix as follows :
$$\begin{bmatrix} 2 & 2 & 0 \\ 2 & 0 & -3 \\ 0 & -3 & -7 \end{bmatrix}$$

which is the required matrix of the given quadratic form.

Example 2. *Find the matrices of the following quadratic forms and verify that they can be written as matrix products $X'AX$.*
(i) $x_1^2 - 18x_1x_2 + 5x_2^2$
(ii) $x_1^2 + 2x_2^2 - 5x_3^2 - x_1x_2 + 4x_2x_3 - 3x_3x_1$

Solution. Let $Q = x_1^2 - 18x_1x_2 + 5x_2^2$
The given quadratic form Q can be written as
$$Q = x_1^2 + (-9 - 9)x_1x_2 + 5x_2^2$$
$$= x_1^2 + (-9)x_1x_2 + (-9)x_1x_2 + 5x_2^2$$

Therefore, the matrix corresponding the given quadratic form is
$$A = \begin{bmatrix} 1 & -9 \\ -9 & 5 \end{bmatrix}$$

Let $X = \begin{bmatrix} x_1 \\ x_2 \end{bmatrix}$, then $X' = [x_1 \quad x_2]$

Now
$$X'AX = [x_1 \quad x_2]\begin{bmatrix} 1 & -9 \\ -9 & 5 \end{bmatrix}\begin{bmatrix} x_1 \\ x_2 \end{bmatrix}$$
$$= [x_1 \quad x_2]\begin{bmatrix} x_1 - 9x_2 \\ -9x_1 + 5x_2 \end{bmatrix}$$
$$= x_1(x_1 - 9x_2) + x_2(-9x_1 + 5x_2)$$
$$= x_1^2 - 9x_1x_2 - 9x_2x_1 + 5x_2^2$$
$$= x_1^2 - 18x_1x_2 + 5x_2^2$$

(ii) Let $Q = x_1^2 + 2x_2^2 - 5x_3^2 - x_1x_2 + 4x_2x_3 - 3x_3x_1$

The given quadratic form Q can be written as

$$Q = x_1^2 + \left(-\frac{1}{2} - \frac{1}{2}\right)x_1x_2 + \left(-\frac{3}{2} - \frac{3}{2}\right)x_1x_3 + 2x_2^2 + (2+2)x_2x_3 + (-5)x_3^2$$

$$= x_1^2 + \left(-\frac{1}{2}\right)x_1x_2 + \left(-\frac{3}{2}\right)x_1x_3 + \left(-\frac{1}{2}\right)x_2x_1 + 2x_1^2 + 2x_2x_3 + \left(-\frac{3}{2}\right)x_3x_1$$

$$+ 2x_3x_2 + (-5)x_3^2$$

Therefore, the matrix corresponding to the given quadratic form Q is

$$A = \begin{bmatrix} 1 & -\dfrac{1}{2} & -\dfrac{3}{2} \\ -\dfrac{1}{2} & 2 & 2 \\ -\dfrac{3}{2} & 2 & -5 \end{bmatrix}$$

Let $X = \begin{bmatrix} x_1 \\ x_2 \\ x_3 \end{bmatrix}$, then $X' = [x_1 \quad x_2 \quad x_3]$

Now $X'AX = [x_1 \quad x_2 \quad x_3] \begin{bmatrix} 1 & -\dfrac{1}{2} & -\dfrac{3}{2} \\ -\dfrac{1}{2} & 2 & 2 \\ -\dfrac{3}{2} & 2 & -5 \end{bmatrix} \begin{bmatrix} x_1 \\ x_2 \\ x_3 \end{bmatrix}$

$$= [x_1 \quad x_2 \quad x_3] \begin{bmatrix} x_1 - \dfrac{1}{2}x_2 - \dfrac{3}{2}x_3 \\ -\dfrac{1}{2}x_1 + 2x_2 + 2x_3 \\ -\dfrac{3}{2}x_1 + 2x_2 - 5x_3 \end{bmatrix}$$

$$= x_1\left(x_1 - \frac{1}{2}x_2 - \frac{3}{2}x_3\right) + x_2\left(-\frac{1}{2}x_1 + 2x_2 + 2x_3\right) + x_3\left(-\frac{3}{2}x_1 + 2x_2 - 5x_3\right)$$

$$= x_1^2 - \frac{1}{2}x_1x_2 - \frac{3}{2}x_1x_3 - \frac{1}{2}x_2x_1 + 2x_2^2 + 2x_2x_3 - \frac{3}{2}x_3x_1 + 2x_3x_2 - 5x_3^2$$

$$= x_1^2 + 2x_2^2 - 5x_3^2 - x_1x_2 + 4x_2x_3 - 3x_3x_1.$$

Example 3. *Obtain the matrices corresponding to the following quadratic forms :*

(i) $x^2 + 2y^2 + 3z^2 + 4xy + 5yz + 6zx$

(ii) $ax^2 + by^2 + cz^2 + 2fyz + 2gzx + 2hxy$

(iii) $a_{11}x_1^2 + a_{22}x_2^2 + a_{33}x_3^2 + 2a_{12}x_1x_2 + 2a_{23}x_2x_3 + 2a_{31}x_3x_1$

(iv) $x_1^2 - 2x_2^2 + 4x_3^3 - 4x_4^4 - 2x_1x_2 + 3x_1x_4 + 4x_2x_3 - 5x_3x_4$

(v) $x_1^2 - 2x_2x_3 - x_3x_4$

(vi) $d_1x_1^2 + d_2x_2^2 + d_3x_3^2 + d_4x_4^2$

(vii) $x_1x_2 + x_2x_3 + x_3x_1 + x_1x_4 + x_2x_4 + x_3x_4$

Solution.   (i)   The given quadratic form can be written as

$$x^2 + (2+2)xy + (3+3)xz + 2y^2 + \left(\frac{5}{2} + \frac{5}{2}\right)yz + 3z^2$$

$$= x^2 + 2xy + 3xz + 2yx + 2y^2 + \frac{5}{2}yz + 3zx + \frac{5}{2}zy + 3z^2$$

Therefore, the matrix corresponding to the given quadratic form is

$$\begin{bmatrix} 1 & 2 & 3 \\ 2 & 2 & \dfrac{5}{2} \\ 3 & \dfrac{5}{2} & 3 \end{bmatrix}$$

(ii)   The given quadratic form can be written as

$$ax^2 + (h+h)xy + (g+g)xz + by^2 + (f+f)yz + cz^2$$

$$= ax^2 + hxy + gxz + hyx + by^2 + fyz + gzx + fzy + cz^2$$

Therefore, the matrix corresponding to the given quadratic form is

$$\begin{bmatrix} a & h & g \\ h & b & f \\ g & f & c \end{bmatrix}$$

(iii)   The given quadratic form can be written as

$$a_{11}x_1^2 + (a_{12} + a_{12})x_1x_2 + (a_{31} + a_{31})x_1x_3 + a_{22}x_2^2 + (a_{23} + a_{23})x_2x_3 + a_{33}x_3^2$$

$$= a_{11}x_1^2 + a_{12}x_1x_2 + a_{31}x_1x_3 + a_{12}x_2x_1 + a_{22}x_2^2 + a_{23}x_2x_3 + a_{31}x_3x_1 + a_{23}x_3x_2$$
$$+ a_{33}x_3^2$$

Therefore, the matrix corresponding to the given quadratic form is

$$\begin{bmatrix} a_{11} & a_{12} & a_{31} \\ a_{12} & a_{22} & a_{23} \\ a_{31} & a_{23} & a_{33} \end{bmatrix}$$

(iv)   The given quadratic form can be written as

$$x_1^2 + (-1-1)x_1x_2 + (0+0)x_1x_3 + \left(\frac{3}{2} + \frac{3}{2}\right)x_1x_4 + (-2)x_2^2 + (2+2)x_2x_3$$

$$+ (0+0)x_2x_4 + 4x_3^2 + \left(-\frac{5}{2} - \frac{5}{2}\right)x_3x_4 + (-4)x_4^2$$

$$= x_1^2 - x_1x_2 + 0x_1x_3 + \frac{3}{2}x_1x_2 + (-1)x_2x_1 + (-2)x_2^2 + 2x_2x_3 + 0.x_2x_3 + 0.x_3x_1$$

$$+ 2x_3x_2 + 4x_3^2 + \left(-\frac{5}{2}\right)x_3x_4 + \frac{3}{2}x_4x_1 + 0.x_4x_2 + \left(-\frac{5}{2}\right)x_4x_3 + (-4)x_4^2$$

Therefore, the matrix corresponding to the given quadratic form is

$$\begin{bmatrix} 1 & -1 & 0 & \dfrac{3}{2} \\ -1 & -2 & 2 & 0 \\ 0 & 2 & 4 & -\dfrac{5}{2} \\ \dfrac{3}{2} & 0 & -\dfrac{5}{2} & -4 \end{bmatrix}$$

(v) The given quadratic form can be written as

$$x_1^2 + (0+0)x_1x_2 + (0+0)x_1x_3 + (0+0)x_1x_4 + 0.x_2^2 + (-1-1)x_2x_3$$
$$+ (0+0)x_2x_4 + 0.x_3^2 + \left(-\frac{1}{2}-\frac{1}{2}\right)x_3x_4 + 0.x_4^2$$

$$= x_1^2 + 0.x_1x_2 + 0.x_1x_3 + 0.x_1x_4 + 0.x_2x_1 + 0.x_2^2 + (-1)x_2x_3 + 0.x_2x_4 + 0.x_3x_1$$
$$+ (-1)x_3x_2 + 0.x_3^2 + \left(-\frac{1}{2}\right)x_3x_4 + 0.x_4x_1 + 0.x_4x_2 + \left(-\frac{1}{2}\right)x_4x_3 + 0.x_4^2$$

Therefore, the matrix corresponding to the given quadratic form is

$$\begin{bmatrix} 1 & 0 & 0 & 0 \\ 0 & 0 & -1 & 0 \\ 0 & -1 & 0 & -\frac{1}{2} \\ 0 & 0 & -\frac{1}{2} & 0 \end{bmatrix}$$

(vi) The given quadratic form can be written as

$$d_1x_1^2 + 0.x_1x_2 + 0.x_1x_3 + 0.x_1x_4 + 0.x_2x_1 + d_2x_2^2 + 0.x_2x_3 + 0.x_2x_4 + 0.x_3x_1$$
$$+ 0.x_3x_2 + d_3x_3^2 + 0.x_3x_4 + 0.x_4x_1 + 0.x_4x_2 + 0.x_4x_3 + d_4x_4^2$$

Therefore, the matrix corresponding to the given quadratic form is

$$\begin{bmatrix} d_1 & 0 & 0 & 0 \\ 0 & d_2 & 0 & 0 \\ 0 & 0 & d_3 & 0 \\ 0 & 0 & 0 & d_4 \end{bmatrix} = diag.\left(d_1,d_2,d_3,d_4\right)$$

(vii) The given quadratic form can be written as

$$0.x_1^2 + \left(\frac{1}{2}+\frac{1}{2}\right)x_1x_2 + \left(\frac{1}{2}+\frac{1}{2}\right)x_1x_4 + 0.x_2^2 + \left(\frac{1}{2}+\frac{1}{2}\right)x_2x_3 + \left(\frac{1}{2}+\frac{1}{2}\right)x_2x_4$$
$$+ \left(\frac{1}{2}+\frac{1}{2}\right)x_3x_1 + 0.x_3^2 + \left(\frac{1}{2}+\frac{1}{2}\right)x_3x_4 + 0.x_4^2$$

$$= 0.x_1^2 + \frac{1}{2}x_1x_2 + \frac{1}{2}x_1x_3 + \frac{1}{2}x_1x_4 + \frac{1}{2}x_2x_1 + 0.x_2^2 + \frac{1}{2}x_2x_3 + \frac{1}{2}x_2x_4 + \frac{1}{2}x_3x_1$$
$$+ \frac{1}{2}x_3x_2 + 0.x_3^2 + \frac{1}{2}x_3x_4 + \frac{1}{2}x_4x_1 + \frac{1}{2}x_4x_2 + \frac{1}{2}x_4x_3 + 0.x_4^2$$

Therefore, the matrix corresponding to the given quadratic form is

$$\begin{bmatrix} 0 & \frac{1}{2} & \frac{1}{2} & \frac{1}{2} \\ \frac{1}{2} & 0 & \frac{1}{2} & \frac{1}{2} \\ \frac{1}{2} & \frac{1}{2} & 0 & \frac{1}{2} \\ \frac{1}{2} & \frac{1}{2} & \frac{1}{2} & 0 \end{bmatrix}$$

Example 4. *Find the matrix of the following quadratic form :*

$$(x_1 - x_2 + x_3)^2$$

Solution. The given quadratic form can be written as

$$(x_1 - x_2 + x_3)^2 = x_1^2 + x_2^2 + x_3^2 - 2x_1x_2 + 2x_1x_3 - 2x_2x_3$$

$$= x_1^2 + (-1-1)x_1x_2 + (1+1)x_1x_3 + x_2^2 + (-1-1)x_2x_3 + x_3^2$$

$$= x_1^2 - x_1x_2 + x_1x_3 - x_2x_1 + x_2^2 - x_2x_3 + x_3x_1 - x_3x_2 + x_3^2$$

Therefore, the matrix corresponding to the given quadratic form is

$$\begin{bmatrix} 1 & -1 & 1 \\ -1 & 1 & -1 \\ 1 & -1 & 1 \end{bmatrix}$$

Example 5. *Find the quadratic form of the real symmetric matrix*

$$A = \begin{bmatrix} 2 & 2 & 0 \\ 2 & 0 & -3 \\ 0 & -3 & -7 \end{bmatrix}$$

Solution. Let $X = \begin{bmatrix} x_1 \\ x_2 \\ x_3 \end{bmatrix}$, then $X' = [x_1 \quad x_2 \quad x_3]$

Now $$X'AX = [x_1 \quad x_2 \quad x_3] \begin{bmatrix} 2 & 2 & 0 \\ 2 & 0 & -3 \\ 0 & -3 & -7 \end{bmatrix} \begin{bmatrix} x_1 \\ x_2 \\ x_3 \end{bmatrix}$$

$$= [x_1 \quad x_2 \quad x_3] \begin{bmatrix} 2x_1 + 2x_2 + 0.x_3 \\ 2x_1 + 0.x_2 - 3x_3 \\ 0.x_1 - 3x_2 - 7x_3 \end{bmatrix}$$

$$= x_1(2x_1 + 2x_2 + 0.x_3) + x_2(2x_1 + 0.x_2 - 3x_3) + x_3(0.x_1 - 3x_2 - 7x_3)$$

$$= 2x_1^2 + 2x_1x_2 + 0.x_1x_3 + 2x_2x_1 + 0.x_2^2 - 3x_2x_3 + 0x_3x_1 - 3x_3x_2 - 7x_3^2$$

$$= 2x_1^2 + 4x_1x_2 - 6x_2x_3 - 7x_3^2$$

$$\therefore \quad X'AX = 2x_1^2 + 4x_1x_2 - 6x_2x_3 - 7x_3^2$$

which is the required quadratic form.

Example 6. *Write down the quadratic forms corresponding to the following symmetric matrices:*

(i) $\begin{bmatrix} 1 & 2 & 3 \\ 2 & 0 & 3 \\ 3 & 3 & 1 \end{bmatrix}$ (ii) $diag.(\lambda_1, \lambda_2, ..., \lambda_n)$

Solution. (i)   Let $A = \begin{bmatrix} 1 & 2 & 3 \\ 2 & 0 & 3 \\ 3 & 3 & 1 \end{bmatrix}$ and let $X = \begin{bmatrix} x_1 \\ x_2 \\ x_3 \end{bmatrix}$, then $X' = [x_1 \quad x_2 \quad x_3]$

Now $X'AX = [x_1 \quad x_2 \quad x_3] \begin{bmatrix} 1 & 2 & 3 \\ 2 & 0 & 3 \\ 3 & 3 & 1 \end{bmatrix} \begin{bmatrix} x_1 \\ x_2 \\ x_3 \end{bmatrix}$

$$= [x_1 \quad x_2 \quad x_3] \begin{bmatrix} x_1 + 2x_2 + 3x_3 \\ 2x_1 + 0.x_2 + 3x_3 \\ 3x_1 + 3x_2 + x_3 \end{bmatrix}$$

$$= x_1(x_1 + 2x_2 + 3x_3) + x_2(2x_1 + 0.x_2 - 3x_3) + x_3(3x_1 + 3x_2 + x_3)$$

$$= x_1^2 + 2x_1x_2 + 3x_1x_3 + 2x_2x_1 + 0.x_2^2 + 3x_2x_3 + 3x_3x_1 + 3x_3x_2 + x_3^2$$

$$= x_1^2 + x_3^2 + 4x_1x_2 + 6x_1x_3 + 6x_2x_3$$

which is the required quadratic form.

(ii) Let $A = \text{diag.}(\lambda_1, \lambda_2, ..., \lambda_n)$. Then A is an $n \times n$ diagonal matrix so let us take

$$X = \begin{bmatrix} x_1 \\ x_2 \\ \vdots \\ x_n \end{bmatrix}$$

then $X' = \begin{bmatrix} x_1 & x_2 & \cdots & x_n \end{bmatrix} \text{diag.}(\lambda_1, \lambda_2, ..., \lambda_n) \begin{bmatrix} x_1 \\ x_2 \\ \vdots \\ x_n \end{bmatrix}$

$$= \lambda_1 x_1^2 + \lambda_2 x_2^2 + ... + \lambda_n x_n^2$$

which is the required quadratic form.

⬗ EXERCISE 9.2

1. Find the matrices corresponding to the following quadratic forms as :

   (i) $ax^2 + 2hxy + by^2$

   (ii) $5x_1^2 - 2x_1x_2 + x_2^2$

   (iii) $x_1^2 + 5x_2^2 - 7x_3^2$

   (iv) $4x_1x_3 + 2x_2x_3 + x_3^2$

   (v) $(x_1 + x_2)^2 - x_3^2$

   (vi) $x_1^2 - 2x_2^2 - 3x_3^2 + 4x_1x_2 + 6x_1x_3 - 8x_2x_3$

   (vii) $2x_1x_2 + 6x_1x_3 - 4x_2x_3$

   (viii) $5x_1^2 + 3x_2^2 + 2x_3^2 - x_1x_2 + 8x_2x_3$

   (ix) $8x_1^2 + 7x_2^2 - 3x_3^2 - 6x_1x_2 + 4x_1x_3 - 2x_2x_3$

   (x) $4x_1x_2 + 6x_1x_3 - 8x_2x_3$

   (xi) $5x_1^2 - x_2^2 + 7x_3^2 + 5x_1x_2 - 3x_1x_3$

2. Compute the quadratic form $X'AX$, when

   (i) $A = \begin{bmatrix} 5 & 1/3 \\ 1/3 & 1 \end{bmatrix}$

   (ii) $A = \begin{bmatrix} 4 & 0 \\ 0 & 3 \end{bmatrix}$

   (iii) $A = \begin{bmatrix} 3 & -2 \\ -2 & 7 \end{bmatrix}$

3. Write down the quadratic forms corresponding to the following matrices :

   (i) $\begin{bmatrix} 2 & 1 & 5 \\ 1 & 3 & -2 \\ 5 & -2 & 4 \end{bmatrix}$

   (ii) $\begin{bmatrix} 1 & 0 & 0 \\ 0 & 2 & 0 \\ 0 & 0 & 3 \end{bmatrix}$

   (iii) $\begin{bmatrix} 1 & 2 & 3 \\ 2 & 2 & 5/2 \\ 3 & 5/2 & 3 \end{bmatrix}$

   (iv) $\begin{bmatrix} 0 & 5 & -1 \\ 5 & 1 & 6 \\ -1 & 6 & 2 \end{bmatrix}$

   (v) $\begin{bmatrix} 2 & 2 & 0 \\ 2 & 0 & -3 \\ 0 & -3 & -7 \end{bmatrix}$

   (vi) $\begin{bmatrix} 1 & -1/2 & -3/2 \\ -1/2 & 2 & 2 \\ -3/2 & 2 & -5 \end{bmatrix}$

   (vii) $\begin{bmatrix} 0 & 1 & 3 \\ 1 & 0 & -2 \\ 3 & -2 & 0 \end{bmatrix}$

   (viii) $\begin{bmatrix} 0 & a & b & c \\ a & 0 & l & m \\ b & l & 0 & p \\ c & m & p & 0 \end{bmatrix}$

Answers

1. (i) $\begin{bmatrix} a & h \\ h & b \end{bmatrix}$

(ii) $\begin{bmatrix} 5 & -1 \\ -1 & 1 \end{bmatrix}$

(iii) $\begin{bmatrix} 1 & 0 & 0 \\ 0 & 5 & 0 \\ 0 & 0 & -7 \end{bmatrix}$

(iv) $\begin{bmatrix} 0 & 0 & 2 \\ 0 & 0 & 1 \\ 2 & 1 & 1 \end{bmatrix}$

(v) $\begin{bmatrix} 1 & 1 & 0 \\ 1 & 1 & 0 \\ 0 & 0 & -1 \end{bmatrix}$

(vi) $\begin{bmatrix} 1 & 2 & 3 \\ 2 & -2 & -4 \\ 3 & -4 & -3 \end{bmatrix}$

(vii) $\begin{bmatrix} 0 & 1 & 3 \\ 1 & 0 & -2 \\ 3 & -2 & 0 \end{bmatrix}$

(viii) $\begin{bmatrix} 5 & -1/2 & 0 \\ -1/2 & 3 & 4 \\ 0 & 4 & 2 \end{bmatrix}$

(ix) $\begin{bmatrix} 8 & -3 & 2 \\ -3 & 7 & -1 \\ 2 & -1 & -3 \end{bmatrix}$

(x) $\begin{bmatrix} 0 & 2 & 3 \\ 2 & 0 & -4 \\ 3 & -4 & 0 \end{bmatrix}$

(xi) $\begin{bmatrix} 5 & 5/2 & -3/2 \\ 5/2 & -1 & 0 \\ -3/2 & 0 & 7 \end{bmatrix}$

2. (i) $5x_1^2 + \dfrac{2}{3}x_1x_2 + x_2^2$

(ii) $4x_1^2 + 3x_2^2$

(iii) $3x_1^2 + 7x_2^2 - 4x_1x_2$

3. (i) $2x_1^2 + 3x_2^2 + + 4x_3^2 + 2x_1x_2 + 10x_1x_3 - 4x_2x_3$

(ii) $x_1^2 + 2x_2^2 + 3x_3^2$ (iii) $x_1^2 + 2x_2^2 + 3x_3^2 + 4x_1x_2 + 6x_1x_3 + 5x_2x_3$

(iv) $x_2^2 + 2x_3^2 + 10x_1x_2 - 2x_1x_3 + 12x_2x_3$ (v) $2x_1^2 - 7x_3^2 + 4x_1x_2 - 6x_2x_3$

(vi) $x_1^2 + 2x_2^2 - 5x_3^2 - x_1x_2 - 3x_1x_3 + 4x_2x_3$ (vii) $2x_1x_2 + 6x_1x_3 - 4x_2x_3$

(viii) $2ax_1x_2 + 2bx_1x_3 + 2cx_1x_4 + 2lx_2x_3 + 2mx_2x_4 + 2px_3x_4$

9.10 CONGRUENCE OPERATION ON A SQUARE MATRIX

Any one of the following three operations is a congruence operation on a square matrix :

(i) The operations $R_i \leftrightarrow R_j$ and $C_i \leftrightarrow C_j$ should be applied simultaneously in such a way that the operations $R_i \leftrightarrow R_j$ is followed by $C_i \leftrightarrow C_j$.

(ii) The operation $R_i \rightarrow kR_i$ is followed by, where k is a non-zero scalar.

(iii) The operation $R_i \rightarrow R_i + kR_j$ is follwed by $C_i \rightarrow C_i + kC_j$.

We know that every elementary row (column) transformation of a matrix can be brought about by pre-multiplication (post-multiplication) with the corresponding elementary matrix. If a matrix B is obtained from A by a finite chain of congruent operations applied on A, then there exist elementary matrices $E_1, E_2, E_3, ..., E_s$ such that

$$B = E'_s E'_{s-1} ... E'_2 E'_1 A E_1 E_2 ... E_{s-1} E_s$$

or $B = (E_1 E_2 ... E_{s-1} E_s)' A (E_1 E_2 ... E_{s-1} E_s)$

or $B = P'AP$

where $P = E_1 E_2 ... E_{s-1} E_s$ (being a product of non-singular matrices) is a non-singular matrix. It follows that B is congruent to A.

Hence, every matrix B obtained from A by applying a finite chain of congruent operations on A is congruent to B. The converse of above result is also true. If B is congruent to A, then

$$B = P'AP$$

where P is a non-singular matrix. Since every non-singular matrix can be expressed as the product of elementary matrices, therefore there exist $E_1, E_2, ..., E_s$ such that

$$P = E_1, E_2, ..., E_s$$
$$B = E'_s ... E'_2 E_1 A E_1 E_2 ... E_s$$

9.11 CONGRUENCE OF QUADRATIC FORMS

Two quadratic forms $Q(X) = X'AX$ and $Q(Y) = Y'BY$ over a field F are said to be congruent or equivalent over F. If A and B are congruent over F.

Therefore, $X'AX$ is equivalent to $Y'BY$ if there exists a non-singular matrix P such that $B = P'AP$.

9.12 EQUIVALENCE OF REAL QUADRATIC FORMS

Two real quadratic forms $X'AX$ and $Y'BY$ are said to be :

(i) real equivalent if there exists a non-singular real matrix P such that $B = P'AP$,

(ii) an orthogonally equivalent if there exists a non-singular orthogonal matrix P such that $B = P'AP$.

(iii) complex equivalent if there exists a non-singular complex matrix P such that

$$B = P'AP$$

9.13 THE LINEAR TRANSFORMATION OF A QUADRATIC FORM

Let $Q(X) = X'AX$ be a quadratic form over a field F and let

$$X = PY$$

be a non-singular linear transformation where P is a non-singular matrix.

Now
$$X'AX = (PY)'A(PY)$$
$$= Y'(P'AP)Y$$
$$= Y'BY, \text{ where } B = P'AP$$

Since $B = P'AP$, therefore, B is congruent to a symmetric matrix A so that B is also symmetric, thus $Y'BY$ is a quadratic form which is called a linear transform of the form $X'AX$ by the non-singular matrix P.

Also $B = P'AP$ which is the matrix of the quadratic form $Y'BY$, hence the quadratic form $Y'BY$ is congruent to $X'AX$.

THEOREM 1. *The ranges of values of two congruent quadratic forms are the same.*

Proof. Let $Q(X) = X'AX$ and $Q(Y) = Y'BY$ be two congruent quadratic forms, then there exists a non-singular matrix P such that

$$B = P'AP$$

Let $X = PY$ be a linear transformation, and let $Q(X) = \alpha$, when $X = X_1$, then

$$Q(X_1) = \alpha$$
$$\Rightarrow \qquad \alpha = X_1'AX_1$$

Now, when $Y = P^{-1} X_1$ we have

$$Q(Y) = Q(P^{-1}X_1)$$
$$= (P^{-1}X_1)'B(P^{-1}X_1)$$
$$= X_1'(P^{-1})'BP^{-1}X_1$$
$$= X_1'(P^{-1})'P'APP^{-1}X_1$$
$$= X_1'AX_1 = \alpha$$

$\therefore$ Each value of $Q(X)$ is equal to some value of $Q(Y)$.

Conversely, let $Q(Y) = \beta$ when $Y = Y_1$ then

$$Q(Y_1) = \beta$$
$$\Rightarrow \qquad Y_1'BY_1 = \beta$$

Now, when $X = PY_1$, we have

$$Q(X) = Q(PY_1)$$
$$= (PY_1)'A(PY_1)$$
$$= Y_1'P'APY_1$$
$$= Y_1'BY_1 = \beta$$

$\therefore$ Each value of $Q(Y)$ is equal to some value of $Q(X)$.

Hence $Q(X)$ and $Q(Y)$ have the same ranges of values.

9.14 CONGRUENT REDUCTION OF A SYMMETRIC MATRIX

THEOREM 1. *If A be an $n \times n$ non-zero symmetric matrix of rank r, over a field F, then there exists an $n \times n$ non-singular matrix P over F such that*

$$P'AP = \begin{bmatrix} A_1 & O \\ O & O \end{bmatrix}$$

where A_1 is an $r \times r$ non-singular diagonal matrix over F and each O is a null matrix of suitable order.

Or

Every symmetric matrix of rank r is congruent to a diagonal matrix with r of its diagonal elements are only non-zero.

Proof. Let $A = [a_{ij}]$ be an $n \times n$ symmetric matrix of rank r over a field F.

We shall prove the theorem by induction hypothesis on n, the order of the matrix A. If $n = 1$, then the theorem is trivally true. Suppose that the theorem is true for all symmetric matrices of order $n - 1$.

First of all we shall show that there exists a matrix $B = [b_{ij}]$ of order n over F which is congruent to A such that $b_{11} \neq 0$.

Case I: If $a_{11} \neq 0$, then we take A as B.

Case II: If $a_{11} = 0$ and some of diagonal elements of A, say, $a_{ii} \neq 0$.

Now applying the operations $R_i \leftrightarrow R_1$ and $C_i \leftrightarrow C_1$ to A, we get a matrix B which is congruent to A such that

$$b_{11} = a_{ii} \neq 0.$$

Case III : If $a_{ii} = 0$ for $i = 1, 2, ..., n$ in A, then some $a_{ij} \neq 0$ in A since A is a non-zero matrix. Let a_{ij} be a non-zero element in A. Since A is symmetric matrix, so that

$$a_{ij} = a_{ji} \neq 0$$

Now applying the operations $R_i \to R_i + R_j$ and $C_i \to C_i + C_j$ on A, we get a matrix $D = [d_{ij}]$ which is congruent to A such that

$$d_{ii} = a_{ij} + a_{ji} = 2a_{ij} \neq 0$$

Again applying the operations $R_i \leftrightarrow R_1$ and $C_i \leftrightarrow C_1$ to D, we get a matrix B which is congruent to D, therefore congruent A such that

$$b_{11} = d_{ii} \neq 0$$

Thus, in all cases there always exists a matrix $B = [b_{ij}]$ which is congruent to A such that $b_{11} \neq 0$.

As B is congruent to A which is symmetric so that B itself is symmetric matrix. Since $b_{11} \neq 0$, then all the elements in first row and first column of B, except b_{11}, can be made zero by applying suitable congruent operations. We, therefore, get a matrix

$$C = \begin{bmatrix} a_{11} & 0 & \cdots & 0 \\ 0 & & & \\ \vdots & & B_1 & \\ 0 & & & \end{bmatrix}_{n \times n}$$

which is congruent to B and B is congruent to A so that C is congruent to A. Since A is a symmetric matrix, therefore C is also symmetric matrix and so that B_1 is symmetric matrix of order $n - 1$. Thus by induction hypothesis B_1 reduced to a diagonal matrix by applying congruent operations on B_1. If the congruent operation applied to B_1 for this purpose be applied to C because these operations can not affect the first row and first column of C. Therefore C can be reduced to a diagonal matrix. Since A is congruent to C, therefore A is congruent to a diagonal matrix say diag. $(\lambda_1, \lambda_2, ..., \lambda_k, 0, 0, ..., 0)$.

Thus there exists a non-singular matix P such that
$$P'AP = \text{diag. } (\lambda_1, \lambda_2, ..., \lambda_k, 0, 0, ..., 0)$$
We know that the rank of a matrix does not change on pre-multiplication and post-multiplication by a non-singular matrix. Since the rank of A is r, therefore rank of $P'AP$ is also r, it follows that exactly r elements of diag. $(\lambda_1, \lambda_2, ..., \lambda_k, 0, 0, ..., 0)$ are non-zero, thus $k = r$.

Hence
$$P'AP = \text{diag. } (\lambda_1, \lambda_2, ..., \lambda_r, 0, 0, ..., 0)$$
$$= \begin{bmatrix} A_1 & O \\ O & O \end{bmatrix}$$
where A_1 is a diagonal matrix of order r.

Theorem 2. *Corresponding to every quadratic form $X'AX$ over a field F, there exists a non-singular linear transformation*
$$X = PY$$
over F, such that the form $X'AX$ transforms to a sum of r square terms
$$\lambda_1 y_1^2 + \lambda_2 y_2^2 + ... \lambda_r y_r^2$$
where $\lambda_1, \lambda_2,..., \lambda_r$ belong to the field F and r is the rank of A and $Y' = [y_1, y_2, ..., y_n]$.

Proof. Since A is a real symmetric matrix of rank r, then there exists a non-singular real matrix P such that
$$P'AP = \text{diag. } (\lambda_1, \lambda_2, ..., \lambda_r, 0, ..., 0)$$
Now
$$X'AX = (PY)'A(PY)$$
$$= Y'(P'AP)Y$$

$$= [y_1, y_2, ..., y_n] \text{ diag. } (\lambda_1, \lambda_2, ..., \lambda_k, 0, 0, ..., 0) \begin{bmatrix} y_1 \\ y_2 \\ \vdots \\ y_n \end{bmatrix}$$

$$= \lambda_1 y_1^2 + \lambda_2 y_2^2 + ... \lambda_r y_r^2$$

9.15 RANK OF A QUADRATIC FORM

Let $Q(X) = X'AX$ be a quadratic form over a field F. Then the rank of $Q(X)$ is the rank of the matrix A.

If the rank of $Q(X)$ is r, then there exists a non-singular matrix P which reduces $Q(X)$ to a sum of r square terms.

i.e.,
$$Q(X) = \lambda_1 y_1^2 + \lambda_2 y_2^2 + ... \lambda_r y_r^2$$

WORKING RULE

Let A be $n \times n$ real symmetric matrix. In order to find a non-singular matrix P such that $P'AP = $ diagonal matrix, we use the following steps :	
STEP 1.	Write $A = I_n A I_n$
STEP 2.	Apply congruent row operations on pre-factor I_n of A on RHS and congruent column operations on post-factor I_n of A . Applying such operations simultaneously on A on LHS till A reduces to a diagonal matrix. When the matrix A reduces to a diagonal matrix, the post-factor I_n ultimately gives the non-singular matrix P such that $P'AP = $ diagonal matix

Solved Examples

Example 1. *Determine a non-singular matrix P such that P´AP is a diagonal matrix, where*

$$A = \begin{bmatrix} 0 & 1 & 2 \\ 1 & 0 & 3 \\ 2 & 3 & 0 \end{bmatrix}$$

Solution. We have $\qquad A = I_3 A I_3$

or $\qquad \begin{bmatrix} 0 & 1 & 2 \\ 1 & 0 & 3 \\ 2 & 3 & 0 \end{bmatrix} = \begin{bmatrix} 1 & 0 & 0 \\ 0 & 1 & 0 \\ 0 & 0 & 1 \end{bmatrix} A \begin{bmatrix} 1 & 0 & 0 \\ 0 & 1 & 0 \\ 0 & 0 & 1 \end{bmatrix}$

Applying $R_1 \to R_1 + R_2$, we get

$$\begin{bmatrix} 1 & 1 & 5 \\ 1 & 0 & 3 \\ 2 & 3 & 0 \end{bmatrix} = \begin{bmatrix} 1 & 1 & 0 \\ 0 & 1 & 0 \\ 0 & 0 & 1 \end{bmatrix} A \begin{bmatrix} 1 & 0 & 0 \\ 0 & 1 & 0 \\ 0 & 0 & 1 \end{bmatrix}$$

Applying $C_1 \to C_1 + C_2$, we get

$$\begin{bmatrix} 2 & 1 & 5 \\ 1 & 0 & 3 \\ 5 & 3 & 0 \end{bmatrix} = \begin{bmatrix} 1 & 1 & 0 \\ 0 & 1 & 0 \\ 0 & 0 & 1 \end{bmatrix} A \begin{bmatrix} 1 & 0 & 0 \\ 1 & 1 & 0 \\ 0 & 0 & 1 \end{bmatrix}$$

Applying $R_2 \to R_2 - \dfrac{1}{2} R_1, R_3 \to R_3 - \dfrac{5}{2} R_1$, we get

$$\begin{bmatrix} 2 & 1 & 5 \\ 0 & -1/2 & 1/2 \\ 0 & 1/2 & -25/2 \end{bmatrix} = \begin{bmatrix} 1 & 1 & 0 \\ -1/2 & 1/2 & 0 \\ -5/2 & -5/2 & 1 \end{bmatrix} A \begin{bmatrix} 1 & 0 & 0 \\ 1 & 1 & 0 \\ 0 & 0 & 1 \end{bmatrix}$$

Applying $C_2 \to C_2 - \dfrac{1}{2} C_1, C_3 \to C_3 - \dfrac{5}{2} C_1$, we get

$$\begin{bmatrix} 2 & 1 & 5 \\ 0 & -1/2 & 1/2 \\ 0 & 1/2 & -25/2 \end{bmatrix} = \begin{bmatrix} 1 & 1 & 0 \\ -1/2 & 1/2 & 0 \\ -5/2 & -5/2 & 1 \end{bmatrix} A \begin{bmatrix} 1 & -1/2 & -5/2 \\ 1 & 1/2 & -5/2 \\ 0 & 0 & 1 \end{bmatrix}$$

Applying $R_3 \to R_3 + R_2$, we get

$$\begin{bmatrix} 2 & 0 & 0 \\ 0 & -1/2 & 1/2 \\ 0 & 0 & -12 \end{bmatrix} = \begin{bmatrix} 1 & 1 & 0 \\ -1/2 & 1/2 & 0 \\ -3 & -2 & 1 \end{bmatrix} A \begin{bmatrix} 1 & -1/2 & -5/2 \\ 1 & 1/2 & -5/2 \\ 0 & 0 & 1 \end{bmatrix}$$

Applying $C_3 \to C_3 + C_2$, we get

$$\begin{bmatrix} 2 & 0 & 0 \\ 0 & -1/2 & 0 \\ 0 & 0 & -12 \end{bmatrix} = \begin{bmatrix} 1 & 1 & 0 \\ -1/2 & 1/2 & 0 \\ -3 & -2 & 1 \end{bmatrix} A \begin{bmatrix} 1 & -1/2 & -3 \\ 1 & 1/2 & -2 \\ 0 & 0 & 1 \end{bmatrix}$$

$\Rightarrow \quad \text{diag.} \left(2, -\dfrac{1}{2}, -12 \right) = P´AP$

where $\qquad P = \begin{bmatrix} 1 & -1/2 & -3 \\ 1 & 1/2 & -2 \\ 0 & 0 & 1 \end{bmatrix}$

Example 2. *Determine a non-singular matrix P such that P´AP is a diagonal matrix, where*

$$A = \begin{bmatrix} 6 & -2 & 2 \\ -2 & 3 & -1 \\ 2 & -1 & 3 \end{bmatrix}$$

Interpret the result in terms of quadratic form.

Solution. We have
$$A = I_3 A I_3$$

or
$$\begin{bmatrix} 6 & -2 & 2 \\ -2 & 3 & -1 \\ 2 & -1 & 3 \end{bmatrix} = \begin{bmatrix} 1 & 0 & 0 \\ 0 & 1 & 0 \\ 0 & 0 & 1 \end{bmatrix} A \begin{bmatrix} 1 & 0 & 0 \\ 0 & 1 & 0 \\ 0 & 0 & 1 \end{bmatrix}$$

Applying $R_2 \to R_2 + \dfrac{1}{3} R_1$, we get

$$\begin{bmatrix} 6 & -2 & 2 \\ 0 & 7/3 & -1/3 \\ 2 & -1 & 3 \end{bmatrix} = \begin{bmatrix} 1 & 0 & 0 \\ 1/3 & 1 & 0 \\ 0 & 0 & 1 \end{bmatrix} A \begin{bmatrix} 1 & 0 & 0 \\ 0 & 1 & 0 \\ 0 & 0 & 1 \end{bmatrix}$$

Applying $C_2 \to C_2 + \dfrac{1}{3} C_1$, we get

$$\begin{bmatrix} 6 & 0 & 2 \\ 0 & 7/3 & -1/3 \\ 2 & -1/3 & 3 \end{bmatrix} = \begin{bmatrix} 1 & 0 & 0 \\ 1/3 & 1 & 0 \\ 0 & 0 & 1 \end{bmatrix} A \begin{bmatrix} 1 & 1/3 & 0 \\ 0 & 1 & 0 \\ 0 & 0 & 1 \end{bmatrix}$$

Applying $R_3 \to R_3 - \dfrac{1}{3} R_1$ we get

$$\begin{bmatrix} 6 & 0 & 2 \\ 0 & 7/3 & -1/3 \\ 0 & -1/3 & 7/3 \end{bmatrix} = \begin{bmatrix} 1 & 0 & 0 \\ 1/3 & 1 & 0 \\ -1/3 & 0 & 1 \end{bmatrix} A \begin{bmatrix} 1 & 1/3 & 0 \\ 0 & 1 & 0 \\ 0 & 0 & 1 \end{bmatrix}$$

Applying $C_3 \to C_3 - \dfrac{1}{3} C_1$, we get

$$\begin{bmatrix} 6 & 0 & 2 \\ 0 & 7/3 & -1/3 \\ 0 & -1/3 & 7/3 \end{bmatrix} = \begin{bmatrix} 1 & 0 & 0 \\ 1/3 & 1 & 0 \\ -1/3 & 0 & 1 \end{bmatrix} A \begin{bmatrix} 1 & 1/3 & -1/3 \\ 0 & 1 & 0 \\ 0 & 0 & 1 \end{bmatrix}$$

Applying $R_3 \to R_3 + \dfrac{1}{7} R_2$, we get

$$\begin{bmatrix} 6 & 0 & 0 \\ 0 & 7/3 & -1/3 \\ 0 & 0 & 16/7 \end{bmatrix} = \begin{bmatrix} 1 & 0 & 0 \\ 1/3 & 1 & 0 \\ -2/7 & 1/7 & 1 \end{bmatrix} A \begin{bmatrix} 1 & 1/3 & -1/3 \\ 0 & 1 & 0 \\ 0 & 0 & 1 \end{bmatrix}$$

Applying $C_3 \to C_3 + \dfrac{1}{7} C_2$, we get

$$\begin{bmatrix} 6 & 0 & 0 \\ 0 & 7/3 & 0 \\ 0 & 0 & 16/7 \end{bmatrix} = \begin{bmatrix} 1 & 0 & 0 \\ 1/3 & 1 & 0 \\ -2/7 & 1/7 & 1 \end{bmatrix} A \begin{bmatrix} 1 & 1/3 & -2/7 \\ 0 & 1 & 1/7 \\ 0 & 0 & 1 \end{bmatrix}$$

$$\Rightarrow \quad \text{diag.}\left(6, \frac{7}{3}, \frac{16}{7}\right) = P'AP \qquad \qquad \dots(1)$$

where
$$P = \begin{bmatrix} 1 & 1/3 & -2/7 \\ 0 & 1 & 1/7 \\ 0 & 0 & 1 \end{bmatrix}$$

The quadratic form of the matrix A is given by

$$X'AX = \begin{bmatrix} x_1 & x_2 & x_3 \end{bmatrix} \begin{bmatrix} 6 & -2 & 2 \\ -2 & 3 & -1 \\ 2 & -1 & 3 \end{bmatrix} \begin{bmatrix} x_1 \\ x_2 \\ x_3 \end{bmatrix}$$

$$X'AX = 6x_1^2 + 3x_2^2 + 3x_3^2 - 4x_1x_2 + 4x_1x_3 - 2x_2x_3 \qquad \text{...(2)}$$

The non-singular transformation corresponding to the matrix P is given by

$$X = PY \qquad \text{...(3)}$$

where

$$Y = \begin{bmatrix} y_1 \\ y_2 \\ y_3 \end{bmatrix} \text{ and } X = \begin{bmatrix} x_1 \\ x_2 \\ x_3 \end{bmatrix}$$

From (3) we have

$$\begin{bmatrix} x_1 \\ x_2 \\ x_3 \end{bmatrix} = \begin{bmatrix} 1 & 1/3 & -2/7 \\ 0 & 1 & 1/7 \\ 0 & 0 & 1 \end{bmatrix} \begin{bmatrix} y_1 \\ y_2 \\ y_3 \end{bmatrix}$$

i.e.,

$$\left. \begin{aligned} x_1 &= y_1 + \frac{1}{3}y_2 - \frac{2}{7}y_3 \\ x_2 &= y_2 + \frac{1}{7}y_3 \\ x_3 &= y_3 \end{aligned} \right\} \qquad \text{...(4)}$$

The transformations given by (4) reduce the quadratic form (2) to the diagonal form

$$X'AX = Y'(P'AP)Y$$

$$= Y' \text{diag.} \left(6, \frac{7}{3}, \frac{16}{7} \right) Y$$

$$= 6y_1^2 + \frac{7}{3}y_2^2 + \frac{16}{7}y_3^2$$

Clearly, $X'AX$ has a sum of 3 square terms, hence the rank of $X'AX$ is 3.

9.16 REDUCTION OF A REAL QUADRATIC FORM OVER REAL FIELD

THEOREM. *If A is an $n \times n$ real symmetric matrix of rank r, then there exists a real non-singular matrix P such that*

$$P'AP = diag. (1, 1, 1, ...1, -1, -1, -1, -1, ...-1, 0, 0, ... 0)$$

where 1 appears p times and -1 appears $r - p$ times.

Proof. Since A is an $n \times n$ real symmetric matrix of rank r, therefore there exists a non-singular matrix Q such that $Q'AQ$ is a diagonal matrix which has exactly r non-zero elements.

Suppose that

$$Q'AQ = \text{diag.} (\lambda_1, \lambda_2, ..., \lambda_r, 0, 0,... 0)$$

where each $\lambda_i \neq 0 (i = 1, 2, ... r)$

Again suppose that out of r non-zero elements, p are positive and $r - p$ are negative. Since in a diagonal matrix the positions of the diagonal elements occurring in i^{th} and j^{th} rows can be interchanged by using the congruent operations $R_i \leftrightarrow R_j$ and $C_i \leftrightarrow C_j$, so that without any loss of generality we may take $\lambda_1, \lambda_2, ..., \lambda_p$ to be positive and $\lambda_{p+1}, \lambda_{p+2}, ..., \lambda_r$ to be negative.

Let

$$S = \text{diag.} \left(\frac{1}{\sqrt{\lambda_1}}, \frac{1}{\sqrt{\lambda_2}}, ..., \frac{1}{\sqrt{\lambda_p}}, \frac{1}{\sqrt{-\lambda_{p+1}}}, ..., \frac{1}{\sqrt{-\lambda_r}}, 0, ..., 0 \right)$$

Then S is a real non-singular symmetric matrix.

Let $P = QS$, then P is a real non-singular matrix.

Now

$$P'AP = (QS)'A(QS)$$
$$= S'(Q'AQ)S$$
$$= S' \text{diag}.(\lambda_1, \lambda_2, ..., \lambda_r, 0, 0, 0, ..., 0) \, S$$
$$= S' \, \text{diag}.\left(\sqrt{\lambda_1}, \sqrt{\lambda_2}, ..., \sqrt{\lambda_p}, \frac{\lambda_{p+1}}{\sqrt{-\lambda_{p+1}}}, ..., \frac{\lambda_r}{\sqrt{-\lambda_r}}, 0, ..., 0 \right)$$
$$= S' \, \text{diag}. (1, 1, 1, ...1, -1, -1, -1, -1, ...-1, 0, 0, ... 0)$$

where 1 appears p times and -1 appears $r - p$ times.

Corollary. *If $X'AX$ is a real quadratic form of rank r in n variables over real field, then there exists a non-singular linear transformation $X = PY$ which reduces $X'AX$ to the form*

$$Y'PAPY = y_1^2 + y_2^2 + ... + y_p^2 - y_{p+1}^2 - y_{p+2}^2 - ... - y_r^2$$

Proof. By above theorem

$$P'AP = \text{diag}.(1, 1, 1, ...1, -1, -1, -1, -1, ...-1, 0, 0, ... 0)$$

where 1 appears p times and -1 appears $r - p$ times.

Let $\quad Y = \begin{bmatrix} y_1 \\ y_2 \\ \vdots \\ y_n \end{bmatrix}$, then

$$X'AX = (PY)'A(PY)$$
$$= Y'(P'AP)Y$$
$$= Y' \, \text{diag}. (1, 1, 1, ...1, -1, -1, -1, -1, ...-1, 0, 0, ... 0)Y$$
$$= y_1^2 + y_2^2 + ... + y_p^2 - y_{p+1}^2 - y_{p+2}^2 - ... - y_r^2$$

9.17 NORMAL (OR CANONICAL) FORM OF A REAL QUADRATIC MATRIX

Let $X'AX$ be a real quadratic form in n variables over the real field, then there exists a real non-singular linear transformation $X = PY$, which reduces the given quadratic form to the form

$$Y'(P'AP)Y = y_1^2 + y_2^2 + ... + y_p^2 - y_{p+1}^2 - y_{p+2}^2 - ... - y_r^2$$

This new form is known as the normal (or canonical) form of $X'AX$.

Theorem 1. *The number of positive terms in any two normal forms of a real quadratic form is the same.*

Proof. Let $X'AX$ be a real quadratic form of rank r in n variables over real field.

Suppose that the real non-singular linear transformations $X = PY$ and $X = QZ$ reduce the given quadratic form to two normal forms.

$$X'AX = Y'P'APY = y_1^2 + y_2^2 + ... + y_p^2 - y_{p+1}^2 - y_{p+2}^2 - ... - y_r^2 \quad ...(1)$$

and $\quad X'AX = Z'Q'AQZ = z_q^2 + z_2^2 + ... + z_q^2 - z_{q+1}^2 - z_{q+2}^2 - ... - z_r^2 \quad ...(2)$

respectively.

We shall prove that $p = q$.

Let us suppose that $p < q$, then $q - p > 0$, therefore $n - (q - p)$ is less than n or $(n - q) + p$ is less than n.

Since $X = PY$ and $X = QZ$ it follows that $y_1, y_2, ..., y_n$ and $z_1, z_2, ..., z_n$ are the homogeneous functions of $x_1, x_2, ... x_n$.

Consider the homogeneous equations in $x_1, x_2, ... x_n$ variables

$$y_1 = 0, \ y_2 = 0, ..., y_p = 0, \ z_{q+1} = 0, \ z_{q+2} = 0, ..., z_n = 0$$

Clearly these are $p + (n - q)$ equation in n variables. Since $p + (n - q)$ is less than n so that these equations will have non-zero solutions.

Let $x_1 = a_1, x_2 = a_2, ..., x_n = a_n$ be a non-zero solution of these equations.

Let $X_1 = \begin{bmatrix} a_1 \\ a_2 \\ \vdots \\ a_n \end{bmatrix}$, then $\quad X_1 \neq O$

Let $Y1 = \begin{bmatrix} b_1 \\ b_2 \\ \vdots \\ b_n \end{bmatrix}$ and $Z_1 = \begin{bmatrix} c_1 \\ c_2 \\ \vdots \\ c_n \end{bmatrix}$ be the values of Y and Z where $X = X_1$, then we have

$$b_1 = 0 = b_2 = ... = b_p$$

and $\qquad c_{q+1} = 0 = c_{q+2} = ... = c_n.$

Putting $Y = \begin{bmatrix} b_1 \\ b_2 \\ \vdots \\ b_n \end{bmatrix}$ in (1) and $Z = \begin{bmatrix} c_1 \\ c_2 \\ \vdots \\ c_n \end{bmatrix}$ in (2) we get

$$-b_{p+1}^2 - b_{p+2}^2 - ... - b_r^2 = c_1^2 + c_2^2 + ... + c_q^2$$

$\Rightarrow \quad b_{p+1} = 0, b_{p+2} = 0, ..., b_r = 0$ and $c_1 = 0, c_2 = 0, ..., c_q = 0$

$\Rightarrow \qquad\qquad\qquad Y_1 = O$ and $Z_1 = O$

Since $\qquad\qquad\qquad X = PY$ and $X = QZ$, then

$$X_1 = PY_1 \text{ and } X_1 = QZ_1$$

$\Rightarrow \qquad\qquad\qquad X_1 = O$

which is a contradiction since $X_1 \neq O$.

Thus, $p \not< q$.

Similarly, $q \not< p$.

Hence $p = q$.

Theorem 2. *The number of negative terms in any two normal forms of a quadratic form is the same.*

Proof. If $X'AX$ is a real quadratic form of rank r in n variables over the real field.

Suppose in one normal form of $X'AX$, there are p positive terms and in other normal form, there are q positive terms, then in first normal form there will be $r - p$ negative terms and in second normal form the negative terms will be $r - q$. By above theorem, we have $p = q$.

$\therefore \qquad\qquad r - p = r - q$

Hence the corollary.

Theorem 3. *The excess of the number of positive terms over the number of negative terms in any two normal forms of a real quadratic form is the same.*

Proof. Since the number positive terms and negative terms in any two normal form of a real quadratic form are the same, therefore, the excess of positive terms over negative terms is the same.

9.18 SIGNATURE AND INDEX OF A REAL QUADRATIC FORM

Let $X'AX$ be a real quadratic form of rank r in n variables over the field of reals and its normal form be

$$y_1^2 + y_2^2 + \dots + y_p^2 - y_{p+1}^2 - y_{p+2}^2 - \dots - y_r^2$$

Then the number p of positive terms in a normal form is called the index of $X'AX$. The excess of the number of positive terms over the number of negative terms in a normal form *i.e.*, $p - (r-p) = 2p - r$ is called the signature *of* $X'AX$.

THEOREM 1. *The signature of a real quadratic form is invariant for its all normal forms.*

Proof. Since the number of positive and negative terms in any two normal forms of a real quadratic form are the same so that their difference are the same *i.e.*, the signature of given quadratic is the same for its all normal forms.

THEOREM 2. *Two real quadratic forms in n variables over the field of reals are real equivalent if and only if they have the same rank and signature (or index).*

Proof. Let $X'AX$ and $Y'BY$ be two real quadratic forms in the same number of variables.

Suppose that $X'AX$ and $Y'BY$ are equivalent then there exists a real non-singular linear transformation $X = PY$ such that

$$B = P'AP$$

Again suppose that the real non-singular linear transformations $Y = QZ$ transforms the quadratic form $Y'BY$ to a normal form $Z'CZ$ such that

$$C = Q'BQ$$

Since P and Q are non-singular matrices so that PQ is a non-singular matrix, therefore the linear transformation $X = (PQ)Z$ transforms the quadratic form $X'AX$ to the form

$$\begin{aligned} (PQZ)'A(PQZ) &= Z'(PQ)'A(PQ)Z \\ &= Z'Q'(P'AP)QZ \\ &= Z'(Q'BQ)Z \\ &= Z'CZ \end{aligned}$$

It follows that $X'AX$ and $Y'BY$ reduce to the same normal form, hence $X'AX$ and $Y'BY$ have the same rank and same signature (or index).

Conversely, $X'AX$ and $Y'BY$ have the same rank r (say) and the same signature s (say), then the index of $X'AX$ and $Y'BY$ is $\dfrac{s+r}{2} = p$.

Therefore, $X'AX$ and $Y'BY$ can be reduced to the same normal form.

$$Z'CZ = z_1^2 + z_2^2 + \dots + z_q^2 - z_{q+1}^2 - z_{q+2}^2 - \dots - z_r^2$$

by non-singular linear transformations

$X = PY$ and $Y = QZ$ respectively. That is, $X'AX$ reduces to $Z'CZ$ by $X = PY$, then $C = P'AP$ and $Y'BY$ reduces to $Z'CZ$ by $Y = QZ$, then $C = Q'BQ$.

$$\therefore \qquad P'AP = Q'BQ$$
$$\Rightarrow \qquad B = (Q')^{-1}(P'AP)Q^{-1}$$
$$\Rightarrow \qquad B = (Q^{-1})'(P'AP)Q^{-1}$$
$$\Rightarrow \qquad B = (PQ^{-1})'A(PQ^{-1})$$

It follows that the real non-singular linear transformation $X = (PQ^{-1})Y$ reduces $X'AX$ to $Y'BY$, therefore the two given quadratic forms are real equivalent. Hence the theorem.

9.19 REDUCTION OF A REAL QUADRATIC FORM OVER THE FIELD OF COMPLEX NUMBERS

THEOREM 1. *Let A be an $n \times n$ real symmetric matrix of rank r, then there exists a non-singular matrix P whose elements may be complex numbers such that*

$$P'AP = diag.(1, 1, 1, ...1, 0, 0, ... 0)$$

where 1 appears r times.

Proof. Since A is a real symmetric matrix of rank r, then there exists a non-singular real matrix Q such that

$$Q'AQ = diag. (\lambda_1, \lambda_2, ..., \lambda_r, 0, 0, ... 0) = D \text{ (say)}$$

where D is a diagonal matrix which has exactly r non-zero elements $\lambda_1, \lambda_2, ..., \lambda_r$ and $\lambda_1, \lambda_2, ..., \lambda_r$ may be positive or negative or both.

Let $S = diag.\left(\dfrac{1}{\sqrt{\lambda_1}}, \dfrac{1}{\sqrt{\lambda_2}}, ..., \dfrac{1}{\sqrt{\lambda_r}}, 1, 1, ..., 1 \right)$

be an $n \times n$ complex diagonal matrix, which is obviously, a complex non-singular diagonal matrix and $S' = S$.

Let us take $P = QS$, clearly P is a complex non-singular matrix.

Now $P'AP = (QS)'A(QS)$

$$= S'(Q'AQ)S$$

$$= S'DS \qquad\qquad\qquad\qquad\qquad [\because Q'AQ = D]$$

$$= SDS \qquad\qquad\qquad\qquad\qquad [\because S' = S]$$

$$= diag.\left(\frac{1}{\sqrt{\lambda_1}}, \frac{1}{\sqrt{\lambda_2}}, ..., \frac{1}{\sqrt{\lambda_r}}, 1, 1, ..., 1 \right) diag.(\lambda_1, \lambda_2, ..., \lambda_r, 0, 0, ...0)S$$

$$= diag\left(\sqrt{\lambda_1}, \sqrt{\lambda_2}, ..., \sqrt{\lambda_r}, 0, 0, ..., 0 \right) S$$

$$= diag\left(\sqrt{\lambda_1}, \sqrt{\lambda_2}, ..., \sqrt{\lambda_r}, 0, 0, ..., 0 \right)$$

$$\qquad\qquad\qquad\qquad diag\left(\frac{1}{\sqrt{\lambda_1}}, \frac{1}{\sqrt{\lambda_2}}, ..., \frac{1}{\sqrt{\lambda_r}}, 1, 1, ..., 1 \right)$$

$$= diag(1, 1, 1, ..., 1, 0, 0, ..., 0)$$

$\therefore$ $P'AP = diag(1, 1, 1, ..., 1, 0, 0, ..., 0)$

where 1 appears r times.

Theorem 2. *Every real quadratic form $X'AX$ is complex equivalent to the form*

$$z_1^2 + z_2^2 + ... + z_r^2$$

where r is the rank of A.

Proof. Let $Z = [z_1, z_2, ..., z_r, z_{r+1}, ... z_n]'$, where z_i is a complex number.

The quadratic form $X'AX$ can be reduced to the form $Z'BZ$ by a complex non-singular linear transformation $X = PZ$ where P is a complex non-singular matrix.

Since A is a real symmetric matrix of rank r, then by above theorem, there is a non-singular complex matrix P such that

$$P'AP = diag. (1, 1, 1, ..., 1, 0, 0, ... 0)$$

where 1 appears r times.

$\therefore$ $X'AX = (PZ)'A(PZ) \quad = Z'(P'AP)Z$

$$= [z_1, z_2, ..., z_r, ... z_n] \text{ diag.}(1, 1, 1, ... 1, 0, 0, ... 0) \begin{bmatrix} z_1 \\ z_2 \\ \vdots \\ z_r \\ \vdots \\ z_n \end{bmatrix}$$

$$= z_1^2 + z_2^2 + ... + z_r^2$$

REMARK

- Two real quadratic forms in n variables are complex equivalent if and only if they have the same rank.

9.20 ORTHOGONAL REDUCTION OF A REAL QUADRATIC FORM

THEOREM 1. *If $X'AX$ be a real quadratic form of rank r in n variables, then there exists a real orthogonal transformation $X = PY$ which transforms $X'AX$ to the form $\lambda_1 y_1^2 + \lambda_2 y_2^2 + ... \lambda_r y_r^2$ where $\lambda_1, \lambda_2, ..., \lambda_r$ are the r non-zero eigenvalues of A and $n - r$ eigenvalues of A being equal to zero.*

Proof. Since A is a real symmetric matrix of order n, then there exists a real orthogonal matrix P such that

$$P^{-1}AP = D$$

where D is a diagonal matrix, whose diagonal elements are the eigenvalues of A. Again, since the rank of A is r, then the rank of $P^{-1}AP$ is also r, therefore D will have exactly r non-zero diagonal elements, hence A has exactly r non-zero eigenvalues and remaining $n - r$ eigenvalues of A are all zero.

So we can take

$$D = \text{diag.}(\lambda_1, \lambda_2, ..., \lambda_r, 0, 0, .., 0)$$

where $\lambda_1, \lambda_2, ..., \lambda_r$ are the r non-zero eigenvalues of A

$$\therefore \qquad P^{-1}AP = \text{diag.}(\lambda_1, \lambda_2, ..., \lambda_r, 0, 0, ... 0)$$

Since P is an orthogonal matrix, then $P^{-1} = P'$

$$\therefore \quad P^{-1}AP = P'AP = \text{diag.}(\lambda_1, \lambda_2, ..., \lambda_r, 0, 0, ... 0) = D$$

It follows that A is congruent to D

Now, let us take a real orthogonal transformation $X = PY$ such that

$$X'AX = (PY)'A(PY)$$
$$= Y'(P'AP)Y$$
$$= Y'DY$$

$\lambda_1 y_1^2 + \lambda_2 y_2^2 + ... \lambda_r y_r^2$, if $Y = [y_1, y_2, ... y_r, ..., y_n]$

Hence the theorem.

THEOREM 2. *Every real quadratic form $X'AX$ in n variables is real equivalent to the form*

$$y_1^2 + y_2^2 + ... + y_p^2 - y_{p+1}^2 - y_{p+2}^2 - ... - y_r^2$$

where r is the rank of A. and p is the number of positive eigenvalues of A.

Proof. Since A is a real symmetric matrix, then there exists a real orthogonal matrix Q such that

$$Q^{-1}AQ = D$$

or $\qquad\qquad Q'AQ = D \qquad\qquad\qquad\qquad [\because Q^{-1} = Q']$

where D is a diagonal matrix whose diagonal elements are the eigenvalues of A. Since the rank of A is r so that D is also of rank r, therefore D has exactly r non-zero elements it follows that A has exactly r non-zero eigenvalues and remaining $n - r$ eigenvalues of A are all zero.

If $\lambda_1, \lambda_2, ..., \lambda_r$ be non-zero eigenvalues of A, then we have

$$Q'AQ = D = \text{diag. } (\lambda_1, \lambda_2, ..., \lambda_r, 0, 0, ... 0)$$

Suppose out of r eigenvalues of A, $\lambda_1, \lambda_2, ..., \lambda_p$ are positive eigenvalues and $\lambda_{p+1}, \lambda_{p+2}, ..., \lambda_r$ are negative eigenvalues of A.

Let $S = \text{diag.}\left(\dfrac{1}{\sqrt{\lambda_1}}, \dfrac{1}{\sqrt{\lambda_2}}, ..., \dfrac{1}{\sqrt{\lambda_p}}, \dfrac{1}{\sqrt{-\lambda_{p+1}}}, ..., \dfrac{1}{\sqrt{-\lambda_r}}, 0, ..., 0\right)$

Then S is a non-singular diagonal matrix, and $S' = S$.

Let us take $P = QS$, clearly P is a real non-singular matrix, then we have

$$
\begin{aligned}
P'AP &= (QS)'A(QS) \\
&= S'(Q'AQ)S \\
&= S'DS && [\because Q'AQ = D] \\
&= SDS && [\because S' = S] \\
&= S \text{ diag. } (\lambda_1, \lambda_2, ..., \lambda_r, 0, 0, ..., 0)S \\
&= \text{diag.}(1, 1, ..., 1, -1, -1, ... -1, 0, 0, ..., 0)
\end{aligned}
$$

where 1 appears p times and -1 appears $r - p$ times.

Consider a real non-singular linear transformation $X = PY$ which reduces $X'AX$ to the form.

$$Y'(P'AP)Y = y_1^2 + y_2^2 + ... + y_p^2 - y_{p+1}^2 - y_{p+2}^2 - ... - y_r^2$$

REMARKS

- Two real quadratic forms $X'AX$ and $Y'BY$ in the same number of variables are real equivalent if and only if A and B have the same number of positive and negative eigenvalues.
- If $X'AX$ is a real quadratic form, the number of non-zero eigenvalues of A is equal to the rank of $X'AX$ and the number of positive eigenvalues of A is equal to the index of $X'AX$.

THEOREM 3. *Two real quadratic forms $X'AX$ and $Y'BY$ are orthogonally equivalent if and only if A and B have the same eigenvalues and these occur with the same multiplicities.*

Proof. Suppose A and B have the same eigenvalues. Let $\lambda_1, \lambda_2, ..., \lambda_n$ be the eigenvalues of A and B. If D is a diagonal matrix with $\lambda_1, \lambda_2, ..., \lambda_n$ as diagonal elements, then there exists orthogonal matrices P and Q such that

$$P'AP = D = Q'BQ$$
$$\Rightarrow \qquad B = (Q')^{-1}(P'AP)Q^{-1}$$
$$\Rightarrow \qquad B = (Q^{-1})'(P'AP)Q^{-1} \qquad [\because (Q^{-1})' = (Q')^{-1}]$$
$$\Rightarrow \qquad B = (PQ^{-1})'A(PQ^{-1})$$

Since P and Q are orthogonal, then PQ^{-1} is orthogonal, therefore $Y'BY$ is orthogonally equivalent to $X'AX$.

Conversely, suppose that $X'AX$ and $Y'BY$ are orthogonally equivalent, then there

exists an orthogonal matrix P such that
$$B = P'AP = P^{-1}AP$$
$\Rightarrow$ A and B are similar matrices.
$\Rightarrow$ A and B have the same eigenvalues with the same multiplicities.
Hence the theorem.

Solved Examples

Example 1. *Reduce each of the following quadratic forms in three variables to real canonical form and find its rank and signature. Also write in each case the linear transformation which reduces to normal form.*

(i) $x^2 + 2y^2 + 2z^2 - 2xy - 2yz + zx$

(ii) $x^2 - 2y^2 + 3z^2 - 4yz + 6zx$

(iii) $2x_1^2 + x_2^2 - 3x_3^2 - 8x_2x_3 - 4x_3x_1 + 12x_1x_2$

(iv) $6x_1^2 + 3x_2^2 + 14x_3^2 + 4x_2x_3 + 18x_3x_1 + 4x_1x_2$

Solution.   (i)   The given quadratic form is
$$X'AX = x^2 + 2y^2 + 2z^2 - 2xy - 2yz + zx$$

$$\therefore \qquad A = \begin{bmatrix} 1 & -1 & \dfrac{1}{2} \\ -1 & 2 & -1 \\ \dfrac{1}{2} & -1 & 2 \end{bmatrix}$$

We write
$$A = IAI$$

or
$$\begin{bmatrix} 1 & -1 & 1/2 \\ -1 & 2 & -1 \\ 1/2 & -1 & 2 \end{bmatrix} = \begin{bmatrix} 1 & 0 & 0 \\ 0 & 1 & 0 \\ 0 & 0 & 1 \end{bmatrix} A \begin{bmatrix} 1 & 0 & 0 \\ 0 & 1 & 0 \\ 0 & 0 & 1 \end{bmatrix}$$

Applying $R_2 \to R_2 + R_1$, we get
$$\begin{bmatrix} 1 & -1 & 1/2 \\ 0 & 1 & -1/2 \\ 1/2 & -1 & 2 \end{bmatrix} = \begin{bmatrix} 1 & 0 & 0 \\ 1 & 1 & 0 \\ 0 & 0 & 1 \end{bmatrix} A \begin{bmatrix} 1 & 0 & 0 \\ 0 & 1 & 0 \\ 0 & 0 & 1 \end{bmatrix}$$

Applying $C_2 \to C_2 + C_1$, we get
$$\begin{bmatrix} 1 & 0 & 1/2 \\ 0 & 1 & -1/2 \\ 1/2 & -1/2 & 2 \end{bmatrix} = \begin{bmatrix} 1 & 0 & 0 \\ 1 & 1 & 0 \\ 0 & 0 & 1 \end{bmatrix} A \begin{bmatrix} 1 & 1 & 0 \\ 0 & 1 & 0 \\ 0 & 0 & 1 \end{bmatrix}$$

Applying $R_3 \to R_3 - \dfrac{1}{2}R_1$, we get
$$\begin{bmatrix} 1 & 0 & 1/2 \\ 0 & 1 & -1/2 \\ 0 & -1/2 & 7/4 \end{bmatrix} = \begin{bmatrix} 1 & 0 & 0 \\ 1 & 1 & 0 \\ -1/2 & 0 & 1 \end{bmatrix} A \begin{bmatrix} 1 & 1 & 0 \\ 0 & 1 & 0 \\ 0 & 0 & 1 \end{bmatrix}$$

Applying $C_3 \to C_3 - \dfrac{1}{2}C_1$, we get
$$\begin{bmatrix} 1 & 0 & 0 \\ 0 & 1 & -1/2 \\ 0 & -1/2 & 7/4 \end{bmatrix} = \begin{bmatrix} 1 & 0 & 0 \\ 1 & 1 & 0 \\ -1/2 & 0 & 1 \end{bmatrix} A \begin{bmatrix} 1 & 1 & -1/2 \\ 0 & 1 & 0 \\ 0 & 0 & 1 \end{bmatrix}$$

Applying $R_3 \to R_3 + \dfrac{1}{2}R_2$, we get

$$\begin{bmatrix} 1 & 0 & 0 \\ 0 & 1 & -1/2 \\ 0 & 0 & 3/2 \end{bmatrix} = \begin{bmatrix} 1 & 0 & 0 \\ 1 & 1 & 0 \\ 0 & 1/2 & 1 \end{bmatrix} A \begin{bmatrix} 1 & 1 & -1/2 \\ 0 & 1 & 0 \\ 0 & 0 & 1 \end{bmatrix}$$

Applying $C_3 \to C_3 + \dfrac{1}{2}C_2$, we get

$$\begin{bmatrix} 1 & 0 & 0 \\ 0 & 1 & 0 \\ 0 & 0 & 3/2 \end{bmatrix} = \begin{bmatrix} 1 & 0 & 0 \\ 1 & 1 & 0 \\ 0 & 1/2 & 1 \end{bmatrix} A \begin{bmatrix} 1 & 1 & 0 \\ 0 & 1 & 1/2 \\ 0 & 0 & 1 \end{bmatrix}$$

Applying $R_3 \to \sqrt{\dfrac{2}{3}}\, R_3$, we get

$$\begin{bmatrix} 1 & 0 & 0 \\ 0 & 1 & 0 \\ 0 & 0 & \sqrt{3/2} \end{bmatrix} = \begin{bmatrix} 1 & 0 & 0 \\ 1 & 1 & 0 \\ 0 & 1/\sqrt{6} & \sqrt{2/3} \end{bmatrix} A \begin{bmatrix} 1 & 1 & 0 \\ 0 & 1 & 1/2 \\ 0 & 0 & 1 \end{bmatrix}$$

Applying $C_3 \to \sqrt{\dfrac{2}{3}}\, C_3$, we get

$$\begin{bmatrix} 1 & 0 & 0 \\ 0 & 1 & 0 \\ 0 & 0 & 1 \end{bmatrix} = \begin{bmatrix} 1 & 0 & 0 \\ 1 & 1 & 0 \\ 0 & 1/\sqrt{6} & \sqrt{2/3} \end{bmatrix} A \begin{bmatrix} 1 & 1 & 0 \\ 0 & 1 & 1/\sqrt{6} \\ 0 & 0 & \sqrt{2/3} \end{bmatrix}$$

$\Rightarrow$
$$D = P'AP$$

where
$$P = \begin{bmatrix} 1 & 1 & 0 \\ 0 & 1 & 1/\sqrt{6} \\ 0 & 0 & \sqrt{2/3} \end{bmatrix}$$

Now the real non-singular linear transformation is
$$X = PY$$

i.e.,
$$\begin{bmatrix} x \\ y \\ z \end{bmatrix} = \begin{bmatrix} 1 & 1 & 0 \\ 0 & 1 & 1/\sqrt{6} \\ 0 & 0 & \sqrt{2/3} \end{bmatrix} \begin{bmatrix} y_1 \\ y_2 \\ y_3 \end{bmatrix}$$

i.e.,
$$x = y_1 + y_2$$
$$y = y_2 + \frac{1}{\sqrt{6}} y_3$$
$$z = \sqrt{\frac{2}{3}} y_3$$

which reduces $X'AX$ to the normal form
$$Y'(P'AP)Y = y_1^2 + y_2^2 + y_3^2$$

The rank of $X'AX$ = number of non-zero terms in normal form
$$= 3$$

The signature of $X'AX$ = the excess of positive terms over the negative terms
$$= 3 - 0 = 3$$

The index of $X'AX$ = the number of positive terms
$$= 3$$

20

(ii) The given quadratic form is

$$X'AX = x^2 - 2y^2 + 3z^2 - 4yz + 6zx$$

$$= \begin{bmatrix} x & y & z \end{bmatrix} \begin{bmatrix} 1 & 0 & 0 \\ 0 & -2 & -2 \\ 3 & -2 & 3 \end{bmatrix} \begin{bmatrix} x \\ y \\ z \end{bmatrix}$$

$$\therefore \qquad A = \begin{bmatrix} 1 & 0 & 3 \\ 0 & -2 & -2 \\ 3 & -2 & 3 \end{bmatrix}$$

We write

$$A = IAI$$

or

$$\begin{bmatrix} 1 & 0 & 3 \\ 0 & -2 & -2 \\ 3 & -2 & 3 \end{bmatrix} = \begin{bmatrix} 1 & 0 & 0 \\ 0 & 1 & 0 \\ 0 & 0 & 1 \end{bmatrix} A \begin{bmatrix} 1 & 0 & 0 \\ 0 & 1 & 0 \\ 0 & 0 & 1 \end{bmatrix}$$

Applying $R_3 \to R_3 - 3R_1$, we get

$$\begin{bmatrix} 1 & 0 & 3 \\ 0 & -2 & -2 \\ 3 & -2 & -6 \end{bmatrix} = \begin{bmatrix} 1 & 0 & 0 \\ 0 & 1 & 0 \\ -3 & 0 & 1 \end{bmatrix} A \begin{bmatrix} 1 & 0 & 0 \\ 0 & 1 & 0 \\ 0 & 0 & 1 \end{bmatrix}$$

Applying $C_3 \to C_3 - 3C_1$, we get

$$\begin{bmatrix} 1 & 0 & 0 \\ 0 & -2 & -2 \\ 0 & -2 & -6 \end{bmatrix} = \begin{bmatrix} 1 & 0 & 0 \\ 0 & 1 & 0 \\ -3 & 0 & 1 \end{bmatrix} A \begin{bmatrix} 1 & 0 & -3 \\ 0 & 1 & 0 \\ 0 & 0 & 1 \end{bmatrix}$$

Applying $R_3 \to R_3 - R_2$, we get

$$\begin{bmatrix} 1 & 0 & 0 \\ 0 & -2 & -2 \\ 0 & 0 & -4 \end{bmatrix} = \begin{bmatrix} 1 & 0 & 0 \\ 0 & 1 & 0 \\ -3 & -1 & 1 \end{bmatrix} A \begin{bmatrix} 1 & 0 & -3 \\ 0 & 1 & 0 \\ 0 & 0 & 1 \end{bmatrix}$$

Applying $C_3 \to C_3 - C_2$, we get

$$\begin{bmatrix} 1 & 0 & 0 \\ 0 & -2 & 0 \\ 0 & 0 & -4 \end{bmatrix} = \begin{bmatrix} 1 & 0 & 0 \\ 0 & 1 & 0 \\ -3 & -1 & 1 \end{bmatrix} A \begin{bmatrix} 1 & 0 & -3 \\ 0 & 1 & -1 \\ 0 & 0 & 1 \end{bmatrix}$$

Applying $R_2 \to \dfrac{1}{\sqrt{2}} R_2$, we get

$$\begin{bmatrix} 1 & 0 & 0 \\ 0 & -\sqrt{2} & 0 \\ 0 & 0 & -4 \end{bmatrix} = \begin{bmatrix} 1 & 0 & 0 \\ 0 & 1/\sqrt{2} & 0 \\ -3 & -1 & 1 \end{bmatrix} A \begin{bmatrix} 1 & 0 & -3 \\ 0 & 1 & -1 \\ 0 & 0 & 1 \end{bmatrix}$$

Applying $C_2 \to \dfrac{1}{\sqrt{2}} C_2$, we get

$$\begin{bmatrix} 1 & 0 & 0 \\ 0 & -1 & 0 \\ 0 & 0 & -4 \end{bmatrix} = \begin{bmatrix} 1 & 0 & 0 \\ 0 & 1/\sqrt{2} & 0 \\ -3 & -1 & 1 \end{bmatrix} A \begin{bmatrix} 1 & 0 & -3 \\ 0 & 1/\sqrt{2} & -1 \\ 0 & 0 & 1 \end{bmatrix}$$

Applying $R_3 \to \dfrac{1}{\sqrt{4}} R_3$, we get

$$\begin{bmatrix} 1 & 0 & 0 \\ 0 & -1 & 0 \\ 0 & 0 & -\sqrt{4} \end{bmatrix} = \begin{bmatrix} 1 & 0 & 0 \\ 0 & 1/\sqrt{2} & 0 \\ -3/\sqrt{4} & -1/\sqrt{4} & 1/\sqrt{4} \end{bmatrix} A \begin{bmatrix} 1 & 0 & -3 \\ 0 & 1/\sqrt{2} & -1 \\ 0 & 0 & 1 \end{bmatrix}$$

Applying $C_3 \to \dfrac{1}{\sqrt{4}} C_3$, we get

$$\begin{bmatrix} 1 & 0 & 0 \\ 0 & -1 & 0 \\ 0 & 0 & -1 \end{bmatrix} = \begin{bmatrix} 1 & 0 & 0 \\ 0 & 1/\sqrt{2} & 0 \\ -3/2 & -1/2 & 1/2 \end{bmatrix} A \begin{bmatrix} 1 & 0 & -3/2 \\ 0 & 1/\sqrt{2} & -1/2 \\ 0 & 0 & 1/2 \end{bmatrix}$$

$\Rightarrow$ $D = P'AP$

where $P = \begin{bmatrix} 1 & 0 & -3/2 \\ 0 & 1/\sqrt{2} & -1/2 \\ 0 & 0 & 1/2 \end{bmatrix}$

Thus the real non-singular linear transformation is

$$X = PY$$

i.e., $\begin{bmatrix} x \\ y \\ z \end{bmatrix} = \begin{bmatrix} 1 & 0 & -3/2 \\ 0 & 1/\sqrt{2} & -1/2 \\ 0 & 0 & 1/2 \end{bmatrix} \begin{bmatrix} y_1 \\ y_2 \\ y_3 \end{bmatrix}$

i.e., $x = y_1 - \dfrac{3}{2} y_3$

$$y = \dfrac{1}{\sqrt{2}} y_2 - \dfrac{1}{2} y_3$$

$$z = \dfrac{1}{2} y_3$$

which reduces $X'AX$ to the normal form

$$Y'(P'AP)Y = \begin{bmatrix} y_1 & y_2 & y_3 \end{bmatrix} \begin{bmatrix} 1 & 0 & 0 \\ 0 & -1 & 0 \\ 0 & 0 & -1 \end{bmatrix} \begin{bmatrix} y_1 \\ y_2 \\ y_3 \end{bmatrix}$$

$$= y_1^2 - y_2^2 - y_3^2$$

The rank of $X'AX$ = number of non-zero terms in normal form
= 3

The signature of $X'AX$ = the excess of positive terms over the negative terms
= 1 − 2 = −1

The index of $X'AX$ = the number of positive terms in the normal form
= 1

(iii) The given quadratic form is

$$X'AX = 2x_1^2 + x_2^2 - 3x_3^2 - 8x_2 x_3 - 4x_3 x_1 + 12 x_1 x_2$$

$$= \begin{bmatrix} x_1 & x_2 & x_3 \end{bmatrix} \begin{bmatrix} 2 & 6 & -2 \\ 6 & 1 & -4 \\ -2 & -4 & -3 \end{bmatrix} \begin{bmatrix} x_1 \\ x_2 \\ x_3 \end{bmatrix}$$

$\therefore$ $A = \begin{bmatrix} 2 & 6 & -2 \\ 6 & 1 & -4 \\ -2 & -4 & -3 \end{bmatrix}$

We write

$$A = IAI$$

or $\begin{bmatrix} 2 & 6 & -2 \\ 6 & 1 & -4 \\ -2 & -4 & -3 \end{bmatrix} = \begin{bmatrix} 1 & 0 & 0 \\ 0 & 1 & 0 \\ 0 & 0 & 1 \end{bmatrix} A \begin{bmatrix} 1 & 0 & 0 \\ 0 & 1 & 0 \\ 0 & 0 & 1 \end{bmatrix}$

Applying $R_2 \rightarrow R_2 - 3R_1$, $C_2 \rightarrow C_2 - 3C_1$, $R_3 \rightarrow R_3 + R_1$, $C_3 \rightarrow C_3 + C_1$ we get

$$\begin{bmatrix} 2 & 0 & 0 \\ 0 & -17 & 2 \\ 0 & 2 & -5 \end{bmatrix} = \begin{bmatrix} 1 & 0 & 0 \\ -3 & 1 & 0 \\ 1 & 0 & 1 \end{bmatrix} A \begin{bmatrix} 1 & -3 & 1 \\ 0 & 1 & 0 \\ 0 & 0 & 1 \end{bmatrix}$$

Applying $R_3 \rightarrow R_3 + \dfrac{2}{17}R_2$, $C_3 \rightarrow C_3 + \dfrac{2}{17}C_2$, we get

$$\begin{bmatrix} 2 & 0 & 0 \\ 0 & -17 & 0 \\ 0 & 0 & -81/17 \end{bmatrix} = \begin{bmatrix} 1 & 0 & 0 \\ -3 & 1 & 0 \\ 11/17 & 2/17 & 1 \end{bmatrix} A \begin{bmatrix} 1 & -3 & 11/17 \\ 0 & 1 & 2/17 \\ 0 & 0 & 1 \end{bmatrix}$$

Applying $R_1 \rightarrow \dfrac{1}{\sqrt{2}} R_1$, $C_1 \rightarrow \dfrac{1}{\sqrt{2}} C_1$, $R_2 \rightarrow \dfrac{1}{\sqrt{17}} R_2$, $C_2 \rightarrow \dfrac{1}{\sqrt{17}} C_2$, $R_3 \rightarrow \sqrt{\dfrac{17}{81}} R_3$, $C_3 \rightarrow \sqrt{\dfrac{17}{81}} C_3$, we get

$$\begin{bmatrix} 1 & 0 & 0 \\ 0 & -1 & 0 \\ 0 & 0 & -1 \end{bmatrix} = \begin{bmatrix} 1/\sqrt{2} & 0 & 0 \\ -3/\sqrt{17} & 1/\sqrt{17} & 0 \\ 11\sqrt{17}/9 & 2\sqrt{17}/9 & 1\sqrt{17}/9 \end{bmatrix} A \begin{bmatrix} 1/\sqrt{2} & -3/\sqrt{17} & 11\sqrt{17}/9 \\ 0 & 1/\sqrt{17} & 2\sqrt{17}/9 \\ 0 & 0 & 1\sqrt{17}/9 \end{bmatrix}$$

$\Rightarrow \qquad\qquad D = P'AP$

where $\qquad\qquad P = \begin{bmatrix} 1/\sqrt{2} & -3/\sqrt{17} & 11\sqrt{17}/9 \\ 0 & 1/\sqrt{17} & 2\sqrt{17}/9 \\ 0 & 0 & 1\sqrt{17}/9 \end{bmatrix}$

Thus the real non-singular linear transformation is

$$X = PY$$

i.e., $\qquad \begin{bmatrix} x_1 \\ x_2 \\ x_3 \end{bmatrix} = \begin{bmatrix} 1/\sqrt{2} & -3/\sqrt{17} & 11\sqrt{17}/9 \\ 0 & 1/\sqrt{17} & 2\sqrt{17}/9 \\ 0 & 0 & 1\sqrt{17}/9 \end{bmatrix} \begin{bmatrix} y_1 \\ y_2 \\ y_3 \end{bmatrix}$

i.e., $\qquad\qquad x_1 = \dfrac{1}{\sqrt{2}} y_1 - \dfrac{3}{\sqrt{17}} y_2 + \dfrac{11}{9}\sqrt{17}\, y_3$

$$x_2 = \dfrac{1}{\sqrt{17}} y_2 + \dfrac{2\sqrt{17}}{9} y_3$$

$$x_3 = \dfrac{1\sqrt{17}}{9} y_3$$

These transformations reduce the given quadratic form $X'AX$ to the normal form

$$Y'(P'AP)Y = Y'DY = \begin{bmatrix} y_1 & y_2 & y_3 \end{bmatrix} \begin{bmatrix} 1 & 0 & 0 \\ 0 & -1 & 0 \\ 0 & 0 & -1 \end{bmatrix} \begin{bmatrix} y_1 \\ y_2 \\ y_3 \end{bmatrix}$$

$$= y_1^2 - y_2^2 - y_3^2$$

The rank of $X'AX$ = number of non-zero terms in normal form
$$= 3$$

The signature of $X'AX$ = the excess of the number of positive terms over the negative terms in the normal form
$$= 1 - 2 = -1$$

The index of $X'AX$ = the number of positive terms in the normal form

$$= 1$$

(iv) The given quadratic form is

$$X'AX = 6x_1^2 + 3x_2^2 + 14x_3^2 + 4x_2x_3 + 18x_3x_1 + 4x_1x_2$$

$$= \begin{bmatrix} x_1 & x_2 & x_3 \end{bmatrix} \begin{bmatrix} 6 & 2 & 9 \\ 2 & 3 & 2 \\ 9 & 2 & 14 \end{bmatrix} \begin{bmatrix} x_1 \\ x_2 \\ x_3 \end{bmatrix}$$

$\therefore$

$$A = \begin{bmatrix} 6 & 2 & 9 \\ 2 & 3 & 2 \\ 9 & 2 & 14 \end{bmatrix}$$

We write

$$A = IAI$$

or

$$\begin{bmatrix} 6 & 2 & 9 \\ 2 & 3 & 2 \\ 9 & 2 & 14 \end{bmatrix} = \begin{bmatrix} 1 & 0 & 0 \\ 0 & 1 & 0 \\ 0 & 0 & 1 \end{bmatrix} A \begin{bmatrix} 1 & 0 & 0 \\ 0 & 1 & 0 \\ 0 & 0 & 1 \end{bmatrix}$$

Applying $R_2 \to R_2 - \dfrac{1}{3} R_1$, $C_3 \to C_2 - \dfrac{1}{2} C_1$ and $R_3 \to R_3 - \dfrac{3}{2} R_1$,

$C_3 \to C_3 - \dfrac{3}{2} C_1$ we get

$$\begin{bmatrix} 6 & 0 & 0 \\ 0 & 7/3 & -1 \\ 0 & -1 & 1/2 \end{bmatrix} = \begin{bmatrix} 1 & 0 & 0 \\ -1/3 & 1 & 0 \\ -3/2 & 0 & 1 \end{bmatrix} A \begin{bmatrix} 1 & -1/3 & -3/2 \\ 0 & 1 & 3/7 \\ 0 & 0 & 1 \end{bmatrix}$$

Applying $R_3 \to R_3 + \dfrac{3}{7} R_2$, $C_3 \to C_3 + \dfrac{3}{7} C_1$, we get

$$\begin{bmatrix} 6 & 0 & 0 \\ 0 & 7/3 & 0 \\ 0 & -1 & 1/14 \end{bmatrix} = \begin{bmatrix} 1 & 0 & 0 \\ -1/3 & 1 & 0 \\ -23/14 & 3/7 & 1 \end{bmatrix} A \begin{bmatrix} 1 & -1/3 & -23/14 \\ 0 & 1 & 3/7 \\ 0 & 0 & 1 \end{bmatrix}$$

Applying $R_1 \to \dfrac{1}{\sqrt{6}} R_1$, $C_1 \to \dfrac{2}{\sqrt{6}} C_1$, $R_2 \to \sqrt{\dfrac{3}{7}} R_2$, $C_2 \to \sqrt{\dfrac{3}{7}} C_2$ and

$R_3 \to \dfrac{1}{\sqrt{14}} R_3$, $C_3 \to \dfrac{1}{\sqrt{14}} C_3$, we get

$$\begin{bmatrix} 1 & 0 & 0 \\ 0 & 1 & 0 \\ 0 & 0 & 1 \end{bmatrix} = \begin{bmatrix} \dfrac{1}{\sqrt{6}} & 0 & 0 \\ -\dfrac{1}{3}\sqrt{\dfrac{3}{7}} & \sqrt{\dfrac{3}{7}} & 0 \\ -\dfrac{23}{14}\dfrac{1}{\sqrt{14}} & \dfrac{3}{7}\dfrac{1}{\sqrt{14}} & 1 \end{bmatrix} A \begin{bmatrix} \dfrac{1}{\sqrt{6}} & -\dfrac{1}{3}\sqrt{\dfrac{3}{7}} & -\dfrac{23}{14}\dfrac{1}{\sqrt{14}} \\ 0 & \sqrt{\dfrac{3}{7}} & \dfrac{3}{7}\dfrac{1}{\sqrt{14}} \\ 0 & 0 & 1 \end{bmatrix}$$

$\Rightarrow$

$$D = P'AP$$

where

$$P = \begin{bmatrix} \dfrac{1}{\sqrt{6}} & -\dfrac{1}{3}\sqrt{\dfrac{3}{7}} & -\dfrac{23}{14}\dfrac{1}{\sqrt{14}} \\ 0 & \sqrt{\dfrac{3}{7}} & \dfrac{3}{7}\dfrac{1}{\sqrt{14}} \\ 0 & 0 & 1 \end{bmatrix}$$

Thus the real non-singular linear transformation is
$$X = PY$$

i.e.,
$$\begin{bmatrix} x_1 \\ x_2 \\ x_3 \end{bmatrix} = \begin{bmatrix} \dfrac{1}{\sqrt{6}} & -\dfrac{1}{3}\sqrt{\dfrac{3}{7}} & -\dfrac{23}{14}\dfrac{1}{\sqrt{14}} \\ 0 & \sqrt{\dfrac{3}{7}} & \dfrac{3}{7}\dfrac{1}{\sqrt{14}} \\ 0 & 0 & 1 \end{bmatrix} \begin{bmatrix} y_1 \\ y_2 \\ y_3 \end{bmatrix}$$

or
$$x_1 = \frac{1}{\sqrt{6}}y_1 - \frac{1}{3}\sqrt{\frac{3}{7}}y_2 - \frac{23}{14}\frac{1}{\sqrt{14}}y_3$$

$$x_2 = \sqrt{\frac{3}{7}}y_2 + \frac{3}{7}\frac{1}{\sqrt{14}}y_3$$

$$x_3 = y_3$$

These transformations reduce the given quadratic form $X'AX$ to the normal form
$$Y'(P'AP)Y = Y'DY = y_1^2 + y_2^2 + y_3^2$$

The rank of $X'AX$ = number of non-zero terms in normal form
$$= 3$$

The signature of $X'AX$ = 2(positive terms in normal form) − rank
$$= 2(3) - 3$$
$$= 6 - 3$$
$$= 3$$

The index of $X'AX$ = the number of positive terms in the normal form
$$= 3$$

Example 2. *Find an orthogonal matrix P that will diagonalize the real matrix*
$$A = \begin{bmatrix} 0 & 1 & 1 \\ 1 & 0 & -1 \\ 1 & -1 & 0 \end{bmatrix}$$

Interpret the result in terms of quadratic form.

Solution. The characteristic equation of A is given by
$$|A - \lambda I| = 0$$

or
$$\begin{bmatrix} -\lambda & 1 & 1 \\ 1 & -\lambda & -1 \\ 1 & -1 & -\lambda \end{bmatrix} = 0$$

Applying $C_1 \to C_1 + C_2$, we get
$$\begin{bmatrix} 1-\lambda & 1 & 1 \\ 1-\lambda & -\lambda & -1 \\ 0 & -1 & -\lambda \end{bmatrix} = 0$$

or $(1-\lambda)\begin{bmatrix} 1 & 1 & 1 \\ 1 & -\lambda & -1 \\ 0 & -1 & -\lambda \end{bmatrix} = 0$

or $(1 - \lambda)[\lambda^2 - 1 + \lambda - 1] = 0$

or $(1 - \lambda)^2(2 + \lambda) = 0$

Therefore, the eigenvalues of A are 1, 1 and –2.

Eigenvector corresponding to the eigen value 1

Let $X = \begin{bmatrix} x_1 \\ x_2 \\ x_3 \end{bmatrix} \neq 0$ be an eigenvector corresponding to the eigen value 1, then X

is a solution of the equation

$$(A - I)X = 0$$

or $\begin{bmatrix} -1 & 1 & 1 \\ 1 & -1 & -1 \\ 1 & -1 & -1 \end{bmatrix} \begin{bmatrix} x_1 \\ x_2 \\ x_3 \end{bmatrix} = \begin{bmatrix} 0 \\ 0 \\ 0 \end{bmatrix}$

Solving these equation, we have

$$-x_1 + x_2 + x_3 = 0 \qquad \qquad \qquad ...(1)$$

The two orthogonal solutions, which satisfy (1) are

$$X_1 = \begin{bmatrix} 1 \\ 0 \\ 1 \end{bmatrix} \text{ and } X_2 = \begin{bmatrix} 1 \\ 2 \\ -1 \end{bmatrix}$$

Thus, the two mutually orthogonal eigenvectors of A are

$$X_1 = \begin{bmatrix} 1 \\ 0 \\ 1 \end{bmatrix} \text{ and } X_2 = \begin{bmatrix} 1 \\ 2 \\ -1 \end{bmatrix}$$

Eigenvector corresponding to the eigen value –2

Let $X = \begin{bmatrix} x_1 \\ x_2 \\ x_3 \end{bmatrix} \neq 0$ be an eigen vector corresponding to the eigen value –2, then X

is a solution of the equation $(A + 2I)X = 0$

or $\begin{bmatrix} 2 & 1 & 1 \\ 1 & 2 & -1 \\ 1 & -1 & 2 \end{bmatrix} \begin{bmatrix} x_1 \\ x_2 \\ x_3 \end{bmatrix} = \begin{bmatrix} 0 \\ 0 \\ 0 \end{bmatrix}$

Applying $R_1 \leftrightarrow R_2$, we get

$\begin{bmatrix} 1 & 2 & -1 \\ 2 & 1 & 1 \\ 1 & -1 & 2 \end{bmatrix} \begin{bmatrix} x_1 \\ x_2 \\ x_3 \end{bmatrix} = \begin{bmatrix} 0 \\ 0 \\ 0 \end{bmatrix}$

Applying $R_2 \rightarrow R_2 - 2R_1, R_3 \rightarrow R_3 - R_1$, we get

$\begin{bmatrix} 1 & 2 & -1 \\ 0 & -3 & 3 \\ 0 & -3 & 3 \end{bmatrix} \begin{bmatrix} x_1 \\ x_2 \\ x_3 \end{bmatrix} = \begin{bmatrix} 0 \\ 0 \\ 0 \end{bmatrix}$

Applying $R_3 \rightarrow R_3 - R_2$, we get

$\begin{bmatrix} 1 & 2 & -1 \\ 0 & -3 & 3 \\ 0 & 0 & 0 \end{bmatrix} \begin{bmatrix} x_1 \\ x_2 \\ x_3 \end{bmatrix} = \begin{bmatrix} 0 \\ 0 \\ 0 \end{bmatrix}$

Solving these equation, we get

$$x_1 + 2x_2 - x_3 = 0$$
$$-3x_2 + 3x_3 = 0$$

Clearly, $x_1 = -1$, $x_2 = 1$, $x_3 = 1$ satisfy these equations, therefore, the eigenvector is

$$X_3 = \begin{bmatrix} -1 \\ 1 \\ 1 \end{bmatrix}$$

Thus, the required matrix P is a matrix whose column vectors are unit vectors which are scalar multiplies of X_1, X_2 and X_3.

$$\therefore \qquad P = \begin{bmatrix} 1/\sqrt{2} & 1/\sqrt{6} & -1/\sqrt{3} \\ 0 & 2/\sqrt{16} & 1/\sqrt{3} \\ 1/\sqrt{2} & -1/\sqrt{6} & 1/\sqrt{3} \end{bmatrix}$$

Now, we have

$$P'AP = D = \text{diag. } (1, 1, -2)$$

The quadratic form of the given symmetric matrix is

$$X'AX = 2x_1x_2 + 2x_1x_3 - 2x_2x_3$$

Thus, the orthogonal linear transformation $X = PY$ reduces the quadratic form $X'AX$ to the diagonal form

$$X'AX = Y'(P'AP)Y = Y'DY = y_1^2 + y_2^2 - 2y_3^2$$

The rank of $X'AX$ = number of non-zero eigenvalues of A

$$= 3$$

The signature of $X'AX$ = the excess of the number of positive eigenvalues over the number of negative eigenvalues of A

$$= 2 - 1 = 1$$

The diagonal form $y_1^2 + y_2^2 - 2y_3^2$ can be reduced to the normal form $z_1^2 + z_2^2 - z_3^2$.

Example 3. *Reduce the following quadratic form in to canonical form and find its rank and signature :*

$$x^2 + 4y^2 + 9z^2 + t^2 - 12yz + 6zx - 4xy - 2xt - 6zt$$

Solution. Let $X = \begin{bmatrix} x \\ y \\ z \\ t \end{bmatrix}$, then the given quadratic form can be written as

$$x^2 + 4y^2 + 9z^2 + t^2 - 12yz + 6zx - 4xy - 2xt - 6zt$$

$$= X' \begin{bmatrix} 1 & -2 & 3 & -1 \\ -2 & 4 & -6 & 0 \\ 3 & -6 & 9 & -3 \\ -1 & 0 & -3 & 1 \end{bmatrix} X = X'AX$$

$$\therefore \qquad A = \begin{bmatrix} 1 & -2 & 3 & -1 \\ -2 & 4 & -6 & 0 \\ 3 & -6 & 9 & -3 \\ -1 & 0 & -3 & 1 \end{bmatrix}$$

We write

$$A = IAI$$

$$\begin{bmatrix} 1 & -2 & 3 & -1 \\ -2 & 4 & -6 & 0 \\ 3 & -6 & 9 & -3 \\ -1 & 0 & -3 & 1 \end{bmatrix} = \begin{bmatrix} 1 & 0 & 0 & 0 \\ 0 & 1 & 0 & 0 \\ 0 & 0 & 1 & 0 \\ 0 & 0 & 0 & 1 \end{bmatrix} A \begin{bmatrix} 1 & 0 & 0 & 0 \\ 0 & 1 & 0 & 0 \\ 0 & 0 & 1 & 0 \\ 0 & 0 & 0 & 1 \end{bmatrix}$$

Applying $R_2 \rightarrow R_2 + 2R_1, C_2 \rightarrow C_2 + 2C_1;$
$R_3 \rightarrow R_3 + 3R_1, C_3 \rightarrow C_3 - 3C_1;$
$R_4 \rightarrow R_4 + R_1, C_4 \rightarrow C_4 + C_1$

We get

$$\begin{bmatrix} 1 & 0 & 0 & 0 \\ 0 & 0 & 0 & -2 \\ 0 & 0 & 0 & 0 \\ 0 & -2 & 0 & 0 \end{bmatrix} = \begin{bmatrix} 1 & 0 & 0 & 0 \\ 2 & 1 & 0 & 0 \\ -3 & 0 & 1 & 0 \\ 1 & 0 & 0 & 1 \end{bmatrix} A \begin{bmatrix} 1 & 2 & -3 & 1 \\ 0 & 1 & 0 & 0 \\ 0 & 0 & 1 & 0 \\ 0 & 0 & 0 & 1 \end{bmatrix}$$

Applying $R_2 \rightarrow R_2 + R_4, C_2 \rightarrow C_2 + C_4$, we get

$$\begin{bmatrix} 1 & 0 & 0 & 0 \\ 0 & -4 & 0 & -2 \\ 0 & 0 & 0 & 0 \\ 0 & -2 & 0 & 0 \end{bmatrix} = \begin{bmatrix} 1 & 0 & 0 & 0 \\ 3 & 1 & 0 & 1 \\ -3 & 0 & 1 & 0 \\ 1 & 0 & 0 & 1 \end{bmatrix} A \begin{bmatrix} 1 & 3 & -3 & 1 \\ 0 & 1 & 0 & 0 \\ 0 & 0 & 1 & 0 \\ 0 & 1 & 0 & 1 \end{bmatrix}$$

Applying $R_4 \rightarrow R_4 - \dfrac{1}{2}R_2, C_4 \rightarrow C_4 - \dfrac{1}{2}C_2$, we get

$$\begin{bmatrix} 1 & 0 & 0 & 0 \\ 0 & -4 & 0 & 0 \\ 0 & 0 & 0 & 0 \\ 0 & 0 & 0 & 1 \end{bmatrix} = \begin{bmatrix} 1 & 0 & 0 & 0 \\ 3 & 1 & 0 & 1 \\ -3 & 0 & 1 & 0 \\ -1/2 & -1/2 & 0 & 1/2 \end{bmatrix} A \begin{bmatrix} 1 & 3 & -3 & -1/2 \\ 0 & 1 & 0 & -1/2 \\ 0 & 0 & 1 & 0 \\ 0 & 1 & 0 & 1/2 \end{bmatrix}$$

Applying $R_2 \rightarrow \dfrac{1}{2}R_2, C_2 \rightarrow \dfrac{1}{2}C_2$, we get

$$\begin{bmatrix} 1 & 0 & 0 & 0 \\ 0 & -1 & 0 & 0 \\ 0 & 0 & 0 & 0 \\ 0 & 0 & 0 & 1 \end{bmatrix} = \begin{bmatrix} 1 & 0 & 0 & 0 \\ \dfrac{3}{2} & \dfrac{1}{2} & 0 & \dfrac{1}{2} \\ -3 & 0 & 1 & 0 \\ -\dfrac{1}{2} & -\dfrac{1}{2} & 0 & \dfrac{1}{2} \end{bmatrix} A \begin{bmatrix} 1 & \dfrac{3}{2} & -3 & -\dfrac{1}{2} \\ 0 & \dfrac{1}{2} & 0 & -\dfrac{1}{2} \\ 0 & 0 & 1 & 0 \\ 0 & \dfrac{1}{2} & 0 & \dfrac{1}{2} \end{bmatrix}$$

$\Rightarrow$ $D = P'AP$

where $P = \begin{bmatrix} 1 & 3/2 & -3 & -1/2 \\ 0 & 1/2 & 0 & -1/2 \\ 0 & 0 & 1 & 0 \\ 0 & 1/2 & 0 & 1/2 \end{bmatrix}$

Thus, the real non-singular linear transformation $X = PY$ reduces the given quadratic form $X'AX$ to the normal form

$$Y'(P'AP)Y = Y'DY = y_1^2 - y_2^2 + y_4^2$$

The rank of $X'AX$ = the number of non-zero terms in the normal form

$$= 3$$

The signature of $X'AX$ = 2(positive terms) – rank

$$= 2(2) - 3$$
$$= 4 - 3$$
$$= 1$$

🐝 EXERCISE 9.3

1. Find the rank of each of the following quadratic forms :

(i) $x^2 - 12xy - 4y^2$ (ii) $3x^2 + 2xy + 3y^2$

(iii) $x^2 - 2xy + y^2$

(iv) $x_1^2 - 2x_1x_2 + 2x_2^2$

(v) $4x_1^2 + x_2^2 - 8x_3^2 + 4x_1x_2 - 4x_1x_3 + 8x_2x_3$

2. Find a real non-singular linear transformation $X = PY$ which reduces the given real quadratic form $x^2 + 2y^2 + 3z^2 + 4xy + 4yz$ to real canonical form. Also find the rank and signature of the given quadratic form.

3. Reduce each of the following quadratic forms to real canonical form and find its rank and signature. Also write in each case the linear transformation which reduce the normal form.

(i) $x_1^2 + 2x_2^2 - 7x_3^2 - 4x_1x_2 + 8x_1x_3$

(ii) $(x_1 + x_2 + x_3)^2 + (x_2 + x_3)^2 + 4x_3^2$

(iii) $x_1^2 + 2x_2^2 + 3x_3^2 + 2x_2x_3 - 2x_3x_1 + 2x_1x_2$

(iv) $4x_1^2 + 9x_2^2 + 2x_3^2 + 8x_2x_3 - 6x_3x_1 + 6x_1x_2$

(v) $3x^2 + 3y^2 + 3z^2 - 2yz + 2zx + 2xy$

(vi) $2x^2 + 9y^2 + 2z^2 - 2yz + 2zx + 6xy$

(vii) $x^2 - 4y^2 + 6z^2 + 2xy - 4xz + 2w^2 - 6zw$

(viii) $x_1x_2 - 4x_1x_4 - 2x_2x_3 + 12x_3x_4$

4. Reduce the quadratic form $7x^2 - 8y^2 - 8z^2 - 2yz - 8zx + 8xy$ to canonical form by an orthogonal transformation and hence find the signature of the quadratic form.

───── **Answers** ─────

1. (i) rank = 2,   (ii) rank = 2,   (iii) rank = 1,   (iv) rank = 2,   (v) rank = 3,

2. $X = \begin{bmatrix} 1 & -1/\sqrt{2} & -2/\sqrt{5} \\ 0 & 1/\sqrt{2} & 1/\sqrt{5} \\ 0 & 0 & 1/\sqrt{5} \end{bmatrix} Y$, canonical form is $y_1^2 - y_2^2 + y_3^2$, rank = 3, signature = 1

3. (i) rank = 3, signature = 1   (ii) rank = 3, signature = 3   (iii) rank = 3, signature = 3

(iv) rank = 3, signature = 1   (v) rank = 3, signature = 3   (vi) rank = 3, signature = 3

(vii) rank = 4, signature = 0   (viii) rank = 4, signature = 0

9.21 CLASSIFICATION OF REAL SYMMETRIC MATRICES

Let A be a real symmetric matrix. Then

(i) the matrix A is said to be positive definite if its corresponding quadratic form $X'AX$ is positive definite.

(ii) the matrix A is said to be definite, semi-definite or indefinite if its corresponding quadratic form $X'AX$ is definite, semi-definite or indefinite, respectively.

(iii) the matrix A is said to be non-negative definite if its corresponding quadratic form $X'AX$ is non-negative definite.

THEOREM 1. *All real equivalent real quadratic forms have the same value class.*

Proof. Let $X'AX$ and $Y'BY$ be any two real equivalent real quadratic forms in the same number of variables.

Since $X'AX$ and $Y'BY$ are real equivalent, then there exists a real non-singular matrix P such that

$$P'AP = B$$

$\Rightarrow \qquad A = (P')^{-1}BP^{-1}$

$\Rightarrow \qquad A = (P^{-1})'BP^{-1}$

It follow that the real non-singular linear transformation $X = PY$ transforms the quadratic form $X'AX$ into the quadratic form $Y'BY$ and the non-singular linear transformation $Y' = P^{-1}X$ transforms the quadratic form $Y'BY$ to the quadratic form $X'AX$. The quadratic forms $X'AX$ and $Y'BY$ have the ranges of values. Therefore the vector Y for which $Y'BY$ has the same value as $X'AX$ has for the vector X is given by $Y = P^{-1}X$ and the vector X for which $X'AX$ has the same value as $Y'BY$ has for the vector Y is given by $X = PY$.

Now we have the five cases.

Case 1. *$X'AX$ is positive definite if and only if $Y'BY$ is positive definite.*

Suppose that $X'AX$ is positive definite, then

$$X'AX > 0, \text{ when } X \neq O$$

Since $X'AX$ and $Y'BY$ have the same ranges of values, therefore

$$X'AX > 0, \text{ when } X \neq O$$

$\Rightarrow \qquad\qquad Y'BY > 0 \text{ when } PY \neq O$

$\Rightarrow \qquad\qquad Y'BY > 0 \text{ when } Y \neq O$

$\Rightarrow \quad Y'BY$ is positive definite.

Conversely, suppose that $Y'BY$ is positive definite, then

$$Y'BY > 0 \text{ when } Y \neq O$$

Since $X'AX$ and $Y'BY$ have the same ranges of values, therefore

$$Y'BY > 0, \text{ when } Y \neq O$$

$\Rightarrow \qquad\qquad X'AX > 0 \text{ when } P^{-1}X \neq O$

$\Rightarrow \qquad\qquad X'AX > 0 \text{ when } X \neq O$

$\Rightarrow \quad X'AX$ is positive definite.

Case 2. *$X'AX$ is negative definite if and only $Y'BY$ is negative definite.*

$$X'AX < 0, \text{ when } X \neq O$$

$\Leftrightarrow \qquad\qquad (PY)'A(PY) < 0 \text{ when } PY \neq O$

$\Leftrightarrow \qquad\qquad Y'(P'AP)Y < 0 \text{ when } Y \neq O$

$\Leftrightarrow \qquad\qquad Y'BY < 0 \text{ when } Y \neq O$

Case 3. *$X'AX$ is positive semi-definite if and only if $Y'BY$ is positive semi-definite.*

$$X'AX \geq 0 \text{ for all } X$$

i.e., $\qquad\qquad X'AX = 0 \text{ for some } X \neq O$

$\Leftrightarrow \qquad\qquad (PY)'A(PY) < 0 \text{ for some } PY \neq O$

$\Leftrightarrow \qquad\qquad Y'(P'AP)Y = 0 \text{ for some } Y \neq O$

$\Leftrightarrow \qquad\qquad Y'BY = 0 \text{ for some } Y \neq O$

$\Leftrightarrow \qquad\qquad Y'BY \geq 0 \text{ for all } Y$

Case 4. *$X'AX$ is negative semi-definite if and only if $Y'BY$ is negative semi-definite.*

$$X'AX \leq 0 \text{ for some } X$$

$$X'AX = 0 \text{ for some } X \neq O$$

$\Leftrightarrow \qquad\qquad (PY)'A(PY) = 0 \text{ for some } PY \neq O$

$\Leftrightarrow \qquad\qquad Y'(P'AP)Y = 0 \text{ for some } Y \neq O$

$\Leftrightarrow \qquad\qquad Y'BY = 0 \text{ for some } Y \neq O$

$\Leftrightarrow \qquad\qquad Y'BY \leq 0 \text{ for all } Y.$

Case 5. *X'AX is indefinite if and only if Y'BY is indefinite.*

Since $X'AX$ and $Y'BY$ have the same ranges of values, therefore

$$X'AX > 0 \text{ for some } X$$
and
$$X'AX < 0 \text{ for some other } X$$
$$\Leftrightarrow$$
$$Y'BY > 0 \text{ for some } Y$$
and
$$Y'BY < 0 \text{ for some other } Y$$

Hence the theorem.

THEOREM 2. *Let $Q(X) = X'AX$ be a real quadratic form in n variables, of rank r and signature s, then*

   (*i*) *Q(X) is positive definite if and only if $s = r = n$.*

  (*ii*) *Q(X) is negative definite if and only if $-s = r = n$.*

 (*iii*) *Q(X) is positive semi-definite if and only if $s = r < n$.*

 (*iv*) *Q(X) is negative semi-definite if and only if $-s = r < n$, and*

  (*v*) *Q(X) is indefinite if and only if $|s| \neq r$.*

Proof. Suppose $X'AX$ is reduced to the real canonical form

$$Y'BY = y_1^2 + y_2^2 + \dots + y_p^2 - y_{p+1}^2 - y_{p+2}^2 - \dots - y_r^2$$

where p is the number of positive terms in the canonical form.

Since r and s are the rank and signature of $X'AX$ respectively, then $s = 2p - r$. Also $X'AX$ and $Y'BY$ are real equivalent real quadratic forms, therefore they have the same value class.

  (i) Suppose that $s = r = n$, then we have

$$2p = s + r = 2n$$
$$\Rightarrow \qquad p = n$$
$$\therefore \qquad Y'BY = y_1^2 + y_2^2 + \dots + y_n^2$$
$$\Rightarrow \quad Y'BY \text{ is positive definite.}$$
$$\Rightarrow \quad X'AX \text{ is positive definite.}$$
$$\Rightarrow \quad Q(X) \text{ is positive definite.}$$

Conversely, suppose that $Q(X) = X'AX$ is positive definite, then $Y'BY$ is also positive definite, so we have

$$\Rightarrow \qquad Y'BY = y_1^2 + y_2^2 + \dots + y_n^2$$
$$\Rightarrow \qquad r = n, p = n$$
$$\Rightarrow \qquad s = 2p - r = 2n - n = n$$
$$\Rightarrow \qquad s = r = n$$

 (ii) Suppose that $-s = r = n$, then

$$2p = s + r = 0$$
$$\Rightarrow \qquad p = 0$$
$$\Rightarrow \qquad Y'BY = -y_1^2 - y_2^2 - \dots - y_n^2$$
$$\Rightarrow \quad Y'BY \text{ is negative definite.}$$
$$\Rightarrow \quad X'AX \text{ is negative definite.}$$
$$\Rightarrow \quad Q(X) \text{ is negative definite.}$$

Conversely, suppose that $Q(X) = X'AX$ is negative definite, then $Y'BY$ is also negative definite.

$$\Rightarrow \qquad Y'BY = -y_1^2 - y_2^2 - \ldots - y_n^2$$
$$\Rightarrow \qquad r = n, p = 0$$
$$\Rightarrow \qquad s = 2p - r = 0 - n = -n$$
$$\Rightarrow \qquad -s = r = n$$

(iii) Suppose that $s = r < n$, then

$$2p = r + s = 2r$$
$$\Rightarrow \qquad p = r < n$$
$$\Rightarrow \qquad Y'BY = y_1^2 + y_2^2 + \ldots + y_n^2$$

Since $r < n$, therefore there is a positive semi-definite form in n variable, so that $Y'BY$ is positive semi-definite, thus $Q(X)$ is also positive semi-definite.

Conversely, suppose that $X'AX$ is positive definite, then $Y'BY$ is also positive semi-definite in n variables, therefore, we have

$$Y'BY = y_1^2 + y_2^2 + \ldots + y_r^2, \text{ where } r < n$$

It follows that $p = r < n$ so that

$$s = 2p - r = 2r - r = r$$

thus, $\qquad s = r < n.$

(iv) Suppose that $-s = r < n$, then

$$2p = s + r = 0$$
$$\Rightarrow \qquad p = 0$$

$\therefore \qquad X'AX$ is reduced to a canonical form.

$Y'BY$ in which there is no positive terms.

$$\therefore \qquad Y'BY = -y_1^2 - y_2^2 - \ldots - y_r^2 \qquad\qquad [\because r < n]$$
$$\Rightarrow \qquad Y'BY \text{ is negative semi-definite}$$
$$\Rightarrow \qquad X'AX \text{ is also negative semi-definite.}$$

Conversely, suppose that $X'AX$ is negative semi-definite, then its canonical form $Y'BY$ is also negative semi-definite in n variable.

$$Y'BY = -y_1^2 - y_2^2 - \ldots - y_r^2, \text{ where } r < n$$
$$\Rightarrow \qquad \text{There is no positive term in } Y'BY$$
$$\Rightarrow \qquad p = 0$$
$$\therefore \qquad s = 2p - r = 0 - r = -r$$

thus, $\qquad -s = r < n.$

(v) Suppose that $|s| \neq r$

$$\therefore \qquad |s| \neq r \Rightarrow |2p - r| \neq r$$
$$\Rightarrow \qquad p \neq 0 \text{ and } p \neq r$$
$$\Rightarrow \qquad 0 < p < r$$
$$\Rightarrow \qquad Y'BY \text{ has positive as well negative terms.}$$
$$\Rightarrow \qquad Y'BY \text{ is indefinite.}$$
$$\Rightarrow \qquad X'AX \text{ is indefinite.}$$

Conversely, suppose $X'AX$ is indefinite, then its canonical form $Y'BY$ is also indefinite, so that $Y'BY$ has positive as well negative terms, thus $|s| \neq r$.

Theorem 3. *A real quadratic form $Q(X) = X'AX$ in n variables is positive definite if and only if all the eigenvalues of A are positive.*

Proof. Since the matrix A is real symmetric matrix of order n, then there exists an orthogonal matrix P such that

$$P^{-1}AP = P'AP = D = \text{diag. } (\lambda_1, \lambda_2, ..., \lambda_n)$$

where $\lambda_1, \lambda_2, ..., \lambda_n$ are the eigenvalues of A.

Now the real non-singular linear transformation $X = PY$ reduces $X'AX$ to the canonical form

$$X'AX = (PY)'A(PY) = Y'(P'AP)Y = Y'DY$$

$$= \lambda_1 y_1^2 + \lambda_2 y_2^2 + ... + \lambda_n y_n^2$$

Suppose that $\lambda_1, \lambda_2, ..., \lambda_n$ are all positive;

$$\lambda_1 y_1^2 + \lambda_2 y_2^2 + ... + \lambda_n y_n^2 > 0 \text{ when } Y \neq 0$$

$\Rightarrow \qquad\qquad X'AX > 0 \text{ when } PY \neq 0$

$\Rightarrow \qquad\qquad X'AX > 0 \text{ when } X \neq 0$

$\Rightarrow \quad X'AX$ is positive definite.

Conversely, suppose that $X'AX$ is positive definite

$\therefore \qquad\qquad X'AX > 0 \text{ when } X \neq 0$

$\Rightarrow \lambda_1 y_1^2 + \lambda_2 y_2^2 + ... + \lambda_n y_n^2 > 0 \text{ when } Y \neq 0$

$\Rightarrow \qquad\qquad \lambda_1, \lambda_2, ..., \lambda_n > 0$

$\therefore \quad$ All the eigenvalues of A are positive.

Hence the theorem.

REMARKS

- A real symmetric matrix is positive definite iff all its eigenvalues are positive.
- A real quadratic form $Q(X) = X'AX$ in n variables is,
  (i) negative definite if and only if all the eigenvalues of A are negative,
  (ii) positive semi-definite if and only if all the eigenvalues of A are non-negative.
  (iii) negative semi-definite if and only if all the eigenvalues of A are non-positive.
  (iv) indefinite if and only if A has positive as well negative eigenvalues.

THEOREM 4. *A positive definite real symmetric matrix is non-singular.*

Proof. Let A be a positive definite real symmetric matrix, then its all eigenvalues are positive.

Since A is real symmetric matrix, then there exists an orthogonal matrix P such that

$$P^{-1}AP = D = \text{diag. } (\lambda_1, \lambda_2, ..., \lambda_n)$$

where $\lambda_1, \lambda_2, ..., \lambda_n$ are the eigenvalues of A.

Since $\lambda_i > 0$ $(i = 1, 2, ..., n)$, then D is non-singular.

Now $\qquad\qquad P^{-1}AP = D$

$\Rightarrow \qquad\qquad A = PDP^{-1}$

$\Rightarrow \quad A$ is non-singular.

THEOREM 5. *A real symmetric matrix A is positive definite if and only if there exists a non-singular matrix Q such that*

$$A = Q'Q$$

Proof. Suppose that A is positive definite, then there exists an orthogonal matrix P such that

$$P^{-1}AP = P'AP = D = \text{diag.} (\lambda_1, \lambda_2, ..., \lambda_n)$$

where $\lambda_1, \lambda_2, ..., \lambda_n$ are the eigenvalues of A.

Since A is positive, therefore $\lambda_i > 0$ $(i = 1, 2, ..., n)$

Let
$$D_1 = \text{diag.} \left(\sqrt{\lambda_1}, \sqrt{\lambda_2}, ..., \sqrt{\lambda_n}\right), \text{ then}$$
$$D_1^2 = D \text{ and } D_1' = D_1$$

Now
$$P^{-1}AP = D$$
$$\Rightarrow \qquad A = PDP^{-1}$$
$$\Rightarrow \qquad A = P(D_1^2)P^{-1}$$
$$\Rightarrow \qquad A = PD_1D_1P^{-1}$$
$$\Rightarrow \qquad A = (PD_1)(D_1P') \qquad\qquad [\because P^{-1} = P']$$
$$\Rightarrow \qquad A = (PD_1)(PD_1)' \qquad\qquad [\because D_1' = D_1]$$

Let $Q = (PD_1)'$, then
$$A = Q'Q$$

where Q is non-singular being the product of non-singular matrices P and D_1.

Conversely, suppose that $A = Q'Q$, where Q is non-singular matrix.

Now, let $Y = QX$, then $Y \neq 0$ when $X \neq 0$

$$\therefore \qquad X'AX = X(Q'Q)X \qquad\qquad [\because A = Q'Q]$$
$$= (QX)'(QX)$$
$$= Y'Y > 0 \text{ when } Y \neq 0$$
$$\Rightarrow \quad X'AX > 0 \text{ when } X \neq 0$$
$$\therefore \quad X'AX \text{ is positive definite, hence } A \text{ is positive definite.}$$

THEOREM 6. *Every real non-singular matrix A can be written as a product $A = PS$, where S is a positive definite symmetric matrix and P is orthogonal.*

Proof. Since A is real non-singular matrix, then by theorem 6, $A'A$ is a positive definite, real symmetric matrix.

As $A'A$ is a real symmetric matrix of order n so that there exists an orthogonal matrix Q such that

$$Q^{-1}(A'A) = Q'(A'A)Q = D = \text{diag.} (\lambda_1, \lambda_2, ..., \lambda_n)$$

where $\lambda_1, \lambda_2, ..., \lambda_n$ are the eigenvalues of $A'A$. Since $A'A$ is positive definite, then $\lambda_i > 0$ $(i = 1, 2, ..., n)$

Let
$$D_1 = \text{diag.} \left(\sqrt{\lambda_1}, \sqrt{\lambda_2}, ..., \sqrt{\lambda_n}\right), \text{ then}$$
$$D_1^2 = D \text{ and } D_1' = D_1$$

Let us consider $S = QD_1Q'$, then $S' = S$.

Since S is similar to D_1 therefore S is positive definite so that S will have positive eigenvalues.

Also,
$$S^2 = (QD_1Q')(QD_1Q')$$

$$= QD_1(Q'Q)D_1Q'$$
$$= QD_1^2Q'$$

$$[\because Q \text{ is orthogonal} \Rightarrow Q'Q = I, \text{ the identity matrix}]$$

$$= QDQ'$$
$$= QDQ^{-1} \qquad\qquad\qquad [\because Q' = Q^{-1}]$$
$$= Q(Q^{-1}A'AQ)Q^{-1}$$
$$= A'A$$

Let us take $P = AS^{-1}$,

Now
$$(P'P) = (AS^{-1})'(AS^{-1})$$
$$= (S^{-1})'A'AS^{-1}$$
$$= (S')^{-1}A'AS^{-1}$$
$$= S^{-1}A'AS^{-1} \qquad\qquad\qquad [\because S' = S]$$
$$= S^{-1}S^2S^{-1} \qquad\qquad\qquad [\because S^2 = A'A]$$
$$= (S^{-1}S)(S^{-1}S)$$
$$= (I)(I)$$
$$= I$$

$\therefore$ P is orthogonal matrix.

Now
$$P = AS^{-1}$$
$$\Rightarrow \qquad A = PS$$

Hence A can be written as a product PS where S is a positive definite real symmetric matrix and P is orthogonal.

REMARK

- Here the decomposition $A = PS$ is called the polar factorization of A.

9.22 POSITIVE-DEFINITENESS OF A QUADRATIC FORM X´AX IN TERMS OF LEADING PRINCIPAL MINORS OF A

Let $A = [a_{ij}]_{n \times n}$ be a matrix of order $n \times n$. Then the leading principal minors of A are

$$A_1 = a_{11}, A_2 = \begin{vmatrix} a_{11} & a_{12} \\ a_{21} & a_{22} \end{vmatrix}, A_3 = \begin{vmatrix} a_{11} & a_{12} & a_{13} \\ a_{21} & a_{22} & a_{23} \\ a_{31} & a_{32} & a_{33} \end{vmatrix}$$

$$A_4 = \begin{vmatrix} a_{11} & a_{12} & a_{13} & a_{14} \\ a_{21} & a_{22} & a_{23} & a_{24} \\ a_{31} & a_{32} & a_{33} & a_{34} \\ a_{41} & a_{42} & a_{43} & a_{44} \end{vmatrix}, \dots A_n = \begin{vmatrix} a_{11} & a_{12} & \cdots & a_{1n} \\ a_{21} & a_{22} & \cdots & a_{2n} \\ \vdots & & & \\ a_{n1} & a_{n2} & \cdots & a_{nn} \end{vmatrix}$$

Before providing the main theorem we first prove the following lemma :

Lemma.   *Let $X'AX$ be a positive definite real quadratic form, then*

$$det (A) = |A| > 0$$

Proof.   Since $X'AX$ be a positive definite real quadratic form, then there exists a real non-singular matrix P such that

$$P'AP = I$$

where I is the unit matrix of order $n \times n$.

$$\therefore \qquad |P'AP| = |I| = 1$$

$$\Rightarrow \qquad |P'|\,|A|\,|P| = 1$$

$$\Rightarrow \qquad |A| = \frac{1}{|P'|\,|P|}$$

$$\Rightarrow \qquad |A| = \frac{1}{|P|^2} \qquad\qquad\qquad [\because |P'| = |P|]$$

$$\Rightarrow \qquad |A| > 0$$

THEOREM. *A necessary and sufficient condition for a real quadratic form $X'AX$ to be positive definite is, that the leading principal minors of A are all positive.*

Proof. **Necessary condition :** Suppose that $X'AX$ is positive definite in n variables. Let m be any positive integer such that $m \le n$.

Let us take

$$X = \begin{bmatrix} x_1 \\ x_2 \\ \vdots \\ x_m \\ 0 \\ \vdots \\ 0 \end{bmatrix}_{n \times 1}$$

Then $X'AX$ is a positive definite in m variable $x_1, x_2, ..., x_m$. Therefore, the determinant of the matrix in new quadratic form is the leading principal minor of A of order $m \times m$, thus by above lemma this leading principal minor of order $m \times m$ is positive.

Hence every leading principal minor of A is positive.

Sufficient Condition : Suppose that the leading principal minors of A are all positive.

We shall prove that $X'AX$ is positive definite by induction hypothesis on n.

If $n = 1$, then

$$X'AX = a_{11}x_1^2$$

Since $a_{11} > 0$, then $\qquad X'AX = a_{11}x_1^2 > 0$

Now by induction hypothesis, the theorem is true for quadratic form in k variables and we then shall prove that the theorem is true for quadratic form in $k + 1$ variables.

Let S be any real symmetric matrix of order $k + 1$ and the leading principal minors of S be all positive, then we write

$$S = \begin{bmatrix} B & B_1 \\ B_1' & \alpha \end{bmatrix}$$

where B is a real symmetric matrix of order k and B_1 is an $k \times 1$ column matrix and α is any real number.

By given hypothesis, the leading principal minors of S are all positive, so that $|S|$ and the leading principal minors of B are all positive and B is a symmetric matrix of order k, thus by induction hypothesis the quadratic form corresponding to B

is positive definite, therefore, there exists a real non-singular matrix P such that
$$P'BP = I_k$$
where I_k is the unit matrix of order k.

As $|B| > 0$ so that B is non-singular, therefore, B^{-1} exists.

Let $C = -B^{-1}B_1$, then C is a column matrix of order $k \times 1$, since B_1 is a column matrix of order $k \times 1$.

Now
$$C' = -(B^{-1}B_1)' = -B_1'(B^{-1})' = -B_1'(B')^{-1}$$
But $B' = B$, then we have
$$C' = -B_1'B^{-1}$$

Now we have
$$\begin{bmatrix} P' & O \\ C' & 1 \end{bmatrix} S \begin{bmatrix} P & C \\ O & 1 \end{bmatrix}$$
$$= \begin{bmatrix} P' & O \\ C' & 1 \end{bmatrix} \begin{bmatrix} B & B_1 \\ B_1' & \alpha \end{bmatrix} \begin{bmatrix} P & C \\ O & 1 \end{bmatrix}$$
$$= \begin{bmatrix} P' & O \\ C' & 1 \end{bmatrix} \begin{bmatrix} BP & BC + B_1 \\ B_1'P & B_1'C + \alpha \end{bmatrix}$$
$$= \begin{bmatrix} P'BP & P'(BC + B_1) \\ C'BP + B_1'P & C'(BC + B_1) + B_1'C + \alpha \end{bmatrix}$$
$$= \begin{bmatrix} P'BP & P'BC + P'B_1 \\ C'BP + B_1'P & C'BC + C'B_1 + B_1'C + \alpha \end{bmatrix}$$
$$= \begin{bmatrix} I_K & O \\ O & B_1'C + \alpha \end{bmatrix} \qquad \left[\because C = -B'B_1, C' = -B_1'B^{-1} \right]$$

Taking determinant on both sides, we have
$$|P'| |S| |P| = |I_k| |B_1'C + \alpha|$$
$$\Rightarrow \quad |P|^2 |S| = B_1'C + \alpha \quad [\because |P'| = |P| \text{ and } B_1'C + \alpha \text{ is a matrix of order } 1 \times 1]$$
As $|S| > 0$ and $|P| \neq 0$, then $B_1'C + \alpha$ is poositive.

Let $B_1'C + \alpha = \lambda^2$, λ is a real number.
$$\therefore \quad \begin{bmatrix} P' & O \\ C' & 1 \end{bmatrix} S \begin{bmatrix} P & C \\ O & 1 \end{bmatrix} = \begin{bmatrix} I_K & O \\ O & \lambda^2 \end{bmatrix}$$
Pre-multiplying and post-multiplying both sides by $\begin{bmatrix} I_K & O \\ O & \lambda^2 \end{bmatrix}$, we get

$$\begin{bmatrix} I_k & O \\ O & \lambda^{-1} \end{bmatrix} \begin{bmatrix} P' & O \\ C' & 1 \end{bmatrix} S \begin{bmatrix} P & C \\ O & 1 \end{bmatrix} \begin{bmatrix} I_k & O \\ O & \lambda^{-1} \end{bmatrix} = I_{k+1}$$

Let us take
$$Q = \begin{bmatrix} P & C \\ O & 1 \end{bmatrix} \begin{bmatrix} I_k & O \\ O & \lambda^{-1} \end{bmatrix}$$

Then
$$Q' = \begin{bmatrix} I_k & O \\ O & \lambda^{-1} \end{bmatrix} \begin{bmatrix} P' & O \\ C' & 1 \end{bmatrix}$$

$$\therefore \quad Q'SQ = I_{k+1}$$
$$\Rightarrow \quad \text{S is congruent to } I_{k+1}.$$

$\Rightarrow$ the quadratic form corresponding to S is positive definite.

Hence by induction hypothesis, the theorem is proved.

Solved Examples

Example 1. *Prove that the quadratic form*

$$6x_1^2 + 49x_2^2 + 51x_3^2 - 82x_2x_3 + 20x_3x_1 - 4x_1x_2$$

in three variables is positive definite.

Solution. The matrix of the given quadratic form is

$$A = \begin{bmatrix} 6 & -2 & 10 \\ -2 & 49 & -41 \\ 10 & -41 & 51 \end{bmatrix}$$

The leading princpal minors of A are

$$A_1 = 6,$$

$$A_2 = \begin{vmatrix} 6 & -2 \\ -2 & 49 \end{vmatrix} = 294 - 4 = 290$$

$$A_3 = \begin{vmatrix} 6 & -2 & 10 \\ -2 & 49 & -41 \\ 10 & -41 & 51 \end{vmatrix}$$

Applying $R_1 \to R_1 + 3R_2, R_3 \to R_3 + 5R_2$

$$A_3 = \begin{vmatrix} 0 & 145 & -113 \\ -2 & 49 & -41 \\ 0 & 204 & -154 \end{vmatrix}$$

$$= 2\begin{vmatrix} 145 & -113 \\ 204 & -154 \end{vmatrix} = 2(145 \times -154 + 113 \times 204) = 1444$$

Clearly $A_1 > 0$, $A_2 > 0$ and $A_3 > 0$, the leading principal minors of A are all positive, hence the given quadratic form is positive definite.

Example 2. *Prove that the quadratic form*

$$X'AX = 6x_1^2 + 3x_2^2 + 14x_3^2 + 4x_2x_3 + 18x_3x_1 + 4x_1x_2$$

in three variables is positive definite.

Solution. The matrix of the given quadratic form is

$$A = \begin{bmatrix} 6 & 2 & 9 \\ 2 & 3 & 2 \\ 9 & 2 & 14 \end{bmatrix}$$

We write

$$A = I_3 A I_3$$

or

$$\begin{bmatrix} 6 & 2 & 9 \\ 2 & 3 & 2 \\ 9 & 2 & 14 \end{bmatrix} = \begin{bmatrix} 1 & 0 & 0 \\ 0 & 1 & 0 \\ 0 & 0 & 1 \end{bmatrix} A \begin{bmatrix} 1 & 0 & 0 \\ 0 & 1 & 0 \\ 0 & 0 & 1 \end{bmatrix}$$

Applying $R_2 \to R_2 - \dfrac{1}{3}R_1, C_2 \to C_2 - \dfrac{1}{3}C_1$ and $R_3 \to R_3 - \dfrac{3}{2}R_1, C_3 \to C_3 - \dfrac{3}{2}C_1,$ we get

$$\begin{bmatrix} 6 & 0 & 0 \\ 0 & 7/3 & -1 \\ 0 & -1 & 1/2 \end{bmatrix} = \begin{bmatrix} 1 & 0 & 0 \\ -1/3 & 1 & 0 \\ -3/2 & 0 & 1 \end{bmatrix} A \begin{bmatrix} 1 & -1/3 & -3/2 \\ 0 & 1 & 0 \\ 0 & 0 & 1 \end{bmatrix}$$

Applying $R_3 \to R_3 + \dfrac{3}{7} R_2, C_3 \to C_3 + \dfrac{3}{7} C_2$, we get

$$\begin{bmatrix} 6 & 0 & 0 \\ 0 & 7/3 & 0 \\ 0 & 0 & 1/14 \end{bmatrix} = \begin{bmatrix} 1 & 0 & 0 \\ -1/3 & 1 & 0 \\ -23/14 & 3/7 & 1 \end{bmatrix} A \begin{bmatrix} 1 & -1/3 & -23/14 \\ 0 & 1 & 3/7 \\ 0 & 0 & 1 \end{bmatrix}$$

Applying $R_1 \to \dfrac{1}{\sqrt{6}} R_1, C_1 \to \dfrac{1}{\sqrt{6}} C_1; R_2 \to \sqrt{\dfrac{3}{7}} R_2, C_2 \to \sqrt{\dfrac{3}{7}} C_2$ and

$R_3 \to \dfrac{1}{\sqrt{14}} R_3, C_3 \to \dfrac{1}{\sqrt{14}} C_3$, we get

$$\begin{bmatrix} 1 & 0 & 0 \\ 0 & 1 & 0 \\ 0 & 0 & 1 \end{bmatrix} = \begin{bmatrix} \dfrac{1}{\sqrt{6}} & 0 & 0 \\ -\dfrac{1}{3}\sqrt{\dfrac{3}{7}} & \sqrt{\dfrac{3}{7}} & 0 \\ -\dfrac{23}{14}\dfrac{1}{\sqrt{14}} & \dfrac{3}{7}\dfrac{1}{\sqrt{14}} & 1 \end{bmatrix} A \begin{bmatrix} \dfrac{1}{\sqrt{6}} & -\dfrac{1}{3}\sqrt{\dfrac{3}{7}} & -\dfrac{23}{14}\dfrac{1}{\sqrt{14}} \\ 0 & \sqrt{\dfrac{3}{7}} & \dfrac{3}{7}\dfrac{1}{\sqrt{14}} \\ 0 & 0 & 1 \end{bmatrix}$$

Clearly the given quadratic form $X'AX$ is reduced to the normal form

$$Y'(P'AP)Y = y_1^2 + y_2^2 + y_3^2$$

∴   The rank of $X'AX = r = 3$

The signature of $X'AX = s = 3$

∴ $s = r = n = 3$

Hence the given quadratic form is positive definite.

Example 3. *Prove that the quadratic form*

$$X'AX = 2x_1^2 + x_2^2 - 3x_3^2 - 8x_2x_3 - 4x_3x_1 + 12x_1x_2$$

in three variables is indefinite.

Solution. The matrix of the given quadratic form is

$$A = \begin{bmatrix} 2 & 6 & -2 \\ 6 & 1 & -4 \\ -2 & -4 & -3 \end{bmatrix}$$

We write

$$A = I_3 A I_3$$

or

$$\begin{bmatrix} 2 & 6 & -2 \\ 6 & 1 & -4 \\ -2 & -4 & -3 \end{bmatrix} = \begin{bmatrix} 1 & 0 & 0 \\ 0 & 1 & 0 \\ 0 & 0 & 1 \end{bmatrix} A \begin{bmatrix} 1 & 0 & 0 \\ 0 & 1 & 0 \\ 0 & 0 & 1 \end{bmatrix}$$

Applying $R_2 \to R_2 - 3R_1, C_2 \to C_2 - 3C_1$ and

$R_3 \to R_3 + R_1, C_3 \to C_3 + C_1$, we get

$$\begin{bmatrix} 2 & 0 & 0 \\ 0 & -17 & 2 \\ 0 & 2 & -5 \end{bmatrix} = \begin{bmatrix} 1 & 0 & 0 \\ -3 & 1 & 0 \\ 1 & 0 & 1 \end{bmatrix} A \begin{bmatrix} 1 & -3 & 1 \\ 0 & 1 & 0 \\ 0 & 0 & 1 \end{bmatrix}$$

Applying $R_3 \rightarrow R_3 + \dfrac{2}{17}R_2, C_3 \rightarrow C_3 + \dfrac{2}{17}C_2$, we get

$$\begin{bmatrix} 2 & 0 & 0 \\ 0 & -17 & 0 \\ 0 & 0 & -81/17 \end{bmatrix} = \begin{bmatrix} 1 & 0 & 0 \\ -3 & 1 & 0 \\ 11/17 & 2/17 & 1 \end{bmatrix} A \begin{bmatrix} 1 & -3 & 11/17 \\ 0 & 1 & 2/17 \\ 0 & 0 & 1 \end{bmatrix}$$

Applying $R_1 \rightarrow \dfrac{1}{\sqrt{2}}R_1, C_1 \rightarrow \dfrac{1}{\sqrt{2}}C_1; R_2 \rightarrow \dfrac{1}{\sqrt{17}}R_2, C_2 \rightarrow \dfrac{1}{\sqrt{17}}C_2$ and

$$R_3 \rightarrow \sqrt{\dfrac{17}{81}}R_3, C_3 \rightarrow \sqrt{\dfrac{17}{81}}C_3 \text{, we get}$$

$$\begin{bmatrix} 1 & 0 & 0 \\ 0 & -1 & 0 \\ 0 & 0 & -1 \end{bmatrix} = P'AP$$

where

$$P = \begin{bmatrix} \dfrac{1}{\sqrt{2}} & -\dfrac{3}{\sqrt{17}} & \dfrac{11}{17}\sqrt{\dfrac{17}{81}} \\ 0 & \dfrac{1}{\sqrt{17}} & \dfrac{2}{17}\sqrt{\dfrac{17}{81}} \\ 0 & 0 & \sqrt{\dfrac{17}{81}} \end{bmatrix}$$

Clearly the given quadratic form $X'AX$ is reduced to the normal form
$$y_1^2 - y_2^2 - y_3^2$$

$\therefore$ The rank of $X'AX = r = 3$

The signature of $X'AX = s = -1$

$\therefore$ $|s| \neq r$

Hence the given quadratic form is indefinite.

Example 4. *Classify the following quadratic forms in three variables as definite, semi-definite and indefinite*

(i) $2x^2 + 2y^2 + 3z^2 - 4yz - 4zx + 2xy$

(ii) $26x^2 + 20y^2 + 10z^2 - 4yz - 16zx - 36xy$

(iii) $x_1^2 + 4x_2^2 + x_3^2 - 4x_2x_3 + 2x_3x_1 - 4x_1x_2$

Solution.   (i) The matrix of the given quadratic form is

$$A = \begin{bmatrix} 2 & 1 & -2 \\ 1 & 2 & -2 \\ -2 & -2 & 3 \end{bmatrix}$$

We write
$$A = I_3 A I_3$$

or
$$\begin{bmatrix} 2 & 1 & -2 \\ 1 & 2 & -2 \\ -2 & -2 & 3 \end{bmatrix} = \begin{bmatrix} 1 & 0 & 0 \\ 0 & 1 & 0 \\ 0 & 0 & 1 \end{bmatrix} A \begin{bmatrix} 1 & 0 & 0 \\ 0 & 1 & 0 \\ 0 & 0 & 1 \end{bmatrix}$$

Applying $R_2 \rightarrow 2R_2, C_2 \rightarrow 2C_2$ we get
$$\begin{bmatrix} 2 & 2 & -2 \\ 2 & 8 & -4 \\ -2 & -4 & 3 \end{bmatrix} = \begin{bmatrix} 1 & 0 & 0 \\ 0 & 2 & 0 \\ 0 & 0 & 1 \end{bmatrix} A \begin{bmatrix} 1 & 0 & 0 \\ 0 & 2 & 0 \\ 0 & 0 & 1 \end{bmatrix}$$

Applying $R_2 \rightarrow R_2 - R_1, C_2 \rightarrow C_2 - C_1$ and $R_3 \rightarrow R_3 + R_1, C_3 \rightarrow C_3 + C_1$, we get

$$\begin{bmatrix} 2 & 0 & 0 \\ 0 & 6 & -2 \\ 0 & -2 & 1 \end{bmatrix} = \begin{bmatrix} 1 & 0 & 0 \\ -1 & 2 & 0 \\ 1 & 0 & 1 \end{bmatrix} A \begin{bmatrix} 1 & -1 & 1 \\ 0 & 2 & 0 \\ 0 & 0 & 1 \end{bmatrix}$$

Applying $R_3 \rightarrow 3R_3, C_3 \rightarrow 3C_3$, we get

$$\begin{bmatrix} 2 & 0 & 0 \\ 0 & 6 & -6 \\ 0 & -6 & 9 \end{bmatrix} = \begin{bmatrix} 1 & 0 & 0 \\ -1 & 2 & 0 \\ 3 & 0 & 3 \end{bmatrix} A \begin{bmatrix} 1 & -1 & 3 \\ 0 & 2 & 0 \\ 0 & 0 & 3 \end{bmatrix}$$

Applying $R_3 \rightarrow R_3 + R_2, C_3 \rightarrow C_3 + C_2$, we get

$$\begin{bmatrix} 2 & 0 & 0 \\ 0 & 6 & 0 \\ 0 & 0 & 3 \end{bmatrix} = \begin{bmatrix} 1 & 0 & 0 \\ -1 & 2 & 0 \\ 2 & 2 & 3 \end{bmatrix} A \begin{bmatrix} 1 & -1 & 2 \\ 0 & 2 & 2 \\ 0 & 0 & 3 \end{bmatrix}$$

$$\Rightarrow \quad \begin{bmatrix} 2 & 0 & 0 \\ 0 & 6 & 0 \\ 0 & 0 & 3 \end{bmatrix} = P'AP$$

$$\Rightarrow \quad \text{diag.}(2, 6, 3) = P'AP$$

Clearly the given quadratic is reduced to the diagonal form $2y_1^2 + 6y_2^2 + 3y_3^2$.

$\therefore$ The rank of given quadratic form $r = 3$

The signature of $s = 2$(positive terms) $-$ rank

$$= 2(3) - 3 = 3$$

$\therefore$ $s = r = n$

Hence the given quadratic form is positive definite.

(ii) The matrix of the given quadratic form is

$$A = \begin{bmatrix} 26 & -18 & -8 \\ -18 & 20 & -2 \\ -8 & -2 & 10 \end{bmatrix}$$

The characteristic equation of A is

$$|A - \lambda I| = 0$$

or

$$\begin{bmatrix} 26 - \lambda & -18 & -8 \\ -18 & 20 - \lambda & -2 \\ -8 & -2 & 10 - \lambda \end{bmatrix} = 0$$

Applying $C_1 \rightarrow C_1 + C_2 + C_3$ we get

$$\begin{bmatrix} -\lambda & -18 & -8 \\ -\lambda & 20 - \lambda & -2 \\ -\lambda & -2 & 10 - \lambda \end{bmatrix} = 0$$

or

$$(-\lambda) \begin{bmatrix} 1 & -18 & -8 \\ 1 & 20 - \lambda & -2 \\ 1 & -2 & 10 - \lambda \end{bmatrix} = 0$$

Applying $R_2 \rightarrow R_2 - R_1, R_3 \rightarrow R_3 - R_1$, we get

$$(-\lambda) \begin{bmatrix} 1 & -18 & -8 \\ 0 & 38 - \lambda & 6 \\ 0 & 16 & 18 - \lambda \end{bmatrix} = 0$$

$$(-\lambda) \left[(38 - \lambda)(18 - \lambda) - 96 \right] = 0$$

or

$$(-\lambda) \left[\lambda^2 - 56\lambda + 588 \right] = 0$$

The eigenvalues of A are
$$\lambda = 0, \lambda = 42, \lambda = 28$$
Clearly, the eigenvalues of A are all non-negative and A has one zero eigenvalue, hence, the given quadratic form is positive semi-definite.

(iii) The matrix of the given quadratic form is
$$A = \begin{bmatrix} 1 & -2 & 1 \\ -2 & 4 & -2 \\ 1 & -2 & 1 \end{bmatrix}$$

The characteristric equation of A is
$$|A - \lambda I| = 0$$
or
$$\begin{bmatrix} 1-\lambda & -2 & 1 \\ -2 & 4-\lambda & -2 \\ 1 & -2 & 1-\lambda \end{bmatrix} = 0$$

Applying $C_1 \to C_1 + C_2 + C_3$ we get
$$\begin{bmatrix} -\lambda & -2 & 1 \\ -\lambda & 4-\lambda & -2 \\ -\lambda & -2 & 1-\lambda \end{bmatrix} = 0$$

or
$$(-\lambda) \begin{bmatrix} 1 & -2 & 1 \\ 1 & 4-\lambda & -2 \\ 1 & -2 & 1-\lambda \end{bmatrix} = 0$$

Applying $R_1 \to R_1 - R_3$, we get
$$(-\lambda) \begin{bmatrix} 0 & 0 & \lambda \\ 1 & 4-\lambda & -2 \\ 1 & -2 & 1-\lambda \end{bmatrix} = 0$$

or
$$(-\lambda)(\lambda) \begin{bmatrix} 0 & 0 & 1 \\ 1 & 4-\lambda & -2 \\ 1 & -2 & 1-\lambda \end{bmatrix} = 0$$

or
$$(-\lambda)(\lambda)(-2-4+\lambda) = 0$$

or
$$-\lambda^2(\lambda - 6) = 0$$

The eigenvalues of A are 0, 0, 6.

Clearly, the eigenvalues of A are all non-negative, hence the given quadratic form is positive semi-definite.

Example 5. *Write the matrix A of the following quadratic form*
$$6x^2 + 65y^2 + 11z^2 + 4zx$$

Find the eigenvalues of A and hence determine the value class of the given quadratic form.

Solution. The matrix A of the given quadratic form is
$$A = \begin{bmatrix} 6 & 0 & 2 \\ 0 & 65 & 0 \\ 0 & 0 & 11 \end{bmatrix}$$

The characteristic equation of A is

$$|A - \lambda I| = 0$$

or

$$\begin{bmatrix} 6-\lambda & 0 & 2 \\ 0 & 65-\lambda & 0 \\ 2 & 0 & 11-\lambda \end{bmatrix} = 0$$

or $\quad (6-\lambda)(65-\lambda)(11-\lambda) - 4(65-\lambda) = 0$

or $\quad\quad (65-\lambda)[(6-\lambda)(11-\lambda) - 4] = 0$

or $\quad\quad (65-\lambda)(\lambda^2 - 17\lambda + 62 = 0$

The eigenvalues of A are

$$\lambda = 65, \lambda = \frac{1}{2}(17 \pm \sqrt{41})$$

Clearly ,the eigenvalues of A are all positive, hence the given quadratic form is positive definite.

Example 6. *Prove that the quadratic form*

$$6x_1^2 + 3x_2^2 + 3x_3^2 - 4x_1x_2 - 2x_2x_3 + 4x_3x_1$$

in three variables is positive definite.

Solution. The matrix A of the given quadratic form is

$$A = \begin{bmatrix} 6 & -2 & 2 \\ -2 & 3 & -1 \\ 2 & -1 & 3 \end{bmatrix}$$

The leading principal minors of A are

$$A_1 = 6$$

$$A_2 = \begin{vmatrix} 6 & -2 \\ -2 & 3 \end{vmatrix} = 18 - 4 = 14$$

$$A_3 = \begin{vmatrix} 6 & -2 & 2 \\ -2 & 3 & -1 \\ 2 & -1 & 3 \end{vmatrix}$$

$$= 6(9-1) + 2(-6+2) + 2(2-6)$$
$$= 48 - 8 - 8 = 32$$

Clearly, $A_1 > 0, A_2 > 0, A_3 > 0$

The leading principal minors of A are all positive, hence the given quadratic form is positive definite.

Example 7. *Show that the quadratic form*

$$6x^2 + 17y^2 + 3z^2 - 20xy - 14yz + 8zx$$

in three variables is positive semi-definite and find a non-zero set of values of x, y, z which makes the form zero.

Solution. The matrix of the given quadratic form is

$$A = \begin{bmatrix} 6 & -10 & 4 \\ -10 & 17 & -7 \\ 4 & -7 & 3 \end{bmatrix}$$

We write

$$A = I_3 A I_3$$

or

$$\begin{bmatrix} 6 & -10 & 4 \\ -10 & 17 & -7 \\ 4 & -7 & 3 \end{bmatrix} = \begin{bmatrix} 1 & 0 & 0 \\ 0 & 1 & 0 \\ 0 & 0 & 1 \end{bmatrix} A \begin{bmatrix} 1 & 0 & 0 \\ 0 & 1 & 0 \\ 0 & 0 & 1 \end{bmatrix}$$

Applying $R_2 \to 3R_2, C_2 \to 3C_2, R_3 \to 3R_3, C_3 \to 3C_3$ we get

$$\begin{bmatrix} 6 & -30 & 12 \\ -30 & 153 & -63 \\ 12 & -63 & 27 \end{bmatrix} = \begin{bmatrix} 1 & 0 & 0 \\ 0 & 3 & 0 \\ 0 & 0 & 3 \end{bmatrix} A \begin{bmatrix} 1 & 0 & 0 \\ 0 & 3 & 0 \\ 0 & 0 & 3 \end{bmatrix}$$

Applying $R_2 \to R_2 + 5R_1, C_2 \to C_2 + 5C_1$ and $R_3 \to R_3 - 2R_1, C_3 \to C_3 - 2C_1$, we get

$$\begin{bmatrix} 6 & 0 & 0 \\ 0 & 3 & -3 \\ 0 & -3 & 3 \end{bmatrix} = \begin{bmatrix} 1 & 0 & 0 \\ 5 & 3 & 0 \\ -2 & 0 & 3 \end{bmatrix} A \begin{bmatrix} 1 & 5 & -2 \\ 0 & 3 & 0 \\ 0 & 0 & 3 \end{bmatrix}$$

Applying $R_3 \to R_3 + R_2, C_3 \to C_3 + C_2$, we get

$$\begin{bmatrix} 6 & 0 & 0 \\ 0 & 3 & 0 \\ 0 & 0 & 0 \end{bmatrix} = \begin{bmatrix} 1 & 0 & 0 \\ 5 & 3 & 0 \\ 3 & 3 & 3 \end{bmatrix} A \begin{bmatrix} 1 & 5 & 3 \\ 0 & 3 & 3 \\ 0 & 0 & 3 \end{bmatrix}$$

$\Rightarrow$ diag.$(6, 3, 0) = P'AP$

where

$$P = \begin{bmatrix} 1 & 5 & 3 \\ 0 & 3 & 3 \\ 0 & 0 & 3 \end{bmatrix}$$

The non-singular linear transformation

$$X = PY$$

i.e.,

$$\begin{bmatrix} x \\ y \\ z \end{bmatrix} = \begin{bmatrix} 1 & 5 & 3 \\ 0 & 3 & 3 \\ 0 & 0 & 3 \end{bmatrix} \begin{bmatrix} y_1 \\ y_2 \\ y_3 \end{bmatrix}$$

or

$$\left. \begin{array}{l} x = y_1 + 5y_2 + 3y_3 \\ y = 3y_2 + 3y_3 \\ z = 3y_3 \end{array} \right\} \qquad \ldots(1)$$

reduces the given quadratic form to the diagonal form

$$X'AX = Y'(P'AP)Y = Y' \text{ diag.}(6, 3, 0)Y$$
$$= 6y_1^2 + 3y_2^2 + 0.y_3^2$$

$\therefore$ The rank of given quadratic form $= r = 2$

The signature of the given quadratic form $= s = 2$

The number of variables $n = 3$

$\Rightarrow$ $\qquad\qquad\qquad\qquad s = r < n$

$\therefore$ The given quadratic form is positive semi-definite.

If we put $y_1 = 0, y_2 = 0$ and $y_3 = 1$, the given quadratic form reduces to zero.

Putting $y_1 = 0, y_2 = 0$ and $y_3 = 1$ in (1), we get $x = 3, y = 3, z = 3$

Hence $x = 3, y = 3, z = 3$ is a non-zero set of values of x, y, z which makes the given quadratic form zero.

Example 8. *Show that the quadratic form*

$$x_1^2 + 2x_2^2 + 3x_3^2 + 2x_2x_3 - 2x_3x_1 + 2x_1x_2$$

in three variables is indefinite and find two sets of values of x_1, x_2, x_3 for which the form assumes positive and negative values.

Solution. The matrix of the given quadratic form is

$$A = \begin{bmatrix} 1 & 1 & -1 \\ 1 & 2 & 1 \\ -1 & 1 & 3 \end{bmatrix}$$

We write

$$A = I_3 A I_3$$

or

$$\begin{bmatrix} 1 & 1 & -1 \\ 1 & 2 & 1 \\ -1 & 1 & 3 \end{bmatrix} = \begin{bmatrix} 1 & 0 & 0 \\ 0 & 1 & 0 \\ 0 & 0 & 1 \end{bmatrix} A \begin{bmatrix} 1 & 0 & 0 \\ 0 & 1 & 0 \\ 0 & 0 & 1 \end{bmatrix}$$

Applying $R_2 \to R_2 - R_1, C_2 \to C_2 - C_1$ and $R_3 \to R_3 + R_1, C_3 \to C_3 + C_1$, we get

$$\begin{bmatrix} 1 & 0 & 0 \\ 0 & 1 & 2 \\ 0 & 2 & 2 \end{bmatrix} = \begin{bmatrix} 1 & 0 & 0 \\ -1 & 1 & 0 \\ 1 & 0 & 1 \end{bmatrix} A \begin{bmatrix} 1 & -1 & 1 \\ 0 & 1 & 0 \\ 0 & 0 & 1 \end{bmatrix}$$

Applying $R_3 \to R_3 + R_2, C_3 \to C_3 + C_2$, we get

$$\begin{bmatrix} 1 & 0 & 0 \\ 0 & 1 & 0 \\ 0 & 0 & -2 \end{bmatrix} = \begin{bmatrix} 1 & 0 & 0 \\ -1 & 1 & 0 \\ 3 & -2 & 1 \end{bmatrix} A \begin{bmatrix} 1 & -1 & 3 \\ 0 & 1 & -2 \\ 0 & 0 & 1 \end{bmatrix}$$

$\Rightarrow$ diag.$(1, 1, -2) = P'AP$

where

$$P = \begin{bmatrix} 1 & -1 & 3 \\ 0 & 1 & -2 \\ 0 & 0 & 1 \end{bmatrix}$$

Thus, the non-singular linear transformation

$$X = PY$$

i.e.,

$$\begin{bmatrix} x_1 \\ x_2 \\ x_3 \end{bmatrix} = \begin{bmatrix} 1 & -1 & 3 \\ 0 & 1 & -2 \\ 0 & 0 & 1 \end{bmatrix} \begin{bmatrix} y_1 \\ y_2 \\ y_3 \end{bmatrix}$$

or

$$\left. \begin{array}{l} x_1 = y_1 - y_2 + 3y_3 \\ x_2 = y_2 - 2y_3 \\ x_3 = y_3 \end{array} \right\} \qquad \ldots(1)$$

reduces the given quadratic form to the diagonal form

$$X'AX = y_1^2 + y_2^2 - 2y_3^2$$

The rank of given quadratic form $= 3 (= r)$

The signature of the given quadratic form $= 2(2) - 3 = 2 (= s)$

$\Rightarrow \qquad |s| \neq r$

Hence the given quadratic form is indefinite.

If we put $y_1 = 0, y_2 = 0$ and $y_3 = 1$, the given quadratic form reduces to negative value and if put $y_1 = 0, y_2 = 1$ and $y_3 = 0$, the given quadratic form reduces to positive value.

Substituting these values of y_1, y_2 and y_3 in (1) respectively, we get two sets of non-zero values of x_1, x_2, x_3 as $x_1 = 3, x_2 = -2, x_3 = 1$ and $x_1 = -1, x_2 = 1, x_3 = 0$, which make the given quadratic from zero.

Example 9. *If A is a positive definite real symmetric matrix, show that there exists a positive definite real symmetric matrix B such that*

$$B^2 = A$$

Solution. Since A is a positive definite real symmetric matrix, then there exists an orthogonal matrix P such that
$$P^{-1}AP = D = \text{diag. } (\lambda_1, \lambda_2, ..., \lambda_n)$$
where $\lambda_1, \lambda_2, ..., \lambda_n$ are the eigenvalues of A and each $\lambda_i > 0$ $(i = 1, 2, ..., n)$.

Let $D_1 = \text{diag. } \left(\sqrt{\lambda_1}, \sqrt{\lambda_2}, ..., \sqrt{\lambda_n}\right)$, then
$$D_1^2 = D \text{ and } D_1' = D$$

Also each
$$\sqrt{\lambda_i} > 0 \ (i = 1, 2, ..., n)$$

Let us take $B = PD_1P^{-1} = PD_1P'$ $[\because P^{-1} = P']$
$\Rightarrow$ $B' = (PD_1P')' = PD_1'P'$
$\Rightarrow$ $B' = PD_1P'$ $[\because D_1' = D_1]$
$\Rightarrow$ $B' = B$

$\therefore$ B is a real symmetric matrix and B is similar to D_1.

Thus, B and D_1 have the same eigenvalues, therefore the eigenvalues of B are all positive, hence B is a positive definite real symmetric matrix.

Also, we have $B^2 = (PD_1P^{-1})^2$
$$= (PD_1P^{-1})(PD_1P^{-1})$$
$$= PD_1(P^{-1}P)D_1P^{-1}$$
$$= PD_1^2P^{-1} = PDP^{-1} = A$$

Example 10. *Show that every real non-singular matrix A can be expressed as*
$$A = QDR$$
where Q and R are orthogonal and D is real diagonal.

Solution. Since A is a real non-singular matrix, then $A'A$ is a positive definite real symmetric matrix, therefore there exists an orthogonal matrix P such that
$$P^{-1}(A'A)P = P'(A'A)P = D = \text{diag. } (\lambda_1, \lambda_2, ..., \lambda_n)$$
where $\lambda_1, \lambda_2, ..., \lambda_n$ are the eigenvalues of $A'A$ and each $\lambda_i > 0$ $(i = 1, 2, ..., n)$.

Let $D = \text{diag. } \left(\sqrt{\lambda_1}, \sqrt{\lambda_2}, ..., \sqrt{\lambda_n}\right)$, then
$$D^2 = D \text{ and } D' = D$$

Now we have
$$D'D = D^2 = \text{diag. } (\lambda_1, \lambda_2, ..., \lambda_n)$$
$\Rightarrow$ $D'D = P'(A'A)P$
$\Rightarrow$ $(P')^{-1}D'DP^{-1} = A'A$
$\Rightarrow$ $(P^{-1})'D'DP^{-1} = A'A$ $[\because P \text{ is orthogonal} \Rightarrow P' = P^{-1}]$
$\Rightarrow$ $PD'DP^{-1} = A'A$
$\Rightarrow$ $(A')^{-1}PD'DP^{-1}A^{-1} = I$
$\Rightarrow$ $(A^{-1})'PD'DP^{-1}A^{-1} = I$
$\Rightarrow$ $(A^{-1})'PD'DP'A^{-1} = I$ $[\because P' = P^{-1}]$
$\Rightarrow$ $(DP'A^{-1})'(DP'A^{-1}) = I$
$\Rightarrow$ $DP'A^{-1}$ is orthogonal.

Let $Q = (DP'A^{-1})^{-1}$, then Q is orthogonal.

Again, let $R = P'$, then R is orthogonal.

Now we have $QDR = (DP'A^{-1})^{-1}DP'$
$$= A(P')^{-1}D^{-1}DP'$$
$$= A(P')^{-1}IP' = A(P')^{-1}(P')$$
$$= AI = A$$

☕ EXERCISE 9.4

1. Show that the quadratic form $x^2 - 2y^2 + 3z^2 - 4yz + 6zx$ in three variables is indefinite.

2. Show that the quadratic form $y^2 + 2z^2 - 2yz + 2zx - 2xy$ in three variables is indefinite.

3. Show that the quadratic form $2x^2 + 6y^2 + 8z^2 + 6yz + 3zx - 4xy$ in three variables is positive definite.

4. Determine the value class of the form $6x^2 + 12y^2 + 8yz + 4zx$ in three variables.

5. Determine the value class of the form $-y^2 + 2yz - 2xy$ in three variables.

6. Show that the quadratic form $5x_1^2 + 26x_2^2 + 10x_3^2 + 4x_2x_3 + 14x_3x_1 + 6x_1x_2$ in three variables is positive semi-definite and find a non-zero set of values of x_1, x_2, x_3 which makes the form zero.

7. Prove that every positive semi-definite symmetric matrix A has a positive semi-definite square root B such that $B^2 = A$.

Answers

4. Indefinite   5. Indefinite

9.23 HERMITIAN FORMS

In this section we assume that V is a vector space over the complex field C.

Definition. *If $A = [a_{ij}]$ is an $n \times n$ matrix over C, then we write $\overline{A}$ for the matrix obtained by taking the complex conjugate of every entry of A i.e., $\overline{A} = [\overline{a_{ij}}]$. We also write $A^\theta = (\overline{A})^t = (\overline{A^t})$. Thus A^θ is the* conjugate transpose *of A.*

For Example: If $A = \begin{bmatrix} 2+3i & 5+4i \\ 6-7i & 1+9i \end{bmatrix}$, $A^\theta = (\overline{A})^t = \begin{bmatrix} 2-3i & 6+7i \\ 5-4i & 1-9i \end{bmatrix}$

Definition 2. *A matrix A is said to be Hermitian if $A^\theta = A$.*

Definition 3. *A matrix A is said to be skew-Hermitian if $A^\theta = -A$.*

Definition 4. *Let V be a vector space over the complex field C, then a mapping $f : V \times V \to C$ is said to Hermitian form which satisfies the following properties :*

(i) $f(a\alpha + b\beta, \gamma) = af(\alpha, \gamma) + bf(\alpha, \gamma)$

(ii) $f(\alpha, \beta) = \overline{f(\beta, \alpha)}$, *where $a, b \in C$ and $\alpha, \beta \in V$.*

If f is Hermitian on V, then $f(\alpha, a\beta + b\gamma) = \overline{a} f(\alpha, \beta) = \overline{b} f(\alpha, \gamma)$

or

If $A = [a_{ij}]_{nxn}$ is a Hermitian matrix of order $n \times n$, then

$$X^\theta A X = \sum_{i=1}^{n} \sum_{j=1}^{n} a_{ij} \overline{x_i} x_j$$

is called a Hermitian form in n variables $x_1, x_2, ..., x_n$. The matrix A is called the matrix of this Hermitian form.

Definition 5. *Let f be a Hermitian form on V. Then a mapping $q : V \to R$ defined by $q(\alpha) = f(\alpha, \alpha)$ is called the Hermitian form or* complex quadratic form *associated with the Hermitian form f. Moreover, one can obtain f from q by the following identity, called polar form of f :*

$$f(\alpha, \beta) = \frac{1}{4} (q(\alpha + \beta) - q(\alpha - \beta)) + (q(\alpha + i\beta) - q(\alpha - i\beta)).$$

Definition 6. *A Hermitian form f and its quadratic form q are said to be non-negative semi-definite if $q(\alpha) = f(\alpha, \alpha) > 0$ for every $\alpha \in V$, and are said to be positive definite if $q(\alpha) = f(\alpha, \alpha) > 0$ for every $\alpha \neq 0$.*

9.24 MATRIX REPRESENTATION OF A HERMITIAN FORM

Let V be an n-dimensional vector space over C and let $B = \{\alpha_1, \alpha_2, ..., \alpha_n\}$ of V. Then the matrix $A = [a_{ij}]$, where $a_{ij} = f(\alpha_i, \alpha_j)$ is called the matrix representation of f relative to the basis B.

Since $f(\alpha_i, \alpha_j) = \overline{f(\alpha_i, \alpha_j)}$, thus A is Hermitian and in particular the diagonal entries of A are all real.

THEOREM 1. *Let f be a Hermitian form on V. Let A be the matrix of f relative to the basis $B = \{\alpha_1, \alpha_2, ..., \alpha_n\}$ of V. Then $f(\alpha, \beta) = [\alpha]_B^t A [\bar{\beta}]_B$, for all $\alpha, \beta \in V$.*

Proof. Since $\alpha, \beta \in V$ so that

$$\alpha = a_1 \alpha_1 + a_2 \alpha_2 + ... + a_n \alpha_n \text{ and } \beta = b_1 \alpha_1 + b_2 \alpha_2 + ... + b_n \alpha_n, \text{ for } a_i, b_i \in C$$

Then, $f(\alpha, \beta) = f(a_1 \alpha_1 + a_2 \alpha_2 + ... + a_n \alpha_n, b_1 \alpha_1 + b_2 \alpha_2 + ... + b_n \alpha_n)$

$$= \sum_{i=1}^{n} \sum_{j=1}^{n} a_i . \bar{b}_j . f(\alpha_i, \alpha_j) = (a_1, a_2, ..., a_n) A \begin{bmatrix} \overline{b_1} \\ \overline{b_2} \\ \vdots \\ \overline{b_3} \end{bmatrix} = [\alpha]_B^t A [\bar{\beta}]_B$$

THEOREM 2. *Let P be the change of basis matrix from a basis B of V to a new basis B'. Let A be the matrix of a Hermitian form f in the original basis B. Then $P'AP = Q^\theta AQ$, where $Q = \bar{P}$ is the matrix of f in the new basis B'.*

Proof. Let $\alpha, \beta \in V$. Since P is the change of basis matrix from B to B', we have

$$P[\alpha]_{B'} = [\alpha]_B \qquad \text{and} \qquad P[\beta]_{B'} = [\beta]_B$$

$$\therefore \qquad [\alpha]_B^t = [\alpha]_B^t P' \qquad \text{and} \qquad [\bar{\beta}]_B = \overline{P[\beta]_B}$$

Then by above theorem 1, we have

$$f(\alpha, \beta) = [\alpha]_B^t = A[\bar{\beta}]_B = [\alpha]_{B'}^t P'A\overline{P}[\beta]_B$$

Since α, β are arbitrary elements of V. Hence $P'A\bar{P}$ is the matrix of f in the new basis B'.

REMARK

- Let f be a Hermitian form on V. Then there exists a basis $\{\alpha_1, \alpha_2, ..., \alpha_n\}$ of V in which f is represented by a diagonal matrix i.e., $f(\alpha_i, \alpha_j) = 0$ for $i \neq j$. Moreover, every diagonal representation of f has the same number p of positive entries and the same number N of negative entries. The difference $p - N$ is called *signature of f*.

Solved Examples

Example 1. *If f is a Hermitian form on V. Show that*

$$f(\alpha, a\beta + b\gamma) = \bar{a} f(\alpha, \beta) + \bar{b} f(\alpha, \gamma)$$

Solution. By definition of Hermitian form

$$f(\alpha, a\beta + b\gamma) = \overline{f(a\beta + b\gamma, \alpha)} = \overline{af(\beta, \alpha) + bf(\gamma, \alpha)}$$

$$= \overline{af(\beta, \alpha)} + \overline{bf(\gamma, \alpha)} = \bar{a}f(\alpha, \beta) + \bar{b}f(\alpha, \gamma)$$

Example 2. *Let C^n be the set all square matrices of order $n \times n$ and A be a Hermitian matrix. Then show that f is Hermitian form on C^n where f is defined by $f(X, Y) = X'A\bar{Y}$.*

Solution. For $a, b \in C$ and all $X_1, X_2, Y \in C^n$,

$$f\left(aX_1 + bX_2, Y\right) = \left(aX_1 + bX_2\right)^t A\overline{Y} = \left(aX_1' + bX_2'\right)A\overline{Y}$$

$$aX_1'A\overline{Y} + bX_2'A\overline{Y} = af\left(X_1, Y\right) + bf\left(X_2, Y\right)$$

Also
$$\overline{f(X,Y)} = \overline{\left(XA\overline{Y}\right)} = \left(\overline{XAY}\right)^t$$

$$= \overline{\left(Y^t AX\right)} = Y^t A^\theta \overline{X} = Y^t A\overline{X} \qquad [\because A^\theta = A]$$

$$\therefore \qquad \overline{f(X,Y)} = f(Y,X) \text{ . Hence } f \text{ is Hermitian form.}$$

Example 3. Let $A = \begin{bmatrix} 1 & 1+i & 2i \\ 1-i & 4 & 2-3i \\ -2i & 2+3i & 7 \end{bmatrix}$, *a Hermitian matrix. Find a non-singular matrix P*

such that $P'A\overline{P}$ *is diagonal.*

Solution. First form the matrix $M = (A : I)$ as

$$M = \begin{bmatrix} 1 & 1+i & 2i & \vdots & 1 & 0 & 0 \\ 1-i & 4 & 2-3i & \vdots & 0 & 1 & 0 \\ -2i & 2+3i & 7 & \vdots & 0 & 0 & 1 \end{bmatrix}$$

Apply the row operation $R_2 \to R_2 + (-1+i)R_1$ and $R_3 \to R_3 + 2iR_1$ to M and corresponding Hermitian column operations $C_2 \to C_2 + (-1-i)C_1$ and $C_3 \to C_3 - 2iC_1$ to A we get

$$\begin{pmatrix} 1 & 1+i & 2i & \vdots & 1 & 0 & 0 \\ 0 & 2 & -5i & \vdots & -1+i & 1 & 0 \\ 0 & 5i & 3 & \vdots & 2i & 0 & 1 \end{pmatrix}$$

and then
$$\begin{pmatrix} 1 & 0 & 0 & \vdots & 1 & 0 & 0 \\ 0 & 2 & -5i & \vdots & -1+i & 1 & 0 \\ 0 & 5i & 3 & \vdots & 2i & 0 & 1 \end{pmatrix}$$

Now apply the row operations $R_3 \to 2R_3 - 5iR_2$ and the corresponding Hermitian column operation $C_3 \to 2C_3 + 5iC_2$, we get

$$\begin{pmatrix} 1 & 0 & 0 & \vdots & 1 & 0 & 0 \\ 0 & 2 & -5i & \vdots & -1+i & 1 & 0 \\ 0 & 0 & -19 & \vdots & 5+9i & -5i & 2 \end{pmatrix}$$

and then
$$\begin{pmatrix} 1 & 0 & 0 & \vdots & 1 & 0 & 0 \\ 0 & 2 & 0 & \vdots & -1+i & 1 & 0 \\ 0 & 0 & -38 & \vdots & 5+9i & -5i & 2 \end{pmatrix}$$

Thus A is diagonalized.

Setting $P = \begin{bmatrix} 1 & -1+i & 5+9i \\ 0 & 1 & -5i \\ 0 & 0 & 2 \end{bmatrix}$. Then, $P^t A \overline{P} = \begin{bmatrix} 1 & 0 & 0 \\ 0 & 2 & 0 \\ 0 & 0 & -38 \end{bmatrix}$.

☙ EXERCISE 9.5

1. Find the quadratic form $q(x,y,z)$ corresponding to the following matrices:

(i) $A = \begin{bmatrix} 3 & 0 & 0 \\ 0 & -4 & 0 \\ 0 & 0 & 6 \end{bmatrix}$

(ii) $A = \begin{bmatrix} 2 & -5 & 1 \\ -5 & -6 & -7 \\ 1 & -7 & 9 \end{bmatrix}$

(iii) $A = \begin{bmatrix} 2 & 1 & 5 \\ 1 & 3 & -2 \\ 5 & -2 & 4 \end{bmatrix}$ (iv) $A = \begin{bmatrix} 1 & 2 & 3 \\ 2 & 0 & 3 \\ 3 & 3 & 1 \end{bmatrix}$

2. Find the real symmetric matrices corresponding to the following quadratic forms:

  (i) $q(x,y) = ax^2 + 2hxy + by^2$

  (ii) $q(x,y,z) = 2xy + 6xz - 4yz$

  (iii) $q(x,y) = 4x^2 + 5xy - 7y^2$

  (iv) $q(x,y,z) = 2x^2 - 10xy - 6y^2 + 2xz - 14yz + 9z^2$

  (v) $q(x,y,z) = x^2 - 2yz + xz$.

3. Let $q(x, y, z) = x^2 + 4xy + 3y^2 - 8xz - 12yz + 9z^2$. Find a non-singular linear substitution expressing the variables x, y, z in terms of the variables r, s, t so that $q(r, s, t)$ is diagonal. Also find the signature of q.

4. Show that $q(0) = 0$ for any quadratic q on V.

5. Suppose $q(\alpha) = 0$ for a quadratic form q on $V(F)$. Show that $q(k\alpha) = 0$ for any $k \in F$.

6. Show that rank $(f) = \text{rank}(q) = p + N$, where p is the number of positive entries and N the number of negative entries in any diagonal representation of f and q.

7. Let $q(x,y,z) = x^2 + y^2 + 2xz + 4yz + 3z^2$. Is q positive definite?

8. Let $q(x,y) = x - 4xy + 5y^2$. Show that q is positive definite.

9. Let $q(x,y) = x^2 - 6xy + 3y^2$. Show that q is not positive definite.

10. Let $q(x,y) = x^2 + 4xy + y^2$. Find an orthogonal change of co-ordinates which diagonalizes q. Also find its signature.

11. Let $q(x,y) = 3x^2 - 6xy + 11y^2$. Find an orthogonal change of co-ordinates which diagonalizes q. Also find its signature and rank.

12. The following expressions define quadratic forms q on $\mathbb{R}^2$. Find the symmetric bilinear form corresponding to each q.

  (i) cx_2^2 (ii) $3x_1 x_2 - 2x_2^2$

  (iii) $2x_1^2 - \dfrac{1}{3}x_1 x_2$

13. Suppose f is a Hermitian form on V. Show that $f(\alpha, \alpha)$ is real for any $\alpha \in V$.

14. Define a non-negative semi-definite and a positive definite Hermitian form.

15. Let f be the dot product on $\mathbb{C}^n$. Let $f(\alpha, \beta) = x_1 \bar{y}_1 + x_2 \bar{y}_2 + \ldots\ldots + x_n \bar{y}_n$ for $\alpha = (x_1, x_2, \ldots, x_n)$ $\beta = (y_1, y_2, \ldots, y_n) \in \mathbb{C}^n$. Is f a Hermitian? Is f positive definite?

16. Show that the Hermitian form assumes one real values for all complex n-vectors X.

17. Every Hermitian form $X^\theta A X$ is unitarily equivalent to the form

$$\lambda_1 \bar{y}_1 y_1 + \lambda_2 \bar{y}_2 y_2 + \ldots + \lambda_n \bar{y}_n y_n$$

where $\lambda_1, \lambda_2, \ldots, \lambda_n$ are the eigenvalues of the Hermitian matrix A, and $X = PY$, P the unitary.

Hint to Selected Problems

1. (ii) Let $X = \begin{bmatrix} x_1 \\ x_2 \\ x_3 \end{bmatrix}$. Then we have,

$$q(X) = X'AX = [x_1, x_2, x_3] \begin{pmatrix} 2 & -5 & 1 \\ -5 & -6 & -7 \\ 1 & -7 & 9 \end{pmatrix} \begin{bmatrix} x_1 \\ x_2 \\ x_3 \end{bmatrix}$$

$$= [2x_1 - 5x_2 + x_3, -5x_1 - 6x_2 - 7x_3, x_1 - 7x_2$$

$$+ 9x_3] \begin{bmatrix} x_1 \\ x_2 \\ x_3 \end{bmatrix}$$

$$= x_1(2x_1 - 5x_2 + x_3) + x_2(-5x_1 - 6x_2 - 7x_3)$$

$$+ x_3(x_1 - 7x_2 + 9x_3)$$

$$= 2x_1^2 - 10x_1 x_2 + 2x_1 x_3 - 6x_2^2 - 14x_2 x_3 + 9x_3^2$$

$$= 2x_1^2 - 6x_2^2 + 9x_3^2 - 10x_1 x_2 + 2x_1 x_3 - 14x_2 x_3$$

2. (iv) $q(x, y, z) = 2x^2 - 10xy - 6y^2 + 2zx - 14yz + 9z^2$

$q(x,y,z)$
$= 2x^2 - 5xy + xz - 5yx - 6y^2 - 7yz + zx - 7zy + 9z^2$
$= x(2x - 5y + z) + y(-5x - 6y - 7z)$
$\quad + z(x - 7y + 9z)$

$$= [x, y, z] \begin{pmatrix} 2 & -5 & 1 \\ -5 & -6 & -7 \\ 1 & -7 & 9 \end{pmatrix} \begin{bmatrix} x \\ y \\ z \end{bmatrix}.$$

Hence the matrix corresponding to the given quadratic form is $\begin{pmatrix} 2 & -5 & 1 \\ -5 & -6 & -7 \\ 1 & -7 & 9 \end{pmatrix}$

4. Since we have
$q(\alpha) = f(\alpha, \alpha)$,
$q(0) = f(0, 0) = f(0\alpha, 0) = 0 f(\alpha, 0) = 0$.

7. $q(x,y,z) = x^2 + y^2 + 2zx + 4yz + 3z^2$.

$$A = \begin{bmatrix} 1 & 0 & 1 \\ 0 & 1 & 2 \\ 1 & 2 & 3 \end{bmatrix}$$

Reduce A to diagonal form as follows :

$$A = \begin{bmatrix} 1 & 0 & 0 \\ 0 & 1 & 2 \\ 0 & 2 & 2 \end{bmatrix} = \begin{bmatrix} 1 & 0 & 0 \\ 0 & 1 & 0 \\ 0 & 0 & -2 \end{bmatrix}.$$

There is a negative entry -2 in the diagonal representation of q. Hence q is not positive definite.

10. $q(x, y) = x^2 + 4xy + y^2$, the matrix of q is $A =$ $\begin{bmatrix} 1 & 2 \\ 2 & 1 \end{bmatrix}$, then

$$\Delta(t) = |tI - A| = \begin{vmatrix} t-1 & 2 \\ -2 & t-1 \end{vmatrix} = t^2 - 2t - 3$$

$$= (t-3)(t+1)$$

Thus eigenvalues of A are 3 and –1. Substitute into the matrix tI–A to obtain the corresponding homogeneous system of linear equations $2x - 2y = 0$, $-2x + 2y = 0$. A non-zero solution is $\alpha_1 = (1,1)$.

Next substitute $t = -1$ into the matrix tI–A to obtain the corresponding homogeneous system of linear equation $-2x - 2y = 0$, $-2x - 2y = 0$. A non-zero solution is $\alpha_2 = (1,-1)$. Normalize α_1, α_2 to obtain orthonormal basis.

$$\begin{cases} \beta_1 = \left(1\sqrt{2}, 1/\sqrt{2}, 1/\sqrt{2}\right), \\ \beta_2 = \left(-1/\sqrt{2}, 1/\sqrt{2}\right) \end{cases}$$

Let P be the matrix whose columns are β_1 and β_2 respectively, then

$$\begin{bmatrix} x \\ y \end{bmatrix} = P \begin{bmatrix} x' \\ y' \end{bmatrix}$$

or $\quad x = \dfrac{x' - y'}{\sqrt{2}}$, $y = \dfrac{x' + y'}{\sqrt{2}}$.

Under this change of co-ordinates, q is transformed into the diagonal form

$q(x', y') = 3x'^2 - y'^2$
$\mathrm{sig}(q) = 1 - 1 = 0$.

13. Since f is a Hermitian form, then by definition of Hermitian form, $f(\alpha, \alpha) = \overline{f(\alpha, \alpha)}$. This shows that $f(\alpha, \alpha)$ is real.

15. Let $\alpha = (x_1, x_2,..., x_n)$, $\beta = (y_1, y_2,...,y_n)$, $z = (z_1, z_2,..., z_n) \in C^n$

For $a, b \in C$, we have

$a\alpha + b\beta$
$= (ax_1 + by_1, ax_2 + by_2,...,ax_n + by_n)$
$f(a\alpha + b\beta, \gamma)$
$= (ax_1 + by_1)\,\overline{z}_1 + (ax_2 + by_2)\,\overline{z}_2 + ...$
$\qquad\qquad + (ax_n + by_n)\,\overline{z}_n$
$= a(x_1\,\overline{z}_1 + x_2\,\overline{z}_2 + ... + x_n\,\overline{z}_n) + b(y_1\,\overline{z}_1$
$\qquad\qquad + y_2\,\overline{z}_2 + ... + y_n\,\overline{z}_n)$
$= af(\alpha, \gamma) + bf(\beta, \gamma)$

Also, $f(\alpha, \beta) = x_1\overline{y}_1 + x_2\overline{y}_2 + ... + x_n\overline{y}_n$

$= \overline{(y_1\overline{x}_1 + y_2\overline{x}_2 + ... + y_n\overline{x}_n)} = \overline{f(\beta, \alpha)}$

Thus, f is Hermitian.
Let $\qquad \alpha \neq 0$
$f(\alpha, \alpha) = x_1\overline{x}_1 + x_2\overline{x}_2 + ... + x_n\overline{x}_n$
$\qquad\quad = |x_1|^2 + |x_2|^2 + ... + |x_n|^2 > 0$

Hence f is positive definite.

Answers

1. (i) $q(x,y,z) = 3x^2 - 4y^2 + 6z^2$ (ii) $q(x,y,z) = 2x^2 - 6y^2 + 9z^2 - 10xy + 2xz - 14yz$

(iii) $q(x,y,z) = 2x^2 + 3y^2 + 4z^2 + 2xy + 10xz - 4yz$ (iv) $q(x,y,z) = x^2 + z^2 + 4xy + 6xz + 6yz$

2. (i) $A = \begin{bmatrix} a & h \\ h & b \end{bmatrix}$ (ii) $A = \begin{bmatrix} 0 & 1 & 3 \\ 1 & 0 & -2 \\ 3 & -2 & 0 \end{bmatrix}$ (iii) $A = \begin{bmatrix} 4 & 5/2 \\ 5/2 & -7 \end{bmatrix}$

(v) $A = \begin{bmatrix} 2 & -5 & 1 \\ -5 & -6 & -7 \\ 1 & -7 & 9 \end{bmatrix}$ (vi) $A = \begin{bmatrix} 1 & 0 & 1/2 \\ 0 & 0 & -1 \\ 1/2 & -1 & 0 \end{bmatrix}$

3. $x = r - 2s$, $y = s + 2t$, $z = t$, $\mathrm{sig}.(q) = -1$

7. q is not positive definite, change of co-ordinates is $x = \dfrac{x'}{\sqrt{2}} - \dfrac{y'}{\sqrt{2}}$, $y = \dfrac{x'}{\sqrt{2}} + \dfrac{y'}{\sqrt{2}}$.

$\therefore \qquad q(x', y') = 3x'^2 - y^2$, $\mathrm{sig}.(q) = 0$.

11. Orthogonal change of co-ordinates is $x = \dfrac{3x' - y'}{\sqrt{10}}$, $y = \dfrac{x' + 3y'}{\sqrt{10}}$.

so $\qquad q(x', y') = 3x'^2 - y'^2$, $\mathrm{sig}.(q) = 2$, $\mathrm{rank}(q) = 0$.

12. (i) $\qquad f((x_1, x_2),(x_1, x_2)) = cx_2^2$ (ii) $f((x_1, x_2),(x_1, x_2)) = 3x_1x_2 - 2x_2^2$

(iii) $f((x_1, x_2),(x_1, x_2)) = 2x_1^2 - \dfrac{1}{3}x_1x_2$

CHAPTER REVIEW : A COMPETITIVE APPROACH

Selected Terms and Results

TERMS

- **Bilinear Form:** A bilinear form on a vector space $V(f)$ is a mapping $f : V \times V \to f$ which satisfies the following properties:
  (i) $f(a\alpha+b\beta,\gamma)=af(\alpha,\gamma)+bf(\beta,\gamma)$
  (ii) $f(\alpha,a\beta+b\gamma)=af(\alpha,\beta)+bf(\alpha,\gamma)$
  for all $a,b \in F$, $\alpha,\beta,\gamma \in V$.

- **Bilinear Space :** A vector space V together with a bilinear form f defined above is called a bilinear space.

- **Rank of a Bilinear Form:** Let f be a bilinear form on V, then rank of a bilinear form f, denoted by rank(f) is defined to be the rank of any matrix representation of f.

- **Degenerate and Non-degenerate Bilinear Form:** A bilinear form f on V is said to be degenerate if rank $(f) < \dim V$ and it is said to be non-degenerate if rank $(f) = \dim V$.

- **Quadratic Mapping:** A mapping $q:V \to f$ is called a quadratic mapping if $q(\alpha)=f(\alpha,\alpha)$, where f is a symmetric bilinear form on V.

- **Quadratic form:** A mapping $q:V \to f$ is called a quadratic form if $q(\alpha)= f(\alpha, \alpha)$ for some bilinear form f on V.

- **Signature:** Let q be a real quadratic form and f be a bilinear form then the signature of f and of q are defined by $\text{sig}(f) = \text{sig}(q) = p-N$ where p is the number of positive entries and N is the number of negative entries in any diagonal

representation of f and q.

- **Hermitian Form :** A matrix A is said to be Hermitian if $A^\theta=A$ and skew-hermitian if $A^\theta = -A$.

- **Similar Martices:** For two square matrix A and B, the matrix B is said to be similar to A if there exists an invertible matrix P such that $B= P^{-1}AP$.

- **Invariant Subspace:** Let $T:V \to V$ be a linear transformation. Then a subspace W of V is said to be invariant under T if $T(W) \subset W \ \forall \ \alpha \in W$, $T(\alpha) \in W$.

- **Congurence of a Quadratic Forms :** Two quadratic forms $Q(X) = X'AX$ and $Q(Y) = Y'BY$ over a field F are said to be congruent or equivalent to over F if A and B are congruent over F.

- **Equivalence of Real Quadratic Forms :** Two real quadratic forms $X'AX$ and $Y'BY$ are said to be real equivalent if there exists a non-singular real matrix P such that $B = P'AP$.

- **Orthogonally Equivalent :** Two real quadratic forms $X'AX$ and $Y'BY$ are said to be orthogonally equivalent if there exists a non-singular orthogonal matrix P such that $B = P'AP$.

- **Rank of a Quadratic Form :** Let $Q(X) = X'AX$ be a quadratic form over a field F. Then rank of $Q(X)$ is the rank of matrix A.

RESULTS

- A bilinear form f is called symmetric if $f(\alpha,\beta)=f(\beta,\alpha)$ for all $\alpha, \beta \in V$.

- A bilinear form f is called skew-symmetric if $f(\alpha,\beta)= -f(\beta, \alpha)$ for all $\alpha,\beta \in V$.

- A bilinear form f is called alternating if $f(\alpha,\alpha)=0$ for all $\alpha \in V$.

- A bilinear form is reflexive if and only if it is either symmetric or alternating.

- A bilinear form is reflexive if it is either symmetric or skew-symmetric.

- A symmetric bilinear form on a finite

dimensional vector space V over a field F of characteristic not equal to 2 is diagonalizable.

- A mapping $q:V \to F$ is called a quadratic form if $q(\alpha)=f(\alpha,\alpha)$ for some bilinear form f on V.

- If dim. $V = n$ and if $T \in A(V)$ has all its characteristic roots in F, then T satisfy a polynomial of degree n over F.

- The ranges of values of two congruent quadratic forms are same.

- Every symmetric matrix of rank r is congruent to a diagonal matrix with r of its diagonal elements are only non-zero.

- The number of positive terms in any two normal forms of a quadratic form are same.
- The number of negative terms in any two normal forms of a quadratic form are the same.
- The excess of the numbers of positive terms over the number of negative terms in any two normal forms of a real quadratic form is the same.
- The signature of a real quadratic form is invariant for its all normal norms.
- Two real quadratic forms in n variables over the field of reals are real equivalent if and only if they have the same rank and signature (or index).
- Two real quadratic forms $X'AX$ is complex equivalent to the form
$$z_1^2 + z_2^2 + \ldots + z_r^2$$
where r is the rank of A.

- Two real quadratic forms in n variables are complex equivalent if and only if they have the same rank.
- Two real quadratic forms $X'AX$ and $Y'BY$ are orthogonally equivalent if and only if A and B have the same eigenvalues and these occur with the same multiplicities.
- A real quadratic form $Q(X) = X'AX$ in n variables is positive definite if and only if all the eigenvalues of A are positive.
- A positive definite real symmetric matrix is non-singular.
- Every real non-singular matrix A can be written as a product of positive definite symmetric matrix and orthogonal matrix.
- A necessary and sufficient condition for a real quadratic form $X'AX$ to be positive definite is that the leading principal minors of A are all positive.

Review Questions and Project Work

1. If V is an n-dimenstonal vector space over the field F, then show that dim. $L[V,V,F]=n^2$

2. Let F be a subfield of complex numbers and A be a symmetric $n \times n$ matrix over F. Show that there exists an invertible $n \times n$ matrix P over F such that $P'AP$ is diagonal.

3. Describe explicity all symmetric bilinear form S on R^3.

4. Let V be an n-dimensional vector space over the field of complex numbers and f be a skew-symmetric bilinear form on V. Prove that rank of f is even.

5. Let A be $n \times n$ matrix over the field f, show that
$$f(x,y)=X'AY$$
is a bilinear form on F^n.

6. Show that the symmetric matrix of a real quadratic form is unique.

7. Let $\alpha = (x_1, x_2, x_3)$, $\beta = (y_1, y_2, y_3) \in V_3(C)$
Define a function $f : V_3(C) \times V_3(C) \to C$ by
$$f(\alpha, \beta) = i x_1 \bar{y}_2 - i x_2 \bar{y}_1 + x_3 \bar{y}_3$$
Show that f is a Hermitian form on $V_3(C)$.

8. Find all invariant subspace of $A = \begin{pmatrix} 2 & -4 \\ 5 & -2 \end{pmatrix}$ viewed as a linear operator R^2.

9. Let $T:V \to V$ be linear and suppose $T = T_1 \oplus T_2$ w.r.t. a T-invariant direct sum decomposition $V = V \oplus W$. Let $m(t)$, $m_1(t)$ and $m_2(t)$ denote respectively the minimum polynomials of T, T_1

and T_2. Show that $m(T)$ is the least common multiple of $m_1(t)$ and $m_2(t)$.

10. Show that congruent matrices have the same rank.

11. Let $A = \begin{pmatrix} 0 & 1 & 1 \\ 1 & -2 & 2 \\ 1 & 2 & -1 \end{pmatrix}$ be a symmetric matrix. Find a non-singular matrix P such that $P^T A P$ is diagonal. Also find the matrix $P^T A P$.

12. Show that the quadratic form $q(x,y,z)$ corresponding to $\begin{pmatrix} 2 & -5 & 1 \\ -5 & -6 & -7 \\ 1 & -7 & 9 \end{pmatrix}$ is given by
$$q(x,y,z)=2x^2-10xy-6y^2+2xz-14yz+9z^2.$$

13. Consider the quadratic form $q(x,y,z)=x^2 + 4xy + 3y^2 - 6xz + 10yz + 7z^2$. Find a non-singular linear substitution expressing the variables x, y, z in terms of variables r, s, t such that $q(r,s,t)$ is diagonal.

14. Diagonalize $q(x,y) = 3x^2 - 12xy + 7y^2$ by completing the square.

15. Let $q(x,y,z)=x^2+y^2+2xz+4yz+3z^2$. Show that q is not positive definite.

16. Show that $q(x,y)=ax^2+bxy+cy^2$ is positive definite iff $b^2 - 4ac < 0$.

Objective Type Questions

FILL IN THE BLANKS

1. A bilinear form f is reflexive if for $\alpha, \beta \in V, f(\beta, \alpha) = 0$ whenever _____.

2. A bilinear form f is _____ if $f(\alpha, \beta) = f(\beta, \alpha)$ $\forall \alpha, \beta \in V$.

3. A bilinear form f is _____ if $f(\alpha, \beta) = -f(\beta, \alpha)$ $\forall \alpha, \beta \in V$.

4. A bilinear form f is _____ if $f(\alpha, \alpha) = 0$ $\forall \alpha \in V$.

5. A bilinear form f is non-degenerate if rank $(f) =$ _____.

6. A bilinear form f is alternating if $f(\alpha, \alpha) =$ _____ $\forall \alpha \in V$.

7. Let f be a bilinear form on a vector space V over F. Then the quadratic form on V associated with the bilinear form f is the function $q:V \to F$ defined by $q(\alpha) =$ _____ $\forall \alpha \in V$.

8. Every bilinear form on the vector space V over the field of complex numbers can be uniquely expressed as the _____ of a symmetric and skew-symmetric bilinear forms.

9. If f is a Hermitian form on a vector space V over the complex field C. Then $f(\alpha, a\beta + b\gamma)$ _____.

10. The sum of Hermitian form is _____.

TRUE/FALSE

Write 'T' for true and 'F' for false statement.

1. A bilinear form f is alternating if $f(\alpha, \alpha) = 0$ $\forall \alpha \in V.$ **(T/F)**

2. A symmetrical bilinear form f is positive definite for $0 \neq \alpha \in V, f(\alpha, \alpha) > 0$ **(T/F)**

3. Let F be a symmetric form on V, then signature$(f) = \dim.V^+ + \dim.V^-.$ **(T/F)**

4. A quadratic form f is negative semi-defintie if $q(\alpha) \neq 0$ $\forall \alpha \in V.$ **(T/F)**

5. The rank of a Hermitian form F on a vector space V if definite to be the rank of the matrix representation of $f.$ **(T/F)**

MULTIPLE CHOICE QUESTIONS

Choose the most appropriate one :

1. Let f be a symmetric bilinear form on V, then sig(f) is equal to :
   (a) $\dim.V^+ + \dim.V^-$ (b) $\dim.V^+ - \dim.V^-$
   (c) $\dim.V^+$ (d) none of the above

2. Let f be a symmetric bilinear form on V, then rank(f) is equal to :
   (a) $\dim.V^+ + \dim.V^-$ (b) $\dim.V^+ - \dim.V^-$
   (c) $\dim.V^+$ (d) none of these

3. A bilinear form f on a vector space $V(F)$, is called degenerate if:
   (a) for each $0 \neq \alpha \in V; f(\alpha, \beta) = 0$ $\forall \beta \in V$
   (b) for each $0 \neq \beta \in V; f(\alpha, \beta) = 0$ $\forall \alpha \in V$
   (c) both (a) and (b) are true
   (d) none of the above

4. If a linear transformation $X = BY$ is applied to a quadratic form $A(x, x)$, then the resultant is a quadratic form $C(y, y)$ whose matrix C can be expressed as :
   (a) $C = B'AB$ (b) $A = B'CB$
   (c) $A = B^{-1}CB$ (d) none of the above

5. When a quadratic form $q = X'AX$ is transformed to a quadratic form $Y'CY$ by a non-singular transformation $X = BT$, then :
   (a) $\rho(C) < \rho(A)$ (b) $\rho(C) > \rho(A)$
   (c) $\rho(C) \leq \rho(A)$ (d) none of these

6. The quadratic form corresponding to the diagonal matrix.$(\lambda_1, \lambda_2, ..., \lambda_n)$ is :
   (a) $x_1^2 + x_2^2 + ... + x_n^2$
   (b) $\lambda_1 x_1^2 + \lambda_2 x_2^2 + ... + \lambda_n x_n^2$
   (c) $\lambda_1^2 x_1^2 + \lambda_2^2 x_2^2 + ... + \lambda_n^2 x_n^2$
   (d) none of the above

7. The corresponding matrix of the quadratic form $q = ax^2 + 2hxy + by^2$ is :
   (a) $\begin{bmatrix} a & h \\ -h & b \end{bmatrix}$ (b) $\begin{bmatrix} a & h \\ h & b \end{bmatrix}$
   (c) $\begin{bmatrix} a & -h \\ -h & b \end{bmatrix}$ (d) none of the above

8. The corresponding matrix of the quadratic form

$$q = ax^2 + by^2 + cz^2 + 2fyz + 2gzx + 2hxy \text{ is :}$$

(a) $\begin{bmatrix} a & h & g \\ h & b & f \\ g & f & c \end{bmatrix}$

(b) $\begin{bmatrix} a & h & -g \\ -h & -b & f \\ g & f & c \end{bmatrix}$

(c) both (a) and (b) are true

(d) none of the above

Answers

FILL IN THE BLANKS

1. $f(\beta, \alpha) = 0$
2. symmetric
3. skew-symmetric
4. alternating
5. $\dim.V^+ + \dim.V^-$
6. 0
7. $f(\alpha, \alpha)$
8. sum
9. $\bar{a}f(\alpha, \beta) + \bar{b}f(\alpha, \gamma)$
10. Hermitian

TRUE/FALSE

1. T
2. T
3. F
4. T
5. T

MULTIPLE CHOICE QUESTIONS

1. (b)
2. (b)
3. (c)
4. (a)
5. (b)
6. (b)
7. (b)
8. (a)

COMPETITION CORNER
for JRF, NET/SET, GATE Aspirants

SOME FASCINATING FACTS

- If U and V are two vector spaces over the field F and $L(U, V, F)$ denote the set of all bilinear forms on $U \times V$, then $L(U, V, F)$ is also a vector space over the field F with respect to addition and multiplication of bilinear forms on $U \times V$.

- $L(U, V, F)$ is closed with respect to scalar multiplication defined on it.

- If V is a vector space of dimension n over F then there exists a one-one correspondance between the set of bilinear form on V and the set $n \times n$ matrices over F.

- If F is a subfield of the complex numbers the symmetric bilinear form according to the polarization identity.

- Every bilinear form can be expressed as the sum of a symmetric and skew-symmetric bilinear forms.

- If V is finite dimensional vector space over the field F then a bilinear form F on V is symmetric or non-symmetric as the matrix F in any ordered basis is symmetric or non-symmetric.

- Any bilinear form on V is the sum of a symmetric bilinear form and skew-symmetric bilinear form.

- Symmetric matrix of a real quadratic form is unique.

- If V is an n-dimensional vector space over the field F of complex numbers and f be a skew-symmetric bilinear form on V then rank of f is even.

- Let q be the quadratic form associated with the symmetric bilinear form f then

$$f(\alpha, \beta) = \frac{1}{2}\left[q(\alpha + \beta) - q(\alpha) - q(\beta)\right]$$

- The rank of a bilinear form f on a vector space F is defined as the rank of any matrix representation of f.

- If V is an n-dimensional vector space over the field F then dim.$(V, V, F) = n^2$.

- Every positive semi-definite symmetric matrix A has a positive semi-definite square root B such that $B^2 = A$.

SOME IMPORTANT ILLUSTRATIONS

- Let A be $n \times n$ matrix over K. Then the mapping f defined by $f(X, Y) = X^TAY$ is a bilinear form on K^n.

- Let P be change of basis matrix from a basis S to a basis S^1. Let A be the matrix representing a bilinear form in the basis S, then $B = P^TAP$ is the matrix representing f in the basis S^1.

- $q(x, y) = ax^2 + bxy + cy^2$ is positive definite if and only if $a > 0$ and the discriminant $D = b^2 - 4ac < 0$.

- Let A be a Hermitian matrix then f is a Hermitian form on C^n where f is defined by $f(X, X) = X^TAY$.

- Let A be a complex non-singular matrix then $H = A^*A$ is Hermitian and positive definite.

- Let f be a Hermitian form on V then there is a basis S of V in which f is represented by a diagonal matrix and every such diagonal representation has the same number p of

positive entries and the same number n of negative entries.

- The vector space $L(V, V, F)$ of all bilinear forms on V is isomorphic to vector space of all $n \times n$ matrices.

- Let V be a vector space of all ordered n-tuples over the field F such that $\alpha = (a_1, a_2, ..., a_n)$, $\beta = (b_1, b_2, ..., b_n)$ are any two elements of V. If f is a function from $V \times V$ into F defined by $f(x, y) = a_1b_1 + a_2b_2 + ... + a_nb_n$, then f is a bilinear form on V.

- Let T be a linear operator on $V(F)$ and f a bilinear form on V then the function $g : V \times V \to F$ defined by $g(\alpha, \beta) = f(T\alpha, T\beta)$ is a bilinear form on V.

- Let L_1 and L_2 be linear functionals on $V(F)$ then the function $f : V \times V \to F$ defined by $f(\alpha, \beta) = L_1(\alpha).L_2(\beta)$ is a bilinear form on V.

Self Assessment Test

1. Let V be any vector space over K and let $f : V \times V \to K$ be the zero function *i.e.*, $f(u, v) = 0$ for every $u, v \in V$. Show that f is bilinear.

2. Show that $f(0, v) = f(v, 0) = 0$ for every $v \in V$.

3. Find the matrix representation of f in the usual basis $E = \{e_1 = (1, 0), e_2 = (0, 1)\}$ of R^2.

4. Find the symmetric matrix A which corresponds to the quadratic form $q(x, y, z) = 3x^2 + 4xy - y^2 + 8xz - 6yz + z^2$.

5. Diagonalize $q(x, y) = 3x^2 - 12xy + 7y^2$.

6. Show that $q = x^2 + y^2 + 2xz + 4yz + 3z^2$ is not positive definite.

7. Let A be a Hermitian matrix. Show that f is a Hermitian form on C^n where f is defined by $f(x, y) = X^T A Y$.

8. Write the matrix A of the quadratic form $6x^2 + 35y^2 + 11z^2 + 4zx$. Find the eigenvalues of A and hence determine the value class of the given quadratic form.

9. Prove that the quadratic form $6x_1^2 + 3x_2^2 + 3x_3^2 - 4x_1x_2 - 2x_2x_3 + 4x_3x_1$ in three variables is positive definite.

10. Let V be a vector space of dimension 3 over a field F. Then show that dim $B(V)$ is 9.

11. Let f be the bilinear form on R^3 defined by
$f((x_1, x_2, x_3),(y_1, y_2, y_3)) = 3x_1y_1 - 2x_1y_2$

$+ 5x_2y_1 + 7x_2y_2 - 8x_2y_3 + 4x_3y_2 - x_3y_3$ and $A = \begin{bmatrix} 3 & -2 & 0 \\ 5 & 7 & -8 \\ 0 & 4 & -1 \end{bmatrix}$ is a matrix of f relative to the basis $\{e_1 = (1, 0, 0), e_2 = (0, 1, 0), e_3 = (0, 0, 1)\}$ then show that matrix B of f relative to the basis $\{e_1' = (1, 1, 0), e_2' = (1, 0, 1), e_3' = (0, 0, 1)\}$ is given by $\begin{bmatrix} 13 & 0 & -8 \\ 5 & 2 & -1 \\ 4 & -1 & -1 \end{bmatrix}$.

12. Show that the matrix of quadratic form q on R^3 given by
$$q(x_1, x_2, x_3) = x_1^2 - x_3^2 + 3x_1x_2 - 6x_2x_3$$
is given by $\begin{bmatrix} 1 & 3/2 & 0 \\ 3/2 & 0 & -3 \\ 0 & -3 & -1 \end{bmatrix}$.

13. Let f be a bilinear form on V over F. If characteristic of $F \neq 2$ *i.e.*, $1 + 1 \neq 0$ then $f(u, v) = -f(v, u)$ if and only if $f(v, v) = 0 \ \forall \ v \in V$.

14. A real quadratic form in three variables is equivalent to the diagonal form $6y_1^2 + 3y_2^2 + 0y^3$. Show that quadratic form is positive definite.

15. A real quadratic from $X^T A X$ in three variables is equivalent to the diagonal form $3y_1^2 - 4y_2^2 + 5y_3^2$. Show that the quadratic form $X^T A X$ is indefinite.

10 CANONICAL FORMS

10.1 INTRODUCTION

The relation of similarity (in matrices) arise when we study the various matrix representation of linear transformation of vector space into itself. Under similarity, the rank of matrix A is invariant since two similar matrices are certainly equivalent and rank is even invariant under equivalence.

Now, to check the similarity of two linear transformation, we have to compute a particular canonical form for each and check if these are same.

10.2 SIMILARITY OF LINEAR TRANSFORMATIONS

Definition 1. *Let $V(F)$ be the n-dimensional vector space over the field F and $A(V)$ be the set of all linear transformation from V to V. Then two linear transformation S, $T \in A(V)$ are said to be similar if $\exists$ an invertible linear transformation $C \in A(V)$ such that*

$$T = CSC^{-1}$$

The similarity of linear transformation is also an equivalence relation. Thus, we can decompose $A(V)$ into equivalence classes, each such classes is called similarity class.

Definition 2. *The special form of matrix representation (in some basis of V) of linear transformation in each similarity class is called canonical forms.*

Therefore, as we earlier said, in order to check the similarity of two linear transformations, we have to form a particular canonical form for each matrix representation of linear transformation and then we have to verify if these are the same.

10.3 INVARIANT SUBSPACE

Let $T:V \to V$ be a linear transformation. Then a subspace W of V is said to be invariant under T if $T(W) \subset W$ i.e., $\forall \alpha \in W$, $T(\alpha) \in W$.

THEOREM 1. *If W is a subspace invariant under $T \in A(V)$, then T induces a linear transformation T_q on quotient space V/W defined by $T_q(\alpha + W) = T(\alpha) + W$.*
Further, if T satisfies the polynomial $q(x) \in F[x]$, then so is T_q. Thus the minimal polynomial of T_q divide the minimal polynomial of T.

Proof. We have to show that T_q is well defined and also T_q is linear.

(i) T_q is well defined.

Take two elements $\alpha + W$ and $\beta + W$ of $V|W$ such that

$$\alpha + W = \beta + W, \text{ this implies } \alpha - \beta \in W$$

Now, $\qquad T(\alpha - \beta) = T(\alpha) - T(\beta) \in W \qquad\qquad$ [W is T-invariant.]

so, $\qquad T(\alpha)+W = T(\beta) + W$

$\Rightarrow \qquad T_q(\alpha + W) = T_q (\beta + W)$

Thus, T is well-defined.

(ii) T_q is linear transformation. For $(\alpha + W)$, $(\beta + W) \in V/W$

We have

$$T_q\{(\alpha + W) + (\beta+W)\} = T_q(\alpha+\beta+W)$$

$$= T(\alpha+\beta)+W$$

$$= T(\alpha)+T(\beta)+W \qquad [\because T \text{ is linear.}]$$

$$= T(\alpha)+W+T(\beta)+W$$

$$= T_q(\alpha+W)+T(\beta+W)$$

Also, $\qquad T_q\{C(\alpha+W)\} = T_q(C\alpha+W)$

$$= T(C\alpha)+W = CT(\alpha)+W \qquad [\because T \text{ is linear.}]$$

$$= C\{T(\alpha)+W\}$$

$$= CT_q(\alpha+W)$$

So, T_q is linear.

Again $\alpha + W \in V/W$, then

$$T_q{}^2(\alpha+W) = T^2(\alpha)+W = T(T(\alpha))+W$$

$$= T_q(T(\alpha)) + W = T_q (T_q(\alpha+W))$$

$$= T_q^2(\alpha + W)$$

Similarly, we can prove $(T_q{}^n) = (T_q)^n; \forall\ n \ge 0$

Now, for a polynomial $q(x) \in F[x]$ where

$$q(x) = a_n x^n + a_{n-1}x^{n-1} + \ldots\ldots + a_0$$

$$q(T_q)(\alpha+W) = q(T)(\alpha)+W$$

$$= a_n T^n(\alpha) + a_{n-1}T^{n-1}(\alpha) + \ldots\ldots + a_0 I(\alpha)+W$$

$$= \Sigma a_i T^i(\alpha)+W = \Sigma a_i(T^i(\alpha)+W)$$

$$= \Sigma a_i T_q{}^i(\alpha+W)$$

$$= a_i(T_q)^i(\alpha+W)$$

$$= q(T_q)(\alpha+W)$$

Hence T_q satisfy the polynomial $q(x)=0$ *i.e.,* $q(T_q)=0$. Thus, T_q is root of $q(x)=0$.

10.4 INVARIANT DIRECT-SUM DECOMPOSITIONS

Let $T:V\rightarrow V$ be a linear transformation such that V is the direct sum of T-invariant subspaces $W_1, W_2, \ldots\ldots, W_r$, *i.e.,* $V = W_1 \oplus W_2 \oplus \ldots\ldots \oplus W_r$, where $T(W_i)\subset W_i$ for $i\in N$ of T_i in the linear transformation restricted to W_i. Then T is said to be decomposable into operator T, so T can be written as $T = T_1 \oplus T_2 \ldots\ldots \oplus T_r$ and subspace $W_1, W_2, \ldots\ldots W_r$ are said to be T-invariant direct sum decomposition of V.

THEOREM 1. *If $V = W_1 \oplus W_2 \oplus \ldots\ldots \oplus W_r$ where n_i is dimension of each subspace W_i and every subspace is invariant under $T \in A(V)$, then a basis of V can be found so that the matrix of T in this basis is of the form*

$$\begin{bmatrix} A_1 & 0 & \cdots & 0 \\ 0 & A_1 & \cdots & 0 \\ \vdots & \vdots & \cdots & \vdots \\ 0 & 0 & \cdots & A_r \end{bmatrix}$$

where each A_i is an $n_i \times n_i$ matrix of linear transformation induced by T on W_r.

Proof. Let $\left\{\alpha_1^{(1)}, \alpha_2^{(1)}, ..., \alpha_n^{(1)}\right\}, \left\{\alpha_1^{(2)}, \alpha_2^{(2)}, ..., \alpha_n^{(2)}\right\} ... \left\{\alpha_1^{(r)}, \alpha_2^{(r)}, ..., \alpha_n^{(r)}\right\}$ be the basis of $W_1, W_2, ..., W_r$ respectively.

Since $V = W_1 \oplus W_2 \oplus ...... \oplus W_r$, therefore

$$\left\{\alpha_1^{(1)}, \alpha_2^{(1)}, ..., \alpha_n^{(1)}, \ \alpha_1^{(2)}, \alpha_2^{(2)}, ..., \alpha_n^{(2)}, ..., \alpha_1^{(r)}, \alpha_2^{(r)}, ..., \alpha_n^{(r)}\right\}$$

form a basis of V. Also, each W_i is T-invariant, so that $T(\alpha_j^i) \in W_i$ and it is linear combination of $\alpha_1^{(i)}, \alpha_2^{(i)}, ..., \alpha_n^{(i)}$ *i.e.,*

$$T\left(\alpha_j^{(i)}\right) = a_1^{(i)}\alpha_1^{(i)} + a_2^{(i)}\alpha_2^{(i)} + ... + a_n^{(i)}\alpha_n^{(i)} \qquad .....(1)$$

for every $i = 1, 2, ..., r$ and $j = 1, 2, ..., n$.

Thus, matrix representation of T with respect to basis V is obtained by (1) which is

$$\begin{bmatrix} A_1 & 0 & \cdots & 0 \\ 0 & A_1 & \cdots & 0 \\ \vdots & \vdots & \cdots & \vdots \\ 0 & 0 & \cdots & A_r \end{bmatrix}$$,where A_i is the matrix of T_i induced on W_i by T.

Now, we are in position to discuss some important canonical form for checking the similarity of two linear transformation. They are of following types:

  (i)   Normal form

  (ii)   Triangular form

  (iii)   Jordan form

  (iv)   Rational form

10.5 NORMAL FORM

A matrix A is said to be in *normal form* if it can be written as $A = \begin{pmatrix} I_r & 0 \\ 0 & 0 \end{pmatrix}$, where I_r is square identity matrix of order r.

THEOREM 1. *Let $T: U \rightarrow V$ be a linear transformation and $rank(T) = r$. Then there exists bases of U and V such that matrix representation of T has the form $A = \begin{pmatrix} I_r & 0 \\ 0 & 0 \end{pmatrix}$, $I_r \rightarrow$ represents the identity matrix of order r.*

Proof. Let the dim. $U = m$ and dim. $V = n$. Let W be kernel of T. Now rank of T is r so the dimension of kernel space of T is $m-r$. Consider $\{\alpha_1, \alpha_2, ..., \alpha_{m-r}\}$ be the basis of W. By extension theorem, it can be extended to form the basis of U. Let this extension be $\{v_1, v_2, ..., v_n, \alpha_1, \alpha_2, ..., \alpha_{m-r}\}$. By setting a transformation $T(v_i) = u_i$, the set $\{u_1, u_2, ..., u_r\}$ form a basis of image (T) and thus base can be extended to form the basis of V. Let this base be

$$\{u_1, u_2 ... u_r, u_{r+1} .... u_n\}$$

Here, we can observe that every v_i under T can be written as the linear combination of u_i as

$$T(v_1)=u_1=1u_1+0u_2+...+0u_r+0u_n$$
$$T(v_2)=u_2=0u_1+1u_2+...+0u_r+0u_n$$
$$...\quad...\quad...\quad...\quad...\quad...\quad...$$
$$T(v_r)=u_r=0u_1+0u_2+...+1u_r+0u_n$$
$$T(\alpha_1)=0=0u_1+0u_2+...+0u_r+0u_{r+1}+...+0u_n$$
$$T(\alpha_2)=0=0u_1+0u_2+...+0u_r+0u_{r+1}+...+0u_n$$
$$...\quad...\quad...\quad...\quad...\quad...\quad...$$
$$T(\alpha_{m-r})=0=0u_1+0u_2+...+0u_r+0u_{r+1}+...+0u_n$$

so the matrix representation of T is given by

$$A=\begin{bmatrix} 1 & 0 & 0 & \cdots & 0 & 0 & \cdots & 0 \\ 0 & 1 & 0 & \cdots & 0 & 0 & \cdots & 0 \\ \cdots & \cdots & \cdots & \cdots & \cdots & \cdots & \cdots & \cdots \\ 0 & 0 & 0 & \cdots & 1 & 0 & \cdots & 0 \\ 0 & 0 & 0 & \cdots & 0 & 0 & \cdots & 0 \\ \vdots & \vdots & \vdots & \cdots & \vdots & \vdots & \cdots & \vdots \\ 0 & 0 & 0 & \cdots & 0 & 0 & \cdots & 0 \end{bmatrix}_{n\times n}$$

Hence, $\quad A=\begin{pmatrix} I_r & 0 \\ 0 & 0 \end{pmatrix}$.

10.6 TRIANGULAR FORM

Let $T:V\to V$ be a linear transformation on V over F, then the matrix of T in the basis $\{\alpha_1,\alpha_2...,\alpha_n\}$ of V is triangular if

$$T(\alpha_1)=a_{11}\alpha_1$$
$$T(\alpha_2)=a_{21}\alpha_1+a_{22}\alpha_2$$
$$T(\alpha_3)=a_{31}\alpha_1+a_{32}\alpha_2+a_{33}\alpha_3$$
$$...\quad...\quad...\quad...\quad...\quad....$$
$$T(\alpha_n)=a_{n1}\alpha_1+a_{n2}\alpha_2+a_{nn}\alpha_n$$

THEOREM 1. *If $T \in A(V)$ has all its characteristic root in F, then there is a basis of V in which matrix representation of T is triangular.*

Proof. This result can be proved by induction on the dimension of V

(i) If dim. $V=1$, then every matrix representation is a matrix of order 1×1 which is trivially triangular.

(ii) Let this result hold good for all vector space over F of dimension $n-1$. Let dim $V=n > 1$. If $\lambda_1 \in F$ be a characteristic root of T, then $\exists$ a non-zero eigen-vector α_1 corresponding to λ_1 such that $T(\alpha_1)=a_{11}\alpha_1$. It is due to the fact that T has all its characteristic roots in F. Let W be the one dimensional subspace of V spanned by α_1 and T invariant, then quotient space $V_q =V/W$, and

$$\text{dim. } V_q = \text{dim. } V - \text{dim. } W = n-1.$$

Now, T induces a linear transformation T_q on V_q, whose minimal polynomial divides the minimal polynomial or T so all roots of minimal polynomial of T_q are also the roots of minimal polynomial of T and hence all roots lie in F.

Thus V and T satisfy the hypothesis of the theorem.

Now, dim. of $V_q = n-1$. Then by the hypothesis of induction, there is a basis for V_q as $\{\bar{\alpha}_2, \bar{\alpha}_3, ..., \bar{\alpha}_n\}$ of V_q such that

$$T_q(\bar{\alpha}_2) = a_{22}\bar{\alpha}_2$$
$$T_q(\bar{\alpha}_3) = a_{32}\bar{\alpha}_2 + a_{33}\bar{\alpha}_3$$
$$\cdots \quad \cdots \quad \cdots \quad \cdots$$
$$T_q(\bar{\alpha}_n) = a_{n2}\bar{\alpha}_2 + a_{n3}\bar{\alpha}_3 + ... + a_{nn}\bar{\alpha}_n$$

Now, elements $\{\alpha_2, \alpha_3, ..., \alpha_n\}$ being the elements of V, also belong to cosets $\bar{\alpha}_2, \bar{\alpha}_3, ..., \bar{\alpha}_n$ respectively $i.e.$, $\bar{\alpha}_i = \alpha_i + W$.

Now, $\qquad T_q(\bar{\alpha}_2) = a_{22}\bar{\alpha}_2$

$\Rightarrow \qquad T_q(\alpha_2 + W) = a_{22}(\alpha_2 + W) \Rightarrow T(\alpha_2) + W = a_{22}\alpha_2 + W$

$\Rightarrow \qquad T(\alpha_2) - a_{22}\alpha_2 \in W$

But W is spanned by α_1 so

$$T(\alpha_2) - a_{22}\alpha_2 = a_{21}\alpha_1$$

$\Rightarrow \qquad T(\alpha_2) = a_{21}\alpha_1 + a_{22}\alpha_2$

Similarly, for $\bar{\alpha}_3, \bar{\alpha}_4, ..., \bar{\alpha}_n$ we have $T(\alpha_i) = a_{i1}\alpha_1 + a_{i2}\alpha_2 + ... + a_{in}\alpha_n$.

In this way, we get

$$T(\alpha_1) = a_{11}\alpha_1$$
$$T(\alpha_2) = a_{21}\alpha_1 + a_{22}\alpha_2$$
$$\cdots \quad \cdots \quad \cdots \quad \cdots \quad \cdots \quad \cdots$$
$$T(\alpha_n) = a_{n1}\alpha_1 + a_{n2}\alpha_2 + a_{nn}\alpha_n$$

Hence the matrix of T in the basis $\{\alpha_1, \alpha_2 ...... \alpha_n\}$ is triangular.

REMARKS

- The above theorem can be restated as : "If a square matrix A has all its characteristic roots in F, then A is similar to a triangular matrix, $i.e.$, there exists an invertible matrix P such that $P^{-1}AP$ is triangular."

- If any linear transformation T is represented by a triangular matrix

$$A = \begin{bmatrix} a_{11} & a_{12} & \cdots & a_{1n} \\ 0 & a_{22} & \cdots & a_{2n} \\ \vdots & \vdots & \cdots & \vdots \\ 0 & 0 & \cdots & a_{nn} \end{bmatrix}$$

the characteristic polynomial of T is a product of linear factors and is given by
$$\Delta(x) = |A - x| = (x - a_{11})(x - a_{22}) ...... (x - a_{nn}).$$

THEOREM 2. *If dim. $V = n$ and if $T \in A(V)$ has all its characteristic roots in F, then T satisfy a polynomial of degree n over F.*

Proof. Let $\lambda_1, \lambda_2, ..., \lambda_n$ be the characteristic roots of T. Since T has all its characteristic root in F so there exists a basis $\{\alpha_1, \alpha_2, ..., \alpha_n\}$ of V such that
$$T(\alpha_1) = \lambda_1\alpha_1$$

$$T(\alpha_2) = a_{21}\alpha_1 + \lambda_2\alpha_2$$
$$T(\alpha_3) = a_{31}\alpha_1 + a_{32}\alpha_2 + \lambda_3\alpha_3$$
$$... \quad ... \quad ... \quad ... \quad ... \quad ....$$
$$T(\alpha_n) = a_{n1}\alpha_1 + a_{n2}\alpha_2 + ... + \lambda_n\alpha_n$$

The all above relations can be rewritten as

$$(T-\lambda_1 I)(\alpha_1) = 0$$
$$(T-\lambda_2 I)(\alpha_2) = a_{21}\alpha_1$$
$$(T-\lambda_3 I)(\alpha_3) = a_{31}\alpha_1 + a_{32}\alpha_2$$
$$... \quad ... \quad ... \quad ... \quad ... \quad ....$$
$$(T-\lambda_n I)(\alpha_n) = a_{n1}\alpha_1 + a_{n2}\alpha_2 + ... + \alpha_{n(n-1)}\alpha_{n-1}$$

Now, $(T-\lambda_2 I)(T-\lambda_1 I)(\alpha_2) = (T-\lambda_1 I)(T-\lambda_2 I)(\alpha_2)$

$$= (T-\lambda_1 I)a_{21}\alpha_1 \qquad \text{[From above relations]}$$
$$= a_{21}(T-\lambda_1 I)(\alpha_1)$$

and $(T-\lambda_3 I)(T-\lambda_2 I)(T-\lambda_1 I)(\alpha_3)$

$$= (T-\lambda_2 I)(T-\lambda_1 I)(T-\lambda_3 I)(\alpha_3)$$
$$= (T-\lambda_2 I)(T-\lambda_1 I)(a_{31}\alpha_1 + a_{32}\alpha_2) \qquad \text{[From above relations]}$$
$$= (T-\lambda_2 I)(T-\lambda_1 I)(a_{31}\alpha_1) + (T-\lambda_2 I)(T-\lambda_1 I)(a_{32}\alpha_2)$$
$$= a_{31}(T-\lambda_2 I)(T-\lambda_1 I)(\alpha_1) + a_{32}(T-\lambda_2 I)(T-\lambda_1 I)(\alpha_2)$$
$$= 0 + 0 = 0.$$

Proceeding in the same way, we get

$$(T-\lambda_n I)(T-\lambda_{n-1} I)(T-\lambda_{n-2} I)...(T-\lambda_1 I)(\alpha_n) = 0$$

If $(T-\lambda_n I)(T-\lambda_{n-1} I)...(T-\lambda_1 I)$ is represented by S, then we have $S(\alpha_1) = S(\alpha_2) = ... S(\alpha_n) = 0$. Thus S, being the annihilator of base of V, annihilates all elements of V.

i.e., $\qquad S = 0 \Rightarrow (T-\lambda_n I)(T-\lambda_{n-1} I),...(T-\lambda_1 I) = 0$

Hence T satisfy the polynomial $q(x) = (x-\lambda_n)(x-\lambda_{n-1}),...(x-\lambda_1)$ in $F[x]$ of degree n.

THEOREM 3. *If a subspace W of V is T-invariant, then T has a matrix representation $\begin{pmatrix} A & B \\ 0 & C \end{pmatrix}$, where A is matrix representation of the restricted T_q of T to W.*

Proof. The proof of this result can be showed simply by matrix representation of T_q. Let $\{\beta_1, \beta_2, ..., \beta_r\}$ be the basis of W then by extension theorem, it can be extended to the basis $\{\beta_1, \beta_2, ..., \beta_r, \alpha_1, \alpha_2, ..., \alpha_s\}$ of V.

Since W is T-invariant, so

$T_q(\beta_i) = T(\beta_i)$, for $i = 1, 2, ..., r$. Now we have

$$T_q(\beta_1) = T(\beta_1) = a_{11}\beta_1 + ... + a_{1r}\beta_r$$
$$T_q(\beta_2) = T(\beta_2) = a_{21}\beta_1 + ... + a_{2r}\beta_r$$
$$... \quad ... \quad ... \quad ... \quad ... \quad .... \quad ....$$
$$T_q(\beta_r) = T(\beta_r) = a_{r1}\beta_1 + a_{r2}\beta_2 + ... + a_{rr}\beta_r$$

and $\qquad T(\alpha_1) = b_{11}\beta_1 + ... + b_{1r}\beta_r + c_{11}\alpha_1 + ... + c_{1s}\alpha_s$

Continuing in this way, we get

$$T(\alpha_1) = b_{s1}\beta_1 + ... + b_{sr}\beta_r + c_{s1}\alpha_1 + ... + b_{ss}\alpha_s$$

$$\begin{bmatrix} a_{11} & a_{21} & \cdots & a_{r1} & b_{11} & b_{21} & \cdots & b_{s1} \\ \vdots & \vdots & \cdots & \vdots & \vdots & \vdots & \cdots & \vdots \\ a_{1r} & a_{2r} & \cdots & a_{r2} & b_{1r} & b_{2r} & \cdots & b_{sr} \\ \cdots & \cdots & \cdots & \cdots & \cdots & \cdots & \cdots & \cdots \\ 0 & 0 & \cdots & 0 & C_{11} & C_{21} & \cdots & C_{s1} \\ \vdots & \vdots & \cdots & \vdots & \vdots & \vdots & \cdots & \vdots \\ 0 & 0 & \cdots & 0 & C_{1s} & C_{2s} & \cdots & C_{ss} \end{bmatrix}$$

$$= \begin{pmatrix} A & B \\ 0 & S \end{pmatrix}$$

where, $A = \begin{bmatrix} a_{11} & a_{21} & \cdots & a_{r1} \\ \cdots & \cdots & \cdots & \cdots \\ a_{1r} & a_{2r} & \cdots & a_{rr} \end{bmatrix}_{r \times r}$ $B = \begin{bmatrix} b_{11} & b_{21} & \cdots & b_{s1} \\ \cdots & \cdots & \cdots & \cdots \\ b_{1r} & b_{2r} & \cdots & b_{sr} \end{bmatrix}_{r \times s}$

$C = \begin{bmatrix} c_{11} & c_{21} & \cdots & c_{r1} \\ \cdots & \cdots & \cdots & \cdots \\ c_{1s} & c_{2s} & \cdots & c_{ss} \end{bmatrix}$ $O = \begin{bmatrix} 0 & 0 & \cdots & 0 \\ \cdots & \cdots & \cdots & \cdots \\ 0 & 0 & \cdots & 0 \end{bmatrix}_{s \times r}$

10.7 NILPOTENT TRANSFORMATION

A linear transformation $T : V \to V$ is said to be nilpotent if $T^n = 0$ for some least positive integer n.

or

Any $T \in A(V)$ is nilpotent then for some $k \in Z^+$, $T^k = 0$ but $T^{k-1} \neq 0$, where k is index of nilpotency.

REMARK

- The characteristic root of nilpotent transformation are zero, So they belong to F and hence all characteristic roots of nilpotent transformation belong to F. Then, we can say that nilpotent linear transformation can always be brought to triangular form over F.

10.8 JORDAN CANONICAL FORM

The matrix of the form

$$J = \begin{bmatrix} \lambda & 1 & 0 & \cdots & 0 & 0 \\ 0 & \lambda & 1 & \cdots & 0 & 0 \\ \cdots & \cdots & \cdots & \cdots & \cdots & \cdots \\ 0 & 0 & 0 & \cdots & \lambda & 1 \\ 0 & 0 & 0 & \cdots & 0 & \lambda \end{bmatrix}$$

is called Jordan block matrix belonging to λ. In this matrix A, λ's are on the diagonal and 1's are on the superdiagonal and other elements are zero.

THEOREM 1. *Let $T : V \to V$ be a linear operator whose characteristic and minimal polynomial are respectively given by*

$$\Delta(x) = (x - \lambda_1)^{n_1} (x - \lambda_2)^{n_2} \ldots (x - \lambda_r)^{n_r}$$

and $$m(x) = (x - \lambda_1)^{m_1} (x - \lambda_2)^{m_2} \ldots (x - \lambda_r)^{m_r}$$

where, λ_i are different scalars, then T has a block diagonal matrix representation

$$J = \begin{bmatrix} J_1 & & & \\ & J_2 & & \\ & & \cdots & \\ & & & J_r \end{bmatrix}$$

where

$$J_i = \begin{bmatrix} \lambda_i & 1 & 0 & \cdots & 0 & 0 \\ 0 & \lambda_i & 1 & \cdots & 0 & 0 \\ \cdots & \cdots & \cdots & \cdots & \cdots & \cdots \\ 0 & 0 & 0 & \cdots & \lambda_i & 1 \\ 0 & 0 & 0 & \cdots & 0 & \lambda_i \end{bmatrix}$$

For each λ_i, the corresponding block J_i have the following properties :

1. There is at least one J_i of order m, and all other J_i are of order less than or equal to m_i.

2. The sum of the orders of J_i is n_i.

3. The number J_i equals the geometric multiplicity of λ_i.

4. The number J_i of each possible order is uniquely determined by T.

Proof. We can write T by primary decomposition theorem as
$$T = T_1 \oplus T_2 \oplus \ldots \ldots \oplus T_r.$$
where $(x - \lambda_i)^{m_i}$ is the minimal polynomial of T_i.

Since the minimal polynomial is satisfied by the operator, therefore we have

$$\left(T_i - \lambda_i I\right)^{m_i} = 0 \qquad \text{for } i = 1, 2, 3, \ldots, r$$

Now taking $\quad N_i = T_i - \lambda_i I$ for $i = 1, 2, 3, \ldots, r$

$\Rightarrow \qquad T_i = N_i + \lambda_i I$ and $N_i{}^{m_i} = 0$

This implies that N_i is nilpotent of index m_i and T_i is the sum of N_i, and scalar operator $\lambda_i I$.

Now, N_i being the nilpotent of index m_i, we can select a basis in which N_i is represented by a canonical form as

$$\begin{bmatrix} 0 & 1 & 0 & \cdots & 0 & 0 \\ 0 & 0 & 1 & \cdots & 0 & 0 \\ \cdots & \cdots & \cdots & \cdots & \cdots & \cdots \\ 0 & 0 & 0 & \cdots & 0 & 1 \\ 0 & 0 & 0 & \cdots & 0 & 0 \end{bmatrix}$$

In this basis $T_i = N_i + \lambda_i I$ can be reduced to a block diagonal matrix whose block are block Jordan matrix J_i. T is direct sum of $T_1, T_2, \ldots \ldots T_r$ therefore the direct sum of matrix representation of T_i gives the block diagonal matrix representation of T whose diagonal block are matrices J_i which have the following properties:

1. N_i is the nilpotent of index m_i so there is at least one J_i of order m_i.

2. Since T and the block diagonal matrix representation of T have the same characteristic polynomial so that the sum of orders of J_i is n_i.

3. Since the nullity of N_i is equal to the geometric multiplicity of eigenvalue λ_i because characteristic equation of N_i is $(x - \lambda_i)^{m_i} = 0$. Hence the number of J_i is equal to the geometric multiplicity of λ_i.

4. Since T_i and N_i are uniquely determined by T. Hence number of J_i of each possible order is uniquely determined by T.

10.9 RATIONAL CANONICAL FORM

The Jordan canonical form is exerted when the minimal polynomials cannot be factored into linear polynomial while in rational canonical form, the minimal polynomial is taken as the product of linear polynomial.

THEOREM 1. *Let $T : V \to V$ be a linear operator with minimal polynomial*

$b_1(x) = q_1(x)^{l_1} . q_2(x)^{l_2} ... q_k(x)^{l_k}$, *where $q_1(x), q_2(x), ... q_k(x)$ are distinct monic irreducible polynomial. Then T has a unique block diagonal matrix representation*

$$\begin{bmatrix} C_1 & & & & \\ & C_2 & & & \\ & & C_3 & & \\ & & & ... & \\ & & & & C_k \end{bmatrix}$$

where C_i is the companion matrix of polynomial $q_i(x)^{l_{ij}}$ where

$$l_1 = l_{11} \geq l_{12} \geq ... \geq l_{1r}...$$
$$l_k = l_{k_1} \geq l_{k_2} \geq .... \geq l_{k_r}$$

Proof. By primary decomposition theorem, V can be decomposed as $V = V_1 \oplus V_2 \oplus ... \oplus V_k$, where each V_i is T-invariant and the minimal polynomial of T_i, linear transformation induced by T on V_i has minimal polynomial $q_i(x)^{l_{ij}} ...$

The matrix representation of T_i in some of V_i is companion matrix C_i. But $V = V_1 \oplus V_2 \oplus ...... \oplus V_k$.

Thus the matrix of T is

$$\begin{bmatrix} C_1 & & & & \\ & C_2 & & & \\ & & C_3 & & \\ & & & ... & \\ & & & & C_k \end{bmatrix}$$

This matrix representation is called rational canonical form and polynomial

$$q_1(x)^{l_{11}}, q_1(x)^{l_{12}}, ..., q_1(x)^{l_{1r}} ... q_k(x)^{l_{k_1}} ... q_k(x)^{l_{k_2}} ... q_k(x)^{l_{k_r}}$$

are called the elementary divisor of T.

10.10 RAW AND COLUMN SPACE OF A MATRIX

Definition 1. Let A be an arbitrary $m \times n$ matrix over a field K and

$$A = \begin{pmatrix} a_{11} & a_{12} & ... & a_{1n} \\ a_{21} & a_{22} & ... & a_{2n} \\ ... & ... & ... & ... \\ a_{m1} & a_{m2} & ... & a_{mn} \end{pmatrix}$$

The rows of A i.e., $R_1 = (a_{11}, a_{12}, ..., a_m)$, ..., $R_m = (a_{m1}, a_{m2}, ..., a_{mn})$, viewed as vectors in K^n, span a subspace of K^n called the row space of A. That is,

$$\text{rowsp } (A) = \text{span } (R_1, R_2, ..., R_m).$$

Definition 2. The columns, C_1, C_2, ..., C_n of an $m \times n$ matrix A over a field K, viewed as vectors in K_n, span a subspace of K_m called the column space of A. That is,

$$colsp~(A) = span~(C_1, C_2, ..., C_n).$$

THEOREM 1. *Row equivalent matrices have the same row space.*

Proof. Suppose we apply elementary row operations on a matrix A : (i) $R_i \leftrightarrow R_j$, (ii) $R_i \to kR_i$, $k \neq 0$, or (iii) $R_i \to kR_j + R_i$ and obtain a matrix B. Then each row of B is clearly a row of A or a linear combination of rows of A. Hence the row space of B is contained in the row space of A. On the other hand, we can apply the inverse elementary row operation on B and obtain A; hence the row space of A is contained in the row space of B. Therefore, A and B have the same row space. Thus any sequence of elementary row operations produces a matrix with the same row space. Hence, row equivalent matrices have the same row space.

THEOREM 2. *Let $A = (a_{ij})$ and $B = (b_{ij})$ be echelon matrices in row canonical form. Then A and B have the same row space if and only if they have the same non zero rows.*

Proof. Clearly, if A and B have the same non zero rows then they have the same row space. Thus we only have to prove the converse.

Suppose A and B have the same row space and suppose $R \neq 0$ is the ith row of A. Then there exist scalars $c_1, c_2, ..., c_s$ such that

$$R = c_1 R_1 + c_2 R_2 + ... + c_s R_s \qquad ...(1)$$

where the R_i are the non-zero rows of B.

The theorem is proved if we show that $R = R_i$, that is, $C_i = 1$ but $c_k = 0$ for $k \neq i$.

Let a_{ij_i} be the leading non zero entry of R. By (1) and we know that

$$a_{ij_i} = c_1 b_{1j_i} + c_2 b_{2j_i} + ... + c_s b_{sj_i} \qquad ...(2)$$

Since, b_{ij_i} is a leading non zero entry of B and, since B is row reduced, it is the only nonzero entry in the j_ith column of B. Thus from (2) we obtain $a_{ij_i} = b_{ij_i}$. However, $a_{ij_i} = 1$ and $b_{ij_i} = 1$ since A and B are row reduced; hence $c_i = 1$.

Now suppose $k \neq i$, and $b_{k j_k}$ is the distinguished entry in R_k. By (1) and we know that

$$a_{i j_k} = c_1 b_{1 j_k} + c_2 b_{2 j_k} + ... + c_s b_{sj_k} \qquad ...(3)$$

Since B is row reduced, $b_{i k_k}$ is the only non zero entry in the j_k^{th} column of B; so by (3), $a_{i j_k} = c_k b_{k j_k}$. Since, a_{kj} is a leading non zero entry of A and, since A is row reduced, $a_{i j_k} = 0$. Thus $c_k b_{k j_k} = 0$ and, since $b_{kj} = 0$, $c_k = 0$.

Hence, $\qquad R = R_i$.

THEOREM 3. *Let A be any matrix. Then A is row equivalent to a unique matrix in row canonical form.*

Proof. Suppose A is row equivalent to matrices A_1 and A_2 where A_1 and A_2 are in row canonical form. By theorem 1,

rowsp (A) = rowsp (A_1) and rowsp (A) = rowsp (A_2); hence
rowsp (A_1) = rowsp (A_2)

Since A_1 and A_2 are in row canonical form, then by theorem 2

$$A_1 = A_2$$

THEOREM 4. *Matrices A and B have the same row space iff their row canonical forms have the same non zero rows.*

Proof. Let A_1 and B_1 be the row canonical forms of A and B, respectively. Suppose A and B have the same row space. Then

$$\text{rowsp } (A_1) = \text{rowsp } (A) = \text{rowsp } (B) = \text{rowsp } (B_1).$$

By theorem 2, A_1 and B_1 have the same nonzero rows.

Then $\text{rowsp } (A) = \text{rowsp } (A_1) = \text{rowsp } (B_1) = \text{rowsp } (B).$

THEOREM 5. *Let A and B be matrices such that the product AB is defined. Then the row space of AB is contained in the row space of B.*

Proof. The rows of AB are R_iB where R_i is the ith row of A. Hence by the above result each row of AB is in the row space of B. Thus the row space of AB is contained in the row space of B.

THEOREM 6. *colsp (AB) ⊆ colsp (A).*

Proof. Using theorem 4, we have

$$\text{colsp}(AB) = \text{rowsp } [(AB)^T]$$
$$= \text{rowsp } (B^TA^T) \text{ rowsp } (A^T) = \text{colsp}(A).$$

Solved Examples

Example 1. *Suppose W is invariant under S:V→V and T : V→V. Show that W is also invariant under S+T and ST.*

Solution. Since W is S-invariant and T-invariant, if $\alpha \in W$, then $S(\alpha) \in W$ and $T(\alpha) \in W$. Now, if $\alpha \in W$, then

$$(S+T)(\alpha) = S(\alpha) + T(\alpha)$$

Since $S(\alpha), T(\alpha) \in W$, so $S(\alpha) + T(\alpha) \in W$

$$(S + T) (\alpha) \in W$$

∴ W is invariant under $S + T$.

Also, for $\alpha \in W$.

$$(ST)(\alpha) = S(T(\alpha))$$

Since $T(\alpha) \in W$ and W is invariant under S, so that

$$S(T(\alpha)) \in W$$

$\Rightarrow$ $(ST) (\alpha) \in W$

∴ W is invariant under ST.

Example 2. *Determine all invariant subspaces of $A = \begin{bmatrix} 2 & -5 \\ 1 & -2 \end{bmatrix}$ viewed as an operator on R^2.*

Solution. Since R^2 and $\{0\}$ are invariant subspace of A. If A has any other invariant subspaces, then it must be one-dimensional. The characteristic polynomial of A is

$$\Delta(x) = |xI - A|$$
$$= \begin{vmatrix} x-2 & 5 \\ -1 & x+2 \end{vmatrix}$$
$$= (x-2) (x+2) + 5$$
$$= x^2 + 1$$

Clearly, A has no eigenvalue in R so A has no eigenvectors in R^2. Hence, $R^2 + \{0\}$ are the only subspace invariant under A.

Example 3. *Suppose $T: V \rightarrow V$ is linear and suppose $T = T_1 \oplus T_2$, with respect to a T-invariant direct sum decomposition $V = V_1 \oplus V_2$, Show that*

    (i) *$m(x)$ is the least common multiple of $m_1(x)$ and $m_2(x)$ where $m(x), m_1(x)$ and $m_2(x)$ are the minimal polynomials of T, T_1 and T_2, respectively.*

    (ii) *$\Delta(x) = \Delta_1(x)\Delta_2(x)$ where $\Delta(x), \Delta_1(x)$ and $\Delta_2(x)$ are the characteristic polynomials of T, T_1 and T_2 respectively.*

Solution.   (i)   Since T_1 is induced of T on V_1 and T_2 is induced of T on V_2, therefore the minimal polynomials $m_1(x)$ of T_1 and $m_2(x)$ of T_2 each divides $m(x)$.

         Suppose $p(x)$ is a multiple of both $m_1(x)$ and $m_2(x)$, then

$$(P(T_1))(V_1) = 0 \quad \text{and} \quad (P(T_2))(V_2) = 0$$

         Let $\alpha \in V$, then $\alpha = \alpha_1 + \alpha_2$, where $\alpha_1 \in V_1$ and $\alpha_2 \in V_2$

         Now, $\quad (P(T))(\alpha) = (P(T))(\alpha_1) + (P(T))(\alpha_2)$

$$= 0 + 0$$
$$= 0$$

         T is the root of $P(x)$. Hence $m(x) \mid p(x)$. So that $m(x)$ is the least common multiple.

  (ii)   Since $T = T_1 \oplus T_2$, then the matrix representation of T is

$$M = \begin{bmatrix} A & 0 \\ 0 & B \end{bmatrix}$$

         where A and B are the matrix representations of T_1 and T_2 respectively.

$$\Delta(x) = |xI - M|$$

$$= \begin{vmatrix} xI - A & 0 \\ 0 & xI - B \end{vmatrix} = (xI - A)(xI - B)$$

$$= \Delta_1(x)\Delta_2(x)$$

Example 4. *Let $T: V \rightarrow V$ be linear and let W be the eigenspace belonging to an eigenvalue λ of T. Show that W is T-invariant.*

Solution. By the definition of eigenspace, we have

$$W = \text{kernel}(T - \lambda I)$$

If $\alpha \in W$, then

$$(T - \lambda I)(\alpha) = 0 \Rightarrow T(\alpha) - \lambda I(\alpha) = 0$$

$\Rightarrow \qquad\qquad\qquad T(\alpha) = \lambda \alpha \qquad\qquad\qquad\qquad\qquad \text{...(1)}$

Since W is a subspace of W, so for any scalar $\lambda \in F$ and $\alpha \in W$, we have

$$\lambda \alpha \in W \Rightarrow T(\alpha) \in W \qquad\qquad \text{[Using (1)]}$$

Hence W is T-invariant.

Example 5. *If $\{W_i\}$ is a collection of T-invariant subspaces of a vector space V. Show that the intersection $W = \cap W_i$ is also T-invariant.*

Solution. Since each W_i is T-invariant, then for $\alpha \in W_i$, we have $T(\alpha) \in W_i$ for every i.

$\therefore \qquad\qquad T(\alpha) \in \cap W_i$ for every i.

$\Rightarrow \qquad\qquad T(\alpha) \in W.$

Hence W is T-invariant.

Example 6. *Let A be a square matrix over the complex field* C. *Suppose* λ *is an eigenvalue of* A^2. *Show that* $\sqrt{\lambda}$ *or* $-\sqrt{\lambda}$ *is an eigenvalue of A.*

Solution. Since, A is similar to a triangular matrix

$$B = \begin{bmatrix} u_1 & b_{12} & \cdots & b_{1n} \\ 0 & u_2 & \cdots & b_{2n} \\ \cdots & \cdots & \cdots & \cdots \\ 0 & 0 & \cdots & u_n \end{bmatrix}$$

Thus, A^2 is similar to the matrix.

$$B^2 = \begin{bmatrix} u_1^2 & u_1 b_{12} + b_{12} u_2 & \cdots & u_1 b_{1n} + \ldots + u_n b_{1n} \\ 0 & u_2^2 & \cdots & u_2 b_{2n} + \ldots + u_n b_{2n} \\ \cdots & \cdots & \cdots & \cdots \\ 0 & 0 & \cdots & u_n^2 \end{bmatrix}$$

Since similar matrices have the same eigenvalues, so $\lambda = u_i{}^2$ for some i.

$$\Rightarrow \qquad u_i = \sqrt{\lambda} \quad \text{or} \quad u_i = -\sqrt{\lambda} .$$

Hence $\sqrt{\lambda}$ or $-\sqrt{\lambda}$ is an eigenvalue of A.

Example 7. *Show that similar matrices have the same eigenvalues.*

Solution. If a matrix A is similar to B, then there exists an invertible matrix P such that
$$A = P^{-1}BP.$$

So, $|xI - A| = |xI - P^{-1}BP|$
$$= |P^{-1}(xI - B)P| = |P^{-1}||xI - B||P|$$
$$= |xI - B| \qquad\qquad [|P^{-1}||P| = 1]$$

This shows that similar matrices have the same characteristic polynomials. Hence they have the same eigenvalues.

Example 8. *Let the matrix A is given by*

$$A = \begin{bmatrix} 0 & 1 & 1 & 0 & 1 \\ 0 & 0 & 1 & 1 & 1 \\ 0 & 0 & 0 & 0 & 0 \\ 0 & 0 & 0 & 0 & 0 \\ 0 & 0 & 0 & 0 & 0 \end{bmatrix}$$

Show that it is nilpotent and find its index of nilpotency. Also, find the nilpotent matrix m in canonical form, which is similar to A.

Solution. We have, $A^2 = \begin{bmatrix} 0 & 0 & 1 & 1 & 1 \\ 0 & 0 & 0 & 0 & 0 \\ 0 & 0 & 0 & 0 & 0 \\ 0 & 0 & 0 & 0 & 0 \\ 0 & 0 & 0 & 0 & 0 \end{bmatrix} \neq 0$

$$A^3 = A^2 A$$

$$= \begin{bmatrix} 0 & 0 & 1 & 1 & 1 \\ 0 & 0 & 0 & 0 & 0 \\ 0 & 0 & 0 & 0 & 0 \\ 0 & 0 & 0 & 0 & 0 \\ 0 & 0 & 0 & 0 & 0 \end{bmatrix} \begin{bmatrix} 0 & 1 & 1 & 0 & 1 \\ 0 & 0 & 1 & 1 & 1 \\ 0 & 0 & 0 & 0 & 0 \\ 0 & 0 & 0 & 0 & 0 \\ 0 & 0 & 0 & 0 & 0 \end{bmatrix}$$

$$= \begin{bmatrix} 0 & 0 & 0 & 0 & 0 \\ 0 & 0 & 0 & 0 & 0 \\ 0 & 0 & 0 & 0 & 0 \\ 0 & 0 & 0 & 0 & 0 \\ 0 & 0 & 0 & 0 & 0 \end{bmatrix} = 0$$

Thus, A is a nilpotent matrix of index 2.

Since A is nilpotent of index, thus we can say that M contains the diagonal block matrix of order less than or equal to 2.

Clearly, the rank of $A = 2$ and the matrix A is of order 3. So that nullity of $A = 3$.

Thus, M will contain 3 diagonal block matrices, in which 2 diagonal block of order 2 each and 1 diagonal block of order 1.

$$M = \begin{bmatrix} M_2 & & \\ & M_2 & \\ & & M_1 \end{bmatrix}$$

where, $\qquad M_2 = \begin{bmatrix} 0 & 1 \\ 0 & 0 \end{bmatrix}, \quad M_1 = [0]$

Thus, we have

$$M = \left[\begin{array}{cc|ccc} 0 & 0 & 0 & 0 & 0 \\ 0 & 0 & 0 & 0 & 0 \\ \hline 0 & 0 & 0 & 1 & 0 \\ 0 & 0 & 0 & 0 & 0 \\ \hline 0 & 0 & 0 & 0 & 0 \end{array} \right]$$

Example 9. *If T is nilpotent of index k, show that $T^n, n > 1$ is nilpotent of index k.*

Solution. Since $T^k = 0$ but $T^{k-1} \neq 0$, then

$$(T^n)^k = (T^k)^n = 0^n = 0 \qquad\qquad [\because T^{k-1} \neq 0]$$

$$(T^n)^{k-1} = (T^{k-1})^n \neq 0$$

and thus T^n is nilpotent of index $\leq k$.

Example 10. *Suppose S and T are nilpotent operators which commutes i.e., $ST = TS$, Show that $S+T$ and ST are also nilpotent.*

Solution. Since S and T are nilpotent, so we have $S^m = 0$ and $T^n = 0$ for some positive integers m and n. Since S and T commutes, then

$$(S+T)^{m+n} = \sum_{r=0}^{m+n} C_r T^{m+n-r} . S^r \qquad\qquad \dots (1)$$

(i) If $r \geq m$, then $S^r = 0$. So from (1), we get
$$(S+T)^{m+n} = 0$$

(ii) If $r < m$, so $m+n-r \geq n$, then $T^{m+n-r} = 0$
Therefore, from (1), we get
$$(S+T)^{m+n} = 0$$
$\Rightarrow \qquad S+T$ is nilpotent.

Now to show that ST is nilpotent.

Suppose $S^m = 0$ and $T^n = 0$ and $m \leq n$.

Then, $(ST)^n = S^n T^n = S^n = 0 = 0$

If $m \geq n$, then

$$(ST)^m = S^m T^m = 0.T^m = 0$$

Thus ST is nilpotent.

Example 11. *Determine all possible Jordan canonical forms for a linear operator $T: V \to V$ whose characteristic polynomial is $\Delta(x) = (x-2)^3 (x-5)^2$.*

Solution. Since $x - 2$ has exponent 3 in $\Delta(x)$ and $x-5$ has exponent 2 in $\Delta(x)$, therefore Jordan canonical form will be the matrix of order 5×5.

We may write $\Delta(x)$ as

$$\Delta(x) = (x - \lambda_1)^3 (x - \lambda_2)^2, \text{ where } \lambda_1 = 2 \text{ and } \lambda_2 = 5.$$

Now, $\lambda_1 = 2$ must appear a three times on the main diagonal and $\lambda_2 = 5$ must appear two times. Hence the possible Jordan canonical forms are

$$(i)\ \begin{bmatrix} 2 & 1 & 0 & 0 & 0 \\ 0 & 2 & 1 & 0 & 0 \\ 0 & 0 & 2 & 0 & 0 \\ 0 & 0 & 0 & 5 & 1 \\ 0 & 0 & 0 & 0 & 5 \end{bmatrix} \qquad (ii)\ \begin{bmatrix} 2 & 1 & 0 & 0 & 0 \\ 0 & 2 & 1 & 0 & 0 \\ 0 & 0 & 2 & 0 & 0 \\ 0 & 0 & 0 & 5 & 1 \\ 0 & 0 & 0 & 0 & 5 \end{bmatrix}$$

$$(iii)\ \begin{bmatrix} 2 & 1 & 0 & 0 & 0 \\ 0 & 2 & 1 & 0 & 0 \\ 0 & 0 & 2 & 0 & 0 \\ 0 & 0 & 0 & 5 & 1 \\ 0 & 0 & 0 & 0 & 5 \end{bmatrix} \qquad (iv)\ \begin{bmatrix} 2 & 1 & 0 & 0 & 0 \\ 0 & 2 & 1 & 0 & 0 \\ 0 & 0 & 2 & 0 & 0 \\ 0 & 0 & 0 & 5 & 1 \\ 0 & 0 & 0 & 0 & 5 \end{bmatrix}$$

$$(v)\ \begin{bmatrix} 2 & 1 & 0 & 0 & 0 \\ 0 & 2 & 1 & 0 & 0 \\ 0 & 0 & 2 & 0 & 0 \\ 0 & 0 & 0 & 5 & 1 \\ 0 & 0 & 0 & 0 & 5 \end{bmatrix} \qquad (vi)\ \begin{bmatrix} 2 & 1 & 0 & 0 & 0 \\ 0 & 2 & 1 & 0 & 0 \\ 0 & 0 & 2 & 0 & 0 \\ 0 & 0 & 0 & 5 & 1 \\ 0 & 0 & 0 & 0 & 5 \end{bmatrix}$$

Example 12. *If A is a complex 5×5 matrix with characteristic polynomial*

$$\Delta(x) = (x - 2)^3 (x + 7)^2$$

and minimal polynomial $m(x) = (x - 2)^2 (x + 7)$. What is the Jordan form for A?

Solution. Here, in $\Delta(x)$, the exponent of $(x - 2)$ is 3 and that of $(x + 7)$ is 2. Thus, A will be of order 5×5. Also, in $m(x)$, the exponent of $(x - 2)$ is 2 and that of $(x + 7)$ is 1. Therefore, Jordan form will have one block of order 2 and other two blocks must be of the order 2 or 1. Hence the required Jordan form of A is

$$\begin{bmatrix} 2 & 1 & 0 & 0 & 0 \\ 0 & 2 & 0 & 0 & 0 \\ 0 & 0 & 2 & 1 & 0 \\ 0 & 0 & 0 & 2 & 0 \\ 0 & 0 & 0 & 0 & -7 \end{bmatrix}$$

Example 13. *Determine all possible Jordan canonical forms for a matrix of order 5 whose minimal polynomial is* $m(x) = (x-2)^2$.

Solution. Since the minimal polynomial of a matrix of order 5×5 is $(x-2)^2$, therefore, its characteristic polynomial will be $(x-2)^5$.

Thus, Jordan canonical form must have an Jordan block matrix of order 2 and other must be of order 2 or 1.

Hence the possible Jordan canonical forms are :

(i) $$\begin{bmatrix} 2 & 1 & 0 & 0 & 0 \\ 0 & 2 & 0 & 0 & 0 \\ 0 & 0 & 2 & 1 & 0 \\ 0 & 0 & 0 & 2 & 0 \\ 0 & 0 & 0 & 0 & 2 \end{bmatrix}$$
(ii) $$\begin{bmatrix} 2 & 1 & 0 & 0 & 0 \\ 0 & 2 & 0 & 0 & 0 \\ 0 & 0 & 2 & 1 & 0 \\ 0 & 0 & 0 & 2 & 0 \\ 0 & 0 & 0 & 0 & 2 \end{bmatrix}$$

Example 14. *Suppose* $T : V \to V$ *has characteristic polynomial*

$$\Delta(x) = (x+8)^4(x-2)^3$$

and minimal polynomial

$$M(x) = (x+8)^3(x-1)^2$$

Find the Jordan canonical form of the matrix representation of T.

Solution. Since degree of $\Delta(x)$ is 7 so that Jordan form will be a matrix of order 7×7, in which -8 will be repeated 4 times on the diagonal and 1 will be three times on the diagonal. Also, $(x+8)$ has the exponent 3 in $m(x)$, therefore Jordan form will have one block of order 3×3 belonging to -8 and $(x-1)$ has the exponent 2 in $m(x)$. So, there must be one block of order 2×2 belonging to 1.

Hence the required Jordan canonical form is

$$\begin{bmatrix} -8 & 1 & 0 & 0 & 0 & 0 & 0 \\ 0 & -8 & 1 & 0 & 0 & 0 & 0 \\ 0 & 0 & -8 & 0 & 0 & 0 & 0 \\ 0 & 0 & 0 & -8 & 0 & 0 & 0 \\ 0 & 0 & 0 & 0 & 1 & 1 & 0 \\ 0 & 0 & 0 & 0 & 0 & 1 & 0 \\ 0 & 0 & 0 & 0 & 0 & 0 & 1 \end{bmatrix}$$

Example 15. *Find all possible rational canonical forms for* 6×6 *matrices with minimal polynomial* $m(x) = (x+1)^3$.

Solution. Let $T : V \to V$ be a linear with minimal polynomial $m(x) = (x+1)^3$ and dim. $V = 6$, then T is one of the following direct sum of companion matrices:

(i) $C((x+1)^3) \oplus C((x+1)^3)$

(ii) $C((x+1)^3) \oplus C((x+1)^2) \oplus C(x+1)$

(iii) $C((x+1)^3) \oplus C(x+1) \oplus C(x+1) \oplus C(x+1)$

Now, $C((1+x)^3) = C(x^3+3x^2+3x+1) = \begin{bmatrix} 0 & 0 & -1 \\ 1 & 0 & -3 \\ 0 & 1 & -3 \end{bmatrix}$

$C((1+x)^2) = C(x^2+2x+1) = \begin{bmatrix} 0 & -1 \\ 1 & -2 \end{bmatrix}$

$C(1+x) = [-1]$.

Thus, the rational canonical form of T is one of the following matrices :

(i) $\left[\begin{array}{ccc:ccc} 0 & 1 & -1 & 0 & 0 & 0 \\ 1 & 0 & -3 & 0 & 0 & 0 \\ 0 & 1 & -3 & 0 & 0 & 0 \\ \hdashline 0 & 0 & 0 & 0 & 0 & -1 \\ 0 & 0 & 0 & 1 & 0 & -3 \\ 0 & 0 & 0 & 0 & 1 & -3 \end{array}\right]$
(ii) $\left[\begin{array}{ccc:ccc} 0 & 0 & -1 & 0 & 0 & 0 \\ 1 & 0 & -3 & 0 & 0 & 0 \\ 0 & 1 & -3 & 0 & 0 & 0 \\ \hdashline 0 & 0 & 0 & 0 & -1 & 0 \\ 0 & 0 & 0 & 1 & -2 & 0 \\ 0 & 0 & 0 & 0 & 1 & -1 \end{array}\right]$

(iii) $\left[\begin{array}{ccc:ccc} 0 & 0 & -1 & 0 & 0 & 0 \\ 1 & 0 & -3 & 0 & 0 & 0 \\ 0 & 1 & -3 & 0 & 0 & 0 \\ \hdashline 0 & 0 & 0 & -1 & 0 & 0 \\ 0 & 0 & 0 & 0 & -1 & 0 \\ 0 & 0 & 0 & 0 & 0 & -1 \end{array}\right]$

Example 16. *Let A be a 4×4 matrix with minimal polynomial $m(x) = (x^2+1)(x^2-3)$. Find the rational canonical form for A if A is a matrix over (i) the rational field F, (ii) the real field R, (iii) the complex field C.*

Solution. (i) If the field is rational Q. The rational canonical form of A is the following direct sum of companion matrices :

(a) $C(x^2+1) \oplus C(x^2-3)$

Now, $C(x^2+1) = \begin{bmatrix} 0 & -1 \\ 1 & 0 \end{bmatrix}$

$C(x^2-3) = \begin{bmatrix} 0 & 3 \\ 1 & 0 \end{bmatrix}$

Thus the rational canonical form is

(i) $\left[\begin{array}{cc:cc} 0 & -1 & 0 & 0 \\ 1 & 0 & 0 & 0 \\ \hdashline 0 & 0 & 0 & 3 \\ 0 & 0 & 1 & 0 \end{array}\right]$

(ii)   If the field is the field of reals R, then the rational canonical form of A is the direct sum of the companion matrices

$$C(x^2+1) \oplus C(x-\sqrt{3}) \oplus C(x+\sqrt{3})$$

Now, $C(x-\sqrt{3}) = [\sqrt{3}]$ and $C(x+\sqrt{3}) = [-\sqrt{3}]$

Thus the required rational canonical form is

$$\begin{bmatrix} 0 & -1 & 0 & 0 \\ 1 & 0 & 0 & 0 \\ 0 & 0 & \sqrt{3} & 0 \\ 0 & 0 & 0 & -\sqrt{3} \end{bmatrix}$$

(iii) If the field is field of complex numbers C, then the rational canonical form of A is the direct sum of the companion matrices

$$C(x-i) \oplus C(x+i) \oplus C(x-\sqrt{3}) \oplus C(x+\sqrt{3})$$

Thus the required rational canonical form is

$$\begin{bmatrix} i & 0 & 0 & 0 \\ 0 & -i & 0 & 0 \\ 0 & 0 & \sqrt{3} & 0 \\ 0 & 0 & 0 & -\sqrt{3} \end{bmatrix}$$

Example 17. *Let V be a vector space of dimension 6 over R and let T be a linear operator whose minimal polynomial is $m(x) = (x^2 - x + 3)(x - 2)^2$. Find the rational canonical form of T.*

Solution. Since dim.$V = 6$, then T is one of the following direct sums of companion matrices:

(i) $C(x^2 - x + 3) \oplus C(x^2 - x + 3) \oplus C((x-2)^2)$

(ii) $C(x^2 - x + 3) \oplus C((x-2)^2) \oplus C((x-2)^2)$

(iii) $C(x^2 - x + 3) \oplus C((x-2)^2) \oplus C(x-2) \oplus C(x-2)$

where $C(q(x))$ is the companion matrix of $q(x)$.

Now $C(x^2 - x + 3) = \begin{bmatrix} 0 & -3 \\ 1 & 1 \end{bmatrix}$

$$C((t-2)^2) = C(t^2 - 4t + 4) = \begin{bmatrix} 0 & -4 \\ 1 & 4 \end{bmatrix}$$

$$C((x-2)) = [2]$$

Thus the canonical forms of T is one of the following matrices :

(i) $\begin{bmatrix} 0 & -3 & 0 & 0 & 0 & 0 \\ 1 & 1 & 0 & 0 & 0 & 0 \\ 0 & 0 & 0 & -3 & 0 & 0 \\ 0 & 0 & 1 & 1 & 0 & 0 \\ 0 & 0 & 0 & 0 & 0 & -4 \\ 0 & 0 & 0 & 0 & 1 & 4 \end{bmatrix}$

(ii) $\begin{bmatrix} 0 & -3 & 0 & 0 & 0 & 0 \\ 1 & 1 & 0 & 0 & 0 & 0 \\ 0 & 0 & 0 & -4 & 0 & 0 \\ 0 & 0 & 1 & 4 & 0 & 0 \\ 0 & 0 & 0 & 0 & 0 & -4 \\ 0 & 0 & 0 & 0 & 1 & 4 \end{bmatrix}$

(iii) $\begin{bmatrix} 0 & -3 & 0 & 0 & 0 & 0 \\ 1 & 1 & 0 & 0 & 0 & 0 \\ 0 & 0 & 0 & -4 & 0 & 0 \\ 0 & 0 & 1 & 4 & 0 & 0 \\ 0 & 0 & 0 & 0 & 2 & 0 \\ 0 & 0 & 0 & 0 & 0 & 2 \end{bmatrix}$

✿ EXERCISE 10.1

1. Prove that the relation of similarity is an equivalence relation in $A(V)$.

2. If A is triangular $n \times n$ matrix with entries λ_1, $\lambda_2, \ldots \lambda_n$ on the diagonal, then
$(A - \lambda_1 I)(A - \lambda_2 I) \ldots (A - \lambda_n I) = 0$

3. Suppose $T : V \to V$ is linear. Show that each of the following is invariant under T :

  (i) kernel of T (ii) {0}

  (iii) image of T (iv) V

4. Determine the invariant subspace of $A = \begin{bmatrix} 2 & -4 \\ 5 & -2 \end{bmatrix}$ viewed as linear operator on:

  (i) R^2 (ii) C^2

5. Suppose A is super triangular matrix (all entries below the main diagonal are 0). Show that A is nilpotent.

6. What is the minimal polynomial of nilpotent matrix A of index k?

7. Show that following matrix are nilpotent :
$$A = \begin{bmatrix} -2 & 1 & 1 \\ -3 & 1 & 2 \\ -2 & 1 & 1 \end{bmatrix}; A = \begin{bmatrix} 1 & -3 & 2 \\ 1 & -3 & 2 \\ 1 & -3 & 2 \end{bmatrix}$$

Find also the index of nilpotency in each matrix.

8. Find the canonical nilpotent form of the matrix
$$A = \begin{bmatrix} -2 & 1 & 1 \\ -3 & 1 & 2 \\ -2 & 1 & 1 \end{bmatrix}$$

9. If matrix A and B are similar, then show that A is nilpotent of index k, if and only if B is nilpotent of index k.

10. If F is a field of characteristic zero and if S and T in $A(V)$ are such that $ST - TS$ commutes with S, then $ST - TS$ is nilpotent.

CHAPTER REVIEW : A COMPETITIVE APPROACH

Selected Terms and Results

TERMS

- **Canonical Forms :** The special form of matrix representation (in some basis of V) of linear transformation in each similarity class is called canonical form.

- **Invariant Subspace :** Let $T : V \to V$ be a linear transformation. Then a subspace W of V is said to be invariant under T if $T(W) \subset W$ i.e., for all $\alpha \in W$, $T(\alpha) \in W$.

- **Normal Form :** A matrix A is said to be in normal form if it can be written as $A = \begin{pmatrix} I_r & 0 \\ 0 & 0 \end{pmatrix}$, where I_r is square identity matrix of order r.

- **Triangular Form :** Let $T : V \to V$ be a linear transformation on V over F, then the matrix of T in the basis $\{\alpha_1, \alpha_2 \ldots \ldots \alpha_n\}$ of V is triangular if
$$T(\alpha_1) = a_{11}\alpha_1$$

$$T(\alpha_2) = a_{21}\alpha_1 + a_{22}\alpha_2$$
$$T(\alpha_3) = a_{31}\alpha_1 + a_{32}\alpha_2 + a_{33}\alpha_3$$
$$\ldots \quad \ldots \quad \ldots \quad \ldots \quad \ldots \quad \ldots$$
$$T(\alpha_n) = a_{n1}\alpha_1 + a_{n2}\alpha_2 + \ldots + a_{nn}\alpha_n$$

- **Nilpotent Transformation :** A linear transformation $T : V \to V$ is said to be nilpotent if $T^n = 0$ for some least positive integer n.

- **Jordan Canonical Form :** The matrix of the form

$$J = \begin{bmatrix} \lambda & 1 & 0 & \cdots & 0 & 0 \\ 0 & \lambda & 1 & \cdots & 0 & 0 \\ \cdots & \cdots & \cdots & \cdots & \cdots & \cdots \\ 0 & 0 & 0 & \cdots & \lambda & 1 \\ 0 & 0 & 0 & \cdots & 0 & \lambda \end{bmatrix}$$

is called Jordan block matrix belonging to λ.

RESULTS

- If W is a subspace invariant under $T \in A(V)$, then T induces a linear transformation T_q on quotient space V/W defined by $T_q(\alpha + W) = T(\alpha) + W$.

- If $V = W_1 \oplus W_2 \oplus \ldots \oplus W_r$ where n_i is the dimension of each subspace W_i and every subspace is invariant under $T \in A(V)$, then a basis of V can be found so that the matrix of T in this basis is of the form

$$\begin{bmatrix} A_1 & 0 & \cdots & 0 \\ 0 & A_1 & \cdots & 0 \\ \vdots & \vdots & \cdots & \vdots \\ 0 & 0 & \cdots & A_r \end{bmatrix}$$

where each A_i is an $n_i \times n_i$ matrix of linear transformation induced by T on W_r.

- If $T \in A(V)$ has all its characteristic root in F, then there is a basis of V in which matrix representation of T is triangular.

- If $\dim . V = n$ and if $T \in A(V)$ has all its characteristic roots in F, then T satisfy a polynomial of degree n over F.

- The characteristic root of nilpotent transformation are zero.

- All characteristic roots of nilpotent transformation belong to F.

- The nilpotent linear transformation can always be brought to triangular form over F.

- If W is invariant under $S : V \to V$ and $T : V \to V$ then W is also invariant under $S + T$ and ST.

- If $T : V \to V$ is linear and W be eigenspace belonging to an eigenvalue λ of T, then W is T-invariant.

- Similar matrices have the same eigenvalues.

- If T is nilpotent of index k, then T^n, $n > 1$ is also nilpotent of index k.

Review Questions and Project Work

1. If $T : V \to V$ is linear operator on V over a field F of characteristic zero such that tr. $T^k = 0$ for all $k \geq P$, then show that T is nilpotent.

2. Find all possible Jordan canonical forms for those matrix whose characteristic polynomial $\Delta(x)$ and minimal polynomial $m(x)$ are as

follows :

(i) $\Delta(x) = (x - 2)^4 \cdot (x - 3)^2,$
  $m(x) = (x - 2)^2 (x - 3)^2$

(ii) $\Delta(x) = (x - 2)^7, m(x) = (x - 2)^2$

(iii) $\Delta(x) = (x - 3)^4 (x - 5)^4,$
  $m(x) = (x - 3)^2 (x - 5)^2$

3. How many possible Jordan forms are there for 6×6 complex matrix with characteristic polynomial $\Delta(x) = (x + 2)^4 (x- 1)^2$?

4. Find all possible Jordan forms 8×8 matrices having $x^2 (x - 1)^3$ as minimal polynomial.

5. Show that every complex matrix is similar to its transpose.

6. Prove that the matrix $\begin{bmatrix} 1 & 1 & 1 \\ -1 & -1 & -1 \\ 1 & 0 & 0 \end{bmatrix}$ is nilpotent and find its invariant and Jordan form.

7. Determine the Jordan canonical form for the matrices

(i) $\begin{bmatrix} 1 & 1 & 1 & 1 \\ 0 & 1 & 1 & 1 \\ 0 & 0 & 1 & 1 \\ 0 & 0 & 0 & 1 \end{bmatrix}$ (ii) $\begin{bmatrix} 1 & 0 & 1 & 1 \\ 0 & 1 & 0 & 1 \\ 0 & 0 & 1 & 0 \\ 0 & 0 & 0 & 1 \end{bmatrix}$

8. Show that all complex matrices of order n for which $A^n = I$ are similar.

9. Find all possible rational canonical forms for 6×6 matrix with minimal polynomial $m(x) = (x^2 + 2)^2 (x + 3)^2$.

Objective Type Questions

FILL IN THE BLANKS

1. N is nilpotent if there is some integer r such that $N^r =$ _____ .

2. An operator T can be put into _____ if its characteristic and minimal polynomials fields into linear polynomials.

3. T can be decomposed into operator each the sum of a scalar and a _____ operator.

4. If C is a characteristic value of T, then $\det(T - CI) =$ _____ .

TRUE/FALSE

Write 'T' for true and 'F' for false statement.

1. The operator $(T - CI)$ is singular $\Rightarrow \det(C - \lambda I) = 0.$ **(T/F)**

2. $\det(C - \lambda I) = 0 \Rightarrow C$ is a characteristic value of T. **(T/F)**

3. The characteristic polynomial $f(t)$ and minimal polynomial $m(t)$ of matrix A have the same irreducible factors. **(T/F)**

4. k is called the index of nilpotency if $T^K = 0$, but $T^{k-1} \neq 0$. **(T/F)**

MULTIPLE CHOICE QUESTIONS

Choose the most appropriate one :

1. Let $T : V \rightarrow V$ be linear, then which of the following is invariant under T ?

   (a) $\{0\}$ (b) V

   (c) Ker(T) (d) All are true

2. The subspace of $A = \begin{bmatrix} 2 & -5 \\ 1 & -2 \end{bmatrix}$ which are invariant under T is/are ?

   (a) $\{0\}$ (b) R^2

   (c) Both (a) and (b)   (d) None of these

3. Let $T : V \rightarrow V$ be linear. If $v \in V$ be such that $T^k(v) = 0$ but $T^{k-1}(v) \neq 0$, then which of the following is/are true ?

   (a) The set $S = \{v, T(v), ..., T^{k-1}(v)\}$ is linearly independent

   (b) The subspace W generated by S is

   T-invariant.

   (c) Both (a) and (b) are true.

   (d) None of the above

4. Which of the following is not true ?

   (a) If W is invariant under $T : V \rightarrow V$, then W is invariant under $f(t)$.

   (b) Every subspace of V is invariant under identity and zero operator.

   (c) The eigen spaces E_λ is T-invariant.

   (d) All are true.

5. If T_1, T_2 are nilpotent operators that commutes i.e., $T_1 T_2 = T_2 T_1$ then :

   (a) $T_1 + T_2$ is nilpotent.

   (b) $T_1 T_2$ is nilpotent.

   (c) Both (a) and (b) are true.

   (d) None of the above.

Answers

FILL IN THE BLANKS

1. 0   **2.** Jordan canonical form   **3.** nilpotent   **4.** 0

TRUE/ FALSE

1. T   **2.** T   **3.** T   **4.** T

MULTIPLE CHOICE QUESTIONS

1. (d)   **2.** (c)   **3.** (c)   **4.** (d)   **5.** (c)

SOME FASCINATING FACTS

- If E_1, E_2, ..., E_k are the projections associated with the primary decomposition of T, then each E_i is a polynomial in T and accordingly if a linear operator U commute with T then U commutes with each of the E_i i.e., each subspace W_i is invariant under U.

- Let T be a linear operator on the finite dimensional vector space V over the field F. If for the polynomial, there is a diagonalizable operator D on V and a nilpotent operator N on V such that

$$T = D + N \text{ and } DN = ND$$

- The characteristic polynomial $f(t)$ and minimal polynomial $m(t)$ of a matrix A have the same irreducible factors.

- The minimal polynomial $m(t)$ of a matrix divides every polynomial that A as a zero. In particular $m(t)$ divides the characteristic polynomial $f(t)$ of A.

- The minimal polynomial of lowest degree over the field F that annihilates A is called minimal polynomial of A.

SOME IMPORTANT ILLUSTRATIONS

- Let T be a linear operator on C^2, the characteristic polynomial for T is either $(x - c_1)(x - c_2)$, where c_1 and c_2 are distinct complex numbers $(x - c)^2$. T is diagonalizable and is represented in the basis of $\begin{pmatrix} c_1 & 0 \\ 0 & c_2 \end{pmatrix}$.

- If a_0, a_1, ..., a_{n-1} are complex numbers and let V be the space of all n times differential functions f on the interval of the real time which satisfy the differential equation

$$a_n \frac{d^n f}{dx^n} + a_{n-1} \frac{d^{n-1} f}{dx^{n-1}} + \ldots + a_1 \frac{df}{dx} + a_0 f = 0$$

and if D be the differential operator. Then V is invariant under D.

- Let T be the linear operator on R^2, which represents the standard ordered basis by the matrix $A = \begin{bmatrix} 0 & -1 \\ 1 & 0 \end{bmatrix}$ then T has no characteristic values.

- If T is a linear operator on the finite dimensional space V then T is diagonalizable if there is a basis for V of each vector, which is the characteristic vector of T.

- The minimal polynomial of $A = \begin{bmatrix} 2 & 2 & -5 \\ 3 & 7 & -15 \\ 1 & 2 & -4 \end{bmatrix}$ is given by $m(t) = t^2 - 4t + 3$.

- The minimal polynomial of $A = \begin{bmatrix} 4 & -2 & 2 \\ 6 & -3 & 4 \\ 3 & -2 & 3 \end{bmatrix}$ is given by $m(t) = (t - 1)(t - 2)$.

- The minimal polynomial of $A = \begin{bmatrix} 5 & 1 \\ 3 & 7 \end{bmatrix}$ is given by $m(t) = (t - 4)(t - 8)$.

- The minimal polynomial of the matrix $A = \begin{bmatrix} 3 & -2 & 2 \\ 4 & -4 & 6 \\ 2 & -3 & 5 \end{bmatrix}$ is given by $m(t) = (t - 1)^2(t - 2)$.

Self Assessment Test

1. Find all invariant subspaces of $A = \begin{pmatrix} 2 & -5 \\ 1 & -2 \end{pmatrix}$ viewed as a linear operator on R^2.

2. Show that :
   (i) $\{0\}$ is invariant under T.
   (ii) V is invariant under T.
   (iii) Kernel of T is invariant under T.

3. Show that every subspace of V is invariant under identity and zero operator.

4. Show that there are two non-similar canonical nilpotent matrices of order 4 and index 2.

5. Show that the matrix $A = \begin{pmatrix} 0 & 1 & 1 & 0 & 1 \\ 0 & 0 & 1 & 1 & 1 \\ 0 & 0 & 0 & 0 & 0 \\ 0 & 0 & 0 & 0 & 0 \\ 0 & 0 & 0 & 0 & 0 \end{pmatrix}$ is nilpotent of index 3.

6. Find all Jordan matrices with characteristic polynomial $\Delta(t) = (t-7)^4$.

7. Find the minimal polynomial of the matrix $A = \begin{pmatrix} 7 & 1 & 0 & 0 \\ 0 & 7 & 1 & 0 \\ 0 & 0 & 7 & 1 \\ 0 & 0 & 0 & 7 \end{pmatrix}$.

8. Find all possible Jordan canonical forms for a linear map $T : V \rightarrow V$ whose characteristic polynomial is $\Delta(t) = (t-7)^5$ and whose minimal polynomial is $m(t) = (t-7)^2$.

9. Let W be a subspace of a vector space V. Show that following are equivalent :
   (i) $u \in v + W$
   (ii) $u - v \in W$
   (iii) $v \in u + W$.

10. Find all rational canonical forms with minimal polynomial $m(t) = (t-1)^3$ and characteristic polynomial $\Delta(t) = (t-1)^7$.

11. Find the rational canonical form M of A is a matrix over the rational field Q.

12. Find the rational canonical form for A if A is a matrix over the real field R.

MODULES

11.1 INTRODUCTION

Module is a generalized concept of vector space. Basically vector space is generalization of abelian group. By this fact, the Module can be considered as the generalization of abelian group. In module, the scalar element can be taken from a ring while in vector space scalar element belong to the field.

11.2 MODULES

Definition 1. *Let R be a ring. An R-module (left) is an additive abelian group M together with a function $R \times M \to M : \forall\ r,s \in R$ and $a,b \in M$ with following conditions*

(i) $r(a + b) = ra + rb$

(ii) $(r + s)a = ra + sa$

(iii) $r(sa) = (rs)a$

Definition 2. (Unitary R-module). *If R has an identity element, i.e., $1 \in R$ and $1a = a\ \forall a \in M$, then M is said to be a unitary R-module. If R is a field then unitary R-module is called a vector space over the field R.*

Similarly we can define a right R-module by putting the elements of ring, in right to the element of abelian group in the all conditions mentioned above. If R is commutative then every left R-module M can be given the structure of a right R-module by defining $ar = ra$ for $r \in R$ and $a \in M$ the difference between left R-module and right R-module is merely that of notations therefore theory of both module (left of right) can be developed in the same manner. Here, we shall develop the theory of left R-modules and omit the replaced use of adjective (left) in this way left R-module is written now, simply R-module.

☛ ILLUSTRATIONS

(1) *Every abelian group G is a module over the ring of integers Z.*

To prove that G is module over the ring of integer Z, we have to show that, $\forall\ m,n \in Z$ and $a,b \in G$

(i) $m(a + b) = ma + mb$ (ii) $(m + n)a = ma + na$ (iii) $m(na) = (mn)a$

All these conditions can be proved by using the properties of elements in vector space. In fact concept of module is generalisation of that of a vector space.

(2) *Every left ideal M of ring R is an R-module.*

M , being left ideal, is an additive abelian group. Also, by definition of left ideal of R , $rm \in M$, $\forall\ r \in R$ and $m \in M$, further

(i) $r(m_1 + m_2) = rm_1 + rm_2$, $\forall\ r \in R$ and $m_1, m_2 \in M$.

This is hold by left distributions in R.

(ii) $(r_1 + r_2)m = r_1 m + r_2 m \ \forall \ r_1, r_2 \in R$ and $m \in M$ [By right distribution in R]

(iii) $r(sm) = (rs)m \ \ \forall \ r, s \in R$ *and* $m \in M$. This is hold by associativity.

        Hence, M is an R-module.

(3) *Every ring is a R module over itself.*

       (if we take $M = R$ then this result is a special case of example 2)

(4) *Let R be a ring then set of R^n of all n-tuples is an R-module with an internal and external composition defined by*

$$(a_1, a_2, ..., a_n) + (b_1, b_2, ..., b_n) = (a_1 + b_1, a_2 + b_2, ..., a_n + b_n)$$

and $\qquad\qquad a(a_1, a_2, ..., a_n) = (aa_1 + aa_2, ..., aa_n)$

11.3 COSET R-MODULE

Let R be a ring and S be a left ideal of R. Let, $M = \{a + S : a \in R\}$ be the set of all cosets of S in R. Then, M is a R-module with the composition defined by

$$(a + S) + (b + S) = (a + b) + S; \text{ and } r(a + S) = (ra + S)$$

11.4 GENERAL PROPERTIES OF MODULES

Let M be a module over a ring R. Then

(i) $r \, 0 = 0, \ \forall \ r \in R$

(ii) $0a = 0, \ \forall \ a \in M$

(iii) $(-r)a = r(-a) = -(ra), \ \forall \ r \in R$ and $a \in M$

(iv) $(-r)(-a) = ra, \ \forall \ r \in R$ and $a \in M$

(v) $r(a-b) = (ra - rb), \ \forall \ r \in R$ and $a, b \in M$

(vi) $(r-s)a = (ra - sa), \ \forall \ r, s \in R$ and $a \in M$

The proof of all these properties may be given in the same way as in vector space.

11.5 SUBMODULES

A non empty subset S of a R-module M is said to be its submodule if

            (i) S is an additive subgroup of M;

and   (ii) $r \in R$; and $a \in S = r \, a \in S$

11.5.1 IMPROPER SUBMODULES OF M

If M is module over a ring R then M and $\{0\}$ are improper submodules of M.

11.5.2 PROPER SUBMODULES OF M

Any submodule of M other than M and $\{0\}$ is called the proper submodule of M.

11.5.3 IRREDUCIBLE R-MODUIE

A R-module of M is said to be irreducible if its only submodule are M and $\{0\}$.

THEOREM 1.   **(Intersection of two submodules).** *If A and B one two submodules of module M over a ring R, the $A \cap B$ is also a submodule of M.*

Proof.   We have

        (i) $A \cap B$ is an additive subgroup of M [$\because$ A and B are subgroups.]

(ii) For $r \in R$ and $a \in A \cap B$ $\Rightarrow$ $ra \in A \cap B$

Now, since A and B are the submodule of M.

Therefore, $r \in R, a \in A$ $\Rightarrow$ $ra \in A$

Also, $r \in R, a \in B$ $\Rightarrow$ $ra \in B$

So $ra \in A \cap B$ $[\because$ since $ra \in A$ and $ra \in B]$

Hence $A \cap B$ is a submodule of M.

THEOREM 2. *Arbitrary intersection of submodules is a submodule.*

Proof. Let $\pi = \{\cap A_\lambda : A_\lambda$ is a submodule for $\lambda = 1,2,...,n\}$. Then we have

(i) $\pi = \cap A_\lambda$ is an additive subgroup of M $[\because$ each λ is a subgroup of M. $]$

(ii) For $r \in R$ and $a \in \pi$

$\Rightarrow$ $r \in R$ and $a \in A_\lambda$ for each $\lambda = 1,2,...,n$

Now since each A_λ is submodule of M

Therefore; $r \in R, a \in A_\lambda \Rightarrow ra \in A_\lambda$ for each $\lambda = 1,2,...,n$

$\Rightarrow ra \in \pi \cap A_\lambda$ for each $\lambda = 1,2,...,n$ $\Rightarrow ra \in \pi$

Hence π (intersection of n submodules) is a submodule.

11.5.4 SUBMODULE GENERATED BY A SET

Let M be a module over a ring R and A be a non empty subset of M. Then a submodule S of M containing A is called the submodule generated by A; if S is contained in every submodule of M containing A.

THEOREM 3. *Let S be a non-empty subset of M and M be the unital R-module, then show that the set of all linear combinations of elements of S is a submodule of M, generated by S.*

Proof. Let $a_1, a_2, a_3, ..., a_n$ be arbitrary finite subsets of S, then take $L(S) = r_1 a_1 + r_2 a_2 + ... + r_n a_n$ where $L(S)$ denote the set of all linear combination of elements of S and $r_1, r_2,...,r_n$ is any arbitrary finite subset of the ring R. Now, if a,b be two elements of S then

$$a = r_1 a_1 + r_2 a_2 + ... + r_n a_n; \quad b = s_1 a_1 + s_2 a_2 + ... + s_n a_n$$

where $r_i's, s_i's \in R; a_i's$ and $b_i's \in S$.

So, $a-b = r_1 a_1 + r_2 a_2 + ... + r_n a_n + (-s_1)b_1 + (-s_2)b_2 + ... + +(-s_n)b_n$

which is linear combination of some element of S.

So, $a-b \in L(S)$. Here $a,b \in L(S) \Rightarrow a-b \in L(S)$. Hence, $L(S)$ is an additive subgroup of M.

Again for $r \in R$, we have

$$ra = r(r_1 a_1 + r_2 a_2 + ... + r_n a_n)$$
$$= r(r_1 a_1) + r(r_2 a_2) + ... + r(r_n a_n)$$
$$= (rr_1)a_1 + (rr_2)a_2 + ... + (rr_n)a_n$$

$\Rightarrow$ $ra \in L(S)$ $\begin{bmatrix} \text{Since}\quad rr_1, rr_2 ... rr_n \in R \\ \text{so, } ra \text{ is linear combination or some element or } S \end{bmatrix}$

Therefore $L(S)$ is submodule of M.

Also each $a_i \in S$ can be expressed as $a_i = 1.a_1 \in L(S)$. So, $S \subset L(S)$.

So $L(S)$ is submodule containing S.

Now we have to show $L(S) \subset T$ (any submodule of m containing S)

Let $a = r_1 a_1 + r_2 a_2 + ... + r_n a_n \in L(S)$, where $a_i's \in S$ so $a_i's \in T$ since $S \subset T$

Then $a \in T$ (because T is also a module). Therefore, $L(S) \subset T$

Hence, $L(S)$ is the smallest submodule of M containing S of $L(S)$ is submodule generalised by S.

11.6 LINEAR SUM OF TWO MODULES

Let M_1 and M_2 be two submodules of an R-module then linear sum of M_1 and M_2 is denoted by $M_1 + M_2$ and is defined as

$$M_1 + M_2 = [a+b : a \in M_1, b \in M_2)$$

Since M_1 and M_2 is submodule of M so $M_1 + M_2$ is a subset of M.

THEOREM 1. *The linear sum of two submodules of an R-module is also a submodule of the R module.*

Proof. Let M_1 and M_2 be the two submodules of a R-module then $a = M_{11} + M_{21}$, $b = M_{12} + M_{22}$ be the two elements of $M_1 + M_2$ where $m_{11}, m_{12} \in M_1$ and $m_{12}, m_{22} \in M_2$

Now $\qquad a - b = (m_{11} + m_{21}) - (m_{12} + m_{22}) = (m_{11} - m_{12}) + (m_{21} - m_{22})$

Since M_1 is an additive subgroup so $m_{11} + m_{21} \in M_1$

Similarly $m_{21} - m_{22} \in M_2$.

Therefore $(m_{11} - m_{12}) + (m_{21} - m_{22}) \in M_1 + M_2$

Therefore $M_1 + M_2$ is additive subgroup of M.

Now for any $r \in R$

$$ra = r((m_{11} + m_{21}) = rm_{11} + rm_{21} \qquad \qquad \dots(1)$$

Since M_1 is submodule so $r \in R$, $m_{11} \in M_1 \Rightarrow rm_{11} \in M_1$

Similarly $rm_{21} \in M_2$ thus $rm_{11} + rm_{21} \in M_1 + M_2$, also from (1) $ra \in M_1 + M_2$

Hence, $M_1 + M_2$ is a submodule of M.

11.6.1 DIRECT SUM OF SUBMODULES

A R-module M is a direct sum of two of its submodules M_1 and M_2 if each element of M can be uniquely expressed as sum of an element of M_1 and element of M_2. It is denoted as $M = M_1 \oplus M_2$.

This definition of direct sum can be extended to n submodules of M. Let $M_1, M_2 \dots M_n$ be n submodule of M.

Then M is said to be the direct sum of $M_1, M_2 \dots M_n$ of each element of $a \in M$ can be uniquely written as $a = a_1 + a_2 + \dots + a_n$ where $a_1 \in M_1, a_2 \in M_2 \dots a_n \in M_n$

THEOREM 2. *The necessary and sufficient conditions for a modules to be a direct sum of its two submodules M_1 and M_2 are that*

(i) $M = M_1 + M_2$ \qquad (ii) $M_1 \cap M_2 = \{0\}$

Proof. Let us first suppose M is the direct sum of M_1 and M_2, i.e., $M = M_1 \oplus M_2$. By definition of direct sum, every element of M can be uniquely expressed as the sum of the element of M_1 and the element of M_2. If $m \in M$ be arbitrary, then

$$m = \text{element of } M_1 + \text{element of } M_2.$$

$\Rightarrow \qquad\qquad M = M_1, + M_2$

Further, we shall show that $M_1 \cap M_2 = (0)$

Let if possible $0 \neq x \in M_1 \cap M_2$. Then clearly x is a non-zero element common to both M_1 and M_2. Therefore, we can write

$$x = x + 0 \in M_1 + M_2, \quad x \in M_1 \text{ and } 0 \in M_2$$

and

$$x = 0 + x \in M_1 + M_2, \quad 0 \in M_1 \text{ and } x \in M_2$$

$\Rightarrow$ $x \in M$ may be expressed as the sum of element of M_1 and M_2 in two ways which contradicts the fact that each element of M is uniquely expressed as sum of an element of M_1 and an element of M_2.

Therefore, 0 is the only element common to both M_1 and M_2. Hence, $M_1 \cap M_2 = \{0\}$.

Conversely, suppose that (i) and (ii) conditions are satisfied. We have to prove that $M = M_1 \oplus M_2$. For this we shall prove that each element of M is uniquely expressed as the sum of an element of M_1 and an element of M_2. Now from (i) we have $M = M_1 + M_2$

$\Rightarrow$ Every element of M is expressed as the sum of an element of M_1 and an element of M_2.

Let $x \in M$ be arbitrary then $x = x_1 + y_1, x_1 \in M, y_1 \in M_2$

We shall show that the representation $x = x_1 + y_1$ is unique.

Let if possible $\qquad x = x_1 + y_1$ with $x_1 \in M_1, y_1 \in M_2$

and $\qquad\qquad\qquad x = x_2 + y_2$ with $x_2 \in M_1, y_2 \in M_2$ be two representations.

Then, clearly we have $x = x_1 + y_1 = x_2 + y_2$

$\Rightarrow \qquad\qquad\qquad x_1 - x_2 = y_2 - y_1 \in M_1 \cap M_2$

$\qquad\qquad\qquad [\because x_1 - x_2 \in M_1 \text{ and } y_2 - y_1 \in M_2 \text{ and } x_1 - x_2 = y_2 - y_1]$

But, we have assumed that $M_1 + M_2 = \{0\}$

Therefore $\qquad\qquad x_1 - x_2 = y_2 - y_1 = 0$

$\Rightarrow \qquad\qquad\qquad x_1 = x_2 \text{ and } y_1 = y_2$

$\Rightarrow$ The presentation $x = x_1 + y_1$ is unique.

$\Rightarrow \qquad\qquad\qquad M = M_1 \oplus M_2$

Hence, M is the direct sum of M_1 and M_2.

11.7 HOMOMORPHISM OF MODULES (LINEAR TRANSFORMATIONS)

A mapping f from a R-module M to other R-module N; is said to be module homomorphism (R-homomorphism) if

$\qquad$ (i) $f(m_1 + m_2) = f(m_1) + f(m_2), \forall m_1 m_2 \in M$

and $\qquad$ (ii) $f(rm) = r f(m), \forall r \in R$ and $m \in M$

The R-module N is said to be homomorphic image of M under f.

11.7.1 ISOMORPHISM OF MODULES

A mapping f from a R-module M to other R-module N is said to isomorphism (R-isomorphism) if

$\qquad$ (i) f is one-one and onto

$\qquad$ (ii) $f(m_1 + m_2) = f(m_1) + f(m_2), \forall m_1, m_2 \in M$

$\qquad$ (iii) $f(rm) = rf(m), \forall r \in R$ and $m \in M$

REMARK

- From the definitions of homomorphism and isomorphism it is clear that isomorphism is a particular case of homomorphism.
- If f be a homomorphism of a R-module M into this R-module N. Then,
  - (i) $f(0) = 0$
  - (ii) $f(-m) = -f(m), \forall\ m\ \in M$
  - (iii) $f(m_1 - m_2) = f(m_1) - f(m_2)\ \forall m_1, m_2 \in M$

11.7.2 KERNEL OF A HOMOMORPHISM

Let f be a homomorphism of an R-module M into an R-module N then Kernel of f denoted by $K(f)$ is set of all those elements of M which are mapped to 0 (identity element of N), *i.e.,*

$K(f) = \{m \in M : f(m) = 0\}$ where 0 is identity element of additive group $N.\}$

THEOREM 1. *The Kernel of a homomorphism is a submodule.*

Proof. Let f be a homomorphism from R-module M into an R-module N then kernel of f is given by $K(f) = \{m \in M : f(m) = 0\}$. Since $f(0) = 0$.

Therefore at least $0 \in K(f)$. So $K(f)$ is non empty subset of M

Now for any m_1, $m_2 \in K(f)$.

Then $f(m_1) = 0$ and $f(m_2) = 0$ and $f(m_1 - m_2) = f(m_1) - f(m_2) = 0 - 0 = 0$ therefore $m_1 - m_2 \in K(f)$. Thus $K(f)$ is an additive subgroup of M.

Again for second condition $r \in R$ and $m \in K(f)$ such that $f(m) = 0$ We have $f(rm)$ = $rf(m) = rf(m) = r0 = 0$ therefore $rm \in K(f)$. Hence $K(f)$ is submodule of M.

THEOREM 2. *If f be a module homomorphism. Then f is an isomorphism iff $K(f) = 0$, i.e., $Kr(f) = 0$.*

Proof. Let f be a homomorphism of an R-module M onto a R-module N. First we shall prove that if $K(f) = 0$ then f is an isomorphism. For this take any two elements $m_1, m_2 \in M$ such that

$$f(m_1) = f(m_2) \qquad \Rightarrow f(m_1) - f(m_2) = 0$$
$$\Rightarrow \qquad f(m_1 - m_2) = 0 \qquad \Rightarrow m_1 - m_2 \in K(f)$$
$$\Rightarrow m_1 - m_2 = 0 \qquad [\ker (f) = 0 \text{ gives}]$$
$$\Rightarrow \qquad m_1 = m_2$$

Therefore f is one-one. Hence f is isomorphism.

Conversely. Suppose f be an isomorphism. Then f is one-one.

Now for any $m \in \ker(f)$, we have

$$\ker(f) = f(m) = 0 = f(0) \Rightarrow m = 0.$$

Hence, $K(f)$ or $\ker(f) = 0$.

THEOREM 3. *The range of module-homomorphism is submodule.*

Proof. Let f be a homomorphism of an R-module M into a R-module M and let $I(f)$ denote the range of T. Then $I(f) = \{f(m): m \in M\}$. To prove that $I(f)$ is an additive subgroup of N. Let $f(m_1)$ and $f(m_2)$ be the element of $I(f)$ then $f(m_1) - f(m_2) = f(m_1 - m_2) \in I(f)$ therefore $I(f)$ is an additive subgroup of N.

Again for any $r \in R$ and $f(m) \in I(f)$ we have

$$rf(m) = f(rm) \in I(f) \text{ since } rm \in M.$$

Hence, $I(f)$ is a submodule of N.

11.8 QUOTIENT MODULES

Let M be an R-module and A be its any submodule. Then by definition A is a subgroup of the additive group M for a $m \in M$, $A + m$ is a coset of A in M. Such set of all cosets of A in M is denoted by M/A and the structure M/A is called quotient modules of M relative to the submodule A such that $M/A = \{A+m : m \in M\}$, the set of all coset of A in M with respect to the composition defined by

$$(A + m_1) + (A + m_2) = A + (m_1 + m_2), \forall r \in R \text{ and } m \in M$$

REMARK

- Let M be the R-module and A be its submodule then the mapping $f : m \to M/A : f(m) = W + m$, $\forall m \in M$ is a homomorphism of M onto (M/W) with $\ker(f) = W$.

THEOREM 1. *(Fundamental theorem on homomorphism of modules).*

If f be is a homomorphism of a R-module M onto a R-module N with ker (f) = A then N is isomorphic to M/A, i.e., $N \cong M/A$.

 Proof. Let f be a homomorphism of an R-module onto R-module N then $\ker(f)$ is submodule of M. But $\ker(f) = A$ (Given) so A is submodule of M and M / A is a quotient module which is given as

$$M/A = \{m+A : m \in M\}$$

Also, for all $m_1 \in A$, we have $f_1(m_1) = 0$ and range of $T(N) = N$ (Since f is onto). Let us define a mapping $T : N \to M/A$ such that

$$T(f(m)) = (m + A), \forall \in N, f(m) \in N \text{ when } m \in M.$$

Obviously T is well defined mapping. To show $N \cong M/A$. First, take $m_1 / m_2 \in M$ so that $f(m_1), f(m_2) \in N$ such that

$$T(f(m_1)) = m_1 + A, T(f(f(m_2))) = m_2 + A$$
$$T(f(m_1)) = T(f(m_2)) \Rightarrow \quad m_1 + A = m_2 + A$$
$$\Rightarrow \quad m_1 - m_2 \in A \Rightarrow f(m_1 - m_2) = 0$$
$$\Rightarrow \quad f(m_1) + f(m_2) = 0 \Rightarrow f(m_1) = f(m_2)$$

Hence $\quad T(f(m_1)) = T(f(m_2)) = T(f(m_1 + m_2)) = (m_1 + m_2) + A$
$$= (m_1 + A) + (m_2 + A) = T(f(m_1)) + T(f(m_2))$$

Let $r \in R$ then

$$T(rf(m)) = T(f(rm)) \qquad \text{[By definition of } f]$$
$$= rm + A \qquad \text{[By definition of } f]$$
$$= r(m + A) = rT(f(m))$$

Hence T is also linear. Thus T is isomorphism.

11.9 CYCLIC MODULE

A R-module M is said to be a cyclic module generated by an element $a \in M$ if each $m \in M$ is expressible as $m = ra$ for some $r \in R$.

The element a is called generator of M and we write it as $M = (a)$

REMARK

- Let M be a unital R-module and $W = \{ra : r \in R \text{ and } a \in M\}$. Then W is a cyclic submodule of m, generated by a.

THEOREM 1. *An irreducible R-module is cyclic.*

Proof. Let M be a unital, irreducible R-module. Then only submodule of M are $\{0\}$ and M itself.

Case (i) If $M = \{0\}$, it is clearly cyclic.

Case (ii) If $M \neq \{0\}$ Take $a \in M$ such that $a \neq 0$.

Let $W = \{ra : r \in R\}$, Also $a = 1. a = 1 (R$ is unital$) = a \in W$

So $W \neq \{0\}$ this W is non zero submodule of m.

Hence $W = m$ [$\because$ only non zero submodule of M is m.]

Also W is generated by $a \in M$ so W is cyclic. Hence M is cyclic.

11.9.1 FINITELY GENERATED MODULE

A R-module M is called finitely generated if there exist finite number of elements $m_1, m_2, ..., m_n \in M$ such that every $m \in M$ is expressible as

$$m = \sum_{i=1}^{n} r_i m_i, \text{ for some } r_i \in R.$$

THEOREM 2. **(Fundamental theorem of finitely generated modules over Euclidean Rings).** *If R is an Euclidean Ring, then any finitely generated R-module M is the direct sum of a finite number of cyclic modules.*

Proof. Let M be a R-module which is finitely generated and R be a Euclidean rings whose generating sets have the minimum element as minimal generating sets. The number of element in such minimal generating is said to be the rank of M. We shall prove the theorem by method of mathematical induction on rank of M. If $n = 1$, then M is generated by single element which is generator of M and M is cyclic. Hence the result is true. Now let the result is true for all module of rank $n = k - 1$.

Let M be R-module of rank k, we shall prove that the result is true for this M. For any given minimal generating set $\{a_1, a_2,...,a_n\}$ of M and relation of the form $r_1 a_1 + r_2 a_2 + ... + r_k a_k$ for $r_i \in R$ implies that $r_1 a_1 = r_2 a_2 = ... = r_k a_k = 0$ then M clearly is the direct sum of $M_1, M_2,...M_k$, where each M_i is cyclic submodule of M generated by a_i. Thus

$$M = M_1 \oplus M_2 \oplus ... \oplus M_k.$$

Hence by mathematical induction the theorem is true for every n.

So for any given minimal generated set $\{b_1, b_2,...,b_k\}$ of M, there must be elements $r_1, r_1, ...r_k \in R$ such that

$$r_1 b_1 + r_2 b_2 + ... + r_k b_k = 0$$

in which not all of $r_1 b_1 + r_2 b_2 + ... + r_k b_k$ are equal to zero.

Among all such possible relations, for all minimal generating sets, there is an element $s_i \in R$ whose d-value $d(s_i)$ is minimal. Let the set $\{C_1, C_2, ,...,C_k\}$ be the generating set such that

$$\sum_{i=1}^{n} s_i C_i = 0 \qquad ... (1)$$

Now R is an Euclidean ring so for $s_1, r_1 \in R, \exists m, t \in R$ such that $r_1 = ms_1 + t$ $\qquad ... (2)$

where $t = 0$ or $d(t) < d(s_1)$.

Let $\sum\limits_{i=1}^{k} r_i c_i = 0$ for some $r_1 \in R$ ...(3)

Multiply (1) by m and subtracting from (3) we get

$$\sum_{i=1}^{k} (r_i - ms_i)c_i = 0$$

$$\Rightarrow \quad (r_1 - ms_1)c_1 + \sum_{i=1}^{k} (r_i - m s_i)c_i = 0$$

$$\Rightarrow \quad tc_1 + \sum_{i=1}^{k} (r_i - m s_i)c_i = 0 \quad \text{[from (2)]} \quad ...(4)$$

If $t \neq 0$, then $d(t) < d(S_1)$ which contrary to the fact $d(s_1)$ is minimal. Hence $t = 0$ which implies $r_1 = ms_1$ by (2). This means s_1 is a divisor of r_1. Now we shall show that s_1 is divisor of s_i for $i = 2, 3, ... k$.

For $s_1, s_2 \in R \, \exists \, m_2, t_2 \in R$ such that
$$s_2 = m_2 s_1 + t_2$$
where either $t_2 = 0$ or $d(t_2) < d(s_1)$.

Now if $c_1^1 = c_1 + m_2 c_2$, then $\{c_1^1, c_2, c_3 .... ck\}$ is also a generating set of M

So $\quad s_1 c_1^1 + t_2 c_2 + s_3 c_3 + ... + s_k c_k = s_1(c_1 + m_2 c_2) + t_2 c_2 + s_3 c_3 + ... + s_k c_k$
$$= s_1 c_1 + (s_1 m_2 + t_2)c_2 + s_3 c_3 + ... + s_k c_k$$
$$= s_1 c_1 + (s_2 c_2 + ... + s_k c_k) = 0 \quad ... (5)$$

If $t_2 \neq 0$ then $d(t_2) < d(s_1)$ and in this case solution (5) contradict for our choice of s_1 so $t_2 = 0$ then $s_2 = ms_1$, which means s_1 is divisor of s_2.

If we proceed in this manner, we can show that s_1 is a divisor of s_i for $i = 3, 4, ...,$ k. Therefore $\quad s_2 = m_2 s_1, s_3 = s_2 = m_2 s_1 .... m_k s_1$

Again let

$$c^*_1 = c_1 + m_2 s_2 = m_2 c_2 + ... + m_k c_k \quad ... (A)$$

Let M_1 (generted by $c_1, c_2, ... c_k$) be cyclic module of M generated by c_1 and M_2 be generated by $\{c_1, c_2, ..., c_k\}$

$\therefore \quad\quad\quad\quad M = M_1 + M_2$ ...(6)

Now for any $\alpha \in M_1 \cap M_2 \Rightarrow \alpha \in M_1, \alpha \in M_2$

$\Rightarrow \quad\quad\quad \alpha = r_1, c^*_1, \alpha = r_2 c_2 + r_3 c_3 + ... + r_k c_k$

Since M is generated by a_1, therefore
$$r_1 c^*_1 = r_2 c_2 + r_3 c_3 + ... + r_k c_k$$
$$\Rightarrow \quad r_1(c_1 + m_2 c_2 + ... + m_k c_k) - r_2 c_2 - r_3 c_3 .... - r_k c_k = 0$$
$$\Rightarrow \quad r_1 c_1 + (r_1 m_2 - r_2) c_2 + ... + (r_1 m_k - r_k)c_k = 0 \quad ...(7)$$

In this relation coeffcient of c_i is r_i. Therefore s_1 is divisor of r_1 and hence $r_1 = ps_1$, where $p \in R$.

Thus $\quad\quad\quad\quad \alpha = r_1 c^*_1 = ps_1 c_1 = p(s_1 c_1)$

$\therefore \quad\quad\quad\quad \alpha = p[s_1(c_1 + m_2 + c_2 + ... m_k c_k)] \quad\quad\quad \text{[By (A)]}$
$$= p[s_1 c_1 + s_1 m_2 c_2 + ... s_1 m_k c_k]$$
$$= p[s_1 c_1 + s_2 c_2 + ... s_k c_k] \quad [\because s_i = m_i c_i + 1 = 2, 3, ... k]$$

$$= p \cdot 0 = 0 \qquad \text{[By (3)]}$$

Thus $\qquad \alpha \in M_1 \cap M_2$ and $\alpha = 0$

$$M_1 \cap M_2 = \{0\}$$

Finaly, we have $\quad M = M_1 \oplus M_2$ and $M_1 \cap M_2 = \{0\}$

Therefore $\qquad M = M_1 \oplus M_2$

Now $\qquad M_1 = \{a_1{}^*\} \Rightarrow$ rank of $M_1 = 1 \Rightarrow M_1$ is cyclic.

$\qquad\qquad M_2 = \{a_2, \ldots\ldots a_k\} \Rightarrow$ rank of M_2 is at most $k - 1$.

Hence by induction, M_2 is direct sum of cyclic modules.

Therefore, M is direct sum of cyclic modules.

EXERCISE 11.1

1. Show that the polynomial ring $R(x)$ over ring R is a R–module.
2. If M and N are R – modules and if $f : M \rightarrow N$ is homorphism them
   (i) $f(0) = 0$
   (ii) $f(-m) = -f(m), \forall, m \in M$
3. Show that a subring S of a ring R is a module over the whole ring only if S is an ideal of R.
4. If M is the set of $m \times n$ matrices over a ring R. Then show that M is a module over R.
5. If M, N, Q are R-module of an R-module M such that $M \supset N$, Show that
   $$M \cap (N + Q) = N + (M \cap Q)$$
6. If R is a non-zero commutative ring with unity.

   Show that $R(x)$ is not a finitely generated R- module.
7. Show that, the set of rational numbers Q is not finitely generated Z-module.
8. If K is a submodule of a finitely generated R-module M, then M/K is also a finitely generated R-module.
9. Let $f : M \rightarrow N$ be and R-module homomorphism and let the kernel contains a submodule M_1 of M. Show that the mapping $\phi : M / M_1 \rightarrow N$ defined by $\phi(x + M_1) = f(x)$ is an R-module homomorphism.
10. If R is a ring with unity. Show that a R-module M is cyclic if and only is $M \cong R/I$ for some left ideal I or R.

11.10 SIMPLE AND SEMI-SIMPLE MODULES

NOTATION

(1) In this section, all rings will be rings with unity and all R-modules will be unital.

(2) R-modules means left "R-module."

Definition 1. *An R-module M is said to be simple if it has no submodules other than itself and the zero submodule*

For Example :

(1) An abelian group $G \neq 0$ is a simple Z-module if and only if it is cyclic of prime order.

(2) A vector space over a division ring R is a simple R-module if and only if it has dimension 1.

(3) If M is the left R-module formed by the addtive group of a ring R, then a submodule $N \neq 0$ of M is simple if and only if it is a minimal left ideal of M.

Definition 2. *A homomorphism of a ring R into the ring $End_R(M)$ of all endomorphism of an abelian group M is a called representation of R.*

Definition 3. *A left R-module M is said to be faithful if and only if the representaion of R associated with M is injective.*

Definition 4. *An R-module M is said to be semi-simple if it can be expressed as a direct sum of simple submodules.*

REMARK

- If M is a left R-module, then a mapping $f : R \rightarrow \text{End}_R (M)$, $f \rightarrow I_r$ is the representation of R associated with M, where for each $r \in R$, the mapping $I_r(x) = rx$ is an endomorphism of M.

THEOREM 1. **(Schur's Theorem).** *If M is a simple R-module and N is any R-module then*

(i) *every non-zero homomorphism $f : M \rightarrow N$ is injective (Monomorphism)*

(ii) *every non-zero homomorphism $f : M \rightarrow N$ is surjective (Epimorphism)*

(iii) *End_R is a division ring, where $\text{End}_R (M) \text{ Hom}_R (M, M)$*

Proof.

(i) Since, f is a homomorphism, therefore, ker(f) is a submodule of M. Now, sinec M is simple, thus either Ker $(f) = [0]$ or Ker $(f) = M$. But $f \neq 0$, therefore

$$\text{Ker} (f) \neq M$$

$\Rightarrow \qquad \text{Ker} (f) = [0]$

Henec, f is surjective.

(ii) Since M is a non-zero R-module homomorphism from M to N. But Im (f) is also a non-zero submodule of M and M is simple. Thus

$$\text{Im} (f) = M$$

Hence, f is surjective.

(iii) Suppose, f is a nor-zero R-module homomorphism from M to M. Now, since, M is simple, then using (i) and (ii), f is an automorphism.

$\Rightarrow \qquad f$ is unit in the ring. Hence, $\text{End}_R (M)$ is a division ring.

THEOREM 2. *If R is a ring and I is a minimal left ideal of R. Also, if M is a faithful simple left R-module, then M is isomorphic to the left R-module I.*

Proof. Let R be a ring and I be a minimal left ideal of R.

Let $a_0 \in I$, $a_0 \neq 0$

Since M is faithful, thus there is some $x \in R$ such that $a_0 x \neq 0$ and the mapping $f_x : I \rightarrow R$ defined by $f_x (a) = ax$, $a \in I$ is a non-zero homomorphism. By Schur's lemma, f_x is an epimorphism because the R-module M is simple.

Hence, f_x is an isomorphism.

THEOREM 3. *Let M be an R-module and $\{ M_i : i \in \Delta\}$ be a family of simple submodules of M such that $M = \sum_{i \in \Delta} M_i$, then for each submodule N of M, there is a Δ' of Δ such that*

$$M = \underset{i \in \Delta'}{\oplus} M_i \oplus N .$$

Proof. In case of $M = N$, $\Delta' = \phi$, thus therom is trivially true. Suppose $M \neq N$, then for some $k \in \Delta$, $M_K \not\subset N$. Now, since M_k is simple $M_k \cap N = [0]$ and $N + M_k$ is a direct sum Therefore, the set of all families $F = \{M_i : i \in \Delta'' : \Delta'' \subset \Delta\}$ for which the sum is direct, $\sum_{i \in \Delta^n} M_i + N$ is non empty. By Zorn's lemma, F contains a maximal subfamily $\{M_i : i \in \Delta'\}$.

Now, suppose $\underset{i \in \Delta'}{\oplus} M_i \oplus N . \neq M$, then for some $j \in \Delta, M_j \not\subset \underset{i \in \Delta'}{\oplus} M_i \oplus N .$

Because M_j is simple, therefore $M_j \cap [\underset{i \in \Delta'}{\oplus} M_i \oplus N] = 0$ and $M_j \cap [\underset{i \in \Delta'}{\oplus} M_i \oplus N]$ is a direct sum. Then, we have

$$[M_i : i \in \Delta' \cup [J]] \in F$$

which contridicts the maximality of $\{M_i : i \in \Delta'\}$

Hence, we have

$$M = \underset{i \in \Delta'}{\oplus} M_i \oplus N$$

THEOREM 4. *Let M be an R-module. Then following conditions are equivalent :*

    (i) *M is a sum of simple submodules.*

    (ii) *M is a semi-simple module.*

    (iii) *Every submodule of M is a direct summand of M.*

Proof. (i) $\Rightarrow$ (ii) Let us suppose $\{M_i : i \in \Delta\}$ be a family of simple submodule of M and if $M = \sum\limits_{i \in \Delta} M_i$, then using previous theorem 3, with $N = 0$, there is a subset Δ' of Δ such that

$$M = \underset{i \in \Delta'}{\oplus} M_i$$

Hence, M is a sum of simple submodule.

(ii) $\Rightarrow$ (iii) Let N be a submodule of M and let

$$M = \underset{i \in \Delta'}{\oplus} M_i$$

$$M = \underset{i \in \Delta}{\oplus} M_i \qquad\qquad [\because M \text{ is semi simple}]$$

Then again by theorem-3, there is a subset Δ' of Δ such that

$$M = \underset{i \in \Delta'}{\oplus} M_i \oplus N.$$

(iii) $\Rightarrow$ (i) Let us suppose that every submodule of M is a direct summand of M. Firstly, we shall prove that M has simple submodule.

Let $a \in M$, $a \neq 0$ and let $N = Ra$. Then clearly, N is a finitely generated R-module, therefore N has a maximal submodule L. Thus, N/P is a simple R- module. By our hypothesis, L has a complementary submodule L' such that $\qquad M = L \oplus L'$

Thus $\qquad\qquad N = \text{L} \oplus (N \cup L')$.

$\Rightarrow$ $N \cup L'$ is a submodule of M which is isomorphic to N/L .

$\Rightarrow$ $N \cup L'$ is simple.

Now, since every non-zero submodule of M contains the cyclic submodule generated by each of its non-zero elements. Thus, every non-zero submodule contains a simple submodule.

Further, let P be the sum of the simple submodules of M. Then, $M = P \oplus P'$ for some submodule P . If P' is non-zero, then it contains a simple submodule which is not possible, since all the simple submodules are contained in P.

Therefore $\qquad M = P.$

Hence, M is a sum of simple submodules.

THEOREM 5. *Every submodule and every homomorphic image of a semi-simple module is semi-simple.*

Proof.   Let M be a semi-simple R-module and let N be a submodule of M and let L be a submodule of N. Then, there exists a submodule L' of M such that $M = L \oplus L'$. In this case, $\qquad L \oplus (L' \cap N) = N$.

Then, by previous theorem, every submodule of N is direct summand of N.

Hence, N is semi-simple.

Finally, since every homomorphic image of R-module M is isomorphic to M / N for some submodule N of M. But $M / N \cong N'$.

Hence, M is semi-simple.

THEOREM 6. *Let M be a semi simple R-module and it is equal to the direct sum of a family $[M_i : i \in \Delta]$ of simple submodules, then every simple submodule of M is isomorphic to one of the M_i.*

Proof. Let M be a semi-simple R-module and N is a simple submodule of M, then N is a direct summand of M. Therefore, there is an epimorphism

$$\phi : M \to N \text{ and } N = \sum_{i \in \Delta} \phi(M_i).$$

Now, since $N = \phi(M_K)$ for $K \in \Delta$. But M_K is simple. Hence, by Schur's lemma, ϕ is an isomorphism.

11.11 FREE MODULES

Definition 1. *Let M. be a module over a ring R with unity and let S be a subset of M, then S is said to be basis for M. if*

(i) *S generates M.*

(ii) *S is linearly independent.*

Definition 2. *An R-module M is said to be free module if there exists a subset S of M such that S generates M and S is linearly independent set.*

For Example. If $M = (0)$, then the set ϕ is its basis. Therefore, it is free module.

REMARK

- If S is a basis for M, then, in particular $M \neq [0]$ and if $R \neq [0]$, then every element of M can be expressed uniquely as a linear combination of the elements of S.

THEOREM 1. *Let R be a ring and M be a module over R. Also let S be a non-empty set and let $\{x_i : i \in S\}$ be a basis of M. If N is an R-module and $\{y_i : i \in S)$ be a family of elements of N, then, there exists a unique homomorphisrn $f : M \to N$ such that $f(x_i) = yi, \forall i \in S$.*

Proof. It is given that $\{x_i : i \in S\}$ is a basis of M, then every element of M can be expressed uniquely as linear combination of elements of the set $\{x_i : i \in S\}$. Let $x \in M$. Then, there exists a unique family $\{r_i : i \in S \}$ of elements of R such that

$$x = r_1 x_1 + r_2 x_2 + ... + r_k x_k ... = \sum_{i \in S} r_i x_i \qquad ...(1)$$

Define a mapping $f : M \to N$ such that

$$f(x) = \sum_{i \in S} r_i y_i, \forall x \in M, \forall y_i \in N \qquad ...(2)$$

Clearly, f is a homomorphism and it is unique, because

$$f(x) = \sum_{i \in S} r_i f(x_i) \qquad ...(3)$$

From (2) and (3), we conclude that $f(x_i) = y_i, \forall i \in S$.

THEOREM 2. *Let M be a free R-module with a basis $[e_i : i \in N]$, where e_i is n-tuples in which the ith entry is 1 and rest are zero. Then $M \cong R^n$.*

Proof. Let us define a mapping $\phi : M \rightarrow R^n$ such that

$$\phi(x) = \sum_{i=1}^{n} r_i y_i \text{ where } x \in M \text{ and } y_i \in R^n.$$

Thus,

$$x = \sum_{i=1}^{n} r_i e_i$$

Let x, y be any two elements of M and suppose that

$$x = y \Rightarrow \sum_{i=1}^{n} r_i e_i = \sum_{i=1}^{n} s_i e_i$$

Now, since $\{e_1, e_2, ..., e_n)$ is linearly independent.

$\Rightarrow \qquad r_i - s_i = 0$

$\Rightarrow \qquad r_i = s_i$

$\Rightarrow \qquad \sum_{i=1}^{n} r_i y_i = \sum_{i=1}^{n} s_i y_i$

$\Rightarrow \qquad \phi(x) = \phi(y)$

Thus, ϕ is well defined.

Again, if

$$x = \sum_{i=1}^{n} r_i e_i, y = \sum_{i=1}^{n} s_i e_i \text{ and } r \in R. \text{ Then}$$

$$\phi(x+y) = \phi\left(\sum_{i=1}^{n} r_i e_i + \sum_{i=1}^{n} s_i e_i \right) = \phi\left(\sum_{i=1}^{n} (r_i + s_i) e_i \right)$$

$$= \sum_{i=1}^{n} (r_i + s_i) y_i = \sum_{i=1}^{n} r_i y_i + \sum_{i=1}^{n} s_i y_i$$

$$= \phi(x) + \phi(y)$$

Also

$$\phi(rx) = \phi\left(r \sum_{i=1}^{n} r_i e_i \right) = \phi\left(r \sum_{i=1}^{n} r_i e_i \right)$$

Therefore, ϕ is a homomorphism.

Finally, if $\phi(x) = 0$, then $\sum_{i=1}^{n} r_i \cdot y_i = 0$

$\Rightarrow \quad r_1 = 0 = r_2 = ... = r_n \Rightarrow x = 0$

$\Rightarrow \quad \phi$ is one-one

Also, ϕ is onto.

Hence, ϕ is an isomorphism. Thus, $M \cong R^n$.

THEOREM 3. *Every finitely generated module is a homomorphic image of a finitely generated free module.*

Proof. Let M be an R-module generated by the set $[x_1, x_2, ..., x_n]$, i.e., $M = \{x_1, x_2, ..., x_n\}$. Now, let $e_i = (0, 0, ..., 1, 0... 0\}$, then the set $\{e_1, e_2, ..., e_n\}$ is linearly independent over R and generates a free module R^n.

Now, define a mapping $\phi : R^n \to M$ given by

$$\phi(x) = \sum_{i=1}^{n} r_i\, x_i, \ x \in R^n \text{ so that } x = \sum_{i=1}^{n} r_i e_i$$

Now, since every element of R^n has a unique representation as $x = \sum_{i=1}^{n} r_i e_i$. Thus, ϕ is well defined.

Let have $x = \sum_{i=1}^{n} r_i e_i$ and $y = \sum_{i=1}^{n} s_i e_i$ be any two elements of R^n . Then, we have

$$\phi\,(x+y) = \phi\left(\sum_{i=1}^{n} (r_1 + s_i)e_i \right) = \sum_{i=1}^{n} (r_i + s_i)x_i$$

$$= \sum_{i=1}^{n} r_i\, x_i + \sum_{i=1}^{n} s_i\, x_i = \phi(x) + \phi(y)$$

Also $\phi\,(rx) = \phi = \left(\sum_{i=1}^{n} rr_i e_i \right) = \sum_{i=1}^{n} r_i r x_i = r \sum_{i=1}^{n} r_i x_i = r\phi(x)$

Hence, ϕ is a homomorphism of R^n into M.

THEOREM 4. *Let R be a ring and M be a free module with basis B. If N is any R-module and $F : B \to N$ is any mapping, then there exists a unique R-module homomorphism $\phi : M \to N$ such that $\phi_{iB} = f$.*

Proof. Let $B = \{x_i : i \in \Lambda\}$ be a basis of M. For any $x \in M$,
we have $x = \sum_{i \in \wedge} r_i x_i, r_i \in R$ [By definition of basis]

Define a mapping $\phi : M \to N$ be $\phi\,(x) = \sum_{i \in \Delta} r_i f(x_i)$

If $x, y \in M$, then for $r_i \in R$ and $s_i \in R$, we have

$$x = \phi \sum_{i \in \Delta} r_i x_i, \ y = \sum_{1 \in \Delta} s_i x_i$$

Consider, $\phi\,(x+y) = \phi\left(\sum_{i \in \Delta} r_i x_i + \sum_{i \in \Delta} s_i x_i \right)$

$$= \phi\left(\sum_{i \in \Delta} (r_i + s_i)x_1 \right) = \sum_{i \in \Delta} (r_i + s_i)f(x_i)$$

$$= \sum_{i \in \Delta} r_i f(x_i) + \sum_{i \in \Delta} s_i f(x_i) = \phi(x) + \phi(y)$$

Now, for $r \in R, x \in M$,

$$\phi(rx) = \sum_{i \in \Delta} r_i f(x_i) = r \sum_{i \in \Delta} r_i f(x_i) = r\phi(x)$$

Hence, ϕ is an R-module homomorphism and $\phi_{1B} = f$.

THEOREM 5. *Let M be a finitely generated free module over a commutative ring R. Then, all basis of M are finite..*

Proof. Since M is finitely generated free module, let $M = (x_1, x_2 ..., x_M)$ and $\{e_i : i \in s\}$ be a basis for M.

Then, for any $x_j \in M$, we have

$$x_j = \sum_i a_{ij} \text{ where } a_{ij} \in R \qquad \text{...(1)}$$

Now, since finite number of a_{ij} are zero, therefore the set of those e_i's which occur in all expression of the type (1) for $j = 1, 2, \dots n$ is finite.

THEOREM 6. *If R is a commutative ring and M is a finitely generated free module over R, then all basis of M have the same number of elements.*

Proof. It is known that a free module with a basis $[e_1, e_2, \dots, e_n]$ is isomorphic to R^n. So, let R^m and R^n be any two basis of M. We have to prove that if $R^m \cong R^n$, then $m = n$.

Let us suppose that $m < n$. Define a homomorphism $\phi : R^m \to R^n$ such that $\phi = \phi^{-1}$

Suppose $[e_1, e_2 \dots e_m]$ and $[f_1, f_2, \dots f_n]$ be the ordered basis of R^m and R^n respectively. Now, since ϕ is an isomorphism from R^m to R^n and g is an isomorphism from R^n to R^m. Thus, we have

$$\left.\begin{aligned}
\phi(e_1) &= a_{11}f_1 + a_{21}f_2 + \dots\dots + a_{n1}f_n \\
\phi(e_2) &= a_{12}f_1 + a_{22}f_2 + \dots\dots + a_{n2}f_n \\
\dots\quad &\quad \dots\quad\quad \dots\quad\quad\quad \dots \\
\phi(e_m) &= a_{1m}f_1 + a_{2m}f_2 + \dots\dots + a_{nm}f_n
\end{aligned}\right\} \qquad \dots\dots(1)$$

and

$$\left.\begin{aligned}
g(f_1) &= b_{11}e_1 + b_{21}e_2 + \dots\dots + b_{m1}e_m \\
g(f_2) &= b_{12}e_1 + b_{22}e_2 + \dots\dots + b_{m2}e_m \\
\dots\quad &\quad \dots\quad\quad \dots\quad\quad\quad \dots \\
g(f_n) &= b_{1n}e_1 + b_{2n}e_2 + \dots\dots + b_{mn}e_m
\end{aligned}\right\} \qquad \dots\dots(2)$$

Let $A = [a_{ij}]_{n \times m}$ and $B = [b_{kj}]_{m \times n}$ be two matrices. Then, we can write

$$g(\phi(e_i)) = g(a_{11}f_1 + a_{21}f_2 + \dots + a_{n1}f_n)$$

$$= a_{11}g(f_1) + a_{21}g(f_2) + \dots + a_{n1}g(f_n)$$

Similarly

$$\left.\begin{aligned}
g(\phi(e_1)) &= a_{11}b_{11}e_1 + a_{21}b_{21}e_2 + \dots + a_{n1}b_{m1}e_m \\
g(\phi(e_2)) &= a_{12}b_{12}e_1 + a_{22}b_{22}e_2 + \dots + a_{n2}b_{m1}e_m \\
\dots\quad &\quad \dots\quad \dots\quad\quad \dots\quad \dots\quad \dots\quad \dots \\
g(\phi(e_m)) &= a_{1m}b_{1m}e_1 + a_{2m}b_{2m}e_2 + \dots + a_{nm}b_{mn}e_m
\end{aligned}\right\} \qquad \dots(3)$$

Now, equation (3) can be rewritten as

$$g(\phi(e_i)) = \sum_{k=1}^{m} \sum_{j=1}^{n} b_{kj} a_{ji} a_{ji} e_k \text{ for } 1 \leq i \leq m$$

Also, since e_i's are linearly independent and $g = \phi^{-1}$, then

$$\sum_{k=1}^{m} \sum_{j=1}^{n} b_{kj} a_{ji} = \delta_{ki} = \begin{cases} 1 & ; \quad k = i \\ 0 & ; \quad k \neq i \end{cases} \qquad \dots(4)$$

which gives

$$\begin{bmatrix} b_{11} & b_{12} & \cdots & b_{1n} \\ b_{21} & b_{22} & \cdots & b_{2n} \\ \vdots & \vdots & \vdots & \vdots \\ b_{m1} & b_{m2} & \cdots & b_{mn} \end{bmatrix} \begin{bmatrix} a_{11} & a_{12} & \cdots & a_{1n} \\ a_{21} & a_{22} & \cdots & a_{2n} \\ \vdots & \vdots & \vdots & \vdots \\ a_{m1} & a_{m2} & \cdots & a_{mn} \end{bmatrix} = \begin{bmatrix} 1 & 0 & \cdots & 0 \\ 0 & 1 & \cdots & 0 \\ \vdots & \vdots & \vdots & \vdots \\ 0 & 0 & \cdots & 1 \end{bmatrix}$$

$\Rightarrow \quad BA = I_m$, where I_m is the unit matrix of order $m \times n$.

Similarly, we may get
$$AB = I_n$$

Now, let $A_1 = [A : 0]$ and $B_1 = \begin{bmatrix} B \\ \vdots \\ 0 \end{bmatrix}$

be agumented matrices, where 0 blocks are zero matrices.
Then, we have

$$A_1B_1 = I_n \text{ and } B_1A_1 = \begin{bmatrix} i_m & 0 \\ 0 & 0 \end{bmatrix}$$

$\Rightarrow \qquad |A_1B_1| = |I_n| = 1 \text{ and } |B_1A_1| = 0 \qquad \qquad \dots(5)$

Also, since R is commutative, therefore $|A_1.B_1| = |B_1A_1|$, which is a contraduction by virtue of (5).

Thus $m \not\subset n$ so that $m \geq n$. If the role of $[e_i]$ and $[f_i]$ are interchanged, then we get $n \geq m$. Hence, $m = n$.

☞ EXERCISE 11.2

1. Show that every ideal of Z is free as Z-module.
2. Show that every left ideal is an integral domain R with unity is free as a left R-module.
3. Show that the Z-module Q is not free.
4. If M and N are free modules over a commutative ring with unity such that both can be freely generated by n elements, show that $M = N$.
5. Let R be a non-zero commutative ring with

unity. Consider R as an R-module. Show that every submodule of R is free if and only if R is PID.

6. Let M be a semi-simple module and let $K \neq M$ be a submodule of M. Show that M/K is semi-simple module.
7. Show that, the polynomial ring $R[x]$ is a free R-module with basis $[X^n : n \in N]$.
8. Show that a simple R-module M is free if and only if R is a division ring and rank $(M) = 1$.

11.12 NOETHERIAN AND ARTINIAN MODULES

Definition1. (Modules with Chain Conditions)

Let R be a ring and M be the R-module. Also $S = [M_\alpha : \alpha. \in \Delta]$ is the set of all submodules of M. Then S is said to satisfy

(i) *the ascending chain condition if and only if every increasing chain of submodules*
$$M_1 \subseteq M_2 \subseteq M_3 \subseteq \dots\dots$$
in S is stationary, i.e., there exists a positive integer n such that
$$M_n = M_{n+1} = M_{n+2} = \dots\dots$$
(ordered by the inclusion relation $\subseteq$)

(ii) *the descending chain condition if and only if every decreasing chain of submodules*
$$M_1 \supseteq M_2 \supseteq M_3 \supseteq \dots\dots$$
in S is stationary, i.e., there exists a positive integer n such that
$$M_n = M_{n+1} = \dots\dots$$
(S is ordered by $\supseteq$)

Definition 2. *An R-module is said to be Noetherian module if it satisfy the ascending chain condition.*

Definition 3. *An R-module is said to be Artinian if it satisfies the descending chain condition.*

Definition 4. *An R-module M is said to be co-generated if for each family $[M_\alpha : \alpha \in A]$ of submodules of M $\bigcap_{\alpha \in \Delta} M_\alpha = 0$ implies that $\bigcap_{\alpha \in \Delta'} M_\alpha = 0$ for some finite subset Δ' of Δ.*

REMARK

- The finitely co-generated modules satisfies the certain chain condition on their submodules.

Definition 5. (Uniform Module). *A non-zero R-module is said to be uniform if any two non-zero submodules of M have non-zero intersection.*

Definition 6. (Subisomorphic submodule). *Let M and N be two uniform modules, then M is said to be subisomorphic to N, divided by M ~ N if M find N contain non-zero isomorphic submodules.*

REMARK

- Clearly ~ is an equivalence relation and [M] denotes the equivalence class of M, i.e., [M] = [N ~ M , for all uniform modules N].

Definition 7. (Primary Module). *An R-module M is said to be primary if each non-zero submodule of M has uniform submodule and any two uniform submodules of M are subisomorphic.*

Definition 8. *Let R be a commutative Noetherian ring and P be the prime ideal of R. Then, P is said to be associated with the module M if P = r(x), for some $x \in M$, where r(x) =[a $\in$ R: xa= 0] is the annihilator of x or if R/P embeds in M.,*

Definition 9. *An R-module M is said to be P-primary for some prime ideal P if P is the only prime ideal associated with M.*

THEOREM 1. *Let M be the R-module. Then following are equivalent :*

   (i) *M is Noetherian.*

   (ii) *Every submodule of M of finitely generated.*

   (iii) *Every non-empty set S of submodules of M has a maximal element.*

Proof. (i) $\Rightarrow$ (ii) Let M be Noetherian module. Let us suppose that every submodule N of M is finitely generated.

Let if possible N is not finitely generated. Then for any positive integer r if
$$a_1, a_2, ..., a_r \in N.$$
Then, $N \neq (a_1, a_2, ..., a_r)$

Now, let $a_{r+1} \in N$, then $a_{r+1} \notin (a_1, a_2, ..., a_r)$.

Thus we obtain an infinite proper ascending chain.

$(a_1) \subset (a_1, a_2) \subset (a_1, a_2, a_3) \subset ...... \subset (a_1, a_2, ..., a_r) \subset (a_1, a_2, ..., a_{r+1}) \subset ...$ of submodules of M. Since, M satisfies a finite ascending chain of its submodules, therefore we arrive at a contradiction, because N is taken as not to be finitely generated. Hence, N is finitely generated.

(ii) $\Rightarrow$ (iii) Let us suppose every submodule of M is finitely generated. We have to show that every non-empty set of submodules of M has a maximal element.

Let N_0 be an element of S. Suppose that N_0 is not maximal. In this case, there is a submodule N_1 in S such that
$$N_0 \subset N_1 \qquad\qquad ...(1)$$

If N_1 is not maximal, then similarly as above, there is a submodule N_2 in S such that

$$N_1 \subset N_2 \qquad \qquad ...(2)$$

From (1) and (2), we conclude that

$$N_0 \subset N_1 \subset N_2$$

Proceeding in the similar manner, we obtain an infinite property ascending chain of submodules

$$N_0 \subset N_1 \subset N_2 ...... \subset N_i \subset N_{i+1} \subset ...... \qquad \qquad ...(3)$$

Let us write $N = \bigcup_i N_i$, then N is a submodule of M, being the arbitrary union of property containing submodules. But by our assumption, every submodule of M is finitely generated, therefore N is finitely generated. Thus, there exist $a_1, a_2, ..., a_n \in N$ such that

$$N = (a_1, a_2, ..., a_n)$$

Now, since $N = \bigcup_i N_i$, then there must exist $N_k \subset N$, such that $a_1, a_2, ..., a_n \in N_k$. Further, N is the smallest submodule containing $a_1, a_2, ..., a_n$.

Then, we get $N_k = N$.

Therefore, $N_k = N_{k+1} = ...$

which is a contradiction because we have assumed that S has no maximal element. Hence, S must have a maximal element.

(iii) $\Rightarrow$ (i)   Let us suppose that every non-empty set of submodules of M has maximal element. We have to show that M is Noetherian. Let us consider an ascending chain

$$N_1 \subset N_2 \subset N_3 \subset ... \qquad \qquad ...(4)$$

of submodule of M, but we have assumed that every non-empty set of submodules has a maximal element. Therefore, the chain (4) has a maximal element say N_i, but then

$$N_i = N_{i+1} = N_{i+2} = ...$$

$\Rightarrow$ (4) is a finite ascending chain of submodules of M. Hence, M is Noetherian module.

THEOREM 2. *Let M be a R-module. Then following are equivalent :*

   (i)   *M is Artinian.*

  (ii)   *Every quotient module of M is finitely co-generated.*

 (iii)   *Every non-empty set of submodules of M has a minimal element.*

Proof.   The proof is consequence of Theorem 1.

THEOREM 3. *Every submodule of a Noetherian (Artinian) module is Noetherian (Artinian).*

Proof.   Let M be a Noetherian (Artinian) module and N be its submodule. We have to prove that N is Noetherian (Artinian). Since M is Noetherian (Artinian), then it satisfying ascending (descending) chain condition of submodules. Also, every non-empty set of submodules of M has a maximal (minimal) element.

$\Rightarrow$ N is Noetherian (Artinian).

Hence, every submodule of M is Noetherian (Artinian).

THEOREM 4. *Every homomorphic image of a Noetherian (Artinian) module is Noetherian (Ertinian).*

Proof. Let f be a homomorphism of a Noetherian (Artinian) module M. We have to show that $f(M)$ is Noetherian (Artinian).

Since M is Noetherian, then we have an ascending chain of submodules of M.

$$M_1 \subset M_2 \subset M_3 \subset \ldots \subset M_n \subset \ldots$$

such that $M_k = M_{k+1} = \ldots$ for some positive integer k. Also, in case of Artinian, we have a descending chain of submodules of M.

$$N_1 \supset N_2 \supset N_3 \supset \ldots \supset N_n \supset \ldots$$

such that $N_r = N_{r+1} = \ldots$, for some positive integer r.

Now, in case of Noetherian

$$f(M_1) \subset f(M_2) \subset \ldots f(M_n) \subset \ldots \text{ such that } f(M_k) = f(M_{k+1}) = \ldots$$

Also, in case of Artinian, we have

$$f(N_1) \supset f(N_2) \supset \ldots \ldots \supset f(N_n) \supset \ldots \ldots$$

such that $f(N_r) = f(N_{r+1}) = \ldots$

$\Rightarrow \qquad f(M)$ is Noetherian (Artinian).

Hence, every homomorphic image of a Noetherian (Artinian) module is Noetherian (Artinian).

THEOREM 5. *Let M be an R-module and N be an R-submodule of M, then M is Noetherian (Artinian) if and only if both N and M/N are Noetherian (Artinian).*

Proof. Let us first suppose that N and M/N are both Noetherian. Also, let L be a submodule of M, then $(L + N)/N$ is a submodule of M/N. But M/N is Noetherian, therefore $(L + N)/N$ is finitely generated. Therefore, there exist $\bar{a}_1, \bar{a}_2, \ldots, \bar{a}_n \in L/N \cap L$ such that

$$L/N \cap L = (\bar{a}_1, \bar{a}_2, \ldots, \bar{a}_n)$$

which implies that

$$L = (a_1, a_2, \ldots, a_n) + (N \cap L), \forall a_i \in L$$

Since, N is Noetherian and $N \cap L$ is a submodule of N. Therefore, $N \cap L$ is finitely generated. Then, there exist $b_1, b_2, \ldots, b_m \in N \cap L$ such that

$$N \cap L = (b_1, b_2, \ldots, b_m)$$

$\Rightarrow \quad L = (a_1, a_2, \ldots, a_n) + (b_1, b_2, \ldots, b_m)$

$\Rightarrow \quad L$ is finitely generated and L is any submodule of M.

$\Rightarrow \quad M$ is Noetherian.

Conversely, let M be Noetherian. Then, its every submodule is Noetherian.

$\Rightarrow \quad N$ is Noetherian.

Now, let $f : M \to M/N$ be a canonical homomorphism and let $\overline{M_1} \subset \overline{M_2} \subset \ldots$ be an ascending chain of submodule of M/N.

If $M_i = f^{-1}(\overline{M_i})$, then we have $M_1 \subset M_2 \subset M_3 \subset \ldots$ is an ascending chain of submodules of M.

Since, M is Noetherian, therefore there exists a positive integer r such that

$$M_r = M_{r+1} = \ldots \text{ Then, we have}$$

$$f(M_i) = \overline{M_i}, \forall i \geq r \Rightarrow \overline{M_r} = \overline{M_{r+1}} = \ldots$$

$\Rightarrow \quad M/N$ is Noetherian.

The proof for Artinian module is similar as above.

THEOREM 6. *Let M be a module and N, N' be submodules. If M = N + N' and if both N, N' are Noetherian, then M is Noetherian. Further, a finite direct sum of Noetherian module is Noetherian.*

Proof. We observe that the direct product $N \times N'$ is Noetherian because it contains N as a submodule whose factor module is isomorphic to N'. Then, applying previous theorem, we obtain a surjective homomorphism $N \times N' \to M$ such that the pair (x, x'), $x \in N$, $x' \in N'$ maps on $x + x'$.

Therefore, M is Noetherian.

Further, by using induction, we can prove the result of finite direct sum.

REMARK

- A ring is called Noetherian if it is Noetherian as a left module over itself, *i.e.,* every left ideal is finitely generated.

THEOREM 7. *Let A be a Noetherian ring and M be a finitely generated module. Then, M is Noetherian.*

Proof. Let $x_1,..., x_n$ be the generators of M.

Then, there exists a homomorphism $f : A \times A \times ... \times A \to M$ of the product of A with itself n times such that

$$f(a_1,...,a_n) = a_1 x_1 + ... + a_n x_n$$

Since, this homomorphism is surjective, then by previous theorem, the product is Noetherian. Also, we know that every homomorphic image of a Noetherian module is Noetherian. Hence, M is Noetherian module.

THEOREM 8. *Let M be a Noetherian module. Then, each non-zero submodule of M contains a uniform module.*

Proof. Let M be a Noetherian R-module. We have to show that each non-zero submodule of M contains a uniform module. For this, we shall prove that μR contains a uniform submodule for non-zero element μ of M.

If R is Noetherian ring, then μR is also Noetherian. Also, if M is Noetherian, then the submodule μR is also Noetherian. Further, let N be a non-zero submodule of M and if $N \cap L \neq [0]$ for all non-zero submodules L of M.

Now, let S be a family of all submodules of μR which are not large. Since, μR is Noetherian, then every non-empty set of submodules of μR has a maximal element, thus S has a maximal element, say L.

But L is not large, so $L \cap L = [0]$ for some non-zero submodule K of μR.

We shall prove that K is uniform.

Suppose K_1 and K_2 be any two non-zero submodules of K. Let if possible,

$$K_1 \cap K_2 = [0]$$

If $u \in (L \oplus K_1) \cap K_2$, then we have

$$u \in (L \oplus K_1 \quad \text{and} \quad u \in K_2$$

$\Rightarrow$ $\qquad u = a + a_1 = a_2$, for some $a \in L$, $a_1 \in K_1$, $a_2 \in K_2$

$\Rightarrow$ $\qquad a = u - a_1 \in L \cap K$

$\Rightarrow$ $\qquad a = 0 \quad \text{and} \quad a_1 = a_2$

$\Rightarrow$ $\qquad a_1 = a_2 \in K_1 \cap K_2 = [0]$

$$\Rightarrow \qquad a_1 = a_2 = 0$$
$$\Rightarrow \qquad u = 0$$

which is a contradiction, therefore $K_1 \cap K_2 \neq [0]$. Hence, K is uniform.

THEOREM 9. *Let R be a commutative Noetherian ring and M be a uniform R- module. Then, M contains a submodule isomorphic to R/P for exactly one prime ideal of P.*

Proof. Let $x \in M$, $x \neq 0$ and S be the family of annihilators ideals $r(x)$. Further, since R is Noetherian, S will contain a maximal element say $r(x)$. We have to show that $P = r(x)$ is prime ideal.

Suppose $a.b \in r(x)$ with $a \notin r(x)$.

By definition of $r(x)$, $\quad xa \neq 0$ and $r(x) \subset r(xa)$.

But $r(x)$ is maximal in S, therefore $r(x) = r(xa)$.

Therefore, $\qquad xab = 0$, then $b \in r(x)$

$\Rightarrow \qquad\qquad P$ is prime,

and xR is isomorphic to $R/r(x) = R/P$. Therefore R / P embeds in M.

Now, it remains to prove that P is unique.

Let if possible, Q be an other prime ideal of R such that R/Q embeds in M.

Also, $yR \simeq R/Q$ and $[R/P] = [R/Q]$.

Then, there exist two cyclic modules xR of R/P and yR of R/Q such that $xR \simeq yR$.

Therefore, $\qquad\qquad R / P \simeq R/Q$

$\Rightarrow \qquad\qquad P = Q$

REMARK

- Let M be a non-zero finitely generated module over a commutative Noetherian ring R, then there are only a finite number of prime associated with M.

THEOREM 10. (**Noether-Lasker Theorem**). *Let M be a finitely generated module over a commutative Noetherian ring R. Then, there exists a finite family $\{N_1, N_2, ..., N_m\}$ of submodules of M such that*

(i) $\displaystyle\bigcap_{j=1}^{m} N_j = 0$ and $\displaystyle\bigcap_{\substack{j=1 \\ j \neq r}}^{m} N_j \neq 0$ for all $1 \leq r \leq m$

(ii) *Each quotient module M / N_j is a P_j-primary for some prime ideal P_j associated with M.*

(iii) *The Pj are all distinct $1 \leq j \leq m$.*

(iv) *The primary components N_f is unique if and only if P_j does not contain P_k for any $k \neq j$.*

11.13 FILTERED AND GRADED MODULES

Definition 1 (Filtration). *Let R be a commutative ring and E is a module. By a filtration of E, we mean a sequence of submodules.*

$$E = E_0 \supset E_1 \supset E_2 \supset ... \supset E_n \supset$$

For example: Let I be an ideal of a ring R and E an R-module. Also, let $E_n = I^n E$. Then, the sequence of submodules $< E_n >$ is a filtration.

REMARK

- The filtration defined above is known as descending filtration.

Definition 2. *Let* $< E_n >$ *be any filtration of a module E, It is said to be an I-filtration of* $IE_n \subset E_{n+1}$ *for all n.*

Definition 3. *An l-filtration is said to be I-stable or stable if* $IE_n = E_{n+1}$ $\forall n$ *, where n is sufficiently large.*

Definition 4 (Graded Ring). *A ring R is called graded (by the natural numbers) if we can write R as a direct sum (an abelian group).*

$$A = \overset{\infty}{\underset{n=0}{\oplus}} A_n$$

such that for all integers, m,n $\geq$ 0, we have $R_n R_m \subset R_{n+m}.$

REMARK

- R_0 is a subring and each component R_n is an R_0-module.

Definition 5 (Graded Module). *Let R be a graded ring. A module E is called a graded module if E can be expressed as a direct sum (as abelian groups)*

$$E = \overset{\infty}{\underset{n=0}{\oplus}} E_n$$

such that $R_n E_m \subset E_{n+m}.$

In particular, E_n *is an* R_0 *-module and elements of* E_n *are then called homogeneous of degree n.*

REMARK

- Using above definition, any element E can be expressed uniquely as a finite sum of homogeneous elements.

THEOREM 1. *Let* $<E_n>$ *and* $< E'_n >$ *be stable I-filtration of E, then there exists a positive integer d such that* $E_{n+d} \subset E'_n$ *and* $E'_{n+d} \subset E_n$ *for all n $\geq$ 0.*

Proof. Here, it is sufficient to prove the proposition when $E'_n = I^n E$. Since

$IE_n \subset E_{n+1}$ for all n, we have $I^n E \subset E_n$.

By stability hypothesis, there exists d such that

$$E_{n+d} = I^n E_d \subset I^n E$$

$$\Rightarrow \qquad E_{n+d} \subset E'_n \, .$$

THEOREM 2. *Let R be a graded ring. Then R is Noetherian if and only if* R_0 *is Noetherian and R is finitely generated as* R_0-*Algebra.*

Proof. Clearly, a finitely generated algebra over a Noetherian ring is Noetherian because it is a homomorphic image of the polynomial ring in finitely many variables and we can then apply Hilbert's theorem.

Conversely, suppose R is Noetherian, the sum

$$p_n^+ = \bigoplus_{n=0}^{\infty} R_n$$

is an ideal of R, whose residue class ring is R_0, which is thus a homomorphic image of R, and is therefore Noetherian. Also, R^+ has a finite number of generators $x_1, \ldots x_s$ by hypothesis. Expressing each generator as a sum of homogeneous elements, we may assume without loss of generality that these generators are homogeneous, say of degrees $d_1, \ldots\ldots d_s$ respectively with all $d_i > 0$. Let B be the subring of R generated over R_0 by $x_1, \ldots, x_s$. We claim that $R_n \subset B$ for all n. This is certainly true for $n = 0$. Let $n > 0$. Let x be homogeneous of degree n, then there exists some elements $a_i \in R_{n-d}$ such that

$$x = \sum_{i=1}^{s} a_i x_i$$

Since $d_i > 0$, by induction, each a_i is in $R_0 [x_1, \ldots, x_s] = B$.

$\Rightarrow \qquad\qquad\qquad x \in B$

Hence, the theorem.

11.13.1 CONSTRUCTION OF GRADED RINGS FROM FILTRATION

Let R be a ring and I is an ideal. We view R is a filtered ring by the powers I^n. We define the first associated graded ring to be

$$S_I(A) = S = \bigoplus_{n=0}^{\infty} I^n$$

Similarly, if E is an R-module and E is filtered by an I-filtration, we define

$$E_s = \bigoplus_{n=0}^{\infty} E_n$$

Then, we can easily verify that E_s is a graded S-module.

REMARK

- If R is Noetherian and I is generated by elements $x_1, \ldots, x_s$, then S is generated as an R-algebra also by $x_1, \ldots, x_s$ and is therefore also Noetherian.

THEOREM 3. *Let R be a Noetherian ring and E a finitely generated module with an I-filtration. Then E_s is finite over S if and only if the filtration of E is I-stable.*

Proof. Let $F_n = \bigoplus_{i=0}^{n} E_i$ and let $G_n = E_0 \oplus \ldots \oplus E_n \oplus IE_n \oplus I^2E_n \oplus I^3E_n \oplus \ldots$. Then, G_n is a S-submodule of E_s and is finite over S, since F_n is finite over R.

We have $G_n \subset G_{n+1}$ and $\cup G_n = E_s$

Since S is Noetherian, we get

$\qquad E_s$ is finite over S

$\Rightarrow \quad E_s = G_N$ for some N.

$\Rightarrow \quad E_{N+M} = I^m E_n$ for all $m \geq 0$.

Hence, filtration of E is I-stable.

THEOREM 4 (**Artin-Ress**). *Let R be a Noetherian ring and I an ideal E a finite R-module with a stable I-filtration. Let F be a submodule and let $F_n = F \cap E_n$. Then $<F_n>$ is a stable I-filtration of F.*

Proof. We have $I[F \cap E_n] \subset IF \cap IE_n \subset F \cap E_{n+1}$

Therefore, $< F_n >$ is an I-filtration of F. Then, we can form the associated graded S-module F_s, which is a submodule of E_s and is finite over S, since S is Noetherian. Finally, using the previous theorem, we conclude that $< F_n >$ is a stable I-filtration of F.

REMARK

- Artin-Ress theorem can be restated as follows : "Let R be a Noetherian ring, E is a finite R-module and F a submodule. Let I be an ideal. There exists an integer i such that for all integers $n \geq i$, we have $I^n E \cap F = I^{n-i}(a^i E \cap F)$

THEOREM 5 (**Krull Theorem**). *Let R be a Noetherian ring and I be an ideal contained in every maximal ideal of R. Let E be a finite R-module. Then*

$$\bigcap_{n=1}^{\infty} I^n E = 0$$

Proof. Write $F = \cap I^n E$. Then apply Nakayama's lemma, whose statement is given below.

Nakayama's Lemma

Let I be an ideal of R which is contained in every maximal ideal of R. Let E be a finitely generated R-module. Suppose that $IE = E$. Then $E = [0]$.

11.14 SMITH NORMAL FORM OVER A PID AND RANK

In this section, we shall consider the ring R as a principal ideal domain with unity, so in particular, R may be a field.

Definition 1. (Row Module and Column Module). *Let R be a ring and A be $n \times n$ matrix of order $n \times n$ over R. The submodule of R^n generated by the m rows of A is called the rows module of A and the submodule of R^m generated by the n columns of A and the submodule of R^m generated by the n columns of A is known as column module of A.*

Definition 2. (Rank). *Let A be an $m \times n$ matrix over R. The rank of the row module (column module) is known as the row rank (column rank) of A.*

Definition 3. *Let A be an $m \times n$ matrix over R. The following types of operations on the rows (column) of A are known as elementary row (column) operations*

   (i)   *interchanging two rows (columns)*

   (ii) *multiplying the element of one row (column) by a non-zero element of R*

   (iii) *adding to the elements of one row (column) x times the corresponding elements of a different row (column) where $x \in R$.*

Definition 4. We define the following matrices

   (1) E_{ij} = *the matrix obtained from the identity matrix by interchanging i^{th} and i^{th} rows (columns).*

   (2) $L_i(x)$ = *the matrix obtained from the identity matrix by multiplying the ith row (column) by a non-zero $\alpha \in R$.*

(3) $M_{ij}(\alpha)$ = the matrix obtained from the identity matrix by adding to the elements of the ith row (column) α times the corresponding elements of the jth row (column).

Then the matrices E^{ij}, $L_i(\alpha)$ and $M_{ij}(\alpha)$, obtained from identity matrix by elementary operations are known as *elementary matrices.*

Definition 5. *Two m×n matrices A and B over R are said to be equivalent if there exists an inuertible matrix P∈R^m and an invertible matrix Q∈R^n such that B = PAQ.*

Definition 6. (Smith Normal Form). *Let A be m×n matrix over a principal ideal domain R, then A is equivalent to a matrix of the diagonal form*

$$\begin{bmatrix} q_1 & & & & & \\ & q_1 & & & & \\ & & \ddots & & & \\ & & & q_1 & & \\ & & & & 0 & \\ & & & & & \ddots \\ & & & & & & 0 \end{bmatrix}_{m \times n} ,\text{with}\, q_i \neq 0 \,\text{and}\, q_1 \,|\, q_2 \,|\, q_3 \dots$$

This form of matrix A is called Smith normal form, denoted by S(A).

Definition 7. *The non-zero diagonal elements of Smith normal form are called the invariant factors of A, where S(A) is the Smith normal form.*

Definition 8. *Let A be an m× n matrix over a PID R. The common value of the row rank of A and the common rank of A is called the rank of A.*

Definition 9. *Let R be a PID, and an invariant factor*

$$q_k = u_k p_1^{\alpha_{k_1}} p_2^{\alpha_{k_2}} ,\dots, p_m^{\alpha_{k_m}} , k = 1, 2,\dots,r$$

where $p_1, p_2, \dots, p_m$ are distinct primes in the complex system of non-associates in R, α_{kj} are non-negative.

Integers j = 1, 2, …, m, k = 1, 2, …. r and u_k are units.

From $q_k | q_{k+1}$ it follows that $\alpha_{ki} \subseteq \alpha_{(k+1)}$, k = 1, 2, …, r – 1, j = 1, 2, …,m.

A prime factor $P_i^{\alpha_{kj}}$ in which $\alpha_{kj} > 0$ is called an elementary divisor of h.

Solved Examples

Example 1.   *Find the Smith normal form and rank of the matrix*

$$A = \begin{bmatrix} 1 & 2 & 3 \\ 4 & 5 & 0 \end{bmatrix}$$

Solution.   The given matrix is $A = \begin{bmatrix} 1 & 2 & 3 \\ 4 & 5 & 0 \end{bmatrix}$

Performing elementary row and column operation
Performing $R_2 \to R_2 - 4R_1$, we get

$$\begin{bmatrix} 1 & 2 & 3 \\ 0 & -3 & -12 \end{bmatrix}$$

Performing $C_2 \to C_2 - 2C_1$ and $C_3 \to C_3 - 3C_1$, we get

$$\begin{bmatrix} 1 & 0 & 0 \\ 0 & -3 & -12 \end{bmatrix}$$

$C_3 \to C_3 - 4C_2$

$$\begin{bmatrix} 1 & 0 & 0 \\ 0 & -3 & 0 \end{bmatrix}$$

$R_2 \to (-1)R_2$

$$A \sim \begin{bmatrix} 1 & 0 & 0 \\ 0 & 3 & 0 \end{bmatrix}$$

Therefore Smith normal form

$$S(A) = \begin{bmatrix} 1 & 0 & 0 \\ 0 & 3 & 0 \end{bmatrix}$$

with $q_1 = 1$, $q_2 = 3$
Clearly $q_1 | q_2$. Hence, rank $(A) = 2$.

Example 2. *Reduce the matrix*

$$A = \begin{bmatrix} -x & 4 & -2 \\ -3 & 8-x & 3 \\ 4 & -8 & -2-x \end{bmatrix}$$

Over the ring $Q(x)$ to smith normal form. Also, find the rank.

Solution.
$$A = \begin{bmatrix} -x & 4 & -2 \\ -3 & 8-x & 3 \\ 4 & -8 & -2-x \end{bmatrix} \sim \begin{bmatrix} 4 & -8 & -2-x \\ -3 & 8-x & 3 \\ -x & 4 & -2 \end{bmatrix} \quad [\text{Performing } R_1 \leftrightarrow R_3]$$

$$\sim \begin{bmatrix} 4 & -8 & -(2+x) \\ 0 & 2-x & \dfrac{1}{4}(6-3x) \\ 0 & 4-2x & -\dfrac{1}{4}(x^2+2x+8) \end{bmatrix} \quad [R_2 \to R_2 + \dfrac{3}{4}R_1, R_3 \to R_3 + \dfrac{x}{4}R_1]$$

$$\sim \begin{bmatrix} 4 & -8 & -(2+x) \\ 0 & 2-x & \dfrac{1}{4}(6-3x) \\ 0 & 0 & -\dfrac{1}{4}(x^2+4x+20) \end{bmatrix} \quad [R_3 \to R_3 \to 2R_2]$$

$$\sim \begin{bmatrix} 4 & 0 & -(2+x) \\ 0 & 2-x & \dfrac{1}{4}(6-3x) \\ 0 & 0 & -\dfrac{1}{4}(x^2+4x+20) \end{bmatrix} \quad [C_2 \to C_2 + 2C_1]$$

$$\sim \begin{bmatrix} 4 & 0 & -(2+x) \\ 0 & 2-x & 0 \\ 0 & 0 & -\dfrac{1}{4}(x^2+4x+20) \end{bmatrix} \qquad [C_3 \to C_3 - \dfrac{3}{4}C_2]$$

$$\sim \begin{bmatrix} 1 & 0 & -(2+x) \\ 0 & 2-x & 0 \\ 0 & 0 & -\dfrac{1}{4}(x^2+4x+20) \end{bmatrix} \qquad [C_1 \to \dfrac{1}{4}C_1]$$

$$\sim \begin{bmatrix} 1 & 0 & 0 \\ 0 & 2-x & 0 \\ 0 & 0 & x^2+4x+20 \end{bmatrix} \qquad [C_3 \to C_3 + (2+x)\, C_3 \to -4C_3]$$

In a similar manner, we apply some more transformation, we get

$$\sim \begin{bmatrix} 1 & 0 & 0 \\ 0 & 1 & 0 \\ 0 & 0 & (x-2)(x^2+4x+20) \end{bmatrix}$$

Then, we have $q_1 = 1$, $q_2 = 1$,
$$q_3 = (x-2)\,(x^2 - 4x + 20)$$
$$q_1 \neq 0,\ q_2 \neq 0,\ q_3 \neq 0 \text{ if } x \neq 2$$

Also $q_1 \mid q_2$ and $q_2 \mid q_3$

Therefore, Smith normal form is given by

$$\begin{bmatrix} 1 & 0 & 0 \\ 0 & 1 & 0 \\ 0 & 0 & (x-2)\left(x^2+4x+20\right) \end{bmatrix}$$

Hence, rank $(A) = 3$.

Example 3. *If $D = diag\ (3^2 - 7^3.\ 24)$, then find its Smith normal form overr Z.*

Solution. Clearly, the elementary divisors of D are $2^3.\ 3.\ 3^2.\ 7^3$.

Then, the list of invariant factors is given by
$$q_3 = 2^3.3^2.7^3 = 24696$$
$$q_2 = 3$$
$$q_1 = 1$$

Hence, ths Smith normal of D is given by
$$S\ (D) = diag.\ (1, 3, 24969)$$

Example 4. *Let $2^3,\ 2^4,\ 2^4,\ 3,\ 3^3,\ 3^3,\ 5^2,\ 5^5, 7,\ 7^2$ be the complete list of elementary divisors of $A \in Z^5 \times Z^7$, i.e., A is 5×7 matrix over Z and rank of $A = 4$. Find the Smith normal form of A.*

Solution. It is given thar rank of A is 4.

Thus, there are four invariant factors q_1, q_2, q_3, q_4.

Also, q_4 must be the product of all the highest powers of the distinct primes.

Therefore, $q_4 = 2^4 . 3^3 . 5^5 . 7^2$.

Further, q_3 must be the product of all the highest powers of the distinct primes in the list after q_4 has been constructed, *i.e.*, $q_3 = 2^4 . 3^3 . 5^2 .7$

In a similar manner, we get $q_2 = 2^3 . 3$ and q_1 must be a unit, so $q_1 = 1$.

Hence, Smith normal form is given by

$$S(A) = \begin{bmatrix} q_1 & 0 & 0 & 0 & 0 & 0 & 0 \\ 0 & q_2 & 0 & 0 & 0 & 0 & 0 \\ 0 & 0 & q_3 & 0 & 0 & 0 & 0 \\ 0 & 0 & 0 & q_4 & 0 & 0 & 0 \\ 0 & 0 & 0 & 0 & 0 & 0 & 0 \end{bmatrix}$$

11.15 FINITELY GENERATED MODULES OVER A PID

Definition 1. (Torsion and Torsion Free Element). *Let R be a principal ideal domain and M be an R-module. An element $x \in M$ is called* torsion element *if $rx = 0$ for some non-zero $r \in R$. If x is not a torsion element, then it is known as torsion free element.*

Definition 2. (Torsion and Torsion Free Module). *Let R be a principal ideal domain and M be an R-module. If T(M) is a set of all torsion element of M, then T(M) is a submodule of M. Also, if T(M). Also, if T(M) = M, then M is known as torsion module. On the other hand, if T(M) = 0, then M is called torsion free module.*

THEOREM 1. **(Decompositon Theorem).** *If M is a non-zero finitely generated module over a PID R. If n is the minimal number of elements required to generate M, then M is the direct sum of cylic submodules.*

$$M = Ra_1 \oplus Ra_2 \oplus .... \oplus Ra_n = \sum_{i=1}^{n} Ra_i$$

such that Ann. $(a_{i+1}) \subseteq$ Ann.(a_i) for all i= 1, 2, ..., n–1, where Ann. $(a_1) \neq R$ and $A_{nn}. (a_n) = A_{nn}. (M)$

Proof. As per given, M is a non-zero finitely generated module over a *PID R*. Since M is finitely generated, therefore, there exist $\alpha_1, \alpha_2, ..., \alpha_n \in R$ such that

$$M = (\alpha_1, \alpha_2, ..., \alpha_n)$$

Define a mapping $f : R^n \to M$ such that $f(c_1, c_2, ..., c_n) = c_1\alpha_1 + c_2\alpha_2 + ... + c_n\alpha_n$

Clearly, f is an R-module epimorphism.

Now, if $K = \text{Ker}(f)$, then K is a free submodule of rank m (say) $m \leq n$.

Further, choose a basis $[\beta_1, \beta_2, ..., \beta_n]$ of R^n and non-zero elements $d_1, d_2, ..., d_m \in R$ such that $[d_1\beta_1, d_2\beta_2, ..., d_m\beta_m]$ form a basis of K and for each i, $d_i | d_{i+1}$.

Now, since f is epimorphism, then for each i=1, 2, ..., n.

$$a_i = f(\beta_i)$$

Thus, $[a_1, a_2, ..., a_n]$ will generate M.

Since, for any $a \in M$, there is $b \in R^n$ such that $a = f(b)$ but

$$b = c_1\beta_1 + c_2\beta_2 + ... + c_n\beta_n \text{ for some } c_1, c_2, ..., c_n \in R$$

such that $\qquad a = c_1a_1 + c_2a_2 + ... + c_na_n = \sum_{i=1}^{n} c_ia_i$

We want to show that $M = Ra_1 \oplus Ra_2 \oplus ... \oplus Ra_n$.

Now, suppose that $a = 0$, i.e., $\sum_{i=1}^{n} c_i a_i = 0$, for $a_i \in R$.

Then, $\qquad\qquad f(b) = 0$

$\Rightarrow \qquad\qquad b \in \text{Ker}\,(f) = K$

$\Rightarrow \qquad\qquad c_1\beta_1 + c_2\beta_2 + ... + c_n\beta_n \in K$

$\Rightarrow \qquad\qquad \sum_{i=1}^{n} C_i\beta_i \in K$

But $[d_1\beta_1, d_2\beta_2, ..., d_m b_m]$ is the basis of K so that

$$\sum_{i=1}^{n} c_i\beta_i = \sum_{j=1}^{m} e_i d_j\beta_j, \text{ for some } e_i \in R$$

$\Rightarrow \qquad\qquad \sum_{i=1}^{m} (c_i - e_i d_i)\beta_i = \sum_{i=m+1}^{n} c_i\beta_i = 0$

Now, since $[\beta_i : i = 1, 2, ..., m, m+1, ..., n]$ is linearly independent so that $c_i - e_i d_i = 0$ for $i = 1, 2, ..., m$.

and $\qquad\qquad c_i = 0$ for $i = m+1, m+2, ..., n$.

For $\qquad\qquad i = 1, 2, ..., m,$

we have $\qquad\qquad a_i = f(\beta_i)$

So that $\qquad\qquad c_i a_i = f(c_i \beta_i) = f(e_i d_i \beta_i) = 0$

Therefore, $c_i a_i = 0$, for all i. Thus, we get

$$M = Ra_1 \oplus Ra_1 \oplus .... \oplus Ra_n$$

We have to show that

Ann. $(a_{i+1}) \subseteq$ Ann. (a_i) for $i = 1, 2, ..., n-1$

For $\alpha \in$ Ann. (a_i) $\alpha a_i = 0$ and $a_i = f(\beta_i)$ so that

$$f(\alpha\beta_i) = \alpha a_i = 0$$

$\Rightarrow \qquad\qquad \alpha\beta_i \in K$

If $i > m$, then $\alpha = 0$, thus Ann. $(a_i) = [0]$. If $i \leq m$, then $\alpha\beta_i = \sum_{j=1}^{m} e_j d_j\beta_j$ for some $e_1, e_2, ..., e_m \in R$, so that $\alpha = e_i d_i$

$\Rightarrow \qquad\qquad d_i \mid a$

Therefore, $\qquad$ Ann. $(a_i) \subseteq d_i$

But $\qquad\qquad d_i a_i = d_i f(\beta_i) = f(d_i \beta_i) = 0$

because $d_i \beta_i \in K$ so that $d_i \in Ann.\,(a_i)$

$\Rightarrow \qquad\qquad (d_i) \subseteq$ Ann.$[a_i]$

Thus, we get Ann. $(a_i) = d_i$

Since $d_i \mid d_{i+1}$ for each i, therefore

$$\text{Ann. } (a_{i+1}) \subseteq \text{Ann } (a_i)$$

It remains to prove that Ann. $(a_n) =$ Ann. (M).

If $m < n$, then Ann. $(a_n) = [0] =$ Ann. $[M]$

If $m = n$, then M is torsion module and $d_i \mid d_n$ for all i

$\Rightarrow \qquad\qquad d_i M = [0]$

Therefore, $\qquad$ Ann. $(a_n) = (d_n) =$ Ann. (M)

THEOREM 2. **(Uniqueness of Decompositon Theorem).** *Let M be a finitely generated module over a principal ideal domain R. If M has cyclic decompositions given by*

$$M = Ra_1 \oplus Ra_2 \oplus \ldots \oplus Ra_m$$

and $$M = Rb_1 \oplus Rb_2 \oplus \ldots \oplus Rb_m$$

such that   Ann. $(a_{i+1}) \subseteq$ Ann. (a_i), for $i = 1, 2, \ldots, m-1$

With   Ann. $(\alpha_i) \neq R$ and Ann. $(a_n) =$ Ann. (M)

and   Ann. $(b_{j+1}) \subseteq$ Ann. (b_j), for $j = 1, 2, \ldots, n-1$

With   Ann. $(b_1) \neq R$ and Ann. $(b_n) =$ Ann. (M), then $m = n$.

and   Ann. $(a_i) =$ Ann. (b_j) for all $i = 1, 2, \ldots, m$.

Proof. We have   Ann. $(a_m) =$ Ann. $(M) =$ Ann. (b_n)

So that,   Ann. $(a_m) =$ Ann. (b_n)

Let us assume that $m \geq n$.

Let Ann.$(a_i) = (r_i)$ for all $i = 1, 2, \ldots, m$, $r_i \in R$. Also, let p be a prime element in R such that $p \mid r$.

Then,   Ann.$(a_i) \subseteq (p) \forall i = 1, 2, \ldots, m$

Clearly, pM is a submodule of R and M/pM is a module of R/pR with scalar multiplication defined by

$$(r + pR)(a + pM) \Rightarrow ra + pM, \text{ for } r \in R, a \in M \qquad \ldots(1)$$

But R/pR is a field, so that $M|pM$ can be treated as a vector space over R/pR. Thus, M/pM is a free module. Now, we shall prove that

$$S = [a_1 + pM, a_2 + pM, \ldots a_m + pM] \text{ is a basis of } M/pM$$

Since, M is finitely generated. Therefore, $M = \{a_1, a_2, \ldots, a_m\}$ and if $x \in M$, then

$$x = \alpha_1 a_1 + \alpha_2 a_2 + \ldots + \alpha_m a_m$$

for $\alpha_1, \alpha_1, \ldots, \alpha_m \in R$.

Therefore,

$$x + pM = (\alpha_1 a_1 + \alpha_2 a_2 + \ldots + \alpha_m a_m) + pM$$
$$= (\alpha_1 a_1 + \alpha_2 a_2 + \ldots + \alpha_m a_m) + pM$$
$$= (\alpha_1 a_1 + pM) + (\alpha_2 a_2 + pM) + \ldots + (\alpha_m a_m + pM)$$

Thus, $$x = \sum_{i=1}^{m} (\alpha_i + pR)(a_i + pM) \qquad \text{[Using (1)]}$$

Therefore S spans M/pM because $a_i + pM \neq 0$. Further, we shall show that S is linearly independent. Suppose

$$\sum_{i=1}^{m} (\alpha_i + pR)(a_i + pM) = pM \qquad [\because pM \text{ is the zero element of } M/pM.]$$

$$\Rightarrow \qquad \sum_{i=1}^{m} (a_i \alpha_i + pM) = pM, \qquad\qquad i.e., \sum_{i=1}^{m} \alpha_i a_i \in pM, \forall a_i \in R$$

$$\Rightarrow \qquad \sum_{i=1}^{m} \alpha_i a_i = p \sum_{i=1}^{m} \beta_i a_i, \text{ for some } \beta_i \in R$$

$$\Rightarrow \qquad \sum_{i=1}^{m} (\alpha_i - p\beta_i) a_i = 0$$

$$\Rightarrow \qquad (\alpha_i - p\beta_i)a_i = 0 \ \forall \ i = 1, 2,...., m$$

$$\Rightarrow \qquad \alpha_i - p\beta_i \in A_{nn}(a_i) \subseteq (p) \Rightarrow a_i \in (p)$$

$$\Rightarrow \qquad \alpha_i + pR = pR \qquad [\because pR \text{ is the zero element of } R/pR]$$

$\Rightarrow \quad S$ is linearly independent.

Thus, S form a basis of M/pM over R/pR of dimension m.

Also, $\qquad\qquad\qquad M = (b_1, b_2,....., b_n).$ Then

$$S_1 = (b_1 + pM, b_2 + pM,....., b_n + pM)$$

gernerates M/pM, thus we get $m \le n$.

$$\Rightarrow \qquad\qquad\qquad m = n.$$

It remains to prove that

Ann.$(a_i) =$ Ann.(b_i), for $i = 1, 2, 3,.....m$.

We know that if I is an ideal of R, then $I = (x)$. Then, $l(x)$ denotes the primes appearing in the prime factorization of X.

Now, we apply induction on $l(\text{Ann.}(M))$. If $l(\text{Ann.}(m)) = 1$, then Ann.$(M) = (p)$.

But $\quad$ Ann.$(a_m) =$ Ann.$(M) =$ Ann.$(a_m) = (p)$ and p is the maximal ideal of R.

Therefore, $\qquad$ Ann.$(a_i) = (p) =$ Ann.$(b_i) \ \forall \ i = 1, 2,...., n$

$\Rightarrow \quad$ Theorem is true for $l(A_{nn}, (M)) = 1$

Further, assume that theorem is true for all finitely generated module N over R with

$$l(A_{nn}, (N)) < l(A_{nn}, (M))$$

Thus, we have

$$pM = pRa_1 \oplus pRa_2 \oplus ....\oplus pRa_m = pRa_{s+1} \oplus pRa_{s+2} \oplus ....\oplus pRa_m$$

and $\qquad\quad pM = pRb_1 \oplus pRb_2 \oplus ....\oplus pRb_m = pRb_{t+1} \oplus pRb_{t+2} \oplus ....\oplus pRb_m$

where, Ann.$(a_1) =$ Ann.$(a_2) = ...=$ Ann.$(a_s) = (p)$

Also, Ann.$(a_{s+1}) \ne (p)$ and Ann.$(b_1) =$ Ann. $(b_2) = ...=$ Ann.$(b_k) = (p)$

$\qquad$ Ann.$(b_{k+1}) \ne (p)$

Now, Ann.$(pM) = (r|p)$, if Ann.$(M) = (r)$

Therefore,

$$l(\text{Ann.}(p(M)) = l(\text{Ann.}(p(M)) - 1$$

Then by induction

$$m - s = m - t \qquad \Rightarrow \quad s = t$$

and $\quad$ Ann.$(pa_i) =$ Ann.(pb_i) for $i = s+1,......,m$

Hence, Ann.$(a_i) =$ Ann. $(b_i) \ \forall \ i = 1, 2,...., n$

⮞ EXERCISE 11.3

1. Let M be a finitely generated module over a principal ideal domain R, then show that $M = \text{Tor}(M) \oplus F$, where F is a free module of finitely rank.

2. Find the Smith normal form and rank of the following matrices over a principal ideal domain,

$$(i) \begin{bmatrix} 0 & 2 & -1 \\ -3 & 8 & 3 \\ 2 & -4. & -1 \end{bmatrix}, R = Z.$$

(ii) $\begin{bmatrix} -x-3 & 2 & 0 \\ 1 & -x & 1 \\ 1 & -3 & -x-2 \end{bmatrix}$, R = Q

$\begin{bmatrix} 5-x & 1 & -2 & 4 \\ 0 & 5-x & 2 & 2 \\ 0 & 0 & 5-x & 3 \\ 0 & 0 & 0 & 4 \end{bmatrix}$

3. Obtain the invariant of the following matrix over Q(x).

Answers

2.(i) $\begin{bmatrix} 1 & 0 & 0 \\ 0 & 1 & 0 \\ 0 & 1 & 10 \end{bmatrix}$, rank=3

(ii) $\begin{bmatrix} 1 & 0 & 0 \\ 0 & 1 & 0 \\ 0 & 0 & (x+1)^2(x+3) \end{bmatrix}$, rank=3

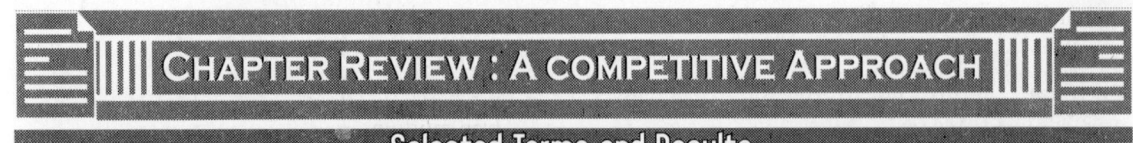

CHAPTER REVIEW : A COMPETITIVE APPROACH

Selected Terms and Results

TERMS

- **Module :** Let R be a ring. A R-module (left) is an additive abelian group M together with a function $R \times M \to M$ $\forall r, s \in R$ $a, b \in M$ with the following conditions.

  (i) $r(a+b) = ra+rb$

  (ii) $(r+b)(a) = ra + sa$

  (iii) $r(sa) = (rs)(a)$

- **Unitary R-module :** If R has identily element, *i.e.*, $I \in R$ and $1.a = a, \forall a \in m$ Then M is said to be a unitary R-module.

- **Coset R-module :** Let R be a ring and S be a left ideal of R. Let $m = \{a+S : a \in R\}$ be the set of all cosets of S in R. Then M is a R-Module with the composition defined by

  $(a+s) + (b+s) = (a+b)+s$ and $r(a+s) = ra+s$

- **Submodules :** A non-empty subset S of a R-module M is said to be submodule if

  (i) S is an additive subgroup of M

  and (ii) $r \in R, a \in S \Rightarrow ra \in S$

- **Module homomorphism :** A mapping f from a R-module M to other R-module N is said to be module homomorphism if

  (i) $f(m_1+m_2) = f(m_1) + f(m_2) \forall m_1 m_2 \in M$

  (ii) $f(rm) = r \cdot f(m) \forall r \in R, m \in M.$

- **Modules Isomorphism :** A mapping f from a R-module M to other R-module N is said to be an isomorphism if

  (i) f is a homomorphism.

  (ii) f is one-one and onto.

- **Kernal of a Module Homomorphism:** Let f be homomorphism of an R-module M into an R-module N then kernal of f denoted by $K(f)$ is the set of all those elements which are mapped to o(identity element of N),

  i.e., $K(f) = \{m \in M : f(m) = 0$, where 0 is the identity element of additive group $N.\}$

- **Quotient Module :** $M/A = \{A+m : m \in M\}$

- **Cyclic Module :** A R-module M is said to be simple if it has no submodules other than it self and zero submodule.

- **Semi-simple Module :** An R-module M is said to be semi-simple if it can be expressed as a direct sum of simple submodules.

- **Basis of Module:** Let M be a module over a ring R with unity and let S be a subset of M then S is said to be basis for m if

- (i) S generates M

- (ii) S is linearly independent.

- **Free Module :** An R-module M is said to be free modul if there exists a subset S of M such that S generates M and S is linearly independent set.

- **Noetherian Module:** An R-module is said to be Noetherian module if it satisfy the ascending chain condition.

- **Artinian Module:** An R-module is said to be Artinian if it satisfies the descending chain condition.

- **Uniform Module:** A non-zero R-module is said to be uniform if any two non-zero submodules of M have non-zero intersection.

- **Primary Module:** An R module M is said to be primary if each non-zero submodule of M has uniform submodule and any two uniform submodules of M are subisomorphic.

- **Row and Column Module:** Let R be a ring and A be $n \times n$ matrix of order $n \times n$ over R. The submodule of R^n generated by the m rows of A is called the row module of A and the submodule of R^m generated by the n-columns of A and the submodule of R^m generated by the n-columns of A and the submodule of R^m generated by the n-columns of A is known as column module of A.

- **Rank of a Module:** Let A be a $m \times n$ matrix over R. The rank of the row module (column module) is known as the row rank (column rank) of A.

- **Invariant Factor of a Smith Normal form:** The non-zero diagonal elements of Smith normal form are called the invariant factors of A.

- **Torison and Torsion free Element:** Let R be a principal ideal domain and M be an

R-module. An element $x \in M$ is called torsion element if $rx = 0$ for some non-zero $r \in R$. If x is not a torsion element, then it is called torsion free element.

- **Torsion and Torsion free module:**

Let R be a principal ideal domain and M be an R-module. If $T(m)$ is a set of all torsion element of M, then $T(M)$ is a submodule of M. Also if $T(M) = M$, then M is called as torsion module and if $T(m)$, then M is called torsion free module.

RESULTS

- Module is a generalised concept of vector space.
- Every abelian group is a module over the ring of integers.
- Every left ideal M of a ring R is an R-module.
- Every ring is a R-module itself.
- Arbitrary intersection of submodules is again a submodule.
- The linear sum of two submodules of an R-module is also a submodule of the R-module.
- The kernel of a homomorphism is a submodule.
- Let f be a module homomorphism. Then f is an isomorphism iff $\text{Ker}(f) = 0$.
- The range of a module homomorphism is a submodule.
- An irreducible R-module is cyclic.
- If R is an Euclidean ring, then any finitely generated R-module M is the direct sum of a finite number of cyclic modules (Fundamental theorem).
- Polynomial ring $R(x)$ over a ring R is a R-module.
- A subring S of a ring R is a module over the whole ring only if S is an ideal of R.
- The set of rational numbers Q is not finitely generated Z-module.
- If R is a ring with unity, then R-module M is cyclic if and only if $M \cong R / I$ for some left ideal I of R.
- Every submodule and every homomorphic image of a semi-simple module is semi simple.
- If R is a ring and I is a minimal left ideal of R. Also, if M is a faithful simple left R-module, then M is isomorphic to the left R-module I.
- Every finitely generated module is a homomorphic image of a finitely generated free module.
- If M is a finitely generated free module over a commutative ring R. Then, all basis of M are finite.
- If R is a commutative ring and M is a finitely generated free module over R, then all basis of M have the same number of elements.
- The finitely co-generated modules satisfies the certain chain condition on their submodules.
- Every submodule of a Noetherian (Artinian) module is Noetherian (artinian).
- Every homomorphic image of a Noetherian (artinian) module is Noetherian (artinian).
- A ring is called Noetherian if it is Noetherian as a left module over itself, *i.e.*, every left ideal is finitely generated.
- Every free module M over a ring R with identity is protective.

Review Questions and Project Work

1. Let M be the set of all $m \times n$ matrix over a ring R. Show that M is a module over R.

2. Let G be a multiplication abelian group. Define $ng = g^n$ for $g \in G$ and $n \in Z$. Show that G is a Z-module.

3. Let M be an additive abelian group. Show that there is one way of making it a Z-module.

4. If M is an R-module and $x \in M$, prove that the set
   $$K = \{rx + nx : r \in R, n \in Z\}$$
   is an R-module of M containing x. Further if R has unity then prove that $K = Rx$.

5. Let R be a commutative ring over I, a finitely generated ideal in R with $I = I^2$. Show that I is a direct summand of R. Also, give an example to show that if I is not finitely generated then I need not be a summand.

6. If A and B are R-submodule of an R-module M. Show that $(A + B) | B \cong A | A \cap B$.

7. Let M be a completely reducible module and let K be a non-zero submodule of M. Show that K is completely reducible. Also show that K is a direct summand of M.

8. Show that every finitely generated module is a homomorphic image of a finitely generated free module.

9. Show that every principal left ideal is an integral domain R with unity is free as a left R-module.

10. Let R be a commutative noetherian ring and let S be a multiplicative subset of R. Show that the ring of fraction R_S is also Noetherian.

11. Let R be a left Artinian integral domain with more than one element. Show that R is a division ring.

12. Let R be a prime left Artinian ring with unity. Show that R is isomorphic to the $n \times n$ ring over a division ring. Hence, show that a prime ideal is an Artinian ring is maximal.

13. Show that an infinite dimensional vector space is neither artinian nor noetherian.

14. Show that the abelian group generated by x_1 and x_2 subject to $x_1 + x_2 = 0$ is isomorphic to Z.

Objective Type Questions

FILL IN THE BLANKS

1. Every abelian group G is a module over the _____ .

2. Let M be an additive abelian group. Then there is _____ way of making it a Z-module.

3. Let $f:M \to N$ be a R-module of N. Then the range of homomorphism f is an _____ of N.

4. An anti-homomorphism which is one-one and onto is called _____ .

5. Any minimal R-submodule N of an R-module M is _____ .

6. An R-submodule N of M is _____ in M if and only if M/N is simple.

7. Any one-dimensional vector space is _____ .

8. A R-module M is called _____ if it is a direct sum of a family of minimal submodule.

9. An R-module M is called _____ if M admits a basis.

10. Let M be a finitely generated free module over a commutative ring R. Then all basis of M are _____ .

TRUE/FALSE

Write 'T' for true and 'F' for false statement.

1. If R is a field then the rank of M is called dimension of the vector space.   **(T/F)**

2. Every ideal of Z is not necesarily is free as Z-module.   **(T/F)**

3. Every principal left ideal is an integrated domain R with unity is free as a left R-module.   **(T/F)**

4. Every finite abelian group is free as a module over Z.   **(T/F)**

5. A module which has only finitely many submodule is Artinian.   **(T/F)**

6. Infinite cycle group are Artinian.   **(T/F)**

7. If M is Artinian then every quotient module of M is finitely generated.   **(T/F)**

8. The ring of integers is Noetherian and as well as Artinian.   **(T/F)**

9. Every homomorphic image of a Noetherian module is a Noetherian.   **(T/F)**

10. The subring of a Noetherian ring is Noetherian.   **(T/F)**

MULTIPLE CHOICE QUESTIONS

Choose the most appropriate one.

1. The ring of integers is :
   (a) Noetherian
   (b) Artinian
   (c) both (a) and (b) are true
   (d) none of the above

2. The subring of Artinian ring is :
   (a) Artinian
   (b) need not be Artinian
   (c) never Artinian
   (d) none of the above

3. If R is a Noetherian ring then $ab=1, a,b \in R$ if and only if :
   (a) $a=1$ (b) $b=1$
   (c) $ba=1$ (d) none of the above

4. If R is right Noetherian then :
   (a) maximum condition holds for right ideal of R
   (b) every right ideal of R is finitely generated
   (c) both (a) and (b) are true
   (d) none of the above

5. The set of integers Z as a Z-module is :
   (a) uniform module
   (b) primary module
   (c) both (a) and (b) are true.
   (d) none of the above
6. Let R be a PID with unity then :
   (a) every irreducible element is prime in R.
   (b) every non-zero prime ideal is maximal.
   (c) both (a) and (b) are true.
   (d) none of the above
7. If $rx = 0$, $r \in R$ $r = 0$ then element x is called :
   (a) Torsion element
   (b) Torsion free element
   (c) Free element
   (d) None of the above
8. Which one of the following is not module ?

(a) An additive abelian group
(b) A ring R with property $a, m \in R \Rightarrow am \in R$
(c) The polynomial ring $R[x]$ over a ring R
(d) All are true.

9. An R-module M is called semi-simple if :
   (a) it is a direct sum of a family of minimal submodule.
   (b) it is a direct sum of simple submodules.
   (c) both (a) and (b) are true.
   (d) none of the above

10. A boolean Noetherian ring is :
   (a) finite
   (b) a finite direct product of fields with two elements
   (c) both (a) and (b) are true.
   (d) none of the above

Answers

FILL IN THE BLANKS
1. ring of integers 2. one 3. R-submodules 4. anti-isomorphism 5. simple
6. semi-simple 7. simple 8. semi-simple 9. free module 10. finite

TRUE/FALSE
| 1. T | 2. F | 3. T | 4. T | 5. T |
| 6. F | 7. T | 8. F | 9. T | 10. F |

MULTIPLE CHOICE QUESTIONS
| 1. (a) | 2. (b) | 3. (c) | 4. (c) | 5. (c) |
| 6. (c) | 7. (b) | 8. (d) | 9. (c) | 10. (c) |

Bibliography

1.	**Asha Rani Singhal**	*Algebraic Structures* Rastogi and Company, Meerut
2.	**A. Ganesh**	*Linear Algebra and its Applications* CBS Publishers and Distributors Pvt. Ltd. New Delhi
3.	**E.Artin**	*Galois Theory* University of Notre Dame Press
4.	**F.M. Hall**	*An Introduction to Abstract Algebra* Cambridge University Press
5.	**F.Stewart**	*Introduction to Linear Algebra* East-West Press, New Delhi
6.	**G. Briefkhoff and S.K. Lane**	*A Survey of Modern Algebra* Macmillan Company, New York
7.	**I.N. Herstein**	*Topics in Algebra* Vani Educational Books
8.	**J.B. Frayleigh**	*A First Course in Abstract Algebra* Addison-Wesley Publishing Company
9.	**J.T. Moore**	*Introduction to Abstract Algebra* Academic Press
10.	**L.R. Goldstein**	*Abstract Algebra: A First Course* Prentice Hall India
11.	**N. Jacobson**	*Lectures in Abstract Algebra* East-West Press Pvt. Ltd., New Delhi
12.	**P.L. Bhatnagar**	*Introductory Lessons in Modern* Mathematical Concepts East-west Press, New Delhi
13.	**P.R. Halmos**	*Finite Dimensional Vector Spaces* Springer Verlag
14.	**S.Lang**	*Algebra* Addison-Wesley Longman
15.	**Surjeet Singh and Qazi Zameeruddin**	*Modern Algebra* Vikas Publishing House, New Delhi
16.	**Sudhir K. Pundir**	*A Competitive Approach to Modern Algebra* CBS Publishers and Distributors (Pvt) Ltd. New Delhi
17.	**Seymour Lipschutz and Marc Lipson**	*Linear Algebra Schaum's Outline Series* Mcgraw Hill Education (India) Private Ltd. New Delhi
18.	**Vijay K. Khanna and S.K. Bhambri**	*A Course in Abstract Algebra* Vikas Publishing House, New Delhi

Index

A

Abelian or commutative group,236
Absolute value of a real number, 3
Addition and multiplication of two cosets, 318
Addition of matrices, 65
Additive identity, 67
Additive inverse, 68
Algebra of linear operators, 356
Algebra of linear transformations, 352
Algebra of subspaces, 254
Algebraic and geometric multiplicity of an eigenvalue, 222
Algebraic structure, 48
Annihilator of an annihilator, 431
Annihilator, 429
Associative law, 67
Augmented matrix, 148

B

Basis of a vector space, 290
Bilinear forms and matrices, 557
Bilinear forms, 555
Binary operation, 47

C

Cancellation law, 68
Cardinal number of a set, 7
Cartesian product of two sets, 23
Change of basis, 375
Characteristic vectors or eigenvectors of a matrix, 178
Characterization of spectra, 526
Classification of real symmetric matrices, 616
Classification of relations, 30
Commutative law, 67
Complement of a set , 9
Complex matrix, 81
Composition of relations , 32

Composition table for binary operation on finite sets, 55
Concept of sets, 3
Condition for consistency, 148
Congruence modulo 'm' , 31
Congruence of quadratic forms, 593
Congruence operation on a square matrix, 593
Congruent operators, 498
Congruent reduction of a symmetric matrix, 594
Conjugate of a complex matrix, 81
Consistency and inconsistency, 148
Construction of graded rings from filtration, 692
Conversion of a symmetrtc matrix into quadratic form, 587
Coordinate vector, 370
Coset R-module, 670
Cosets, 317
Cyclic module, 675

D

Determinant of a linear transformation on a finite dimensional vector space, 381
Determinant of a square matrix, 75
Diagonalisation of a matrix, 214
Diagonalization, 453
Difference of two sets, 15
Dimension of subspace of a vector space, 294
Direct sum of submodules, 672
Direct sum of vector subspaces, 265
Disjoint sets, 15
Dual basis, 423
Dual spaces, 420

E

Echleon from of a matrix, 88
Eigen values and Eigen vectors of a matrix   177
Eigenvalues and eigenvectors of a linear transformation,   445

Eigenvalues of special type of matrices, 180

Elementary matrices, 93

Elementary properties of vector spaces, 251

Elementary properties of vector subspaces, 253

Elementary transformations
(or e-transformations) of a matrix, 92

Equality of matrices , 66

Equivalence of matrices, 98

Equivalence of real quadratic forms , 593

Equivalence relations, 31

Equivalent sets, 7

Evolution of a determinant by sarrus diagram, 76

External composition , 235

F

Field, 236

Filtered and graded modules, 690

Finite and infinite group, 236

Finite dimensional vector space, 290

Finitely generated module, 676

Finitely generated modules over a PID, 697

Free modules, 681

Functions, 39

G

Gauss elimination method , 155

General properties of modules, 670

H

Hermitian and skew-Hermitian matrices, 82

Hermitian forms, 634

Homogeneous linear equations, 140

Homomorphism of modules, 673

I

Identity element, 52

Identity relation , 29

Improper submodules, 670

Inner product spaces, 471

Inner product vector space isomorphism, 499

Internal composition , 235

Interval, 2

Invariance and reducibility in inner product space, 518

Invariance of linear operator, 449

Invariance of rank under E-transformations, 95

Invariant direct-sum decompositions, 646

Invariant subspace, 645

Inverse of a matrix by elementary transformations, 121

Inverse of a matrix, 116

Inverse of a relation, 29

Inverse of an element, 53

Invertible linear transformation, 361

Irreducible r-module, 670

Isomorphism of modules, 673

Isomorphism, 321

Isotropic subspace, 562

J

Jordan canonical form, 651

K

Kernel of a homomorphism, 674

L

Law of excluded middle and law of contradiction, 16

Length of an interval, 2

Linear combination of vectors, 138

Linear combination of vectors, 273

Linear dependence and independence of any matrix, 140

Linear dependence and independence of vectors, 138

Linear dependence and independence of vectors, 275

Linear dependence of the rows and columns of a square matrix, 139

Linear functionals, 419

Linear operator, 356

Linear sum of two modules, 672

Linear sum of two subspaces, 264
Linear transformations, 349-418
LU decomposition method, 163

M

Matrix representation of a Hermitian form, 635
Matrix representation of a linear transformation, 370
Minimal polynomial, 448
Minor and cofactors, 77
Minors of a matrix , 87
Modules, 669
Monic polynomials, 446
Multiplication of a matrix by a scalar , 66
Multiplication of a vector by a scalar , 138
Multiplication of matrices, 72

N

Natural mapping, 429
Nature of the solution of the equation AX = 0, 142

Nilpotent transformation, 651
Noetherian and artinian modules, 685
Non-homogeneous equations, 148
Non-singular linear transformations, 362
Non-trivial solution, 140
Normal (or canonical) form of a real quadratic matrix, 600
Normal form, 96
Normal form, 647
Normal operators , 507
Notation, 678
Notations for E-transformations, 93
Number of binary operations, 47
Number of subsets of a set, 8
Number system, 1

O

Operation on matrices, 65
Operations on sets, 13
Order of a group, 236
Ordered pair, 23

Ordered triplet, 23
Orthogonal and unitary matrices, 82
Orthogonal diagonalization of the quadratic form, 578
Orthogonal expansion, 480
Orthogonal projections, 521
Orthogonal reduction of a real quadratic form, 604
Orthogonality and orthonormality, 474

P

Perpendicular projection, 514
Polynomials in a linear operator, 360
Positive definite, 566
Positive operators, 509
Positive-definiteness of a quadratic form x´ax in terms of leading principal minors of a, 622
Product of linear transformations, 358
Prologue to vector space, 235
Proper submodules of m, 670
Proper subset, 8
Properties of a perpendicular projection, 516
Properties of binary operation, 47
Properties of determinants, 76
Properties of Hermitian and skew-Hermitian matrices, 82
Properties of linear transformations, 350
Properties of matrix addition, 67
Properties of matrix multiplication, 73
Properties of multiplication of matrix by a scalar, 68
Properties of symmetric and skew-symmetric matrix, 80
Properties of the adjoint, 490
Properties of transpose conjugate of matrix, 82
Properties of transpose of a matrix, 78

Q

Quadratic forms and matrices, 585
Quadratic forms, 575
Quotient modules, 675
Quotient spaces, 318

R

Range and domain of a function, 39

Range and null space of a linear transformation, 357

Rank and signature, 567

Rank of a matrix, 87

Rank of a quadratic form, 596

Rational canonical form, 653

Raw and column space of a matrix, 653

Real quadratic form , 585

Real symmetric bilinear and quadratic forms: law of inertia, 576

Reducibility, 519

Reduction of a real quadratic form over real field, 599

Reduction of a real quadratic form over the field of complex numbers, 603

Relation between eigenvalues and eigenvectors, 178

Relation, 28

Relations other than equivalence, 37

Representation of a set, 4

Row and column equivalence of matrices, 99

S

Scalar transformation, 381

Second dual space : bidual space, 426

Self-adjoint transformation, 491

Signature and index of a real quadratic form, 602

Similarity of linear transformations, 378, 645

Similarity of matrices, 377

Simple and semi-simple modules, 678

Singular and non-singular matrix, 77

Skew-symmetric matrix, 79

Smith normal form over a PID and rank, 693

Some results on venn diagram, 21

Spectral form of t (normal), 533

Spectral form of t (self-adjoint), 537

Spectral theorem for self- adjoint operators, 535

Subfield, 237

Submatrix of a matrix, 86

Submodule generated by a set, 671

Submodules, 670

Subset, 7

Subtraction of matrices, 66

Sum of two vectors, 138

Symmetric difference of two sets, 16

Symmetric matrix, 79

T

The adjoint of a linear transformation, 489

The Cayley-Hamilton theorem, 202

The characteristic equation of a matrix, 177

The linear transformation of a quadratic form, 594

The spectral theorem for normal operators, 530

Total number of relations, 29

Trace of a linear transformation on a finite dimensional vector space, 382

Trace of a matrix, 381

Transpose conjugate of a matrix, 81

Transpose of a matrix, 78

Triangular form, 648

Trivial solution, 140

Type of functions, 40

Type of matrices, 63

Type of sets, 6

U

Union and intersection operations, 13

Unitary operators, 502

Universal set, 9

V

Vector spaces, 237

Vector subspaces: vector space within vector space, 253

Vectors and their dependence and independence, 137

Venn diagram, 13